P9-CPZ-983

ENCYCLOPEDIA OF EVOLUTION

Editorial Board

Encyclopedia of

Evolution

Mark Pagel

EDITOR IN CHIEF

VOLUME 2

OXFORD

UNIVERSITY PRESS

2002

OXFORD
UNIVERSITY PRESS

Oxford New York
Auckland Bangkok Buenos Aires Cape Town Chennai
Dar es Salaam Delhi Hong Kong Istanbul Karachi Kolkata
Kuala Lumpur Madrid Melbourne Mexico City Mumbai Nairobi
São Paulo Shanghai Singapore Taipei Tokyo Toronto

and an associated company in Berlin

Published by Oxford University Press, Inc.
198 Madison Avenue, New York, New York 10016
www.oup.com

Library of Congress Cataloging-in-Publication Data

(The Oxford) encyclopedia of evolution / Mark Pagel, editor in chief.
p. cm.
Includes bibliographical references (p.).
ISBN 0-19-512200-3 (set)
ISBN 0-19-514864-9 (vol. 1)
ISBN 0-19-514865-7 (vol. 2)
1. Evolution (Biology)—Encyclopedias. I. Pagel, Mark D.
QH360.2.O83 2002
576.8'03—dc21 2001021588

3 5 7 9 8 6 4 2

Printed in the United States of America
on acid-free paper

ENCYCLOPEDIA OF EVOLUTION

I–J

IDEAL FREE DISTRIBUTION

The ideal free distribution refers to the idea that individuals will distribute themselves among areas, or patches, in such a way that the average returns to all individuals are equal. The concept was invented independently by Parker (1970) to account for the distribution of male dung flies searching for mates, and Fretwell and Lucas (1970) to account for the settlement patterns of birds. It represents one of the earliest applications of game theory to evolution or ecology, but it was devised independently of game theory.

The assumption is that animals are "ideal" (i.e., they will go where their mating success or intake rate is highest and are not constrained by a lack of information about alternatives or by dispersal) and "free" (they are not constrained by, for example, territoriality or dominance). Individuals settle in the best-quality patch (e.g., the one with the highest density of prey) until the competition within that patch reduces the rate of intake to the extent that it equals that of the next best patch (with one fewer prey but also less competition). After this point, both patches will be occupied. It thus follows that the expected mean gain for each individual must be equal: if individuals can obtain a higher expected mean intake elsewhere, then they are expected to do so. The expected mean gain will thus be the same across sites. The ideal free distribution is usually applied in one of two ways: (1) as the distribution of foraging individuals in relation to both the food supply and the levels of interference or depletion or (2) as the behavior of individuals searching for mates.

Although the ideal free distribution provides a useful framework for considering how individuals might distribute themselves, it is extremely unlikely that it ever precisely describes any given field situation. A range of factors makes it unlikely that population will be ideal or free, and much of the research examines these discrepancies.

The nature of competition (whether territoriality, interference, or depletion) greatly affects the expected distribution. If territoriality is the main factor, then the required concept is the "ideal despotic distribution," in which individuals defend resources. Thus in the ideal free distribution, all experience a reduced gain as the number of competitors increases, while in the ideal despotic distribution, the first to arrive in a patch defend the best territory so that subsequent individuals have a lower gain. Thus in the ideal despotic distribution, it is the potential gain from settling in each patch that is equalized.

Interference is the decline in intake rate as a result of the presence of other individuals—for example from fighting or disturbance of the resource. In this case, equal gains across patches are theoretically achieved by high densities of competitors in the better patches resulting in increased interference, thus offering the same gains as for a poorer patch with less competition. Thus the stronger the interference, the greater the expected use of the poorer patch.

Depletion is the actual removal of resources. If depletion is the major source of competition, the theoretical expectation is that individuals will accumulate in the patches of higher prey density, deplete these, and then occupy patches of lower densities. The expectation is that the density of available prey becomes equalized across patches. This resulting prediction of the distribution and numbers of individuals was tested by examining the distribution of black-tailed godwits *Limosa limosa* across patches, mudflats, and estuaries (Gill et al., 2001); the models provided remarkably accurate predictions across all scales.

Many of the tests of the ideal free distribution apply to the situation in which there is rapid short-term depletion combined with rapid and repeated renewal. The earliest example was Parker's (1970) study of dung flies searching for mates, in which females arrived at a given rate and soon mated with males. The arrival rate of the females and the density of competing males could then explain the male distribution. Similarly, the most common way of testing the ideal free distribution is to create two patches and provision these at different rates: the distribution should then be in proportion to the provisioning rate. The usual tests are for waterfowl on a water body or fish in a tank. Although most of the tests and theory apply to such situations, in reality, for foraging, this pattern of depletion and rapid renewal is relatively rare in the real world.

A major discrepancy between reality and the basic concept of the ideal free distribution is the assumption that all individuals are equal. The models can be extended to consider individual differences, for example in competitive abilities of foraging efficiencies. The conclusions of these models are that differences in foraging efficiencies will not affect the choice of patch, but differences in the intensity of interference will (Sutherland and Parker, 1992). Using the concept of phenotype-limited evolutionary stable strategies, it is possible to derive the game theoretical solution for each class of individuals, while considering the decision of members of other classes. The solution is similar, with the ideal free distribution applying to all members of a phenotype class so that the mean gains should be the same across sites. If an individual of a given class could have higher gains elsewhere, then it should move. The mean gains and optimal decisions may, however, differ between classes.

Another useful application has been to consider how distributions of individuals are modified by factors such as predation risk. Milinski (1985) showed that the levels of depletion within different patches could predict the distribution of sticklebacks *Gasterosteus aculeatus* avoiding a predator. Grand and Dill (1997) studied juvenile coho salmon to measure the relationship between preferences for patches that differed in food availability and in the extent of cover they provided against a predator. In the presence of a model avian predator, cover was even more strongly preferred at the cost of reduced intake. From this it was possible to determine the amount of food that had to be applied to the patch with reduced cover to overcome the predation risk, and tests showed that this prediction was correct. Gill et al. (1996) used a similar approach to quantify the extent to which pink-footed geese *Anser brachyrhynchus* are affected by human disturbance. They measured the extent to which sites were not equally depleted by the geese and found that depletion was strongly correlated with distance from the road. The same approach found no evidence that black-tailed godwits *Limosa limosa* were affected by disturbance.

The ideal free distribution and ideal despotic distribution can be used as a component of population models to answer applied questions. The ideal free (or despotic) distribution can predict the patch choice of individuals and the fitness (e.g., intake rate, reproductive success, or breeding output) of each. From this it is possible to assess how these components of fitness change with the density of individuals both as a result of competition and of differential use of different patches. It is then possible to determine how birth rate or death rate change with density, information that can be used to predict the consequences of habitat loss or habitat deterioration on populations (Goss-Custard et al., 1995; Sutherland, 1996).

A further application of the ideal free distribution is in understanding the distribution of hunters or fishers in relation to the abundance of prey. Gillis and Frank (2001), studying the fishing of cod *Gadus morhua* show that simple measures of changes in catch per unit effort are unlikely to be understandable unless the factors affecting the distribution of fishing boats is also understood.

It is likely that future research will concentrate on three issues: (1) the mechanisms by which individuals can gather information on patch quality and the consequences for decision making, (2) the factors that influence decisions made in the field, and (3) the applied implications of the ideal free and ideal despotic distribution in predicting the consequences of local and global environmental change.

[*See also* Optimality Theory, *article on* Optimal Foraging; Territoriality.]

BIBLIOGRAPHY

Fretwell, S. D., and J. H. J. Lucas. "On Territorial Behaviour and Other Factors Influencing Habitat Distribution in Birds." *Acta Biotheoretica* 19 (1970): 16–36.

Gill, J. A., K. Norris, P. Potts, T. Gunnarsson, P. W. Atkinson, and W. J. Sutherland. "Large-scale Population Regulation in Migratory Birds." *Nature.*

Gill, J. A., W. J. Sutherland, and K. Norris. "Depletion Models Can Predict Shorebird Distribution at Different Spatial Scales." *Proceedings of the Royal Society,* Series B, 246 (2001): 369–376.

Gill, J. A., W. J. Sutherland, and A. R. Watkinson. "A Method to Quantify the Effects of Human Disturbance on Animal Populations." *Journal of Applied Ecology* 33 (1996): 786–792.

Gillis, D. M., and K. T. Frank. "Influence of Environment and Fleet Dynamics on Catch Rates of Eastern Scotian Shelf through the Early 1980s." *ICES Journal of Marine Science* 58 (2001): 61–69.

Goss-Custard, J. D., R. G. Caldow, R. T. Clarke, S. E. A. le V dit Durell, and W. J. Sutherland. "Deriving Population Parameters from Individual Variations in Foraging Behaviour I. Empirical Game Theory Distribution Model of Oystercatchers *Haematopus ostralegus* Feeding on Mussels *Mytilus edulis.*" *Journal of Animal Ecology,* 64 (1995): 265–276.

Grand, T. C., and L. M. Dill. "The Energetic Equivalence of Cover to Juvenile Coho Salmon (*Oncorhynchus kisutch*): Ideal Free Distribution Theory Applied." *Behavioural Ecology* 8 (1997): 437–447.

Parker, G. A. "The Reproductive Behaviour and the Nature of Sexual Selection in *Scatophaga stercoraria* L. (Diptera: Scatophagia). IX Spatial Distribution of Fertilization Rates and Evolution of Male Searching Strategy within the Reproductive Area." *Evolution* 28 (1970): 93–108.

Sutherland, W. J. *From Individual Behaviour to Population Ecology.* Oxford, 1996.

Sutherland, W. J., and G. A. Parker. "The Relationship between Continuous Input and Interference Models of Ideal Free Distributions with Unequal Competitors." *Animal Behaviour* 44 (1992): 345–355.

— WILLIAM J. SUTHERLAND

IMMUNE SYSTEM

[*This entry comprises two articles. The first article describes the structure and function of the vertebrate immune system and discusses the immune system as a microcosm of evolution by natural selection; the second article explains the challenges and mechanisms that microorganisms such as bacteria, viruses, and protozoa encounter as they seek to colonize a vertebrate host and evade its immune system. For related discussions, see articles on* Acquired Immune Deficiency Syndrome *and* Disease.]

Structure and Function of the Vertebrate Immune System

Discrimination between self and invader—the most basic component of an immune system—is a ubiquitous problem for multicellular organisms. The identification and elimination of the barrage of parasites, bacteria, and viruses from the outside, or cancerous cells on the inside, is essential for the survival of the simplest as well as the most complex animals.

Overview. A given pathogen, such as a bacterium that penetrates the physical barriers of the body, or an attenuated one injected as a vaccine, encounters two kinds of immune response. Immediately, the incriminating features of the pathogen, such as lipopolysaccharides common to many bacterial cell walls, alert components of the *innate* immune system, machinery of which exists prior to exposure to the pathogen and responds the same way in subsequent encounters. After several days, if the infection has not already been eliminated, an *adaptive* immune response, tailored to the specifics of the pathogen, mounts an attack that improves with time and "remembers" features of the pathogen for the next encounter. Innate immune components, evidenced by their presence in all extant metazoan phyla, are as old as multicellular animals themselves (Figure 1). Adaptive immunity, the biologically costly, time-consuming, custom approach, exists only in vertebrates (excluding the most primitive jawless fish).

The adaptive immune response relies on two types of cell-surface receptors—B cell and T cell receptors—found on white blood cells called lymphocytes (Figure 2). B cell receptors, or *Immunoglobulin* (Ig) receptors, recognize molecular signatures on extracellular pathogens such as virus particles or parasitic worms. The reactive region of the pathogen, referred to as the *antigen*, binds the Ig receptor, signalling cellular uptake of the pathogen; this induces maturation of the B lymphocyte into a plasma cell that secretes its Ig receptor as a soluble *antibody*. T cell receptors (TCR) bind to peptide fragments of pathogens inside cells, such as viruses or endocytosed bacteria. T cells survey the peptides as they are cradled by the major histocompatibility complex (MHC) receptor, also called the human leukocyte antigen (HLA) receptor in humans. If a TCR fits snugly with the peptide:MHC complex, the T cell divides and differentiates into an army that promotes the killing of infected cells. The potential repertoire of 10^{14} different Igs and 10^{18} different TCRs enables detection of nearly any pathogen. The extraordinary diversity of these receptors could not possibly be coded as individual genes. Instead, variations are generated throughout life by rearrangement of genomic DNA, directed by the recombinase-activated genes (RAG).

Successful identification of Ig, TCR, MHC, and RAG homologues in all classes of jawed vertebrates but not in agnathans, protochordates, or invertebrates implies that the basic molecules that function together to generate and modulate the adaptive immune response arose in a lineage that evolved some 500 million to 550 million years ago, after the radiation of jawless fish and before the divergence of cartilaginous fish (see Figure 1). Inquiry into the origins of adaptive immunity must involve a reassembly of the past through comparison of species in the present or a search for the molecular raw material that might survive in jawless fish today.

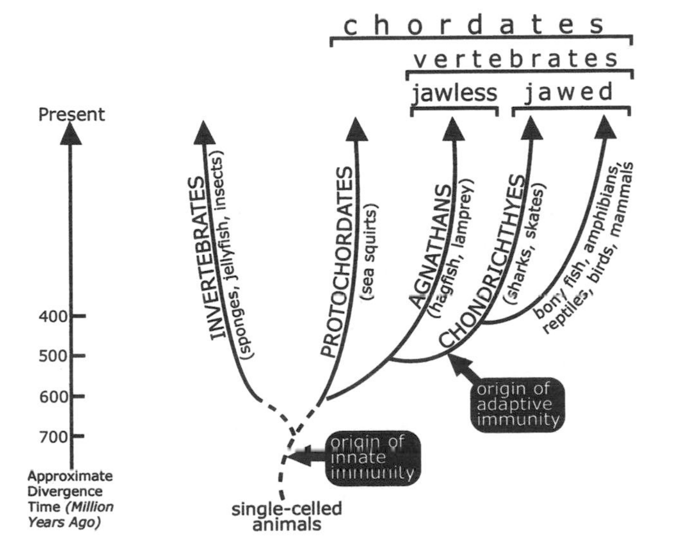

FIGURE 1. Evolution and Taxonomy of Metazoans. Shown are the probable origins of innate and adaptive immunity. Drawing by Diana J. Laird.

Structures of the Vertebrate Immune System. Despite the quantity of immune receptors, signaling molecules, and serum factors that carry out antigen recognition and the accompanying inflammatory response, these structures share much in common. Protein motifs such as the immunoglobulin fold, which appears in Ig, TCR, natural killer (NK) cell receptors, MHC receptors, complement proteins, and adhesion molecules, imply extraordinary recycling and even a common origin during evolution. The hypothesis that a primordial immunoglobulinlike receptor gave rise to the rearranging receptors of lymphocytes has inspired the study of Ig, TCR, and other immune receptors of primitive vertebrates. This section focuses on the evolution of the main components of the adaptive immune system, unique to jawed vertebrates.

Immunoglobulins (Igs). Igs are modular proteins with different domains performing different roles. Antigen binding by the "variable" region of Ig receptors helps

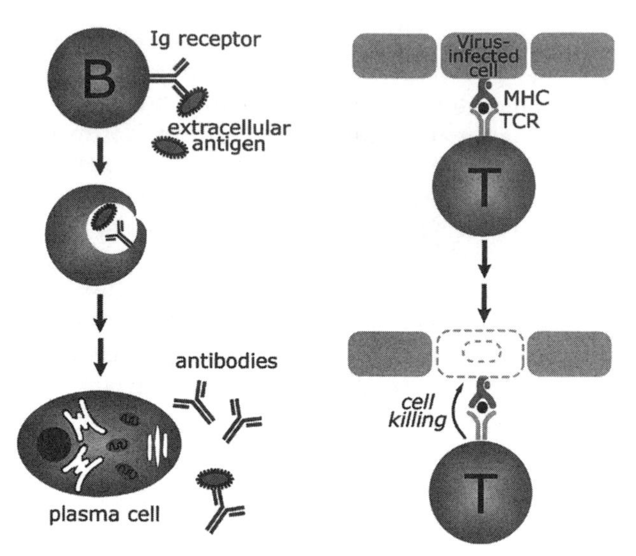

FIGURE 2. Lymphocytes and their Receptors.
Left: Binding of extracellular antigens to immunoglobulin (Ig) receptors on B cells induces uptake of antigen and leads to the division and differentiation of the B cell into a plasma cell that secretes an antibody. Right: Antigens from within the cell, such as viral peptides, are presented in MHC receptors to T cells. T cell receptor (TCR) recognition of the peptide:MHC complex stimulates expansion and maturation of the naïve T cell into killer cells that destroy infected target cells. Drawing by Diana J. Laird.

activate B cells and, later in the immune response, secreted Igs, or antibodies, aggregate and neutralize pathogens. The "constant" region elicits other immune responses, such as uptake of the antibody-antigen complex by antigen-presenting cells and activation of inflammatory response by an enzymatic cascade of serum proteins called the complement cascade. Igs are comprised of a symmetrical pair of two subunits, the heavy chain (HC) and light chain (LC). Each chain is encoded by gene segments called V, J, and C, plus an additional segment, D, in the HC. Ig genes break the cardinal rule of DNA immutability by rearranging V, D, and J segments at the DNA level to generate unique antigen-binding sites. V(D)J recombination is initiated by enzymes specific to B and T lymphocytes, the recombinase-activating genes (RAG). The outcome of this irreversible genetic splicing is a very high diversity of antigen-binding receptors that could not fit in the genome as separately coded genes.

During vertebrate evolution, immunglobulin genes have undergone a striking genetic reorganization (see Figure 3). Chondrichthyians and bony fish LCs are arranged into over 200 clusters of single V, J, and C segments (plus two D's in the HC). By contrast, most other vertebrate Ig loci consist of tandem arrays of many Vs, followed by many D's, J's, and C's. RAG genes act on both arrangements to bring a single combination of V, D, J, and C segments in proximity to enable the expression of a single Ig receptor. However, these differences in gene organization produce vastly different antigen repertoires. The predetermined arrangements of shark Ig segments are subject to some diversity in different HC-LC combinations and imprecise joining of V, D, and J segments, or junctional diversity. The array organization increased the number of different antibodies that could be produced by enabling random combinations of different V, D, and J segments; despite other mechanisms of diversity generation that subsequently evolved, combinatorial diversity remains the most significant contributor to the antibody repertoire.

Later in evolution, another very different mechanism for diversity generation arose, which is found in birds and some mammals, including rabbits. RAG-directed recombination creates identical Ig receptors by splicing a single set of functional V, (D), and J segments. Diversity results from subsequent gene conversion events that exchange short sequences of the functional Ig receptor with otherwise unexpressed (pseudogene) V segments. Speculations that the RAG genes also played a role in the Ig locus reorganizations during early vertebrate evolution have been bolstered by the recent finding of already rearranged V, D, and J segments in the germ-line DNA of sharks and some mammalian neonates. The "natural antibodies" that result from these pre-rearrangements tend to be cross-reactive against many antigens, suggestive of a prototypical intermediate to innate and adaptive immune receptors.

Immunoglobulin evolution may be examined on the basis of sequence divergence, genetic structure, or function. Sequence comparison of antibody splice variants across species suggests that IgM, the only isotype present throughout jawed vertebrates, remains closest to a primordial template Ig, which gave rise, through duplication and point mutation, to the other known C gene cassettes (Figure 4). Disparate organization of the Ig loci between Chondrichthyes and other vertebrates suggests that a major genome-wide event must have occurred to create and arrayed V_n-D_n-J_n structure from a repeated cluster arrangement. During avian radiation, deletion and reorganization mechanisms reduced the pool of functional V and J segments considerably. Recently discovered antibody functions argue for their origin as effective antigen-neutralizing proteins that later acquired immunostimulatory mechanisms. The production of catalytic antibodies that function as enzymes relegated immunoglobulins to the realm of ordinary proteins with recognition capability. Demonstration that antibodies, as well as other types of proteins, generate free radicals that can directly kill target cells lends support to the hypothesis that antigen neutralization preceded lymphocyte stimulation.

Recombinase-activating genes (RAG). Two recombinase activating genes (RAG 1 and RAG 2) initiate rearrangement of Ig and TCR loci by binding conserved sequences flanking each V, D, and J segment, and inducing double-stranded breaks in the DNA. In the absence of either RAG gene, Ig and TCR rearrangement and expression is abolished, and lymphocytes die early in their development. RAG homologues have been isolated in all jawed vertebrates, but either do not exist or are unrecognizable in jawless fish. The sudden appearance of RAG in Chondrichthyes, coupled with their peculiar genetic features (they lack introns and are adjacent in most species) suggests their origin as a retrotransposon, or a foreign DNA element introduced by a virus some 500 million years ago. The question remains whether this genetic "intruder" initiated the development of the entire adaptive immune system by interrupting an ancient gene between what became V and J segments.

T cell receptors (TCR). T cells play a central role in activating macrophages and B cells, as well as directly killing virus-infected cells. T cell receptors, like their B lymphocyte sisters, are generated by RAG-mediated recombination of subgenic segments within a single locus. Four separate loci (α, β, γ, δ) provide the genetic materials for the receptor chains on two different classes of T lymphocyte. γδ T cells originate in the lymphoid tissue of the gut and, much like membrane Igs, γδ TCRs directly bind to epitopes on protein and phospholipid antigens. The development of αβ T cells is tightly regulated, because their receptors uniquely survey intracellular antigens displayed in MHC receptors and the consequence of unleashing self-reactive cells is potentially fatal. αβ T cells arise from precursors in the thymus and proceed through a series of selective steps termed *thymic education*, which eliminate those cells with defective or self-reactive receptors.

Errors in the development and regulation of T and B cells can result in cancers of the immune system, such as leukemia and lymphoma. Double-strand breaks of the DNA during V(D)J recombination increase the opportunity for unintentional chromosomal translocations that disrupt expression of cell-cycle proteins, causing transformation of the cell to a malignant state. High rates of proliferation in stimulated T and B cell clones also increase the odds of replication errors that can cause a lymphoma.

The study of TCR evolution, like that of many immune system genes, is complicated by high rates of protein divergence and, for αβ TCR, by the additional selective constraint of binding MHC. Comparison of α,β,γ and δ genes reveals greater sequence divergence between vertebrate species than seen for Ig, but more conservation in the overall structure of TCR loci. In the search for an ancestral Ig/TCR gene, phylogenetic trees suggest different relationships depending on the region of comparison. Analysis of constant regions shows γδ sequences to be most ancient and supports the hypothesis that γδ T cells are closest to the primordial type of lymphocyte. The finding that γδ T cells recognize intact antigen suggests that antigen processing and MHC restriction were later acquired features of αβ TCR. Their origin in the gut, their response to a wider variety of antigens (especially conserved mycobacterial molecules and heat shock proteins), and their immediate effector response without B cell co-stimulation suggest that γδ T cells are a first line of defense—closer to innate mechanisms than part of the highly specific adaptive response. γδ T cells as the primordial lymphocyte may thus displace IgM as the candidate ancestral prototype and suggest that secretion is a derived feature of Igs.

Major histocompatibility complex (MHC). Major histocompatibility complex (MHC) denotes both the antigen-presenting receptors and the four-megabase genetic locus (named for its association with graft rejection or acceptance) that encodes these receptors. Two structurally related forms of MHC receptors present peptides to separate subsets of αβ T cells and so specify distinct immune-response pathways. Nearly all somatic cells display fragments of endogenously produced proteins in MHC I receptors to cytotoxic (CD4) T cells. Recognition of peptides in MHC I by TCR induces direct killing of the cell, since the presence of foreign peptides indicates viral infection. Specialized antigen-presenting cells (such as B cells, macrophages, and dendritic cells) possess a second type of MHC receptor that displays peptide fragments from endocytosed antigen. MHC II peptide complexes bound by helper (CD8) TCR initiate a cytokine-mediated B cell and macrophage response, which is effective against microorganisms and parasites.

The evolutionary constraints on MHC receptor evolution differ from those acting on Ig and TCR. MHC must be at once conservative and promiscuous—conservative enough to recognize invariant domains of many different TCRs and promiscuous enough to engage a multitude of peptides in 10^{18} different possible TCRs. Failure to preserve this balance results in the predisposition of certain MHC alleles toward autoimmune disease, such as type 1 diabetes. MHC is the most polymorphic genetic locus in vertebrates, both in the number and diversity of alleles. Multiple selective forces appear to maintain polymorphism of the MHC. Given the high rates of viral and bacterial evolution, MHC allelic diversity is crucial for preventing decimation of a host population by a pathogen. Evidence for associations between MHC type and mating preference or reproductive compatibility argues that there is also selection against homozygosity in the MHC. The genetic mechanisms generating MHC diversity in jawed vertebrates include gene conversion, mutation, and recombination.

The structure of the MHC locus documents a rich his-

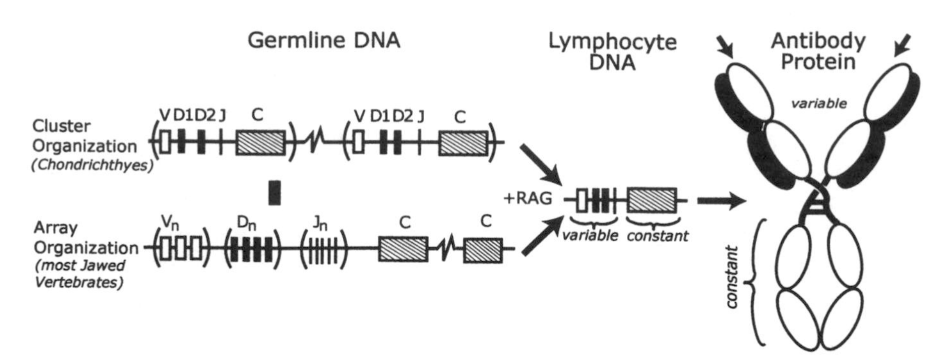

FIGURE 3. Genomic Organization of Immunoglobulin Genes.
The cluster organization of sharks and skates (top left) consists of preselected combinations of V, D, J, and C gene segments in the heavy chain (shown) and V, J, and C in the light chain (not shown). By contrast, in jawed vertebrates, tandem arrays of heavy chain Vs, Ds, Js, and Cs are found in a single genetic locus (bottom left) and Vs, Js, and Cs for the light chain (not shown). Both organizations are rearranged in B lymphocytes by RAG to produce a single Ig receptor (middle). The resulting secreted Ig, or antibody (right) is comprised of multiple immunoglobulin domains (denoted by ovals), heavy chain (white) and light chain (shaded), constant domain and variable domain, with antigen-binding sites denoted by arrows. Drawing by Diana J. Laird.

tory of the vertebrate genome and the adaptive immune system. Genes encoding class I and II receptors each span several hundred kilobases of a single chromosome in all jawed vertebrates except fishes; woven between receptor genes are many related immune genes plus other genes without known immune function. Sequence similarity and linkage between MHC class I and II receptors argues for their common origin from a primordial Ig-domain gene. However, the assembly of other genes in the MHC locus and the reasons for their continued linkage remain unresolved issues. Before the genomic era, Ohno posited two duplications of the entire genome during vertebrate evolution, based on gene number alone. Consistent with this, several paralogous regions were recently discovered in the human genome, each containing a stretch of genes with sequence similarity to many in the MHC, which is located on chromosome 6. The paralogous regions on chromosomes 1, 9, and 19 imply two rounds of partial or complete chromosomal duplication after the divergence of jawless fish and before the emergence of Chondrichthyes, although sequence analysis of each individual set of paralogs suggests a wider range of duplication times. The discovery of a candidate precursor of the MHC locus in the roundworm, *C. elegans*, which encodes homologues of nonimmune genes in the vertebrate MHC, may push back the existence of this region by several hundred million years. Furthermore, it suggests that MHC receptor genes arose within and became associated with an already ancient linkage group. Our nascent ability to study the structures of entire genomes promises to catapult the understanding of MHC evolution.

Architecture of the Adaptive Immune System. The molecular and cellular innovations that apparently arose in Chondrichthyian ancestors would be useless without a highly structured lymphoid system capable of sequestering different stages of cellular development and facilitating interactions between antigens and rare lymphocytes. It is therefore no coincidence that lymphoid tissue exists in all jawed vertebrates but is absent in Agnathans. The lymphoid system comprises specialized organs and tissues connected by a network of lymphatic capillaries that collect interstitial fluid and concentrate antigens in regional nodes. Primary lymphoid organs provide microenvironments that support immune cell development. The primary site of blood cell formation (the bone marrow in mammals, birds, and reptiles; the heart or kidney in amphibians and fish) supports continuous production of lymphoid and myeloid cell progenitors by stem cells. B lymphocyte development continues in the bone marrow in mice and humans, or proceeds in gut-associated lymphoid tissue in some mammals and birds, where self-reactive B cells are eliminated. T lymphocyte progenitors migrate to the thymus, a primary lymphoid organ next to the heart, where separate regions facilitate each stage of thymocyte development. Naive but mature lymphocytes then congregate in secondary lymphoid organs, such as the spleen, the lymph nodes, and follicles throughout the gut and mucosal tissues. As a "meeting place" for antigen and lymphocytes, secondary lymphoid tissues overcome the otherwise dismal odds of a lymphocyte encountering a reactive antigen. Selected lymphocyte clones become effector cells following co-stimulation by other cells in secondary lymphoid sites, then migrate to sites with contact to the outside world, sometimes called tertiary lymphoid tissues. Infected cells or invading pathogens are thereby destroyed at points of entry, such as wounds to the skin and mucous membranes, and memory cells are dispersed to patrol these sites. Trafficking of cells during develop-

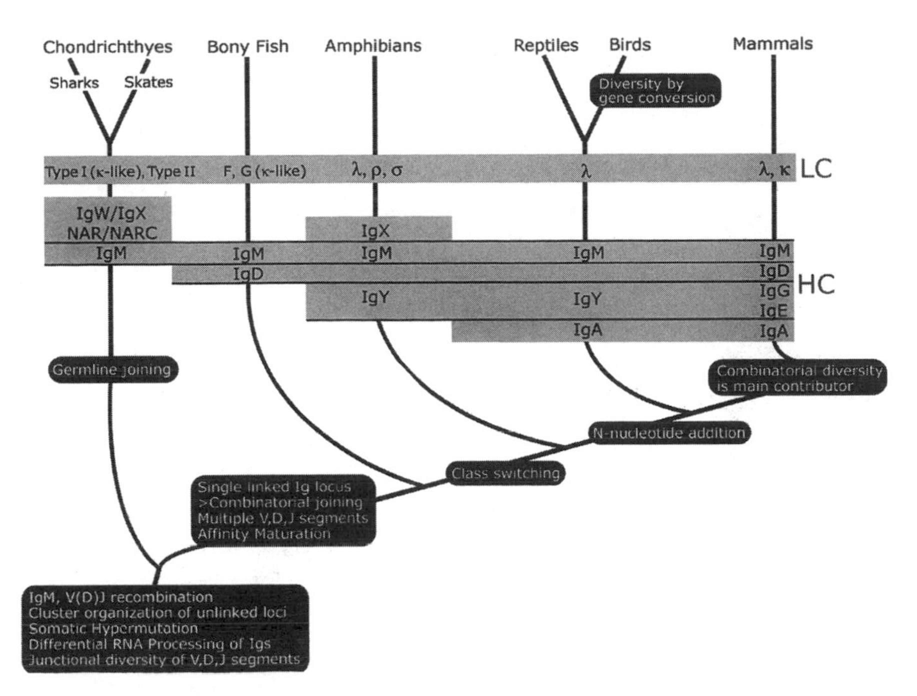

FIGURE 4. Immunoglobulin Evolution in Vertebrates.
Heavy chain (HC) constant-region splice variants, or isotypes, are compared between vertebrate subphyla, with homologues on the same line. Light chain (LC) genes are shown above, with homologues indicated in parentheses. Black boxes denote molecular innovations and their probable origin during phylogeny. Drawing by Diana J. Laird.

ment and immune responses is directed by a complex system of homing receptors that interact with vessel walls, chemotactic signals, and tissue components.

Immunology and the Study of Evolution. The idea of the immune system as a microcosm of evolution emerged early in the study of immunology, exerting influence over development of the field and seeding new paradigms. In a prescient 1955 paper, Niels Jerne applied the theory of natural selection to antibodies. Departing from previous models of antibody formation from antigen templates, Jerne proposed that organisms possess a priori a repertoire of antibodies complementary to all possible antigens. "Selection" of lymphocytic clones with reactive receptors by antigen then induces replication and differentiation. Clonal selection theory explains the observed improvements in the speed and quality of the antibody response between primary and secondary exposures to antigen. Competition among cell lineages within an individual is not limited to the immune system, but exemplifies an evolutionary mechanism at work in development, cancer, and cell migration. The study of coevolving pathogen and host immune response brings together the fields of evolution, immunology, and microbiology.

The origin of adaptive immunity can be approached from many angles. The genomic approach was first employed by Masanori Kasahara, who hypothesized that genome duplications in vertebrate ancestors paved the way for the molecular experimentation that gave rise to this system. Comparison of MHC loci across vertebrate species provides clues about how and when this chromosomal region of immune and nonimmune genes was assembled. The study of RAG structure and mechanism across vertebrates fuels the theory that receptor genes of the adaptive immune system first arose by transposon-induced shuttling of duplicated segments of DNA. Phylogenetic data estimate the appearance of RAG in a vertebrate ancestor about 500 million years ago, approximately coincidental with the last hypothesized genome duplication and the origins of TCR, Ig, and MHC receptors. Another strategy is the reconstruction of the hypothetical ancestral receptor through discovery of surviving remnants, such as nonrearranging immunoglobulin domain receptors in vertebrates and invertebrates alike. Candidates include the paired Ig-like receptors, whose role as both inhibitory and activating receptors for MHC I and MHC I-like molecules in mammals suggests their involvement in ancient self-nonself recognition. A number of other Ig-type nonrearranging antigen receptors, such as the NAR in sharks and the NITR in pufferfish, contribute to our understanding of receptor evolution. Finally, a different angle on the problem of immune system origin involves the broader study of histocompatibility systems, including MHC in vertebrates. The existence of allorecognition (the molecular identification of genetically different individuals of the same species) in plants, sponges, and fungi confirms it as a general phenomenon that has probably evolved multiple times. The study of allorecognition in colonial organisms seeded the idea that histocompatibility systems arose not for immunity to infection, but to protect the individual from invasion and outcompetition by allogenic cell lines. Given the possibility that early vertebrates were colonial, as are many extant protochordates, it is possible that the MHC survives as a vestige of ancestral colonialism.

A major force shaping the evolution of multicellular animals has been the need to protect the individual from the incessant barrage of pathogens, its own dangerously altered cells, and even parasitic lineages of cells from other members of the species. The work emerging from the confluence of immunology and evolution attests to the importance of cross-disciplinary thinking. The study of biology will be enriched by exploring the origin of adaptive immunity in vertebrates.

[*See also* Overview Essay *on* Darwinian Medicine; Acquired Immune Deficiency Syndrome, *article on* Dynamics of HIV Transmission and Infection; Developmental Selection; Disease, *articles on* Hereditary Disease *and* Infectious Disease; Emerging and Re-Emerging Diseases; Influenza; Malaria; Myxomatosis; Plagues and Epidemics; Transmission Dynamics; Vaccination; Virulence; Vertebrates.]

BIBLIOGRAPHY

Buss, L. W. *The Evolution of Individuality*. Princeton, 1987. Extensive discussion of competition among cell lineages in the immune system, cancer, development, and colonial organisms.

Buss, L. W., and D. R. Green. "Histoincompatibility in Vertebrates: The Relict Hypothesis." *Developmental and Comparative Immunology* 9 (1985): 191–201. A concise treatment of graft rejection in an evolutionary context.

DuPasquier, L., and G. W. Litman, eds. *Origin and Evolution of the Vertebrate Immune System*. Berlin and New York, 2000. See especially chapters by M. Kasahara; J. D. Hansen and J. F. McBlane; M. F. Flajnik and L. L. Rumfelt.

Janeway, C., P. Travers, M. Walport, and J. D. Capra. *Immunobiology: Immune System in Health and Disease*. 4th ed. London and New York, 1999. Comprehensive textbook with evolutionary and integrative perspectives on immunology.

Jerne, N. K. "The Natural-Selection Theory of Antibody Formation." *Proceedings of the National Academy of Science* 41 (1955): 849–857.

Laird, D. J., A. W. DeTomaso, M. D. Cooper, and I. L. Weissman. "Fifty Million Years of Chordate Evolution: Seeking the Origins of Adaptive Immunity." *Proceedings of the National Academy of Science* 97 (2000): 6924–6926.

Litman, G. W., M. K. Anderson, and J. P. Rast. "Evolution of Antigen Binding Receptors." *Annual Review of Immunology* 17 (1999): 109–147.

McDevitt, H. O. "Discovering the Role of the Major Histocompatibility Complex in the Immune Response." *Annual Review of Immunology* 18 (2000): 1–17.

Ohno, S. *Evolution by Gene Duplication*. Berlin and New York, 1970.

Parham, P., ed. Genomic Organization of the MHC: Structure, Origin and Function. *Immunological Reviews*. Vol. 167, 1999. Collection of articles provides a comprehensive view of MHC evolution.

Picker, L. J., and M. H. Siegelman. "Lymphoid Tissues and Organs." In *Fundamental Immunology*, edited by W. E. Paul. 4th ed. Philadelphia, 1999. Detailed review of the field of lymphocyte homing. Section on "Compartmentalization of the Immune System," which draws the connection between cellular development and lymphoid structures, is essential reading for any serious student of immunology.

— DIANA J. LAIRD

Microbial Countermeasures to Evade the Immune System

Pathogens—the viruses, bacteria, parasites (protozoa and helminths), and fungi that cause disease—represent a minority of the life-forms that make up the microbial world. They possess intrinsic and evolved characteristics to facilitate their potential to infect vertebrate hosts. The population sizes of pathogenic microbes exceed that of their vertebrate hosts by several orders of magnitude; an adult human harbors more microbes than there are people on the planet. In addition, the doubling time (the time required to double the size of the population) of most pathogens is orders of magnitude less than that of their hosts. Therefore, evolution of microbes is more rapid and is characterized by greater population diversity than is the case for their vertebrate hosts. The prevalence of pathogens accounts for the fact that any animal host encounters scores or perhaps hundreds of infections during a lifetime. The very existence and persistence of pathogens indicates their success in countering the innate and acquired host clearance mechanisms. Mutation and genetic recombination within large microbial populations that undergo rapid replication is a key factor in the evolution of pathogen fitness. Global control by humans of specific infections has been achieved (e.g., the eradication of smallpox), but in overall terms, it is hard to envisage how any major, sustained reduction in the totality of infections could occur, as new pathogens continually emerge over time.

The interaction of pathogens with their hosts can be considered in the context of several stages of the infectious process: transmission, colonization, dissemination, and tissue damage. The characteristics of particular pathogens depend on specific genes (virulence factors) that interact with the host. This mutuality of host–pathogen interactions is an essential consideration. To persist in a host population, the average number of new infections caused by any microbe must exceed one. This average is often referred to as the basic reproductive rate (R_0). Thus, a pathogen must not inflict too great a burden of damage or death upon its natural hosts, lest it be threatened with their extinction. To an extent, pathogen strategies tend toward balanced parasitism, in which the costs of infection do not impair host fitness and fecundity to the detriment of the microbe. However, because evolution is blind, within and between host selection may select for short-term microbial survival even if this leads to an evolutionary dead end for the pathogen.

Transmission and Colonization. Pathogens exploit inanimate and living vectors, through a mixture of opportunism and evolved mechanisms. For example, the propensity of pathogens to infect hosts via water supplies or to be transmitted by air currents or wind is probably opportunistic. However, some factors favoring transmission have evolved, for example, spore formation by bacteria or factors that facilitate pathogen spread, such as the induction of massive diarrhea by *Vibrio cholerae* or the induction of coughing and sneezing by rhinoviruses (the common cold). To take up residence in a host, a pathogen must negotiate the structural or anatomical barriers that sustain the integrity of its host. The interaction of a pathogen with the skin or mucous membranes of animal hosts involves opportunism and highly evolved tropism. Opportunistic (accidental) infection may arise following damage to a host integument (e.g., skin) through injury (e.g., the introduction of spores of *Clostridium* to cause tetanus). However, many pathogens have evolved specific determinants that target natural receptors of host cells. Epstein-Barr virus targets epithelial cells and B lymphocytes by engaging a specific human glycoprotein receptor, CR2. *Neisseria meningitidis*, a pathogenic bacterium, possesses filamentous proteins (pili) on its surface. Pili bind through a specific protein adhesin (PilC) to a human receptor (CD46) of respiratory tract epithelia. *Plasmodium falciparum* (one of the malaria parasites) expresses surface molecules encoded by genes (*var*) that engage with a variety of human receptors including ICAM-1. These ligand-receptor interactions exemplify not only evolved strategies for facilitating interactions with the host but define the basis of host and tissue tropism. In many instances, the targeting of a particular host receptor has evolved because it transduces signals to host tissues that favor microbial persistence, for example, endocytosis, so that the pathogen is taken up into the cell where it can multiply in a protected site. Entry into cells is an obligatory requirement for all viruses, but it is also crucial to the life cycle of cell wall–deficient organisms, such as *Mycoplasma* spp., and obligate intracellular bacterial pathogens, such as *Listeria monocytogenes* and *Mycobacterium tuberculosis*. Parasites such as *Leishmania* spp. must invest in strategies that accommodate the dual requirements of a complex life cycle. This requires separate mechanisms of tissue tropism for the digestive tract of their insect vector (sandflies) and monocytic cells of their mammalian hosts. Some pathogens have evolved mechanisms for subverting the functions of host cells to facilitate their entry. Pathogenic *Salmonella* spp. elaborate tyrosine phosphatases, structural mimics of host enzymes, that the bacterium injects into host cells and that paralyze the antibacterial activity of macrophages.

Evasion of Immunity. Once pathogens have negotiated epithelial surfaces, they encounter a myriad of host defenses. These include the innate immune mechanisms, such as phagocytic cells, complement proteins, and NK (natural killer) cells. Pathogens must also contend with sophisticated acquired repertoires of T (thymus-dependent) and antibody (B) cells. There are many examples of specific pathogen factors that induce or exploit deficiencies in the specialized functions that make up the innate immune responses that have evolved to counter pathogens. These are not the result of a global impairment of immune responses and a consequent lowering of resistance to all microbes. Rather, species-specific antigens have evolved functions that promote persistence of a particular pathogen. Pathogenic bacteria are particularly prone to attack by host phagocytes, primarily macrophages and granulocytes. These host cells, aided by host proteins that facilitate the engulfment of invading microbes (opsonization), are thwarted by a number of countermeasures. For example, many pathogenic bacteria possess surface polysaccharide capsules. These are typically polymers of repeating sugar units, for example, sialic acid. The negative charge and hydrophilic properties of capsules combat efficient phagocytosis. In addition, capsular polysaccharides interfere with opsonization in which host proteins, such as the complement pathway, facilitate phagocytosis. In the case of sialic acid, there is selective binding of a key protein (factor H) that hinders the assembly of the active complement components. The parasite *Toxoplasma gondii* has a different strategy. It prevents the orderly process whereby the engulfed parasite is killed within specialized compartments (endosomes) of the phagocyte. The parasites then take advantage of the cell by multiplying in it. More ruthless still, *Yersinia enterocolitica*, a bacterium closely related to the infamous plague pathogen, injects proteins (Yops) into nearby phagocytes to inhibit and even kill them. The specialized apparatus or injectosome used is a striking example of a sophisticated virulence factor. Many different bacterial pathogens (e.g., salmonella, shigella, and bordetella) have acquired similar versions of this large region of DNA (enough to code for about fifty genes), known as the type 3 system. Similar to a flagellar basal body, this secretion apparatus makes contact with the host cell, fuses with it, and injects preformed pathogen molecules through a pore.

Once established in the short term, pathogens are confronted with host clearance mechanisms that have been triggered by their presence in the form of antibodies and specialized effector lymphocytes (T cells) that specifically target them. Pathogens once again have evolved numerous mechanisms to prevent their elimination. For example, rubella virus infects the human fetus at a time when the T-cell response is especially weak, thereby enabling the virus to persist during pregnancy and for long periods postnatally. Some antigens fail to elicit a strong immune response by exploiting genetically determined deficiencies in the repertoire of germline genes encoding the antigen binding sites of immunoglobulin molecules or T-cell receptors. When a special category of virulence factors, for example, the bacterial toxin superantigens of *Staphylococcus aureus*, encounter thymic lymphocytes during development, there is elimination of specific clones resulting in an absence of T cells with specific V beta clones. During infections with the cryptococcus, antigens of this pathogenic fungus induce a state of anergy in which the T-cell responses are diminished, despite a rigorous antibody response, so that elimination of the fungus is impaired. Many microbial surface antigens are structurally identical to host molecules, and it has been proposed that this similarity marks them as "self," thereby deceiving the immune response into overlooking the pathogen. An example of this mechanism is the scrapie agent in sheep (the protein that cause the spongiform encephalopathy disease called scrapie) that shares a similar amino acid sequence to the host prion protein, PrP. Some pathogens suppress responses to its antigens, typically through effects on dendritic cells, macrophages, or lymphocytes. For example, certain cell surface phenolic glycolipids of *Mycobacterium leprae* induce suppressor cells that in turn inhibit the responses of other T cells that specifically recognize these antigens.

These examples of pathogen strategies should not detract emphasis from an appreciation of the dynamic nature of host–microbe interactions. This mutuality has been captured in the metaphor of the "gene for gene arms race." To survive and propagate, all organisms must maintain their fitness in diverse and changing environments, and pathogens have therefore evolved mechanisms for responding to such changes. The capacity of bacterial pathogens to run the gauntlet of the differing environments of their hosts is remarkable, especially because infections may involve clonal expansion of a single bacterial cell. This adaptive potential ultimately depends on gene regulation and gene variation.

Pathogens respond to host environmental factors by regulating virulence determinants and by controlling their growth rate. When the animal bacterial pathogen *Salmonella typhimurium* is present in the intestinal lumen, it regulates coordinately six different genes required for invasion, each of which is dependent on specific environmental cues, including oxygen, osmolarity, and pH. This coordination of virulence genes in different locations is achieved through evolving regions of DNA that recognize the same regulatory proteins (regulon). Expression of virulence factors may depend on the microbial population attaining a critical density to trigger the elaboration of bacterial cell-to-cell signaling molecules such as N-acetyl-L-homoserine lactone in *Pseudomonas aeruginosa. Staphylococcus aureus* uses a global regulator, which is activated by secreted autoinducing peptides, to control the expression of its major virulence genes. The autoinducing peptides show sequence variation that affects their specificity. These peptides are thiolactones that either activate or inhibit gene expression depending on the conformation of the ligand-receptor interaction.

Antigenic Variation. Microbes encounter selective landscapes of such diversity that the prescriptive mechanisms of gene regulation may be inadequate to encompass the plethora of host factors that have evolved to eliminate them. How, then, do the relatively small numbers of organisms that make up a potentially infectious inoculum generate the necessary diversity to adapt and thereby to evade these multifarious host clearance mechanisms? One strategy is through the evolution of an increased mutation rate of those genes that are involved in critical interactions with their hosts. In many RNA viruses, high mutation rates are an intrinsic property and are a major factor in the extraordinary diversity that is characteristic, for example, of the human immunodeficiency virus. In parasites such as trypanosomes and malaria, this capacity for variation is highly evolved and is mediated through gene conversion of hypermutable DNA sequences located in telomeres. In many bacterial pathogens, antigenic switching has also evolved and is mediated by hypermutable loci through DNA transposition, inversion, recombination, or slippage during replication. The combinatorial phenotypic effects of the independent switching of just a few such loci in a pathogen genome, given that mutation occurs at random, can generate substantial diversity within a pathogen population. These genetic elements have been called contingency loci to emphasize their potential to enable at least a few organisms in a given pathogen population to adapt to unpredictable and precipitous changes in the host environment, while minimizing deleterious effects on pathogen fitness, that would arise if mutation rates were elevated throughout the genome.

The key points can be summarized as follows. The large population sizes and rapid replication of microbes generate enormous diversity within microbial populations. This diversity and its rapid evolution, shaped by selection, is critical to the potential of microbes to thrive in virtually every environmental niche on our planet, including vertebrate hosts. Host-microbial interactions are characterized by a dynamic, coevolutionary state in which the fittest characteristics of host and microbe typically prevail through natural selection. Among microbes, population diversity is such that every opportunity for exploitation of the host has had a high probability of occurring at least once, if not many times. Thus, strategies of evasion, stealth, mimicry, and sequestration have evolved to counter the powerful deterrents and clearance mechanisms of immunity (innate and acquired) and the genetically determined variation of hosts. These microbial strategies for perpetuating themselves sometimes result in damage to host tissues. Microbes with this propensity to cause disease are referred to as pathogens. Although symbiotic microbial-host interactions are the rule, diseases caused by microbes are the most important cause of death and disability in animals on our planet.

[*See also* Acquired Immune Deficiency Syndrome, *article on* Dynamics of HIV Transmission and Infection; Bacteria and Archaea; Basic Reproductive Rate (R_0); Coevolution; Disease, *article on* Infectious Disease; Influenza; Malaria; Mutation, *article on* Evolution of Mutation Rates; Protists; Vaccination; Virulence; Viruses.]

BIBLIOGRAPHY

Anderson, R. M., and R. M. May, eds. *Population Biology of Infectious Diseases.* Berlin, 1982. A review of the extent to which pathogens regulate natural populations (including human) under the general headings of impact, transmission, control, and coevolution.

Bajaj, V., R. L. Lucas, C. Hwang, and C. A. Lee. "Coordinate Regulation of *Salmonella typhimurium* Invasion Genes by Environmental and Regulatory Factors Is Mediated by Control of *hilA* Expression." *Molecular Microbiology* 22 (1996): 703–714. A description of experiments on cultured cells to demonstrate that coordinate regulation of several virulence genes occurs in response to environmental variables.

Berry, J. D., and R. A. McCulloch. "Antigenic Variation in Trypanosomes: Enhanced Phenotypic Variation in Eukaryotic Parasite." *Advances in Parisitology* 49 (2001): 1–55. A comprehensive review of antigenic variation in parasites, with a major emphasis on trypanosomes.

Ewald, P. W. *Evolution of Infectious Disease.* New York, 1994. A stimulating book that focuses on how evolutionary principles must be applied to an understanding of pathogens and their control.

Fuqua, C., S. C. Winans, and E. P. Greenberg. "Census and Consensus in Bacterial Ecosystems: The *luxR–LuxI* Family of *Quorum-sensing* Transcriptional Regulators." *Annual Review of Microbiol.* 50 (1996): 727–751. A review of the mechanisms by which bacterial populations at high density coordinate their behavior through specific cell-to-cell signaling mechanisms.

Garrett, L. *The Coming Plague.* New York, 1995. A comprehensive and scholarly book outlining the dangers posed by past, present, and emerging infectious diseases.

Groisman, E. A., ed. *Principles of Bacterial Pathogenesis.* New York, 2001. An excellent multiauthor volume providing in-depth chapters on the molecular basis of infectious diseases.

Mims, C. A., A. Nash, and J. Stephen. *Mims' Pathogenesis of Infectious Disease.* 5th ed. London, 2001. A classic reference book on the myriad ways in which pathogens cause infection with particular emphasis on the diversity of strategies used by both host and microbes.

Moxon, E. R., and P. A. Murphy. "*Haemophilus influenzae* Bacteraemia and Meningitis Resulting from Survival of a Single Organism." *Proceedings of the National Academy of Science USA* 75 (1978): 1534–1536. An experimental study that demonstrates how all of the infecting organisms involved in the infectious process were derived from one founder bacterium.

Moxon, E. R., P. B. Rainey, M. A. Nowak, and R. E. Lenski. "Adaptive Evolution of Highly Mutable Loci in Pathogenic Bacteria." *Current Biology* 4 (1994): 24–33. A review of how bacteria evolve hypermutable loci (contingency genes) to facilitate adaptation to their host environment.

Moxon, R., and C. Tang. "Challenge of Investigating Biologically Relevant Functions of Virulence Factors in Bacterial Pathogens." *Philosophical Transcripts of the Royal Society of London* 255 (2000): 643–656. A review of the molecular microbiology of pathogens with emphasis, on *in vivo* mechanisms.

Rosebury, T. *Life on Man.* New York, 1969. An instructive and popular account of the invisible world of microbes that live on and in human body.

Weiss, R. A. "How Does HIV Cause AIDS?" *Science* 260 (1993): 1273–1279. A concise but authoritative version of the key biological characteristics of the pathogenicity of human immunodeficiency virus.

— E. RICHARD MOXON

INBREEDING

Inbreeding is mating between relatives. This produces offspring that are more likely to be homozygous (i.e., to have the same allele at both copies of a locus) than individuals produced by random mating. This is because inbred offspring are likely to receive two alleles descended from the same ancestor in the recent past. Inbreeding has two main direct effects: an increase in the homozygosity (the probability that the two alleles at the same genetic locus are identical) and, commonly, a reduction in fitness of inbred relative to outbred individuals called inbreeding depression. Inbreeding depression is one of the oldest observations related to evolutionary biology; perhaps as a result, most human societies have prohibited close inbreeding. Charles Darwin spent a lot of effort on experiments documenting inbreeding depression, motivated in part by the health difficulties of his own children. (Darwin, the grandson of Josiah Wedgwood, had married Emma Wedgwood, his first cousin.)

Inbreeding depression is usually thought to arise from two mechanisms termed dominance and overdominance (Charlesworth and Charlesworth, 1987). Alleles with low fitness can be maintained in a population by recurrent mutation. If these deleterious alleles are recessive, then the mean fitness of inbred individuals with a greater proportion of homozygous loci will be less than the mean fitness of outbred individuals. This is the dominance model of inbreeding depression. If, instead, the heterozygote is more fit than either homozygote, then both alleles can be maintained in the population by selection alone. Again, inbreeding will increase the proportion of homozygous loci, and the mean fitness would be reduced relative to outbred individuals. This is the overdominance model of inbreeding depression. Both of these mechanisms have been shown to occur, but in general the dominance model is thought to cause a greater proportion of inbreeding depression (Charlesworth and Charlesworth, 1987).

Inbreeding is measured by means of Sewall Wright's inbreeding coefficient, F. F measures the probability that two alleles in the same individual are identical by descent, that is, the probability that they derive from the same ancestral allele in the recent past. Figure 1 shows a simple example. In general, the larger the inbreeding coefficient of an individual, the more homozygous the genotype of that individual is expected to be. In randomly mating populations, the frequencies of the three genotypes at a locus with two alleles at frequencies p and q are expected to be p^2 and q^2 for the two homozygotes and $2pq$ for the heterozygote. (These are often called the Hardy–Weinberg frequencies, after their co-discoverers.) In a population that has inbred individuals, these frequencies are expected to be $p^2(1 - F) + pF$ and $q^2(1 - F) + qF$ for the two homozygotes and $2pq(1 - F)$ for the heterozygote. Inbreeding alone does not change the allele frequency but can do so indirectly because it changes the nature of selection.

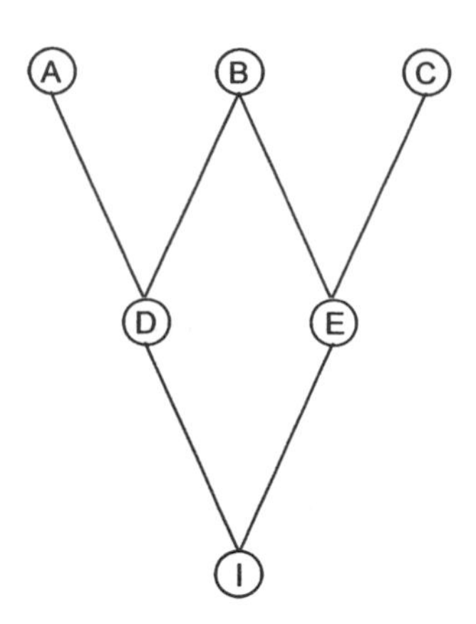

FIGURE 1. Estimating the Inbreeding Coefficient (F) from the Pedigree of a Set of Individuals.
Individual I at the bottom of this pedigree is inbred, having parents who are half-siblings (they share a parent marked B). If A, B, and C are unrelated to each other, the probability that the two alleles in individual I are identical by descent is $F_I = 1/8\ (1 + F_B)$, where F_B is the inbreeding coefficient of individual B. This can be calculated as follows: there is a probability of 1/2 that the allele that I gets from E is the allele that E received from B, the chance that D received the same allele from B as did E is $1/2 + F_B/2$, and the chance that I received the same allele from D as D got from B is also 1/2. Multiplying these together gives the answer above. Drawing by Michael C. Whitlock.

There are multiple ways to estimate the inbreeding coefficient of a natural population. First, F can be calculated directly from the pedigree of a set of individuals (Wright, 1969). This is most accurate, but it requires information that is not always available. Second, F can be inferred from the data about homozygosity of marker loci, by use of the equations in the preceding paragraph. Though much easier, this method is subject to great error when applied to estimating inbreeding coefficients of individuals because the homozygosity of a small set of loci does not predict very well the homozygosity of an individual (Chakraborty, 1981).

If individuals are likely to remain near their birthplace, then a spatial genetic pattern will emerge in which individuals near each other are more likely than average to be related to each other. Thus, genetic differences can emerge by chance between different subpopulations of the same species. Furthermore, the inbreeding coefficient can be subdivided into a term that describes the nonrandom mating that occurs as a result of local mating between individuals within the same population and a term that describes the inbreeding within local populations. The deviation from homozygosity that results can be described using the term F_{ST}, and the deviations from the Hardy–Weinberg genotype frequencies of p^2, $2pq$, and q^2 within a subpopulation are described by F_{IS}.

The main importance of inbreeding to evolution is inbreeding depression. Inbreeding depression can affect the persistence time of a population (Mills and Smouse, 1994; Saccheri et al., 1998) and therefore is potentially important in conservation biology. Inbreeding depression is also very important in the evolution of mating systems; the evolution of selfing (relatively common in plants) may depend on the inbreeding depression being less than a critical value. Finally, the avoidance of inbreeding may drive the evolution of dispersal.

[*See also* Genetic Drift.]

BIBLIOGRAPHY

Barrett, S. C. H., and D. Charlesworth. "Effect of a Change in the Level of Inbreeding on the Genetic Load." *Nature* 352 (1991): 522–524.

Chakraborty, R. "The Distribution of the Number of Heterozygous Loci in an Individual in Natural-Populations." *Genetics* 98 (1981): 461–466.

Charlesworth, D., and B. Charlesworth. "Inbreeding Depression and Its Evolutionary Consequences." *Annual Review of Ecology and Systematics* 18 (1987): 237–268.

Darwin, C. *The Effects of Cross and Self Fertilisation in the Vegetable Kingdom*. London, 1876.

Falconer, D. S. *Introduction to Quantitative Genetics*. Essex, England, 1981.

Keller, L. F., P. Arcese, J. N. M. Smith, et al. "Selection against Inbred Song Sparrows during a Natural-Population Bottleneck." *Nature* 372 (1994): 356–357.

Mills, L. S., and P. E. Smouse. "Demographic Consequences of Inbreeding in Remnant Populations." *American Naturalist* 144 (1994): 412–431.

Saccheri, I., M. Kuussaari, M. Kankare, P. Vikman, W. Fortelius, and I. Hanski. "Inbreeding and Extinction in a Butterfly Metapopulation." *Nature* 392 (1998): 491–494.

Thornhill, N. W., ed. *The Natural History of Inbreeding and Outbreeding: Theoretical and Empirical Perspectives.* Chicago, 1993.

Wright, S. *Evolution and the Genetics of Populations*, vol. 2, *The Theory of Gene Frequencies.* Chicago, 1969.

— MICHAEL C. WHITLOCK

INCLUSIVE FITNESS

Why do the plants in a certain populations grow to a particular height (on average) and not to another? What determines the relative numbers of male and female offspring in a population? How should an individual allocate resources between growth and fecundity, or between foraging and defense, or between her own offspring and those of a sister? What proportion of an individual's offspring should disperse and attempt to breed far from home? These are examples of the questions we consider here, and the answers we seek are evolutionary—that is, we want to argue that "deviant" individuals with alternative traits must have lower fitness than "normal" individuals who exhibit the established trait. For example, we might try to find evidence that a plant that was shorter than average would receive less sunlight or disperse less pollen, whereas a plant that was taller than average would consume too much energy in growth or would take too long to attain maturity and would have less reproductive success.

There is an extra complication that enters here. It is possible that by growing a bit taller, the individual would get more offspring, but, overall, the trait (of increased height) would be at a disadvantage. The reason is that the neighbors of the taller individual would have slightly less fitness, and, because of limited dispersal, they might be related to the taller individual and thus share the "tallness" genes. As a result, this would contribute a negative fitness force to such genes. In short, if what is good for the individual is bad for the group, and if there is some within-group genetic similarity, the evolutionarily stable level of the trait will be a compromise between the two. We can look at it another way too—at the evolutionarily stable state (ESS), a shorter individual could have less personal fitness, if its relatives had more. In this context, tallness would be called "selfish" and shortness would be called "altruistic," and the ESS represents a balance.

The central problem in modeling the evolution of social characteristics, those whose behavior affects the fitness of neighbors, is to find a good way to account for the effect of these interactions. Here we will describe a remarkable approach to this problem known as the inclusive fitness method. This method gives us an expression that can tell us whether or not a deviant behavior will increase in frequency. The analysis applies to a wide range of characteristics, but for concreteness, we will work here with our example of plant height.

As a simple illustration, suppose that the height deviation is caused by a mutation at a certain locus that alters its owner's height, and thereby changes the fitness of the owner and of exactly one neighbor by the amounts b_0 and b, respectively. To calculate the resulting change in the frequency of the mutant allele in the next generation, we must weight these two fitness changes by the "effectiveness" of the owner and the neighbor in propagating the mutant allele, that is, by its frequency in their gametes (p_0 and p, respectively). It would seem that the allele should increase in frequency exactly when $p_0b_0 + pb$ is positive. This is the simple idea behind the inclusive fitness method, and with some elaboration, it works.

If we take the above alteration to be a decrease in height, then the fitness change to the focal individual is actually a cost and could be written as $b_0 = -c$. Substituting $b_0 = -c$ into the formula and solving for c, the condition $p_0b_0 + pb > 0$ for the mutant allele to increase in frequency can be written $rb > c$, where $r = p/p_0$ is called the relatedness between owner and neighbor. This formula is known as Hamilton's rule. Hamilton's rule tells us that an altruistic act is selectively favored if the benefit to the recipients, weighted by the relatedness of the altruist to the recipients, exceeds the cost to the altruist. If relatedness is high, even costly, altruistic acts can be favored. But if relatedness is low, altruism is less likely to spread through a population.

We now look more carefully at the above calculation. Let us take a simple population structure, and suppose that individuals occur in patches of five, and that an individual's behavior affects only those on the same patch. Suppose that a single deviant individual changes its own fitness by b_0 and changes the fitness of its ith patchmate by b_i ($1 \leq i \leq 4$), where we use some suitable fitness measure such as contributions to the effective gamete pool of the next generation. Let this gamete pool contain a total of T gametes, D of which are mutant.

Now select a random copy of the mutant allele, which we will suppose causes its owner, the focal individual, to be deviant. We wish to calculate the change in frequency of the mutant allele brought about by this one deviant individual. To do this, we need to know the frequency of the mutant allele in each of the five inhabitants of the patch. Let these frequencies be p_i ($0 \leq i \leq 4$) and let $\bar{p} = D/T$ be the average frequency of the mutant allele in the population as a whole. Then the changes in D and T are

$$\Delta D = \Sigma p_i b_i \quad (1)$$

$$\Delta T = \Sigma b_i, \quad (2)$$

where Δ means "change in" and Σ is "the sum of" and the sums are over all five inhabitants of the patch ($0 \leq i \leq 4$). We have made an assumption here of "fair meiosis"—that the gametes produced by any individual contain its alleles in the same proportion as found in its genotype.

Next, we take the average of equations (1) and (2) over all copies of the mutant allele. We use E to mean expectation or average of a quantity. To get the final expression in equation (3), we employ a technical assumption that the b_i and p_i variables are uncorrelated.

$$\mathrm{E}(\Delta D) = \mathrm{E}(\Sigma p_i b_i) = \Sigma \mathrm{E}(p_i)\, \mathrm{E}(b_i) \qquad (3)$$

$$\mathrm{E}(\Delta T) = \Sigma \mathrm{E}(b_i). \qquad (4)$$

For the average in equations (3) and (4) to make sense, there must be a natural way of assigning to each focal individual an $i = 1$ patchmate and an $i = 2$ patchmate, and so on. This is essentially an assumption of homogeneity of the population, that each individual has the same spatial cluster of patchmates, or neighborhood configuration of relatives. For example, the five individuals might be equally spaced around a circle, (I_0, I_1, I_2, I_3, I_4) so that each individual (I_0) has two "near" neighbors (I_1 and I_4) and two far neighbors (I_2 and I_3).

If the mutant allele happens to be rare (as is often assumed), an increase in numbers will correspond to an increase in frequency, and we can expect the sign of $\mathrm{E}(\Delta D)$ to predict frequency change. When the mutant allele is not rare, the average change in frequency must be calculated as

$$\begin{aligned}\Delta \bar{p} &= \frac{D + \mathrm{E}(\Delta D)}{T + \mathrm{E}(\Delta T)} - \frac{D}{T} = \frac{D + \Sigma \mathrm{E}(p_i)\, \mathrm{E}(b_i)}{T + \Sigma \mathrm{E}(b_i)} - \bar{p} \\ &= \frac{\Sigma[\mathrm{E}(p_i) - \bar{p}]\, \mathrm{E}(b_i)}{T + \Sigma \mathrm{E}(b_i)} \\ &= \frac{\mathrm{E}(p_0) - \bar{p}}{T + \Sigma \mathrm{E}(b_i)} \left[\Sigma \frac{\mathrm{E}(p_i) - \bar{p}}{\mathrm{E}(p_0) - \bar{p}}\, \mathrm{E}(b_i) \right].\end{aligned} \qquad (5)$$

The expression in the square brackets is called the inclusive fitness effect of the deviant behavior. It can be expressed as

$$\Delta W_{\mathrm{IF}} = \Sigma r_i\, \mathrm{E}(b_i), \qquad (6)$$

where

$$r_i = \frac{\mathrm{E}(p_i) - \bar{p}}{\mathrm{E}(p_0) - \bar{p}} \qquad (7)$$

is the average relatedness of the focal individual to its ith neighbor. Note that $i = 0$ is one of the summands in equation (4) (as the focal individual is one of those affected by the deviant behavior), and the relatedness r_0 of the focal individual to itself is 1.

According to equation (5) (assuming $\mathrm{E}(p_0) > \bar{p}$), the inclusive fitness effect has the same sign as the average allele frequency change due to the deviant behavior of a focal individual, where an individual is selected to be focal according to its mutant allele frequency p_0. For example, in a diploid population, an individual with two mutant alleles will have twice the probability of being selected as an individual with one mutant allele.

Now recall that we have assumed that only one mutant copy acts at a time. But what happens if we allow them all to act together? First of all, we note that there is no reason to expect the inclusive fitness effect necessarily to represent the resulting change in allele frequency, because it contains no information about the fitness effect of simultaneous deviant behavior from several individuals, and that can certainly make a difference to allele frequency change. For example, for a trait such as a warning call, it might be that an expression by several individuals serves the group no better (and might even do worse) than a single call. On the other hand, a complex task requiring cooperation might have little effect if performed by one, but a large effect if performed by two. Some assumption is required and the right one turns out to be additivity of effect. For example, if a deviant height in one individual increases the fitness of the actor by 5 percent and decreases the fitness of the four others in the patch by 3 percent each, then a deviant height in two individuals should increase the fitness of each actor by 2 percent ($2 = 5 - 3$) and decrease the fitness of the three others in the patch by 6 percent each.

The Fundamental Theorem of Inclusive Fitness. Suppose we are studying the effect of selection on a behavioral or physical trait determined at a single locus. Suppose we want to measure the fitness of a mutant allele at that locus which codes for a deviant behavior. With this assumption of additive fitness effects between actors, the mutant allele will increase in frequency when the inclusive fitness effect defined in equation (6) is positive, and will decrease when it is negative.

This is a remarkable result. It tells us that we can calculate whether the deviant behavior will increase or decrease in frequency by adding up the effects of an "average" deviant actor on the fitnesses of all affected individuals, each affected individual weighted by its relatedness to the actor. Its applicability is much wider than our simple example might suggest. The genius behind this result is that of the late British biologist W. D. Hamilton in a 1964 paper. It is no exaggeration that the appearance of this result revolutionized the evolutionary modeling literature, and it is a measure of the depth of the result that this revolution has unfolded gradually over the past forty years, each insight into the power of the method giving rise in due course to others. This "power" is twofold—conceptual and computational. First of all, expression (6) provides a heuristic that gives us a powerful way to analyze behavior and make intuitive guesses. Second, it gives us a way to make calculations when a straightforward accounting of all the different possible local (neighborhood) configurations in which an allele might find itself is completely infeasible.

To implement the fundamental theorem we have to be able to calculate the fitness effects b_i and the relatedness coefficients r_i, and we look at these one at a time.

The Fitness Effects. Typically, there are different kinds of interactions between actors and potential recipients, some with a greater effect than others. To model this situation, we often group interactions into categories and express fitness in terms of average behavior in a category. As a simple example, suppose the fitness $W = W(x, y, z)$ of a plant depends on its own height x, the average height y of its two near neighbors, and the average height z of its two far neighbors. Here we have three categories, the focal individual and its near and far neighbors. Note that we expect W to increase with x and decrease with y and z.

Now, W will not usually be linear in x, y, and z, but we can always get the additive effects required by the inclusive fitness method (at least approximately) by assuming that the deviations in height are small enough that a differential approximation is valid, that is, W can be approximately by a linear expression. If we let x^* be the established height in the population, then an individual whose x, y, and z values are $x^* + dx$, $x^* + dy$, and $x^* + dz$ will, for small deviations, dx, dy, and dz, have approximate fitness

$$W = W^* + \frac{\partial W}{\partial x}\,dx + \frac{\partial W}{\partial y}\,dy + \frac{\partial W}{\partial z}\,dz$$

where W^* is fitness in a x^* population. Here the partial derivatives are evaluated at $x = y = z = x^*$. For example, $\partial W/\partial y$ provides the rate at which a plant's fitness changes with the average height of its two near neighbors.

Now fasten attention on a focal individual I_0 with height $x^* + dx$, all others having normal height x^*. Applying the above formula for W, its fitness effects on the five individuals of the neighborhood are

on itself:

$$b_0 = \frac{\partial W}{\partial x}\,dx \tag{8}$$

on each near neighbor:

$$b_y = \frac{\partial W}{\partial y}\,dy = \frac{\partial W}{\partial y}\frac{dx}{2} \tag{9}$$

on each far neighbor:

$$b_z = \frac{\partial W}{\partial z}\,dz = \frac{\partial W}{\partial z}\frac{dx}{2} \tag{10}$$

For example, to get (9), we want to calculate the fitness of a near neighbor, say I_1. Of the five individuals in the neighborhood of I_1, only I_0 is deviant, and its deviation is dx. Since I_0 is one of two near neighbors of I_1, the average near-neighbor deviation for I_1 is $dy = dx/2$. The inclusive fitness effect is:

$$\begin{aligned}\Delta W_{IF} &= b_0 + 2r_y b_y + 2r_z b_z \\ &= \left[\frac{\partial W}{\partial x} + \frac{\partial W}{\partial y}\,r_y + \frac{\partial W}{\partial z}\,r_z\right]dx,\end{aligned} \tag{11}$$

where r_y and r_z are the relatednesses of the focal individual to near and far patchmates. Note that equation (6) has five summands but equation (11) has only three—as it is over categories of individuals grouped according to like effects, with two near neighbors, $2r_b b_y$, and two far neighbors, $2r_z b_z$.

Interestingly enough, the expression in the square brackets reminds us of the expression we would get for dW if we treated y and z as functions of x and used the standard chain rule of calculus to expand the derivative:

$$dW = \frac{dW}{dx}\,dx = \left[\frac{\partial W}{\partial x} + \frac{\partial W}{\partial y}\frac{dy}{dx} + \frac{\partial W}{\partial z}\frac{dz}{dx}\right]dx. \tag{12}$$

This provides a useful heuristic for the inclusive fitness calculation—treat the average phenotype y of each category as a function of focal phenotype, take the total derivative of the fitness function with respect to focal phenotype, substitute $dy/dx = r_y$ and $dz/dx, = r_z$ as the category-specific relatednesses, and we get an expression for inclusive fitness.

Example: Parasite Virulence. Completely analogous to the plant height model is the question of the level of aggressiveness of a parasite within a host—too low, and it cannot compete against others; too high, and the life of the host and the group of parasites as a whole is endangered. A simple model capturing this tension between within-group competition and prudent use of resources is

$$W(x,y) = \frac{x}{y}\,G(y),$$

where x determines individual competitiveness, y is the average value of competitiveness within the group, and $G(y)$ measures the resources available to the group, which is assumed to be a decreasing function of y. Taking $dx = 1$, equation (11), when simplified, gives us

$$\Delta W_{IF} = \frac{\partial W}{\partial x} + \frac{\partial W}{\partial y}\,r = \frac{G(x^*)}{x^*}(1 - r) + G'(x^*)r. \tag{13}$$

Here, r is the average relatedness of any individual to all members of the group as a whole including to focal individual, and everything is evaluated at $x = y = x^*$. The first term of equation (13) treats the total resources G as fixed and measures the gain to the focal individual of its increased competitiveness, less the amount that must therefore be lost by a random group member, and the second term measures the reduction in total resources available (note that G' is negative). A simple model for G is $G(z) = 1 - cz$, where c is the rate at which average group competitiveness reduces available resources. If we put this into equation (13), and set the expression to zero, we get the ESS to be $x^* = (1 - r)/c$. Thus, the model predicts that higher relatedness decreases the competitiveness or virulence of individuals.

Relatedness. Sometimes we may be able to measure r empirically. We can then use the measured value directly in a formula such as $x^* = (1 - r)/c$. Alternatively, we may wish to calculate r based on theoretical assumptions about population structure. The theoretical calculation of relatedness (7) requires that we find the $E(p_i)$, the average mutant frequency in the genotype of the ith neighbor of a random focal allele. Note that, because this average is taken over all mutant alleles and a focal individual is deviant in proportion to its number of mutant alleles, we can interpret $E(p_i)$ as the probability that the ith neighbor of an individual of deviant height is deviant.

The calculation of the $E(p_i)$ may seem to require complete knowledge of the local distribution of the mutant allele. It is remarkable that we can usually manage with a simple recursive argument based on the concept of identity by descent. Two genes are said to be identical by descent (IBD) if they have a common ancestor (with no intervening mutation) some number of generations back. It turns out that, under mild technical assumptions, two genes chosen in a pair of interactants will be either IBD or probabilistically independent.

Next, choose a random mutant copy and choose a copy in the ith neighbor of its owner ($0 \le i \le 4$). Note that $E(p_i)$ is the average probability that the second copy is mutant. Let g_i be the probability that the two copies are IBD. If they are not IBD, they are independent, so that $E(p_i) = g_i + (1 - g_i)\bar{p}$. If we put these into (7), we get

$$r_i = \frac{g_i}{g_0}. \tag{14}$$

For example, (14) tells us that the relatedness between sibs in an outcrossed diploid population is $R = (1/4)/(1/2) = 1/2$. A striking feature of equation (14) is that it displays relatedness as independent of allele frequency. The definition of r in equation (7) does not lead us to expect this as all the terms in the formula there depend on the frequency of the allele in the population. As an example, suppose we are using equation (6) to assess the selective advantage of an altruistic interaction between siblings. We might well need to know the frequency of the behavior in the population to calculate the fitness effects b_i, but equation (14) tells us that we will not need this to calculate the relatedness r_i.

To calculate the g_i in our plant population, we must specify the amount of gene flow between patches. As an overly simplified example, assume that the population is effectively infinite and that all pollen disperses widely, but fertilized seeds always remain on their native patch. The force of the word *widely* is that uniting gametes are never IBD. We assume also that the four patchmates of a focal individual are indistinguishable, so that the $E(p_i)$ have a common value for $1 \le i \le 4$.

We calculate the g_i recursively, that is, we obtain g_0 and then find each g_i in terms of the previous value. First of all, g_0 is clearly 1/2, as uniting gametes are never IBD. For $1 \le i \le 4$, $g_i = g$ is the probability that random genes in two different individuals are IBD. This can be the case only if they both came from previous generation ovules, and that has probability 1/4. In this case, they were random alleles in "two" plants in the parental patch that might have been the same plant (with probability 1/5). If they were the same plant, they are IBD with probability 1/2; if they were different plants, they are IBD with probability g. This gives us the recursion

$$g_{i+1} = \frac{1}{4}\left[\frac{1}{5}\frac{1}{2} + \frac{4}{5}g_i\right].$$

If we set $g_{i+1} = g_i = g$ and solve, we get $g = 1/32$. Hence,

$$r = g/g_0 = 1/16.$$

This possibility of calculating r with a simple recursive argument transforms inclusive fitness from being just an interesting concept to being an important analytic tool for the study of evolutionary stability.

Reproductive Value. There is one final aspect of inclusive fitness we have not touched upon, and this arises when the population consists of individuals of different classes, for example, age classes (adults and juveniles), sex classes (male and female), and role classes (resident and satellite males, breeder and helper females). In this case, we have the problem of how to compare fitness effects on individuals in different classes. That is what reproductive value is designed to accomplish.

For example, in our plant population, we may have different age classes (for a perennial species) or sexes (for a dioecious species). In this case, a mutant allele may have a different effect in different classes, and a deviant individual of a certain class may affect the fitness of neighbors of different classes differently, so that the analysis can become complex. We handle this by first calculating the inclusive fitness effect ΔW_j of the mutant allele in a class j individual and then ΔW_{IF} is the average of these weighted by the class reproductive values, the relative contribution of each class to the future gene pool of the population. Second, for each class j, we generalize equation (6) as

$$\Delta W_j = \Sigma v_i\, r_i\, E(b_i), \tag{15}$$

where v_i is the individual reproductive value of neighbor i, defined as its relative contribution to the future gene pool of the population. Here, to obtain a common fitness yardstick, the b_i are typically measured as percentage increases. Examples of these calculations are found in the literature.

Comments and Notes.

Evolutionary Stability. Inclusive fitness plays an important role in the analysis of evolutionary stability (ESS), where we typically must determine whether alleles for a certain trait will increase in frequency. In our patch example, for the established height to be evolutionarily stable, it must be the case that alternative heights have reduced fitness; for small deviations, that implies a negative inclusive fitness effect. From equation (11),

$$\left[\frac{\partial W}{\partial x} + \frac{\partial W}{\partial y} R_y + \frac{\partial W}{\partial z} R_z\right] dx \leq 0 \quad (16)$$

for small dx both positive and negative. This can be the case only when we have the equilibrium condition:

$$\frac{\partial W}{\partial x} + \frac{\partial W}{\partial y} R_y + \frac{\partial W}{\partial z} R_z = 0. \quad (17)$$

In practice, we would begin the analysis of the model with this equation evaluated at $x = y = z = x^*$ and solve it to obtain the ESS value of x^*.

Direct fitness. Our approach here was to calculate the fitness effect of a random mutant allele on all neighboring copies, then add these up over all copies of the allele. This is the standard inclusive fitness approach. It can be contrasted with an equivalent "direct fitness" approach, which calculates the fitness effect on a random mutant allele by all neighboring copies and adds these up. Sometimes one approach works better than the other, but the difference is really in how the accounting is done.

Approximations. The calculations of inclusive fitness are approximate for several reasons. First of all, the assumptions of additive genetic effects on fitness are crucial to the argument. These are of two kinds: within and between individuals. Within individuals, we have assumed a "semidominance," that a mutant homozygote is twice as likely to be deviant as a mutant heterozygote. Between individuals, we have assumed that the effects of several deviant individuals acting together can be added. These are, of course, unlikely to hold in practice, but they may hold to a good approximation for small effects. Second, the action of the deviant allele will usually disturb the distribution of the allele, but relatedness is calculated in the "neutral" distribution in which the deviant allele is not permitted to act. Again, this approximation can be expected to be good when the effects of the deviant behavior are small. Typically, we use these calculations to determine the equlibrium, or ESS. Classical methods of dynamics tell us that nonlinear (nonadditive) effects often become effectively linear (additive) near equilibria. So we usually get additivity automatically where we need it—in the calculation of ESS.

Haplodiploidy. Many of the early applications of inclusive fitness concern the phenomenon of haplodiploidy, which induces very unusual patterns of relatedness whereby, for example, a female is more closely related to her sister than to her own daughter. This is one explanation for the evolution of eusociality, a phenomenon in which some individuals help to raise their siblings rather than having any reproductive output of their own. The roots of this hypothesis lie in Hamilton's rule, which, as mentioned earlier, implies that we expect greater altruism to be directed toward more closely related individuals, in this case toward sisters rather than daughters.

The Covariance Form of Relatedness. The power of the inclusive fitness method lies in its ability to reduce a complex allelic distribution to simple relatedness coefficients r, which can be calculated by a recursive argument. In our approach, the expectation E has been taken over the set of mutant alleles. An alternative approach makes the same calculation, but takes the expectation over all individuals. For this to give the same answer as before, an individual must be counted in proportion to the number of mutant alleles it possesses. We obtain that by using the focal frequencies p_0 as weights. For example, $E(\Delta D) = \Sigma E(p_i)E(b_i)$ in equation (3) would become $E(p_0 \Delta D) = \Sigma E(p_0 p_i)E(b_i)$, where E now denotes the average over all individuals, mutant or not. What we get instead of equation (7) is

$$R_i = \frac{E(p_0 p_i) - \bar{p}^2}{E(p_0{}^2) - \bar{p}^2} = \frac{\mathrm{Cov}(p_0, p_i)}{\mathrm{Cov}(p_0, p_0)} \quad (18)$$

which is the statistical regression, or slope, of allele frequencies p_i (recipient) on p_0 (actor).

Relatedness is a measure of genetic similarity, and this covariance form nicely reflects that.

Connections with group selection. When the population is divided into patches or groups, a fruitful point of view is to juxtapose individual fitness with the fitness of the group as a whole. It is often the case that what is good for one is bad for the other, and evolution seeks a right balance. Inclusive fitness tells us that high within-group relatedness should shift the balance toward the needs of the group.

An intuitive interpretation of high within-group relatedness is that individual similarity within a group should be much greater than individual similarity between groups. If we let p be the average allele frequency within each group, then $\mathrm{Var}(p)$ is called the between-group variance. If we take the average of equation (18) over all members of the patch (i.e., over all i $0 \leq i \leq 4$), then the numerator becomes the covariance between two individuals chosen at random (with replacement) from the same patch and that equals the within-patch variance. That is,

$$E(R_i) = \frac{\text{Var}(p)}{\text{Var}(p_0)}. \quad (19)$$

This says that the average relatedness between two individuals chosen at random (with replacement) from the same patch equals the ratio of within-group variance to total variance. Thus, relatedness captures the distribution of genetic variances within and between groups. There is no formal distinction between a model that measures the distribution of variances by r or by the variance between groups.

[*See also* Altruism; *articles on* Eusociality; Fitness; Group Selection; Kin Recognition; Kin Selection; Social Evolution.]

BIBLIOGRAPHY

Bulmer, M. G. *Theoretical Evolutionary Ecology.* Sunderland, Mass., 1994.

Frank, S. A. *Foundations of Social Evolution.* Princeton, 1998.

Grafen, A. "Natural Selection, Kin Selection and Group Selection." In *Behavioural Ecology*, edited by J. R. Krebs and N. B. Davies, 2d ed., pp. 62–84. Oxford, 1984.

Hamilton, W. D. *The Narrow Roads of Gene Land.* vol. 1. *Evolution of Social Behaviour.* New York, 1996.

Hamilton, W. D., and R. M. May. "Dispersal in Stable Habitats." *Nature* 269 (1977): 578–581.

Michod, R. E., and W. D. Hamilton. "Coefficients of Relatedness in Sociobiology." *Nature* 288 (1980): 694–697.

Queller, D. C. "A General Model for Kin Selection." *Evolution* 46 (1992): 376–380.

Taylor, P. D., and S. Frank. "How to Make a Kin Selection Argument." *Journal of Theoretical Biology* 180 (1996): 27–37.

— Peter D. Taylor and Troy Day

INFANTICIDE

Infanticide is the killing of an infant. There is no general definition of *infant.* Some animal behaviorists refer to any killing of a victim who is not yet sexually mature as an infanticide, but others restrict the term to a much briefer life stage. (A common convention in discussion of human cases is to define an infanticide as the killing of a child before its first birthday.)

The term *infanticide* is generally restricted to killings by members of the same species. Interspecific killings of infants are usually to be understood as predation, with the victim's infancy being of no relevance to the predator except insofar as it affects vulnerability. Deaths that occur as accidental byproducts of other motivated pursuits, such as when battling bull seals crush pups that get in their way, would also not normally be called "infanticides," although deciding whether a case is accidental can be difficult. In the human case, accidents are contrasted with intended actions, but intent is not readily imputed to other animals; instead, the issue is largely whether the infant's death is likely to benefit the perpetrator and appears to be the adaptive function for which the infanticidal action evolved. In many species of fish, for example, cannibalizing eggs or fry confers nutritional benefits, and is performed not only by unrelated fish—who also eliminate competitors of their own young in the process, and sometimes gain a nest site—but also by the victims' own parents, who can avoid the necessity of leaving an entire brood unguarded and vulnerable while feeding elsewhere, by staying at the nest and eating just a few of their own young. Infanticidal behavior has been observed and studied in many animals, including numerous insects, fish, birds, and mammals.

Infanticide perpetrated by the victim's genetic parent often represents a failure to fully engage parental care mechanisms for a particular offspring or brood. Proposed explanations for such failures typically invoke either pathology or the adaptive regulation of reproductive effort. Pathology seems relevant in rare and aberrant cases, as well as when animals are housed in artificial environments such as research laboratories or are subject to disturbance and overcrowding. However, "explaining" infanticide as mere pathology tends to foreclose investigation and prevent a better understanding of the mechanisms that normally engage parental investment. Pathologies are failures of anatomical, physiological, and psychological mechanisms and processes such that they exhibit reduced effectiveness in achieving the adaptive functions for which they evolved. Identification of the nature of the failures can help elucidate their proper functioning.

There are several distinct reasons why cessation of investment in an individual offspring or a whole brood may be an adaptive course of action for a parent, and the factors influencing parental divestment "decisions" should be intelligible in terms of the circumstances and cues associated with ancestral fitness benefits. Consumption of its own eggs or fry by a fish, for example, or the resorption of embryos or partial consumption of a newborn litter by a mammalian mother may occur in conditions of reduced food availability and reflect benefits of reallocating or postponing parental effort. Adaptive parental divestment is typically seen at early stages, rather than after much parental effort has been expended, because the future costs and risks required to get an offspring to independence shrink, and the offspring's potential contribution to parental fitness increases, as time passes, making infanticide less and less likely to promote parental fitness. One reason why the term *infanticide* commonly includes victims beyond early infancy is that as long as young are dependent on their parents, the adaptive rationales for parental termination are likely to be similar: continued investment in the current offspring promises to cost the parent more future fitness than it will provide.

The following categories of adaptive rationales account for the majority of known cases of infanticide perpetrated by avian and mammalian parents:

1. The parent does not have adequate material or social resources to successfully rear the young, so continued efforts are futile and may compromise the parent's future fitness. Many mammals kill or abandon nursing litters if the mother's food supply suddenly fails. In some biparental bird species, a widowed parent cannot raise unfledged young alone and will abandon them.
2. In those relatively rare species in which older young are not yet fully independent when the new young are conceived or born, investment in a newborn can compromise the survivorship of an older (or healthier) offspring. This is another reason why a parent may be overburdened. Several mammalian species, such as hamsters and kangaroos, have the capacity to delay implantation of embryos while still nursing older young. In species with an extended period of parental care, there is an optimal interval between successive young such that too short an interval would reduce the likelihood of both the older and the younger siblings surviving.
3. Parental efforts are likely to be wasted because the infant is deformed, injured, seriously ill, or otherwise unresponsive to parental care. The rationale for parental divestment from a moribund infant is obvious, but a lack of responsiveness is an ambiguous cue, and natural selection on mothers must have balanced the costs of continuing to invest in hopeless cases against the costs of giving up on those who might recover. There are many descriptions of nonhuman mothers, including chimpanzees, baboons, and elephants, continuing to carry, or try to arouse, their dead infants. These mothers' persistent efforts may take days to disappear entirely.

When maternal overburdening is a reason for terminating investment, one may wonder why parents would ever kill their young rather than try (or at least hope) to have them adopted by others. In fact, adoption does occur in a number of species, but unless the psychological mechanisms by which parents normally identify and prefer their own offspring have been somehow circumvented, the fitness benefits for the adoptive parents would have to exceed the costs for adoption to persist under natural selection. In modern human adoption, adoptive parents are unrelated to the child, but in several species, those who make "alloparental" (parentlike) investments in young are typically relatives of the infant who lack alternative breeding opportunities for themselves, or are helping their parents while hoping to gain a scarce and valuable breeding opportunity. The care of younger siblings by older siblings in humans is analogous in some respects.

Parental care by a stepparent, the new mate of a widowed or abandoned genetic parent, is a different sort of adoption, for which the fitness benefits of tolerance and alloparenting would again have to offset the costs in foregone alternative reproductive options for the practice to be maintained under selection. If the new partner is more likely to establish a mateship with the stepparent, then stepparental tolerance and care of the young may be expected, and such adoption as a sort of courtship effort occurs in a variety of birds and mammals, including humans. In other species, such as group-living lions and langur monkeys, nursing infants are routinely killed by "stepfathers" after defeating and replacing resident males who had sired the victims. This has been called "sexually selected infanticide" because it is a form of competition between males for access to the reproductive "resource" of females. Its benefits derive from destroying the offspring of a rival and from the fact that the lactating females of the group he has taken over will be able to conceive sooner than if the females had continued nursing. Similarly, in some biparental birds such as acorn woodpeckers and tree swallows, new mates are likely to kill the dependent young of the prior mate, but in many other species, such as American kestrels and black-billed magpies, tolerance and care of "step-offspring" are common outcomes. Various combinations of factors, such as sex ratio imbalances in the population, opportunities to nest again in the same breeding season, the season-to-season persistence of pair bonds, and the costs imposed by defenders of the infants, affect the utility of killing versus investing in a new mate's dependent young. [*See* Sex Ratios.]

Human infants, like many other species, do not typically reside with stepparents, but when they do, they are at elevated risk of being killed. However, these events do not exhibit the hallmark features of sexually selected infanticide, as observed in lions and langurs. Homicides by human stepparents are much more likely to be the culmination of a pattern of episodic neglect and abuse interspersed with periods of adequate care, and are probably not fitness promoting. Human steprelationships are not a modern novelty. In recent centuries in Europe, for example, mortality was substantial among parents of dependent children, and if one parent died, then the child's mortality risk was elevated; according to one study, if the surviving parent remarried, the child's mortality risk was elevated further. In the face-to-face societies of our ancestors, social services beyond kin assistance were nonexistent, and the situation for stepchildren was probably worse than in peasant societies. According to one study of contemporary South

American hunter-gatherers, the Ache of Paraguay, 43 percent of children brought up by a mother and stepfather died before their fifteenth birthday, compared to 19 percent of those brought up by the two genetic parents. In light of these facts, an obvious question concerns the fates of children raised only by unrelated adoptive parents; where such adoptions are highly regulated by agencies, which select ideal parents who very much want to adopt, the risks of maltreatment and death are not elevated as they are in stepfamilies.

Selective infanticide of daughters has been the object of considerable study and debate, although it is by no means typical of societies in which infanticide occurs. In complex, stratified societies, the practice has been status-graded, with the upper classes more likely to eliminate daughters while concentrating investments in sons. In many societies, there has been a general tendency to prefer sons over daughters, and possible explanations continue to be debated.

Maternal youth is another risk factor for infanticide in both traditional and modern societies. The most detailed anthropological study of the incidence and determinants of infanticide found that during a period of adverse social and economic circumstances, Ayoreo mothers of South America were less and less likely to dispose of newborns with increasing age. In modern industrialized nations such as Canada, England, and the United States, mothers are also decreasingly likely to commit infanticide with increasing age. The decreased risk with older mothers has been interpreted as reflecting a life span developmental change in women's valuation of their newborns as their capacity to produce additional future children declines.

In many legal codes, infanticide by the mother carries a lesser penalty than other homicides, bespeaking a moral sentiment that such crimes should be distinguished from killings of infants by others. Indeed, maternally perpetrated infanticide is not universally criminalized at all. In traditional nonstate societies, it is widely considered an unfortunate but appropriate recourse in some circumstances. These circumstances accord with the functional rationales described above: maternal incapacity to cope with the demands of child rearing because of illness, famine, lack of paternal assistance, a still-nursing older sibling, or the birth of twins; inappropriate paternity, perhaps as a result of sexual coercion; and low probability of the infant surviving because of illness or deformity.

BIBLIOGRAPHY

Blaffer Hrdy, S. *Mother Nature.* New York, 1999.

Borries, C., K. Launhardt, C. Epplen, J. T. Epplen, and P. Winkler. "DNA Analyses Support the Hypothesis that Infanticide Is Adaptive in Langur Monkeys." *Proceedings of the Royal Society of London B* 266 (1999): 901–904.

Bugos, P. E., and L. M. McCarthy. "Ayoreo Infanticide: A Case Study." In *Infanticide*, edited by G. Hausfater and S. B. Hrdy, pp. 503–520. New York, 1984.

Daly, M., and M. Wilson. *Homicide.* New York, 1988.

Daly, M., and M. Wilson. *The Truth about Cinderella.* New Haven, 1999.

Dickemann, M. "Female Infanticide, Reproductive Strategies, and Social Stratification: A Preliminary Model." In *Evolutionary Biology and Human Social Behavior*, edited by N. A. Chagnon and W. Irons. North Scituate, Mass., 1979.

Ebensperger, L. A. "Strategies and Counterstrategies to Infanticide in Mammals." *Biological Reviews* 73 (1998): 321–346.

Hausfater, G., and S. B. Hrdy, eds. *Infanticide.* New York, 1984.

Hill, K., and A. M. Hurtado. *Ache Life History.* New York, 1996.

Parmigiani, S., and F. S. vom Saal, eds. *Infanticide and Parental Care.* Chur, Switzerland, 1994.

Rohwer, S., J. C. Herron, and M. Daly. "Stepparental Behavior as Mating Effort in Birds and Other Animals." *Evolution and Human Behavior* 20 (1999): 367–390.

Scheper-Hughes, N. "Culture, Scarcity, and Maternal Thinking: Maternal Detachment and Infant Survival in a Brazilian Shantytown." *Ethos* 13 (1985): 291–317.

Van Schaik, C. P., and C. H. Janson, eds. *Infanticide by Males and Its Implications.* Cambridge, 2000.

— Margo Wilson and Martin Daly

INFECTIOUS DISEASE. *See* Disease, *article on* Infectious Disease.

INFLUENZA

Influenza causes almost yearly epidemics of respiratory infection in humans. We contract influenza repeatedly throughout our lives, making it unlike other diseases such as measles, to which we typically become immune after a single childhood infection. Continual reinfection occurs because the influenza virus evolves so quickly; within a few years our immune systems no longer recognize the new mutant strains. This continual evolutionary change is called antigenic drift. On rare occasions, viruses bearing genes from influenza viruses native to birds become established in humans—a process called antigenic shift. We have little or no immune protection against those foreign strains, and particularly severe and widespread outbreaks of disease called pandemics result. This article describes aspects of influenza biology that make antigenic drift and shift possible, and illustrates how these processes affect the evolution of influenza in humans and other animal hosts.

Centuries-old historical records describe winter respiratory epidemics that appear suddenly, persist for about two weeks, then abruptly disappear—the pattern we see in influenza today. The cause of those historical outbreaks was discovered with the isolation of the influenza virus in 1933. Although we typically perceive little difference in our symptoms from one bout of influenza to the next, subsequent research has revealed a wealth of variation among influenza viruses.

Types of Influenza. The influenza viruses compose three genera in the viral family *Orthomyxoviridae.* In practice, the genera are referred to as the three "types" of influenza: types A, B, and C. Influenza type C, a distant relative of types A and B, typically causes mild upper respiratory infections in children. The symptoms and epidemiology of influenza A and B are similar in that both cause winter epidemics, with influenza B infections being somewhat less severe. However, types A and B have distinct evolutionary histories and pose very different threats to human health.

Only influenza A undergoes antigenic shift, the potentially deadly transfer of influenza genes from viruses in birds to viruses infecting humans. Antigenic shift does not occur in influenza B because nonhuman hosts do not carry type B. By contrast, influenza A viruses infect humans, swine, horses, and, most importantly, birds. Wild aquatic birds harbor a wide diversity of influenza A strains and are thought to be the original host of the influenza A viruses that infect other species.

Type A influenza strains have been classified into subtypes based on variation in the two major antigenic proteins on the surface of the virus, hemagglutinin and neuraminidase. Subtypes are defined as antigenically distinct strains; that is, infection by one subtype does not provide immunity against infection by other subtypes. To date, fifteen hemagglutinin and nine neuraminidase alleles have been isolated from birds. Subtypes are named according to the particular hemagglutinin and neuraminidase alleles they carry (e.g., subtype H3N2).

Examples of Pandemic Strains. Two recent pandemics (in 1957 and 1968) originated through the process of reassortment—genetic mixing between different viruses. The genome of influenza A consists of only ten genes, one or two on each of eight separate RNA segments. When two or more influenza A subtypes infect a host cell simultaneously, reassortment of segments from different parental viruses may occur. If a reassortant progeny contains a hemagglutinin or neuraminidase allele that is novel to humans, and the virus is capable of being spread easily from person to person, a worldwide epidemic, or pandemic, may occur.

The influenza A virus isolated in 1933 descended from the H1N1 strain that caused the deadly "Spanish flu" epidemic of 1918. In 1957, this strain obtained avian H2 and N2 alleles through reassortment. The progeny caused the H2N2 ("Asian flu") pandemic, then continued circulating in humans until the acquisition of an avian H3 hemagglutinin allele in 1968. The resulting reassortant strain caused the H3N2 ("Hong Kong flu") pandemic, and subsequent H3N2 outbreaks that continue to this day.

With the emergence of the H2N2 and H3N2 pandemic strains (in 1957 and 1968, respectively), the previously circulating human influenza A strains became extinct. These serial replacements of one influenza A subtype by another suggested that only a single influenza A subtype could circulate in humans at one time. However, in 1977, an H1N1 influenza A virus virtually identical to an H1N1 strain that infected humans in 1950 reappeared from an unknown source, possibly from a laboratory. This strain has cocirculated with both the H3N2 subtype of influenza A and influenza B ever since.

The reemergence of subtype H1N1 in humans in 1977 did not cause a pandemic. Transmission may have been low because of preexisting host immunity; many older members of the human population in 1977 had been infected by H1N1 three decades earlier. Another explanation could be reduced virulence; like other pandemic strains observed to date, the deadly H1N1 pandemic strain evolved to a much lower level of virulence within a few years after its introduction to humans in 1918.

The decay of long-term immunity to influenza is not well understood. However, it is known that the first variant of a subtype encountered by an individual typically causes the strongest antibody response. Subsequent infections by related variants tend to reinforce the antibody response to the first variant. Thus, the highest antibody levels in an age class are against the virus responsible for the childhood infections of that group. This phenomenon, curiously called "original antigenic sin," allowed historical reconstruction of the strains that infected people prior to 1933. These "seroarchaeological" studies suggested that viruses carrying H2 and H3 alleles infected humans in the late 1800s, and that these disappeared during or prior to the H1N1 pandemic of 1918.

Together with modern observations, these historical observations suggest that the H1, H2, and H3 hemagglutinin alleles have each been introduced into humans, in that order, at least twice in the recent past. The apparent recirculation of these three alleles does not guarantee, however, that viruses carrying other hemagglutinin alleles cannot infect humans. In 1997, a limited outbreak of H5N1 influenza A occurred in Hong Kong, killing six of the eighteen people known to be infected. This outbreak was caused by an H5N1 virus of entirely avian origin; it did not arise by reassortment with a human-adapted influenza virus.

Routes of Transmission. Influenza viruses bind to different host cell receptors in birds and humans, and this difference was thought to prevent direct transmission of avian viruses to humans. Swine, which have both avianlike and humanlike cell receptors, have been proposed as necessary intermediate hosts for the transmission of avian viruses to humans. This hypothesis was consistent with the observation that many epidemics and pandemics appear to originate in Southeast Asia, where agricultural practices put ducks, swine, and humans in close contact. However, the direct infection of humans by avian H5N1 viruses established that receptor specificity is not strictly limiting to host range.

Even rare infection of humans by avian strains poses a severe danger. It provides the avian virus an opportunity to reassort with human-adapted viruses, and subsequently to acquire the ability to transmit between humans during replication in this new host. The H5N1 virus in Hong Kong apparently did not transmit between humans, and no further human infections were seen after the elimination of infected local poultry. However, the H5N1 virus remains a threat because of recurrent infections of domestic poultry by this virus in Southeast Asia.

Variability in Virulence. Another puzzle concerns variability in virulence among influenza strains. The H1N1 virus that caused the 1918 pandemic was particularly lethal, killing tens of millions of people worldwide. Most deaths occurred among young adults instead of the usual victims, the elderly. Researchers have amplified and sequenced certain genes of H1N1 viruses preserved in the tissues of victims of the 1918 pandemic. However, the gene sequences of these viruses have not yet revealed why this strain was so deadly.

Evolution of Strains. Antigenic shifts occur rarely—we have not experienced a shift event since 1968. In contrast, antigenic drift occurs continually. Drift results from the accumulation of amino acid replacements. Eventually those replacements alter the shape and the electric charge of the hemagglutinin and neuraminidase surfaces enough to allow viral escape from existing antibodies. Antigenic drift can sometimes proceed so quickly that we risk becoming reinfected yearly.

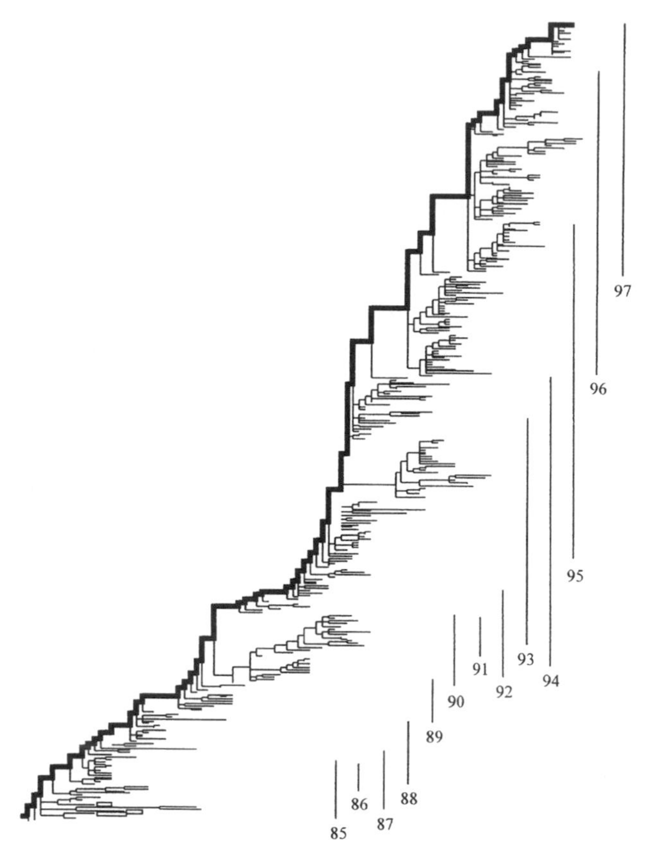

FIGURE 1. Human Influenza Evolution.
Human influenza evolution is characterized by the rapid emergence and extinction of many new lineages over time (H3N2 subtype of influenza A shown from 1985 to 1997).
Drawing by Robin M. Bush.

Gene sequencing has recently produced new insights into the evolutionary processes responsible for antigenic drift. Figure 1 shows that new mutant lineages of the H3N2 subtype of influenza A arise continually. At any time a number of closely related lineages cocirculate, but most, shown as numerous short side branches, die out within a few years. The hemagglutinins of influenza B and the H1N1 subtype of influenza A show similar patterns of evolution, except that influenza B has split into two deep lineages that presently cocirculate.

The phylogeny in Figure 1 shows dramatic variation in long-term success among strains. Host immune pressure probably causes this variation. Studies of human H3N2 influenza A support this hypothesis. A small subset of positions in the antibody binding sites of the H3 hemagglutinin appear to have been under selection to change repeatedly in the past. Such "positive selection" causes a much higher than expected rate of nonsilent (amino acid changing) as opposed to silent nucleotide substitutions over time. Lineages with new amino acid replacements at these positions usually leave more descendants than other lineages. If replacements at key amino acid positions truly explain differential success, then additional replacements in these positions may also be advantageous in the future. Screening new strains for additional replacements at these particular sites may provide an early warning of potentially successful viral strains, and thus prove a useful adjunct to current methods of influenza surveillance and strain selection for vaccine production.

Influenza is among the best studied of human diseases. Nonetheless, much remains to be learned. For example, epidemics caused by drift variants account for substantial amounts of mortality over time, yet we know very little about the types and amounts of genetic change needed to produce a new epidemic strain. Changing distribution of immunity in the human population probably contributes to the temporal order in which H1N1, H3N2, and influenza B have caused recent epidemics, but this remains an untested hypothesis. Current efforts toward understanding host immunity in both humans and birds, along with expanded influenza surveillance, will surely increase our understanding of antigenic drift. It is hoped that these efforts will also produce knowledge that will help us survive the next pandemic.

[*See also* Disease, *article on* Infectious Disease; Immune System, *articles on* Microbial Countermeasures to Evade the Immune System *and* Structure and Function of the Vertebrate Immune System; Plagues and Epidemics; Vaccination; Viruses.]

BIBLIOGRAPHY

Bush, R. M., C. A. Bender, K. Subbarao, N. J. Cox, and W. M. Fitch. "Predicting the Evolution of Human Influenza A." *Science* 286 (1999): 1921–1925. Use of molecular techniques to study antigenic drift.

Cox, N. J., and C. A. Bender. "The Molecular Epidemiology of Influenza Viruses." *Seminars in Virology* 6 (1995): 359–370.

Cox, N. J., and K. Subbarao. "Global Epidemiology of Influenza: Past and Present." *Annual Review of Medicine* 51 (2000): 407–421.

Dowdle, W. R. "Influenza A Virus Recycling Revisited." *Bulletin of the World Health Organization* 77 (1999): 820–828.

Glezen, W. P., and R. B. Couch. "Influenza Viruses." In *Viral Infections of Humans*, edited by A. S. Evans and R. A. Kaslow, 4th ed., 473–505. New York, 1997. General overview of influenza.

Murphy, B. R., and R. G. Webster. "Orthomyxoviruses." In *Fields Virology*, edited by B. N. Fields, D. M. Knipe, and P. M. Howley, 3d ed., 1397–1445. Philadelphia, 1996. General overview of influenza.

Nicholson, K. G., R. G. Webster, and A. K. Hay (eds). *Textbook of Influenza*. Oxford, 1998. In-depth chapters covering many areas of influenza biology.

Reid, A. H., J. K. Taubenberger, and T. G. Fanning. "The 1918 Spanish Influenza: Integrating History and Biology." *Microbes and Infection* 3 (2001): 81–87.

— Robin M. Bush

INSECTS

Insects are the most diverse group of organisms on earth and seem to have been diverse since at least the Permian period, about 250 million years ago. This qualifies them as the most successful animals ever to have lived on earth. Scientists recognize at least 750,000 species of insects and place them in the order Insecta. They estimate, however, that the total number of living insect species is ten million or more. The number of beetle species alone (order Coleoptera) at about 500,000, is roughly twice that of the nearest other major group (green plants). This amazing richness is evidence that evolution has taken varied paths to fill or subdivide niches. Larger (and smaller) organisms seem not to have exploited niches to this extent. Many researchers have posed the question Why are there so many insect species? This question may be rephrased as What key adaptations have allowed insects to be so species rich?

Insects as Arthropods. Insects are the most species-rich class in the phylum Arthropoda. Arthropods include major groups such as the extinct trilobites, chelicerates, crustaceans, myriapods, and insects. The phylum Arthropoda contains roughly three-quarters of the species of animals on earth. Insecta alone accounts for about two-thirds of the animal species (Hammond, 1992). Researchers have diagnosed the phylum Arthropoda in different ways. The most generally accepted features are a chitinous exoskeleton and jointed appendages. Other commonly cited synapomorphies (shared, derived homologous features) include an open circulatory system, Malpighian tubules used for nitrogenous waste concentration and excretion, the hemocoel body cavity, the lack of cilia on any body cells, the lack of nephridia (the excretion organ of segmented worms), and separate sexes (as opposed to a near sister group, segmented worms, which can be hermaphroditic). Some experts have included the lobopods, *Peripatus*, for example, as arthropods but most recognize those interesting creatures as a separate phylum, the Onychophora. The exact topology of relationships among major groups of the phylum Arthropoda is a contentious issue. The phylogenetic tree in Figure 1 summarizes the relationships among major groups of arthropods. Note that the sister group of insects is myriapods. Other recent molecular considerations list Crustacea as the sister group. Support for this arrangement with Crustacea is sparse and does not overwhelm the purported synapomorphies of Insecta and Myriapoda when morphological and molecular lines of evidence are evaluated together (Wheeler et al., 2001). Whether myriapods form a monophyletic group (a group consisting of a common ancestor and all of its descendants) is unclear. Myriapods could be merely a paraphyletic group (an evolutionary grade consisting of some, but not all, descendants of a common ancestor). Both points of view have propo-

FIGURE 1. Relationships Among Classes of Arthropoda.
C. Riley Nelson.

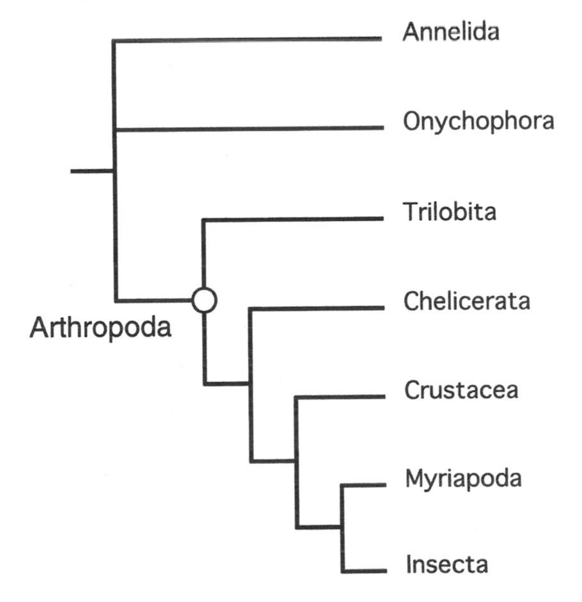

nents. In any event, the ancestral insect (hexapod) is thought to have evolved from some member of this myriapod group. The controversy, in part, relates to exactly which myriapod, whether millipede, centipede, symphylan, or pauropod, shares a common ancestor with insects.

Insects as Hexapods. A well-supported tree showing phylogenetic relationships among living six-legged arthropods is shown in Figure 2. To produce this arrangement, Wheeler and colleagues (2001) reviewed the vast morphological literature and added characters given by sequences from two genes. It has become common practice in recent years to list the class to which insects belong as Hexapoda. But a rationale for recognition of all six-legged arthropods as class Insecta is supportable as well. A primary reason for recognizing class Hexapoda (rather than class Insecta) is generally given as the lack of resolved phylogenetic relationships among Collembola, Diplura, and Protura. Advocates maintain that a named basal node will clarify the relationships with taxa further up the tree. Also, a need is articulated to recognize individual names for nodes from which Archaeognatha, Zygentoma, and winged insects emanate. The five groups of wingless animals in question here have been included in the "Apterygota" in the past. Apterygota, in this sense, is clearly a paraphyletic group, and its use as a formal clade name should be discouraged. Most (but not all) recent reviews of the classification status of these five orders list Collembola, Diplura, and Protura as a monophyletic group, Entognatha. This status is based on the presence of mouthparts that can retract into a facial pouch, as well as several other characteristics. Archaeognatha and Zygentoma each form monophyletic groups (Figure 2). The rationale for use of the taxon name Hexapoda does not hinge on attempts to avoid paraphyletic groups. The second rationale, naming of every node in a phylogenetic sequence, has been criticized regularly in the past as redundant and unnecessary. On a more positive note, several characters supporting monophyly of Hexapoda (major body regions of head, thorax, and abdomen; six legs) have a long traditional use as defining characteristics for Insecta in the general scientific community and for the public as well. Although "tradition" alone is not a valid argument for a particular naming scheme, stability of use is desirable. Here, class Insecta is used for the node connecting the Entognatha with the remainder of six-legged arthropods.

Insect Phylogeny and Key Innovations. We can begin to answer questions such as "Why are there so many kinds of insects?" by considering a few key adaptations. Six breakthroughs in morphological adaptation are largely responsible for the success of insects. (A few others have been important for insects on a smaller, but significant, species-richness scale). The six major morphological innovations of insects are having six legs; tagmatization, having the body divided into three regions; mandibles with two condyles; wings; the ability to compactly fold these wings; and complete metamorphosis.

The oldest insect fossils are pieces of Collembola from the Rhynie Chert of Scotland. Several specimens including the described *Rhyniella precursor* have been found in these lower Devonian deposits dated to approximately 400 million years ago. These specimens are the first in the fossil record to show the six-legged condition and having the body segmentation coalesced into three main body regions: head, thorax, and abdomen. Thus, the minimum age of the insect clade is 400 million years.

Numerous authors have speculated on the advantage that six legs and three body regions might have conferred on these species. The dual tripod gait with six legs allows significant stability with few contact points on the substrate. Functional morphologists have studied this phenomenon, as have mechanical robot designers. A metachronal gait is employed by most terrestrial arthropods. It is typified by lifting one leg at a time from the substrate while keeping the leg behind it down to bear the animal's weight. This allows good stability while decreasing net energy use. By decreasing the number of legs to six (from the myriapod condition of many), mechanical simplification is achieved. Although it is intuitive that simplification with no loss of stability or speed could give an organism a selective advantage, we must note that other gait systems work. Is the six-legged condition a breakthrough adaptation regarding diversity? Not necessarily, but it could have allowed channeling of energy and integration resources elsewhere. Resources needed for multiple limbs could be diverted to other important features or functions necessary for survival and reproduction. Additionally, limbs freed from use during locomotion could take on new roles as mouthparts for enhanced food acquisition and manipulation. Ancestral legs could also be modified into complex genitalia and enhanced reproduction.

Tagmatization, the specialization of different regions of the body for different functions, is not unique to insects, arthropods, or protostomes for that matter. What is unique for the head, thorax, and abdomen arrangement seen in insects, however, is a significant simplification of the annelidlike or myriapod model. The head in insects is largely responsible for environmental sensory perception and food acquisition. The thorax bears legs and wings in insects. It can be considered the locomotion center of the animal. In contrast, the abdomen's particular functions of food processing and waste removal can be considered more as retained ancestral conditions. The complex genitalia composed in large part of modified legs, however, can clearly confer reproductive advantage (Eberhard, 1985).

A second condyle or contact point in the mandibles

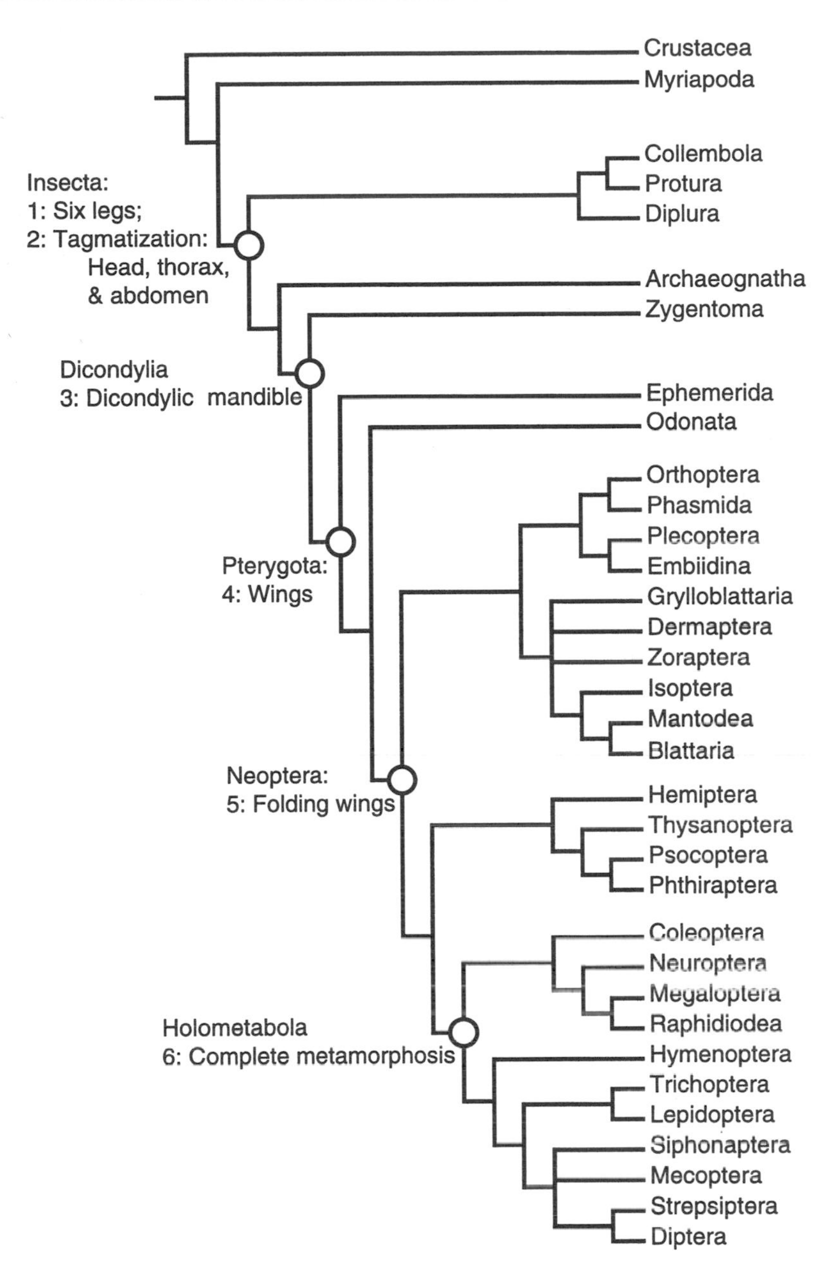

FIGURE 2. Relationships Among Orders of Insecta, with Key Innovations.
C. Riley Nelson.

of Zygentoma and the remainder of insects (Dicondylia) conferred increased mechanical advantage with similar muscle mass. This was a key adaptation because it allowed exploitation of tougher food sources.

Wings and active flight confer habitat location, food-finding, dispersal, and mate-locating advantages to the animals that have them. Active flight has evolved at least four times in animals: insects, pterosaurs, birds, and bats. Each of these groups is species rich when compared to their nonflying immediate ancestors. The fossil history of wings begins with the insects of the Carboniferous period, about 320 million years ago. The winged clade of insects is by far its most species rich, with at least 750,000 extant species capable of active flight. Insects are the only arthropod group employing active flight, and they are the most species rich as well. Of all the key adaptations of insects, flight is the feature that is probably most responsible for diversification and success.

Where did these wings come from? In vertebrates, clearly the wings are modified forelimbs. Pterosaurs, birds, and bats all fly (or flew) by modifying these limbs into broad, lift-generating surfaces. This is not true in insects. The wings of insects cannot easily be associated with any preexisting features. One of the most interesting, and at times acrimonious, debates in insect evolution centers on the origin of their wings. A recent summary of wing origin theories is given by Dudley (2000). He also discussed theories related to function at intermediate stages of wing development in insects. Two basic theories, each supported by extensive conjecture and limited data, are most commonly discussed in the origin of the actual wing structure. The paranotal extension hypothesis alleges that the wings originated as thin dorsolateral extensions of the individual segments of the thorax. The greatest evidential support for the paranotal hypothesis comes from the existence of extensive flanges in some large Palaeodictyoptera of the Carboniferous period, approximately 300 million years ago. The location of wings and protowings at the appropriate near-head location with respect to center of mass also lends credence to this hypothesis. The most apparent problem with this hypothesis is that the extensions would bear neither articulation nor musculature necessary for flapping flight. The second major summary, the pleural hypothesis, claims that the wings are modified portions of leg appendages. In one representation of this hypothesis, the wings are thought to be modified gills enlarged to take on flight capabilities. Support for this hypothesis is greatest because of the readily apparent blade, articulation, musculature, and venation that exists in gills of extant Ephemerida and some extinct basal insect orders. One difficulty with this hypothesis is that these types of gills are never on the thorax, where the necessary moment to center of mass would need to be for flight. Additionally, this hypothesis would require wings to have originated in aquatic insects, despite the best evidence pointing to the wingless ancestor being terrestrial. Another permutation of the pleural theory, championed by Kukulova-Peck (1983), notes the existence of exite lobes on the basal parts of legs of some insects of the Carboniferous period. She suggested that these exites have broken away from the main leg support structure and migrated dorsally in the body wall of the insect. These parts, with associated articulation and musculature, went on to become wings. Detractors of this hypothesis cite the lack of direct evidence of actual structures in the fossils. They also note that the immense physiological and ontogenetic changes necessary to accomplish such a radical migration of tissue makes the exite hypothesis virtually untenable.

Besides speculations about the physical origin of wings, many have considered the functional history of wing development. The crux of the issue centers on how short protowings of limited flight enhancement capability could be modified by natural selection to produce the magnificent flight organs we see today. Again, two basic hypotheses are evident. The first is that protowings and wing have always (or nearly always) been under direct selection for aerodynamic function. Second, protowings initially had other more important functions. Later they were adapted for aerodynamic function.

The aerodynamic hypothesis maintains that gliding or flight was the selective issue throughout the wing elongation process. That even short wings confer benefit from an aerodynamic perspective on the way to powered flight has been substantiated experimentally: it could have happened. A related variation of this hypothesis purports that ancestral preflying insects could have used protowings as sails to skim across water surfaces.

The second set of hypotheses, grouped together as preadaptation hypotheses, insists that a purpose other than flight and locomotion initiated wing development and elongation. Flight, according to adherents, only became important in wing evolution at later stages. Various ideas for the initial value of protowings have been proposed: thermoregulation, courtship display, warning display, jump escape benefits, and apparent size increase for predator deterrence.

Note that many of these hypotheses are not necessarily mutually exclusive. This may explain the lack of consensus on the issue of wing origin and function. The diversity in form and function that we see in current insect wings may arise from a diversity of function historically. In any case, wings are obviously important for aerodynamic reasons (movement, dispersal, mate finding, food acquisition, and predatory escape) as well as for a variety of other purposes.

Despite the advantages of wings and flight, they come at a cost. The wing surface necessary to generate lift must be large relative to the organism's total size and weight. The beating surface can be made lighter through thinning but must remain expansive to generate enough lift. These expansive beating surfaces then can become a burden when predators attempting to evade, even while simply maneuvering as the insect feeds or mates. One solution to this problem would be to fold the wings neatly out of the way when not in use. This is what has occurred. Another solution would be to dispense with wings. This has also occurred repeatedly. Thus, we note that a variety of insects, scattered all over the phylogenetic tree, have become flightless.

The final key innovation to be discussed is that of holometaboly or complete metamorphosis. The Holometabola, holometabolous insects, contains about 85 percent of the species in the group. If diversification is a measure of success, then holometaboly could be the single most important factor because it is used by so many species. The postembryonic developmental regime that holometaboly follows consists of the passage from egg through several stages of larvae, then a quiescent pupa followed by the adult. Such a complex pattern is energetically, morphologically, and ecologically unparsimonious. It comes at great cost. Why, then, would such a system evolve? What are the benefits of holometaboly? The key function that holometaboly accomplishes is segregation of immatures from adults. This is accomplished by insertion of the pupal stage between other immature stages and the adult. This juvenile/adult segregation correlates with several ecological incentives: decreased intraspecific competition, predator avoidance, separation of feeding and reproducing stages, and allowance for massive reorganization from juvenile to adult.

Intraspecific competition can be the most severe type of competition. Complete metamorphosis may reduce it by segregating the juvenile and adult resources, such as food and shelter. This decrease in competition hypothesis is generally cited as the most important reason for the success of insects having holometabolous development. However, many hemimetabolous insects also segregate these stages in an analogous way that can reduce intraspecific competition. Most prominent in this regard are those that have aquatic immatures and aerial adults. This simply indicates that several pathways to segregation have been used with success. A second professed reason for segregation of immatures and adults is that, by selecting different forms for the two stages, predators cannot key in on a uniform search image. One or the other stages may avoid detection and escape. The third concept of functional segregation of feeding and reproduction into two life stages allows for specialization of morphological features more advantageous in each stage. This requires, then, a radical reorganization of form, the fourth idea given above. Again, these hypotheses are not mutually exclusive but may all contribute to the success of the organisms.

Other Innovations and Relationships. Whether the first insects lived in water or air is a point of contention among evolutionary biologists. This controversy and the available evidence were summarized by Pritchard and colleagues (1993). They considered evidence as diverse as osmoregulation, locomotion, fossils, phylogeny, gills, life history, and the morphological arrangement of the tracheal system. The researchers concluded that a terrestrial origin of hexapods, and thus insects, is more reasonable.

The rise to dominance of flowering plants is regularly suggested as causing the diversification of the insect lineage. Many insects are closely tied to plants by morphologies related to feeding guilds such as defoliators, phloem suckers, and pollinators. In considering the diversity of life on earth, two forms dominate: green plants and insects. A coevolutionary system in which plants and insects both diversify in response to the other must thus be seriously considered. Ehrlich and Raven (1964) proposed a coevolutionary "arms race" between butterflies and plants. They used this idea to explain the diversity of secondary compounds found in plants and the close relationships many insects have with plants. The family-level diversity of fossil insects, however, was already growing exponentially before the angiosperms appeared in the fossil record (Labandeira and Sepkoski, 1993). In fact, the rate of family-level diversification decreased when angiosperms came on the scene. Taken together, this information from the fossil record actually shows that diversity of insects and angiosperms is decoupled. Other analyses, however, have shown that many herbivorous lineages of insects have diversified when compared to their nonherbivorous sister groups. The support for a close insect/plant coevolutionary trend in diversification therefore is equivocal. At a minimum it has not had as direct an influence on the diversification of insects as is often proposed.

Social insects, including ants and termites, are dominant biological forces on earth, especially in tropical regions. The numbers, biomass, and density of tropical ants and termites are nothing less than astounding. For example, ant and termites compose one-third of the animal biomass in an Amazonian rain forest, and by adding two other social groups of insects this number reaches 75 percent. Ants alone in these systems are diverse, with up to 128 species in 250 square meters. However, on a worldwide scale, considering the amazing diversity of all insects, ants and termites are not outrageously species rich. They consist of about 11,500 species worldwide. Still we must note that this is roughly equivalent to the 9,000 species of known birds. In any event, these social insects are extremely successful. Whether the evolution of societies is directly responsible for that success is uncertain.

Insects are not important components in marine ecosystems. This is enigmatic when one considers how important they are in all other major ecosystems. Many insect groups have evolved the morphological and physiological adaptations needed to exploit these habitats, but it is thought that the exquisite range of adaptations to the terrestrial habitat has hindered them in attempts to colonize the seas. Crustaceans, trilobites, and other arthropods radiated in Devonian and Carboniferous seas. This gave them a preemptive competitive advantage over the insects, which continues to this day. So the reasons why there are no marine insects may be that they did not reach the seas initially and were burdened with terrestrial adaptations when returning later.

Conservation. The terminal Permian extinction (240 million years ago) devastated insect diversity. Of the twenty orders of insects in the Permian era, only thirteen survived into the Triassic period. Equally devastating losses at family and generic levels were sustained. Diversity worldwide in most other animal groups paralleled this most prominent extinction event the world has ever known. Insects passed unscathed through all other extinction events, such as the terminal Cretaceous event that eliminated large dinosaurs. But insects today may be facing the greatest extinction force of their 400 million–year history. The indications are that insect species are disappearing from our planet in numbers that overshadow those of the end-Permian event. Causes cited for their disappearance are habitat destruction and indiscriminate use of pesticides. Insects have a long history of evolving key innovations to deal with novel challenges. Already we see evolution occurring in the form of pesticide resistance over a very few years. Is this enough? Will insects survive these new human-induced challenges? Some species, but probably not most, will survive. Are we willing to lose the incredible beauty in form, function, and strategy that insects in their hyperdiversity present to us? Are we willing to replace this diversity with resilient cockroaches, house flies, and fire ants?

[*See also* Animals.]

BIBLIOGRAPHY

Dudley, R. *The Biomechanics of Insect Flight: Form, Function, Evolution.* Princeton, 2000. An up-to-date summary of the insect wing origin and early function controversy.

Eberhard, W. G. *Sexual Selection and Animal Genitalia.* Cambridge, Mass., 1985. An introduction to the the incredible diversity of animal genitalia with an interpretation that it is the product of sexual selection.

Ehrlich, P. R., and P. H. Raven. "Butterflies and Plants: A Study in Coevolution." *Evolution* 18 (1964): 586–608. The early introduction of coevolution and arms race ideas.

Hammond, P. M. "Species Inventory." In *Global Biodiversity, Status of the Earth's Living Resources*, edited by B. Groombridge, pp. 17–39. London, 1992. A compilation and enumeration of living species.

Holldobler, B., and E. O. Wilson. *The Ants.* Cambridge, Mass., 1990. Contains an introduction to the diversity, abundance, and ecological importance of ants, termites, and other social insects.

Kukulova-Peck, J. "Origin of the Insect Wing and Articulation from the Arthropodan Leg." *Canadian Journal of Zoology* 61 (1983): 1619–1669. An interpretation of structures from the side (pleuron) of Carboniferous fossils.

Labandeira, C. C., and J. J. Sepkoski. "Insect Diversity in the Fossil Record." *Science* 261 (1993): 310–315. A discussion of family-level insect diversity in the fossil record. Clarifies that insect fossils are neither rare nor necessarily poorly preserved.

Pritchard, G., et al. "Did the First Insects Live in Water or in Air?" *Biological Journal of the Linnean Society* 49 (1993): 31–44. A particularly clear summary of the issue, with a ranking of the value of different evidence germane to the aquatic/terrestrial origin question.

Wheeler, W. C., et al. "The Phylogeny of Extant Hexapod Orders." *Cladistics* (2001). A summary of morphological and molecular data used to support a phylogeny for the insects. Contains a complete bibliography of the subject.

— C. Riley Nelson

INTERSPECIFIC COEVOLUTION. *See* Coevolution.

INTRONS

One of the biggest surprises in the history of molecular biology came in 1977, when it was discovered that genes in eukaryotes are interrupted by noncoding regions known as introns. Introns are transcribed along with the coding region during the synthesis of messenger RNA and have to be removed, or spliced, by a protein-RNA complex called a spliceosome, before the transcript can be translated intro a protein (see Vignette). The regions of the gene that are spliced together to create a functional transcript are known as exons. Sequences that interrupt genes are also found in bacteria and the genomes of organelles such as mitochondria and chloroplasts. However, such introns are self-splicing parasitic elements, and it is the evolution of eukaryotic, or spliceosomal, introns that is the focus of this article.

The number and size of introns varies enormously between genes and organisms. In yeast, most genes are uninterrupted, whereas mammalian genes typically contain multiple introns, which together make up the vast majority of DNA within a gene. For example, the gene coding for dystrophin in humans, mutations in which cause Duchenne muscular dystrophy, consists of seventy-nine exons spread over about two million base pairs of DNA. Less than 1 percent of the gene, which takes up to sixteen hours to transcribe, is present in the spliced transcript. Even more remarkable, the introns of some genes are so large that other unrelated genes are located within them.

Introns and Gene Regulation. The location of introns in coding sequences is associated with certain short sequence motifs, known as splice sites, which are recognized by the spliceosomal machinery. Typically, introns begin with the nucleotide sequence GU and end with the pair AG, but there are other patterns, both in the intron and the adjacent coding regions, associated with splice sites (see Vignette). Regulatory sequences may also be found in introns. For example, expression of the *Igf2r* gene in mice is prevented by the binding of a repressor to a site within the second intron.

One type of gene regulation, called alternative splicing, depends critically on the presence of introns. Alternative splicing is the production of different transcripts from a single gene, by the splicing of alternative sets of exons. Alternative transcripts may be produced in different tissues or at different stages of development. Many transcripts differ only in the first exon, which is usually untranslated but contains regulatory elements. Some genes, however, can produce completely different proteins through alternative splicing. For example, two of the key genes involved in the sex-determination pathway of the fruit fly (*Drosophila melanogaster*), *doublesex* and *transformer*, have different forms expressed in developing male and female embryos. Another remarkable example of the use of introns occurs during the generation of antibody diversity in vertebrates, in which gene rearrangements within the loci coding for immunoglobulin genes create a vast diversity of sequences from a library of exons.

However, for most introns, the majority of the nucle-

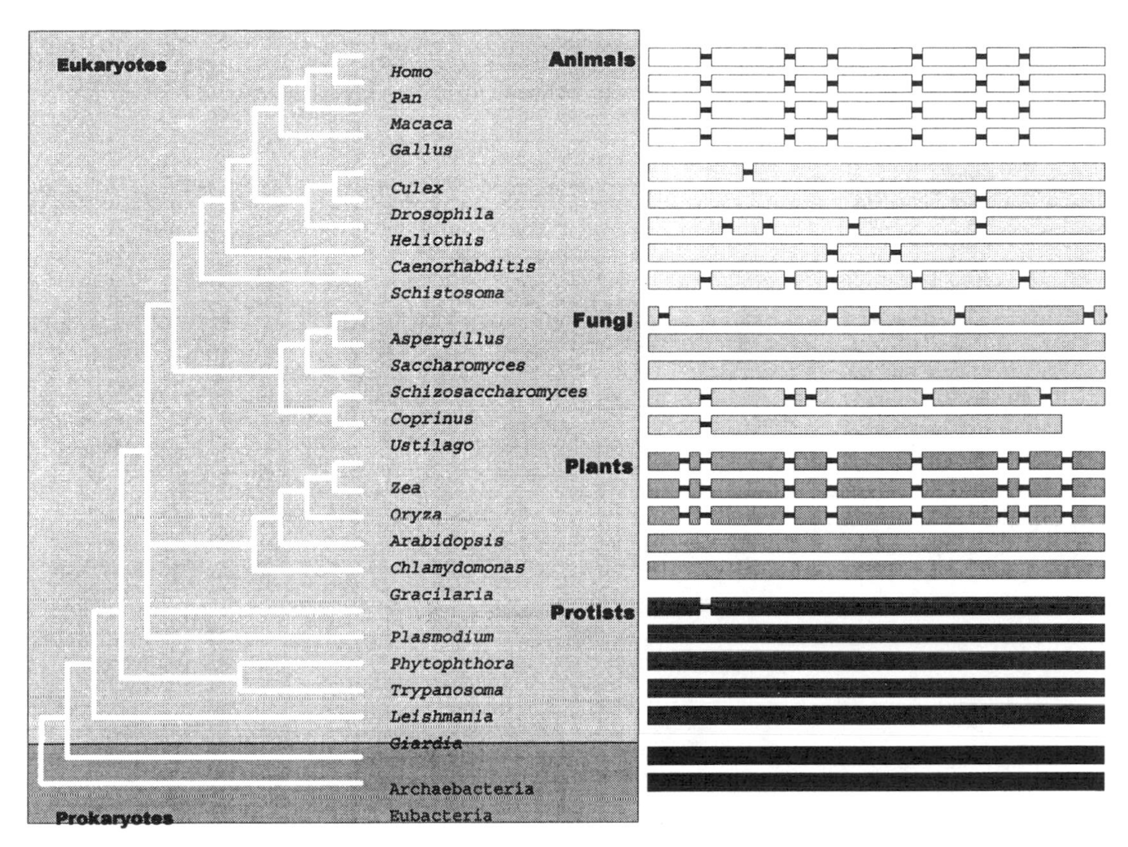

FIGURE 1. Exon Structure of the Triose-Phosphate Isomerase Gene in Eukaryotes and Prokaryotes.
Gaps in the solid bars indicate the presence of an intron. No intron has been found in any of the thirty-eight bacterial genomes that have been sequenced completely. *The Proceedings of the National Academy of Sciences, USA* 92 (1995): 8507–8511.

otide sequence is irrelevant to gene function. In agreement with the notion of little or no functional constraint, the rate of sequence evolution in introns is typically much higher than in the surrounding coding regions. In addition, they accumulate insertions and deletions at a high rate, which make it difficult to establish sequence homology, even between closely related species. Although the nucleotide sequence of introns tend to evolve rapidly, the location of introns within genes can be highly conserved between distantly related species or between duplicated (paralogous) genes within genomes. One example is the globin gene superfamily, members of which are involved in oxygen binding in animals, plants, and even bacteria. In vertebrates, the alpha- and beta-globin genes, thought to have been the result of a gene duplication event some 250 million years ago, both contain two introns located at identical positions relative to the coding sequence. Introns are also found in equivalent positions in the related myoglobin gene, which binds oxygen in muscle. Even the distantly related leghemoglobin, which binds oxygen in the root nodules of leguminous plants, is divided similarly, though an extra intron separates the central exon.

The Antiquity of Introns. The conservation of intron location across large phylogenetic distances has been documented for several genes present in both bacterial and eukaryotic genomes. This has led to the suggestion that the exon structure of genes is an ancient condition and represents the relic of a time when novel genes were formed by the shuffling of smaller, functional domains, each encoded by a separate exon. Under this scenario, prokaryotes and many unicellular eukaryotes have secondarily lost introns because of strong selective pressures to "streamline" their genomes.

The exon theory of genes, or introns-early hypothesis, is attractive because it provides a plausible means by which complex protein structures could have been assembled in the early stages of life and gives an adaptive explanation for the presence of what are perceived to be useless stretches of DNA. As a result, the theory has gained widespread acceptance in textbooks on molecular biology. The alternative, the introns-late hypothesis, is that the original genes were uninterrupted, and that introns have been inserted subsequently into eukaryotic genes.

Evidence for exon shuffling on an evolutionary times-

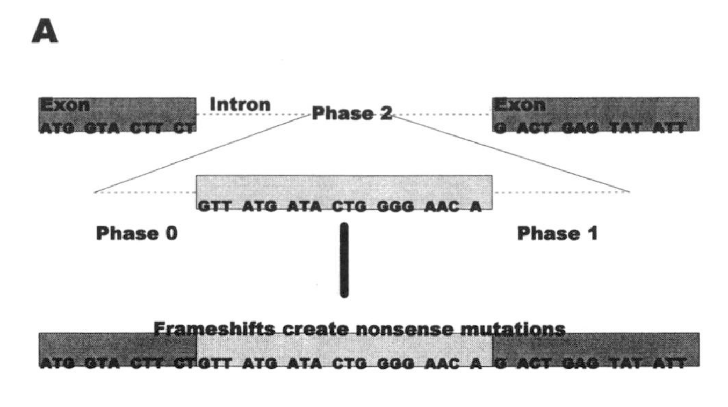

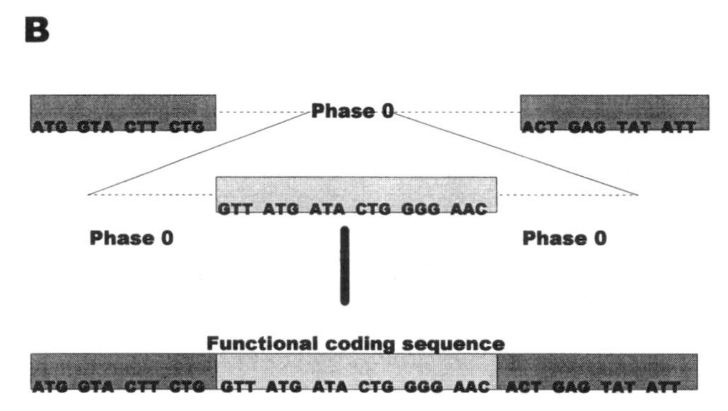

FIGURE 2. Intron Phase and Exon Shuffling.
Insertion of whole exons into the introns of existing genes can create novel functions, but it depends critically on intron phase (the position within a codon that an intron interrupts). If the phases of the combined introns differ, the resulting coding sequence will contain nonsense frameshift mutations (A). The resulting sequence will only be free of frameshifts if the introduced exon both begins and ends in the same phase as the intron into which it is inserting (B). Gilean McVean.

cale comes from the chimeric nature of genes such as the human low-density lipoprotein (LDL) receptor. The first six exons of the LDL receptor are related to the blood protein complement factor C9, and the next eight exons are related to the exons of the gene for the precursor of epidermal growth factor. Other genes show evidence for internal exon duplication. For example, the gene encoding collagen in chickens contains several repeats of a fifty-four base pair exon. But evidence for recent exon shuffling does not prove that introns are ancient. For the introns-early hypothesis to be credible, two things must be proved. First, putatively ancient introns should define functional subunits. Second, the distribution of introns in modern genes should be best explained by the loss of ancient introns. Both these points have been the focus of heated debate, and the gene Triose Phosphate Isomerase (TPI) has been at the center of the argument. The crystal structure of this enzyme is virtually identical in bacteria and mammals, but although yeast and bacterial genes are uninterrupted, there are multiple introns in both plants and animals, several of which are in identical locations. At first inspection, the shared introns appear to divide the protein into functional domains, occurring between major structural elements, such as alpha helices and beta sheets. This led Walter Gilbert to make the prediction in 1986 that there would have been an intron between the sequences encoding two lobes of a bilobed domain in the ancestral sequence. Eight years later, such an intron was found within the gene of the mosquito *Culex*.

However, triumph for the introns-early hypothesis was short-lived. The discovery in *Culex* prompted other groups to sequence TPI from a wider phylogenetic range of organisms. The picture that emerged was that the intron in *Culex* is one of the clearest examples of intron gain. An homologous intron is present in the close relative *Aedes*, but it is not found in the more distantly related mosquito *Anopheles* or any other organism. Although it is theoretically possible that *Culex* and *Aedes* are the only two lineages to retain an ancient intron, this hypothesis would require ten or more independent losses. Furthermore, these studies identified a large number of introns in novel positions within the TPI gene (Figure 1). A diversity of intron locations has also been found for other genes sequenced from a broad taxonomic range. Diversity of intron location is more parsimoniously explained by the frequent insertion of introns, rather than by the loss of ancient ones.

The claim that introns subdivide proteins into functional domains has also faced strong criticism. The original observation in TPI of a correlation between intron location and protein structure became obscured after the discovery of the novel intron locations in different species. Likewise, reanalysis of four ancient proteins, in which exon structure was proposed to correlate with protein domains, failed to find any evidence for nonrandom locations of any of the introns. However, analyses of larger data sets have identified other interesting patterns. First, introns are not randomly distributed with respect to position within codons that they interrupt. There is an excess of introns that neatly separate two codons, known as phase 0 introns. There is also an excess of symmetrical exons, such that an exon that starts within a given reading frame will tend to finish in the same reading frame. Both features facilitate exon shuffling, by increasing the probability that a novel protein will be free from frame-shift mutations (Figure 2). In addition, there is evidence that introns of phase 0 tend to separate proteins into modules of closely situated residues. However, because introns that interrupt codons do not show this pattern, even advocates of the introns-early hypothesis have to acknowledge that intron insertion does occur, and furthermore that the majority of introns are of recent origin.

The admission that intron insertions have occurred frequently in the history of many genes seriously weakens the argument for the antiquity of any intron. The process by which insertion occurs is unknown, but current evidence suggests that it is far from random. For example, the coding sequence immediately flanking introns tends to contain specific motifs (related to the motifs within introns known to be important in splicing), suggesting that they are targets for intron insertion. Nonrandom distribution of such motifs throughout genes may perhaps explain some of the correlations between intron phase and protein structure listed above. In addition, if the motifs are conserved between species, introns could independently insert into identical sites in independent lineages. One of the few cases where there is good evidence for a recent intron insertion is in the globin genes of the chironomid midge (*Chironomus thummi*). All members of the genus have three duplicated copies of the globin genes, but *C. thummi* is unique in that each gene contains an intron at an identical position (which is flanked by the putative insertion sequence). High levels of sequence similarity between the introns suggest that one of the genes acquired an intron in the *C. thummi* lineage, which subsequently spread to the other paralogues through gene conversion.

FIGURE 3. Intron Splicing in Yeast.
Based on Figure 31.8 of *Genes V*, written by Benjamin Lewin
© Oxford University Press and Cell Press 1994.

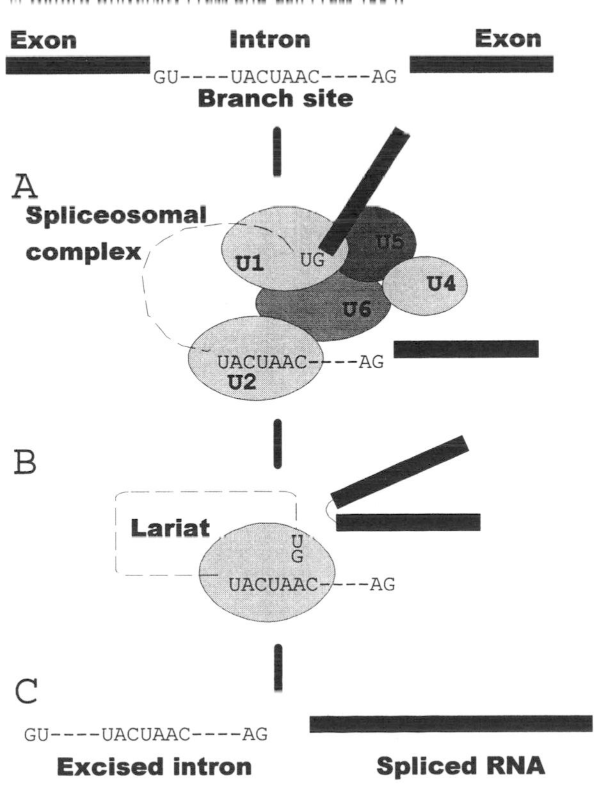

SPLICEOSOMAL REMOVAL OF INTRONS

Introns are stretches of DNA that interrupt the coding sequence of many eukaryotic genes. They are transcribed along with the coding sequence and are removed from the RNA by a molecular structure called the spliceosome. The spliceosome consists of several small nuclear RNAs, associated with protein complexes in ribonucleoprotein particles (snRNPs). During the splicing reaction, snRNPs bind to three specific motifs located within the intron sequence. These are the 5′ GU and 3′ AG doublets, which start and finish all eukaryotic introns, and the branch site, which is located about thirty bases upstream of the 3′ end of the intron. In yeast, the branch site is highly conserved and has the consensus sequence UACUAAC. Higher eukaryotes have a much greater diversity of branch site sequences, but they can be recognized by the preference for purines or pyrimidines at each position.

During the splicing reaction, an snRNP called U1 binds the 5′ splice site while U2 snRNP binds the branch site (see Figure 3). Subsequently, a complex of U4, U5, and U6 snRNPs bring the two sites together (A), allowing the 5′ splice site to be cleaved and forming a specific structure from the intron called a lariat (B). Finally, the 3′ splice site is cleaved, the spliced RNA is released, and the intron is linearized and rapidly degraded (C). Many of the features of the spliceosome, including the formation of the lariat intermediate, are also found in the self-splicing group II introns of bacterial and organellar genomes. However, whereas group II intron splicing is autocatalytic and requires no energetic input, the spliceosomal reaction requires the presence of adenosine triphosphate (ATP).

—GILEAN MCVEAN

Gene conversion can also provide a mechanism by which members of ancient gene families acquired introns in identical locations.

Without a clear understanding of the mechanism of intron insertion, perhaps the most critical evidence in the debate over the antiquity of introns is their phylogenetic distribution. If introns (and the genomic machinery required for their removal) were found in prokaryotes, this would be very strong evidence in favor of the introns-early hypothesis. However, no spliceosomal intron has been identified in any of the diverse eubacterial and archaebacterial genomes that have been completely sequenced. Within the eukaryotes, introns have been found in some of the lineages thought to represent early radiations such as microsporidians and amoebas. But other groups, currently thought to represent even earlier radiations, such as the diplomonad *Giardia* and the parabasalid *Trichomonas*, do not have any known introns. However, their genomes have not been sequenced completely, and there is uncertainty over the antiquity of several of these supposedly primitive eukaryotes. For example, the presence of mitochondria-derived genes in the nucleus of *Giardia* indicates that this group has lost mitochondria, rather than never having had them. As our understanding of the eukaryotic radiation improves, the inferred antiquity of spliceosomal introns may well increase. But there is no evidence to date that they predate the origin of the eukaryotes.

One piece of evidence, however, has continued to puzzle proponents of the introns-late theory. Organelles, such as mitochondria and chloroplasts, are descendants of eubacteria that became permanent residents (endosymbionts) within the cells of early eukaryotes. Comparison of organelle genomes with free-living relatives (the alpha-proteobacteria and cyanobacteria, respectively) shows that most of the ancestral symbiont's genes have been lost, and many others have been transferred to the nuclear genome of their hosts. Although organelle genomes do not contain spliceosomal introns, a test of the introns-early hypothesis would be to ask whether organelle genes that have been transferred to the nucleus share introns with ancient diverged homologues. There is one such pair of genes, those encoding the cytoplasmic and chloroplast forms of glyceraldehyde 3-phosphate dehydrogenase, which share five intron positions. The similarities are clearly no coincidence, and the gene pair was for a long time a cornerstone of the introns-early hypothesis. But the pattern could also be explained by gene conversion. In addition, across eukaryotes the density of introns in (nuclear-encoded) organelle-derived genes correlates with the density of introns in the rest of the genome, which suggests that the intron structure of organelle-derived genes reflect recent, rather than ancient, dynamics of intron evolution.

The Origin of Introns. The majority of evidence favors the introns-late hypothesis that spliceosomal introns arose early in the eukaryotic lineage, and therefore played no role in the early evolution of proteins. This conclusion raises two pertinent questions: where did introns come from, and what are they doing? The most likely scenario is that they are derived from a type of mobile genetic element found in bacteria and the organelle genomes of several eukaryotes. Although these genomes do not possess spliceosomal introns, some contain apparently functionless stretches of DNA within genes, which can excise themselves from messenger RNA transcripts without the aid of any cellular machinery. These elements are known as self-splicing introns and are classified into several different groups on the basis of the splicing mechanism. One such class, known as group II introns, bears a striking resemblance to the spliceosomal introns of eukaryotes in terms of the reaction sequence involved in splicing. In particular, both systems use a two-step transesterification scheme that generates an intermediate structure known as a lariat (see Vignette). The key difference is that the spliceosomal process requires a large number of proteins and RNAs, encoded for by nuclear genes, whereas group II introns are genuinely autocatalytic. Many features of the spliceosome are probably derived from elements within group II introns.

Self-splicing introns are almost certainly selfish genetic elements. Like transposable elements, they are genetic parasites that are of no benefit to the genome in which they are found, but promote their own survival through replication outside the cell cycle. Experiments have shown that group II introns excised through self-splicing produce a DNA copy that can insert into DNA free from the intron, a mechanism by which they can spread through a genome. Spliceosomal introns are therefore likely to be the descendants of selfish genetic elements that invaded the nucleus, perhaps from the genomes of bacterial endosymbionts. The loss of the self-splicing ability must have occurred after the host genome evolved the mechanism for their efficient removal, perhaps through co-option of the group II intron's machinery. The evolutionary forces that favored this transition are obscure, but because it shifts control of the splicing process (and therefore transposition) to the host genome, it may have led to the restraint, if not eradication, of the genomic parasites.

[*See also* Gene Families; Genes.]

BIBLIOGRAPHY

Blake, C. C. F. "Do Genes-in-Pieces Imply Proteins-in-Pieces?" *Nature* 273.5660 (1978): 267–268.

Cavalier-Smith, T. "Intron Phylogeny: A New Hypothesis." *Trends in Genetics* 7.5 (1991): 145–148. A discussion of the evolutionary origin of the different classes of intron; suggests that spliceosomal introns are the descendants of group II introns introduced into eukaryotes by the endosymbiotic ancestors of mitochondria.

Doolittle, W. F. "Genes in Pieces: Were They Ever Together?" *Nature* 272.5654 (1978): 581–582.

Gilbert, W. "Why Genes in Pieces?" *Nature* 271.5645 (1978): 501. The first suggestion that introns might speed the rate of evolution. The article sparked a series of comments, most importantly the suggestions that introns may be ancient (Doolittle, 1978) and that exons may correspond to protein domains (Blake, 1978).

Gilbert, W., M. Marchionni, and G. McKnight. "On the Antiquity of Introns." *Cell* 46.2 (1986): 151–154. Walter Gilbert's famous prediction about the location of an intron in TPI.

Gö, M. "Correlation of DNA Exonic Regions with Protein Structural Units in Haemoglobin." *Nature* 291.5810 (1981): 90–92. Important for the influential graphical method the article presents for relating intron location to protein structure.

Lewin, B. *Genes V*. Oxford, 1995. Chapter 23 of this established undergraduate textbook presents a completely introns-early view, with little reference to any counter arguments. Chapter

31 is an accessible summary of the mechanisms of intron splicing.

Logsdon, J. M., Jr. "The Recent Origins of Spliceosomal Introns Revisited." *Current Opinion in Genetics and Development* 8.6 (1998): 637–648. An excellent survey of the various battles between proponents of the introns-early and intron-late theories, from the viewpoint of the latter.

Sambrook, J. "Adenovirus Amazes at Cold Spring Harbor." *Nature* 268.5616 (1977): 101–104. A meeting report that captures the excitement surrounding the discovery of introns. The leaders of two key groups, Phillip Sharp and Richard Roberts, were later awarded the Nobel Prize in medicine and physiology.

Stoltzfus, A., D. F. Spencer, M. Zuker, J. M., Jr. Logsdon, and W. F. Doolittle. "Testing the Exon Theory of Genes: The Evidence from Protein Structure." *Science* 265.5169 (1994): 202–207. The first attempt to statistically analyze the claim that introns divide proteins into meaningful subunits.

— Gilean McVean

INVERTEBRATES. *See* Animals.

ISLAND BIOGEOGRAPHY. *See* Biogeography, *article on* Island Biogeography.

JELLYFISH. *See* Cnidarians.

KEY INNOVATIONS. *See* Novelty and Key Innovations.

KIMURA, MOTOO

Motoo Kimura was born in Okazaki, Japan, on 13 November 1924 and died in Mishima in 1994 on his seventieth birthday. Kimura is best known for his espousal of the neutral mutation, random drift theory of molecular evolution, but he also made many other important contributions to theoretical population genetics.

As a youth, Kimura wanted to be a plant taxonomist. With this goal in mind, he studied plant cytogenetics at Kyoto University. Ultimately, he began to dream of developing a rigorous mathematical approach to biology. As an undergraduate student, he taught himself mathematics and decided to become a theoretical population geneticist when he realized that the field was already quantitative. In 1949, a position was offered to him at the National Institute of Genetics in Mishima, and he remained at the institute until his death. Kimura and Gustave Malécot (1911–1998) were two successors of the three founders of population genetics, J. B. S. Haldane, R. A. Fisher, and Sewall Wright. They both focused on stochastic, or random, processes and thus brought about rebirth of the mathematical theory of population genetics.

Kimura worked on diverse problems, most of them in either population genetics or molecular evolution. The former includes problems related to random drift, natural selection and genetic load, population structure, fluctuating selection intensity, linkage and recombination, genetics of quantitative characters, and inbreeding systems. [*See* Linkage; Sex, Evolution of.] Early on in his research career, Kimura became interested in the debate between Fisher and Wright concerning the importance of random drift in evolution and strove to obtain a complete solution of the diffusion equation, which he hoped would settle the debate. A diffusion equation describing Brownian motion was developed independently by Albert Einstein and Marian von Smoluchowski. Fisher and Wright were the first to use it in genetics. Kimura succeeded in applying the diffusion method to significant problems in population genetics. In particular, he formulated the probability and the time of fixation (reaching a frequency of 100 percent) or extinction of mutant genes in a population. He was able to demonstrate that in the absence of natural selection, the fixation probability of a mutant gene becomes equal to the frequency in the population: thus, when a new mutation first appears, it typically has a very low chance of ever going to fixation. He also showed that the average time until fixation is in accord with Fisher's earlier conjecture that the number of individuals having a gene derived from a single neutral mutation cannot greatly exceed the number of generations since its occurrence. Kimura opened up the field of retrospective analysis of gene frequency changes. He was particularly interested in the age of a mutant gene and used this type of information to analyze the nature of existing genetic variation. His retrospective analysis later culminated in the rigorous mathematical treatment by J. F. C. Kingman of the genealogical relationships of genes at a locus—the coalescent.

Kimura was a theoretical biologist and constantly sought biologically interesting problems. This characteristic is reflected in many of the mathematical models he proposed. One of them is the stepping-stone model, named for a characteristic feature in Japanese gardens. In this model, each stepping-stone represents a breeding group, and the whole population of groups is united by migrations that are permitted only between neighboring groups. Topological relationships among locally breeding groups are essential for the understanding of how genetic variation correlates with geographic distance. Kimura also developed the model of infinitely many alleles, a widely used simplification in studying molecular evolution and polymorphism. In a paper coauthored with James F. Crow (1964), Kimura's mentor at the University of Wisconsin and a lifelong collaborator, Kimura analyzed the population consequences of highly mutable multiple alleles. All these models were developed in an attempt to interpret the emerging data produced by the then new technologies of gel electrophoresis and DNA sequencing.

Kimura was interested in evolutionary mechanisms of intraspecific variation as well as interspecific differences in homologous proteins and DNA segments. He never treated these two aspects separately; rather, he regarded intraspecific variation as a transient phase of interspecific differences and believed that a single theory must explain both. Comparative studies of hemoglobin and other molecules among different groups of animals led Kimura to the discovery that the rate of amino acid substitutions is too high to be consistent with Haldane's well-known estimate. This conclusion led him to propose in 1968 the neutral theory of molecular evolution, which states that most mutations are almost selectively neutral. Because of Kimura's forceful arguments over the years, the theory became widely accepted and revolutionized the way we think about evolution.

The neutral theory also fostered the development of molecular evolution and molecular taxonomy from the viewpoint of evolutionary mechanisms. Molecular evolution contrasts sharply with phenotypic evolution in rates and modes. Evolutionary rates are irregular at the phenotypic level but are much more regular at the molecular level. This regularity or constancy is referred to as the molecular clock hypothesis and is regarded as evidence of the neutral theory. The mode of phenotypic evolution is opportunistic, whereas that of molecular evolution is conservative. Yet random drift dominates at the molecular level, and the fates of new mutations are not predetermined. The neutral theory thus revealed an important aspect of nature, namely, that evolution at the molecular level is not an unconsciously creative process, but is largely dependent on a succession of accidents.

[*See also* DNA and RNA; Genetic Drift; Molecular Clock; Molecular Evolution; Neutral Theory.]

BIBLIOGRAPHY

Kimura, M. "Solution of a Process of Random Genetic Drift with a Continuous Model." *Proceedings of the National Academy of Science USA* 41 (1955): 144–150.

Kimura, M. "Evolutionary Rate at the Molecular Level." *Nature* 217 (1968): 624–626.

Kimura, M. "Preponderance of Synonymous Changes as Evidence for the Neutral Theory of Molecular Evolution." *Nature* 267 (1977): 275–276.

Kimura, M. *The Neutral Theory of Molecular Evolution*. Cambridge, 1983.

Kimura, M., and J. F. Crow. "The Number of Alleles that Can Be Maintained in a Finite Population." *Genetics* 49 (1964): 725–738.

— Naoyuki Takahata

KINGDOMS. *See* Classification.

KIN RECOGNITION

Kin recognition is the ability to identify relatives. More precisely, it is the differential treatment of members of the same species in a way that depends on their genetic relatedness to the discriminating individual. Generally, we can ask two kinds of questions about kin recognition. First, what function does it serve? Second, how do individuals recognize relatives? Before addressing these questions, consider the following examples of kin recognition.

Mexican free-tailed bats live in caves in colonies of millions. After giving birth to a single pup, a mother bat leaves her offspring clinging to the roof of the cave in a "crèche" that may contain four thousand pups per square meter. When the female returns, she flies back to where she last nursed her pup. Although mothers are besieged by dozens of unrelated babies, they recognize their own pup through vocal and olfactory communication.

Tunicates begin life as planktonic larvae. They eventually settle on a firm substrate, multiply asexually, and form an interconnected colony of genetically identical animals. Because large colonies survive better than small ones, colonies often attempt to fuse. If a tunicate attempts to join an unrelated colony, however, the colony emits chemicals that repel the unrelated individual.

Many plants possess self-sterility alleles that promote outcrossing. These alleles prevent the growth of pollen tubes in styles with which they have an allele in common by producing an antibody-like substance. Some species use this mechanism to avoid breeding not only with close relatives but also with plants that are extremely different genetically.

Certain amphibian larvae occur as two distinct types: a small-headed morph that eats invertebrates and a large-headed "cannibal" morph that feeds on other larvae. All larvae are born as the small-headed morph. They typically stay that way if they grow up among siblings, but often transform into cannibals if they grow up among nonkin. Moreover, when confronted with kin and nonkin, cannibals prefer to feed on the latter. In tiger salamanders, cannibals can even discriminate between nonkin and cousins, with whom they share, on average, only 12.5 percent of their alleles.

As these examples illustrate, kin recognition occurs in diverse species, and it may be expressed through behavior, chemical production, or morphological development. Given this diversity, what function does kin recognition serve?

Benefits of Kin Recognition. The main benefits of kin recognition are nepotism (i.e., helping relatives) and choosing an optimally related mate. Nepotism is important, because natural selection favors those individuals that are most successful at propagating their own alleles (alternative forms of the same gene). Personal reproduction achieves this goal in direct fashion, and this explains why selection often favors parental care. However, helping nondescendant kin (e.g., cousins) provides an indirect route to the very same end, because individuals also share alleles with such relatives through descent from a common ancestor (e.g., cousins have the same grandparents). Thus, individuals can propagate their alleles not only by reproducing themselves but also by helping relatives to reproduce. Kin recognition thereby facilitates actions that may help propagate the discriminating individual's alleles, a process known as kin selection.

Another important benefit of kin recognition is that it enables organisms to avoid inbreeding and perhaps choose an optimally related mate. All organisms possess a few deleterious alleles that are normally not expressed, and such alleles are likely to be carried by close relatives owing to recent common ancestry. With inbreeding (or selfing), progeny can inherit multiple copies of these alleles, leading to the outward expression of their harmful effects. Conversely, extreme outbreeding can also produce detrimental effects, by breaking up allele combinations that are adaptive for a given environment. Kin recognition enables organisms to select mates that are neither too closely nor too distantly related.

Mechanisms of Kin Recognition. The discussion above illustrates why kin recognition is adaptive. Indeed, kin recognition is so beneficial that it has evolved independently in many taxa. Moreover, the varied mechanisms used to identify kin reflect this diversity. To examine these mechanisms, it is useful to partition the process of kin recognition into three components: a recognition label is produced (e.g., a small-headed salamander larva produces a chemical signal as a metabolic byproduct), and an individual perceives and interprets this label (e.g., a cannibal morph uses the signal to assess its relatedness to the other larva); consequently, that individual acts appropriately (e.g., the cannibal does or does not eat the small-headed larva, depending on its hunger state). Below, we examine each component in turn.

Production component. Kin recognition labels may be produced by recipient organisms themselves (phenotypic recognition), or, alternatively, may lie in clues related to time or place that do not involve recipient phenotypes (nonphenotypic recognition). The latter works only when relatives occur at predictable times or locations. For example, before engaging in cannibalistic behavior, mother spiders leave the area into which their young disperse, thereby decreasing the probability of killing their own offspring.

Phenotypic recognition cues can be any aspect of an organism's phenotype that signifies kinship reliably. Chemical cues in particular are widely used, possibly because they potentially contain much information while requiring little energy to obtain or produce, as when they are metabolic byproducts. Moreover, chemical cues are readily interpreted by the body's immune system.

Phenotypic recognition labels can be of genetic or environmental origin. Examples of the former are glycoproteins, which are encoded by genes in the major histocompatibility complex (MHC). These proteins reside in cell membranes where their function is to recognize invader organisms. Glycoproteins react differently toward genetically identical cells of the body in which they reside than toward genetically dissimilar cells from different individuals (e.g., bacteria).

Although MHC genes may have evolved initially in the context of fighting pathogens, they now mediate kin recognition in many animals, including tunicates and humans. Because each MHC locus has as many as fifty alleles, each individual often has a unique MHC genotype. Relatives tend to be more similar than nonrelatives, however.

The MHC somehow affects the odor of its carrier, enabling other individuals to use smell to detect the carrier's genotype. For example, female house mice prefer males with dissimilar MHC genotypes, apparently to avoid inbreeding. However, when pregnant or lactating, females form communal nesting and nursing associations with other females that have similar MHC genotypes. Presumably, such nepotism helps the female mice propagate alleles they share in common with their pregnant relatives. Interestingly, human females also prefer odors of males with dissimilar MHC genotypes, except when they are taking birth control pills; in that case, they prefer odors of males with similar genotypes. The function of the former choice may be optimal outbreeding, whereas the latter may indicate nepotistic kin recognition by (pseudo-) pregnant females, that is, females may prefer to associate with kin when pregnant. (See Vignette for another example of kin recognition and nepotism.)

KIN RECOGNITION AND NEPOTISM IN BEE-EATER SOCIETIES

Kin recognition plays a vital role in facilitating nepotism. Consider, for example, the white-fronted bee-eater, a small bird native to East and Central Africa. These birds breed in colonies by excavating nesting chambers in sandy riverbanks. The 40 to 450 individuals in a colony are subdivided into "clans" consisting of from three to seventeen individuals. Each clan may contain several sets of parents and their offspring, which defend their own feeding territory.

Young that are old enough to breed on their own often remain and help breeding pairs. These helpers assist with nest building, nest defense, and help feed the nestlings. Helpers usually have a choice of assisting nestlings that vary in degree of relatedness to them.

Emlen and Wrege (1989) found that bee-eater helpers assist the most closely related individuals available (see Figure 1). This assistance is vital to the recipients: on average, the presence of each helper results in an additional 0.47 offspring being reared to fledging. Moreover, because helpers are related to recipients, helpers benefit indirectly by promoting the spread of alleles they share with recipients.

—David W. Pfennig

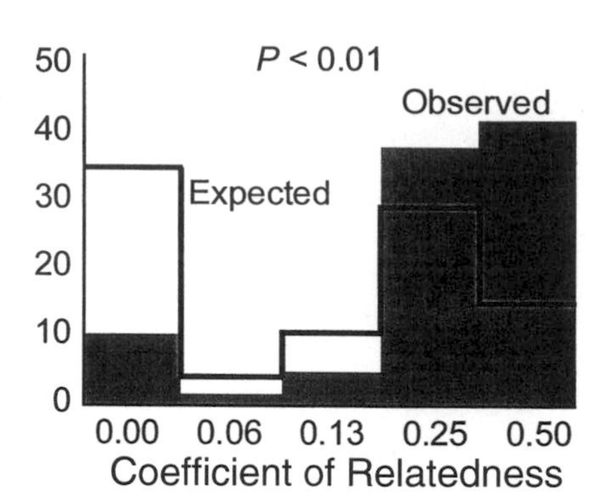

FIGURE 1. Probability of Helping as a Function of Relatedness. Courtesy of David W. Pfennig.

Other organisms, such as paper wasps, use labels that may be environmental in origin. Paper wasps distinguish nestmates, which are often kin, from unrelated non-nestmates by smell. Each wasp assimilates a colony-specific odor from its nest, which the queen and workers construct by mixing saliva with plant fibers. This colony-specific odor, which serves as the recognition cue, is locked into each enclosing wasp's epicuticle (skin) before it hardens. The source of the cue is odoriferous hydrocarbons. These compounds are derived from the plant fibers that make up the nest paper as well as from secretions produced by the wasps that constructed the nest. The relative importance in kin recognition of odors derived from plant fibers versus those applied by the wasps themselves is uncertain. However, environmental cues potentially could rival genetic cues in diversity, and therefore, in reliability as recognition cues, if different colonies use different mixtures of plants in nest construction.

Generally, the relative importance of genetically based versus environmentally acquired labels depends on which type of cue is less likely to cause a recognition "error." Relying solely on environmentally acquired cues might cause individuals to mistakenly assist nonrelatives that inhabit the same environment. These nonrelatives could reap the fitness rewards of beneficence without reciprocating, and so increase in frequency in the population. Relying solely on gene products also might cause individuals to mistakenly assist nonrelatives that carry "outlaw alleles"—alleles that encode the recognition cue, and which spread by helping themselves, but not the rest of their bearer's genome.

The likelihood that these errors will occur depends on each species' genetic system and ecology. Organisms can minimize the chance that two nonrelatives will share similar gene products by using cues encoded in highly polymorphic genetic loci, such as MHC loci. Such genetically based labels are most useful for organisms that occur in relatively homogeneous environments. For organisms that occur in more heterogeneous habitats, environmentally acquired labels permit accurate discrimination of kin without the potential risks of outlaw alleles.

Perception component. Once a recognition label is produced, how do individuals use such labels to identify kin? Interestingly, no organism appears to possess a genetically encoded internal representation or "template" of its kin. This is probably because as a result of meiotic shuffling of genetic cues and spatial variation in environmental cues, the characteristics of kin will differ for different individuals, rendering genetically encoded templates unreliable.

Organisms form kin recognition templates through learning. In particular, individuals learn labels from their nestmates, their environment, or themselves, depending on which of these referents is available and which is least likely to cause a recognition error. For example, young animals often attend to the cues associated with early companions, which are often parents or siblings. Thus, female great tit chicks learn their fathers' songs and, as adults, females avoid pairing with males whose songs match this template. In many species, parents also imprint on their newborn offspring.

Other species, such as paper wasps, learn cues from their nest. Use of the nest as a referent may provide a more complete record of kin recognition labels than would a subset of nestmates. Still other species, such as honeybees, learn characteristics of kin by imprinting on their own phenotype. Such self-inspection enables kin recognition when available relatives would yield error-prone templates as, for example, when nestmates are full and half siblings due to multiple mating by their mother, or when opportunities to assess relatives' phenotypes are limited. Regardless of the referent used, the process of template formation resembles classical imprinting in that learning occurs during a sensitive period, often early in life, and memories of the labels are durable.

Once its template is formed, an individual discriminates among others on the basis of how similar the recipient individual's phenotype is to the previously learned cues. If the match is close enough, then the recipient is identified as kin. Such a mechanism—dubbed phenotype matching—causes individuals to behave differently toward others in a manner correlated with their degree of relatedness.

Action component. Given a sufficiently close match between the recipient individual's phenotype and the discriminator's template, the outward manifestation of kin recognition takes place only when the latter decides what action to take. This decision often depends upon the context. In general, individuals are expected to treat relatives and nonrelatives differently only when the fitness benefits of discrimination exceed the costs.

For example, cannibalistic tadpoles are less likely to avoid eating siblings when they are hungry than when they are satiated. Presumably, cannibals cease to discriminate kin when the cost of passing up a meal increases, and, thus, when their own survival is threatened. This flexibility in kin preference makes sense, because a cannibal is always more closely related to itself than to its sibling. This example also illustrates why kin recognition is not always expressed: the fitness costs of expression might exceed the benefits.

Significance of Kin Recognition. The ability to identify relatives can affect an individual's fitness profoundly. The diversity of kin recognition mechanisms reflects the underlying diversity among species in their ecology, their genetic systems, and the selective context in which kin recognition is used. Such variety in kin recognition mechanisms might even trace back to the origins of life. In their book *The Major Transitions in Evolution*, John Maynard Smith and Eörs Szathmáry speculate that kin selection was critical in facilitating cooperation among genes, chromosomes, cells, and organisms, which in turn enabled the diversity we see in nature. Thus, kin recognition may have been vital in the origin of life.

[*See also* Alarm Calls; *articles on* Eusociality; Kin Selection; Social Evolution.]

BIBLIOGRAPHY

Bateson, P. "Optimal Outbreeding." In *Mate Choice*, edited by P. Bateson, pp. 257–277. Cambridge, 1983. Discusses how kin recognition might help individuals choose an optimally related mate.

Dawkins, R. *The Selfish Gene*. Oxford, 1976. A highly accessible book by an author who argues that natural selection acts on genes through their phenotypic effects.

Emlen, S. T., and P. H. Wrege. "A Test of Alternative Hypotheses for Helping Behavior in White-Fronted Bee-Eaters of Kenya." *Behavioral Ecology and Sociobiology* 25 (1989): 303–319.

Gamboa, G. J., H. K. Reeve, and W. G. Holmes. "Conceptual Issues and Methodology in Kin-Recognition Research: A Critical Discussion." *Ethology* 88 (1991): 109–127.

Grafen, A. "Do Animals Really Recognize Kin?" *Animal Behaviour* 39 (1990): 42–54. According to Grafen, most examples of kin recognition are epiphenomenal, meaning that they are not maintained by kin selection, but rather by selection for species or group member recognition. Grafen also asserts that kin recognition favored by kin selection must be mediated by genetic cues. For an opposing view, see Sherman et al. (1997).

Hamilton, W. D. *The Narrow Roads of Gene Land*, vol. 1, *Evolution of Social Behaviour*. Oxford, 1996. Contains the seminal writings of W. D. Hamilton, the developer of kin selection theory.

Holmes, W. G., and P. W. Sherman. "The Ontogeny of Kin Recognition in Two Species of Ground Squirrels." *American Zoologist* 22 (1982): 491–517. Describes a classic series of experiments.

Manning, C. J., E. K. Wakeland, and W. K. Potts. "Communal Nesting Patterns in Mice Implicate MHC Genes in Kin Recognition." *Nature* 360 (1992): 581–583.

Maynard Smith, J., and E. Szathmáry. *The Major Transitions in Evolution*. Oxford, 1995.

Pfennig, D. W., J. P. Collins, and R. E. Ziemba. "A Test of Alternative Hypotheses for Kin Recognition in Cannibalistic Tiger Salamanders." *Behavioral Ecology* 10 (1999): 436–443. Surprisingly, the fitness consequences of kin recognition have rarely been documented. This study examines the function of kin recognition in cannibalistic species.

Reeve, H. K. "The Evolution of Conspecific Acceptance Thresholds." *American Naturalist* 133 (1989): 407–435. Provides a theoretical framework for understanding successes and failures of kin recognition.

Sherman, P. W., H. K. Reeve, and D. W. Pfennig. "Recognition Systems." In *Behavioural Ecology: An Evolutionary Approach*, edited by J. R. Krébs and N. B. Davies, 4th ed., pp. 69–96. Oxford, 1997. Discusses principles common to all recognition systems.

Waldman, B. "Kin Recognition in Amphibians." In *Kin Recognition*, edited by P. G. Hepper, pp. 162–219. Cambridge, 1991. An excellent overview of the evidence for, and mechanisms and functions of, kin recognition in amphibians and other organisms.

Waser, N. M., and M. V. Price. "Crossing Distance Effects on Prezygotic Performance in Plants: An Argument for Female Choice." *Oikos* 68 (1993): 303–308. Describes how kin recognition might facilitate mate complementarity in plants.

— David W. Pfennig

KIN SELECTION

Kin selection theory has its origins in attempts to understand why some organisms have evolved to help other organisms of the same species. Such helping behavior has puzzled researchers, because an organism that helps another will likely incur some reproductive cost, such as a loss of resources to allocate to its own offspring, increased mortality risk, or reductions in other components of Darwinian fitness (i.e., reproductive success). This means that any gene that increases the probability of helping behavior will reduce its frequency in future generations and ultimately will be lost, unless there is some compensating reproductive benefit—some additional path by which helping behavior causes a net increase in the underlying gene's frequency.

William D. Hamilton (1964) provided a pivotal insight into the evolution of altruistic behavior by pointing out that altruistic acts directed toward relatives produce an important kind of reproductive compensation. By enhancing the reproduction of relatives, a gene that increases the probability that its bearer provides help indirectly propagates copies of itself in those same relatives. When the relatives are the helper's offspring, the help is referred to as "parental care" and the enhancement of propagation of the helping genes as a positive "direct effect" of those genes. But Hamilton's theory showed that help-inducing genes benefit in an essentially identical way when the help is directed toward nondescendant relatives, such as siblings. When the latter "indirect effect" is sufficiently strong, the helping gene can spread (increase in frequency in the population) despite decreases in its propagation through the offspring of the individual in which it resides (i.e., the positive indirect effects can outweigh negative direct ef-

COMPUTING RELATEDNESS FROM GENETIC DATA

Relatedness can be computed from genetic data as follows (Queller and Goodknight, 1989; Reeve et al., 1992): Let A be the average frequency of the altruism gene in altruists, p be the population-average frequency of the altruism gene, and R the average frequency of the altruism allele in the kin group within which an altruist interacts. Then it can be shown that $Ar + (1 - r')p = R$ (Grafen 1985, 1991), where r is the "regression measure of relatedness" (so called because it is equal to the slope of the regression of the gene frequency of an individual's group on the gene frequency within the individual). Rearranging the latter equation, $r = (R - p)/(A - p)$. The regression measure of relatedness is essentially identical to the genealogical measure of relatedness r when selection is not too strong (Grafen, 1985, 1991). This measure of relatedness nicely illustrates that relatedness depends on above-background genetic similarity ($R - p$), not only on the absolute genetic similarity (R) between altruists and recipients.

—LAURENT KELLER AND KERN H. REEVE

fects). Despite the terminology, there is no sense in which indirect effects, when measured properly, are less important than direct effects (Dawkins, 1979).

There are two distinct, mutually consistent, fitness-accounting methods in quantitative kin selection theory: neighbor-modulated fitness and inclusive fitness (Hamilton, 1964; Grafen, 1982; Maynard Smith, 1982a; Reeve, 1998). Neighbor-modulated fitness, which is closely allied to the concept of fitness used in population genetics theory, focuses on the reproductive costs or benefits *received* by an altruist from others (the term *neighbor-modulated* is intended to connote fitness influences from the social environment). In contrast, inclusive fitness focuses on the reproductive benefits and costs *dispensed* by an altruist to others (the term *inclusive* is used to emphasize that effects of altruism on all kin, not just on oneself, are to be combined).

Hamilton showed that both methods of accounting lead to exactly the same condition for the spread of altruism. This was an important achievement, because inclusive fitness is generally much easier to compute than is neighbor-modulated fitness, but there are pitfalls (Grafen, 1982) as the latter but not the former depends on the population frequency of the altruism allele.

Inclusive Fitness. In the inclusive-fitness accounting approach, the focus is on reproductive effects dispensed by the individual bearing the altruism-producing gene. The idea is to combine all the dispensed reproductive effects of an individual into an "inclusive fitness" for that individual in a way that correctly predicts the evolution of the altruism (Hamilton, 1964). Hamilton's verbal prescription for calculating inclusive fitness was essentially as follows: Take the baseline personal reproductive output of the individual and add to this the possibly costly effect of the individual's behavior on itself (this sum is called the "personal component" of inclusive fitness), and also add the sum of the effects of the individual's behavior on the reproduction of others, weighted by the individual's genetic relatedness to those other individuals (this weighted sum is called the "kin component" of inclusive fitness). The behavior that maximizes average inclusive fitness is predicted to spread through natural selection.

Before computing inclusive fitness, a rigorous understanding of the concept of genetic relatedness must be reached. Relatedness is a measure not of the absolute genetic similarity between two individuals (a common misconception), but of the degree to which this similarity exceeds the "background" similarity between individuals randomly drawn from the population. One way to compute relatedness is to calculate the probability that a given allele in the donor of altruism (the altruist) is present in the recipient of altruism via common descent (Hamilton, 1964; Grafen, 1985); this is termed *genealogical relatedness*. (Another way to compute relatedness from genetic data is explained in the Vignette.) Having obtained the value of the genetic relatedness, r, between an altruist and a recipient, we can now compute inclusive fitness. Let the average reproductive output of nonaltruists be equal to x offspring. This is also the baseline reproductive output for altruists, and thus the total personal reproductive component for altruists is $x - c$, where c is the cost of altruism. The recipient of altruism receives a mean increment b to its reproductive output. It then follows from the definition that the average inclusive fitness for altruists is

$$x - c + rb \tag{1}$$

Altruism will begin to spread when the inclusive fitness for altruists exceeds the inclusive fitness for nonaltruists; that is when $x - c + rb > x$, which simplifies to

$$rb > c \tag{2}$$

The latter condition is referred to as Hamilton's rule for the evolution of altruism. Hamilton's rule is easily generalized to encompass multiple interactions with different kinds of kin as

$$\sum_{i-1}^{N} r_i \, b_i > 0 \tag{3}$$

where N is the number of kin plus self, relatedness to self equals one and the b_i's associated with the *ith* kin classes can be either positive or negative (i.e., either benefits or costs to the recipients). Importantly, this rule can be applied to the evolution of any phenotype, not just behavioral altruism.

However, the generalized rule has important limitations (Grafen, 1985). In particular, its use assumes that (1) costs and appropriately weighted benefits can be added together (as opposed to multiplied or otherwise combined) to determine the overall fitness effect of altruism, (2) selection is weak enough that the genetic relatedness can be treated as constant and equivalent to the genealogical relatedness, and (3) altruism is random according to whether recipients of a given relatedness r actually possess the gene promoting altruism (Grafen, 1984, 1991). When these assumptions are not met, other methods for computing the evolutionary outcomes of kin selection must be sought (Queller, 1992; Frank, 1998). Importantly, Hamilton's rule is more likely to be valid when the expression of altruism is so strongly context dependent that recipients of altruism do not themselves express the same altruistic behaviour (Parker, 1989).

Altruism among relatives. Hamilton's theory of kin selection predicts that an individual will tend to behave differently toward conspecifics of different degrees of relatedness, and there are numerous examples demonstrating that this is indeed the case. Here we take the example of alarm calls to illustrate how variation in kin structure can influence the expression of altruism within groups.

In many social groups of vertebrates, individuals give alarm calls when predators approach. For example, when groups of Belding's ground squirrels are threatened by a coyote or a weasel, some of the squirrels stand up on their hind legs and produce a high-pitched screech. Callers most likely suffer a cost because calling makes them conspicuous and presumably more likely to be attacked by the predator. Two lines of evidence suggest that individuals are more likely to give alarm calls when surrounded by a higher proportion of relatives. First, there is sex-biased dispersal, and females are more likely than males to stay close to their natal area, resulting in a higher relatedness of females than males to neighbors. Accordingly, females were found to call much more frequently than males. Second, females that had close relatives nearby called more frequently than females without (Sherman, 1977), again as predicted by kin selection.

In another species of rodent, the black-tailed prairie dog, Hoogland (1983) further investigated whether individuals behaved differently depending on whether kin were offspring or nondescendant relatives. He studied alarm calling by presenting individuals with a stuffed badger, the badger being a natural predator. He found that the proportion of times individuals gave an alarm call significantly increased when they were surrounded by kin, but that it made no difference whether these were offspring or nondescendant kin. These results are in line with the prediction that the warning of offspring and the warning of siblings (direct and indirect effects of altruism, respectively) are in fact equivalent ways of increasing inclusive fitness.

Reproductive altruism in vertebrate societies. In at least 220 bird and 120 mammal species, young are reared not only by their parents, but by other individuals as well. Typically, these helpers are young individuals that help their parents to rear younger siblings. The occurrence of such cooperative breeding raises two interconnected questions. Why do offspring remain with their parents rather than disperse and attempt to breed independently on their own? And why do the offspring that remain at home engage in costly tasks, for example collecting food to feed their parents' offspring?

Numerous studies, mainly in birds, suggest that offspring stay home because the opportunities for successful dispersal and independent breeding are limited relative to the payoff for staying at home. There are several, ways in which helping might be beneficial given that offspring stay at home. First, helpers may obtain direct benefits, for example by gaining experience and thus increasing their fitness as parents when they breed on their own later. Second, helpers may gain indirect benefits by increasing the numbers of relatives produced, thereby increasing their inclusive fitness.

Several lines of evidence demonstrate that kin selection is an important force favoring the evolution of cooperative breeding. In most of the cooperative-breeding bird and mammal species in which it has been investigated, individuals were found to preferentially assist their closest relatives (Emlen, 1997). For example, in white-fronted bee-eaters, kinship is a strong predictor of both whether a given individual becomes a helper and to whom it provides help (Emlen and Wrege, 1988). Nonbreeders are more likely to become helpers when the breeding pairs in their family are close genetic relatives, and when faced with a choice of potential recipient nests, they preferentially help the pair to whom they are most closely related.

Interestingly, kinship does not only determine whether or not an individual will help, but also the amount of help provided. In pied kingfishers, for instance, helpers may be related or unrelated to the breeding pair. Related helpers work as hard as the breeding pair, but unrelated ones work less (Reyer, 1984). Similarly, cooperative-breeding Seychelles warblers exhibit significantly higher helping efforts (food provisioning and period of helping) when rearing full sibs than when rearing half sibs (Komdeur, 1994). Overall, these results demonstrate that, as predicted by kin selection, individuals are more likely to help kin than nonkin and that the level of altruism increases with relatedness.

Eusociality in insects. Of all the cases of altruism to be found in the animal kingdom, surely the most extreme is the behavior of the workers of social insects. Some ants, for example, form colonies comprising up to one million sterile workers specializing in tasks such as building the nest, collecting food, rearing the young, and defending the colony. In these colonies reproduction is restricted to one or a few individuals, the queens. The term *eusociality* refers to such societies that are characterized by reproductive division of labor, cooperative brood care, and (generally) overlapping of generations.

There is currently no doubt that kin selection has been the all important selective force responsible for the evolution of eusociality and reproductive altruism by workers (Bourke and Franks, 1995; Crozier and Pamilo, 1996). Numerous genetic studies have revealed that eusociality evolved within groups of highly related individuals, such as one mother and her offspring. There are a few ant species in which the relatedness between nest mates is close to zero, but this low relatedness stems from an increase in queen number that occurred long after the evolution of morphological castes and reproductive division of labor. It is still unclear whether such societies are of recent origin and evolutionarily unstable, or whether the benefits of worker helping in such societies is large enough so that Hamilton's rule is still satisfied despite the low relatedness of workers to the helped brood (Bourke and Franks, 1995; Keller, 1995).

Eusociality has evolved independently many times among the insects and most frequently in the Hymenoptera (wasps, bees, and ants). Interestingly, Hymenoptera have a haplodiploid mechanism of sex determination that generates peculiar patterns of relatedness. Because unfertilized eggs give rise to males and fertilized eggs to females, sisters always receive the same set of paternal genes and therefore share 75 percent of their genes identical by descent. (This assumes, as is often the case, that the queen has mated a single father. Because males are haploid, all of the father's sperm are identical.) As a consequence, Hymenopteran females are more related to their full sisters than to their own offspring, and it has been suggested that this may explain the prevalence of eusocial origins in the Hymenoptera (Hamilton, 1964). The haplodiploidy hypothesis would also explain another interesting feature, that workers are exclusively females in social Hymenoptera but not in the termites (which are always diploid).

Although the haplodiploid hypothesis is appealing, it turns out that it is not so simple, because the high relatedness between sisters is balanced by the low relatedness of females to brothers ($r = 0.25$). Hence, workers gain by rearing siblings rather than offspring only if workers lay the male eggs or if workers can concentrate on raising sisters while males are produced by solitary females (Seger, 1983). These conditions do not always apply in contemporary species and it is possible that haplodiploidy has not been a very important factor responsible for the maintenance and possibly even the origin of eusociality. The fact that most cases of extreme reproductive altruism are found in Hymenoptera might be attributable to other features of this group, for example the unusually high frequency of parental care (Alexander, 1974), a preadaption for caring for nondescendant relatives, or the enhanced protection from chance loss that rare alleles for worker altruism receive in haplodiploid as compared with diploid systems (Reeve, 1993).

Conflicts within Kin Groups. Although high relatedness favors high levels of cooperation, potential conflicts persist in groups composed of kin, whether they are cooperative breeding birds or eusocial insects. Potential conflicts arise because, in contrast to cells of an organism, group members are not genetically identical. Hence, kin selection predicts that individuals with partially divergent genetic interests may attempt to favor the propagation of their own genes, possibly to the detriment of other group members. Group members can compete over direct reproduction or over how to allocate group resources to various relatives, and the potential conflict may translate into actual conflict or may remain unexpressed (Ratnieks and Reeve, 1992). Interestingly, the study of conflicts provides some of the best tests of kin selection, as illustrated by the patterns of allocation of resources to the production of male and female reproductives in social Hymenoptera.

Kin selection theory predicts that the value of new queens and males is influenced by their relatedness to other colony members (Hamilton, 1964). The haplodiploid sex-determination system in social Hymenoptera results in asymmetries in the relatedness of workers to females and males with, in colonies headed by a single queen, workers being three times more related to new sister queens than they are to their brothers. Thus, if workers control the sex ratio, they should invest three times more in females than males. In contrast, because queens are equally related to their sons and daughters, they should invest equal amounts in males and females if queens control colony sex ratios. Therefore, kin selection predicts a conflict between queens and workers over sex-investment ratios (Trivers and Hare, 1976).

Cross-species comparison of population-level sex-investment ratios in ants that generally have a single queen per colony showed that the population sex-investment ratio is globally female-biased (1.7:1). This indicates that workers have some control over colony sex ratios and that, as predicted by kin selection, they bias sex investment toward females. Unfortunately, it is very difficult to assess reliably the relative cost of production of males and queens, and a proportion of the ant species in this comparative study probably depart from the simple family structure expected when the colony is headed by one queen mated with a single male. As a result, interspecific comparisons of this type do not allow precise determination of the theoretically expected and observed population-wide investment sex ratios and thus the relative power of queens and workers in biasing colony sex ratios to their advantage.

The most complete demonstration of queen-worker conflict over sex allocation and the all-important role of kin selection comes from a study in the ant *Formica exsecta* (Sundström et al., 1996). The study population consists of colonies headed by single queens mated with either one or multiple males. Multiple mating by queens decreases the relatedness between workers and the new queens to be raised but does not influence the relatedness of workers to males. Theory shows that under such conditions workers maximize their inclusive fitness by producing the sex to which they are most related compared to the population average, that is new queens in nests headed by a singly-mated queen and males in colonies with a multiple-mated queen (Boomsma and Grafen, 1990). The queen controls the primary proportion of males and females by regulating the proportion of haploid and diploid eggs she lays. However, workers may subsequently modify the sex-investment ratio by selective rearing one sex or the other. Comparison of sex ratio at the egg and adult stage showed that workers eliminated a high proportion of males in colonies headed by single-mated queens, leading these colonies to produce mostly females. By contrast, males were kept alive in colonies headed by a multiple-mated queen (Sundström et al., 1996). Hence, in this population, workers win the conflict against the queen and bias colony sex ratio so as to maximize their inclusive fitness. Similar patterns of split sex ratios have been documented in sixteen other species of social Hymenoptera (Queller and Strassmann, 1998). In some other species, however, there is apparently no association between sex ratio and colony level relatedness, suggesting that queens may also have some means to achieve their colony sex ratio interests, for example by limiting the number of female eggs produced (Pamilo, 1982; Reuter and Keller, 2001).

Overall, studies of sex allocation in social Hymenoptera and other within-group conflicts (e.g., Pfennig, 1993) demonstrate that the nature and expression of many conflicts depends on the genetic structure of the group, as predicted by kin selection theory. Paradoxically, the outcomes of within-group conflicts strongly support kin selection theory, a theory which was first proposed to explain the evolution of cooperation. More generally, these conflicts also reflect the most basic principle of Darwinian evolution, namely that organisms are selected to maximize the number of copies of their own genes transmitted to the next generation.

BIBLIOGRAPHY

Alexander, R. D. "The Evolution of Social Behavior." *Annual Review of Ecology and Systematics* 5 (1974): 325–383.

Boomsma, J. J., and A. Grafen. "Intraspecific Variation in Ant Sex Ratios and the Trivers-Hare Hypothesis." *Evolution* 44 (1990): 1026–1034.

Bourke, A. F. G., and N. R. Franks. *Social Evolution in Ants*. Princeton, 1995.

Crozier, R. H., and P. Pamilo. *Evolution of Social Insect Colonies: Sex Allocation and Kin-Selection*. Oxford and New York, 1996.

Dawkins, R. *The Selfish Gene*. Oxford and New York, 1976.

Dawkins, R. "Twelve Misunderstandings of Kin Selection." *Zeitschrift für Tierpsychologie* 51 (1979): 184–200.

Emlen, S. T. "Predicting Family Dynamics in Social Vertebrates." In *Behavioural Ecology: An Evolutionary Approach*, edited by J. R. Krebs and N. B. Davies, 4th ed., pp. 228–253. Oxford, 1997.

Emlen, T. S., and P. H. Wrege. "The Role of Kinship in Helping Decisions among White-Fronted Bee-Eaters." *Behavioural Ecology and Evolution* 323 (1988): 305–315.

Frank, S. A. *Foundations of Social Evolution*. Princeton, 1998.

Grafen, A. "How Not to Measure Inclusive Fitness." *Nature* 298 (1982): 425.

Grafen, A. "Natural Selection, Kin Selection, and Group Selection." In *Behavioural Ecology: An Evolutionary Approach*, edited by J. R. Krebs and N. B. Davies, 2d ed., pp 62–89. Sunderland, Mass., 1984.

Grafen, A. "A Geometric View of Relatedness." In *Oxford Surveys in Evolutionary Biology*, edited by R. Dawkins and M. Ridley, vol. 2, pp. 28–89. Oxford and New York, 1985.

Grafen, A. "Modelling in Behavioural Ecology." In *Behavioural Ecology: An Evolutionary Approach*, edited by J. R. Krebs and N. B. Davies, 3d ed., pp. 5–31. Oxford, 1991.

Hamilton, W. D. "The Genetical Evolution of Social Behaviour, I and II." *Journal of Theoretical Biology* 7 (1964): 1–52.

Hoogland, J. L. "Nepotism and Alarm Calling in the Black-Tailed Prairie Dog (Sciuridae: Cynomys spp.)." *Animal Behaviour* 31 (1983): 472–479.

Keller, L. "Social Life: The Paradox of Multiple-Queen Colonies." *Trends in Ecology and Evolution* 10 (1995): 355–360.

Keller, L., and H. K. Reeve. "Dynamics of Conflicts within Insect Societies." In *Levels of Selection in Evolution* edited by L. Keller, pp. 153–175. Princeton, 1999.

Komdeur, J. "The Effect of Kinship on Helping in the Cooperative Breeding Seychelles Warbler (Acrocephalus sechellensis)." *Proceedings of the Royal Society of London, B*, 256 (1994): 47–52.

Maynard Smith, J. "The Evolution of Social Behaviour: A Classification of Models." In *Current Problems in Sociobiology*, edited by King's College Sociobiology Group, pp. 29–44. Cambridge, 1982a.

Maynard Smith, J. *Evolution and the Theory of Games*. Cambridge, 1982b.

Pamilo, P. "Genetic Evolution of Sex Ratios in Eusocial Hymenoptera: Allele Frequency Simulations." *American Naturalist* 119 (1982): 638–656.

Parker, G. A.. "Hamilton's Rule and Conditionality." *Ethology, Ecology, and Evolution* 1 (1989): 195–211.

Pfennig, D. W., and J. P. Collins. "Kinship Affects Morphogenesis in Cannibalistic Salamanders." *Nature* 362 (1993): 836–838.

Queller, D. C. "A General Model for Kin Selection." *Evolution* 46 (1992): 376–380.

Queller, D. C., and K. F. Goodknight. "Estimating Relatedness Using Genetic Markers." *Evolution* 43 (1989): 258–275.

Queller, D. C., and J. E. Strassmann. "Kin Selection and Social Insects." *Bioscience* 48 (1998): 165–175.

Ratnieks, F. L. W., and H. K. Reeve. "Conflict in Single-Queen Hymenopteran Societies: The Structure of Conflict and Processes

That Reduce Conflict in Advanced Eusocial Species." *Journal of Theoretical Biology* 158 (1992): 33–65.

Reeve, H. K. "Haplodiploidy, Eusociality and Absence of Male Parental and Alloparental Care in Hymenoptera: A Unifying Genetic Hypothesis Distinct from Kin Selection Theory." *Philosophical Transactions of the Royal Society of London*, Series B: Biological Sciences, 342 (1993): 335–352.

Reeve, H. K. "Acting for the Good of Others: Kinship and Reciprocity with Some New Twists." In *Handbook of Evolutionary Psychology: Ideas, Issues, and Applications*, edited by C. Crawford and D. L. Krebs, pp. 43–86. Mahwah, N. J., 1998.

Reeve, H. K., D. W. Estneat, and D. C. Queller. "Estimating Average within Group Relatedness from DNA Fingerprints." *Molecular Ecology* 1 (1992): 223–232.

Reuter, M., and L. Keller. "Sex Ratio Conflict and Worker Production in Eusocial Hymenoptera." *American Naturalist* 158 (2001): 166–177.

Reyer, H. U. "Investment and Relatedness: A Cost/Benefit Analysis of Breeding and Helping in the Pied Kingfischer (Ceryle rudis)." *Animal Behaviour* 32 (1984): 1163–1178.

Seger, J. "Partial Bivoltinism May Cause Sex-Ratio Biases That Favour Eusociality." *Nature* 301 (1983): 59–62.

Sherman, P. W. "Nepotism and the Evolution of Alarm Calls." *Science* 197 (1977): 1246–1253.

Sundström, L., M. Chapuisat, and L. Keller. "Conditional Manipulation of Sex Ratios by Ant Workers—A Test of Kin Selection Theory." *Science* 274 (1996): 993–995.

Trivers, R. L., and H. Hare. "Haplodiploidy and the Evolution of the Social Insects." *Science* 191 (1976): 249–263.

— Laurent Keller and Kern H. Reeve

L

LAMARCK, JEAN BAPTISTE PIERRE ANTOINE DE MONET

Naturalist Jean Baptiste Pierre Antoine de Monet, Chevalier de Lamarck (1744–1829) was born in Picardy, France. He grew up during the Enlightenment, a period characterized by great confidence in reason, criticism of traditional institutions, dissemination of knowledge, and, in particular, confidence in the explanatory power of science. The Enlightenment attempted to liberate human thought from mythical and metaphysical concepts and construct knowledge of the world based on scientific evidence. Lamarck subscribed to the Enlightened perspective of his times, in which living beings represented a progression, from the simplest organisms to the most complex ones, with humans as the highest life form.

Lamarck's major conceptual works included the first extensive theory of evolution and the notion of the "living being," according to which both animals and plants have characteristics in common and should not be studied separately as botany and zoology, but instead as a new science: biology. Lamarck aimed to go beyond the description of organisms outlined by Carolus Linnaeus, passing from the historical and descriptive to the explanatory and philosophical. He is perhaps best known for his belief that evolution and adaptation proceed via the inheritance by offspring of traits acquired by the parents during their lifetimes, even though this was a relatively minor part of his wider thinking.

Lamarck promoted two complementary theories of Nature. On the one hand, Nature has a plan and a final goal. On the other hand, Nature undergoes unpredictable changes. Evolution occurs as the result of two fundamental causes. One cause is the inherent capacity of life for change, a capacity that is led by Nature's plan for the creation of different organisms. The other cause is the environmental change that obliges organisms to adapt themselves to new situations.

Lamarck, like his mentor, Georges-Louis Leclerc Buffon, was a nominalist and maintained that Nature had not formulated classes, orders, families, genera, or even species. Instead, there are only individuals who supersede one another. This was a controversial view in the eighteenth century. For example, creationists such as Linnaeus understood adaptation as a God-given match of the organism to its environment.

By contrast, Lamarck understood adaptation as an ongoing process based on the interaction between an organism and its environmental circumstances. Although Lamarck did not use the term *adaptation*, his fundamental belief was that organisms change to suit their environment. He understood organismal change not as a direct action by, for example, a creator, but rather as the intervention of necessity. Considering that organisms have needs to satisfy and that they do so in an environment, when this environment changes, the organisms find themselves obliged to change to continue satisfying their needs. In a word, changes in circumstances entail changes in needs, and changes in needs entail changes in the ways an organism lives its life.

Lamarck believed that where these changes are long lasting, the use or nonuse of certain organs provokes either enhancement or atrophy. For example, organisms that need to climb trees will have, over time, developed strong muscles for climbing. The traits that are acquired are then inherited by offspring provided that the new morphological feature is common to both sexes of the species. This is Lamarck's inheritance of acquired characteristics and is now often referred to as Lamarkian inheritance.

Lamarck believed in the uniformitarian view that the same causes that at present transform organisms little by little have, over long periods of time, caused major changes (gradualism). He was opposed to positions, such as those of Georges Cuvier, that explained changes in the history of life by rare catastrophes rather than the gradual accumulation of small changes.

During the seventeenth and eighteenth centuries, many theorists maintained that the parts of animals were designed for particular functions. Thus, Linnaeus and the English theologian and philosopher William Paley (1743–1805) spoke of a perfect adaptation to the environment in which living beings exist, given that each organ was specifically constructed to carry out certain tasks and not others. In the Linnaeus–Paley view, the adapted form was externally created to match the function. Lamarck proposed the opposite: the functions of organs (which arise as responses to a changing environment) have led to the form we observe in each animal. Lamarck's functionalist concept was highly criticized by Cuvier, who emphasized that the internal laws of development determine the structure of each organ rather than organs being modified to meet the external challenges posed by the environment. Darwin's vision includes both perspectives.

Lamarck developed an instructionist theory of adaptation, whereas modern evolutionary ideas are based on a selectionist theory of adaptation. According to Lamarck, the environment provides information directly to each organism about the characteristics needed to survive. The organism changes directly in response to the environmental information. According to modern theory, the environment selects only those organisms that, from a population of variants already present, are endowed with the characteristics necessary for survival. The population changes because the environment selects those organisms that are best adapted.

[*See also* Lamarckism.]

BIBLIOGRAPHY

Burkhardt, R. W., Jr. *The Spirit of System: Lamarck and Evolutionary Biology.* Cambridge, Mass., 1977. A seminal book on Lamarck that covers the tenets of Lamarck's methodology, especially his inclination for setting up complete explanations based on general and unusual principles.

Corsi, Pietro. *The Age of Lamarck.* Berkeley, 1988. Makes an important contribution to the comprehension of the fundamental change that Lamarck proposed for the classification of living beings, from a linear to a branching system: the "series."

Madaule, M. B. *Lamarck ou le mythe du précurseur.* Paris, 1979. Analyzes Lamarck's life, his distinguished career and conceptions, in a historical context. Here, the relationship between Lamarck and other contemporary and earlier thinkers is well documented.

— ROSAURA RUIZ

LAMARCKISM

Lamarckism has come to mean the theory that evolutionary change can occur through the inheritance of acquired characters. However, this notion is only part of the complex set of ideas about evolution that were developed by the French biologist Jean Baptiste Lamarck (1744–1829), and it was not original to him. Over two thousand years earlier, the Hippocratic physicians had suggested that differences between human populations—for example, differences in skin color and stature—are the result of inherited environmental effects, especially the effects of climate and diet. They reasoned that "semen," which comes from all parts of the body of both parents and mingles in the embryo, contains substances representing the parents' features—their skin color, limb shape, temperament, and so on. Whatever changes the parents undergo are reflected in their semen and are therefore inherited by their offspring. Although later thinkers found flaws in the theory, the idea that acquired characters are inherited was accepted by the majority of naturalists and philosophers until the early twentieth century, and was incorporated into nineteenth-century evolution theories.

Lamarck's Lamarckism. Lamarck first wrote about the transmutation of species in 1800, but he described his theory more fully in *Philosophie zoologique*, published in 1809. Earlier, in the mid-1790s, Charles Darwin's grandfather Erasmus Darwin had suggested something similar in his *Zoonomia*, in which he described how, through the cumulative effects of the inheritance of features that arise in response to the animal's way of life, one species is transformed into another. However, it was Lamarck who first suggested a coherent and systematic theory of evolution, based on the then available knowledge of anatomy, paleontology, physiology, and psychology.

Lamarck saw in existing animals a continuous progressive series from the simple to the most complex. He believed that this series is formed because, during successive generations, spontaneously generated simple organisms undergo transformations. He assumed two causes of change. The first and most important was inherent in the organization of living matter, which he believed differs from nonliving matter in that it is self-organizing, self-maintaining, and self-complicating. The tendency for self-complication produces a progressive increase in complexity, which, if it acted alone, would lead to a linear evolutionary series.

Lamarck's evolutionary series are not completely linear, however. They are affected by his second cause of evolutionary transformation, the ability of organisms to adapt to their conditions physiologically and to transmit this acquired adaptation to their offspring. For example, a giraffe develops a long neck and forelegs through the habit of browsing on tall trees, constantly having to stretch upward. In cave-dwelling organisms, the eyes often degenerate as a result of many generations of lack of use. Lamarck believed that slight changes resulting from use and disuse are transmitted to the progeny, and very slowly, over many generations, they accumulate and become stable. Individual physiological adaptations become evolutionary adaptations.

Darwin's Lamarckism. Darwin assigned the major role in evolution to natural selection, but even in the first edition of *On the Origin of Species* (1859) he acknowledged that the inherited effects of use and disuse also play a part. In later editions and in subsequent books, as he struggled to understand the origin and maintenance of variation, he gave the inheritance of acquired characters a more prominent role in evolution.

Darwin's theory made understanding heredity very important, because for natural selection to work, organisms must have a continuous supply of heritable variations. In *The Variation of Animals and Plants under Domestication* (1868), Darwin put forward his "provisional hypothesis of pangenesis" to explain the mass of information about heredity that he had collected. It was a version of the ancient Hippocratic theory: each "organic unit" of the body continuously throws off minute "gemmules," some of which end up in the reproductive organs, where they form the hereditary material. During embryonic development, gemmules from the parents join together, and each type forms the corresponding part of the embryo's body. Darwin's theory thus accommodated the inheritance of acquired characters, because modified body parts throw off modified gemmules.

Neo-Lamarckism. Although the idea of organic evolution was accepted by most biologists soon after the publication of *On the Origin of Species*, Darwin's main mechanism of evolution, natural selection, was not. The main problem besetting natural selection was the difficulty in seeing how it would work when, as then assumed, inheritence occurred through a process of blending. Alternative mechanisms were sought. The most popular was the inheritance of acquired characters, with or without an internal driving force. Prominent evolutionary theorists such as Herbert Spencer and Ernst Haeckel believed that the inheritance of acquired characters was as important as natural selection. The coordinated nature of adaptations, which required correlated changes in many parts of the organism, seemed to demand it, and the parallels between embryology and phylogeny were seen as evidence of it.

In the United States, Lamarckism flourished. Paleontological findings indicating long-term, linear evolutionary trends, and seemingly nonadaptive geographic clines in size and color, were marshaled as proofs of Lamarckian mechanisms. In 1885 the American naturalist Alpheus Packard coined the term *neo-Lamarckism* to describe the set of evolutionary theories that incorporated the inheritance of acquired characters and internal driving forces. These ideas were in marked contrast to the neo-Darwinian views that were being championed by August Weismann, who denied any role for the inheritance of acquired characters. Weismann attacked Lamarckism on the following grounds: first, the empirical evidence was weak; second, there was no known mechanism for the transfer of somatic (bodily) adaptations to the germ line; and third, it was extremely unlikely that adaptive changes at the organ level could be translated to the chromosomal level. Weismann demanded experimental proof of the inheritance of acquired somatic characters and a plausible mechanism for its operation.

As Mendelian genetics developed and pangenesis-type theories were abandoned, neo-Darwinians also expected Lamarckians to show that acquired characters affected genes. However, the experimental evidence Lamarckians provided was beset with problems. First, it was not conclusive—usually it was possible to explain the results through selection of preexisting hidden variations or new mutations. Second, the acquired characters were frequently unstable, or did not segregate in a Mendelian fashion. Third, the results were often peculiar to a particular group of organisms and hence deemed nonrepresentative. For example, strange patterns of inheritance in protozoa, especially ciliates, were assumed to be a peculiarity of this group. Fourth, it was sometimes difficult or impossible to repeat the results, and some results were fraudulent.

The case of the midwife toad is the most notorious example of fraud. Paul Kammerer, a Viennese biologist, conducted a study in which midwife toads, which normally mate on dry land, mated in moist conditions. The males developed dark swellings on their forelimbs, which resembled the nuptial pads that help males grasp females in water-mating species. According to Kammerer, the adaptive pads in his experimental toads became heritable. Most of Kammerer's experimental material was destroyed during World War I, but a single old and tatty pickled toad remained. When other scientists examined it in 1926, they found that the pad area had recently been injected with ink. Soon after this, Kammerer shot himself, thereby reinforcing the suspicion that he was responsible for the forgery, and casting doubts on his other evidence for Lamarckian inheritance. The author and maverick evolutionist Arthur Koestler tried to salvage Kammerer's reputation in the 1970s but with limited success.

Just before his suicide, Kammerer had been offered a position in the Soviet Union, where during the Stalinist era Mendelian genetics and its practitioners would be repressed. Under the direction of Trofim Lysenko, Lamarckian ideas flourished and were applied to agriculture. The consequences were disastrous, and Western scientists saw the approach as at best methodologically unsound and at worst fraudulent. As the cold war intensified, Lamarckism, which had little experimental evidence to support it and no plausible mechanisms to explain it, became more and more suspect in the West. Its last stronghold in the 1940s was in bacteriology. The discovery that after a lengthy lag period, bacteria acquired and transmitted the ability to use new foodstuffs or resist new antibiotics was at first interpreted in Lamarckian terms. Later it was recognized that these adaptations are not genetic, but physiological, involving the turning on of genes and the transmission of gene products to daughter cells. Other cases of induced changes and non-Mendelian inheritance, found especially in plants and fungi, were also at first interpreted as instances of Lamarckian inheritance, but were later reinterpreted in genetic terms when it was recognized that genes are present in cytoplasmic organelles.

Lamarckism Today. Since Weismann's day, five major theoretical arguments against Lamarckism have been reiterated in various guises. The first is that it is impossible to see how the hereditary material, which can preserve and transmit characters for many generations, can also respond to and be changed by environmental influences. The Lamarckians' answer is that induced heritable changes are small, and only persistent environmental influences produce the cumulative effects that lead to visible stable inheritance.

A second objection, made in recent years by John Maynard Smith and Richard Dawkins, is that many acquired variations are the harmful results of old age, injuries, and disease, and it would be detrimental for organisms to pass on these "acquisitions." The Lamarckians' answer is that, as with DNA changes, most of which are also detrimental, evolved internal selective filters eliminate many harmful variants, and most of those that remain are removed by the external filter, natural selection.

A third argument, voiced by Weismann and others, is that because many aspects of evolution cannot be explained in terms of the inheritance of acquired characters, Lamarckism is an inadequate and unnecessary concept. The evolution of characters that are not associated with use and disuse (for example, warning colors in butterflies) are impossible to explain in Lamarckian terms. Moreover, although Lamarckism depends on adaptability, it cannot explain the evolution of the capacity to adapt. Lamarckians have responded by acknowledging that natural selection is indeed sometimes the only explanation for the evolution of an adaptation, and it alone can explain the evolution of adaptive plasticity. However, this is no reason to deny the possibility of the inheritance of acquired physiological adaptations. The inheritance of acquired characters is not an alternative to natural selection, it is complementary.

A fourth argument, again stemming from Weismann, is that continuity between generations is through the germ line, not the soma, and germ-line cells cannot be influenced by somatic events. The Lamarckians' response is that many organisms, particularly plants, do not have a separate germ line, and some plants and invertebrates reproduce through fragmentation or budding. This certainly allows the inheritance of acquired characters. Moreover, modifications in the soma can affect germ cells, as discussed below.

The fifth argument also originated with Weismann. The modern version, stressed by Dawkins, is that since development is epigenetic, there is no one-to-one correspondence between genes and parts of the body; therefore phenotypic changes cannot be back-translated into genotypic changes. There are two responses to this objection. One is that the assumption that phenotypes can be inherited only via genes is invalid, and therefore Lamarckism does not require the back-mapping of phenotypes into genotypes. The other is that molecular biology has revealed mechanisms through which genomes can be modified in response to changed environmental conditions.

In the late 1970s Ted Steele proposed a plausible mechanism through which adaptive somatic changes could cause changes in germ-line DNA. He suggested that if somatic mutations exist that produce diversity among cells and this is followed by selection (as happens in the immune system), copies of the messenger RNA in the selectively favored variant cells could be picked up by retroviral vectors and carried to the germ line. There they could be integrated into DNA through reverse transcription. In this way the selected somatic character could be transmitted to descendants.

A different type of genomic change in response to external factors has been found in microorganisms. Recurring environmental conditions, such as those that produce moderate starvation, seem to have led to the evolution of bacterial systems that preferentially generate DNA changes in relevant genes. The result is that new mutations are not completely blind; they are biased by external stimuli.

It is accepted that external conditions bias the generation of variation in the non-DNA heredity systems—those that transmit differences in cell states and in behavior. In cell lineages, epigenetic variations in gene expression or cell structures are inherited through transmissible differences in chromatin structure (such as the state of DNA methylation), protein complexes (such as prions), and self-sustaining regulatory loops. Such heritable, non-DNA differences, which may be accidental but are frequently induced, can sometimes also be passed from one generation of organisms to the next. At a higher level of organization, environmentally induced parental effects—for example those mediated by hormones—and learned behavior are often transmitted through several generations.

For most of the twentieth century, biological heredity was identified with genetics, which was based on genes that change rarely and randomly. Lamarckism was difficult to reconcile with this view of heredity. However, now that molecular studies have revealed that some genetic variations are not completely blind, and it has been recognized that epigenetic and behavioral inheritance are part of biological heredity, there is a renewed interest in the inheritance of acquired characters. Lamarckism is once more being reexamined and reassessed, but in a much more sober manner, based on facts and experiments. It is clear that evolutionary adaptations predominantly arise from natural selection, but there is probably also a role for Lamarckism, not as a rival to Darwin's evolutionary theory but as part of it.

[*See also* Lamarck, Jean Baptiste Pierre Antoine de Monet.]

BIBLIOGRAPHY

Blacher, L. I. *The Problem of the Inheritance of Acquired Characters*. New Delhi, India, 1982. A translation by F. B. Churchill of an account of the rise and fall of Lamarckian ideas in the Soviet Union, written in 1971, soon after Lysenko was denounced. Difficult to find, even in good libraries.

Bowler, P. J. *The Eclipse of Darwinism.* Baltimore, 1983. An excellent review of neo-Lamarckism and other late-nineteenth- and early-twentieth-century evolution theories.

Dawkins, R. *The Blind Watchmaker.* Harlow, Essex, England, 1986. One of the strongest statements of modern neo-Darwinism; criticizes alternative explanations of adaptive evolution, including various forms of Lamarckism.

Jablonka, E., and M. J. Lamb. *Epigenetic Inheritance and Evolution: The Lamarckian Dimension.* Oxford, 1995. Gives an historical background before discussing Lamarckism in the light of modern cell and molecular biology.

Jablonka, E., M. J. Lamb, and E. Avital. "Lamarckian Mechanisms in Darwinian Evolution." *Trends in Ecology and Evolution* 13 (1998): 206–210. A review of the evolution and the evolutionary effects of heredity systems that allow the inheritance of induced and learned variations.

Koestler, A. *The Case of the Midwife Toad.* London, 1971. Describes the life and work of Paul Kammerer, focusing on the famous deception.

Lamarck, J. B. *Zoological Philosophy.* London, 1914. English translation by H. Elliot of *Philosophie zoologique,* Paris, 1809. Reprinted, with a useful historical introduction, Chicago, 1984. Chapter 7 of part 1 summarizes Lamarck's evolution theory, but the entire book, including the rarely read parts 2 and 3, is worth reading.

Steele, E. J., R. A. Lindley, and R. V. Blanden. *Lamarck's Signature.* St. Leonards, New South Wales, Australia, 1998. A summary of the somatic selection hypothesis of the inheritance of acquired characters.

Weismann, A. *The Evolution Theory.* 2 vols. London, 1904. J. A. Thomson and M. R. Thomson's translation of *Vorträger über Descendenztheorie* (2d ed., Jena, 1904), Weismann's masterly exposition of neo-Darwinism at the turn of the twentieth century.

— EVA JABLONKA AND MARION J. LAMB

LANGUAGE

[*This entry comprises two articles. The first article focuses on the issue of language versus nonhuman primate communication; the companion article discusses the evolution of linguistic diversity and the relationship of genetic diversity to linguistic diversity. For related discussions, see the* Overview Essay *on* Human Genetic and Linguistic Diversity *at the beginning of Volume 1;* Hominid Evolution, *article on* Neanderthals; *and* Primates, *article on* Primate Societies and Social Life.]

An Overview

Although the ancestral and closely related hominid species that preceded contemporary human beings are extinct, comparative studies of the communicative abilities of humans, chimpanzees, and other primates suggest that some of the biological bases of human language have a long evolutionary history, while other attributes may have more recent origins. Theory and models consistent with the process of evolution by natural selection applied to several lines of evidence provide the most promising tools to account for the origin and subsequent development of human language.

Human speech has a central role in language, allowing us rapidly to transmit information vocally. It is not possible to present a detailed timetable for the evolution of the anatomy and neural mechanisms that regulate human speech. Some aspects of speech are "primitive," occurring in many species that are ancestrally distant from human beings. For example, virtually all mammals and many other species (for example, frogs) communicate by means of laryngeally generated vocal signals. The larynx generates audible signals through the process of phonation. The vocal cords, or folds, complex assemblages of muscle, cartilage, and other soft tissues, rapidly open and close, producing a sequence of puffs of air. We perceive the rate at which these puffs of air occur—the fundamental frequency of phonation (F0), as the pitch of a person's voice; childrens' voices have high F0s, adult males low F0s. Many human languages, such as the Chinese languages, use controlled variations in F0 to signify different words. In English, the end of a spoken declarative sentence is signaled by an abrupt falling F0, and yes–no questions conclude with a rising F0. Because chimpanzees and many other species, including aquatic mammals, modulate F0 to produce meaningful calls, it is apparent that this capacity existed in the earliest stages of human evolution, five million years ago.

Fully developed human speech depends on specialized anatomy that appears to have evolved more recently in anatomically modern humans. The melding of individual sounds into syllables that yields the rapid transmission rate of human speech is an automatic consequence of the manner in which the airway above the larynx, the supralaryngeal vocal tract (SVT), modulates sound energy. The SVT acts as an acoustic filter, letting maximum energy through at particular frequencies that are called formants. Different vowels and consonants are characterized by particular formant frequency patterns that are generated by forming different SVT shapes. The shape of the SVT, independent of laryngeal activity, determines the formant frequency pattern much as the filtering properties of a pair of sunglasses are determined by its lenses. The formant frequency patterns of individual phonemes are automatically merged, or encoded, into syllables as we speak. For example, when the word "do" is uttered, the SVT initially produces the shape that yields the formant pattern of the consonant /d/. However, the SVT cannot instantly change in shape from the /d/ sound to the shape that is necessary to produce the formant frequencies of the vowel that follows the /d/. As the SVT shape gradually changes, the formant frequency pattern melds the consonant and vowel into a syllable. We "hear" the individual sounds that constitute a syllable by means of a speech mode of perception that unconsciously accounts for the physics of speech production.

Natural selection directed at enhancing the reliability of speech mode perception appears to account for the evolution of the human supralaryngeal airway, which differs from that of all other primates. The relationship between formant frequency patterns and phonemes depends on the length of a speaker's SVT. A long SVT produces lower formant frequency patterns for the same speech sound than does a short SVT, and a particular formant frequency pattern can signify different words spoken by speakers having longer or shorter supralaryngeal airways. As we listen to speech, we must unconsciously adjust for the length of a talker's SVT. The vowel /i/ (in phonetic notation, the vowel of English "see") provides an optimal signal for determining the length of a speaker's SVT. Studies of the fossil record and of living primates show that the human skull and supralaryngeal airway have evolved to allow the production of the vowel /i/.

The supralaryngeal airways of apes and monkeys cannot produce the formant frequency patterns of the vowel /i/ because their tongues are contained entirely in their mouths. The larynx in nonhuman primates is positioned close to the passage opening to the nose and can move upward, locking into the nasal airway to form a sealed respiratory pathway that allows the animal to drink or swallow small pieces of food while breathing. Human newborn infants retain this anatomy until about three months of age; however, the human mouth migrates backward relative to the base of the skull from birth to one year of age, while the larynx moves down into the throat, reaching an adultlike position by age six. This unique human growth process yields a supralaryngeal airway that allows us to produce the vowel /i/, as well as other sounds that are inherently more distinct than the sounds other primates can produce. The relatively thin tongues of other primates are positioned entirely in their mouths. In contrast, the back of the human tongue is almost round in profile, and half of the tongue is positioned in the mouth and half in the pharynx of the throat, oriented at a right angle to the mouth. The human mouth and pharynx have about the same length; this allows the production of the vowel /i/ as well as the vowels of "father" (/a/) and "who" (/u/).

Nonhuman primates follow an opposite growth pattern. Their faces resemble newborn humans at birth but gradually project forward as they mature. The fossil remains of the skulls of early hominids bear a striking resemblance to those of living primates in this regard. Their faces project ahead of their foreheads; their long mouths preclude the production of the vowel /i/ and other speech sounds. The skulls of fossil Neanderthals, who diverged from modern human beings within the last half million years, appear to follow the nonhuman primate growth pattern. In contrast, anatomically modern human skulls dated to between 100,000 and 150,000 years before the present appear to have supported vocal tracts capable of the full modern range of speech sounds. The biological cost of this restructuring of human speech anatomy was noted by Charles Darwin: food can lodge in the human larynx, causing death by asphyxiation. Moreover, in contrast to Neanderthals, our molar teeth are crowded and can become impacted and infected. The utility of speech apparently offsets these negative attributes of the human supralaryngeal airway. Nonetheless, it is evident that some form of speech must have existed in the variants of *Homo erectus* ancestral to both Neanderthals and modern humans; otherwise, there would have been no reason for the retention of the otherwise negative biological characteristics of the human SVT that enhanced speech. The skulls of chimpanzees and earlier hominids appear to support similar vocal anatomy that would have precluded their producing sounds like /i/.

Studies of the linguistic abilities of chimpanzees provide a window on the earliest stage of hominid evolution. Present-day chimpanzees probably are not equivalents of the common ancestors of apes and humans who lived five million years ago; however, the brain volumes of living chimpanzees are similar to those of the earliest fossil hominids who diverged from the common ancestor. When chimpanzees are exposed to manually transmitted languages (American Sign Language and computer keyboards that key artificial speech) and spoken human languages in an environment similar to that of young human children, they acquire limited lexical and syntactic ability. In these circumstances, chimpanzees can productively use and comprehend 150 to 200 words. Moreover, they can comprehend simple spoken English sentences such as "Put the apple in the refrigerator." Language-using chimpanzees coin new words; they communicate and argue with humans and each other, using language. One young chimpanzee acquired sign language from other chimpanzees without formal instruction, as human children do. Therefore, it is quite unlikely that early hominids, who possessed somewhat larger brains than contemporary chimpanzees, lacked the biological substrate that confers the ability to form and use words that encode concepts, or to produce and comprehend short sentences that employ simple syntax. The quantitative distinctions that mark human syntactic and lexical ability could have evolved gradually in the past five million years. Apes, however, cannot talk; therefore, human speech appears to have evolved at a later date.

Although the vocal anatomy of apes prevents them from producing the full range of human speech sounds, segments of ape vocalizations correspond to sounds that signify spoken words. However, ape vocalizations are tied to particular emotional states or situations. Apes are unable to alter the sequences of articulatory gestures that generate their stereotyped signals to produce novel vocalizations, or approximations to the words of any human language. The human brain clearly is specialized to regulate speech production. As we talk, the sequence of phonemes (meaningful sounds, roughly approximated by the letters of the alphabet) transmits words at rates of 20 to 30 phonemes per second. In contrast, the human auditory system fuses discrete nonspeech sounds into a buzz at rates exceeding 15 per second. The rapid transmission rate of speech is achieved by encoding—that is, melding phonemes into syllables transmitted at a slower rate that accommodates the constraints of human hearing. If humans were limited to the slow nonspeech rate, they would forget the first word of a typically complex sentence before the sentence ended. Listeners recover the phonemic sequences that spell words by means of speech mode perception, which is distinct from other aspects of auditory perception. The speech mode appears to entail neural operations that internally model the motor commands that generate speech. Studies of sentence comprehension reveal another role for speech: words heard are silently "rehearsed," by means of the neural mechanisms that regulate overt speech, to maintain them in a neural buffer or verbal working memory in which sentence comprehension takes place.

We now know that the neural mechanisms involved in speech reach beyond the neocortex. The traditional brain-language theory, which dates back to the last quarter of the nineteenth century and localizes language in Broca's and Wernicke's areas of the cortex, is incorrect. Converging evidence from studies of humans suffering permanent language loss (aphasia) or neurodegenerative diseases such as Parkinson's, along with studies of the brains of other species, show that neural circuits linking activity in regions of the cortex and subcortical structures regulate complex behaviors such as comprehending the meaning of a sentence, picking up a cup, solving a puzzle, or planning a day's activities.

Many subcortical structures of the brain, such as the basal ganglia, cerebellum, and thalamus, support these neural circuits. In fact, studies of humans who have suffered brain damage show that permanent aphasia does not occur in the absence of subcortical damage. The subcortical basal ganglia, which have been intensively studied, confer cognitive and motor flexibility. During motor tasks, they sequence individual gestures, interrupting a sequence when environmental stimuli indicate that another course of action is appropriate. They perform similar functions when people think or attempt to comprehend the meaning of a sentence—for example, they interrupt the sequence of syntactic operations necessary for comprehension when the hearer encounters a relative clause in a sentence such as "The boy who was fat sat down." Basal ganglia also are implicated in learning. In short, the basal ganglia that regulate overt motor activity also form part of the neural systems that confer human cognitive and linguistic ability.

The mark of evolution is often evident when the human brain or body is studied. In this instance, it is probable that brain structures adapted for motor control took on cognitive roles in the course of evolution. Complex, learned motor activities, such as upright bipedal walking, may have played a part in triggering the evolution of the neural bases of complex human language around five million years ago. The process must have been complete before the migration of anatomically modern humans from Africa between 200,000 and 100,000 years ago, because any normal human child, native to any region of the world, can easily acquire any language on Earth.

[*See also* Consciousness; Hominid Evolution, *article on* Neanderthals; Primates, *article on* Primate Societies and Social Life.]

BIBLIOGRAPHY

Gardner, R. A., B. T. Gardner, and T. A. Van Cantfort. *Teaching Sign Language to Chimpanzees*. Albany, N.Y., 1989. A detailed presentation of the evidence for chimpanzees acquiring the ability to use words and simple phrases.

Hauser, M. D. *The Evolution of Communication*. Cambridge, Mass., 1996. A comprehensive survey of animal communication.

Lieberman, P. *The Biology and Evolution of Language*. Cambridge, Mass., 1984. A detailed treatment of the anatomy and physiology of speech production, language acquisition in children, and other aspects of language.

Lieberman, P. *Human Language and Our Reptilian Brain: The Subcortical Bases of Speech, Syntax, and Thought*. Cambridge, Mass., 2000. An introduction to current neurophysiologic studies of brain systems and brain-language models in an evolutionary framework.

Negus, V. E. *The Comparative Anatomy and Physiology of the Larynx*. New York, 1949. The classic work on the evolution of the larynx.

— PHILIP LIEBERMAN

Linguistic Diversity

Human language is remarkably diverse: at least 6,500 different languages are currently spoken, and many more were spoken in the past. This diversity is of interest to the evolutionist for several reasons. First, languages evolve by a process analagous to organic evolution, and studying their diversification sheds light both on the biological background of language and on evolutionary processes in general. Second, languages can be placed in phylogenetic relationships and thus used as markers for the recent population history of the groups who speak them. Third, the distribution of languages is related to ecological factors in somewhat the same way

PIDGIN AND CREOLE LANGUAGES

Pidgin languages are communication systems that arise where large groups of people with no common language are thrown together by major social upheavals, such as enslavement and the plantation system, the colonial encounter, or changing patterns of trade. Pidgins are characterized by a basic vocabulary drawn largely from the variety with highest status at its birth (the "lexifier" language, often English, French, or Portugese), with some words from other sources. They lack the full range of grammatical devices of a normal language.

When children grow up in a pidgin-speaking environment, they transform it gradually, over several generations, into a creole. This is a language with its origins in a pidgin, but with a full grammatical structure. The structure is produced by the process of generations of children learning the language and using it as their mother tongue. As they do this, they generate solutions to such problems as tense and negation that demonstrably do not come from the lexifier language. Creolization has been taken as evidence of the human instinct for language in action. Creoles have been argued to show strong similarities to one another in phonology and grammar, regardless of where they were born. This may reflect the recurrence of the simplest solutions to the functional needs of grammatical communication, or perhaps more specific universal properties of what language acquisition mechanisms generate. Either way, it can be argued that creoles are not members of any language family. Though many of their words have an obvious descent, the grammatical system is created anew by speakers using their own cognitive and acquisitional resources.

There are at least 100 pidgins and creoles spoken in the world today, with lexifier languages as varied as Tupi, Malay, and Arabic. The largest—such as Tok Pisin (New Guinea), Krio (Sierra Leone), and Haitian (Haiti)—are spoken by millions of people; Tok Pisin and Haitian are official national languages.

—DANIEL NETTLE

the distribution of species. Finally, the current rapid loss of diversity in language is analogous to that seen in biodiversity.

Origins of Diversity. The acquisition of language is believed to be genetically guided to a significant extent. This is inferred from the speed at which children acquire linguistic competence, the characteristic age-related stagelike pattern, and the finding that all languages can be described in terms of similar processes and templates both the set of forms and the precise rules for combining words into phrases and sentences vary from language to language. This diversity is the result of two factors. First, language is genetically underspecified, relying on learning to fill in the details; second, learning results in imperfect replication of the linguistic system. Genetic underspecification is probably adaptive: because guided learning mechanisms are evidently so good at homing in on the language of one's immediate social group, there is unlikely to be any selective advantage in a genetic code for language becoming entirely innate, even if that were possible. Imperfect replication stems both from individual idiosyncrasies and from the continual reanalysis that a learner performs in extracting a productive linguistic system from the messy, variable, error-prone set of utterances to which he is exposed. Differences between individuals become amplified into social dialects by the uneven way individuals assort with and identify with one another; as a result, class-based dialectal differences exist in all large, complex societies. Where groups become socially or geographically isolated, their dialects eventually diverge to the point of mutual unintelligibility; at that point, they are called "languages," though this distinction can be a difficult one to make objectively, much as biological species are.

The evolutionary processes of languages and of species have been seen as significantly analogous. The linguistic equivalent of the genotype is the system of words and rules, or grammar, that makes up a language; the equivalent of the phenotype is the actual utterances that are produced by speakers. There are, however, several major differences between organic evolution and that of languages, or indeed of any culturally transmitted system. First, replication of the genotype in biological species is direct from gene-to-gene copy, whereas grammars reproduce themselves by giving rise to utterances, from which new grammars are inferred by the next generation of speakers. This means that linguistic evolution is Lamarckian (features acquired during a parent's life can be transmitted to offspring), and that grammars can undergo much more drastic transformation in a single generation than genes can (an example of such drastic transformation is seen in creole languages; *see* the vignette). Second, the number of "parents" of a grammar is not limited to one or two but is as large as the speech community that surrounds the growing child, with peer groups thought to be particularly important. Third, and because of the first two factors, the potential for horizontal transmission or admixture between languages is very high. This is most commonly observed in the borrowing of words; thousands of words in English are derived from French, not from the Germanic ancestor from which English's rules of word and sentence formation come. This process can also result in more profound examples of linguistic hybridization. There are a few much-discussed cases of languages, like Ma'a (or Mbugu,

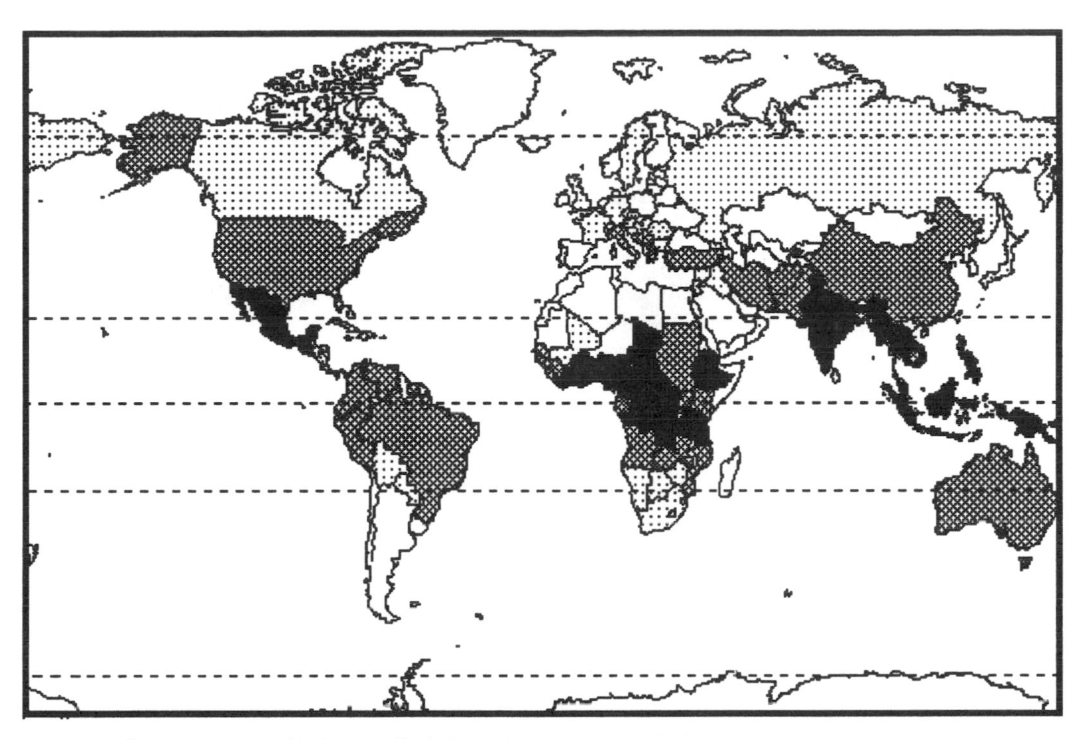

MAP 1. Language Diversity. The relative language diversity of the major countries (from Nettle, 1999). The denser the shading, the more the diverse the country for its size.

in Tanzania) and Copper Island Aleut, whose words and rules are so equally drawn from different sources that they cannot be placed into any single family.

Linguistic Diversity and Recent Human Evolution.

Phylogenetics of language. The idea that languages could be placed in phylogenies (or "family trees," as they are known in linguistics) stems back to Sir William Jones. He asserted in 1786 that Sanskrit, parent to many of the languages of India, bore such a resemblance to Greek and Latin that they could only have sprung from a common source. That source was the long-disappeared Proto-Indo-European, and the tree connecting it to its many descendants across Europe and Asia was soon sketched. The result influenced both linguistics and biology; Darwin himself acknowledged the parallel with his proposals on species in *The Descent of Man.*

Modern work in historical linguistics has succeeded in assigning most of the 6,500 extant languages to around 250 language families (the exceptions are pidgins and creoles—*see* the vignette—and the few cases of mixed languages). These families vary in size from a single member ("linguistic isolates") to nearly one thousand in the cases of Austronesian and Niger-Congo. The internal trees of these families have been resolved to a greater or lesser extent, depending on the data available, the existence of historical texts, and the amount of borrowing and convergence that has gone on in the family. The work of classification depends on what is known as the comparative method: the establishment of a core of vocabulary and, particularly, of inflectional patterns that show not just similarity but regular patterns of sound correspondence. These forms are identified as "cognate" (deriving from a common ancestor), and an ancestral form may be reconstructable by deduction from the structure of its descendants. Regularity in sound correspondence allows the elimination of borrowing and chance resemblance from language comparison. The internal topography of the tree can then be deduced from structural innovations and shifts that are shared by some branches and not found in others.

The quantitative comparison of vocabularies is known as *lexicostatistics.* It was long hoped that the mass, objective comparison of languages made possible by lexicostatistics, particularly with the advent of computers, would facilitate the construction of linguistic phylogenies. In fact, lexicostatistics has neither allowed deeper (that is, with older branching points) phylogenies to be made nor caused decisive reinterpretation of any existing ones, and it remains at best a secondary tool. This is because, first, lexicostatistics depends on the accurate identification of cognates, which requires the comparative method; second, borrowing, particularly between related languages, can produce incorrect classification when the criteria depend on number rather than the structure of shared forms; and third, lexicostatistics becomes unreliable at a depth similar to the point where more traditional methods do. Thus, the lack of an agreed method with more power than traditional qualitative comparison to detect phylogenetic history is still a barrier.

Linguistic phylogenies have never yet been pushed back much beyond the depth of Indo-European (perhaps 5,000 years), though some families, such as Afroasiatic, may be a little deeper. This situation probably reflects linguistic reality more than methodological limitations. Linguistic structure erodes at such a rate that, beyond a certain "ceiling," the resemblances that remain are too few, too tenuous, and too slight too allow the emphatic rejection of the hypotheses of chance resemblance or ancient borrowing. Furthermore, borrowing and convergence between the small-scale, mobile societies of the past was probably rather common, and this obscures phylogenetic relationship. Thus, despite a continual stream of proposals and putative cognates, no firmly established reduction of the 250 established families to fewer macro-families has been published. There are attempts, notably those summarised by Merritt Ruhlen (1987), but the consensus among linguists is that these groupings are loosely geographical rather than demonstrably phylogenetic. (A global language phylogeny can of course be drawn, trivially, but it becomes star-shaped for lack of resolution beyond the level of the 250 families.)

Distribution of language families. Though there are no absolute methods for dating splits on linguistic criteria alone, it is thought that the maximum time depth of a recognizable family is no more than 10,000 years, and usually less. Thus, if today's languages had all been evolving *in situ* since the Pleistocene, they would appear to be linguistic isolates because they would have diverged so much from their neighbors as to be unrelatable. This is true of a few languages, such as Japanese and Basque; however, most languages clearly belong to the radiation of some substantial family. This indicates that an ancestor spread out, pushed by some important demographic or cultural process, probably within the last 10,000 years. Thus, examining the distribution of language families is a window into the major population movements of the Holocene, the period since the end of the last glaciation.

Perhaps the most significant transformations in human activity in that period have been the transitions in several parts of the globe to agriculture. Agriculture allows a great increase in rates of population growth, causing a "wave of advance" as farmers spread out into lands at their periphery. Precisely such a process could account for the spread of proto-languages. The case was argued by the British archaeologist Colin Renfrew (1987) for the Indo-European family, which is hypothesized to have been pushed outward from the Anatolian centers of cereal cultivation by the spread of farmers, beginning around 8,000 years ago. It has also been noted that the centers of radiation of several other major families coincide with centers of agricultural origin, such as the Yellow and Yangtze river basins in China (Sino-Tibetan family), South and East China (Austronesian, spreading into the Pacific), and central West Africa (Niger-Congo, spreading both west to the coast and southeast throughout tropical Africa). Thus, it seems likely that a major factor accounting for today's linguistic map is the spread of agriculture.

Where such large-scale homogenous farming spread did not occur, the level of diversity in language families is strikingly high. The Americas, Australia, and New Guinea have among them more than 200 of the 250 global families, all fairly localized. This gives some insight into the enormous diversity that must have characterized the whole world 10,000 years ago. With the exception of a few outliers like the Khoisan languages of southern Africa and Basque in Europe the distribution of language families we see today in Africa and Eurasia does not reflect the primordial diversity of those continents but is rather the result of homogenization and extinction processes operating in the relatively recent past.

Linguistic and genetic diversity. Most of the little genetic diversity there is within the human species varies between individuals rather than between populations; however, there is some regional or ethnic variation, and the extent to which this is congruent with the classification of groups on linguistic grounds has been investigated. Linguistic barriers are often, though not always, zones of rapid change in allele frequencies. This may be true because the same geographical boundaries impede communication of both types. Linguistic boundaries themselves can act as barriers to gene flow. This is clearly not universally true; in areas where ethnolinguistic groups are very small, such as the northwestern Amazon or central Nigeria, linguistic exogamy (marriage outside the group) is normal or even prescribed, and it may also be common in larger groups.

Whether there is a relationship between the structure of genetic variation and deeper linguistic entities such as families is still a matter of debate. It is clear that speakers belonging to the same language family do not generally share private polymorphisms, nor are they pure in terms of haplotypes on any major genetic system. There are reports of frequencies of haplotypes, or of other genetic markers, correlating with language families. Evaluation of the significance of such findings is confounded by geography. Most language families are geographically continuous, and the main factor affecting allele frequencies is simply distance. The conclusions are perhaps not very surprising; people both interact and marry close by, and thus there is strong spatial autocorrelation in both language and genes. However, ethnolinguistic groups have been fluid and have absorbed many migrants, so none is unique in molecular terms.

A well-known study by L. L. Cavalli-Sforza and his colleagues (1988) purported to show a correlation between the worldwide tree of inter-population differences in allele frequencies, and the tree of linguistic macrofamilies prepared by Merritt Ruhlen. The apparent correspondence is in fact an illusion. Many of Ruhlen's groupings, such as "Australian," "Papuan," and "Amerind," are not accepted by many linguists, who believe they are not true phylogenetic entities but simply loose geographical groupings. This would make the tree essentially a geographical construct. Because allele frequencies are known to vary primarily with geographical distance, the apparent correspondence of the two trees is to be expected. On the other hand, to the extent that differences in allele frequencies indicate reduced contact among human groups, we would also expect differences in language. These two processes would give rise to the tree proposed by Cavalli-Sforza.

The lumping of languages into regional groups actually masks interesting facts about the distribution of diversity. There are more language families in the Americas than anywhere else (150 of 250 families), and rather few in Africa (perhaps 20 of 250 families). This is the inverse of genetic diversity, where—at least on fast-evolving systems like mitochondrial DNA—African populations are more diverse because of their greater time depth. In the Americas, the most recently settled continents, the diversity of language families does not indicate great time depth, as genetic diversity would. The explanation must be that the tempo of evolution is different in the two cases. For language families, it is quite plausible that the initial radiation of diversity can be very fast, whereas the predominant force in recent millennia has been extinction, particularly in continents where there have been major agricultural spreads. Thus, the diversity of the Americas reflects not great age but relative lack of extinction, and the uniformity of Africa not recent settlement but great age with sustained homogenization.

Distribution and Abundance of Languages. Diversity in terms of language families and diversity in terms of languages are independent variables. Zambia, for example, has around 30 indigenous languages, which all belong to the same family, while Chile's 11 languages belong to at least 7 families. Whereas diversity of language families reflects radiations of groups in prehistory, diversity of languages is a reflection of the socio-spatial arrangement of people on the ground in the recent past.

Map 1 shows the language diversity of all the major countries of the world, relative to their areas. The darker the shading, the more relatively diverse the country. Note that the resolution of the map is only at the level of the political; countries like India and Mexico that have very diverse and less diverse regions are shaded uniformly according to the country-wide average.

As the map shows, language diversity is generally greatest near the Equator and decreases as one moves north or south away from it. Species diversity decreases in a similar way as one moves away from the Equator. There are more specific associations between language and biological diversity, as well. In the Old World, there are two great belts of very high language diversity, which correspond almost exactly to the two great belts of equatorial forest that harbor so many of the Old World's species. One runs from Ivory Coast across West Africa into Zaïre. The other runs from South India and peninsular Southeast Asia down through the Indonesian islands into New Guinea and the islands of the Western Pacific. At least 60 percent of all languages are spoken in these two belts. As well as being plentiful where species are numerous, languages are few where species are few. Large areas of white on the map correspond to the Sahara and Arabian deserts, which, though in the tropics, are arid and poor in species.

The fact that the geographical range of each language tends to be smaller the closer one gets to the Equator is the linguistic equivalent of the phenomenon known in ecology as Rapoport's rule. The reason for Rapaport's rule in the case of species is still being debated, though it obviously relates to the biological productivity of the environment. But how can it be explained in the linguistic case? Human groups that share a common mother tongue generally do so because there are close, usually face-to-face, social bonds between them. In subsistence societies, which until very recently included almost all societies, the important social bonds are also key economic ones, and it is through such bonds that foodstuffs, labor, and gifts are exchanged. Subsistence economies are characterized by zero- or low-growth trajectories and limited diversity of available goods. People thus have little incentive to enlarge their exchange networks beyond what is required for secure food supply, or to accumulate surpluses beyond those required for local cultural success. The key determinant of the scale of the economy, then, is the size of the exchange network required to ensure an adequate food supply at all times.

This is the link with biological productivity. Near the Equator, temperatures are high all year round, as is rainfall, which is generally abundant, and day length is nearly constant. Thus, food production is nearly constant from month to month and from year to year. This does not mean that food is abundant, of course; the equatorial zones are chronically hostile disease environments, and average productivity is low. However, that productivity is at least fairly constant and dependable. This means that the size of community required to be a securely viable food-producing unit near the Equator is small. In contrast, communities living farther from the Equator experience greater seasonal and annual fluctuations in the food supply, which can be extremely dangerous. In general, the greater the temporal fluctuations, the larger the network of individuals needs to be ensure that there will always be access to food supplies. This is a dynamic parameter, and social networks wax and wane in response to changing situations, often leading to the spread or shift of new languages. In the extreme case of desert-fringe pastoralists, people cope with the variability of their environment by large circuits of migration and vast networks of kin and stock partners, which bring them into contact with individuals dispersed over thousands of square kilometers. Consequently, the languages of the desert fringes all cover vast areas and large numbers of people.

This rather simplified picture leaves out political factors, social and religious ties which may bind communities together, and other geographical factors which may set them apart. In principle, however, it explains how diversity of species and of languages can be so strongly correlated.

Loss of Diversity. If species and language diversity are parallel in their abundance, they are also parallel in their current rapid disappearance. In both cases, the greatest diversity is found in tropical countries and undeveloped areas that are now undergoing rapid change toward more intensive, more globalized exploitation. A significant proportion of languages are not now being passed on to a new generation of children, and linguists estimate that at least half are in serious danger of disappearance within a human lifetime. This is on a par with the worst estimates of the threat to biological diversity.

Linguistic and biological extinction are obviously different because the former involves human choices; however, many of the conditioning factors are the same in the two cases. Where previously self-sufficient areas come into the orbit of a distant center, their pristine habitats are removed to make way for low-diversity cultivation of that center's economically valued species (usually wheat, rice, and livestock). Metropolitan languages like Spanish and English replace indigenous varieties as part of the same process.

Many linguists feel that languages are inherently worth supporting, though the fact that language use is the outcome of human choices means that the issues involved are somewhat different from those for species. In both cases, though, it is likely that allowing local communities control over the management and organization of their social and natural surroundings will be beneficial for the sustaining of diversity.

[*See also* Hominid Evolution, *article on* Archaic *Homo sapiens*.]

BIBLIOGRAPHY

Campbell, L. *Historical Linguistics: An Introduction.* Edinburgh, 1998. A general introduction to language change and family trees.

Cavalli-Sforza, L. L., A. Piazza, P. Menozzi, and J. Mountain. "Reconstruction of Human Evolution: Bringing Together Genetic, Archaeological and Linguistic Data." *Proceedings of the National Academy of Sciences of the USA* 85 (1988): 6002–6006. Famous paper that shows correspondence between a language-family diagram and a global tree of gene frequencies.

Dixon, R. M. W. *The Rise and Fall of Languages.* Cambridge, 1997. Provocative and readable discussion of the tempo and mode of language evolution, including consideration of the roles of borrowing and convergence.

Grimes, B. F. *Ethnologue: The World's Languages.* 14th ed. Dallas, 2000. Comprehensive list of all the world's languages by country and language family.

Nettle, D. 1999. *Linguistic Diversity.* Oxford, 1999. Models and data on the mechanisms of linguistic evolution as well as the distribution of languages and language families.

Nettle, D. submitted. Human genetic andlinguistic diversity: Continental patterns and time depth. *Nature.* Comparison of geographical distributions of genetic diversity and language families.

Nettle, D., and S. Romaine. *Vanishing Voices: The Extinction of the World's Languages.* New York, 2000. Discussion of linguistic extinction, with special attention to the parallels with biodiversity.

Nichols, J. *Linguistic Diversity in Space and Time.* Chicago, 1992. Seminal work on the geographical distribution of language families.

Renfrew, C. *Archeology and Language: The Puzzle of Indo-European Origins.* London, 1987.

Renfrew, C. "Before Babel: Speculations on the Origin of Linguistic Diversity." *Cambridge Archaeological Journal* 1 (1991): 3–23. This and the previous work develop the model of agricultural expansion as the key to understanding the spread of language families.

Romaine, S. *Pidgin and Creole Languages.* London, 1988. On pidgins and creoles.

Ruhlen, M. *A Guide to the World's Languages. Volume 1: Classification.* London, 1987. Ambitious scheme to reduce all world linguistic diversity to just 17 or so macro-families.

Thomason, S. G., and T. Kaufman. *Language Contact, Creolization and Genetic Linguistics.* Berkeley, 1988. Considers cases of mixed languages within a phylogenetic framework.

— Daniel Nettle

LEKS

A lek is a cluster of male territories, together with their occupants, visited by females for courtship and mating. The term *lek* refers to both the males and the site (or arena) where they display, and derives from a Scandinavian word meaning "play." Lekking is an uncommon mating system, first described in birds, and subsequently in a variety of other taxa including mammals and insects. Well-known avian examples include black grouse and the ruff (a sandpiper) in Eurasia, sage grouse in North America, and the greater birds of paradise of New Guinea. Mammalian examples include hammer-headed bats, fallow deer, and Uganda kob. In these "classical" lek breeders, displaying males are highly aggregated, their display territories contain no significant resources required by females (e.g., food and roosting or nesting sites), and males play no role in rearing their offspring (Bradbury, 1981). In "quasi" lek breeders, such as the North American ruffed grouse, males are more evenly spaced. Additional leklike mating systems, in which courting males are aggregated but other defining features of classical lek breeders may be absent, have been described in mammals, birds, lizards, amphibia, fish, and insects. Some authors consider frog choruses and insect mating swarms to be leks. However, the focus here will be on those classical lek breeding vertebrates to which the term was originally applied and whose behavior inspired the interest of evolutionary biologists.

Figure 1. Peacock Pair with the Male Displaying His Plumage. © William J. Weber/Visuals Unlimited.

The Lek Paradox. As Charles Darwin was aware, lek-breeding species are often strikingly sexually dimorphic (males and females differ) in morphological and behavioral components of courtship used by males to attract females, indicating that these traits have been subject to sexual selection. [*See* Sexual Dimorphism.] Field studies of lek-breeding birds and mammals in the 1960s and 1970s confirmed the potential for strong sexual selection by showing that a small percentage of males at a lek typically perform most of the observable mating activity (Wiley, 1973). The same studies also indicated that in classical lek breeders, particularly birds, females are often relatively free to choose their mates. Together, these observations suggested that female choice for sexually dimorphic courtship traits may be highly developed in lekking species. However, as argued by George Williams (1975) and others, this inference appears paradoxical from the standpoint of equilibrium population genetic theory. The reasoning is as follows. Lek males appear to offer their mates little except genes. If fitness were heritable, a female might increase the fitness of her offspring by mating with a phenotypically fitter male. However, population genetic theory suggests that the benefit of doing so should be small. This is because natural selection should fix alleles that increase fitness and eliminate alleles that decrease fitness. This process can be accelerated if females choose fitter males. Under this scenario, what is called the additive genetic variance for fitness disappears. The additive genetic variance is variability that can be attributed to additive effects of alleles. The additive effect of an allele is the average effect of substituting it for alternative alleles. Once additive genetic variance for fitness is exhausted, phenotypic differences in fitness among males are no longer heritable.

When this occurs, females will no longer obtain heritable benefits for their offspring. If this is so, how can we explain the evolution of highly developed mate choice in a mating system where the benefits of such choice should be slight? This is the paradox of the lek.

This dilemma made leks the "poster child" for two decades of research into the evolution of mating preferences in nonresource-based mating systems and, particularly, into mechanisms by which mating preferences for apparently costly "sexual ornaments" or "handicaps" (traits that attract mates but otherwise impair fitness) might evolve.

The results of this research have had implications for sexual selection in general, and for leks in particular. An important outcome includes the demonstration that in at least some species with nonresource-based mating systems, females do indeed prefer males with more exaggerated development of conspicuous sexual ornaments (Andersson, 1994). Also important are the demonstration of the conditions under which the expression of a costly secondary sexual trait (handicap) may reveal variation in individual fitness (Grafen, 1990) and the development of plausible models for the evolution of mating preferences and sexual ornaments, including those based on "good genes," R. A. Fisher's principle of "self-reinforcing choice" (runaway sexual selection), immediate fitness consequences of choice, and selection occurring on sensory mechanisms in contexts other than mating (sensory bias) (Andersson, 1994). A final outcome is the recognition that traits closely related to fitness often exhibit substantial additive genetic variation, even if their heritabilities are low, because they are also strongly affected by environmental variation (Houle, 1992).

For some biologists, these results have resolved the lek paradox as a problem in evolutionary biology. But do they really explain sexual selection in lek-breeding animals? Recent studies indicate that some initial perceptions about lek breeders were not well founded. For example, lek-breeding birds are not necessarily more sexually dimorphic than closely related species that do not lek (Hoglund and Alatalo, 1995). In those classical lek breeders in which sexual selection has been investigated in detail, it has proved to be far more complex than was initially suggested. Leks may involve nested levels of male–male competition and female choice based on multiple, substitutable criteria, including aspects of male display performance or morphology, prior experience, and the mating decisions of other females (Gibson et al., 1991; Hoglund and Alatalo, 1995). Field studies also show that lekking females make choices in social environments where they may be under predation risk, in which individuals of both sexes are prone to interfere aggressively in each other's mating attempts, and in which a male might transfer pathogens or ectoparasites during copulation. It would be surprising if these contingencies did not also exert selection on mating preferences. These observations need not mean that females fail to choose males of high fitness. However, they indicate that both the choice process and its selective consequences are likely to be more complex than is assumed in evolutionary models inspired by the lek paradox.

Explaining the Evolution of Leks. Leks also have been of interest to students of social behavior because they appear to challenge the idea that ecological variables such as the kind and spatial distribution of food, cover, or nesting habitat drive the evolution of mating systems through their effects on the spatial and temporal dispersion of breeding females (Emlen and Oring, 1977). Interspecific comparisons have not revealed consistent ecological correlates of lek breeding versus alternative mating systems, although they have revealed that females of lek-breeding birds and mammals characteristically move over much larger areas than closely related and otherwise similar species in which males do not aggregate (Bradbury, 1981).

High female mobility may have favored aggregated display in these taxa for at least two reasons. First, where females range widely their movements, interacting with topographic features, should generate "traffic hot spots" at which males seeking to encounter females should aggregate (Bradbury et al., 1986). Although this process on its own does not predict the formation of classical leks, it could initiate clustering. It also explains where leks are located in two species of birds that have been studied in detail (Gibson, 1996; Westcott, 1997). High mobility also provides females with opportunities to visit and compare multiple males, creating a situation in which females may favor aggregated males if this reduces the costs of mate choice or increases the chance of finding a good mating partner. This process has the potential to drive males into groups. It is supported by evidence that females are less likely to visit or mate with males in very small groups in some species of lekking birds and swarming insects (Hoglund and Alatalo, 1995).

One difficulty with the idea that males join leks solely to increase mating opportunities is that the rewards are unequally shared. Less successful males have been suggested to participate in leks because proximity to more attractive or dominant "hotshot" males increases their mating opportunities. This does not explain the participation of males that fail to mate at all. Such individuals could be queueing for breeding status (McDonald and Potts, 1994), obtaining covert matings around the lek periphery, or, if females prefer to mate in larger leks, their presence may increase the breeding success of a close relative within the group (Petrie et al., 1999). However, it is also likely that factors other than breeding opportunities promote lek display. Because sexual advertisement commonly attracts predators as well as mates, the antipredator benefits of grouping, which have been studied extensively in the context of foraging groups, are a likely candidate. Lack (1968) regarded predation as the principal factor selecting for lek behavior in birds, but its potential role has since been largely neglected.

Gaps in our understanding of leks reflect the practical difficulties of studying many lekking species when they are away from their display grounds. In some species, breeding activity takes place away from as well as at leks. Our understanding of leks is likely to be advanced by analysis of the behavioral mechanisms and fitness consequences (for both sexes) of participation in lekking versus alternative reproductive tactics in such species.

[*See also* Male–Male Competition; *articles on* Mate Choice; Mating Systems, *article on* Animals; Optimality Theory, *article on* Optimal Foraging.]

BIBLIOGRAPHY

Andersson, M. *Sexual Selection*. Princeton, 1994. An authoritative review of empirical and theoretical work on this topic.

Bradbury, J. W. "The Evolution of Leks." In *Natural Selection and Social Behavior*, edited by R. D. Alexander and D. W. Tinkle, pp. 138–169. New York, 1981. A stimulating review that inspired many modern field studies.

Bradbury, J. W., R. M. Gibson, and I.-M. Tsai. "Hotspots and the Dispersion of Leks." *Animal Behaviour* 34 (1986): 1694–1709.

Emlen, S. T., and L. W. Oring. "Ecology, Sexual Selection and the Evolution of Mating Systems." *Science* 197 (1977): 215–223. An important review of ecological influences on animal mating systems.

Gibson, R. M. "A Re-evaluation of Hotspot Settlement in Lekking Sage Grouse." *Animal Behaviour* 52 (1996): 993–1005.

Gibson, R. M., J. W. Bradbury, and S. L. Vehrencamp. "Mate Choice in Lekking Sage Grouse Revisited: The Roles of Vocal Display, Female Site Fidelity and Copying." *Behavioral Ecology* 2

(1991): 165–180. A detailed analysis of mate choice cues in a lekking bird.

Grafen, A. "Biological Signals as Handicaps." *Journal of Theoretical Biology* 144 (1990): 517–546. A particularly clear exposition of the handicap principle.

Höglund, J. and R. V. Alatalo. *Leks.* Princeton, 1995. A survey of lek behavior, featuring the authors' research on black grouse and ruffs.

Houle, D. "Comparing Evolvability and Variability of Quantitative Traits." *Genetics* 130 (1992): 195–204.

Lack, D. *Ecological Adaptations for Breeding in Birds.* Oxford, 1968. An early attempt to explore the relationship between ecology and social systems.

McDonald, D. W., and W. K. Potts. "Cooperative Display and Relatedness among Males in a Lek-Mating Bird." *Science* 266 (1994): 1030–1032.

Petrie, M., A. Krupa, and T. Burke. "Peacocks Lek with Relatives Even in the Absence of Social and Environmental Cues." *Nature* (London) 401 (1999): 155–157.

Westcott, D. A. "Lek Locations and Patterns of Female Movement and Distribution in a Neotropical Frugivorous bird." *Animal Behaviour* 53 (1997): 235–247.

Wiley, R. H. "Territoriality and Non-random Mating in the Sage Grouse, *Centrocercus urophasianus.*" *Animal Behaviour Monographs* 6 (1973): 87–169. A classic study demonstrating skewed mating success in a lekking bird.

Williams, G. C. *Sex and Evolution.* Princeton, 1975. Contains the first, and one of the clearest, discussions of the lek paradox.

— ROBERT M. GIBSON

LEVELS OF SELECTION

In scientific disagreements over the levels at which natural selection operates in biological evolution, semantic confusion has played a significant role. On the surface, such disputes seem straightforward. At what level or levels does natural selection take place? Does it act only at the level of individuals or can it also take place at the level of groups? The trouble is that the terms delineating the debate are used in highly ambiguous ways. This essay seeks to present a unified and coherent discussion of these terms so that the empirical issues can be stated clearly. The issue of levels of selection is important primarily because of its connection to social evolution and altruism. Societies are an instance of higher-level organization that needs explaining, and altruism is one mechanism that can promote higher-level organization (Boyd and Richerson, 1985; Cavalli-Sforza and Feldman, 1981; Sober and Wilson, 1998; Sterelny and Griffiths, 1999).

Individuals and Groups. The "levels" at issue in the controversy over levels of selection make up the traditional organizational hierarchy, with nucleotides, genes, and cells at one end and organisms and kinship groups (such as beehives), demes, and species at the other end. The contrast between individuals and groups is usually presented in terms of a sharp dichotomy. Either something is an *individual* or it is a *group*, but as these terms are used in the traditional organizational hierarchy, individuals gradate imperceptibly into groups. Individuals are a good deal more coherent and internally organized than are groups. Genes in the simplest case are segments of DNA. Although genetic material cannot be divided into discrete genes as easily as some introductory texts imply, genes do form reasonably unproblematic individuals. Most organisms exhibit a high degree of cohesiveness and internal organization. They comprise cells that have a common origin and by and large retain the same genetic makeup. A Portuguese man-of-war may seem like a single organism; it is cohesive and internally well organized. But it does not form a single organism. Instead, it is a colony (Buss, 1987; Wilson, 1998).

Groups pose the same sort of difficulties. A beehive possesses some of the properties of individuals (it is cohesive and exhibits internal organization), but not all the bees that are part of a particular beehive exhibit the same genetic makeup. Demes exhibit even fewer characteristics of individuals. A deme is made up of organisms that belong to the same species, but demonstrates very little in the way of cohesiveness and internal organization. Instead, the organisms that make up a deme are held together by an external force. Species exhibit population structure and inhabit a range—analogous to inhabiting the same haystack—but that is about it. Thus, a continuum exists between genes at one end of the spectrum and species at the other end. The failure to treat terms like *individual* and *group* in a clear and precise way leads to numerous irritating confusions. For example, many biologists insist that natural selection operates only at the level of genes. Because genes are individuals of a sort, they are considered individual selectionists. Other biologists argue that the organism is the primary focus of selection. Because organisms are individuals, they are also termed individual selectionists.

With respect to species, one confusion must be avoided. In the above discussion, individuals and groups are distinguished as being at opposite ends of a single continuum. Traditionally, species have been treated as "classes," as if they do not belong on the individual-group continuum at all. Philosophers commonly distinguish between individuals (or particulars) and classes (or sets), the contrast between one particular atom of gold and all the gold atoms in the universe. Particulars are localized in space and time and exhibit reasonably discrete boundaries both at any one time and through time. The entities that comprise a class can occur anywhere and at any time. They belong to the classes that they do because of the characteristics that they exhibit. Both philosophers and biologists have argued that species, as the things that evolve, are better treated as belonging to the individual-group continuum, not set off from it as abstract classes. It is this distinction that is at issue in the claim that species are individuals and not classes. They belong at the group end of the continuum between individuals and groups. Even though groups are less well-organized and continuous than individuals, they are not classes (Ghiselin, 1974; Hull, 1976).

In sum, biological entities can be ordered into a series of increasingly inclusive individuals, from genes, cells, and organisms at one end of the spectrum to kinship groups, demes, and entire species at the other end. All of these entities are related by the part-whole relation. Genes are part of cells, cells are part of organisms, some organisms are also part of kinship groups such as hives, some are also part of demes, while all organisms are part of one species or other. Thus, debates over levels of selection concern the level or levels in this organizational continuum at which natural selection takes place. Only at the genetic level? Primarily at the level of organisms? Possibly at the level of entire species?

Selection: Replication and Environmental Interaction. One problem with respect to selection that has emerged through the years is that selection is not *one* process but *two* precisely related processes. Certain entities are replicated successively through time, forming lineages. Chief among these entities are genes. They are the primary replicators. They form gene lineages. If by "selection" all one means is "replication," then genes are the primary units of selection. However, a second sort of process also contributes to selection. Periodically, genes as well as more inclusive entities interact with their environments in such a way that replication is differential,—that is, some entities replicate more than others. Chief among these entities are organisms. They interact with their environments as cohesive wholes forming organism lineages. A process counts as a selection process if owing to its particular environmental interaction a replicator leaves more copies of itself than other comparable replicators. Thus, gene selectionists and organism selectionists need not be read as disagreeing with each other. When gene selectionists claim that genes are the primary units of selection, they mean replication, and replication is concentrated at the genetic level. When organism selectionists claim that organisms are the units of selection, they mean environmental interaction, and environmental interaction is concentrated at the organismic level (Keller, 1999; Lloyd, 1992; Sterelny, 2001; Sterelny and Griffiths, 1999; Williams, 1992).

Each of these processes—replication and environmental interaction—is necessary for selection. For example, differential replication in the absence of environmental interaction is commonly termed "drift." A third element is also needed if selection is going to result in the evolution of lineages. Variation must be introduced into the system. The character of this introduction determines whether the process is Darwinian or Lamarckian. If there is no correlation between those variations that an organism might need and those that it gets, then inheritance is "Darwinian." If there is such a correlation, then it is "Lamarckian": modifications in an organism's phenotype must somehow find their way to its genotype to be passed on in heredity.

Once the preceding distinctions are made, the general answer to the questions about levels of selection can be given quite succinctly. Replication occurs primarily at the level of the genetic material, while environmental interaction can occur at a wide variety of levels, from genes, chromosomes, and cells to organisms, demes, and possibly entire species. The preceding answer is general. Particular cases are highly problematic. Perhaps fission in single-celled organisms can count as replication. In response to this suggestion, some authors have complained that organisms are not replicated. They simply die. But on this same line of argument, genes are not replicated, either. At meiosis they lose half of their constituent material. All that literally gets passed on is the information incorporated in their structure. Kin selection can be treated quite naturally as individual selection, but kinship *groups* are another matter. Perhaps beehives are sufficiently well organized that they can function as individuals in environmental interaction. Perhaps trait groups can perform this same function. Some authors consider any appropriately related pairs of organisms "groups" (Sober and Wilson, 1998).

The traditional organizational hierarchy can also be viewed in terms of increasingly more inclusive lineages. Entities at various levels form lineages, and the existence of such lineages is central to the selection processes. Genes, organisms, and even entire species form lineages. The lower-level lineages are contained in the higher-level lineages. In fact, the latest solution to the species problem involves treating species as lineages (de Queiroz, 1999). Although treating the living world in terms of a hierarchy of organizational levels gets complicated at times, one thing is clear: as we proceed up the organizational hierarchy, environmental interaction of the relevant sort becomes increasingly difficult, because selection tends to occur much more rapidly at lower levels of organization than at higher levels.

Replicators and environmental interaction. *Selection* has been defined in a variety of ways, but all these definitions include roughly the same items—reproduction of some sort (heredity), the transmitting of information from one generation to the next, the formation of lineages, the interaction of various entities with their environments so that not all of these entities survive to reproduce, and the introduction of variation. In biological evolution, genes are the entities that have evolved to contain information and pass it on from generation to generation. In fact, *generation* is defined in terms of this process. Other entities may also play this role, but genes evolved to perform it. One frequently hears that genes are self-replicating. If by this it is meant that a stretch of DNA can replicate in the absence of all else, including the necessary cellular machinery, then genes do not replicate themselves—but then no one holds such a view about self-replication, anyway. Of course, much else besides genes are necessary for the replication of these genes, but genes contain the relevant information and are the things that end up getting replicated.

Genes replicate (autocatalysis); they also produce organisms to aid in coping with their environments (heterocatalysis). As it turns out, the relation between genotypes and phenotypes is extremely complicated and idiosyncratic. So far, developmental geneticists have found little in the way of regularities to help in understanding the selection process, but some may be in the offing. In a few cases, biologists have been able to trace the molecular pathways that connect a certain set of genes to a phenotypic trait, but in most cases the existence of such pathways must be taken on faith. We may not know in detail what is going on at the molecular level, but we can assume that something is taking place. Adaptations are those characteristics that evolved to aid in environmental interaction.

Development is one way in which entities are produced that can function in environmental interaction, but it is a special case of a more general process—one that is defined in terms of the selection process itself. Any environmental interaction that results in replication being differential is good enough for selection to occur, even if it is not a developmental process. Development is one way in which interactors are produced, but it is not the only way. For example, genes are paradigm replicators. They also exhibit phenotypic traits of their own. The order of bases constitute the information that gets passed on in replication. The weak interactions between paired nucleotides are part of the machinery that allows this information to be replicated. The relation between the order of nucleotides and the rest of the structure of DNA is not one of development. At the other extreme, species might well function as entities with respect to environmental interaction. Some species have highly convoluted peripheries. If speciation in sexual organisms occurs primarily at the peripheries of species through peripheral isolation, then those species with more convoluted peripheries are likely to speciate more frequently than those having less convoluted peripheries. The problem is heredity. Is the tendency to produce highly convoluted peripheries passed down from generation to generation?

Reaction of the Immune System to Antigens. The structure of selection can be seen even more clearly in the reaction of the immune system to antigens. Selection in the immune system takes place at the cellular level, but its influence is primarily at the organismic level—either the organism survives or it dies. The chief complication with respect to the immune system is that it is a selection process taking place within more inclusive selection processes—selection within selection. Numerous sorts of cells are produced that are interrelated in highly complex ways. One class of cells that play a central role in the highly complicated reaction of the immune system to antigens are T cells. T cells are produced in large numbers in the thymus gland. On the surface of each of these cells is a reaction site. The genes that produce the structure of these reaction cites are so constituted that they introduce massive variations that are as close to random as is physically possible. These reaction cites in turn can match up to some portion of invading antigens.

When this happens, the T cells, as part of a larger immune reaction, proliferate and kill off the antigens. If they succeed and the organisms survive, this defense system is then shut down. Only a relatively few "memory" cells remain laying in wait for any future invasion by this same antigen. Every time that the same antigen invades an organism, the better the immune system of the organism is in reacting to it. Of course, antigens also change their structure, resulting in arms races between the T cells and the antigens that they attack.

The most striking feature of the immune system is the presence of such huge numbers. The thymus produces millions of T cells, but only a small percentage of these cells ever leave the thymus. Most are reabsorbed, possibly because they would attack the body of the organism itself. Of the millions of T cells that leave the thymus, only a tiny fraction ever come across an antigen that matches its reaction cite. If anything is apparent with respect to selection in the immune system, it is huge numbers and massive waste. Parallel observations hold for gene-based selection in biological evolution. In most organisms, millions of germ cells are produced. Only a tiny percentage of these germ cells ever fuse to form zygotes. Of the millions of zygotes produced, only a tiny percentage ever succeed in developing to sexual maturity. In selection processes, huge numbers and massive waste take the place of intentional behavior.

The chief difference between gene-based selection in biology and the reaction of the immune system is the level of organization at which they occur. In the first, gene lineages are contained within organism lineages. The relevant generations are those of organisms, and the result of selection processes are passed on from one generation of organisms to the next. In the immune system, some parental immunity is passed on to offspring, but it is soon replaced by the offspring's own immune reactions. By and large, the immunity that an organism develops in its lifetime is confined to that organism. The immune system functions at the cellular level with effects at the organismal level.

Selection and Reductionism. In any one case, deciding the level or levels at which environmental interaction is taking place is extremely difficult. Any evolutionary phenomenon can be explained with a bit of tweaking by more than one evolutionary process. For example, if ordinary individual fitness is partitioned into within- and between-group components, detailed predictions can be made, but these models are mathematically equivalent to inclusive fitness models. One would think that distinguishing between replication and environmental interaction would have resolved disputes over what is the ultimate unit of selection. There are no units of selection, only units of replication and units of environmental interaction. However, at this juncture an additional issue is introduced—reduction. In one common usage, classical thermodynamics is in principle reducible to statistical mechanics and Mendelian genetics is reducible to molecular biology. Once spelled out in sufficient detail, the higher-level theory is derivable from the lower-level theory. In practice, however, such derivations are at best extremely complicated. In the case of levels of selection, the issue is whether environmental interaction can be ignored.

The work of Richard Dawkins (1976, 1982) is famous for two reasons. First, Dawkins discusses particular adaptations and how they might have evolved. His opponents dismiss these narratives as "just-so stories," the sort of fictitious adaptationist stories told by Rudyard Kipling. Some just-so story or other must be correct, but deciding which is more than a little difficult. Second, Dawkins acknowledges the role of environmental interaction in selection—what he terms *vehicular selection*—but only to reduce it to gene selectionism. To use Dawkins's own terminology, he introduced vehicles only to bury them. After all, the only vehicular selection that matters is that which influences gene frequencies. Although selection acts on vehicles (structures that carry genes), its evolutionary effects are on gene frequencies, and only indirectly through these on phenotypes. Hence, they can be ignored. Because of this indirect transmission and the relative ease of individuating genes, textbooks and technical papers are filled with references to gene frequencies as if they were all that matters in selection (see Dawkins in Wilson and Sober, 1994).

Dawkins's opponents insist that downplaying environmental interaction amounts to ignoring the mechanisms that produce adaptations. Changes in gene frequencies provide understanding of sorts, but what makes biological evolution interesting is adaptation, and adaptations are the phenomena that need explaining more than any others. Strangely, the very biologists who object to just-so stories are among those biologists who complain most vehemently that gene selectionists are ignoring the mechanisms that give rise to the adaptations that these stories are designed to explain. Conversely, those biologists who spend the most time telling highly ingenious just-so stories ultimately abandon this part of the selection process for changes in gene frequencies. The point to notice, however, is that everyone agrees on the relevance of the organizational hierarchy to the question of levels of selection. The chief bone of contention is whether or not higher levels in this hierarchy are reducible to lower levels, and ultimately to genes (Sarkar, 1998, Sterelny and Griffiths, 1999).

[*See also* Adaptation; Group Selection; Macroevolution; Selfish Gene.]

BIBLIOGRAPHY

Boyd, R., and P. J. Richerson. *Culture and the Evolutionary Process*. Chicago, 1985.

Buss, L. *The Evolution of Individuality*. Princeton, 1987.

Cavalli-Sforza, L. L., and M. W. Feldman. *Cultural Transmission and Evolution: A Quantitative Approach*. Princeton, 1981.

Dawkins, R. *The Selfish Gene*. Oxford, 1976.

Dawkins, R. *The Extended Phenotype*. Oxford, 1982.

De Queiroz, K. "The General Lineage Concept of Species and the Defining Properties of the Species Category." In *Species: New Interdisciplinary Essays*, edited by R. A. Wilson, pp. 49–89. Cambridge, Mass., 1999.

Ghiselin, M. "A Radical Solution to the Species Problem." *Systematic Zoology* 25 (1974): 536–544.

Hull, D. L. "Are Species Really Individuals?" *Systematic Zoology* 25 (1976): 174–191.

Keller, L., ed. *Levels of Selection in Evolution*. Princeton, 1999.

Lloyd, L. "Unit of Selection." In *Keywords in Evolutionary Biology*, edited by E. Fox Keller and E. A. Lloyd, pp. 334–340. Cambridge, Mass., 1992.

Maynard Smith, J. "How to Model Evolution." In *The Latest on the Best: Essays on Evolution and Optimality*, edited by J. Dupré, pp. 119–131. Cambridge, Mass., 1987.

Sarkar, S. *Genetics and Reductionism*. Cambridge, 1998.

Sober, E., and R. C. Lewontin. "Artifact, Cause and Genic Selection." *Philosophy of Science* 49 (1984): 147–176.

Sober, E., and D. S. Wilson. *Unto Others: The Evolution and Psychology of Unselfish Behavior*. Cambridge, Mass., 1998.

Sterelny, K. *The Evolution of Agency and Other Essays*. Cambridge, 2001.

Sterelny, K., and P. E. Griffiths. *Sex and Death: An Introduction to Philosophy of Biology*. Chicago, 1999.

Williams, G. C. *Natural Selection: Domains, Levels, and Challenges*. Oxford, 1992.

Williams, G. C., ed. *Group Selection*. Chicago, 1971.

Wilson, D. S., and E. Sober. "Reintroducing Group Selection to the Human Behavioral Sciences." *Behavioral and Brain Science* 17 (1994) 585–654. Includes open peer commentary and authors' response.

Wilson, J. *Biological Individuality*. Cambridge, 1998.

— David L. Hull

LICHENS

Lichens are obligate mutualistic ectosymbioses between fungi (mycobionts) and either or both of the photobionts green algae and cyanobacteria. Usually the photobiont (inhabitant) lives in the supporting tissue of the mycobiont (exhabitant), which receives carbohydrates from the photobiont. The symbionts have been grown separately in axenic (sterile) culture. The lichen symbiosis has been resynthesized in vitro, but late developmental stages with typical complex phenotypic features have been most successfully reached in nonsterile or near-nature experiments.

In general, the mycobiont seems the most dependent of the two types of symbionts found in lichens. The lichenized mycobiont species have never been found growing without their photobiontic partner in nature and are often difficult to grow in axenic culture. Many algae found in lichens (e.g., *Trentepohlia* and *Myrmecia*) are facultative photobionts, known to occur independently as epiphytes, endoliths, or soil algae. They often have growth rates in culture that are similar to those of axenic cultures of free-living algal species that are never found in lichens. However, the free-living ability of some of the other most commonly found photobionts in lichens is still unclear (e.g., *Trebouxia*).

Symbionts grown separately do not exhibit the complex growth forms (e.g., crustose, foliose, and fruticose thalli) and structures (e.g., cyphellae, rhyzines, fibrils, cilia, and soralia) resulting from their symbiotic interactions. In contrast, some of the so-called lichen substances that are often uniquely found in lichens (e.g., depsides and depsidones) can still be synthesized by the mycobiont when grown in axenic culture. The number of lichen substances with known structures is about 700, many of which have important biological properties, such as antibiotic, antitumor, antimutagenic, allergenic, plant growth inhibitory, and enzyme inhibitory activity, as well as UV protection. One compound is reported to inhibit the cytopathic effect of HIV in vitro. Little is known about the role these substances may play in the biology, ecology, and evolution of lichens.

The development of the symbionts into one lichen thallus is so integrated that lichens have been perceived and studied as single organisms until very recently. Although the symbiotic nature of lichens was first revealed in 1867, it took more than 60 years before an integrated classification of lichenized and nonlichenized fungi was attempted, and only since 1971 have lichen-forming fungi been included in the *Index of Fungi* and *Dictionary of the Fungi*.

Because many of the photobiont species are associated with many different mycobionts, the symbiotic lichen association is named after the mycobiont. For example, the lichen *Parmelia sulcata* is the name of the ascomycete species forming this lichen. Photobionts found in lichens have their own taxonomic names. The photobiont associated with *Parmelia sulcata* belongs to the green alga genus *Trebouxia*.

Sexual reproduction is restricted, most often, to the lichen mycobiont, which produces typical fungal fruitbodies (ascomata or basidiomata), such as apothecia and perithecioid ascomata, or typical mushrooms. With the exception of the green alga *Trentepohlia*, the algae do not reproduce sexually when part of the lichen symbiosis. For example, asexual flagellated spores (zoospores) of the green alga *Trebouxia* are often observed in axenic culture, but sexual reproduction was never observed. Therefore, it is believed that most green algae found in lichens are part of asexual lineages. Because a large proportion of lichen photobionts (e.g., *Trebouxia*) are very rarely found outside the lichen symbiosis, it is a longstanding question how sexually produced fungal spores from the mycobiont manage to find the appropriate photobionts.

Various lines of evidence have provided at least partial explanations for this enigma. Because distantly related lichen-forming species can share the same photobiont species, it has been proposed that some of these fungi can acquire the photobiont from other lichens or from symbiotic propagules. It has also been reported that at very early ontogenic stages, the mycobiont mycelium can temporarily associate with ultimately incompatible algal cells, possibly increasing its chances of eventually finding the appropriate alga. Recent physiological studies have revealed that *Trebouxia* is better adapted to desiccation than its mycobiont *Cladonia*, suggesting that individual or small groups of algal cells might be widely distributed and readily available in a dormant state in nature. Finally, the production of asexual propagules (e.g., soredia and isidia) containing both symbionts is another way lichens can bypass this hurdle. It is rare that the same mycobiont individual will produce post-meiotic spores and specialized asexual symbiotic propagules simultaneously. As for many other Basidiomycota and Ascomycota, many lichen-forming fungi are known only (or mostly) in their asexual form.

Because of their slow growth (often <1 mm per year), most lichens need substrates that are sufficiently stable to support their development until reproductive maturity. Despite these growth and reproductive constraints, lichens are virtually omnipresent in nature from the poles to the tropics, and they range from extremely dry to aquatic habitats (fresh and salt water) where they can be submerged for most of the year. They grow on, or sometimes in, the bark of trees and shrubs, on wood, on or in rocks (endolithic lichens), on long-lived leaves (mostly in subtropical and tropical habitats), on soils, on mosses, on bones and antlers lying on the ground, and even on other lichens. Some lichens are narrowly substrate-specific and are confined to one or two tree species or a particular rock type, while others are found not only as epiphytes on various tree species but sometimes also on wood, soil, or rock. Lichens can also colonize manmade substrates such as concrete, various types of roof tiles, plastics, or glass. The facts that lichens colonize a broad range of substrates (mostly nutrient-poor substrates where few other organisms can grow), and that closely related species often colonize the same types of substrates, strongly suggest that substrate availability and colonization ability greatly contributed to their impressive diversification.

Species diversity is believed to be highly unequal between lichen mycobionts and photobionts. The number of known extant lichen-forming fungus species is usually estimated to be about 13,500, representing about one-fifth of all known extant species in the Fungi kingdom. Lichenization is among the most successful ways fungi fulfill their requirement for carbohydrates. Because many areas of the world have been poorly sampled and many of the lichenized fungi are in need of modern systematic revisions, the total world lichen diversity has been estimated to be between 17,000 and 20,000 species. The evidence currently available strongly suggests that the photobiont diversity found in lichens is remarkably lower than its mycobiont counterpart, although recent DNA studies promise to reveal a much higher number of photobiont species than previously expected in lichens.

Although a remarkable one out of five fungal species is lichenized, 98 percent of lichen diversity is highly concentrated within the largest of the four fungal phyla, the Ascomycota, so that more than 40 percent of Ascomycota species are lichenized. A few lichen-forming fungal species (<2 percent, about 50 species) are found within the Basidiomycota (sister phylum to the Ascomycota and second to it in size), accounting for less than 0.5 percent of the basidiomycete diversity. The great majority of lichen-forming fungi species (roughly 85 percent) are associated with green algae, forming bimembered lichen symbioses. About 10 percent of the lichenized fungi are associated only with cyanobacteria, and a small portion (3–4 percent) are associated with both photobionts, forming tri-membered lichen symbioses.

In the mid-1980s, the mycological and lichenological communities accepted that obligate mutualistic and ectosymbiotic associations between basidiomycetes and photobionts should be recognized as lichens, establishing for the first time that the lichen symbiosis had separate origins. Within the Basidiomycota, three independent origins have been documented (*Dictyonema*, *Multiclavula*, and *Omphalina*). Within the Ascomycota, where almost all lichens are found, the evolution of lichens is not as clear. Although ascolichens are found in 15 of the 46 orders of Ascomycota reported in the *Dictionary of the Fungi*, only 4 of these 15 orders of euascomycetes are comprised strictly of lichen-forming species. The remaining 11 orders encompass a mixture of lichenized and nonlichenized species, representing 61 percent of the total ascomycete diversity. Because of this high diversity and mixed occurrence of lichenized and non-lichenized ascomycete species, lichens were widely believed to have arisen independently on multiple occasions within the Ascomycota.

More recently, molecular studies have suggested that lichens evolved earlier than formerly believed and that, in fact, gains of lichenization have been infrequent during the evolution of the ascomycetes. Instead, the evolution of lichens may have been characterized by multiple independent losses of the lichen symbiosis. This suggests that the lichen symbiosis has been a relatively long-standing relationship, and consequently, that it has played a larger role in the evolution of the Ascomycota than previously believed. Some of the major Ascomycota lineages, including the non-lichenized Ostropales, the Chaetothyriales, and the Eurotiales (which include the antibiotic *Penicillium* and the pathogenic *Aspergillus*) may have been derived from lichen-forming ancestors. Distinguishing between secondarily derived and ancestrally nonlichenized Ascomycota is of fundamental importance for any study of the ascomycetes.

The comparative phylogenetic study of the model genus *Omphalina* (nonlichenized and lichenized mushrooms), where a recent transition to the lichen symbiotic state took place (with the green alga *Coccomyxa*), was used to identify potential predispositions, stresses, and consequences associated with the acquisition of the lichenized state. Phenotypic plasticity/broad ecological amplitude, low fungal virulence/high photobiont infection resistance, and desiccation and sun irradiation tolerance/efficient DNA repair mechanisms were identified as three critical predispositions for a successful transition to the lichen habit. Slow growth in axenic culture, basidiomata with uninucleate hyphae, a highly variable number of spores per basidium, and an accelerated rate of nucleotide substitutions were four potential consequences revealed by the study of this model system. These consequences strongly suggest that a high level of stress is associated with a transition to the lichenized state, providing a possible explanation for the low rate of gains of lichenization observed within the Ascomycota.

Possible clues for how mycobionts "escape" from the lichenized condition can be found in lichenicolous fungi, which comprise more than 1,000 species of nonlichenized mycobionts obligately dwelling on or in lichens. Based on morphological evidence, it seems that many lichenicolous species belong to nonlichenized ascomycotic lineages derived from lichen ancestors. By colonizing lichens, these fungi thereby continue to obtain directly, or indirectly from the lichenized fungus, carbohydrates generated by the algae and cyanobacteria found in lichens. Hence, these species derive many of the benefits of lichenization without having to find a photobiont to form the obligate lichen symbiotic association. This lichenicolous state may act as a fungal "halfway house" that could facilitate further transitions to different substrates and types of interactions, such as pathogens on humans and crops.

Currently, about 100 species of cyanobacteria and algae, classified in less than 40 genera, have been found in lichen symbioses. However, the photobionts have been identified to the species level for less than 5 percent of lichen species. *Trebouxia*, *Trentepohlia* (green algae), and *Nostoc* (cyanobacteria) are the most common. Molecular analyses reported *Trebouxia* species as being a paraphyletic lineage interspersed with coccoid lichen and soil algae. Molecular studies suggest that the cyanobacteria *Nostoc* associated with all lichen-forming fungi species are derived from a single common ancestor nested at the base of the nostocalean clade. The genetic variation within this symbiotic *Nostoc* lineage, which also includes all *Nostoc* associated with plants (except *Azolla* and *Gunnero*), corresponds to the level of variation expected within a single cyanobacterium species. Similarly, molecular phylogenetic analyses also suggest that the *Coccomyxa* associated with lichen-forming ascomycetes and basidiomycetes form a single monophyletic species. Therefore, a higher rate of speciation appears to be taking place within the genus *Trebouxia* than in other lichen photobionts. Cyanobacteria and green algae found in lichens are often shared extensively across closely and distantly related lichen-forming fungi. Multiple photobiont species (e.g., *Trebouxia*) have also been isolated from different lichen thalli belonging to the same lichen species.

The nature of the mycobiont–photobiont coevolution has only been recently explored for lichens. It is believed that lichenized ascomycetes have been coevolving longer with photobionts than have lichenized basidiomycetes. In terms of cospeciation, mycobiont and photobiont specificity is rather high, so that there is a certain degree of cospeciation that can be detected when broad phylogenetic comparisons are made. Most major groups of lichen-forming ascomycetes (genera to supraordinal levels) are associated with specific groups of photobionts. However, within each of the lichenized fungal lineages (at the genus and species levels) there is usually little phylogenetic evidence of cospeciation (or colineage sorting) between mycobionts and photobionts, although this may change when rapidly evolving molecular markers are used to study the slowly evolving photobionts.

The rate of nucleotide substitution for the same coding gene in the mycobiont and photobiont growing within the same thallus has been estimated to be between three times faster for RNA coding genes in the mycobiont, and up to 25 times higher for noncoding DNA. A major increase in rates of nucleotide substitution has been detected in lichenized *Omphalina* as a result of a recent transition to lichen mutualism in this genus. The impact of this transition on the photobiotic lineage (accelerated vs. decelerated, vs. status quo) is unknown.

[*See also* Algae; Coevolution; Fungi; Mutualism; Symbiosis.]

BIBLIOGRAPHY

Ahmadjian, V., and M. E. Hale. *The Lichens*. New York, 1973. A classic overview of the lichen symbiosis.

Beck, A., T. Friedl, and G. Rambold. "Selectivity of Photobiont Choice in a Defined Lichen Community: Inferences from Cultural and Molecular Studies." *New Phytology* 139 (1998): 709–720. The most detailed study on lichen symbiont specificity.

Brodo, I. M., S. Duran Sharnoff, and S. Sharnoff. "Lichens of North America." New Haven, 2001. One of the most comprehensive introductions to lichenology with spectacular illustrations. This is a must.

Culberson, C. F. *Chemical and Botanical Guide to Lichen Products*. Chapel Hill, N.C., 1969. A classic overview of compounds found in lichens, including an excellent section on the biogenesis of lichen substances and their relationship to the products found in nonlichen-forming fungi.

Friedl, T. "The Evolution of the Green Algae." In *The Origin of Algae and Plastids*, edited by D. Bhattacharya, *Plant Systematics and Evolution*, Suppl. 11 (1997): 87–101. The best integrated phylogeny of lichenized and nonlichenized green algae.

Galun, M. *Handbook of Lichenology*. Volumes I–III. Boca Raton, Fla., 1988. One of the best overviews of lichenology. All topics highlighted here are addressed in much more detail in this series of three volumes.

Hawksworth, D. L., P. M. Kirk, B. C. Sutton, and D. N. Pegler. *Ainsworth and Bisby's Dictionary of the Fungi*. Cambridge, 1995. One of the most complete and most cited books for fungi (including lichens) diversity and classification.

Huneck, S., and I. Yoshimura. *Identification of Lichen Substances*. New York, 1996. A recent and extensive overview of compounds found in lichens.

Kranner, I., and F. Lutzoni. "Evolutionary Consequences of Transition to a Lichen Symbiotic State and Physiological Adaptation to Oxidative Damage Associated with Poikilohydry." In *Plant Responses to Environmental Stresses: From Phytohormones to Genome Reorganization*, edited by H. R. Lerner. Basel, 1999. A recent synthesis of micro- and macroevolution of the lichen symbiosis.

Kroken, S., and J. W. Taylor. "Phylogenetic Species, Reproductive Mode, and Specificity of the Green Alga *Trebouxia* Forming Lichens with the Fungal Genus *Letharia*." *The Bryologist* 103 (2000): 645–660. The most detailed coevolutionary study of lichens.

Lutzoni, F., and M. Pagel. "Accelerated Evolution as a Consequence of Transitions to Mutualism." *Proceedings of the National Academy of Sciences USA* 94 (1997): 11422–11427. The first comparative phylogenetic study of the lichen symbiosis using statistical and analytical tools.

Lutzoni, F., M. Pagel, and V. Reeb. "Major Fungal Lineages are Derived from Lichen Symbiotic Ancestors." *Nature* (in review), 2001. The most extensive reconstruction of the evolution of the lichen symbiosis within the Ascomycota. First phylogenetic comparative study where uncertainty associated with phylogenetic and ancestral state reconstructions was accounted for.

Nash, T. H. *Lichen Biology*. Cambridge, 1996. One of the best recent and condensed overviews for lichenology. A large part of the information presented here was synthesized from this book.

Rambold, G., and D. Triebel. "The Inter-Lecanoralean Associations." *Bibliotheca Lichenologica* 48 (1992): 1–201. A major contribution to our understanding of the biology and evolution of lichenicolous fungi.

Rambold, G., T. Friedl, and A. Beck. "Photobionts in Lichens: Possible Indicators of Phylogenetic Relationships?" *The Bryologist* 101 (1998): 392–397. The broadest and most exhaustive compilation of photobionts associated with specific lineages of lichen-forming fungi Lecanorales.

— FRANÇOIS LUTZONI

LIFE HISTORY STAGES

Life history stages are the various forms in which an organism expresses or manifests itself beginning with its gametes and ending with its adult stage (when it reproduces). A description of an organism's life cycle consists of these ontogenetic, developmental stages and usually includes information about the timing and habitat of each stage. The form and existence of life history stages are the result of environmental and historical influences, including natural selection, on the developmental processes and genetic information that control the expression of the stages.

A key way in which organisms and life cycles evolve is through changes in their life history stages. When one considers life history stages, the focus is on how and why that part of the life cycle is present. How does it work—functionally, ecologically, and evolutionarily? How are life history stages added, modified, or removed from life cycles? Traditional approaches to these questions include comparative methods that seek taxonomic or biogeographic patterns and theoretical modeling to predict life cycle composition or changes based on a set of biological, usually ecological, conditions. Most recently there is growing interest in the developmental and genetic basis of evolutionary changes in life history stages. These studies analyze changes in gene expression as organisms progress through different life history stages or differences in genetic and developmental processes of a particular life history stage that differs between related taxa.

Collectively, living organisms assume a myriad of life history stages that may differ in size, form, genetic architecture (haploid-diploid), reproductive mode (sexual-asexual), and representation within the life cycle (dominant-transient). Many organisms, including the fungi, most macroalgae (e.g., kelps and red algae), the lower land plants (e.g., bryophytes and ferns), most invertebrate animals (e.g., especially marine phyla and the holometabolus insects), and some vertebrate animals (e.g., amphibians), have multiphasic or complex life cycles in which sequential life history stages have distinctly different forms or occupy different habitats. By changing form or habitat, a particular stage can occupy an ecological niche or environmental setting different from an earlier or later stage of the life cycle. The significant increase in size (covering several to many orders of magnitude) that is associated with progression through a life cycle from egg to adult may force changes in ecological niches and thus lead to formation of complex life cycles (see Werner, 1988).

The most common life cycle among animals is indirect development in which the zygote (fertilized egg) develops into a free-living larval stage that transforms through a metamorphosis into a juvenile that is morphologically similar to an adult. Metamorphosis necessarily includes the loss of specific larval structures (and form) and usually is associated with a shift in habitats. The dramatic differences in form between a tadpole and a frog, a caterpillar and a butterfly, or a sea urchin larva and adult draw attention to the structure and function of the different body plans present in these different life history stages. By analyzing morphological traits of larvae, biologists begin to understand them as devices adapted for their environment or the niche they occupy. Most benthic and many pelagic invertebrates have planktonic larval forms that are microscopic and show limited resemblance to the macroscopic adult forms. The small size and aquatic environment of these larvae demand a different body organization for effective feeding and locomotion compared to what works for the large adult form. Larval instars of holometabolous insects occupy distinct sedentary habitats, before entering the stationary pupal stages and emerging as winged adults. Amphibian larvae have tails, feeding structures, and gills required in their aquatic habitats. These structures are modified or lost at metamorphosis to juvenile stages.

Other animal life cycles (e.g., mammals and many invertebrates) show direct development, in which a zygote transforms into the juvenile form without an intermediate larva, with a larval stage that occurs inside the mother or as an unhatched developmental stage. The term *direct development* has also been used to refer to life cycles that have larvae that are highly modified from an ancestral pattern of indirect development.

Many closely related groups of organisms show wide variation in the types of life history stages present in their life cycles. Morphological variation among equivalent life history stages is particularly well documented within many groups of marine invertebrates, including echinoderms, crustaceans, and urochordates. Amphibians also show remarkable variation in the form, size, and periods of life history stages among related taxa. For example, among frogs, a life cycle with a free-living aquatic tadpole is believed to be the ancestral or primitive condition. However, this free-living tadpole stage has been lost from the life cycle and replaced by a directly developing embryo at least ten times. This conclusion is based on the fact that each of ten family-level taxonomic groupings of frogs contains species with tadpoles and species without tadpoles (see Hanken in Hall and Wake, 1999). A similar survey of living sea urchins indicates that the indirect development via a feeding larval form has been lost at least fourteen times and replaced by direct development via nonfeeding larval forms or brooded embryos. Changes such as these are also often associated with changes in duration of the larval period, size of eggs and embryonic stages, and number of offspring during a given bout of reproduction. This remarkable evolutionary flexibility in life cycles offers opportunities to study morphological adaptation of the different stages, ecological and evolutionary reasons for shifts in development, and possible influences of larval body form on subsequent adult form.

Other groups of animals (e.g., placental mammals) are relatively invariant in the form of the life history stages present in the life cycle. Variation appears to be restricted to the length of time spent in a given stage.

Studies on the origin of multicellular animals (the Metazoa) have also examined life histories and life cycles. These comparative surveys consider whether the first metazoan was pelagic or benthic, whether it had a simple or a complex life cycle (e.g., Nielsen, 2001). Some authors go on to consider how variation in life histories and life cycles may have changed to give rise to modern phyla by differences in emphasis given to the larval and the adult phases.

One way that life history stages can evolve is by changes in the timing of when features are expressed. This change in timing is called heterochrony and is recognized by comparison with known or inferred life cycles of (putative) ancestors. Heterochrony can lead to a reordering of expression of traits or loss of expression of traits found in the ancestral condition. Correctly identifying the ancestral condition is essential for recognizing and interpreting heterochrony.

[*See also* Cell Lineage; Heterochrony; Metamorphosis, Origin and Evolution of; Phenotypic Plasticity.]

BIBLIOGRAPHY

Hall, B. K., and M. H. Wake, eds. *The Origin and Evolution of Larval Forms.* San Diego, Calif., 1999.

McEdward, L. *Ecology of Marine Invertebrate Larvae.* Boca Raton, Fla., 1995.

Nielsen, C. *Animal Evolution: Interrelationships of the Living Phyla.* Oxford, 2001.

Raff, R. A. *The Shape of Life.* Chicago, 1996.

Werner, E. E. "Size and Scaling and the Evolution of Complex Life Cycles." In *Size-Structured Populations*, edited by B. Ebenman, pp. 60–81. Berlin, 1988.

— Richard B. Emlet

LIFE HISTORY THEORY

The first article serves as an overview of the scope of life history theory and its relationship with demography; the second article provides an introduction to life history issues in humans and how we have adjusted the various things we do for good adaptive reasons; the third article specifically attempts to understand the unusual phenomenon of an extended period of nonreproductive life in humans; the fourth article is a case study of the life history of guppies and focuses on the tradeoffs between offspring size and offspring number, age of first reproduction, and the experimental evidence of life-history evolution in response to predation pressure. For a related discussion, see Life History Stages.]

An Overview

Life history theory analyzes the evolutionary significance of the many varied aspects of organisms' life histories. An organism's life history includes how long it lives, at what age it matures, and how many offspring it has. The theory shows how natural selection acts on life history traits. It provides the evolutionary underpinning of behavioral and physiological ecology and aims to explain why particular life history strategies are adaptive in particular environments. The theory is essentially mathematical in character. A mathematical model of a life history is constructed—usually all the individuals in the population are assumed to have the same life history—and the success or failure of mutant genes producing variant life histories is analyzed. Any such mathematical model is necessarily a simplification of reality, and informed decisions have to be taken as to how best to construct such models, as described here. The main areas in which simplications are made are in describing the life history (see Figure 1) and in constructing an index of fitness.

Historical Background. Perhaps the father of life history theory is R. A. Fisher, the English geneticist and statistician who, in addition to being a founder of statistical theory, wrote *The Genetic Theory of Natural Selection* in 1930. Here, Fisher not only outlined one of the main methods of measuring the fitness of genes affecting life histories but also indicated a central area of interest: "It would be instructive to know not only by what physiological mechanism a just apportionment is made between the nutriment devoted to the gonads and that

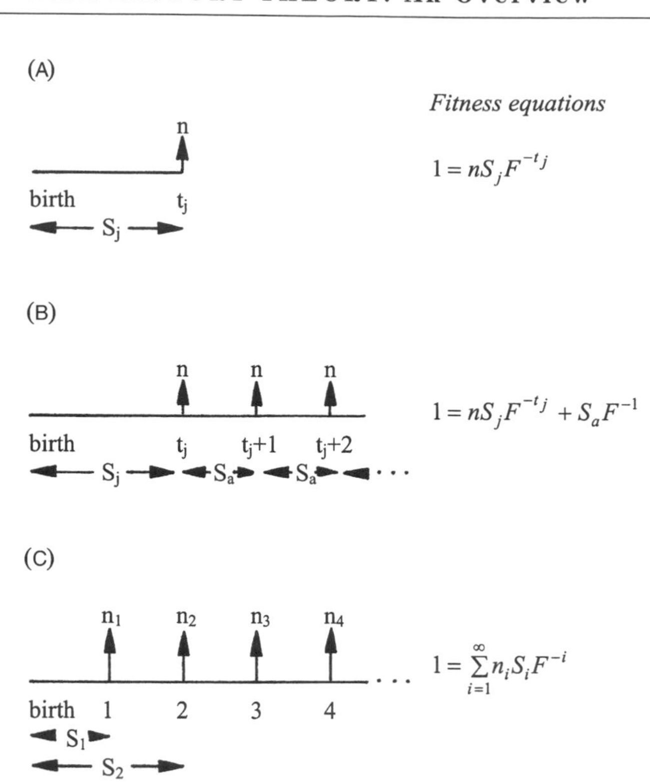

FIGURE 1. The Main Types of Animal Life-History Models. In each case the x-axis is age. In (A) females breed for the first time at age tj years, when they produce n daughters, then die. This type of life history, in which animals breed once then die, is referred to as *semelparity.* Alternatively animals may breed repeatedly, as in (B) and (C), this is referred to as *iteroparity.* In (B) females start as in (A) but do not die after first reproduction, thereafter females breed at yearly intervals, producing *n* daughters each time they breed. There are two stages in this life history: juvenile and adult. In (C) females breed each year, producing n_1 daughters at age 1, n_2 at age 2 and so on. In general not all animals survive all the stages, and so mortality schedules are also needed, these are here given in terms of survivorships, *S.* The equations on the right give for each life history the fitness, *F,* of the genes that code for it. The fitness of a gene is here measured as its *per copy* population growth rate. Drawing by Richard M. Sibly.

devoted to the rest of the parental organism, but also what circumstances in the life history and environment would render profitable the diversion of a greater or lesser share of the available resources towards reproduction." He suggested the possibility of physiological variants that would alter the allocation of resources between reproduction and body maintenance and explored the features in the environment that selected for particular life histories. Postwar interest was kindled by L. C. Cole's 1954 paper on life history phenomena, which caught the imagination with the following apparent paradox: "For an annual species, the absolute gain in intrinsic population growth which could be achieved by changing [from a single bout of reproduction] to the perennial reproductive habit would be exactly equivalent to adding one individual to the average litter size." Following Fisher, Cole used population growth rate as a measure of fitness and suggested that parental self-sacrifice would be adaptive if it led to the production of one additional surviving offspring. At first glance, this seems paradoxical, because it appears relatively easy to achieve an extra offspring from the relatively large resources of the parent.

Selecting a Fitness Measure. A fitness index measures how fast genes that affect the life histories of their carriers in specified ways spread in the population (see Figure 1). The two main ways to consider fitness are outlined below. For simplicity, this is done only for females.

Lifetime reproductive success (LRS) calculates the number of daughters a female can expect to produce in her lifetime. It is a good fitness measure if it is reasonable to model the population's size as constant; this is justified on the grounds that over evolutionary time, short-term population fluctuations can be ignored. In the long term, the rate of increase of any population that persists is very close to 0. Otherwise, the population would expand unchecked. In populations with growth rates of 0, the LRS is equal to 1, because each individual expects to replace itself. Genes whose carriers have LRS less than 1 would be selected against, whereas those with LRS greater than 1 would spread in the population. It is true that the spread of such genes might increase population size short term, but it is assumed that the ecological process of density dependence would eventually bring population size down again.

An alternative approach, adopted here, is to use a gene's population growth rate (pgr) as a fitness measure. Genes decreasing the pgr of their carriers would be selected against, whereas those with increased pgr would spread in the population. This approach allows the evolutionary process to be studied in periods when pgr is nonzero.

Equations are widely available for both fitness measures for a large number of life histories and are given in Figure 1. The equation in Figure 1C, known as the Euler–Lotka equation, was derived independently by the Swiss mathematician Leonhard Euler in 1760 and by the American population biologist Lotka in 1907.

The Evolutionary Process. The simplest case to consider is that in which a mutant gene affects just one of the life history traits. It is intuitively obvious that genes that increase fecundity or survival will spread in the population. This is demonstrated mathematically by showing that fitness is increased by increasing fecundity or survival. If age-specific death rates are fixed, then fitness is also increased by breeding earlier, so an increased growth rate is also always favored by selection. Much of behavioral and physiological ecology is concerned with the study of adaptations that increase birth and growth rates or decrease death rates.

Formulas calculating the sensitivity of fitness to increases in fecundity are widely available. An important application of these methods is to show that the strength of natural selection decreases with age. For example, genes that increase the reproductive competence of the young are under much stronger selection than equivalent genes affecting older individuals. This can be understood intuitively because as a result of natural mortality, only a few of the individuals affected when young survive to be affected by late-acting genes. Consequently, the number of gene copies affecting the young is greater than the number affecting old individuals. This translates into greater selection pressure on young than old individuals and is thought to be one of the evolutionary mechanisms producing senescence.

Trade-Offs. Although selection always favors increases in birth or growth rates or decreases in death rates, these improvements cannot continue indefinitely, because eventually organisms come up against limitations in, for example, the availability of resources. At that point, improvements in one trait may be achievable only at the expense of others—there is a trade-off between the traits. For example, increasing fecundity may entail a mortality cost, so that there is a cost of reproduction.

Many trade-offs have their basis in the laws of physics and chemistry. Some arise because time, space, or energy allocated to one function is not available to others. Thus, resources allocated to defense, lowering mortality rate, may not be available for somatic growth. Defense may be achieved by physical structures, such as spines or shells, or by vigilance that is achieved at the expense of feeding. All such defenses absorb resources that could otherwise go to increasing productivity.

The importance of such trade-offs is that they represent the limits of what is genetically possible for individuals of a given species. Trade-offs may thus have an important role in the evolutionary process. Figure 2 shows what happens in a constant environment. During the evolutionary process, some alleles are selected and others are eliminated (Figure 2A). The alleles that remain at the end of the evolutionary process are those with the highest fitness. These alleles occur in a thin band on or beside the trade-off curve, around the optimal strategy (Figure 2B). These alleles affect more than one life history character and are called pleiotropic.

One of the most important trade-offs involves a cost of reproduction, in which increases in fecundity can only be obtained at a cost in terms of adult survival. Using the life history model of Figure 1B, it is usual to suppose the trade-off has a convex up shape, as in Figure 3. If all animals breed first at age 1, then fitness is given by the simple equation $fitness = nS_j + S_a$ (this comes from the equation in Figure 1B). This equation

Figure 2. The Operation of the Evolutionary Process in a Constant Environment.
It is assumed that genetic variation affects just two life-history characters, the average phenotypes of a number of genetic possibilities are shown as dots. The thick curve represents the trade-off limiting what is genetically possible. Light lines represent fitness contours, fitness increases toward top right. (A) and (B) show hypothetical situations before and after some generations of selection. * indicates the *optimal strategy* maximizing fitness. Peter Calow, ed. *The Encyclopedia of Ecology and Environmental Management.* Blackwell Science Ltd., 1998. Reprinted with permission.

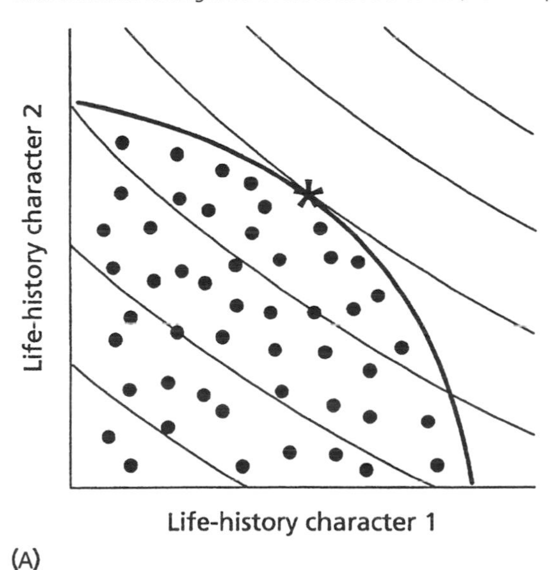

Life-history character 2

Life-history character 1

(B)

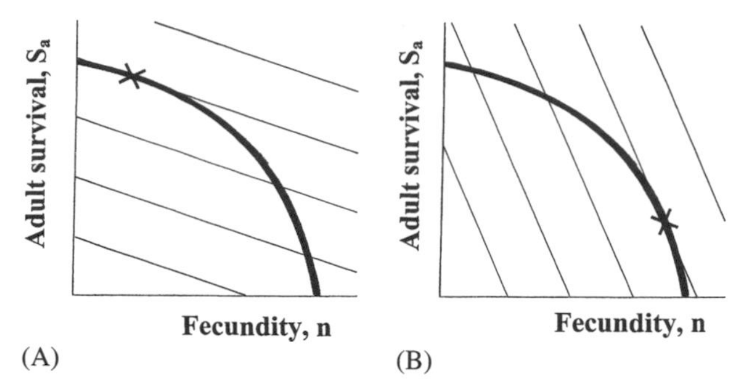

FIGURE 3. Analysis of the Cost of Reproduction Trade-off. The form of the trade-off between fecundity *n* and adult survivorship *Sa*- is given by the thick convex curve (variables defined as in Fig. 1b). The straight lines indicate fitness contours, the slope of the lines depends on juvenile survival, S_j (see text). (A) and (B) show the cases that juvenile survival is poor or good respectively. * indicates the *optimal strategy* maximizing fitness. Drawing by Richard M. Sibly.

can be rewritten as $S_a = fitness - nS_j$. Note that if juvenile survival S_j and *fitness* are fixed, then S_a decreases linearly with fecundity n. It follows that contours of equal fitness are straight lines, as shown in Figure 3. The slope of these lines is given by S_j. If juvenile survival is low, as in Figure 3A, then the fitness contours are shallow; if juvenile survival is high, then the fitness contours are steep. Inspection of Figure 3 reveals that the optimal strategies are to breed repeatedly (iteroparity) if juvenile survival is poor (Figure 3a), but to breed once, then die (semelparity), if juvenile survival is high (Figure 3B).

This analysis of the cost of reproduction trade-off shows how the optimal strategy may depend on a factor extrinsic to the trade-off (juvenile survival), determined perhaps by the local environment. If this factor varies between environments, then the optimal strategy may also vary. This shows how one of the goals of life history theory may be achieved: demonstrating which strategies are adaptive in which environments.

The cost of reproduction trade-off also crops up in an intriguing way in discussions of senescence. The trade-off occurs because resources allocated to fecundity are not available for adult maintenance. If the terms of the trade-off are like those in Figure 3, it follows that the optimal strategy will involve submaximal adult survival. Thus, organisms could be designed for immortality, in which case the trade-off curve would go through the y axis where $S_a = 1$. However, this has never been favored by selection because it is always advantageous to sacrifice some personal survival to enhance breeding success.

Trade-offs may occur between any life history characteristics. The number and identity of life history characteristics vary between species, but many life histories can be satisfactorily described by development period, fecundity, and juvenile and adult survival, as in Figure 1B. Between these four life history characteristics there are six possible trade-off pairs, each produced in one or more of the following ways. Growth rate and survival may be traded off because resources allocated to defense (increasing survival) are not available for production; thus, development rate and fecundity are both reduced. If more offspring are produced, the survival of each may be lower if offspring rely on parents for food and this is in short supply. Adult size and egg size may both affect more than one life history characteristic, another source of trade-offs. For example, larger adults are often more fecund, but it takes longer to grow to be bigger, so fecundity is traded off against development period. The same trade-off can occur through variation in egg size. Thus, if eggs are smaller, more can be produced, but each takes longer to grow to reach adult size, so development period is increased. It is also sometimes found that smaller juveniles incur increased mortality costs.

Trade-offs may also occur between related individuals because copies of the same gene may occur in both individuals. Thus, parents may take risks (decreasing adult survival) to protect their offspring (increasing juvenile survival). Alternatively, parents may feed juveniles (speeding development) instead of feeding themselves; this has been shown to result in reduced subsequent fecundity for the adults. When trade-offs occur between individuals in this way, the calculation of fitness is complicated by the need to take account of the chance that the gene is present in the helped individual. For example, a rare allele affecting parental care is transmitted to offspring with chance 0.5, given by Mendel's laws. This uncertainty affects whether or not the genes increase in the population, according to Hamilton's rule. [*See* Cooperation.] The evolution of altruism between related individuals may also be analyzed in this way.

It has also been shown that suites of life history characteristics (r strategies) that are advantageous in periods of rapid population growth may be incompatible with those needed at carrying capacity (K strategies). Under some assumptions, it can be shown that r strategists should be characterized by semelparity, producing huge numbers of tiny offspring at an early age when the adults are relatively small. In contrast, K strategists should be iteroparous, putting limited resources into reproduction and producing large, competitive offspring. Adults, too, need to be large to compete effectively, and age at first reproduction is therefore likely to be delayed.

Recently, some have questioned whether any individual species have the suite of traits thought to designate one or the other strategies designated by the r/k distinction, and whether these traits are found only in the ecological niches after which they are named. In place of this, life history theorists now prefer the notion of a fast/slow continuum of life history traits. On this view some species, quite independently of whether they are in a population at carrying capacity, tend to reproduce early, have large litters, short gestation and weaning times (where these traits apply), and short life spans, whereas other species are characterized by the reverse. Intriguingly, a relative fast/slow continuum seems to hold even when differences in body size among species are controlled. One finds some large-bodied animals with relatively fast life histories for their size, and small-bodied animals with relatively slow life histories. Moreover, there are exceptions even to these trends. Humans, for example, have quite late ages at maturity, but then reproduce at a fast pace given their size.

Over the last three decades a great deal of effort has gone into comparing life history strategies in contrasting environments, interpreting differences as adaptations to local environments. Delineating the shape of trade-off curves, however, has not proved easy. Five methods are used.

1. Between-individual phenotypic correlations might reflect genetic correlations, and thus provide evidence of trade-offs, as in Figure 2B. However, individual life histories are greatly affected by the environments individuals grew up in, and the effects of developmental environment generally swamp genetic effects. Phenotypic correlations are therefore considered to give unreliable evidence of trade-offs.
2. Genetic correlations can be measured using the methods of quantitative genetics, but these generally involve large-scale breeding experiments requiring that the organisms be individually housed. Not only are the experiments difficult to perform, but they also can be difficult to interpret, because the study environment generally differs from the environment in which the animals evolved. The remaining methods are more useful.
3. Transplant experiments consist of transplanting individuals from a number of habitats to a common environment in which their life histories are measured. Provided that no maternal or other nongenetic effects are carried over from the previous environment, this can give direct evidence of the genetic options set.
4. Experimental manipulation changes individuals' life histories directly. This may be done by genetic engineering, which has had some startling successes. Previously, manipulation was achieved by altering the animals' breeding opportunities or implanting hormones; those methods, however, need supporting evidence that what was achieved by the experimenter could also have been produced genetically.
5. Selection experiments alter life histories genetically over several generations by selecting for change in one life history characteristic. These last two methods both manipulate one life history characteristic and observe the correlated responses of others. If other life history characteristics change, this is evidence that they were involved in a trade-off with the manipulated characteristic.

In summary, life history theory provides the mathematical tools with which to calculate the selection pressures acting on life history traits. To achieve this, the complete life history must be known. Generally, some aspects depend on the local environment; for example, resource availability affects productivity, and predation affects mortality rate. A major aim is to explain the differences between species in terms of the differences in the trade-offs that constrain them, together with the differences in the selection pressures that act upon the trade-offs.

BIBLIOGRAPHY

Caswell, H. *Matrix Population Models*. 2d ed. Sunderland, Mass., 2001. The standard reference for a very powerful method that uses matrices to describe life histories. Uses the population growth rate approach to assessing fitness. Very mathematical.

Charlesworth, B. *Evolution in Age-Structured Populations*. 2d ed. Cambridge, 1994. The standard reference for the population growth rate approach to assessing fitness in age-structured populations. Very mathematical.

Charnov, E. O. *Life History Invariants*. Oxford, 1993. Argues for the existence of invariant relationships between life history traits, a possible source of trade-offs.

Cole, L. C. "The Population Consequences of Life History Phenomena." *Quarterly Review of Biology* 29 (1954): 103–137.

Fisher, R. A. *The Genetic Theory of Natural Selection* (1930). Variorum edition. New York, 1999.

Roff, D. A. *The Evolution of Life Histories*. London, 1992. An excellent treatment with comprehensive reviews of many important areas.

Sibly, R. M. "An Allelocentric Analysis of Hamilton's Rule for Overlapping Generations." *Journal of Theoretical Biology* 167 (1994): 301–305. Life history treatment of Hamilton's rule.

Sibly, R. M., and J. Antonovics. "Life-History Evolution." In *Genes in Ecology*, edited by R. J. Berry, T. J. Crawford, and G. M. Hewitt, 87–122. Oxford, 1992. Extended discussion of the topic.

Sibly, R. M., and R. N. Curnow. "An Allelocentric View of Life-History Evolution." *Journal of Theoretical Biology* 160 (1993): 533–546. Provides the theoretical basis for the population growth rate version of the fitness index. Very mathematical.

Stearns, S. C. *The Evolution of Life Histories*. 1992. Important work by charismatic leader in the field.

— RICHARD M. SIBLY

Human Life Histories

Human life histories are extreme in several ways. Compared to other land mammals, humans have lower infant mortality, depend on parents for a longer time, mature later, and are much longer lived. In addition to these specialized life history characteristics, human brains grow faster and continue to grow longer after birth. Humans also exhibit plasticity in growth rate, age of maturation, birthrate, age-specific mortality, and senescence. For example, in the developed world during the past century and a half, average age at menarche among girls has decreased by about four years, completed family size has decreased from six to two children, and the number of people living to over 100 years of age has increased manyfold. The study of human life history evolution must explain both those extreme values and the pattern of flexible response. Although the field has not achieved consensus regarding those explanations, and there are many lively debates, research into human life history evolution is progressing rapidly, with many exciting results to report.

Human Life Histories in a Comparative Perspective. Some life history features can be indexed in the fossil record, but many vital statistics are not directly visible in paleontological or archaeological remains. We use observations of living populations of humans and other modern primates to help understand how life histories vary, to reconstruct the life histories of our ancestors, and to develop hypotheses about the selection pressures acting on them. Modern hunter-gatherers are not living replicas of our Stone Age past; all have been affected by global socioeconomic forces. Yet, in spite of the wide variety of historical, ecological, and political circumstances in which different groups live, there is remarkable similarity in the life histories displayed by our species, and even the variation often makes adaptive sense. Modern chimpanzees, our closest living relatives, are not replicas of the first australopithecine radiation of our lineage, yet those australopithecines have been called "bipedal apes" because they were similar in body size, brain size, and maturation rate to living chimpanzees. That makes comparisons between modern chimpanzees and modern humans who live by foraging an especially instructive way to characterize the evolutionary changes in human life history.

The age-specific mortality profile among chimpanzees is relatively V-shaped, decreasing rapidly after infancy to its lowest point (in real terms, about 3 percent per year) at about 13, the age of first reproduction for females, and increasing sharply thereafter. In contrast, mortality among human foragers decreases to a much lower point (in real terms, about 0.5 percent per year) and remains low with no increase between about 15 and 40 years of age. Mortality then increases slowly, until there is a very rapid rise in the sixties and seventies. The pattern is much more "block U-shaped." The strong similarities in the mortality profiles of the foraging populations suggest that this pattern is an evolved life history characteristic of our species.

As a result of these differences in mortality pattern, 60 percent of hunter-gatherer children survive to age of first reproduction, about 19 years of age, while only 35 percent of chimpanzee offspring survive to age of first reproduction, which is about 13 years in that species. Chimpanzees have a much shorter adult lifespan than humans as well. At first reproduction, chimpanzee life expectancy is an additional 15 years, compared to 38 more years among human foragers. Importantly, women spend more than a third of their adult life in a post-reproductive phase, whereas very few chimpanzee females survive to reach the post-reproductive phase. Less than 10 percent of chimpanzees ever born survive to age 40, but more than 15 percent of hunter-gatherers survive to age 70.

The long dependency of human offspring is illustrated by comparing the age when chimpanzees are self-sufficient to the age when young humans are able to support themselves. Before age 5, young chimpanzees are entirely, then partially, dependent upon mother's milk. Sometime before, but certainly after, weaning they are able to feed themselves; and by sexual maturity, females, but not males, gather sufficient calories to feed themselves and support a nursing infant. In contrast, young humans are dependent upon food produced by others for much longer, and adult humans of both sexes are capable of producing more calories than they consume. Kaplan and colleagues have termed the difference between food produced and food consumed *net production* and have estimated net production at each age for chimpanzees and humans. This value is negative much longer for humans and then increases to a much higher peak at a later age than it does for chimpanzees. The long period during which humans produce less than they consume could not exist if humans had the same mortality profile as chimpanzees. Only 30 percent of chimpanzees ever born reach the age when humans produce what they consume on average, and less than 10 percent reach the age when human production peaks.

These features of human life histories raise several fundamental questions. Why is age of first reproduction so late? Why is minimum mortality so low? Why does it remain low for so long and not rise quickly after first reproduction? Why do human females live so long after menopause? Why do mortality rates rise sharply in the 60s and 70s among both men and women? And, finally, why is production so low early in life and so high late in life? I discuss two alternative explanations of these patterns.

Two Models of Human Life History Evolution. One model, the "grandmother hypothesis" (Hawkes et al., 1998; O'Connell et al., 1999), derives a set of life history shifts from an ancestral increase in mother–child food sharing. Among modern low-latitude foragers, women gather foods—such as nuts, roots, and seeds—that are hard for children to acquire, but not for older women, especially those who are unencumbered by child care. Ancestral human females were able to depend on these foods only by provisioning their young offspring. The increased mother–child food sharing provided a novel way for older females with declining fertility to increase their fitness. Without infants of their own, they could help their daughters feed weanlings. The childbearing-aged women were able to have babies at a faster rate without lowering offspring survival because grandmothers helped feed weaned grandchildren. Females who were more vigorous as they neared menopause passed this vigor on to more grandchildren. As a result, our ancestors maintained the same pattern of age-specific fertility decline still seen in chimpanzees, but they evolved slower rates of aging and resulting longer lifespan. The greater longevity favored delaying maturity because, following Charnov (1993), lower adult mortality rates reduce the fitness costs of prolonged growth (Hawkes et al., 1998).

A second model proposes that the timing of life events is best understood as an "embodied capital" investment process (Kaplan et al., 2000; Kaplan, 1996). In a physical sense, embodied capital is organized somatic tissue—muscles, brains, and so on—and the strength, skill, and knowledge required to exploit the suite of high-quality, difficult-to-acquire resources humans consume. Those abilities require a large brain and a long time commitment to development. This extended learning phase, during which productivity is low, is compensated by higher productivity during the adult period. Since productivity increases with age, the time investment in skill acquisition and knowledge leads to further selection for physiological and behavioral changes that lower the mortality rate and result in greater longevity. Because the returns on the investments in development occur at older ages, the long human lifespan coevolved with the lengthening of the juvenile period, increased brain capacity for information processing and storage, and intergenerational food sharing.

Both models propose that foraging ecology is the principal driving force in the evolution of human life histories. Humans focus on resources that are difficult for juveniles to acquire but that are acquired by adults in sufficient quantities to feed themselves and others. However, the models differ in the reasons why food production increases with age. According to the grandmother hypothesis, most of the increase in productivity with age is due to strength (adults vs. children). In contrast, the embodied capital theory proposes that it is the learning-intensive nature of human foraging that results in both brain and lifespan expansion. The theories also differ in the role of male parental investment in the evolution of the human life course. The grandmother hypothesis focuses on the problem of feeding young children and the role older women play in solving it. According to the brain capital theory, human hunting is the most skill-intensive foraging strategy, requiring decades to learn; it is protein- and lipid-rich meat that both supports and requires the extended juvenile period. The feeding niche specializing particularly on hunting promotes cooperation between men and women and high levels of male parental investment. (For an example, see the Vignette.)

Future research will resolve these issues. It is possible that both views are partially correct. It may be that initial extension of the lifespan among early hominids was due to a shift to plant resources difficult for young juveniles to handle, the consequent increase in maternal provisioning, and a novel role for older female kin. Hunting and male provisioning may have played an important

SEX AND AGE DIFFERENCES IN PRODUCTIVITY

The embodied capital hypothesis proposes that human life history traits evolved in a feeding niche specializing on large, valuable food packages, particularly hunting. This promotes cooperation between men and women and high levels of male parental investment, because it favors sex-specific specialization in skills and generates a complementarity between male and female inputs. The extension of the lifespan operates through selection on both men and women. Supporters of the alternative hypothesis note that strength seems to be more important than practice in determining return rates for some resources, such as shellfish collecting (Bleige et al., in press) and the archery component of hunting (Blurton Jones et al., in press). On the other hand, even though strength peaks at about age 22, return rates for hunting do not peak until about 35 to 40 years of age among the Ache and Hiwi (Kaplan et al., 2000), one Hadza data set (Marlowe, unpublished data), the Machiguenga and Piro of Peru (Kaplan, unpublished data), and the Gidra of Papua (Ohtsuka, 1989). There is also evidence showing how variable meat acquisition can be, and that meat given is often not balanced with meat received at the individual level (Hawkes, 1991). This evidence is used to reject the argument that hunting is paternal provisioning. By contrast, there is evidence that men provide the bulk of the excess calories to support the deficits of unproductive children. Averaging estimated age- and sex-specific annual consumption and production for a ten-group sample of foragers for which quantitative data are available showed that after adult consumption is subtracted, women supplied only 3 percent of the calories remaining for children, with men providing the remaining 97 percent (Kaplan et al., 2001).

—Hillard S. Kaplan

role in the subsequent expansion of the brain and lifespan with the emergence of *Homo sapiens.* It is also possible that hunting supports juvenile dependence and favors increased investment in the brain and longevity, but is motivated by mating effort rather than by parental investment. Finally, it may be the case that the relative importance of strength, skill, and knowledge in determining the age profile of production varies across time, space, and ecology, with each playing a selective role in human life history evolution.

Reproduction. Like other primates, humans usually bear only a single baby at a time and invest substantially in each one. However, there is variation both within and among populations in the amount of resources that parents invest and the way that parental effort is allocated—relatively less to many children or relatively more to few. Broadly speaking, there are two different fertility regimes exhibited by women. One emerged with what has been called "the demographic transition" and is discussed below. The other and more ancient, often termed "natural fertility" (Henry, 1961), depends on physiological design to produce adaptive responses to the energy demands of childbearing. The physiological regulation of ovulation, fertilization, implantation, and maintenance of a pregnancy is highly responsive to energy stores in the form of fat, energy balance (calories consumed minus calories expended), and energy flux (rate of energy turnover per unit time). Low body fat, weight loss resulting from negative energy balance, and extreme energy flux (either very low intake and very low expenditure, or very high intake and very high expenditure) all lower monthly probabilities of conceiving a child that will survive to birth. Seasonal variation in work loads and diet have been shown to affect female fecundity and fertility, and variation across groups in both age of menarche and fecundity has been linked to differences in food intake and work load. Peter Ellison's book *On Fertile Ground: A Natural History of Human Reproduction* (Cambridge, 2001) is an elegant synthesis that provides the original sources of evidence.

Behavior and the underlying psychological processes that govern parental investment also affect fertility indirectly via maternal physiology. The longer children are provisioned and the greater the proportion of their food needs that is subsidized, the lower the reproductive rate that can be supported with a given adult income. Thus, natural selection appears to have acted upon both the psychology of parental investment and maternal physiology to produce a flexible system of fertility regulation.

The understanding of fertility regulation in traditional settings provides insights into the dramatic lowering of fertility accompanying modernization. With the demographic transition, a second fertility regime was born, in which completed family sizes are low (2–3 children), women adjust the likelihood of having the next child on the basis of how many they already have in order to achieve a "target" family size, and observed fertility does *not* maximize reproductive success (Kaplan, 1996). Education is the best and most consistent predictor of fertility variation, both within and among nations. This is associated with the changing payoffs for education, which have increased for two reasons. First, more investment in education for children can sharply improve their position in education-based labor markets. Second, changing medical technology and public health have greatly reduced mortality rates for all age groups, increasing the expected years of return from educational investments. Lower mortality rates have a further consequence: they increase the expected costs per child born, and so favor further increases in offspring quality (Kaplan, 1996).

These factors result in fertility being regulated by a consciously determined fertility plan realized through birth control technology and/or controlled exposure to sex. Reproductive physiology sets the broad age limits within which a fertility plan can be realized. In "natural fertility" regimes, fertility and fecundity are low during late teen years, increasing to their highest levels between 20 and 29 years of age, and declining at an increasing rate after age 35 to near zero by age 45. These physiological age limits may actually lower fertility when family planning includes significant delays to first reproduction (Kaplan et al., in press).

A great deal of further research is necessary, however, before we can understand why these changes in socioecology have resulted in such low levels of fertility and high levels of parental investment and wealth consumption. Clearly there is much to learn, but an understanding of our evolutionary past will surely assist in explaining the present and predicting the future.

[*See also* Demographic Transition; Demography; Life History Theory, *article on* Grandmothers and Human Longevity; Primates, *articles on* Primate Classification and Phylogeny *and* Primate Societies and Social Life.]

BIBLIOGRAPHY

Bleige Bird, R., and D. W. Bird. "Constraints of Knowing or Constraints of Growing: Fishing and Collecting Among the Children of Mer." *Evolution and Human Behavior* (in press). Examines age profiles of foraging returns for fishing and shell-fish collection among children and adults in Mer, Australia. The authors argue that for some methods, increases in returns with age are due to strength rather than skill.

Blurton Jones, N. B., and F. Marlowe. "The Forager Olympics: Does It Take 20 Years to Become a Competent Hunter-Gatherer?" *Human Nature*, (in press). Experiments conducted with Hadza foragers to determine the impacts of age, strength and school attendance on foraging abilities found no impact of school attendance on archery skill and suggest that skill-building may not determine the length of the human juvenile period.

Blurton Jones, N. B., K. Hawkes, and J. O'Connell. "The Antiquity of Post-reproductive Life: Are There Modern Impacts on

Hunter-Gatherer Post-Reproductive Lifespans?" *Human Biology*, in press. Evidence from foragers is presented to show that long post-reproductive lifespans among humans are not an artefact of recent historical conditions, but an evolved human trait.

Charnov, E. L. *Life History Invariants: Some Explanations of Symmetry in Evolutionary Ecology*. Oxford, 1993. Examines fundamental tradeoffs in the evolution of life histories; shows how those tradeoffs result in relationships that are invariant to many transformations.

Hawkes, K. "Showing Off: Tests of an Hypothesis about Men's Foraging Goals." *Ethology and Sociobiology* 12 (1991): 29–54. Evidence that hunting of large game is an unreliable way to provision families, because of high temporal variance in rates of return and large scale redistribution of animals; suggests that men hunt in order to "show off," as a form of mating effort, rather than to provision their children.

Hawkes, K., J. F. O'Connell, N. G. Blurton Jones, H. Alvarez, and E. L. Charnov. "Grandmothering, Menopause, and the Evolution of Human Life Histories." *Proceedings of the National Academy of Science USA* 95 (1998). The most thorough presentation of the grandmother hypothesis also relates the model to existing life history theory, especially as developed by Charnov (1993).

Henry, L. "Some Data on Natural Fertility." *Eugenics Quarterly* 8 (1961): 81–91. A classic paper in demography that first makes the distinction between natural and parity-specific fertility, as defined in the text.

Hill, K., C. Boesch, J. Goodall, A. Pusey, J. Williams, and R. Wrangham. "Mortality Rates among Wild Chimpanzees." *Journal of Human Evolution* 39 (2001): 1–14. The most complete analysis of mortality patterns among wild-living chimpanzees; it produces a synthetic life table, using data from all field sites with systematic mortality records.

Hill, K., and A. M. Hurtado. *Ache Life History: The Ecology and Demography of a Foraging People*. Hawthorne, N.Y., 1996. This book examines virtually all aspects of the demography of Ache hunter-gatherers prior to, during, and after regular contact with Western society. Mortality, age of first reproduction, and fertility rates are all examined in the light of evolutionary theory.

Kaplan, H. K., K. Hill, J. B. Lancaster, and A. M. Hurtado. "A Theory of Human Life History Evolution: Diet, Intelligence, and Longevity." *Evolutionary Anthropology* 9 (2000): 156–185. Presents a theory of human origins based upon the co-evolution of foraging niche, brain size and longevity; summarizes the cross-cultural and comparative evidence from studies of modern hunter-gatherers and wild-living chimpanzees.

Kaplan, H. S. "A Theory of Fertility and Parental Investment in Traditional and Modern Human Societies." *Yearbook of Physical Anthropology* 39 (1996): 91–135. The embodied capital theory of life history evolution is presented formally, with specific emphasis on the evolution of human fertility and parental investment.

Kaplan, H. S., K. Hill, A. M. Hurtado, and J. B. Lancaster. "The Embodied Capital Theory of Human Evolution." In *Reproductive Ecology and Human Evolution*, edited by P. T. Ellison. Hawthorne, N.Y., 2001. Applies embodied capital theory to understanding primate brain evolution and longevity, extreme intelligence and long lifespans among humans, and the sexual division of labor.

Kaplan, H. S., J. B. Lancaster, K. G. Anderson, and W. T. Tucker. "An Evolutionary Approach to Below Replacement Fertility." *American Journal of Human Biology*, (in press). An evolutionary theory of extremely low fertility is developed, with evidence that in the United States below replacement fertility is due primarily to long delays in initiating reproduction and is confined largely to the most educated sector of the society.

Kaplan, H. S., J. B. Lancaster, J. A. Bock, and S. E. Johnson. "Does Observed Fertility Maximize Fitness among New Mexican Men? A Test of an Optimality Model and a New Theory of Parental Investment in the Embodied Capital of Offspring." *Human Nature* 6 (1995): 325–360.

O'Connell, J. F., K. Hawkes, and N. G. Blurton Jones. "Grandmothering and the Evolution of *Homo erectus*." *Journal of Human Evolution* 36 (1999): 461–485. Analyzes the initial evolution of long hominid lifespans in the early Pleistocene. On the basis of archeological and modern behavioral evidence, argues that root foraging, as opposed to big game hunting, set the stage for grandmothering.

Ohtsuka, R. "Hunting Activity and Aging Among the Gidra Papuans: A Biobehavioral Analysis." *American Journal of Physical Anthropology* 80 (1989): 31–39. The first quantitative analysis of male hunting returns by age, showing that men in early adulthood achieve much worse returns than hunters in their 30s and 40s, who, in turn, have higher rates than old men.

— HILLARD S. KAPLAN

Grandmothers and Human Longevity

Longevity varies enormously among animals. Schoolchildren know a dog can live longer than a hamster, and most can calculate the age of their dog in human years. Given that life span is an obvious feature of all plant and animal life histories it is surprising that many investigators believe our life span is an artifact of recent events. Widespread inferences about ancient life spans rely on the fact that human life expectancies have nearly doubled in many places over the last century as public health systems improved and medical interventions reduced mortality risk from many sources. As a consequence the fraction of old people is rising quickly with large social and economic consequences. Some therefore see human longevity as a recent novelty, and propose that we live longer because we are better fed, better clothed, better housed and more caring for each other. Parallels are drawn to living in a zoo, as husbandry of captive populations delays the deaths of fragile, aging animals. Captive macaques for example often live through menopause displaying marked senescence in all other physiological systems, quite unlike menopausal women.

Those who think long human life spans are recent point not only to aging modern populations but also to paleodemographic models. The geneticist Kenneth Weiss derived eight model mortality schedules for paleolithic populations from age distributions in cemetery populations and the limited fertility and mortality sched ules of hunter gatherers then available to him. His models produced survival schedules that were more like those of chimpanzees than modern people. Some paleoanthropologists who estimate ages from fossils, especially the sizable sample of Neanderthal specimens, conclude that there is no evidence for unusual longevity in pre-modern populations. The survivorship curves constructed for the Sima de los Huesos and for the Krapina samples show zero survivorship past age thirty-five. Among the problems that plague reconstructions from cemetery remains and hominid fossils are the the large errors in estimating adult ages and the differential preservation of the bones of both infants and older individuals. Additional problems in using cemetery assemblages to reconstruct the age specific survival of populations have drawn critical attention from paleodemographers.

When deciding whether ancestral life spans exceeded those of other great apes two variables, life expectancy at birth and longevity, must be distinguished. Life expectancy at birth is a misleading index of longevity because it is an average of all the life spans expected for a cohort. It can vary widely with changes in infant and juvenile mortality rates even when age-specific adult mortality remains unchanged. In typical mammalian populations juvenile mortality starts out high, so life expectancy increases for some years after this initial high rate of death. The force of mortality curve for human populations is the shape of a broad valley with the mortality rate falling rapidly from birth to five years then flattening through the middle of the life span, generally reaching its lowest level at reproductive maturity, after which it begins a moderate rise, accelerating toward the end of the life span but slowing at very late ages in modern populations (Figure 1). Because life expectancy at birth is so sensitive to juvenile mortality rates, some investigators have focused attention on "maximum life span," and found no credible evidence that it has changed over the time of historic record keeping.

An especially important source of information about human longevity is the increasingly better demographic data, on modern people hunting and gathering for a living, that has accumulated since the mid 1960s. In these cases ancient causes of mortality apply as public health measures and scientific medicine have little effect. Ethnographer Richard Lee reported that 10 percent of the population of !Kung Bushmen he studied was over sixty. An observation confirmed by Nancy Howell's classic demography in which she showed that 30 percent of the adult women were over forty-five. Among Ache foragers of eastern Paraguay ethnographers Kim Hill and Magdelena Hurtado estimated a life expectancy at birth of thirty-seven years, but as with the !Kung, mortality rates decline through childhood so that a women who survived to age at last birth, forty-two years, had an expectation of living twenty-four more years. Careful analyses from hunter-gatherer populations demonstrate that even though life expectancies at birth are around thirty-five, most of those who make it to adulthood live past menopause and 30 percent or more of the population of adult women at any one time are past the age of last birth. The human survival curves are similar among populations living in different environments and under different recent historical circumstance, and very unlike those of chimpanzees, our closest living relatives.

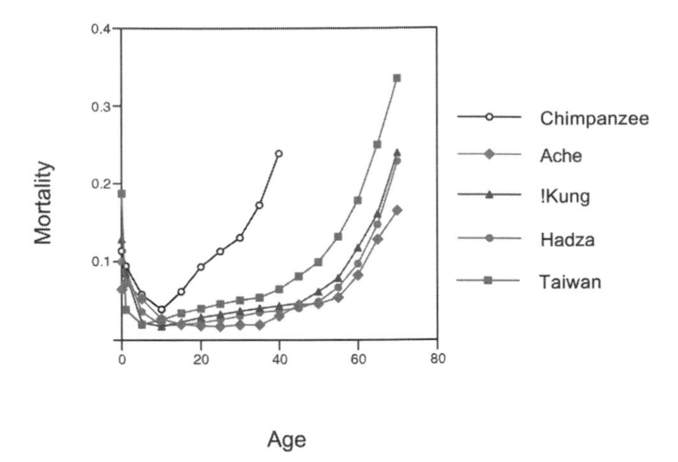

Figure 1. Mortality Curves.
Shown are the mortality curves for the populations calculated using the method of Hamilton (1966). Chimpanzee data from five study sites: Hill et al. (2001); !Kung data from Howell (1979); Ache data from Hill and Hurtado (1996); Hadza data from Blurton Jones et al. (2002); Taiwan data for rural populations about 1906 as reported in Hamilton (1966).
Drawing courtesy of Helen Alvarez.

One set of longstanding ideas links our distinct longevity to our relatively large brains. The high correlations, across mammalian species, between body size, brain size, and life span noted by biologist and gerontologist George Sacher have been cited by others to support the notion that the selective advantage of large brains favored later ages of maturity, growth to large body size and consequent long lives. Sacher suggested two hypothesis for the relationship between brain size and longevity; (1) the brain is responsible for the maintenance of physiological stability which promotes longevity, or (2) the added metabolic and developmental demands of the large brain are accommodated by a lower rate of reproduction over a longer life span. The high correlations demonstrated by Sacher prompted others to undertake similar investigations, segregating the data by sex and by taxonomic order, using other body organs, and taking advantage of larger data sets than those available to Sacher. These more recent studies have failed to find high levels of correlation between brain size and life span, but brain size often emerges as a significant independent contributor in multiple regression models.

A separate, although overlapping suite of ideas builds on the influential hunting hypothesis of biological anthropologist Sherwood Washburn, who proposed that the drying environments of the Late Tertiary favored increased hunting which in turn promoted tool use, bipedalism, and big brains. But bigger brains and bipedalism created an obstetrical dilemma requiring babies to be born earlier and more dependent on maternal care, forcing females to rely on provisioning by hunting mates. Recent refinements and additions add long life spans to this cluster, proposing that gains in hunting effectiveness with age increase the benefit of behavioral adjustments and physiological investments that lower mortality risks and lengthen the life span.

Another recent hypothesis takes advantage of developments in mammalian life history evolution to explain human longevity. Mammalian life spans vary by fifty fold. There is a broad correlation between body size and life span but with many exceptions, so much so that life history features are more strongly correlated with each other than they are with body size. These correlations led evolutionary biologist Paul Harvey and his colleagues to propose that life histories fall along a "fast/slow" continuum of variation. In relative terms, species on the "fast track" are characterized by short gestation periods, small brained neonates, fast rates of birth, early age of maturity, and short life spans.

The most basic ideas about life history evolution deal with variation in aging and life span. Peter Medawar initially showed that the force of selection declines with age. George Williams and William Hamilton elaborated the implications of changing selection pressure for the evolution of senescence. Their population genetics models showed that variation in environmentally imposed mortality risk has consequences throughout life histories. If the risk of dying goes up, selection favors maturing earlier and putting more into current reproduction. In his disposable soma model, gerontologist Tom Kirkwood modeled this decision as an optimal tradeoff between somatic investment in a more durable soma and the physiological proccesses of somatic repair on one hand and investment in current reproduction on the other. When mortality risks go down selection can favor postponing reproduction and allocating more to building and maintaining a soma. In populations subject to low extrinsic risks more individuals are expected to die at later ages from causes incident to ageing. Strategies of increasing allocation to maintenance and repair can have large effects on fitness in these ecologies.

Within a given population natural selection is expected to drive allocation tradeoffs to an evolutionarily stable strategy, the ESS. In mammals these tradeoffs are reflected in such traits as size at birth, age and size at weaning, growth rate, age and size at maturity, birth rate and length of the life span. Evolutionary biologist Eric Charnov reasoned that evolution to the ESS should produce characteristic relationships between certain traits, their products should be approximately constant across transformations of size and phylogeny. As noted above, selection pressure on life history traits arises from mortality risks in the environment. Where mortality risks from predation, disease, or unpredictable environmental variability are high, selection favors early maturity against the risk of dying before reproducing. Where average adult mortalities are low, natural selection favors larger investments in cells that are expensive to build and maintain resulting in slower growth, later age of maturity, longer life spans, and low rates of birth. Slow growing mothers produce fewer, larger, more expensive babies.

Charnov's approach highlights the fast-slow variation both within and between taxa. Primates as an order have slower life histories than other mammals, and humans have the slowest of primate life histories. Two of the most conspicuous features of our slow life history are our late age of first birth and long postmenopausal survival. The tradeoffs modeled by Charnov predict that an age of maturity as late as ours depends upon reproductive benefits accruing throughout a long life span. If postmenopausal women are contributing to reproduction we should also have higher annual fecundity than a "grandmotherless" primate with our age at maturity. These predictions are confirmed by comparisons across primates. Charnov's invariants in combination with a grandmother hypothesis, provide the framework for an argument about the evolution of our extreme longevity.

The grandmother hypothesis proposes that an ancestral population with a chimpanzee-like life history, an efficient bipedal foraging range, genetic variation for somatic robusticity, and a primate proclivity to share some food with infants encountered socioecological circumstances with a novel fitness opportunity for senior females. The novelty arose from an increase in mother–child food sharing.

All other primate mothers are limited to habitats with food resources that weaned offspring can access directly. Mother-child food sharing allows human mothers to occupy habitats with few foods that young children can acquire themselves. Mother-child food sharing also imposes a cost on mothers that allomothers can readily reduce. Aging females whose fertility has dropped (so they have no youngsters of their own to feed) can provision the weaned offspring of their daughters, shortening interbirth intervals, with less reduction in child survival. In this scenario vigorous older females have large effects on their own fitness. There is no selection pressure to lengthen the fertile span because that would reduce the help for younger mothers and so lower early fecundity. As a consequence we continue to have fertile spans similar to those of chimpanzees. Our marked longevity—slowed senescence in other aspects of physiology—is the derived trait.

This grandmother hypothesis was stimulated in part

by observations among the Hadza, hunter gatherers of the lake Eyasi region of Tanzania. Among these modern people, women collect plant foods which provide predictable daily resources for provisioning themselves and their children even though Hadza children as young as five years old are able to provide a substantial portion of their own food in certain seasons. When these resources are not in season children are dependent upon foods acquired by mothers and grandmothers. When the welfare of Hadza children is estimated from changes in body weight, positive effects are correlated with the foraging productivity of their mothers as long as the mothers are not nursing. When mothers focus on newborns the welfare of their dependent children is closely correlated to the foraging returns of their grandmothers.

These observations and arguments from life history models have been linked to paleoanthropological evidence in a hypothesis that our unique life histories originated with the evolution of *H. erectus*. Body size and age at maturity in this taxon shift toward the modern range. Because the life history model links are at maturity and adult life span, longevity more like moderns may be implied as well. Our grandmothering life history might explain the expansion of our genus out of Africa and the successful occupation of much of the temperate and tropical old world.

[*See also* Demography; Human Families and Kin Groups; Life History Theory, *article on* Human Life Histories; Reproductive Physiology, Human.]

BIBLIOGRAPHY

Alvarez, H. P. "Grandmother Hypothesis and Primate Life Histories." *American Journal of Physical Anthropology* 113 (2000): 435–450. A comparative examination of life history invariants across sixteen primate species, including humans, that tests predictions of the Grandmother hypothesis, and finds that, as expected, humans are within the 95 percent confidence interval for the relationship between age at maturity and life span, but outside it for age at maturity and fecundity.

Austad, S. N. *Why We Age*. New York, 1997. Entertaining and informative summary of the work of researchers who ask both ultimate and proximate questions.

Blurton Jones, N. G., K. Hawkes, and J. O'Connell. "The Antiquity of Post-Reproductive Life: Are There Modern Impacts on Hunter-Gatherer Post-Reproductive Life Spans?" *American Journal of Human Biology* (2002). A careful look at the demography of the Hadza, Ache, and !Kung with an evaluation of the claims that the reported numbers of aged individuals arise from errors in the demography and/or modern influences on longevity.

Charnov, E. L. *Life History Invariants: Some Explorations of Symmetry in Evolutionary Ecology*. Oxford, 1993. This important work demonstrates the general nature of life history tradeoffs in organisms across transformations of body size and phylogeny. Chapter 5 is especially informative for anthropologists interested in primate evolution.

Charnov, E. L. "Evolution of Mammal Life Histories." *Evolutionary Ecology Research* 3 (2001): 521–535. The most recent statement of Charnov's life history evolution relates the cost of building expensive cells to the classic life history tradeoffs.

Finch, C. E. *Longevity, Senescence, and the Genome*. Chicago, 1990. A massive compilation of the proximate causes of ageing in many different species by someone who understands the relevance of the empirical evidence to ultimate questions.

Hawkes, K., J. F. O'Connell, and N. G. Blurton Jones. "Hadza Women's Time Allocation: Offspring Provisioning and the Evolution of Long Post Menopausal Life Spans." *Current Anthropology* 38 (1997): 551–577. The authors develop the relationship between the grandmother hypothesis and empirical evidence from the Hadza on changes in childrens' welfare and maternal and grandmother foraging and distinguish between early hypotheses of the evolution of menopause and their own hypothesis which focuses attention on a phylogenetically conserved fertile span within a novel life span.

Hawkes, K., J. F. O'Connell, and N. G. Blurton Jones. "The Evolution of Human Life Histories: Primate Tradeoffs, Grandmothering Socioecology, and the Fossil Record." In *The Role of Life Histories in Primate Socioecology*, edited by P. Kappeler and M. Pereira. Chicago, in press. A very careful and detailed treatment of evolutionary ideas about life history variation, and the link between Charnov's model, the ethnographic evidence from the Hadza, the grandmother hypothesis, and the archeological evidence linking modern life history traits to Homo erectus.

Holliday, R. "Understanding Ageing." *Philosophical Transactions of the Royal Society (London), series B* 352 (1997): 1793–1797. A short version of the book of the same name reviewing the problems faced by all organisms in the allocation of energy and the proximate solutions to those problems as they relate to the evolution of the life span.

Kaplan, H., K. Hill, J. Lancaster, and A. M. Hurtado. "A Theory of Human Life History Evolution: Diet Intelligence, and Longevity." *Evolutionary Anthropology* 9 (2000): 156–185. The authors propose that the unique constellation of human life history traits arose from selection pressure for expanding brains and long juvenile periods in which children learn skills required for successful adult foraging, especially hunting.

Kirkwood, T. B. L., and M. R. Rose. "Evolution of Senescence: Late Survival Sacrificed for Reproduction." *Philosophical Transactions of the Royal Society (London), series B* 332 (1991): 15–34. A valuable comparison of two models of senescence, the antagonistic pleiotropy model and the disposable soma model with a clear discussion of how these models relate to the evolution of the life span and the relevant experimental evidence for testing the evolutionary origins of senescence and longevity.

O'Connell, J. F., K. Hawkes, and N. G. Blurton Jones. "Grandmothering and the Evolution of Homo erectus." *Journal of Human Evolution* 36 (1999): 461–485. Develops the predictions from the grandmother hypothesis that can be tested in the archeological record.

Ricklefs, R. E. "Evolutionary Theories of Aging: Confirmation of a Fundamental Prediction with Implications for the Genetic Basis and Evolution of the Life Span." *The American Naturalist* 152 (1998): 24–44. An excellent discussion of life span evolution with a model designed to tease apart extrinsic and intrinsic sources of mortality using life tables.

Stearns, S. C. *The Evolution of Life Histories*. New York, 1992. The classic work combining descriptions of the models and empirical evidence from many species clearly written and well organized.

— Helen Alvarez

Guppies

Life history theory proposes that the relative survival rates of juveniles and adults can influence many life history traits (Charlesworth, 1994). Natural populations of guppies (*Poecilia reticulata*), a small freshwater fish, co-occurring with different communities of predators were thought to differ in their rates of survival of juveniles and adults making them an excellent study organism. Guppies are live bearers; that is, their young develop in the female and are free-swimming at birth. They have a promiscuous breeding system in which the tan-colored females choose among males exhibiting a wide range of color patterns that are primarily genetically determined. The natural distribution of guppies includes streams and rivers on two small tropical islands in the West Indies, Trinidad and Tobago, and adjacent areas of South America.

In some populations of guppies in Trinidad, the main predator of guppies is a small killifish, *Rivulus hartii*, which can swallow and hence can prey only upon small, mostly immature guppies (hereafter called "low-predation" sites). In other populations, there are several larger fish species, some of which are serious predators of guppies ("high-predation" sites). One of these is the pike cichlid, *Crenicichla alta*, which prefers to prey on adult guppies. High and low predation sites are often found in the same river, separated by a waterfall that excludes the larger predators.

In high-predation localities, selection should act to reduce the age at maturity, to increase energy allocation to reproduction and to increase the number of offspring produced, usually at the expense of reduced offspring size. Each of these effects makes it more likely that an individual will produce several offspring before being eaten by a predator. In a comparison of the phenotypes of field-caught individuals, guppies from high-predation localities matured at smaller sizes than guppies from low-predation localities (Table 1). High-predation females had more, smaller offspring per brood (corrected for female body size) than those from low-predation localities. One of the convincing aspects of this research was that guppies from five low-predation localities were compared with those from seven high-predation localities.

However, to demonstrate that these life history differences had evolved, it was essential to show that there was a genetic basis for them. To do this, guppies were collected from several sites and reared for two generations under controlled conditions in the laboratory. Differences between high- and low-predation guppies emerged just as in the wild, indicating that there is indeed a genetic basis for the life history differences (Reznick, 1982). Guppies from high-predation sites matured earlier and at smaller sizes than guppies from low-predation sites. High-predation females allotted more energy to their broods, composed of many, small offspring borne at shorter intervals than guppies from low-predation localities (Reznick, 1982).

Table 1. Comparison of the Mean Life History Traits of Guppies from High- and Low-Predation Localities

	Predation Locality	
	High	*Low*
Trait		
Female		
Size at maturity (mm)	14.6	17.4
Age at maturity (days)	49	57
Offspring number/brood	6.6	3.0
Offspring size (mg)	0.9	1.5
Reproductive effort (%)	25.0	19.2
Interbrood Interval (days)	22.8	25.0
Male		
Size at maturity (field) (mm)	14.9	16.4
Age at maturity (days)	51	59

From Reznick and Endler, 1982; Reznick, 1982.

As further evidence for the role of predation in the life history differentiation of guppies, the researcher David Reznick took advantage of a transplant experiment that John Endler had performed in 1976. Endler had moved guppies upstream from a high-predation population to a low-predation site that had previously had only the small predator, *Rivulus*—but no guppies (Endler, 1980). Reznick repeated this experiment in another stream in 1981. Two decades later, for both experimental populations, Reznick and his colleagues found evidence for the rapid evolution of most, but not all, life history traits (Reznick and Bryga, 1987; Reznick et al., 1990); the descendants of the transplanted guppies had life histories more similar to low-predation guppies than to their own ancestors. A quantitative genetic study showed how rapidly the life history traits had evolved in those experiments (Rodd et al., 1997).

An underlying assumption of these studies was that there was a difference in the relative mortality rates of juvenile and adult guppies in low- and high-predation sites. Two-week mark-recapture studies in 14 populations showed that guppies in high-predation sites suffered higher mortality rates and that the probability of surviving to maturity was lower in those sites (Reznick et al., 1996a). New, long-term mark-recapture assays indicate that the mortality rate of adult female guppies is two to three times greater in high-predation localities. This projects to an approximately thirtyfold difference in the probability of survival for six months. This means that it is not unusual to find females in low-predation sites that are one and one-half to two years old, but it would be very unusual to find a female in a high-predation site that was more than six months old.

There are other theories that make predictions about life history patterns that could be relevant to the guppy system (e.g., Charlesworth, 1994). An ecological study suggested that differences in resource availability associated directly with a predator community and/or mediated through an indirect effect of the predators may be contributing to the interpopulation differences in life histories (Reznick et al., 2001). Density-dependent selection can also play a role in the evolution of the life history strategies (Charlesworth, 1994). Density manipulations of guppies in natural streams indicate that this form of selection may also be contributing to the life history differences.

Intriguing results of Reznick's studies are the discrepancies between the phenotypes of the field-caught and lab-reared guppies. For example, newborn offspring from field-caught, low-predation females are nearly twice as large as those from high-predation females, while the difference is only 10 percent when lab-reared offspring are compared. Food availability appears to be one environmental factor contributing to these differences. Female guppies produce larger, but fewer, offspring in response to low or varying food availability (Reznick and Yang, 1993). Food availability is lower in the low-predation sites for a number of reasons, including lower productivity because of lower light levels, and also because of higher guppy densities (an indirect result of lower mortality rates) (Reznick et al., 2001). Therefore, lower levels of food availability in the low-predation sites apparently contribute to the larger offspring size in the field-caught individuals. Phenotypic plasticity in response to interactions with conspecifics also appears to contribute to phenotypic, interpopulation differences in male size at maturity (Rodd et al., 1997); in low-predation populations, male size at maturity increases with the density of males. The fact that males in high-predation sites do not respond in this way to interactions with conspecifics suggests that plasticity in the life history traits has itself evolved (Rodd et al., 1997).

The many natural populations of guppies, with their well-defined differences in community structure and risk of mortality, have thus provided a natural laboratory for testing predictions of life history theory.

BIBLIOGRAPHY

Charlesworth, B. *Evolution in Age-Structured Populations.* 2d ed. New York, 1994.

Endler, J. A. "Natural Selection on Color Pattern in *Poecilia reticulata*." *Evolution* 34 (1980): 76–91.

Endler, J. A. "Multiple-Trait Coevolution and Environmental Gradients in Guppies." *Trends in Ecology and Evolution* 10 (1995): 22–29. An excellent survey of the traits that differ between guppies from high- and low-predation sites.

Houde, A. E. *Sex, Color and Mate Choice in Guppies.* Princeton, 1997. An excellent review of studies on the ecology and evolution of guppies, including natural and sexual selection.

Reznick, D. N. 1982. "The Impact of Predation on Life History Evolution in Trinidadian Guppies: Genetic Basis of Observed Life History Patterns." *Evolution* 36 (1982): 1236–1250.

Reznick, D. N. "Life-History Evolution in Guppies: 2. Repeatability of Field Observations and the Effects of Season on Life Histories." *Evolution* 43 (1989): 1285–1297.

Reznick, D. N., and H. Bryga. "Life-History Evolution in Guppies (*Poecilia reticulata*): 1. Phenotypic and Genetic Changes in an Introduction Experiment." *Evolution* 41 (1987): 1370–1385.

Reznick, D. N., and H. A. Bryga. "Life-History Evolution in Guppies (*Poecilia reticulata*: Poeciliidae). V. Genetic Basis of Parallelism in Life Histories." *American Naturalist* 147 (1996): 339–359.

Reznick, D. N., and J. A. Endler. "The Impact of Predation on Life History Evolution in Trinidadian Guppies (*Poecilia reticulata*)." *Evolution* 36 (1982): 160–177.

Reznick, D. N., and A. P. Yang. "The Influence of Fluctuating Resources on Life History: Patterns of Allocation and Plasticity in Female Guppies." *Ecology* 74 (1993): 2011–2019.

Reznick, D. N., M. J. Butler IV, and F. H. Rodd, "Life-History Evolution in Guppies. VII. The Comparative Ecology of High- and Low-Predation Environments." *American Naturalist* 157 (2001): 126–140.

Reznick, D. N., M. J. Butler IV, F. H. Rodd, and P. Ross, "Life-History Evolution in Guppies (*Poecilia reticulata*): 6. Differential Mortality As a Mechanism for Natural Selection." *Evolution* 50 (1996a): 1651–1660.

Reznick, D. N., F. H. Rodd, and M. Cardenas. Life-History Evolution in Guppies (*Poecilia reticulata*: Poeciliidae). IV. Parallelism in Life-History Phenotypes. *American Naturalist* 147 (1996b): 319–338.

Rodd, F. H., D. N. Reznick, and M. B. Sokolowski. "Phenotypic Plasticity in the Life History Traits of Guppies: Response to Social Environment." *Ecology* 78 (1997): 419–433.

— F. Helen Rodd

LINKAGE

Linkage describes the nonindependent inheritance of genes located on the same chromosome. The phenomenon of linkage was discovered by William Bateson and Reginald C. Punnett in the first years of the twentieth century, while they were studying the genetics of the sweet pea plant. They worked with two loci; one that determined flower color and that had two alleles (*P*, purple, and *p*, red), and another that influenced the shape of pollen grains, also with two alleles (*L*, long, and *l*, round). In their experiments, they crossed plants of genotypes *PPLL* (purple, long) and *ppll* (red, round). The first-generation progeny, *PpLl* heterozygotes, were self-pollinated to obtain a second-generation sample. The proportions of each phenotype observed (see Table 1) in the second generation differed dramatically from the expected 9:3:3:1 Mendelian ratio. Instead, there was an excess of parental phenotypes purple-long and red-round. Bateson and Punnett hypothesized that the first-generation plants had produced more *PL* and *pl* gametes than predicted under Mendelian independent assortment be-

TABLE 1. The Sweet Pea Phenotypes Observed by Bateson and Punnett in Second-Generation Plants

The dashes in the genotypes indicate that both heterozygous and homozygous genotypes may be present for the dominant alleles, but they cannot be distinguished on the basis of phenotype.

	Number of progeny	
Phenotype and Genotype	*Observed*	*Expected from 9:3:3:1 ratio (independent Mendelian assortment)*
Parental: purple, long (*P-L-*)	284	215
Recombinant: purple, round (*P-ll*)	21	71
Recombinant: red, long (*ppL-*)	21	71
Parental: red, round (*ppll*)	55	24
Total	381	381

(From Griffiths et al., 1996.)

cause of physical coupling between the dominant *P* and *L* alleles and between the recessive *p* and *l* alleles.

Further insights came from the work of Thomas Hunt Morgan, who beginning in 1909 used the fruit fly *Drosophila* as a model organism for genetic investigation. Morgan proposed that coupling occurred when the genes determining both phenotypes were located on the same chromosome. Morgan went on to suggest that when homologous chromosomes pair during meiosis, the chromosomes sometimes reciprocally exchange segments through a process called crossing over. The original allelic arrangements on the chromosomes are the parental combinations, while the products of crossing over lead to nonparental combinations of alleles, or recombinant genotypes. In the sweet pea example, recombinant genotypes would be the occurrence of alleles *Pl* or *pL* together on the same chromosome.

Morgan noticed that the proportion of recombinant progeny among his fruit flies varied according to which pairs of linked genes were being studied, and he thought that this may reflect the physical distance separating the genes on the chromosome. At the end of 1911 one of Morgan's students, Alfred Sturtevant, realized that variations in the strength of linkage could be used to determine the relative sequence of genes along the chromosome.

These ideas lead to the development of linkage mapping, a fundamental technique in genetics. The basic principle behind linkage mapping is that the greater the distance between genes on a chromosome, the greater the proportion of recombinants produced. In other words, recombination frequency varies in proportion to the distance between genes. Therefore, by measuring the recombination frequency, a measure of distance can be obtained. One genetic map unit (m.u.; also known as a centimorgan [cM], after Morgan) is defined as the distance between genes for which 1 product of meiosis out of 100 is recombinant; that is, a recombination frequency of 1 percent equals one genetic map unit. In the sweet pea example, the map distance between the two loci is equivalent to the percentage of recombinant genotypes among the second-generation progeny: ([Recombinant progeny/Total progeny] × 100) = ([PurpleRound (*P-ll*) + RedLong (*ppL-*)/Total progeny] × 100) = ([21 + 21/381] × 100) = 11 m.u. (The dashes in the genotypes indicate that both heterozygous and homozygous genotypes may be present for the dominant alleles, but they cannot be distinguished on the basis of phenotype.)

One property of map distances is that generally they are additive. Thus if genes A and B are separated by 7 m.u. and A and C by 3 m.u., the distance between B and C could be either 10 m.u. or 4 m.u. The linear order of the three genes can be determined by measuring the recombination frequency between B and C. Recombination frequencies significantly exceeding 50 percent are never observed. This is because double crossovers, which reconstitute parental genotypes, cause the true recombination frequency to be underestimated. If genes are located on separate chromosomes, nonparental classes will always make up 50 percent of the progeny due to independent assortment of the chromosomes. Therefore, recombination frequencies of 50 percent indicate either that genes are located far apart on the same chromosome or that they are located on separate chromosomes.

Genes, or combinations of genes between which the recombination frequency is significantly less than 50 percent, are lumped together in units known as linkage groups, which correspond to individual chromosomes. Linkage maps and linkage groups are abstract concepts based on genetic data, which can be derived without knowledge of physical chromosomes. Linkage maps place loci relative to each other using map units. They cannot determine where on a physical chromosome the loci reside or to which physical chromosome different linkage groups correspond. This requires experiments that involve observing or manipulating the chromosome

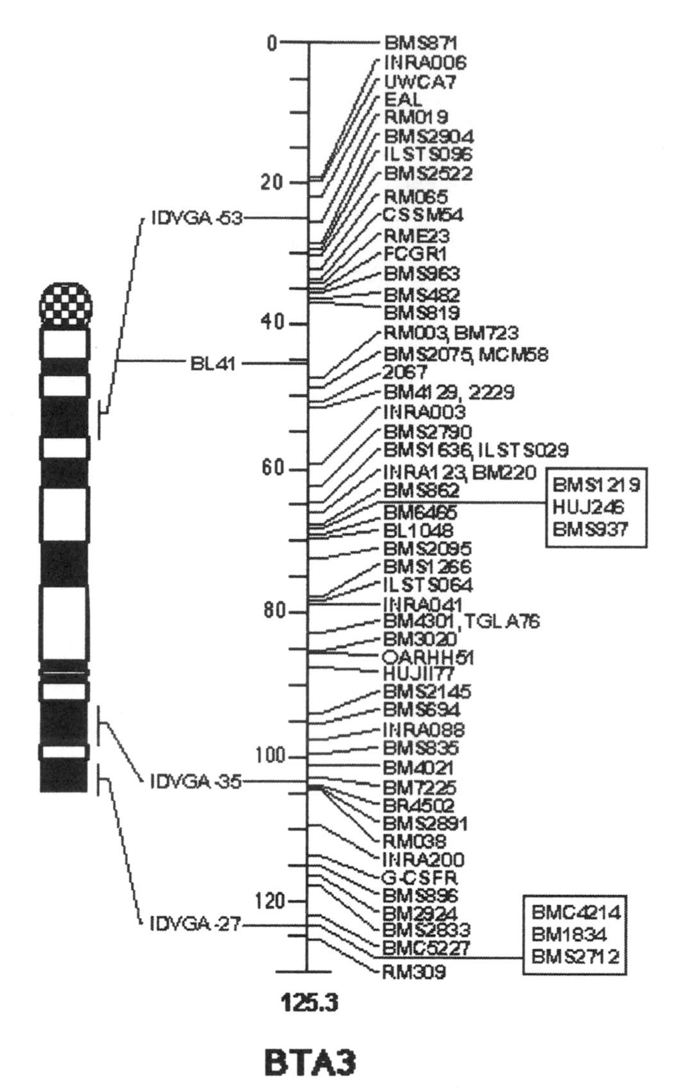

FIGURE 1. Linkage Map of Bovine Chromosome 3. Linkage map based on microsatellite DNA repeat length polymorphisms. On the left is a diagram of the actual chromosome showing the relative physical position of the markers, and on the right the linkage map. Text gives the names of individual microsatellite markers. Numbers refer to distance in map units from the upper end of the chromosome. Large values are calculated as the sum of smaller intervals since recombination frequency cannot exceed 50 percent. S. M. Kappes et al. 1997, *Genomics Research* Mar 7(3):235–49. http://sol.marc.usda.gov/.

content of cells. Modern linkage-mapping typically uses many thousands of markers based on neutral DNA polymorphisms to accurately locate genes of interest. Figure 1 gives an example of a modern linkage map of a cow chromosome constructed using microsatellite loci.

Sex Linkage. *Sex linkage* is a term applied to genes located on sex chromosomes. Sex linkage has important consequences when gender is determined by sex chromosomes with large size differences, such as the mammalian X and Y sex chromosomes. Females have two X chromosomes, one derived from each parent, while males carry a single X chromosome exclusively inherited from the mother and a Y chromosome inherited from the father. Y chromosomes typically carry only a few genes involved in male sex determination and spermatogenesis, while X chromosomes carry a full complement of genes. Because males have only a single X chromosome inherited from the mother, most genes located on the X chromosome are present in males only as a single copy. Therefore the mothers X-linked genes are the sole determinant of the son's phenotype. If the mother is heterozygous for a recessive X-linked mutation, and the son inherits the X chromosome carrying that allele, the son will express a sex-linked phenotype not observed in the mother. Females may also express recessive X-linked phenotypes, but only if they inherit an X chromosome from both parents carrying the recessive mutation. This is much less likely, making many X-linked mutant phenotypes more common in male offspring.

X-linked loci cause many disease phenotypes in humans including Duchenne muscular dystrophy (muscle wastage), hemophilia A (a failure of blood clotting) as well as other traits such as red-green color blindness.

Linkage Disequilibrium. Often genes interact with each other rather than behaving as unconnected units. When genes interact, changes at one locus will have consequences for other genes. This introduces the problem of how to think about evolution at more than one locus at the same time. Hardy–Weinberg equilibrium describes the expected genotype frequencies for a single locus in the absence of selection (differences in reproduction between genotypes) and with random mating. The concept of linkage equilibrium is the multilocus equivalent, and gives the expected frequencies of combinations of alleles at different loci (haplotypes) under the same conditions. Deviation from independent association between loci is termed *linkage disequilibrium.* A key point of linkage disequilibrium is that loci which show deviation from random association do not have to be physically linked on the same chromosome.

In an example in which there are two alleles A, a at one locus and B, b at another (with frequencies p_a, p_b, p_A, p_B), there would be four possible combinations in the haploid gametes: AB, Ab, aB, ab. The frequencies of the gametes can be represented by p_{AB}, p_{Ab}, and so on. When there are no associations between genes, and alleles at the two loci occur independently (i.e., the probability that a gamete carries allele a is independent of whether it also carries allele B) they are said to be in linkage equilibrium. In this case, the frequency of gamete $p_{ab} = p_a p_b$ and so on for the remaining gamete types. However, if this assumption does not hold and the occurrence of specific alleles at the two loci in a gamete is not independent, then there is linkage disequilibrium.

The gamete frequency is now given by $p_{ab} = p_a p_b + D$, where D is the linkage disequilibrium, which quantifies the departure from linkage equilibrium, or degree of association between alleles. Overall, $p_{ab} = p_a p_b + D$; $p_{aB} = p_a p_B - D$; $p_{Ab} = p_A p_b - D$; $p_{AB} = p_A p_B + D$; and $D = p_{ab} p_{AB} - p_{aB} p_{Ab}$. In very large populations and with no selection; linkage disequilibrium for unlinked loci halves each generation because of random assortment of chromosomes at meiosis. For linked loci, recombination breaks down linkage disequilibrium, causing it to decay at an exponential rate equal to the recombination frequency between loci.

Several processes can generate and maintain linkage disequilibrium, either singularly or in combination:

1. *Suppression or absence of recombination.* Chromosomal rearrangements called inversions suppress effective recombination and maintain linkage disequilibrium among the loci on the inverted segment. Inversions occur by chance when a segment of chromosome is excised and reattached in the opposite orientation. In individuals heterozygous for an inversion and a normal chromosome, recombination within the inversion generates a range of abnormal chromosomes or DNA segments with duplicated or deleted loci. Gametes that carry recombinant products from an inversion do not usually give rise to viable progeny.

In some parts of the genome, recombination is rare or absent; these include much of the Y chromosome (excluding the pseudo-autosomal region), and the whole mitochondrial genome. For these examples, all loci on the chromosomes are in linkage disequilibrium and effectively behave as a single locus. New haplotype variants arise only through mutation. This makes these parts of the genome very informative for studies of phylogeny, genetic population history, and patterns of migration.

2. *Selection.* Selection can act in several ways to produce and maintain linkage disequilibrium. One example is a phenomenon called epistasis, where the fitness of an individual due to one gene depends on what alleles are present at other loci. Linkage disequilibrium is generated when selection favors some specific allele combinations over others, increasing the association of those alleles in the population.

Directional selection, which increases the frequency of a favored allele, generates linkage disequilibrium between the favored allele and alleles at other tightly linked loci that flank it. As the favored allele increases in frequency, so do alleles at closely linked loci, even if they are selectively neutral. This effect is known as "hitchhiking."

3. *Drift and mutation in a finite population.* Genetic drift can generate linkage disequilibrium owing to the random sampling of haplotypes transmitted to the next generation. If any haplotypes are overrepresented by chance, then linkage disequilibrium will arise. In each generation, drift causes the extent of disequilibrium to change randomly, because it is equally likely to move haplotype frequencies toward equilibrium or disequilibrium.

In small populations, a novel mutation at a specific site is likely to be unique. It will occur on a particular chromosome carrying a certain set of alleles. Therefore, the novel mutation will initially be in linkage disequilibrium with the other markers on the chromosome.

4. *Nonrandom mating.* Mating preferences with respect to genotype generate linkage disequilibrium. For instance, if individuals of genotype A at locus 1 prefer genotype B individuals over b carriers at locus 2, then there will be an excess of AB haplotypes over the expectation for a randomly mating population.

5. *Population admixture and hybridization.* If two genetically distinct populations start to mix and interbreed, linkage disequilibrium may exist between the genes from each population for a few generations before being broken down by random assortment and recombination. Linkage disequilibrium can be used to estimate the number of interacting populations and measure the rate of gene flow, and to identify migrants or their recent descendants. Continued dispersal of parental genotypes into an area of mixing, mating preferences and selection on hybrid genotypes can maintain linkage disequilibrium in the face of recombination and random assortment. This is a key component of hybrid zone theory.

Linkage disequilibrium has implications for many areas of genetics and evolutionary biology. Patterns of linkage disequilibrium are of practical importance in using data from the human genome project to identify genes associated with disease susceptibility. Linkage disequilibrium can also be used to gain insights into many evolutionary processes. Mitochondrial and Y-chromosome haplotypes are used to construct phylogenies and population histories. For example, in humans, the distributions of different types of Y-chromosome haplotypes indicate that the ancestors of modern Europeans are derived from two paleolithic and one neolithic wave of colonization from the Near East. Linkage disequilibrium can also be used to infer past population demography. In many genes across the human genome, linkage disequilibrium extends over shorter chromosome distances in African populations than in northern Europeans. This suggests that, in the past, African populations expanded, while northern Europeans experienced a population bottleneck, either as a result of the last ice age or during the movement of modern humans out of Africa. Analyses of population structure employing linkage disequilibrium between nuclear markers such as microsatellites have been used for many species, to estimate the number of different populations contributing to wild caught samples of individuals of unknown origin. This type of information is often of ecological or conservation interest or important for stock management. Linkage-disequilibrium-based analyses are central to studies of hybrid zones. They are used to estimate parameters such as the rate of hybridization and introgression, or to identify individuals having hybrid ancestry.

Linkage disequilibrium can be used to assess the strength and type of selection acting in natural populations. Female swallowtail butterflies exhibit Batesian mimicry (a palatable mimic species resembles the warning coloration of a distasteful model species, which predators avoid eating). Mimics therefore gain the protection from predation enjoyed by model species. Females of the African swallowtail species *Papilio memnon* often show genetic polymorphism within a single region, with several distinct forms mimicking different models. Different mimics are specified by multiple loci that are tightly linked in a supergene complex. Crosses between different mimic types generate occasional recombinants. Recombinants resemble neither model species and are not recognized as distasteful, and therefore are removed by predators, leaving just parental combinations of alleles in the population. This is an example of epistatic selection, as genes for specific color patterns are selected against if they are combined with genes for other color patterns.

Linkage disequilibrium attributable to selection can be found at work within the human lymphocyte antigen (HLA) system. The HLA system is a set of linked genes on chromosome 6 that are involved in immune-system function. In different populations, specific combinations of genes occur at greater-than-expected frequencies. Among northern European populations, a combination called A1B8 typically shows an excess. Many genes in the HLA are associated with the recognition of antigens from pathogens. Such patterns of disequilibrium may reflect selection on favorable gene combinations in relation to interactions with pathogens.

The human glucose-6-phosphate-dehydrogenase (G6PD) locus provides another example of selection generating linkage disequilibrium. Mutations that reduce the activity of this enzyme are associated with increased resistance to malaria. In Africa, southern Europe, the Middle East, and India, resistance mutations are strongly associated with specific alleles at neutral flanking loci. Different resistance mutations in G6PD from sub-Saharan Africa and the Mediterranean are associated with different haplotype groups, and there are fewer haplotypes associated with the resistance mutations compared with many for normal versions of the gene. This suggests that the resistance mutations have rapidly increased in frequency after arising independently in sub-Saharan Africa and the Mediterranean only 2,000 to 12,000 years ago.

Although only a few examples of linkage disequilibrium are given here, it is central to theories in evolutionary biology, such as the evolution of trait and preference genes in sexual selection, the evolution of recombination in sexual reproduction, the behavior of hybrid zones, genetics of reproductive isolation, speciation, and the spread of adaptations.

[*See also* Mendelian Genetics.]

BIBLIOGRAPHY

Barton, N. H., and K. S. Gale. "Genetic Analysis of Hybrid Zones." In *Hybrid Zones and the Evolutionary Process*, edited by R. G. Harrison. Oxford, 1993. Explains many fundamental aspects of hybrid zones and their evolutionary genetics, including the role of linkage disequilibrium.

Griffiths, A. J. F., J. H. Miller, D. T. Suzuki, R. C. Lewontin and W. M. Gelbart. *An Introduction to Genetic Analysis*. New York, 1996. An accessible introduction to many of the areas covered in this article, suitable for advanced high school or undergraduate students.

Hedrick, P. W. "Evolutionary Genetics of the Major Histocompatibility Locus." *American Naturalist* 143 (1994): 945–964. A good account, though somewhat dated, of the evolutionary explanations for the patterns of variation seen in MHC (HLA) genes.

Kohn, M. H., H. J. Pelzand, and R. K. Wayne. "Natural Selection Mapping of the Warfarin Resistance Gene." *Proceedings of the National Academy of Science* 97 (2000): 7911–7915. A nice example of how linkage disequilibrium can be used to identify selection acting on the genome of a natural population.

Maynard Smith, J. *Evolutionary Genetics*, New York, 1998. A good introduction to evolutionary biology, with clear explanations of the theoretical background to linkage disequilibrium and evolutionary phenomena in which linkage disequilibrium plays an important role.

Pritchard, J. K., M. Stephens, and P. Donnelly. "Inference of Population Structure Using Multilocus Genotypes." *Genetics* 155 (2000): 945–959. An important contribution to the statistical methodology used to analyze genetic-maker data for population structure and gene flow. Assumes a basic knowledge of statistics.

Reich, D. E., M. Cargill, S. Bolk, J. Ireland, P. C. Sabeti, D. J. Richter, T. Lavery, R. Kouyoumrian, S. Farhadian, R. Ward, and E. S. Lander. "Linkage Disequilibrium in the Human Genome." *Nature* 401 (2001): 199–204. One of the first studies to describe patterns of linkage disequilibrium at a full genome level in humans. Illustrates the types of modern analyses possible with the large genetic data sets now available.

Ridley, M. *Evolution*. Cambridge, Mass., 1998. An introduction to the subject of evolution for undergraduate-level students; includes many examples of linkage and linkage disequilibrium as important in evolution.

Semino, O., G. Passarino, P. J. Oefner, A. A. Lin, S. Arbuzova, L. E. Beckman, G. De Benedictis, P. Francalacci, A. Kouvatsi, S. Limborska, M. Marcikiae, A. Mika, B. Mika, D. Primorao, A. S. Santachiara-Benerecetti, L. L. Cavalli-Sforza, and P. A. Underhill. "The Genetic Legacy of Paleolithic Homo Sapiens Sapiens in Extant Europeans: A Y Chromosome Perspective." *Science* 290 (2000): 1155–1159. An example of the use of haplotypic markers to reconstruct population history.

Tishkoff, S. A., R. Varkonyi, N. Cahinhinan, S. Abbes, G. Argyropoulos, G. Destro-Bisol, A. Drousiotou, B. Dangerfield, G. Lefranc, J. Loiselet, A. Piro, M. Stoneking, A. Tagarelli, G. Tagarelli, E. H. Touma, S. M. Williams, and A. G. Clark. "Haplotype Diversity and Linkage Disequilibrium at Human G6PD. Recent Origin of Alleles That Confer Malarial Resistance." *Science* 293 (2001): 455–462. Provides modern insight into the well-known association between G6PD mutations and resistance to malaria in humans. It uses patterns of linkage disequilibrium to show how selection is acting on this locus.

Turner, J. R. G. "Mimicry: The Palatability Spectrum and Its Consequences." *The Biology of Butterflies*, edited by R. I. Vane-Wright and P. R. Ackery. London, 1984. A comprehensive reference on mimicry in butterflies.

Weir, B. S. *Genetic Data Analysis II*. Sunderland, Mass., 1996. An advanced text that describes the population genetics underlying disequilibrium measures and statistical approaches for testing for disequilibrium in genetic data.

White, R., M. Leppert, D. T. Bishop, D. Barker, J. Berkowitz, C. Brown, P. Callahan, T. Holm, and L. Jerominski. "Construction of Linkage Maps with DNA Markers for Human-Chromosomes." *Nature* 313 (1985): 101–104. Describes the modern approach to constructing linkage maps.

— SIMON J. GOODMAN

LINNAEAN CLASSIFICATION. *See* Classification.

LINNAEUS, CAROLUS

Carolus Linnaeus (1707–1778), the founder of modern systematics and nomenclature, was born in Råshult, Sweden, on 23 May 1707. He studied medicine in Lund and Uppsala before earning his medical degree in the Netherlands. Linnaeus combined his interest in medicine with that of natural history, and with a grant from the Swedish Royal Academy, he traveled for five months throughout Lapland collecting animals, plants, and minerals. Later in Amsterdam he published a botanical account of the trip, the *Flora Lapponica* (1737).

As a consequence of Linnaeus's journey to the Netherlands, he met an influential set of Dutch naturalists and physicians who were greatly impressed with his knowledge of natural history. They persuaded him to stay in the Netherlands for three years. During this period, Linnaeus became familiar with the extensive Dutch natural history collections that were some of the most impressive of the day. Part of that time he served as superintendent of the garden to one of the major collectors of Holland, George Clifford (1685–1760), a wealthy financier and director of the Dutch East India Company. In Clifford's gardens (and private zoo) Linnaeus studied living specimens from southern Europe, Asia, Africa, and the New World.

It was a time of great excitement in natural history, as Europeans were encountering thousands of new species of plants and animals, plus many new rocks and minerals. The Dutch, along with the French and the British, explored vast regions of the globe during the eighteenth century with an eye toward future trade and colonization. Governments were interested in documenting the natural history of exotic places in hopes of finding commercially valuable products. The influx of so much new material caused serious problems for naturalists. Many of the new species did not fit easily into accepted classification systems, and the lack of a standard system of nomenclature compounded the problem.

FIGURE 1. Carolus Linnaeus.
Visuals Unlimited.

While in the Netherlands, Linnaeus published a short work (twelve printed pages), the *Systema naturae* (1735), which proposed a new system of classification for plants, animals, and minerals. The most original part of this work contained a sexual system of classification that hierarchically arranged plants into twenty-four classes according to the number and relative positions of their stamens (male parts). The classes were further broken down into sixty-five orders, mostly on the basis of the number and position of the pistils (female parts). Then, using other criteria, he distinguished particular genera and individual species. The simplicity and relative ease of application made Linnaeus' system appealing, and its use spread throughout much of Europe.

Linnaeus also revolutionized the naming of organisms by proposing a binomial nomenclature (in Latin) based on genus and species which he first used in his *Species plantarum* (1753), a work that recorded all the known species of plants. Like his hierarchical classification, the binomial nomenclature caught on with the scientific public, and today all species are still assigned a Linnean binomial classification (e.g., *Homo sapiens*).

After Linnaeus returned to Sweden in 1738, he lived for a short period in Stockholm, then moved to Uppsala, where he remained as professor at the university until his death in 1778. He carried on correspondence to bring the editions of his classification systems up to date. Linnaeus also inspired many of his students to travel abroad to add to the existing stock of knowledge of natural history.

Linnaeus's career greatly expanded the realm of natural history. He served as a general clearinghouse for information, much of it finding its way into successive editions of his works. His system gave naturalists and enthusiasts a simple tool that made identification relatively easy, and his nomenclature provided a common language for naming. Linnaeus considered natural history a subject closely related to religion. In naming, describing, and ordering, he hoped to document the Creation. He viewed God's Creation as a static, balanced, and harmonious system. Classification reflected the harmony of nature, and in his writings on biogeography he described a general balance in nature. Every plant and animal fills a place in a grand system and helps maintain that network. The reciprocal relationship of predator and prey, for example, linked each in an overall, harmonious, and static plan. Linnaeus held that the original species created by God had the same relationships in nature that they currently have, even after dispersing from the place of creation to their assigned location where they have remained ever since. Similarly, Linnaeus stated in his early writings that species had not changed since their creation (later he modified this view to accept the notion that hybridization had produced some new species in time).

Linnaeus' system and perspective stressed a static picture of nature. His works inspired naturalists to explore and expand the dimensions of natural history, which led some, like Charles Darwin, to question the assumptions of his work and ultimately to transform them.

[*See also* Classification.]

BIBLIOGRAPHY

Blunt, W. *The Compleat Naturalist.* New York, 1971. A broad and accessable view of Linnaeus's life and work.

Frängsmyr, Tore, ed. *Linnaeus: The Man and His Work.* Berkeley, 1983. A set of essays exploring the major features of Linnaeus's science.

Koerner, Lisbet. *Linnaeus: Nature and Nation.* Cambridge, Mass., 1999. A study of Linnaeus that situates him in the context of his contemporary Sweden.

Larson, J. *Reason and Experience: The Representation of Natural Order in the Work of Carl von Linné.* Berkeley, 1971. A careful study of Linnaeus's system of classification.

— PAUL LAWRENCE FARBER

LYELL, CHARLES

Charles Lyell (1797–1875), Scottish geologist, was a major contributor to both the content and the methodology of geology during the nineteenth century. His early training as a lawyer is apparent in his skillful marshaling of the evidence in his scientific writings. Lyell traveled extensively in Europe and North America. Field work in 1828 convinced him that geological change takes place gradually, over long periods of time, and that the conditions of life have been relatively constant. His book *Principles of Geology, being an Attempt to Explain the Former Changes of the Earth's Surface, by Reference to Causes now in Operation*, began to appear in 1830 and was repeatedly revised. It profoundly influenced Charles Darwin, who read it during his famous voyage on the HMS *Beagle* (1831–1836).

Lyell's position, called uniformitarianism, is encapsulated in the saying "The present is the key to the past." However, several different notions were at issue, and they tended to be confused from the outset. The ideas that the laws of nature are true irrespective of time and place and that reconstructed histories must not contradict them seems all too obvious and has never been seriously contested. But Lyell had additional points in mind, most notably his rejection of the widely held view called catastrophism, according to which the fossil record was thought to represent a series of faunas and floras that had been successively destroyed as the result of catastrophes such as floods, which were often associated with the biblical Deluge, then replaced by new assemblages of organisms. Lyell replaced the idea of repeated catastrophes with the notion that change was always gradual and caused by processes that could be observed in the present. Methodologically, it seemed appropriate to rely on causes that were currently observed before proceeding to invoke additional factors that no one had ever witnessed.

Lyell also advocated a steady-state concept of the solar system, in which a limited amount of change did occur, but it was cyclical (like the seasons of the year) rather than directional. There would be uplift and erosion, but these would balance each other out, with continents rising in some places and sinking in others.

Extrapolated to biology, a steady state meant that organisms might be adapted to different circumstances, but there would be no directional, or "progressive," changes in the earth's biota. Lyell reasoned that if the number of species remains constant, new species must somehow replace those that have become extinct. He did not profess to know how new forms of life have come into being. Upon reading Jean Baptiste Lamarck's *Philosophie Zoologique* in 1827, he much admired that author's argument for evolution, but he remained unconvinced. Lyell's respectful but somewhat distorted account of Lamarck's views in his *Principles* had a considerable influence, especially on Darwin.

As a geologist, Darwin was one of Lyell's most enthusiastic supporters, and he took Lyell's denial of progressive change in the fossil record very seriously. However, Lyell began to change his mind as evidence accumulated that better adapted organisms have replaced the older ones. He was particularly impressed when, on 26 November 1855, he read a paper on biogeography by Alfred Russel Wallace. Lyell's field work in Madiera and the Canary Islands the previous year had affected his thinking about biogeography and its evolutionary implications.

On 16 April 1856, Darwin explained natural selection to him. Lyell urged Darwin to publish his theory so as not to lose priority. When a manuscript on natural selection by Wallace reached Darwin in 1858, Lyell was instrumental in arranging for joint publication.

Lyell did not immediately embrace the concept of evolution. On 6 February 1863, his book *Geological Evidences of the Antiquity of Man with Remarks on the Origin of Species by Variation* was published. In it, Lyell argued that during the ice ages humans had coexisted with very different assemblages of animals and therefore had a long history. Although he maintained that evolution was the only scientific explanation for the facts, he held back from offering his support of the theory, and Darwin was much disappointed. It was only on 30 November 1864, during a meeting at which Darwin was awarded the Copley Medal of the Royal Society of London, that Lyell publicly endorsed evolution by natural selection. The later editions of *Principles* reflect his new position. Lyell's ability to change his mind rather late in life has often been interpreted as indicating strong commitment to the ideals of science. Be that as it may, supporting Darwin's theory kept Lyell a leader to the very end of his career.

[*See also* Darwin, Charles; Geology.]

BIBLIOGRAPHY

Bailey, E. *Charles Lyell.* London, 1962. A brief and readable biography.

Blundell, D. J., and A. C. Scott, eds. *Lyell: The Past Is the Key to the Present.* London, 1998. Recent essays.

Lyell, C. *Principles of Geology, Being an Attempt to Explain the Former Changes of the Earth's Surface, by Reference to Causes Now in Operation.* 2 vols. London, 1830–1832: Reprint: Chicago. The reprint has an excellent introduction by Martin Rudwick.

Lyell, K. M. *Life, Letters and Journals of Sir Charles Lyell, Bart.* London, 1881. Standard work, with reprint available.

Wilson, L. G. *Sir Charles Lyell's Scientific Journals on the Species Question.* New Haven, 1970.

Wilson L. G. *Charles Lyell: The Years to 1841—The Revolution in Geology.* New Haven, 1972.

Wilson, L. G. "Lyell, Charles." In *Dictionary of Scientific Biography,* edited by C. C. Gillispie. New York, 1981. Authoritative like Wilson's other works, this is concise and covers Lyell's whole career.

Wilson, L. G. *Lyell in America: Transatlantic Geology, 1841–1853.* Baltimore, 1998.

— MICHAEL T. GHISELIN

M

MACROEVOLUTION. *See* Overview Essay *on* Macroevolution *at the beginning of Volume 1.*

MAJOR HISTOCOMPATIBILITY COMPLEX. *See articles on* Modern *Homo sapiens.*

MAJOR TRANSITIONS IN EVOLUTION. *See* Overview Essay *on* Major Transitions in Evolution *at the beginning of Volume 1.*

MALARIA

Malaria is a disease caused in vertebrates by infection with parasitic protozoa of the genus *Plasmodium*. The genus is classified within the complex protist phylum Apicomplexa (order Haemosporida, class Hematozoa), which consists of nearly 5,000 described species. In addition to *Plasmodium*, notable apicomplexans include *Cryptosporidium parvum*, *Babesia bigemina*, and *Toxoplasma gondii*. All these single-celled organisms have the characteristic "apical complex" structure from which the phylum derives its name. The taxonomy and phylogeny of the phylum have been the subject of controversy and frequent revision. At issue is whether *Plasmodium* evolved directly from monogenetic (one-host) parasites of the ancient marine invertebrates from which the chordates evolved, or whether they originated by lateral transfer from other, already digenetic (two-host) vertebrate parasites (Barta 1989; Garnham 1966; Huff 1938; Maxwell, 1955). It is clear that the digenetic lifestyle has evolved independently several times in the various apicomplexan classes.

As is the case for most microorganisms, there are no apicomplexan fossils (Margulis et al., 1993). The evolutionary history of apicomplexans is best discerned by molecular phylogenetic methods. A conclusion of the molecular investigations is that the phylum is very ancient, perhaps as old as the Cambrian period, or about 500 million to 600 million years old. The genus *Plasmodium*, which consists of nearly 200 named species that parasitize various reptiles, birds, and mammals, arose sometime in the last 55 million to 129 million years.

Four species—*P. falciparum*, *P. ovale*, *P. malariae*, and *P. vivax*—are parasites of humans and are the causative agents of human malaria. The phylogeny of the numerous species has been elucidated using several genes, such as the small subunit ribosomal RNA gene (SSUrRNA, Figure 1; see Escalante and Ayala, 1994, 1995). These phylogenetic analyses have shown that the four human malarial parasites are distantly related to each other, so that each has a distinct nonhuman parasite as its closest relative. The most parsimonious conclusion from this observation is that all four species became human parasites independently. This is likely to have occurred by lateral transmission of parasites to the human ancestral lineage from nonhuman hosts. These conclusions are consistent with the diversity of physiological and epidemiological characteristics of the human *Plasmodium* species.

The most deadly of the human malaria parasites, *Plasmodium falciparum*, is more closely related to *P. reichenowi*, the chimpanzee parasite, than to any other *Plasmodium* species. The time of divergence between these two *Plasmodium* species is estimated to be eight million to 11 million years ago, which is roughly consistent with the time of divergence between the lineages that eventually led to the two contemporary host species. A parsimonious interpretation of this situation is that *P. falciparum* has been associated with human ancestors since the divergence of the hominids from the great apes. If so, then this would put to rest the long-held notion that *P. falciparum's* extreme virulence derives from its becoming only recently associated with its human host. Parasitologists often assume that parasites gradually evolve toward a benign association with their hosts, a view referred to as Fahrenholz's rule. Population biologists, based largely on theoretical grounds, have shown that decreased virulence is not the only plausible outcome of parasite–host coevolution.

There are some 500 million clinical cases of *P. falciparum* each year; at least one million people die from the infection. Despite clear indications that *falciparum* malaria has an ancient association with its human hosts, strong evidence shows that the current global distribution of the parasite is recent. This conclusion is based on analyses of silent nucleotide polymorphisms. Silent-site polymorphisms are nucleotide substitutions to silent sites in genes; that is, they do not alter the amino acid that is encoded and so are thought to be largely neutral with respect to fitness and natural selection. Accordingly, nucleotide substitutions at these sites occur at a steady rate through geological time periods, as a function of the mutation rate and time elapsed. The rate of such substitutions can be obtained empirically by

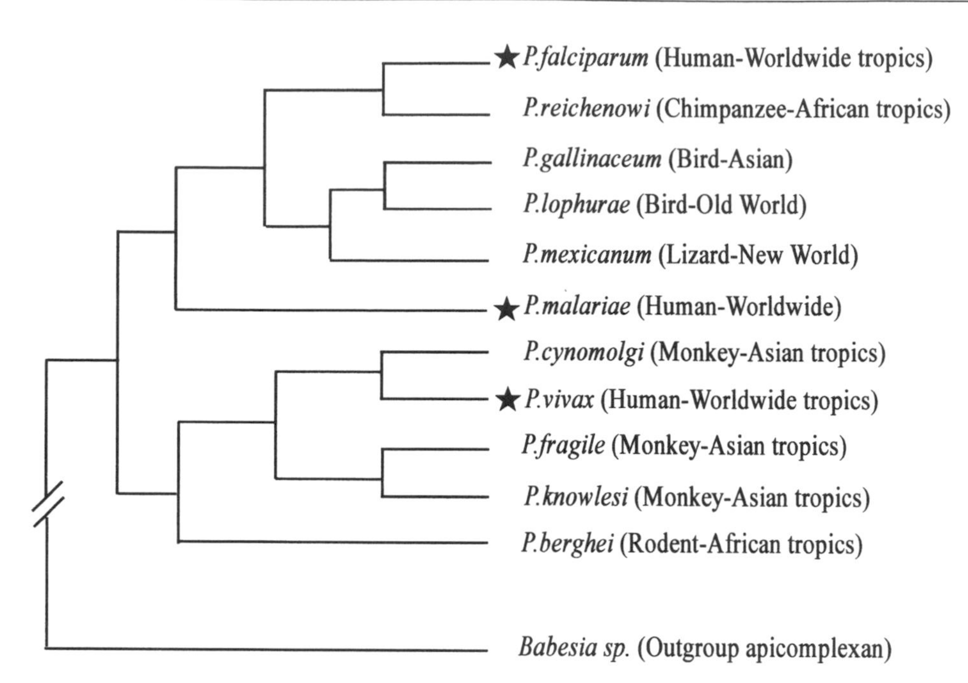

Figure 1. Phylogeny of Various *Plasmodium* Species Based on SSUrRNA.
The vertebrate host and geographic range are given after each species name. Three of the four human malarias are present in this tree (indicated by a star ★). Courtesy of author.

counting differences among gene sequences from species for which the divergence time is known, for example, from the fossil record. *P. falciparum* manifests few silent-site polymorphisms, even among globally sampled populations. This is true for several nuclear-encoded genes (Escalante et al., 1998; Rich et al., 1998, 2000) as well as for mitochondrial genomes (Conway et al., 2000). Based on these observations, the current distribution of *P. falciparum* throughout the world's tropical regions is believed to have derived from a small ancestral population within the past 20,000 to 60,000 years.

Evidence exists to show that natural selection acts strongly on both the *falciparum* and its hosts. The human response is evident in the various hemoglobinopathies (e.g., sickle-cell trait) that have arisen to combat the *Plasmodium*. The *P. falciparum* genotype itself shows evidence of positive selection among several genes encoding immunogenic surface proteins as well as drug resistance factors. Consequently, *falciparum* malaria affords an excellent system in which to study genetic variation and population structure of the parasite, genetic responses of the host, and host–parasite coevolution. Such investigations have considerable clinical relevance because the rate of evolution of certain genes, such as those encoding drug resistance factors, helps to determine means for intervening in the transmission cycle.

P. falciparum, like all members of the genus, relies on a two-host life cycle that involves a vertebrate host and a mosquito vector. Additional complexity of the parasite's life cycle derives from obligatory sexual and asexual stages. All *Plasmodium* have a life cycle that includes asexual proliferation of haploid parasites, known as merozoites, which are found in the blood cells of the vertebrate host (Figure 2). A subset of the asexual forms differentiate into gametocytes, which are imbibed with the mosquito's blood meal and subsequently develop into gametes inside the mosquito. In the mosquito midgut, the gametes fuse to form a diploid, zygotic ookinete. After migrating through the midgut wall, the ookinete undergoes meiotic reduction division into haploid nuclei that eventually develop into sporozoites. The sporozoites migrate through the mosquito's body into the salivary glands and are passed to a vertebrate host at the next blood meal. In mammalian malarias, these sporozoites make their way to the liver, where they enter hepatocytes, proliferate, and upon liberation from the ruptured hepatocyte invade red blood cells as merozoites, thereby completing the cycle.

This complex life cycle requires a finely balanced regulation of gene expression that involves stage-specific expression of various factors involved in parasite survival. These have been most studied and are best understood in *P. falciparum*. Particular attention has been paid to genetic loci that encode immunogenic surface proteins, many of which are involved in cytoadherence and invasion of host tissue. One example is the *pfEMP* gene family. The protein encoded by a *pfEMP* sequence

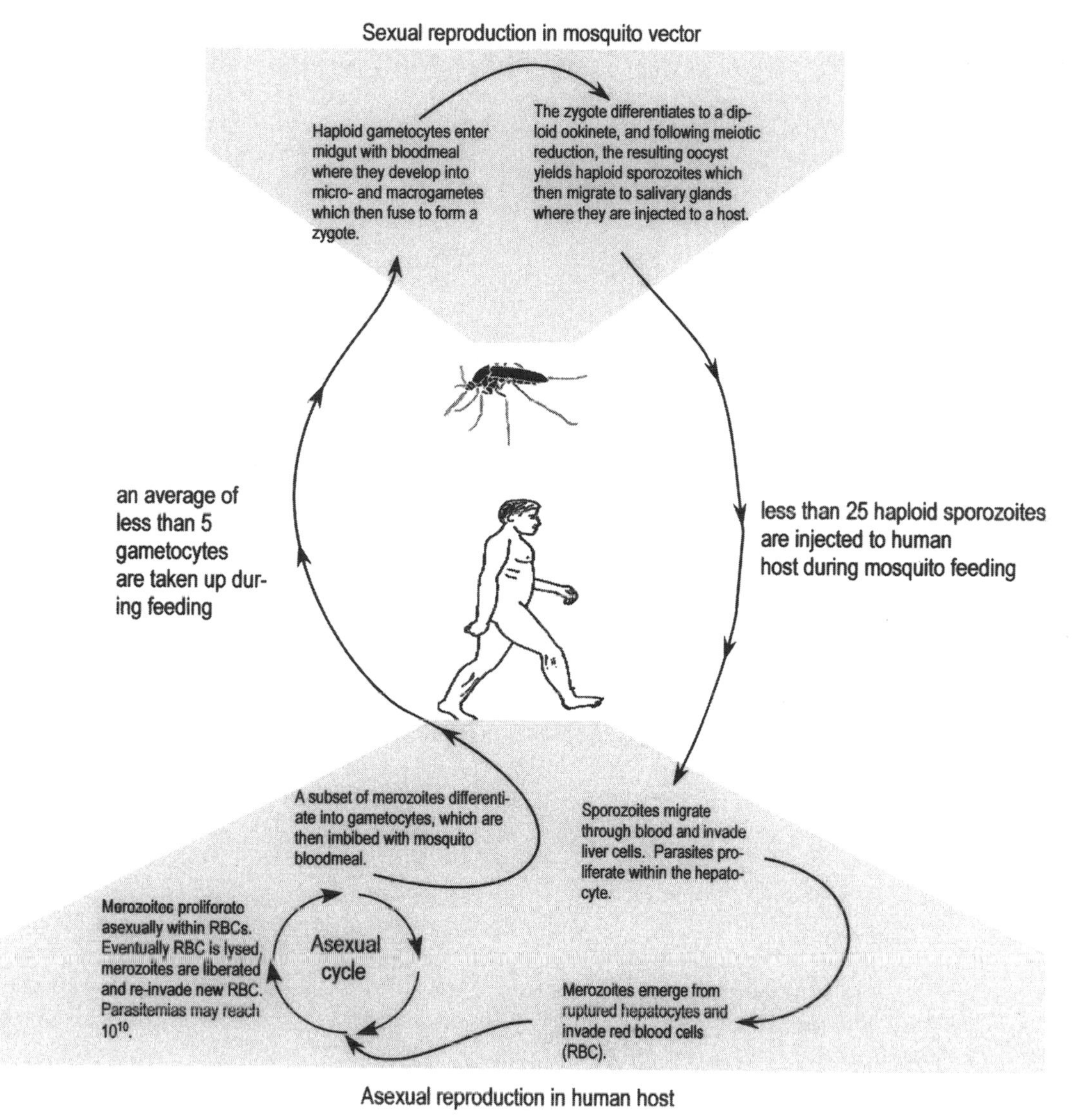

FIGURE 2. Life Cycle of *Plasmodium falciparum*.
Courtesy of author.

is expressed on the erythrocyte (red blood cell) surface, thus facilitating adherence to receptors in vasculature, but it must avoid appearing as antigen to the ever-vigilant host immune system. *P. falciparum* has evolved an extraordinary genomic innovation that falls in a general category of phenomena known as antigenic variation. The parasite maintains multiple copies of the *pfEMP* gene sequences within its genome and serially expresses these polymorphic gene sequences to avoid immune recognition. Antigenic variation is an adaptation that allows a single parasite to take different forms, as mediated by transcription of different gene sequences, in the course of its development. As a result, even a single clone of the parasite carries the requisite variability for evading host response.

However not all surface proteins of *P falciparum* exhibit antigenic variation. Examples of proteins that are likely targets for vaccine interventions include the product of the circumsporozoite gene *Csp*, which is expressed solely on sporozoites, and the merozoite surface proteins *Msp-1* and *Msp-2*, which are expressed on merozoites. Each gene has its own stage-specific function, but unlike *pfEMP*, each parasite genome has only one copy of each gene sequence. Nonetheless, among the various clones that make up a *P. falciparum* population, there is an abundance of amino acid polymorphism in the various single-copy surface proteins. In order to develop sustainable vaccines, it is essential to understand the origin of these polymorphisms and the mechanisms by which they are maintained.

Evolutionary biologists who study malaria also seek to understand the role that sex plays in maintaining diversity among *P. falciparum* genotypes. The advantages and disadvantages of sexual recombination are discussed elsewhere in this volume. Because all *Plasmodium* species have an obligatory sexual phase in their life cycle, the question is how much recombination actually occurs in natural populations. Among various parasitic protozoa and nematodes, it is known that matings in nature often occur between genetically similar, even identical, individuals. The population structure of these organisms may therefore be effectively clonal such that a few multilocus genotypes are maintained with little genetic exchange between them.

The relevance of sexual recombination for maintaining genotypic variance in populations of *P. falciparum* can be ascertained by means of molecular assays for detecting genetic variation. It is becoming increasingly clear that relative rates of sexual recombination vary locally and continuously, from highly clonal to largely outbreeding population structures. The general pattern is that outcrossing among strains is common in areas of highest malaria transmission, whereas clonality prevails where transmission rates are low. This is not surprising given the parasite's life cycle, wherein high transmission rates indicate numerous mosquito bites, and hence opportunity for genetically dissimilar gametocytes to be present and fuse within an individual mosquito.

Nevertheless, the population structure of these parasites shows that they are not solely dependent on sexual recombination for generating and maintaining new genotypes. Even among populations that show high degrees of inbreeding and linkage disequilibrium among loci, there is still a tremendous diversity of parasite genomes. This is not surprising, because at certain times in the species' history the opportunities for gametic fusion among distinct clones were limited, for example, during the demographic reduction within the last 20,000 to 60,000 years. Moreover, the ontogeny of individual parasites and passage among mosquito and human hosts involve a concomitant occurrence of bottlenecks. A feeding mosquito imbibes an average of fewer than five *P. falciparum* gametocytes per infected blood meal (Taylor and Read, 1997); even when a mosquito feeds on a human with mixed genotype infection, there is thus a high probability that all of the gametocytes taken up in the blood meal will be genetically identical. Among the gametocytes ingested, not all will differentiate into gametes, and the gametes that actually form zygotes may be fewer still. Furthermore, from a fully mature oocyst, only 25 percent of the sporozoites successfully invade the salivary glands, and only 2 percent of all infected mosquitoes transmit more than twenty-five sporozoites to their human target (Vaughan et al., 1994). Even when one considers extreme instances of entomological inoculation rate, such that a single person may receive 1,000 infectious bites per year, the number of sporozoites transmitted is vanishingly small when compared to the 10^{10} parasites that will result from asexual proliferation within a single human host.

The role of natural selection in maintaining genetic diversity in *P. falciparum* is also of interest. The effect of selection for antigenic diversity may well give the impression that recombination is frequent, whereas instead it may be selection that has preserved the result of extremely rare events. Because many of the selective pressures on the parasite are imposed by the human immune response, one might suspect that some of the variation arises in the human host where parasites proliferate strictly asexually. Because a single human infection may yield up to 10^{10} asexual blood stage parasites (McConkey et al., 1990), within-host variation can arise by mutation and selections. Moreover, the genes encoding surface proteins of the parasite have certain features that may make them highly mutable. A characteristic of nearly every *P. falciparum* single-copy surface protein gene is the presence of simple DNA repeats. These gene regions evolve at rates several orders of magnitude greater than those of genes that lack repeats. The most likely mechanism driving this rapid evolution is the slipped-strand mutation process, such as is associated with microsatellite DNA, so that the number of repeats greatly varies among individual parasites. The repeat regions within the *P. falciparum* surface proteins may therefore be an adaptive mechanism for overcoming the losses of variation arising from transmission bottlenecks and inbreeding.

At the time of this writing, the entire genome sequencing of *P. falciparum* is nearing completion. This important accomplishment follows a century of research on parasite biology. An important next step will be to understand the processes that alter the genes and genomes of these organisms, facilitating their continued interaction with their hosts, and to focus on the evolutionary potentiality that they present.

[*See also* Clonal Structure, *article on* Population Structure and Clonality of Protozoa; Disease, *article on* Infectious Disease; Heterozygote Advantage; *article on* Sickle-Cell Anemia and Thalassemia; Immune System, *article on* Microbial Countermeasures to Evade the Immune System; Plagues and Epidemics; Protists; Transmission Dynamics; Vaccination; Virulence.]

BIBLIOGRAPHY

Barta, J. R. "Phylogenetic Analysis of the Class Sporozoea (phylum Apicomplexa Levine, 1970): Evidence for the Independent Evolution of Heteroxenous Life Cycles." *Journal of Parasitology* 75 (1989): 195–206.

Conway, D. J., C. Fanello, J. M. Lloyd, B. M. Al-Joubori, A. H. Baloch, et al. "Origin of *Plasmodium falciparum* Malaria Is Traced by Mitochondrial DNA." *Molecular and Biochemical Parasitology* 111 (2000): 163–171.

Escalante, A. A., and F. J. Ayala. "Phylogeny of the Malarial Genus *Plasmodium*, Derived from rRNA Gene Sequences." *Proceedings of the National Academy of Sciences USA* 91 (1994): 11373–11377.

Escalante, A. A., and F. J. Ayala. "Evolutionary Origin of *Plasmodium* and Other Apicomplexa Based on rRNA Genes." *Proceedings of the National Academy of Sciences USA* 92 (1995): 5793–5797.

Escalante, A. A., A. A. Lal, and F. J. Ayala. "Genetic Polymorphism and Natural Selection in the Malaria Parasite *Plasmodium falciparum*." *Genetics* 149 (1998): 189–202.

Garnham, P. C. C. *Malaria Parasites and Other Haemosporidia*. Oxford, 1966.

Huff, C. G. "Studies on the Evolution of Some Disease-producing Organisms." *Quarterly Review of Biology* 13 (1938): 196–206.

Maxwell, R. "Some Evolutionary Possibilities in the History of the Malaria Parasites." *Indian Journal of Malariology* 9 (1955): 247–253.

Margulis, L., H. McKhann, and L. Olendzenski. *Illustrated Guide of Protoctista*. Boston, 1993.

McConkey, G. A., A. P. Waters, and T. F. McCutchan. "The Generation of Genetic Diversity in Malaria Parasites." *Annual Review of Microbiology* 44 (1990): 479–498.

Rich, S. M., M. U. Ferreira, and F. J. Ayala. "The Origin of Antigenic Diversity in *Plasmodium falciparum*." *Parasitology Today* 16 (2000): 390–396.

Rich, S. M., M. C. Licht, R. R. Hudson, and F. J. Ayala. "Malaria's Eve: Evidence of a Recent Population Bottleneck throughout the World Populations of *Plasmodium falciparum*." *Proceedings of the National Academy of Sciences USA* 95 (1998): 4425–4430.

Taylor, L. H., and A. F. Read. "Why So Few Transmission Stages? Reproductive Restraint by Malaria Parasites." *Parasitology Today* 13 (1997): 135–140.

Vaughan, J. A., B. H. Noden, and J. C. Beier. "Sporogonic Development of Cultured *Plasmodium falciparum* in Six Species of Laboratory-reared Anopheles Mosquitoes." *American Journal of Tropical Medicine and Hygiene* 51 (1994): 233–243.

— STEPHEN M. RICH AND FRANCISCO J. AYALA

MALE–MALE COMPETITION

Evolution arising from sexual selection may be defined as changes in population gene frequencies that result from selection on traits influencing mating success or, ultimately, fertilization success of the individual. Sexual selection is traditionally divided into competition among the members of one sex (usually males) for mating access to the other sex and mate choice. Mate choice involves an interaction between the sexes and is called intersexual selection; competition for mates involves an interaction between members of the same sex and is called intrasexual selection. Male–male competition is intrasexual selection in which males compete to enhance their mating success. However, it should be noted that the mental association between intrasexual selection and male–male competition is arbitrarily established by the convenience; normally, males are the competing sex and females the choosing sex, but sexual roles may be reversed. Likewise, the distinction between intrasexual and intersexual selection is not as clear as was initially thought.

Why is sexual selection stronger among males than females? In general, the number of offspring sired by a male increases linearly with the number of females he mates with, whereas female fecundity does not increase with multiple mating. This bias in reproductive potential between sexes was formalized by Robert Trivers in 1972. Typically, females cannot enhance their reproductive success by further copulations because (1) females are limited by their egg production and (2) there may be a long delay between mating and giving birth. However, in sex-role reversed species such as sea horses and jacana birds, males care for the offspring, enabling females to search for new mates immediately after egg laying. In such mating systems, sexual selection is stronger in females than in males. Finally, in monogamous species where the need for biparental care limits the reproductive potential of both sexes, intrasexual competition may occur in both sexes.

Intrasexual versus intersexual selection as concepts. Since the early 1980s, the major emphasis in sexual selection has focused on understanding the evolution of female choice for ornamented males that otherwise offer no tangible benefits (such as paternal care of offspring) for the females. In such cases, females are assumed to choose for indirect benefits in terms of good genes for their offspring or arbitrary sexiness of their sons. This form of intersexual selection is an intriguing theoretical and empirical problem, whereas intrasexual selection with direct benefits in terms of increased mating success is simpler to understand. However, for a comprehensive understanding of evolution through sexual selection, we must consider both. Males may compete for females in three ways: fighting for direct access to females, fighting for resources that are essential for female reproductive success (territories, food, and nest sites), and attracting females with sexual signals. The first case refers to intrasexual selection among males. The second case is intrasexual selection among males, but intersexual selection in the form of female choice for direct benefits is also involved. The third case represents intersexual selection, where female choice for indirect benefits plays a role. Several of the mechanisms may operate simultaneously, and to understand the evolution through sexual selection, all of them must be considered.

Forms of Male–Male Competition. Competition

occurs whenever the use of a resource by one individual makes the resource harder to come by for others. Thus, to have competition, there is no need for actual contact between the contestants. Two categories of competition in the context of sexual selection are contest competition and scramble competition. There may also be competition to be chosen as a mate. However, this is equivalent to intersexual selection, and was discussed earlier.

Contest competition. For competition to be classified as contest, some form of direct signaling or physical contact is necessary. Physical fights of horned beetles and ungulate mammals are examples of contest competition. However, contest competition may also be ritualized without physical contact in which weapons or other traits are used in assessing the fighting ability or resource-holding potential of the contestants. Contest competition is similar to interference competition that is used in ecological literature to define aggressive physical struggles over the resource. Contest competition may result in territoriality, males holding female harems, or males defending resources essential for reproduction of females.

Scramble competition. In scramble competition, rapid location and exploitation of the resource is important. Early emergence or maturation in order to gain an edge on competitors serves as an example of scramble competition. Scramble over early emergence may lead to evolution of protandry, in which males become sexually mature before females. The probability of mating may also increase with endurance or persistence, in which case it is called endurance rivalry. Scramble competition is related to the ecological term *exploitation competition*, in which rivals exploit the same resource but do not necessarily meet each other. An extreme example of scramble competition is reproduction in sessile marine organisms, such as coral, in which gametes are released in the water and fertilization happens by scramble.

Male–male competition as a female strategy. Because females are the limiting resource, there are more opportunities for females to manipulate males than vice versa. Therefore, it may be female strategy to make males fight for them. In this scenario, females manipulate males to sort out the ranks among each other, then mate with the winner, having paid few costs themselves to be choosy. For example, in three-spined stickleback fish, male–male competition reinforces male dominance by increasing the difference in their expression of red coloration. Increased difference in coloration in turn facilitates female choice of dominant males. In lekking black grouse, males fight for central territories, and females choose dominant males using centrality of the territory as a cue. However, if territories are unstable, as in leks on ice without landmarks, dominant males will encircle females, leaving little opportunity for female choice. In all cases, females end up mating with dominant males, which may be the best strategy to ensure the most viable sire for the offspring. However, for researchers, it is troublesome to separate male–male competition from sophisticated female choice, in particular, because they may both cause similar evolution in male traits.

Sexual conflict as a result of competition. Reproductive interests of males and females may not be mutual, resulting in sexual conflict. For example, if the most dominant males have the best territories for female reproduction but are not providing parental care, because they are able to become polygynous, a conflict between the sexes will result. In many insects, males harass females continuously with mating attempts, and there may arise evolutionary arms races for male and female reproductive characters. For the male, it is almost invariably disadvantageous if the female with whom he mated remates with a rival male. Thus, males that are able to inflict a cost of mating on females and thus discourage remating are favored by selection. Sexual conflict over reproduction and mating appears common and deserves further consideration as a possible outcome from intrasexual competition.

Theoretical Treatments of Male–Male Competition. In the early 1970s, John Maynard Smith and Geoff Parker, theoreticians of evolution, introduced the theory of games into evolutionary biology. Ever since, game theory has set the framework in which many biological interactions between individuals have been modeled, and male–male competition makes no exception. According to game theoretical models, the assessment of relative fighting ability and motivation to fight are vital for settling disputes. In the most advanced game, the sequential assessment game, a fight consists of a series of costly assessment rounds, and a dispute is settled when one opponent has gained reliable information about the other opponent's superior fighting ability and withdraws. Game theoretical models help us to understand the decision making in animal conflicts.

Evolution of Weaponry. We easily recognize as weapons traits that we feel threatened by ourselves. These traits include antlers, horns, tusks, teeth, spurs, and claws. However, there may be other kinds of weapons that we have difficulty perceiving as such. Weaponry may have naturally selected advantages, such as defense against predators, but here we concentrate on weapons that may have evolved through sexual selection in the context of male–male competition.

If a larger trait bestow a benefit in male–male competition, directional sexual selection for larger traits will result. Directional selection for larger size in turn will result in evolution of larger traits, until the costs of growing or having an even larger trait will balance the benefit and an optimal trait size will evolve. The evolutionary maintenance of weapons is easier to understand than the maintenance of ornamental traits. This is because weaponry bestow benefits in terms of increased mating success directly to the male carrying the weapon. In contrast, ornamental traits bestow benefits through female preference, and female preference for the ornamental traits is often relying on indirect benefits.

Some weapons such as very complex antlers seem unnecessarily elaborate and are often less than perfect for fighting. In these instances, the evolution of the complexity may have risen as a response to some other function of the antlers. There are two likely candidates. Complex and elaborate weapons may have a dual function and be used in male–male contests but also in female choice. If fighting becomes highly costly, such that males are selected to avoid physical contact, then selection on the function of the traits as weapons may be relaxed. As a result, female choice may select weapons to be less perfect for fighting but better for revealing male quality. Another possibility is that weapons are used as shields in defending the male against the attacks of rivals. Weapons that are complex may not be optimal for attacking and harming the rivals, but the loss as attack weapons may be compensated by the better function as shields.

Fighting without weaponry. Not all male–male competition and fighting involve weapons. In particular, badges of status (color patterns, tufts, etc.), vocal signaling, or chemical scent marking may be used in ritualized assessment of the fighting ability. These signals are often used in male–male competition of territories. Territoriality can be understood in game-theoretical contexts, and it is possible that ownership is used as an arbitrary convention in addition to reflecting resource-holding potential of males.

Evolution of Body Size Dimorphism. Body size often differs between males and females. There are several causes for such dimorphism, such as fecundity benefit of larger females or foraging competition between the sexes, but one major factor is male–male competition. There is abundant evidence that larger male size gives an advantage in disputes over territories, in dominance contests, and in fights over females. In these cases, sexual selection leads to evolution of size dimorphism between sexes. However, we must be careful when we interpret the evolution of body size dimorphisms from the current state of affairs. The traditional view is that males evolve larger body size because of the advantages that size bestows in male–male competition. Nevertheless, an alternative that has received much less attention exists: it may be that females evolve smaller body size. An example can be found in species in which the body size of both sexes is dependent on the length of time they have to grow, and this time in turn is related to the age at first breeding. Consider that a later age at first breeding reduces female fitness but has less effect on males. Then, selection acting on the age at first breeding may favor females that are smaller than males. In such cases, females may evolve to be smaller despite the fact that male–male competition may favor larger males.

Alternative Mating Strategies. In addition to size dimorphism between the sexes, dimorphism occurs within one sex. As in everyday life, evolutionary problems may be overcome in several ways. Where the problem is one of reproductive success, male–male competition over females may lead to the evolution of alternative male mating strategies through differences in morphology, behavior, or, as is often the case, both. The interesting feature of male dimorphisms is that they represent alternative selective regimes within a sex and thus serve as a powerful tool in testing the theory of evolution through adaptation.

Sperm Competition. Sexual selection, in particular, male–male competition, is normally understood as competition for access to mates. However, it is very common for females to mate with more than one mate (polyandry) within one reproductive cycle. Once this is realized, it becomes clear that simple competition for access to mates may not be an adequate determinant of competitors' reproductive success. This is because in those cases where females are polyandrous and ejaculates from multiple mates overlap in time in the female reproductive tract, there is a possibility for further competition mechanism to take place: sperm competition. Geoff Parker, the founder and developer of the modern sperm competition theory, defines sperm competition as "competition between the sperm from two or more males for the fertilization of a given set of ova." This definition includes the external fertilizers as found in many fish and frogs. Sperm competition is now recognized as a powerful selective force in evolution, which may lead to a variety of adaptations at the level of behavior, physiology, or morphology.

Infanticide. Infanticide or even induced abortion may be seen as a form of male–male competition in cases where the infanticidal male has fitness benefits from the act of killing, evicting, or inducing the abortion of the young. To gain these fitness benefits, a few requirements must be met. First, the infant that is being killed must not be related to the infanticidal male. Second, the period of time elapsing from the death of the infant to the time when the mother of the infant is receptive and ready to reproduce again must be shorter after the infant has been killed than it would be otherwise. Finally, owing to the death of the previous young, infanticidal male must have an increased probability of mating with the mother and siring her forthcoming offspring. When these requirements are fulfilled, infanticide is adaptive for males and may evolve through sexual selection.

Male–Male Competition in Plants. Whether we like it or not, the view of the behavioral ecologists is often biased because of our animal perspective. Perhaps for this reason, plants have received much less attention than animals among the students of sexual selection. However, plants behave and compete for mates. Male–male competition in plants is most likely to take a form of scramble competition. In animal-pollinated plants, males may compete for the attraction of pollinators with conspicuous flowers that signal the nectar reward the plant has to offer. Thus, male–male competition for pollinators may play a role in the evolution of florescences. Male–male competition may also occur after the pollen arrives in the stigma of the female flower. Here it is a scramble over reaching the ovules first, and selection may favor pollen that has rapid germination and pollen tube growth. Traditionally, the evolution of conspicuous florescences has been addressed to naturally selected factors such as advantages of cross-fertilization and avoidance of inbreeding, but it is becoming increasingly obvious that sexual selection may have a pronounced role in the evolution of plant mating systems.

[*See also* Cryptic Female Choice; Mate Choice, *article on* Human Mate Choice; Sexual Dimorphism; Sperm Competition.]

BIBLIOGRAPHY

Ahnesjö, I., C. Kvarnemo, and S. Merilaita. "Using Potential Reproductive Rates to Predict Mating Competition among Individuals Qualified to Mate." *Behavioral Ecology* 12 (2001): 397–401.

Andersson, M. *Sexual Selection.* Princeton, 1994.

Berglund, A., A. Bisazza, and A. Pilastro. "Armaments and Ornaments: An Evolutionary Explanation of Traits of Dual Utility." *Biological Journal of the Linnean Society* 58 (1996): 385–399.

Borries, C., K. Launhardt, C. Epplen, J. T. Epplen, and P. Winkler. "DNA Analyses Support the Hypothesis That Infanticide Is Adaptive in Langur Monkeys." *Proceedings of the Royal Society of London B* 266 (1999): 901–904.

Candolin, U. "Male–Male Competition Facilitates Female Choice in Sticklebacks." *Proceedings of the Royal Society of London B* 266 (1999): 785–789.

Enquist, M., O. Leimar, T. Ljungberg, Y. Mallner, and N. Segerdahl. "A Test of the Sequential Assessment Game: Fighting in the Cichlid Fish *Nannacara anomola.*" *Animal Behavior* 40 (1990): 1–14.

Gross, M. R. "Alternative Reproductive Tactics: Diversity within Sexes." *Trends in Ecology and Evolution* 11 (1996): 92–98.

Höglund, J., and R. V. Alatalo. *Leks.* Princeton, 1995.

Kemp, D. J., and C. Wiklund. "Fighting without Weaponry: A Review of Male–Male Contest Competition in Butterflies." *Behavioral Ecology and Sociobiology* 49 (2001): 429–442.

Kotiaho, J. S., R. V. Alatalo, J. Mappes, and S. Parri. "Honesty of Aonistic Signalling and Effects of Size and Motivation Asymmetry in Contests." *Acta Ethologica* 2 (1999): 3–21.

Maynard Smith, J. *Evolution and the Theory of Games.* Cambridge, 1982.

Parker, G. A. "Sperm Competition and its Evolutionary Consequences in the Insects." *Biological Reviews* 45 (1970): 525–567.

Qvarnström, A., and E. Forsgren. "Should Females Prefer Dominant Males." *Trends in Ecology and Evolution* 13 (1998): 498–501.

Trivers, R. L. "Parental Investment and Sexual Selection." In *Sexual Selection and the Descent of Man*, edited by B. Campbell. Chicago, 1972.

— JANNE S. KOTIAHO AND RAUNO V. ALATALO

MAMMALS

Modern mammals are characterized by a number of attributes: a four-chambered heart, a diaphragm separating the thoracic and abdominal cavities, the ability to raise the body temperature by internal metabolic processes (endothermy), the appearance of hair or bristles at some stage of their development, a dentition that if present is usually divisible into functional groups, a replacement of the first dentition by a set of permanent teeth, and the early nourishment of the young provided by secretions from the mammary glands of the mother.

The pathway toward the evolution of these attributes is based on a fossil record extending over 200 million years, and almost all evidence is preserved as teeth, jaws, skull fragments, and partial skeletons. Thus, the evolutionary history of the inner ear, tooth structure, and the articulation of the jaw with the skull dominate the discussion of early mammal evolution.

The first modern effort to classify the fossil and living mammals was attempted by George Gaylord Simpson (1945). He listed 2,864 genera, of which 1,932, or 67 percent, were extinct. Discoveries in the past 50 years have raised the number of genera of living and extinct genera above 5,000 (McKenna and Bell, 1997). Of these, about 1,135 are living and contain 4,692 species considered recently extant (Wilson and Reeder, 1993). Thus, the history of the class Mammalia includes major extinction events and periods of intense adaptive radiation. A major survival event for mammals occurred 65 million years ago at the end of the Cretaceous, when the dinosaurs and many other taxa went extinct, but the major taxa of mammals survived (Archibald, 1982). The surviving mammalian lineages gave rise to the Age of Mammals, which includes the Cenozoic epochs. Before this time of dominance lies a 148-million year history of mammals; these forms are referred to as the Mesozoic mammals (Lillegraven et al., 1979).

Mammals evolved from mammal-like reptiles, the Therapsida, which flourished from 250 to 180 million years ago. One derivative lineage, the Moganucodontids, began to evolve two of the most distinctive features of the mammals, the single bone in each half of the lower jaw that articulates directly with the skull, and the three articulated bones of the inner ear—the malleus, incus, and stapes (Jenkins and Crompton, 1979). The transitional forms leading to the Mammalia are becoming better known (Hotton et al., 1986), but it must be appreciated that early mammals were small—about the size of a mouse or shrew. The recovery of a complete skull or postcranial skeleton is a rare event.

The therapsids that gave rise to the mammals were the Cynodontia. This group exhibited the beginnings of tooth modifications so that functional groups of teeth such as incisors, canines, and postcanine teeth were discernible. Tooth specializations continued in the mammalian lineages so that, by the Jurassic, some 150 million years ago, the molar teeth had evolved a combination of cusp and plane surfaces to increase the efficiency of mastication. A final derivative, the tribosphenic molar, became the one defining feature of the stocks that led to modern mammals. In the Jurassic, three major lines of mammals can be distinguished: the monotremes, the multituberculates, and the so-called therians. It is postulated that these taxa exhibited the following characters: hair, young nourished by secretions from mammary glands, adults feeding principally on invertebrates, and the capacity to raise their body temperatures by their own body heat, thus being partially independent of solar heat. These inferences are based not only on certain anatomical features but also on the possesion of all the characters by the living descendants of the monotremes and the early therians.

The living mammals exhibit three modes of reproduction. The monotremes (platypus and echidna) reproduce by laying eggs. The marsupials, or metatherians, bear live young, but the major nourishment of the intrauterine fetus is by means of a yolk sac placenta, and the lactation phase is prolonged, with milk sustaining the phase of extra-uterine fetal development. The reproductive tracts of the marsupials are distinctly different from the rest of the living mammals (see below). The remainder of the living mammals (the Placentalia) nourish the young during the uterine phase by means of a chorio-allantoic placenta, and the lactation phase may be of variable length. There is tantalizing anatomical evidence in the pelvic structure of fossil multituberculates suggesting that they laid eggs (Kielan-Jaworowska and Gambaryn, 1994). The multituberculates survived the extinction events at the end of the Cretaceous together with other mammalian stocks, but they went extinct in the Oligocene.

The Cretaceous was a time when the mammals were small but diverse in habits and ecology. Some were arboreal, others terrestrial. The multituberculates may have been feeding on fruits and seeds. One hundred million years ago the marsupials were already a separate lineage from the Placentalia. This was also a time when the breakup of the world continent of Pangaea was in full development. Gondwanaland (Africa, Madagascar, India, South America, Antarctica, and Australia) was separating into East Gondwanaland (Africa) and West Gondwanaland (South America, Australia and Antarctica). This breakdown of connections may have had an early influence on the faunal compositions of the continents, but the major radiations of the mammalian stocks occurred after the Cretaceous, and subsequent geological events had more important consequences for the modern distribution of the mammals. These events include very low sea levels during the Oligocene, reestablishment of connections between Africa and Asia during the Miocene, the completion of the Panamanian land bridge connecting North America and South America during the Pliocene, and the intermittent connections between North America and Asia throughout the Cenozoic.

Molecular studies of mammalian relationships have been very controversial, but two recent findings, based on very large nuclear protein-coding datasets deserve mention. First, although whole mitochondrial DNA sequences supported the surprising sister group relationship of monotremes and marsupials, the latest nuclear gene data (approximately 100,000 million years BP [before present]) strongly support the traditional view that the egg-laying monotremes branched off before the divergence of marsupials and eutherians (placental mammals). Two recent studies of the controversial relationships within the eutherian mammals, both based on independent large datasets, agree almost perfectly on the major relationships of eutherian mammals (Murphy et al., 2001; Madsen et al., 2001). In particular, these studies agree that the "Afrotheria," composed of elephants, aardvarks, manatees, elephant shrews, anteaters, and sloths, are the sister taxon to the remaining eutherians, and that primates fall in a clade that includes rodents.

Sixty-five million years ago, the early mammalian lineages that had survived the Cretaceous extinctions began a dramatic adaptive radiation. In the course of 10 million years, all orders of mammals that are alive today, as well as numerous extinct orders, became defined. This has led to much research to determine the relationships among the orders. Fossil evidence and the use of the techniques of molecular evolution have led to interesting hypotheses. Liu and Miyamoto (1999) have compared the morphological and molecular approaches and find a fundamental conformity, but the techniques of DNA sequencing have also raised important questions. The assertion that a fundamental relationship exists among the old, endemic orders of African mammals that reflect isolation events surrounding the formation of East Gondwanaland have rocked the field of paleontology (Waddel et al., 1999; Springer et al., 1997).

The recent mammals have been surveyed in a number of books. Eisenberg (1981) reviews the trends in adaptation displayed by the living species. Nowak (1999) surveys the biology of each genus, with lists of the species. Wilson and Reeder (1993) present an exhaustive taxo-

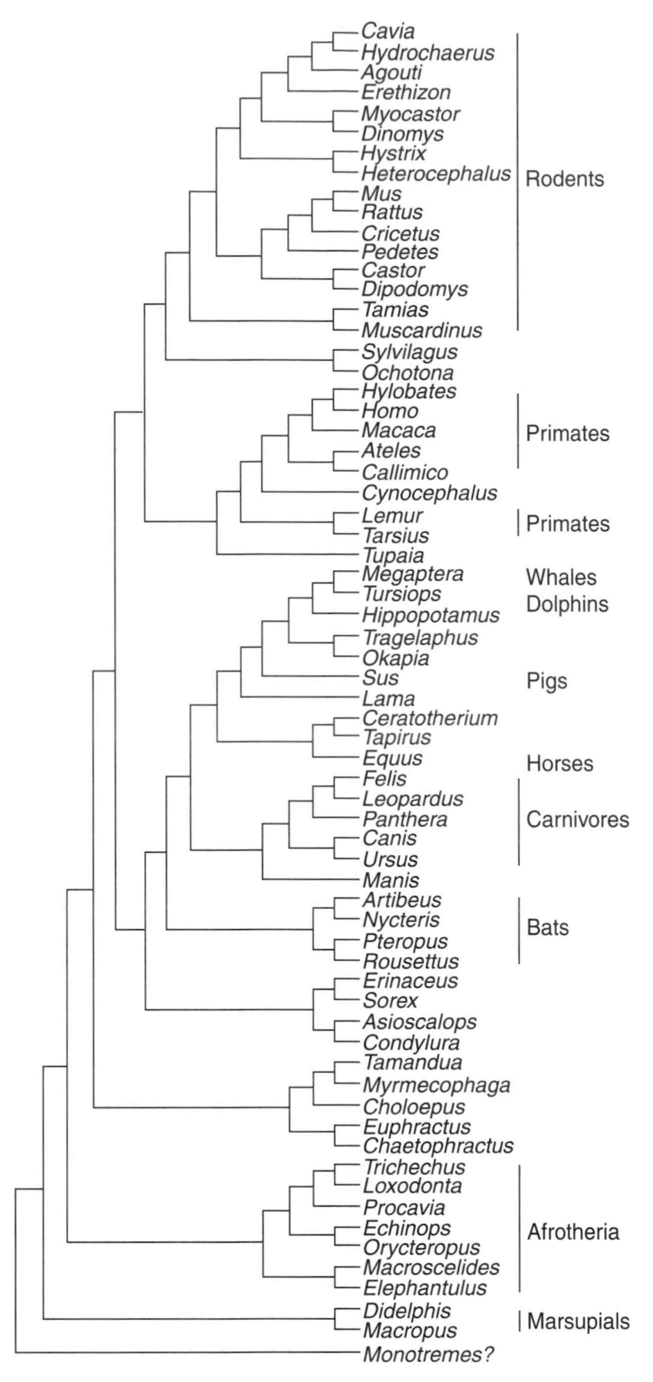

FIGURE 1. Phylogenetic Tree of the Mammals Derived from Gene Sequence Data.
After Murphy et al., 2001.

nomic and geographical reference at the species level. The remainder of this article will discuss briefly some aspects of the biology of the living orders.

Monotremes. The Monotremata have a sparse fossil record from the Early Cretaceous of Australia, the Pliocene of South America, and the Miocene of Australia. The recent forms are found only in Australia, New Guinea, and Tasmania. They include the spiny anteaters or echidnas and the platypus. The females reproduce by laying rather small eggs, and the young upon hatching are nourished by milk. The echidnas are specialized for feeding on ants and termites with long sticky tongues and large foreclaws for digging. The platypus is aquatic, with the skull specialized into a beaklike structure, and the fore and hind feet with webbed toes. They feed on aquatic invertebrates and nest in burrows constructed on riverbanks. Since the monotremes are survivors carrying into the present a rather conservative body plan, they have been the subject of great scientific interest since their discovery by Europeans in the eighteenth century (Griffiths, 1978).

Marsupials. The Metatheria or Marsupialia have traditionally been classified as a single order, but the antiquity of the group and its considerable adaptive radiation have led workers to subdivide them into several orders. The earliest fossil marsupials appear in the Cretaceous of North and South America. Although Cretaceous therians are known from Australia (Rich et al., 1997), it is not yet possible unequivocally to assign the skeletal fragments to the marsupials. Marsupial fossils have also been found in the Eocene of peninsular Antarctica, Australia, Europe, North Africa, and Asia. Since the Oligocene, the major centers of marsupial evolution have been confined to the continents where they currently exhibit their greatest diversity: Australia and South America. At present, more than 270 species of marsupials have been described that were recently extant. Seventy of these inhabit Central and South America, with one species (the opossum) occurring north of Mexico in the United States and Canada.

The marsupials exhibit unique features in the anatomy of the reproductive tract. The female has three vaginas, two laterally placed and the median vagina open only at the time of birth. The male has a bifurcate penis and a prepenial positioning of the scrotum. The prominent role of the yolk sac placenta during the uterine phase of embryonic development, coupled with the prolonged lactation and obligate teat attachment of the young to the nipple of the female during early lactation are unique to the marsupials. In some female marsupials, the teat area is enclosed by folds of skin, often termed a pouch (Tyndale-Biscoe and Renfree, 1987).

The living New World marsupials are grouped into three orders, of which the Didelphimorphia or opossums are the most diverse, including about 70 species. The marsupials of the New World exhibit a variety of lifestyles, including an aquatic form, the yapok; arboreal forms, the wooly opossums; and terrestrial forms such as the short-tailed opossums. All living New World marsupials are omnivores, insectivores, or frugivores. Some extinct marsupials of South America were carnivorous, and some are believed to have been plant feeders. Much of the marsupial diversity in South America declined during the Pliocene.

The surviving marsupials of Australia and New Guinea are grouped into four orders and represent a full spectrum of adaptation that parallels the extant placental radiations of the mammals on the more contiguous continental land masses. The Diprotondtia include herbivorous marsupials such as kangaroos, wombats, the koala, phalangers, honey possums, sugar gliders, and other well-known forms. Both terrestrial and arboreal specializations are well represented. The Notoryctemorphia includes only two surviving species, the curious marsupial moles. The Peramelemorphia are represented by the bandicoots, terrestrial insectivorous forms. The Dasyuromorphia, with 63 species, include insectivorous shrewlike species (*Planigale*), small carnivores (*Dasyurus*), a recently extinct doglike form (*Thylacinus*), a specialized ant and termite feeder (*Myrmecobius*), and the scavenging Tasmanian devil (*Sarcophilus*). It is clear that the marsupials, in relative isolation on the island continent of Australia, accomplished an adaptive radiation that, when studied in detail, can offer insights into the forces of natural selection that molded the radiations of the Placentalia.

Placental Mammals. The remaining orders of living mammals are referred to as the Eutheria, or placentals. The relationships among their orders is an active area of research. A good introduction to some of the problems and recent advances is given in Szalay et al. (1993).

Sloths, anteaters, and armadillos. The order Xenarthra includes the living tree sloths, New World anteaters, and armadillos. Fossil forms include the ground sloths and glyptodonts. This is a group of archaic mammals that first appeared in the Paleocene of South America. A movement into North America occurred in the late Miocene when ground sloths arrived, followed by armadillos. Ground sloths appear in the West Indies during the Miocene. Xenarthrans or edentates represent such an ancient radiation that the living forms at the family level seem to bear little resemblance to one another, yet certain features of the skeleton and dentition tie them into a single taxon. The armadillos, with their dorsal surfaces covered with a leathery skin embedded with dermal bone, present a distinctive appearance. There are 20 living species of armadillos, one of which reaches as far north as Oklahoma in the United States. Some species of armadillo are omnivorous, but others are specialized for feeding on ants and termites. The tree sloths are specialized for browsing on leaves in tropical forests. The anteaters, as the name implies, are specialized for feeding on ants and termites.

Insectivores. The order Insectivora (Lipotyphla) includes seven recent families. There is doubt that this grouping reflects a true, natural taxon. The Solenodontidae and the recently extinct Nesophontidae were confined to the larger islands of the West Indies. The solenodons remain of great interest because they are true living fossils, surviving on Hispaniola and Cuba. The Chrysochloridae (golden moles) and tenrecoid insectivores (Tenrecidae) are African, with most of the Tenrecidae confined to the island of Madagascar. These two families are thought to be the last survivors of an early placental group confined to East Gondwanaland. The remaining families include the northern hemisphere moles (Talpidae), the African and holarctic shrews (Soricidae), and the African and Eurasian hedgehogs (Erinaceidae). All these families carry into the present a suite of conservative dental and skeletal features, as well as other specializations associated with feeding on small vertebrates and soil invertebrates. Most species are terrestrial; only a few are semiaquatic or arboreal.

There are 428 species of recent insectivores, 312 of which belong to the family Soricidae (shrews). This family first appears in the Eocene and throughout its history has exemplified a specialization for small size and occupancy of the terrestrial, invertebrate feeding niche within the northern continents (Wojcik and Wolson, 1998). The family invaded Africa in the Miocene and northern South America in the Pliocene. They never entered Australia. The moles appear in the Mid-Eocene and have maintained a distribution in North America and northern Asia. The hedgehogs have a long history of occupancy in Europe and Asia, with an appearance in North America during the Oligocene and persistence until the Pliocene. The hedgehogs entered Africa in the Miocene and have remained.

The squirrel-sized tree shrews, order Scandentia, represent a small taxon with only 19 living species confined to South and Southeast Asia, including Sumatra, Borneo, Java, and Mindanao. The pentailed tree shrew is nocturnal, but all other species are diurnal or crepuscular. All species have climbing ability, but many forage mainly on the ground or in low shrubs. Fruits and invertebrates form the bulk of their diet. The fossil history is confined to Asia. Certain morphological resemblances to early fossil primates focused much attention on this group as a possible source of information on the behavior and ecology of early primates (Emmons, 2000). The Dermoptera is an order with an origin in the Paleocene, with fossil representation in North America and subsequently in the Eocene of Asia. At present there are only two living species, found in Southeast Asia including Borneo and Mindanao. The living forms are cat-sized and possess a gliding membrane between the fore and hind limbs. They are arboreal and feed on a variety of leaves and fruit in the forest canopy.

Bats. The Chiroptera (bats) are distinctive in the modification of the hands to support a membrane attached to the body to form a wing. They are the only mammals to have evolved true flight. Early fossil bats show complete adaptation for flight in the Paleocene and Eocene of North America, Europe, and Asia. The earlier forms of bats remain unrecognized in the fossil record. At present, bats include over 930 species and comprise 20 percent of the species richness of the mammals. The bats are divided into the Old World fruit bats (Megachiroptera: Pteropodidae) and the 16 other families grouped in the suborder Microchiroptera. This division reflects genuine anatomical and physiological differences such as the ability to echolocate by producing high-frequency vocalizations, an attribute of the Microchiroptera. The Megachiroptera include around 167 species confined to the Old World tropics. Some are of considerable size, weighing up to 800 grams. They feed primarily on fruit and flowers and forage at night relying on excellent night vision, although some species use tongue clicks for crude echolocation.

The Microchiroptera are relatively small, most species weigh less than 20 grams. They are distributed over all the continents in suitable habitats. The range of dietary adaptation is remarkable: fish-eating, insectivory, frugivory, carnivory, nectar-feeding, and feeding on the blood of living vertebrates are all part of the spectrum displayed by this group. The role of bats in the control of insect populations is enormous. In the tropical forests, bats may be primarily dispersers of fruit seeds and have a significant role in pollination (Findley, 1993).

Primates. The order Primates includes about 233 species. Primates are currently found on all continents, if we include the distribution of humans. Nonhuman primates are currently found in the tropical and subtropical regions of all continents except Australia and Antarctica. Primate origins trace to the Paleocene of North America, but by the Eocene primates are distributed in Europe, Asia, and North Africa. Primates first appear in the Oligocene in South America. The most conservative living primates belong to the Strepsirrhini and include the galagos, lorises, and lemurs. Most are nocturnal, arboreal omnivores and frugivores ranging in size from the mouse lemur at 100 grams to the indri of Madagascar, which weighs up to 7 kilograms. Galagos and lorises are tropical African and Asian in distribution, but the lemurs and their relatives are currently confined to Madagascar, where more than 30 living species are found. The tarsiers are a relic group of five species currently confined to the larger islands of Southeast Asia.

The ceboid primates include the marmosets and tamarins as well as more familiar New World monkeys such as the howler, spider, and capuchin monkeys. There are more than 80 species ranging in the forests from southern Mexico to northern Argentina. The Old World monkeys belong to the family Cercopithecidae, including more than 80 species such as the macaques, baboons, guenons, and the leaf-eating colobus and langur monkeys. The remaining living primates are grouped into two families: the Hylobatidae or gibbons, and the Hominidae, which includes the three great apes (chimpanzees, gorillas, and orangutans) as well as the human species *Homo sapiens*.

Carnivores. The order Carnivora appears in the Paleocene and by the Eocene is represented in North America, Europe, and Asia. Early on the lineage divides into the Feliformia and the Caniformia. The feliforms are represented today by the mongooses, cats, civets, and hyenas. The caniforms include the weasels and allies, raccoons and allies, bears, canids, and an early offshoot, the Pinnipedia (seals, sea lions, elephant seals, and walruses).

The early history of the carnivores takes place in North America, Europe, and Asia. In Africa the carnivores appear in the Late Oligocene and Miocene. Carnivores are absent in South America until the Late Miocene. Until humans transported the dog to Australia, its only carnivorous mammals were marsupials. Today the terrestrial carnivores include about 236 species; the aquatic pinnipeds include 34 species. As the ordinal name implies, this taxon has adapted to a carnivorous way of life, but this is misleading. The weasels, felines, some canines, and the fish-eating pinnipedia are carnivorous, but the giant panda is almost entirely herbivorous.

Scaly anteaters. The Pholidota includes the scaly anteaters or pangolins. It is a small order with one family and seven living species, found in tropical Africa and South Asia. The group has its origin in the Paleocene of North America and Europe, spreading to Asia and later in the Late Miocene to Africa. As the common name implies, the dorsal surface is covered with overlapping epidermal scales. The anatomy—long sticky tongue, long claws, and loss of the teeth—is modified for feeding on ants and termites. The pangolins represent a remarkable example of convergence in form and function when compared with the Xenarthran anteaters of the New World.

Elephant shrews. The elephant shrews are included in the order Macroscelidea. There are only 15 surviving species. They occur as fossils in the Late Eocene of Africa and remain as African endemics. The living species range in weight from 30 to 700 grams and have long snouts. They are insectivorous and terrestrial. Some species are diurnal. Their distinctive morphology and habits have fostered some interesting studies on their relationship to other mammalian groups (Rathbun, 1979). Many consider them to be related to the other ancient mammals of Africa to be discussed below.

Aardvarks, elephants, hyraxes, and sea cows. The Tubulidentata (aardvarks), Proboscidea (elephants, mastodonts and allies), Hyracoidea (hyraxes), and Sirenia (sea cows) all had their origins in northern Africa. The hyraxes and aardvarks have remained for the most part examples of African endemism. The aquatic sea cows (manatees and dugongs) have spread to many continental coastal areas. The Proboscidea moved widely and at one time occupied all the continents with the exception of Antarctica and Australia.

The aardvarks are large (40–80 kg), long-snouted mammals adapted for digging and feeding on ants and termites. There is only one living species inhabiting much of tropical and subtropical Africa. The six species of hyraxes are small and tailless, with blunt toes that are hooflike. Weighing less than 6 kilograms they are semi-arboreal and terrestrial browsers. Although mainly African they extend to the Middle East. The four species of manatees and dugongs are fully aquatic, with the body streamlined and the hind limbs lost. They are herbivores, feeding for the most part on aquatic plants. The proboscideans have a very complete fossil record, and their spread from Africa to the other continents is well documented. Today there are only two living species—the African and Asian elephants—although a recent proposal suggests that the forest elephants of west Africa are a third distinct species. Their large size and possession of a mobile trunk have made elephants universally recognizable (Shoshani and Tassy, 1996).

Odd-toed ungulates. The Perissodactyla or odd-toed ungulates are characterized in part by having hooflike toes, but the major weight-bearing axis is through the middle toe. Loss of lateral toes as they evolved digitigrade locomotion is a common theme in the evolution of the group. The group was once very diverse, but today there are only three living families: the Tapiridae (tapirs) with 4 species, the Rhinocerotidae (rhinoceroses) with 5 species, and the Equidae (horses, zebras, and asses) with 9 species. Rhinoceroses, horses, and tapirs appear in the Eocene of North America and Asia and subsequently in the Miocene of Africa. Migrations by tapirs and equines to South America did not occur until the Pliocene. The living perissodactyls have all adapted to a cursorial (running) form of locomotion and are specialized for feeding on plants. The perissodactyls, together with the sirenians, proboscideans, and hyraxes, represent in their morphology the mechanism of hindgut fermentation, in which plant structural carbohydrate is broken down by bacterial symbionts in the large intestine and a blind pouch of the hind gut, the caecum. This trait is shared by some herbivorous marsupials and rodents. The Artiodactyla evolved foregut fermentation in a capacious stomach as did some marsupials, the leaf-eating primates, and the tree sloths, but in the ruminant artiodacyls, eructation and remastication together with a chambered stomach raise digestive efficiency to a new level.

Extinctions have plagued the perissodactyl lineages; the equines of North and South America disappeared in the Pleistocene (MacFaden, 1992). At present, most wild equines and tapirs are threatened and the rhinoceros species hover on the verge of extinction.

Whales, dolphins, and porpoises. The Cetacea (whales, dolphins, and porpoises) include 78 living species distributed worldwide in the oceans and some of the great river systems. These are fully aquatic mammals and carry out their entire life cycle in the water. With a streamlined body shape and loss of the hind limbs, they have modified the basic body plan of mammals in their adaptation for aquatic life. The fossil record for Asian and African waters demonstrates a sequence of limb loss leading to the modern forms and commencing in the Eocene (Gingerich et al., 1983; Thewissen and Fish, 1997). The cetaceans exhibit two major living taxa: the Mysticeti or baleen whales have lost their teeth and have epidermally derived horny plates in the roof of the mouth that function as filters for collecting swarms of aquatic crustaceans which serve as their primary food source; the Odontoceti (toothed whales, porpoises, and dolphins) retain the teeth and are adapted for feeding on squid and fish, or in the case of killer whales, other aquatic mammals. Odontocetes have also evolved echolocation for finding their prey in the dark or turbid waters they feed in. Research on cetaceans and their behavior has generated an intense effort because of their complex communication systems and rich social behavior. Although early research on the protein chemistry of cetaceans suggested a link with the Artiodactyla, recent research clarifies the relationship (Gatesy and Arctander, 2000).

Even-toed ungulates. The Artiodactyla or even-toed ungulates are classically defined by their limb structure. In the living forms, the weight-bearing axis lies between the third and fourth toes, and the astragulus of the ankle bones has a grooved joint at either end. It has long been recognized that the taxon has a fundamental division between the Suiformes (swine, peccaries, and hippopotami) and the remaining artiodactyls (camels, deer, antelopes, and bovines). The whales and hippopotami seem to have shared a common ancestor 65 million years ago (Gatesy et al., 1999; Nikaida et al., 1999).

The suiform artiodactyls (pigs and allies) include 23 living species distributed on all continents with the exception of Australia, where they were introduced. In common with the remaining artiodactyls, their origin appears to have been in Asia and Europe, with subsequent dispersals. The camels (Camelidae) and their allies have had a long history since the Oligocene in North America, with exchanges into Asia and a Pliocene passage to South America. The Tragulidae or mouse deer include only five living species found in Africa and South Asia. The giraffes and okapi (Giraffidae) have been an Asian and African radiation, surviving today in Africa. The pronghorn antelopes (Antilocaridae) were an important

Miocene radiation in North America, where they survive as a single species.

The true deer or Cervidae originated in Asia and subsequently moved out to Europe and North America. In the Pliocene they entered South America. Today there are about 44 species of deer found in Europe, Asia, and the Americas. They have been introduced to some oceanic islands and Australia. The Bovidae (sheep, goats, true antelopes, bovines, and allies) appeared in the Oligocene of Asia and moved to Europe. They appeared in the Miocene of Africa and made several movements to North America beginning in the Pliocene. Today there are 137 species of extant wild bovids. Until humans transported domestic forms, bovids were absent from Australia and South America. The digestive tracts of the Tragulidae, Cervidae, Giraffidae, Antilocapridae, and Bovidae are sufficiently similar in form and function that they are often grouped as the Ruminantia. The ruminant digestive system, with a four-chambered stomach and foregut fermentation of structural carbohydrate, was a key adaptation. Apparently the Camelidae independently evolved a similar system. Early in their history, artiodactyls evolved cursorial adaptations including lengthening the limbs and reducing the number of toes. The evolution of horns and hornlike structures has been a recurrent feature of the evolution of large body size in the ungulates. The evolution of large body size and gigantism seems to accompany the adaptation for feeding on plants and especially the processing of structural carbohydrates; a discussion of the phenomenon is presented by Owen-Smith (1988).

Rabbits, hares, and pikas. The Lagomorpha include the rabbits, hares, and pikas. In their dentition they show ancient affinities with rodents. They appear in the fossil record during the Late Eocene of Asia. In their long evolutionary history they have occupied all the continents except Australia and Antarctica. They were recently introduced by Europeans into Australia, where they became a serious pest. There are about 80 species of lagomorphs, of which 26 are pikas. Twenty-five species of pika occur in the mountains of Asia, and the remaining species is found in North America.

Rodents. The Rodentia living today include more than 2,300 species; almost one-half of the living species of mammals are rodents. With their single pairs of upper and lower incisors separated from the molars and premolars by a gap, rodents early on exhibit a trait that has served them well: the ability to gnaw into hard surfaces. Their fossil history begins in the Paleocene. In Africa the earliest rodents belong to an ancient suborder, the Hystricognathi (porcupines, guinea pigs, agoutis, and relatives). Hystricognathous rodents appear next in the Oligocene of South America, where they have persisted and radiated, sending one species (the porcupine) to North America. In Europe, Asia, and North America, the squirrellike rodents appear in the Paleocene and Eocene. Mouselike and ratlike rodents appear somewhat later in the Oligocene but radiate rapidly. Australia received its rodent fauna in several invasions from island-hopping immigrants via New Guinea. South America received squirrels and mouselike rodents in the Late Miocene and Pliocene.

Rodents exhibit stunning diversity. At least twice they evolved gliding forms, including the flying squirrels and the scaly-tailed flying squirrels of Africa. Separate stocks adapted for underground life (tuco tucos, pocket gophers, mole rats) (Nevo, 1999); bipedal desert-adapted rodents evolved several times (jerboas, kangaroo rats, jumping mice) (Prakash and Ghosh, 1975); semiaquatic adaptations are exemplified by the Australian water rats, muskrats, beaver, and water voles. Arboreal adaptations include the squirrels and numerous other groups such as dormice and climbing rats. Convergence in form and function have led to numerous studies of rodent morphology, ecology, and behavior. The range of social systems evolved by rodents is marvelous in its variety. Of special interest are the cursorial, large rodents of South America, which occupy niches similar to small ungulates on other continents—for example, the paca, agouti, Patagonian cavy, and capibara (Rowlands and Weir, 1974). The blind naked mole-rat is the only mammal known to have evolved social groupings based on a single fertile pair and their sterile offspring, reminiscent of social insect societies (Sherman et al., 1991)

[*See also* Animals; Vertebrates.]

BIBLIOGRAPHY

Archibald, J. D. "A Study of Mammalia and Geology across the Cretaceous-Tertiary Boundary in Garfield County, Montana." *University of California Publications in Geological Science* 122 (1982).

Eisenberg, J. F. *The Mammalian Radiations*. Chicago, 1981.

Emmons, L. *Tupai: A Study of Bornean Tree Shrews*. Berkeley, 2000.

Ewer, R. F. *The Carnivores*. Ithaca, 1973. An admirable review of the terrestrial carnivores.

Findley, J. *Bate: A Community Perspective*. Cambridge, 1993.

Gatesy, J., and P. Arctander. "Molecular Evidence for the Phylogenetic Affinities of the Ruminantia." In *Antelopes, Deer and Relatives*, edited by E. S. Vrba and G. B. Schaller, pp. 143–155. New Haven, 2000.

Gatesy, J., et al. "Stability of Cladistic Relationships between Cetacea and Higher Level Artiodactyl Taxa." *Systematic Biology* 48 (1999): 6–20.

Gingerich, P. D., et al. "Origin of Whales in Epicontinental Remnant Seas." *Science* 220 (1983): 403–406.

Griffiths, M. *The Biology of the Monotremes*. New York, 1978.

Hotton, N., P. D. MacLean, J. J. Roth, and E. C. Roth, eds. *The Ecology and Biology of Mammal-like Reptiles*. Washington, D.C., 1986.

Jenkins, F. A., and A. W. Crompton. "Origin of Mammals." In *Mesozoic mammals*, edited by J. A. Lillegraven, et al., eds., pp. 59–73, 1979.

Kielan-Jaworowska, Z., and P. P. Gambayran. "Postcranial Anatomy and Habits of Asian Multituberculate Mammals." In *Fossils and Strata*. Oslo, Copenhagen, Stockholm, 1994.

Killian, J., T. Buckley, N. Stewart, B. Munday, and R. Jirtle. "Marsupials and Eutherians Reunited: Genetic Evidence for the Theria Hypothesis of Mammalian Evolution." *Mammalian Genome* 12 (2001): 513–517.

Lillegraven, J. A., Z. Kielan-Jaworowska, and W. A. Clemens, eds. *Mesozoic Mammals: The First Two-thirds of Mammalian History*. Berkeley, 1979.

Liu, F-G, R., and M. Miyamoto. "Phylogenetic Assessment of Molecular and Morphological Data for Eutherian Mammals." *Systematic Biology* 48 (1999): 54–64.

Madsen, O., M. Scally, C. J. Douady, D. J. Kao, R. W. DeBry, R. Adkins, H. M. Amrine, M. J. Stanhope, W. W. D. Jong, and M. S. Springer. "Parallel Adaptive Radiations in Two Major Clades of Placental Mammals." *Nature* 2001.

Martin, R. D. *Primate Origins and Evolution*. Princeton, 1990.

MacFaden, B. J. *Fossil Horses: Systematics, Paleobiology and Evolution of the Family Equidae*. Cambridge, 1992.

McKenna, M., and S. K. Bell. *Classification of Mammals above the Species Level*. New York, 1997.

Mann, J. R., C. Conner, P. L. Tyack, and H. Whitehead, eds. *Cetacean Societies*. Chicago, 2000.

Montgomery, G. G., ed. *The Evolution and Ecology of the Armadillos, Sloths and Vermilinguas*. Washington, D.C., 1985.

Murphy, W. J., E. Eizirik, W. E. Johnson, Y. P. Zhang, O. A. Ryder, and S. J. O'Brien. "Molecular Phylogenetics and the Origins of Placental Mammals." *Nature* 409 (2001): 614–618.

Nevo, E. *Mosaic Evolution of Subterranean Mammals*. Oxford, 1999.

Nikaida, M., A. P. Rooney, and N. Okada. "Phylogenetic Relationships among the Certartiodactyla." *Proceedings of the National Academy of Sciences* 96 (1999): 10,261–10,266.

Owen-Smith, R. N. *Megaherbivores*. Cambridge, 1988.

Prakash, I., and P. K. Ghosh, eds. *Rodents in Desert Environments*. The Hague, 1975.

Rathbun, G. "The Social Structure and Ecology of Elephant Shrews." *Journal of Comparative Ethology* 20 (1979).

Rich, T. H., et al. "A Tribosphenic Mammal from the Mesozoic of Australia." *Science* 278 (1997): 1438–1442.

Rowlands, I. W., and B. J. Weir, eds. *The Biology of Hystricomorph Rodents*. Symposium of the Zoological Society of London no. 34. London, 1974.

Sherman, P. W., J. U. M. Jarvis, and R. Alexander, eds. *The Biology of the Naked Mole Rat*. Princeton, 1991.

Shoshani, J., and P. Tassy, eds. *The Proboscidea: Evolution and Paleoecology of Elephants and Their Relatives*. Oxford, 1996.

Simpson, G. G. "The Principles of Classification and a Classification of Mammals." *Bulletin of the American Museum of Natural History* 85 (1945).

Springer, M. S., et al. "Endemic African Mammals Shake the Phylogenic Tree." *Nature* 388 (1997): 61–63.

Szalay, F. S., M. Novacek, M. McKenna, eds. *Mammal Phylogeny*. Berlin and New York, 1993.

Thewissen, J. G., and F. E. Fish. "Locomotor Evolution in the Earliest Cetaceans." *Paleobiology* 23 (1997): 482–490.

Tyndale-Biscoe, C. H., and M. B. Renfree. *Reproductive Physiology of Marsupials*. Cambridge, 1987.

Waddel, P. J., N. Okada, and M. Hasegawa. "Towards Resolving the Interordinal Relationships of Placental Mammals." *Systematic Biology* 48 (1999): 1–5.

Wilson, D. E., and D. M. Reeder, eds. *Mammal Species of the World*. 2d ed. Washington, D.C., 1993.

Wojcik, J. M., and M. Wolson, eds. *Evolution of Shrews*. Bialoieza, 1998.

— JOHN F. EISENBERG

MARGINAL VALUES

The concept of marginal values, borrowed from microeconomics, provides a powerful tool for evolutionary biologists to find the optimal strategy of how to allocate efforts to different options. The marginal value of an option is defined as the additional utility obtained (generally, a token of Darwinian fitness) from a resource by investing an additional, infinitesimally small, amount of effort into its procurement. In mathematical terms, this can be best visualized by a function that reflects utility and that describes how the cumulative gain, $g(t)$, from a given resource relates to the amount of effort, t, already invested in its exploitation. The marginal value at point t is given by the derivative, $dg(t)/dt$. Typically, the biological question of interest is to identify the strategy that maximizes the (inclusive) Darwinian fitness of the individual adopting it, thereby turning the economist's concept of a descriptive utility into an a priori specification of what the organism should do. The scope of strategies that can be analyzed in this way is virtually unlimited and can refer to behavior, resource allocation during life history, developmental strategies, genetic conflicts, and so forth.

Allocation problems can be solved when the set of available strategies, the resulting consequences for utility (fitness), and the constraints operating on the system (e.g., costs and individual state) are known. In particular, the marginal value theorem states that the optimal allocation strategy reduces all options to the same marginal value. This marginal value maximizes the utility under scrutiny if total costs and benefits of the strategy are taken into account. The theorem makes intuitive sense, because under these conditions each option has been exploited to a degree that could not be bettered by switching to another option. These concepts have been most widely used in optimal foraging theory (Charnov, 1976; Stephens and Krebs, 1986), where it is convenient to think of $g(t)$ as the cumulative gain of net energy (energy collected minus cost of foraging) and of t as the time needed to exploit a resource. Typically, an animal has to decide on the best strategy to maximize its long-term net rate of energy gain. This rate is assumed to be a token for the animal's fitness.

Imagine a bee that collects nectar from an inflorescence with many flowers, such as when bees visit foxglove or monkshood. After the bee has already visited ten flowers on the stalk, the marginal value of nectar on this inflorescence is the additional nectar the bee gets

MARGINAL VALUES IN OPTIMAL FORAGING THEORY

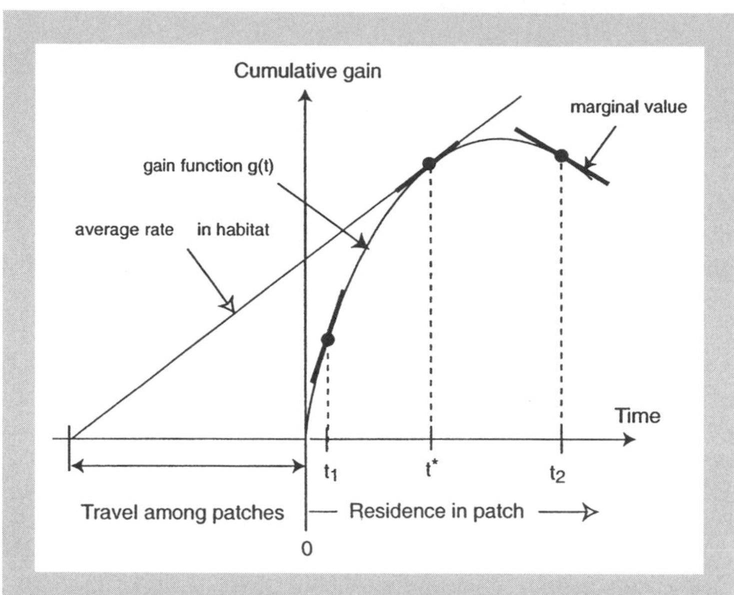

The most often used application of marginal values in optimal foraging theory. A forager enters, at time 0, a patch that contains resources. As it starts to consume the resources, gains accumulate according to a gain function, *g(t)*. Because the resources in the patch deplete, it becomes harder to extract additional resources the longer the forager continues; thus, *g(t)* shows decreasing returns to scale and eventually decreases. Mathematically, at any one point along the gain curve, the derivative with respect to time, *dg(t)/dt*, is the marginal value of the resource in the patch. The graphical equivalent is the tangent (thick lines) to *g(t)* at a given point (t_1, t_2, or t^*). With travel time τ between patches, the marginal value theorem states that a forager maximizes its long-term net rate of gain, γ^*, by leaving the patch at a point where the marginal value of the patch drops to the maximum possible rate in the habitat, γ (thin line). In the graph, the rate γ is given by the tangent to *g(t)* that intersects the *x* axis at point $-\tau$. The graphical solution to this maximization problem is the identity of the tangent representing the marginal value at point t^* (thick line) and the maximum possible rate, γ^* (thin line), in the habitat.

—PAUL SCHMID-HEMPEL

when it is visiting just one more flower relative to the additional cost needed to do so. The marginal value theorem states that the forager should stay on the inflorescence just as long as to reduce its marginal value to a value that is characteristic for the habitat and that represents the maximum possible long-term net rate of energy gain, $\gamma^* = \gamma^*(t)$. More formally, assume that the bee visits a (large) number, n, of inflorescences in the habitat and leaves each of them at time t, having collected amount of resource $g(t)$. In microeconomic terms, leaving time, t, is the consumer's strategy, and $g(t)$ represents utility relative to price, t. The time needed to complete the foraging is

$$T \approx \mathrm{n} \cdot (t + \tau)$$

where τ is the time to travel from one inflorescence to another. The expected net gain from this strategy is

$$G = \mathrm{n} \cdot g(t).$$

Hence, the average long-term net rate of gain, provided strategy t is adhered to, is

$$\gamma(t) = \frac{n \cdot g(t)}{n \cdot (t + \tau)}.$$

To find the maximum ($\gamma = \gamma^*$) we differentiate with respect to strategy, t, and set the term to 0, to find

$$\frac{d\gamma(t)}{dt} = \frac{d}{dt}\left(\frac{g(t)}{t + \tau}\right) = 0.$$

This simplifies to

$$\left.\frac{dg(t)}{dt}\right|_{t=t^*} = \gamma^*,$$

which defines the optimal leaving time, t^*. A graphical equivalent of this solution is shown in Figure 1. (Technically, flowers on stalks occur in discrete steps, whereas the formal treatment assumes continuous time.) Should inflorescences differ in their gain functions, all should be reduced to the same marginal value, but the leaving times would be different, so that stalks with steeper gain functions are visited longer.

Marginal values can also be used to test for the kind of utility used by a forager and thus to gain an insight into an animal's behavior. For example, honeybees visit a large number of flowers from which they collect nectar that is delivered to the hive. The marginal value theorem predicts how many flowers a bee should visit during a single trip to maximize nectar delivery to the colony. However, different numbers are predicted depending on how the utility is defined. Taking into account the extra cost of carrying the accumulating nectar load, the behavior of individual bees was found to be compatible with efficiency maximization (maximizing the ratio of net energy gain and foraging energy cost) but not with the maximization of the long-term net rate of energy gain as postulated above (Schmid-Hempel et al., 1985). In other words, the honeybees were using a utility that evaluated how much net energy they gained per unit of energy spent, whereas time was not of prime importance.

The use of marginal values goes far beyond foraging theory. For example, male dung flies transfer sperm to the female and guard it against rivaling males as long as they remain in copula. A gain function with decreasing returns to scale results from the decreasing likelihood, with time, that an additional minute spent in copula will actually lead to sire offspring. In contrast, it takes the male a certain time to search for the next female (Parker, 1978)—a similar situation as for a bee that travels from one inflorescence to the next. The same kind of analysis has been applied to parasitoid wasps that have to decide on how many eggs to lay per host caterpillar before moving on to the next (Hubbard and Cook, 1978). Marginal value thinking also underlies the concept of evolutionary stable strategies. Marginal values and the marginal value theorem are general tools for an idealized world and thus help to explain the general structure of problems. In any single case, this general approach is bound to lead to predictions that are not perfectly matched by the observations. Such deviations can often be used to generate new insights into the underlying assumptions and processes.

[*See also* Mathematical Models.]

BIBLIOGRAPHY

Charnov, F. L. "Optimal Foraging: The Marginal Value Theorem." *Theoretical Population Biology* 9 (1976): 129–136.

Hubbard, S. F., and R. M. Cook. "Optimal Foraging by Parasitoid Wasps." *Journal of Animal Ecology* 53 (1978): 283–299.

Parker, G. A. "Searching for Mates." In *Behavioural Ecology: An Evolutionary Approach*, edited by J. R. Krebs and N. B. Davies, pp. 214–244. Oxford 1978.

Schmid-Hempel, P. A. Kacelnik, and A. I. Houston. "Honeybees Maximize Efficiency by Not Filling Their Crop." *Behavioral Ecology and Sociobiology* 17 (1985): 61–66.

Stephens, D. W., and J. R. Krebs. *Foraging Theory*. Princeton, 1986.

— PAUL SCHMID-HEMPEL

MARSUPIALS. *See* Mammals.

MASS EXTINCTIONS

A mass extinction is an extinction of a significant proportion of the world's biota in a geologically brief period of time. The vagueness about what is "significant" can be dealt with fairly satisfactorily in particular cases by giving the percentages of taxa that go extinct, but the vagueness about time is more difficult to deal with. An important question about mass extinctions is how catastrophic they were, so a definition is required of *catastrophe* in this context. A useful one is that it is a biospheric perturbation that appears instantaneous when viewed at the level of resolution provided by the geological record. This is never likely to be more precise than a few thousand years, but within a time frame of many millions of years virtually all geologists would concur with a mass extinction of this duration being catastrophic. As regards geographic extent, more than half

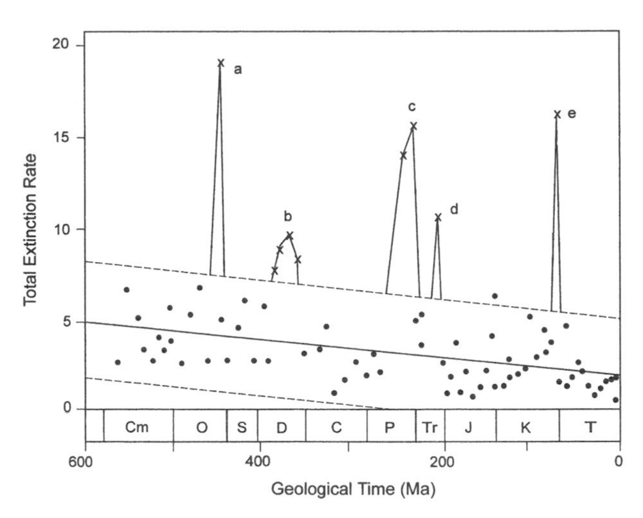

FIGURE 1. Extinction Rates, in Families Per Million Years, of Marine Animals During the Phanerozoic.
The "big five" mass extinctions are shown as clear peaks with crosses standing along the enveloped background extinction level. The term *Phanerozoic*, derived from the ancient Greek, means "evident life" and refers to the time elapsed since the start of the Cambrian period, about 540 million years ago. The periods are (A) Late Ordovician; (B) Late Devonian; (C) Late Permian; (D) Late Triassic; (E) Late Cretaceous. "Cm" represents Cambrian; "O," Ordovician; "S," Silurian; "D," Devonian; "C," Carboniferous; "P," Permian; "Tr," Triassic; "J," Jurassic; "K," Cretaceous; "T," Tertiary. Oxford University Press.

of the earth's surface must be environmentally affected to produce mass extinctions of the magnitude found in the fossil record.

Intensive paleontological research since the 1970s has established that there were five major episodes of marine mass extinction, at the end of the Ordovician, Permian, Triassic, Cretaceous, and near the end of the Devonian. This is shown in Figure 1, with a statistical analysis of extinction rates expressed as families going extinct per stratigraphic stage. The "big five" events are portrayed as clear peaks standing above the background extinction level, which shows a clear decline of extinction rate through time. The mass extinctions also show up as abrupt decreases in diversity (Figure 2). Table 1 indicates the relative importance of each event. That at the end of the Permian was by far the largest, and appropriately divides the Palaeozoic from the Mesozoic era.

The fossil record of marine invertebrates has the advantage of abundant specimens, good stratigraphic control, closely spaced samples, uniform preservation quality, and broad geographic distribution. That for terrestrial animals and plants is less satisfactory, but good progress has been made at least for the tetrapods. Four mass extinctions and two minor Cenozoic extinctions

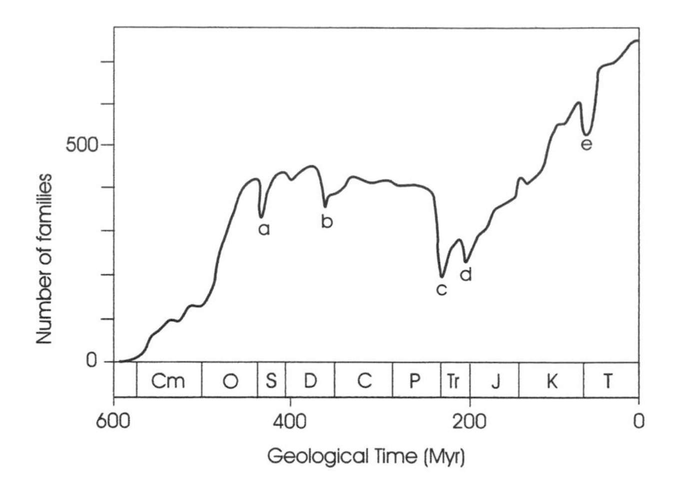

FIGURE 2. Marine Family Diversity During the Phanerozoic. The long-term increase punctuated by diversity crashes caused by the "big five" mass extinction events. Oxford University Press.

have been recognized, but only the end-Permian and end-Cretaceous events correspond with major extinctions in the marine record. The terrestrial plant record does not reveal such a clear-cut pattern of global mass extinctions as the animal record. This could at least partly be due to problems of data quality.

The pattern shown in Figure 1 has led to the suggestion that mass extinctions are different in kind from background extinctions, and that organisms that were adapted to survive the latter were vulnerable to the far more significant environmental perturbations that led to mass extinctions. This suggestion has proved to be rather controversial, and a consensus view has been emerging among paleontologists that the big five mass extinctions are merely the most extreme of a whole spectrum of extinction events of varying magnitude, with extinction intensity varying inversely with frequency according to a power law. Thus a larger number of lesser mass-extinction events are now widely recognized. Taking into account some of these lesser events, the claim has been made by the American paleontologists David Raup and Jack Sepkoski of a 26-million-year cycle in mass extinction events within the last 250 million years. This claim has proved very controversial, with criticisms being made of the validity of the statistical methods chosen and of the reality of some of the lesser extinction events. Raup and Sepkoski also used a particular timescale based on dating by radioactive decay of elements in minerals. This timescale has been modified recently as a result of the collection of more accurate data, and this has not supported the suggested periodicity. Today, the periodicity hypothesis is given little credence. Mass extinctions seem to be episodic rather than periodic events.

The typical sequence of events in a mass extinction is initiated by an extinction phase, during which diversity falls rapidly, followed by a survival or lag phase of minimal diversity and then a recovery phase of rapid diversity increase (Figure 3). In practice, the distinction between the extinction phase and the survival phase is rather arbitrary as extinctions can continue into the survival phase. Those taxa that outlive the majority of their clade are known as *holdover taxa*, though their survival is commonly brief. *Progenitor taxa*, on the other hand, appear during the extinction or survival phase and rapidly radiate during the recovery phase. The survival phase is also characterized by *disaster taxa*, usually widely distributed species of opportunists, whose presence in low numbers of species but huge numbers of individuals is a sure sign of elevated environmental stresses. Disaster taxa, and opportunists generally, are typically small, morphologically simple forms.

The recovery interval is marked by the radiation of both progenitors and the taxa that survived the mass extinction. It is also marked by the reappearance of taxa that had disappeared from the fossil record during the extinction phase several millions of years earlier. This significant phenomenon of many mass extinctions has been called the Lazarus effect, after the biblical character who was brought back from the dead by Jesus. The absence of Lazarus taxa from the fossil record, known as their *outage*, is one of most intriguing aspects of the survival phase. A simple explanation for Lazarus taxa might lie in the reduction of population sizes at the time of extinction. While populations of many species may decline below the minimum viable population sizes

FIGURE 3. Phases of a Mass Extinction Crisis. The extinction phase expressed by fall in diversity appears to be extended through time, but in some cases this can be a statistical artifact. Oxford University Press.

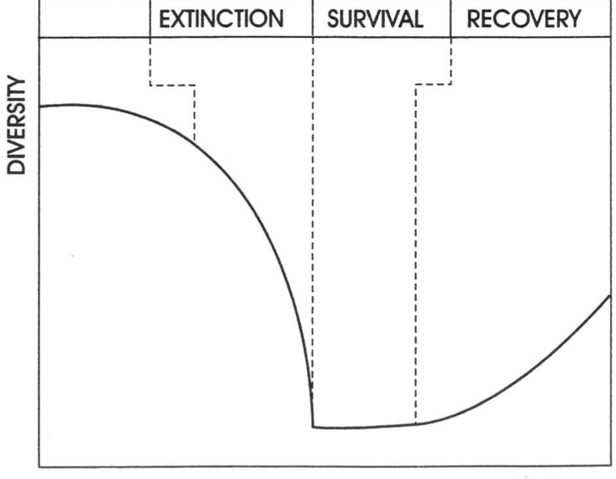

TABLE 1. Extinction Intensities at the Five Major Mass Extinctions in the Fossil Record
Species-level estimates are based on a rarefaction technique.

	Families		*Genera*	
Mass extinction	*Observed Extinction (%)*	*Calculated Species Loss (%)*	*Observed Extinction (%)*	*Calculated Species Loss (%)*
End Ordovician	26	84	60	85
Late Devonian	22	79	57	83
End Permian	51	95	82	95
End Triassic	22	79	53	80
End Cretaceous	16	70	47	76

and thus go extinct, a certain proportion of species will presumably become very rare but just manage to avoid extinction. They will thus become correspondingly rare fossils that are unlikely to be collected. Accordingly, the proportion of Lazarus taxa becomes an important measure of the extinction crises. That this is the case is indicated by the fact that the most significant Lazarus effects occur after two of the biggest extinctions, the end-Ordovician and end-Permian. The key factor seems to be the relationship between population size and habitat suitability and how the latter recovers after the perturbation.

The precise pattern of extinction during the geologically brief crisis has proved to be a major subject of debate, since it has an important bearing on interpretation; accordingly, statistical analysis has been used to analyze stratigraphic range charts of species. Three distinct extinction patterns have been recognized, abrupt or catastrophic, stepped, and gradual (Figure 4). Clearly there can be overlap between these patterns, but distinguishing among them is important in identifying the underlying "kill mechanism." Thus stepped extinction has been predicted for a multiple cometary impact kill mechanism, while an abrupt extinction is held to be characteristic (but by no means diagnostic) of a single, large asteroid or comet impact. Unfortunately, extinction patterns are as much influenced by the data quality and sampling methodology as by the nature of the event itself—hence the need for statistical rigor. Range charts record the series of point occurrences of fossil species against a stratigraphic column. Such charts give a feel for the relative frequency of occurrence of species; more common species will have shorter gaps between their occurrences than rarer ones. An important point to be remembered when interpreting extinctions in range charts is that the highest occurrence of a fossil species is unlikely to represent the very last individual of that species. Thus, all species' ranges are artificially truncated to some extent. This will tend to make mass extinctions appear more gradual and slightly earlier in the stratigraphic record, an effect known as backsmearing. Thus gradual extinction patterns prior to mass extinction do not necessarily eliminate catastrophic extinction hypotheses. Geologists use the term *condensation* for strata that are unusually thin compared with correlative strata elsewhere, as a result of low rates of sediment accumulation. Stratigraphic condensation related to low sedimentation rates may make a gradual extinction event appear abrupt. The main contribution of various statistical assessments has been to demonstrate that when the ex-

FIGURE 4. Three End-Members for the Extinction Phase. The vertical axis represents time, with vertical lines representing the stratigraphic ranges of species. Oxford University Press.

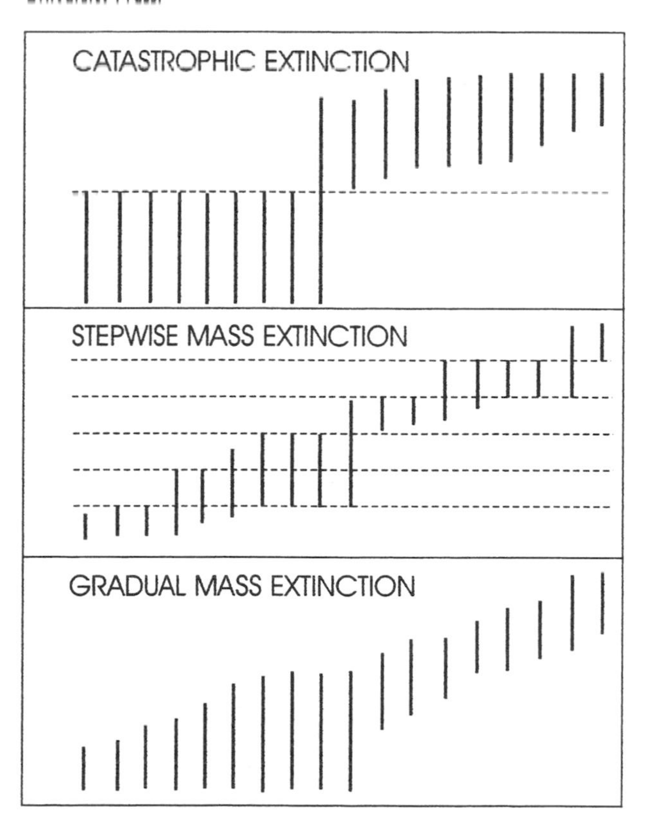

tinction phase is of short duration (i.e., less than one million years), it is not possible to distinguish the three types of extinction portrayed in Figure 4. The major biotic changes produced by the big five mass extinctions are presented below.

End Ordovician. The first of the major mass extinctions occurred at the end of the Ordovician period, approximately 438 million years ago. All parts of the marine realm were severely affected. Thus the dominant early Paleozoic planktonic group, the graptolites, underwent the worst crisis in its history, with only a few species surviving. Subsequently, in the Silurian, the monograptid graptolites underwent a spectacular radiation. Among the actively swimming organisms, the conodonts, an important Paleozoic group now considered to be small vertebrates, suffered a crisis almost as severe as that for the graptolites, while the dominant large bottom-dwellers (macrobenthos) the brachiopods and trilobites, underwent two distinguishable phases of extinction. Other benthic groups, such as the bivalves and rugose corals, together with other reef organisms, were less severely affected.

Late Devonian. The most important mass extinction phase during this time was at the boundary of the last two Devonian stratigraphic stages, the Frasnian and the Famennian. For a number of groups, however, the extinctions were more extended in time. This is true, for instance, for the brachiopods, whose extinction rate remained high through the Frasnian but was balanced by a high origination rate until toward the end of the stage, resulting in a sharp fall in species diversity. Reef faunas were severely affected, with the rugose corals suffering one of their most severe crises, and the tabulate corals and stromatoporoids also suffering heavy losses; it took a long time, until well into the Carboniferous, before reef systems recovered. The ammonites underwent a number of severe crises extending over several million years. Among the chordates, the conodonts suffered a severe setback, while there was a virtual elimination of primitive jawless fish, notably the heavily armored forms. There were also major extinctions among the jawed fish, the placoderms and arthrodires. Terrestrial plants experienced the first major crisis in their history, with diversity declining to a prolonged low point from the early middle Frasnian to the mid-Famennian.

End Permian. Although it has received less attention than the mass extinction at the end of the Cretaceous, that at the end of the Permian was by a wide margin the biggest such event known in the history of life. Although an important late Permian extinction event has been recognized in the excellent stratigraphic record in China, the most significant event took place at the end of the period. Radiometric dating of volcanic ash layers in the Permian-Triassic boundary at Meishan, China, suggest strongly that the marine extinction at least was a catastrophic event, taking place probably in less than 500,000 years. Reef systems suffered badly, the extinctions involving both calcareous sponges and the rugose and tabulate coral groups, which finally disappeared. It took reefs no less than 7–8 million years to recover from the extinction, one of the longest "reef gaps" in the Phanerozoic record. The other major colonial group, the Bryozoa, also underwent a severe crisis, with the four major stenolaemate Paleozoic orders undergoing major generic-level extinctions, though only one group, the fenestrates, disappeared completely. The echinoderms suffered an almost complete annihilation, with the crinoids losing two subclasses and the blastoids going extinct. The Foraminifera experienced the disappearance of the dominant late Paleozoic group the fusulinids, while the event marked the end of the dominance of the most important benthic macroinverebrate group, the brachiopods. It was indeed the most severe crisis in their history, with 90 percent of families and 95 percent of genera going extinct. Unlike these marine invertebrate groups, however, neither conodonts nor fishes suffered much extinction.

Among the terrestrial biota, the insects suffered the most severe crisis in their history, although their fossil record is not good enough to pinpoint stratigraphic horizons with much accuracy. The record is much better for the tetrapods, for which it has become clear that the Permian-Triassic boundary marks an event even more severe than that for the marine biota, with the loss of twenty-one families, or 63 percent of the total. Those going extinct included most of the herbivores, including the dominant pareiasaurs, gliding reptiles, and two-thirds of amphibian families. Ninety percent of assemblages of the Lower Triassic consist solely of the dicynodont *Lystrosaurus*. While terrestrial floras did not suffer mass extinctions comparable to marine or terrestrial animals, there were nevertheless some notable changes. The most striking was the catastrophic disappearance of the glossopterids, a taxonomically problematic group probably related to seed ferns. This had been a dominant group in the Gondwanan Province. Elsewhere in the world there was a significant disappearance of the dominant large plants, such as tree-forming cordaitids in the Angara and *Gigantopteris* in the Cathaysia Province, with the Early Triassic flora being dominated by small, weedy survivors. A major consequence of this was a so-called coal gap in the stratigraphic record that lasted until well into the Middle Triassic.

End Triassic. Once again there was a spectacular demise of reef faunas, affecting both calcareous sponges and corals. Among the invertebrates, the most striking event was among the ammonites, with no fewer than six superfamilies going extinct; perhaps no more than one genus survived into the Jurassic. Important Paleozoic groups among the brachiopods and gastropods, which had survived the end-Permian crisis, went extinct at this time. The same is true of the conodonts. The now dominant bivalves also underwent significant extinctions, but, as with other extinction events, they were less severe than for many other groups. Among terrestrial organisms, the record at the critical stratigraphic horizons is poor, and there remains some controversy about whether or not there was a major extinction event among the tetrapods. At any rate, the major biotic turnover had been accomplished in the late Triassic, with the earliest dinosaurs and mammals replacing other reptile groups as dominant land organisms. As regards plants, a mass extinction event has been claimed in eastern North America and Greenland, but there are no strong grounds as yet for accepting an event on a global scale.

End Cretaceous. For the general public the only event of note at this time was the extinction of the dinosaurs (see Vignette) but the Cretaceous-Tertiary boundary is much more clearly marked in most parts of the world by a marine mass extinction. This is strongly shown in the calcareous plankton, namely planktonic foraminifera and coccolithophorid algae. Both groups underwent catastrophic mass extinctions in what were low latitudes at that time, leaving only a small percentage of survivors, probably within a few thousand years at the most. Molluscan groups characteristic of the Mesozoic, such as inoceramid and rudist bivalves, ammonites and belemnites, finally disappeared, but the first two groups probably went extinct a million or so years before the end of the Cretaceous, while the other two had been in slow decline, though for the ammonites there seems to have been a catastrophic final disappearance. Extinctions among other marine invertebrates were somewhat variable in intensity. Among the marine vertebrates, the ichthyosaurian and plesiosaurian reptiles finally disappeared, as did the giant "sea lizards" the mosasaurs. The first two groups had long been in slow decline, but the mosasaurs died out after a spectacular radiation in the late Cretaceous.

On land, besides the dinosaurs a number of marsupials went extinct, but the extinction rate among placentals was not exceptionally high, while freshwater groups such as crocodiles and turtles passed into the Tertiary more or less unscathed. There is no evidence to support a high extinction rate either among birds or insects, but the fossil record across the Cretaceous-Tertiary boundary for these groups is rather poor. Among plants, a severe phase of mass extinction is recorded only in North America, and in the southern hemisphere there is little evidence of any notable change.

Causes. The causes of mass extinction events have been hotly debated by both scientists and nonscientists, but only a limited number are supported by evidence generally accepted as satisfactory today. Among a multiplicity of possible causes, only a limited number merit consideration as having the potential to cause significant environmental change on a global scale and affect a wide range of biota simultaneously. Information on paleoenvironments can be sought from a wide range of information derived from the stratigraphic record, based on sedimentology, paleoecology, mineralogy, and geochemistry.

Bolide (asteroid or comet) impact has been widely

EXTINCTION OF THE DINOSAURS

Dinosaurs continue to appeal to people's imaginations and sense of wonder more than any other fossil organisms, and numerous theories about their extinction 65 million years ago have been proposed. Apart from conventional ones usually involving climate change, it has been speculated that the newly emergent mammals stole their eggs, or even that they died of an AIDS epidemic introduced by viruses from outer space. The trouble with all these more or less fanciful theories is not so much the lack of supporting evidence but their ad hoc character, ignoring the fact that the dinosaur went extinct contemporaneously with numerous other organisms both on land and in the sea, suggesting strongly that all were adversely affected by some drastic deterioration of the physical environment.

Since the late twentieth century, the popular theory, which does have some respectable evidential support, is that the dinosaurs were killed off by an asteroid or comet impact in Mexico. Some cautionary observations are required, however. The fossil record of dinosaurs is rather poor compared to many other organisms and the only really good one at the critical horizon comes from the state of Montana in the western United States. While a consensus has emerged that there was a catastrophic event at the end of the Cretaceous, a number of experts have maintained that the dinosaurs were in decline for some time before the end of the period. If this is so (and the point is disputed), then it is difficult to argue that an impact was the sole cause of the extinction.

Large organisms such as dinosaurs are always relatively vulnerable to extinction because of their relatively small population sizes and low reproductive rates. There was indeed a considerable turnover in time of dinosaurs through their 150 million years history, with many significant extinction phases. Jurassic dinosaurs are quite different from their Cretaceous relatives. Thus, whatever finally killed them off need not necessarily have been strikingly different in character, as opposed to intensity, from events earlier in their history.

—ANTHONY HALLAM

	Bolide impact	Volcanism	Cooling	Warming	Regression	Anoxia/ transgression
Late Precambrian						●
Late Early Cambrian					●	●
Biomere boundaries			○			●
Late Ashgill			●	●	●	●
Frasnian-Famennian			○		○	●
Devonian-Carboniferous			○			●
Late Maokouan					●	
End-Permian		●		●	○	●
End-Triassic		●		○	●	○
Early Toarcian		○		○		●
Cenomanian-Turonian			○			●
End-Cretaceous	●	●	●		●	○
End-Palaeocene		●		●		●
Late Eocene			●			

● strong link ○ possible link

FIGURE 5. Summary of the Proposed Causes of the Main Phanerozoic Mass Extinction Events.
Some of the stratigraphic terms represent stages or series rather than periods. The biomere boundaries relate to the Late Cambrian. Ashgill is the last series of the Ordovician period; Maokouan is Late Permian, Toarcian is Early Jurassic, Cenomanian and Turonian Mid-Cretaceous, and Palaeocene Early Tertiary. Oxford University Press.

discussed ever since Luis Alvarez and colleagues published their classic paper on the Cretaceous-Tertiary boundary event. An extreme view has been put forward by David Raup that not just mass extinctions but virtually all extinctions are probably due to this cause. For this view to be more than an act of faith, good evidence must be brought forward in support. The presence of a significant anomaly in abundance of the platinum-group element iridium was the evidence originally put forward by Alvarez, and this, together with the evidence of so-called shocked quartz (quartz with distinctive planar laminae diagnostic of the imposition of huge pressures) and the discovery of what appears to be a huge impact crater in the Yucatán Penninsula of Mexico, have persuaded a consensus of scientists that this is the likeliest cause of extinction of dinosaurs and other groups during this event. There are several ideas about how the extraterresteral impact caused mass extinction. The extinction scenarios are wide ranging. They embrace one or more of several factors: severe reduction of photosynthesis by an earth-enveloping dust cloud, intense cooling associated with consequent reduction of sunlight reaching the earth's surface, intense heating associated with the release of CO_2 from sedimentary carbonates, sulphur dioxide gases expelled from gypsum deposits, extensive acid rain, and wildfires on land. No adequate resolution of conflicting arguments has yet been reached, nor has there been achieved a satisfactory determination of the role of other purely terrestrial factors such as cooling climate or sea-level fall.

What has become much better established is that the end-Cretaceous bolide impact was unique in causing a major mass extinction, because evidence for bolide impact at other mass extinction horizons is either weak, ambiguous, or absent. Furthermore, there is only a poor correlation between undoubted impact craters in the geological record and mass extinctions.

Among other factors, volcanism on a massive scale seems to be implicated in a number of greater and lesser mass extinction events, though by no means all. Thus, if there is no causal correlation, it is a remarkable coincidence that two of the greatest series of eruptions of continental flood basalts, the Siberian Traps and Decan Traps of India, coincide in time respectively with the end-Permian and end-Cretaceous events. The most obvious long-term environmental change produced by such volcanism would be global warming induced by the increased amount of CO_2 in the atmosphere. Climatic cooling has also been invoked as a cause of mass extinctions, but here the evidence is weaker. So has reduction in habitat area of epicontinental seas (or shallow seas covering parts of the continents) produced by marine regression, that is, retreat of the sea from the continents related to sea-level fall. Although this does seem to be involved in some extinction events, the evidence is stronger for marine extinctions being produced by the spread of oxygen-deficient bottom waters associated with marine transgression (which is the converse of regression and is associated with rise of sea level), and sometimes also by global warming (Figure 5). The evidence for a significant phase of global warming and increased oxygen deficiency (anoxia) in the oceans is strongest for the biggest mass extinction event of all, at the end of the Permian.

A great deal of interest has been shown in extinctions of large mammals and birds in the late Quaternary, within the last few tens of thousands of years, most notably in Australasia and North America. That this corresponds in time with the expansion of human populations is unlikely to be coincidental, though the climatic change alternative has not yet been decisively ruled out. Destruction of habitat area as a result of the activities of our species seems to be the main cause of the increasing rate of extinction of a wide variety of animals and plants, together with other human activities such as overhunting, pollution, and introduction of alien species. The corresponding depredation of the world's biota has led some to make a case for this being the sixth big mass extinction. The somber implication of the geological record extending back thousands of years is that mankind has never lived in harmony with nature.

Evolutionary Significance. The role of catastrophic events has until recently been unappreciated by evolutionary biologists. Darwin was firmly of the opinion that biotic interactions, such as competition for food and space—the "struggle for existence"—were of considerably greater importance in promoting evolution and extinction than changes in the physical environment, but this is not in accord with the geological record of extinctions. Simultaneous extinctions of groups of different biology and habitat, on a global or regional scale, clearly implies adverse changes in the physical environment, on a scale that is not consistent with normal rates of evolutionary process. In other words, to use Raup's laconic phrase, extinctions could have been more a matter of bad luck than bad genes. Much of evolution has depended on a succession of historical contingencies and opportunistic responses. Mass extinctions have not operated randomly among different kinds of organisms but have been selective to some degree. Thus both reef ecosystems and large animals have been relatively vulnerable. Features evolved adaptively during normal times have proved insufficient to prevent extinction at times of catastrophic change.

It has become increasingly apparent that in general the fossil record of both vertebrates and invertebrates favors the view that the prime motor of evolutionary change is the physical environment. This is most evident from the record of mass extinctions and subsequent radiations, for example the end-Mesozoic extinction of dinosaurs and early-Cenozoic radiation of mammals, but it is clear also from a host of lesser extinction and radiation events. Some believe that the most significant phenomenon for long-term evolutionary progress may be what has been called incumbent replacement. New clade species acquire a key adaptation that gives them a higher competitive speciation than the old clade sources of replacement of extinct species. Incumbent replacement proceeds at a rate limited by the extinction rate and often seems to be related to mass extinction events. It has been maintained that the turnover of plant taxa with time exhibits a fundamentally different pattern from that of animals, with plants being less vulnerable to mass extinction episodes and with displacive competition playing a major role. This has been disputed, however, in some instances, such as the Carboniferous-Permian vegetational transition, which is considered to be replacive rather than displacive. Taxa that originated in peripheral, drier habitats in the Late Paleozoic tend to be subgroups of seed plants. The life histories of seed plants suggests a priori a greater resistance to extinction than most groups of "lower" vascular plants, such as ferns, lycopsids, and sphenopsids. This prediction is confirmed by the nearly continuous expansion of seed-plant diversity since the Paleozoic, at the expanse of more primitive vascular plants.

[*See also* Extinction; Red Queen Hypothesis.]

BIBLIOGRAPHY

Alvarez, W. *T. Rex and the Crater of Doom*. Princeton, 1997. A popular account of the impact theory of the end-Cretaceous extinction, written by one of the original team that discovered the iridium anomaly in Italy.

Archibald, J. D. *Dinosaur Extinction and the End of an Era*. New York, 1996. While not denying that bolide impact had an influence, this book relates longer term extinctions to changes of habitat induced by marine regression.

Chaloner, W. G., and A. Hallam, eds. *Evolution and Extinction*. London, 1989. The proceedings of a joint symposium of the Royal Society and the Linnean Society held in November 1989.

Courtillot, V. *Evolutionary Catastrophes: The Science of Mass Extinctions*. Cambridge, 1999. A popular account promoting the volcanism theory.

Donovan, S. K., ed. *Mass Extinctions: Processes and Evidence*. London, 1989. A multiauthor volume containing some valuable and informative articles.

Eldredge, N. *The Miner's Canary*. New York, 1991. A popular account of mass extinctions promoting climatic change as a major influence.

Erwin, D. H. *The Great Paleozoic Crisis: Life and Death in the Permian*. New York, 1993. A well-written review, now outdated in some respects but still useful, of the greatest of all mass extinctions.

Hallam, A., and P. B. Wignall. *Mass Extinctions and Their Aftermath*. Oxford, 1997. The only comprehensive scholarly monograph on the subject.

Hart, M. B., ed. *Biotic Recovery from Mass Extinction Events*. London, 1996. A multiauthor volume, the only book to date focusing on biotic recovery.

Jablonski, D. "Extinctions in the Fossil Record." *Philosophical Transactions of the Royal Society*, B, 344 (1994): 11–17. A concise review of the facts.

Larwood, G. P., ed. *Extinction and Survival in the Fossil Record*. London, 1988. Accounts by a number of experts of all the major fossil groups.

MacLeod, N. "Extinctions: Causes and Evolutionary Significance."

In *Evolution of the Earth*, edited by L. Rothschild and A. Lister London, 2002. The author promotes the thesis that the primary controls on long- and intermediate-term taxonomic extinction patterns at all scales are tectonic and probably mediated through the waxing and waning of lineages that occupy the base of marine ecological hierarchies.

Martin, P. S., and R. G. Klein, eds. *Quaternary Extinctions: A Prehistoric Revolution*. 1984. A multiauthor volume presenting the facts and rival theories concerning the geologically recent extinction of large vertebrates.

Raup, D. M. *Extinction: Bad Luck or Bad Genes?* New York, 1991. A lucidly written, intellectually rigorous popular account, with a pronounced partiality toward bolide impact.

— ANTHONY HALLAM

MATE CHOICE

[*This entry comprises three articles:*

An Overview
Mate Choice in Plants
Human Mate Choice

The first article offers a broad and thorough discussion of the fundamental reasons why mate choice arises, what it is that males and females derive from mate choice, and how theories of sexual selection account for the ways that males (and females) signal their characteristics to each other; the second article provides an overview of female choice and male competition in plants; the third article discusses human cross-cultural mating preferences and generalizations. For related discussions, see Mating Systems; Sexual Selection; *and* Signalling Theory.]

An Overview

In many animal species, males and females have strikingly different appearances; male birds of paradise, for example, have elaborate colorful plumage that females lack. Charles Darwin was greatly intrigued by these traits, called secondary sexual characteristics, which differ between the sexes but are not directly used in reproduction. In addition to ornaments such as the colorful feathers, they include horns and antlers found on many male ungulates and beetles.

History of Mate Choice Studies. Darwin said that secondary sexual characteristics could evolve in one of two ways. First, they could be useful to one sex, usually males, in fighting for access to members of the other sex. They are advantageous because better fighters get more mates and have more offspring. The second way was more problematic. Darwin noted that females often pay attention to traits such as long tails and elaborate plumage during courtship, and he concluded that the traits evolved because the females preferred them. Female peafowl, for example, find males with long tails attractive. Sexual selection, or selection for traits that improve mating success, consists of two components: male–male competition, which results in weapons, and female choice, which results in ornaments. [*See* Male–Male Competition.]

Although competition among males to mate with females seemed reasonable enough to Darwin's Victorian contemporaries, virtually none of them could swallow the idea that females could possibly do anything as complex as discriminating between males with different degrees of development of a characteristic like colorful plumage. Alfred Russel Wallace, who also independently arrived at some of the same conclusions about evolution and natural selection that Darwin did, was particularly vehement in his objections. [*See* Wallace, Alfred Russel.] He, and many others, simply found it absurd that females could make the sort of complex decision required by Darwin's theory; it would require the female to possess an aesthetic sense like that of humans, an idea they were unwilling to accept. Besides, what would be the point of choosing one male over another? If the only difference between them was the secondary sexual trait, why should the female bother?

Largely because of this opposition to the idea of female choice, sexual selection as a theory lay dormant for several decades. Even after genetics became incorporated with Darwin's ideas on evolution in the early decades of the twentieth century to form the new synthesis, the major evolutionary biologists of that time were largely uninterested in sexual selection. When they discussed extravagant traits at all, these were suggested to have arisen to allow females to find a mate of the right species. Choosing a male of a different species could have disastrous consequences, because hybrid offspring, if they can develop at all, are often infertile. Thus, species recognition was the only form of sexual selection that was recognized as important for a long time.

These ideas, and the lack of interest in sexual selection, continued through the whole first half of the twentieth century. It was not until the 1960s that evolutionary biologists began to reconsider the portrait they had painted of animal social life. Instead of being for the good of the species, behaviors were now seen as selfish acts that promoted an individual's fitness.

Modern Views on Mate Choice. Probably the most important new insight came from a paper written by Robert Trivers in 1972. He pointed out that females and males inherently differ because of how they put resources and effort into the next generation, which he termed parental investment. [*See* Parental Care.] Females are limited by the number of offspring they can successfully produce and rear. Because they are the sex that supplies the nutrient-rich egg, and often the sex that cares for the young, they leave the most genes in the next generation by having the highest quality young they can. Which male they mate with could be very important, because a mistake in the form of poor genes or no help with the young could mean that they have lost their whole breeding effort for an entire year. Mating with many males will not increase a female's reproductive success; it would not increase the number of eggs she produces.

Males, in contrast, can leave the most genes in the next generation by fertilizing as many females as possible. Because each mating requires relatively little investment from him, a male who mates with many females sires many more young than a male mating with only one female. Hence, males are expected to compete among themselves for access to females, and females are expected to be choosy and to mate with the best possible male they can. Trivers (1972) thus gave Darwin's idea a more modern rationale, and he eliminated the need for invoking a complex conscious process by females by pointing out that as long as females could discriminate among different individuals, and benefited by choosing particular types, female choice would be favored.

Evolutionary biologists now generally agree that mate choice, usually by females (for exceptions see the Vignette on Sex Role Reversal and Male Choice) is an important part of sexual selection. Female preference for exaggerated traits has been demonstrated in a variety of animals, both by observing what individuals do in nature and by performing controlled experiments in the field or laboratory. For example, male African long-tailed widowbirds have, as the name implies, elongated tail feathers that are used in a sexual display. When Swedish biologist Malte Andersson (1994) cut off the tails of some male widowbirds and used the pieces to augment the tails of others, he found that the exceptionally long-tailed males attracted more females than the control males with tails of normal length, and that these males in turn were more successful than those with shortened tails.

Mechanisms and Function of Mate Choice. But how does mate choice operate? What circumstances have led to the evolution of mate choice? Three basic hypotheses have been proposed.

First and most straightforward is the direct benefits hypothesis. When males give females something they need for themselves or their offspring, females might be expected to discriminate among males on the basis of the resources the males have. For example, male scorpion flies and hanging flies studied by Randy Thornhill catch a prey item and give it to the females when they mate. The larger this so-called nuptial gift, the longer a pair stays in copula and the more of the male's sperm fertilize the eggs. Thus, the male "trait" of acquiring food, as well as the nuptial gift itself, evolves via female choice. Another example is seen in moorhens in England, studied by Marion Petrie. Here, males do much of the incubation of eggs. In addition, females prefer to mate with small males that have a higher proportion of body fat, because those males have greater reserves to last them during the incubation period. Again, a trait in males is selected via female choice; thus, it makes sense that females prefer the males that give them what they need.

Another direct benefit a female can gain is parental care. The good parent hypothesis suggests that females choose males that will be good fathers. But the problem there is that females do not see males being parental during courtship. How does a female tell whether a male is going to be a good parent? It has been suggested that

SEX ROLE REVERSAL AND MALE CHOICE

Although in most animals females are limited by the offspring they can produce and rear, whereas males are limited by the number of fertilizations they obtain, this is not always the case. In several groups of animals, males actually contribute a great deal to each mating. Katydids, also called bush crickets (family Tettigoniidae), have a mating system in which males produce a nutritive gelatinous mass called a spermatophylax when they mate. The spermatophylax is joined to the sperm package or sperm ampulla, and both are attached to the female's genital opening during mating. As the sperm flow into the female's body, she begins to consume the spermatophylax. When she is finished, she often removes the ampulla. The larger the spermatophylax is, the longer she takes to eat it, and the more eggs are fertilized by the male, so males benefit by making as large a spermatophylax as possible. From the male's perspective, such contributions are costly, because each may weigh up to 30 percent or more of his body weight; therefore, it should be advantageous for a male to mate only with the best females. Larger females lay more eggs, and in many katydid species males will refuse to mate with females that are too small, whereas females will compete for access to gift-bearing males, a reversal of the usual situation.

Similar role reversal is seen in sea horses and their relatives, the pipefish. Females deposit eggs into a male pouch, where they are fertilized and harbored by the male until they hatch, when the male "gives birth" to them. In at least some species of pipefish, females are also ornamented, and males prefer more elaborately decorated females. Again, these are the exceptions that prove the rule; in both cases, the sex that is limited by the number of mates is chosen by the sex that is limited by the number of offspring it can care for.

—Marlene Zuk

male birds with elaborate songs also tend to have good physiological reserves, rather like the moorhens, and so that trait is linked to parental behavior. In some species, males give females bits of food during courtship, an activity called courtship feeding. Some scientists have suggested that if a male does a lot of courtship feeding, he is more likely to be diligent about feeding the young.

In a large number of species, however, males do not give females anything other than the sperm to fertilize their eggs, and it is in these species that female choice, along with the evolution of elaborate secondary sexual ornaments, is the most puzzling. In many of the animal species with the most showy traits, for example, birds of paradise and lyrebirds, females do not see the males after copulation, but instead rear the chicks by themselves. There are two major ideas about the evolution of female choice under these circumstances, and the argument is far from settled.

First is the suggestion, originated by R. A. Fisher in 1930 and elaborated by several other scientists, including Russ Lande, Steve Arnold, and Mark Kirkpatrick, that female choice in these species is arbitrary. [*See* Fisher, Ronald Aylmer.] Imagine an ancestral population in which some females have a slight preference for a male with, say, a longer tail than average. Perhaps the length gives the male a slight advantage in flight. The offspring resulting from males with long tails and females with the preference for them will inherit both the preference and the trait. The male offspring will grow to be preferred, and the daughters have the preference. The trait and the preference for it thus become genetically correlated, and selection will proceed to exaggerate the characteristic, such as the tail on a peacock, until natural selection stops it, for example, because it makes the male too conspicuous to predators. This snowballing effect is called runaway sexual selection or sometimes Fisherian selection. According to this view, the trait does not "mean" anything other than that it happens to have been preferred a long time ago. In other words, the ornament does not indicate that the male carrying it has greater survival ability or would be a better parent.

In contrast to this idea, the good genes or adaptive sexual selection hypothesis suggests that males with elaborate traits are indicating their viability and fitness to females, so that a female mating with a male who has an exceptionally long tail has offspring that not only have long tails themselves but are also more fit than offspring of less ornamented males. The secondary sexual characteristics are therefore indicators and are not arbitrary. This hypothesis differs from Fisherian sexual selection because here the trait actually is linked to higher survival, whereas in runaway sexual selection, the trait is neutral or even negative with respect to survival. Under Fisherian selection, females do not "know" anything else about a male save for the length of his tail. According to the good genes hypothesis, females "know" that a long-tailed male is better able to find food, escape predators, or has some other trait related to fitness.

Theoretically, both of these notions are plausible. Is one or the other the more likely basis of female choice? It has been difficult to test these hypotheses conclusively, in part because if an advantage to a trait is not found, it is difficult to know if that is because the advantage is absent, which would support runaway selection, or was simply not detected. A recent example of a test of the good genes hypothesis, by Marion Petrie, is discussed in the Mate Choice in Peafowl Vignette. Evidence such as Petrie's suggests that at the very least, arbitrary processes cannot always be responsible for mate choice in nature.

A few versions of good genes hypotheses deserve special mention. Amotz Zahavi, an Israeli biologist, suggested in 1975 that ornamental traits were "handicaps," which indicated to females that the male carrying them was so fit that he could cope with the cost of the ornament. Zahavi did not formally develop his ideas into a model; when other researchers attempted to do so, it became clear that handicaps could work only under some circumstances. Specifically, if males express the trait only when they are in a good condition, then handicaps can evolve. Such facultative models of handicaps appear to work the best, and numerous versions have been developed. For example, male guppies have orange patches that are attractive to females. The color comes from carotenoids in the foods that males eat, and it is likely that only males in good condition can sequester enough carotenoids to produce large, attractive orange patches. Several other animals, including house finches and other birds with red, orange, or yellow feathers, exhibit similar condition-dependent ornamentation.

W. D. Hamilton and Marlene Zuk developed a good genes model in 1982 that linked parasite resistance to female choice. Their model also addressed the problem that arises with continued directional female choice. In time, if only the highest quality males mate, all of the poor quality males (and their genes) will have been selected out of the population. The only variation among individuals in their secondary sexual characteristics that remains will thus be environmentally caused, so that some males who may have had poor nutrition, for example, became stunted as adults, but genetically they are no different from larger, better fed individuals. If this is the case, why should female choice for particular males matter?

This puzzle has been called the lek paradox, because it appears to be most prominent in species that display communally on traditional areas called leks. Several different solutions have been proposed. Most of these solutions involve changes in the kinds of genes that are advantageous in a particular time and place; if the "good genes" change, female choice can keep introducing variation into the population. Hamilton and Zuk suggested that the genes for resistance to parasites and pathogens may be especially subject both to this kind of change in time and to scrutiny by females looking for good genes. The reason for the advantage to females is simple: parasites are a continual problem for most animals in the wild, and whatever else is happening, it is always better to be able to resist disease. The reason for parasite–host interactions being especially subject to genetic change is also fairly simple: because parasites have genes of their own and have short generation times, they are able to respond to changes by the host, requiring ever more refined counterattacks. The result is what has been called an "evolutionary arms race." To illustrate, imagine a parasite that is lethal in one type of host, for example, one with a vulnerable gut mucosa, but not so harmful in another. Selection would favor the type of host that is not harmed by the parasite. If the parasite in response uses another avenue of attack, for example, the lungs, then the type of host with the good gut is not at an advantage anymore. Instead, the "good genes" are good lung genes. If this kind of varying selection can continue indefinitely, the parasite poses a continually moving target to the host.

From a female's point of view, it does not matter whether it is the lungs, gut, or legs—she merely has to find a male with apparent resistance to parasites. Hamilton and Zuk proposed that a female could do this best by looking at secondary sexual characteristics. They suggested that female choice for traits that revealed a male's ability to resist parasites had been important in the evolution of these ornaments.

Hamilton and Zuk predicted that, because of this pressure on males to look resistant and on females to scrutinize them, species that had been most heavily subjected to parasites in their evolutionary history should also be those with elaborate ornamentation. Within a species, however, females are trying to find the individual male who is most resistant to pathogens, so an individual male with brighter ornaments ought to have fewer parasites than a duller male. The ornaments can be seen as the female's cue to a male's ability to fight off parasites. Scientists have tested these predictions in many kinds of animals, and although results vary, parasites often have an impact on sexually selected traits that are important in mate choice.

How did the process of mate choice, and the accompanying exaggeration of secondary sexual traits, get started? Although R. A. Fisher speculated that the ornament was originally functional, this question has received relatively little attention. A recent suggestion is that male traits exploit the sensory biases of females that are already present in the nervous system. For example, if a female's visual system is particularly sensitive to the color red because prey items are often red, a male with a red appendage will automatically attract attention because of the preexisting bias for the trait. This preference will lead to selection of traits already attractive to females. Mike Ryan and other researchers have suggested that an appropriate test of this idea is to examine preferences for traits that do not occur in a particular species or population; if a sensory bias exists, females will prefer traits that have not evolved in their branch of the phylogenetic tree. Alexandra Basolo has performed such a study in a group of fish, including the swordtails (see the Vignette on Mate Preference and Sensory Bias in Swordtails). A related notion is that females respond to ornamental traits because they form supernormal stimuli, or exaggerated versions of every-

MATE CHOICE IN PEAFOWL

Marion Petrie has spent many years studying peafowl (strictly speaking, only males are called peacocks and females are peahens) that range freely in a large park in England. As in their native India, males gather in groups to display to females with their magnificent tails, but they do not provide females with any resources. Petrie first showed that males with more elaborate trains had higher mating success than other males, which suggests that the trait evolved because of female choice. She then experimentally altered the number of eyespots on the males' trains and found that peacocks with reduced eyespots had lower mating success than they had before the manipulation, whereas those with increased numbers of eyespots had higher mating success.

To see what females might be getting out of their choice for ornamented males, Petrie put groups of randomly chosen females in pens that each had one male. The males differed in the number of eyespots they had, so females were in effect getting preferred or unpreferred males as mates. Petrie then took the eggs from each female and reared them all in an incubator. By controlling for the parental behavior of the mothers in this way, she could see how well offspring of different fathers survived and grew. She found that the average weight of the offspring from males with larger eyespots was greater than that of offspring from less ornamented males. When the chicks were older, Petrie released a sample of them into the park where she had been working, then monitored their survival rates. It turned out that the survival of chicks was positively correlated to the area of eyespots on the tails of their fathers, which supports the good genes hypothesis.

—MARLENE ZUK

MATE PREFERENCE AND SENSORY BIAS IN SWORDTAILS

Swordtails are fish in the genus *Xiphophorus*, which also contains some species that do not have long swordlike appendages on their tails. Alexandra Basolo discovered that females of one such species, the platyfish *X. maculatus*, nevertheless prefer males with artificial swordtails attached to their caudal fins. If, as some versions of the *Xiphophorus* phylogeny suggest, the ancestor of the platyfishes arose before the species that gave rise to the swordtails, Basolo thought it was reasonable to conclude that a preference for long tails preceded the origin of the trait itself, supporting the sensory bias hypothesis.

Alternatively, the platyfish may be derived from a species with the sword on its tail. In this case, female platyfish prefer sworded males because they have retained the preference that arose when males also possessed the trait. To further examine the origin of the female preference, Basolo studied female preference for swords in another relative of the swordfish, *Priapella olmecae*, which also lacks the ornament. Females of this species also preferred males with an artificial sword, further bolstering the idea that the trait evolved after the preference for it was already present. Why preference for a long tail, rather than some other character, is maintained in the population is unclear.

—MARLENE ZUK

day objects. Perhaps because of the architecture of the nervous system, reactions to larger or brighter than normal signals are common, as in the goose that is more likely to attempt to incubate an egg many times normal size than she is to incubate a natural one.

Still another idea about the function of female choice comes from W. G. Eberhard (1996), who suggests that many forms of courtship behavior represent cryptic female choice, a term originally coined by Randy Thornhill. [*See* Cryptic Female Choice.] According to this idea, females can influence which male fertilizes their eggs by continuing to exert choice after copulation has started. Eberhard points out that many processes, including sperm transport, sperm storage, ovulation, and latency to remating, occur after copulation and may be under female control. By biasing the operation of these processes, females can increase or decrease the proportion of their offspring fathered by a particular male. The extremely complex genitalia of many animals, especially insects, may indicate an evolutionary history of male attempts to overcome the female gauntlet requiring more and more effort to fertilize the eggs. Eberhard also suggests that prolonged and complicated courtship rituals, such as those occurring during copulation of many flies and beetles, may have evolved via cryptic female choice. The advantage to females of using such courtship details to discriminate among males is unclear, although it is likely that Fisherian rather than good genes processes prevail.

Finally, it is important to remember that sexual selection via female choice may be opposed by natural selection. Traits that make a male conspicuous to prospective mates may also make him easily detectable by predators and parasites. In tungara frogs from Central America, for example, the same portion of the male's call that is most attractive to females is also used by predatory bats to find their prey. A group of parasitoid flies from the family Tachinidae localizes cricket and other calling insect hosts by homing in on the male's song. In some cases, the evolution of the male sexual signal is constrained by pressure from these natural enemies; as mentioned above, male guppies exhibit bright orange patches that attract female attention. These patches are reduced in areas with relatively high predation levels, where the sexual signal becomes more risky. Mate choice is only part of the explanation for the variation found in sexually dimorphic characteristics.

[*See also* Assortative Mating; Leks; Sexual Dimorphism; Sperm Competition.]

BIBLIOGRAPHY

Andersson, M. *Sexual Selection*. Princeton, 1994. A good, even-handed treatment of most topics in sexual selection, with several chapters on mate choice, including mate choice in plants.

Basolo, A. "Female Preference Predates the Evolution of the Sword in Swordtail Fish." *Science* 250 (1990): 808–810.

Bateson, P., ed. *Mate Choice*. Cambridge, 1983. Contains articles on general aspects of mate choice and on its operation in particular species.

Cronin, Helena. *The Ant and the Peacock: Altruism and Sexual Selection from Darwin to Today*. Cambridge, 1991. A popular and well-written discussion of two of the most perplexing ideas in Darwin's writings, including a consideration of mate choice.

Darwin, C. *The Descent of Man, and Selection in Relation to Sex* (1871). Princeton, 1981. Fascinating to read, even if the English is a bit archaic, as Darwin painstakingly explores the foundation of sexual selection and the function of mate choice.

Eberhard, W. G. *Female Control*. Princeton, 1996. Explanation of cryptic female choice and an argument for its prominence in nature, with many examples from both invertebrates and vertebrates.

Hamilton, W. D., and M. Zuk. "Heritable True Fitness and Bright Birds: A Role for Parasites?" *Science* 213 (1982): 384–387.

Houde, A. E. *Sex, Color, and Mate Choice in Guppies*. Princeton, 1997. Detailed discussion of mate choice and the evolution of color patterns in the guppy, a model system for sexual selection studies. Contains a particularly good treatment of how experiments to test mate choice can be constructed, along with good suggestions for future research.

Ligon, J. D. *The Evolution of Avian Breeding Systems*. Oxford and

New York, 1999. A thorough consideration of bird mating behavior, with several sections on mate choice; note particularly the discussion of the role of phylogeny in the evolution of mating systems.

Low, B. S. *Why Sex Matters*. Princeton, 1999. Written for a general audience, the book focuses on humans and discusses the evolution of mate choice along with other topics in human evolution.

Petrie, M. "Female Moorhens Compete for Small Fat Males." *Science* 220 (1983): 413–415.

Ridley, M. *The Red Queen: Sex and the Evolution of Human Nature*. New York, 1995. A popular book examining both the evolution of sexual reproduction and its repercussions, including mate choice.

Ryan, M. J., and A. S. Rand. "The Sensory Basis of Sexual Selection in the Túngara Frog *Physalaemus pustulosus* (Sexual Selection for Sensory Exploitation)." *Evolution* 44 (1990): 305–314. An explanation of sensory bias and how it may operate in a tropical frog.

Thornhill, R. "Sexual Selection and Nuptial Feeding Behavior in *Bittacus apicalis* (Insecta: Mecoptera)." *American Naturalist* 110 (1976): 529–548. One of the first demonstrations of nuptial gifts and their costs and benefits.

Trivers, R. L. "Parental Investment and Sexual Selection." In *Sexual Selection and the Descent of Man 1871–1971*, edited by B. Campbell, 136–179. London, 1972. The classic explanation of male and female reproductive strategies.

Williams, G. C. *Sex and Evolution*. Princeton, 1975. Concise, well-written explanation of the basic differences between males and females and how these are manifested in various species. Contains some mathematical treatments, but these mainly require only algebra to understand.

Zahavi, A. "Mate Selection—A Selection for a Handicap." *Journal of Theoretical Biology* 53 (1975): 205–214. The original explication of the handicap principle.

— Marlene Zuk

Mate Choice in Plants

Competition among males for access to females, as well as choice by females among potential mates, is often responsible for striking traits such as the bright plumage of male pheasants, a process known as sexual selection. Although most discussion of sexual selection concerns animals, these kinds of selective pressures may be important in plant mating as well. Whenever more pollen grains arrive on a stigma than are necessary to fertilize all of the ovules, there is an opportunity for selection among pollen grains during mating. Pollen deposition frequently greatly exceeds the number of ovules. The opportunity for selection is further increased if the pollen grains are from more than one pollen donor. Field studies show that pollinators are likely to carry mixtures of pollen from several different plants, so it is likely that stigmas receive mixtures of pollen from different plants.

Once a mixture of pollen from two or more pollen donors arrives at a stigma, pollen donor success may be nonrandom for a variety of reasons. Many plants have self-incompatibility systems that prevent or reduce fertilization of ovules by their own pollen or by the pollen of some close relatives. Some plants also show reduced fertilization by pollen from too distant relatives, and most discriminate against pollen from other species. None of these patterns of nonrandom mating fit with models of sexual selection. These patterns are based on degree of relatedness among plants and will not cause some pollen donors to be more successful than others across a number of maternal plants.

However, another pattern of nonrandom mating may also be common in plants and may indicate the potential for sexual selection. For a variety of crop and noncrop species, artificial pollination using mixtures of pollen from more than one compatible, unrelated donor resulted in differential pollen donor success. This differential pollen donor success is usually measured by using genetic markers to determine which pollen donor sired each seed. For example, pollen donors carrying genes for different seed phenotypes sired different numbers of seeds on ears of corn. In wild radish (*Raphanus sativus*), use of genetic markers showed that differential pollen donor performance is very common. Mixtures of pollen from groups of two to six donors have been studied, and virtually all result in unequal siring of seeds by the pollen donors.

When pollen donors show differences in the number of seeds sired, there are at least two different explanations: (1) pollen donors may have varying abilities to compete for access to ovules or (2) the maternal plant may be able to influence the outcome of pollen tube growth and ovule fertilization, effectively choosing its mates. Of these possibilities, competition among pollen donors has been less controversial and more frequently studied.

Competition among Pollen Donors. Competition among pollen donors for access to ovules can occur by a variety of means. During pollination, pollen donors that make bigger flowers, have greater rewards for pollinators, or make more pollen may simply get more of their pollen to other flowers. This may allow them to fertilize more ovules because they have more opportunities to do so. Once pollen arrives on a stigma, pollen from different donors is usually thought of as racing for ovules. Pollen that germinates faster and grows more quickly through the stigma would reach and fertilize more ovules. In truth, only a few studies actually measure both pollen tube growth rate and ability to sire seeds. An excellent example of such a study comes from work done on *Hibiscus moscheutos*. For this species, pollen tube growth varies among pollen donors. These differences in pollen tube growth are very similar across maternal plants. In addition, plants with faster pollen tube growth sire more seeds, as determined by genetic markers (Snow and Spira, 1996).

More frequently, effects of pollen competition have

been studied by making a fundamental assumption about the correlation between pollen tube growth rate and offspring growth rate. Many of the genes expressed during pollen tube growth are also expressed during seedling growth. Thus, it is assumed that increasing the opportunity for pollen competition will cause the fastest growing pollen tubes to sire seeds. Because there is overlap in gene expression, seeds sired by fast-growing pollen tubes are also assumed to produce highly vigorous progeny. Attempts to manipulate the amount of pollen competition and measure offspring growth rate have produced mixed results. Sometimes an increase in pollen load size results in improvements in offspring growth, and sometimes it does not. This discrepancy may occur because (1) pollen load sizes were not varied sufficiently to affect differential pollen donor success, (2) nonrandom mating does not occur at any pollen load size, (3) there is no correlation between the genes that affect mating and the genes that affect offspring growth, or (4) mate choice is important and occurs at all of the pollen load sizes used.

Demonstrating Mate Choice in Plants. Although it is logical to assume that mate choice will occur by some means if pollen is abundant and the paternity of seeds affects offspring quality, it has been difficult to demonstrate and very controversial in plants. The controversy comes from several sources. First, some scientists have been uncomfortable with the use of the word *choice* because it seems to imply consciousness. However, models of mate choice in animals do not assume conscious choice. Rather, they refer to mechanisms by which females consistently alter the outcome of mating. Second, there has been concern about applying terms developed for animal mating to plant mating systems. However, this use of similar terms can aid us in understanding conceptual similarities among these mating systems. Third, mate choice in plants has been very difficult to demonstrate.

The difficulty in demonstrating mate choice in plants rests largely on the problem of separating mate choice from pollen competition. To test for mate choice, a diversity of pollen tubes must be present. However, under these circumstances, pollen competition may also occur. Three kinds of evidence have been used to tease apart the potential for mate choice from that for pollen competition.

First, it was suggested that variability in pollen donor performance across maternal plants must be a result of effects of maternal tissue. Thus, finding such variation could be taken as evidence for mate choice. Unfortunately, variation in pollen donor performance across maternal plants also suggests that pollen donor performance is affected largely by the genetic relationships among maternal plants and pollen donors. Although this may be construed as choice for mates of a suitable degree of relatedness, variable pollen donor success across maternal plants will not produce selection for particular male characters.

Second, attempts have been made to measure pollen donor performance across maternal plants of varying condition. If maternal plant condition affects the outcome of mating, then the maternal plant must play some role in regulating pollen donor performance. Several experiments using wild radish show that pollen donor performance is different on maternal plants that have been subjected to stress than on other maternal plants. The pattern of change is that when maternal plants are in poor condition, seed siring success is more equal among pollen donors than when maternal plants are in good condition. This suggests that the tissues of healthy maternal plants are more competent to regulate the outcome of mating.

Detailed examination of pollen tube growth and ovule fertilization support this interpretation. For peach flowers (*Prunus persica*), pollen tubes grow to the base of the style, where their further development is arrested by the maternal tissue. A change in the maternal tissue releases the pollen tubes to grow the remainder of the distance to the ovule. Thus, the maternal tissue sets up a temporary barrier. In wild radish, initial pollen tube growth and ovule fertilization are slower for plants grown in optimal conditions than for plants given low water. This suggests that maternal tissue also sets up barriers that regulate mating and that the barriers are reduced when maternal plants are under stress.

Further work on changes in mating patterns in wild radish over the length of the flowering season also supports a role for the maternal tissue in regulating mating. Early in the flowering season, there is little difference in seed-siring success among pollen donors. A few weeks later, variation in seed-siring success is maximized. Over the period of time involved, maternal plant condition is changing because both the number of previous pollinations and the number of developing fruits are increasing. However, over the same period of time, neither pollen size nor number of pollen grains per flower changes in wild radish.

A third observation also suggests a role for the maternal tissue in regulating mating. For wild radish, fruits are filled to greater mass when several seeds within the fruits are sired by a variety of pollen donors than when all of the seeds are sired by a single pollen donor. Thus, maternal plants allocate more resources to fruits when genetic diversity of the seeds is highest. This pattern is difficult to explain solely in terms of differential pollen donor ability. Further data showing that the mass of fruits can be affected by the identity of neighboring fruits and that fruits are filled to larger mass when they are sired by pollen donors that did not sire nearby fruits support this interpretation. Once again, maternal plants of wild radish appear to be allocating resources such that offspring diversity is increased. Interestingly, this is very similar to a rare male advantage in animal mating systems.

All of the examples presented have dealt with choice by the maternal tissue among the pollen grains that arrive on stigmas. However, maternal plants may affect the prepollination aspects of mating. Changes in the number of flowers produced or the timing of flower production may alter the amount and diversity of pollen received, which would increase the opportunity for mate choice. Likewise, it has been suggested that maternal plants may be selected to alter the length of stigma receptivity or the period of time during which pollen will germinate. For example, if pollen will adhere to stigmas for a period of time (several hours or days) before germination can occur and if pollen germination is then permitted by a change in condition of the stigma, the amount of competition among pollen grains that arrived over a period of time would be maximized. Regulation of the amount of competition could be a substantial maternal effect on mating.

Finally, it is important to consider the most prevalent objection to the possibility of mate choice in plants. Because physiological self-incompatibility is widely known and widely studied, it has frequently been suggested that the only mate choice that occurs in plants is through sorting among compatible and incompatible pollen. Under this hypothesis, the observations of plant mating that have invoked sexual selection would be misinterpretations. Rather, the data would represent unusual manifestations of the self-incompatibility system. However, for at least one species, wild radish, unexpected effects of the self-incompatibility system cannot explain the results. Pollen donors show consistent patterns of unequal seed-siring success across groups of maternal plants known to have different incompatibility genotypes. Furthermore, differential seed-siring success of pollen donors can occur after pollinations that are designed to circumvent the incompatibility system (bud pollinations).

Nonetheless, the relationship between a variety of processes that occur during plant mating remains a significant unanswered question. For many plants, physiological self-incompatibility, choice among compatible mates, and avoidance of inbreeding, too distant outbreeding, and hybridization may occur simultaneously on the same plant if not the same flower. The degree to which these mating processes are related through similar function or through similar selective events is unknown.

[*See also* Male–Male Competition; Mate Choice, *article on* Human Mate Choice; Sperm Competition.]

BIBLIOGRAPHY

Charlesworth, D., D. W. Schemske, and V. L. Sork. "The Evolution of Plant Reproductive Characters: Sexual vs. Natural Selection." In *Evolution of Sex*, edited by S. C. Stearns, pp. 317–335. Basel, 1987. Criticizes of much of the previous work on sexual selection in plants and points out some of the difficulties in interpretation of such data.

Delph, L. F., and K. Havens. "Pollen Competition in Flowering Plants." In *Sperm Competition and Sexual Selection*, edited by T. R. Birkhead and A. P. Moller, pp. 149–173. New York, 1998. Considers male–male competition in plants and contrasts these studies with work on sperm competition in animals.

Herrero, M., and J. L. Hormaza. "Pistil Strategies Controlling Pollen Tube growth." *Sexual Plant Reproduction* 9 (1996): 343–347. Considers several mechanisms by which maternal tissue may control mating.

Lyons, E. E., N. M. Waser, M. V. Price, J. Antonovics, and A. F. Motten. "Sources of Variation in Plant Reproductive Success and Implications for Concepts of Sexual Selection." *American Naturalist* 134 (1989): 409–433. Points out some of the problems in previous studies of sexual selection in plants and suggests types of experiments that might be informative.

Marshall, D. L., and P. K. Diggle. "Mechanisms of Differential Pollen Donor Performance in Wild Radish, *Raphanus sativus* (Brassicaceae)." *American Journal of Botany* 88 (2001): 242–257. Covers patterns of pollen tube growth and ovule fertilization and strongly supports a role for the female tissue. It references most of the earlier work on wild radish.

Marshall, D. L., and M. W. Folsom. "Mate Choice in Plants: An Anatomical to Population Perspective." *Annual Reviews of Ecology and Systematics* 22 (1991): 37–63. Literature review, concentrating on material published since 1983.

Snow, A. A. "Postpollination Selection and Male Fitness in Plants." *American Naturalist* 144 (1994): S69–S83.

Snow, A. A., and T. P. Spira. "Pollen-Tube Competition and Male Fitness in *Hibiscus moscheutos*." *Evolution* 50 (1996): 1866–1870.

Stephenson, A. G., and R. I. Bertin. "Male Competiton, Female Choice, and Sexual Selection in Plants." In *Pollination Biology*, edited by L. Real, pp. 109–149. Orlando, Fla., 1983. Reviews both pre- and post-pollination possibilities for male-male competition and female choice.

Willson, M. F., and N. Burley. *Mate Choice in Plants: Tactics, Mechanisms, and Consequences*. Princeton, 1983. This monograph set the stage for debate on the possibility of mate choice in plants. It was regarded as controversial.

— Diane L. Marshall

Human Mate Choice

Although human mating patterns vary from culture to culture, patterns of parental investment are less variable: mothers everywhere do most of the work of taking care of children. This trait, which we share with most other species, has a predictable set of reproductive consequences. Maternal investment is, for males, a scarce resource worth competing over. To the extent that a male does not devote time and energy to caring for offspring, his reproductive success depends on his ability to succeed in competition with other males in mating with as many females as possible. A female, on the other hand, constrained as she is by a long period of intensive nurture, does not gain additional conceptions from mating with a large number of males. On the contrary, she can lose much valuable time and energy on the consequences of a bad mating. Human females, therefore, resemble other mammals in being more discriminating than males in their choice of mates. The role of male–male competition in our species is manifested both physiologically and behaviorally: the variance in reproductive success is larger among men than among women, especially in economies where a few men are able to monopolize a large fraction of the resources; and men are larger than women, more physically aggressive, more risk-prone, and later to mature. In all these respects we are typical mammals, although less sexually dimorphic and less polygynous than most.

While this pattern of sex differences is true in general outline, humans (and most other primates) also have a range of tactics and strategies. Although women do not gain additional conceptions by having a variety of mates, they, like other female primates, can sometimes gain resources, protection, and good genes by engaging in short-term matings with more than one male. Among human males, conversely, the intensity of male–male competition is mitigated to some extent by long-term (although not necessarily exclusive) pair bonds and by the importance of paternal provisioning in most human societies. Being solicitous toward a mate and her offspring also helps males of many primate species (including humans) gain mating advantages. The importance of pair bonds and the fact that men frequently control economic resources adds complexity to the usual mammalian picture. Human females may compete for resource-holding males, and human males may be quite selective in their choice of long-term mates.

Given all this, it is not surprising that women and men are similar in most of their mate choice criteria. In addition to those familiar attributes consciously valued by people in all cultures (kindness and intelligence are high on the list for both sexes), human mate choice for men and women involves (1) finding someone genetically different, but not too different, and (2) finding someone in good health and with a robust genetic constitution.

Optimal Outbreeding. Sociologists have long observed that people mate with people similar to themselves. This produces a moderate degree of inbreeding, which some biologists have suggested may have beneficial genetic consequences. We know little about these consequences for humans (with isolated exceptions such as Rh incompatibility). We know more about the deleterious genetic consequences of close inbreeding (i.e., mating between siblings), and there are at least two mechanisms by which people avoid it. One is a form of early learning whereby people develop an aversion to mating with people they were reared with. This appears to be a kind of imprinting, with the critical period being the first few years of life.

More recently, it has been discovered that we even prefer the smell of people who are genetically different from ourselves. The major histocompatibility complex (MHC) is a highly polymorphic group of genes that code for immunological proteins, and heterozygosity in these genes is thought to be advantageous in promoting resistance to a diversity of pathogens. Like rats and some other species, people appear to prefer the smell of those whose MHC genes differ from their own, presumably because the offspring are better able to resist infectious disease.

Fluctuating Asymmetry. Both men and women value good health and a robust genetic constitution in a mate. But how is one to assess it? Certain obvious indicators, such as clear skin, are valued the world over, but other indicators may be less obvious. Because perfect bodily symmetry is difficult to attain, fluctuating asymmetry—asymmetry in normally symmetrical body parts—is thought by some to be a good indicator that an individual was exposed to stressors during development and did not have a genotype robust enough to withstand them perfectly. In some bird species, males with low fluctuating asymmetry (FA) also have traits (such as longer tails) that make them attractive to females. The same may be true for humans. A number of studies purport to show that women prefer men with low fluctuating asymmetry and that such men have reproductive advantages. Men with low FA have more sexual partners and more extra-pair sex, their mates have more sperm-retaining orgasms (discussed below), and they are judged by women to be more physically attractive. Evidence that men prefer symmetrical women is weaker, which is not surprising given the greater choosiness of females.

Although there is a clear preference for mates with low FA, it is not known how people are assessing it. It is unlikely to be the asymmetry itself, because FA measurements, which are taken with calipers, are probably not perceptible to the naked eye. In one recent study, women preferred the faces of symmetrical men even though they were unable to judge which faces were most symmetrical. More symmetrical individuals had more masculine features (wide cheekbones and longer lower face), and the women may have been responding to that. If so, it suggests that masculine facial features are an honest signal of good physical condition in a man. This possibility is discussed further in a later section. Fascinating as these studies of FA are, however, there is a growing perception that the results of much work on FA are difficult to replicate.

What Women Want: Resources and Investment. A large body of literature, including a questionnaire study conducted in 37 societies, shows that women everywhere value resources and financial prospects in a mate more than men do. Women also value dependability and emotional stability in a mate—traits that would be important in a long-term partner. These preferences are clearly adaptive, because paternal investment and protection are important to child survivorship. Two fathers may be even better than one: the Ache and the Bari (indigenous South American societies) believe that biological fatherhood is partitioned among all the men that a woman has sex with while she is pregnant, and in both societies, children with two fathers have the highest survival rates.

Belief in "partible paternity" is unusual, but female interest in extrapair matings is not. What might women gain from this? One possibility is that females solicit matings from other males in order to confuse paternity and thereby procure more male investment, or to reduce the danger to infants that males otherwise may pose. This strategy has been shown to be adaptive for female dunnocks (a bird species) and langurs (a species of monkey) and seems plausible for a wide array of other primates, including our closest living relatives, the chimpanzees, but it is not known how widely it applies to people.

Does Human Polygyny Result from Female Choice? In all societies, wealthy, high-status men have more mates. This robust association could arise from female choice; a woman who shares a very wealthy man with other women might still have more resources left for her offspring than if she had a poor man all to herself. Her children might also benefit from his social connections, as has been suggested for both Ache hunter-gatherers and nineteenth-century Mormons. However, the association of wealth with polygyny in men could also stem from the ability of a wealthy, powerful man to coerce other males and prospective wives to do as he wishes. It is difficult to distinguish which is most important in explaining human polygyny. The female-choice argument (formalized as the "polygyny threshold" model when it was first applied to birds) seems to explain polygyny among the Kipsigis (an agro-pastoral people of East Africa), since Kipsigis women choose men who can provide them with the most land after it has been divided and appear to benefit reproductively from their choice. Among the Dogon of Mali, on the other hand, children in polygynous households have higher morbidity and mortality than do children in monogamous households. Other cases of human polygyny, such as the large harems maintained by ancient despots, seem more obviously a case of male coercion than female choice. In any event, both the degree of polygyny and the importance of material resources to women are probably greater today than they were during most of our evolutionary past, since the accumulation and control of large stores of wealth is a post-agricultural development.

What Women Want: Immediate Resources. Like many bird species, humans form pair bonds and males invest a lot in offspring. But, again as with many bird species, it has recently been recognized that there are a lot of extrapair matings in these supposedly monogamous relationships, at least some of which are initiated by females. One reason—procurement of additional investment—was discussed above. There are two other plausible reasons for a female to seek extrapair matings. If long-term investment is not likely to be forthcoming, a woman might still benefit from a short-term mating if she can gain immediate resources. This might involve expensive gifts in modern industrial societies but is sometimes claimed to involve meat among hunter-gatherer societies. Ache hunters, like foragers everywhere, share their kills widely with members of the band, yet successful Ache hunters have more mates than poor hunters do. Since they are sharing the meat widely, perhaps their motive is less the exchange of meat for sex than it is the display of qualities that make men desirable friends and dangerous competitors.

What Women Want: Good Genes. If a woman is unlikely to get either immediate or long-term resources, she may still benefit by getting "good genes" from a short-term mate. Those good genes will be beneficial if they provide her with healthy, attractive, dominant offspring. But how can a female get an honest signal of good genes? Some evolutionary theorists have suggested that "costly signals" (traits that are energetically expensive or that reduce resistance to disease) are honest precisely because only highly fit individuals can produce them and still survive. Many of the traits that women find attractive in men, particularly when they are interested in protection or in short-term matings, appear to fit this criterion. The tendency of human males to show off through risky activities and extravagant production may be costly signals of this sort. So might masculine morphological traits (enhanced muscle mass, wide shoulder-to-hip ratio, masculine and mature facial features), all of which are considered attractive by women, particularly when they are seeking a short-term mate. Testosterone facilitates competitive behavior in men and promotes masculine morphology, but it does so at an immunological cost. The cost may be worth it to males not only in direct male–male status competition but also as an honest signal to women of high mate value.

There is growing evidence that women's mate preferences are responsive to their reproductive status and their particular goals. Women seeking short-term mates (hence presumably more interested in genetic quality) place a greater value on attractiveness than do women seeking long-term mates. We would also expect that women seeking genetic quality in a mate would act so as to enhance the chance of conceiving with such a male. Such an expectation is consistent with the finding that only fertile women prefer men with low FA, that women are more likely to seek extrapair partners when they are in the fertile phase of their cycle, and that single women in their fertile phase are more likely to engage in such mate-seeking behaviors as wearing revealing clothes. One (as yet unreplicated) study has claimed that female orgasms at or after ejaculation retain more sperm, and that women having extrapair copulations have such orgasms more frequently with their extrapair than with their regular partner. Such "cryptic female choice" would be favored if a woman is trying to get paternal investment from one male while getting genes from a male whose high mate quality makes him an unlikely long-term partner. Men, obviously, have evolved counter-strategies to such behavior.

Sperm Competition. Throughout the animal kingdom, one male response to female promiscuity has been sperm competition, one form of which involves outnumbering the competition through sheer volume of sperm. There is some dispute about whether sperm competition is important in humans. The best evidence for the magnitude of sperm competition in a species is probably the size of the testes, adjusted for body weight. Humans are average primates by this measure, with a relative testis size smaller than chimpanzees, who live in multi-male groups, but larger than gorillas, whose groups usually contain only one breeding male.

What Men Want: Assurance of Paternity. Males in a number of species respond to the threat of female infidelity by guarding their mate to keep her from mating with other males. A woman always knows that her child is her own, but a man can never be as certain. Men, therefore, value chastity and sexual fidelity in a mate more than women do. Mate guarding can reach extreme forms in some human societies, involving such things as secluding women to keep them from other men and clitoridectomy (removing the clitoris) to reduce their sexual desire. In many such societies, the "sexual purity" of a family's women is essential to its honor—and to a woman's likelihood of attracting a desirable husband. The importance of chastity to men varies considerably across cultures, probably in part because of differences in the degree to which men invest in offspring.

What Men Want: Beauty and Reproductive Potential. Men value physical attractiveness in a mate more than women do, and this sex difference, while not large, appears consistently across societies. The reason for this sex difference is apparent when we consider what people consider beautiful in a woman. If people are asked to rank photos of female faces (or drawings, or experimentally altered photos), the results indicate a preference for youthful features such as small chin and full lips. These preferences are found in nearly all societies. A youthful physique, with firm breasts and a low waist-to-hip ratio, are also considered attractive. This preference for youthful features is mirrored by the universal male preference for younger women as mates (women universally express the opposite preference, presumably because older males are more likely to have more resources and higher status). We take this preference for granted because of its familiarity, but both young and old male chimpanzees do not share it: old female chimpanzees, who have demonstrated fertility and experience as mothers, retain their sexual attractiveness. Why are humans different?

The reason may lie in our life history. Human females come to the end of their fertility at about the same age chimpanzees do but are much longer lived. Unlike other primates, we usually live beyond our childbearing years and have a long and active postmenopausal life. Humans are also unusual among mammals in forming pair bonds that often last many years. Any man who found older women more sexually attractive and more desirable as mates might lead a happy life, but probably not one favored with many offspring. Reproductive success is the coin of evolution, and human mate preferences, conscious or not, show clear evidence of its effects.

[*See also* Fluctuating Asymmetry; Human Sociobiology and Behavior, *article on* Evolutionary Psychology; Sexual Selection, *article on* Bowerbirds; Sperm Competition.]

BIBLIOGRAPHY

Baker, M., and M. Bellis. *Human Sperm Competition: Copulation, Masturbation, and Infidelity.* New York, 1995. A controversial book summarizing fascinating but unreplicated evidence for the existence of sperm competition and cryptic female choice in humans. Their suggestion that different sperm morphs have different functions in human males has been criticized, but their findings about the function of female orgasm have been widely cited.

Borgerhoff Mulder, M. 1992. "Women's Strategies in Polygynous Marriage: Kipsigis, Datoga, and Other East African Cases." *Human Nature* 192 (1992): 45–70. Shows, among other things, that polygyny among the Kipsigis is a consequence of female choice.

Buss, D. *The Evolution of Desire: Strategies of Human Mating.* New York, 1994. Discusses the findings from his cross-cultural questionnaire studies in 37 societies and extends these findings with other relevant data.

Cashdan, E. "Women's Mating Strategies." *Evolutionary Anthropology* 5 (1996): 134–143. A review of women's mate choice aims and tactics, with an emphasis on explaining cross-cultural variation.

Hrdy, S. B. *The Woman That Never Evolved.* Cambridge, 1981. One of the first books to refute the stereotype of coy and chaste primate females; reviews and explains differences in primate mating patterns and suggests that females in multi-male primate groups solicit copulations from a variety of males in order to confuse paternity and get better treatment for their offspring; speculates on the human implications of these findings.

Mealey, L. *Sex Differences: Developmental and Evolutionary Strategies.* San Diego, 2000. A comprehensive textbook that reviews all the human mate choice issues covered in this entry.

Scheib, J., S. Gangestad, and R. Thornhill. "Facial Attractiveness, Symmetry and Cues of Good Genes." *Proceedings of the Royal Society Biological Sciences Series B* 266 (1999): 1913–19. Replicates earlier findings about female preference for symmetrical men, and considers what it is about these men that women find attractive.

Singh, D. "Adaptive Significance of Female Attractiveness: Role of Waist-to-Hip Ratio." *Journal of Personality and Social Psychology* 65 (1993): 293–307. Low waist-to-hip ratio (WHR), with fat deposits on the hips rather than the abdomen, is widely, if not universally, found attractive in women. Singh argues that this is an adaptive preference since low WHR is associated with higher fecundity and lower rates of certain degenerative diseases.

Wedekind, C., and S. Furi. "Body Odor Preferences in Men and Women: Do They Aim for Specific MHC Combinations or Simply Heterozygosity?" *Proceedings of the Society of London B* 264 (1997): 1471–1479. Reports that body odor preferences ensure greater offspring heterozygosity; results based on experimental data.

Wilson, M., and M. Daly. "The Young Male Syndrome: An Analysis of Male-Male Competition for Mates." *Ethology and Sociobiology* 6 (1985): 59–62. Young men take more risks than women or older men, and are disproportionately represented as both perpetrators and victims of homicide. Wilson and Daly interpret the evidence in terms of status and reproductive competition.

— Elizabeth Cashdan

MATERNAL CYTOPLASMIC CONTROL

The regulation of postfertilization development by cytoplasmic messages of maternal origin is referred to as maternal cytoplasmic control (MCC). Its initiation and cessation are critical for embryo survival and development. The first experimental evidence for MCC was obtained in studies of sea urchin hybrids in the late 1800s and early 1900s.

Fertilization brings the maternal and paternal genomes together in a common maternally derived cytoplasm and initiates a cascade of events leading to the temporal and spatial expression of genes prerequisite for embryonic and fetal development. However, the earliest stages of development are largely dependent on maternally derived messages stored in the oocyte prior to fertilization. This is an intriguing time of development because during the period of MCC, the maternal lineage in effect dominates the genes of paternal origin, and the maternal lineage charts the course of early development. Maternal messages are eventually depleted or modified, and embryo-derived messages become key controlling factors. The transition from dependence on oocyte-derived messages to embryo-produced messages is referred to as the maternal-to-embryonic transition (MET) or embryonic genome activation (EGA). The timing of this crucial transition varies among species and ranges from a few hours to several days after fertilization, as follows: *Drosophila*—two hours and ten minutes postfertilization during the fifteenth cell cycle (about 6,000 cells); *Caenorhabditis elegans*—two hours and thirty minutes during the fourth to sixth cell cycle (about twenty-eight cells); *Xenopus*—seven to eight hours after fertilization during the twelfth cell cycle (about 4,000 cells); mouse—one day postfertilization during the second cell cycle (two-cell stage); humans—two to three days postfertilization during the third or fourth cell cycle (four- to eight-cell stage); cattle—three days postfertilization during the fourth cell cycle (eight- to sixteen-cell stage).

The so-called zygotic clock governs these precisely timed sequences of events. However, the mechanism by which this clock operates is a matter of some speculation. Possibilities include time elapsed from fertilization, number of rounds of DNA replication, nucleo-cytoplasmic ratio, and the dilution or titration of transcriptionally repressive maternal factors. The zygotic clock may not be universal and may differ between those species with rapid cell turnover, lack of G1 and G2 phases in early cell cycles of development, and short time to MET and those with slower developmental kinetics, longer cell cycles, including G1 and G2, and longer time to MET.

Transition from Maternal to Embryonic Control. The transition from maternal to embryonic control of development is a multistep process that initially involves the repression of transcription of the embryonic genome followed by the acquisition of a transcriptionally permissive state and the coordinated activation of groups of genes. The degradation of maternal RNA and protein, sensitivity to transcriptional inhibitors such as alpha-amanitin, and a burst of transcriptional activity from the embryonic genome are consistent with maternal cytoplasmic control of development. The content of polyadenylated and ribosomal RNA declines from the oocyte to the morula stage, after which there is a marked increase in all classes of RNA. Inhibition of polymerase-dependent transcription by alpha-amanitin during the earliest stages of development has shown that embryos can survive in the absence of transcription from the embryonic genome until a specific stage of development. This generally coincides with the point at which EGA and MET occur. Thus, embryo sensitivity to the lethal effects of alpha-amanitin occurs at the two-cell stage in mice and the five- to eight-cell stages in cattle.

Assessment of transcriptional activity from the embryonic genome in mammals has been examined using a variety of techniques, including measurement of incorporation of [S^{35}]UTP and [H^3]uridine into RNA, differential display reverse transcription-polymerase chain reaction (RT-PCR) northern blot analysis, and sequence-specific PCR and two-dimensional electrophoretic analysis of proteins. In mice and cattle, low levels of transcription from the embryonic genome have been identified within hours of fertilization in the male pronucleus of the zygote. Transcription from the embryonic genome appears to occur at a low level in early stages of development when the G1 phase of the cell cycle is short in cycle 10 in *Drosophila* and cycle 1 in cattle and mice, followed by a burst of transcriptional activity during cycles 12 in *Drosophila*, 4 in cattle, and 2 in mice.

EGA occurs against a backdrop of structural changes to the nucleus that prepare the DNA template for transcription and molecular changes to the cytoplasm that regulate the level of activity and abundance of transcription and translation factors. A key event that precedes the transition from maternal cytoplasmic control to embryo genomic control of development is the conversion of the maternal and paternal gametic genomes into a single embryonic genome. This involves the nucleus and the transcriptional and translational machinery. Essential to this process is the replacement of the nonpermeable sperm nuclear envelope with a porous one that facilitates the replacement of sperm-specific protamines by oocyte and cleavage stage–specific histones. Concurrent with formation of the paternal pronucleus is the completion of meiosis and the development of the maternal pronucleus. The two pronuclei are in direct competition for the uptake of specific chromatin proteins, such as hyperacteylated histone H4, which are in limited supply in the oocyte cytoplasm.

Changes to the configuration of nuclear chromatin structure play a major role in creating a genome that is able to change back and forth from a transcriptionally repressive to a transcriptionally permissive state. At the chromatin level, this is regulated by the array of histone subtypes and posttranslational modifications, most notably the changes to the acetylation of histones. These changes affect the nucleosome repeat length and thereby alter the accessibility of DNA binding sites. The changes in histone content and configuration occur in a progressive manner, and the kinetics of these processes is species-dependent and correlated somewhat with the timing of EGA.

In addition to changes to the protein structures surrounding the DNA, changes to the methylation pattern of DNA itself (methylation of the fifth position of cytosine in cytosine-guanine doublets) are capable of modifying chromatin organization to prevent or allow the binding to transcription sites of activators or inhibitors of transcription. The DNA of pronuclei is differentially methylated with housekeeping genes generally unmethylated in both gametic genomes, whereas tissue-specific loci are highly methylated in sperm and less methylated in oocyte pronuclei. The preferential accumulation of acetylated histones and the differential methylation of DNA provide epigenetic mechanisms for establishing long-lasting differential gene expression from the parental genomes in instances of postfertilization imprinting.

Cytoplasmic changes that may be envisioned include a combination of translational and posttranslational modifications to maternal and early embryo transcripts. The basal transcriptional machinery, largely derived from the oocyte, requires a series of modifications to become fully functional. Phosphorylation patterns of RNA polymerase II and translocation to the nucleus are critical components of the regulation of inhibition of transcription. Dephosphorylation of the carboxy terminal domain of the largest subunit of RNA polymerase II has been shown to occur just before the minor burst of transcription in mice. The phosphorylation patterns resemble that of somatic nuclei of later stages of development by the time EGA has occurred.

Factors Involved in the Transfer of Cytoplasmic Control. The transfer of terminally differentiated somatic nuclei into enucleated oocytes has resulted in the birth of live cloned offspring in several species. In this process, maternal cytoplasmic factors in the oocyte promote the dedifferentiation and reprogramming of the genome of the transplanted nucleus. In essence, the oocyte is capable of recreating the totipotent status that the differentiated nucleus had as a zygote. Based on biochemical and morphological studies, it appears that the transcriptionally competent transplanted nuclei are reprogrammed and enter a period of transcriptional suppression, and go through many of the same modifications as nuclei prior to EGA. In mammals, the low success rate currently observed has been attributed to a variety of causes including the failure of transcriptional repression and the inappropriate postnuclear transfer reprogramming of the genome. Embryonic loss during the first few days of development is a consistent feature of normal development in many species. In humans and domestic animals, a high rate of embryonic loss occurs around the time that maternal cytoplasmic control of development switches to EGA. The comparison of normal in vivo developing, in vitro cultured, and cloned embryos promises to further our understanding of the process of early development and the maternal cytoplasmic control of it.

[*See also* Cortical Inheritance; Epigenetics.]

BIBLIOGRAPHY

Latham, K. E. “Mechanisms and Control of Embryonic Genome Activation in Mammalian Embryos.” *International Reviews of Cytology* 139 (1999): 71–124.

Memili, E., and N. L. First. “Zygotic and Embryonic Gene Expression in Cow: A Review of Timing and Mechanisms of Early Gene Expression as Compared with Other Species.” *Zygote* 8 (2000): 87–96.

Thompson, E. M., E. Legouy, and J. P. Renard “Mouse Embryos Do Not Wait for the MBT: Chromatin and RNA Polymerase Remodeling in Genome Activation at the Onset of Development.” *Developmental Genetics* 22 (1998): 31–42.

Zuccotti, M., S. Gargna, and C. A. Redi “Nuclear Transfer, Genome Reprogramming and Novel Opportunities in Cell Therapy.” *Journal of Endocrinological Investigations* 23 (2000): 623–629.

— W. Allan King

MATERNAL-FETAL CONFLICT

As David Haig (1993) pointed out, "The most intimate of human relationships is that between a mother and her fetus." Traditional views of pregnancy have emphasized the harmonious aspects of this intimate relationship. From an evolutionary genetic perspective, however, it is apparent that maternal-fetal interactions involve a complex interplay between mutual and conflicting interests (Spencer et al., 1998). In a sexually reproducing species, typically only half of the autosomal alleles (alternative forms of the same gene) present in mother and offspring are identical by descent (r, the coefficient of relatedness for biparentally inherited nuclear genes, is 0.5). As a consequence of this genomic nonequivalence, mother and fetus are likely to differ in what constitutes the optimal allocation of maternal resources to individual offspring. Whereas selection acting on the mother should favor equal allocation of parental investment across all offspring, any individual offspring is selected to manipulate a bias in maternal investment toward itself. Nonetheless, offspring selfish behavior should be tempered, at least to some extent, by the cost it imposes on siblings, with the degree of restraint being influenced by the coefficient of relatedness between siblings.

Multiple Paternity and Maternal-Fetal Conflict. Maternal-fetal conflict is a special form of parent–offspring conflict that results from a viviparous (live birth) mode of reproduction. [*See* Viviparity and Oviparity.] By providing a direct conduit for manipulation of the mother's reproductive physiology by genes expressed in the embryo, viviparity creates an arena for forms of genetic conflict absent in species that lay eggs (Zeh and Zeh, 2000). In theory, viviparity can give rise to three different forms of conflict: (1) between mother and developing fetus, (2) between siblings competing for maternal resources within the womb, and (3) between maternally and paternally inherited alleles within the fetus. This latter possibility is particularly intriguing because it involves conflict between genetic elements contained within a single individual (intragenomic conflict). Haig and colleagues have proposed that such intragenomic conflict is a driving force in the evolution of maternal-fetal interactions (e.g., Haig, 1993, 1997; Haig and Graham, 1991). The essential feature of Haig's conflict hypothesis is that polyandry (females mating with more than one male) is the key factor generating conflict between fetal maternal and paternal genomes. Half-siblings resulting from polyandry have the same mother but different fathers. This means that the coefficient of relatedness (r_p) between paternally inherited alleles is zero (assuming that fathers are unrelated). By contrast, all the offspring have the same mother, and thus maternally inherited alleles have the expected coefficient of relatedness (r_m) of 0.50. Consequently, maternally derived alleles will be selected to seek resorces from the mother such that their selfish interests are balanced against those of related siblings. By comparison, a given paternally derived allele will not, under polyandry, be highly related to other offspring. Consequently, selection should favor paternally inherited alleles in the fetus that act to extract more resources from the mother than is optimal for their maternally inherited counterparts.

Genomic Imprinting, Fetal Development, and Maternal Behavior. Intragenomic conflict between maternal and paternal genomes in the fetus can be manifested only if a mechanism exists that enables alleles to vary their expression as a function of parental origin. Such a mechanism, known as genomic imprinting or parent-of-origin gene expression, has now been demonstrated to occur at a limited number of autosomal loci in marsupials, placental mammals, and angiosperm plants. [*See* Genomic Imprinting.] In the case of genomically imprinted genes, alleles are marked (methylated) differently in eggs and sperm. Consequently, one of the alleles in the fetus is switched off while its counterpoint is expressed. The mechanisms and consequences of genomic imprinting have been most thoroughly investigated in mice. Nuclear transplantation experiments on mice have provided evidence of a critical role for imprinting during early developmental stages in mammals (McGrath and Solter, 1984; Surani et al., 1984). Comparisons of developmental patterns in mouse embryos possessing only paternal (androgenetic) or maternal (gynogenetic) chromosomes have revealed that growth of the trophoblast and the placenta, the embryonic organs that function in acquiring resources from the mother, is largely the result of expression of paternally inherited alleles. By contrast, morphological differentiation of the embryo proper, at least during early developmental stages, is primarily controlled by expression of maternally inherited alleles. Neither androgenetic nor gynogenetic embryos are capable of completing fetal development.

More recently, advanced molecular techniques, such as messenger RNA phenotyping and targeted gene deletions ("knockouts"), have been used to investigate imprinting at specific gene loci. More than forty genes in the mouse have been found to exhibit parent-of-origin-specific gene transcription, with many of these loci having homologues in humans that display similar imprinting patterns. In nearly all cases, imprinted alleles are either fully expressed or silent, depending on parent of origin. For example, at the *insulin-like growth factor 2 (Igf2)* locus, only the paternal allele is expressed, and it produces a peptide (IGF-2) that increases resource transfer from mother to fetus. At the *insulin-like growth factor 2 receptor (Igf2r)* locus, the pattern of imprinting is reversed. In this case, the paternal allele is silent and the maternal allele codes for a receptor that degrades the IGF-2 peptide. Knockout experiments in mice have demonstrated that normal fetal development depends on a balance between the antagonistic effects of these two reciprocally imprinted genes: loss-of-function mutations in paternal *Igf2* alleles result in a 40 percent reduction in growth, whereas fetuses containing maternal *Igf2r* knockouts are oversized and ultimately inviable. Interestingly, the double mutant is completely viable and of normal size (reviewed in Tilghman, 1999), supporting the interpretation that an important function of *Igf2r* is to counter *Igf2*.

Targeted deletions of two imprinted mouse genes, *Peg1/Mest* and *Peg3*, have revealed that imprinting effects can, in some cases, extend beyond fetal development to maternal physiology and nurturing behavior. At both these loci, only the paternally inherited allele is expressed, and, as expected, knockouts result in fetal growth retardation (Lefebvre et al., 1998; Li et al., 1999). In addition, females with *Peg1* loss-of-function mutations fail to exhibit the normal, hygenic behavior of consuming the placenta shortly after birth. These mothers also show little propensity to feed their pups. In the case of *Peg3*, mutant females (those with no *Peg3* function) engage in defective nest building and pup retrieval behavior and secrete insufficient quantities of milk in response to suckling. This disruption of maternal behavior stems from the fact that *Peg1* and *Peg3* display imprinted expression in both the developing and the adult brain and have a major impact on the differentiation of the central nervous system. Although it may seem paradoxical that maternal behavior and lactation are controlled by paternally inherited allele expression, this is precisely the pattern predicted by Haig's conflict hypothesis (see Li et al., 1999). A female mouse maximizes her reproductive success by producing multiple litters. If she invests too heavily in her current offspring, she risks compromising her prospects for future reproduction. However, given the occurrence of multiple paternity across litters, paternal interest is best served by prolonging care and feeding beyond what would be optimal for maternally inherited alleles. Although the potential effects of *Peg1* and *Peg3* on human maternal behavior have yet to be investigated, it is known that the two loci in humans exhibit imprinted expression in a pattern similar to that in mice (Murphy et al., 2001).

Alternatives to the Conflict Hypothesis? According to Hurst (1997), at least thirteen other theories, ranging from defense against ovarian tumors to chromosome surveillance, have been proposed to explain the evolution of genomic imprinting. None of these alternative hypotheses seems sufficiently general to explain the diversity of patterns associated with parent-of-origin gene expression. By contrast, both whole genome and locus-by-locus analyses of mammalian imprinted gene expression have revealed a pattern that is highly concordant with intragenomic conflict playing a major role in maternal-fetal interactions. Most imprinted loci exert strong effects on fetal growth and development through expression in trophoblast and placenta. In the majority of cases, alleles at paternally expressed loci act to promote maternal resource transfer to the fetus, whereas alleles at maternally expressed loci down-regulate paternal genome activity. The conflict hypothesis also predicts that imprinting should be restricted to live-bearing species because only with viviparity does the embryonic paternal genome have the opportunity to directly manipulate the mother's reproductive physiology. In support of this prediction is the observation that imprinting in plants is limited to angiosperms, whose reproductive mode is analogous to animal viviparity, with embryos being nourished by the endosperm for an extensive period after fertilization (Mora-Garcia and Goodrich, 2000). Moreover, parent-of-origin gene expression does not occur in *Drosophila*, birds (O'Neil et al., 2000), and the platypus (Killian et al., 2000), all of which lay eggs before or soon after fertilization.

Health Implications of Maternal-Fetal Conflict. An increasing number of pregnancy-related health risks, previously viewed as the failure of individual adaptation, are now being linked to imbalances in imprinted gene expression (Haig, 1993). Maternal-fetal conflict may thus provide an evolutionary explanation for many of the complications of pregnancy, several of which are discussed in the following.

Gestational trophoblast disease. Gestational trophoblast disease refers to a set of conditions arising from aberrant growth of the fetal chorion. The chorion serves to attach the placenta to the uterus. This potentially pernicious outcome of pregnancy is caused by unusual fertilization events that result in an imbalance in maternal/paternal genomic contribution to the fetus. Fetuses lacking a maternal genome but containing two paternal genome copies (complete hydatidiform moles, or CHMs) are particularly dangerous to the mother. CHMs exhibit a growth-without-form morphology similar to that of androgenetic mouse embryos, and carry a 15–20 percent risk of persistent trophoblastic tumor (Tham and Ratnam, 1998). There seems little doubt that unrestrained expression of the fetal paternal genome is the cause of this disease.

Preeclampsia. Preeclampsia is a relatively common complication of pregnancy characterized by hypertension, proteinuria, and an excessive inflammatory response in the uterus. The disease is usually considered to be one of first pregnancies (primigravidity), although recent analysis suggests that it can be best understood as a condition of the first pregnancy with a particular male partner (primipaternity). After changing partners, women experience a risk of preeclampsia comparable to that in first pregnancies (Robillard et al., 1999), but the risk is lower if the father is the same in both pregnancies. Haig (1993) considered preeclampsia to be a likely outcome of maternal-fetal conflict, because paternally inherited alleles should be selected to enhance nutrient transfer to the fetus by increasing maternal blood pressure. Although preeclampsia appears to have a genetic basis, the mechanisms involved are complex and poorly understood, with most physicians favoring an immunologically based explanation for the condition. It should be pointed out, however, that maternal immune reactions are directed largely against paternal antigens expressed by genomically imprinted genes in the trophoblast and placenta. The conflict and immunological explanations for preeclampsia are therefore not mutually exclusive.

Fetal-maternal cell trafficking. Fetal-maternal cell trafficking is a complication of pregnancy whose potential significance is only beginning to be appreciated (Bianchi, 2000). Although it has long been known that fetal cells circulate in the mother during pregnancy, the recent discovery that such cells can persist in the mother for up to twenty-seven years postpartum came as a startling revelation. The presence of these foreign cells (microchimerism) has already been implicated in a number of inflammatory or autoimmune diseases of women, such as polymorphic skin eruptions during pregnancy, postpartum scleroderma, and primary biliary cirrhosis. Transfusion of fetal cells to the mother is significantly increased in pregnancy disorders associated with an imbalance in imprinted gene expression, such as preeclampsia and fetal cytogenetic abnormalities. This raises the possibility that microchimerism-based disease may be an epiphenomenon of maternal-fetal conflict.

Morning sickness. The nausea and vomiting experienced by approximately two-thirds of women during the first trimester of pregnancy, commonly referred to as morning sickness, has been hypothesized to occur as a byproduct of maternal-fetal conflict. Evidence suggests, however, that this is not the case, and women who experience morning sickness are significantly less likely to miscarry than women who do not. By protecting against parasites, pathogens, and teratogenic chemicals, morning sickness probably serves an adaptive function for both mother and fetus (Flaxman and Sherman, 2000).

Evolution of Viviparity. The evolution of viviparity has almost invariably been considered from the perspective of the mother. Clearly, egg retention, the first step in this process, is likely to have been favored because it enhanced the lifetime reproductive success of the female in the face of ecological factors such as adverse climatic conditions and predation pressure (Shine, 1985). The evolution of egg retention, however, also opened up the possibility for embryonic manipulation of the mother. After this initial step, the paternal genome, operating within a context of multiple paternity, then became a major and, until recently, an unappreciated player in driving the course of mammalian development and evolution.

[*See also* Overview Essay *on* Darwinian Medicine; Life History Theory, *article on* Human Life Histories; Parental Care; Parent–Offspring Conflict.]

BIBLIOGRAPHY

Bianchi, D. W. "Fetomaternal Trafficking: A New Cause of Disease." *American Journal of Medical Genetics* 91 (2000): 22–28.

Flaxman, S. M., and P. W. Sherman. "Morning Sickness: A Mechanism for Protecting Mother and Embryo." *Quarterly Review of Biology* 75 (2000): 113–148.

Haig, David. "Genetic Conflicts in Human Pregnancy." *Quarterly Review of Biology* 68 (1993): 495–532.

Haig, David. "Parental Antagonism, Relatedness Asymmetries, and Genomic Imprinting." *Proceedings of the Royal Society of London B* 264 (1997): 1657–1662.

Haig, D., and C. Graham. "Genomic Imprinting and the Strange Case of the Insulin-like Growth Factor II Receptor." *Cell* 64 (1991): 1045–1046.

Hurst, Laurence D. "Evolutionary Theories of Genomic Imprinting." In *Genomic Imprinting*, edited by W. Reik and A. Surani, pp. 211–237. Oxford, 1997.

Killian, J. K., J. C. Byrd, J. V. Jirtle, B. L. Munday, M. K. Stoskopf, R. G. MacDonald, and R. L. Jirtle. "M6P/IGF2R Imprinting Evolution in Mammals." *Molecular Cell* 5 (2000): 707–716.

Lefebvre, L., S. Viville, S. C. Barton, F. Ishino, E. B. Keverne, and M. A. Surani. "Abnormal Behaviour and Growth Retardation Associated with Loss of Imprinted Gene Mest." *Nature Genetics* 20 (1998): 163–169.

Li, E., C. Beard, and R. Jaenisch. "Role of Methylation in Genomic Imprinting." *Nature* 366 (1993): 362–365.

Li, L. L., E. B. Keverne, S. A. Aparicio, F. Ishino, S. C. Barton, and M. A. Surani. "Regulation of Maternal Behavior and Offspring Growth by Paternally Expressed *Peg3*." *Science* 284 (1999): 330–333.

McGrath, J., and D. Solter. "Completion of Mouse Embryogenesis Requires Both the Maternal and Paternal Genomes." *Cell* 37 (1984): 179–183.

Mochizuki, A., Y. Takeda, and Y. Iwasa. "The Evolution of Genomic Imprinting." *Genetics* 144 (1996): 1283–1295.

Mora-Garcia, S., and J. Goodrich. "Genomic Imprinting: Seeds of Conflict." *Current Biology* 10 (2000): R71–R74.

Murphy, S. K., A. A. Wylie, and R. L. Jirtle. "Imprinting of *PEG3*, the Human Homologue of a Mouse Gene Involved in Nurturing Behavior." *Genomics* 71 (2001): 110–117.

O'Neill, M. J., R. S. Ingram, P. B. Vrana, and S. M. Tilghman. "Allelic Expression of *Igf2* in Marsupials and Birds." *Development Genes and Evolution* 210 (2000): 18–20.

Robillard, P.-Y., G. A. Deckker, and T. C. Hulsey. "Revisiting the Epidemiological Standard of Preeclampsia: Primigravidity or Primipaternity." *European Journal of Obstetrics, Gynecology, and Reproductive Biology* 84 (1999): 37–41.

Shine, R. "The Evolution of Viviparity in Reptiles: An Ecological Analysis." In *Biology of the Reptilia*, edited by C. Gans and F. Billet, vol. 15, pp. 605–694. New York, 1985.

Spencer, H. G., M. W. Feldman, and A. G. Clark. "Genetic Conflicts, Multiple Paternity and the Evolution of Genomic Imprinting." *Genetics* 148 (1998): 893–904.

Surani, M. A. H., S. C. Barton, and M. L. Norris. "Development of Reconstituted Mouse Eggs Suggests Imprinting of the Genome during Gametogenesis." *Nature* 308 (1984): 548–550.

Tham, K. F., and S. S. Ratnam. "The Classification of Gestational Trophoblastic Disease: A Critical Review." *International Journal of Gynecology and Obstetrics* 60 (1998): S39–S49.

Tilghman, S. M. "The Sins of Mothers and Fathers: Genomic Imprinting in Mammalian Development." *Cell* 96 (1999): 185–193.

Zeh, D. W., and J. A. Zeh. "Reproductive Mode and Speciation: the Viviparity-driven Conflict Hypothesis." *BioEssays* 22 (2000): 938–946.

— David Zeh and Jeanne Zeh

MATHEMATICAL MODELS

Mathematical models have contributed to evolutionary biology from its beginnings. Although Charles Darwin did not write equations or draw graphs, he often tried to reason in quantitative terms about the dynamic outcomes of idealized biological situations. This is modeling: clearly stating a set of assumptions, then using the rules of logic to deduce consequences. Modelers today use a suite of powerful mathematical tools in addition to verbal reasoning, but the aim and general strategy remain essentially the same.

Could a blind process of variation, inheritance, and natural selection explain the adaptedness and diversity of living things? Darwin and his contemporaries understood that the answer hinged on numbers, probabilities, rates, and times, but they were unable to resolve the question because they did not know about genes. The rediscovery of Gregor Mendel's work in 1900 clarified the heritable basis of variation and also suggested how mathematical models of variation and evolution might be constructed. In 1908, G. H. Hardy and Wilhelm Weinberg independently showed that diploid genotypes in a randomly mating population reappear in the same equilibrium proportions generation after generation. Roughly ten years later, R. A. Fisher, J. B. S. Haldane, and Sewall Wright began to sketch the outlines of theoretical population genetics. Collectively, their models (1) showed that a Darwinism based on Mendelian inheritance could in principle do everything Darwin had proposed; (2) provided an explicit genetic basis for the continuous variation that had been studied statistically by Francis Galton and other biometricians; and (3) made mathematical analysis an essential component of genetics, and thereby (implicitly) of all biology.

In the second half of the twentieth century, evolutionary modeling became much more diverse. Ecologists and behaviorists sought to identify general principles governing the evolution of life histories, sex differences, social structures, and other specific kinds of phenotypes; molecular geneticists uncovered a wealth of evolutionary information in DNA sequences; and systematists began to use that information to infer the phylogenetic relationships of populations, species, and higher taxa. Models have played critical roles in all these developments, even though relatively few biologists contribute to (or even read) the mathematical literature.

We begin by contrasting models in evolutionary biology with those in other disciplines. Then we discuss key contributions of models in three areas that correspond to the major elements of Darwin's argument: (1) characterizing variation within populations, (2) explaining how populations change and new species are formed, and (3) analyzing the adaptive evolution of particular kinds of phenotypes.

What Evolutionary Models Do. Any formal device that facilitates "what if" reasoning can be called a model. Architects and airplane designers make miniatures for this purpose, and fashion models illustrate how the viewer might (with dieting) appear to others if wearing the same clothes. Some scientific models are equally simple. For example, J. D. Watson and Francis Crick's double-helical molecular model embodies a specific hypothesis about the chemical structure of DNA. Such models give insight into a hypothesis by allowing some of its implications to be visualized. However, once the hypothesis is known to be correct, models of this kind become uninteresting except as historical artifacts; they are superseded by the factual understanding that they helped to create.

Models in evolutionary biology, by contrast, are not hypotheses to be tested for possible acceptance as descriptions of reality. Indeed, most evolutionary models are known in advance to be false in important respects. A typical model intentionally oversimplifies many aspects of reality, so as to focus on the effects of a few other aspects. The question being asked is not whether the world really works like this, but instead whether a certain process can be better understood when all but a few of the potentially relevant variables are ignored. Evolutionary models abstract the relationships of a few variables (the way z depends on x and y, perhaps) in order to identify general principles that may continue to apply under the full complexity of real life (where z actually depends on hundreds of variables in addition to x and y). A model can be successful scientifically despite being "true" in only a limited sense and in particular situations. Indeed, the aim is often to identify those situations to which the model applies.

For example, the Hardy–Weinberg model ignores everything except two versions of a gene (alleles) at a diploid locus in a randomly mating, infinitely large population not subject to selection. Some of the genes (comprising a proportion p of the total) are allele A, and the others (proportion $q = 1 - p$) are allele a. Hardy and Weinberg reasoned that if individuals mate at random with respect to their genotypes at the *A* locus, then the diploid genotypes of offspring will be formed in proportion to the relative abundances of the alleles in the parents. A proportion p^2 will be homozygous AA, $2pq$ will be heterozygous Aa, and q^2 will be homozygous aa. The allele proportions or "frequencies" will remain p and q in the offspring and in subsequent generations. Thus, the distribution will re-create itself each generation. This may seem obvious to many readers, but it was not obvious to anyone at the time; several geneticists had even published confused alternatives.

As fundamental as it is, the Hardy–Weinberg model rests on highly simplified assumptions. Real organisms have thousands of genetic loci, not one; most loci have many alleles, not two; real populations are often small, and never infinite; individuals do not mate at random; the segregation of alleles during gamete formation is less than perfectly fair; selection is almost certain to be acting (if not at the A locus, then at some nearby linked locus); and so on. Conceptually, these simplifications focus our attention on a fundamental issue; in practice, the neglected factors often have only minor effects, in part because they affect all alleles equally.

Evolutionary models often provide the only way to describe and study processes that are too large, too slow, too abstract, or too far in the past to be approached directly. Even the simple Hardy–Weinberg model lets us see things that would otherwise be invisible. For example, suppose that allele A is dominant (i.e., AA and Aa individuals have identical phenotypes). Then the model allows us to estimate the relative numbers of homozygotes and heterozygotes in the population, even though we cannot distinguish them by inspection.

Models of Variation. Ernst Mayr has argued that Darwinism made "population thinking" central to biology and thereby to modern thought generally. If a species consists only of its individually variable members rather than some unchanging essence, then to describe a species, one must describe its variation. How should this be done? An exhaustive catalog of all the individuals and their characteristics would be accurate but not useful. Francis Galton and other early biometricians recognized that the distributions of many quantitative characters are bell-shaped (approximately normal, or Gaussian), and that for many purposes they can be characterized by two summary statistics: the mean (which describes the distribution's "center") and the variance (which describes the dispersion of individual character states around that center). With this efficient language for describing variation, we can ask questions that would otherwise be difficult or even unthinkable; for example, we can easily compare two populations by comparing their means and variances.

The relationships among different traits can be analyzed in a similar way. Darwin often used the word *correlation* to refer to situations where extreme deviations of one trait tend to be associated with extreme deviations of another (e.g., where individuals with long feathers tend to have long toes). The biometricians developed quantitative methods that apply not only to correlations among traits but also to the similarities among family members (e.g., where some families tend to have longer feathers than average), which is also a type of correlation. The "regression" coefficient of statistics is called that because it was originally used to model the tendency of parents with extreme values of some trait to have offspring with somewhat less extreme values of the same trait; the coefficient quantifies the degree to which offspring phenotypes "regress," on average, toward the population mean.

The resemblance of parents and offspring is central to Darwinian natural selection, but "blending inheritance" (the pre-Mendelian mechanism assumed by Darwin and most of his contemporaries) should rapidly eliminate the heritable component of variation through repeated averaging in offspring. The resulting lack of variation would shut down adaptive evolution because there can be no selection among identical individuals. Fisher, Haldane, and Wright showed that Mendelian assumptions predict observed patterns of parent–offspring resemblance (including regression to the mean), but without a progressive loss of variation from one generation to another. Their models of Mendelian inheritance, enlarged to include many genes and effects of the environment, gave rise to the theory of quantitative genetics, in which effects of genotypes and environments are aggregated into a few simple parameters. For example, the concept of heritability directly relates genetic and environmental influences on a trait to the degree of resemblance between parents and offspring. This approach to modeling variation has proved almost infinitely adaptable. It forms the basis of all agricultural breeding programs, and it has been applied to theoretical problems in development, sex differences, ecological niche breadths, speciation, and many other areas.

Few models in science make something so complicated so simple. The models of quantitative genetics reveal the forest (a trait's genetic and environmental variance components) by merging the trees (huge numbers of individually small effects of genes and experiences) in a biologically appropriate way.

These classic models imply that the phenotypic variation in a population must reflect a balance between forces that increase and forces that decrease the level of genetic variation. Random mutation is clearly the ultimate source of new genetic variation, but even without blending there are processes that tend to reduce variation. Wright was among the first to appreciate that random changes of allele frequencies in finite populations would reduce genetic diversity. Models of this process of genetic drift showed that the rate of loss is inversely proportional to population size (N), and models incorporating mutation (μ) showed that the level of standing variation should be proportional to the product $N\mu$.

When this product is much greater than 1, there will be much variation, and when it is less than 1, there will be little. In other words, if there is more than one new mutation per generation, then a genetic locus will tend to be variable, regardless of the absolute population size. Once revealed by models, these simple relationships seem intuitive, but no one could have guessed them. They have recently acquired increased importance as keys to the analysis of DNA sequence variation within and between populations and species.

Drift had been well understood by population geneticists for nearly half a century when J. F. C. Kingman and others realized that much could be gained by thinking of contemporary gene copies as sitting at the tips of the branches of a genealogical tree that descends from a single ancestral copy at some time in the past. There are simple relationships between the population size and the expected depths of branch points in this "gene tree," and mutations are naturally thought of as random marks on its branches. The classical results about mutation and drift are easily recovered from this genealogical model, which has also been applied to problems involving recombination, selection, and several kinds of population substructuring. The resulting coalescent theory has revolutionized the analysis of genetic variation within populations, transforming and clarifying a subject that had previously been much more difficult and opaque.

Models of Evolution. The simplest models of natural selection are straightforward extensions of the Hardy–Weinberg law. Instead of assuming that individuals of all three genotypes survive and reproduce with equal probabilities, we allow them to have different fitnesses, which are numbers proportional to their average contributions to the next generation. Using these assumptions, it is easy to derive an equation that gives the gene frequency in the next generation as a function of the gene frequency in the present generation. This relationship can be represented in many different ways, each offering different insights into the process of selection. For example, Wright showed that Δp, the *change* in allele frequency, obeys

$$\Delta p = \frac{pq}{\overline{W}} \frac{d\overline{W}}{dp}.$$

Here $\overline{W}$ is the average fitness of all individuals in the population, and $d\overline{W}/dp$ is the derivative of mean fitness with respect to p. An allele increases in frequency ($\Delta p > 0$) when its presence increases mean fitness. This relationship inspired the metaphor of evolution as a fitness-maximizing process that "climbs hills" in an "adaptive landscape."

Even in their simplest forms, these basic models of selection offer unexpected and important insights. For example, selection against harmful recessive alleles becomes extremely weak as those alleles become rare, because almost all of them occur in heterozygotes where they enjoy fitness equal to that of the dominant "wild type" allele. Thus, recessive "disease genes" may persist at surprisingly high frequencies in outbred populations. The same framework can be used to address almost any biological situation that can be represented by quantitative relationships between genotypes, phenotypes, and fitnesses (which may depend on the environment, including other phenotypes).

Selection is also easily added to models of continuous quantitative variation, giving expressions for the change of the mean phenotype as a function of the strength of selection and the trait's heritability. These quantitative-genetic models predict rates of evolution that agree closely with those observed in experimental, agricultural, and natural settings.

R. A. Fisher pointed out that fitness itself is a quantitative trait under selection, and that its rate of evolution is approximately equal to the heritable fitness variance. Because selection tends to increase the frequencies of fitter alleles, the fitness variance should decline as favorable alleles rise to high frequencies, displacing less favorable alleles, and evolution should slow to a crawl. Traits closely related to fitness do tend to have relatively low levels of heritable variation, but those levels are well above zero. Why do such high levels of fitness variance persist? Are new mutations a significant source? Or is the environment continually changing, and thereby changing the relative fitnesses of common and rare alleles? These quantitative questions motivate many lines of ongoing empirical research.

Selectively neutral mutations (those with no effects on fitness) can drift to high frequencies and "fix" (replace all other alleles) in finite populations, and their rate of fixation should be equal to the rate at which they arise by mutation. Motoo Kimura and others have suggested that much evolution at the molecular level may be of this kind. "Nearly neutral" mutations (those that lower fitness by no more than the reciprocal of the population size, $1/N$) may also fix, though at a lower rate. Thus, whether a mutation is effectively neutral, nearly neutral, or firmly opposed by selection will depend on the population size. Small populations are expected to fix many slightly deleterious mutations, potentially degrading the precision of adaptations and leading to further reductions of population size. Deleterious mutations may also impose significant "genetic loads" on large populations, even though the mutations do not reach high frequencies individually. The questions framed by models of these processes directly motivate much current research in population genetics and genomics.

The models mentioned above treat species as populations of individuals that share a common "gene pool." Speciation can occur only if some process leads to subdivision of a formerly well-connected gene pool, for example, through the creation of barriers to migration between parts of a species range.

Models of migration between semi-isolated populations show that even a single migrant per generation ($Nm = 1$) is sufficient to keep the subpopulations' allele frequencies similar and thus to stall phenotypic and ecological divergence, even in the face of moderately strong selection. Because low levels of migration can prevent speciation, theoretical and empirical studies of speciation focus on factors that can give rise to nearly complete reproductive isolation early in the speciation process.

The evolutionary relationships of species and higher taxa result from histories of speciation. We view these histories retrospectively, and when we infer them from DNA sequences, we almost literally "run the models backward." The computer programs that implement these methods use sophisticated models of sequence evolution; they evaluate many possible histories of mutation and speciation, looking for the one best able to explain the present-day sequences. The models need to be sufficiently realistic to account for important features of sequence evolution, but not so complex that they lose all statistical power or cannot be implemented as programs that will run at acceptable speeds. Models also play important roles in studies of gene and genome evolution, where the aim is to understand mutational and recombinational processes of several kinds, and their interaction with selection at the level of gene products.

Models of Adaptation. Darwin was an ecologist. He spent much time thinking about the evolution of adaptations. The main purpose of his *Origin of Species* was to show that natural selection of relatively "fit" variants could produce all the adaptation in the world. This claim drew strong criticism and persuaded relatively few of Darwin's contemporaries because the essential substrate (inherited variation) remained so poorly defined that it could not be modeled. As long as people remained unconvinced about selection in general, they were unlikely to start analyzing the evolution of adaptations in particular. This is probably why the first two-thirds of the twentieth century was devoted largely to answering general questions about how evolution works; during this period, remarkably little attention was given to the kinds of questions that had animated Darwin.

In the last third of the century, with the general issues resolved, there was a flowering of evolutionary ecological studies that implicitly assumed (as Darwin had done) that if a trait was being selected in a certain direction, then it would evolve in that direction. The traits studied in this way are typically ones constrained by trade-offs such that more of one "good" necessarily entails less of some other good. For example, parent birds may lay more eggs and attempt to rear more chicks, but doing so may exhaust them and reduce their own survival. Models have been constructed that describe how fledgling success, adult survival, and other relevant variables depend on the number of eggs laid; such models are analyzed to find the "optimal" clutch size (the one that maximizes parental fitness, typically evaluated as lifetime reproductive success) under a given set of conditions. A knowledge of genes informs such models, but the analyses are purely phenotypic.

Genes appear more explicitly in models for the evolution of mate choice, sex differences, and interactions among relatives whose inclusive fitnesses may depend on each others' reproductive success, owing to the fact that relatives have correlated genotypes. In such situations, the course of evolution cannot be predicted by maximizing the population's average fitness because the actors "prefer" different outcomes and irreconcilable "genetic conflicts of interest" arise between individuals and even between genes within genomes.

The phenotypes studied in this way often have fitnesses that vary as a function of the population-wide distribution of phenotypes. Sex ratios provide the canonical example: if males are more common than females, then females will have higher average reproductive success, and vice versa, other things being equal. But other things are not always equal. The circumstances surrounding allocation to male and female offspring vary enormously in nature and have stimulated the development of elegant mathematical models that explain a great deal about the sex-ratio variation seen in nature.

Direct conflict over resources may also give rise to frequency-dependent payoffs. For example, individuals that escalate immediately to high levels of threat and aggression ("hawks") may do well when most of their competitors are peaceable "doves," but may suffer frequent injuries when most competitors are other "hawks." Evolutionary game theory was developed to study such situations. Often there is no optimal strategy that individuals should express regardless of what others are doing; instead, there is typically an "evolutionarily stable strategy" that is best only in the sense that it is unbeatable if everyone else adopts it.

J. B. S. Haldane ended his book *The Causes of Evolution* (1932) as follows: "The permeation of biology by mathematics is only beginning, but unless the history of science is an inadequate guide, it will continue, and the investigations here summarized represent the beginning of a new branch of applied mathematics." An excellent prediction, both for his time and ours.

[*See also* Game Theory; Hardy–Weinberg Equation; Prisoner's Dilemma Games.]

BIBLIOGRAPHY

Bulmer, M. *Theoretical Evolutionary Ecology*. Sunderland, Mass., 1994. A lucid and accessible survey of models and modeling at the ecological end of evolutionary biology.

Bulmer, M. *Francis Galton: Pioneer of Heredity and Biometry*. [www.francisgalton.com] A history for biologists, not historians, and available free on the Web in PDF format.

Charnov, E. L. *The Theory of Sex Allocation*. Princeton, 1982. Synthesizes and extends sex-ratio research, both theoretical and empirical.

Crow, J. F., and M. Kimura. *An Introduction to Population Genetics Theory*. New York, 1970. The classic modern treatment.

Falconer, D. S., and T. F. C. Mackay. *Introduction to Quantitative Genetics*. 4th ed. Essex, England, 1996. The standard treatment, written for professionals but accessible.

Fisher, R. A. *The Genetical Theory of Natural Selection*. Oxford, 1930. The first and most influential book of evolutionary genetics.

Frank, S. A. *Foundations of Social Evolution*. Princeton, 1998. How to model natural selection, especially with G. R. Price's covariance equation and in social contexts.

Gillespie, J. H. *A Concise Guide to Population Genetics*. Baltimore, 1998. A unique and accessible book for beginners that insults no one's intelligence.

Haldane, J. B. S. *The Causes of Evolution*. New York, 1932. The second and more entertaining book of evolutionary genetics.

Hamilton, W. D. *Narrow Roads of Gene Land*. Oxford, 1996. Hamilton's papers on social behavior, sex ratios, senescence, and dispersal, collected with introductory essays that explain how they came to be written and what the author thought about them years later.

Hudson, R. R. "Gene Genealogies and the Coalescent Process." *Oxford Surveys in Evolutionary Biology* 7 (1990): 1–44. An accessible but thorough introduction.

Kingman, J. F. C. "Origins of the Coalescent: 1974–1982." *Genetics* 156 (2000): 1461–1463. A short, informal history by the central player.

Maynard Smith, J. *Mathematical Ideas in Biology*. Cambridge, 1968. A master modeler at work.

Maynard Smith, J. *Models in Ecology*. Cambridge, 1974.

Maynard Smith, J. *Evolution and the Theory of Games*. Cambridge, 1982. The theory and practice of evolutionary game theory made clear and instructive.

Mayr, E. *The Growth of Biological Thought*. Cambridge, Mass., 1982. A comprehensive but accessible history of ideas in biology, including "population thinking" and other conceptual foundations of modern mathematical models.

Provine, W. B. *The Origins of Theoretical Population Genetics*. Chicago, 1971. Classic history by a historian.

— JON SEGER AND FREDERICK R. ADLER

MATING STRATEGIES, ALTERNATIVE

In many animal populations, individuals adopt mating strategies that differ from the conventional strategy employed in the population. These *alternative mating strategies* are observed mainly in males, and they are widespread in species in which opportunities for multiple mating are greater for males than for females. Thus, there is little doubt that alternative mating strategies evolve when sexual selection is strong. Research is active on the degree to which the expression of alternative mating strategies depends upon discrete genetic differences among individuals, upon interactions between genetic and environmental factors, or upon behavioral and developmental "choices," mediated by individual perceptions of social status. This article will explain the nature of this variation, identify its possible underlying causes, and describe the evolutionary forces that allow alternative mating strategies to persist in nature.

Sexual Selection and the Mating Niche. Charles Darwin, in *The Descent of Man and Selection in Relation to Sex*, (1874) considered the effects of sexual selection to be functionally similar to those that exist in populations with a surplus of males. He noted that "*if each male secures two or more females, many males cannot pair*" (p. 212). This elegant observation explains why sexual selection causes the divergence of male and female phenotypes, as well as why alternative mating strategies so readily evolve. When only a small number of males secure the majority of matings, male characteristics that promote polygamy are disproportionately transmitted to the next generation. Over time, males are expected to become more modified in their appearance than females. Consistent with this expectation, males show greater phenotypic diversity than females in related species, including a greater tendency to become polymorphic.

Why should alternative mating strategies evolve when some males mate and others do not? The answer lies in the *average* and *variance* in mating success among males, particularly among males expressing the most common—that is, the *conventional*—mating phenotype. Suppose, in a species with equal numbers of males and females, that males defend mating territories. If each such male acquires exactly one mate, sexual selection cannot exist because each male contributes equally to the next generation. However, if some territorial males acquire more than one mate, other males attempting to defend territories will be prevented from mating. When this occurs, the *average* harem size, as well as the *variance* in mate numbers among males, both will increase. Even in species that form pairs, when some males are excluded from mating, variance in male fitness will exist.

This condition not only causes sexual selection; it also creates a "mating niche" for males engaging in *unconventional* mating behaviors. Males who avoid direct combat with males defending territories can often invade breeding territories by indirect means. Once inside breeding territories, unconventional males may surreptitiously copulate with receptive females, as occurs in isopods (see Vignette), lizards, beetles, and ruffs. Alternatively, in species that spawn, unconventional males may position themselves near ovipositing females and spread their sperm over unfertilized ova, as occurs in horseshoe crabs, coral-reef fish, salmon, and midshipmen fish. Stolen matings may yield unconventional males only a fraction of the fertilization success gained by males defending harems, but, if unconventional males sire even a few offspring, then on average they will have higher fitness than territorial males who secure no mates at all.

Game theory and population genetic analyses agree on the conditions necessary for the invasion and persistence of evolutionary stable strategies. For males employing alternative mating strategies, these conditions are most easily met when both the average harem size and the variance in mating success among conventional males are large. Increasing harem size among mating conventional males increases the fraction of conventional males who are excluded from mating altogether. Thus, the larger the average harem size becomes, the fewer fertilizations unconventional males need to acquire within harems in order for their average fitness to equal that of all conventional males combined. Although the fitness of unconventional males seems inferior to that of conventional males, in fact the average fertilization success of all unconventional males often equals or exceeds the average success of all conventional males (see Vignette).

Making the Best of a Bad Job. Note that the above description is distinct from the notion that unconventional males, often called *satellite* males, "make the best of a bad job." This latter hypothesis is often used to describe the relative fitnesses of males exhibiting unconventional phenotypes, and it is widely invoked to explain reports that satellite males are less successful at mating than territorial males. However, these apparent tests of the "best of a bad job" hypothesis consider only the mating success of males who actually mate. As explained above, when some territorial males are excluded from mating, average mating success, calculated only for mating territorial males, is certain to exceed the average mating success of satellite males. This is true because average harem size always overestimates the average mating success of all territorial males, unless every male in the population mates once. When mating males, as well as nonmating males, are included in calculations of average mating success, the fitnesses of territorial and nonterritorial males are expected to be, and have been shown to be, equivalent.

Alternative Mating Strategies in Males and Females. Why do alternative mating strategies appear more often in males than in females? The answer is once again found by considering the average and variance in mate numbers, and thus the average and variance in offspring numbers, within each sex. The average number of offspring produced by males who mate is equal to the average number of offspring per female, multiplied by the number of females with which each male mates. Thus, the average fitness of mating males can be many times greater than that of the average female. The fitness of the average unsuccessful male, on the other hand, is zero, and is therefore less than that of the average female. The more females tend to mate only with certain males in the population, the more the average number of offspring sired by mating males increases, and the larger the variance in offspring numbers among all males becomes.

Even in species in which females routinely seek multiple mates, individual female fecundity tends to be more limited than individual male fertility. Thus, on a populational level, the variance in offspring numbers among females is seldom as large as it is among males. Moreover, unlike males, relatively few females are prevented from mating altogether. Smaller fitness variance among females, compared to that among males, limits the size of the "mating niche" available to females attempting to reproduce by unconventional means. Thus, it is only in species in which considerable variance in female fitness exists that alternative female mating strategies are expected to evolve (see below).

The Expression of Alternative Mating Strategies. Three fundamental patterns of phenotypic expression exist for alternative mating strategies: Mendelian strategies, developmental strategies, and behavioral strategies. Each pattern of expression depends, at the most proximate level, on hormonal and neurological factors that regulate the timing and degree with which phenotypic differences appear. The nature of each regulatory mechanism depends, in turn, on its underlying mode of inheritance. Ultimately, the genetic architectures responsible for each mode of inheritance depend on the circumstances in which mating opportunities arise—that is, on the intensity of selection favoring distinct reproductive morphologies, as well as on the predictability or unpredictability of mating opportunities relative to individual lifespan.

Mendelian strategies. Alternative mating strategies controlled by few loci of major effect, which segregate in populations according to Mendelian rules, are well documented in diverse animal species. Examples include marine isopods (Vignette), bulb mites, damselflies, fig wasps, several species of poeciliid fish, side-blotched lizards, and ruffs. In each of these cases, specific allelic combinations produce morphologically and behaviorally distinct male phenotypes.

Mendelian strategies are expected to arise when sexual selection favors specialized mating phenotypes, and when the relative mating success of each phenotype is unpredictable within male lifetimes. By chance, a morph is well or poorly suited for securing mates in a given environment, and accordingly, its relative fitness, as well as its relative population frequency, rises or falls. In such circumstances, genes of major effect are expected to exclude genetic architectures that allow a phenotypic response to environmental cues predicting mating success. When such cues are lacking, the benefits of being

STRATEGIES AND TACTICS

The term *strategy* comes from evolutionary game theory. In this sense, a strategy is a preprogrammed set of behavioral or life history characteristics. Alternative mating strategies can therefore be viewed as functional sets of behaviors or morphologies used by their possessors to acquire mates. An *evolutionary stable strategy* (ESS) is a strategy that persists in a population. Such persistence occurs for one of two reasons. Either the average fitness of individuals expressing the ESS *equals* that of all other strategies existing in the population, or the average fitness of individuals expressing the ESS *exceeds* that of other strategies that might invade the population. If a strategy's average fitness is consistently *less than* that of other strategies, it will be removed from the population by selection.

Strategies are *adaptive.* Implicit in this are two further assumptions. First, genetic variation is presumed to underlie such traits. Heritability is required for *any* trait to change in frequency or be removed from a population, as described above. If genetic variation is lacking, no response to selection is possible. Second, stabilizing selection is presumed to refine trait expression. This is the process by which less fit trait variants are culled, more fit trait variants reproduce, and over time, a trait's function becomes recognizable. Traits with uniformly inferior fitness seldom persist in populations long enough to be shaped by selection.

Recent descriptions of discontinuous variation in mating phenotype distinguish between genetically distinct "strategies" and phenotypes that represent condition-dependent "tactics." The term *tactic* is used to describe behavioral or morphological characteristics whose expression is contingent, either on environmental conditions, or on the "status" of the individuals in which they appear. *Status-dependent selection* (SDS), the term used to describe how selection may operate on such traits, is presumed to allow individuals to assess their potential mating opportunities, and then make a behavioral or developmental "decision" that leads to greater mating success than if the choice had not been made. According to the SDS hypothesis, individuals in dimorphic populations are identical in their ability to choose one or another status-dependent phenotype.

The term *tactic* is clearly useful for describing phenotypes that are flexible in their expression, as opposed to those that segregate according to Mendelian rules. However, considerable evidence now indicates that polymorphisms in mating phenotype, which arise by phenotypic plasticity, in fact represent *mixtures* of evolutionary stable strategies. If this pattern is a general one, genetic architectures that allow phenotypic flexibility, like polymorphisms controlled by Mendelian factors, are expected to persist in populations by frequency-dependent selection. Thus, while "tactic" adequately describes reproductive phenotypes that vary with environmental conditions, "strategy" is appropriate for *all* evolved polymorphisms in reproductive behavior, regardless of how their expression is controlled.

—Stephen M. Shuster

able to change phenotype are low or nonexistent. Because trait heritabilities of the genetically determined strategies approach unity, the frequencies of Mendelian strategies are expected to oscillate over brief intervals with changes in morph fitness. Over longer durations, different morphs are expected to persist in the population because their average fitnesses are equal.

Developmental strategies. Discontinuous phenotypes produced by distinct developmental trajectories, which *do not* segregate in a Mendelian manner, are also well documented among animals. Examples include earwigs, crickets, horned beetles (see below), amphipods, freshwater prawns, salmonids, coral-reef fish, midshipmen fish, and again, side-blotched lizards. Developmental strategies are expected to arise when sexual selection favors specialized mating phenotypes, *and* when environmental cues, detectable by males, predict the type of mating opportunities likely to become available. Genetic architectures sensitive to environmental cues allow males to tune their mating phenotypes in response to changing environments. Such architectures are expected to exclude major genes that do not permit phenotypic plasticity.

In many species, the environmental cue to which males respond appears to be their own growth rate. In certain species, males unable to reach some threshold size within a given duration tend to mature early as satellites, whereas males who cross this threshold continue to grow and mature later as territorials. In other species, rapidly growing males tend to mature as satellites, whereas slower growers tend to become territorial. Genetic variation controlling male growth rate, like most traits influenced by numerous hereditary factors, appears to be normally distributed. Thus, the position of the body size threshold within the distribution of male growth rates determines the proportions of the population likely to consist of satellites and territorials. The relative success of satellites and territorials, in turn, appears to determine where the average male growth rate lies with respect to the body size threshold. Although circumstances favoring satellites or territorials can influence the population frequencies of each morph, directional selection can also change the position of the threshold itself.

For example, males in most species of *Onthophagus* beetles bear impressive horns. In all such species, horns are present only on large males and are used in contests for access to females. Horns vary in size, shape, and location on individual beetles, and a steplike relationship exists between horn length and body size in adult males. Small individuals possess rudimentary horns or lack them entirely, whereas large individuals possess well-developed weaponry. Although smaller males lack effective combat structures, these males are sexually mature and highly mobile and steal matings within the burrows of horned males. Horn size correlates with success in burrow defense, but small body size correlates with success in stealing matings. Thus, males that are intermediate in either of these characteristics are unlikely to mate at all.

Douglas Emlen and his colleagues have used controlled breeding designs and diet-manipulation experiments to explore the genetic basis of intraspecific variation in male horn length in several species of *Onthophagus*. Within lineages, morph frequencies do not segregate in a Mendelian manner. However, Emlen found that he could produce males with unusually large and unusually small horns by selecting in opposite directions on male horn length. This approach changed horn length, as well as the size within each lineage at which males developed horns. These results indicate not simply that horn length is heritable. They also demonstrate that a heritable "threshold" exists, which influences how males respond to their feeding history and thus to their own growth rate.

Threshold models for phenotypic plasticity require that individuals are genetically variable, *not* genetically identical. This is a reasonable assumption, given what is known about genetic variation in natural populations. In the case of growth rate polymorphisms, individuals either commit to an accelerated developmental trajectory or not, depending on the position of their genotype relative to the population threshold. Thus, only *part* of the population, not the *entire* population, must respond to an environmental cue for male polymorphism to appear. Threshold inheritance explains much about variation in male phenotype within and among populations. As the frequency of circumstances favoring one or another male phenotype changes, the proportion of a population likely to respond to environmental cues is also expected to change. Interpopulational variation in the proportion of males exhibiting each phenotype is expected to exist, and it does. Such variation is *not* expected if males are genetically monomorphic with respect to their ability to express one or another mating phenotype.

Behavioral strategies. The most commonly observed alternative mating strategies appear to be those involving discontinuous shifts in mating behavior. Behavioral strategies are expected to evolve when environmental changes influencing mating success occur often within individual lifetimes, *as well as* when the circumstances in which successful matings occur are highly variable. Thus, mobile species that breed more than once are most likely to exhibit behavioral polyphenism in the context of mating. Examples of such variation in mating behavior include dungflies, solitary bees, scorpionflies, many amphibians, songbirds (see below), ruffs, rodents, ungulates, felids, and primates. In each of these cases, males, and often females as well, rapidly change their behavior in ways that allow them to exploit mating opportunities as they arise.

An excellent example of such variation exists in cases of extrapair mating and egg dumping in songbirds. Darwin was intrigued by sexual dimorphism in group-nesting, apparently monogamous birds. He reasoned that sexual selection could operate in such species if females in superior condition nested early with the most attractive males. Elaborate male characters could thus be favored through the enhanced fecundity of their mates. However, an alternative, more powerful hypothesis exists, now that DNA fingerprinting has shown that males and females in a large number of "socially monogamous" species engage in extrapair copulations (EPCs). If multiple females sneak matings with the most attractive males and then rear the resulting offspring with cuckolded males, variance in offspring numbers among males will increase, and sexual dimorphism can evolve.

A surprising number of non-dimorphic, "socially monogamous" animals also engage in EPCs. In these species, both sexes routinely seek multiple mates, although particular individuals do not appear to be favored. Paternity in most nests is mixed, and mixed maternity also exists because females "dump" eggs in other females' nests. If EPCs and egg dumping increase the variance in offspring numbers in either sex, sexual dimorphism could evolve. However, if most individuals tend to mate more than once, although the average and the variance in mate numbers may increase in both sexes, the sex difference in the variance in offspring numbers will be small. In such cases, the species will remain monomorphic.

The underlying genetic architectures responsible for

behavioral variability are not well understood, but they appear to be similar to those described above for developmental strategies. That is, genetic variation underlying quantitative traits is expected to influence the likelihood that individuals will express a particular mating behavior. In a given situation, individuals with phenotypes below the liability threshold express a default set of mating behaviors, whereas individuals with phenotypes above this threshold express another behavioral set. Genetically variable characters influencing behavioral liability are known to include individual sensitivities to crowding and to circulating hormone levels. Other characters likely to influence mating behavior may include heritable sensitivities to pheromones, or to observations of multiple mating by other individuals. In the presence of a strong environmental cue, many individuals are expected to adjust their phenotype, whereas weaker cue intensity may induce few or no individuals to change. Thus, the same female distributions that induce some males to assume satellite behavior are expected to cause other males to persist as territorial males, as is widely observed.

Satellite Males on Leks in Ruffs. Delicate sandpipers known as ruffs (*Philomachus pugnax*) inhabit marshy regions in northern Europe and Asia. Ruffs are named for the mane of feathers borne by adult males. Ruffs breed on clusters of mating courts (leks), and in most populations, males exhibit two color morphs. About 85 percent of males consist of territorial residents, who bear darker plumage and defend mating courts against other court residents. Nonterritorial, satellite males, bearing lighter plumage, make up the remainder of male populations. Satellites are recruited onto residents' courts, where pairs of males form temporary breeding alliances.

David Lank and his colleagues (1999) have shown that plumage differences between residents and satellites are controlled by a single Mendelian locus, or perhaps by a chromosomal inversion that segregates within families. Females do not exhibit plumage variation like males, although females treated with testosterone implants develop male characters. When a female arrives on a mating territory, both males of the pair court her, but the resident often drives the satellite away. However, if a neighboring resident challenges the courting resident, the satellite may return and mate with the female while the residents fight. Females visiting mating courts routinely mate with both males, particularly on smaller leks. The combined displays of residents and satellites on small leks may be more attractive than larger leks containing only residents. Despite temporal variation in ruff behavior on leks, the average fitnesses of resident and satellite males appear to be equal.

Developmental Polymorphism in Pacific Salmon. Two male morphs coexist in southern populations of coho salmon (*Oncorhynchus kisutch*). Fry of this species leave their natal streams and mature in the Pacific Ocean. Larger, *hooknose* males return to spawn after three or more years. Smaller males, or *jacks*, mature early and return after only two years. Each hooknose defends a gravel nest on the stream bottom where females spawn. Hooknoses usually fertilize the ova in their nests. However, jacks lurking nearby may steal fertilizations by darting in and ejaculating when spawning begins. The male closest to the spawning female is most likely to sire young. Large hooknoses do best against other hooknoses, small jacks do best against other jacks, and most males of intermediate size are excluded from mating altogether.

Because jacks must be rare to mate successfully, the relative fitnesses of hooknoses and jacks appear to be frequency-dependent. Mart Gross (1996) demonstrated this relationship by considering the simultaneous influences of survival probability, reproductive tenure, and frequency-dependent spawning efficiency for males of each type. To standardize the units of these diverse fitness measures, Gross estimated the ratio of each jack-to-hooknose fitness estimate, then calculated the product of the three ratios. His result, .95, while not bounded by confidence limits, suggested that hooknoses and jacks experience approximately equal fitness. Such conditions are necessary and sufficient to maintain a genetic polymorphism in male phenotype by frequency dependent selection.

[*See also* Frequency-Dependent Selection; Sexual Selection, *article on* Bowerbirds.]

BIBLIOGRAPHY

Austad, S. N. "A Classification of Alternative Reproductive Behaviors, and Methods for Field Testing ESS models." *American Zoologist* 24 (1984): 309–320. Provides a summary of the forms in which alternative mating strategies appear and discusses early difficulties with using ESS approaches to understand mating polymorphism.

Crow, J. F. *Basic Concepts in Population, Quantitative and Evolutionary Genetics.* New York, 1986. An excellent summary of fundamental quantitative principles in evolutionary genetics; these concepts provide powerful tools for understanding how novel phenotypes may invade and persist in natural populations.

Cook, J. M. S. G. Compton, E. A. Herre, and S. A. West. "Alternative Mating Tactics and Extreme Male Dimorphism in Fig Wasps." *Proceedings of the Royal Society, London, B* 264 (1997): 747–751. A phylogenetic analysis demonstrating that male dimorphism in fig wasps evolves in response to sexual selection.

Darwin, C. R. *The Descent of Man and Selection in Relation to Sex.* New York, 1874. This book not only explains the basis of sexual selection, it also contains a large number of examples of alternative mating strategies that are rarely cited. In this context, Darwin clearly recognized the conditions necessary for the invasion, persistence and elimination of alternative mating strategies in natural populations.

Falconer, D. S. *Introduction to Quantitative Genetics.* 2d ed. New

York, 1989. A detailed yet readable description of the theory and methods necessary to identify quantitative genetic variation in natural and domestic populations.

Gadgil, M. "Male Dimorphism as a Consequence of Sexual Selection." *American Naturalist* 106 (1972). 574–580. An early discussion of how alternative mating strategies evolve in response to sexual selection.

Gross, M. R. "Alternative Reproductive Strategies and Tactics: Diversity within Sexes." *Trends in Ecology and Evolution* 11 (1996): 92–97. A brief summary of alternative mating strategies and a detailed description of the status dependent selection model for the evolution of alternative mating "tactics."

Lank, D. B., M. Coupe, and K. E. Wynne-Edwards. "Testosterone-Induced Male Traits in Female Ruffs (*Philomachus pugnax*): Autosomal Inheritance and Gender Differentiation." *Proceedings of the Royal Society of London, B: Biological Sciences* 266 (1999): 2323–2330.

Levins, R. *Evolution in Changing Environments*. Princeton, 1968. A technical but fundamental source of concepts and theory underlying the evolution of phenotypic plasticity.

Lucas, J., and R. D. Howard. "On Alternative Reproductive Tactics in Anurans: Dynamic Games with Density and Frequency Dependence." *American Naturalist* 146 (1995): 365–397. An application of contingency analysis and game theory to alternative mating tactics in bullfrogs.

Maynard Smith, J. *Evolution and the Theory of Games*. New York, 1982. The classic text for evolutionary game theory as it applies to animal behavior; several examples including horned beetles and the "rock-paper-scissors" model are discussed.

Moore, M. C. "Application of Organization-Activation Theory to Alternative Male Reproductive Strategies: A Review." *Hormones and Behavior* 25 (1991): 154–179. A summary of hormonal and developmental mechanisms responsible for the expression of alternative mating strategies.

Roff, D. A. "The Evolution of Threshold Traits in Animals." *Quarterly Review of Biology* 71 (1996): 3–35. A detailed description of how threshold characters evolve and persist in natural populations. An excellent source of empirical examples, particularly those by Daphne Fairbairn and Roff himself, with convincing discussion of how threshold characters explain complex patterns of development and behavior, including alternative mating strategies.

Schlicting, C. D., and M. Pigliucci. *Phenotypic Evolution: A Reaction Norm Perspective*. Sunderland, Mass., 1998. An encyclopedic work on evolution in changing environments and plasticity in phenotypic expression of traits; a timely and readable book.

Shuster, S. M., and M. J. Wade. *Mating Systems and Alternative Mating Strategies*. Princeton, in press. Uses methods from population genetics and selection theory to explain the evolution and dynamics of mating systems and alternative mating strategies.

Sinervo, B., and C. M. Lively. "The Rock-Paper-Scissors Game and the Evolution of Alternative Male Strategies." *Nature* 380 (1996): 40–243. A now classic, living example of Maynard Smith's "rock-paper-scissors" model for the persistence of polymorphic mating strategies; provides excellent illustrations of orange, blue, and yellow male side-blotched lizards as well as the frequency-dependent dynamics of the three, genetically distinct male morphs.

Wade, M. J. "Sexual Selection and Variance in Reproductive Success." *American Naturalist* 114 (1979): 742–764. Methods for measuring the opportunity for sexual selection; a fundamental concept necessary for measuring the mating niche and understanding how alternative mating strategies evolve.

— Stephen M. Shuster

MATING SYSTEMS

[*This entry comprises two articles. The first article provides a discussion of plant mating systems; the second article is an overview of the major kinds of animal mating systems with a discussion of their implications for mate choice. For related discussions, see* Male–Male Competition; *and* Mate Choice.]

Plant Mating Systems

Mating systems of flowering plants are highly diverse. Part of the reason for this diversity is simple: plants can repeatedly make new reproductive organs throughout their lives. Hence, different plants can, depending on the species, vary whether a flower contains only male reproductive organs (pollen grains—the male gametophytes), only female reproductive organs (ovules—the female gametophytes), or both, and they can even vary this from year to year. Mating systems range from one in which seeds can be produced without fertilization by an asexual process (apomixis), to those in which pollen needs to be transferred from one individual to another in a process known as pollination.

If you look at most flowers, you will see both male and female reproductive organs (or parts) in the same flower ("perfect" flowers). This mating system, hermaphroditism, is the most widespread one in flowering plants. Look even closer and you may notice that the male and female parts actually touch each other or can easily be brought into contact. This proximity makes it possible for plants to fertilize themselves (selfing). The most extreme case of selfing occurs in flowers that never open, but instead always self inside the bud. These are called cleistogamous flowers. However, not all hermaphroditic plant species self or are even capable of selfing. Selfing can lead to inbreeding depression, which occurs when seeds produced from selfing are not as vigorous or die more often than those produced from nonselfing (outcrossed). Hence, mechanisms for reducing or avoiding selfing exist in many hermaphroditic species with perfect flowers. These outcrossing mechanisms include separation of the male and female parts in time (dichogamy), separation of the parts in space (herkogamy), and genetic prevention of selfing (self-incompatibility). Finally, some plants have evolved unisexuality (separation of the sex parts into different flowers or individuals) as a way of reducing or preventing selfing.

As mentioned, even if a plant makes perfect flowers, it can reduce selfing by presenting male and female parts at different times in the life of the flower (dichogamy). For example, some flowers present their male parts early and their female parts later (protandry). An alternative is protogyny (female parts presented first), but this system is less common. Separation in time can be combined with separation in space. Figure 1 shows an example of a flower that presents its stamens first and moves these to the edge of the flower, followed by the elongation of the style and presentation of the stigma for pollen deposition. In addition to reducing selfing, dichogamy and herkogamy help to prevent physical interference between male and female parts.

Although separation in time and space can reduce the level of selfing within perfect flowers, pollinators can still move pollen from one flower on a plant to another flower on the same plant and cause selfing (geitonogamy). Consequently, dichogamy and herkogamy are unlikely to eliminate selfing all together. Selfing is entirely eliminated with self-incompatibility—a genetically based mechanism that prevents pollen grains from germinating when placed on a stigma of a flower from the same plant, or even another plant that shares the same self-incompatibility gene (*S* gene). These *S* genes have evolved many times within the flowering plants. Large numbers of self-incompatibility alleles can be maintained in populations, because a pollen grain with a rare *S* allele will be able to germinate and fertilize ovules on most of the other plants in the population, whereas a pollen grain with a common *S* allele will be more limited in its mating opportunities.

Heterostyly is another type of mating system with perfect flowers. Heterostylous species exhibit both herkogamy and self-incompatibility. One form of heterostyly is called distyly—there are two morphs, pin and thrum, that differ reciprocally in the placement of male and female parts within flowers. The pin morph has a longer style and shorter stamens, whereas the thrum morph has longer stamens and a shorter style. This mating system prevents selfing: pollen grains are incapable of germinating on the plant from which they came; they also cannot germinate on plants of the same morph. In addition to preventing selfing, heterostyly reduces interference between the parts while simultaneously promoting precise cross-pollen transfer because of the reciprocal and complementary placement of the sex parts.

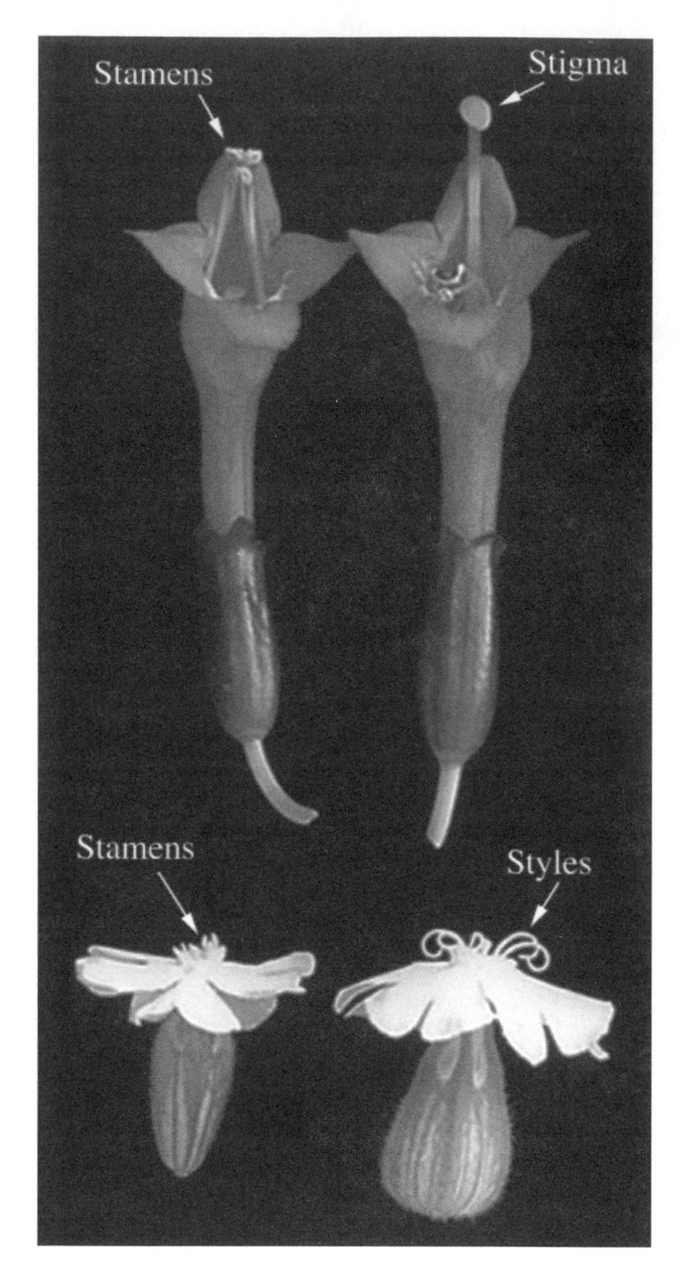

FIGURE 1. Plant Mating Systems.
The top two flowers are from a hermaphroditic species that exhibits separation of the sex parts in time and space. The upper left flower is in the male phase, with pollen being presented for removal. The upper right flower is in the female phase, exposing its stigma for pollen receipt. The bottom two flowers are from a dioecious species (*Silene latifolia*), which has separate male and female individuals. The flower on the lower left is from a male plant and the flower on the lower right is from a female plant. Photograph by Lynda F. Delph.

Not all hermaphroditic species produce perfect flowers. Some species separate the male and female parts into separate, unisexual flowers, but the plants themselves are still considered hermaphrodites, because they produce both types of reproductive organs within the same plant. There are variations on this theme of hermaphrodites producing unisexual flowers, but all have the same suffix of monoecious (meaning "one house"). This term highlights the fact that although different forms of flowers exist, there is only one type of plant—a hermaphroditic one. Monoecious species make two types of flowers: pistillate flowers (containing female parts only) and staminate flowers (male parts only). In some monoecious species, both types of unisexual flowers are made on each plant each year. Others exhibit what is called "sex choosing"—they produce only one type of flower in a given year, but may "change sex" the following year, producing only the other type of flower. Still other variations include gynomonoecious species, which produce pistillate and perfect flowers, and andromonoecious species, which produce staminate and perfect flowers. These terms are easier to remember if you focus on the prefixes, *gyno-* ("female") and *andro-* ("male"). Regardless of whether only perfect flowers are produced or there is some combination of pistillate, staminate, and perfect flower production, plants with these monoecious mating systems achieve half of their evolutionary fitness via female function and half via male function, on average. This is because each seed produced has only one mother and one father. Hence, an average andromonoecious individual does not gain more fitness via male function, even though staminate flowers contain only male parts, any more than a gynomonoecious individual gains more fitness via female function.

Although monoecious mating systems may reduce selfing, it is thought that the primary benefit to making unisexual flowers is the greater flexibility in resource allocation to male and female function that this confers to hermaphroditic plants. For example, a small plant might not have enough resources to produce fruits from all of its flowers and consequently might make some staminate and some perfect flowers. Such a strategy would allow the plant to avoid wasted investment in female parts it could not mature into fruit, while at the same time allowing it to produce flowers with pollen that could fertilize the flowers of other plants. A similar strategy is seen in sex choosers, which often make only staminate flowers early in life and switch to producing pistillate flowers later in life when they are larger and have more resources.

In addition to systems that involve only one gender morph per population, there are mating systems in which two types of plants coexist within a population; that is, the population is dimorphic for gender. These systems have the suffix dioecious (meaning "two houses"). Individual plants of dioecious species make only staminate flowers or only pistillate flowers (see Figure 1). These plants are referred to as males and females, respectively, and this mating system is roughly equivalent to the one found in humans. Adding the prefixes *gyno-* and *andro-* as before gives gynodioecious and androdioecious, two variations on the dioecious theme. Gynodioecious species, containing female plants and hermaphroditic plants, are relatively numerous, whereas androdioecious species, containing male plants and hermaphroditic plants, are extremely scarce.

Gender dimorphism is thought to have evolved from hermaphroditism. Two evolutionary pathways to gender dimorphism have been considered. In one, a unisexual mutant (that is solely male or female) invades a hermaphroditic population and becomes established. Because gynodioecious species are much more common than androdioecious ones, it seems clear that mutations causing plants to be females either arise more frequently than mutations causing plants to be males or, more likely, female mutants can become established and spread more easily. The other pathway involves selection acting on a preexisting dimorphism such that each morph gradually specializes in one sex function to the exclusion of the other. For example, heterostylous species already have two morphs, pin and thrum, and it has been shown that the pins become females and the thrums become males when dioecy evolves from heterostyly.

There are two broad nonexclusive mechanisms underlying why gender dimorphism might evolve. One involves the elimination of inbreeding depression. Unisexual mutants cannot self. Consequently, the fitness of seeds from a female mutant might be higher than that of self-compatible hermaphrodites from the same population. The second mechanism involves reallocation of resources into one of the sex functions following the loss of the other sex function. Again, females might achieve higher fitness through their seeds than the hermaphrodites if they are able to make more seeds using the resources that would otherwise have been used to produce pollen. Hence, if both mechanisms are at work, females might make more, higher-quality seeds as compared to hermaphrodites. Furthermore, the mutation causing a plant to be strictly female is often in the mitochondria, and mitochondria are almost always inherited solely from the mother. Hence, whether or not a particular type of mitochondria will spread in the population depends on its seed fitness compared to other types of mitochondria. This means that a mitochondrial gene for femaleness will spread, and females will be maintained in the population, as long as it confers slightly higher seed fitness as compared to plants whose mitochondria do not have such genes. In contrast, a similar gene in the nucleus would spread only if it conferred twice the seed fitness of the hermaphrodites, because nuclear genes are inherited from both the mother and the father, and the females would therefore have to compensate for not spreading genes via pollen.

Given that the avoidance of selfing is one of the selective factors thought to be responsible for the evolution of dioecism, one might expect that flowering plant genera with self-incompatibility might not have as many dioecious species as those genera without self-incompatibility. If you already have one mechanism to avoid selfing, why would you need another? Although a tendency for a negative association between self-incompatibility and dioecism has been documented, it is difficult to evaluate the strength of this association because we lack sufficient data on the number of species in which self-incompatibility occurs. Self-incompatibility may be underestimated in some cases and overestimated in others. An example of this is when diploid, self-incompatible species give rise to new species via the duplication of chromosome number (polyploidy). Polyploidy disrupts self-incompatibility in some species, allowing selfing to occur, and sometimes triggers the evolution of dioecism in what at first glance appears to be a genus with self-incompatibility. Uncovering such "hidden" compatibility strengthens the hypothesis that the avoidance of selfing is a potent selective force for the evolution of gender dimorphism.

Once gender dimorphism has evolved, it sets the stage for the evolution of differences between the morphs. Just as there are differences between the sexes in animals (e.g., males are often larger), sexes of plants differ in a variety of ways. Such differences, or sexual dimorphism, can arise because of differences in allocation to reproduction, access to mates (i.e., via sexual selection), or character displacement. Two easily noticed differences are the number and size of flowers produced: in temperate species, males typically produce a showier display, by producing both more flowers and larger flowers than females (note that there are exceptions, as shown in Figure 1). This difference is hypothesized to occur, in part, because of competition among males for access to the ovules of females. By being more attractive to pollinators than the male next to him, a particular male might acquire more visits and consequently transfer more pollen to the receptive stigmas of the flowers on females. Sexual dimorphism occurs not only in terms of what the flowers of the two sexes look like, but also for other morphological (e.g., leaf traits), physiological, life history, and ecological traits.

[*See also* Mate Choice, *article on* Mate Choice in Plants.]

BIBLIOGRAPHY

Barrett, S. C. H., ed. *Evolution and Function of Heterostyly*. Berlin, 1992. Contains articles dealing with the forces selecting for heterostyly and its consequences.

Barrett, S. C. H., and L. D. Harder. "Ecology and Evolution of Plant Mating." *Trends in Ecology and Evolution* 11.2 (1996): 73–79. Focuses on how mechanisms of pollen dispersal influence the level of selfing.

Darwin, C. *The Different Forms of Flowers on Plants of the Same Species* (1877). Reprint, Chicago, 1986. One of two books that Darwin wrote on plant mating systems.

Geber, M. A., T. E. Dawson, and L. F. Delph, eds. *Gender and Sexual Dimorphism in Flowering Plants*. Berlin, 1999. Deals with theoretical and empirical studies of why gender dimorphism evolves, and why males and females are different from each other in a variety of ways.

Lloyd, D. G., and C. J. Webb. "The Avoidance of Interference between the Presentation of Pollen and Stigmas in Angiosperms: 1. Dichogamy." *New Zealand Journal of Botany* 24 (1986): 135–162. A classic paper suggesting why dichogamy is adaptive.

Miller, J. S., and D. L. Venable. "Polyploidy and the Evolution of Gender Dimorphism in Plants." *Science* 289 (2000): 2335–2338. A paper showing how polyploidy can lead to the breakdown of self-incompatibility and hence the evolution of gender dimorphism.

Schlessman, M. A. "Gender Modification in North American Ginsengs." *BioScience* 37 (1987): 469–475. A review on sex-choosing plants.

Wyatt, R. "Pollinator–Plant Interactions and the Evolution of Breeding Systems." In *Pollination Biology*, edited by L. Real. Orlando, Fla., 1983. An ecological view of plant mating systems.

— LYNDA F. DELPH

Animals

Mating systems are descriptions of behavior related to the acquisition of mates for reproduction. In their narrowest sense, mating systems describe the numbers of mating partners obtained by members of both sexes during the breeding season, as well as forms of courtship, conflicts within and between the sexes, and resources that are used to attract mates. The term *mating system* is often used in a broader sense to include parental care, as originally described in the classic papers by Gordon Orians in 1969 and Stephen Emlen and Lewis Oring in 1977. However, the term *breeding system* is more appropriate for this broader understanding that includes parental care.

Anisogamy and Potential Reproductive Rates. Males and females are defined by the different sizes of their gametes (anisogamy). The larger gametes produced by females enhance the survival of zygotes, while the numerous small gametes produced by males compete to fertilize the female's egg. From this fundamental difference between the sexes, the stage is set for competition for mates, conflicts between the sexes, and mate choice.

In 1948 A. J. Bateman showed how gametic competition could be scaled up to the level of the individual when gametes were the only investment made into offspring. He showed that reproductive success of male *Drosophila* fruit flies increased with each additional mating, while for females reproductive success was constrained by the number of ova produced. Thus, there is a difference in the potential reproductive rate of each sex—that is, the maximum rate at which animals can produce offspring if given unlimited access to members of the opposite sex. This difference between the sexes generates divergence in reproductive interests that are manifested in mating systems.

Potential reproductive rates are also influenced heavily by parental care. As Robert Trivers identified in the 1970s, females usually invest more into each offspring than do males. This generates mating systems in which males (as the lower investors) are forced into competition with each other for reproductive access to the higher-investing females. Males therefore usually have greater variation in reproductive success.

The Operational Sex Ratio. The vignette shows how differences between the sexes in gametic and parental input to the offspring translate into differences in potential rates of reproduction. To convert these individual differences into population differences, we can translate them into the operational sex ratio (OSR), defined as the ratio of available reproducing females to reproducing males. A male-biased OSR (i.e., more available males than females), which is typical of many species, means that males have more to gain from mate competition and courtship than females, which are in shorter supply. Competition can occur between individual males prior to mating or between the sperm of males in the many species in which females mate with more than one male during a breeding bout. Females are able to be more selective of mating partners than are males, because more receptive males are available.

The OSR is also influenced by the distributions and ratios of males and females in space and time, as well as sex-dependent variance in age at maturity, risk of mortality, and reproductive lifespan. For example, males have often evolved more risky traits through sexual selection (such as conspicuous coloration or behavior), which leads to higher mortality rates. These factors are clearly not fixed but vary at both population and individual levels, and so we find that a species' OSR can change in different environments, with correlated shifts in mating systems.

Ecological Influences. Natural selection can be important in shaping mating systems. For example, if females live in widely dispersed territories, individual males may have trouble visiting them and defending them from rivals. The mating system may therefore involve low variation among males in numbers of mates. Conversely, if males can monopolize sites that are important to large numbers of females, there may be a higher variation among males in mating success. Thus, in red deer (*Cervus elephas*), dominant stags defend popular grazing areas and the groups of hinds that visit them.

A FRAMEWORK FOR ANIMAL MATING SYSTEMS

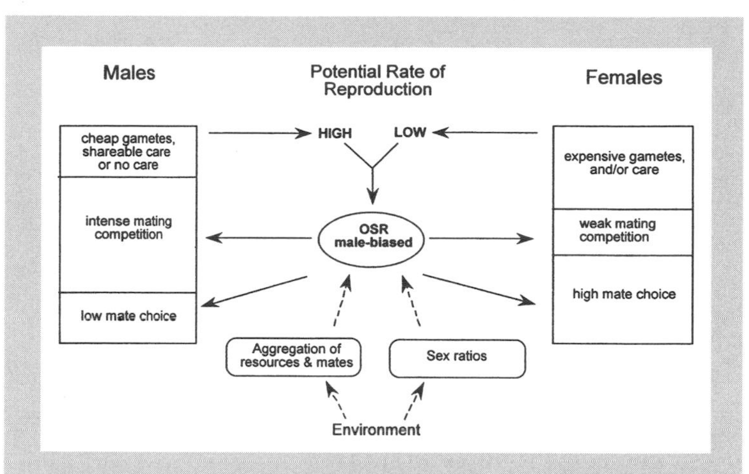

This flow diagram illustrates a "typical" animal, in which males provide less care per brood than females, in terms of time or energy devoted to the young. For courtship-role-reversed species, the terms *males* and *females* at the top can be reversed with females retaining expensive gametes. The sizes of the boxes for each sex represent relative differences in time or energy devoted to each activity. Low parental input per brood by males leads to high potential rates of reproduction, and therefore causes a male-biased OSR (operational sex ratio). This excess of available males selects for mating competition and courtship, and males are less able to afford being less choosy about mates than are females. The OSR will also depend on the environment, such as the extent of aggregation of mates and resources for reproduction, as well as sex ratios of adults.

—JOHN D. REYNOLDS AND MATTHEW J. G. GAGE

The classic theory to explain defense of ecological resources for mating access and mate choice is the polygyny threshold model, first formalized by Gordon Orians in 1969. The theory was inspired by avian mating systems in which males defend territories of variable quality in which females breed. A male crosses the polygyny threshold from monogamy when his territory quality exceeds that of other males to a sufficient degree that additional females choose to breed with him there. The model establishes a simple scenario to explain the interactions between reproductive and ecological resources for males and females, and illustrates the potential for variance in mating systems within a species. Although it is an elegant "null model," it ignores the rich array of female tactics that have come to light in the last two decades, including competition with other females and mating with males outside of the pair bond.

Definitions of Mating Systems. It is impossible to capture all of the elements of mating systems with a simple classification because of individual variation in behavior within so-called "systems." We return to this point with an example after some key terms have been defined.

1. *Monogamy:* One male mates with one female. This is usually the majority pattern in species that have a prolonged pair bond and biparental care. Genetic analyses of parentage have shown this mating system to be much less common than previously thought. Even when apparently stable pair bonds are formed and both parents contribute care to the young, either sex may seek copulations with individuals outside the pair bond. Thus, "social monogamy" (an appearance of monogamy based on pair bonds) is often unmasked as not being "genetic monogamy" (which we here consider to be true monogamy). While the reproductive benefits of extra-pair matings to males are clear, the benefits to females are less obvious, especially if females risk losing parental contributions from males due to lower confidence of paternity. Females may gain fertility assurance or improved genetic quality for their offspring.

2. *Polygyny:* One male mates with several females during a breeding season. May occur with or without pair bonds and defense of resources and territories.

3. *Polyandry:* One female mates with several males during a breeding season. May occur with or without pair bonds and defense of resources and territories.

4. *Promiscuity:* Both sexes have multiple mates and no pair bonds are formed.

5. *Polygynandry:* Both sexes have multiple mates and pair bonds are formed. For example, in the dunnock, *Prunella modularis*, females may bond with two or sometimes three partners. Some of these males will also be paired with other females so that both polygyny and polyandry occur in the same population simultaneously. As reviewed in his monograph *Dunnock Behaviour and Social Evolution* (1992), Nicholas Davies and colleagues have shown that food resource defense is an important predictor of mating pattern in dunnocks. When food is limiting, females may pair polyandrously with multiple males who contribute jointly to nestling provisioning. When food is plentiful the system shifts away from resource limitation, allowing male reproductive potential to expand and polygyny to be manifested.

6. *Lekking:* Males offer nothing to females other than sperm, and perform courtship displays from aggregated arenas. For example, in black grouse (*Tetrao tetrix*), mating success is heavily skewed toward males that retain a central position in the lek site through aggression and dominance. Traits chosen by females may indicate genetic quality. While research has traditionally focussed on the high rates of polygyny shown by some males in lekking mating systems, recent research has indicated a rich array of female behaviors, ranging from monogamy to polyandry in lekking species. Lekking is a rare mating system that occurs in a taxonomically disparate group of species that include various flies, butterflies, cichlid fishes, frogs and toads, ungulates, grouse, hummingbirds, manakins, and birds of paradise.

7. *Courtship-role reversal*, also known as *sex-role reversal*, occurs when females are more competitive than males. For example, in the red-necked phalarope, *Phalaropus lobatus*, females require about eleven days to produce a clutch of eggs, and then abandon males to perform incubation and brood rearing by themselves for forty days. This generates a female-biased OSR because females have time to form pair bonds with two males in one season. Females therefore compete strongly for mates. Additional examples of this rare mating system include some species of orthopterans in which males provide large nuptial gifts to females, some giant water bugs, most species of jacanas, some dendrobatid frogs, and some species of pipefishes. As expected from mating system theory (see Vignette), courtship-role reversal is associated with much greater parental care contributions by males than by females. Note, however, that exclusive paternal care by itself does not guarantee courtship-role reversal, as shown by the many species of fishes such as sticklebacks (*Gasterosteus*) and gobies, in which males are the more competitive sex despite providing exclusive care of the young. Care per brood is "cheaper" in these species than in birds, so males can still process broods faster than females can lay eggs. Courtship role-reversal occurs when male care is so costly or time consuming that the potential rate of reproduction of males is less than that of females.

8. *Hermaphrodite* mating systems are those in which a single individual produces both male and female gametes, either simultaneously or sequentially. In simultaneous hermaphrodites such as many molluscs and hamlet fishes (*Hypoplectrus* spp.) there is potential conflict between individuals during mating when sperm and ova are exchanged. Individuals transferring sperm achieve fertilizations for less investment than when producing eggs. Some species have resolved this conflict using an iterated Tit For Tat sequence in which small quantities of sperm and ova are traded reciprocally. In sequential sex changers, individuals can switch from female to male (*protogyny*) or vice versa (*protandry*). These reproductive strategies have evolved in a variety of fish species, with well-studied examples of protogyny in labrids such as bluehead wrasse (*Thalassoma bifasciatum*) and protandry in anenomefish (*Amphiprion* sp.). Protogyny is associated with mating systems in which large male size is essential for defending mating territories. Thus, individuals are selected to begin reproducing as females when small, and then switch to males when they are large enough to compete effectively. Protandry appears in taxa that have lower size-based competition amoung males and where the fecundity benefits of large female body size can be sustained.

The foregoing terms provide only a start to describing mating systems, because mating systems also encompass the manner in which mates are acquired, and the extent of variation in numbers of mates obtained by both sexes. For example, in the red-winged blackbird (*Agelaius phoeniceus*), males attempt to defend territories in marshes, and some individuals attract more than one female to nest on their territories. Classically, this would be called resource defense polygyny. Note that this term ignores females and focuses on the subject of males that are successful, ignoring males that are unable to defend territories ("floaters") and those that are socially monogamous. As for females, studies of paternity have revealed that they may mate with males from nearby territories as well as with their resident males. This mating system could therefore be described as *polyandry* (from the female's view) or *polygyny* (from the successful male's view).

Sexual Conflict. Conflicts abound both within and between the sexes. Mate desertion is one result of conflict resolution that usually results in a reproductive disadvantage for the deserted partner. Experimental studies have shown that male cichlid fishes are more apt to desert their mates and broods early when they are offered access to unmated females. Similarly, females are more prone to desertion when they have extra mating opportunities. Thus, there is two-way feedback between sexual selection and parental care, and we can expect different individuals within populations to allocate their time and energy budgets differently according to their ability to provide care and compete for mates. Forced copulation is another outcome of sexual conflict in which males attempt to shift the mating system away from the female optimum. This behavior occurs in a wide variety of internally fertilizing species, including primates, turtles, waterfowl, various insects, and livebearing fishes.

A graphic illustration of the importance of sexual conflict was provided by experiments on *Drosophila melanogaster* by Brett Holland and William Rice in 1999. In this species of fruit fly, males compete with one another at the gametic level using toxic seminal fluids that are also harmful to females. The researchers manipulated the mating system of the fruit flies in the laboratory by enforcing monogamy in some populations and comparing them with promiscuous controls. After forty-seven generations, males in the monogamous populations had evolved to become less harmful to their mates, and females had become less resistant to such male-induced damage. Furthermore, the monogamous populations had also evolved a greater net reproductive rate than promiscuous controls. The results show that in the drive for reproductive success, male–male competition has led to strategies that are deleterious to female reproduction.

[*See also* Ideal Free Distribution; Male–Male Competition; Mate Choice, *article on* Human Mate Choice; Parental Care; Sexual Dimorphism; Sexual Selection, *article on* Bowerbirds; Territoriality.]

BIBLIOGRAPHY

Andersson, M. *Sexual Selection.* Princeton, N.J., 1994. An excellent, wide-ranging review of sexual selection and mating systems.

Bateman, A. J. "Intra-Sexual Selection in *Drosophila.*" *Heredity* 2 (1948): 349–368.

Clutton-Brock, T. H., and G. A. Parker. "Potential Reproductive Rates and the Operation of Sexual Selection." *Quarterly Reviews of Biology* 67 (1992): 437–455. Explains links with operational sex ratios.

Emlen, S. T., and L. W. Oring. "Ecology, Sexual Selection, and the Evolution of Animal Mating Systems." *Science* 197 (1977): 215–223. A classic explanation of how mating systems are shaped by the environment.

Höglund, J., and R. V. Alatalo. *Leks.* Princeton, N.J., 1995.

Holland, B., and W. R. Rice. "Experimental Removal of Sexual Selection Reverses Intersexual Antagonistic Coevolution and Removes a Reproductive Load." *Proceedings of the National Academy of Sciences of the USA* 96 (1999): 5083–5088. An empirical selection experiment illustrating the net reproductive costs of sexual conflict.

Johnstone, R. A., J. D. Reynolds, and J. C. Deutsch. "Mutual Mate Choice and Sex Differences in Choosiness." *Evolution* 50 (1996): 1382–1391. Game-theoretic models of interactions between males and females when each chooses the other.

Orians, G. H. "On the Evolution of Mating Systems in Birds and Mammals." *American Naturalist* 103 (1969): 589–603. A classic paper with a graphical model of the "polygyny threshold hypothesis." The underlying concept was later reinvented and used widely in behavioral ecology as the ideal free distribution.

Reynolds, J. D. "Animal Breeding Systems." *Trends in Ecology and Evolution* 11 (1996): 68–72. An update on the theory, incorporating mate choice for genes and resources and multiple mating by females. It provided the flow diagram in the vignette which is updated here.

Trivers, R. L. "Parental Investment and Sexual Selection." In *Sexual Selection and the Descent of Man*, edited by B. Campbell. London: Heinemann, 1972.

— John D. Reynolds and Matthew J. G. Gage

MAXIMUM LIKELIHOOD. *See* Phylogenetic Inference.

MCCLINTOCK, BARBARA

Plant cytogeneticist Barbara McClintock (1902–1992) is most famous as a geneticist, but she considered her primary contributions to have been in development and evolution. Born in Hartford, Connecticut, raised mainly in Brooklyn, New York, McClintock received her bachelor's (1923), master's (1925), and doctoral (1927) degrees from Cornell University. In these early years, she established nearly single-handedly the cytogenetic study of maize, or Indian corn, and ushered in the "golden age of maize genetics" (1930–1935).

A string of superb papers followed. In 1931, with graduate student Harriet Creighton, she proved the long-held assumption that genetic crossing over reflected physical exchange of chromosomal material. She discovered the nucleolar organizer and the telomere, and described the behavior of broken and ring-shaped chromosomes.

In 1936, she accepted a position at the University of Missouri, in Columbia. While there, she described the breakage-fusion-bridge (BFB) cycle, a pattern of repeating chromosome breakages and fusions. Unhappy in academic life, she resigned from Missouri before coming up for tenure and nearly left science altogether. Loyal colleagues, including Lewis Stadler at Missouri, showed their support by nominating her for the National Academy of Sciences, to which she was elected in 1944, young for the honor at age forty-one and only the third woman member.

After a year at Columbia University, in 1941 McClintock was hired by the Carnegie Institution of Washington's Department of Genetics, at Cold Spring Harbor, New York, where she remained until her death. In 1944–1946, she uncovered a pair of genetic loci that appeared to control the mutation rate of other genes. She believed these to be part of a system for controlling gene action during development. In the spring of 1948, she discovered that the two "controlling units" (later, "controlling elements") could transpose, or change position, on the chromosomes. Between 1948 and 1950 she developed a theory in which platoons of controlling elements executed the developmental program by coordinated transpositions, inhibiting or modulating the action of genes near where they landed.

Transposition was rapidly and repeatedly confirmed in maize. But to most geneticists it seemed a random process. One who accepted her developmental argument was Richard Goldschmidt, an enemy of the classical gene concept and champion of physiological genetics and macroevolution. McClintock had known Goldschmidt since 1932, and in 1951 they reinforced each others' positions in widely noticed papers at the Cold Spring Harbor Symposium on Quantitative Biology.

McClintock's interest in evolution grew out of a research project, begun in 1957, to describe the chromosome constitutions of the many cultivated strains of maize in the Americas. She found she could map the migrations and evolution of the races of maize by the patterns of chromosomal swellings called knobs. Knobs harbor controlling elements, and she believed they were involved in macroevolutionary changes, large-scale genomic rearrangements leading to major changes in developmental patterns.

By the beginning of the 1980s, transposition was understood to be a fundamental, universal genomic mechanism with medical implications. McClintock's emphasis on the role of controlling elements in development and evolution were rejected. Controlling elements were recast as transposable elements—evolutionary opportunists, probably of viral origin. The press began to refer to them as "jumping genes." In 1981, she won a string of distinguished awards, including the Lasker Prize in medicine and a lifetime MacArthur fellowship. Two years later, she won the Nobel Prize in physiology or medicine. The honor was bittersweet; though glad for the recognition, she thought it missed her efforts to understand the genetic basis of development and evolution.

[*See also* Overview Essay *on* Macroevolution; Ancient DNA; Development, *article on* Evolution of Development; Transposable Elements.]

BIBLIOGRAPHY

Comfort, N. C. "Two Genes, No Enzyme: A Second Look at Barbara McClintock and the 1951 Cold Spring Harbor Symposium." *Genetics* 140 (1995): 1161–1166.

Comfort, N. C. "'The Real Point Is Control': The Reception of Barbara McClintock's Controlling Elements." *Journal of the History of Biology* 32 (1999): 133–162. An article-length synopsis of the core arguments of *The Tangled Field.*

Comfort, N. C. *The Tangled Field: Barbara McClintock's Search for the Patterns of Genetic Control.* Cambridge, Mass., 2001. Argues that McClintock was neither ignored nor marginalized and details her core science in lay language.

Fedoroff, N. "How Jumping Genes Were Discovered." *Nature Structural Biology* 8 (2001): 300–301. Fedoroff has been the primary scientist in the molecular characterization of McClintock's controlling elements.

Fedoroff, N., and D. Botstein. *The Dynamic Genome: Barbara McClintock's Ideas in the Century of Genetics.* Cold Spring Harbor, N. Y., 1992. A Festschrift assembled for McClintock's ninetieth birthday, with reminiscences of colleagues and reprints of a few of McClintock's major articles.

Keller, E. F. *A Feeling for the Organism: The Life and Work of Barbara McClintock.* New York, 1983. The first major biographical study of McClintock.

McClintock, B. "Significance of Chromosome Constitutions in Tracing the Origin and Migration of Races of Maize in the Americas." In *Maize Breeding and Genetics*, edited by W. D. Walden, pp. 159–184. New York, 1978. Article-length treatment of the races-of-maize story; the basis of McClintock's evolutionary thinking.

McClintock, B. *The Discovery and Characterization of Transposable Elements: The Collected Papers of Barbara McClintock.* Edited by J. A. Moore. New York, 1987. Nearly complete, this collection omits McClintock's early papers and a few later ones. An essential reference.

McClintock, B., T. A. Kato Y., and A. Blumenschein. *Chromosome Constitution of Races of Maize: Its Significance in the Interpretation of Relationships between Races and Varieties in the Americas.* Chapingo, Mexico, 1981. The product of McClintock's major study of the races of maize; the basis of her evolutionary thinking. Hard to find.

— NATHANIEL C. COMFORT

MEIOSIS

Meiosis, or reduction division, is the mechanism by which chromosome numbers are halved once every sexual life cycle. Without it, chromosomes would double in number every sexual generation as the result of the fusion of gametes (syngamy) and eventually their nuclei. Meiosis is best understood by contrasting it with mitosis, the normal way that most nuclei divide in animal and plant bodies as well as during multiplication of single-celled eukaryotes (e.g., protozoa and yeasts).

All living organisms are either eukaryotes, having nuclei that divide by mitosis, or bacteria, which have a chromosome directly attached to the cell surface and lack nuclei and mitosis. Sexual eukaryotes have life cycles involving both types of nuclear division, in addition to cell and nuclear fusion (Figure 1). In thinking about the biological significance and evolution of meiosis, it is important to note that the process evolved not in animals but in a very early single-celled eukaryote. Meiosis is usually considered a specialized derivative of mitosis, but both possibly evolved together and played a key role in the evolutionary origin of the nucleus itself and the conversion of the ancestral circular bacterial chromosome into linear eukaryotic chromosomes.

DNA Replication and the Key Role of Cohesins. Eukaryote chromosomes are immensely long DNA threads wound around histone proteins, normally invisible in growing cells. Virtually all eukaryotes have more than one chromosome per genome. Each chromosome comprises one DNA molecule millions of times longer than the diameter of the nucleus. Chromosomes would suffer breakage if the cell attempted to divide with them in this highly extended state. To avoid this, during the first stage of mitosis (prophase) each chromosome thread tightly winds into a thick, compact but flexible rod suitable for being moved efficiently into daughter cells by the division apparatus—the mitotic spindle. Prior to this condensation of the DNA-histone threads into microscopically visible chromosomes, each is genetically doubled by DNA replication during cell growth. But it still appears as a single rod because cohesin proteins tightly glue the two folded daughter DNA molecules side by side (Figure 2a). Toward the end of prophase, a protease enzyme digests the cohesins except at the centromere and the daughters partially separate, making the "chromosome" visibly double.

Each half of the double chromosome, conventionally called a chromatid but genetically a complete chromosome, remains attached to its partner at one position (rarely more)—its centromere. The centromere contains special histones and other proteins that attach it to microtubules within the mitotic spindle. Typically the chromosomes then move into the equatorial plane of the cell, being held there for a period (metaphase) by a balance of forces pulling them to opposite ends (poles) of the cell. Suddenly the residual cohesins are digested, splitting the centromere and thereby allowing sister chromatids to be pulled to opposite poles (anaphase movement). Once there, the chromosomes cluster and unwind, becoming invisible again within the two daughter nuclei, between which the cytoplasm of the cell divides. In ancestral mitosis and meiosis, as in most fungi and protozoa and many lower plants, the nuclear envelope, a double membrane that surrounds the chromosomes in growing eukaryotic cells, stays largely intact, dividing into two after anaphase chromosome separation—like the cell's surface membrane. In animals, higher plants, and most large-celled protists, the envelope fragments at the end of prophase and is reassembled around the chromosome mass at the end (telophase) of mitosis and meiosis. Several lineages independently

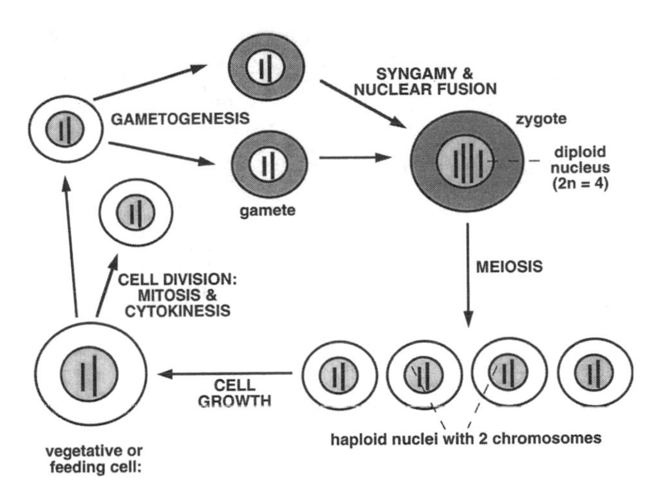

FIGURE 1. A Generalized Sexual Life Cycle.
Shown here is the probable ancestral condition in which chromosome numbers are doubled by sexual fusion of gamete cells and halved by meiosis to produce four daughters without any intervening mitotic divisions. The haploid chromosome number, n, is shown as two: more often it is larger—usually between 4 and 40, but it can be many hundreds or as low as one. Note that diploid cells have twice the volume of haploid ones and predivision cells are twice as large as daughters (with binary fission). Diagrams by geneticists often wrongly show diploid and haploid cells the same size, concealing one of the two key consequences of diploidy. But cell size is often selectively at least as important as genetic duplicity. Drawing by Thomas Cavalier-Smith.

evolved this envelope breakdown and rearrangement of the membrane fragments within the spindle, probably so that their calcium pumps can regulate calcium ion concentrations, a key controller of mitosis, more effectively within larger cells.

Origin of Eukaryotic Chromosomes and Mitosis. Eukaryotic chromosomes originated immediately prior to the formation of the nucleus during the conversion of a bacterium into the first eukaryote about 850 million years ago. Some histones, the main proteins of our chromosomes, had already evolved in the common ancestor of eukaryotes and archaebacteria (Figure 3), perhaps as an adaptation to thermophily; their origin caused marked changes in all enzymes that interact directly with DNA. As in all bacteria, the ancestral chromosome would have been a single replication unit: replication, occurring throughout the cell cycle, started at one point (the origin) and finished at another, the terminus. In bacteria the origin and terminus, as well as the replication machinery that moves the two replication forks bidirectionally from origin to terminus, are attached to the cell surface membrane and wall. As bacterial chromosomes are usually circular, completion of replication requires recombination at specific sites near the terminus to generate two daughter circles, which are actively moved apart into daughter cells by motor proteins: there is no mitotic spindle, and the separation mechanisms are unrelated to those of eukaryotes and depend on chromosome attachment to the rigid cell surface. Following chromosome separation, bacterial cells divide by a contractile ring of FtsZ protein.

The key step in the origin of eukaryotes was the origin of phagotrophy: the ability to engulf other cells by phagocytosis and digest them internally. This process was made possible by evolving a flexible surface coat of novel glycoproteins that first evolved in the immediate common ancestors of eukaryotes and archaàebacteria (Figure 3). A soft surface and the temporary internalization of surface membrane by phagocytosis would have interfered with bacterial DNA segregation and cell division mechanisms. Evolution of an internal cytoskeleton of actin and of the molecular motor myosin allowed these eukaryote ancestors to divide by a novel actomyosin contractile ring. This freed the redundant bacterial FtsZ division protein to evolve into tubulin proteins, to form spindle microtubules and the centrosomes at the poles of the mitotic spindle where microtubules are assembled. The origin of centrosomes, microtubules, and the motors dynein and kinesin that propel membrane vesicles along them, were fundamental innovatory prerequisites for mitosis. The key innovations coupling them to chromosome segregation were the evolution of centromeres to bind microtubules and of cohesins to glue sister DNA molecules together until properly attached to the spindle. All arose as indirect consequences of phagotrophy and the soft surface, explaining why our archaebacterial sisters that retained a rigid cell wall in-

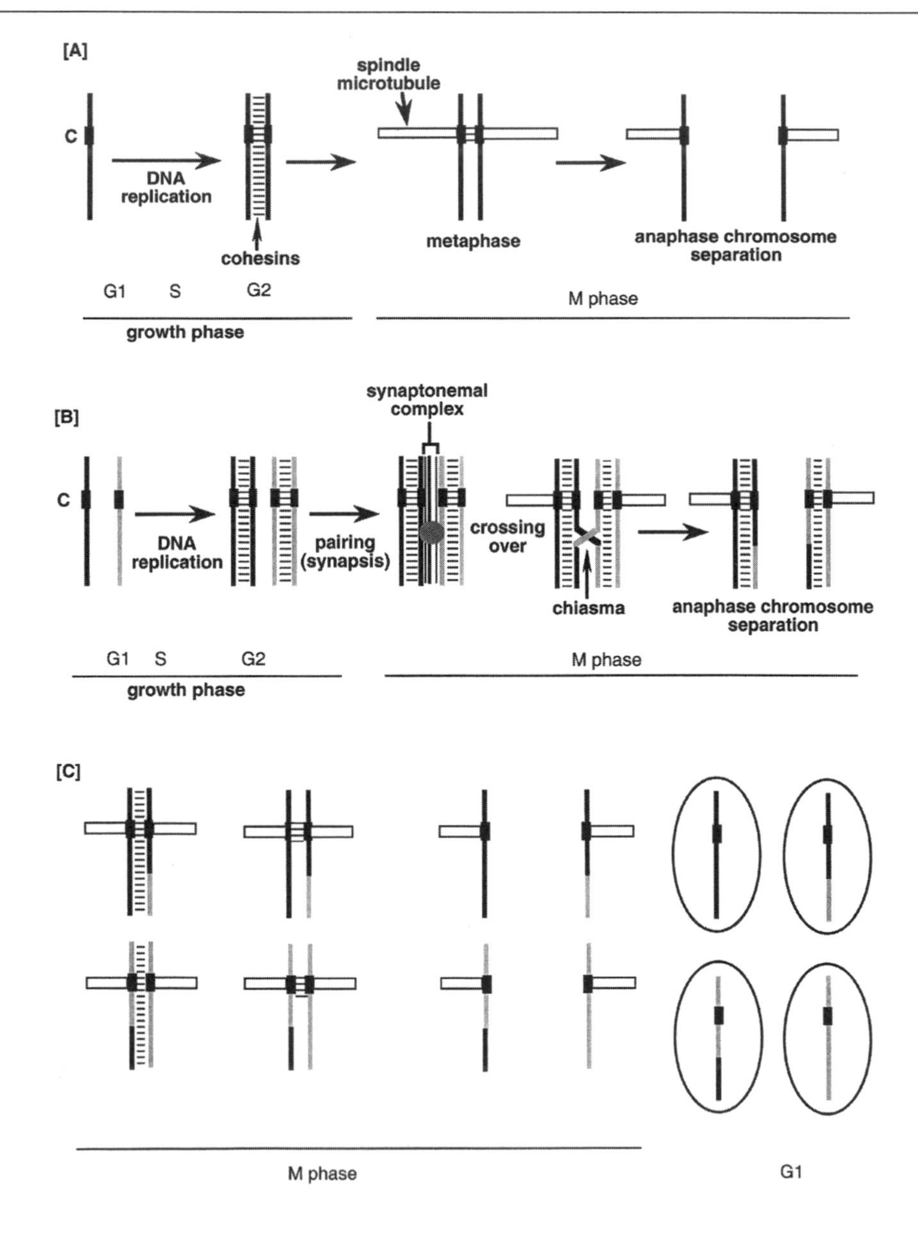

stead (and never evolved phagotrophy) retain typical bacterial chromosomes. Novel DNA-folding mechanisms also evolved, using histone H1 already present in the actinobacterial ancestor of neomura (but lost by archaebacteria) and additional histones that evolved only in the first eukaryotes. They reversibly fold the chromosome more compactly during mitotic chromosome separation, which is more likely to break extended DNA than the bacterial mechanisms, being much more rapid.

The origin of mitosis abolished stabilizing selection maintaining the single chromosome origin and terminus essential for bacterial chromosome separation that had dominated chromosome evolution for 3 billion years. Mutations multiplying replication origins inevitably spread. Thus, eukaryote chromosomes simultaneously initiate replication from numerous points; replication time no longer limits growth rates, removing the key evolutionary constraint keeping bacterial genomes small. This allowed, but did not require, large increases in eukaryotic genome size, which are related to their more greatly varying cell and nuclear volumes. Novel cell cycle controls, mediated by cyclin-dependent kinases (enzymes that phosphorylate proteins), ensured that DNA replicates once only per cell cycle, finishing before mitosis. Thus the eukaryotic cell cycle arose by new mechanical machinery and controls over the timing of chromosome separation and cell division. Without either, high viability could not have been maintained following the cytoskeletal and endomembrane innovations that phagocytosis entailed. The origin of the nuclear envelope was probably to prevent DNA breakage by the novel contractile machinery.

FIGURE 2. How Meiosis Halves Chromosome Numbers.
Each unreplicated chromosome consists of a single linear double-stranded DNA molecule represented by a single thick vertical line, and its centromere (C) by a black oblong. In normal eukaryotic cell division by mitosis (A), every nuclear and cell division is preceded by a DNA replication that occurs during the DNA synthesis period (S) in the preceding growth phase of the cell's reproductive cycle. In meiosis, two successive nuclear divisions (meiosis I [B] and meiosis II [C]) occur without intervening growth and replication, thus halving the nuclear DNA content from 2C to C and chromosome number from 2n to n. For simplicity n is assumed to be 1 and the nuclear envelope and cell membranes are mostly omitted.

(A) Mitosis. During replication, sister DNA molecules (chromatids) become mutually attached by cohesin proteins (horizontal lines) and remain so in the last growth phase (G2). During early nuclear division by mitosis (M phase), the cohesins are digested except at the centromere, where the spindle microtubules attach. At metaphase, molecular motors attempting to push the chromosome/microtubule complexes to opposite poles of the cell are in balance, so align all chromosomes at its equator. Digestion of the centromeric cohesins allows the motors to pull the sister chromosomes to opposite poles (anaphase movement) where they are incorporated into the two daughter nuclei and cells.

(B) Meiosis I uses special meiotic cohesins assembled at the premeiotic S period. Replicated chromosomes with homologous sequences pair at many points, and are stabilized by the synaptonemal complex to form a bivalent. Meanwhile, some chromatids have undergone double-strand DNA breakage at homologous positions and rejoining to nonsister chromatids (crossing over). Initially the break points are marked by a recombination nodule (gray disc) containing the recombination enzymes; when the synaptonemal complexes disassemble the bivalent is held together only by the chimeric junctions (chiasmata). Partial cohesin digestion and the poleward forces on the microtubules allow the chromosomes to separate to opposite poles in daughter nuclei, each of which now has n chromosomes but a 2C DNA content. Meiosis I differs from mitosis in three ways: [1] having homologue pairing and a synaptonemal complex [2] absence of centromere splitting by cohesin digestion and [3] modified centromeres that attach only to spindle fibres directed to a single pole.

(C) Meiosis II is mechanistically identical to mitosis (A), reducing DNA content from 2C to C. If crossing over occurred (not invariable) some chromosomes (DNA molecules) will now be chimera of maternal and paternal genes (here shown in gray and black). In lower eukaryotes all four cells resulting from meiosis survive and can transmit their genes to the next generation—the ancestral condition. In animals, all four become sperm, but in females only one forms the egg nucleus, the others being ejected as polar bodies that die; this death allows meiotic drive (biased allele segregation) to evolve so that a selfish allele may avoid it. In land plants, all four form the haploid gametophyte generation, which in bryophytes and primitive (homosporous) vascular plants is free-living and photosynthetic; in advanced (heterosporous) vascular plants, the haploid male phase is pollen and in females the embryo sac, which is parasitic on the mother plant. One or more embryo sac cells becomes the egg and the others form nutritive tissue for it, which in angiosperms becomes triploid or polyploid by additional nuclear fusions. Drawing by Thomas Cavalier-Smith.

Centromere Behavior and Cell Cycle Controls. Animal meiosis occurs only during the formation of gametes: eggs and sperm. After they merge at fertilization, their haploid nuclei, each with one genome (n chromosomes) fuse to form one diploid nucleus containing two genomes (2n chromosomes). These diploid cells divide by mitosis to build the animal body. Meiosis extends over two successive nuclear and cell divisions, not just one, between which no DNA replication occurs. Consequently, the DNA amount per nucleus is halved. The second meiotic division is identical in mechanism to mitosis. The unique meiotic processes occupy the first division (meiosis I).

The key event is pairing between homologous (genetically related) chromosomes (Figure 2b). Each chromosome of haploid nuclei (twenty-three in humans) differs in DNA sequence and bears different genes. Each of the n egg chromosomes has an evolutionarily related partner (its homologue) in the sperm, sufficiently similar in sequence to be able to pair with its homologue (by complementary nucleotide pairing) in meiosis I. This pairing (synapsis) holds them together in a complex, the bivalent, in which two fully replicated chromosomes (four DNA molecules in all) are held together by a proteinaceous structure unique to meiosis, the synaptonemal complex. The bivalents, not individual replicated chromosomes as in mitosis, attach to spindle microtubules in meiosis I.

At anaphase of meiosis I, sister centromeres remain unsplit and complete replicated chromosomes (not sister chromatids) move to opposite poles. Specific pairing of homologues and blockage of centromere splitting together ensure that daughter cells have n replicated (double) chromosomes, not 2n single ones as in mitosis. In meiosis II, centromeres split as usual at anaphase, giving each of the four daughter cells n unreplicated chromosomes. Thus, halving chromosome number depends on three things: pairing of homologous chromosomes; blocking centromere splitting in meiosis I, and blocking DNA replication between meiosis I and II.

Although chromosome reduction from diploid to haploid is the fundamental function of meiosis, it also effects genetic recombination. All chromosomes that came originally from one parent of the preceding generation will not stay together, because orientation of different bivalents on the spindle at meiosis I is random. Therefore each daughter cell randomly acquires some maternal and some paternal chromosomes—the explanation of Mendel's first law of independent assortment of alleles of unlinked genes. This is the only form of meiotic recombination (e.g., in male flies).

More typically groups of adjacent genes cross over from one homologue to the other by breakage and rejoining of DNA of different chromatids (Figure 2b). When chromatids involved are from different parents this yields chimeric chromosomes bearing genes from each. This physical joining of homologous chromosomes is visible microscopically as junctions called chiasmata, which hold bivalents together after the synaptonemal complex disassembles in late prophase. Breakage is not truly random, but at a possibly random subset of localized hotspots; chiasma interference prevents others nearby, nsuring that only one or a few are on each chromosome arm and reducing chiasma frequency somewhat below crossover frequency.

Evolution of Meiosis. The first eukaryote was a phagotrophic nonphotosynthetic unicellular organism. Recent phylogenetic advances make it likely that the last common ancestor of all eukaryotes was already sexual, with a well-developed two-step meiosis (Figure 3). Most likely it was a flagellate or amoeboflagellate (possibly an amoeba) that multiplied as haploid cells and formed gametes only when food became scarce; their syngamy formed a diploid resting zygospore or cyst that could germinate when conditions improved, undergoing meiosis to form four haploid feeding cells. Early sexual life cycles probably reflected an evolutionary trade-off between the advantages of larger diploid cell volumes and genetic redundancy during hazardous dormancy and of more numerous smaller haploid cells during growth and multiplication in more benign times. At germination, meiosis has a twofold numerical advantage over mitosis

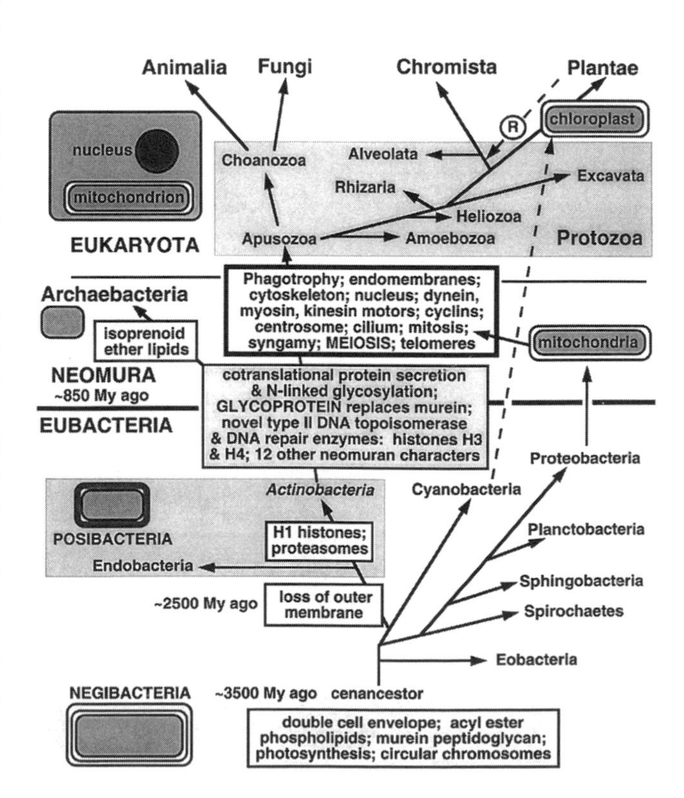

if smaller cells do not reduce viability. The enzymatic machinery for breaking and rejoining DNA was already present in the common ancestor of eukaryotes and archaebacteria. In bacteria its most fundamental role is to repair stalled replication forks in most cell cycles. It is also induced by radiation and rescues cells from otherwise lethal DNA damage by reconstructing an intact chromosome by recombination between two differently damaged sisters; in early eukaryotes its induction perhaps became associated with cyst germination to repair damage incurred during dormancy.

Meiosis is remarkably uniform among the most evolutionarily diverse eukaryotes. Synaptonemal complex structure is universal, though most of its molecules evolve too rapidly for easy detection of homology across the whole tree. Meiotic cohesins are homologous in all eukaryotes and diverged from mitotic cohesins following a very early gene duplication. Almost certainly meiosis evolved only once: subsequently meiosis and sex have been frequently lost in protozoa, fungi, algae, and higher plants, but more rarely in animals.

Speculations that meiosis was originally one-step are unfounded, because no protists with such a mechanism are basal on the best phylogenetic trees and several, possibly all, actually have two-step meiosis. The simplest origin for meiosis necessarily makes it two-step. The normal mitotic cell cycle is a bistable oscillator: cells switch reversibly between a growth phase when DNA replication is allowed but mitosis is inhibited and an M-phase where mitosis is allowed but replication inhibited. These switches are mediated by the alternating synthesis and proteolytic destruction of cyclin-dependent protein kinases. These kinases are not destroyed after meiosis I as they are after mitosis. Thus the interval between meiosis I and II is not a true interphase or growth phase and DNA replication automatically remains blocked. This prolongation of M-phase over two successive divisions could have been caused simply by inhibition of centromere splitting at meiosis I, if centromere splitting is the normal signal for cyclin destruction to terminate M-phase and return cells to interphase growth and replication. Thus the eukaryotic cell cycle may be so constructed that a single change in centromere behavior, probably mediated by the evolution of special meiotic cohesins, could simultaneously evolve two of the three features essential for ploidy reduction. Cohesins must be put in place during DNA replication and numerous mechanisms prevent the switch to M-phase in the absence of DNA replication, so meiotic differentiation necessarily occurs in G1 of the cell cycle, not G2. It would be mechanistically much harder to have evolved a reduction division by initiating division in the absence of replication, necessary for a hypothetical single-step meiosis.

The other key innovation was chromosome pairing; this occurs in two stages. Probably only the first was

FIGURE 3. Origin of Meiosis in the Context of Cell Evolution and the History of Life.
Eukaryotes are sisters of archaebacteria, known jointly as neomura ("new walls") because their common ancestor replaced the ancestral bacterial peptidoglycan cell wall by surface glycoproteins. This happened about 3 million years after life began and enabled the ancestor of eukaryotes to evolve a more flexible cell surface capable of phagocytosis and cell fusion. Perfecting phagotrophy entailed the evolution of an internal skeleton and endomembrane system, leading directly to formation of the nucleus and concomitant evolution of mitosis involving microtubules and cytokinesis (cytoplasmic division) using actin, and to their respective novel molecular motors (dynein/kinesin and myosin). The neomuran common ancestor evolved histones and the novel DNA topoisomerase II that makes the double strand breaks for meiotic crossing over, as well as novel DNA polymerases and repair enzymes, including a derivative of the eubacterial protein RecA that enables a single-stranded DNA end to invade a double strand homologue and pair with it (a likely key to meiotic homologue recognition), and also glycoproteins. Syngamy was caused by fusogenic surface glycoproteins—whether to further the selfish spread of transposons encoding them or for the benefits of ploidy cycles or as the incidental consequence of selection for intracellular membrane fusions may never be known. But because archaebacteria retained the novel glycoproteins for their original function as a rigid cell wall, only eukaryotes evolved syngamy and meiosis: as they evolved neither mitosis nor meiosis, they retained the circular chromosomes that had originally evolved prior to the last common ancestor of all life (the cenancestor), a negibacterium with an envelope of two membranes, as in the proteobacterium and cyanobacterium that were implanted into eukaryote cells to form mitochondria and chloroplasts. Eukaryotes alone evolved linear chromosomes with telomeres, as their novel cell cycle controls and mitotic/meiotic machinery allowed circles to be dispensed with: linearity allows odd numbers of crossovers, unlike circles where dimers or interlocks would form. A third symbiogenesis implanted a red alga within a protozoan cell to form the chromalveolate common ancestor of Chromista and alveolates. The position of the root of the eukaryotic part of the tree is uncertain: it might lie among Amoebozoa or Heliozoa, rather than apusozoan flagellates as shown, or somewhere between these three groups. As all four advanced kingdoms (animals, fungi, plants and chromists [which comprise cryptomonad, heterokont, and haptophyte algae and their secondarily heterotrophic derivatives]) evolved independently from sexual protozoa as shown, sex and meiosis evolved in a nonphotosynthetic, phagotrophic protozoan. Uncertainty over the root of the eukaryotic tree means that a few primitively asexual eukaryotes might exist, but it is more likely that two-step meiosis evolved prior to the last common ancestor of all extant eukaryotes; protists sometimes claimed (mostly mistakenly) to have single-step mitosis all belong to definitely derived, not basal groups (e.g., microsporidia to Fungi and oxymonads and parabasalids to excavates). Drawing by Thomas Cavalier-Smith.

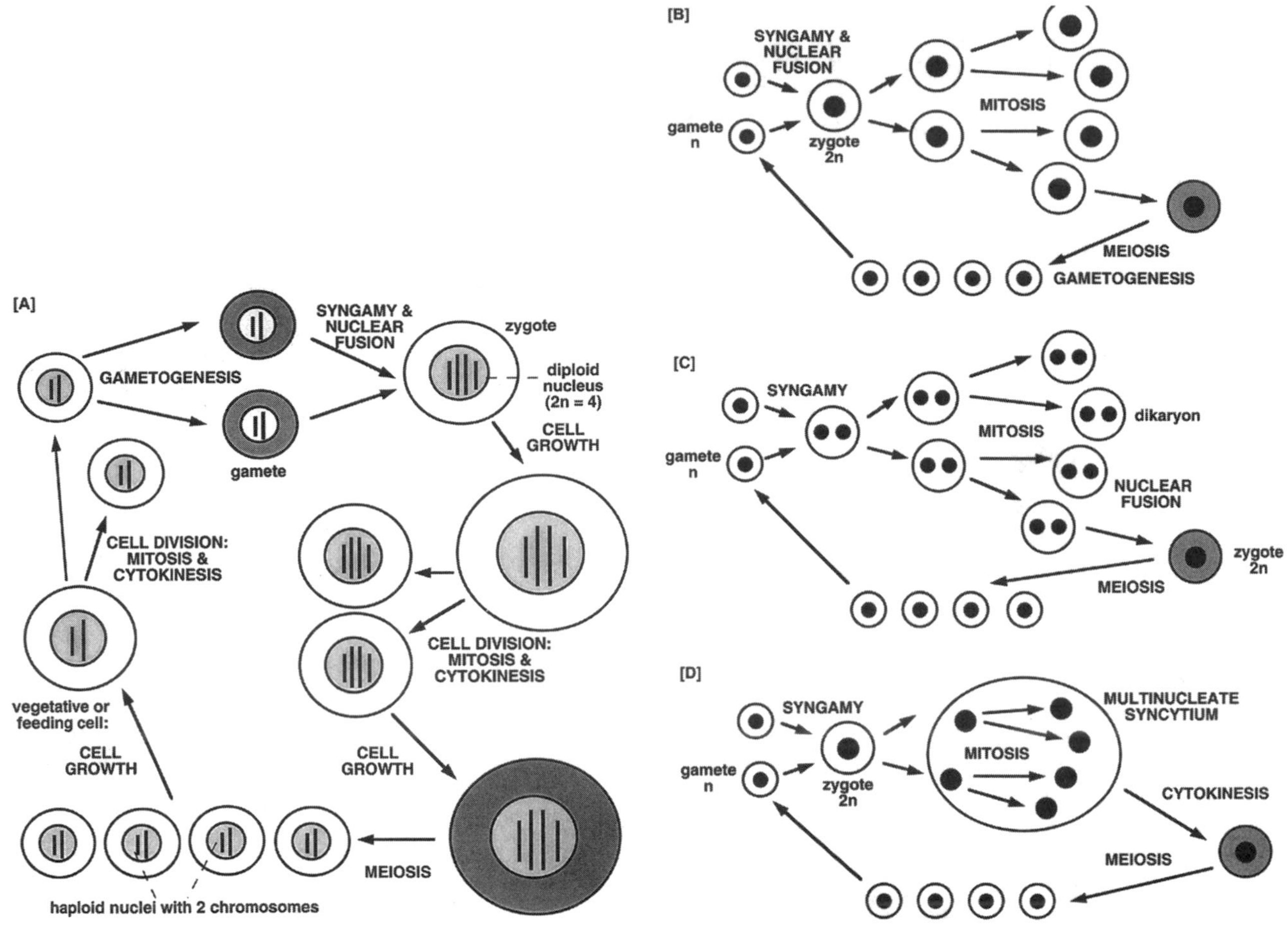

FIGURE 4. Cyclic Variation of Chromosome and Nuclear Numbers in Eukaryotic Sexual Life Cycles.
A great variety of different life cycles and body forms evolved by varying the timing of the division of nuclei (by mitosis or meiosis) and cytoplasm and of syngamy and nuclear fusion. The phases that became multicellular (usually by daughter cell adhesion, rarely by aggregation as in cellular slime moulds, e.g. the amoebozoan protozoan Dictyostelium) also greatly vary; only rarely (most animals and fucoid brown algae) is the diploid phase alone multicellular. Meiosis after syngamy with no intervening mitosis is probably the ancestral condition (Figure 1), found in dinoflagellate and sporozoan alveolate protozoa (e.g., malaria parasites), the slime mould Dictyostelium (Amoebozoa), many green algae (e.g., Chlamydomonas), most lower fungi, and some chromists; all cells except the dormant zygospores or cysts are haploid.

(A) Meiosis during gametogenesis yields diploid bodies, as in most animals, ciliate and probably trypanosome protozoa, and some chromists (e.g., diatoms, fucoid brown seaweeds).

(B) Separation of meiosis from both gametogenesis and syngamy by many mitotic cycles yields an alternation of diploid and haploid generations, as in most plants (all land plants, many green and red algae), some chromists (e.g., the brown algal kelps and some cryptomonads and haptophytes), a few lower fungi (e.g., Allomyces), the yeast Saccharomyces cerevisiae, and numerous protozoa (e.g. myxogastrid Mycetozoa, many foraminifera).

(C) Separating syngamy and nuclear fusion gives a dikaryophase, as in most higher fungi (ascomycetes, basidiomycetes and some microsporidia): delayed nuclear fusion was probably originally an adaptation to keep nuclei smaller to reduce breakage when squeezing through minute pores in hyphal septa.

(D) Delaying cytokinesis after mitosis produces multinucleate coenocytes or plasmodia: an adaptation to make very large (often macroscopic) cells in either diplophase or haplophase, e.g., many green (e.g., Codium) and red algae, some chromists (e.g., the xanthophyte alga Vaucheria, the frog rectal parasite Opalina), and many lower fungi (e.g., Mucor) and protozoa (e.g. myxogastrid slime moulds like Physarum, some cercomonads). Drawing by Thomas Cavalier-Smith.

essential—an initial homology search probably mediated by complementary base-pairing between related sequences of homologues, an inherent capacity of DNA. Subsequent synaptonemal complex assembly may have been a secondary refinement to increase bivalent stability. Even crossing over may have evolved so chiasmata could ensure accurate chromosome segregation: double-strand DNA breaks and rejoining might originally have been a prerequisite for stable synapsis, as in yeast. Some animal and plant chromosomes have other means of pairing, involving heterochromatin, but the rarity of heterochromatin in protozoa suggests that these are derived, not ancestral. Halving chromosome number is the primary function of meiosis. The advantages of such ploidy reduction and avoiding accompanying chromosome loss suggest that meiosis could have originated in basically asexual early eukaryotes to reduce ploidy caused by cell cycle accidents or accidental cell fusions.

Bacterial chromosomes are single, directly attached to the cell surface, and cell division is constrained to occur between each chromosome. Because of this, bacteria cannot accidentally become irreversibly polyploid as eukaryotes can. Polyploidy caused by cell cycle accidents originated as soon as eukaryotic chromosomes detached from the cell surface and became segregated instead by mitosis, allowing the genome to fragment into several dissimilar chromosomes. If meiosis originated to solve this novel problem, it would have simplified the subsequent evolution of sexual life cycles that took place by controlling cell fusion more efficiently.

A key question in understanding the origin of sex is whether it occurred before or after the origin of the cell nucleus. If syngamy preceded the nuclear envelope, the evolution of nuclear fusion must have been the last step; if syngamy followed the origin of the envelope, one could envision an intermediate stage in which reversible cell fusion occurred with no nuclear fusion or genetic recombination. Such life cycles might have preceded the origin of meiosis. Possibly these are false dichotomies. The early naked ancestors of eukaryotes may simultaneously have been evolving the mitotic machinery, a nuclear envelope, an ability to reduce ploidy by a primitive meiosis, and a capacity to impede or stimulate cell and nuclear fusion according to their needs.

Position of Meiosis in Life Cycles. The timing of meiosis relative to syngamy and nuclear fusion varies immensely, being responsible for a great wealth of different behaviors of chromosomes in eukaryotic life cycles (Figure 4). Multicellular animals are typically diploid and meiosis is essential for making gametes. But in bees, wasps, ants, and other Hymenoptera and in thrips, and a few beetles and coccids, males are haploid and make gametes without meiosis. In land plants, diploid and haploid generations alternate, meiosis occurring during formation of spores, not gametes: in mosses and liverworts the photosynthetic plant is haploid, but the capsules and stalks that parasitize it are diploid, whereas heterosporous vascular plant bodies are diploid and only pollen and embryo sacs are haploid. Some algae have separate haploid and diploid plants, sometimes structurally indistinguishable, sometimes entirely different—as also do homosporous fern generations, where prothalli are haploid and the large leafy plant diploid.

Ploidy increases by cell fusion or delayed cytoplasmic division are common in protozoan life cycles, as is polyploidy in somatic development in invertebrates and plants, often simply to increase cell size. Since somatic cells do not contribute to the next generation, they have not evolved ploidy-reduction mechanisms. But ploidy reduction is essential for the germ line of all sexual organisms. Some protozoa undergo cell fusion unconnected with sex to form plasmodia as feeding stages. As soon as the eukaryote cell evolved with an internal cytoskeleton and a soft surface indulging in phagocytosis, cell fusion would have been far easier than in bacteria with external walls, where it is almost unknown. But continued indefinitely, repeated cell and nuclear fusion would inexorably harmfully increase ploidy. Meiosis probably either evolved to counteract this in the earliest eukaryotes capable of cell fusion or even earlier to correct polyploidy caused by misdivision.

Meiosis enables genetic recombination among individuals in eukaryotic populations. Although traditionally viewed as the fundamental function of sex, this is more likely to have been a minor side effect of its origin rather than its driving force. Probably the twin primary roles of syngamy prior to forming resting cysts were to double cellular food stores and provide genetic redundancy to allow recombinational repair of DNA damage incurred during dormancy: both adaptations to maintain viability in life cycles subject to alternating want and plenty. The common timing of meiosis after long periods of dormancy in unicells or after long-extended growth to maturity of multicells with a semidormant germ line may also help eliminate accumulated harmful recessive mutations. Although probably not the primary reason for its origin, the ability of recombination to speed up evolution in some circumstances might also have helped its initial establishment. Origin of the eukaryote cell required rapid incorporation of thousands of synergistic rare novel mutations; if syngamy and meiosis arose during that process, they could have combined ones arising in different cells and been fixed by their joint success.

[*See also* Linkage; Meiotic Distortion; Sex, Evolution of.]

BIBLIOGRAPHY

Cavalier-Smith, T. "R and K-Tactics in the Evolution of Protist Developmental System: Cell and Genome Size, Phenotype Diversifying Selection, and Cell Cycle Patterns." *BioSystems* 12 (1980): 43–59. Discusses the importance of selection for different cell sizes in protist life cycles, significant for evolution of ploidy cycles and the timing of meiosis.

Cavalier-Smith, T. "The Phagotrophic Origin of Eukaryotes and Phylogenetic Classification of Protozoa." *Int. J. Syst. Evol. Microbiol.* In press. Discusses the origin of the eukaryote cell, early eukaryote phylogeny, the nature of the first sexual protozoan and the origins of meiosis and syngamy.

Cavalier-Smith, T. "Origins of the Machinery of Recombination and Sex." *Heredity.* In press. Contrasts the bacterial and eukaryotic recombination machinery and discusses the origins of both in detail.

Donachie, W. D. "Co-ordinate Regulation of the *Escherichia coli* Cell Cycle or the Cloud of Unknowing." *Molecular Microbiology* 40 (2001): 779–785. Shows how bacterial chromosome replication and segregation are coordinated with cell division in a radically different way from the mitotic cell cycle of eukaryotes.

John, B. *Meiosis.* Cambridge, 1990. A good general review.

Kondrashov, A. S. "The Asexual Ploidy Cycle and the Origin of Sex." *Nature* 370 (1994): 213–216. Shows that ploidy reduction may be sufficient selective force for the origin of meiosis.

———. "Evolutionary Genetics of Life Cycles." *Annual Review of Ecology and Systematics.* 28 (1997): 391–435. Excellent review of life cycles emphasizing lower eukaryotes, genetic aspects, and ploidy cycles.

Petes, T. "Meiotic Recombination Hotspots and Coldspots." *Nature Reviews Genetics* 2 (2001): 360–369. Reviews molecular evidence for sequence-dependent nonrandomness of crossing over.

Roeder, G. S. "Meiotic Chromosomes: It Takes Two to Tango." *Genes and Development* 11 (1997): 2600–2621. A general review of the molecular basis of meiosis.

Toth, A., K. P. Rabitsch, M. Galova, A., Schleiffer, S. B. Buonomo, and K. Nasmyth. "Functional Genomics Identifies Monopolin: A kinetochore Protein Required for Segregation of Homologs during Meiosis I." *Cell* 103 (2000): 1155–1168. A novel protein may explain why meiotic chromosomes attach to only one pole, not two as in mitosis. Apparent absence of homologues from animals means that they use an unrelated one or that this protein evolves too rapidly for easy detection.

Watanabe, Y., S., Yokobayashi, M., Yamamoto, and P. Nurse. "Pre-meiotic S Phase Is Linked to Reductional Chromosome Segregation and Recombination." *Nature* 409 (2001): 359–363. Provides evidence for the theory that the delay in centromere splitting until meiosis II by the evolution of specialized meiotic cohesins was central to the origin of meiosis.

— THOMAS CAVALIER-SMITH

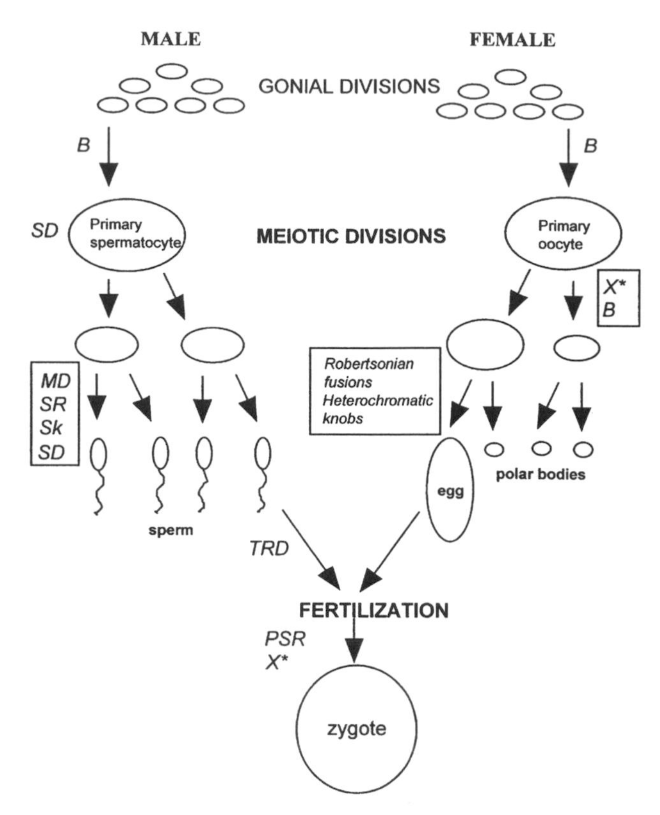

FIGURE 1. Meiotic Distortion.
Points of exploitation in gametogenesis and fertilization by meiotic cheaters. Courtesy of Terrence Lyttle.

MEIOTIC DISTORTION

Meiosis, or reduction division, is the process by which the paired or diploid chromosomes in a cell divide into two equal or haploid sets of chromosomes that form the gametes, for example, the sperm and eggs. For meiosis to be "fair," we require a set of rules whereby the paired diploid chromosomes and genes "agree" to segregate and be represented equally in the haploid gametes. One of the ongoing questions of evolutionary theory asks why a fair meiosis should be stable. Meiosis is obviously exploitable, as selfish "cheaters" can potentially double their net fitness simply by arranging to end up in all rather than the expected half of the gametes. In fact, some of the most interesting examples of intragenomic conflict involve competition between genetic elements that exploit the delicately balanced processes of meiosis and fertilization. Figure 1 illustrates the many points of gametogenesis and fertilization that are susceptible to meiotic distortion, as discussed in detail below.

Most cases of biased meiotic transmission can be collected under the category of meiotic drive. This is broadly defined as exploitation of meiosis by a heterozygous gene or a heteromorphic chromosome, resulting in its inclusion in excess of the 50 percent of gametes expected under Gregor Mendel's law of segregation. For simplicity here, we use the term *driver* for the allele or chromosome recovered in excess; the exploited genetic element correspondingly recovered in deficient numbers is called the target. Crow coined the term *ultraselfish DNA* to distinguish meiotic drive from more familiar categories of "selfish" DNA (i.e., the middle and highly repetitive DNA sequences that employ transposition or unequal crossing over, respectively, to increase their copy number). Ultraselfish elements go further: they not only proliferate but actually eliminate the gametic transmission of their homologues.

Meiotic drive is found widely in natural populations of a number of species (e.g., fungi, *Drosophila*, mice, and mosquitoes), as well as for some cases of laboratory-induced chromosome arrangements. Table 1 presents the details of these systems. The majority of cases of drive depend on mechanisms of gamete killing. For example, the driver Segregation Distorter (SD) causes *SD/SD*$^{+}$ males in *Drosophila* to exhibit spermatogonial cysts with up to 100 percent of the thirty-two developing

TABLE 1. Meiotic Distortion. A Comparison of Drive Systems

+, presence of phenomenon; −, absence of phenomenon; /, differences among-strains or species; ?, no information available; na, not applicable.

	Drive System				
Phenomenon	*SD Dros mel*	*TRD Mouse*	*SR Dros spp*	*MD Mosquito*	*Sk Fungi*
MODE OF ACTION					
X, Y, or autosomal (A) drive	A	A	X	Y	A
Gamete dysfunction (E = early visible gamete abnormalities, L = late)	+ E	+ L	+ E	+ E	+ L
Limited to males	+	+	+	+	−
Sensitive target causes gamete dysfunction		+	+	+	+
Insensitive target rescues gamete	−	−	−	−	+
Target locus *cis*-acting	+	+	+	+	+
Distorter locus *trans*-acting	+	+	?	−	+
Homozygous effects (S = male sterility, R = reduced fertility, L = lethality)	R/S	S, L	R	na	−
Degree of distortion in nature	>99%	90–99%	>99%	50–61%	>99%
GENETIC AND POPULATION STRUCTURE					
DNA sequence information	+	+	−	−	−
Evidence for transcription of target sequences	−	+	?	?	?
Heterochromatic elements involved	+	+	?	+	?
Drive elements linked to centromere	+	+	+	+	+/−
Linked chromosome rearrangements	+	+	+	+	+/−
Drive and target loci recombinationally separable	+	+	+	+	+
Known *trans*-acting modifiers present	+		+/	+	
System widespread in natural populations	+	+	+	+	+/−

SD⁺-bearing spermatids rendered dysfunctional because they fail to properly complete the step of individualization (Figure 2).

There are also cases where distorted gametic ratios from individual loci (non-Mendelian segregation) are a consequence of biased gene conversion, in which one heterozygous allele exploits the processes of genetic recombination and repair to replace its homologue with a DNA copy of itself. Prominent examples are homing endonuclease genes (HEG), widely distributed in fungi, protists, bacteria, and viruses. HEG copies (HEG+) are inserted into introns and encode endonucleases that cut at highly specific and rare sites in the host genome, generally at the homologous HEG− chromosomal location. Repair of the double-stranded break results in insertion of HEG+, which converts the genotype from HEG+/HEG− to HEG+ homozygosity. Subsequent meiosis produces what superficially resembles the distorted gametic ratio caused by meiotic drive, but the population genetic consequences are quite different (see below). Homing endonucleases are essentially a class of transposable elements, and are capable of spreading laterally to other sites in the genome, or indeed, across species barriers.

Categories of Meiotic Drive.

Sex-ratio distortion. Given that meiosis offers such an exploitable opportunity for "cheaters," it is perhaps surprising that drive is not more common in nature than represented by the cases described here. But, to be observed, the driver must not have reached fixation, and genetic or phenotypic markers must exist to measure the non-Mendelian segregation ratios. Sex chromosomal drive will produce the strikingly obvious phenotype of sex-ratio distortion, which may suggest why it is more often discovered in nature than are cases of autosomal meiotic drive. On the other hand, the consequences of an unbalanced sex ratio are so detrimental to fitness that extreme cases of sex chromosome drive cannot long persist in nature unchecked, but will either lead to unisexual populations and extinction or to strong evolutionary pressure to accumulate suppressors that restore the optimal 50:50 ratio, thereby masking the drive system from detection in its natural background. This is

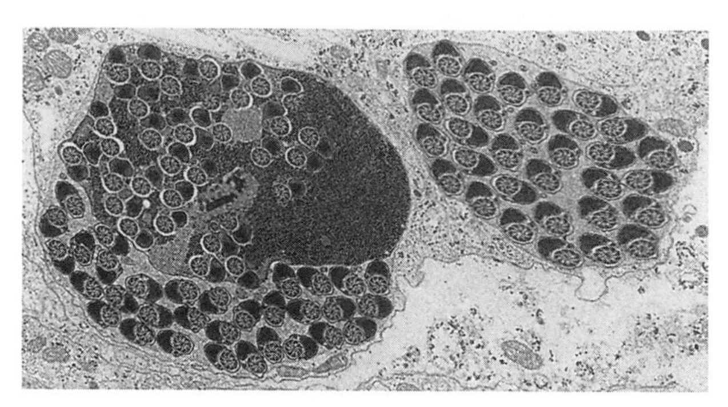

FIGURE 2. Transverse SEM Section of a Spermatogonial Cyst in an *SD/SD*+ Heterozygous Male of *D. melanogaster*.
Shown are two profiles of spermatid tails from the same cyst of 64 developing sperm. On the left, 32 normal (*SD*) and 32 abnormal (*SD*+) sperm tails are visible, while on the right, only the 32 normal tails extend further down the coiled cyst.
Courtsey of Dr. K. T. Tokuyasu.

especially true for Y chromosome drive, because population fecundity is more reduced from the resulting scarcity of females than it would be for a comparable deficiency of males under X chromosome drive. Not surprisingly, the only examples of Y (or W) drive observed in nature are weak drive systems (*MD* in the mosquito, *W* drive in the butterfly).

Cases of X chromosome drive, however, are relatively common among *Drosophila* species. Although details of the genetic mechanisms for these systems are largely unknown, they generally represent cases of postmeiotic gamete competition, with reduced survival of Y-bearing sperm. Drivers are prevented from reaching fixation by a combination of lowered overall fecundity or virility of heterozygous males and reduced viability of females homozygous for the driving X. Well-studied cases of sex-ratio distortion in both *D. pseudoobscura* and *D. simulans* reveal that multiple X-linked loci are involved, held in a coadaptive complex by inversions. This evolutionary pattern is also observed for the *MD* system of male drive in the mosquito and in the cases of drive caused by autosomal gene complexes, described later. In addition, some X-drive systems are held in check by segregation of resistant Y chromosomes.

Some interesting special cases of sex chromosomal drive have been observed. In the lemming *Myopus schisticolor*, males are normally XY and females XX. However, presence of a variant X carrying a specific mutational change causes X*Y chromosomal males to develop as females. Meiotic drive in X*Y transformed females yields exclusively X*-bearing eggs, thus leading to all-daughter broods. It appears that stable persistence of this system of drive is promoted through inbreeding. Finally, sex-ratio distortion can occur as the result of certain types of B chromosome meiotic drive, as discussed in the next section.

Biased chromosome transmission: B chromosomes. Many plants and insects harbor varying numbers of largely genetically inert supernumerary (B) chromosomes. Many of these preferentially accumulate from one generation to the next. There are several mechanisms for this, including extra mitotic replication cycles, mitotic nondisjunction in the gametophyte stage of some plants, and more often orientation on the meiotic spindle so as to preferentially segregate to the oocyte during meiosis. They appear to introduce relatively minor fitness effects in their diploid carriers, although they levy a metabolic load comparable to that incurred by species harboring extra heterochromatin on their A chromosomes. Heterochromatic knobs, which are essentially large segments of B chromosome–like sequences inserted into A chromosomes, also show preferential segregation into oocytes in many species and have been suggested as playing a major role in the evolution of chromosome structure in maize.

However, the most striking case of B chromosomal drive can be found with the paternal sex ratio (PSR) system in the parasitic wasp *Nasonia vitrepennis*, where fertilized eggs normally produce diploid daughters, and unfertilized eggs yield haploid sons (haplo-diplo sex determination) The supernumerary PSR chromosome not only is transmitted in all sperm of PSR males, but in a unique form of genetic imprinting, its presence has the effect of causing all other male-derived chromosomes to be consigned to a densely condensed amorphous chromatin mass, which is lost at the first postfertilization mitotic division. The overall effect of PSR is to cause the production of only haploid sons, including both normal males arising from unfertilized eggs and patroclinous PSR males arising from masculinized fertilized eggs.

Meiotic drive in males: Postmeiotic gamete competition. As discussed above, meiotic drive sensu strictu involves exploitation of the orientation of chromosomes on the spindle during gametogenesis to allow the driving chromosome to achieve preferential inclusion in those meiotic products destined to be functional (i.e., oocyte, megaspore). Such true meiotic drive has no intrinsic effect on gamete number, or indeed on the fecundity of the carrier. Conversely, drive systems in sperm or pollen, where all meiotic products share intrinsically equal expectations for participation in fertilization, are more likely to produce non-Mendelian segregation through gamete competition or gamete killing (Table 1) arising from the action of coadaptive gene complexes. The expected evolutionary impact of such systems (which could potentially cut male fecundity in half) is initially limited in impact to the population dynamics of the specific drive and target loci, and other loci accidentally found in close linkage. Linked alleles (especially if they enhance drive) may enjoy indirect drive through genetic hitchhiking, leading eventually to the establishment of drive haplotypes. For strong drive, this haplotype may spread to fixation, eliminating polymorphism in a meiotic sweep. The haplotype may be further extended by incorporating chromosome rearrangements that reduce recombination and promote additional linkage disequilibrium between the drive locus and more distant enhancer loci. In the extreme, the haplotype becomes coextensive with an entire chromosome, yielding population consequences essentially indistinguishable from the types of drive discussed in the previous section. Thus, although the differences between drive systems are often biologically and mechanistically significant, if the number of successful gametes produced with each system are similar, they converge to allow the same mathematical models to describe both.

Note the quite different evolutionary consequences for linked loci under biased gene conversion. Here, there is no genetic hitchhiking for neighboring loci outside of the few hundred base pairs of the DNA region of conversion, nor is there any selective advantage for enhancers of the process, linked or not. The population effects of biased gene conversion do not extend beyond the individual locus exhibiting it.

Simple meiotic drive can be modeled using a parameter K to denote the proportion of progeny (and, by inference, successful gametes) carrying the driver from heterozygotes. K can vary from 0.5 (Mendelian segregation) to 1.0 (only one gamete class recovered). The minimal requirement for establishment and increase of a driver is:

$$2KW > 1,$$

where W = heterozygous fitness. An ongoing puzzle concerning the evolutionary dynamics of meiotic drive is the observation that most drive systems in nature reach polymorphic equilibrium, despite the fact that the equation represents what should be an unstable equilibrium point; that is, a driver should either proceed to fixation if this inequality is satisfied or become extinct if not.

Although closely linked enhancers are favored to accumulate by "hitchhiking" along with the correspondingly enhanced driver, unlinked suppressors and more resistant target alleles are also favored. The advantage for the latter in escaping killing is obvious. Less intuitive is the reason for the accumulation of suppressors at unlinked genetic loci not themselves the direct targets of meiotic drive. Such alleles are selectively favored because they enhance the overall fitness of the individual organism in the face of secondary effects on viability or fecundity associated with the driver, beyond its primary effect of gamete killing. Overall, a driver will increase in frequency as long as suppressors or insensitive target loci do not accumulate to lower the value of K sufficiently to violate this inequality.

If fecundity is directly related to gamete number, as it is for organisms that broadcast gametes (e.g., most plants, many marine organisms, and spore-producing fungi), then K and W are directly inversely related, and even very efficient gamete killing produces little or no net meiotic drive. If this is the case, drivers should rarely become established. They may persist in small populations, where the elimination of haploid spores or gametes may significantly affect the actual census number of target alleles, and they enjoy a small deterministic advantage inversely correlated with population size. Combined with the effects of genetic drift, this may have allowed drive systems such as Spore killer (*Sk*) in the fungus *Neurospora* to become serendipitously established in isolated populations, but rarely to rise deterministically to a significant frequency.

However, animals with internal fertilization often exhibit nonlinear relationships between gamete number and fecundity, especially when sperm are produced in excess. In such cases, even a reduction of as much as 50 percent in the number of functional gametes may lead to little or no loss in fertilization capacity, allowing drive systems to easily satisfy the above equation. Not surprisingly, the majority of drive cases in nature, such as transmission ratio distortion (*TRD*) for *t* alleles in the house mouse, Sex-Ratio distortion (*SR*) and Segregation Distorter (*SD*) in *Drosophila*, and male drive (*MD*) in the mosquito occur in organisms in which males deliver an excess of sperm during fertilization (Table 1).

The best studied cases of autosomal drive (*TRD* in the mouse, *SD* in *Drosophila*) share superficial features arising from convergent evolution elicited by strong selection for drive enhancers and suppressors. Each represents a binary system comprised of *trans*-acting killer loci and a DNA region serving as a *cis*-acting target that tags sperm for dysfunction. These are held in tight linkage disequilibrium with each other and with drive modifiers through reduced recombination associated with neighboring heterochromatin or linked inversions. *SD*-bearing and *t*-bearing sperm are each transmitted upwards of 99 percent from SD/SD^+ and $t/+$ males, respectively.

Closer examination proves this surface similarity to mask quite distinct mechanisms. For the *SD* system, the main "killer" is a dominant gain-of-function mutation (*Sd*) yielding a radically altered version of the RanGap nuclear transport protein, which is then distributed aberrantly in the nuclei of developing sperm. Some other "killer" loci may also be involved (*E(SD)*, *M(SD)*), but they appear of secondary consequence. The novel form

SEX-RATIO DISTORTION IN DROSOPHILA MELANOGASTER

Strong selection maintains an optimal 50:50 sex ratio. Using a chomosome translocation to link the *SD* system of autosomal drive to the Y chromosome in *Drosophila melanogaster*, one can create an extreme form of sex-ratio distortion, the production of only male offspring. This model system was used by Lyttle (1991) to identify genetic mechanisms that evolve rapidly to suppress meiotic drive via strong selection for restoration of an equal sex ratio. Experimental laboratory populations containing the synthetic Y^D chromosome demonstrated that both polygenic drive suppressors and insensitive target alleles accumulated, as predicted by theoretical models. In comparison to this slow but deterministic accumulation of modifiers of small effect, a more rapid population response involving remodeling of chromosome structure to negate sex-ratio distortion was often observed. Some populations accumulated XXY females, which arise spontaneously in low frequency due to primary nondisjunction errors in meiosis, but are normally removed from the population by purifying selection. These produced significant proportions of XX-bearing eggs, which in turn yielded more XXY daughters after fertilization by Y-bearing sperm produced by Y^D males, reducing the strength of male drive. In one exceptional population, a spontaneous secondary translocation of part of an *SD* chromosome to the tip of the X chromosome allowed the sex-ratio distortion to actually be reversed, providing a startling example of the flexibility of evolutionary response to sex-ratio distortion.

—TERRENCE W. LYTTLE

of RanGap coded by *Sd* is associated with improper chromatin condensation and subsequent dysfunction of sperm carrying a copy of the responder (*Rsp*) target locus. A unique 120-base pair repetitive satellite DNA sequence defines this locus, which is normally located adjacent to an autosomal centromere. The *Rsp* repeats harbor no coding sequences, yet the level of sperm dysfunction associated with sensitive versions of this target allele can be demonstrated to be highly correlated with repeat number. Moreover, sensitivity cotranslocates when these sequences are transposed to other sites in the genome. In particular, their translocation to the X or the Y chromosome of *Drosophila* causes *SD* to become a Y- or X-drive system, respectively, providing an excellent model for studying the evolution of both autosomal and sex chromosomal drive in experimental populations (see Vignette).

For TRD in mice, it appears that a *t* chromosome harbors several quantitatively interacting "killer" (*Tcd*) loci, all involved in sperm motility. In contrast to *SD*, however, the target locus (*Tcr*) appears to be a fusion gene ($Smok^{TCR}$) that includes sequences from a member of a family of sperm motility kinases. This novel *Smok* product is expressed late in spermatogenesis and is limited in its action to the sperm in which it is located. It appears that the "killer" loci in this system act by inducing abnormally high levels of the regular *Smok* activity of phosphorylizing target sites, which in turn causes defective sperm motility. In *t*-bearing sperm, however, the impaired kinase activity of the $Smok^{TCR}$ gene product may serve to down-regulate the *Tcd*-induced phosphorylation effect, allowing normal development.

Other differences between *SD* and TRD include the tendency for the latter to accumulate linked lethal mutations. These may be selectively favored because sterility of *t* homozygotes is a basic property of the *TRD* system. If linked lethal mutations allow *t* homozygotes otherwise destined to be sterile to be eliminated early enough in gestation, reproductive compensation can take place. The resulting excess of *t*/+ heterozygotes (a phenomenon not possible during *Drosophila* reproduction) may promote the spread of *t* chromosomes, paradoxically generating a positive selection for recessive lethal mutations.

Evolutionary Implications of Meiotic Distortion. Recently, there has been considerable debate as as to whether complementary systems of sex chromosome drive fixed in different species can provide an explanation for Haldane's Rule. Haldane's Rule is the empirical observation that many apparently successful species hybridizations result in progeny in which the heterogametic sex (XY or ZW) is sterile. However, this idea remains controversial and probably explains only a few cases. It has also been speculated that drive plays a role in chromosome evolution by promoting fixation of some types of chromosome rearrangements or the evolution of increased rates of recombination. There are also some tantalizing observations that suggest that for some human monogenic disorders, particularly where allele function correlates with the copy number of families of trinucleotide DNA repeats, disease alleles persist because they exhibit meiotic drive. None of these are yet statistically compelling, however.

To summarize, we might emphasize the wide range of organisms displaying meiotic distortion. As information on the best studied systems accumulate, we find that the apparently great similarities among them are largely the result of convergent evolution resulting from common population genetic pressures. Conversely, the biological (and molecular) mechanisms involved in the various cases of meiotic distortion appear to involve quite different aspects of gametogenesis, and range from affecting single loci to whole haploid genomes. This serves to highlight the rather surprising question to which one is inevitably led when considering the great variety of meiotic distortion mechanisms: why is Mendelian segregation evolutionarily stable? As Crow (1988, p. 391): so succinctly notes: "Mendelism is a magnificent invention for fairly testing genes in many combinations, like an elegant factorial experimental design. Yet it is vulnerable at many points and is in constant danger of subversion by cheaters that seem particularly adept at finding such points."

[*See also* Haldane's Rule; Meiosis; Mendelian Genetics; Plasmids; Selfish Gene; Sex Chromosomes; Sex Ratios.]

BIBLIOGRAPHY

Bengtsson, B. O., and M. K. Uyenoyama. "Evolution of the Segregation Ratio: Modification of Gene Conversion and Meiotic Drive." *Theoretical Population Biology* 38 (1990): 192–218.

Buckler, E., T. Phelps-Durr, C. Buckler, R. Dawe, J. Doeble, and T. Holtsford. "Meiotic Drive of Chromosomal Knobs Reshaped the Maize Genome." *Genetics* 153 (1999): 415–426.

Cabot, E., P. Doshi, M.-L. Wu, and C.-I. Wu. "Population Genetics of Tandem Repeats in Centromeric Heterochromatin: Unequal Crossing Over and Chromosomal Divergence at the Responder Locus of *Drosophila melanogaster*." *Genetics* 135 (1993): 477–487.

Cazemajor, M., C. Landre, and C. Montchamp-Moreau. "The Sex-Ratio Trait in *Drosophila simulans*: Genetic Analysis of Distortion and Suppression." *Genetics* 147 (1997): 635–642.

Crow, J. F. "The Ultraselfish Gene." *Genetics* 118 (1988): 389–391.

Fisher, R. A. *The Genetical Theory of Natural Selection*. Oxford and New York, 1930. One of the classic works of population genetics and evolutionary theory. See Pages 158–160 for a discussion of selection for a 1:1 sex ratio.

Fredga, K. "Aberrant Chromosomal Sex-Determining Mechanisms in Mammals, with Special Reference to Species with XY Females." *Philosophical Transactions of the Royal Society of London B* 322 (1988): 83–95.

Goddard, M. R., and A. Burt. "Recurrent Invasion and Extinction of a Selfish Gene." *Proceedings of the National Academy of Sciences USA* 96 (1999): 3880–3885.

Hamilton, W. D. "Extraordinary Sex Ratios." *Science* 156 (1967): 477–488. The seminal work on the evolutionary consequences of sex-ratio distortion.

Hurst, L. D., A. Atlan, and B. O. Bengtsson. "Genetic Conflicts." *Quarterly Review of Biology* 71.3 (1996): 317–364.

Hurst, L. D., and A. Pomiankowski. "Causes of Sex Ratio Bias May Account for Unisexual Sterility in Hybrids: A New Explanation of Haldane's Rule and Related Phenomena." *Genetics* 128 (1991): 841–858.

Johnson, N. A., and C. I. Wu. "An Empirical Test of the Meiotic Drive Models of Hybrid Sterility: Sex-Ratio Data from Hybrids between *Drosophila simulans* and *Drosophila sechellia*." *Genetics* 130 (1992): 507–511.

Jones, R. N., and H. Rees. *B Chromosomes*. New York, 1982.

Lyttle, T. W. "Segregation Distorters." *Annual Review of Genetics* 25 (1991): 511–557. Exhaustive review that includes a discussion of most systems while serving as an umbrella reference describing work by the author on this topic.

Lyttle, T. W., ed. "Proceedings for the Genetics and Evolutionary Biology of Meiotic Drive." *American Naturalist* 137.3 (1991): 281–486. Individual papers summarizing contemporaneous state of understanding of most systems of meiotic drive.

Maynard Smith, J., and E. Szathmary. *The Major Transitions in Evolution*. Oxford: 1995. See pages 171–176 for a discussion of the vulnerability of gametogenesis to meiotic cheaters.

Merrill, C., L. Bayraktaroglu, A. Kusano, and B. Ganetzky. "Truncated RanGAP Encoded by the Segregation Distorter Locus of *Drosophila*." *Science* 283 (1999): 1742–1745.

Schimenti, J. "Segregation Distortion of Mouse *t* Haplotypes: The Molecular Basis Emerges." *Trends in Genetics* 16 (2000): 240–243.

Silver, L. M. "The Peculiar Journey of a Selfish Chromosome: Mouse *t* Haplotypes and Meiotic Drive." *Trends in Genetics* 9 (1993): 250–254.

White, Michael J. D., *Modes of Speciation*. San Francisco, 1978. Classic work outlining mechanisms of speciation, including the role of meiotic drive in establishing chromosome rearrangements.

Wyttenbach, A., P. Borodin, and J. Hausser. "Meiotic Drive Favors Robertsonian Metacentric Chromosomes in the Common Shrew (*Sorex araneus, Insectivora, mammalia*)." *Cytogenetics and Cell Genetics* 83 (1998): 199–206.

— TERRENCE W. LYTTLE

MEME

The term "meme" was coined in 1976 by the Oxford biologist Richard Dawkins in his book *The Selfish Gene*. Dawkins explained the basic principles of universal Darwinism—that when information is copied again and again, with variations and with selection of some variants over others, then evolution must occur. In this process, the information that is copied is called the replicator. The process itself is what the American philosopher, Daniel Dennett, calls the "evolutionary algorithm": a mindless procedure that generates design without a designer.

Dawkins emphasized that this is a general process, and not confined to our most familiar replicator, the gene. To give an example of a different replicator he invented the term *meme*; deriving it from the Greek *mimeme* ("that which is imitated"). Examples of memes are stories, songs, habits, skills, inventions, and ways of doing things. Memes are replicated by copying from person to person, mainly by imitation. Variation arises either through forgetting and distortion during copying, or through recombination of old memes to make new ones. Selection occurs because only some variants are successfully copied while others die out. Memes therefore qualify as replicators, and we should expect them to produce a new evolutionary process: memetic evolution.

An analogy can be drawn between memetic and genetic evolution because memes and genes are both replicators, but the analogy is not close because the two replicators work so differently. Genes are information stored in molecules of DNA and replicated by an extremely high fidelity chemical copying process. Memes, in contrast, are much more variable (including oral histories, written texts, actions, and artifacts), and they are copied by a variety of mostly rather inaccurate copying mechanisms. Therefore, although the meme-gene analogy can be useful, it may be misleading if not treated with care.

Memes are selfish in the same sense that other replicators are. They are either copied or not, and successful copying benefits their own survival and replication, not necessarily that of the people who copy them or those people's genes. For example, the memes of contraception are successful (including the idea of controlling fertility as well as the methods of doing so), even though they reduce the biological fitness of their carriers. The people who use them devote more time and energy to spreading their memes than their genes, including contraception memes.

This distinguishes memetics from the related fields of sociobiology and evolutionary psychology. The founder of sociobiology, E. O. Wilson, famously said that the genes hold culture on a leash, and most evolutionary psychologists assume that the ultimate function of culture is to serve the genes. But in memetic theory memes are also replicators, with their own replicator power. The ultimate function of culture is the replication of culture itself.

Memes may group together for mutual protection and replication as do genes. A *memeplex* (shortened from *co-adapted meme complex*) will form whenever memes can survive better as part of the group than on their own. For example, the memes in a book are replicated together when the book is printed; the different components of a scientific theory are studied together; the doctrines of a political party are promoted in the same speech. Other memeplexes are financial institutions, religions and cults, paintings, poems, websites, and television programs. In larger memeplexes, some memes may act as adaptations that help the whole memeplex to replicate (like the book's title or the picture on the cover), while others ride along for free.

Viruses of the Mind. Memes are often likened to diseases, referred to as "thought contagions," or said to infect or parasitize people. This epidemiological approach has been productive, for example in understanding the spread of body piercing, suicide contagion, and Princess Diana memorabilia, or tracking words and phrases that spread through electronic discussion lists. However, like all analogies it requires caution, and some argue that it is positively misleading.

More specifically, the term *mind virus* is used to refer to harmful memes. These are often emphasized in memetics because they may help distinguish its predictions from those of sociobiology or other gene-centered theories. According to memetic theory, some memes are copied because they are of real value to people, such as clothes, methods of transport, and tasty recipes. Cooperative and altruistic behaviors can often spread successfully, making humans an especially altruistic species. But some memes get selfishly copied in spite of being useless or even positively harmful to the people who adopt them.

The simplest type of selfish meme is the self-replicating sentence. In 1985, the American philosopher Douglas Hofstadter collected examples through his *Scientific American* column *"Metamagical Themas."* These ranged from the simple instruction "Copy me!" through sentences including baits, promises, and curses, to virulent chain letters. E-mail viruses now use similar tricks. Many appear as an honest request to help others by passing on the message warning them of a (usually nonexistent) computer virus or threat to world peace. They may sound urgent, appeal to authority, and raise genuine fears (Table 1). They are easy to copy and many have survived for years. Their basic structure is an instruction to "copy me" backed up with threats and promises. Some have suggested that religions use just this structure, which is why Dawkins has called them "viruses of the mind".

Meme-Gene Coevolution. The established field of gene-culture coevolution includes theories and mathematical modelling based on units called "culturgens" and "cultural traits." Several authors have shown how maladaptive cultural traits can sometimes survive. However, they mostly define "maladaptive" from only the genes' point of view, and treat inclusive fitness as the final arbiter. If memetics differs from these theories it is by emphasizing the meme as a replicator, surviving for its own sake.

Memes are defined as that which is imitated. Although imitation presumably evolved because it was biologically adaptive, once memes appeared a new replicator was let loose that was not necessarily subservient to the old. Human evolution has been shaped for much of its history by the coevolution of genes and memes. Some have suggested that this resulted not only in bigger brains but in brains especially suited to language, music, ritual, dance, and storytelling.

In this process memes are not seen as an aspect of the phenotype but as an evolving system in their own right. As memetic evolution proceeded, the outcome of the competition between memes may have altered the environment for gene selection. Music provides a good example. Imagine simple sounds competing to be copied. If people who are best at copying the winning sounds acquire status and a mating advantage, the successful sounds give an advantage to genes for the ability to copy those particular sounds. If these processes have

TABLE 1. Some Examples of Self-Replicating Texts, with Adaptations That Aid Their Replication

Memeplex	*Adaptations*
Copy me.	None
Copy me or you'll die.	Threat
Pass on this sentence and you'll win $1,000.	Promise
Hurry! Dangerous virus! Aens this warning to all your friends or their hard-drives will be wiped.	Threat, urgency, exploits readers' desire to help their friends
Spread the good news of Jesus's love and you'll go to heaven.	Untestable promise
Those who deny Allah have a garment of fire prepared for them.	Untestable threat

been important in human evolution, then modern brains reveal the past history of memetic evolution.

Might a similar process explain the evolution of language? In general, successful replicators are those with high fidelity, longevity, and fecundity. Digitization of sounds into words may increase fidelity, combining words into novel combinations may improve fecundity, and every improvement leads to increased memetic competition. The people who can best copy the winning sounds have an advantage and pass on the genes that gave them that ability. Gradually, human brains would be driven by the emerging language itself. In most theories of language evolution, the ultimate function of language is to benefit genes. On the memetic theory it is to benefit memes. Underlying these examples is the general principle that replicators coevolve with their replication machinery; just as genes must once have coevolved with their cellular copying machinery, so memes have coevolved with human brains.

Self and Consciousness. Both Blackmore and Dennett have used meme theory to try to understand the nature of self and consciousness. Dennett suggests that a human being is a special sort of ape infested with memes, and that the human mind is an artifact created when memes restructure the brain to make it a better habitat for memes. Human consciousness itself is a huge complex of memes, and the self is a "benign user illusion." Blackmore suggests that the illusory self is a memeplex constructed by the memes to aid their own survival and replication, and, far from being benign, is a major source of human suffering.

These theories raise questions about who or what is in charge of human behavior and decision making, and critics fear a new form of determinism and the undermining of human responsibility and the rule of law. Perhaps the most extreme memetic view is that the replicator power of memes is the ultimate answer. We humans are meme machines and free will is an illusion.

Controversies in Memetics. Culture is not obviously particulate and some have argued that this is a problem for a science of memetics. Dennett suggests that the relevant memetic units are the smallest elements that replicate themselves with reliability and fecundity. This means that the first four notes of Beethoven's Fifth Symphony are a meme because they have spread widely on their own, whereas most four note combinations are not. The whole symphony is a memeplex. More generally, if memes are understood as replicated information there is no need for it to be particulate. However, digitized information can be copied with higher fidelity than analogue information, so we might expect memes to evolve toward greater digitization. This seems to be happening with the appearance of computers, digital broadcasting, and other information technologies.

A more important disagreement concerns whether memes are only inside brains, only outside brains, or both. Dawkins originally defined memes to include both, but in his 1982 book *The Extended Phenotype*, he restricted them to information residing in a brain. These views are sometimes known as Dawkins A and B. Memeticist Aaron Lynch (1996) follows Dawkins B and treats memes as neurally stored information. British biologist Derek Gatherer takes a behaviorist perspective, counting behaviors and artifacts as memes rather than effects inside brains, because it is behaviors and artifacts that are copied.

David Hull, as well as Blackmore (1999) and Dennett (1991, 1995), prefer to count all replicated information as memes, whether in brains, behaviors, or artifacts. This avoids the related problem of whether there are memetic equivalents of vehicles, interactors, and phenotypic effects. Before the term *meme* was coined, American anthropologist F. T. Cloak compared i-culture (cultural instructions stored in people's heads) to the genotype and m-culture (behavior, technology, and social organizations produced by those instructions) to the phenotype, with the ultimate function of both being the maintenance of the i-culture. Later the terms *meme-phenotype*, *sociotype* and *phemotype* appeared but there is simply no agreement about how these terms should be used.

These arguments make it easy to defuse one of the most common objections to memetics, that memetic transmission is Lamarckian. Usually objectors mean that memes are characteristics that are acquired by one generation and then inherited by the next. But this reveals a confusion between the gene's-eye view and the meme's-eye view. From the genes'-eye view the phenotype acquires new characteristics, but there is no Lamarckian process because our memes do not alter the genes we pass on. From the meme's point of view, the human body is not the relevant phenotype and there is no obvious equivalent of the germ line. Therefore, comparisons with Lamarck are simply misplaced.

Within memetics, an interesting disagreement concerns the mechanism of replication. Dawkins originally defined memes as copied by "imitation, in the broad sense." Blackmore follows this, arguing that most other species, because they are incapable of true imitation, do not have memes. Some birds and whales can imitate songs, but chimpanzee cultures such as nut-cracking techniques are probably passed on by simpler forms of social learning that cannot sustain a new evolutionary process and are not memes. Cambridge biologist Kevin Laland disagrees, arguing that memes can be replicated by any kind of social learning. If this is so, meme theory cannot be used to account for the uniqueness of human culture.

At the other end of the scale, some memeticists argue that linking memes to imitation ignores the many rich and diverse ways in which humans share schemas (such as the rules of how to behave in a restaurant) and learn by direct teaching and reading books. This makes memetics too restrictive and oversimplifies how culture changes. Perhaps a new word is needed to describe the replication of memes with variation and selection, but we probably need to understand the process much better before inventing one. Memetics is a fledgling science engaged in lively and largely constructive controversy, but the most important open question is just how useful it will prove to be.

[*See also* Overview Essay *on* The New Replicators; Adaptation; Cultural Evolution, *article on* Cultural Transmission; Levels of Selection; Natural Selection, *article on* Natural Selection in Contemporary Human Society; Selfish Gene.]

BIBLIOGRAPHY

Aunger, R. A., ed. *Darwinizing Culture: The Status of Memetics as a Science*. Oxford, 2000. Useful collection of scholarly chapters by some of the best proponents and critics of memetics.

Blackmore, S. J. *The Meme Machine*. Oxford, 1999. Overview of memetics and meme-gene coevolution.

Brodie, R. *Virus of the Mind: The New Science of the Meme*. Seattle, 1996. Popular book with practical applications and lots of examples.

Cavalli-Sforza, L. L., and M. Feldman. *Cultural Transmission and Evolution*. Princeton, 1981.

Dawkins, R. (1976) *The Selfish Gene*. Oxford, 1976; new ed. with additional material, 1989. A classic, and the origin of the meme meme. The chapter on memes is essential reading.

Dawkins, R. "Viruses of the Mind." In *Dennett and his Critics: Demystifying Mind*, edited by B. Dahlbohm. Oxford, 1993. Elaboration of Dawkins's idea applied especially to religions.

Dennett, D. *Consciousness Explained*. Boston, 1991. Has several short sections on memetics, especially as applied to the nature of mind and consciousness.

Dennett, D. *Darwin's Dangerous Idea*. London, 1995. A long and fascinating book exploring the power of the evolutionary algorithm. Several sections on memes, some repeated from *Consciousness Explained*.

Durham, W. H. *Coevolution: Genes, Culture and Human Diversity*. Stanford, 1991. Not primarily about memetics, but an excellent overview of gene-culture coevolution.

Lynch, A. (1996) *Thought Contagion: How Belief Spreads Through Society*. New York, 1996. A popular book with lots of examples of sex memes, cults, religious beliefs, and taboos.

ONLINE RESOURCES

Journal of Memetics: Evolutionary Models of Information Transmission. An on-line journal with bibliography, overview of memetics, and links to other useful sites. http://www.cpm.mmu.ac.uk/jom-emit/

Memetics publications on the web. An extensive set of high quality papers on memes, as well as links to other sites. http://users.lycaeum.org/~sputnik/Memetics/

— SUSAN BLACKMORE

MENDEL, GREGOR

Gregor Mendel (1822–1884) is credited as the founder of the study of genetics and discovered the fundamental laws of heredity. At the meeting of the Natural Science Society of Brno in 1865, Mendel presented the results of his ten-year experiments into plant hybridization. The paper describing this work was published the following year, and largely forgotten. It was rediscovered in 1900, and his theory was generalized as Mendel's laws of heredity, acknowledged as the foundation of genetics.

Mendel (born Johann) was born on 22 July 1822 in Hynčice in the province Moravia (now Czech Republic). His formal education included six years at the gymnasium in Opava and two years at the Philosophical Institute in Olomouc, where he studied philosophy, mathematics, and physics, In 1843, he entered the Augustinian monastery in Brno, where he took the name Gregor. After being ordained a priest, he was sent to teach Greek and mathematics at the gymnasium in Znojmo. In 1851–1853, he studied physics, mathematics, chemistry, botany, and plant physiology at the University of Vienna. On returning to Brno, he taught physics and natural history at the new Realschule until 1868, when he was elected abbot, retaining this position until his death on 6 January 1884.

In his lecture in 1865, Mendel first explained the motivation for the research arising from fertilization in plant breeding and plant hybridizing experiments by botanists. Mendel aimed to understand how heredity determines the characteristics of offspring produced by parents with different traits. For crossing, he selected varieties of garden peas that differed in easily distinguished characters. First, he crossed plant varieties that differed in whether their seeds were either round or wrinkled. Hybrid progeny from this cross had only round seeds. Thus, the round character was dominant to the wrinkled character, and the wrinkled character was recessive to the round character. Mendel then self-fertilized the hybrid, round-seeded progeny. The resulting offspring had round and wrinkled seeds in a ratio of 3:1.

Mendel realized that the self-fertilized progeny in ratio 3:1 could be expressed as a series of types AA + 2Aa + aa with ratio 1:2:1, in which the pure, dominant round type is AA, the pure, recessive wrinkled types is aa, and the hybrid is Aa. Because A dominates a, the Aa hybrid expresses the dominant character. When Mendelian genetics became the basis of genetic analysis after 1900, the pattern of segregation for dominant and recessive traits became known as Mendel's first law, the segregation of characters.

Crossing plants with two characters, such as round or wrinkled seeds and purple or white flowers, Mendel found that each individual character pair followed the law of segregation independently from the other character. Thus, the ratios of characters among progeny could be obtained by combining the series of types expected independently for each character. This became known as Mendel's second law, the independent assortment of characters. Mendel followed his studies on Pisum (peas) with further studies of segregation and assortment in Phaseolus and other plant species.

Mendel focused on discrete characters that differed between parents because of his interest in the nature of varieties and species. For Mendel, it was difficult and uncertain to draw a sharp line between species and varieties, and equally between hybrids of species and those of varieties. He denoted his experimental plans with the German term *Art*. In English translation, this term is consistently given as *species*. The idea of evolution appeared already in Mendel's geological essay for the university examination in 1851. He read Charles Darwin's books in German translation, and according to his marginal notes, he accepted the theory of natural selection but rejected Darwin's hypothesis of pangenesis, as he himself was explaining the enigma of variability and heredity.

[*See also* Mendelian Genetics.]

BIBLIOGRAPHY

Fisher, R. A. "Has Mendel's work been rediscovered?" *Annals of Sciences* 1 (1936): 115–137.

Iltis, H. *Life of Mendel.* 2d ed. New York, 1966.

Orel, V. *Gregor Mendel, the First Geneticist.* Oxford and New York, 1996.

Stern, C., and E. R. Sherwood. *The Origin of Genetics: A Mendel Source Book.* San Francisco and London, 1966. Included is the English translation of Mendel's *Pisum* paper, published in German in 1866; his *Hieracium* paper, published in German in 1870; and Mendel's letters to C. Naegeli, published in German by C. Correns in 1905.

Wood, R. J., and V. Orel. *Genetic Prehistory in Selective Breeding—A Prelude to Mendel.* Oxford, 2001.

— Vitezslav Orel

MENDELIAN DISORDERS. *See* Disease, *article on* Hereditary Disease.

MENDELIAN GENETICS

The crucial question in the study of inheritance is the nature of the rules that govern the transmission of traits, so that offspring resemble their parents. In the mid-nineteenth century, Gregor Mendel (1822–1884) discovered two fundamental principles governing the flow of inherited information, which are referred to as the law of segregation and the law of independent assortment. These principles apply equally well to plants and animals, and they help us to understand the origin of the variation in traits on which the process of natural selection depends. Mendel discovered these laws through carefully executed experiments with garden peas. In these experiments, he made controlled crosses between certain pea plant varieties and then recorded the traits of each hybrid and the numbers of progeny and statistically analyzed them.

Biologists before Mendel's time failed to discover these laws because they relied on incorrect assumptions and misconceptions. Aristotle in the fourth century BCE, Hippocrates before him, and even Mendel's contemporary Charles Darwin believed in an idea known as pangenesis, which correctly assumed that male animals provided hereditary guidance through information contained in their semen. Some adherents to pangenesis assumed that sperm contained a collection of heritable information from each part of the body that the offspring inherited, and that this information could contain variation caused by interactions between the parent organism and the external environment. Although today pangenesis is a discredited idea, at that time it was an appealing conceptual framework to explain some of the observed range of phenotypic variation on which natural selection could act.

Before Mendel's discovery of the laws of segregation and independent assortment, hereditary information from each parent was not considered to consist of discrete qualities; instead, traits were thought to combine much like two miscible fluids. This "blending theory" of inheritance was based on reproducible observations that progeny usually look like a combination of traits seen in both parents. However, it was incorrectly assumed that the isolated parental traits could not be recovered in further matings of progeny because the genetic information encoding these traits had become blended and thus inextricably merged.

Gregor Mendel's Garden Peas. Gregor Mendel's discoveries were first published in 1865. He studied pea plants (*Pisum sativum*), which offered several important experimental advantages. Peas mature and produce offspring in a single season and can be studied in great numbers, giving statistical power to clear results. Seed merchants offered many varieties with easily observed characteristics. Mendel chose to work with fourteen true-breeding varieties of the pea plant that represented seven pairs of contrasting features, such as height (tall vs. short plants), seed color (green vs. yellow seeds), and seed texture (smooth vs. wrinkled). Pea plants can be cross-fertilized as well as self-fertilized because a pea flower produces both pollen and ova. The stamen and pistil structures on a single plant can be manipulated experimentally to prevent self-pollination and to allow reliable crosses to be performed that produce viable, fertile hybrid plants.

The first of Mendel's crucial strategies involved the selection of true-breeding varieties with simple characteristics. Repeated self-crossings produced strains that contained only a single form of a gene, reliably present on what we now know are both homologous chromosomes carried by the plant. To use one of the traits studied by Mendel as an example, seed colors came in two varieties, green and yellow. Mendel did not begin his experiments until he had obtained a strain of peas with green seeds that, when crossed with themselves, always and only produced offspring with green seeds. Likewise, he worked with another strain with yellow seeds that produced only yellow seeds. Years later, it was understood that these different traits are encoded by different forms of the same gene (in this instance, the "seed color gene"); these forms are now known as alleles of that gene.

F1 Hybrids. Once Mendel was confident that he had pure-breeding strains, he systematically crossed strains with each other, grew the offspring, and carefully enumerated the numbers and types he observed. Mendel's first striking observation was that the offspring resembled one of the parental strains and not the other. There was no blending of the parental traits. The hybrid nature of the progeny was not apparent from observing the plants. These offspring, the first generation from a controlled monohybrid cross, are today referred to as the F1 generation; in the example of green vs. yellow seeds, all plants in the F1 generation displayed yellow seeds. The form of the trait exhibited in the F1 (yellow seed color) is now called the dominant character, while the parental trait not observed in the F1 hybrids (the hidden trait, in this case green seed color) is referred to as the recessive character.

How did Mendel discover that the genetic information encoding the recessive character (green seed) was residing hidden in the F1 plants? He was able to deduce this because he took the F1 plants and crossed them with each other. The offspring of this cross, known as the F2 generation, contained two types of plants with respect to seed color, demonstrating that the green-seed F1 plants were able to give rise to offspring with both yellow and green seed types. In this way, Mendel's data allowed him to distinguish between an organism's genetic makeup, now called its genotype, and its outward appearance, now called its phenotype. A helpful underpinning of genetics is that a phenotype can sometimes be predicted if the genotype is known; however, the genotype cannot always be predicted from the phenotype, because different genotypes can give rise to the same phenotype.

Law of Segregation. Crossing F1 plants with each other gave the same results as crossing these plants to themselves ("selfing"). In the F2 generation, plants with the dominant and recessive phenotypes reappeared. Mendel carefully counted the number of F2 plants displaying the dominant phenotype and those displaying the recessive phenotype. In each case, the numbers were consistent; the ratio of dominant form to recessive form was 3:1. Mendel stated "that among each four plants of this generation three display the dominant character and one the recessive." He reported that there were 6,022 yellow and 2,001 green seeds. Moreover, the green seeds gave rise to true-breeding plants, producing only plants with green seeds. In contrast, only one-third of the yellow seeds bred true. The remaining plants grown from yellow seeds gave rise to both green and yellow seeds when crossed with each other or themselves.

Mendel deduced from these observations that each organism carries two copies of a gene, one from each parent. In a hybrid, the dominant form of the gene (the dominant allele) is always expressed preferentially over the recessive allele. When an organism contains two copies of the dominant allele—when it is homozygous—it displays the dominant phenotype. When it contains one copy of the dominant allele and one copy of the recessive allele (now referred to as heterozygous), it displays the dominant phenotype. Only when it is homozygous for the recessive allele does it display the recessive phenotype.

The 3:1 phenotypic ratio, first observed by Mendel, can be illustrated by a simple mathematical graphic form sometimes referred to as a Punnet square after the scientist who developed it. A simple Punnet square is shown in Figure 1. The four quadrants of the square show three different genotypes of the offspring from gametes of a cross between F1 plants, which are heterozygous for the trait under study. One quarter of the progeny are genotypically homozygous for the dominant allele, and these plants show the dominant phenotype.

(genotype of parent 1) Aa X Aa (genotype of parent 2)

⇓

gametes	A	a
A	AA round	Aa round
a	Aa round	aa wrinkled

FIGURE 1. Segregation of Alleles (A, a) During Meiosis Followed by Random Fertilization of Pollen and Ova in a Monohybrid Cross.
Assume A designates the dominant allele specifying round seeds and a is the recessive allele for wrinkled seeds. The phenotype is indicated for each genotype. The genotypic ratio of the three different genotypes among the F2 progeny in the Punnet square is 1:2:1 while the phenotypic ratio is 3:1.
Courtesy of author.

Two quarters are heterozygous and also show the dominant phenotype. One quarter of the progeny are homozygous for the recessive allele and show the recessive phenotype. Thus, the 1:2:1 genotypic ratio gives rise to the observed 3:1 phenotypic ratio. Mendel's law of segregation can also be understood in the context of the process of meiosis that gives rise to gametes. The two alleles of a gene separate from one another in the gonads during gamete formation, usually with equal numbers of gametes of each allele produced during meiosis.

Law of Independent Assortment. We know today that eukaryotic organisms contain thousands of genes. Humans are estimated to have at least 40,000 genes. *Arabidopsis thaliana*, a flowering weed, has 25,498 genes. It is important, therefore, to understand the rules governing the simultaneous inheritance of multiple independent genes. Again Gregor Mendel's ground breaking work helped in this process. In another series of experiments, Mendel examined the inheritance of two different pairs of traits in combination, referred to as a dihybrid cross. One parental plant was true-breeding for round seeds and yellow cotyledons, while the other plant was true-breeding for wrinkled seeds and green cotyledons. All of the F1 progeny plants exhibited the dominant phenotype for each pair of traits and were round and yellow. The F1 plants then were allowed to self-pollinate. In 1866, Mendel reported that "The plants raised therefrom yielded seeds of four sorts . . . . In all, 556 seeds were yielded by 15 plants, and of these there were: 315 round and yellow, 101 wrinkled and yellow, 108 round and green and 32 wrinkled and green." The relative ratio of these four phenotypes was 9:3:3:1, respectively.

Each parent from this dihybrid cross can produce four types of gametes. Construction of a 4 × 4 Punnet square, with 16 cells, will demonstrate that 9:3:3:1 is the expected phenotypic ratio. Note that embedded in the 9:3:3:1 ratio are two 3:1 ratios, if each phenotype is considered separately. For example, there is a total of 423 round seeds (315 + 108) and 133 wrinkled seeds (101 + 32), which is close to a ratio of 3:1, providing confirmation that the Law of Segregation still applies in a dihybrid cross. This experiment also revealed the second general law of heredity, the Law of Independent Assortment. Mendel observed that, among a large number of progeny, seed shape (round or wrinkled) and cotyledon color (yellow or green) behaved with respect to each other as if they were independently transmitted to the offspring. He did not report data that suggested that one trait was always associated with another. This behavior of independent inheritance of multiple genes is now referred to as the Law of Independent Assortment. Today we understand that different genes assort independently when they reside on different chromosomes, or when they are located on the same chromosome but very far apart. Such genes are said to be unlinked.

The types of experiments done by Mendel have been repeated in numerous plant and animal species, with analogous results. The resulting principles are now clearly established: genetic information is particulate; it undergoes segregation, and it typically shows independent assortment as it is passed from parents to offspring.

Exceptions to Mendel's Law of Independent Assortment. Mendel's choice of traits to work with allowed for the unambiguous elucidation of the basic laws of genetics. As in many explorations to understand a process, however exceptions to the basic laws have been discovered, adding layers of complexity to the original understanding.

Autosomal linkage. Among all pairs of genes examined, a small fraction have been observed to violate Mendel's law of Independent Assortment (see Sturtevant, 2001). One of the first such reports was described in the *Matthiola* plant, regarding two genes, one specifying leaf and petal color (the different alleles specify dark and white forms) and another specifying leaf and stem hairs (hoary and smooth alleles). When crosses were completed, only the original two parental types appeared; no independent assortment had occurred. This co-inheritance is now called linkage. Linkage occurs because the two genes in question reside adjacent to each other on the same chromosome. This means that they are not subject to random independent segregation of chromosomes as they are passed from parent to offspring.

Complex phenotypes. Understanding reproducible

patterns of inheritance that are exceptions to Mendel's two rules informs us of other hereditary phenomena. The reason Mendel's simple laws were not recognized earlier is that some of the most readily observable phenotypes in many organisms arise from the simultaneous action of many different genes. As an example of complex multigenic inheritance of a trait, skin color in humans has been estimated to be caused by the action of four or five different genes (Cavalli-Sforza and Bodmer, 1971). Thus, the offspring of two individuals with markedly different skin color typically appear intermediate in skin color, but since human families are of limited size, the degree of skin coloration does not follow a recognizable pattern.

Sex linkage. Early studies in fruit flies (*Drosophila melanogaster*) revealed that some traits have a highly sex-specific expression. For example, the normal eye color in fruit flies is red. When red-eyed female flies were crossed with males displaying the variant white eye color, male and female progeny had red eyes. However, the reciprocal cross between white-eyed females and red-eyed males produced male progeny with white eyes and female progeny with red eyes. The difference in outcome of reciprocal crosses stems from the fact that this gene (w^+, normal or wildtype allele of the white gene) for eye color resides on a sex chromosome, the X chromosome, which is involved in sex determination. In other words, the white gene is sex-linked. Male *Drosophila* have one X chromosome and one Y chromosome, while females have two X chromosomes. In the fruit fly, the *white* gene resides only on the X chromosome. Therefore, a recessive sex-linked mutant allele of the *white* gene, *w*, is not expressed in heterozygous females (w/w^+), which carry the normal red eye color allele of the *white* gene. Males do not carry a second X chromosome, allowing expression of the recessive *w* allele. In Mendel's experiments with peas, he observed identical results in reciprocal crosses.

Codominance. Certain traits show another inheritance pattern, called codominant or additive inheritance. In these cases, heterozygotes display a phenotype intermediate between that of homozygous dominant and homozygous recessive individuals. In domestic chickens, the length of the comb is under the control of such a system. Homozygotes for the long form have the longest combs observed; homozygotes for the short form have the shortest combs, while heterozygotes have combs intermediate in length. When homozygous long comb is crossed with homozygous short comb and the F1 progeny are then crossed among themselves, three types of progeny appear: long comb, intermediate comb, and short comb, in a 1:2:1 ratio.

Conclusion. Other exceptions to the simple but elegant principles discovered by Mendel are known, such as mitochondrial inheritance, imprinting, X-chromosome inactivation, and new mutations. In general, however, it is now clear that Mendel's laws and the exceptions to them govern the behavior of genes of all higher organisms. Thus, they also govern the expression of genetic variability in individual organisms on which natural selection acts. Although many exceptions to Mendel's laws have been elucidated, they remain the sturdy original constructs on which future complexity could be anchored.

[*See also* Cytoplasmic Genes; Hardy–Weinberg Equation; Linkage; Meiotic Distortion; Mendel, Gregor.]

BIBLIOGRAPHY

The Arabidopsis Genome Initiative. "Analysis of the Genome Sequence of the Flowering Plant *Arabidopsis thaliana*." *Nature* 408 (2000): 796–815.

Cavalli-Sforza, L. L., and W. F. Bodmer. "The Genetics of Human Populations." San Francisco, 1971.

Mendel, Gregor. "Versuche über Pflanzen-Hybriden." *Verhandlungen des naturforschenden Vereines in Brunn*, 4 (1865): 3–47. English translation, *Experiments in Plant Hybridization*, is available at Mendel Web, http://www.netspace.org/MendelWeb/, and at Electronic Scholarly Publishing, http://www.esp.org/.

Sturtevant, A. H. *A History of Genetics*. New York, 2001.

— Thomas B. Friedman and Dennis Drayna

METAMORPHOSIS, ORIGIN AND EVOLUTION OF

Metamorphosis is inextricably linked to the concepts of a larval stage and complex life history. A species has a complex (biphasic) life history or indirect development when it has a larval stage that is different in body form and possibly aspects of its habitat and feeding behavior from its juvenile or adult stage. Larvae require a metamorphosis or period of concentrated development to transform to a juvenile or an adult, usually after the completion of embryonic development and before or synchronously with sexual maturation. A species lacking a larva and metamorphosis is said to be direct developing. Deciding if and when a species exhibits a larval stage and metamorphosis hinges on some rather arbitrary distinctions. Are larval and postlarval morphologies sufficiently different and long lasting that their transformation constitutes a distinct period of concentrated development? Which events delineate the start and end of metamorphosis? Animals exhibit immense variation in the type, number, magnitude, and timing of postembryonic changes. Understanding the origin and evolution of metamorphosis requires consideration of morphological features, cell pathways, and signaling mechanisms.

The Origin of Primary Metamorphosis. Metamorphosis by definition occurs only in metazoans (multicellular animals). The ancestral metazoan stock gave rise to poriferans (sponges) and cnidarians (corals and their relatives) before evolving bilateral symmetry. Larvae and metamorphoses of some form occur in all metazoan phyla except onychophorans, tardigrades, and the pseudocoelomates (although rotifers show traits common to larvae of other forms). Most metamorphoses are primary; they arose in early metazoans and involve small, ciliated larvae that are specialized for planktotrophy (feeding on plankton), dispersal, and/or settlement in marine environments.

Opinions differ on when and how many primary metamorphoses evolved: only once, in the ancestor of all metazoans, and lost in the ancestors of phyla that lack a ciliated larva (Jägersten, 1972) or later, in the common ancestor of deuterostomes and protostomes (Peterson et al., 1997). Others (Nielsen, 1998; Rieger, 1994) point to fundamental differences among existing larval forms to argue that primary metamorphoses evolved two, four, or more times.

There are three hypotheses for how primary metamorphoses evolved. One is that ancestral metazoans were originally small planktonic forms that evolved metamorphosis through the addition of a new benthic adult stage (Nielsen, 1998). This is based on the view introduced by the German scientist and philosopher Ernst Haeckel (1834–1919) that the adult stages of ancestral sponges and cnidarians were small blastula- and gastrula-like forms, respectively. The second hypothesis holds that ancestral metazoans were originally small, direct developing, and lecithotrophic (yolk feeding). They evolved a planktotrophic larval stage only after attaining a sufficiently large adult size to produce large numbers of small eggs (Raff in Hall and Wake, 1999). The adult body plans were in place when metamorphosis emerged to remove newly evolving larval specializations. The third hypothesis holds that the earliest metazoans already had a complex life cycle characterized by a microscopic, motile larva and macroscopic, colonial adult (Rieger, 1994). Metamorphosis became more defined with the appearance of specializations and tissue types that drove the divergence between larvae and adults. Most researchers agree that the earliest metamorphoses were periods of slight and gradual change between planktonic larval and benthic postlarval stages that underwent many diverse modifications in the radiation of metazoan phyla.

Direct Development and Secondary Metamorphoses. Loss of the ciliated larval stage and the readoption of direct development are believed to have evolved separately in the ancestors of panarthropods (arthropods, onychophorans, and tardigrades, correlated with the postembryonic development of a chitinous exoskeleton or cuticle), chordates, and, according to some phylogenetic hypotheses, pseudocoelomates. Direct development also evolved within all other metazoan phyla except sponges, cnidarians, phoronids, brachiopods, and bryozoans. Many phyla also exhibit intermediate forms that lost the larval specializations and some of the metamorphic changes found in metamorphosing relatives, but still pass through a ciliated, lecithotrophic stage, which may be free-living or occur inside a benthic egg mass or brood chamber. In addition, some arthropods and chordates evolved new larval stages and secondary metamorphoses, two of which (those of butterflies and frogs) are usually chosen as the textbook examples of metamorphosis.

Among arthropods, secondary metamorphoses evolved in the ancestors of crustaceans and holometabolous insects. Ancestral insects are thought to have been direct developing or ametabolous. Holometabolous insects (beetles, flies, moths, and bees) hatch as specialized terrestrial feeding larvae (maggots, grubs, or caterpillars), which pass through a number of stages (instars) bordered by sheddings of the cuticle (molting). Upon reaching a critical size, they pass through a metamorphic molt into a pupal stage, during which the cuticle is hardened and the larval body form is reorganized into the winged adult. A final imago molt releases the adult from the pupa. Intermediate to ametabolous and holometabolous forms is a paraphyletic assemblage of hemimetabolous forms. These hatch at or just after a pronymph stage and pass through several nymph stages, during which the wing rudiments appear and become larger, and the body form, bristle pattern, and pigmentation approach those of the adult. These features are completed and the external genitalia formed during the final molt, which is called an incomplete or hemimetabolous metamorphosis. In some cases, though, additional changes in body shape during the final molt may produce almost as great a transformation as a holometabolous metamorphosis. Holometabolous insects are hypothesized to have evolved from a hemimetabolous ancestor whose pronymph and nymph stages were expanded and compressed respectively into larval and pupal stages (Truman and Riddiford, 1999).

Among chordates, dramatic secondary metamorphoses occur in ascidian tunicates, lampreys, elopomorph fishes (e.g., bonefishes and eels), flatfish, stomiiform fishes (e.g., lightfish and viperfishes), lantern fishes, deep-sea anglerfishes, frogs, salamanders, and caecilians. Ascidians metamorphose from a nonfeeding, tailed, and free-swimming larva into a filter-feeding, tailless, and sessile adult. Lampreys metamorphose from a freshwater, filter-feeding ammocoete into a marine adult specialized for parasitic feeding on fish. Teleost metamorphoses usually accompany transitions from one habitat to another. The most dramatic changes are specific to individual taxa, for example, migration of one eye to the other side of the head in flatfish, remodeling of the glass-like leptocephalus larva in elopomorphs, and shortening of the larval eyestocks in stomiiforms. Frogs, salamanders, and caecilians share an aquatic-to-terrestrial transition marked by the loss of external gills, gill-supporting skeleton, gill slits, tail fin, and Leydig cells, and the development of multicellular skin glands, although only frogs exhibit dramatically different larval and adult body forms. Another kind of metamorphosis, exemplified by salmon, involves developmental changes during sexual maturation that transform the unspecialized postlarval skull into a highly specialized adult skull. Ascidian metamorphosis is currently thought to have evolved by addition of the adult stage to the primitive life cycle. Frogs, salamanders, and caecilians are thought to have evolved from a metamorphosing common ancestor. The dissimilarities of other vertebrate metamorphoses and the absence of metamorphosis in phylogenetically intermediate groups (hagfish, chondrichthyans, primitive bony fishes, coelocanths, and lungfish) suggest separate evolutions in most cases.

Direct development has reevolved multiple times in crustaceans, tunicates, salamanders, caecilians, and frogs. Larvae and metamorphosis in these forms are either nonexistent or retained as transient late embryonic stages. Whereas most amphibian direct developers exhibit large yolky eggs, some frogs and caecilians retain the embryo within the oviduct and supplement its yolk supply with maternal nutrition. Certain frogs and salamanders also appear to have reevolved aquatic larval stages and what would be tertiary metamorphoses, although this awaits confirmation by more robust phylogenetic analyses.

Insertion of Additional Larval Stages. Metamorphosing taxa have often evolved additional larval stages that follow or replace the original. Examples include the pelagosphera of sipunculids, veliger of gastropods, and nauplius of crustaceans. The nauplius and some later larval stages may occur before hatching, especially in brooding forms. Transformations between these larval stages and to the juvenile are usually regarded as separate metamorphoses. Whereas some added larval stages are specialized for dispersal or selecting a settlement site, others are intermediate in form and habitat between the original larva and juvenile. Parasitic flatworms may pass through as many as five distinct larval stages, each specialized for invading, inhabiting, or reproducing within a different host species.

Some hemimetabolous insects and certain holometabolous species, especially ones with parasitic larval stages, exhibit heteromorphosis or hypermorphosis, meaning conspicuous changes in body form between instars. In each case, the convention is not to recognize the different instars as distinct larval types, nor their intervening molts as metamorphoses. The same applies for polycheate annelids that continuously acquire trunk segments throughout the trochophore larval stage.

Larval Reproduction. Metamorphosing lineages can escape metamorphosis by evolving the ability to retain the larval morphology to adulthood. This occurs either by neoteny, where metamorphosis fails or is incomplete, so that some or all larval features persist and some or all adult features fail to appear, or by progenesis, where the larva becomes sexually mature and stops developing before the metamorphosis. Sexual maturation at a smaller size or younger age in the larval reproducing form implies progenesis; sexual maturation at the same approximate size or age implies neoteny.

Larval reproduction has been proposed to explain the origin of many major taxa, including the pseudocoelomates, tunicates, and vertebrates. However, the lack of a clearly defined ancestral pattern of development for these groups makes it difficult to test such propositions and impossible to distinguish between progenesis and neoteny. Within taxa that are primitively metamorphosing, larval reproduction by progenesis is found in interstitial annelids, which live between the grains of marine sediment, in parasitic flatworms and repeatedly in holometabolous and hemimetabolous insects, crustaceans, and males of some teleost fishes, again with a prevalence in parasitic forms. Larval reproduction by neoteny is best described in salamanders, where it is the exclusive mode of development in four families and has evolved repeatedly in four of the other six families. Larval reproducing salamanders lie on a spectrum from forms that undergo little postembryonic modification to forms that complete all of the larval changes and most of the metamorphic changes exhibited by their metamorphosing relatives.

Complexity of Metamorphosis. Metamorphoses generally involve three kinds of change at the tissue level: loss of larva-specific tissues, development of juvenile/adult-specific tissues, and transformation of larval into adult tissues. Evolutionary changes in the complexity of metamorphosis are marked by differences in the way that larval cells are organized and instructed to participate in these processes. Metamorphoses range from simple (where most or all larval cells are used to make the adult tissues, and adults body forms are only slightly more complex than larval ones) to complex or radical (where most of the larval body is lost, and a very different, more complex adult body is produced from pockets or undifferentiated or "set aside" larval cells).

Sponge and cnidarian metamorphoses are simple and probably indicative of primitive metazoan metamorphoses. Although some larval cell types are lost at metamorphosis, there is little or no larval cell death. New adult cell types are derived from the differentiation of unipotent and multipotent larval cells, although certain sponge larvae may already possess all the cell types found in the adult. Whether adult cell fates are prespecified in larval cells or specified on the basis of their position during metamorphosis is unclear. Sponges may utilize both mechanisms, and hydrozoans appear to prespecify their adult body plans.

Sea urchins, some nemertean worms, and polycheate annelids exhibit radical metamorphoses. Sea urchins derive all of their adult body from an imaginal rudiment that forms on the left side of the larval body. Heteronemertean worms derive most of their adult body from five to eight imaginal disks that form as inpocketings or invaginations of ectoderm. Both build much of their adult body within the larval body. The emergence of the adult (or overt metamorphosis) coincides with resorption of the larval tissues in sea urchins and ingestion of the larval tissues by the newly emerged juvenile in nemerteans. Polycheate annelids derive their adult tissues from different parts of their trochophore. The larval gut becomes the adult gut, ectodermal and mesodermal stem cells generate the posterior growth zone for making trunk segments, and the apical part of the trochophore forms unsegmented head tissues. The ciliary organs and associated ectoderm are discarded. Circumstantial evidence suggests that adult rudiments of these forms are specified independently of other larval cells. Peterson and colleagues (1997) argue that this specification involves newly evolved and sophisticated genetic mechanisms to pattern the complex adult body plans.

Other metamorphoses lie somewhere between these simple and complex metamorphoses. Like heteronemertean worms, some holometabolous insects develop adult structures from a variable number of imaginal disks that arise as invaginations of ectoderm during the embryonic or larval period. Hemimetabolous insects develop only wings from imaginal disks that arise as outpocketings, rather than inpocketings, of ectoderm. Vertebrates that undergo major changes in body form (e.g., frogs, lampreys, flatfish) exhibit little change in body plan. They utilize extensive cell death, cell division, and recruitment of undifferentiated cells to reposition, reshape, or replace sense organs, nerves, skeletal elements, muscles, and gut regions. However, the basic arrangement of the central and peripheral nervous systems, gut tube, coeloms, and internal organs remains unchanged. In contrast to sea urchins, insects and vertebrates pattern their adult body plans in embryogenesis. This may be a consequence of their ancestors having lost their primary metamorphoses to direct development.

Regulation of Metamorphosis. Metamorphosis in marine invertebrates is usually associated with settlement of dispersing larvae. Larval settlement is triggered by various environmental cues, including light, gravity, and chemical signals from bacteria, conspecific adults, and other species. Chemicals (pheromones) from conspecifics may promote gregarious settling or help determine the sex of metamorphosing individuals. Chemicals from other species may promote or inhibit settlement. Failure to encounter the appropriate signal(s) may prolong dispersal even to the point of death in nonfeeding forms. The mechanisms by which larvae perceive environmental cues and use this information to activate tissue changes are largely unknown. Cnidarians may use neurosensory cells to relay a metamorphic stimulus. Once activated, these cells release a neuropeptide that stimulates neurons to release other neuropeptides. Sea urchins appear to use thyroid hormone (TH) from ingestion of algae to mediate development of the adult rudiment and resorption of larval tissues.

Vertebrate and insect metamorphoses are triggered by a combination of environmental factors and metabolic and developmental changes that control the production of activating hormones. The environmental factors involved vary with species and include food, temperature, light, density, substrate conditions, and the presence of pheromones and certain chemicals in food. Amphibians, flatfish, and many other teleosts produce TH to mediate metamorphosis. Because most vertebrates produce a brief surge in TH to mediate developmental changes at or around hatching or birth, metamorphic vertebrate ancestors appear to have co-opted this signal to mediate programs of larval and metamorphic development. Larval frogs, salamanders, and flatfish exhibit a steady rise in plasma TH concentration, which activates a sequence of concentration-specific tissue responses. Observations on TH production and the TH dependence of metamorphic changes in direct developing frogs and some neotenic salamanders suggest that the evolution of these forms involved changes in the timing and/or amount of TH production.

Generally speaking, insects produce two hormones to mediate their metamorphoses. Larval and nymph stages produce high levels of juvenile hormone (JH) and pulses of ecdysone (Ec), which, in the presence of high JH maintain the larval phenotype and activate the epidermal events involved in molting. During the last larval instar, JH production declines and becomes erratic thereafter. Ec pulses now operate in the midst of fluctuating JH levels and fluctuating tissue sensitivities to JH to sequentially suppress the larval phenotype, commit cells to pupa formation, and activate larval tissue resorption and imaginal disk differentiation. JH production in hemimetabolous forms occurs in two phases, one in early embryogeny and the second during nymph formation. The second promotes maturation of nymph structures from tissues that become responsive to JH at embryonic stages. JH production in holometabolous forms lasts from embryogeny to the start of the final larval instar, although another brief pulse occurs later in this instar. The transition from hemimetabolous to holometabolous development is hypothesized to have involved a shift in the start of the second phase of JH production to embryonic stages (Truman and Riddiford, 1999). The precocious maturation of embryonic tissues would thus convert the pronymph into a functional larval stage. The subsequent appearance and maturation of new nymphal structures would require a new JH-free interval followed by a return to high JH. As the duration of the larval period increased, the JH-free interval and following JH peak became compressed and delayed to produce the JH profile found in the final instar of holometabolous forms.

Although amphibians and insects use unrelated hormones to activate their metamorphoses, they show striking similarities at other levels of regulation. Both use a neural center to activate one gland whose hormone product activates a second gland to produce the primary metamorphic hormones thyroxine and ecdysone, which are converted to their more active forms triiodothyronine and 20-hydroxyecdysone by peripheral tissues. To activate a tissue response, the active hormone binds to a nuclear receptor complex that functions as a ligand-dependent transcription factor. The two proteins that form the receptor complex in amphibians and insects are homologues. One is a member of the nuclear hormone receptor superfamily (thyroid hormone receptor and ecdysone receptor), and the other is a retinoic acid receptor in amphibians and ultraspiracle in insects.

[*See also* Life History Stages.]

BIBLIOGRAPHY

Callery, E. M., and R. P. Elinson. "Thyroid Hormone-dependent Metamorphosis in a Direct Developing Frog." *Proceedings of the National Academy of Sciences USA* 97 (2000): 2615–2620.

Chino, Y., et al. "Formation of the Adult Rudiment of Sea Urchins Is Influenced by Thyroid Hormones." *Developmental Biology* 161 (1994): 1–11. Well-illustrated description of the morphological changes and thyroid hormone dependency of sea urchin metamorphosis.

Gilbert, S. F., and A. M. Raunio, eds. *Embryology, Constructing the Organism.* Sunderland, Mass., 1997. Written by leading developmental biologists, this is the first compendium of detailed developmental descriptions for all major metazoan phyla in over a hundred years.

Gould, S. J. *Ontogeny and Phylogeny.* Cambridge, Mass., 1977. A highly influential overview of the history and concepts of heterochrony, with emphasis on the adaptive significances of neoteny and progenesis.

Hall, B. K., and M. H. Wake, eds. *The Origin and Evolution of Larval Forms.* San Diego, Calif., 1999. Discussions on the morphology and development of invertebrate, fish, amphibian, and insect larvae, the hormonal basis of amphibian and insect metamorphoses, and the origin of invertebrate metamorphosis.

Jägersten, G. *Evolution of the Metazoan Life Cycle: A Comprehensive Theory.* London, 1972. A seminal, though dated, work on the evolution of life history stages across metazoans.

McEdward, L. R., and D. A. Janies. "Life Cycle Evolution in Asteroids: What Is a Larva?" *Biological Bulletin* 184 (1993): 255–268. A concise overview of the concepts and definitions of larvae, metamorphosis, and direct development.

Nielsen, C. "Origin and Evolution of Animal Life Cycles." *Biological Reviews* 73 (1998): 125–155. Comparative summary of larval morphologies to support a classic theory for the origin of metamorphosis.

Nijhout, H. F. *Insect Hormones.* Princeton, 1994. A comparative overview for nonspecialists on the endocrinology, physiology, and morphology of insect development.

Peterson, K. J., R. A. Cameron, and E. H. Davidson. "Set-Aside Cells in Maximal Indirect Development: Evolutionary and Developmental Significance." *Bioessays* 19 (1997): 623–631. Discussion of developmental and genetic evidence for an intriguing new theory of how metamorphosis evolved.

Rieger, R. M. "The Biphasic Life Cycle—a Central Theme of Metazoan Evolution." *American Zoologist* 34 (1994): 484–491. Discussion of tissue-level evidence to support another theory on metamorphic origins.

Rose, C. S., and J. O. Reiss. "Metamorphosis and the Vertebrate Skull: Ontogenetic Patterns and Developmental Mechanisms." In *The Skull*, edited by J. Hanken and B. K. Hall, vol. 1, *Development*, pp. 289–346. Chicago, (1993). Discussion of the concept of metamorphosis and a detailed summary of the morphological changes, remodeling pathways, and tissue interactions involved in skull development in lampreys, fish, and amphibians.

Shi, Y.-B. *Amphibian Metamorphosis.* New York, 2000. Detailed explanation of the molecular biology of frog metamorphosis, with comparisons to nonmetamorphic vertebrates and insects.

Strathmann, R. R. "Hypotheses on the Origins of Marine Larvae." *Annual Review of Ecology and Systematics* 24 (1993): 89–117. A detailed discussion of the diverse modifications involving larval structures and life histories that have evolved in marine invertebrates.

Truman, J. W., and L. M. Riddiford. "The Origins of Insect Metamorphosis." *Nature* 401 (1999): 447–452. Phylogenetic, hormonal, and morphological evidence to support an interesting new theory on the evolution of holometabolous metamorphosis.

Wray, G. A., and A. E. Bely. "The Evolution of Echinoderm Development Is Driven by Several Distinct Factors." *Development Supplement* (1994): 97–106. An insightful discussion of the evolutionary relationships among developmental, morphological, and life history features using echinoderms as the data set.

Youson, J. H. "Is Lamprey Metamorphosis Regulated by Thyroid Hormones?" *American Zoologist* 37 (1997): 441–460. A concise summary of evidence for the role of thyroid hormones in lamprey development.

— Christopher Stewart Rose

METAPOPULATION

Many, if not most, species are subdivided into multiple subpopulations, whose members are more likely to mate locally than would be expected by pure random mating. We say that these species have spatial population structure, and these "populations of populations" are sometimes called metapopulations (Hanski and Gilpin, 1997). The term *metapopulation* is used in population biology in this broad sense (and it is this meaning we use in this article), but it also has a stricter, original meaning as a collection of populations that experience local extinction and recolonization (Levins, 1969).

This confusion in terminology also extends to the words used to mean a local population within a species. For these local populations, the terms *deme*, *subpopulation*, and even *population* are often used interchangeably.

The evolutionary study of population structure involves examining the distribution of allele frequencies over subpopulations, measuring and inferring from these patterns, and predicting the effects on evolution caused by this genetic differentiation.

Measurement of Population Structure. The main population genetic consequence of spatial population structure is that different subpopulations can come to have different allele frequencies. This "genetic differentiation" among populations can be studied and measured in a variety of ways; most of these relate to a quantity introduced by Sewall Wright called F_{ST}. Unfortunately, F_{ST} has a variety of definitions (see Cockerham and Weir, 1987; Rousset, 1997); for a locus with only two alleles at average frequency p and q in the metapopulation and a variance among populations in allele frequency of $Var(p)$, then by definition $F_{ST} = Var(p)/pq$. Thus, F_{ST} can be thought of as a standardized measure of the genetic variance among populations, which for some models is expected to be similar at multiple loci, even if these loci have different allele frequencies.

F_{ST} can be measured for any locus for which allele frequencies can be determined, whether by examination of phenotypes, gene products, or DNA sequence directly. The most widely used method to estimate F_{ST} was proposed by Weir and Cockerham (1984). Other, related statistics estimate quantities approximately equal to F_{ST} for other types of data, for example, G_{ST} for heterozygosity data with multiple alleles (Nei, 1986), R_{ST} (Slatkin, 1995) and ρ_{ST} (Rousset, 1996) for microsatellite data, Φ_{ST} for molecular haplotypes (Excoffier et al., 1992), and $\langle F_{ST} \rangle$ (Hudson et al., 1992) or N_{ST} (Lynch and Crease, 1990) for DNA sequence data (see Michalakis and Excoffier, 1996, for a comprehensive overview).

Sometimes more information can be extracted from data; for example, one might measure patterns of increasing similarity of allele frequencies of nearby populations. This pattern can result from "isolation by distance" (migrants from one population are more likely to move to nearby populations than to more distant ones), local selection, or both. In these cases, the pattern of genetic differentiation as a function of distance can be used to estimate dispersal rates or patterns of selection (Lenormand et al., 1998; Rousset, 1997).

Island Model. Historically, the most common and simplest model of population structure is the island model, in which discrete populations each have N diploid individuals (or $2N$ haplotypes). Each of these populations is assumed to exchange a fraction m of its genes with other populations each generation. This simple model therefore assumes that all populations are equal in size and that the migration rate is equal for all populations (Wright, 1931). A simple starting place in population genetics often is to assume selective neutrality; in other words, there is no difference among alleles in fitness. Assuming an island model with no selection, the probability that two alleles randomly chosen from the same population share a common ancestor in that population (i.e., that they coalesce in that same population) is approximately $1/(4Nm + 1)$, a quantity that is also the predicted F_{ST}. For a neutral locus with two alleles with mean frequencies p and q, the variance among populations in local allele frequency is expected to be $2pq/(4Nm + 1) = 2pq\,F_{ST}$. This variance among populations in allele frequencies is therefore increased by stronger genetic drift (i.e., as N gets smaller) and decreased by migration mixing alleles among populations (i.e., as m gets larger). The resulting pattern is a balance between migration and drift.

Deviations from Island Model. Although the island model is historically important and a simple starting place, it does not do a particularly good job of describing natural populations. The island model assumes that each population is equal in size and contributes equally to migration to every other population. In reality, of course, population sizes and migration rates vary dramatically in space and time, and populations can become extinct, recolonize, grow, or shrink. Migration is usually distance limited and can be asymmetric between locally successful and unsuccessful populations (sources and sinks) (Dias, 1996). These factors can have dramatic effects on the evolutionary properties of structured populations (Whitlock, 2001).

Furthermore, selection can have a strong role in determining the genetic differentiation of populations: strong uniform selection can reduce differentiation, whereas spatially variable selection can increase differentiation (Hedrick et al., 1976).

Importance in Evolution. The importance of population structure to evolutionary biology has many facets, but the reason population structure matters always comes back to the fact that different local populations can have somewhat different allele frequencies and evolutionary trajectories. As a result, the total amount of genetic variation maintained in a species can be substantially affected, adaptation to spatially variable environments becomes more likely, and competition among populations for reproductive success becomes possible.

Genetic variance and its partitioning. Population structure can change both the amount and the partitioning of genetic variation. The amount of genetic variation in a species determines its short- and perhaps long-term rate of evolution. The level of genetic variation is determined by the interactions between genetic drift, selection, mutation, and migration. The effect of these forces can act somewhat differently in subdivided populations, such that the amount of genetic variation can be either much increased or much decreased relative to an undivided species.

Genetic drift tends to reduce the amount of genetic variance. Subdivided species tend to have a lower effective population size than undivided species even if they have the same total number of individuals (Whitlock and Barton, 1997). A lower effective size means that genetic drift is more effective, and therefore subdivided populations are likely to lose genetic variation by drift more rapidly.

In contrast, a subdivided species can more easily adapt to local environmental conditions. This may have two effects on genetic variance, at least: the total population size may be increased due to better local adaptation (with a concomitant reduction in drift), and genetic variation can be directly maintained by balancing selection. Furthermore, local populations can be genetically differentiated even with uniform selection, thus allowing more genetic variation in the species (Goldstein and Holsinger, 1992). It is not yet clear which of these effects is most important in determining the total effect of population subdivision on genetic variance.

The amount of genetic variation within a local population depends heavily on the population structure context it finds itself in. A local population that is entirely isolated from other populations will have properties different from those of a similar population that receives migrants from other populations on a regular basis. Indeed, for most loci, the genetic variance in a small deme connected even occasionally by migration to other demes will have a distribution of alleles similar to that expected of a population about the size of the whole species, rather than to that of an isolated population of the same small size. It will have more genetic variance and fewer fixed deleterious mutations than if it were truly isolated. (The reduction in extinction probability resulting from this input of genetic variance from nearby populations is called the genetic rescue effect.) Thus, the study of a single population often tells us more about the properties of the species as a whole rather than the population itself.

The increasing fragmentation of species into local populations caused by human effects in the environment means that more and more species will be subdivided. The main effect of this fragmentation is its most obvious one: as habitat becomes rarer and of poorer quality, the total population size will become smaller. Therefore, the major genetic effect of fragmentation will be a decrease of effective population size and the concomitant reduction of genetic variance. This will be extreme in species that are poorly adapted for migration between these new fragments, because the dynamics of genetic variance will then depend on the size of the local populations themselves. With such low population sizes, deleterious alleles can fix in populations, and the mean fitness may further decline.

Changes in the effects of selection in subdivided populations. As different local populations encounter different environmental conditions, there is the potential for local adaptation. These effects of selection are countered by migration from other populations, so that the extent of local adaptation is a balance between migration and selection. Local adaptation has several consequences: alleles that are advantageous only in rare environmental conditions have a much higher chance of being maintained in the population, the mean fitness of the species can be increased, and the total number of individuals in the species that can be supported can be larger.

Local adaptation can develop over the space of a few generations, if gene flow is sufficiently low. Some good examples of local adaptation are the local evolution of insecticide resistance (Pasteur and Raymond, 1996) and the adaptation of plants to soils poisoned by heavy metals, especially taxa with restricted gene flow (Antonovics et al., 1971). These local adaptations would not be possible without population structure.

Local adaptation carries the potential for a reduction in fitness in crosses between individuals from different populations, called outbreeding depression. Outbreeding depression can result from deleterious epistatic interactions between variant alleles in different populations, if the average of the optimal phenotypes in two environments is not well adapted anywhere. Local adaptation can therefore be an important first step toward speciation (Schluter, 1999).

A process that can be strongly affected by population structure is host–parasite evolution. An important component of the environment of both the host and the parasite species is determined by the spatial distribution of relevant phenotypic or genetic variability of the other. Variation in the distribution of host or parasite will result in variation in the distribution of the outcome of the host–parasite interaction. This has led to the formulation of the geographic mosaic theory of coevolution (Thompson, 1994) and is still a largely underexplored area of research.

Population structure and the genetic differentiation that results can also strongly affect the evolution of particular traits. One trait whose study is meaningless outside of a spatial context is the dispersal rate. Several factors determine the rate at which individuals move between populations, and spatial structure is important for factors selecting for and against dispersal. If there is local adaptation, then dispersal will be selected against because it will generally lead individuals to environments where they are poorly adapted. Spatial structure can select for dispersal in several ways. First, because populations are differentiated with respect to deleterious allele frequencies, individuals that move to another population are less likely to have offspring that are homozygous for deleterious recessive alleles. In other words, dispersal could be a mechanism to avoid inbreeding depression. Dispersal will also be favored in a temporally variable environment, in particular when disturbances drive some local populations extinct. Finally, dispersal reduces competition between relatives (Clobert and Danchin, 2001).

Evolution at the population level. Another unique feature of subdivided populations is that variance among subpopulations allows for another level of selection. If populations differ in fitness and this difference is to some extent the result of genetic differences between those populations, then allele frequencies will change over generations as a result of this group selection. Although this group selection is theoretically plausible and can be experimentally demonstrated (Wade, 1978), its general applicability to natural populations has not been shown.

[*See also* Genetic Drift.]

BIBLIOGRAPHY

Antonovics, J., A. D. Bradshaw, and R. G. Turner. "Heavy Metal Tolerance in Plants." *Advances in Ecological Research* 7 (1971): 1–85.

Aviles, L. "Sex-Ratio Bias and Possible Group Selection in the Social Spider *Anelosimus eximius*." *American Naturalist* 128 (1986): 1–12.

Clobert, J., and E. Danchin, eds. *Causes, Consequences and Mechanisms of Dispersal at the Individual, Population and Community Level.* Oxford, 2001.

Cockerham, C. C., and B. S. Weir. "Correlations, Descent Measures: Drift with Migration and Mutation." *Proceedings of the National Academy of Sciences USA* 84 (1987): 8512–8514.

Dias, P. C. "Sources and Sinks in Population Biology." *Trends in Ecology and Evolution* 11 (1996): 326–330.

Excoffier, L., P. E. Smouse, and J. M. Quattro. "Analysis of Molecular Variance Inferred from Metric Distances among DNA Haplotypes: Application to Human Mitochondrial DNA Restriction Data." *Genetics* 131 (1992): 479–491.

Goldstein, D. B., and K. E. Holsinger. "Maintenance of Polygenic Variation in Spatially Structured Populations: Roles for Local Mating and Genetic Redundancy." *Evolution* 46 (1992): 412–429.

Hanski, I. A., and M. E. Gilpin. *Metapopulation Biology Ecology, Genetics and Evolution.* New York, 1997.

Hedrick, P. W., M. E. Ginevan, and E. P. Ewing. "Genetic Polymorphism in Heterogeneous Environments." *Annual Review of Ecology and Systematics* 7 (1976): 1–32.

Hudson, R. R., M. Slatkin, and W. P. Maddison. "Estimation of Levels of Gene Flow from DNA Sequence Data." *Genetics* 132 (1992): 583–589.

Lenormand, T., T. Guillemaud, D. Bourguet, and M. Raymond. "Evaluating Gene Flow Using Selected Markers: A Case Study." *Genetics* 149 (1998): 1383–1392.

Levins, D. "Some Demographic and Genetic Consequences of Environmental Heterogeneity for Biological Control." *Bulletin of the Entomological Society of America* 15 (1969): 237–240.

Lynch, M., and T. J. Crease. "The Analysis of Population Survey Data on DNA Sequence Variation." *Molecular Biology and Evolution* 7 (1990): 377–394.

Michalakis, Y., and L. Excoffier. "A Generic Estimation of Population Subdivision Using Distances between Alleles with Special Reference for Microsatellite Loci." *Genetics* 142 (1996): 1061–1064.

Nei, M. "Definition and Estimation of Fixation Indexes." *Evolution* 40 (1986): 643–645.

Pasteur, N., and M. Raymond. "Insecticide Resistance Genes in Mosquitoes: Their Mutations, Migration, and Selection in Field Populations." *Journal of Heredity* 87 (1996): 444–449.

Rousset, F. "Equilibrium Values of Measures of Population Subdivision for Stepwise Mutation Processes." *Genetics* 142 (1996): 1357–1362.

Rousset, F. "Genetic Differentiation and Estimation of Gene Flow from F-Statistics under Isolation by Distance." *Genetics* 145 (1997): 1219–1228.

Rousset, F. "Inferences from Spatial Population Genetics." In *Handbook of Statistical Genetics*, edited by D. Balding, M. Bishop, and C. Cannings. New York, 2001.

Schluter, D. "Ecological Causes of Adaptive Radiation." *American Naturalist* 148 (1996): S40–S64.

Slatkin, M. "Gene Flow in Natural Populations." *Annual Review of Ecology and Systematics* 16 (1985): 393–430.

Slatkin, M. "A Measure of Population Subdivision Based on Microsatellite Allele Frequencies." *Genetics* 139 (1995): 457–462.

Thompson, J. N. *The Coevolutionary Process.* Chicago, 1994.

Thornhill, N. W. *The Natural History of Inbreeding and Outbreeding: Theoretical and Empirical Perspectives.* Chicago, 1993.

Wade, M. J. "A Critical Review of the Models of Group Selection." *Quarterly Review of Biology* 53 (1978): 101–114.

Weir, B. S., and C. C. Cockerham. "Estimating F-Statistics for the Analysis of Population-Structure." *Evolution* 38 (1984): 1358–1370.

Whitlock, M. C. "Dispersal and the Genetic Properties of Metapopulations." In *Causes, Consequences and Mechanisms of Dispersal at the Individual Population and Community Level*, edited by J. Clobert and E. Danchin. Oxford, 2001.

Whitlock, M. C., and N. H. Barton. "The Effective Size of a Subdivided Population." *Genetics* 146 (1997): 427–441.

Wright, S. "Evolution in Mendelian Populations." *Genetics* 16 (1931): 97–159.

— MICHAEL C. WHITLOCK AND YANNIS MICHALAKIS

METAZOA. *See* Animals.

METAZOANS

Metazoans are multicellular animals, from simple organisms such as sponges to architecturally more complex organisms such as squid, insects, and vertebrates. These animals show great morphological variety, from organisms of only a few dozen cells and a few cell types, to humans with a complex nervous system and over two hundred different cell types. A variety of molecular and morphologic evidence establishes that the Metazoa are a single, monophyletic clade. Characteristics that define the Metazoa include the production of collagen, the acetylcholine/cholinesterase system, septate/tight junctions between cells, and several developmental control genes. These features separate animals from their nearest relatives, the choanoflagellates, a group of single-celled eukaryotes. The Metazoa clade traditionally is divided into thirty-two to thirty-six phyla, each recognized on the basis of a common architectural framework, or body plan. Analysis of molecular data since 1990, principally 18S ribosomal RNA gene sequences, has driven a dramatic revision of views on the phylogenetic relationships among the metazoans.

The new views of metazoan phylogeny have revolutionized our understanding of the history of the group. Among the most intriguing insights is the frequency with which similar complex morphological structures such as segmentation have arisen, evidently independently, in very different groups. Even more surprising, to some, is that many seemingly simple animals actually represent forms that have degenerated from more complex ancestors. The frequency of such homology and degeneration explains much of the past difficulty of resolving phylogenetic relationships only with morphological data. Similarly, the use of comparative molecular data on metazoan development has led to some startling discoveries. There appear to be extensive homologies among the transcription factors and signaling transduction molecules that guide the regulation of development. This discovery allows inferences to be made about the nature of the last common ancestor of very disparate groups, shedding insight on events documented by the fossil record of the earliest metazoans.

Theories on the Origins of Metazoans. Since Ernst Haeckel produced the first phylogenetic tree in 1866, metazoan phylogenies have been of two types: model-driven, generated by a hypothesis about the processes of evolution of major animal groups, and character-driven, in which selected morphological, developmental, or, more recently, gene sequence data are employed to define the topology, followed by evolutionary inference. Each approach tends to produce very different views of metazoan relationships, although they blend where character-driven approaches emphasize a particular suite of characters to the exclusion of others. The coelom, a tissue-lined cavity found in many bilaterians, has played a critical role in many discussions of metazoan phylogeny prior to the advent of molecular studies. Sponges and cnidarians are diploblasts, with two thin but well-differentiated cell layers, an outer ectoderm and an inner endoderm separated by a gelatinous layer, the mesoglea. Most other metazoans have three tissue layers (the triploblasts), with a mesodermal layer replacing the mesoglea; most muscles develop from the mesoderm. If no cavity is present within the mesodermal layer, the animals are termed acoelomate, as in flatworms (phylum: Platyhelminthes). A mesodermal cavity not lined with tissue is a pseudocoel, and a tissue-lined cavity is termed a coelom.

Prior to the advent of molecular studies, a common view of metazoan evolution in America and parts of Europe was of a progressive development from two cell layers to three, and from acoelomate to pseudocoelomate and then coelomate. But the coelom forms in very different ways during the development of different groups. Furthermore, some groups may have lost their coelom, becoming pseudocoelomate. In other groups, particularly the molluscs, it has never been entirely clear whether they truly have a coelom. Thus, whether coelomates evolved once or several times was a key point of contention for decades. Similar controversies involved the evolution of segmentation, the ciliary tracts that facilitate feeding in metazoan larvae, and other morphological structures.

Most morphologically based discussions of metazoan phylogeny agree on several points: sponges then cnidarians are the two most basal metazoan lineages; ctenophores are the sister clade to all bilaterian metazoans; the pseudocoelomates are monophyletic and lie either below the protostome/deuterostome divergence or at the base of the protostome clade; annelids and arthropods are closely related; the lophoporate phyla (brachiopods, bryozoans, and phoronids) either are closely related to deuterostomes or are intermediate between deuterostomes and protostomes; and chordates and hemichordates are sister taxa. Recent molecular and morphological analyses have raised questions about all of these conclusions. Only the position of sponges, cnidarians, and the ctenophores basal to bilaterians seems probable, although views of metazoan phylogeny have changed so frequently in recent years that more surprises should be expected.

Metazoan Phylogeny. The view of metazoan relationships shown in Figure 1 contains many significant departures from earlier views. The most recent analyses continue to recognize the sponges (Porifera) and the Cnidaria as the most ancient metazoans, but divide the bilaterally symmetrical animals (the Bilateria) into three large clades: the deuterostomes, including the chordates, hemichordates, and echinoderms; the ecdysozoans, including the arthropods, nematodes, onycophorans, priapulids, and some other groups that shed their exoskeleton during growth; and the lophotrochozoans, including molluscs, annelids, brachiopods, and most other invertebrates. The acoelomate and pseudocoelo-

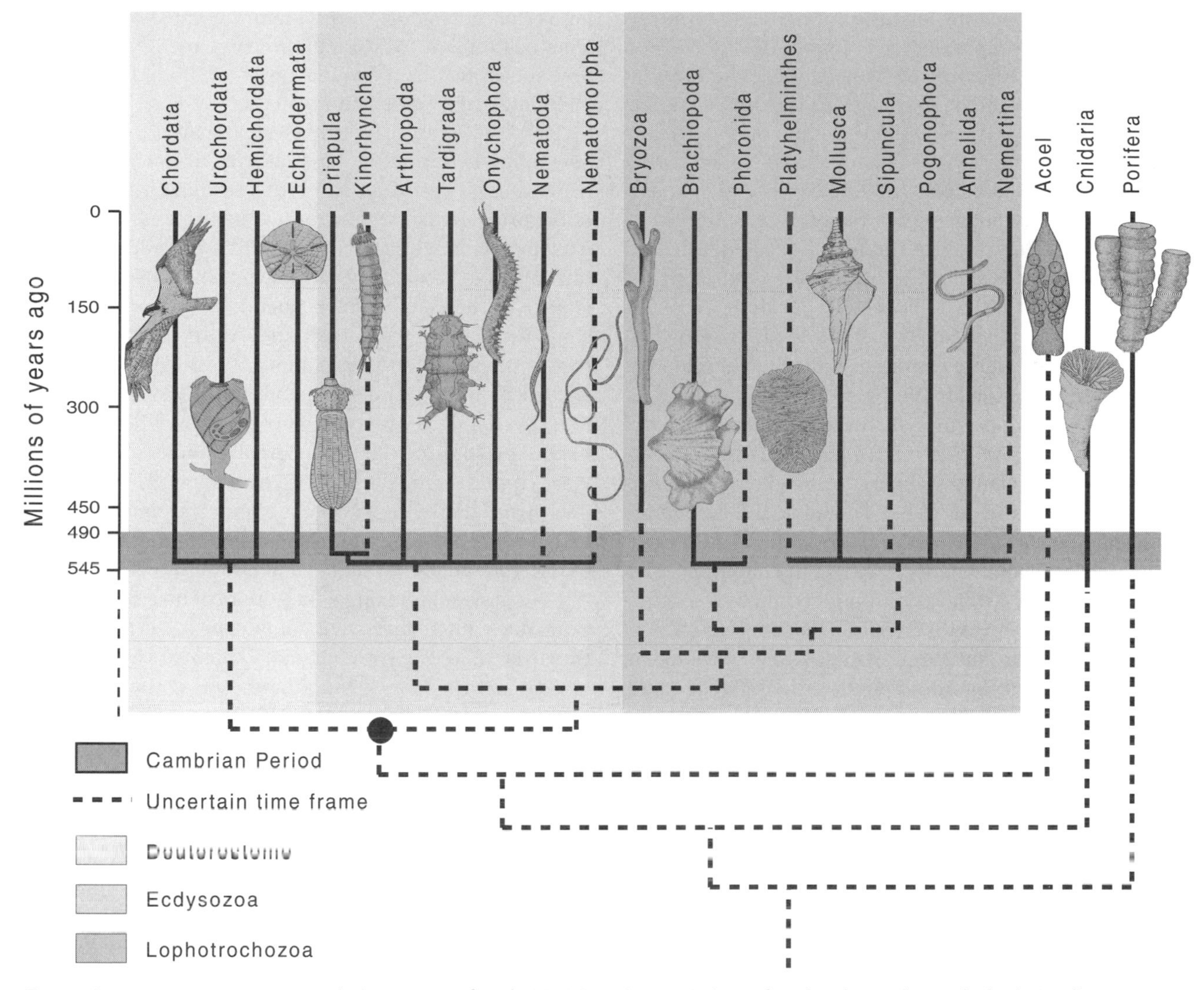

Figure 1. A Consensus Metazoan Phylogeny As of Early 2001 Based on a Variety of Molecular and Morphologic Studies. The bilaterian metazoa divide into three great clades, the deuterostomes, including echinoderms and chordates, and the two protostome clades, the ecdysozoa and the lophotrochozoa. The dot marks the last common ancestor of the protostome and deuterostome lineages. Drawing by Elisabeth Valiulis.

mates have all been distributed among the Ecdysozoa and Lophotrochozoa. Annelids and arthropods anchor the two major subclades of the protostomes, with no particularly close relationship between them, and the lophopohorate phyla nest within the Lophotrochozoa. Such a topology, of course, obliterates the progressive development from acoelomate to pseudocoelomate to coelomate, and the very simple architecture of the clades at the base of the deuterostomes, ecdysozoans and lophotrochozoans, suggests that many complex morphological structures, including the coelom, segmentation, and eyes, must have evolved independently in each group.

An important sidelight of the recent burst of activity on metazoan relationships is the recognition that many highly distinctive clades do not warrant recognition as separate phyla, but nest comfortably within larger clades. Examples include the acanthocephalans within rotifers; pogonophorans, and vestimentiferans within the polychaete annelids; and pentastomids as crustacean arthropods (see Peterson and Eernisse, 2001, for further reading). The monophyly of many other traditional phyla remains unresolved, including the status of sponges, brachiopods, and other groups. Meanwhile, other studies have questioned whether onycophorans belong within Arthropoda, acoel flatworms within Platyhelminthes, and echiurans with Annelida. Some of these uncertainties are due to the lack of adequate molecular data, particularly when some species that have been sequenced exhibit very high rates of molecular evolution. This can cause anomalies in the analysis where high turnover taxa artificially group together.

The Early Evolution of Metazoans. The appearance of metazoans in the fossil record is relatively rapid, beginning with a variety of impressions of soft-bodied organisms known as the Ediacaran fossils from rocks that are from 575 million to 543 million years old. These fossils have been discovered in many parts of the globe and in different marine environments. Some of the Ediacaran fossils are long fronds, often with a central stalk; others are round discs; still others appear superficially segmented. The phylogenetic affinities of these organisms have been widely disputed. Some paleontologists have assigned particular fossils to the arthropods, annelids, and echinoderms, yet these assignments have been questioned by others. Significantly, none of these fossils preserves clear evidence of appendages, a mouth, or internal anatomy. In all but one case, the Ediacaran fossils can be viewed as of the same level of complexity as the Cnidaria, and may well represent cnidarians. The sole exception is *Kimberella*, first described as a cubozoan (a cnidarian) from Australia. The discovery of better preserved material from northern Russia has demonstrated a clear bilaterian affinity for *Kimberella*, and evidence suggests it may be a mollusc. The rocks in which *Kimberella* has been found also contain long, parallel scratches that may be feeding traces.

Such trace fossils are another important clue to the levels of metazoan complexity during this interval. Only very simple horizontal trace fossils have been discovered in rocks between 575 million and 555 million years old, and essentially no verified signs of animal fossils have been found in rocks older than 600 million years. Obvious bilaterian trace fossils are present by 555 million years ago, and with *Kimberella* demonstrate that bilaterian metazoans must have evolved by this time. After 543 million years ago, the base of the Cambrian period, fossils show a rapid increase in complexity and abundance. Simple skeletonized tubes appear first, with an increase in the complexity of trace fossils, followed by the first brachiopods, molluscs, and trilobites. Fortunately, paleontologists have also discovered a number of extraordinary fossil deposits from the Early and Middle Cambrian period where animal soft parts have been preserved. Best known from the Burgess Shale deposit in British Columbia, Canada, these deposits provide a window into the multitude of priapulid and annelid worms, arthropods, sponges, and many other taxa that first appeared during this Cambrian radiation. The Changjiang fauna, a deposit similar to the Burgess Shale from Yunnan Province, China, also includes possible chordates and some fossils that may be very early fish. Thus, in only sixty-five million years, the metazoan radiation progresses from sponges and cnidarians to essentially all readily fossilizable metazoan phyla, including vertebrates. Most of the Cambrian radiation proper, from 543 million to 520 million years ago, involves the first appearance and diversification of bilaterian clades, including arthropods and annelids.

The fossil record is fairly clear: the metazoans diversified rapidly during the interval from about 575 million to 520 million years ago. But is this an accurate depiction of the origin and diversification of metazoans? Evidence from molecular clock analyses paint a very different picture. A molecular clock assumes a fairly even rate of change within molecular sequences. If the dates that some living groups diverged are known from the fossil record, these calibration points and the comparison of sequences from various living groups allow the estimation of other divergences deep in the past. When applied to the question of metazoan origins, molecular clocks yield results from 670 million years ago to over 1.2 billion years ago, and averaging about 800 million years ago. Although the spread of ages is partly due to the use of different genes, methods, and assumptions about the rates of molecular change, the spread of over 500 million years does not inspire great confidence in the technique. However, all of the estimates for both the origin of metazoans and the last common ancestor of the protostomes and deuterostomes considerably predate the estimates from the fossil record. Thus, we must seriously entertain the idea that the origin and initial divergence of metazoan groups is considerably older than suggested by the fossil record.

Identifying the time of origin of the metazoans and some of the early divergences is critical for discriminating between alternative views for the metazoan radiation between 575 million and 520 million years ago. If the origin of the Metazoa group, or more specifically the origin of the Bilateria, is about 600 million years ago, then the metazoan radiation may simply reflect the origin of the group, including the developmental innovations that allowed the evolution of bilaterians. Alternatively, if molecular clock analyses are generally correct and the origin of metazoans and bilaterians predates the Cambrian period by hundreds of millions of years, then much of the early history of metazoans may be missing from the fossil record. In this latter case, the Cambrian metazoan radiation represents the independent appearance and diversification of a number of long-established lineages, and the cause of this event must lie in either changes in the physical environment, such as climate or levels of atmospheric oxygen, or in the development of novel ecological interactions that fueled the evolutionary radiation through a kind of positive feedback.

Comparative developmental studies across the Metazoa have revealed remarkable conservation of regulatory elements, primarily signaling transduction molecules and transcription factors. This information has been interpreted in very different ways. Many developmental biologists have viewed the discovery that similar genes in both mice (deuterostomes) and flies (ecdysozoans) control body patterning, eye and heart formation, segmentation, appendage differentiation, and other features as evidence that the last common ancestor of these two lineages possessed all of these characteristics. If the possession of shared regulatory sequences necessarily implied a conservation of function, this interpretation would be correct. Considerable developmental evidence suggests that this view may be too simplistic. An alternative is that many of these regulatory sequences provide only a very generalized function in the ancestor, and later independently developed the specific functions seen today. Thus, the gene *Pax6* coordinates the development of eyes in flies and vertebrates. But flies and vertebrates have very different eyes, and the most straightforward interpretation is that the ancestor probably had a very simple, light-sensitive structure produced by *Pax6*. Resolving this matter is of considerable interest to paleontologists, for if the common ancestor was as complex as indicated by the first scenario, it would be difficult to hide from the fossil record, and thus unlikely to date to 800 million years ago. The less complex ancestor envisioned by the second scenario, however, is consistent with either a very ancient or a relatively young origin of the metazoans.

[*See also* Body Plans; Multicellularity and Specialization.]

BIBLIOGRAPHY

Aguinaldo, A. M. A., J. M. Turbeville, L. S. Linford, M. C. Rivera, J. R. Garey, R. A. Raff, and J. A. Lake. "Evidence for a Clade of Nematodes, Arthropods and Other Molting Animals." *Nature* 387 (1997): 489–493.

Akam, M. "Arthropods: Developmental Diversity within a (Super) Phylum." *Proceedings of the National Academy of Sciences USA* 97 (2000): 4438–4441.

Budd, G. E., and S. Jensen. "A Critical Reappraisal of the Fossil Record of the Bilaterian Phyla." *Biological Reviews* 75 (2000): 253–295.

de Rosa, R., J. K. Grenier, T. Andreeva, C. E. Cook, A. Adoutte, M. Akam, S. B. Carroll, and G. Balavoine. "*Hox* Genes in Brachiopods and Priapulids and Protostome Evolution." *Nature* 399 (2000): 772–776.

Erwin, D. H. "The origin of bodyplans." *American Zoologist* 39 (1999): 617–629.

Knoll, A. H., and S. B. Carroll. "Early Animal Evolution: Emerging Views from Comparative Biology and Geology." *Science* 284 (1999): 2129–2137. A comprehensive overview of the biological and paleontological aspects of the Metazoan radiation.

Nielsen, C. *Animal Evolution: Interrelationships of the Living Phyla.* 2d ed. Oxford, 2001.

Peterson, K. J., R. A. Cameron, and E. H. Davidson. "Bilaterian Origins: Significance of New Experimental Observations." *Developmental Biology* 219 (2000): 1–17.

Peterson, K. J., and E. H. Davidson. "Regulatory Evolution and the Origin of the Bilaterians." *Proceedings of the National Academy of Sciences USA* 97 (2000): 4430–4433.

Peterson, K. J., and D. J. Eernisse. "Animal Phylogeny and the Ancestry of Bilaterians: Inferences from Morphology and 18S rRNA Gene Sequences." *Evolution and Development* 3 (2001): 1–35. Recent metazoan phylogeny based on both morphology and molecular sequences.

Raff, R. A. *The Shape of Life.* Chicago, 1996. Early introduction to comparative evolutionary developmental biology by one of its founders.

Shubin, N., C. Tabin, and S. B. Carroll. "Fossils, Genes and the Evolution of Animal Limbs." *Nature* 388 (1997): 639–648.

Wilmer, P. *Invertebrate Relationships.* Cambridge, 1990. The last overview of metazoan phylogeny published before the advent of molecular data, but with a good introduction to the major phyla.

— Douglas H. Erwin

METHYLATION. *See* Epigenetics; Genomic Imprinting.

MHC. *See articles on* Modern *Homo sapiens*.

MICROEVOLUTION

Microevolution is a term used to describe genetic change within the lifetimes of populations and species. The complementary term *macroevolution* is used, particularly in paleontology, to describe patterns of diversification at and above the species level, governing issues such as the origin of major evolutionary novelties and rates of species formation in different groups of organisms. It is important to appreciate that the processes that generate microevolution can also account for macroevolution. No unique processes may be required to explain the formation of reproductive isolation or of major changes in morphology, physiology, and behavior that may distinguish higher groups of organisms. Divergence and speciation among organisms is often a gradual process of accumulating microevolutionary change that culminates in final reproductive isolation, although there is disagreement on the extent to which microevolution can explain all macroevolutionary patterns.

In its most general form, microevolution can be characterized as the changing of allele frequencies within populations and species over time. Alleles are alternative copies of the same gene. Frequency change comes about through several processes, and it is helpful to first look at circumstances in which there is no microevolution—that is, in which frequencies do not change. In a large, isolated population of randomly mating individuals, allele frequencies will remain stable as long as there is no mutation and no selection. The Hardy–Weinberg principle states that, after one generation of mating under these circumstances, genotype frequencies in a population of diploid organisms will be $p^2 + 2pq + q^2$, where p and q refer to the properties of two different alleles at any one locus in the population, and $q = 1 - p$, such that $p + q = 1, 0$. Any significant deviation from this equilibrium distribution is an indication that microevolutionary processes are acting on the population.

Five major processes contribute to microevolution: natural selection, genetic drift, gene flow, nonrandom mating, and mutation. Natural selection occurs when environmental conditions act on genetic variation among individuals to cause differences in reproductive success. The alleles of individuals that reproduce at a higher rate will increase in frequency in the next generation. Selection can be *stabilizing* (maintaining the local phenotype), *directional* (selecting for a different mean phenotype), or *diversifying* (favoring two or more different phenotypes). A large body of evidence shows that selection is a major factor in evolution from the level of individual genes to the divergence of populations into distinct species.

Genetic drift is the random change of allele frequencies in small populations. For example, consider two parents, both of whom are heterozygous at a locus; that is, both contain an allele A and an allele a. It is possible just by chance that their offspring will inherit only allele A (or a). This chance fluctuation in gene frequencies is genetic drift. Its effects are small in large populations but increase as population sizes decrease. This is exemplified by the bottleneck effect, in which a population that has been temporarily reduced to a small number of individuals loses many alleles, whereas those remaining increase in frequency. Such loss of genetic variation is of central concern in conservation biology. Another example is the founder effect, where only a few individuals start a new population, for example on an isolated island. Because they represent a small subset of the variation in the source population, and it goes through a bottleneck before population size increases, it will likely experience genetic drift.

Gene flow refers to the movement of alleles into the population through immigrating individuals. As immigrants mate with resident members, new alleles are introduced into the population, changing the frequency of others. Emigration can also lead to changes in gene frequency if certain genotypes are more likely to migrate than others.

Nonrandom mating can have direct and indirect effects on gene frequency. Genes that improve the mating success of their carriers will increase in frequency, a process called natural selection. Other forms of nonrandom mating such as preferential choice of relatives (inbreeding), or individuals with similar phenotype (assortative mating) will increase the frequency of homozygotes, and this may influence the operation of both selection and drift.

Mutations are the source of new alleles, through changes in the genetic material. The most common changes are single nucleotide substitutions and gene duplications. Mutations are a critical source of novelty, but observed rates of mutation are much too low to account in themselves for observed rates of microevolution in organisms.

Only microevolutionary change can be observed directly by evolutionary biologists, but this has now been done on many occasions, providing invaluable insight into the evolutionary process.

[*See also* Natural Selection: An Overview; Neo-Darwinism; Peppered Moth; Pesticide Resistance.]

BIBLIOGRAPHY

Endler, J. A. *Natural Selection in the Wild*. Princeton, 1986. A classic compilation of evidence for the role of natural selection in natural communities.

Futuyma, D. J. *Evolutionary Biology*. 3d ed. Sunderland, Mass., 1998. Comprehensive verbal and mathematical discussion of all aspects of microevolution.

Hendry, A. P., and M. T. Kinnison, eds. *Contemporary Microevolution: Rate, Pattern, Process*. Dordrecht, Boston, and London, in press. Chapters by many authors on different aspects of microevolution.

Lynch, M., and J. S. Conery. "The Evolutionary Fate and Consequences of Duplicate Genes." *Science* 290 (2000): 1151–1155. A good example of how mining genomic databases can contribute to understanding microevolutionary processes.

Weiner, J. *The Beak of the Finch: A Story of Evolution in Our Time*. New York, 1994. A highly readable account of a field study of microevolution on the Galápagos Islands over two decades; winner of the 1995 Pulitzer Prize for nonfiction.

— OLLE PELLMYR

MIMICRY

Mimicry is the parasitic or mutualistic exploitation of a communication channel. More plainly, the term describes the situation in which one organism gets the better of another organism (known as the dupe) by looking, smelling, sounding, or feeling like something else. Mimicry in this sense is very widespread, operating as it can through any of the senses of the dupe and involving all kinds of adaptation, from antipredator defense and predator stealth to courtship and reproduction.

The best understood and most thoroughly studied variety of mimicry is employed by prey against predators that operate by their sense of sight. Very familiar to everyone is the mimetic resemblance of insects to various bits of vegetation—twigs, leaves, even dead and fallen flowers—or to bird droppings. Because insectivorous birds do not feed on plants or dung, this resemblance greatly reduces the rate at which the insect is attacked. This can be called mimetic camouflage. It differs from simple camouflage—resemblance in color to the general background, or optical tricks that reduce the

FIGURE 1. The Patterns of the Six Mimicry Rings Among the Long-Winged Butterflies of the South American Rain Forests. The colors, as well as the patterns, are very different. © John R. G. Turner.

amount of shadow around the animal or break its outline into a meaningless jumble of colors—in the way in which the potential predator processes the information received from the prey. A prey that has simple camouflage is not perceived at all as a distinct object; a mimetic prey, in contrast, is perceived within the visual field, then ignored because it is mistaken for something else. The distinction is quite subtle, and the categories must merge.

Prey also reduce their edibility by containing unpleasant and often poisonous substances (often derived from the host plant). They are then often brightly colored and rather rapidly learned and recognized by predators, and are described as warningly colored or aposematic; they may then be mimicked by palatable species. Familiar examples are the many flies and other harmless insects that resemble bees and wasps, as well as the viceroy butterfly (*Limenitis archippus*) of North America, which resembles the poisonous monarch (*Danaus plexippus*).

Batesian and Müllerian Mimicry. The dynamics and the evolution of defensive mimicry are driven by the psychology of the predators. Predators do not divide their food simply into the palatable and the unpalatable: there are all degrees of palatability in between. This raises interesting questions about the nature of mimicry. Extremely palatable mimics will be parasitic on unpalatable species. If they meet a predator that has learned to avoid the unpalatable model (as the copied organism is called), they will benefit, suffering a lower rate of predation than otherwise similar forms that are not mimetic. The model itself, however, will suffer any time it encounters a predator that has just learned from eating a mimic, that some of the prey with this particular pattern are actually edible. If two equally unpalatable species resemble each other, they will be mutualists: predators encountering either of them will reduce their attacks on both. Conventionally, the mimicry of a defended (i.e., unpalatable) species by a palatable one is called Batesian mimicry, and the mutual resemblance of two or more defended species is known as Müllerian mimicry (after their Victorian discoverers, the naturalists Henry Walter Bates and Fritz Müller).

Batesian mimicry has always been regarded, correctly, as parasitic. The analogous view that Müllerian mimicry was similarly a simple matter of mutual protection has been based on some rather simplistic assumptions, made by Müller himself early in the history of our understanding of learning behavior. It is at least likely that, because of the way in which vertebrates assess the costs and benefits of any action, Müllerian mimicry is only fully mutualistic when the mimics are broadly equal in their unpleasantness. When one is much less unpleasant than the other, then, although both are defended, the

FIGURE 2. The South American Batesian Mimic *Consul fabius* Has a Bright Mimetic Pattern (orange, black and yellow) Only on Its Upper Surface.
The lower surface (shown on the right) is colored like a dead leaf. © John R. G. Turner.

relationship tends to become parasitic. It is uncertain in many cases which kind of mimicry we are encountering: the classic viceroy butterfly, although not poisonous like the monarch (and even some individual monarchs lack their poisons), is somewhat unpalatable and perhaps sometimes much more unpleasant than the Florida queen (*Danaus gilippus*), which it also mimics in some areas. When is the viceroy a parasite, and when is it a mutualist or even the model?

Mimicry Rings. Mimicry is seen to its best advantage in the tropics, with their very high species diversity. Here, defended insects rather seldom have their own independent warning patterns, but are instead grouped into "mimicry rings": most of the warningly colored butterflies in any one area will be found to share only about six patterns, and each of the rings contains butterflies (and moths) from several different families (Figure 1). Many of them are defended and therefore mutualists, but there are also Batesian mimics that act as parasites within the ring. In the temperate zone, mimicry rings are most easily observed among hymenoptera: the many bumble bees of western Europe exhibit only five color patterns.

Two Phase Evolution of Mimicry. Mimicry was discovered shortly after Charles Darwin published *On the Origin of Species* (1859). In addition to being a particularly elegant illustration of adaptation, it has been used repeatedly to investigate, challenge, or support various theories about the mechanism of evolution. Extensive experiments have shown broadly how natural selection favors mimetic patterns: the advantage of mimicry has been demonstrated in the laboratory and in the wild, using both artificial and real prey (butterflies painted over with new patterns), presented to both wild and captive birds.

The evolution of mimetic camouflage seems to present few problems. The evolutionist Stephen Jay Gould famously asked, "What is the advantage of a 5 percent resemblance to a turd?" suggesting that such an adaptation could not evolve gradually. Richard Dawkins responded cogently in *The Blind Watchmaker* that mimetic camouflage could evolve readily from simple camouflage because, although imperfect resemblance was never as good as perfect resemblance, any small change toward resembling an inedible object would be of some small advantage some of the time: 5 percent resemblance would not give 100 percent protection, but it would give more protection than 2 percent resemblance.

This argument seems to break down with Batesian and Müllerian mimicry, because the adoption of bright colors makes the potential mimic conspicuous, and therefore vulnerable, unless the resemblance to the model is also quite accurate from the start. This presents a considerable problem for evolutionists determined to maintain that evolution is always entirely gradual. There are two solutions.

First, on becoming a mimic, an organism does not necessarily increase its conspicuousness, and if it does, mimicry will simply not evolve. Thus, Batesian and Müllerian mimicry are most widespread in organisms that are already conspicuous, such as flying insects. It is much more often reported in butterflies than in caterpillars, and virtually absent in butterfly pupae, which remain motionless. Batesian mimicry especially is often confined to the upper surface of the butterfly's wings: the underneath, which is all that can be seen when the butterfly is at rest, remains camouflaged (Figure 2).

Second, at least some of the time, mimicry does evolve through a relatively large change in pattern: ge-

FIGURE 3. The Parallel Mimicry Between Some of the Races of *Heliconius erato* (right of each pair) and *Heliconius melpomene* (left). © John R. G. Turner.

netic mutations range in their effects from the minute through the striking to the catastrophic. Some North American yellow swallowtail butterflies achieve mimicry of the black pipevine swallowtail (*Battus philenor*) by means of a single mutation that simply floods the wings with black pigment. The use of rather large mutations has allowed various Müllerian mimics in the butterfly genus *Heliconius* to switch their allegiance from one mimicry ring to another in South America, resulting, in *Heliconius erato* and *Heliconius melpomene* (the "postman" butterflies, which are the staple of captive butterfly displays), in a spectacular racial divergence of the patterns within the two species, while they have remained in strict mutual mimicry of each other (Figure 3). Evolution within each species has been through the addition or subtraction of large parts (although not of the whole) of the pattern by single gene mutations. These are able to alter large amounts of pattern simultaneously probably because they affect processes occurring early in the development of the wing.

Mutations of large effect may thus be employed in evolution to cross an "adaptive gap" between one pattern and another. How common such gap jumping is in evolution in general is a matter of debate, as is the possibility that many adaptive changes may commence with a rather large step even when there is no gap to be crossed. It is also possible that new warning patterns sometimes become established through the chance occurrence of mutations when the populations are very small: an example of the efficacy of random genetic drift.

Once this first phase in the evolution of a new mimetic pattern is complete, then the pattern, which is likely only to be an approximate mimic, is free to evolve gradually toward an increasingly refined resemblance. The English physician Sir Cyril Clarke and the geneticist Philip Sheppard demonstrated with great clarity the genes involved in this phase using the Batesian mimic *Papilio dardanus*. Mimicry is thus thought usually to evolve in two phases: a large step, followed by gradual change.

Evolution in Polymorphic Batesian Mimics. A palatable, parasitic mimic is expected to suffer an increased rate of predation the more common it is relative to its model. This implies that if the mimic can produce a new form, mimicking a different model, the new form, being rare, will be at an advantage and will increase its frequency in the population. As it becomes more common, its relative advantage will wane, until at some point both forms are equally fit: this is a stable equilibrium, and both will persist indefinitely in the population. Hence some Batesian mimics are polymorphic, having several forms, each belonging to a different mimicry ring. The most spectacular of these are the African swallowtail *Papilio dardanus*, the southeast Asian swallowtail *Papilio memnon*, and the African *Pseudacraea eurytus*, which can each have up to five or six distinct forms in the same population (Figure 4).

FIGURE 4. Mimetic Forms of the African "Mocker" Swallowtail.
Shown are the four forms of *Papilio dardanus* with their models (left half of each butterfly). The models are (top to bottom): *Amauris niavius, Amauris echeria, Othroeda hesperia,* and *Bematistes poggei.*

The ability of a species to build up spectacular polymorphisms of this kind depends on some special architecture in the genes that control the mimetic patterns. For example, *Papilio memnon* has forms that match both the body and the wing colors of their different models. Wing- and body-color are controlled by different genes, and if these were to be inherited separately, forms would be produced with the wing color of one model and the body of another. Birds are observant enough to detect rearrangements of this kind and would be likely to attack them at an increased rate. This outcome is prevented by the genes being carried close together in the same chromosome. Indeed, almost the whole of the mimetic color of the various forms of *Papilio memnon* and of *Papilio dardanus* is controlled by a bank of around five very closely spaced genes. An interesting evolutionary history must lie behind such a piece of adaptation—not only the patterns of the butterflies but also the arrangement of the appropriate genes has been molded by evolution. It is likely that the location of the genes on the same chromosome arises initially because genes with related functions originate by the duplication of a single original gene, and that the spectacular polymorphisms of the type seen in these few Batesian mimics arise only in those species that happen to have the genes linked in the appropriate way. This is an example of a probably important evolutionary mechanism known as a sieve—only those species that have the appropriate genetic architecture in advance can evolve in a particular way.

Mimicry and Speciation. A new mimetic pattern can lead to the origin of a new species (as was suggested by Bates). *Heliconius melpomene* in all its various forms always has some red marks, which are a signal in courtship. Populations of this butterfly that have changed to mimic a purely blue-and-white model have thus become reproductively isolated from their red ancestors and constitute the separate species *Heliconius cydno.*

BIBLIOGRAPHY

Bates, H. W. "Contributions to an Insect Fauna of the Amazon Valley. Lepidoptera: Heliconidae." *Biological Journal of the Linnean Society* 16 (1981): 41–54. A shortened version of the paper (1861) that first described mimicry. The subsequent literature on mimicry is both enormous and diffuse.

Brower, L. P. "Chemical Defence in Butterflies." In *The Biology of Butterflies (Royal Entomological Society of London Symposium No. 11)*, edited by R. I. Vane-Wright and P. R. Ackery, pp. 109–134. New York, 1984. Relates mimicry and defense to the chemical relations between insects and their host plants.

Dawkins, R. *The Blind Watchmaker.* New York, 1987. Argues that natural selection operates gradually on small changes, using mimetic camouflage as an example.

Endler, J. A. "Interactions between Predators and Prey." In *Behavioural Ecology: An Evolutionary Approach*, edited by J. R. Krebs and N. B. Davis, 3d ed., pp. 169–196. Oxford, 1991. Mimicry is discussed in the context of the ecology and behavior of predation.

Ford, E. B. *Ecological Genetics.* 4 eds. London, 1964–1975 Chapter 13 is a clear summary of the evolutionary genetics of mimicry in swallowtail butterflies.

Gilbert, L. E. "Coevolution and Mimicry." In *Coevolution*, D. J. Futuyma and M. Slatkin, pp. 263–281. Sunderland, Mass., 1983. A thoughtful introduction to some evolutionary questions raised by mimicry and its ecology.

Kimler, W. C. "Mimicry: Views of Naturalists and Ecologists before the Modern Synthesis." In *Dimensions of Darwinism*, edited by M. Greene, pp. 97–127. New York, 1983. The early history of the study of mimicry.

Mallet, J. L. B., and M. Joron. "The Evolution of Diversity in Warning Color and Mimicry." *Annual Review of Ecology and Systematics* 30 (2000): 201–233. Old controversies about mimicry may be solved: this exciting article describes some new ones, presenting the theory that random genetic drift has a major part to play in the diversification of warning patterns and the evolution of mimicry rings.

Mallet, J. L. B., and J. R. G. Turner, "Biotic Drift or the Shifting Balance—Did Forest Islands Drive the Diversity of Warningly Colored Butterflies?" In *Evolution on Islands*, edited by P. R. Grant, pp. 262–280. Oxford, 1997. A detailed debate about the ways the extraordinary mimicry between *Heliconius erato* and *Heliconius melpomene* may have evolved.

Turner, J. R. G. "Butterfly Mimicry: The Genetical Evolution of an Adaptation." In *Evolutionary Biology*, edited by M. K. Hecht, W. C. Steere, and B. Wallace, vol. 10, pp. 163–206. New York and London, 1977. A technical summary of research on the genetics of mimetic butterflies.

Turner, J. R. G. "Mimicry: The Palatability Spectrum and Its Consequences." In *The Biology of Butterflies (Royal Entomological Society of London Symposium No. 11)*, edited by R. I. Vane-Wright and P. R. Ackery, pp. 141–161. New York, 1984. An expanded account of evolution in *Heliconius* and *Papilio* for the general biologist.

Turner, J. R. G. "Fisher's Evolutionary Faith and the Challenge of Mimicry." In *Oxford Surveys in Evolutionary Biology*, edited by R. Dawkins and M. Ridley, vol. 2, pp. 159–196. Oxford, 1985. The great English evolutionist and statistician R. A. Fisher was committed to a view that evolution through natural selection occurred by very small steps and that it represented God's continuing creation of the world. He met the challenge posed by the evolution of mimicry with consummate skill.

Turner, J. R. G. "The Genetics of Adaptive Radiation: A Neo-Darwinian Theory of Punctuational Evolution." In *Patterns and Processes in the History of Life*, edited by D. M. Raup and D. Jablonski, pp. 183–207. Berlin, 1986. Discusses whether the large steps in the evolution of mimicry are an example of evolution by punctuated equilibrium.

Wickler, W. *Mimicry in Plants and Animals.* Translated by R. D. Martin. London, 1968. An elegant introduction to the great diversity of mimicry in all its forms; the best book for the nonspecialist professional reader as well as the general naturalist.

— John R. G. Turner

MISSING LINKS

In its most general sense, the term *missing link* refers to the discovery of an organism, usually a fossil, that occupies an intermediate position in the record of two otherwise separate lineages. Obvious examples are the birdlike reptile (or reptilelike bird) *Archaeopteryx* and the numerous primate fossils filling the gap between the records of modern humans and the rest of the great apes.

Strictly, "missing link" is a metaphor. Used literally, it is an oxymoron, because once the link has been found, it is no longer missing. As a predictive term, it should refer to something that is hypothesized but has not yet been found. In fact, the term is most frequently used to describe the discovery itself and the confirmation of hypothesis. In rare cases the term is used in reverse, with a discovery indicating a pattern of relationship previously unsuspected. There is also possibility for overlap and confusion with the term *living fossil*. For example, the archaic mollusc *Neopilina galathea*, representing a living relic of an ancient group linking annelids and molluscs, is both a living fossil and a missing link, whereas the living fossil *Latimeria chalumnae* turns out only to be another coelacanth.

Missing link is a concept that can be applied to any element, not necessarily ancestral, in a series or group of related entities. The great British geologist Charles Lyell, in his *Manual of Elementary Geology* (1852), referred to the absence of particular fossil species in the sequence of beds of the Lower Greensand, Isle of Wight, as a "break in the chain implying no doubt many missing links in the series. . . ." The term came into its modern usage, however, with the advent of Darwinism. Charles Darwin's theory explicitly proposes a continuity among all organisms (living and fossil), but he well knew that the record of life is flawed. Thus, he introduced the concept in relation to possible objections to his theory: "[I]nnumerable transitional forms must have existed, [so] why do we not find them embedded in countless numbers in the crust of the earth?" (*On the Origin of Species*, 1859).

Gaps due to the imperfection of the fossil record may in part be closed by further research. The discovery of Neanderthal man (1857) and *Pithecanthropus erectus* (1895) gave scientific authority to evolution in one its most sensitive areas—the origins and relationships of *Homo sapiens*. The term *missing link* became closely associated with Thomas Henry Huxley, especially after the publication of his *Man's Place in Nature* (1863). Unfortunately, however, the term even more rapidly gained a wide currency as a term of derogatory humor and ethnic abuse. For example: "It is hoped that the discovery, in the Irish Yahoo, of a Missing Link between Man and the Gorilla, will gratify the benevolent reader, by suggesting the necessity of an enlarged definition of fellow creatures, conceived in a truly liberal and catholic spirit" (*Punch*, 1862).

The successes of modern paleontology are due in no small part to the fascination of the public with the metaphor of the missing link. In addition to the search for human fossils, quests, such as for the origins of birds from dinosaurs and of tetrapods from fishes, have also produced extraordinary results in the past thirty years. Even so, experience usually shows that missing links do not close the relevant gap; they merely narrow or redefine it. Most often, each discovery reveals another gap, presumably filled by some other missing link(s). Thus, one is forced to return to Charles Darwin's question: Why do these gaps in the record exist?

Obviously, the fossil record will always be patchy and imperfect. As Darwin noted, the process of fossilization is itself highly contingent and sporadic, and the vagaries of what will be preserved and eventually found are equally obvious. Thus, even the fossil record of a well-documented group has gaps. It is also widely believed that there is something special about the transitional process between any two lineages of differently adapted organisms with relatively rapid change in adaptive morphology during the occupation of a radically new environmental niche. "Intermediate varieties, from existing in lesser numbers than the forms which they connect, will generally be beaten out and exterminated during the course of further modification and improvement" (Darwin, *On the Origin of Species*, 1859). The transitional forms will be less diverse and less numerous than the ancestral or descendant groups. Not until the new adaptive morphology and physiology have been tried and tested does broad diversification begin again. Although this is an attractive view, the evidence for it is largely negative; the imperfection of the record may simply mask everything else. Meanwhile, missing link remains a powerful metaphor, helping to drive the parallel processes of hypothesis, discovery, and popularization in evolution.

[*See also* Transitional Forms.]

BIBLIOGRAPHY

Desmond, Adrian. *Huxley: From Devil's Disciple to Evolution's High Priest*. Reading, Mass., 1994. A comprehensive overview of Huxley's career, with much attention to the apeman/human evolution debate.

Corbey, Raymond, and Bert Theunissen eds. *Ape, Man, Apeman: Changing Views since 1600*. Leiden, 1993. A study of Eugene duBois's contributions to evolution through the discovery of *Pithecanthropus erectus*, with useful material on the role of the orangutan in early considerations of the relationships of *Homo sapiens*.

— KEITH S. THOMSON

MITOCHONDRIA. *See* Cellular Organelles.

MITOSIS. *See* Meiosis.

MODERN *HOMO SAPIENS*

[*This entry comprises three articles:*

An Overview
Human Genealogical History
Neanderthal–Modern Human Divergence

The introductory article discusses the phylogenies and origins of modern humans, including paleontological, archaeological, and genetic evidence; the second article reviews the recent results from coalescence studies and explains the difference between gene trees and population history; the third article discusses DNA sequence divergence between modern humans and Neanderthals. For related discussions, see the Overview Essay *on* Human Genetic and Linguistic Diversity *at the beginning of Volume 1*; Hominid Evolution, *article on* Neanderthals; *and* Human Evolution.]

An Overview

For many decades, paleoanthropologists have pondered whether modern humans arose simultaneously in Africa and Eurasia in places that nonmodern humans had already occupied or in a more limited area. The debate continues, but fresh fossil, archeological, and genetic findings increasingly point to an exclusively African origin. From Africa, modern humans spread to Eurasia, where they replaced or swamped their nonmodern contemporaries. It is particularly clear that African immigrants replaced the Neanderthals in Europe and western Asia beginning about fifty thousand years ago, but it is less obvious that they replaced nonmodern people in eastern Asia. It is not that eastern Asia suggests a contrary result, but that it presents too few data for any persuasive conclusion.

Many important details remain to be established, including above all how modern humans managed to spread from Africa. Archaeology suggests that the most essential factor was the abrupt development of fully modern behavior in Africa roughly fifty thousand years ago. It is only after about fifty thousand years ago that the archaeological record unequivocally reflects the technological ingenuity, social formations, and ideological complexity of historic hunter-gatherers. Social or demographic changes could explain the sudden African origin of modern human behavior, but the simplest explanation may be a highly advantageous genetic mutation that promoted the fully modern human brain.

Fossils, Genetics, and Modern Human Origins. The genetics of chimpanzees and living people show that their last shared ancestor lived between perhaps eight and four million years ago with different studies giving different ranges of dates. The oldest known people, identified mainly by legs that were shaped for bipedal (habitual two-legged) locomotion, date to about four and a half million years ago. All subsequent humans until roughly two and a half million years ago were essentially bipedal apes, who retained small, ape-sized brains and long arms that were well suited for tree climbing. Brain expansion beyond the ape range first occurred about two and a half million years ago, and enlarged brains help explain the simultaneous appearance of the oldest known archaeological sites. These comprise clusters of stone artifacts that signal a distinctively human reliance on technology. They are accompanied by fragmented animal bones that imply an equally humanlike interest in animal tissue for food. The earliest people lived exclusively in Africa, but sometime between 1.8 million and 1.0 million years ago, a primitive species of the genes *Homo* spread to Eurasia. Fossils show that the far-flung descendants of this species then began to diverge in anatomy. By about 500,000 years ago, the divergence resulted in at least three distinct evolutionary lineages: *Homo sapiens* (the ancestors of modern humans) in Africa, *H. neanderthalensis* (the Neanderthals) in Europe, and *H. erectus* (Java man and Beijing man) in eastern Asia.

The subsequent spread of *H. sapiens* to Eurasia underlies the "Out of Africa" theory of modern human origins. However, the theory in fact depends as much on findings in Europe as in Africa. Between 500,000 and 100,000 years ago, European populations progressively accumulated specialized Neanderthal features, culminating in the appearance of the classic Neanderthals by 100,000 years ago. Classic Neanderthal specializations do not anticipate modern human anatomical features, and Neanderthal anatomy remained unchanged until the Neanderthals were replaced by invading *H. sapiens* beginning about 45,000 years ago.

The African fossil record between 500,000 and 100,000 years ago is much less complete than the European one, but it contains no specimens that anticipate the Neanderthals, and it shows that anatomically modern or near-modern people were widespread in Africa and its immediate Southwest Asian periphery by 100,000 years ago. It is above all the anatomical contrast between Africa and Europe roughly 100,000 years ago that supports the "Out of Africa" theory. The East Asian fossil record is the most sketchy, and it may actually comprise two distinct evolutionary pathways. The first suggests continuity within Southeast Asian *Homo erectus* (Java man) from before 500,000 years ago until perhaps 50,000 years ago. The second may indicate evolution within Chinese *H. erectus* (Beijing man) from populations before 500,000 years ago that resembled Southeast Asian *H. erectus* to people who by 100,000 years ago resembled African *H. sapiens* in some aspects of skull form.

The "Out of Africa" theory depends mainly on the great antiquity of modern or near-modern anatomy in

Africa, but it is also supported by two compelling genetic discoveries. These are, first, the remarkable genetic homogeneity of living humans, which implies that they all shared a common ancestor sometime between 200,000 and 50,000 years ago, and, second, the striking genetic contrast between living humans and Neanderthals, which indicates that they last shared a common ancestor about 500,000 years ago. The second finding is based on the recovery of mitochondrial DNA from Neanderthal fossils in central and eastern Europe. DNA has not yet been extracted from early modern or near-modern African fossils, but if the "Out of Africa" theory is correct, future studies will show that Africans of 100,000 years ago were genetically more similar to living people than to, for example, Neanderthals.

Archaeology and the "Out of Africa" Theory. An obvious objection to the "Out of Africa" theory is the failure of modern or near-modern humans to expand from Africa immediately after they appeared, by about 100,000 years ago. Instead, they seem to have been confined to Africa and to a small sliver of adjacent southwestern Asia—in what is now Israel—until roughly fifty thousand years ago. It is even possible that Neanderthals replaced the Israeli near-moderns when the regional climate turned much cooler after eighty thousand years ago. Archaeology offers a partial solution to the puzzle. The people who inhabited Africa between 100,000 and 50,000 years ago may have been modern or near-modern in anatomy, but in behavior they closely resembled the Neanderthals and other archaic humans.

Admittedly, Africans of 100,000 to 50,000 years ago were superb stone shapers. They often collected naturally occurring iron and manganese compounds that they could have used as pigments; they apparently built fires at will; they buried their dead, at least on occasion; and they routinely acquired large mammals as food. In all these respects, and perhaps others, they may have been advanced over yet earlier, archaic people. However, in common with earlier people and with their Neanderthal contemporaries, they manufactured a relatively small range of recognizable stone tool types; their artifact assemblages varied remarkably little through time and space (despite remarkable environmental variation); they obtained stone raw materials overwhelmingly from local (vs. far distant) sources (suggesting relatively small home ranges or very simple social networks); they rarely if ever utilized bone, ivory, or shell to produce formal artifacts; they left little or no evidence for structures or any other formal modification of their campsites; they were relatively ineffectual hunter-gatherers, who lacked, for example, the ability to fish; their populations were apparently very sparse, even by historic hunter-gatherer standards; and perhaps above all, they left little or no compelling evidence for art or decoration.

If the "Out of Africa" theory is correct, we would expect Africa to provide the oldest secure evidence for art, for the routine manufacture of artifacts from bone and shell, and for other indicators of modern human cognitive abilities. The oldest cultures that reflect these abilities are assigned to the Later Stone Age in Africa and the Upper Paleolithic in Europe and western Asia. There are few reliable dates for the beginning of the Later Stone Age and the Upper Paleolithic, but provisionally, the available dates indicate that modern human behavioral markers appeared first in Africa, probably between fifty thousand and forty-five thousand years ago, that they spread to western Asia and eastern Europe by forty-three thousand years ago, and that they reached western Europe about forty thousand years ago. The geographic sequence is clearly consistent with an expansion of modern humans from Africa beginning about fifty thousand years ago.

The time when modern behavior appeared in eastern Asia remains debatable, and some have argued that modern behavioral markers appear in the Far East only about ten thousand years ago or even later. In fact, however, the East Asian archaeological record is as sketchy as the fossil record, and it is marked more by an absence of evidence than by evidence for an absence. Much fresh archaeological research will be necessary to determine whether the pattern in eastern Asia was truly different than in Africa, western Asia, and Europe.

Range Expansion. The behavioral advances that occurred about fifty thousand years ago allowed fully modern humans to become the first people in northeastern Eurasia, the Americas, and probably Australia. To begin with, between forty thousand and thirty thousand years ago, modern humans settled widely in the harsh continental reaches of northeastern Europe, at and above the latitude of Moscow, where the Neanderthals never lived and where fierce winters challenge people even today. Less than fifteen thousand years later, they had occupied the even more forbidding environments of neighboring Siberia, and by about twelve thousand years ago, they had reached its northeastern corner. Archaeology shows that their spread was promoted by several ingenious innovations, including well-built houses, technologically advanced fireplaces, and tailored clothing.

Once people inhabited northeastern Siberia, they would naturally have spread to Alaska across a broad land bridge that stood where the Bering Strait exists now. They could then have expanded southwards to fill the vast, open expanses of North and South America. The timing of human arrival in the Americas remains controversial, but despite extensive exploration and commercial activity over many decades, the oldest uncontested archaeological sites are less than about twelve thousand years old. Unambiguous older sites may exist, but if so, they are much harder to find than older, often much older sites in Africa and Eurasia. The explanation might be that Americans before about twelve thousand years ago were many times sparser than their African and Eurasian contemporaries, but it is reasonable to conclude on current evidence that humans only arrived in the Americas after about thirteen thousand years ago.

The first Americans encountered a rich mammalian fauna whose diversity resembled that of the historical African savannas. Sometime between twelve thousand and ten thousand years ago, roughly two-thirds of the indigenous American species, including elephants, horses, and camels, abruptly disappeared. The reason may have been climatic change in the transition from the last glaciation to the present interglacial (or Holocene) epoch, but the same kind of change had occurred many times earlier without massive extinctions. The most conspicuous difference about ten thousand years ago was the presence of a powerful new human predator with which native American species had little experience. The sum suggests that the first Americans anticipated much later humans in their ability to mimic a geologic force.

In contrast to the Americas, Australia was never connected to other continents during the entire five- to eight-million-year interval of human evolution. Migrants from mainland Southeast Asia could have island hopped, but even during periods of lower sea level, they would have had to make several voyages of roughly 30 kilometers and at least one of 90 kilometers. The first Australians needed boats that could remain at sea for several days before becoming waterlogged, and the colonization of Australia therefore probably required the same innovative ability that promoted the first human adaptation to northeastern Eurasia. If so, the oldest Australian sites should likewise postdate about forty thousand years ago. Until recently, this appeared to be the case, but there are now claims for sites that antedate fifty or even sixty thousand years ago. These claims are controversial, but if they are sustained, they would indicate that modern humans arrived in eastern Asia before they arrived in Europe. Perhaps there were at least two "Out of Africa" migrations, one through the Sinai Desert to western Asia and Europe, and a second possibly across the southern tip of the Red Sea to southern and eastern Asia. Much fresh research will be necessary to demonstrate this second migration and also to explore the likelihood that in common with the first Americans, the first Australians precipitated a wave of large mammal extinctions. The Australian mammal fauna was far richer before people arrived, but in contrast to the Americas, Australia has yet to provide a site that demonstrates the overlap of people and the extinct species.

The uncertainty surrounding the timing of human arrival in Australia illustrates a fundamental problem in the study of modern human origins: it is far easier to develop explanations for important prehistoric events than to confirm basic facts on which the explanations will survive or fail. A fuller understanding of modern human origins thus depends primarily on field work designed to obtain fresh, well-dated fossils and artifact assemblages.

[*See also articles on* Agriculture; Cultural Evolution, *article on* Cultural Transmission; Globalization; Hominid Evolution, *articles on* Early *Homo and* Archaic *Homo sapiens*.]

BIBLIOGRAPHY

Aitken, M. J., C. B. Stringer, and P. A. Mellars, eds. *The Origin of Modern Humans and the Impact of Chronometric Dating.* Princeton, 1993.

Akazawa, T., K. Aoki, and O. Bar-Yosef, eds. *Neandertals and Modern Humans in Western Asia.* New York, 1998.

Bar-Yosef, O., and D. R. Pilbeam, eds. *The Geography of Neanderthals and Modern Humans in Europe and the Greater Mediterranean.* Cambridge, Mass., 2000.

Clark, G. A., and C. M. Willermet, eds. *Conceptual Issues in Modern Human Origins Research.* New York, 1997.

Klein, R. G. *The Human Career: Human Biological and Cultural Origins.* 2d ed. Chicago, 1999.

Mellars, P. A., and C. B. Stringer, eds. *The Human Revolution: Behavioural and Biological Perspectives on the Origins of Modern Humans.* Edinburgh, 1989.

Nitecki, M. H., and D. V. Nitecki, eds. *Origins of Anatomically Modern Humans.* New York, 1994.

Stringer, C. B., and C. Gamble. *In Search of the Neanderthals.* New York, 1993.

Tattersall, I. *Becoming Human: Evolution and Human Uniqueness.* New York, 1998.

Wolpoff, M. H., and R. Caspari. *Race and Human Evolution: A Fatal Attraction.* New York, 1996.

— RICHARD G. KLEIN

Human Genealogical History

Human genealogical history is the description of the size and structure of the population of human ancestors through time. There are roughly six billion humans alive today, but many studies of diversity in human DNA show a range of variations that would be expected in a population of only twenty to thirty thousand. The estimate from gene differences is a kind of average over the past million years or so, suggesting that human demographic success has been recent and that the founding population of our species was small—perhaps only several thousand individuals—for hundreds of thousands of years.

Because the genus *Homo* occupied much of the temperate Old World for over a million years and because technologies were shared over large areas, the idea of a small deme isolated for hundreds of thousands of years is not plausible. A more likely scenario is that the genetic signature of a bottleneck is a consequence of the way that anatomically modern humans spread and interacted with other archaic populations about one hundred thousand years ago. I will describe how our genetic code bears a signature of ancient population dynamics, review the current evidence from different genetic systems, then describe current thinking about the origin of the supposed bottleneck.

We can think of the genealogies of populations, individuals, and of genes—that is, of segments of DNA. A genealogy of populations would describe the sequence of splits that have occurred since the original separation of the ancestral population. This would be like the description of the history of a group of species. But populations not only separate, they exchange members after separation (called gene flow) and they amalgamate. Although many so-called population trees appear in the literature, they are only suggestive metaphors and never real representations of history. For example, we could not put the Spanish-speaking populations of the Americas into such a tree because they are an amalgamation of long-separated groups from Europe and the New World.

A genealogy of an individual is the list of his ancestors. Each individual has 2 parents, 4 grandparents, and so on. If generations are 25 years long, a newborn today has 16 ancestors born at the turn of the twentieth century. These ancestors in turn have 256 ancestors born in 1800, and they have 65,536 ancestors born in 1600. A newborn has roughly four billion ancestors in the year 1200, many times more than the human population of the earth at that time. This means that the same person must occur at many different places in any human genealogy. We are all multiply inbred, and we each share many ancestors with any other human. Because of these shared ancestors, genealogies of genes are like trees: as we follow a sample of genes backward in time, the lines of descent coalesce in shared ancestors until there is a single most recent common ancestor from which all the genes in a sample are descended. This is called the "coalescent" of the gene tree, while nodes below the coalescent are called "coalescences."

Calculations like those in the Coalescence Theory vignette, from numerous systems, are the basis for the accepted estimate of 10,000 for the genetic effective size of our species. (Effective size is roughly the number of breeding individuals, so an effective size of 10,000 corresponds in humans to a census size of 20,000 to 30,000). It is important to notice that in the discussion in the Vignette, I have implicitly assumed a population of constant size: the theory above is not so relevant when population size changes over time. The key to understanding the effects of size change is to recall that the probability of any pair of genes coalescing at any time is $1/2N$ where N is the current effective size. This means that, in a large population, coalescence events are few; in a small population, they occur rapidly, the rate being inversely proportional to population size.

Figure 1 shows a typical gene genealogy of a sample of $n = 6$ genes from a constant-size population. If the effective size of the population is $N = 10{,}000$ then the average pairwise coalescence time should be $2N$ generations, or half a million years if generations average 25 years. The time to the very top of the tree, the "coalescent," should be nearly $4N$ generations or a million years. (The theory is that the average coalescent should approach $4N$ generations as the sample size becomes large.) Figure 2 shows a gene tree from a population that was initially small and then grew suddenly to a very large size. In a large population, coalescence events are rare, and in this tree no coalescence events occur since the abrupt population growth occurred. Before the population growth the effective size was small and coales-

COALESCENCE THEORY

Reproduction in a random-mating population is statistically equivalent to forming individuals by picking pairs of genes at random with replacement from a pool. If the size of the population is N individuals, there are $2N$ genes in the pool from any locus. If we pick a gene from the pool, then pick another, the probability that we pick the same one is just $1/2N$. This is the pairwise hazard of coalescence of any two genes ancestral to a sample at any generation, and with a constant hazard the probability density function of the time backward to coalescence of a pair of genes is geometric with a mean of $2N$. This means that any pair of genes from a population has a common ancestor, on average, $2N$ generations in the past. Now suppose that the gene we are considering is a DNA sequence in which infrequent mutations have occurred, so infrequent that at any nucleotide position no or only a single mutation has ever occurred. If the average pairwise sequence differences is D, we know that those D mutations occurred on average during the $2N$ generations since the pairs shared a common ancestor. The average time of separation is thus $4N$ generations and, given knowledge of the mutation rate μ of the sequence, we can set $D = 4N\mu$ and solve for N.

As an example, suppose that we study a large collection of 5,000 base DNA sequences and find that the average difference between pairs of sequences is five differences. The mutation rate per base per year is 10^{-9}, so the rate for the sequence of 5,000 bases is 5×10^{-6} per year or, at 25 years per generation, 1.25×10^{-4} per generation. Therefore, we estimate the average number of generations separating pairs of sequences from the population to be 5 mutations divided by 1.25×10^{-4} mutations per generation, or 40,000 generations. Equating this time to $4N$, we conclude that the genetic estimate of the size of the species is 10,000 breeding individuals.

—Henry Harpending

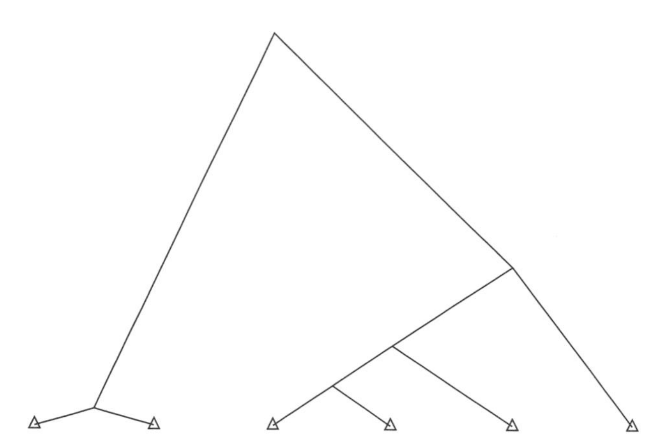

FIGURE 1. Human Genealogical History.
A gene tree of six genes from a population of constant size. Courtesy of Henry Harpending.

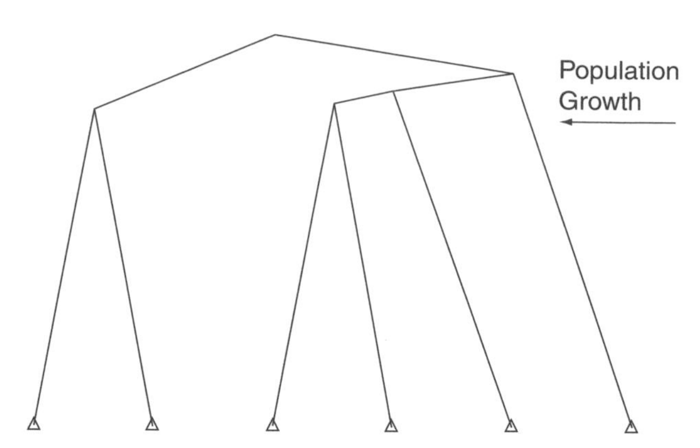

FIGURE 2. Gene Tree from a Population That Grew Rapidly in Size in the Past.
Growth is indicated by the arrow. Courtesy of Henry Harpending.

cence events occurred often, so they are concentrated near the top of the gene genealogy. Compared to the tree in Figure 1, the tree of Figure 2 is "star" or "comb" shaped. Many genetic systems in humans have gene trees like that of Figure 2, suggesting that there was a major episode of expansion from a small founding population in our past. The time of this expansion is estimated to be between fifty thousand and one hundred thousand years ago. The correspondence of this genetic estimate of the time of population expansion with fossil evidence of the first appearance of anatomically modern humans has provided strong support for the "Garden of Eden" model of the evolution of our species, in which we are descended from a small founding population that spread over most of the earth within the past fifty to one hundred thousand years.

The theory of gene genealogies and their relationship to population size is clear and simple, but we cannot observe gene genealogies directly, so our inferences about them must be indirect. There are two simple ways in which a sample of genes can reveal properties of the underlying gene genealogy. First, if we are studying sequences, we can compare all pairwise differences among the genes in our sample. Pairs from the genealogy in Figure 1 would differ greatly among themselves in time back to a common ancestor. Because mutations accumulate along the branches of the genealogy at a roughly constant rate, pairs of genes with a remote common ancestor will differ much more from each other than will pairs with a recent common ancestor. A histogram of the number of pairwise differences among all 6(5)/2 = 15 pairs of genes from Figure 1, called a "mismatch distribution," would be ragged and erratic with widely different numbers of mutations separating the various pairs. Such a histogram from the sample in Figure 2, on the other hand, would be smooth and unimodal. Since most of the coalescence events in the tree are concentrated in a relatively narrow time interval in the past, the total evolutionary timespans along the various branches separating pairs are similar, and the number of mutational differences separating pairs of genes are similar, so the mismatch distribution would be unimodal and smooth. The peak of the mismatch distribution in the case of Figure 2 would reflect the time of the expansion.

Another way to distinguish gene genealogies corresponding to Figures 1 and 2 is through the "frequency spectrum" of mutations in the sample. The branches of a gene genealogy extended backward from the present to the first coalescence event are called "external branches," and any mutation that occurs on an external branch will appear in a single individual in the sample. Old mutations, those occurring high in the tree, are likely to be represented in several individuals in a sample. If we tabulate every variable site according to the frequency of the rarer variant at the site, a gene genealogy like that of Figure 2 will show many more rare variants, especially singletons (a variant that occurs just once in a sample), than will a genealogy like that of Figure 1. Thus population expansion in the past is said to lead to "excess singletons" in a sample.

Many studies have shown that the gene genealogy of human mitochondrial DNA is like that shown in Figure 2. Pairwise sequence differences consistently are unimodal and smooth, and the frequency spectrum always shows an excess of singletons and rare variants. These dramatic patterns in human mitochondrial DNA have been taken as unequivocal support for the "Garden of Eden" model, in which modern humans expanded from a small isolated founding population. The peaks of human mismatch distributions, if current estimates of mutation rate are correct, correspond to expansion times between forty and one hundred thousand years ago.

Although the pattern in human mitochondrial DNA is clear, it is in effect just one genetic locus. Because there is no recombination of mitochondrial DNA, there is only a single history of the molecule in our species, while we need many independent histories for statistical confidence. Other parts of our genome might have different histories. Any credible reconstruction of human population history must be based on a broad sample of the genome, and so far the data are weak to ambiguous. Although some systems, like Alu's and short tandem repeats, do suggest a major expansion in our history, many collections of DNA sequences from the nuclear genome do not show evidence of expansion. One possibility is that there is pervasive natural selection on nuclear genes (most of the nuclear sequences so far are from genes). Another possibility is that parts of our genome were replaced during the expansion of anatomically modern humans one hundred thousand or so years ago, while other parts were retained through admixture between the expanding modern and the resident archaic humans, like the Neanderthals of Europe, throughout the world.

In order to understand how some portions of our genome might reflect a dramatic expansion from a small number of founders while others might reflect a long history with little population size change, it is necessary to review some ideas of the population geneticist Sewall Wright. Wright's idea of an adaptive topography was meant to account for the appearance of new, better-adapted variants and their spread in a population subdivided into partially isolated demes.

Once a new and better adapted form appears, there are several possibilities for how it interacts with other demes as it expands. If the advantage is simply a single locus or quantitative trait, selection will favor the trait and it will spread by interbreeding with other demes. If it is a single locus, the result would be reduced genetic diversity in the chromosomal neighborhood of the new advantageous gene, but over time recombination would erase much of this signature. If the advantage is very complex, involving a number of genes, then hybrids will be highly unlikely to inherit the entire advantageous combination and will face an insurmountable fitness barrier. Thus, the new population will expand to replace other demes with little genetic admixture being seen in the finally emergent population. The new population is essentially a new species. In this case, the whole genome of the new species will carry the bottleneck-and-expansion signature of the founding event, except for loci where old diversity is preserved by balancing selection. (The genealogies of human HLA genes extend tens of millions of years into the past because of such balancing selection.)

But what if the complexity of the advantage is intermediate? Then hybrids between the new population and other demes will suffer a fitness cost, but they might still occasionally inherit the entire gene combination that characterizes the new type, so that their offspring, hav-

WRIGHT'S ADAPTIVE TOPOGRAPHY

Wright considered how new advantageous complexes of traits arise in a population. The problem is that selection may disfavor changes in individual traits without simultaneous changes in others, so there is no selectively favored path from one state to another. Imagine, for example, that in a population of archaic humans there was selection for efficient locomotion and hence a smaller pelvis, but also selection for a larger brain and hence a larger pelvis that could accommodate larger-brained babies. Because of the need to balance opposing constraints, either of these traits—a smaller pelvis or a larger brain—would be individually selected against, but it is possible that co-adapted changes in both pelvic and brain sizes might yet be advantageous. Furthermore, there may be not one but many "optimum" pelvic/brain size combinations (with some being more optimal than others), such that small variations from any optimum size combination would lead to a reduction of fitness.

Now think of a plot of a population's mean values on each of the two traits. With each point in this two-dimensional space there is an associated population fitness; the whole can be pictured as a three-dimensional topography. This is Wright's "adaptive landscape." The several optimal combinations will be represented, on this landscape, as peaks surrounded by valleys of lowered fitness. Selection will move populations from valleys toward peaks of high fitness. But how can a population move from an old peak to a higher new one, when in between the peaks are valleys of lowered fitness? Wright envisioned a species divided into reproductively semi-isolated demes in which there was a lot of genetic drift, so that population genotypes would "wander" at random over the adaptive topography. If a deme by accident happened to wander near the new peak, selection would act to push it up the fitness surface and fix the new trait complex. The mean fitness of this new population would be greater than that of its neighbors and it would expand at their expense. This process of occasional random "jumps" to new fitness optima, and the subsequent spread of new types, is analogous to the punctuated equilibria of sudden changes thought by paleontologists to underlie what is viewed in much of the fossil record. However, unlike the latter, Wright's process is not accompanied by speciation.

—Henry Harpending

ing the new advantageous trait complex, become part of the new population. For example, an individual with a small pelvis and with a small brain could, by continued interbreeding, have descendants with small pelves and large brains. With a suite of traits of moderate complexity, and interbreeding, genes from other demes in the species can thus be introduced into the populations of the new type. The degree of such admixture at neutral loci is related inversely to the complexity of the advantageous gene combination. However, any functional genes that were selectively adapted to the local environment would be likely to be incorporated into the new population, perhaps accounting for morphological evidence of regional continuity in the fossil record.

Recently Vinayak Eswaran of the Indian Institute of Technology at Kanpur has worked out the consequences of Wright's model for gene genealogies and the apparent size of an ancestral population. The founding population expands, and, because of its fitness advantage, net gene flow is outward to neighboring areas. In these areas, the new types have a selective advantage and their frequency increases through time by natural selection. This process is repeated until the new type has swept through the species. As the new type spreads from area A to B to C and so on, the colonizers at each step are a small sample from the previously colonized area: in other words, the colonizers of B are a small sample of the colonizers of A, the colonizers of C a small sample of those from B, and so on; in each case, the colonizers expand to become the entire population of the new area (perhaps with some admixture with the previous inhabitants). At each step of the process there is a founding event, and the result is a rolling bottleneck caused by the expansion of the new advantageous type. Eswaran called this expansion a "diffusion wave of modernity"; it is akin to the "wave of advance" of a single advantageous gene studied by the famous evolutionary biologist and statistician R. A. Fisher in 1937.

The Wright–Eswaran model explains two outstanding puzzles about the human gene genealogies. First, mismatch distributions from human mitochondrial DNA almost always show a clear smooth peak, but there is not the single world peak that we would expect if there had been a single expansion of modern humans. Instead, the peaks correspond to expansion times of nearly a hundred thousand years in African populations, fifty to seventy thousand years in many Asian populations, and as recent as thirty-five thousand years in European populations. These differences are precisely what a slow-moving rolling bottleneck caused by demic diffusion and selection of a new favored type would generate.

Second, even though hybrids between the old and new type are selected against, there is still some possibility of incorporation of local archaic genes into the new population. Some loci may surf the wave of advance over the whole world and show no evidence of incorporation of genes from archaics; at these loci there would be evident signs of population expansions—for only the genes of the founding population, and their subsequent mutations, would now appear in the entire global population. However, other loci would have genes from both the old and the new populations and would accordingly show no evidence of population expansion. This is just what is appearing as more parts of the human genome are studied: a mosaic of loci that show both patterns. According to the Wright–Eswaran model, this suggests that at least some admixture with archaic humans occurred as modern humans spread across the world. This recently proposed model of modern human orgins shows promise, but this is a field in which new data and new theories appear at a high rate and perspectives change rapidly.

[*See also* Molecular Clock; Wright, Sewall.]

BIBLIOGRAPHY

Haigh, J., and J. Maynard Smith. "Population Size and Protein Variation in Man." *Genetical Research, Cambridge* 19 (1972): 73–89. This was the first clear statement that genetic data suggest a population size bottleneck in human history.

Harpending, H. C., and A. R. Rogers. "Genetic Perspectives on Human Origins and Differentiation." *Annual Review of Genomics and Human Genetics* 1 (2000): 361–385. This is a review of genetic evidence about human history.

Hudson, R. R. "Gene Genealogies and the Coalescent Process." In *Oxford Surveys in Evolutionary Biology*, edited by D. Futuyma and J. Antonovics, pp. 1–44. Oxford, 1990. This classic paper is an accessible introduction to coalescence theory.

Wright S. "Character Change, Speciation, and the Higher Taxa." *Evolution* 36 (1992): 427–443. Wright is one of the great founders of evolutionary theory.

— HENRY HARPENDING

Neanderthal–Modern Human Divergence

The relationship between anatomically modern humans and Neanderthals has been debated since the discovery of the first recognized Neanderthal remains in 1856. These bones were found as a result of quarrying at Feldhofer Cave in the Neander Thal (or valley) of western Germany. These bones belonged to an adult male with features that are now recognized as "classic" Neanderthal features, such as a low receding forehead, little or no chin, a long, flat and low cranial vault, strong browridge, large nose, and a robust, muscular body. This suite of features may be an adaptation to a cold environment and is common among Neanderthals dating approximately 120,000 to 30,000 years ago. Approximately 35,000 years ago, anatomically modern humans migrated into Europe. Researchers who think that Neanderthals did not mate with the new immigrants typically classify Neanderthals as *Homo neanderthalensis*, whereas those who think either that anatomically modern humans evolved from Neanderthals or that admixture did occur, and thus that Neanderthals did contribute genes to modern humans, would classify Neanderthals as *Homo sapiens neanderthalensis*. In 1997, mitochondrial DNA sequences were recovered from the individual discovered in 1856. This was the first time that DNA from a Neanderthal was brought to bear on the long-standing debate about the relationship between modern humans and Neanderthals. The genetic data provides a new and independent means of assessing these questions that derive from the fossil record.

To date, ancient DNA has been analyzed from three Neanderthals: the type specimen from the Neander valley, an infant from Mezmaiskaya Cave in the Northern Caucasus dating to 29,000 years ago, and the Vindija 75 sample from Vindija Cave, Croatia, dating to 42,000 years ago (Krings et al., 1997, 1999, 2000; Ovchinnikov et al., 2000). The analysis of DNA from ancient bone and tissue is feasible because of advances in molecular genetics during the last fifteen years. It has been estimated that under appropriate conditions, DNA could be preserved for 100,000 to 150,000 years, which is within the time frame that Neanderthals occupied Europe and western Asia. After careful DNA extraction from bone or tooth roots, the polymerase chain reaction (PCR) is used to copy the DNA fragment of interest millions of times so that there is sufficient DNA for analysis. Damage to the DNA (both chemical and physically fragmenting) and the low number of starting molecules (often less than 1 percent of the original amount) make PCR amplification of ancient DNA difficult (Höss et al., 1996; Pääbo, 1989). Ancient DNA analysis also requires great care because of the high risk of contamination from modern DNA sources. Several precautions must be taken to ensure the authenticity of the results (Handt et al., 1996; Richards et al., 1995; Stoneking, 1995), including using dedicated equipment and reagents, a separate clean laboratory space, and reproducing results multiple times from independent extractions. For unique and potentially controversial samples, such as Neanderthal samples, it is also common to rely on an independent laboratory to confirm the results.

Most analyses of ancient DNA, including the analyses of Neanderthals, have examined mitochondrial DNA (mtDNA). Mitochondrial DNA is found in the mitochondria, small energy-producing organelles within the cell. MtDNA has advantages for ancient DNA analysis because of its high copy number in cells, lack of recombination, rapid mutation rate, maternal transmission, and the large amount of published data for modern humans that can be used for comparison. In particular, the hypervariable region of the mtDNA genome surrounding one of the origins of replication is often used in studies of human population history because of its rapid mutation rate. The first hypervariable region (HVI) was sequenced in small overlapping fragments for all three Neanderthal samples; the second hypervariable region (HVII) was also examined in the Feldhofer and Vindija Cave individuals. Quantification of the amount of mtDNA remaining in the Feldhofer sample indicated that only ten molecules (at least 100 base pairs in length) of template mtDNA per microliter were present. This translates to 1,000–1,500 molecules per 0.4 gram of bone, which effectively rules out the possibility that nuclear DNA can be recovered because it is from 100 to 1,000 times less abundant than mtDNA. Such an amount of DNA would be impossible to amplify consistently if at all using PCR.

The results of HVI and HVII sequencing showed that the two Neanderthal sequences differed by approximately thirty-five nucleotides from modern human sequences, whereas modern humans differ from each other by only eleven mutations on average (Krings et al., 1999, 2000). By comparison, chimpanzees (*Pan troglodytes*) and pygmy chimpanzees or bonobos (*Pan paniscus*) differ from each other by an average of seventy-six nucleotide bases, whereas chimpanzee subspecies differ from each other by an average of forty-seven bases, although the eastern and central subspecies (*P. t. schweinfurthii* and *P. t. troglodytes*) have an average of only twenty-seven differences between them. This suggests that Neanderthals and modern humans are on par with chimpanzee subspecies in terms of amount of diversity between them; however, this does not resolve the issue of whether or not they did interbreed. Chimpanzee subspecies can and do interbreed in zoo environments (they are spatially separate in the wild), and behaviorally they are similar in many ways such that this is not a barrier to mating. How similar were Neanderthals and modern humans in terms of behavior? This issue is highly debated among paleoanthropologists and archaeologists.

To address the question of whether or not Neanderthals were ancestral to modern Europeans, the HVI and HVII sequences of the Feldhofer and Vindija 75 samples were compared with 472 contemporary Europeans, 151 Africans, and 41 Asians. The Neanderthal sequences have an average of 35.3, 33.9, and 33.5 differences from Europeans, Africans, and Asians, respectively. Thus, Neanderthals were not closer to Europeans as might be expected if they were their ancestors. Similar results are found if the HVI sequences of modern humans and all three Neanderthals are compared, leading researchers to conclude that Neanderthals did not contribute mtDNA to the modern human gene pool.

These results, however, do not exclude the possibility that interbreeding took place because genetic drift could result in the loss of specific Neanderthal mtDNA lineages, particularly if the Neanderthal genetic contribution to the modern human gene pool was small. Nordborg (1998) used computer simulations to examine the possibility that Neanderthals and modern humans interbred and concluded that Neanderthals and modern humans were not a single randomly mating population. However, he also noted that some interbreeding is not ruled out because the power to detect inbreeding between a small Neanderthal population and a modern human population that subsequently expands is limited. These results also do not rule out the possibility that other genes were contributed to modern humans by Neanderthals. MtDNA is passed solely from mother to child, whereas Y chromosome DNA is passed from father to son and nuclear DNA is passed on by both parents. It is possible then, though unlikely, that gene flow occurred only between Neanderthal men and modern females, resulting in the transmission of nuclear DNA but not mtDNA from Neanderthals to modern humans. Finally, the possibility of a selective sweep of a particular lineage of anatomically modern mtDNA such that it replaced all other mtDNA types, including Neanderthal types, must be considered. A selective sweep is the process in which a favorable mutation becomes fixed in the population. Because there is no recombination in the mtDNA genome, it is completely linked, and the favorable mutation would drag everything else, including the hypervariable sequence, to fixation with it. Such a selective sweep would result in the removal of all Neanderthal mtDNA lineages but not Neanderthal nuclear DNA, if it were present in the population where the sweep occurred.

The Neanderthal mtDNA data can also be used to estimate the divergence time between modern humans and Neanderthals. Using the data from the Mezmaiskaya and Feldhofer samples, Ovchinnikov and colleagues (2000) estimated the time to the most recent common ancestor (MRCA) of the mtDNA of humans and Neanderthals to be between 365,000 and 853,000 years ago, and the MRCA of the two Neanderthal sequences to date to 151,000 to 352,000 years ago. The latter date coincides with the time that Neanderthal morphological traits begin to appear in the fossil record. The earlier date of human–Neanderthal divergence reflects the fact that DNA begins to diverge prior to population splitting. These estimates use the divergence time between chimpanzees and humans of approximately four million to five million years ago as a calibration point. Estimates of the time to the MRCA for mtDNA of modern humans fall between 100,000 and 150,000 years, about four times smaller than the time to the MRCA for Neanderthal and human mtDNA. This also indicates that the Neanderthal sequences are not ancestral to modern human mtDNA sequences.

To date, no mtDNA has been retrieved from ancient anatomically modern Europeans older than roughly 15,000 years. Recently, however, Hawks and Wolpoff (2001) have suggested that the Mezmaiskaya infant is in fact anatomically modern rather than Neanderthal. They point to the date and the poor definition of stratigraphic layers within the cave as evidence to suggest that the burial is intrusive into the Mousterian layer. The original excavators contend, however, that the infant is Neanderthal based on skull and postcranial morphological characteristics. If the Mezmaiskaya infant is indeed anatomically modern, this supports the hypothesis that at least some Neanderthal and modern human admixture did occur. To ascertain the extent of possible admixture, additional mtDNA sequences from Neanderthals as well as sequences from early anatomically modern humans from Europe need to be examined.

[*See also* Hominid Evolution, *article on* Neanderthals.]

BIBLIOGRAPHY

Handt, O., M. Krings, R. H. Ward, and S. Pääbo. "The Retrieval of Ancient Human DNA Sequences." *American Journal of Human Genetics* 59 (1996): 368–376.

Hawks, J., and M. H. Wolpoff. "Paleoanthropology and the Population Genetics of Ancient Genes." *American Journal of Physical Anthropology* 114 (2001): 269–272.

Höss, M., P. Jaruga, T. H. Zastawny, M. Dizdaroglu, and S. Pääbo. "DNA Damage and DNA Sequence Retrieval from Ancient Tissues." *Nucleic Acids Research* 24 (1996): 1304–1307.

Krings, M., C. Capelli, F. Tschentscher, H. Geisert, S. Meyer, A. von Haeseler, K. Grossschmidt, M. Paunovic, and S. Pääbo. "A View of Neandertal Genetic Diversity." *Nature Genetics* 26 (2000): 144–146.

Krings, M., H. Geisert, R. W. Schmitz, H. Kranitzki, and S. Pääbo. "DNA Sequence of the Mitochondrial Hypervariable Region II from the Neandertal Type Specimen." *Proceedings of the National Academy of Sciences USA* 96 (1999): 5581–5585.

Krings, M., A. Stone, R. W. Schmitz, H. Krainitzki, M. Stoneking, and S. Pääbo. "Neandertal DNA Sequences and the Origin of Modern Humans." *Cell* 90 (1997): 19–30.

Nordborg, M. "On the Probability of Neanderthal Ancestry." *American Journal of Human Genetics* 63 (1998): 1237–1240.

Ovchinnikov, I. V., A. Götherström, G. P. Romanova, V. M. Kharitonov, K. Liden, and W. Goodwin. "Molecular Analysis of Neanderthal DNA from the Northern Caucasus." *Nature* 404 (2000): 490–493.

Pääbo, S. "Ancient DNA: Extraction, Characterization, Molecular Cloning, and Enzymatic Amplification." *Proceedings of the National Academy of Sciences USA* 86 (1989): 1939–1943.

Richards, M. B., B. C. Sykes, and R. E. M. Hedges. "Authenticating DNA Extracted from Ancient Skeletal Remains." *Journal of Archaeological Science* 22 (1995): 291–299.

Stoneking, M. "Ancient DNA: How Do You Know When You Have It and What Can You Do with It?" *American Journal of Human Genetics* 57 (1995): 1259–1262.

— Anne C. Stone

MODERN SYNTHESIS. *See* Overview Essay *on* History of Evolutionary Thought *at the beginning of Volume 1*; Neo-Darwinism.

MOLECULAR CLOCK

The hypothesis of a molecular clock asserts that the rate of DNA or protein sequence evolution is constant over time or among evolutionary lineages. In the early 1960s, when protein sequences became available, it was observed that the rate of evolution for proteins, such as hemoglobin, were relatively constant among different orders of mammals. The observation led to the proposal of the molecular clock hypothesis by Emile Zuckerkandl and Linus Pauling in 1965. The proposal had an immediate impact on the development of the field of molecular evolution. First, the utility of the molecular clock was obvious from the beginning. If proteins evolved at constant rates, they can be used to reconstruct phylogenetic relationships among species and to estimate the dates of species divergences. Second, the accuracy of the clock and the mechanism of molecular evolution have been a focus of controversy. At the time, the synthetic theory of evolution or neo-Darwinism, which maintains that the rate of evolution is determined by environmental changes and natural selection, was generally accepted by evolutionists. A constant rate of evolution among species as different as elephants and mice was unthinkable. For example, morphological characteristics are well known to have markedly different rates of evolution among lineages. Motoo Kimura proposed the neutral theory of molecular evolution in the late 1960s, and the molecular clock was immediately solicited as a major piece of supporting evidence (Kimura, 1983). This theory maintains that most molecular evolution is dominated not by natural selection but by random fixation of neutral mutations, that is, mutations whose effects on fitness are too small for natural selection to play a role in determining their fate. The rate of molecular evolution is then equal to the neutral mutation rate, independent of factors such as environmental changes and population sizes. If the mutation rate was similar and the function of the protein remains the same among lineages so that the same proportions of mutations are neutral, a constant evolutionary rate is predicted by the neutral theory.

MAXIMUM LIKELIHOOD ESTIMATION AND LIKELIHOOD RATIO TEST

Maximum likelihood is a general methodology for estimating statistical parameters. Suppose the probability of observing the data (D) is $P(D; \vartheta)$, where ϑ are parameters under the model. Because the data are observed, we view P as a function of the unknown parameters and write it as $L(\vartheta; D) = P(D; \vartheta)$. L is known as the likelihood function. The values of ϑ that maximize L, or equivalently its logarithm, $l = \ln\{L\}$, are the maximum likelihood estimates.

The likelihood ratio test plays a central role in hypothesis testing. Suppose the more general (alternative) model has p parameters with log likelihood l_1, and the simpler (null) model has q parameters with log likelihood l_0. Then twice the log likelihood difference, $2\Delta l = 2(l_1 - l_0)$, is approximately χ^2 distributed with d.f. $= p_1 - p_0$, if H_0 is true. If the observed value of the test statistic $2\Delta l$ is greater than the χ^2 critical value, we reject H_0.

—Ziheng Yang

The concept of molecular clock has a long history of controversy. It is often a focus of hot debate even today. Early controversies were about whether the clock held, and when it did not, what factors might be responsible for the rate differences among lineages. For example, rodents were suggested to evolver faster than primates; one hypothesis proposed that rodents have a shorter generation time, more germ-line cell divisions per calendar year, and thus a higher mutation rate. Since the 1980s, DNA sequences have become widely available, and these reveal that the molecular clock is violated for most genes or species groups, except for sequences from very closely related species.

More recent controversies have been focused on the most common use of the clock assumption, that is, the dating of divergence times. Dating using molecules has produced a steady stream of controversies, as the molecular dates are often at odds with the fossil records or with the current interpretations of the fossil and morphological data. Two particular examples have attracted much attention, the Cambrian "explosion" about 545 million years ago and the mammalian "radiation" about 65 million years ago. In each case, molecular studies produced dates much older than indicated by the fossil data, sometimes twice as old. Part of the discrepancy probably arises from the incompleteness of fossil data; fossils represent the time when species developed diagnostic morphological characters and were fossilized, and molecules represent the time when the species stopped intermingling, so fossil dates have to be younger

FIGURE 1. A Phylogeny of Three Species to Explain the Relative Rate Test.
Drawing by Ziheng Yang.

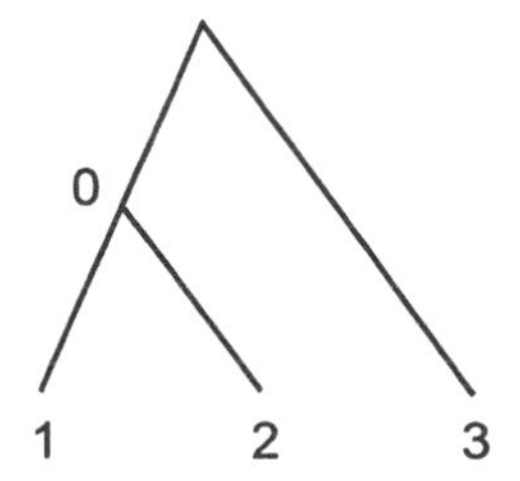

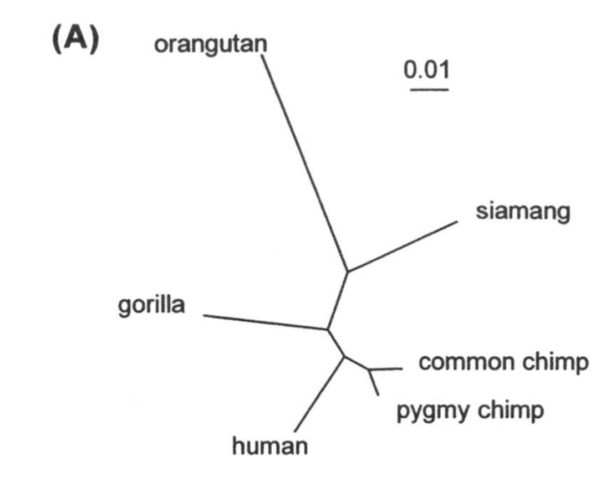

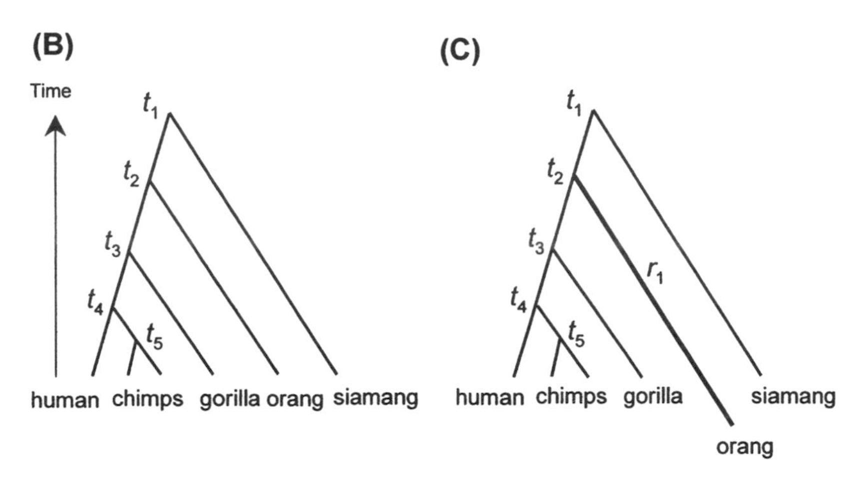

FIGURE 2. Phylogenies of Six Hominoid Species for Testing the Clock Assumption and Estimating Divergence Dates. (A) The unrooted tree without assuming the clock. This is the alternative model, and the unknown parameters include the nine branch lengths. (B) Under the (global) clock assumption, the parameters are the five node distances, the distances of the five internal nodes from the present time, measured by the expected number of nucleotide substitutions per site. (C) A local clock model that assigns a different evolutionary rate to the orangutan lineage. From Yang, 1996.

than molecular dates. Part of the discrepancy seems to arise from inaccuracies in molecular date estimation. In particular, molecular date estimation is very sensitive to violation of the molecular clock (see below).

This article discusses statistical tests of the clock assumption, as well as methods for using the global and local clock models to estimate divergence dates.

A few clarifications are in order. First, the molecular clock was envisaged as stochastic. Molecular changes accumulate at random according to a statistical Poisson process, so that random fluctuations are expected, although the underlying rate is constant over time. Second, different genes or proteins, or different regions of the same gene, may have very different evolutionary rates, and their clocks tick at different rates. The interpretation offered by the neutral theory for this observation is that different genes are under different selective constraints, and those under stronger constraints will have a smaller proportion of neutral mutations and a lower evolutionary rate. Third, the molecular clock is not expected to be universal and is usually applied to a group of species. For example, we might say that the clock holds for a gene within mammals.

Tests of the Molecular Clock. A number of statistical tests have been designed to examine the molecular clock hypothesis. The simplest is the relative rate test based on estimates of pair-wise sequence distances. To test whether species 1 and 2 have the same rate, we get an outgroup species 3, which is more distantly related (see Figure 1). If the clock hypothesis is true, the distance (expected amount of change) from ancestral node 0 to species 1 will be the same as the distance from node 0 to species 2; that is, $d_{01} = d_{02}$. Because we do not have data for the ancestral node 0, we test the clock hypothesis by testing $d_{13} = d_{23}$, using the statistic $d = (d_{13} - d_{23})$. We can use the variances of the estimated pair-wise

TABLE 1. Maximum Likelihood Estimates of Node Distances (*d*) and Node Ages (*t*) under Global and Local Clock Models

	Global clock		*Local clock*	
Node	*Distance*	*Time*	*Distance*	*Time*
1	$d_1 = 0.063$	$t_1 = 15.77 \pm 1.55$	$d_1 = 0.060$	$t_1 = 17.16 \pm 1.86$
2	$d_2 = 0.052$	$t_2 = 13$	$d_2 = 0.045$	$t_2 = 13$
3	$d_3 = 0.031$	$t_3 = 7.73 \pm 1.03$	$d_3 = 0.029$	$t_3 = 8.38 \pm 1.17$
4	$d_4 = 0.020$	$t_4 = 5.04 \pm 0.86$	$d_4 = 0.019$	$t_4 = 5.60 \pm 0.96$
5	$d_5 = 0.009$	$t_5 = 2.35 \pm 0.64$	$d_5 = 0.009$	$t_5 = 2.65 \pm 0.72$
l	-1796.12		-1795.35	
r_1	1		1.36	

Note. The data consist of the transfer RNA genes from five hominoid species: human, chimpanzee, gorilla, orangutan, and siamang. Alignment gaps are removed, with 759 nucleotides in the sequence. The model of nucleotide substitution accounts for different rates of transitions (T ↔ C or A ↔ G) and transversions (T, C ↔ A, G), as well as different frequencies for the four nucleotides. The local clock model assumes an independent rate r_1 for the orangutan lineage.

sequence distances to work out the standard error of the test statistic, σ_d, then compare d/σ_d with the standard normal distribution to test whether d is different from zero. This test is limited to three species only, and it does not test whether the outgroup species has a different rate from the two ingroup species.

A second test of the clock assumption is the likelihood ratio test (see Vignette). This test is applicable to data of any number of species. Figure 2 shows an example of testing the molecular clock using data of eleven transfer RNA genes in the mitochondrial genome from six hominoid species. The null model H_0 assumes the clock, and the parameters of the model are the $n - 1 = 5$ distances from the internal nodes of the tree to the present time, where n is the number of species in the tree (Figure 2B). The more complex model H_1 uses one independent rate parameter for each branch in the tree. Under this model, it is generally impossible to identify the root of the tree, so the "unrooted" tree is used, with $2n - 3$ branches (Figure 2A). Thus, model H_1 involves $(2n - 3) = 9$ parameters. The log-likelihood values under the clock and no-clock models are $l_0 = -1796.12$ and $l_0 = -1794.49$. We then compare $2\Delta_1 = 2(l_1 - l_0) = 3.26$ with a χ^2 distribution with d.f. $= (2n - 3) - (n - 1) = n - 2 = 9 - 5 = 4$. The P value is 0.52, with d.f. $= 4$, and the clock is not rejected.

We should note that failure to reject the clock assumption does not necessarily mean that the evolutionary rate is constant over time. First, the null hypothesis tested by the likelihood ratio test is weaker than the assumption of a constant rate over time. For example, if the evolutionary rate has been accelerating over time in all lineages, the tree will look clocklike, although the rate is not constant. Furthermore, neither the likelihood ratio nor the relative rate tests can distinguish a variable from a constant rate within a lineage. Finally, failure to reject the clock might simply be because of a lack of information in the data rather than the correctness of the clock assumption. These observations suggest that the clock assumption should be accepted with caution when we estimate divergence dates from molecular data.

A third test is based on the index of dispersion, that is, the variance to mean ratio of the number of substitutions over lineages. When the rate is constant, the number of substitutions should follow a Poisson distribution, which has the mean equal to the variance. If all the species diverged at the same time and accumulate substitutions at the same rate since their divergence, the variance to mean ratio of the number of nucleotide substitutions among lineages should be close to one. Most real data sets generate dispersion indices much higher than one, an observation referred to as the overdispersed clock. Early studies by Kimura and J. Gillespie used gene sequences from different orders of mammals and assumed that they diverged in a radiation. This was later shown to be unacceptable as the phylogenetic relationship among species has much effect on the test. Recent analyses used only three lineages (primates, artiodactyls, and rodents) to avoid the problem of phylogeny, but the analysis is prone to errors as the variance calculation using only three lineages is unreliable. Tests based on the dispersion index are out of date and can be performed more rigorously using the likelihood ratio test.

Application of Molecular Clock to Estimate Divergence Dates. The most important practical application of the molecular clock hypothesis is to estimate divergence dates between species, populations, or even viral strains. Molecular sequence data allow estimation of distances only. Under the assumption of a constant rate over time, the distance is a linear function of time. To convert distances into absolute times, an external time (called a calibration point) is used, obtained from fossil data or geological events that mark the separation of species. In the example of Figure 2B, the clock assumption is used to estimate the divergence dates among the hominoid species, and the divergence time of the orangutan is fixed at thirteen million years ago for calibration. There are currently two major approaches to estimating the node distances, the distances from the internal nodes of the tree to the present time measured by the expected number of substitutions per site (the d's). The first approach estimates the sequence distance between each pair of species, then uses least squares to fit the node distances to the pair-wise distance matrix. The second approach is maximum likelihood (Vignette), which is applied to the original sequence alignment rather than estimated pairwise distances. Estimates given in Table 1 were obtained using maximum likelihood. After the node distances are estimated, it is an easy matter to obtain the node times using the calibration point. The substitution rate can then be calculated as $\mu = 0.0517/(13 \times 10^6) = 3.976910^{-9}$ substitutions per site per year. We then use this rate to convert other distances into times, for example, $t_4 = 0.0201/(3.9769 \times 10^{-9}) = 5.04 \times 10^6$ years ago for the separation of humans and chimpanzees. Equivalently, $t_4 = 0.0201/0.0517 \times 13 = 5.04$ million years ago.

Note that both the calculation of pair-wise sequence distances and the likelihood joint analysis of all sequences rely on a model of nucleotide substitution, and it is important to use an adequate substitution model. A simplistic model does not correct for hidden changes (multiple substitutions at the same site) properly; as a result, the estimated distance will not be linear with time. For example, the proportion of differences between two sequences is not a linear function of time. It underestimates the distance, and the underestimation is more serious for large distances than for small ones, because more multiple substitutions are expected the longer the sequences have been diverged. This nonproportional underestimation of distances generates systematic biases in divergence date estimation.

Because the molecular clock hypothesis is often violated, especially for data sets of divergent sequences, much effort has been taken to estimate divergence dates even though the clock does not hold. Recent work has suggested it might be possible to estimate dates without assuming a global molecular clock. Two approaches have been taken. The first uses a random process to describe the change of the evolutionary rate over branches in the phylogeny and then use the Bayes method to estimate the posterior distribution of rates and dates. The second approach uses the likelihood method and assigns specific rate parameters to branches that are assumed to have different rates from other branches. For example, if we assume an independent rate for the orangutan lineage in Figure 2C, the maximum likelihood estimate of the human–chimpanzee divergence became 5.60 ± 0.96 million years ago, older than the estimate under the global clock model (Table 1). Although the clock assumption was not rejected by the likelihood ratio test, we note that date estimation is quite sensitive to assumptions about the clock.

BIBLIOGRAPHY

Cooper, A., and R. Fortey. "Evolutionary Explosions and the phylogenetic Fuse." *Trends in Ecology and Evolution* 13 (1998): 151–156. Discusses discrepancies between molecular dates and fossil records concerning two important periods in the evolutionary history: the Cambrian explosion and the origin and divergence of birds and mammals at the Cretaceous–Tertiary boundary.

Felsenstein, J. "Evolutionary Trees from DNA Sequences: A Maximum Likelihood Approach." *Journal of Molecular Evolution* 17 (1981): 368–376. Introduces maximum likelihood estimation of branch lengths and divergence dates as well as the likelihood ratio test of the clock assumption.

Kimura, M. *The Neutral Theory of Molecular Evolution*. Cambridge, 1983. Summarizes evidence and controversies concerning the molecular clock in relation to the neutral theory.

Thorne, J. L., H. Kishino, and I. S. Painter. "Estimating the Rate of Evolution of the Rate of Molecular Evolution." *Molecular Biology and Evolution* 15 (1998): 1647–1657. This paper Introduces the Bayes approach to date estimation using local molecular clocks.

Wu, C.-I., and W.-H. Li. "Evidence for Higher Rates of Nucleotide Substitution in Rodents Than in Man." *Proceedings of the National Academy of Sciences USA* 82 (1985): 1741–1745. Introduces the relative-rate test.

Yang, Z. "Among-Site Rate Variation and Its Impact on Phylogenetic Analyses." *Trends in Ecology and Evolution* 11 (1996): 367–372. Discusses date estimation using the molecular clock and the importance of evolutionary models.

Yoder, A. D., and Z. Yang. "Estimation of Primate Speciation Dates Using Local Molecular Clocks." *Molecular Biology and Evolution* 17 (2000): 1081–1090. Implements maximum likelihood models of local molecular clocks, and demonstrates the effects of substitution model, clock assumption, and calibration points on date estimation.

Zuckerkandl, E., and L. Pauling. "Evolutionary Divergence and Convergence in Proteins." In *Evolving Genes and Proteins*, edited by V. Bryson and H. J. Vogel, pp. 97–166. New York, 1965. Proposes the molecular clock hypothesis, among many important contributions.

— ZIHENG YANG

MOLECULAR EVOLUTION

Molecular evolution is a discipline of biology that utilizes molecular data to address evolutionary questions. The molecular data are usually DNA or protein sequences but may also include other types of data, such as the three-dimensional structure or some biochemical properties of a protein. DNA and proteins are referred to as *macromolecules* because they are much larger than molecules such as oxygen or ethanol. The field addresses diverse questions, ranging from traditional questions of life science to new questions driven by recently emerged data, such as whole genome sequences. Some examples of these issues will be described in this article.

The evolutionary questions considered in molecular evolution can be divided into two categories. The first one concerns the process and mechanisms of evolution—in other words, how and why evolution of macromolecules has occurred. For example, we may want to know how and why hemoglobin, the oxygen carrier in animal blood, has evolved through time. The second one concerns the evolutionary history of organisms, genes, or genomes. In particular, the molecular data have become a powerful tool to elucidate the differences and relatedness between species. For example, we may ask, "Is the traditional view true that humans are very distantly related to chimpanzees and gorillas and deserve to be classified in a family well separated from the family of chimpanzees and gorillas?" Humans used to think that we were very different from the apes, but DNA sequence data have shown that human and chimpanzee are actually closer to each other than either of them is to gorilla. When the study of molecular evolution deals with this category of questions, it is known as *evolutionary* or *molecular systematics* or *phylogenetics*. This topic will not be discussed in detail here.

To use molecular data to address evolutionary questions, we need statistical and mathematical methods. Because the amounts of data used are often large and the statistical methods involved may be complicated, we also need to develop computer programs to carry out the computations of data analyses. Moreover, the interpretation of the results of an analysis requires an understanding of evolutionary theory. Therefore, molecular evolution is a multidisciplinary subject, requiring training in molecular biology, evolution, statistics, mathematics, and computer science. However, there are computer software packages available for doing statistical analysis of data, so one does not have to be a computer science expert or a statistician.

Birth of Molecular Evolution. Although molecular evolution is a relatively young branch of science, attempts to use molecular techniques to study phylogenetic relationships among mammals were made before the dawn of the twentieth century. The results were summarized in a 1904 book by G. H. F. Nuttall. Nuttall and others found that serological cross-reactions were stronger for more closely related species than for less related ones. For example, he determined that humans' closest relatives are the apes, followed by the Old World monkeys, and then the New World monkeys. However, the field started in earnest only in the early 1960s, when substantial amounts of protein sequence data and electrophoretic data had been accumulated. In particular, based on the analyses of protein sequence data of hemoglobins from different mammals, Zuckerkandl and Pauling (1965) proposed the molecular clock hypothesis, postulating that for any given protein the rate of molecular evolution is approximately constant in time and across evolutionary lineages. This proposal stimulated much interest in the use of macromolecules in evolutionary studies. Indeed, if macromolecules evolve at constant rates, they can be used to estimate the dates of species divergence and to reconstruct phylogenetic relationships among organisms. However, the concept of rate constancy is diametrically opposite to the observed "erratic tempo of evolution" at the morphological and physiological levels. Therefore, the hypothesis also provoked a great controversy. The controversy and the enthusiasm in the new approach to evolutionary study were the impetus for the development of molecular evolution.

Major Issues in Molecular Evolution. Molecular evolution is now a broad field of scientific pursuit, thanks to the rapid development in molecular biology, statistical theory of sequence evolution, and computational tools. It is therefore impossible to describe the field in a limited space. Here we shall describe a number of major issues that have been pursued since the birth of the field. In the last section of the article, we shall mention some future research directions.

Models of DNA sequence evolution. A basic process of molecular evolution is the evolution of a DNA sequence over time. In other words, we need to know how the nucleotides in a DNA sequence change with time. To describe this process, we need a model. The simplest model of nucleotide change at a site was proposed by Jukes and Cantor (1969) and can be represented by Figure 1A, in which A, C, G and T denote the four types of nucleotide. In this figure, α denotes the

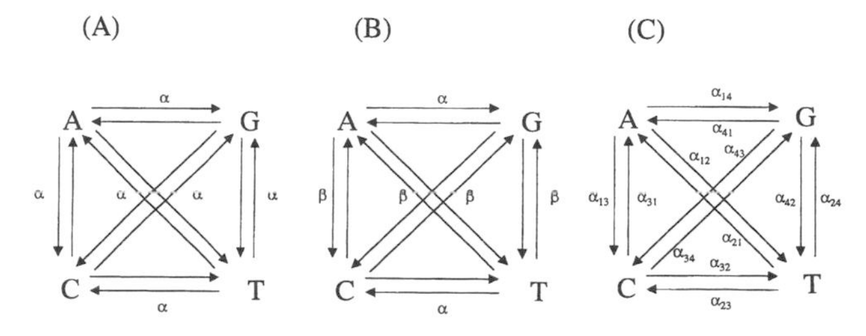

FIGURE 1. One-Parameter Model of Nucleotide Substitution. (A) In this model, the rate of substitution in each direction is α and the total rate of substitution is 3α. (B) Two-parameter model of nucleotide substitution. In this model, the rate of transition (α) may not be equal to the rate of each of the two types of transversion (β). (C) A general model of nucleotide substitution. In this model the rate of change from nucleotide i to nucleotide j may not be equal to the rate of change in any other directions. The rate of change between two nucleotides is depicted as α_{ij}, with nucleotide A corresponding to 1, T to 2, C to 3, and G to 4. Courtesy of Wen-Hsiung Li and Soojin Yi.

probability of change in a unit time (i.e., the rate of change) from one specific nucleotide to another specific nucleotide. Each nucleotide can change to three other nucleotides, and the total rate of change for a nucleotide is 3α. In this model, the four nucleotides have the same rate of change. Since the model is described by a single parameter α, it is known as the one-parameter model.

Figure 1B represents a more sophisticated model of nucleotide change. Here changes are classified into two types: transitions and transversions. There are four types of transition (A ↔ G, C ↔ T) and eight types of transversion (A ↔ T, A ↔ C, G ↔ C, G ↔ T); in animal nuclear DNA, transitions occur more often (~67 percent) than transversions (~33 percent). Therefore, it is more reasonable to distinguish between the transition and transversion rates, being denoted by α and β, respectively, in Figure 1B. This model was proposed by Kimura (1980) and is known as Kimura's two-parameter model.

In Figure 1B as in Figure 1A, the rate of change is the same for the four types of nucleotide. To change this unrealistic assumption, we need to have a more sophisticated model. In general, we can make the model more and more sophisticated by increasing the number of parameters, but the mathematical characterization of the model becomes more and more difficult to do (Li, 1997). Figure 1C denotes the most general model, in which the rates of change from one nucleotide to the three other nucleotides can be different. This model has 12 parameters.

We have assumed that all nucleotide sites on a sequence are equivalent—that is, they evolve at the same rate. Such a sequence is said to be *homogeneous*. We can also remove this assumption and allow the rate of change to vary among sites. The most widely used methods assume that a certain proportion of sites are invariant, or that the distribution of possible rates of change follows a gamma distribution. In summary, one important effort of molecular evolutionists is to improve evolutionary models so that various situations can be treated realistically.

Sequence alignment. When two DNA sequences diverge from a common ancestral sequence, they will gradually accumulate nucleotide differences. In the models above, we considered only the possibility of changes from one nucleotide to another, so the length of the two sequences will always be the same. In practice, however, deletions and insertions of nucleotides can occur, and the two sequences may become different in length. When this occurs, the nucleotide positions of the two sequences will no longer correspond exactly to each other. To compare two such sequences, we will then need to find out which nucleotide positions in one sequence correspond to which nucleotide positions in the other sequence. This process is known as *sequence alignment*.

Aligning two sequences is not a simple task, because there may exist too many possible ways to align them. For example, let us consider the following two sequences (Li, 1997):

```
A:  TCAGACGAGTG
B:  TCGGAGCTG
```

One possible alignment is as follows:

```
     TCAGACGAGTG
(I)  ::  ::    ::
     TCGGA--GCTG
```

In this alignment, there are six pairs of matched nucleotides (denoted by :), three pairs of mismatched nucleotides, and one gap of two nucleotides (denoted by - -). It can be shown that among all the possible alignments with one gap, this alignment has the largest number of matched pairs.

By contrast, among those alignments with two gaps of one nucleotide each, the following two alignments have the largest number of matches (seven):

```
      TCAGACGAGTG
(II)  ::  :: :   ::
      TCGGA-GC-TG

      TCAGACGAGTG
(III) ::  :: :   ::
      TCGGA-G-CTG
```

Which of the three alignments is the best? Obviously, we are facing a choice between having more mismatches or having more gaps. Since comparing mismatches with gaps is like comparing apples with oranges, we must find a common denominator with which to compare gaps with mismatches. This common denominator is called the *gap penalty* (see later).

The basic principle for sequence alignment is either to maximize the number of matched pairs between the two sequences (Needleman and Wunsch, 1970) or to minimize the number of mismatched pairs (Sellers, 1974; Waterman et al., 1976), while keeping the number of gaps as small as possible (Li, 1997). As an illustration, we consider maximizing the number of matched pairs between two sequences.

Let w_k be the penalty for a gap of k nucleotides. Then the Needleman-Wunsch similarity measure between two aligned sequences is defined as

$$S = x - \Sigma w_k z_k$$

where x = number of matches, z_k = the number of gaps of length k, and w_k = the penalty for gaps of length k. The most commonly used penalty function is $w_k = a + bk$, where a and b are nonnegative parameters. However, even with this system, how to choose a and b is a difficult problem. For example, if $a = 0.5$ and $b = 2$, then the similarity for alignment I is $S = 6 - (0.5 + 2 \times 2)$

= 1.5, because there are six matches and one gap of two nucleotides. Since $S = 2$ for both alignments II and III, they are favored over alignment I. (This example shows that there may exist more than one optimal alignment under the same criteria.) However, if $a = 1.5$ and $b = 1$, $S = 2.5$ for alignment I, whereas $S = 2$ for alignments II and III, then alignment I is favored. Therefore, which alignment is preferable depends on the penalty weights used.

This example shows the need for studying the penalty function. More important, the computation is tedious even for such short sequences, and there is a strong need for developing computational tools. When more than two sequences are to be aligned, the task is even more difficult. Therefore, sequence alignment is a very challenging problem for computer scientists and mathematicians.

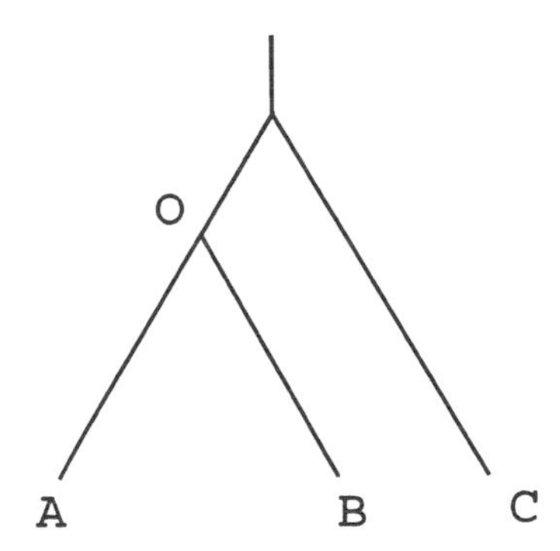

FIGURE 2. Relative Rate Test.
The rooted tree for species A, B, and C, assuming that C is the known outgroup. O denotes the common ancestor of species A and B. Courtesy of Wen-Hsiung Li and Soojin Yi.

Estimation of evolutionary distances. To describe the dissimilarity or divergence between two DNA (or protein) sequences, we need to have a measure. One simple measure is the proportion (p) of different nucleotides between the two DNA sequences. Let L be the length (number of nucleotides) of the two sequences compared (excluding gaps) and n be the number of different nucleotides between the two sequences. Then $p = n/L$. We consider p instead of n because n depends on the sequence length, whereas p does not and is more convenient for comparing the degrees of divergence among different pairs of sequences.

It should be noted that n is the observed number of changes between the two sequences compared and may not represent the true number of changes, because multiple changes may have occurred at the same position. For example, at a certain position, if the nucleotide has changed from A to G in sequence I and from A to T in sequence II, then only one difference will be observed at that position between the two sequences, though in reality two changes have occurred. As another example, at a certain position, in sequence I no change has occurred, but in sequence II A changed to T and then changed back to A. In this case, no change will be observed at this position. One can imagine other situations in which the actual number of changes cannot be fully recovered. The undetected changes are called *hidden changes*, and it requires a statistical theory to estimate the number of hidden changes. Let x be the estimate of the total number of changes between the two sequences. Then, $K = x/L$ is the number of substitutions (changes) per site between the two sequences. Usually, $K \geq p$ because K includes hidden changes.

Both p and K are *evolutionary distances* because they represent certain measures of the differences between the sequences compared and are likely to increase with the time of divergence between the two sequences. There are other types of evolutionary distances (Li, 1997), but K is the most commonly used because it represents an estimate of the average actual number of changes that have occurred since the initial divergence between the two sequences. Evolutionary distances are important quantities in molecular evolution. For example, K is required for estimating the rate of evolution in a sequence or a gene; if the two sequences have diverged t years ago, then $K/(2t)$ is the rate of nucleotide substitution in this sequence. K is also used in many methods of tree reconstruction (Li, 1997).

The estimation of K or other evolutionary distances requires the use of a model of DNA sequence evolution and the development of statistical methods. This is a difficult task when a sophisticated model of sequence evolution is used and when the assumption of sequence homogeneity is removed.

Molecular clocks. The molecular clock hypothesis has been an abiding interest of molecular evolutionists. Therefore, various methods to test the validity of this hypothesis have been proposed. A commonly used one is the so-called *relative rate test* (Figure 2). This test does not require knowing the date of divergence between the two sequences (A and B), which is usually unknown and often controversial, but uses an outgroup (C) as a reference; an outgroup is a related species known to have split before the divergence between A and B. Let O be the common ancestor of A and B. Then, the number of substitutions between sequences A and C, K_{AC}, is equal to the sum of substitutions that have occurred between point A and O and O and C, leading to $K_{AC} = K_{OA} + K_{OC}$. Similarly $K_{BC} = K_{OB} + K_{OC}$ and $K_{AB} = K_{OA} + K_{OB}$. The numbers K_{AC}, K_{BC} and K_{AB} can be directly estimated from the nucleotide sequences as discussed above. From the three equations, we can show that

$$K_{OA} = (K_{AC} + K_{AB} - K_{BC})/2$$
$$K_{OB} = (K_{AB} + K_{BC} - K_{AC})/2$$
$$K_{OC} = (K_{AC} + K_{BC} - K_{AB})/2.$$

The molecular clock hypothesis proposes that $K_{OA} = K_{OB}$. This can be tested using the relation $K_{OA} - K_{OB} = K_{AC} - K_{BC}$. If $K_{AC} - K_{BC}$ is statistically different from zero, then the substitution rates in lineages A and B are different, and the molecular clock hypothesis is rejected.

Employing such tests has established several interesting results (Li, 1997). Within the muroid rodents such as rat and mouse, there appears to be an approximate molecular clock, at least for the synonymous sites (sites that do not cause amino acid substitutions in the proteins). Among the anthropoids, the rate in the human lineage appears to be slower than in the Old World monkey lineage, supporting the so-called hominoid rate-slowdown hypothesis, which postulates that the rate of molecular evolution has become slower in human and apes. This hypothesis was originally proposed from immunological distance and protein sequence data (Goodman, 1961). There are also well-known rate differences between the human and rodent lineages; rodents appear to have a faster clock for both synonymous and nonsynonymous sites. Thus, there may exist local molecular clocks—clocks that hold for a limited group of species such as the muroid rodents—but there exists no molecular clock applicable to a large group of organisms such as the mammals.

One important area of innovation in the field of molecular clocks is to relax the assumption that all lineages must evolve at the same rate, allowing different rates to be estimated for different lineages within the same phylogeny. Both nonparametric (Sanderson, 1997) and Bayesian (Kishino et al., 2001) methods are available for more realistic estimates of divergence times.

Naturally, there has been much speculation as to why there should be differences between lineages in the rates of molecular evolution. One of the hypotheses worth mentioning in this limited space is the generation-time hypothesis. This hypothesis postulates a higher rate of evolution in organisms with a short generation time than in organisms with a long generation time. The underlying assumption is that substitutions in evolution must occur in the germ line, and a primary source of mutation is probably errors in DNA replication. A species with a shorter generation time goes through more rounds of germ cell divisions than a species with a longer generation time. However, whether DNA replication is the primary source of mutation is a controversial issue (Hurst and Ellegren, 1998).

Causes of molecular evolution. A central issue of evolution is how evolution occurs. Mutation, random genetic drift, and natural selection are known as the three causes of evolution, but their relative roles in evolution have been a longstanding question. In the traditional Darwinian view, natural selection plays the dominant role in evolution; although mutation is recognized as the ultimate source of genetic variability, its role in evolution is minor; and random drift, which is the cumulative effect of chance events, plays a negligible role. This view has been well accepted for morphological evolution, but its validity for molecular evolution has been challenged. Kimura (1968) and King and Jukes (1969) proposed that at the molecular level, the majority of differences between species have occurred largely by neutral or nearly neutral mutations and random drift; a mutation is said to be nearly neutral if its effect on selection coefficient is smaller than $1/(2N_e)$, where N_e is the effective size of the population. This proposal is known as the *neutral mutation hypothesis* or the neutral theory of molecular evolution.

Because the neutral mutation hypothesis challenged the traditional view, it provoked a great controversy and some issues soon arose. One was whether certain types of mutation are really selectively neutral. For example, synonymous mutations, which cause no amino acid changes, were then considered by neutralists as good candidates for neutral mutations. However, this view was challenged by Clarke (1970) and Richmond (1970), who argued that synonymous codons for an amino acid may have distinct fitnesses because the tRNAs recognizing different codons may have unequal concentrations in the cell, and because tRNA recognizing multiple synonymous codons may have different binding affinities for different codons. Another was whether the approximate rate-constancy observed in protein evolution was really true and could be used to support the neutral mutation hypothesis (Li, 1997). A third development was the emphasis on slightly deleterious mutations: given the fact that most mutations are deleterious, one would expect slightly deleterious mutations to be frequent and to contribute significantly to polymorphism and evolution.

The controversy over the neutral mutation hypothesis has had two strong impacts on molecular evolution and population genetics. First, it has led to the general recognition that the effect of random drift cannot be neglected when we consider the evolutionary dynamics of molecular changes. Second, it accelerated the fusion of molecular evolution and population genetics; because in the neutralist view molecular evolution and polymorphism are two facets of the same phenomenon (Kimura and Ohta, 1971), an adequate theory of evolution at the molecular level must be able to explain both between-species molecular differences and within-species polymorphisms. At the present time, there is considerable agreement that evolutionary changes at noncoding regions of eukaryotic genomes are due largely to neutral or nearly neutral mutations. In protein-coding regions, however, it remains hotly debated to whether selection or mutation is the major cause of evolution.

Some Future Directions. Active adoption of new

techniques from other fields of life science constantly adds new dimensions to the field of molecular evolution, and the list of research topics in molecular evolution is ever increasing. It is therefore difficult to predict the future directions of this field, but here we suggest that the following three areas will receive much attention in the immediate future. First, detecting positive Darwinian selection in a DNA or protein sequence is an actively pursued subject (Yang and Bielawski, 2000; Anisimova et al., 2001). Positive selection has traditionally been considered the most important type of selection because it leads to adaptive evolution. This subject is theoretically challenging. In this regard, the huge amounts of available sequence data provide wealth of material for data analysis.

Second, the evolution of duplicate genes has received much attention recently (for example, see Lynch and Conery, 2000). Gene duplication is considered the most important step in the emergence of genetic novelties because duplication creates a redundant copy, which can serve as the raw material for the evolution of new function. However, how duplicate genes survive and evolve has not been well understood. Now the completion of several eukaryotic genomes provides unprecedented opportunity for studying the evolution of duplicate genes at the genomic scale. It will be interesting to see the extent of gene duplication in a genome, which types of duplicate genes are common among different genomes and which types of duplicate genes are unique to an evolutionary lineage, how often acceleration in rate of evolution has occurred, and how often divergence in expression pattern has occurred among duplicate genes.

Third, evolution of genomic sequences is an emerging topic. Investigating genomic organization has long been of interest to molecular evolutionists, particularly with regard to the origin of the "genes in pieces" found in eukaryotes, with coding exons being separated by noncoding introns. The debate over whether the uninterrupted genes (few if any introns) found in most prokaryotes represent the ancestral condition goes back two decades, and it has been fueled by the discovery of phenomena such as self-splicing introns (reviewed in Li, 1997).

The new wealth of whole-genomic sequences demands that we examine the evolution of heterogeneous sequences that may contain coding sequences, noncoding sequences, repetitive elements, regulatory elements, and so on. It also demands that we take a large-scale view of evolution (e.g., do the pattern and rate of evolution vary among genomic regions? And if so, why?) The longstanding question of the evolution of genome size may also be answered by comparing whole genome sequences from diverse organisms. Analyzing genomic sequence data requires different methods than traditional ones. For example, the alignment is much trickier because of the length and heterogeneity of the sequences to be aligned. Indeed, new algorithms and theories to accommodate this new type of data are eagerly developed.

There are many potential future developments of molecular evolution. For example, molecular scientists are increasingly recognizing the need for an evolutionary interpretation of newly acquired data. As a result, many of the computational genomic tools borrow principles of molecular evolution to understand the conservation and divergence of genomes. Another promising development with great bearing upon molecular evolution is the field of functional genomics. Testing functional evolutionary hypotheses at the molecular level has been a difficult task because of the long timescale of molecular evolution. However, it is now possible to conceive experimental approaches to test certain functional evolutionary hypotheses. For instance, we can test an evolutionary hypothesis regarding the role of a newly evolved regulatory sequence through a comparative analysis of transcription levels in different tissues or under different conditions by using expression microarrays.

In conclusion, the field of molecular evolution is itself an ever-evolving discipline, often coevolving with other disciplines of biology. It has greatly benefited from the rapidly advancing technology of molecular biology and the unmatched breadth of evolutionary biology. Both traditional and novel questions of evolution can now be investigated at many different biological levels, including genomes. It is becoming clear that without understanding the principles of molecular evolution, a wholistic appreciation and understanding of the evolution of life would be impossible.

[*See also* Codon Usage Bias; DNA and RNA; Phylogenetic Inference, *article on* Methods.]

BIBLIOGRAPHY

Anisimova, M., J. P. Bielawski, and Z. Yang. "Accuracy and Power of the Likelihood Ratio Test in Detecting Adaptive Molecular Evolution." *Molecular Biology and Evolution* 18 (2001): 1585–1592.

Clarke, B. "Selective Constraints on Amino Acid Substitutions During the Evolution of Proteins." *Nature* 228 (1970): 159–160.

Goodman, M. "The Role of Immunochemical Differences in the Phyletic Development of Human Behavior." *Human Biology* 33 (1961): 131–162.

Hurst, L. D., and H. Ellegren. "Sex Biases in the Mutation Rate." *Trends in Genetics* 14 (1998): 446–452.

Jukes, T. H., and C. R. Cantor. "Evolution of Protein Molecules." In *Mammalian Protein Metabolism*, edited by H. N. Munro, pp. 21–123. New York, 1969.

Kimura, M. "Evolutionary Rate at the Molecular Level." *Nature* 217 (1968): 624–626.

Kimura, M. "A Simple Method for Estimating Evolutionary Rates of Base Substitutions through Comparative Studies of Nucleotide Sequences." *Journal of Molecular Evolution* 16 (1980): 111–120.

Kimura, M., and T. Ohta. "Protein Polymorphism as a Phase of Molecular Evolution." *Nature* 229 (1971): 467–469.

King, J. L., and T. H. Jukes. "Non-Darwinian Evolution." *Science* 164 (1969): 788–798.

Kishino, H., J. Thorne, and W. Bruno. "Performance of a Divergence Time Estimation Method under a Probabilistic Model of Rate Evolution." *Molecular Biology and Evolution* 18 (2001): 352–361.

Li, W-H. *Molecular Evolution*. Sunderland, Mass., 1997.

Lynch, M., and J. S. Conery. "The Evolutionary Fate and Consequences of Duplicate Genes." *Science* 290 (2000): 1151–1155.

Needleman, S. B., and C. D. Wunsch. "A General Method Applicable to the Search of Similarities in the Amino Acid Sequence of Two Proteins." *Journal of Molecular Biology* 48 (1970): 443–453.

Nuttall, G. H. F. *Blood Immunity and Blood Relationship*. Cambridge, 1904.

Richmond, R. C. "Non-Darwinian Evolution: A Critique." *Nature* 225 (1970): 1025–1028.

Sanderson, M. J. "A Nonparametric Approach to Estimating Divergence Times in the Absence of Rate Constancy." *Molecular Biology and Evolution* 14 (1997): 1218–1232.

Sellers, P. H. "On the Theory and Computation of Evolutionary Distances." *SIAM Journal of Applied Mathematics* 26 (1974): 787–793.

Waterman, M. S., T. F. Smith, and W. A. Beyer. "Some Biological Sequence Metrics." *Advances in Mathematics* 20 (1976): 367–387.

Yang, Z., and J. P. Bielawski. "Statistical Methods for Detecting Molecular Adaptation." *Trends in Ecology & Evolution* 15 (2000): 496–503.

Zuckerkandl, E., and L. Pauling. "Evolutionary Divergence and Convergence in Proteins." In *Horizons in Biochemistry*, edited by V. Bryson, and H. J. Vogel, pp. 97–166. New York, 1965.

— WEN-HSIUNG LI AND SOOJIN YI

MOLECULAR SYSTEMATICS

Systematists seek to detect, describe, and explain biological diversity. Systematics is thus one of the oldest areas of biological research, a primary field of biology for as long as there have been biologists. Up until the second half of the twentieth century, however, almost all systematic biological research was based on morphological (or, less commonly, behavioral) data. Specimens were collected, and their morphological traits were used to determine species boundaries, assess phylogenetic relationships, study hybridization, and analyze geographic variation. Although morphology and behavior continue to be important for systematic studies today, there is now increasing reliance on molecular characteristics to study biodiversity. The use of protein and nucleic acid data to study variation among biological organisms comprises a field known as *molecular systematics*.

Although the earliest efforts to use protein data in systematic biology date back to at least 1904 (Nutall, 1904), the field of molecular systematics did not begin to develop in any significant way until the 1960s. Until then, the only techniques that were widely used to compare homologous proteins in different species were based on immunological comparisons (see Maxson and Maxson, 1990). These techniques involved producing antibodies (often in a rabbit) to a protein from one species, and then measuring the degree of reaction of these antibodies to the homologous protein from a second species. The more similar the two proteins were (i.e., the fewer amino acid replacements that had occurred between the two), the stronger the reaction. Thus, immunological techniques produced an indirect measure of the similarity of the two related proteins in different samples (usually, different species). Although the techniques were relatively expensive and technically demanding, and they produced only indirect estimates of amino acid differences in the proteins, they were one of the principal means for making comparisons among relatively distantly related species before the advent of DNA sequencing. Various immunological techniques were used through the 1980s, but these have rarely been used for systematic purposes since then.

During the late 1950s and 1960s, techniques of protein electrophoresis were developed and popularized (see Murphy et al., 1996, for background and history). Amino acid replacements in proteins often produce changes in the net charge and/or shape of the protein. Thus, allelic variants of proteins move at different rates through an electric field in a porous medium (typically a gel). The development of histological staining techniques for many enzymes allowed biologists to separate thousands of proteins in crude tissue extracts simultaneously in an electrophoretic field, and then to visualize the location of up to 100 or more specific enzymes, using enzyme-specific staining. Using these techniques, alleleic variants of enzymes (called allozymes) can be identified within and among individual organisms, and variation at dozens of different genetic loci can be examined. Early studies of allozyme electrophoresis revealed a vast and initially unexpected level of genetic variation in many populations and species (e.g., Hubby and Lewontin, 1966). This technique remains one of the fastest and least expensive methods for comparing genetic variation across large numbers of independent loci, although use of the method has dropped precipitously since the early 1990s in favor of nucleic-acid-based methods.

Although the structure of DNA was deduced in the early 1950s and its importance as the genetic basis of all biological variation was understood soon thereafter, it was not until the 1960s that the first systematic studies of variation at the DNA level were undertaken. The first method used to assess variation in homologous regions of DNA was DNA hybridization (Wetmur and Davidson, 1968). The technique was widely applied to systematic problems in the 1970s and 1980s (Werman et al., 1996), but DNA hybridization was largely replaced by DNA sequencing in systematic investigations by the 1990s. DNA hybridization involves separating the two strands of DNA (breaking the hydrogen bonds by heating), and then hybridizing the homologous strands between two different species. Nucleotide substitutions between homologous regions reduce the number of hydrogen bonds that can form within hybrid DNA molecules, and this in turn reduces the temperature at which the two strands will separate. After the hybrid DNA molecules are formed, they are slowly heated, and the temperature at which half the strands separate is measured. Hybridized homologous regions of DNA from different species will separate at low temperature if they differ by many nucleic acid substitutions, but at higher temperature if they differ by fewer nucleic acid substitutions. Thus, melting temperature of the hybrid molecules provides an indirect measure of divergence between the two homologous regions. The advantage of DNA hybridization is that it can provide a measure of divergence across the entire single-copy component of the genome. The primary disadvantages are that all the variation is reduced to a single, average measure of divergence, and thus the level of possible resolution is minimal. Other disadvantages—the technical difficulties of separating repeated gene sequences in the DNA hybridization experiments, the analytical difficulty of distinguishing homologous from homoplastic similarity in DNA sequences, and the need to make all pairwise comparisons—have greatly reduced the application of the technique in systematic investigations.

Enzymes that cleave DNA at specific recognition sequences were first isolated from bacteria in 1970 (see review by Nathans and Smith, 1975). These enzymes are known as restriction endonucleases (REs). By the early 1980s, more than 400 REs had been isolated and characterized (Roberts, 1984). When a particular sequence of DNA is cleaved by a particular RE, a reproducible array of fragments of varying sizes is produced. Analysis of these fragments (or the mapped restriction sites) results in a rapid method of identifying and assessing DNA variation in specific genes. Restriction-site analysis rapidly became an important method of molecular systematics in the 1980s. The methods are still widely used for studying genetic variation across many individuals or many loci (Dowling et al., 1996). Other methods for comparing particular DNA fragments across individuals and species have also been developed. In particular, it is now common to examine variation in copy number of short, tandem-repeated DNA sequences called microsatellites (Tautz, 1989), as well as to use to the genomic location of insertions of short or long interspersed elements (SINES and LINES, respectively) as systematic characters (e.g., Nikaido et al., 1999).

Despite the diversity of molecular techniques that are applied to systematic problems, today molecular systematics is dominated by studies of DNA sequence variation. Practical methods for sequencing long, specific regions of DNA were developed in 1977 (Sanger et al., 1977; Maxam and Gilbert, 1977), and these were routinely used in systematic studies beginning in the early 1980s. However, at first most target sequences of DNA had to be cloned before they could be sequenced, which required considerable time, effort, and expense. The development of the polymerase chain reaction (PCR) to amplify specific regions of DNA in vitro using a thermally stable DNA polymerase removed this obstacle (Mullis and Faloona, 1987). By the late 1980s, DNA sequencing was becoming the method of choice for many systematic studies (Hillis et al., 1996), and new applications of DNA sequencing led to an exponential increase in studies of molecular systematics. Today, DNA sequencing is the most widely used method in molecular systematic investigations. The primary advantages of DNA sequencing are that it provides a large number of characters (potentially equal to the total number of independent genetically determined characters in an organism, if whole genomes are sequenced), and that DNA sequences can now be collected rapidly and easily from virtually any organism. Thus, it is possible to make comparisons at almost any level of biological hierarchy, from recently evolved allelic variants within individuals to homologous genes that are conserved across all of life.

Major Applications of Molecular Systematics. Almost nothing in biology is completely independent; genes, individuals, populations, and species are all linked to one another through their histories. This fact makes biology both complex and interesting, but it also means that any study in biology that concerns more than one thing (e.g., gene, individual, species) requires information on the interrelationships of the units of study to understand and interpret the underlying variation. For this reason, studies of molecular systematics have become nearly ubiquitous in biology. Although molecular systematics had its roots in taxonomy, today it is hard to find a biological journal in any field that does not publish phylogenetic trees in support of its published research articles. Background molecular systematic results have become almost required in fields as diverse as viral epidemiology, molecular evolution, and comparative behavioral ecology.

At the level of the genome, molecular systematic studies inform biologists about the relationships among genes, how the genes are distributed across populations and species, and which alleles are ancestral and derived. Studies at this level have concerned diverse questions: for example, how do new alleles arise, and why do new alleles often appear in hybrid zones (Bradley et al., 1993)? How do repeated gene arrays evolve in concert (Hillis et al., 1991)? What amino acid substitutions were involved in particular adaptations, such as the evolution of foregut fermentation (Stewart and Wilson, 1987; Jermann et al., 1995)?

Molecular systematic studies have also been critical for understanding and analyzing populational variation. For instance, molecular systematic data have been used to distinguish between two distinct hypotheses that have been formulated to account for the evolution of lactose tolerance in adults of a few human populations. In most human populations, adult humans cannot digest lactose. Adult lactose tolerance is broadly correlated with high latitudes (it is common in many European populations, for instance), but it is also correlated with cultures that keep dairy animals. Thus, one hypothesis posits that lactose tolerance is functional related to reduced exposure to sunlight. Sunlight promotes vitamin D production, which promotes calcium deposition; this hypothesis suggests that the enzyme lactase (which both allows lactose digestion and promotes calcium deposition) evolved to become expressed in some northern adult human populations to make up for the lack of exposure to sunlight (Durham, 1991). However, a study of relationships among human populations (Holden and Mace, 1997) showed that lactose tolerance has evolved independently as much as three times in dairy cultures, and also that the latitudinal correlation is spurious. The study further supported the evolution of lactose tolerance *after* the adoption of dairying, so it does not appear that preexisting lactose-tolerant populations were simply more likely to start keeping milk-producing animals.

Molecular systematic methods have also revolutionized epidemiological studies of human (and other) pathogens. New viral diseases are now often identified from phylogenetic studies. For instance, in 1993, there was an outbreak of an unknown disease in the southwestern United States that caused a number of deaths. Prior to the advent of molecular systematic methods, the identification and characterization of the causative pathogen would have taken many years of painstaking investigation. However, PCR amplification, sequencing, and phylogenetic analysis allowed rapid identification of a new virus, now known as the Sin Nombre virus, which was further identified as being closely related to Asian viruses that caused hantavirus pulmonary syndrome. Furthermore, the hosts of the virus—deer mice—were also quickly identified, and additional systematic studies revealed a large number of related viruses in other species of rodents (Morzunov et al., 1998). Once the virus and its hosts had been identified, it was relatively straightforward to implement measures to reduce human infections.

Even if the identity of a pathogen is already established, molecular systematic studies are critical for understanding various aspects of the pathogen's biology that are important for epidemiological investigations. In the case of human immunodeficiency viruses (HIVs), for instance, phylogenetic studies have been used for such diverse purposes as inferring the places and hosts of origin, dating the start of the HIV epidemic, understanding global diversity of HIV for planning vaccine programs, and tracing HIV infections among individuals in health care settings (see reviews in Crandall, 1999, and Rodrigo and Learn, 2001). Because HIVs evolve so quickly, the study of HIV/AIDS would be greatly hampered without molecular systematic studies. Phylogenetic analyses have proved useful for the study of many other viruses as well, and they have even been used to predict the future diversity of influenza viral strains (Bush et al., 1999).

The study of geographic variation of species has long been of interest to biologists and has taken on renewed interest to biogeographers and population biologists who wish to link biological and Earth history, study the origin of new species, or study interactions and hybridization between species. Molecular systematic studies have had major influence in these fields (e.g., Avise, 2000). Of particular interest to humans has been the question of our own geographical origin. Molecular systematic studies of many genes now indicates that modern humans arose in Africa and subsequently moved out of Africa and largely replaced *Homo* populations that were living elsewhere around the globe (see Stoneking, 1997).

Another application of molecular systematic methods involves the identification of previously unknown species. To date, biologists have discovered and described about 1.8 million species, but they estimate that there are many more undescribed species than described species (current estimates range from about 10 to 100 million total species). There are efforts in place to describe all the world's biodiversity (that is, to discover and describe all the species of planet Earth) within the next 25 years (for instance, see the goals of the All Species Foundation at http://www.all-species.org/). This requires a speed-up in species discovery and description that will require automated species identification. These methods will almost certainly include molecular systematic comparisons.

In addition to simply discovering and describing all species in the world, biologists are working to build a complete Tree of Life that depicts the relationships of all the world's biota (see Hillis and Holder, 2000). This would provide a framework for biology similar in nature (but much more complex and detailed) to that provided to chemistry through the Periodic Table of Elements. Molecular systematics contributes to this endeavor by providing virtually the only data that can be compared and analyzed across all of life. Very few morphological or behavioral features can be compared among elephants, oak trees, and *E. coli* bacteria, for instance, but it is straightforward to compare highly conserved genes (such as the ribosomal RNA genes) among all these groups (and indeed, across all of life; see Pace et al., 1986). Obviously, any estimate of the Tree of Life will need to combine data and analyses from many sources and many genes, but clearly molecular analyses will play an important role. There are many computational, logistic, and even political obstacles to completing a Tree of Life, but the benefits for the study of biology are great enough that biologists are willing to tackle these challenges. Already, there are major efforts under way by various individuals to collect and maintain information on the relationships across all of life (e.g., see http://phylogeny.arizona.edu/tree/phylogeny.html).

Why are phylogenies of such interest to biologists? All units of study in biology are connected in varying degrees to all other biological units (through their historical connections), so phylogenetic information is needed to interpret and analyze any variation of interest to biologists. The use of phylogenies to interpret and analyze biological variation has become known as the comparative method (see review by Harvey and Pagel, 1991). The realization of the importance of accounting for phylogenetic information throughout biological studies has been one of three primary driving forces in the increased emphasis in phylogenetic inference. The other two driving forces have been the development and implementation of explicit methods for inferring phylogenies (see Swofford et al., 1996) and the development of methods of molecular biology (in particular, DNA sequencing). Molecular systematics is the intersection of all these areas of evolutionary investigation.

[*See also* Molecular Evolution; Phylogenetic Inference.]

BIBLIOGRAPHY

Avise, J. C. *Phylogeography: The History and Formation of Species.* Sunderland, Mass., 2000.

Bradley, R. D., J. J. Bull, A. D. Johnson, and D. M. Hillis. "Origin of a Novel Allele in a Mammalian Hybrid Zone." *Proceedings of the National Academy of Sciences USA* 90 (1993): 8939–8941.

Bush, R. M., C. A. Bender, K. Subbarao, N. J. Cox, and W. M. Fitch. "Predicting the Evolution of Human Influenza A." *Science* 286 (1999): 1921–1925.

Crandall, K. A. ed. *The Evolution of HIV.* Baltimore, 1999.

Dowling, et al. "Nucleic Acids III: Analysis of Fragments and Restriction Sites." In *Molecular Systematics*, 2d ed., edited by D. M. Hillis, B. K. Mable, and C. Moritz, pp. 249–320. Sunderland, Mass., 1996.

Durham, W. *Coevolution: Genes, Culture, and Human Diversity.* Stanford, 1991.

Harvey, P. H., and M. D. Pagel. *The Comparative Method in Evolutionary Biology.* Oxford, 1991.

Hillis, D. M., C. Moritz, C. A. Porter, and R. J. Baker. "Evidence for Biased Gene Conversion in Concerted Evolution of Ribosomal DNA." *Science* 251 (1991): 308–310.

Hillis, D. M., B. K. Mable, A. Larson, S. K. Davis, and E. A. Zimmer. "Nucleic Acids IV: Sequencing and Cloning." In *Molecular Systematics*, 2d ed., edited by D. M. Hillis, B. K. Mable, and C. Moritz, pp. 321–381. Sunderland, Mass., 1996.

Hillis, D. M., and M. T. Holder. "Reconstructing the Tree of Life." In *New Technologies for the Life Sciences: A Trends Guide* (Dec. 2000): 47–50.

Holden, C., and R. Mace. "A Phylogenetic Analysis of the Evolution of Lactose Digestion." *Human Biology* 69 (1997): 605–628.

Hubby, J. L., and R. C. Lewontin. "A Molecular Approach to the Study of Genic Heterozygosity in Natural Populations. I. The Number of Alleles at Different Loci in *Drosophila pseudoobscura.*" *Genetics* 52 (1966): 203–215.

Jermann, T. M., J. G. Opitz, J. Stackhouse, and S. A. Benner. "Reconstructing the Evolutionary History of the Artiodactyl Ribonuclease Superfamily." *Nature* 374 (1995): 57–59.

Maxam, A. M., and W. Gilbert. "A New Method for Sequencing DNA." *Proceedings of the National Academy of Sciences USA* 74 (1977): 560–564.

Maxson, L. R., and R. D. Maxson. "Proteins II: Immunological Techniques." In *Molecular Systematics*, edited by D. M. Hillis and C. Moritz, pp. 127–155. Sunderland, Mass., 1990.

Morzunov, S. P., J. E. Rowe, T. G. Ksiazek, C. J. Peters, S. C. St. Jeor, and S. T. Nichol. "Genetic Analysis of the Diversity and Origin of Hantaviruses in *Peromyscus leucopus* Mice in North America." *Journal of Virology* 72 (1998): 57–64.

Mullis, K. B., and F. A. Faloona. "Specific Synthesis of DNA *in vitro* via a Polymerase Catalyzed Chain Reaction." *Methods in Enzymology* 155 (1987): 335–350.

Murphy, R. W., J. W. Sites, Jr., D. G. Buth, and C. H. Haufler. "Proteins: Isozyme Electrophoresis." In *Molecular Systematics*, 2d ed., edited by D. M. Hillis, B. K. Mable, and C. Moritz, pp. 51–120. Sunderland, Mass., 1996.

Nathans, D., and H. O. Smith. "Restriction Endonucleases in the Analysis and Restructuring of DNA Molecules." *Annual Review of Biochemistry* 44 (1975). 273–293.

Nikaido, M., A. P. Rooney, and N. Okada. "Phylogenetic Relationships among Cetartiodactyls Based on Insertions of Short and Long Interpersed Elements: Hippopotamuses Are the Closest Extant Relatives of Whales." *Proceedings of the National Academy of Sciences USA* 96 (1999): 10261–10266.

Nutall, G. H. F.. *Blood Immunity and Blood Relationship.* Cambridge, 1904.

Pace, N. R., G. J. Olsen, and C. R. Woese. "Ribosomal RNA Phylogeny and the Primary Lines of Evolutionary Descent." *Cell* 45 (1986): 325–326.

Pagel, M. "Inferring the Historical Patterns of Biological Evolution." *Nature* 401 (1999): 877–884.

Roberts, R. J.. "Restriction and Modification Enzymes and Their Recognition Sequences." *Nucleic Acids Research* 12 (1984): r167–r204.

Rodrigo, A. G., and G. H. Learn, Jr., eds. *Computational and Evolutionary Analysis of HIV Molecular Sequences.* Norwell, Mass., 2001.

Sanger, F., S. Nicklen, and A. R. Coulson. "DNA Sequencing with Chain-Terminating Inhibitors." *Proceedings of the National Academy of Sciences USA* 74 (1977): 5463–5467.

Stewart, C.-B., and A. C. Wilson. "Sequence Convergence and Functional Adaptation of Stomach Lysozymes from Foregut Fermenters." *Cold Spring Harbor Symposium on Quantitative Biology* 52 (1987): 891–899.

Stoneking, M.. "Recent African Origin of Human Mitochondrial DNA: Review of the Evidence and Current Status of the Hypothesis." In *Progress in Population Genetics and Human*

Evolution, edited by P. Donnelly and S. Tavaré, pp. 1–13. New York, 1997.

Swofford, D. L., G. J. Olsen, P. J. Waddell, and D. M. Hillis. "Phylogenetic Inference." In *Molecular Systematics*, edited by D. M. Hillis, B. K. Mable, and C. Moritz, pp. 407–514. Sunderland, Mass., 1996.

Tautz, D. "Hypervariability of Simple Sequences as a General Source for Polymorphic DNA Markers." *Nucleic Acids Research* 17 (1989): 6463–6471.

Werman, S. D., M. S. Springer, and R. J. Britten. "Nucleic Acids I: DNA-DNA Hybridization." In *Molecular Systematics*, 2d ed., edited by D. M. Hillis, B. K. Mable, and C. Moritz, pp. 169–203. Sunderland, Mass., 1996.

Wetmur, J. G., and N. Davidson. "Kinetics of Renaturation of DNA." *Journal of Molecular Biology* 31 (1968): 349–370.

— DAVID M. HILLIS

MOLLUSCS

The great phylum Mollusca is morphologically the most diverse in the animal kingdom and is second only to the Arthropoda in number of known species and in the range of environments it successfully exploits on earth. Estimates vary widely, but the number of living mollusc species probably exceeds 60,000. Moreover, their durable shells provide an excellent fossil record that extends back to the beginning of Phanerozoic time (545 million years ago) and comprises some 25,000 additional species. Modern molluscs occur almost everywhere, from montane forests to the greatest depths of the sea (>10 km). The phylum almost certainly originated in the sea, and the majority of its extant species are marine. The largest class, the Gastropoda or univalves, has also radiated extensively into both terrestrial and freshwater environments, and the second largest class, the Bivalvia, is also well represented in freshwater.

Molluscs vary in size from 1 millimeter to 16 meters in length. They may be extremely abundant (up to 40,000 small snails on intertidal flats in Europe) and ecologically important as herbivores, predators, suspension and deposit feeders, prey of other animals, and hosts and vectors of parasites. Many molluscs are of direct importance to humans, as food, jewelry and other decoration, agricultural pests, and intermediate hosts in schistosomiasis and other parasitic diseases.

Anatomy: What Makes a Mollusc? Molluscs are complex animals; that is, they are at the organ-system grade of biological organization. All of the organ systems that any animal group has evolved are present and well developed in the molluscs, which share the anatomical features of other animals of that grade. These include (1) three fundamental cell layers, (2) bilateral symmetry with many paired structures, and (3) a coelom or secondary body cavity lined by a peritoneum. However, the coelom is small and the major body cavity that houses the internal organs is a haemocoel, as in the Arthropoda.

Although great structural diversity characterizes the Mollusca, in contrast to such other major groups as the arthropods and chordates, molluscs lack a hallmark or phylotypic character—they have no single key feature that is found at some stage of the life cycle in all members of the phylum. Rather, molluscs are recognized by an array or combination of traits, here termed the molluscan body plan.

The molluscan body plan. The molluscan body is soft—the word *mollusc* comes from the Latin *mollis*, *molluscus*, meaning soft—and its component parts are not sharply demarcated from each other. Most molluscs completely lack segmentation, but some body parts in some molluscs, both living and fossil, show hints of segmentation. The molluscan body may be subdivided into a head, a foot, and a visceral mass covered by a skin, called the mantle, that secretes a shell made of calcium carbonate ($CaCO_3$).

These main molluscan body parts and their spatial relation to one another differ markedly from the other animal phyla that are related to Mollusca by the shared possession of spiralian early development, the trochophore larva, and molecular-genetic characteristics such as 18S rRNA and *Hox* gene sequences (Aguinaldo et al., 1997; Rosa et al., 1999). The other main phyla in this group, sometimes now termed the Subkingdom Lophotrochozoa, are the Sipuncula, Annelida, and Platyhelminthes. In these phyla, metamorphosis of the trochophore larva to the juvenile maintains the anterior–posterior axis as the only major body axis. In the Mollusca, however, new structures, especially the visceral mass and mantle-shell, are arranged along a new or secondary body axis that extends from posterodorsal to anteroventral. This is not to say that molluscs have lost their primary body axis. Sometimes called the somatic axis, their primary anterior–posterior axis includes the head and foot and governs the animal's locomotion, sensory and feeding activities, and its muscles power the major functional mechanisms. The head bears sense organs, the brain, and the mouth, although it has been lost in one major group. The molluscan foot is usually a single, broad, creeping structure with a flat sole that has been variously modified (e.g., for digging, for secretion of byssal attachment threads, or as a sucker); it is rarely lost. The visceral mass comprises the internal organs that are concentrated under the mantle, which secretes the shell. The shell is made of $CaCO_3$ crystals with a very thin scleroprotein matrix between them. Its form is extremely variable; it may consist of one, two or eight pieces, often called valves, depending on the class. The shell is usually outside the mantle but may be enveloped within it, and in some groups it is reduced to calcareous spicules or completely absent.

The secondary axis is not an axis of symmetry; rather, it is an axis of differentiation, with the structures of the visceral mass, mantle, and shell arranged along it. The functions of these structures include respiration, digestion, cleansing, and reproduction, and their major power source is ciliary rather than muscular.

Mantle, shell, and mantle cavity. The mantle is a skin covering the visceral mass and continuous with that of the head and foot. Typically it extends laterally beyond the visceral mass as a double-walled skirt, leaving a space between the mantle and visceral mass and the foot, created by the "eaves" or extensions of the mantle. Although, strictly speaking, this space, the mantle cavity, is outside the body, it is extremely important and characteristic of the phylum. Within it are the paired gills, whose specific structure is also an important phylum characteristic. Each gill, called a ctenidium, is formed of a double row of ciliated leaflets arranged like the pages in a book. Some of the cilia beat so as to draw respiratory water currents into the mantle cavity, others filter particles out, and others transport the caught particles so they can be swept out of the mantle cavity to prevent fouling the gills. In some molluscs, the trapped particles of organic matter are transferred to the mouth for food. Typically the inhalent respiratory current enters the mantle cavity ventrally and at the sides, passes through the gills—that is between the pages—and passes out as an exhalent flow dorsally, past the excretory, anal, and reproductive openings, and out of the mantle cavity. The gills thus divide the mantle cavity into inhalent and exhalent chambers, and the ciliary-directed flow pattern ensures separation of inhalent from exhalent flows.

Internal anatomy. The alimentary system begins with the mouth, usually located subterminally on the head. One or a pair of jaws sometimes is present within the mouth, which also contains the closest thing to a hallmark in the phylum: the radula, present in all classes but one. This complex structure is a mechanical food processor, typically a combined rasp and conveyor belt, that is mounted rather like the human tongue in the back of the mouth. It is supported by stiff cartilages and powered by strong muscles. It can be extended out of the mouth to remove bits of food and convey them backward to the esophagus. This leads to a stomach, where ducts of the usually large digestive glands open and where chemical processing of food occurs. The molluscan intestine typically functions in molding feces, and the alimentary system ends with the rectum and anus.

Most molluscs have a central nervous system with the nerve cell bodies concentrated in swellings called ganglia, particularly on a nerve ring around the anterior region of the gut. These ganglia are typically paired left and right and, particularly in the gastropods and cephalopods, several are concentrated in a brain in the head. The different ganglia serve the different body regions described above—the head, foot, mantle, and visceral mass. Nerve commissures join the left and right members of each pair, and nerve connectives join the different ganglia. The most important sensory modality in most molluscs is chemoreception, a combination of smell and taste. Most molluscs have eyes, those of cephalopods being among the most acute in the animal kingdom, and they are also sensitive to vibrational, tactile, and positional stimuli.

The molluscan heart is typically three-chambered, with two auricles each receiving blood from the gill or gills on one side, and one ventricle pumping oxygenated blood via anterior and posterior arteries to the body organs via large blood spaces that comprise the hemocoel. The system is open; that is, capillaries and veins are rare, except in the gills. The hemocoel also often functions as a hydrostatic skeleton, providing both rigidity and plasticity or body shape change. The typical molluscan respiratory pigment is the copper-containing protein hemocyanin, dissolved in the blood, but a few molluscs have the more efficient hemoglobin.

In most complex animals such as vertebrates, the coelom surrounds most of the viscera and is the most important body cavity. In molluscs, however, the coelom is reduced to three parts that usually interconnect: the pericardium that surrounds the heart and provides space for its muscle to expand and contract, and the excretory and reproductive systems. Molluscan excretory organs or kidneys are basically a pair of ducts that receive a filtrate of blood from the pericardium and empty into the mantle cavity.

The sexes of most molluscs are separate, and fertilization may be either internal or external. Molluscan embryonic developmental patterns reveal basic similarities with other animal phyla that elucidate their evolutionary relationships. Early cleavages of the fertilized egg are unequal and result in a specific and complex spiral pattern of daughter cells. This pattern, as well as the subsequent ontogeny summarized below, is strikingly similar to the other phyla now grouped as Lophotrochozoa—the flatworms, annelids, sipunculans, echiurans, and nemerteans—with cell lineages corresponding exactly cell-for-cell in some cases. Development is determinate; that is, separation of cells at the two-cell stage results in two half-embryos rather than twins. Following gastrulation, molluscs pass through a trochophore stage in which the mouth appears at the site where the blastopore of the gastrula disappeared. This condition is called protostomous, and the group of listed phyla characterized by it are usually known as the Protostomia, differentiating them from the deuterostomous phyla (chordates, hemichordates, and echinoderms). This classification is now strongly supported by molecular-genetic evidence, as is the further classification of the Mollusca with the other Lophotrochozoan phyla (see Aguinaldo et al., 1997; Erwin, 1999).

Following the trochophore stage of development, basic molluscan ontogeny diverges markedly from that in the other lophotrochozoan phyla. The major difference is posterodorsal enlargement of the trochophore—that is, expansion along the secondary axis that is lacking in the other groups. The ectoderm of this enlargement forms the shell gland initially and, later, the mantle that secretes the shell. The prototroch, a band of heavily ciliated cells around the trochophore anterior to the mouth, enlarges to form the velum, a more substantial organ for swimming, feeding, and supporting the added weight of the growing shell, foot, and other structures. Hatching of the swimming larva, now called a veliger, typically occurs at this stage. Typical veliger larvae feed and grow, eventually becoming too heavy for the velum to support, and then settling to the sea floor to metamorphose and complete development of juvenile and adult organs.

The Molluscan Classes. The extant members of the Mollusca are usually classified in seven or eight classes, as follows:

Class Aplacophora. Members of this second smallest class (about 150 species) are all marine, occurring from the shallow subtidal region to a depth of 4,000 meters (13,000 feet). The animals are wormlike, and the extant species lack shells (the class name means "without plates"). Rather, the mantle is thickly invested with many $CaCO_3$ spicules. A recently discovered Silurian fossil has seven dorsal shell plates (Sutton et al., 2001). The head and foot are poorly developed in Aplacophora, and there is no visceral mass or secondary axis. The mantle cavity is small and posterior, and may lack gills, but the radula is well developed. The class is divided into two subclasses, each of which is sometimes raised to class status because a recent phylogenetic analysis indicates Aplacophora to be paraphyletic. Members of the larger Class or Subclass Solenogastres (about 125 species) are parasitic predators on Cnidaria. They lack ctenidia but have a more distinct head than the Caudofoveata, which are infaunal burrowers feeding on mud with the head down and posterior end up. According to recent morphologically based cladistic analyses (Salvini-Plawen and Steiner, 1996; Haszprunar, 2000), the Class Solenogastres is the sister taxon of all other molluscan classes, while the Caudofoveata is the sister taxon of the six remaining classes.

Class Monoplacophora or Tryblidia. Known only as Lower Cambrian-to-Devonian (c.525–350 million years ago) fossils until 1957, when the first living representative was discovered, this class remains the smallest in number of extant species, with about 20. More than 100 Paleozoic and Mesozoic fossil genera are known. All are marine, and the known living members occur on hard substrates at depths of 180–4,000 meters (600–13,000 feet). Superficially, living monoplacophorans (the name means "one plate") resemble limpetlike gastropods. However, they are bilaterally symmetrical, with the anterior head directed ventrally, a broad, circular, flat but weak foot, a low, conical visceral mass covered by the shell, and the mantle cavity lateral. The radula is large and strong, but little is known of the diet except that muddy sediment has been found in the guts. Monoplacophora have more serially repeated structures than other molluscs: living members have 5–8 pairs of shell muscles, a ladderlike nervous system without ganglia but with two pairs of longitudinal nerves joined by 10 pairs of connectives, 6 pairs of kidneys, 5–6 pairs of gills each consisting of several fingerlike filaments, two pairs of gonads, and two pairs of auricles. Fossil monoplacophorans are much more diverse in shell form, and many bear scars on the inside of the shell indicating placement of the muscles that attached the body to the shell.

Class Polyplacophora. Commonly called chitons, this is one of the two most morphologically uniform molluscan classes. Its oldest known fossils probably date from the Upper Cambrian, and increasing numbers throughout the later Paleozoic are being documented (Hoare, 2000). The class comprises about 600 extant and about 350 fossil species, most of which occur on hard substrata in shallow water, where they graze on algae, using a radula with teeth hardened by iron compounds. However, some chitons occur in the deep sea, to depths of 7,000 meters (23,000 feet). Several aspects of anatomy resemble the Monoplacophora: the head is directed ventrally in front of a broad, flat foot, and the visceral mass is depressed. The mantle cavity is lateral to the foot and shelters many pairs of gills. However, unlike the Monoplacophora, the gills are typical molluscan ctenidia, and the mantle secretes a shell divided into eight plates or valves. The shell valves differ from those in other classes in having a rather soft outer layer, perforated by extensions of the mantle, that bears photoreceptors and tactile sense organs. The inner layer or layers of each valve extend under the valve in front of it, providing attachment for muscles that give the body strength and flexibility. Internally, the nervous system resembles that of Monoplacophora, as nerve cell bodies are distributed throughout, and ganglionation is minimal.

Class Gastropoda. This is by far the largest and most heterogeneous class of molluscs, with about 35,000 living and perhaps 25,000 fossil species, dating back to the Early Cambrian. They occur in the sea and in freshwater, and they are the only molluscs to have successfully invaded the land. Probably more than half the extant species are terrestrial. Gastropods may be defined as molluscs that have undergone torsion, a 180° twist of the visceral mass on the head and foot, around the secondary axis, resulting in an anterior mantle cavity, a U-shaped gut, and a nervous system twisted into a figure 8. Torsion is an ontogenetic process in some gastropods,

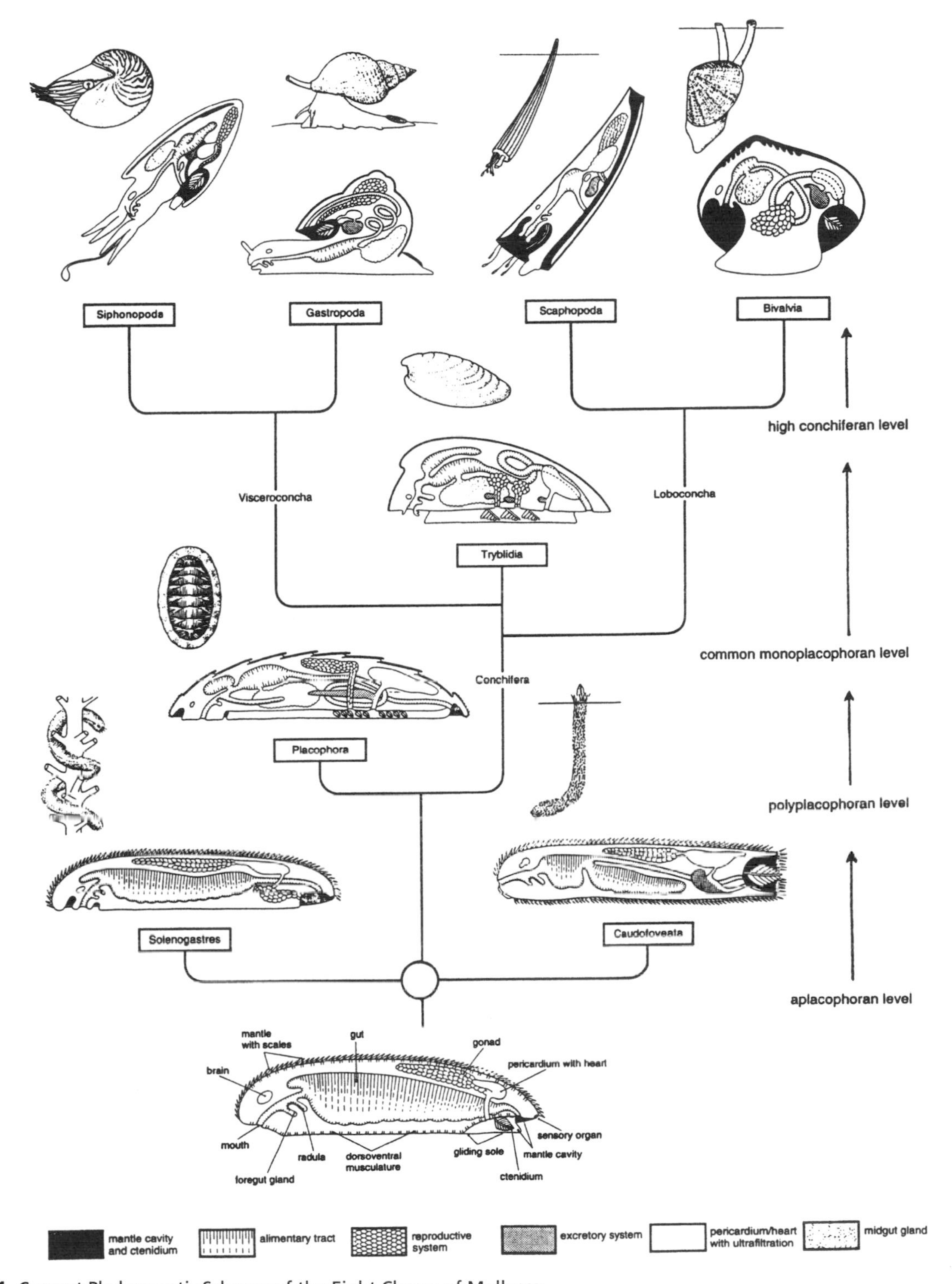

FIGURE 1. Current Phylogenetic Scheme of the Eight Classes of Mollusca.
L. v. Salvini-Plawen and G. Steiner. "Synapomorphies and plesiomorphies in higher classification of Mollusca." In J.D. Taylor (ed.), *Origin and Evolutionary Radiation of the Mollusca,* 1996. By permission of Oxford University Press.

occurring during development of the veliger (Wanninger et al., 2000), but heterochronic developmental changes have resulted in the organs forming in their torted positions during development in most members of the class. The gastropod head is equipped with more sensory structures than those of the classes described above, and the radula has undergone remarkable evolutionary radiation, related to feeding habits that range from collection of small organic particles, to grazing on plants, to predation on other invertebrates and even vertebrates. The foot in most gastropods remains broad and flat, but the visceral mass and mantle-shell are much enlarged along the secondary axis and are usually helically coiled. Most gastropods have a single, large external shell into which the body can withdraw, but a few have bivalved shells, and evolutionary trends toward reduction and loss of the shell have occurred independently in numerous lineages.

Class Bivalvia. The bivalves, comprising clams, oysters, scallops, and so on, are the second largest molluscan class, with about 10,000 extant and more than 40,000 described fossil species. Most occur in the sea, where they are sedentary or sessile on or in sand, mud, and rock everywhere from the intertidal zone to more than 10,000 meters (30,000 feet); they have also successfully invaded freshwater rivers and lakes. Bivalves are the most important molluscan class economically, and they are among the oldest molluscs, dating from the Lower Cambrian.

The secondary axis is much enlarged in bivalves, as the visceral mass is expanded dorso-ventrally and the mantle and shell extend to enclose it. As the class name indicates, two valves comprise the shell, which usually encloses the entire body. The valves are lateral in position, covering the left and right sides; they are separated dorsally by an uncalcified hinge region, and they are open ventrally but may be appressed by one or two adductor muscles. The foot is also usually enlarged along both dorsal–ventral and anterior–posterior axes. In many bivalves, it functions in digging into soft substrata, but in others it secretes tough threads that attach the animal to the substrate, or it forms a sucker to aid mechanical boring by the shell valves in rock or wood. The gills also expand longitudinally and dorso-ventrally as many long lamellae or filaments, and they may occupy much of the mantle cavity. Their main function is not respiration but food collection: most bivalves pump water through the mantle cavity and filter out fine particles of organic matter for food, using cilia on the gills to catch the particles and to move them toward the simple mouth. Water flow in the mantle cavity is often directed by tubular extensions of the mantle called siphons, one for inhalent and the other for exhalent flow. Unlike other molluscs, the bivalves lack a radula as well as a head. Sensory structures are typically located around the periphery of the mantle, and the ganglia of the nervous system are widely separated throughout the body.

Class Scaphopoda. This is the third smallest (or fourth largest) class, with about 300 Recent and 200 fossil species, dating back to the Ordovician (about 450 million years ago). Like the Polyplacophora, it is morphologically conservative. All scaphopods (the name means "spade-footed") occur in soft sediments in the sea, from the intertidal zone to depths exceeding 3,000 meters (9,000 feet). The scaphopod body is elongate along the secondary axis. The foot is very large and pistonlike, thrusting into the sand to enable rapid burrowing. The visceral mass is elongate, and the mantle secretes a tubular shell with the mantle cavity open at both ends. Water is circulated in and out the postero-dorsal opening, but scaphopods lack ctenidia. The head lacks major apparent sensory organs, but it is equipped with two groups of many long, thin, extensile tentacles called captacula, which stick to and collect the foraminifera and small invertebrates on which scaphopods feed. These are drawn into the mouth by contraction of longitudinal muscles of the captacula and passed to a buccal mass containing the largest radula in proportion to body size of any mollusc.

Class Cephalopoda. This class includes the biggest, smartest, most muscular, and fastest marine invertebrates; squid, octopus, cuttlefish, and nautilus are its main living representatives. All 600 living species are marine. The class originated in the Late Cambrian and probably was maximally diverse in the Mesozoic, but most of the 7,000 known fossil species had become extinct by the end of the Cretaceous (65 million years ago). As in the two previous classes discussed, the secondary body axis predominates, to such an extent that the morphologically primary axis is much reduced and the morphologically secondary axis has become the functionally main anterior–posterior axis. The visceral mass is elongate, and it is coiled in the one subclass containing *Nautilus*, which has a large external shell. The other extant cephalopod groups have independently undergone evolutionary reduction of the shell; it is usually internal and may be completely absent, as in the octopus. Concomitantly, the mantle has become very muscular and functions in swimming locomotion by jet propulsion. The head is highly developed, with a large brain and very large, complex eyes, and a predatory feeding mechanism involving a pair of powerful jaws as well as the radula. Cephalopods are primarily visual animals, and their large size, strong musculature, large head, and highly centralized nervous system adapt them for active lifestyles.

Origin and Early Evolution of Mollusca. The durable shells of molluscs persist long after death if conditions are favorable, but the fossil record provides no direct evidence of the origin of the phylum nor of the phylogenetic relationships of the classes. Members of three modern classes, Monoplacophora, Gastropoda, and Bivalvia, appear abruptly in the earliest Cambrian deposits that have yet been discovered and on five modern continents. This diverse Early Cambrian molluscan fauna must have had an earlier, Precambrian origin, perhaps when their ancestors were first evolving shells. This could have been as long ago as 1.0–1.2 billion years ago (Wray, Levinton, and Shapiro, 1996), or less than 630 million years ago (Lynch, 1999), depending on estimates of nucleotide substitution rates in mitochondrial and nuclear genes.

The earliest fossil molluscs were all small (mainly 2–5 mm), and the crystal architecture of their shells was remarkably similar to modern forms (Runnegar and Pojeta, 1985). This is true of other shelled invertebrate groups as well, and the low likelihood of preservation of such small, soft-bodied animals may explain why no Late Precambrian fossil molluscs are known.

The fossil record indicates that molluscan diversity remained low throughout the Cambrian, despite the occurrence of several class-level taxa that went extinct either within the Cambrian or shortly thereafter. Of these, the best documented as true molluscs are the Helcionelloida, Stenothecoida, and Rostroconchia (Runnegar, 1996; Waller, 1998). Three other extinct Paleozoic classes, all originating in the Cambrian, are variously considered to be either molluscs or members of different but related phyla (Runnegar, 1996). Both the size and diversity of molluscs increased dramatically during the Ordovician period (440–500 million years ago). This major radiation appears to have involved seven separate lineages, two each of Bivalvia and Gastropoda, and Monoplacophora, Cephalopoda, and Rostroconchia (Runnegar and Pojeta, 1985).

Most recent students of molluscan phylogeny consider Aplacophora to be the closest living forms to the ancestor of the phylum (Runnegar, 1996; Waller, 1998), specifically the class (or subclass) Solenogastres (Salvini-Plawen and Steiner, 1996; Haszprunar, 2000). These authors suggest that the major apomorphies of the Mollusca included at least one pair of ctenidia, the radula, the spiral cleavage pattern known as the molluscan cross, a dorsal mantle with calcareous epidermal spicules and a mantle cavity, a flat, ciliated ventral foot, and a gonopericardial connection. The extant outgroups most closely related to the Mollusca appear to be the Sipuncula and the free-living flatworms of Class Turbellaria, Phylum Platyhelminthes. The latter is probably the best extant model of an animal with ancestral characteristics of the Mollusca (Haszprunar, 1996).

Fossil Aplacophora were unknown at the time the studies cited above were made; the recently discovered representative, *Acaenoplax*, is of Silurian age (425 million years ago) (Sutton et al., 2001), more than 100 million years later than known fossils of several other molluscan classes. Some recent morphology-based cladistic analyses place the two aplacophoran groups as each other's closest relatives, the Aplacophora as the sister group to the Polyplacophora, and that entire clade as the sister group of the remaining molluscan classes (Scheltema, 1996). Others (Salvini-Plawen and Steiner, 1996; Haszprunar, 2000) place the Solenogastres as the sister group of a clade comprising all the other molluscan classes. The apomorphies of the latter cluster include one pair of ctenidia in the mantle cavity, separate sexes, and a radular membrane supporting the teeth. In this scheme, Caudofoveata is the sister group of the remaining six classes. Prior to the discovery of *Acaenoplax*, a major apomorphy of these classes was the presence of one or more shell plates. Apomorphies that remain valid include 16 or fewer pairs of dorso-ventral muscles, with the lateral nervous system lateral to them, separate gonoducts for gamete exit rather than a gonopericardial connection, radula with serially repeated rows of teeth, and esophageal and midgut glands. The phylogenetic relationships of the remaining classes are given as (Polyplacophora (Monoplacophora (Bivalvia, Scaphopoda) (Gastropoda, Cephalopoda))) by Salvini-Plawen and Steiner (1996), and as (Polyplacophora (Monoplacophora (Bivalvia (Scaphopoda (Gastropoda, Cephalopoda))))) by Haszprunar (2000). Scheltema's (1996) scheme is ((Aplacophora, Polyplacophora) (Monoplacophora, Other classes)).

Evolution Within the Major Molluscan Classes.

Gastropoda. Almost certainly monophyletic despite its vast size and diversity, the Class Gastropoda has attracted a great deal of evolutionary attention, and many alternative phylogenetic schemes have been proposed. The most thorough morphology-based phylogenetic analysis (Ponder and Lindberg, 1997) divides the class into five major clades. The most distinct and probably most primitive is the Patellogastropoda, comprising limpet-like snails with a characteristic docoglossan radula. Examples are *Patella* and *Lottia*. In Ponder and Lindberg's cladistic analysis, Patellogastropoda is the sister group of all other gastropods. The second major clade, the Vetigastropoda, includes the abalones (*Haliotis*), keyhole limpets (*Fissurella*), and top shells (*Trochus*). The first two genera have a pair of gills each sending oxygenated blood to its own auricle, while the top shells and other members of the group have only one gill but retain the vestigial second auricle. Both groups, which comprise the former taxon Archaeogastropoda, are marine, and most are grazers on algae in shallow water. The third group, the Neritopsina, is predominantly marine and tropical (e.g., *Nerita*) but has successfully invaded brackish water and rivers (e.g., *Neritina*). The largest gastropod clade is the Caenogastropoda, comprising most of the former Mesogastropoda plus the Neogastropoda. This group is predominantly marine but includes some freshwater and terrestrial taxa. The univalve shells are helically coiled, and most caenogastropods have a radula with rather few (2–7) teeth in each row. Caenogastropodan feeding types are extremely diverse, including grazing (e.g., *Littorina*), suspension and deposit feeding on organic particulate matter (*Turritella*, *Cerithium*), and predation on a wide variety of other invertebrates such as bivalves (*Natica*), sea urchins (*Cassis*), sipunculans (*Mitra*), polychaetes, and in some cases other gastropods and fishes (*Conus*). The final major gastropod clade, the Heterobranchia, is perhaps the most heterogeneous, including some groups formerly included in the Mesogastropoda (e.g., parasitic predators such as *Architectonica*), as well as the Opisthobranchia (e.g., nudibranchs such as *Tritonia*), and the air-breathing Pulmonata (e.g., the snails *Helix* and *Limnaea*, and the slug *Arion*).

Class Bivalvia. Bivalves are among the oldest fossil molluscs known, but they remained small and low in diversity throughout the Cambrian, although the split between the major subclasses Protobranchia and Autobranchia occurred then (Waller, 1998). Bivalves underwent several radiations in the Lower Ordovician and again in the Silurian and Devonian (Pojeta, 2000; Babin, 2000). By the end of the Paleozoic, all of the modern higher bivalve taxa had originated, as had the major feeding types present in the class. Moreover, some Devonian species reached lengths of 60 centimeters. "In the Mesozoic, this array of animals was to radiate yet further and the stamp of modernity would be placed upon their descendants" (Morton, 1996). Major Mesozoic bivalve diversifications involved deep burrowing into sediments by bivalves that developed siphons—elongate, tubular extensions of the mantle that serve to direct and separate inhalent and exhalent water flows in the mantle cavity. Waller (1998) has synthesized the evolutionary patterns of the major bivalve groups based on the fossil record.

Most bivalves feed on small organic particles, but they do so in different ways. Most collect suspended or deposited particles on the ctenidia from the water pumped through the mantle cavity by cilia on the ctenidia, but the foot and extensions of the labial palps probably preceded the specialized ctenidia as more primitive food-collecting devices in the class (Morton, 1996). A few bivalves have evolved predatory habits, using either the foot or a muscular septum whose contraction forcibly sucks prey into the inhalent siphon.

Class Cephalopoda. Cephalopods, muscle molluscs with a lot of nerve, have evolved along very different paths from the other molluscan classes—paths that more closely resemble the evolutionary trends of vertebrates. Their very well developed sensory and central nervous systems, neuromuscular coordination, reduction of the shell in favor of a muscular body wall, and reliance of behavior on prior individual experience rather than "hard-wiring" all resemble vertebrates, and competition with fishes, particularly during Mesozoic time, may well have selected for these convergent features (Packard, 1972). Cephalopods also anticipated several military weapons developed much later by man: the submarine, smoke screen, and jet propulsion.

The earliest fossil cephalopods are Late Cambrian, and all six subclasses arose in the Palaeozoic. Two became extinct before its end, and two others died out before the end of the Mesozoic (Pojeta and Gordon, 1987). The two very different extant subclasses also nearly perished by the Cretaceous-Tertiary boundary; only one species of the Nautiloidea may have survived, to give rise to the handful of modern species in this group's single genus, the chambered nautilus, the only modern cephalopod with a large, external shell. The history of the other extant subclass, the Coleoidea, is marked by several parallel evolutionary trends toward loss of the shell and the more active lifestyle this permits. The modern squids (e.g., *Loligo*), cuttlefish (*Sepia*), spirula (*Spirula*), and octopus (*Octopus*) represent these independent trends. The first cladistic phylogenetic hypotheses for the subgroups of Coleoidea, based on both morphological and molecular characters, have recently appeared (Vecchione et al., 2000).

All cephalopods are predatory and marine. They include the largest known invertebrate, the giant squid *Architeuthis*, which reaches a length of 16 meters and is captured by sperm whales, but as yet never by man. Despite their variety of size and form, large eyes and brains, and strong and complex neuromuscular systems, cephalopods have never succeeded in invading the terrestrial realm to compete directly with terrestrial vertebrates.

[*See also* Animals.]

BIBLIOGRAPHY

Aguinaldo, A. M. A, J. M. Turbeville, L. S. Linford, M. C. Rivera, J. R. Garey, R. A. Raff, and J. A. Lake. "Evidence for a Clade of Nematodes, Arthropods and Other Moulting Animals." *Nature* 387 (1997): 489–493. The first cladistic analysis (based on 18s rRNA) to provide evidence for two major groups of protostomous animals, the Lophotrochozoa (including the Mollusca) and the Ecdysozoa.

Babin, C. "Ordovician to Devonian Classification of the Bivalvia." *American Malacological Bulletin* 15 (2000): 167–178. New data on bivalve diversification, indicating all subclasses present by end of the Ordovician.

Erwin, D. E. "The Origin of Bodyplans." *American Zoologist* 36 (1999): 617–629. Fossil evidence for the origin of complex invertebrates; Erwin's data support Lynch's (1999) estimate rather than that of Wray et al. (1996).

Haszprunar, G. "The Mollusca: Coelomate Turbellarians or Mesenchymate Annelids?" In *Origin and Evolutionary Radiation of the Mollusca*, edited by J. D. Taylor. Oxford, 1996. Cladistic analysis and interpretation of molluscan ancestry based on morphology of extant invertebrates.

Haszprunar, G. "Is the Aplacophora Monophyletic? A Cladistic Point of View." *American Malacological Bulletin* 15 (2000): 115–130. A morphology-based cladistic analysis indicating that Aplacophora is paraphyletic and that Solenogastres is the sister taxon of all the other molluscan groups.

Hoare, R. D. "Considerations on Paleozoic Polyplacophora Including the Description of *Plasiochiton curiosus* n. gen. and sp." *American Malacological Bulletin* 15 (2000): 131–137.

Lynch, M. "The Age and Relationships of the Major Animal Phyla." *Evolution* 53 (1999): 319–325. A "molecular clock" analysis supporting origin of the Mollusca less than 627 million years ago. Cf. Wray et al., 1996.

Morton, B. "The Ecolutionary History of the Bivalvia." In *Origin and Evolutionary Radiation of the Mollusca*, edited by J. D. Taylor, pp. 337–359, 1996. Descriptive analysis of major features of bivalve evolution, based on the fossil record and functional morphology.

Packard, A. "Cephalopods and Fish: The Limits of Convergence." *Biological Reviews* 47 (1972): 241–207.

Pojeta, J., Jr. "Cambrian Pelecypoda (Mollusca)." *American Malacological Bulletin* 15 (2000): 157–166.

Pojeta, J., Jr., and M. Gordon, Jr. "Class Cephalopoda." In *Fossil Invertebrates*, edited by R. S. Boardman, A. H. Cheetham, and A. J. Rowell, pp. 329–358. Palo Alto, 1987.

Ponder, W. F., and D. R. Lindberg. "Towards a Phylogeny of Gastropod Molluscs—An Analysis Using Morphological Characters." *Zoological Journal of the Linnean Society of London* 119 (1997): 83–265. A cladistic phylogenetic hypothesis based on more than 100 anatomical, shell and ontogenetic characters.

Portmann, A.. "Géneralités sur les mollusques." In *Traité de Zoologie*, 5 (part 2), edited by P.-P. Grassé, pp. 1625–1654. Paris, 1960. General introduction to molluscan morphology, with explanation of the secondary body axis.

Rosa, R. de, J. K. Grenier, T. Andreeva, C. E. Cook, A. Adoutte, M. Akam, S. B. Carroll, and G. Balavoine. "*Hox* Genes in Brachiopods and Priapulids and Protostome Evolution." *Nature* 399 (1999): 772–776. Support of the Lophotrochozoa-Ecdysozoa classification from a second gene family. Cf. Aguinaldo et al., 1997.

Runnegar, B. "Early Evolution of the Mollusca: The Fossil Record." In *Origin and Evolutionary Radiation of the Mollusca*, edited by J. D. Taylor, pp. 77–87. Oxford, 1996.

Runnegar, B., and J. Pojeta, Jr. "Origin and Diversification of the Mollusca." In *The Mollusca, Vol. 10. Evolution*, edited by E. R. Trueman and M. R. Clarke, pp. 1–57. Orlando, Fla., 1985.

Salvini-Plawen, L. v., and G. Steiner. "Synapomorphies and Plesiomorphies in Higher Classification of Mollusca." In *Origin and Evolutionary Radiation of the Mollusca*, edited by J. D. Taylor, pp. 29–51. Oxford, 1996. Morphology-based cladistic analysis of molluscan phylogeny, primarily at the class level.

Scheltema, A. H. "Phylogenetic Position of Sipuncula, Mollusca and the Progenetic Aplacophora." In *Origin and Evolutionary Radiation of the Mollusca*, edited by J. D. Taylor, pp. 53–58. Oxford, 1996.

Sutton, M. D., D. E. G. Briggs, D. J. Siveter, and D. J. Siveter. "An Exceptionally Preserved Vermiform Mollusc from the Silurian of England." *Nature* 410 (2001): 461–463. The first reported fossil aplacophoran, from about 425 million years ago.

Vecchione, M., R. E. Young, and D. B. Carlini. "Reconstruction of Ancestral Character States in Neocoleoid Cephalopods Based on Parsimony." *American Malacological Bulletin* 15 (2000): 179–188. Molecular- and morphology-based cladistic analysis of modern cephalopod groups.

Waller, T. R. "Origin of the Molluscan Class Bivalvia and a Phylogeny of Major Groups." In *Bivalves: An Eon of Evolution*, edited by P. A. Johnson and J. W. Haggart, pp. 1–45. Calgary, 1998. Morphology-based, class-level cladistic phylogeny of the phylum as well as shell morphology-based cladistic analyses of major bivalve taxa.

Wanninger, A., B. Ruthensteiner, and G. Haszprunar. "Torsion in *Patella caerulea* (Mollusca, Patellogastropoda): Ontogenetic Process, Timing, and Mechanisms." *Invertebrate Biology* 119 (2000): 177–187.

Wray, G. A., J. S. Levinton, and L. H. Shapiro. "Molecular Evidence for Deep Precambrian Divergences among Metazoan Phyla." *Science* 274 (1996): 568–573. Comparative "molecular clock" analysis supportiong origin of Mollusca approximately 1,200 mybp. Cf. Lynch, 1999.

— Alan J. Kohn

MONOGAMY. *See* Mating Systems, *article on* Animals.

MORGAN, T. H.

U.S. geneticist Thomas Hunt Morgan (1866–1945) pioneered studies in classical genetics. He was the first to work on the fruit fly (*Drosophila melanogaster*) which has since become a major focus of genetic research.

When the Nobel committee selected Morgan for its 1933 prize in medicine, they cited "his discoveries concerning the role played by the chromosome in heredity." In particular, they noted his combination of methods from Mendelian genetics with microscopic chromosomal studies, his choice of *D. melanogaster* as a research subject, and his genius at assembling a powerful team of collaborators. Although the committee realized that it might seem odd to award the prize in medicine for work on flies, they cited the value for understanding the "hereditary diseases of man."

With hindsight, this recognition still seems right. We retain our hopes for unlocking the hereditary codes to cure human diseases. Yet Morgan always recognized that heredity does nothing by itself and that it is wrong to isolate genetics from the embryology and evolution that also shape organisms. Throughout his career, Morgan saw genetics as just one part of a very complex study of life.

Born into prominent families, Morgan explored the natural history of his native Kentucky. He obtained a doctorate in 1890 from Johns Hopkins University for his study of sea spiders. Morgan spent his career at Bryn Mawr College (1891–1904), Columbia University (1904–1928), and the California Institute of Technology (Cal Tech) (1928–1945). Morgan's study of embryology was inspired by his 1894–1895 visit to the Naples Stazione Zoological. That study of organismal development remained his focus, even though he is more widely known for his contributions to genetics. He saw heredity as part of individual development. Evolution shaped both heredity and development, but Morgan remained skeptical about contemporary interpretations of natural selection's power to effect species change.

Bryn Mawr shaped Morgan as a teacher and a researcher, and he married former Bryn Mawr student Lilian Vaughan Morgan, a fine biologist in her own right. Yet it was at Columbia that he developed his style and new research programs. Always an opportunist in trying different things and following the ones that worked, Morgan worked with various animal species and asked diverse questions. He expressed dissatisfaction with the current state of evolution, for example, in his *Evolution and Adaptation* (1903), and asked how variations arise and are preserved through heredity and evolution. The year 1910 brought two important papers: one critiquing the Mendelian-chromosomal theory of heredity and arguing for epigenetic interpretations of development, the other stimulating the rush of work on the fly and providing key evidence in favor of the Mendelian-chromosomal theory. This apparent contradiction reveals much about Morgan and his time.

Morgan sought empirical evidence to support his theories, and he did not see evidence in favor of the Mendelian-chromosomal theory of heredity. Something was inherited, certainly, but that did not explain development. Among his experimental organisms, however, was *Drosophila*. He realized that it was useful for research—easy to grow, taking little space, satisfied to eat bananas. Among other questions, Morgan was seeking variations sufficient for evolution of new species. In 1910, he observed the white-eyed male fly that changed the way he did biology.

At first, Morgan thought that the white-eyed fly might be the sort of mutation that the Dutch botanist Hugo De Vries (1848–1935) had suggested could give rise to a new species. He did breeding experiments to see whether he could create a new species. He could not, but he did find something at least as interesting. In the first cross between the white-eyed male and red-eyed female flies, the offspring were red-eyed like the mother. Yet the second crossed generation brought some white eyes. Because the white eyes appeared predominantly in males, no matter how the crosses were conducted, there seemed to be a link. And because maleness was connected with chromosomes, white eyes were linked with chromosomes. This raised many possibilities for future research and interpretations.

Morgan did two important things: he recognized the significance of that white-eyed male fly and followed the research path that revolutionized our thinking about heredity and development. Second, he changed his mind. Rather than sticking stubbornly with his skepticism about Mendelism, Morgan followed his evidence.

Furthermore, Morgan created one of the most effective collaborative research teams in the history of biology. Undergraduates Alfred Sturtevant and Calvin Bridges and graduate student Hermann J. Muller became indispensable players in the "fly room" culture. [*See* Muller, Hermann Joseph.] Under Morgan's leadership, the others often took the intellectual lead and directed innovations in laboratory techniques, mapping chromosomes and offering many of the major generalizations that underlie modern genetics. At Columbia and Cal Tech, Morgan directed the team. Increasingly, however, his own focus remained on individual development and the relations of heredity, development, and evolution. Rather than seeing him as a struggling geneticist who could not understand the latest theories of population genetics (as Muller later charged), Morgan should be seen as exhibiting unusual flexibility and recognition of the complexities of biology.

[*See also* Mutation, *article on* Evolution of Mutation Rates.]

BIBLIOGRAPHY

Allen, G. E. *Thomas Hunt Morgan: The Man and His Science*. Princeton, 1978. The best scholarly biography of Morgan.

Kohler, R. E. *Lords of the Fly: Drosophila Genetics and the Experimental Life*. Chicago, 1994. Important interpretation of the culture of the "fly room" at Columbia and Cal Tech.

Maienschein, J. *Transforming Traditions in American Biology, 1880–1915*. Baltimore, 1991. Study of Morgan and three contemporaries during the formative years of American biology, focusing on development.

Morgan, T. H. *Evolution and Adaptation*. 1903.

Morgan, T. H. "Chromosomes and Heredity." *American Naturalist* 64 (1910): 449–496. Morgan's rejection of the Mendelian-chromosomal theory of heredity.

Morgan, T. H. "Sex Limited Inheritance in *Drosophila*." *Science* 32 (1910): 120–122.

Sturtevant, A. H. "Thomas Hunt Morgan." *National Academy of Sciences: Biographical Memoirs* 33 (1959): 283–325.

ONLINE RESOURCES

www.nobel.se/medicine/laureates/1933/press.html. Nobel Prize presentation speech by F. Henschen for the Royal Caroline Institute.

— JANE MAIENSCHEIN

MORPHOMETRICS

Morphometrics is the study of covariances with organismal shape. Traditionally, morphometrics is less concerned with the analysis of the shape per se (the domain of analytic geometry), than with testing hypotheses concerning how biological processes affect shape. Contemporary morphometrics represents a synthesis of two distinct approaches to the quantitative analysis of variation in shape that can be traced back to the late 1800s and early 1900s. One of these approaches is the descriptive and highly statistical discipline of univariate and multivariate biometrics, whereas the other is represented by D'Arcy Thompson's geometric methods devised to represent shape deformation. These distinct approaches were fused into a "geometric morphmetrics" in the 1980s by F. L. Bookstein, Colin Goodall, D. G. Kendall, and F. J. Rohlf. These authors discovered ways of combining the methods of shape description (the biometric tradition) with assessments of shape change (the geometric tradition). This approach allows any comparably sampled shape—real or imaginary—to be located within a finite-dimensional space in which each vector connecting any two points (or shapes) within that space corresponds to a unique pattern of shape deformation.

Morphometric analysis is based on the concept of a landmark. A landmark is any corresponding, relocatable point on a biological structure of interest. Sets or constellations of landmarks, located in either two or three dimensions, represent the structure during the analysis. Three types of landmark classes are recognized: type 1 landmarks (located at junctions between different tissues or structures), type 2 landmarks (maxima of curvature along the boundary or outline of a specimen), and semilandmarks (a miscellaneous category that currently includes extremal points and the entire set boundary coordinates that comprise the trace of a specimen's or structure's outline).

Morphometric data are usually collected with the aid of physical or electronic digitizing devices. In both cases the specimen, or an image of the specimen, is placed within a two- or three-dimensional Cartesian coordinate system such that any point on the specimen's surface can be assigned a unique Cartesian location. Particular points are selected for measurement by placing a stylus or a remote-controlled cursor on the desired point, then instructing the measurement system to record the coordinates of that location. Each system has a spatial resolution that determines the degree to which two adjacent points can be recognized as having different coordinate values. At present, resolutions of tenths to hundredths of a millimeter are common in morphometric measuring systems. Boundary coordinates may be collected by either manually or algorithmically locating a specimen's or a structure's periphery, then employing interpolation methods to sample the boundary. This boundary is then represented as a series of (usually) equally spaced coordinate locations (or semilandmarks).

Once a data set of coordinate locations from a series of specimens has been assembled, the analysis phase of a morphometric investigation may begin. If the purpose is to conduct a traditional multivariate morphometric investigation, for example, to make reference to or compare results with older analyses, interlandmark distances may be computed. The resultant data matrix may then be submitted to a principal component or factor-type analysis procedure (e.g., singular value decomposition) or to a discriminant analysis (e.g., linear discriminant analysis, canonical variates analysis). Factor component analyses are used to characterize variation within the data set as a whole, whereas discriminant analyses are used to find the features that discriminate between a priori determined subgroups within the data set.

A more recently proposed morphometric analysis technique that uses interlandmark distances is Euclidean distance matrix analysis (EDMA). This technique summarizes the shape as a "form matrix" consisting of the Euclidean distances between all pairs of landmarks. This form matrix has the desirable properties of being invariant relative to the specimen's position and orientation at the time it was measured and of being able to preserve a record of localized shape changes that might involve highly localized shape changes. Two specimens' measurements form matrices may be compared by computing their form-difference matrix of ratios between corresponding interlandmark distances. If both specimens are identical, all off-diagonal elements of the form-difference matrix will be 1.0. If all off-diagonal element are equal but differ from 1.0, the specimens have the same shape but differ in size. If the off-diagonal elements exhibit a range of values, the specimens differ in shape. Modes of shape variation among specimens within a sample can be summarized by comparing the patterning of their form-difference matrices, and a global measure of shape dissimilarity (summed over all elements of the form-difference matrix) can be calculated between any two specimens. The EDMA approach has proven popular within the anthropological community, where the method was first applied. However, the method has also been criticized on the grounds that the nature of the geometric shape space implied by EDMA can limit the range of possible results (thereby artificially influencing the ordination of shapes along various trend-based trajectories) and because the results of some EDMA-based analysis imply the existence of biologically impossible shapes. It should be noted that these same criticisms can be made of other multivariate morphometric approaches.

A different approach is represented by the class of methods collectively referred to geometric morphometrics. These methods make use of the geometric data inherent in the Euclidean coordinates of the landmarks themselves and do not require calculation of interlandmark distances. Fundamental to these methods is the scaling and matching of landmark constellations to a reference shape through the method of Procrustes superposition. This method conceptually overlays the two- or three-dimensional centroid of a specimen's landmark constellation with that of the reference specimen's centroid, then isometrically grows (or shrinks) and rotates the specimen's landmarks until the mean root square of the difference between the specimen's landmark locations and those of the reference specimen are minimized. (Note: There are various algorithmic approaches to the iterative procedure whereby this matching is made.) Once all specimens have been matched in this way, the residual root mean square difference between any two Procrustes superposed shapes can be used as a measure of shape dissimilarity. This measure is termed the Procrustes distance.

Procrustes distances are geodesic distances between shapes located on the surface of a high-dimensional, non-Euclidean sphere. The primary drawback in using Procrustes methods to quantify shape differences is that highly localized shape differences (e.g., single-landmarks displacements) will be spread over all landmark posi-

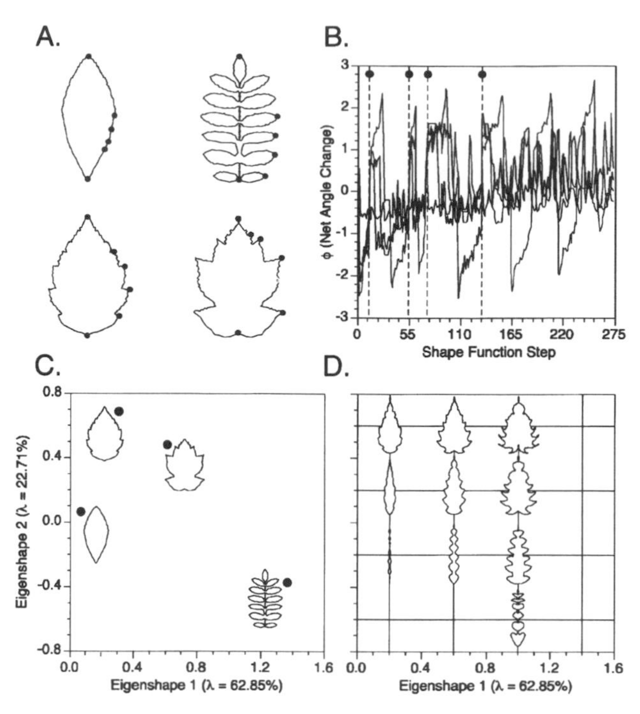

FIGURE 1. Steps in a Typical Combined (landmarks + outlines) Morphometric Analysis. (A) Digitized representations of leaves from four *Sorbus* species. (B) Plot of half-leaf shape functions (= ∅, net angular change between outline points) for the four leaves in A. Dashed vertical lines mark the positions of the landmark points. (C) Plot of eigenshape analysis (= relative warp analysis) results showing scatter of shapes in the space of the two most important shape factors. These results provide an ordination of shapes that can be used for testing a variety of biological hypotheses. (D) Representations of hypothetical shapes that exist at grid nodes within the eigenshape 1-2 plane. [Note: nodes without leaf icons represent regions of the shape space at which biologically realistic morphologies are not possible.] Shape model diagrams like these aid in interpreting the overall shape space and understanding the geometric rationale for the shape ordination shown in C. Comparison of figures C and D indicates that the primary shape distinction within these data represents a contrast between (1) leaves that exhibit a continuous margin (low scores on ES-1) and those characterized by leaflets (high scores on ES-1) and (2) leaves that exhibit a narrow, symmetrical profile (low scores on ES-2) and those that are triangular with a broad base (high scores on ES-2). Drawing by Norman MacLeod.

tions by the matching process. Its advantage is that the surface of the analysis space (the Kendall shape manifold) on which the matched shapes reside is not constrained by asymmetrical irregularities and so is a preferable basis to use for subsequent shape-based ordinations. In addition to these features, the shape differences implied by the comparison of any two Procrustes-matched shapes can be conveniently and rigorously summarized in the form of a Thompsonian deformation grid via use of a pure mathematical object known as a thin-plate spline.

Although an ordination of Procrustes matched shapes could be based on a standard multivariate analysis of the pairwise-Procrustes distances, (provided that the total range of shape variation was sufficiently small so that the positions of species on the surface of the curved Kendall shape manifold could be projected onto a linear plane without engendering significant distortion of the inter-specimen distances), the more typical procedure for analyzing such ordinations is to employ the methods of principal warps or relative warps. The term "warps" in these contexts refers to eigenanalysis-based spatial decompositions of the thin-plate spline or of the set of between-specimen Procrustes distances. In the case of the former, a landmark-referenced metric—the "bending energy"—quantifies shape dissimilarity by representing corresponding landmark displacements as a function of the hypothetical amount of "energy" required to bend a rigid but elastic plate from an initial state (no bends, no shape change) to a comparison state (bent as a result of shape change) in which the reference and comparison states represent the configurations of reference and comparison landmark constellations.

Principal warps are the eigenvectors of the bending energy matrix, calculated between all pairs of landmarks, for a reference and comparison shape. Their orientation is strictly dependent on the reference shape which is traditionally chosen as the grand mean landmark configuration for the sample. Principal warps may be further decomposed into partial warps by projecting the x and y (and z, if available) coordinates of the Procrustes-matched specimens separately onto the principal-warp vectors. These partial-warp vectors can then be used as hierarchically arranged, geometrically (though not statistically) orthogonal shape variables to ordinate specimens and test hypotheses of shape similarity or difference relative to the reference shape. Principal and partial warps are primarily used to examine group mean shape differences.

Relative warps (in the full shape space) are eigenvectors of the covariance matrix calculated between all pairs of landmark coordinates for shapes included within a sample. These differ from principal warps in that they do not contain a uniform component of shape change (the 0ht principal warp) and are strictly sample dependent. Relative warps form an additional series of hierarchically arranged, geometrically and statistically orthogonal shape variables that can be used to ordinate specimens and test hypotheses of intra-sample shape similarity or difference. Relative warps are primarily used to characterize the shape distribution of single samples, though such samples may be composed of different groups (e.g., populations) and the results used to determine whether shape is distributed continuously between the groups. If the correspondence between semilandmarks is accepted as a reasonable approximation of shape correspondence, then the landmark-based methods of geometric morphometrics can be extended to include considerations of interlandmark outlines represented as boundary coordinate series (Figure 1).

Morphometric analysis can be used to investigate any comparison between shapes. Thus far, these methods have been used successfully in systematic, ecological, biogeographical, functional morphological, ontogenetic, and phylogenetic contexts. The general form of such analyses is to postulate a null hypothesis of no systematic relationship between shape variation and an external parameter of interest (e.g., ecological gradient, geographical position, functional group) and use morphometric methods to test the null hypothesis. There is a current need for morphometricians to become more aware of the phylogenetic dimensions of their analytic systems (e.g., via integration of the comparative method into morphometric research strategies) and for systematists to become more familiar with the tools of postsynthesis morphometrics. However, if the historical barriers between the fields of evolutionary biology and morphometrics can be overcome, morphometrics can make a substantial contribution to the understanding of the origin, maintenance, and evolutionary significance of shape variation in organic form.

BIBLIOGRAPHY

Bookstein, F. L. "Size and Shape Spaces for Landmark Data in Two Dimensions." *Statistical Science* 1 (1986): 181–242. An important article documenting the transition between previous morphometric approaches and (the then nascent) geometric morphometrics. First presentation of the concept of shape coordinates.

Bookstein, F. L. *Morphometric Tools for Landmark Data: Geometry and Biology.* Cambridge, 1991. The only book-length treatment of geometric morphometrics by a single author currently available in English. A classic. Referred to as "the orange book" by morphometricians.

Bookstein, F. L. "Can Biometrical Shape Be a Homologous Character?" In *Homology: The Hierarchical Basis of Comparative Biology*, San Diego, edited by B. K. Hall. 1994. An early and extended consideration of the place of morphometrics in systematics in general and phylogenetic systematics in particular.

Goodall, C. R. "Procrustes Methods in the Statistical Analysis of Shape." *Journal of the Royal Statistical Society, Series B* 53 (1991): 285–339.

Kendall, D. G. "Shape Manifolds, Procrustean Metrics and Complex Projective Spaces." *Bulletin of the London Mathematical Society* 16 (1984): 81–121.

Lele, S., and T. Cole. "Euclidean Distance Matrix Analysis: A Statistical Review. In *Proceedings of the International Conference on Current Issues in Statistical Shape Analysis*, edited by K. V. Mardia. Leeds, England, 1995. Most complete description of Euclidean distance matrix analysis.

MacLeod, N. "Generalizing and Extending the Eigenshape Method of Shape Visualization and Analysis." *Paleobiology* 25 (1999): 107–138. An attempt to forge a link between geometric morphometrics and one outline-based method. Includes an extended discussion of the use of morphometrics in systematic and evolutionary contexts.

MacLeod, N., and P. Forey. *Morphometrics, Shape, and Phylogenetics.* London, in press. Collection of essays on the relation between morphometrics and phylogenetic analysis, with contributions by Bookstein, Felsenstein, MacLeod, Rohlf, and Swiderski et al.

Marcus, L. F., E. Bello, and A. García-Valdecasas. *Contributions to Morphometrics.* Madrid, 1993. A collection of essays on various technical and applied topics in geometric morphometrics, including a history of the morphometric synthesis by Bookstein and an important article on relative warps by Rohlf. Referred to as "the black book" by morphometricians.

Marcus, L. F., M. Corti, A. Loy, G. J. P. Naylor, and D. E. Slice. *Advances in Morphometrics.* New York, 1996. A collection of essays on various technical and applied topics in geometric morphometrics. Referred to as "the white book" by morphometricians.

Reyment, R. A. *Multidimensional Paleobiology.* Oxford, 1991. A textbook in multivariate morphometric analysis containing extended description of geometric morphometric methods.

Rohlf, F. J. "On Applications of Geometric Morphometrics to Studies of Ontogeny and Phylogeny." *Systematic Biology* 47 (1998): 147–158. Article detailing criticisms of the Zelditch et al. (1995) use of partial warps in a phylogenetic systematic context.

Rohlf, F. J. "Statistical Power Comparisons among Alternative Morphometric Methods." *American Journal of Physical Anthropology* 111 (2000): 463–478. Article detailing criticisms of Euclidean distance matrix analysis.

Rohlf, F. J., and F. L. Bookstein. *Proceedings of the Michigan Morphometrics Workshop.* Ann Arbor, 1990. A collection of essays on various technical and applied topics in geometric morphometrics from a time when these methods were first being explained to the biological community. Referred to as "the blue book" by morphometricians.

Zelditch, M. L., W. L. Fink, and D. L. Swiderski. "Morphometrics, Homology, and Phylogenetics: Quantified Characters as Synapomorphies." *Systematic Biology* 44 (1995): 179–189. Article detailing a proposed use of partial warps analysis to search for phylogenetic characters.

— Norman MacLeod

MULLER, HERMANN JOSEPH

Hermann Joseph Muller (1890–1967), a pioneer in the study of genetic mutations introduced X rays as a tool for the analysis of mutations, work that led to the Nobel Prize for Medicine in 1946. Muller was also a founding member of the "fly lab" or Morgan school of fruit fly genetics.

Muller was born in New York City. He studied under Thomas Hunt Morgan and received his doctorate in 1915. Muller accepted a Darwinian view of natural selection, in which subtle differences among individuals were selected over long periods of time. His mentor, Morgan, was doubtful that evolution worked in this way, and he preferred the theory of saltations (sudden origins) promoted by the Dutch botanist Hugo De Vries (1848–1935) in his mutation theory of the origin of species. Muller rejected this discontinuous method of evolution. Between 1912 and 1921, his analysis of two genes, beaded wings and truncate wings, supported the genetic basis of evolution along Charles Darwin's original premise. In both the beaded and truncate studies, Muller demonstrated that a major gene caused a mutant phenotype, but there were also numerous genetic modifiers of other loci. These genetic modifiers by themselves had no effect on wing morphology, but they could intensify the mutant trait or normalize the mutant trait. Muller also found that environmental modifiers played a role (the higher temperature favored the mutant expression). Muller used genetic recombination, balanced lethals, mapping of modifiers, and complex breeding methods to isolate the factors involved. He compared the process to anatomical dissection, but he used genetic tools instead to "dissect" a complex hereditary pattern.

A second genetic contribution to evolutionary thought was Muller's advocacy of "the gene as the basis of life." This was a radical concept thirty years before James Watson and Francis Crick describe the structure of DNA. Muller believed that the gene was the only known class of molecules that had the capacity to replicate their errors. He argued that all genes arise from preexisting genes, except for the first genelike molecule that emerged with this property. His "gene doctrine" stemmed from a study of the mutation bar eyes, shown by Sturtevant to have arisen from a duplication of a gene into two similar genetic loci. When salivary chromosome banding became available, Muller's laboratory identified the bar region as a visible gene duplication under the microscope. Muller argued that bar eyes arose by unequal crossing over between chromosomes, establishing a tandem duplication with an adjacent pair of nearly identical genes. Muller believed gene evolution had its primary origin through accidents of pairing and crossing over during meiosis.

Muller combined his interests in crossing over and mutation when developing his now famous theory known as Muller's ratchet. He noted that a particular chromosome will carry different numbers of mutations in different individuals of the population. Suppose that one of those chromosomes carries M mutations and all others have more than M mutations. In a sexual population, the chromosome with M mutations can recombine with another chromosome and gain a wild-type allele in exchange for one of its mutants. This lowers the minimum mutation number to $M - 1$, allowing the sexual population to improve with regard to the mutations carried by its best chromosome. Asexual populations can never improve because they lack the potentially restorative effects recombination. When the chromosome with M mutations gains another mutation, it then has $M + 1$. Thus, each mutation added to the best nonrecombining chromosome causes the ratchet to turn as the asexual population slowly but steadily deteriorates in the quality of its best individuals.

Muller was the first to establish mutation rates in living organisms. With Edgar Altenburg, he measured the spontaneous mutation rate in fruit flies. He later discovered that the mutation rate could vary by an order of magnitude. He also showed it could be increased by temperature, suggesting that most mutations arose from a chemical origin—an important insight at a time when the gene was an abstract concept. Muller's radiation-induced mutations were dramatic in their high frequency, and they quickly eclipsed interest in other methods to induce mutations. However, one of his postdoctoral students, Charlotte Auerbach, in her studies in the 1940s on nitrogen mustard and other alkylating agents, confirmed Muller's suspicion that chemicals would be good mutagens.

Muller applied his knowledge of genetics to advocate radiation protection in industry, medicine, and the military. Cold war politics and Muller's earlier history as a Communist sympathizer made it difficult for him to convince professionals that radiation safety was more than political propaganda. No one disputed the dangers of high doses of radiation. At issue were smaller doses. Muller's chief experimental evidence came from the work of his student S. P. Ray-Chaudhuri, who showed that the same frequency of induced mutations occurred whether a total radiation dose of 400 roentgens was applied in thirty minutes or spread out over a month. The protracted dose suggested to Muller that even X rays for medical imaging could induce mutations. He advocated using lead shields to protect the patient and the physician, an argument that was rejected by medical professionals at the time. After the atomic bomb attacks on Hiroshima and Nagasaki, Japan, in August 1945, Muller had the ear of the public and fought hard for stringent safety standards in international commissions on radiation protection.

Less successfully, Muller applied his knowledge of genetics to human betterment. He was a foe of the American eugenics movement and denounced it at the third, and last, International Congress of Eugenics in 1932. However, he did favor humans directing their own evolution, and he advocated a policy of germinal choice, a voluntary system that would use sperm banks and retain frozen samples of semen from people with unusual talents, such as leadership, altruism, genius, and creativity. In the end, he scuttled an effort to establish such a sperm bank when he found its backers favored aspects of eugenics he had opposed more than a generation earlier.

[*See also* Mutation.]

BIBLIOGRAPHY

Altenburg, E., and H. J. Muller. "The Genetic Basis of Truncate Wing: An Inconstant and Modifiable Character in *Drosophila*." *Genetics* 5 (1920): 1–59.

Carlson, E. A. *Genes, Radiation, and Society: The Life and Work of H. J. Muller.* Ithaca, N.Y., 1982.

Muller, H. J. "Genetic Variability, Twin Hybrids, and Constant Hybrids, in a Case of Balanced Lethal Factors." *Genetics* 3 (1918): 422–499.

Muller, H. J. "The Gene as the Basis of Life." *Proceedings of the International Congress of Plant Science* 1 (1926): 897–921.

Muller, H. J. "Artificial Transmutation of the Gene." *Science* 66 (1927): 84–87.

Muller, H. J. "Bar Duplication." *Science* 83 (1936): 528–530.

Muller, H. J. "Report on Experiments with Gamma Rays." *British Journal of Radiology* 14 (1941): 157–158.

Muller, H. J. *Studies in Genetics.* Bloomington, Ind., 1962.

Muller, H. J., and E. Altenburg. "The Rate of Change of Hereditary Factors." *Proceedings of the Society for Experimental Biology and Medicine* 17 (1919): 10–14.

Muller, H. J. "The Relation of Recombination to Mutational Advance." *Mutation Research* 1 (1964): 2–9.

— ELOF CARLSON

MULTICELLULARITY AND SPECIALIZATION

The evolution of life is punctuated by changes that result in new levels of organismal complexity. Some of these evolutionary innovations include the origin of the cell, the origin of eukaryotic symbiosis, and the origins of multicellularity. Each of these innovations allowed life to evolve in new and very different directions, but at the same time created new evolutionary conflicts that had to be resolved in order for the innovation to be maintained. This is most easily seen with the evolution of multicellularity, an invention that allowed life to achieve levels of cellular specialization never seen before, but at the same time an invention theoretically difficult to keep because of the inherent conflict between the individual organism and the cells comprising that organism. The solution to the conflict may be the very thing that allows for multicellularity in the first place, genetic control over the process of development.

Phylogenetic and Temporal Considerations. The evolution of multicellularity among eukaryotes is actually a fairly common phenomenon: of some twenty-three protist groups, seventeen have multicellular representatives. Furthermore, there are cases among the bacteria (e.g., myxobacteria) that seem to fit almost any definition of multicellularity. Nonetheless, only three groups of eukaryotes have developed multicellularity in more than just a few species: the plants, the fungi, and the animals. Discussions on the origin of multicellularity thus usually, and justifiably, focus on these groups. All three of these multicellular kingdoms are built very differently. Primitively animals consist of single cells without cell walls, plants consist of single cells with cell walls, and fungi are syncytial (multinucleate) with a cell wall. These structural differences indicate that multicellularity was achieved independently in each group.

The independent acquisition of multicellularity is further supported by recent inquiries into the phylogeny of eukaryotes based primarily on ribosomal DNA sequence analysis. Various research groups have shown that each kingdom arose from a different unicellular protist group. Animals and fungi arose from a group of protists represented today by the ancyromonadids, and the most basal animal branches are the choanoflagellates and an enigmatic group of protists known as the Mesomycetozoa. Although the protist sister taxon of Fungi remains elusive, given that there are several protist groups at the base of the Animalia, multicellularity in both groups must have arisen independently. Plants clearly arose among the green algae, so the notion of independent acquisition of multicellularity at the base of the Plantae is not questioned.

Although the independent evolution of multicellularity in the three "kingdoms" is fairly clear, when these evolutionary events transpired is an altogether different and much harder question. There are two interrelated questions: when the lineage arose and when the first multicellular representative of that lineage arose. The earliest record of eukaryotes are biomarkers—geologically stable molecules of known biosynthetic origin—derived from chemicals found exclusively in eukaryotes from 2,700 million-year-old rocks in Australia. However, the first fossil of an unequivocal multicellular eukaryote is not found until 1,500 million years later, in the 1,200 million-year-old Hunting Formation of arctic Canada. Nick Butterfield (2000) from the University of Cambridge described a red algae in the Hunting Formation (Rhodophyta, a diverse group of protists with both uni- and multicellular species) that is clearly multicellular; it shows cellular differentiation with specialization, including the possession of a holdfast, somatic and germ cells, and sexually differentiated whole organisms.

The occurrence of this fossil algae has two immediate effects on our understanding of the origins of multicellularity. First, because red algae are the closest relatives of the green algae and plants, the appearance of a true red algae indicates that the latest common ancestor of green algae/plants must also have evolved. Second, because the red and green algal groups are the closest relatives of the animal/fungi group, then the latest common ancestor of fungi and animals must also have evolved by 1,200 million years ago.

The appearance of the multicellular kingdoms by around 1,200 million years is entirely consistent with molecular clock analyses that estimate that plants, animals, and fungi diversified from one another around

1,500 million years ago. (Molecular clocks estimate divergence times by measuring evolutionary distances between two molecules using at least one time point from the fossil record for calibration.) What is curious is that multicellular representatives of each of the three kingdoms are not seen for another billion years. The oldest animal remains are from rocks in China estimated to be around 560 million years old. A variety of animals forms are found including sponges, cnidarians, and possibly triploblasts (i.e., bilaterian animals). Hence, animal diversification was occurring long before 560 million years ago. Plants make their appearance in the Late Ordovician period, around 450 million years ago, although green algae are known from rocks estimated to be 750 million years old. Unequivocal fungal remains are first seen in the Middle Ordovician period, about 460 million years ago. The disparity seen between the molecular clock estimates and the first appearance in the fossil record could be due to several factors, but the most important is the fact that for much of the billion-year history between their origin and their first appearance, these organisms were unicellular. For example, because multicellular animals are known at 560 million years ago, members of the kingdom Fungi must also have been present, but they may not yet have evolved multicellularity. Therefore, their absence from Precambrian rocks is probably because they were still unicellular and thus much harder to preserve and be recognized. It is unclear just how much time separates the origin of the lineage versus the origin of true multicellularity in each group, and the amount of time will remain unclear until detailed molecular clock estimates testing for the divergence between the multicellular kingdoms and their respective unicellular relatives are undertaken.

Benefits and Costs. Multicellularity confers many benefits to the individual. For example, it allows for an increase in size, which may be advantageous not only for capturing food but also for avoiding being food. An increase in size increases harvesting capabilities because the high surface-to-volume ratio in unicellular organisms makes uptaking nutrients difficult. Moreover, larger size makes it easier to move through an aqueous environment and thus reduces travel costs. However, these advantages are also achieved in colonial organisms, which retain cellular associations but do not specialize any of the constituent cells. Multicellularity, in contrast, reflects a "division of labor" among cell types. This division of labor confers further selective advantages not seen in merely colonial forms. One immediate advantage is that cellular differentiation allows for specialization: rather than being the cellular equivalent of the "jack of all trades but the master of none," multicellularity allows for "master" cells, cells that specialize for one or a few selective jobs. Thus, the individual organism has cells dedicated to very specific jobs. Think of almost any multicellular organism—there are cells for communication, support, digestion, and a myriad of other functions.

Therefore, true multicellularity is recognized by this division of labor, but it is this very invention that is problematic. For example, how do cells communicate with one another? How does a cell "know" whether it is going to be a support cell or a digestive cell? In other words, how does the organism develop? This is really *the* question because multicellularity, unlike mere colonially, requires an ontogeny.

Ontogeny, or development, is the process by which a multicellular individual is formed from a fertilized egg. The cells forming the fertilized egg are haploid and arose via meiosis, whereas the cells of the embryo are diploid and thus are the mitotically derived daughter cells of the zygote. These daughter cells eventually become differentiated to form the various cell types of the individual. One of the most profound questions in all of biology is how cells that are genetically identical can take on such radically different and often terminal cell fates. A large part of the answer to this question lies in the fact that development is essentially the differential expression of the appropriate genes at the appropriate times in the appropriate cells. Thus, the specificity of development is not caused by different cells having different genes, but by differentially expressing a subset of the same genes in different cells.

How is this specificity controlled? The answer lies in understanding genetic regulation, the heritable basis of developmental control. Although complex in its details, for our purposes it is relatively simple and elegant. Any given gene in the genome is "told" to be turned on or to be turned off at the right time and at the right place by either one of two mechanisms: (1) The cell inherits some "factor" from the mother cell that the other daughter cell does not inherit; this maternal factor results in activation or repression of select genes, making the cell distinct with respect to its sister cell. (2) One cell signals to another cell, and this signal results in a differential activation/repression of a subset of genes in the receiving cell. The former mechanism is called "autonomous" specification and the latter "conditional" specification. Both result in the transportation of proteins to the nucleus, which then bind DNA with high specificity either by themselves or with other proteins called cofactors. These DNA-binding proteins, also known as transcription factors, are what actually tell a gene to turn on or to turn off by activating or repressing transcription, the process by which a gene makes a messenger RNA molecule. This combinatorial activation/repression, a process called *cis*-regulation, is responsible for the specificity of gene expression and thus for the process of development.

The importance of these two processes for a multicellular organism, cell signaling and transcriptional regulation, can be seen in the comparative genomics of multi- versus unicellular organisms. The recent completion of bacterial, fungal (yeast; note that this is essentially a unicellular eukaryote), plant (the mustard weed *Arabidopsis*), and animal (the nematode *Caenorhabditis elegans*, the fruit fly *Drosophila melanogaster*, and human) genomes allows for some interesting comparisons between the unicellular (bacterial, yeast) and the multicellular (plant, animal) genomes. First, the eukaryotes seem to have a core "proteome" (rather than comparing numbers of genes, or genome, we are comparing numbers of unique proteins, or proteome) about three to four times that of most prokaryotes: yeast has a core proteome of 4,383 proteins, whereas the bacterium *Haemophilus influenzae* has 1,425. The fly, worm, and plant have core proteomes of 8,065, 9,453, and 11,601 proteins, respectively. The amazing fact to come out of these numbers is that the seemingly complex fly or plant has a core proteome of only two or three times that of the simple unicellular yeast.

Two protein sets seem to be expanded in the genomes of the multicellular representatives with respect to yeast: cellular communication proteins and DNA-binding proteins (i.e., transcription factors). This relative expansion makes sense because these types of proteins underlie the process of development and allow for cellular specialization and cell-to-cell communication. Interestingly, although both plant and animal genomes show expansions of gene types in these two areas, they appear to be independent of one another, in accord with the multiple-origins hypothesis discussed above. With respect to cell signaling, animals use several novel signaling systems, including the Wingless/Wnt, Hedgehog, Notch, TGF-β, JAK/STAT, and receptor tyrosine kinase, just to name a few. Plants have none of these systems; instead, they seem to have evolved unique signaling pathways by combining conserved signaling cascades with new receptor types. Moreover, given that about 30 percent of the *Arabidopsis* proteins are of unknown function, the possibility exists that there are plant-specific systems waiting to be discovered. Animal cells also have evolved a number of distinct cell junctions that allow the cells not only to "stick" together but also to facilitate communication between the cells. Some of these cellular innovations include desmosomes, simple points of attachment between two cells, and gap junctions, or actual passageways between cells. One fascinating example is β-catenin, an adherens junction protein which also serves as a signaling molecule. β-catenin is associated with the Wingless/Wnt signaling pathway in animals, but it has recently been found in the slime mold *Dictyostelium*, a relative of the amoebae (see discussion below). Whether the Wingless/Wnt pathway is present in slime molds is unknown at this time, but it is curious that in the slime mold, another example of the independent acquisition of multicellularity, β-catenin is required for cellular differentiation just like it is in animals. Finally, plants retain cytoplasmic continuations among cells (called plasmodesmata) and thus do not possess the many different types of cell junctions found in animals.

The number of transcription factors found in the respective genomes mirrors what we find with cell-to-cell communication genes. The yeast genome has only 209 transcriptional regulators, whereas the fly, worm, and plant sequence have 635, 669, and 1,533 transcriptional regulators, respectively. Again, animals have evolved several new transcriptional regulatory families, including the *Hox*, *Paired box*, *T-box*, and *ETS* families. Plants also have many unique transcription factors, including the *AP2*, *NAC*, and *WRKY* families. Of interest is the unique amplification of specific transcriptional regulators in each lineage for analogous roles: even though MADS and homeoboxes are found throughout eukaryotes, plants have differentially amplified the MADS family for developmental regulation of pattern formation, whereas animals have differentially amplified the homeoboxes for similar pattern formation roles.

Levels of Selection: Conflict and Compromise. One is used to thinking about natural selection upon the individual, but because most of the constituent cells of that individual are able to divide (i.e., reproduce) independent of the individual, selection can act upon the constituent cells as well. In a typical multicellular organism, cells act altruistically for the benefit of the individual and do not act for the benefit of themselves. However, any cell that can outcompete another cell for resources, and thus increase its relative number of progeny cells, should be selected for a multicellular system. This is in fact one way to look at mammalian cancers, which are simply the proliferation of mutant cells favored at the level of the cell lineage, a selection event quite independent of the consequences to the individual housing those cells. The cancer cells "succeed" because they have a greater replication rate than other "normal" cells, despite the often eventual death of the individual. Thus, the evolution of multicellularity has a problem: how to keep cells from acting on their own behalf, and instead acting on behalf of the individual housing the cells.

If there is an inherent conflict between individuals and constituent cells, then situations should arise whereby cells act upon their own behalf and not for the benefit of the individual. In other words, "cheaters" should arise in which heritable changes confer a reproductive advantage to a cell or a group of cells at the cost of the reproductive viability of not only other cells but possibly the individual as well. Recently, a cheater mutation was described in the slime mold *Dictyostelium* by Herbert Ennis and colleagues (2000) from Columbia University. The slime mold is related to the amoebae and like its cousins is ordinarily a unicellular organism. However, during times of stress (e.g., starvation), individual cells aggregate to form a fruiting body that consists of stalk cells and spore cells. But here is the dilemma: only spore cells are capable of giving rise to the next generation; stalk cells die. Thus, stalk cells are sacrificing themselves for the benefit of the individual. The cheater mutation results in cells that preferentially give rise to spore cells and actively signal wild-type cells to become stalk cells. Whether this mutation is viable in the wild is not known, but several other mutant strains have been found in the wild that behave in a similar fashion. Nonetheless, it does validate the proposition that potential conflict does exist between cells and individuals. Multicellularity would seem to require some sort of control mechanisms to keep individual cells in check, but what are the control mechanisms?

Figure 1. Sea Urchin on Algae Covered Surface. Gordon Rocks, Galapagos.

Leo Buss (1987) of Yale University, in his important and stimulating book *The Evolution of Individuality*, discusses this problem in great detail and proposes a novel solution. He argues that essentially the individual, in this case the mother, maintains control over the process of development so that no one cell can control its own fate. In other words, autonomous specification, whereby substances in the egg control the expression of genes, determines the differentiation of the constituent cells. Thus, selection at the level of the individual opposes selection at the level of the cell lineage by controlling the very process of development itself. Although some animals, including the most familiar developmental model systems, do show some form of maternal control over development, many other animals do not. A sea urchin, for example, does not rely on maternal control over the process of development; the inheritance of a maternal substance merely sets up an initial asymmetry, which is then translated into differential gene expression in different regions of the embryo. This does not obviate the need, however, for the individual to control the process of development. The device used is not autonomous specification, but *cis*-regulation. Thus, the very device that controls the process of development, as well as the device responsible for translating cell signaling events, whether autonomous or conditional, into differential gene expression manages the conflict between the cell and the individual embryo. Under normal circumstances, cells simply do not have the inputs that would allow them to undergo unchecked proliferation and differentiation. They must make the correct number and type of cell because only the inputs required for building that type and number of cell are present in the nucleus. There is simply no other option for the cell.

How does the individual reproduce? Cells must be present that can undergo meiosis and form the germ cells for the next generation. In animals, often this is maternally dictated such that cells that will eventually became the germ cells are sequestered early in development. This sequestration has two effects. One is that it prevents them from responding to any signals they might receive and thus from differentiating into a somatic cell. Similarly, it prevents other cells from attempting to become germ cells and thereby cheating on the individual. In other animals, it is simply not well understood how this decision is made, although it is known that it is not made maternally. Most plant and fungal cells, as well as some animals and many other minor multicellular systems, retain the ability for any somatic cell to become a germ cell. How the decision is made to become a germ cell is again not well understood. Nonetheless, somehow the potential conflict must be mediated to retain the status of the system.

If multicellularity is simply the division of labor into somatic and reproductive roles, and this division is controlled by the genetic regulatory system, then the genetic regulatory system is the device used to manage the conflict between the needs of the individual and the needs of cell lineages. Therefore, the possession of a complex genetic regulatory machinery may give us insight into the origins of multicellularity itself. The traditional problem can simply be posed as follows: in the initial stages of the evolution of a multicellular system, did one cell divide to give rise to two genetically identical cells attached to one another, with one cell performing a somatic function and the other a germinal function, or did two individual (i.e., genetically heterogeneous) cells come together to form a simple two-celled individual, with one cell taking on the somatic function and the other the germinal function? The former is called the "fission" model, the latter the "fusion" model. Given that the slime mold is a fusion system (as are other systems not described here where cheater mutations are known), and aside from cancers, cheaters are rare to nonexistent in animals and plants; thus, the origins of true multicellularity probably fall within the "fission" school of thought. This makes sense in that not only is there genetic homogeneity among the cells (and hence the evolution of altruism is easier to understand and model), but the very process that allows for the substitution of a somatic cell fate is the same mechanism that prevents the cell from trying to evolve its own germinal fate. Slime molds do not really have an ontogeny, and the prediction would be that the complex genetic regulatory controls found in animals, for example, will be absent in slime molds and other simple fusion systems. Where Fungi will fall remains to be seen, but a detailed understanding of their development, if there is one, will go a long ways toward understanding the origin and maintenance of multicellularity in this group.

[*See also* Cell-Type Number and Complexity; Cellular Organelles; Germ Line and Soma; Metazoans.]

BIBLIOGRAPHY

Atkins, M. S., A. G. McArthur, and A. P. Teske. "Ancyromonadida: A New Phylogenetic Lineage among the Protozoa Closely Related to the Common Ancestor of Metazoans, Fungi, and Choanoflagellates (Opisthokonta)." *Journal of Molecular Evolution* 51 (2000): 278–285.

Buss, L. W. *The Evolution of Individuality*. Princeton, 1987. A classic book on the problem of the conflict between individuals and their constituent cells.

Butterfield, N. J. "*Bangiomorpha pubescens* n. gen., n. sp.: Implications for the Evolution of Sex, Multicellularity, and the Mesoproterozoic/Neoproterozoic Radiation of Eukaryotes." *Paleobiology* 26 (2000): 386–404.

Davidson, E. H. *Genomic Regulatory Systems: Development and Evolution*. San Diego, Calif., 2001.

Ennis, H. L., D. N. Dao, S. U. Pukatzki, and R. H. Kessin. "*Dictyostelium* Amoebae Lacking an F-Box Protein Form Spores Rather Than Stalk in Chimeras with Wild Type." *Proceedings of the National Academy of Sciences USA* 97 (2000): 3292–3297.

Grimson, M. J., J. C. Coates, J. P. Reynolds, M. Shipman, R. L. Blanton, and A. J. Harwood. "Adherens Junctions and β-Catenin-mediated Cell Signaling in a Non-Metazoan Organism." *Nature* 408 (2000): 727–731. Describes the system in *Dictyostelium*.

Michod, R. E., and D. Roze. "Cooperation and Conflict in the Evolution of Multicellularity." *Heredity* 86 (2001): 1–7.

Pál, C., and B. Papp. "Selfish Cells Threaten Multicellular Life." *Trends in Ecology and Evolution* 15 (2000): 351–352. A commentary on the Ennis et al. (2000) paper, as well as other cases of "cheaters" known in simple fusion systems.

Philippe, H., A. Germot, and D. Moreira. "The New Phylogeny of Eukaryotes." *Current Opinion in Genetics and Development* 10 (2000): 596–601.

Ragan, M. A., C. L. Goggin, R. J. Cawthorn, L. Cerenius, A. V. C. Jamieson, S. M. Plourde, T. G. Rand, K. Söderhäll, and R. R. Gutell. "A Novel Clade of Protistan Parasites Near the Animal-Fungal Divergence." *Proceedings of the National Academy of Sciences USA* 93 (1996): 11907–11912. Describes the Mesomycetozoa, referred to in this paper as the "DRIP" group.

Riechmann, J. L., J. Heard, G. Martin, L. Reuber, C.-Z. Jiang, J. Keddie, L. Adam, O. Pineda, O. J. Ratcliffe, R. R. Samaha, R. Creelman, M. Pilgrim, P. Broun, J. Z. Zhang, D. Ghanderhari, B. K. Sherman, and G.-L. Yu. "*Arabidopsis* Transcription Factors: Genome-wide Comparative Analysis among Eukaryotes." *Science* 290 (2000): 2105–2110.

Rubin, G. M., et al. "Comparative Genomics of the Eukaryotes." *Science* 287 (2000): 2204–2215. Compares the genomes of yeast, *Caenorhabditis*, and *Drosophila*.

Wang, D. Y.-C., S. Kumar, and S. B. Hedges. "Divergence Time Estimates for the Early History of Animal Phyla and the Origin of Plants, Animals and Fungi." *Proceedings of the Royal Society of London B: Biological Sciences* 266 (1999): 163–171.

— KEVIN J. PETERSON

MUTATION

[*This entry comprises two articles. The first article is a summary overview of the causes and types of mutation; the second article describes how the mutation rate is itself a phenotypic trait that can vary genetically. For related discussions, see* Chromosomes; *and* DNA and RNA.]

An Overview

Mutation is the heritable alteration of a genome (consisting of nucleic acids: DNA in most organisms, but RNA in some viruses) and is the source of new allelic variation in populations. Ultimately, adaptive evolution requires mutation. Paradoxically, however, the overwhelming majority of mutations with phenotypic effects are likely to be deleterious to fitness. This conclusion was argued on theoretical grounds in 1930 by R. A. Fisher, who noted by analogy that a random change in a well-adjusted mechanism is unlikely to improve its function; it is supported empirically by studies estimating that deleterious mutations arise at least 100,000 times more frequently than beneficial mutations in experimental populations of bacteria.

The full distribution of deleterious mutational effects on fitness is unknown in detail in any organism. Some mutations are lethal or confer sterility; others (the majority of those with phenotypic effects) have less deleterious effects, probably including a large class of mildly deleterious mutations. The distribution of beneficial mutational effects is likewise unknown in detail. In an elaboration of the argument cited above, Fisher reasoned that mutations with very slight effects on fitness (analogous to extreme "fine tuning" of a well-adjusted mechanism) are the most likely to be beneficial, and this conclusion is to some extent supported by studies of newly arisen mutations in laboratory organisms. Certainly, major mutations that affect many aspects of the phenotype simultaneously are unlikely to be beneficial, because they are likely to disrupt the integrated functioning of the organism.

Some mutations, however, are effectively neither beneficial nor deleterious: they have no fitness effects, or their slight fitness effects are overwhelmed by random genetic drift. Because beneficial mutations are rare and deleterious mutations are removed from populations by natural selection, substitution of such neutral mutations by genetic drift is the predominant evolutionary force affecting genomic sequences.

In multicellular organisms, mutation affects fitness both by the transmission of changes in the germline to offspring, and by changes in somatic cells. The potential impact of somatic mutations is well illustrated by cancers, which arise as a consequence of somatic mutations that impair controls on cell division and tissue development. In organisms (such as most animals) with separate germline and somatic cell populations, somatic mutations are not transmitted to offspring, whatever their effects on the parent phenotype. In plants, multicellular protists, and multicellular fungi, however, germ cells arise from terminal differentiation of somatic cells, and somatic mutations can be transmitted to offspring.

Mutation Rates. Mutations arise as a consequence of spontaneous errors in genome replication and of damage to nucleic acids caused by chemical and physical factors, such as oxidizing agents and radiation. (An additional major source of mutations in many organisms is the activity of transposable genetic elements, sequence entities that encode properties for their own replication and movement within genomes; this is discussed further below.) Replication errors and damage occur at high rates under normal physiological circumstances, and they are corrected by a variety of elaborate molecular proofreading and repair mechanisms whose evident purpose is to ensure genetic fidelity. These mechanisms are very effective in most organisms: the typical rate of spontaneous mutation per base pair in eukaryotes is extremely low—somewhere in the range of 10^{-10}–10^{-9} per generation. Like other aspects of the phenotype, however, the mutation rate is affected by the environment; environmental extremes (for example, high temperatures, high incident radiation, or the presence of certain mutagenic chemicals), can induce elevated mutation rates.

Total genomic rates of mutation are extremely difficult to measure with accuracy, and measurements made with different methods often disagree substantially. Table 1 lists average mutation rates per nucleotide (base or base pair, as appropriate) and average genomic spontaneous mutation rates, estimated by direct genetic methods in a broad range of taxa. Despite the uncertainty in these values, it is interesting to note that genomic rates of mutation tend to be similar within major groups of organisms. Riboviruses (RNA viruses exclusive of retroviruses) have the highest per-nucleotide and genomic mutation rates, perhaps even approaching a value at which their population viability is threatened, as discussed below. Retroviruses (such as HIV, the cause of AIDS in humans), which have RNA genomes that are reverse transcribed and inserted into host DNA, have genomic mutation rates approximately an order of magnitude higher than those of riboviruses. Among microbes with DNA chromosomes and a wide range of genome sizes (DNA viruses, bacteria, and unicellular fungi), estimated genomic mutation rates are remarkably constant around a mean value of about 0.0032 mutations per generation. Even an organism that lives in a potentially mutagenic environment, the thermophilic archaeon *Sulfolobus acidocaldarius*, has a genomic mutation rate (0.0018) comparable to that of other DNA-based microbes. Finally, genomic mutation rates in the germlines of higher eukaryotes are at least two orders of magnitude higher than those in DNA-based microbes. As discussed below, these rates for higher eukaryotes are almost certainly underestimates, and perhaps even gross underestimates, because the way in which they are measured does not take into account all types of mutational changes.

The influx of deleterious mutations reduces the average fitness of a population below what it would be without mutation; this reduction is termed *mutational load*. In the absence of factors that act to reduce mutational load (such as sexual recombination), extremely high genomic deleterious mutation rates (much above approximately 1) are incompatible with population viability, because they lead to mutational erosion of vital sequence information from generation to generation. Most organisms' genomic deleterious mutation rates lie well below the critical threshold value for population viability, but mutation rates in RNA viruses may be close to this value, given that their genomes consist almost entirely of coding sequence. As shown in Table 1, total genomic mutation rates estimated in some sexual higher eukaryotes are also in the vicinity of 1. Indeed, other methods of estimating genomic mutation rates suggest much higher values in higher eukaryotes. For example, in humans, a phylogenetically based analysis of DNA sequence evolution puts the per generation genomic mutation rate at approximately 200, which almost certainly implies a genomic deleterious rate above 1. There is considerable debate, however, over whether genomic deleterious mutation rates exceed 1 in other higher eukaryotes, and this is a very active area of current research in evolutionary genetics.

Mutation rates are by no means constant across genomic regions. Specific local regions with elevated mutation rates, known as *mutational hotspots*, are known in many organisms. In some cases, hotspots are known to be caused by local sequence features; for example,

TABLE 1. Per Generation Mutation Rate Estimates for a Range of Organisms

Organism	*Mutation rate per nucleotide*	*Genomic mutation rate*
Riboviruses		
Poliovirus[a]	1.97×10^{-5}	0.15
	1.10×10^{-4}	0.82
Measles virus	6.29×10^{-5}	1.00
Human rhinovirus	9.40×10^{-5}	0.67
Vesicular stomatitis virus[b]	9.94×10^{-5}	1.11
Retroelements		
Spleen necrosis virus	5.1×10^{-6}	0.04
Murine leukemia virus[c]	7.2×10^{-6}	$\geq$0.06
	3.1×10^{-5}	0.26
Rous sarcoma virus	4.6×10^{-5}	0.43
Bovine leukemia virus	3.2×10^{-6}	0.027
Human immunodeficiency virus type 1[d]	2.1×10^{-6}	0.19
S. cerevisiae Ty1 retrotransposon	1.9×10^{-5}	0.11
DNA based microbes		
Bacteriophage M13	7.2×10^{-7}	0.0046
Bacteriophage λ	7.7×10^{-8}	0.0038
Bacteriophages T2 and T4	2.4×10^{-8}	0.0040
Escherichia coli	5.4×10^{-10}	0.0025
Sulfolobus acidocaldarius	7.8×10^{-10}	0.0018
Saccharomyces cerevisiae	2.2×10^{-10}	0.0027
Neurospora crassa	7.2×10^{-11}	0.0030
Higher eukaryotes[e]		
C. elegans	2.3×10^{-10}	0.018
Drosophila	3.4×10^{-10}	0.058
Mouse	1.8×10^{-10}	0.49
Human	5.0×10^{-11}	0.16

All of the genomic estimates shown were made by extrapolating from measured mutation rates at individual loci; other methods have yielded different (usually higher) per locus and genomic mutation rates. In particular, estimates made by phylogenetically based analyses of DNA sequence evolution suggest much higher genomic mutation rates in humans and some other higher eukaryotes, as discussed in the text.

Data in the table are taken from the following sources: Drake, J. W. 1993. Rates of spontaneous mutation among RNA viruses. *Proceedings of the National Academy of Sciences USA* 90, 4171–4175. Drake, J. W., Charlesworth, B., Charlesworth, D., and J. F. Crow. 1998. Rates of spontaneous mutation. *Genetics* 148, 1667–1686; Drake, J. W., and J. J. Holland. 1999. Rates of mutation among RNA viruses. *Proceedings of the National Academy of Sciences USA* 96, 13910–13913; Grogan, D. W., Carver, G. T. and J. W. Drake. 2001. Genetic fidelity under harsh conditions: Analysis of spontaneous mutation in the thermoacidophilic archaeon *Sulfolobus acidocaldarius*. *Proceedings of the National Academy of Sciences USA* 98, 7928–7933.

No estimates for plants are shown in the table because the data are presently of insufficient quality. The available data suggest that mutation rates in higher plants are similar to those in other higher eukaryotes.

[a]The lower rate is an average of three similar estimates; the higher rate is an average of two similar estimates.

[b]An average of two similar estimates.

[c]The two divergent estimates are based on different mutational targets.

[d]An average of two similar estimates.

[e]The data for higher eukaryotes are subject to greater uncertainty than those for the other taxa and are likely to be underestimates (perhaps even gross underestimates).

simple (one to a few base pairs) tandem repetitive DNA sequences undergo insertion and deletion of repeat units at high rates as a result of strand slippage during DNA synthesis. Hotspots can also result from preferential insertion and deletion of transposable elements into certain sequences. In some other cases, the mechanistic basis for hotspots remains obscure.

Variation in mutation rates can also encompass large fractions of the genome. For example, the pattern of substitution of neutral mutations in the mouse and rat genomes indicates that the mutation rate on the X chromosome in these species is lower than that on the autosomes. The mechanistic basis for this difference is unknown, but its existence is consistent with the idea that there has been greater selection against deleterious recessive mutations on the X chromosome, which is present in only one copy (hemizygous) in mammalian males.

Comparisons of rates of synonymous substitution between organellar and nuclear DNA have also revealed some interesting differences. In animals, the rate of point mutation appears to be about 10 times higher in the mitochondrial genome than in the nuclear genome. In plants, however, mitochondrial and chloroplast genomes appear to have lower rates of mutation than the nuclear genome.

Types of Mutation. Mutations may be divided into two broad physical categories based on the nucleotide sequence changes they entail: (1) substitution of one nucleotide for another (Figure 1B–1D); and (2) insertion, deletion, or rearrangement of nucleotides (Figure 1E and 2). Single nucleotide substitutions and very short (one to a few nucleotides long) insertions and deletions are collectively termed *point mutations*. Substitutions are further classified as *transversions* or *transitions*. A transversion is the replacement of a purine (adenine or guanine) with a pyrimidine (thymine or cytosine), or vice versa; a transition is the replacement of a pyrimidine with a pyrimidine, or a purine with a purine. Studies of sequence evolution in many organisms have indicated that transitions arise more frequently than transversions.

Insertions, *deletions*, and *rearrangements* collectively range in size from single nucleotides (e.g., Figure 1E) to entire chromosome segements comprised of millions of nucleotides (Figure 2). Many of the larger-scale genomic changes in this category are caused by the activities of transposable genetic elements. These elements cause mutations by inserting into and disrupting gene sequences, and by mediating ectopic recombination events that rearrange chromosome segments. Studies in the fruit fly *Drosophila melanogaster* have suggested that a substantial percentage of newly arising mutations in this species (and perhaps in many other species as well) are the result of transposable element activity. Indeed, the majority of the morphologically detectable mutations studied in classical *Drosophila* genetics are known to be the result of transposable element insertions into genes or into adjacent regulatory regions. Interestingly, the *wrinkled* pea mutant originally studied by Gregor Mendel is also the result of a transposable element insertion.

Phenotype Effects. Whether a mutation affects the phenotype, and hence whether it can affect fitness, depends on where in the genome it arises and on the nature of the sequence change it entails. Mutations of all types arising within noncoding regions, such as introns and intergenic sequences, are likely to have no phenotypic effects unless the mutated sequence contributes to the phenotype in some other manner—for example, by regulating gene expression. Furthermore, because the genetic code is degenerate (more than one codon specifies the same amino acid, in some cases), even mutations occurring in protein coding sequences need not have phenotypic effects. Approximately 70 percent of nucleotide substitutions in the third position and 4 percent of substitutions in the first position of codon triplets do not alter the encoded amino acid and hence leave the protein unchanged (Figure 1B). Such substitutions are termed *silent* or *synonymous*, as opposed to changes that alter the encoded amino acid and are termed *meaningful* substitutions, *nonsynonymous* substitutions, or *missense* mutations (Figure 1C).

Not all synonymous substitutions are neutral. A pattern of biased codon usage has been detected in many organisms in which certain codons within synonymous sets are used more frequently than the random expectation. This bias is stronger in continually expressed

(A) Pro Val Thr His Tyr Gly Arg
CCA GTA ACA CAT TAT GGA AGA

(B) Pro Val Thr His Tyr Gly Arg
CCA GT**G** ACA CAT TAT GGA AGA
↑

(C) Pro **Leu** Thr His Tyr Gly Arg
CCA **C**TA ACA CAT TAT GGA AGA
↑

(D) Pro Val Thr His **STOP**
CCA GTA ACA CAT TA**A** GGA AGA
↑

(E) Pro Val **His** **Thr** **Leu** **Trp** **Lys**
CCA GTA **C**AC ACA TTA TGG AAG A
↑

FIGURE 1. Types of Mutation in Coding Sequences. A hypothetical section of gene sequence is shown, along with the encoded amino acids. The DNA shown corresponds to the sense (coding) strand of DNA; the equivalent mRNA sequence would have uracil (U) in place of thymine (T). (A) Original gene sequence; (B) a synonymous (silent) substitution leaves the amino acid sequence unchanged; (C) a nonsynonymous (meaningful) substitution changes the encoded amino acid; (D) a substitution creating a stop codon truncates the translated amino acid sequence; (E) a frameshift insertion alters the translated amino acid sequence downstream of the insertion site. Drawing by Paul D. Sniegowski.

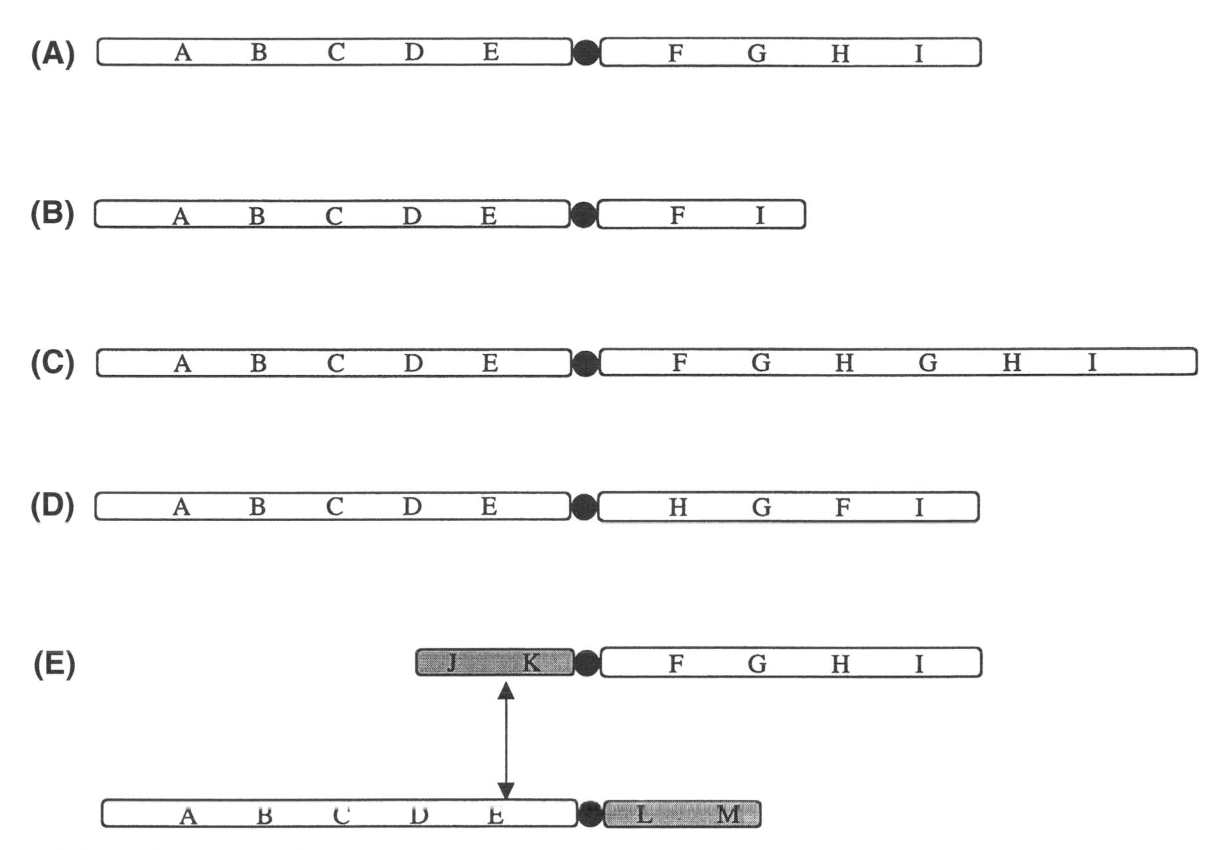

FIGURE 2. Types of Chromosomal Mutation.
Chromosome arms are shown in outline, and centromeres are represented in black. Capital letters represent unique chromosome segments. (A) Original chromosome; (B) Deletion of the HI segment; (C) Duplication of the GH segment; (D) inversion of the FGH segment; (E) reciprocal translocation of whole chromosome arms with another chromosome, shown in darker gray. (Other types of translocations are possible.) Drawing by Paul D. Sniegowski.

genes than in rarely expressed genes, suggesting that it reflects the action of natural selection on translation efficiency.

The effect of a nonsynonymous nucleotide substitution on the phenotype depends on the way in which it alters protein structure and function. Replacement of one amino acid by another with similar chemical properties may have minor effects on protein structure and function, whereas the substitution of a dissimilar amino acid is likely to have profound effects. A classical illustration of the ramifying phenotypic effects of a single amino acid substitution is provided by the example of human sickle-cell disease, in which the replacement of valine with glutamic acid at position 6 of the β-globin polypeptide results in an abnormal hemoglobin. In homozygotes, this abnormal hemoglobin causes sickling and destruction of red blood cells, blockage of circulation, and associated serious medical problems that greatly reduce expected lifespan.

Mutations within coding regions can also alter the size of the expressed polypeptide chain or render it meaningless. Certain nucleotide substitutions, termed *nonsense mutations*, change an amino acid codon into a stop codon (Figure 1D) or vice versa, terminating translation prematurely or allowing it to continue beyond the proper end of the protein sequence; the resultant truncated or elongated protein is unlikely to function correctly. Nucleotide insertions and deletions not in multiples of three (the length of a codon) alter the translational reading frame of the protein. Downstream from such *frameshift* mutations, the expressed amino acid sequence bears no relation to the correct sequence, and protein function is obliterated (Figure 1E). Nucleotide insertions and deletions in multiples of three add to or delete amino acids (or stop codons) from the chain without altering the reading frame, usually with deleterious consequences for the protein. Deletions, insertions, and substitutions can also disrupt gene regulatory regions, thereby altering or eliminating gene expression.

Deletions, duplications, and rearrangements can also span whole genes or multiple genes. Very large deletions may even remove multiple gene sequences, eliminating the expression of many proteins simultaneously. Duplications obviously do not eliminate genes, but they may unbalance the dosage of gene products expressed from the duplicated region, with major phenotypic effects. Despite this possibility, however, duplication and subsequent divergence of coding sequences plays an important role in genome diversification. Rearrangements such as inversions or translocations can disrupt coding or regulatory sequences or change the relationship between coding and regulatory sequences. In addition, some chromosomal rearrangements can cause abnormal synapsis during meiosis, leading to aneuploid gametes that are likely to be nonviable.

Many mutations, as discussed above, result in a *loss of function* at the phenotypic level. Some mutations, however, result in a *gain of function*. For example, the inactivation of a repressor protein by mutation results in constitutive (continual) expression of the repressed gene, which is a gain of function at the phenotypic level. In classical genetic terms, loss-of-function mutations are typically recessive: that is, two copies of the mutation are required in a diploid organism for the phenotype to be substantially affected. Gain-of-function mutations, on the other hand, are usually dominant, affecting the phenotype substantially when present in only one copy. From an evolutionary standpoint, however, dominance and recessivity are best considered as quantitative rather than qualitative aspects of mutations. Many "recessive" loss-of-function mutations nonetheless have slight deleterious effects on fitness in heterozygotes that reflect subtle quantitative alterations of the phenotype.

Conclusion. Mutation and recombination are not always easily distinguished from each other at the molecular level. Some of the molecular machinery that mediates recombination events is also involved in DNA repair and hence contributes to mutagenesis. In addition, mutations involving chromosomal rearrangement occur by reordering of DNA segments using the recombinational machinery of the cell. From a population and evolutionary perspective, however, the distinction between mutation and recombination is clear: mutation creates new allelic variation by altering the contents of a single genome, whereas recombination shuffles previously existing variants that have been contributed by different genomes.

The term "mutation" was not used in its current precise genetic sense until the early twentieth century. Before then, "mutation" was used broadly to refer to any major morphological change from generation to generation. This was the manner, for example, in which Darwin used it in the *Origin of Species* (e.g., Chapter 14, pp. 480–481 in the first edition). Darwin's views on the possible sources of heritable variation (what we now recognize as mutation and recombination) included a minor role for such Lamarckian factors as the "effects of use and disuse" on morphological characters. Unlike Lamarck, however, Darwin made a clear logical distinction between the origin of heritable variation and the origin of adaptation:

> Whatever the cause may be for each slight difference in the offspring from their parents—and a cause for each must exist—it is the steady accumulation, through natural selection, of such differences, when beneficial to the individual, that gives rise to all the more important modifications of structure, by which the innumerable beings on the face of this earth are enabled to struggle with each other, and the best adapted to survive. (*Origin of Species*, first edition, p. 170)

This logical distinction has become a central tenet of modern evolutionary theory: mutations (and recombinants) arise at random with respect to their consequences for cell or organism fitness, and evolutionary adaptation is solely a consequence of the sorting of such genetic variants by natural selection. This viewpoint was supported by the observation in the early twentieth century that visible mutations in higher eukaryotes, such as *Drosophila*, are rare and almost always deleterious, which strongly suggested that mutation alone was unable to account for adaptive evolution. Further support came from the results of several mid-twentieth-century experiments that showed that adaptation of laboratory bacterial populations to lethal selective agents (such as antibiotics or bacteriophages) could be wholly accounted for by the presence of resistant mutants that had arisen at random before selection was imposed.

In the 1980s and 1990s, however, new experiments on bacterial populations challenged the conclusion that mutations arise at random, without regard to their utility. Proponents of the "directed mutation" hypothesis argued that specific mutations could be produced at higher rates when required for growth. However, careful reexamination has led to a rejection of claims for directed mutation. Instead, apparent directed mutation has been shown to result from artifacts and improper controls in the original experiments, or from previously unsuspected pathways of random mutagenesis in bacterial cells (and possibly eukaryotic cells) that act in the absence of cell division. Darwin's logical distinction between the origin of variation and the origin of adaptation remains intact: mutation is the ultimate source of genetic variation, and natural selection is the sole cause of evolutionary adaptation.

[*See also* Cellular Organelles; Codon Usage Bias; Forensic DNA; Genetic Code; Mutation, *article on* Evolution of Mutation Rates; Transposable Elements; Viruses.]

BIBLIOGRAPHY

Fisher, R. A. *The Genetical Theory of Natural Selection*. Oxford, 1930. Chapters 1 and 2 of this classic work present Fisher's fundamental ideas about the role of mutation in evolution and the probability that a new mutation is beneficial.

Foster, P. L. "Adaptive Mutation: Implications for Evolution." *BioEssays* 22 (2000): 1067–1074. A recent review of mechanisms and evolutionary implications of transiently elevated mutation in starving bacterial cells, by an early proponent of "directed mutation."

Friedberg, E. C., G. C. Walker, and W. Siede. *DNA Repair and Mutagenesis*. Washington, DC, 1995. As of 1995, the general state of knowledge of mechanisms of repair and mutagenesis in prokaryotes and eukaryotes.

Gerrish, P. J., and R. E. Lenski. "The Fate of Competing Beneficial Mutations in an Asexual Population." *Genetica* 102/103 (1998): 127–144. See the comment on Schlotterer and Imhof, below.

Griffiths, A. J. F., J. H. Miller, D. T. Suzuki, R. C. Lewontin, and W. M. Gelbart. *An Introduction to Genetic Analysis*. New York, 2000. An excellent college-level introductory genetics text, with a section of chapters on "Generation of Genetic Variation" that covers mechanisms of mutation (including transposable element activity) and recombination.

Keightley, P. D., and A. Eyre-Walker. "Terumi Mukai and the Riddle of Deleterious Mutation Rates." *Genetics* 153 (1999): 515–523. History, methods and results in the effort to measure deleterious mutation rates. Total genomic and deleterious mutation rates estimated by various methods are discussed; many values given are higher than those shown in Table 1.

Kibota, T. T., and M. Lynch. "Estimate of the Genomic Mutation Rate Deleterious to Overall Fitness in *E. coli*." *Nature* 381 (1996): 694–696. The authors estimate the genomic rate of deleterious mutation in *E. coli* as approximately 0.0002 per generation. This is in all likelihood an underestimate, because the method used does not take into account all types of deleterious mutation. This work and other work on experimental populations of *E. coli* allows an estimate of the relative rates of deleterious and beneficial mutation as cited in the text; see the comment on Schlötterer and Imhof, below.

Li, W.-H. *Molecular Evolution*. Sunderland, Mass., 1997. A comprehensive review of molecular evolution, with a chapter devoted to rates and patterns of nucleotide substitution across taxa.

Maynard Smith, J., and E. Szathmáry. *The Major Transitions in Evolution*. New York, 1995. Chapter 4, "The evolution of templates," discusses the evolution of genome size and repair mechanisms.

McVean, G. T., and L. D. Hurst. "Evidence for a Selectively Favorable Reduction in the Mutation Rate of the X Chromosome." *Nature* 386 (1997): 388–392. Comparisons of synonymous substitution rates in mouse and rat X chromosomes and autosomes indicate a lower mutation rate on the X chromosome.

Ridley, M. *The Cooperative Gene: How Mendel's Demon Explains the Evolution of Complex Beings*. New York, 2001. The influence of deleterious mutations on the evolution of genetic systems and complex organisms.

Schlötterer, C., and M. Imhof. "Fitness Effects of Advantageous Mutations in Evolving *Escherichia coli* Populations." *Proceedings of the National Academy of Sciences USA* 98 (2001): 1113–1117. This and Gerrish and Lenski, above, estimate the genomic rate of beneficial mutation in evolving *E. coli* populations as approximately 10^{-9} per genome per generation. Together with the work of Kibota and Lynch, this allows an empirical estimate of the ratio of deleterious to beneficial mutation rates in this organism of at least 10^5.

Sniegowski, P. D., and R. E. Lenski. "Mutation and Adaptation: The Directed Mutation Controversy in Evolutionary Perspective." *Annual Review of Ecology and Systematics* 26 (1995): 553–578. A critical review of claims for directed mutation, including a brief history of ideas about the origin of genetic variation and adaptation.

— PAUL D. SNIEGOWSKI

Evolution of Mutation Rates

Mutation is the ultimate source of the genetic variation required for evolution. Mutation rates are influenced by environmental factors, but heritable alteration of the enzymatic machinery that replicates and repairs genomes can also affect mutation rates, as can variation in the activity of transposable genetic elements. Thus, a population can harbor selectable genetic variation for the mutation rate in the same way that it harbors variation for other phenotypic traits.

Because most mutations of any phenotypic effect are deleterious to fitness, selection usually must favor a decrease in mutation rates. What trade-offs prevent the reduction of genomic mutation rates below the levels that have already evolved? Two possibilities have been identified: (1) Below a certain minimal genomic mutation rate, further reduction might be constrained by the individual-level fitness cost of increased physiological investment in replication, proofreading, and repair systems or by physicochemical limits to accuracy in these systems. (2) Below a certain optimal genomic mutation rate, further reduction might be disfavored by the population-level adaptive cost of failing to produce an adequate supply of new beneficial mutations.

The comparison of mutation rates from a wide variety of species reveals interesting patterns (Figure 1). It is evident that per nucleotide mutation rates are very low in most organisms. It is certainly possible for per nucleotide mutation rates to be higher: defective alleles that can elevate mutation rates slightly to manyfold have been identified in many species. In contrast, almost no alleles that reduce the mutation rate have been identified, suggesting that selection has exhausted most of the available means of reducing muutation.

Per nucleotide mutation rates vary inversely with genome size across a wide range of DNA-based microbes, yielding an almost constant genomic mutation rate. This indicates that lower per nucleotide mutation rates could be achieved in the species with smaller genomes, and it hints at the common operation of a general selective trade-off in setting the mutation rate across all of these species. Whether genomic mutation rates in these species are minimal or optimal, however, remains unresolved.

The mutation rate bottoms out at a constant value of 10^{-10} mutations per nucleotide per generation or lower in species with genome sizes greater than about 10^8 base

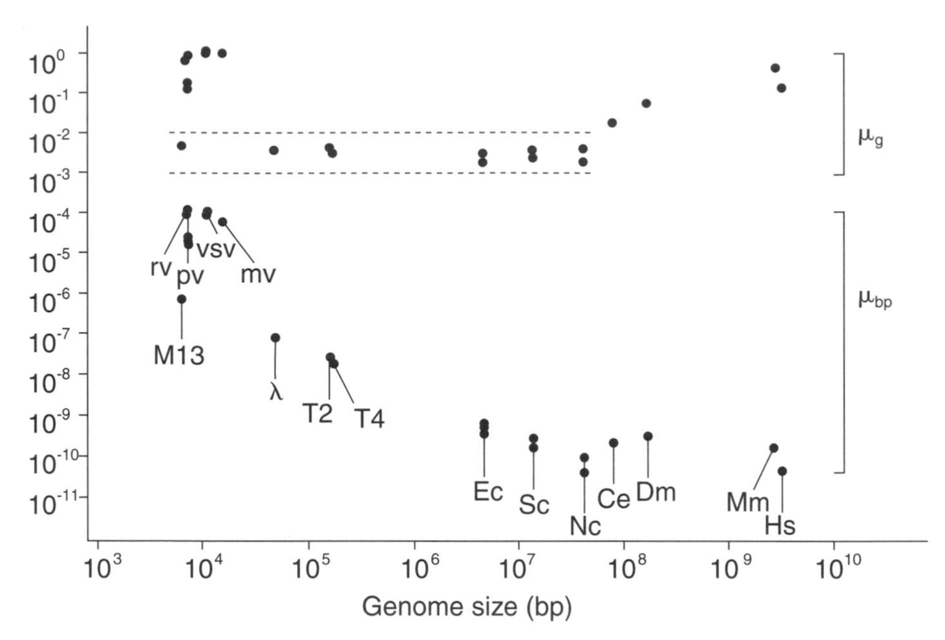

FIGURE 1. Estimates of Mutation Rates per Genome (μ_g, Upper Points) and per Base Pair (μ_{bp}, Lower Points) for a Wide Variety of Species, Plotted Against Genome Size on a Log-Log Scale.
Multiple symbols are drawn where independent estimates from different loci in the same organism were available. No error bars are shown, but the errors are probably large. RNA viruses are as follows: mv = measles virus; pv = poliovirus; rv = rhinovirus; vsv = vesicular stomatis virus. Bacteriophages M13, T2, T4, and λ (DNA-based) are shown in red. *Escherichia coli* (Ec) is shown in orange; *Saccharomyces cerevisiae* (Sc) and *Neurospora crassa* (Nc) are shown in green. Higher eukaryotes are shown in blue: Ce = *Caenorhabditis elegans*; Dm = *Drosophila melanogaster*; Mm = *Mus musculus*; Hs = *Homo sapiens.* Outliers thought to reflect estimates from nonrepresentative loci are not shown. The observation of a conserved genomic rate in DNA-based microbes is highlighted with dashed lines. Data sources are given in and the figure is reprinted from "Evolution of Mutation Rates: Separating Causes from Consequences," by P. D. Sniegowski, P. J. Gerrish, T. Johnson, and A. C. Shaver, in *BioEssays* 22 (2000): 1057–1067. Copyright 2000, Wiley-Liss, Inc., a subsidiary of John Wiley and Sons.

pairs, and thereafter the genomic mutation rate rises with increasing genome size. This suggests that the per nucleotide mutation rate has been minimized in species with large genomes. It is even conceivable that the per nucleotide mutation rate has reached a physicochemical minimum in these species, although some evidence suggests that further reductions are possible but entail excessive fitness costs. Whatever the answer, theoretical studies indicate that this minimal mutation rate is probably higher than the optimal rate required for long-term adaptive success.

Selection can act directly or indirectly to change the frequencies of alleles that modify the mutation rate. Direct selection, for example, might disfavor an allele that reduces the mutation rate because its expression entails an excessive amount of time or energy, even though this would reduce the rate of accumulation of deleterious mutations. Indirect selection depends on the persistence of associations between mutation rate modifier alleles and fitness alleles at other loci. In an asexual population, linkage is absolute, and a modifier that raises the mutation rate can ride to high frequency along with a beneficial allele at another locus; this is a special instance of the phenomenon known as genetic hitchhiking. In a sexual population, the association between a modifier that raises the mutation rate and a beneficial allele is quickly broken down by recombination unless the two are very tightly linked. In both sexual and asexual populations, a modifier that decreases the mutation rate is likely to be favored by indirect selection because it will generally be associated with a lower than average number of deleterious mutations.

The evolution of high mutation rates by indirect selection has been observed in experimental populations of bacteria. In the earliest of these experiments, populations were seeded with variable frequencies of cells bearing a "mutator" allele that raised the mutation rate. Mutator hitchhiking occurred when the ratio of the numbers of mutator to wild-type cells in the population was such that more beneficial mutations arose in the mutator subpopulation. Below this critical threshold ratio, however, the mutator cells declined in frequency in the population. This result seemed to suggest that a mutator allele would have to attain an improbably high frequency in order to hitchhike. However, more recent experimental results have shown that even populations initially free of mutators can evolve high mutation rates via hitchhiking. For example, in a recent experiment, three of twelve populations of *Escherichia coli* founded from a single ancestral cell evolved 100-fold elevations in their genomic mutation rates during 10,000 generations of propagation in the laboratory. In each of these three populations, a defective DNA repair allele rose to fixation as a consequence of hitchhiking. These alleles can only have arisen as a consequence of rare recurrent mutation of ancestrally functional loci in these populations; thus, they were probably present at extremely low frequencies when they began to hitchhike. Mutator phenotypes have also been observed at frequencies of a few percent among natural bacterial isolates, indicating that the hitchhiking phenomenon is not an artifact of laboratory conditions.

It is unclear whether the evolution of high mutation rates in asexual populations should be regarded as pathological or adaptive. A high mutation rate potentially can increase the rate of adaptation by increasing the supply of new beneficial mutations. However, the rate of adaptation in an asexual population is constrained by competition between clones with different beneficial mutations, which cannot be recombined into a single best genotype as they would be in a sexual population. This may limit adaptive benefits from increased mutation to situations in which population sizes are small (and hence very few beneficial mutations arise each generation), or populations are so well adapted that beneficial mutations are very rare. It seems likely that the former situation applies in microbial pathogens that invade hosts in small numbers and must evolve away from the host immune response to survive; a possible association between mutator status and pathogenicity has been noted in natural bacterial isolates. Over the longer term, however, a mutator phenotype seems likely to be disfavored because it increases the influx of deleterious mutations. This probably explains why high mutation rates are not more common in asexual populations.

There are two situations in which a high mutation rate might be advantageous without causing large increases in the load of deleterious mutations. The first of these involves restricting elevation of the mutation rate only to certain loci. Hypermutable "contingency loci" in pathogenic bacteria provide evidence for such locus-specific evolution of high mutation rates. These loci interact with the environment in unpredictable ways, for example, via surveillance by the host immune system, that often favor variation per se rather than conservation of specific function; this further decreases the deleterious effect of elevating the mutation rate.

A mechanism that allowed the transient elevation of genomic mutation rates only when new variation was at a premium might also facilitate adaptation while avoiding the buildup of deleterious mutations. Recent evidence indicates that at least some cells in cultures of bacteria and yeast under starvation stress have elevated genomic mutation rates. This phenomenon has acquired the suggestive name adaptive mutation, and there is much interest in its possible evolutionary significance. To date, however, there has been little analysis of the evolutionary processes by which the capacity for transiently elevated mutation rates might arise and be maintained, and alternative explanations such as unavoidable increases in error rates due to the physiological effects of environmental stress have not been ruled out.

[*See also* Cancer; Experimental Evolution, *article on* A Long-term Study with *E. coli*; Immune System, *article on* Microbial Countermeasures to Evade the Immune System; Sex, Evolution of.]

BIBLIOGRAPHY

Chao, L., and E. C. Cox. "Competition between High and Low Mutating Strains of *Escherichia coli*." *Evolution* 37 (1983): 125–134. This paper provided the first demonstration of mutator hitchhiking in an asexual population.

Drake, J. W. "A Constant Rate of Spontaneous Mutation in DNA-based Microbes." *Proceedings of the National Academy of Science USA* 88 (1991): 7160–7164. Documents the strikingly constant genomic mutation rate of DNA-based microbes.

Drake, J. W., B. Charlesworth, D. Charlesworth, and J. F. Crow. "Rates of Spontaneous Mutation." *Genetics* 148 (1998): 1667–1686. A general review of the literature on genomic mutation rates, with a section on the evolution of mutation rates.

Foster, P. L. "Adaptive Mutation: Implications for Evolution." *BioEssays* 22 (2000): 1067–1074. A review of mechanisms and evolutionary implications of transiently elevated mutation in starving bacterial cells.

Kimura, M. "On the Evolutionary Adjustment of Spontaneous Mutation Rates." *Genetical Research* 9 (1967): 23–34. A seminal theoretical paper that discusses the cost of fidelity and physicochemical limits to the reduction of mutation rates.

Leigh, E. G. "Natural Selection and Mutability." *American Naturalist* 104 (1970): 301–305. An important theoretical paper exploring the consequences of sexual recombination on mutation rate evolution.

Maynard Smith, J. *The Evolution of Sex* Cambridge, 1978. Clearly outlines the concepts of minimal and optimal rates.

Metzgar, D., and Wills. C. "Evidence for the Adaptive Evolution of Mutation Rates." *Cell* 101 (2000): 581–584. Provides a discussion of the evidence for adaptively high mutation rates.

Moxon, E. R., P. B., Rainey, M. A. Nowak, and R. E. Lenski. "Adaptive Evolution of Highly Mutable Loci in Pathogenic Bacteria." *Current Biology* 4 (1994): 24–33. A paper that presents theoretical and empirical evidence for the existence of hypermutable contingency loci that facilitate avoidance of host immune responses in pathogenic bacteria.

Sniegowski, P. D., P. J., Gerrish, T. Johnson, and A. C. Shaver. "Evolution of Mutation Rates: Separating Causes from Consequences." *BioEssays* 22 (2000): 1057–1067. A review of the theoretical and empirical literature on the evolution of mutation rates.

Sturtevant, A. H. "Essays on Evolution: I. On the Effects of Selection on Mutation Rate." *Quarterly Review of Biology* 12 (1937):

467–477. A classic essay in which Sturtevant posed the question Why does the mutation rate not evolve to zero?

— PAUL D. SNIEGOWSKI

MUTUALISM

Mutualisms are interactions between two species that benefit both of them. Individuals who interact successfully with a mutualist experience greater success than those who do not. Behaving mutualistically is therefore of direct benefit to the individual itself. As Charles Darwin first pointed out in *Origin of Species* (1859), it does not require any special concern for the well-being of the partner.

At one time, mutualisms were thought to be rare curiosities of nature, but ecologists now believe that virtually every species on earth is involved in one or more of these interactions. For example, in tropical rain forests, the large majority of plants depend on animals for pollination and seed dispersal. Over 80 percent of all flowering plants are involved in mutualisms with mycorrhizae, fungi that live on and in their roots and that increase their access to some soil nutrients. In the ocean, both coral reef communities and deep-sea vents are exceptionally rich with mutualisms; in fact, corals themselves depend on the photosynthetic algae that inhabit them. Key events in the history of life have been linked to mutualism as well, including the origin of the eukaryotic cell and the invasion of the land by plants. The study of mutualism, therefore, has a major role to play in illuminating both the diversity and the evolutionary diversification of life on earth.

The Benefits of Mutualism. In many ways, mutualisms resemble cooperative interactions that involve individuals of the same species. One major difference, however, is that mutualisms usually involve an exchange of two quite different kinds of benefit. In fact, mutualisms can be thought of as an economic exchange that takes place in a "biological marketplace." Organisms offer their mutualists commodities that are cheap for them to acquire or produce; in exchange, they receive commodities that would otherwise be difficult or impossible for them to acquire.

Although most ecological and evolutionary questions about mutualism still remain open, biologists do have a relatively deep understanding of the kinds of exchanges that take place within mutualisms. In many of these interactions, mutualists make limited but essential nutrients available to their partners. For instance, *Rhizobium* bacteria found in nodules on the roots of many legume (bean) species fix atmospheric nitrogen into a form (NH_3) that can be taken up by plants; in return, the bacteria receive carbon fixed by photosynthesis by their hosts. Consequently, legumes often thrive in nitrogen-poor environments such as deserts. Many plant species also feed fixed carbon to another group of root associates, the mycorrhizal fungi. The primary benefit of mycorrhizae is to vastly increase plants' access to soil phosphorus, a nutrient severely limiting to plant growth. This mutualism is so crucial that many researchers believe that its evolution was a critical step allowing plants to invade land around 400 million years ago.

A second benefit of mutualism comes in the form of transportation. One prominent example is biotic pollination, in which certain animals visit flowers to obtain resources (usually food, such as nectar), and return a benefit by transporting pollen between the plants they visit (Figure 1). As an indication of the importance of plant–pollinator mutualisms, Stephen Buchmann and Gary Nabhan (1996) estimate that half the food humans consume results from one of these interactions. Biotic seed dispersal, in which animals disperse the seeds of the plants whose fruits they consume, is another example of a mutualism in which nutrition is exchanged for transportation. In this case, seeds are transported away from the parent plant, toward more suitable sites for germination and growth.

Protection from biological enemies is a third benefit that mutualists can provide their partners. For example, ants commonly feed on sugar-rich excretions (honeydew) produced by aphids and related insects; they vigorously defend their food source, providing those insects with protection against predators and parasites. Ants similarly guard certain caterpillars (Lycaenidae) as well as diverse plants, both of which have evolved rewards that serve to attract and then keep loyal a core of defenders. Many other fascinating forms of "biological warfare" have now been documented. For example, certain hermit crabs carefully place stinging anemones on

FIGURE 1. Benefits of Mutualism. This bat's nighttime visit to an agave flower enables it to obtain nectar and, at the same time, to return a benefit by transporting pollen among other plants whose flowers it may visit.

their shells. When the crabs are attacked, the anemones effectively come to their defense. The anemones, in turn, receive transport among rich feeding sites.

In the preceding discussion, mutualisms are grouped according to the benefits that participants receive from them. An alternative way to divide up mutualisms is according to whether or not they are symbiotic. Two species found in intimate physical association for most or all of their lifetimes are considered to be in symbiosis. Although symbiosis is often taken to mean the same thing as mutualism, it is more accurate scientifically to say that symbioses may benefit neither, one, or both of the two partner species; only in the latter case is a symbiosis mutualistic. Conversely, only a subset of mutualisms are symbiotic. Of those mutualisms mentioned above, only plant–*Rhizobium* and plant–mycorrhizae associations are generally considered to be symbiotic. In some cases, symbionts become so closely integrated that their individuality becomes difficult and ultimately impossible to distinguish. Most notably, the eukaryotic cell is now believed to have originated as a symbiosis between a primitive cell and bacteria that were ancestors to modern-day mitochondria.

Costs and Variation in Mutualism. Although the benefits that mutualists confer upon one another have been known for some time, it has only recently become recognized that there are substantial costs associated with these interactions as well. Most of these goods and services that organisms provide their mutualists involve some kind of initial investment. For example, up to 20 percent of a plant's total carbon budget can be allocated to support mycorrhizae, and over 40 percent of their total energy investment may be devoted to producing nectar for pollinators. Furthermore, mutualists themselves can inflict costly damage on their partners. For instance, ant defenders sometimes consume aphids rather than protecting them from predators and parasites. Somewhat more abstract costs are associated with an organism's inability to succeed in the absence of its mutualists. Plants may reproduce poorly in parts of their range where their most effective pollinators are unable to persist, for example.

An interaction is mutualistic only when the benefits each of the two species receives from the interaction exceed the costs it experiences. Yet the sizes of both the costs and benefits of mutualism are known to be highly variable in space in time. In some circumstances, a single interaction may vary all the way from mutualism (i.e., benefiting both partners), to commensalism (benefiting one partner and neutral to the other), to antagonism (benefiting one partner and harmful to the other). Buckley and Ebersole (1994) describe an interaction involving a hermit crab and a small hydroid that it frequently places onto its shell. Hydroids are able to exclude most other organisms that attach to and damage hermit crab shells; this behavior clearly benefits the hermit crab. However, shells with hydroids are preferentially colonized by one common marine worm. These worms weaken the shells so much that they can easily be crushed by the hermit crab's own predators (blue crabs). The net effect of the association with hydroids thus varies greatly for the hermit crab: it is mutualistic where the worms or the predatory blue crabs are scarce, antagonistic when both worms and blue crabs are common. This case provides a good example of just how complex mutualisms can be and how dependent their outcomes are on the particular environment in which they occur.

Variation like this can affect both distributions and

THE FIG AND YUCCA POLLINATION MUTUALISMS

Mutualisms between fig trees and fig wasps and between yuccas and yucca moths are highly unusual in their intricacy, specificity, and long histories of coevolution. Most of the 750 fig (*Ficus*) species are pollinated exclusively by a single fig wasp species (family Agaonidae); that wasp in turn is associated exclusively with that fig. Female fig wasps are attracted to the trees by species-specific plant odors. They enter the enclosed fig inflorescences, spread pollen onto the flowers, then lay eggs within some of them. Their offspring feed for the next few weeks, each destroying a single fig seed. When the wasps and intact seeds are mature, the wasps mate, still within their natal fig. Females then collect pollen and depart through an exit hole chewed by the males, in search of figs in which to lay their eggs. Wasp life span is brief, and survival during the dispersal phase is exceedingly low. However, the wasps are evidently successful enough to have made figs ecologically dominant species in tropical and subtropical habitats worldwide.

Striking similarities have long been noted between the fig–fig wasp and yucca–yucca moth mutualisms. Like figs, each yucca species is associated with by a single, short-lived insect species (in this case, a moth in the family Incurvariidae) that exhibits elaborate pollination behaviors. Like fig wasps, yucca moths lay enough eggs to destroy many of the seeds that their actions have initiated in their only possible host plant. Both mutualisms appear to have originated when highly specific seed predators acquired pollination behaviors, guaranteeing that their offspring would not starve during development. Ancient fig and yucca flowers presumably were accessible to diverse pollinators, but both now exhibit features that act to exclude all but one, extremely costly species. Why and how this happened remain intriguing questions.

—JUDITH L. BRONSTEIN

population sizes of species that rely on mutualists. It is also likely to have important implications for how these interactions evolve, although work on this question is just beginning. John Thompson (1994) has argued that variation in outcome is the raw material for the evolution of interactions, just as variation in traits in populations is the raw material for the evolution of species.

Why Cooperate? Natural selection can be expected to act to increase the benefits each partner receives from mutualism, while reducing the costs it experiences. This straightforward observation turns out to have fascinating consequences for the evolution of mutualism. In general, unless an organism has an immediate interest in the well-being of its partner—something thought to be extremely rare, at least in nonsymbiotic mutualisms—natural selection can be expected to favor individuals that obtain what they require from their partners while investing as little as possible into them. If this process continues, mutualisms can be expected to shift progressively toward one-way exploitation. In light of this selection pressure, how can mutualisms possibly persist over the long term? This is an open question currently under very active debate, and no single answer has yet proved satisfying. Three main ideas have been put forth.

First, traits allowing organisms to effectively "cheat" one's partners simply might not arise very often. This does not appear to be the case, however: cheating has been identified in almost every kind of mutualism studied so far. For example, one well-known marine interaction involves a few fish species that consume the external parasites of larger fish, effectively "cleaning" them. Yet cleaners also take bites from the host fish itself, sometimes inflicting considerable damage; in turn, the hosts will on occasion consume the cleaners. As another example, many plants are exploited by animals that collect nectar from flowers but do not pollinate them. Small ants that crawl inside flowers to feed, and certain bees and birds that obtain nectar by boring holes through corollas, rarely if ever pick up or deposit pollen. Conversely, certain plants, including many orchids, entice pollinators to visit but provide them with no nectar once they are there. Historical studies using phylogenetic techniques are beginning to show that some mutualisms have been inflicted with cheaters for most of their evolutionary history. Hence, the presence of cheaters must be considered a central feature of mutualisms.

A second possibility for how mutualism can persist in the face of cheating is that organisms can selectively reward individual partners that cooperate and selectively punish those that do not. A variety of mathematical models suggest that behavioral strategies like these could effectively keep mutualists "honest." However, only a handful of real interactions have been identified in which such strategies may actually exist. The best example involves yucca plants and their obligate yucca moth mutualists. These insects both pollinate and lay eggs in yucca flowers; the caterpillars feed upon developing seeds (see Vignette). Some yucca plants appear able to recognize developing fruits that contain particularly high numbers of the destructive caterpillars and to drop those fruits off the plant before the caterpillars have matured. This mechanism permits only relatively cooperative moths, that is, those that lay few eggs per yucca flower, to survive to reproduce.

A final possibility for why cheating does not extinguish mutualism is that adaptations are present that effectively protect mutualisms from exploitation by cheaters in the first place. For example, traits such as long, thick corollas are thought to protect flowers from attack by nectar robbers. Yet we can often identify common cheaters able to cope quite well with these apparent adaptations. In fact, certain cheater species show elaborate adaptations for getting around apparent defenses against them. One tropical plant, a species of *Piper*, produces oil-rich food rewards on its leaves only when it detects that a specific ant-defender species has colonized it. However, one particular beetle species sometimes invades the plant. The beetle destroys the ant colony, then somehow fools the plant into continuing to produce the expensive and highly nutritious food rewards. It does not defend the plants from its enemies.

Evolutionary Origins of Mutualism. Regardless of the exact nature of their benefits and costs, the large majority of mutualisms are rather generalized. That is, each species can obtain what it requires from a fairly wide range of partner species. Furthermore, most mutualisms are facultative: at least some of the benefit mutualists provide can be obtained from nonbiological sources as well, allowing organisms a degree of success even when mutualists are totally absent. For example, many plants can be pollinated by a wide variety of small flying insects, and many can set at least a few seeds even if no pollinators at all come to the flowers. However, many extremely specialized mutualisms do exist. They are species-specific (i.e., only one mutualist species can provide the necessary commodity), and may be obligate as well (i.e., individuals fail totally in the absence of mutualists). The fig–fig wasp and yucca–yucca moth interactions (see Vignette) are the most thoroughly documented of the highly specialized pollination mutualisms.

The evolutionary origins of specialized and generalized mutualisms are probably somewhat different. Generalized mutualisms are thought to have evolved from interactions in which one species originally inflicted a low level of harm on its partner by feeding on it. This led to natural selection for traits of the prey that reduced the costs of being eaten, in some cases turning a negative interaction into a positive one. For example, ants historically were major predators of lycaenid caterpillars. Today, lycaenids have diverse adaptations that serve to direct ant aggression outwards toward lycaenids' other predators, rather than inwards toward the lycaenids themselves. Thus, the ant–lycaenid interaction has evolved into a protection mutualism. Most species-specific obligate mutualisms, on the other hand, appear to have originated from species-specific, obligate, and highly detrimental host–parasite interactions. For example, traits of the species-specific, obligate pollinating yucca moths and their relatives suggest that the yucca moths' ancestors were species-specific, obligate seed predators on the same group of plants. In general, evolutionary studies indicate that the degree of specificity and obligacy of an interaction may often persist for long periods of evolutionary time, even as its net effects for one partner shift dramatically, from negative to positive.

The preceding discussion should not be taken to imply that all mutualisms are the result of evolutionary pressures acting on one or both participants. Consider the range of evolutionary histories represented by plant–pollinator mutualisms. The obligate, species-specific relationship between figs and fig wasps (see Vignette) clearly has undergone extensive coevolution: unique traits have evolved in both species that increase the likelihood that the mutualism will succeed. In contrast, most plant–pollinator interactions are the product of natural selection on the plant only. In essence, plants have evolved traits, including rewards (e.g., nectar) and attractants (e.g., colorful petals), that allow them to take advantage of preexisting animal behaviors for their own benefit. Other plant–pollinator interactions have involved no detectable evolutionary change on the part of either partner. Some of the most interesting are cases in which one partner is an invasive or introduced species. Honeybees, introduced to the Western Hemisphere from Africa around four hundred years ago, now visit the flowers of diverse native North American plants; they generally do transfer some pollen between flowers, although they tend to be relatively poor mutualists compared to native bees. Mutually beneficial interactions can also arise between two invaders from different regions of the world, as occurs in North America when honeybees pollinate weeds of European origin. These phenomena may hold serious implications for biologists' growing efforts to conserve native species and native interactions.

[*See also* Coevolution; Cooperation; Reciprocal Altruism; Symbiosis.]

BIBLIOGRAPHY

REVIEWS OF PARTICULAR MUTUALISMS

Anstett, M. C., M. Hossaert-McKey, and F. Kjellberg. "Figs and Fig Pollinators: Evolutionary Conflicts in a Coevolved Mutualism." *Trends in Ecology and Evolution* 12 (1997): 94–99.

Buckley, W. J., and J. P. Ebersole. "Symbiotic Organisms Increase the Vulnerability of a Hermit Crab to Predation." *Journal of Experimental Marine Biology and Ecology* 182 (1994): 49–64.

Janzen, D. H. "Co-evolution of Mutualism between Ants and Acacias in Central America." *Evolution* 20 (1966): 249–275. This classic study of mutualism was influential in showing how mutualisms could be investigated using simple field experiments.

Letourneau, D. K. "Code of Ant–Plant mutualism Broken by Parasite." *Science* 248 (1990): 215–217.

Poulin, R., and A. S. Grutter. "Cleaning Symbioses: Proximate and Adaptive Explanations." *Bioscience* 46 (1996): 512–517.

Proctor, M., P. Yeo, and A. W. Lack. *The Natural History of Pollination.* Portland, Ore., 1996. This book is particularly valuable for its excellent illustrations.

OVERVIEW OF MUTUALISM

Boucher, D. S., ed. *The Biology of Mutualism.* New York, 1985. Chapters in this excellent compiled volume span a diversity of topics, including the natural history, ecology, population dynamics, and evolution of mutualisms. Some but not all chapters require a technical background.

Bronstein, J. L. "Our Current Understanding of Mutualism." *Quarterly Review of Biology* 69 (1994): 31–51. This article focuses on what is currently known and not known about mutualism, based on a survey of the published literature.

Buchmann, S., and G. Nabhan. *The Forgotten Pollinators.* Washington, D.C., 1996.

Sapp, J. *Evolution by Association: A History of Symbiosis.* New York, 1994. Although this work focuses on the history of research on symbiosis, much of the discussion revolves around interactions whose outcomes are mutualistic.

Thompson, J. N. *The Coevolutionary Process.* Chicago, 1994. This is a readable overview of the evolution of species interactions, including mutualisms. Current ideas about the evolution of many interesting mutualisms, including the yucca–yucca moth interaction, are clearly described.

— Judith L. Bronstein

MYXOMATOSIS

Myxomatosis, a viral disease of rabbits, apparently originated in the South American rabbit species *Sylvilagus brasiliensis* but also infects the European rabbit *Oryctolagus cuniculus*. In *S. brasiliensis*, myxomatosis is a mild disease, but in *O. cuniculus*, it causes severe skin lesions and is often fatal. The evolutionary significance of myxomatosis is that it was introduced in the 1950s into *O. cuniculus* populations in Australia and Europe that had never before experienced the disease, and, over the next few decades, dramatic evolutionary changes occurred in both host and pathogen. The myxomatosis–*O. cuniculus* system is one of the few cases in which host–pathogen coevolution has been studied as it happened, rather than retrospectively.

Preliminary studies of these changes were carried out by the Australian virologist Frank Fenner and his colleagues. A key feature of their work was the establishment of reference strains of both host and pathogen, which allowed them to track changes in the rabbit and the virus independently of each other. To do this, Fenner and his colleagues injected virus isolates from the field into groups of laboratory rabbits, and tested groups of rabbits from the field with a laboratory strain of the virus. Because rabbits that have recovered from the disease are immune, they used only rabbits that had never been exposed to the disease. To measure rabbit resistance, they had to use the offspring of field-collected rabbits.

To simplify their data, Fenner and his colleagues classified virus strains into what they called "grades," using probability of host death as the measure of virulence. Grade I strains are the most virulent, causing greater than 99 percent mortality in laboratory rabbits; grade V strains are the least virulent, causing less than 50 percent mortality (see Table 1). The strain used to introduce the virus into Australia was in grade I, but within a few years, the most common virus strains from the field in Australia were classified as grades III or IV. Similarly, although rabbits from the field had originally experienced more than 90 percent mortality when infected with the laboratory virus strain, within a few years field-collected rabbits showed only 40 percent mortality when infected with that strain. Similar changes occurred in Great Britain; changes on the European continent have been less well documented.

The increase in rabbit resistance can be explained easily as a response to natural selection on survival, but explaining the drop in virus virulence is not so simple. Fenner and his colleagues showed that part of the explanation is that the myxoma virus is transmitted mechanically on the mouthparts of biting insects, principally mosquitoes and fleas. Because neither mosquitoes nor fleas bite dead rabbits, virus strains that kill rapidly, like the grade I strain used to introduce the virus, are less likely to be transmitted than are strains that kill more slowly. Slower-killing strains, however, produce lower concentrations of virus in the skin of their hosts, and low concentrations of virus make it less likely that a vector insect will become infectious. Finally, because speed of kill is closely correlated with the probability of host mortality, changes in speed of kill lead to changes in host mortality rates. The outcome of natural selection on the virus has thus been the predominance of strains of intermediate virulence. This explanation for the changes in the myxoma virus has inspired theories (see Anderson and May) about the evolution of disease virulence that have suggested that the outcome of natural selection on the virulence of other diseases will also depend on trade-offs between virulence and transmissibility.

Much of the existing theory, however, has focused exclusively on the evolution of virulence without considering the evolution of resistance in the host. Part of the problem is that coevolutionary models have only recently been studied in much depth; *see* Bergelson et al. (2001) for a review. In the case of myxomatosis, an additional problem is that there has been little in-depth testing since the early 1980s of virus and rabbit populations in either Australia or Great Britain. A key issue in host–pathogen coevolution concerns whether host and pathogen will undergo a coevolutionary "arms race" in which virulence and resistance increase rapidly in response to each other, or whether virulence and resistance will reach a stable equilibrium. Some data from Australia in the late 1970s to mid-1980s suggested that the virulence of the virus was indeed increasing in response to increasing rabbit resistance. In contrast, a modest number of samples from Great Britain from the late 1970s and early 1980s suggested that virus strains there had changed little from earlier decades. Over all, though, the amount of data is not enough to be conclusive.

TABLE 1. System Used to Classify Myxoma Virus Strains of Different Levels of Virulence

	Virulence Grade				
	I	*II*	*III*	*IV*	*V*
Average speed of kill	<13	14–16	17–28	29–50	>50
Percent mortality	>99	95–99	70–95	50–70	<50

A full understanding of coevolution in this interaction would require a large-scale effort to sample contemporary host and pathogen populations. An additional difficulty is that the data were originally analyzed in a crude way. The most important problem was that, because accurate measurements of the fraction dying require larger sample sizes than do measurements of the time between infection and death, the original classifications relied on a regression between speed of kill and fraction dying in order to assign virus strains to grades. Individuals that survived infection, however, were arbitrarily assigned a survival time of 60 days. Correcting this and other problems revealed that, in fact, most of the virus strains collected in the field in Australia after 1959 killed nearly as high a percentage of infected rabbits as do Grade I strains, but they allowed infected rabbits to survive for as long as do Grade III strains (Parer 1995). This reanalysis suggests that one of the principal effects of natural selection on the myxoma virus has been to alter the correlation between probability of host mortality and speed of kill. That is, the strains that eventually predominated did indeed kill more slowly than the original strain—thereby prolonging the amount of time during which they could be transmitted—but the chance that their host would die did not change, or at least not by much.

Although field strains of the myxoma virus caused higher mortality in laboratory rabbits than was originally realized, this does not necessarily mean that they had a more severe effect on field rabbits. In fact, because European rabbits are an important agricultural pest, there is concern that the virus will eventually cease to be an effective control, and total mortality attributed to myxomatosis in the field in both countries is indeed often below 25 percent. Nevertheless, measurements by Parer, Ross, and colleagues of the survival rates of experimentally infected wild rabbits in Australia from the 1970s through the 1990s, and in Great Britain in the late 1970s, showed that field strains of the virus often still yield mortality rates of more than 50 percent. Moreover, experimental reductions in myxomatosis incidence on both continents in the late 1970s and early 1980s, either through vaccination of susceptible rabbits or through insecticidal control of vectors, have led to dramatic increases in rabbit populations. It therefore appears that myxomatosis is still an important part of rabbit management, but only a better understanding of coevolution will tell us whether it will continue to be important in the future.

[*See also* Coevolution; Disease, *article on* Infectious Disease; Immune System, *article on* Structure and Function of the Vertebrate Immune System; Transmission Dynamics; Virulence; Viruses.]

BIBLIOGRAPHY

Anderson, R. M., and R. M. May. "Coevolution of Hosts and Parasites." *Parasitology* 85 (1982): 411–426. A clearly written introduction to mathematical modeling of coevolution.

Bergelson, J., G. Dwyer, and J. J. Emerson. "Models and Data on Plant-Enemy Coevolution." *Annual Reviews of Genetics* 35 (2001): 469–499. An overview of recent coevolutionary models.

Fenner, F. "Biological Control, as Exemplified by Smallpox Eradication and Myxomatosis." *Proceedings of the Royal Society (London)*, Series B, 218 (1983): 259–285. Includes an overview of several decades of evolutionary changes in the myxomatosis virus.

Fenner, F., and R. N. Ratcliffe. *Myxomatosis*. London, 1965. The classic work on myxomatosis; although now out of date, the broad perspective is priceless.

Parer, I. "Relationship between Survival Rate and Survival-Time of Rabbits, Oryctolagus cuniculus (L.), Challenged with Myxoma Virus." *Australian Journal of Zoology* 43 (1995): 303–311. A technical inquiry into problems associated with understanding data on virulence.

— Greg Dwyer

N

NAKED MOLE-RATS

Mole-rat is the vernacular name for thirty-seven species of fossorial rodents in ten genera that inhabit subterranean burrows like moles but whose cylindrical body shape, slender tail, and large incisors resemble those of rats. This entry focuses on *Heterocephalus glaber* (literally 'otherheaded smooth'), the most unusual-looking and social of the African mole-rats (family Bathyergidae).

Naked mole-rats resemble overcooked sausages with buck teeth. They are 9–12 cm long and weigh 30–60 g. They have loose, wrinkled, pink-colored skin, minute eyes, and tiny ear pinnae. They dig with chisel-like, procumbent incisors that are powered by muscular jaws (containing 25 percent of the animal's muscle mass). Their skin is thin, and they have little subcutaneous fat, so their body temperature varies with the ambient temperature (i.e., they are poikilothermic and ectothermic). However, they are not completely "naked"—they do have scattered body hairs, long, tactile whiskers, and hair-fringed lips and hind feet. The latter are used like brooms as the animals sweep debris from their tunnels, working backward. Hairs on their lips help keep loose earth out of their throat and windpipe as they dig. Also, when the animals excavate materials that break into fine particles, they often place a flat piece of root bark or tuber husk between their teeth and throat. This serves as an oral dam to keep them from choking.

Naked mole-rats inhabit arid areas of Kenya, Ethiopia, and Somalia. They seldom come above ground, but their presence is revealed by volcano-shaped mounds of earth ejected onto the surface as byproducts of tunneling. They are vegetarians, feeding primarily on subterranean parts of plants. Their specialized gut protozoa aid digestion by breaking down plant cellulose. Curiously, the animals are unable to detect large geophytes through the soil—often they burrow right past a large tuber, only to strike another of the same species a few centimeters later.

Snakes are their main natural predators. At least eight species have been observed preying on naked mole-rats, and African "mole snakes" (genus *Pseudapsis*) specialize on them. The snakes gain entry to burrows through open volcano mounds, which they find by following the scent of freshly turned earth.

Social Behavior. Naked mole-rats have achieved notoriety not just because of their looks, but also because of their complex and fascinating social behavior. They exhibit all three characteristics that define eusociality (true sociality) in insects: overlapping generations, reproductive division of labor, and cooperative care of breeders' offspring by nonbreeders. [*See* Eusociality: Eusociality in Mammals.]

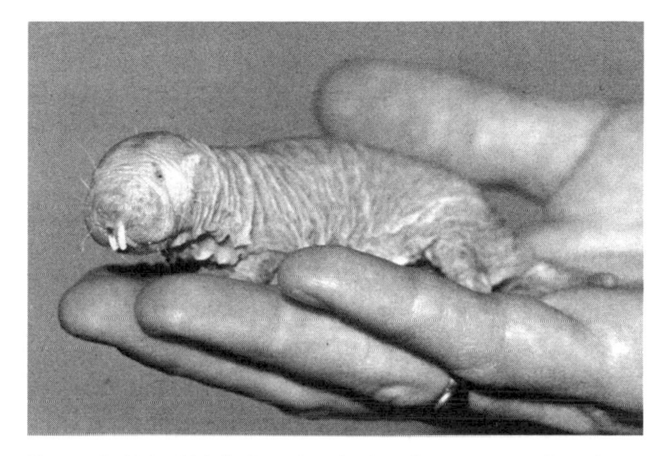

FIGURE 1. Naked Mole-Rat. A naked mole-rat cannot be mistaken for any other mammal. Photo courtesy of Paul W. Sherman.

Like many bees, termites, and ants, naked mole-rats live in colonies that are extended families and within which only one female and one to three males reproduce. Wild colonies contain from ten to three hundred individuals, with a mean of seventy-five to eighty animals per group. Colony mates are very closely related both because young usually do not disperse and because, when a breeder dies, a replacement comes from within the colony. Thus, breeders sometimes are littermates or parents and offspring. Molecular genetic analyses confirm that the mean coefficient of relatedness (denoted by r) among colony mates is $r > 0.80$, which is much higher than among outbred siblings ($r = 0.50$). The inbreeding coefficient (F) is 0.45—the highest F value known for free-living mammals.

However, a small amount of genetic heterogeneity consistently is present within and among wild colonies, indicating that successful dispersal must occasionally occur. Indeed, a few isolated pairs have been discovered, each of which apparently emigrated from a different, nearby colony. These isolated individuals had a specialized morphology that also was noted among putative dispersers in lab colonies (i.e., large size and extra fat reserves). It is unknown how often colonies founded by such pairs of dispersers survive, but probably rarely.

The genetic data suggest, however, that the more successful route of new colony formation may be group fissioning. Indeed, small groups of animals have been captured near larger colonies, and the presence of old volcanos and abandoned tunnels in the intervening area suggest that the smaller groups recently split off. Also, a colony fissioning event was observed in a captive colony. The split occurred right after the breeding female died, and the subcolonies were composed of the most closely related individuals from the parent colony.

Breeding versus nonbreeding colony members. Within a colony, only the breeding female exhibits behavioral estrus and mates. Nonbreeding females are not physiologically sterile, but their reproductive development is suppressed by physical harassment from the dominant breeder. If a breeder dies, she is quickly replaced by a nonbreeder of the same sex. After a female becomes a breeder, her body shape changes, becoming dorsoventrally deep and noticeably elongated, due to lengthening of individual vertebrae. This increases the size of her body cavity, enabling her to gestate extremely large litters.

Female naked mole-rats may bear up to twenty-eight pups in a single litter, although the average litter size is a more modest eleven or twelve pups. A breeder can bear four or five litters and more than one hundred young in one year. Although the litter size often exceeds the number of mammae, the breeding female can rear them because pups rotate between teats (they "take turns" nursing) rather than aggressively defending one. An adequate milk supply is assured because the breeding female is provided with food and protection by nonbreeders. When the pups are about two weeks old, they begin feeding on partially digested fecal pellets provided by nonbreeders. This further eases the physiological burden of lactation for their mother and facilitates the juveniles' transition to independent foraging.

The behavior of nonbreeders is influenced by the breeding female. Colonies are busy around the clock and do not exhibit day–night cycles typical of surface-dwelling rodents. Activity is stimulated by the breeding female, the most aggressive individual in a colony. As she moves through the tunnel system, the breeding female frequently shoves nonbreeding animals. Temporarily removing the breeding female results in decreased colony activity.

Nonbreeding males and females contribute equally to pup care and to maintenance and defense of their colony. They handle, groom, and feed pups, retrieve them if they wander out of the nest, and evacuate them if the colony is disturbed. Nonbreeders also expand, defend, and maintain their family's vast burrow system (2–3 km for an average-sized colony). They dig new tunnels, harvest and bring back food, clear tunnel cave-ins and debris, build nests, and attack predatory snakes. Individuals communicate (e.g., about food sources and danger) using eighteen types of vocalizations—the largest vocal repertoire known among rodents.

The behavior of nonbreeders varies with body size and age. Small individuals are the primary maintenance workers, whereas larger (usually older) nonbreeders defend the colony using their massive jaws and teeth. Body size and associated behaviors vary continuously—there are no morphologically distinct worker and soldier "castes" of mole-rats, as occurs in some ants. Individual mole-rats usually switch from maintenance to defense activities when they are about three years old. However, this "age polyethism" is frequently obscured by behavioral and size changes associated with variations in colony composition and competition to fill reproductive vacancies.

In nature and in captivity, naked mole-rats live a long time. Mark and recapture studies have revealed that, in the field, nonbreeding individuals and breeders can live more than a decade. In laboratory colonies, nonbreeders and breeders live eleven to twelve years on average and at least twenty-five years at the maximum. This is longer than 99 percent of all rodents that have ever been kept in captivity (including much larger species).

Species Differences. Eusociality has evolved twice in the bathyergids: both naked and Damaraland mole-rats (*Cryptomys damarensis*) exhibit all three diagnostic characteristics. In both species, group living appears to be an evolutionary response to the distribution of food, coupled with high energetic costs of burrowing. Both species inhabit deserts where food is dispersed and patchy and rainfall is unpredictable. During prolonged droughts, the mole-rats cannot efficiently excavate the hard soil, nor can they disperse on their own. In brief periods after torrential rains, they must cooperate and dig furiously if they are to locate sufficient food to enable their family to survive through the next dry season.

The mole-rats' "blind" mode of foraging, when combined with limited and unpredictable opportunities to locate food, has had important implications for sociality. By cooperating, arid-zone mole-rats take full advantage of windows of opportunity when conditions are right for burrowing. Working together, they also can repel reptilian predators. Solitary or uncooperative mole-rats simply cannot exist under such harsh ecological conditions. Indeed, all the nonsocial bathyergids inhabit more benign (mesic) environments.

Although some large naked mole-rats attempt to leave home and establish their own colony, for the majority, successful dispersal seldom is possible. Their best reproductive option is to wait for a breeding opportunity in the natal colony. While waiting, individuals can promote their genetic success by helping to raise close kin. This is because rearing a full sibling is equivalent to rearing one's own offspring, because both relatives share the same proportion of genes identical by common descent ($r = 0.50$). Thus, through alloparental behavior, nonbreeders gain reproductive benefits.

In summary, both ecological factors and kin selection are implicated in the social evolution of mole-rats. Variations in aggression and cooperation among colony members reflect the delicate balance between selfish reproductive interests and self-sacrifice on behalf of their family. Studies of naked mole-rats have revealed remarkable parallels with societies of small-colony eusocial insects. Although much remains to be learned about naked mole-rats, they already helped unify theoretical and empirical studies of social evolution in invertebrates and vertebrates.

[*See also* Altruism; Cooperation; Kin Selection; Social Evolution.]

BIBLIOGRAPHY

Bennett, N. C., and C. G. Faulkes. *African Mole-Rats: Ecology and Eusociality*. Cambridge, 2000. A timely monograph discussing the ecology and phylogeny of sociality in the family Bathyergidae.

Braude, S. H. "Dispersal and New Colony Formation in Wild Naked Mole-rats: Evidence against inbreeding as the system of mating." *Behavioral Ecology* 11 (2000): 7–12. Documents formation of nascent colonies by pairs of unrelated dispersers.

Jarvis, J. U. M. "Eusociality in a Mammal: Cooperative Breeding in Naked Mole-rat Colonies." *Science* 212 (1981): 571–573. The classic article on mole-rat social organization.

Jarvis, J. U. M., M. J. O'Riain, N. C. Bennett, and P. W. Sherman. "Mammalian Eusociality: A Family Affair." *Trends in Ecology and Evolution* 9 (1994): 47–51. Explains and tests the aridity–food distribution hypothesis for the evolution of sociality in the Bathyergidae.

Lacey, E. A., and P. W. Sherman. "Cooperative Breeding in Naked Mole-rats: Implications for Vertebrate and Invertebrate Sociality." In *Cooperative Breeding in Mammals*, edited by N. G. Solomon and J. A. French, pp. 267–301. Cambridge, 1997. Explains the pivotal position of this animal in uniting studies of social evolution in vertebrates and invertebrates.

O'Riain, M. J., J. U. M. Jarvis, and C. G. Faulkes. "A Dispersive Morph in the Naked Mole-rat." *Nature* 380 (1996): 619–621. Provides evidence that dispersers are morphologically and behaviorally specialized.

Reeve, H. K. "Queen Activation of Lazy Workers in Colonies of the Eusocial Naked Mole-rat." *Nature* 358 (1992): 147–149. Provides evidence that breeding females focus aggression on less related, larger nonbreeders.

Reeve, H. K., D. F. Westneat, W. A. Noon, P. W. Sherman, and C. F. Aquadro. "DNA 'Fingerprinting' Reveals High Levels of Inbreeding in Colonies of the Eusocial Naked Mole-rat." *Proceedings of the National Academy of Sciences* 87 (1990): 2496–2500. Documents high levels of inbreeding and relatedness within wild colonies.

Sherman, P. W., S. Braude, and J. U. M. Jarvis. 1999. "Litter Sizes and Mammary Numbers of Naked Mole-Rats: Breaking the One-Half Rule." *Journal of Mammalogy* 80 (1999): 720–733.

Sherman, P. W., J. U. M. Jarvis, and S. H. Braude. "Naked Mole-rats." *Scientific American* 267 (1992): 72–78. A brief, nontechnical summary of the animal's ecology and social behavior.

Sherman, P. W., J. U. M. Jarvis, and R. D. Alexander, eds. *The Biology of the Naked Mole-Rat*. Princeton, 1991. The most complete source of information on this remarkable animal.

Shuster, G., and P. W. Sherman. "Tool use by naked mole-rats." *Animal Cognition* 1 (1998): 71–74. Describes how the animals use pieces of bark and tuber husks as oral barriers while digging to keep fine particles of debris out of their throat and windpipe.

— Paul W. Sherman

NATURALISTIC FALLACY

The naturalistic fallacy with respect to morality is reasoning from what *is* the case to what *ought* to be the case. For example, in all societies throughout the history of the human species, men have held most of the power. To conclude from that fact that such patriarchal societies are morally preferable to matriarchal societies, or to societies in which males and female share power equally, is an instance of the naturalistic fallacy. The naturalistic fallacy is a special case of naturalism in general. Naturalism is the view that the only things that exist in nature are natural objects, properties, relations, and processes, and that the same methods can be used to study all those phenomena—the methods of science. If, as naturalists assume, there is nothing over and above the descriptive "is," then reasoning from "is" to "ought" is not a fallacy, but perfectly legitimate. "Ought," in this view, implies nothing that is not already incorporated in "is."

The opponents of naturalism admit that scientists can describe the lay of the land, and they can explain why people find it easier to behave in certain ways than in other ways, but these thinkers contend that the methods used by scientists are inadequate to distinguish between what is morally right and morally wrong. There is something more contained in moral claims. People who think that the naturalistic fallacy *is* a fallacy agree that there has to be something more to moral claims than simply what is the case; however, they disagree dramatically about what this "something more" actually is, and how it can be justified.

Ethical systems must provide some means for people to decide what is right and what is wrong. In addition, they must provide some justification for the criteria used, an activity termed "meta-ethics." With the advent of evolutionary theory, a new form of naturalistic ethics arose—evolutionary ethics. In this line of thinking, good and bad are determined by the relative effects of an act on the evolution of the human species. Any act performed by a human being that contributes to the survival of the human species is good, while any act that in-

EVOLUTIONARY ETHICS

A belief common among citizens of the United States is that everyone deserves all the medical care that they need, regardless of cost. Of course, if more money is spent on health care, then less money is available to be spent elsewhere. Politicians are forced to balance the various needs of society, and frequently medical care for the general public is further down the list than defense, education, and of course medical care for politicians and their families. In the United States an extremely large proportion of money spent on medical care is spent on the aged during their last few months of life. Both medical practitioners and our courts of law are committed to dragging out the process of dying as long as possible. One result of this policy is that too little money is left for the medical care of younger citizens, in particular children.

Ordinary people are of several minds on this issue, as are professional philosophers. What would our morals be like if they were influenced by evolutionary considerations? For example, some species of organisms produce hundreds of thousands of offspring but provide little in the way of parental care. All that matters is that a few of their offspring make it to sexual maturity. At the other extreme, some organisms produce very few offspring but expend considerable effort in raising them. Human beings are of the second sort. We not only have children, but also care for them. Attempting to raise more children than one's resources allow, such that all of them die before having children of their own, is not positive from an evolutionary perspective. Having fewer offspring or producing offspring that aid in the rearing of their siblings can result in more children reaching sexual maturity.

In addition, we are a social species. The rearing of our children takes place in both a biological and social context. The biological context includes extended as well as nuclear families. We tend to expend most of our resources on close kin. If children in our extended family are orphaned, we may come to their aid. But societies include groups whose members are not all that closely related to each other. We also introduce social programs designed to help family groups raise their offspring (e.g., reduced taxes and free education).

If we restrict ourselves just to the calculus of having offspring, people who have just reached their sexual maturity are most valuable because they have the highest potential for reproductive success. This potential steadily decreases thereafter until the aged have almost no reproductive potential at all. With respect to children, reproductive potential increases from birth until it reaches its peak at early sexual maturity. The reason that children are not deemed to have the same reproductive potential as young adults is that they may well die before reaching sexual maturity. Until quite recently, death rates among infants and children were quite high.

On the basis of reproductive considerations alone, our practice of expending huge amounts of resources on the very old is counterproductive. Of course, the same is true in providing extensive health care for babies and very young children, though less so. But the need to rear children to reproductive age and the social practices that we introduce in order to promote these goals complicate the calculus of reproduction. All the education and support that follow consume even more of society's resources. After reaching maturity, the tables are turned. These former children must now invest heavily in their own children. Even older humans play important roles. After menopause, women cannot produce additional children but they still are in a position to rear their own children, not to mention their grandchildren.

According to one common moral perspective, all people are of equal worth. From an evolutionary perspective, they are not. Young adults have the highest reproductive potential. In addition, society has invested a significant amount of its resources in young adults. Aging members of a society are still valuable. They are experienced in maintaining society, but as they near death, their value to society decreases. They consume just as much as when they were younger but contribute less and less to society at large. The rest of society might well owe them care and concern because of their past contributions, but because their reproductive potential is extremely curtailed, they are worth much less from the perspective of reproduction. Of course, according to this same calculus, newborns are also worth more than the aged but less than young adults.

If evolutionary theory is taken seriously, health care should be allotted first and foremost to young adults, next to children, and finally to the aged. According to one study, the elderly agree, at least with respect to children. They would be willing to have less health care available to them if the savings actually went to the aid of children.

—David L. Hull

creases the likelihood that we will go extinct is bad. The overall logic of such inferences seems unproblematic enough, but particular applications in the past have been highly questionable. Founding morality in the evolutionary process gave rise successively to Social Darwinism, eugenics, sociobiology and evolutionary psychology. The first two schools are best known for the blatant racism that they incorporated; in addition, their biology was more than a little questionable. Similar claims, very vigorously denied by adherents, have been made with respect to sociobiology and evolutionary psychology.

The arguments for the claim that the survival of the human species is a necessary condition for ethics and morals are fairly persuasive. Without people, the issue of ethics does not even arise. When confronted with the choice of X or the extinction of the human species, we would all look long and hard at X. For example, the inevitable result of the explosion in the number of human beings here on Earth is the extinction of numerous other species. We are well aware that such an increase in the extinction of other species is dangerous for us. We should hold down our numbers for our own good, but what if the only way to save life here on Earth is for *Homo sapiens* to go extinct? Would contributing to the extinction of the human race be moral or immoral?

Even those who think that the survival of the human species is a necessary condition for ethics and morals draw back at considering it sufficient. Many moral issues are closely connected to the survival of the human species—for example, the equality of the sexes, abortion, childcare, or a purported decrease in sperm count in men—but just as many moral issues have only the most tenuous connection to biological evolution. For example, the final six Hebrew commandments quite obviously deal with matters relevant to survival and reproduction, but the first four deal with God. Advocates of naturalism, of course, reject the notion of a supernatural being, but beliefs about a nonexistent god can nevertheless contribute to group cohesion and other social practices. No belief is better calculated to make warriors fight more bravely than the conviction that God is on their side.

Another example of evolutionary ethics as naturalistic ethics is the relative roles of men and women in rearing children. Someone must rear children; the question is who. For most of human history, children were raised in large extended families. Females did most of the rearing—sometimes maternal grandmothers, sometimes mothers. As the nuclear family evolved, the task of raising offspring fell increasingly to mothers. Now, in first-world countries, these roles are changing once again. Females, usually mothers, must do most of the rearing of children while at the same time working outside the home. Different sorts of arrangements for raising children have characterized people in different environments, but we must raise them if we are to survive. The human species is plastic; the issue is how plastic. If equality of men and women detracts significantly from the successful rearing of children, then from the perspective of evolutionary ethics, the equality of men and women is morally wrong.

All of the preceding discussion emphasizes survival over all else, as evolutionary ethics requires. That this discussion seems to be leaving something out indicates that few of us are willing to settle for naturalistic ethics, including evolutionary ethics. For a significant number of people, equal rights for men and women are good, regardless of the evolutionary effects: "Better that our species go extinct than the oppression of women continue." But even if we conclude that systems of naturalistic ethics are not good enough, we are nevertheless left with the task of the setting out an alternative that is, and thus far no system of ethics has proven to be sufficiently seductive to become prevalent, let alone universally accepted.

[*See also* Altruism; Group Selection; Social Darwinism.]

BIBLIOGRAPHY

Nitecki, M. H., and D. V. Nitecki. *Evolutionary Ethics*. Albany, N.Y., 1993. Biologists and philosophers evaluate various theories of evolutionary ethics.

Richards, R. J. *Darwin and the Emergence of Evolutionary Theories of Mind and Behavior*. Chicago, 1987. A detailed history of evolutionary ethics and a sophisticated justification for such a naturalistic ethics.

Ruse, M. *Taking Darwin Seriously: A Naturalistic Approach to Philosophy*. Oxford, 1986. Ruse defends both evolutionary ethics and evolutionary epistemology in this highly readable book.

Ruse, M., and E. O. Wilson. "Moral Philosophy as Applied Science: A Darwinian Approach to the Foundation of Ethics." *Philosophy* 61 (1986): 173–192. A leading philosopher of biology joins forces with the father of sociobiology to defend evolutionary ethics.

— DAVID L. HULL

NATURAL SELECTION

[*This entry comprises two articles. The overview focuses on how genetic variation coupled with differential success in a particular environment leads to evolutionary change; the companion article considers how natural selection operates when both juvenile and adult mortalities are very low. For related discussions, see* Adaptation; Fitness; Group Selection; Sex Ratios; Sexual Selection; *and* Species Selection.]

An Overview

Natural selection is a mechanism that causes evolution. It occurs when some kinds of organisms in a population leave more offspring than other kinds of organisms; the result is that the kinds of organisms who leave more offspring will increase in frequency over time. The theory was first developed by Charles Darwin in the late 1830s and independently by Alfred Russel Wallace in the 1850s. Darwin and Wallace jointly published the idea in short papers in 1858, but the theory did not become widely known until Darwin addressed it in the book *On the Origin of Species by Means of Natural Selection* (1859).

The Conditions for Natural Selection. In general, natural selection can be understood as a logical argument in which a conclusion follows from a set of premises (or a result follows from a set of initial conditions). Natural selection will operate on anything that has the following three properties:

reproduction
inheritance
variation

Living creatures possess all these properties: they reproduce and parental attributes are inherited, via DNA, in their offspring. Offspring tend to resemble their parents: tall parents, on average, have taller offspring than do short parents. Variation here refers to interindividual differences within a local population (all belonging to the same species) at a particular time. For instance, if we sample several individuals from a population, we could measure their sizes in some way. Some individuals would be larger than others; this would be an example of variation.

For natural selection to work, it is not enough that a population shows variation. The population also must show variation with respect to an attribute that influences reproductive success—or, to be exact, what biologists call fitness. [*See* Fitness.] The fitness of an organism, in its simplest evolutionary sense, equals its chance of survival multiplied by the number of offspring it produces (if it survives). (Notice that evolutionary fitness has nothing to do with its athletic meaning; someone could be athletically very fit but have zero evolutionary fitness if he or she leaves no offspring.)

All entities that show reproduction, inheritance, and variation (in attributes that influence fitness) will inevitably evolve by natural selection. Individuals with attributes that produce high fitness will leave more offspring; those offspring (because of inheritance) will also tend to possess those attributes. The attributes that produce high fitness therefore increase in frequency. As an example, consider the evolution of drug resistance in HIV (human immunodeficiency virus, the cause of AIDS).

HIV uses RNA (ribose nucleic acid) as its hereditary molecule rather than the DNA used by higher organisms. Because HIV uses the cells of human bodies to reproduce itself, the virus has to make a DNA version of itself. It does this by means of a special enzyme called reverse transcriptase, which reproduces the RNA of the virus into DNA code. The human cell, which, like most cells, has the capability to produce RNA, will then run off multiple copies of the HIV from it. Human cells do not use reverse transcriptase in their normal, healthy functioning, and the reverse transcriptase enzyme is one of the favorite targets of scientists who devise antiviral drugs against HIV. DNA is built from a sequence of the four nucleotides symbolized by the letters *A*, *C*, *G*, and *T*. Reverse transcriptase has to recognize these four molecules, as distinct from all the other molecules inside a cell, and bind them together to make the DNA version of HIV. One class of antiviral drugs are called nucleoside inhibitors: they are molecules that closely resemble one of A, C, G, or T, and reverse transcriptase binds them like a real A, C, G, or T. Once the nucleoside inhibitor is added to the DNA chain, it prevents further reproduction. The drug molecule is shaped in such a way as to prevent the chain from growing any longer, and the reproduction of HIV is blocked.

What happens when a nucleoside inhibitor is given to a human AIDS patient? The immediate effect is that the population of HIV inside the person declines precipitously. But then, within a few days, drug-resistant versions of HIV will be detectable inside that individual. Over the next few weeks, the frequency of the drug-resistant version of HIV increases. Within a month or two of the commencement of drug treatment, the HIV population in the patient will have evolved such that 100 percent of the HIV is drug resistant.

The example illustrates natural selection in action, over an exceptionally short time period. (It takes about two days for HIV to reproduce.) HIV shows inheritance (by the RNA that is copied, via a DNA intermediary, into offspring viruses). It also shows variation. The HIV individuals in a human being differ from one another because of copying accidents (called mutations), most of which occur when reverse transcriptase makes a DNA version of the RNA form of HIV. The RNA of one HIV is about 10,000 nucleotides long. Reverse transcriptase has an error rate of about 1 error per 10,000 nucleotides. The result is the extraordinarily high rate at which variant forms of HIV are generated.

When the AIDS patient is treated with a drug, some of the HIV variants can influence fitness. Most versions of HIV in the body initially will be susceptible to the drug, but some rare mutant versions will be resistant. The drug-resistant versions have high fitness. Drug resistance can be due to a change in the reverse transcriptase enzyme. The resistant forms of HIV have reverse transcriptase that is more discriminating. It binds only real A, C, G, or T and does not bind the nucleoside inhibitor that resembles (but is not identical to) them. The change to the more discriminating enzyme is caused by a mutation in HIV, and the change appears rapidly because of HIV's high mutation rate. The exact form of the reverse transcriptase is, in the presence of the drug, an attribute that influences fitness. If an HIV has a discriminating enzyme, it will be drug resistant and reproduce successfully; if it has a normal, slap-dash enzyme, it will be susceptible to the drug and fail to reproduce. The drug-resistant HIV will increase in frequency over time. The term *natural selection* can be used to refer either to the event (biologists would call it an ecological event) in which some individuals reproduce and others do not or to the increase in frequency over time of the successful kinds of individual. Natural selection is acting when one drug molecule terminates the reproduction of one drug-resistant version of HIV. It is also acting when drug resistance increases in frequency between one generation and the next.

The Ecological Struggle for Existence. Natural selection is not limited to situations as extreme as drug resistance in HIV. Natural selection is continually operating in all populations of all species on earth. We can deduce this, not because natural selection has been measured in all species (indeed, it has been measured only in a few hundred species) but because within every species there is a "struggle for existence." Ecological resources are finite, but the reproductive capacity of every species is infinite. Consider, as an extreme example, the giant puffball *Lycoperdon giganteum* (puffballs are fungi, related to mushrooms). The giant puffball is probably the most fecund creature on earth, and a large specimen has been estimated to contain 2×10^{13} spores. If they all grew up and were put in a line, it would circle the earth (at the equator) fifteen times—and that is from just one parental puffball. Clearly, not all the offspring can survive; this rule applies in all species, not just extremes such as puffballs. There is an ecological competition for resources, and the unsuccessful competitors die prematurely. The intensity of ecological competition in nature is the reason to think that natural selection is an almost universal process in life: reproduction and inheritance are known to be universal; variation, because of mutation and other factors, also is universal. Ecological competition will mean that some variants within a species will be more successful than others: variation tends to become, in a competitive environment, variation in fitness.

Natural selection can act on either the survival or the reproduction of organisms, or on both. In technical terms, these correspond to the two components of fitness. One kind of organism may have a higher fitness than another kind because it has a higher chance of survival from egg stage to adulthood or because it has the same chance of survival but produces more offspring. These are referred to as natural selection by differential survival and differential fertility. The ecological environment determines whether selection is by one factor or the other, or both. In the example of drug resistance in HIV, selection was by differential fertility: drug-resistant and drug-susceptible versions of HIV both had the same chance of survival in the presence of the drug, but the drug resistant versions were more fertile. In other examples, selection works by differential survival. In some birds, for example, beak shape influences feeding efficiency; in times of food shortage, birds with inefficient beak shapes are more likely to starve, as will be discussed further below.

Natural selection via differential fertility often differs in males and females. Female reproduction is usually limited by the resources in the natural environment, not by access to males. But male reproduction is usually limited by access to females, and this introduces a second round of competition, one of competition for mates, in addition to the ecological struggle for existence. This second round of competition is called sexual selection and probably caused the evolution of attributes that are confined to males within a species—attributes such as the peacock's tail and the antlers of deer. [*See* Sexual Selection.]

Directing Forces. In evolution by natural selection, the processes generating variation are not the same as the process that directs evolutionary change. Variation is undirected. When HIV individuals are subjected to antiviral drugs, a drug-resistant variation of HIV is no more likely to arise than in the absence of the drug. The variants arise because of copying accidents in the DNA and RNA. Copying accidents happen independently of the environments: some produce an HIV variant that is drug resistant, others (probably the majority) make no difference to drug resistance; some may even make HIV more susceptible to drugs. Mutations are therefore often described as random, although "accidental" is a more accurate descriptive term. What matters is that the mutations are undirected relative to the direction of evolution. Evolution is directed by the selective process, in which resistant variants reproduce better in the presence of the drug. This feature of Darwin's theory has often been misunderstood by its critics. Critics notice that natural selection depends on a random process (mutation) and criticize it for being a random process. Clearly, no random process alone could have generated the nonrandom state of nature that we recognize as life. But in Darwin's theory, mutation only provides the raw material for evolution. The directing process is natural selection, and selection is nonrandom. It is definitely not a random matter whether a drug-resistant or susceptible form will succeed in reproducing in the presence of a nucleoside inhibitor drug.

Natural selection differs from most alternative theories of evolution in the independence between the processes that direct variation and that direct evolution. In the theory of the inheritance of acquired characters, as proposed by the naturalist Jean Baptiste de Lamarck, for example, individual variants tend to occur in the direction of improvement. [*See* Lamarckism.] Evolution is then an extension of the "mutational" process. Darwin's theory is peculiar in that evolution is not an extension of the mutational process.

Explaining Adaptation. Natural selection explains two of the biggest facts about life: adaptation and evolution. In relation to human thought as a whole, natural selection is probably more important as an explanation of adaptation. Adaptation refers to the way living creatures are adjusted (or designed) for life in their environments. Adaptation is a relatively obvious fact about life and was noticed and discussed by the ancient Greeks. It has probably been noticed as a problem in need of explanation for as long as human beings have used rational thought. Adaptation was often explained by the action of God, and indeed the existence of adaptation became one of the most powerful theological arguments for the existence of God, in what was known as the "argument from design." Darwin's theory of natural selection thus undermined a long-standing and important religious argument, and that is one reason why it has been so controversial. Natural selection is a natural, mechanical process that automatically produces adaptation. In an environment containing antiviral drugs, for example, it is adaptive for HIV to be able to resist those drugs. When natural selection favors drug-resistant HIV variants, it results in a population of well-adapted viruses. [*See* Adaptation.]

Natural selection also produces evolution: when the environment changes, natural selection will favor a change in the organisms, too. However, evolution is a less obvious fact about life than adaptation is, and evolution only became a problem in need of an explanation in the nineteenth century.

Natural selection is the only known explanation for adaptation. Other explanations have been suggested, but they have all been scientifically ruled out. Lamarck's theory, for instance, is ruled out by the science of genetics. Whenever an adaptation has evolved, we can infer that it evolved by natural selection. However, natural selection is not the only explanation for evolution as a whole. Another cause of evolution is genetic drift. Genetic drift is random and explains only nonadaptive evolution. It may be that, at the molecular level, the majority of evolutionary change consists of random changes between different, and equally well-adapted, sequences of DNA. If we look at the organismic level, it is less likely that evolution has been by random drift. For example, birds have wings. Wings are adaptations, used for flight. It is practically impossible that wings could have evolved by random drift; they evolved by natural selection.

Biologists differ among themselves in how much of evolution they think is adaptive. Practically all biologists accept that complex adaptations, such as eyes, wings, and legs, evolved by natural selection. They increasingly accept that random genetic drift has been important in molecular evolution and may even account for most molecular evolution. These two claims are consistent, even though eyes and wings are coded for by genes at the molecular level. There may be, at the DNA level, many different sequences of nucleotides that all code for the same kind of wing at the organism level. Wings would then evolve by natural selection, and subsequent evolution at the molecular level would consist of a random drift among all the molecular ways of coding for the same wing.

The main controversy about the importance of natural selection comes in the gray area between complex adaptations such as wings, which undoubtedly evolved by natural selection, and random molecular sequences, which undoubtedly evolved by drift. For instance, the snail *Cepaea nemoralis* has a yellow background color with a number of dark rings. It might easily be thought that a snail would survive and reproduce equally well whether it had one, two, or five rings. But detailed research has shown that natural selection operates on the number of snail rings. Snails with more rings, for example, are less visible to birds in dark environments, such as soil. The pattern and color of a snail shell has turned out to be an adaptation, controlled by natural selection. Many biologists therefore suspect that natural selection will act on almost any detectable attribute of the whole organism. It is only the subtle molecular differences that make no detectable difference to the organism that evolve by drift. Other biologists, however, suggest that some organismic properties are too unimportant for natural selection to act on. It takes careful research to find out whether or not selection is acting on a minor character, such as ring numbers on snails.

The Logic of Natural Selection. Natural selection, as we saw above, can be understood as a deductive argument, in which adaptive evolution results when certain preconditions are met. Two of the preconditions (reproduction and inheritance) are immediately obvious properties of life—so obvious that they might be taken for granted. The third condition concerns a scarcely less obvious property of life: some kinds of organisms survive and reproduce better than others. Given these three conditions, the successful kinds of organisms will increase in frequency. The argument is simple, and for Darwinians this simplicity is part of the power and beauty of the theory. But its simplicity has led some critics to dismiss it as trivial, and they have made two related criticisms—one that natural selection is an empty tautology and the other that it is irrefutable.

The criticism that natural selection is an empty tautology runs something like this. Herbert Spencer, a younger contemporary of Darwin, described natural selection as "the survival of the fittest." But what are the "fittest"? We saw that fitness is defined as the chance of survival (combined with reproduction, but that detail does not matter here). Spencer's phrase seems to reduce to "the survival of those that survive"—which is tautological. If Darwin's theory did simply say that survivors survive, it would be trivial, but there is more to it than that. The problem is an ambiguity in the word *fittest*. In modern biology, fitness has come to mean survival (and reproduction). But Spencer was using the term to refer to adaptation, in the sense that organisms are well adapted (or "fitted") for life. Adaptation is not recognized simply by measuring survival. If we take an adaptation such as the eye, we can observe that it is designed to form visual images. We do so by studying its internal structures. We know that it is an adaptation by our understanding of optical engineering. No biologist has ever tested whether eyes are adaptive by measuring the survival of organisms with and without eyes, or with different kinds of eyes. We have a knowledge of adaptation that is independent of measurements of survival. Adaptation is an independently known fact that is explained, rather than defined, by differential survival.

Likewise, if the theory of natural selection were a trivial tautology, it would be irrefutable. It would no more be possible to disprove the theory of natural selection than to disprove 2 + 2 = 4. However, consider an observation such as the evolutionary increase in the frequency of drug resistance in the population of HIV occupying an AIDS patient, over a period of one or two months after drug treatment begins. The explanation of this observation by natural selection could be disproved if (1) HIV did not reproduce or (2) drug-resistant and drug-susceptible variations of HIV were equally likely to produce drug-resistant offspring. It could be that the observed increase in drug resistance evolved because each individual virus somehow transformed itself into a drug-resistant form. In fact, the HIV do reproduce; drug resistance is inherited, and the evolution is driven by selection rather than individual transformation. Natural selection is a perfectly refutable theory. It has yet to be refuted, as an explanation for adaptive evolution.

Genetic Studies of Natural Selection. The simplest studies of natural selection in action use a species in which there are two forms, and the difference between the two forms is controlled by a single allelic difference at one Mendelian locus. [*See* Mendelian Genetics.] For instance, the peppered moth (*Biston betularia*) has a dark form that is well camouflaged in polluted areas. [*See* Peppered Moth.] The difference between the two forms is mainly controlled by one locus, with the melanic form determined by a dominant allele. If the dominant (dark) allele is labeled *C*, the melanic form is coded for by genotypes *CC* and *Cc*, and the light form

TABLE 1. Simple Population Genetic Model of Natural Selection

The frequency of genotypes in surviving adults (after selection) is calculated from the relative probability of survival (or fitness) and the frequency at birth, which is given by the Hardy–Weinberg theorem.

Moth Color	*Dark*		*Light*
Genotype	*CC*	*Cc*	*cc*
Frequency at birth	p^2	$2pq$	q^2
Fitness	1	1	$1 - s$
Frequency in survivors	$\frac{p^2}{1 - sq^2}$	$\frac{2pq}{1 - sq^2}$	$\frac{q^2(1 - s)}{1 - sq^2}$

p = frequency of *C* allele
q = frequency of *c* allele
$p + q = 1$

by genotype *cc*. In reality, the genetic control of coloration in these moths is not as simple as this. More than two alleles are known of the main locus, the heterozygote may be distinctive, and other genetic loci can also influence coloration; however, the main features of this example can reasonably be understood in terms of a one-locus, two-allele model.

The dark form increased in frequency in nineteenth-century Britain, after the industrial revolution. The rate of increase can be estimated from collections of the moths during this period. These observations can then be used to estimate the relative fitness of the genotypes, as shown in Table 1.

The frequencies of the three kinds of genotype at birth are given by the Hardy–Weinberg equation, $p^2 + 2pq + q^2 + 1$. The fitness, in this simplified example, is the relative chance of survival. Conventionally, the fitness of the best genotype is written as 1, and the fitness of inferior genotypes is written as $(1 - s)$: s is called the selection coefficient and is the fraction by which the chance of survival is reduced. The frequencies in the survivors equal the frequencies at birth multiplied by their fitness; the denominator $(1 - sq^2)$ is needed because without it, the frequencies would not add up to 1. (Notice that $p^2 + 2pq + q^2(1 - s) = 1 - sq^2$.)

In one generation, the frequency of dark moths increases from $(p^2 + 2pq)$ to $(p^2 + 2pq)/(1 - sq^2)$. (It is an increase, because $(1 - sq^2)$ is less than 1.) We can then use the observed rate of increase in dark moths to estimate s, the selection coefficient. It turns out that a figure of about 0.33 explains the observations: light moths in polluted areas had about two-thirds the fitness of dark moths. Traditionally, this was attributed to predation by birds, but subsequent research showed that the dark moths had some other advantage in addition to superior camouflage. Migration also contributed to the observed changes in frequency. The full story, therefore, is complex, but it is in principle possible to study natural selection in a simple Mendelian system.

Most attributes of living creatures are not like the dark and light forms of peppered moths. Attributes such as size and coloration are often continuously distributed rather than coming in discrete forms. Also, the genetic factors at work usually are not as well understood, or as simple, as in the moth example. Size is influenced by genes at many loci. Other techniques are used for these more complex cases. A classic study of natural selection on beak shape in one of Darwin's finches (*Geospiza*) in the Galapagos Islands found that beaks are adapted to the local food supply. [*See* Darwin's Finches.] When many small seeds are available, natural selection favors finches with smaller beaks. From year to year, depending on the weather, the seed supply fluctuates, and in some years large, tough seeds are abundant; then natural selection favors finches with large beaks.

Beak size is inherited: parents with relatively large beaks produce offspring with relatively large beaks. But the exact genetic control is unknown. In this case, natural selection can be described not in terms of the fitness of individual genotypes but in terms of a "selection gradient": a graph that relates relative beak size to chance of survival.

Another recent approach, in cases like beak size in which the genetics is complicated, is to estimate the number of genes that are at work. Things like beak size that vary continuously are called quantitative traits, and the genes that influence them are called quantitative trait loci (QTL). QTLs can be recognized by modern molecular marker techniques, and it is then possible to measure the relationship between which QTLs an organism possesses and its fitness.

Natural selection can also be studied experimentally, both in the field and in the laboratory. For example, *Phlox drummondii* and *P. cuspidata* are two species of flower found in the United States. They mainly occupy separate ranges and have pink flowers. But the two species coexist in parts of Texas, and there *P. drummondii* has red flowers. It has been suggested that the red flowers evolved to prevent cross-pollination by butterflies between the two species. (The two species can hybridize, but the hybrid offspring are inferior; cross-pollination is therefore disadvantageous.) To test the idea, Donald Levin experimentally introduced large numbers of both red- and pink-flowered *P. drummondii* into a region where *P. cuspidata* lives. He found that 38 percent of the seeds of pink *P. drummondii* were inferior hybrids, whereas only 13 percent of the seeds of red plants were. The experiment supports the hypothesis that red flowers have evolved to prevent cross-pollination.

Most experimental work on natural selection has been done in the laboratory, or under artificial conditions. Historically, most work of this kind has consisted of artificial selection experiments. In agriculture, for instance, it may be desirable to produce cows of enhanced milk yield or chickens that lay more eggs. An experimenter can then selectively breed, over a number of generations, from cows with a higher than average milk yield. In a typical experiment, there is an initial large response, but over time selective breeding has less and less effect. This is because there is initially genetic variation for the trait, but over time the genetic variation is used up.

A more recent trend in research is to use bacteria to study natural selection experimentally. Bacteria reproduce fast, and an observable amount of evolution can take place in a convenient time period, such as a few months. In this way, experimenters have studied the efficiency of different forms of selection in producing change and the interaction of selection and drift during evolution. Thus, although natural selection is in many cases an unobservably slow process, and it has operated in the past to produce the adaptations now observable in the world, it is nevertheless possible to study it experimentally, by the normal procedures of science, by picking appropriate organisms and conditions as the subjects of experiments.

Modes of Selection. We can distinguish several different kinds of selection in the simplest case of one genetic locus with two alleles (*A* and *a*) and three genotypes (*AA*, *Aa*, and *aa*). We have looked so far at the case of the peppered moth in which one allele is favored over another; this is called directional selection, and the result is that the favored allele will rise to a frequency of 100 percent (when the allele is said to be "fixed"). Thus, if *AA* and *Aa* have higher fitness than *aa*, natural selection is directional and will fix the *A* allele. Alternatively the *Aa* genotype, known as the heterozygote, may have a higher fitness than *AA* or *aa*. This is called "heterozygous advantage." Natural selection in this case may not favor one allele over the other, and both alleles may be maintained in the population.

A third mode is called frequency-dependent selection. It means that the fitness of a genotype depends on its frequency. The ratio of males to females in a population (the "sex ratio") provides an example (although the genetics of gender is not usually of the *AA*, *Aa*, *aa* type). The reproductive success of an average male increases as males become rarer relative to females. If the sex ratio is one male to two females, the success of an average male will be double what it is if the sex ratio is one to one. Fitness is here negatively frequency dependent: the fitness of a genotype goes down as its frequency goes up. In other cases, fitness may go up as frequency goes up, an example of positive frequency dependence. [*See* Sex Ratios.]

Frequency-dependent selection is related to, but different from, number- or density-dependent selection. Many insects are poisonous and have bright "warning" coloration. Warning coloration is advantageous, because birds have learned from experience not to attack insects of that appearance. Yet the bright warning coloration was not advantageous for the insects that gave the birds the experience: those insects were probably killed. Warning coloration becomes more advantageous when the local density of those colored insects is higher. The higher the density, the less likely it is that any one insect will be killed to educate the local birds. The advantage of warning coloration depends on the absolute numbers of insects, not simply on their relative frequency.

Levels of Selection. Natural selection favors attributes that are of immediate benefit to the organism: attributes that cause the survival and reproductive success of an individual to increase relative to other individuals in the population. In some cases, this may compromise the long-term survival of the population. For instance, natural selection may make the population as a whole more vulnerable to starvation. Some biologists have suggested that natural selection can also act at a higher level, between populations. This process is called group selection (where group means local population). Populations that exploit their food resources recklessly will be more likely to become extinct, whereas populations that preserve their food supply may survive better. There could be a conflict between what is advantageous for the group (preserving food resources) and for the individual (maximizing its own food intake). V. C. Wynne-Edwards, in his book *Animal Dispersion in Relation to Social Behaviour* (1963), suggested that, in such conflicts, between levels of selection, the group advantage will prevail, and individuals reduce their own reproduction in order to preserve the group from extinction. Some biologists still support this view, but most maintain that natural selection between individuals is more powerful than between groups. [*See* Group Selection.]

In most cases, natural selection will not favour individuals that sacrifice their short-term interest for the long-term benefit of the population they live in. Natural selection is a short-term process, and does not operate with long-term strategic plans. Some kinds of adaptation therefore cannot evolve by natural selection, or can only evolve in special conditions. Individual sacrifice for the good of the group is one example. Another kind of example consists of adaptations that are disadvantageous when rare and would only become advantageous after they have fully evolved. Warning coloration, for instance, may initially be disadvantageous in evolution, because the first brightly colored individual would be eaten. It only becomes advantageous after it has become common. Warning coloration has evolved in some cases, however, just as individual self-sacrifice has evolved, but their evolution may be relatively difficult because of the short-term advantages that natural selection requires.

The short-term outlook of natural selection matters for understanding the cause of long-term evolutionary patterns, over a multimillion-year timescale. Some lineages show long-term trends. For example, many mammal lineages have shown a net increase in body size over the past fifty million years; fifty million years ago, the ancestors of modern horses were the size of a modern dog. We can deduce that the cause of the long-term trend was not the goal of producing a modern-sized horse: natural selection does not look ahead or have foresight. One possibility is that there was a sustained short-term advantage to horses of more than average size. When they were a meter long, there was an advantage to being a little more than a meter; when, a few million years later, they were one and a half meters long, there was an advantage to being a little more than a meter and a half; and so on, until modern horses evolved. An alternative view is that, within the ancestral horse population at any one time, larger individuals were not favored by natural selection. Perhaps those species of horse with larger average body size persisted for longer in evolutionary time; then the average size of horses over time would increase. This second process is called species selection. It is not the same as group selection because there is no conflict between individual and group advantage: individual horses are not sacrificing themselves for the long term advantage of the horse lineage. Detailed research is needed to work out the causes of long-term evolutionary patterns. However, we can deduce from the short-term operation of natural selection that natural selection alone will not be working to cause any long-term evolutionary results. The grand patterns of evolution do not contradict the theory of natural selection, but they cannot simply be deduced from it. A short-term study of HIV evolution, for example, can tell us how natural selection is working in those viruses, but we need to know about many other factors to understand long-term viral evolution. [*See* Species Selection.]

BIBLIOGRAPHY

Bell, G. *Selection: The Mechanism of Evolution.* New York, 1997a. A treatise on selection, including the results of artificial selection experiments.

Bell, G. *The Basics of Selection.* New York, 1997b. A shorter and more elementary companion to Bell (1997a).

Dawkins, R. *The Extended Phenotype.* Oxford, 1981. Readable, but aimed at a slightly more knowledgeable reader than *The Selfish Gene;* it defends the idea that genes are units of selection.

Dawkins, R. *The Selfish Gene,* 2d ed. Oxford, 1989. Popular book, based on the idea that the gene is the unit of selection.

Dunbar, R. I. M. "Adaptation, Fitness, and the Evolutionary Tautology." In *Current Problems in Sociobiology,* edited by Kings College Research Group, pp. 9–28. Cambridge, 1982. Thorough account of the problem of whether natural selection is a tautology.

Endler, J. *Natural Selection in the Wild.* Princeton, 1986. Clear conceptual discussion of natural selection (including the tautology issue) and review of research on selection in nature.

Hartl, D. L., and A. G. Clark. *Principles of Population Genetics,* 3d ed. Sunderland, Mass., 1997. Standard text on population genetics, including the theory of natural selection.

Keller, L., ed. *Levels of Selection in Evolution*. Princeton, 1999. Authoritative chapters on topics within the "levels of selection" debate.

Levin, D. A. *The Origin, Expansion, and Demise of Plant Species*. Oxford and New York, 2000. Research monograph, including accounts of experimental studies of natural selection.

Ridley, M. *Mendel's Demon*. London, 2000. Popular book about how natural selection acts against deleterious mutation (also published as *The Cooperative Gene*. New York, 2001).

Weiner, J. *The Beak of the Finch*. New York, 1994. Popular book about Peter Grant and Rosemary the Grant's research on natural selection in Darwin's finches on the Galapagos Islands.

Williams, G. C. *Adaptation and Natural Selection*. Princeton, 1966. Critique of group selection.

Williams, G. C. *Natural Selection*. Oxford, 1992. Update of Williams's thinking about units of selection and reflections on other problematic topics.

Wynne-Edwards, V. C. *Animal Dispersion in Relation to Social Behaviour*. Edinburgh, 1963.

— Mark Ridley

Natural Selection in Contemporary Human Society

Natural selection is the process by which gene frequencies in populations change by virtue of the effects those genes have on the fitness of their bearers. Researchers often wish to know the magnitude and direction of selective forces in populations in order to understand and predict evolutionary change. It is difficult to measure selective forces with sufficient accuracy for that purpose, because even very weak selection can lead to appreciable change when it accumulates over many generations. Consider, for example, pedestrian deaths in traffic. These may be random and thus not related to natural selection; however, if individuals with alleles (alternative copies of the same gene) that lead them to be less fearful are disproportionately victims, then these deaths are selective, and the alleles that suppress fear will decrease in frequency through time.

To what extent is natural selection still operating in modern human society, now that many of the historically important sources of mortality have been reduced? We begin by discussing three common ways in which natural selection operates. We then identify ways of measuring the potential for it in human populations. This article concludes with a discussion of several traits on which natural selection has operated in the recent past and may still be operating in human society.

Examples. Three important categories of natural selection are balanced polymorphism in which selection maintains genetic diversity, selection in response to environmental change, and selection that removes deleterious mutations. We will discuss an example of each category.

Falciparum malaria is one of the most important causes of premature death in much of the tropical and subtropical Old World. Several genes are known to affect the fitness of individuals through their response to falciparum malaria. A particularly well understood case is that of the sickle-cell gene, *HbS*, which causes production of a modified form of hemoglobin. Individuals with a single copy of the sickle-cell gene and a copy of the normal allele (heterozygotes) enjoy better survival and reproduction in environments where infection by falciparum malaria is common than do individuals with two copies of the normal allele (*HbA* homozygotes). Individuals with two copies of the sickle-cell gene (*HbS* homozygotes) ordinarily die before the age of reproduction unless aided by modern medical technology.

In malaria-ridden environments, the fitness advantage of heterozygotes is 10 to 20 percent. This means that a cohort of heterozygotes leaves, one generation later, 10 to 20 percent more offspring than a cohort of normal homozygotes. Using the lower figure of 10 percent and assuming that sickle-cell homozygotes never survive to reproduce, natural selection will lead to population frequencies of about 90 percent for the wild-type (*HbA*) allele, and ten percent for the sickle-cell (*HbS*) allele. At this polymorphic evolutionary equilibrium, the average fitness of the population is only 1 percent greater than the average fitness of a population without the polymorphism.

The sickle-cell gene is the best-understood example of balanced polymorphism in our species: heterozygote advantage leads to the persistence of both alleles in the population. The sickle-cell allele does not spread through the whole population because a higher frequency of this gene results in higher numbers of *HbS* homozygotes, who have very low fitness. It is not known how many such self-destructive pathogen defense polymorphisms are present in our species, but almost all known human polymorphisms causing illness in homozygotes are pathogen defenses. The sickle-cell polymorphism is particular costly in personal terms because homozygotes sicken and die soon after birth as fetal hemoglobin is replaced by normal or adult hemoglobin. If the pathology in affected homozygotes were manifested early in intrauterine life, there would be few or no personal consequences, and we would likely be unaware of the system (Melanesian ovalocytosis, another malaria defense polymorphism, is an example of this: no surviving homozygote has ever been observed). The personal cost of the polymorphism also remains high in emigrant descendants of African and Mediterranean populations with the polymorphism. Where malaria has been eliminated rapidly through technology, illness and premature death from the polymorphism persist for many generations. Its victims are just as much victims of malaria as are people suffering directly from the disease.

The gene *CCR5* codes for a cell surface receptor. In European populations there is a faulty allele that, in homozygotes, causes the absence of this receptor (Schliekelman et al., 2001). Such individuals infected with HIV do not progress to AIDS, while infected heterozygotes enjoy delayed progression to AIDS. Other variant alleles of the gene are known in Africa; some delay and some accelerate the progression to AIDS. In environments with high levels of exposure to HIV, as in much of sub-Saharan Africa, natural selection favors alleles that delay AIDS and disfavors those that accelerate the onset of the disease. Recent projections suggest that the population frequency of resistant genes will increase from 40 percent to 53 percent in the next century, while the frequency of susceptible alleles will decrease from 20 to 10 percent. The frequency of the faulty, AIDS-resistant allele in Europe is so high that it is thought that it increased in response to some other major pathogen, perhaps plague, within the last few thousand years.

Neurofibromatosis is a severely damaging dominant disorder: people who inherit the mutation usually die early in life. Its incidence is determined entirely by the balance between the rate of the mutation that causes new cases and selection against bearers of the mutation. Approximately one or two births per ten thousand have this disorder, one of the highest mutation rates known for a deleterious gene; it reflects the large size of the affected gene and hence the length of DNA susceptible to mutational damage. As in the case of the sickle-cell disorder, we know about this disorder because its effects occur well after birth. Similar disorders that caused very early fetal wastage would have few or no personal consequences and would not be known to us.

In all three of these cases, natural selection has essentially been maintaining the status quo, except for the recent increase in *CCR5* in response to the AIDS pandemic. In general, we would like to know, first, how many individual differences in health and fertility are caused by gene differences; and second, how many of these differences are balanced polymorphisms, how many are responses to environmental change, and how many represent mutational damage. We have almost no solid information about the first question, and very little about the second. Extensive studies of fruit flies suggest that the removal of deleterious mutations has a larger contribution to fitness than does the maintenance of balanced polymorphism. It is not known whether this finding can be extrapolated to mammals.

Opportunity for Selection. With some simplifying assumptions, it is possible to estimate the opportunity for natural selection in human populations. A statistic developed by James Crow (1958) describes the variance in fitness among individuals. If the variance is low, then there is not much difference in number of successful offspring between people and, hence, not much opportunity for selection. If the variance is large, then there are large differences, and the opportunity for selection is great. Crow's index I can be interpreted as follows: if all variation in reproductive performance (fitness) is attributable to additive gene differences, then the fitness of the next generation will change by $1 + I$. However, there are three reasons why we cannot use this index directly: first, the fraction of variation in fitness that is related to gene differences is unknown; second, much of the fitness variation attributable to gene differences must be used up in eliminating deleterious mutations, many of which are newly arisen; and third, this demographic method cannot detect very early deaths. For example, an individual bearing a new dominant lethal mutation that causes death within the first few weeks after conception would lose zygotes bearing the mutation, but these early losses of half her conceptions would mean only a slight, probably undetectable, fertility reduction.

Nevertheless, in humans Crow's index shows that there is enough fitness variation among individuals to accommodate high levels of selection and evolution. In high-mortality populations, the index reaches values of 2 to 3, so that under the simplest assumptions fitness could triple or quadruple each generation. Much of the variation arises from pre-reproductive mortality, but even in contemporary low-mortality populations the index hovers around 0.5, a value that would more than double fitness in two generations under the simplest assumptions. Among Utah Mormons early in the nineteenth century, the index for males was greater than that of females, reflecting the effect of polygyny and acccompanying higher variance in male fitness; after polygyny ended, the values in two sexes became similar.

Quantitative Traits and Behavior. Crow's work shows that there is ample opportunity for ongoing natural selection in modern human populations. The examples of the *HbS* and *CCR5* loci suggest that a substantial fraction of ongoing natural selection in human populations occurs in response to changing environments—movement away from malarial areas in the case of selection against *HbS*, and the advent of widespread HIV infection in the case of *CCR5*. What other changes in human populations are occurring or will occur in the future in response to environmental change? There are many ongoing social changes in industrial societies that will, if continued, result in significant changes in the genetics of these populations. Among the most interesting are those affecting partially heritable quantitative traits, especially behavioral traits.

A quantitative trait is a metric trait that is variable in a population, with part of the variation determined by gene differences. For example, height is known to be influenced both by genes and by environmental factors like diet. Geneticists use heritability as a statistic describing the fraction of variability in a trait that is genetically transmitted in a population. If height in humans were entirely determined by diet and other environmental agents, then its heritability would be zero. In contrast, if differences in height were determined entirely by additive gene differences, the heritability would be unity. In reality, height is affected by both, and similarities among relatives can be used to estimate heritability. It should be clear that there is nothing fundamental about the heritability of a trait. In a sample of people from a population with large differences in nutritional status, we would find that some individuals suffered marginal nutrition and much chronic infectious disease during development, while others were well fed and well supplied with antibiotics; then, much of the variation in height would reflect early environmental conditions. In contrast, height in a sample of people who were all raised in a wealthy community with plentiful food and medical treatment would be highly heritable; all individuals were exposed to a similar highly favorable environment, so phenotypic differences among them would reflect mostly genetic differences.

Empirical studies of twins and other categories of relatives demonstrate overwhelmingly that any trait we examine is partially heritable. Even such traits as political liberalism or conservatism, or the tendency to divorce, are correlated with differences that are transmitted in large part genetically.

The response of quantitative traits to selection is particularly simple. The change in one generation is the product of the trait's heritability and the difference between the population mean and the mean of those who reproduce (called the selection differential).

Intelligence quotient (IQ), the score obtained on any of various standard written tests of cognitive ability, is a familiar example of a quantitative genetic trait (Herrnstein and Murray, 1994). We do not know what genes are involved, but we are sure from a variety of natural experiments, such as identical twins reared apart, that the obtained score has a heritability greater than 0.5. Consider a social system that favored IQ in a population such that those individuals with IQ scores greater than the mean had 5 percent more offspring. If the initial population mean were 100 with standard deviation of 15 points, then the mean IQ of parents would be 100.6, so that the selection differential is 0.6 points. The mean IQ of the next generation would be the product of the differential 0.6 and the heritability 0.5, or 0.3. In a single generation, there would be little change in IQ—from 100 to 100.3. If the social system were to persist for a thousand years, or 40 generations, the net change at the end of the millennium would be 40 × 0.3, or 12 IQ points: the population would then have a mean IQ score of 112. There has been speculation that the high IQ test scores of Ashkenazi Jews reflect a social history like this.

Because different social systems reward different abilities and proclivities, and because almost everything so far examined is more or less heritable, we expect social systems to cause genetic changes in their members over time scales of centuries to millennia. A possible example is that of the D4 dopamine receptor locus, at which long alleles are differentially distributed among human populations (Chen et al., 1999). There is a correlation between long alleles at this locus and novelty-seeking and perhaps activity levels in individuals. Populations that have migrated long distances since the end of the Pleistocene have a higher frequency of long alleles, perhaps because bearers of long alleles were more likely to join population movements in the past. For example, the world's highest frequencies of long alleles, over 50 percent, are in South American populations, furthest from the African home of our species.

We can only speculate about the consequences for genetic change of contemporary social patterns in industrial societies, but there are several obvious candidates. First, the easy availability of birth control technology should lead to selection in females for maternalism and a positive desire to have children. In the past, women rarely had such precise control over conception; now, there is often conscious decision-making involved. Second, human females around the world obtain provisioning from males, but in varying degrees. In dense agricultural societies such as those of Europe, male provisioning has been crucial for female fitness, and this has favored social systems with durable marriages. In many subsistence horticultural societies—notably those in Amazonia, Southeast Asia and Oceania, and central Africa—females essentially feed themselves and their children; provisioning by fathers is less essential to female fitness, and marriage bonds are fragile. As male provisioning becomes less and less relevant to the health and survival of children in industrial societies, we expect selection to favor females who do not seek a provisioning mate and males who are less willing to provision offspring. Finally, new reproductive technologies, if they become widespread, have the potential for drastically altering selection. Currently the wealthiest individuals have the fewest children, but if reproduction could be carried out by surrogates, this negative correlation between wealth and fitness might be reversed.

BIBLIOGRAPHY

Cavalli-Sforza, L. L., and W. F. Bodmer. *The Genetics of Human Populations*. San Francisco, 1971. A classic text on human population genetics, written before the explosion of molecular biology but unexcelled in its treatment of demography and quantitative traits.

Chen, C., et al. "Population Migration and the Variation of Dopamine D4 Allele Frequencies around the Globe." *Evolution and Human Behavior* 20 (1999): 309–324. Describes worldwide variation in population frequency of long alleles at a dopamine receptor locus that may be related to population differences in personality and behavior.

Crow, J. "Some Possibilities for Measuring Selection Intensities in Man." *Human Biology* 30 (1958): 1–13. Crow's classic paper is a description of how to relate demographic patterns to the potential intensity of natural selection.

Herrnstein, R. J., and C. Murray. *The Bell Curve: Intelligence and Class Structure in American Life.* New York, 1994. A standard work on IQ, its transmission, and its social correlates in the United States.

Schliekelman, P., et al. "Natural Selection and Resistance to HIV." *Nature* 411 (2001): 545. Describes the very rapid evolutionary change at the *CCR5* locus that must be occurring in sub-Saharan Africa.

— HENRY HARPENDING AND GREGORY COCHRAN

NEANDERTHALS. *See* Hominid Evolution, *article on* Neanderthals; Modern *Homo sapiens*, *article on* Neanderthal–Modern Human Divergence.

NEO-DARWINISM

The term "neo-Darwinism" is typically used for the kind of Darwinism practiced by today's evolutionary biologists. It is based on the so-called neo-Darwinian synthesis (also called the evolutionary synthesis or modern synthesis), which developed roughly between 1920 and 1950. This synthesis, based on a union between the Darwinian theory of natural selection and Mendelian genetics, resulted in the reformulation of basic Darwinian principles in the new language of mathematical population genetics, and in the subsequent extension of this framework to a number of biological fields under the common name "evolutionary biology."

An earlier use of the term "neo-Darwinism" (coined by British zoologist G. J. Romanes) referred to the views of the German biologist August Weismann and others in the late nineteenth century. As a counter-argument to widespread "neo-Lamarckism," Weismann postulated that the germ line can never be affected by the "soma"—that is, acquired characteristics cannot be inherited—and he declared selection the only evolutionary force.

During the second half of the twentieth century, neo-Darwinism increasingly became identified with new "gene-selectionist" ideas within the framework of adaptation by natural selection. Critics of this version of neo-Darwinism—who also regarded themselves as neo-Darwinists—saw this focus as too restricted and emphasized the plurality of evolutionary forces and levels of selection. At the beginning of the twenty-first century, though this is not widely recognized, two views of neo-Darwinism (and the meaning of the synthesis) exist simultaneously: a "narrow" and a "broader" interpretation.

After Darwin's death, the theory of natural selection was in deep trouble, and alternative evolutionary theories flourished. Weismann's dismissal of neo-Lamarckism had only fortified views about environmentally influenced "soft inheritance." One great problem for natural selection was Darwin's view of "blending inheritance": How would new variation be prevented from being "swamped" by existing variation? Another question was whether the origin of new species is gradual (by isolation) or saltational (literally, in "leaps")? Finally, is variation continuous or discontinuous, and which type is inherited? Darwin's belief that selection acted primarily on small, continuous variations was supported by the population studies of biometricians such as Francis Galton and Karl Pearson, while the champions of discontinuity led by William Bateson (later called Mendelians) also had data that validated their position. The conflict was compounded by the lack of distinction between genotype and phenotype, made by Johannsen in 1911.

After the rediscovery of Mendel in 1900 by Hugo De Vries and others, Mendelians and Darwinians split into two hostile camps. One reason was that Mendelian particulate inheritance (today, we call the "particles" genes) was originally identified with De Vries's "mutation theory," according to which new variations (or species) originated in large jumps, or macromutations, and evolution was exclusively explained by mutation pressure. Darwinian naturalists, believing that Mendelism was synonymous with mutation theory, held on to theories of soft inheritance, while they considered selection a weak force at best. They did not know of the new findings in genetics that would have supported Darwinism.

The theory of particulate inheritance did seem to make sense of the rule-of-thumb methods of practical breeders, but a serious problem was the lack of physical proof of Mendelian "factors." The missing evidence came from cytological studies by Thomas Hunt Morgan (1866–1945) and his co-workers A. H. Sturtevant (1891–1970) and Hermann Muller (1890–1967), who in 1910 demonstrated that the "factors" are physical units located at specific chromosomal sites. Their breeding experiments on *Drosophila* showed that genes for specific traits are inherited as discrete units that remain unchanged over generations. The chromosomal theory was soon to solve the long-standing problem of the source of new genetic variation: ultimately, it comes from such things as (small-scale) spontaneous mutations and chromosomal recombinations, and further, there are large reservoirs of "hidden" (not phenotypically expressed) variability in a population. The discovery of micromutations, in particular, helped refute the theory of soft inheritance.

Now some of the obstacles to a synthesis between Mendelism and Darwinian selection had been eliminated. The final fusion came with the development of quantitative population genetics. Independent contributors to this field were the Russian School under Sergei S. Chetverikov (1880–1959), Ronald A. Fisher (1890–1962), and J. B. S. Haldane (1892–1964) in England, and Sewall Wright (1889–1988) in the United States.

In 1908, G. H. Hardy and Wilhelm Weinberg, working from Mendelian reasoning but independently of each other, had mathematically shown that the original gene frequencies in a random-breeding population would remain constant throughout successive generations unless affected by immigration, mutation, selection, nonrandom mating, or errors of sampling. The Hardy–Weinberg equilibrium principle was to inform population genetics for decades to come. In the late 1920s, Chetverikov, stimulated by a visit from Muller, started applying Mendelian genetics to the study of natural, rather than mathematical, populations. The work of the Russian School (which employed such terms as "gene pool," "microevolution" and "macroevolution") was later introduced in England by Haldane and continued in America in Morgan's laboratory by Chetverikov's student Theodosius Dobzhansky (1900–1975).

Fisher's book *The Genetical Theory of Natural Selection* (1930) is considered the foundation of evolutionary population genetics. It conclusively demonstrated what nobody would have believed 30 years earlier: that Mendelism was not only mathematically compatible with the theory of natural selection, but because of its particulate nature, it actually provided the missing link in Darwin's theory. Already in 1918, Fisher had shown that supposedly non-Mendelian continuous variation could be analyzed in a Mendelian fashion by treating the inheritance of this kind of variation as a result of the additive small effects of many genes, each individually following Mendelian rules.

Fisher's famous Fundamental Theorem of Natural Selection was modeled after the Second Law of Thermodynamics in physics, reflecting the effort to make Darwinian theory more scientific in a time of the "eclipse of Darwinism" and the rise of experimental biology. (Darwinism was also purged of metaphysical-sounding terminology.) Fisher worked on problems of selection, such as the role of dominance, the rate of spread of favorable genes, and the conditions for maintaining balanced polymorphism in a population. Haldane and Wright often studied other factors, such as mutation, mating systems, isolation, migration, and genetic recombination. Haldane's most important contributions are collected in *The Causes of Evolution* (1932). The predictions of his equations, which focused on natural selection and dominant and recessive genes, often turned out to be testable in the field (e.g., in regard to the peppered moth, later confirmed by Kettlewell). During his career, Haldane tackled many and diverse problems, including rates of mutation and genetic change, variation in flower colors, and mapping of the human X chromosome.

Sewall Wright's studies of color inheritance and inbreeding in guinea pigs had convinced him that effective breeding populations (called demes) often are small enough to make sampling errors (the accidental loss or increase of certain traits in small, isolated groups) important, even to the point of giving rise to new species without natural selection. His famous "genetic drift" theory was developed in the late 1920s. As part of his lifelong debate with Fisher, who thought selection works best in large populations, Wright formulated his "shifting balance" theory, which proposes that evolution happens in three phases: first, random drift of gene frequencies in subpopulations, then increase of the preferred gene combination, and finally, its spread throughout the population.

Fisher, Haldane, and Wright were all in contact with real field and experimental research. Fisher, a biometrician and eugenicist, made major advances in statistics while at the Rothamsted agricultural research station; later, he collaborated closely with E. B. Ford (1901–1988), an Oxford ecological geneticist. Haldane, a Marxist and polymath whose interests included physiology and genetics, was for 10 years connected with experimental agricultural research. Wright, a self-taught statistician, was employed as a senior animal breeder for the U.S. Department of Agriculture and later worked closely with Dobzhansky.

With Fisher, Haldane, and Wright, the core of the neo-Darwinian synthesis was firmly in place. Still, the scope of their work was limited, and their mathematical language not easily accessible to biologists. What was now needed to carry the synthesis further was articulators of the synthetic theory, and multidisciplinary bridge-builders between fields and individuals. The most active architects of this stage (who typically had taught themselves population genetics) were Theodosius Dobzhansky, Ernst Mayr (b. 1904), George Gaylord Simpson (1902–1984), G. Ledyard Stebbins (1906–2000) in the United States, Julian Huxley (1887–1975; grandson of T. H. Huxley) in England, and Bernhard Rensch (1900–1990) in Germany.

Dobzhansky's enormously influential *Genetics and the Origin of Species* (1937) would serve as the "textbook" for the new field of evolutionary biology. Huxley's *Evolution: The Modern Synthesis* (1942; dedicated to Morgan) introduced an international audience to evolutionary biology as the new integrative scientific foundation for Darwinism, and a potential unifier of the rest of science as well. Huxley was extremely active as a multidisciplinary scholar and a world traveler. Originally an embryologist, he edited *The New Systematics* in

1940, making available the work of an international group of naturalists.

Dobzhansky's 1937 book brought together a broad range of field studies and experimental work within a population-genetic framework. It also discussed speciation, which was explained as caused by geographic isolation; in general, it presented a continuity between microevolution (evolution below the species level) and macroevolution (evolution at the species level and above). In his own work, Dobzhansky fused the populational approach of Chetverikov with the experimental methods of the Morgan group, combining field sampling studies with follow-up laboratory experiments and cytogenetic analysis of *Drosophila pseudoobscura*.

Dobzhansky's synthesis, presented in a series of books published by Columbia University, was extended in the same series to the field of systematics by Mayr, to paleontology by Simpson, and to botany by Stebbins. Mayr's *Systematics and the Origin of Species* (1942) discussed the contributions by naturalists and proposed additional mechanisms for speciation, while firmly dismissing the idea of speciation by macromutation, reintroduced by Richard Goldschmidt (1878–1958) in 1940. Simpson's influential *Tempo and Mode in Evolution* (1944) made paleontology, too, compatible with the synthesis, and thereby microevolution with macroevolution. Simpson also demonstrated how the fossil record provided evidence for evolution. Discontinuities were explained through "quantum evolution"—periods of rapid, major changes from one adaptive level to another (an idea inspired by Wright). Stebbins employed the new framework to make sense of the mass of often puzzling data on plants, using C. D. Darlington's notion of evolving "genetic systems" (recombinations) to account for phenomena such as polyploidy. His huge *Variation and Evolution in Plants* was the last contribution to the synthesis (1950), after Rensch's independently conceived *Neuere Probleme der Abstammungslehre* (*Evolution above the Species Level*, 1959).

Meanwhile, Edmund Brisco "Henry" Ford at Oxford, who had worked with both Julian Huxley and Fisher, pioneered "ecological genetics," bringing together the study of genetic and ecological variation in natural populations. The work of his colleague geneticist Bernard Kettlewell (1901–1979) on industrial melanism in the 1950s was widely seen as the first serious confirmation of the efficiency of natural selection; Kettlewell even captured the process of selection on film.

The neo-Darwinian synthesis was an international achievement. A list of those who Simpson in 1949 thought had "made outstanding contributions" to the modern synthesis includes, in England, R. A. Fisher, J. B. S. Haldane, Julian Huxley, C. H. Waddington, and E. B. Ford; in the United States, Sewall Wright, Hermann Muller, Theodosius Dobzhansky, Ernst Mayr, and G. Ledyard Stebbins; in Germany, N. W. Timofeeff-Ressovsky and Bernhard Rensch; in the Soviet Union, S. S. Chetverikov and N. P. Dubinin; in France, G. Teissier; and in Italy, Buzzati-Traverso. Mayr's later list is largely similar, although he substituted C. D. Darlington for Waddington and added more naturalists—among others, Francis Sumner in the United States and Erwin Baur in Germany. By 1950, the architects declared the synthesis complete.

Embryology, developmental biology, and physiology, however, remained "unsynthesized." In 1952, Conrad Waddington, supported by Haldane, criticized the narrow focus on mathematical genetics. He later published his own views in *The Strategy of the Genes* (1957). In 1959, at a Darwin celebration at Cold Spring Harbor, Mayr followed this up, pleading with those he called "bean-bag" geneticists to pay attention also to gene interaction, the organism, and its environment, in face of the acute threat against evolutionary biology from reductionist molecular biology. In the early 1960s, this crisis was warded off with the help of Simpson and Dobzhansky. Dobzhansky integrated the molecule, too, as a new level of evolution (coining the slogan "from molecules to Man"), and he argued for the importance of studying both the unity and the diversity of life.

One reason for the popularity of the neo-Darwinian synthesis was surely that Dobzhansky, Huxley, and Simpson conveyed certain moral and political lessons in their scientific and popular books (and Huxley in his BBC radio programs). Man, too, was integrated in the synthesis—as a rational, responsible being, the pinnacle of evolution. Evolution was connected to progress, and knowledge of evolution would be crucial for the future of man. Mayr presented the new "population thinking" as bringing new respect for the uniqueness of the individual, as opposed to earlier essentialist, typological thinking. In the post–World War II climate, the modern synthesis represented optimism and hope.

In the last quarter of the twentieth century, the extension of the synthesis to social behavior around new concepts such as kin selection (with the ensuing "sociobiology controversy") demonstrated the continuing tension between a narrow, selectionist (or "adaptationist") and a broader, "pluralist" conception of neo-Darwinism. It also showed the difficulties of applying neo-Darwinian models directly to humans. Debates about macroevolution (especially "punctuated equilibria"), "neutral molecular evolution," and other theories indicated that, for some evolutionists of the next generation, the synthesis was clearly unfinished or in need of reassessment. Meanwhile, progress in genomics and proteomics promised to extend the neo-Darwinian synthesis toward the field of development.

[*See also* Overview Essay *on* History of Evolutionary

Thought; Fisher, Ronald Aylmer; Haldane, John Burdon Sanderson; Wright, Sewall.]

BIBLIOGRAPHY

Allen, Garland. *Life Science in the Twentieth Century*. (Cambridge History of Science Series.) Cambridge, 1985. Good discussion of conflicting camps before the synthesis and developments leading to the synthesis. Interesting discussion of the Russian School of population genetics.

Dawkins, Richard. *The Extended Phenotype*. Oxford and New York, 1982. Excellent defense of the research program of genetic selectionism against epistemological and moral attacks on "the adaptationist program." Valuable, abundant references and glossary.

Eldredge, Niles. *The Unfinished Synthesis: Biological Hierarchies and Modern Evolutionary Thought*. Oxford and New York, 1985. Complains that the modern synthesis gives an ontologically unsatisfactory picture of the actual levels of evolution and ignores ecology; cites and discusses views of original architects.

Eldredge, Niles. *Reinventing Darwin: The Great Debate at the High Table*. New York, 1995. Argues that there is an opposition in current neo-Darwinism between "Ultra-Darwinians" and "pluralists," and that the latter are the true heirs of Darwin.

Gayon, Jean. "Critics and Criticisms of the Modern Synthesis: The Viewpoint of a Philosopher." *Evolutionary Biology* 24 (1990): 1–49. Presents the debates among philosophers of biology about the status of the modern synthesis and about evolution as a scientific theory.

Gould, Stephen Jay. "The Hardening of the Synthesis." In Dimensions of Darwinism, edited by Marjorie Grene, pp. 71–93. Cambridge, 1983. Argues that the synthesis "hardened" in the 1940s and 1950s, with an increasing emphasis on selection as opposed to its original more "pluralist" recognition of evolutionary forces.

Hamburger, Victor. "Embryology and the Modern Synthesis in Evolutionary Theory." In Mayr, Ernst and William B. Provine, *The Evolutionary Synthesis: Perspectives on the Unification of Biology* edited by Ernst Mayr and William B. Provine, pp. 97–112. Cambridge, Mass., 1980. Presents as "the missing chapter" in the evolutionary synthesis I. I. Schmalhausen's and C. H. Waddington's views of selection's effects on the gene-controlled variability of developmental processes.

Mayr, Ernst. *The Growth of Evolutionary Thought*. Cambridge, Mass., 1982. A magisterial historical overview by one of the architects of the evolutionary synthesis. Emphasizes role of naturalists, and "population thinking."

Mayr, Ernst, and William B. Provine. *The Evolutionary Synthesis: Perspectives on the Unification of Biology*. Cambridge, Mass., 1980. Based on the symposium Mayr and Provine organized in 1974 for the surviving architects and historians and philosophers; detailed "who was who," valuable comments and notes.

Provine, William B. *The Origins of Theoretical Population Genetics*. Chicago, 1971. Discusses Fisher, Haldane and Wright as the founders of the modern synthesis, the origin of their different views, and their disagreements with one another.

Provine, William B. *Sewall Wright and Evolutionary Biology*. Chicago, 1986. Analyzes Wright's role and views.

Ridley, Mark. *Evolution*. Oxford, 1996. A comprehensive and readable overview from a current "stricter" neo-Darwinian perspective.

Ruse, Michael. *Darwinism Defended: A Guide to the Evolutionary Controversies*. Reading, Mass., 1982. An overview of recent debates relevant to the synthetic theory, including those surrounding macromutation, cladistics, the neutral theory of molecular evolution, and the revival of Lamarckism.

Segerstråle, Ullica. *Defenders of the Truth: The Battle for Science in the Sociobiology Debate and Beyond*. Oxford and New York, 2000. Interview-based account and analysis of the epistemological, scientific and moral debates about sociobiology and neo-Darwinism, seen as a later generation's dispute about the modern synthesis in a new socio-historic context; detailed index, glossary.

Smocovitis, Vassiliki Betty. *Unifying Science: The Evolutionary Synthesis and Evolutionary Biology*. Princeton, 1996. Interesting history of the evolutionary synthesis as an effort to unify biology in a climate of physics-dominated positivism; valuable footnotes.

— ULLICA SEGERSTRÅLE

NEOGENE CLIMATE CHANGE

The Neogene and Recent are geological periods of extreme climate change that have had a profound effect on the evolution of humans as well as other animals. At the beginning of the Cenozoic, about 65 million years ago, global climate was warm and equitable, with little difference in temperature between the equator and poles. Beginning in the Eocene, about 50 million years ago, the benthic $\delta^{18}O$ record from Atlantic deep-sea cores (see Deep Sea Cores Vignette) documents a gradual cooling of the global climate punctuated by periods of very rapid cooling (Figure 1). Superimposed on this general trend are high-frequency climate fluctuations reflecting Milankovitch cycles (see Orbital Geometry Vignette). The very rapid periods of cooling occurred first in the Paleogene, about 33 million years ago (mya), then in the Miocene, between 15.6 and 12.5 mya, and again in the Plio-Pleistocene, between 2.95 and 2.52 mya.

The first period of rapid cooling in the Neogene occurred in the middle of the Miocene period and is associated with modern continental geography and the appearance of the modern mode of ocean circulation. One major influence may have been a change in topography of the Greenland-Scotland Ridge in the North Atlantic Basin that reduced the northward flow of warm, saline surface water. Another might have been the closure of the Tethys Sea, which had connected the Atlantic and Indian oceans throughout the Paleogene. This would have considerably reduced the area of tropical and subtropical oceans, restricting a major source of water vapor and thereby contributing to the general cooling trend.

The closure of the Tethys Sea also resulted in the formation of land bridges between Africa and Eurasia and consequent faunal migrations, including the spread of apes, or early hominoids, from Africa into more northern latitudes. By later Miocene times, however, the vast proportion of apes in Africa and Eurasia disappear from the record. Their disappearance coincides with increased aridity and the spread of grassland and savanna environments and is followed by the spread of cercopithecid monkeys and the radiation of the hominins, including the ancestors of humans. The main cause of this marked environmental change was probably the uplift of landmasses, particularly of the Tibetan Plateau.

The uplift of the Tibetan Plateau began in the Paleogene about 50 million years ago, when the Indian subcontinent made initial contact with southern Asia. This may be reflected in the general global cooling trend since that time. However, the main Neogene uplift thrust of between 1 and 2.5 kilometers occurred in the later Miocene at about 8 mya and had two major climate consequences. First, there was a fundamental change in air circulation patterns, resulting in the Asian monsoons. Summer heating of the Tibetan Plateau produces a powerful low-pressure cell over southern Asia and a consequent strong flow of warm, dry air from east central Asia into northeastern and east central Africa. This would have replaced moist, low-level oceanic winds and resulted in aridification throughout eastern Africa. There also would have been additional drying effects over the central and northern Sahara.

The second consequence of the uplift of the Tibetan Plateau was a marked reduction in the CO_2 content of the atmosphere resulting from chemical weathering of silicate rocks. This is associated with general cooling and drying that ultimately caused northern latitude winter precipitation to fall as snow rather than as rain. This is an important precondition for the onset of the northern latitude glaciations that began about 2.95–2.52 mya. Another important effect was the global increase of plants using C_4 photosynthesis (typical savanna/grassland grasses). Particularly in warm climates, C_4 plants have a higher quantum yield and would spread under conditions of relatively low atmospheric CO_2. Based on analysis of the diet of fossil equids, the change from C_3 to C_4 plants occurred by 8.0–7.5 mya in East Africa and slightly later in Pakistan (7.8–6.0 mya), with a generally cooler mean annual temperature.

FIGURE 1. Averaged Oxygen Isotope Curve for the Past 70 Million Years.
Graph based on composite data from benthic foraminifera compiled from Deep Sea Drilling Program sites in the Atlantic Ocean. After K. G. Miller, R. G. Fairbanks, and G. S. Mountain "Tertiary Oxygen Isotope Synthesis, Sea Level History, and Continental Margin Erosion." *Paleoceanography* 2 (1987): 1–19.

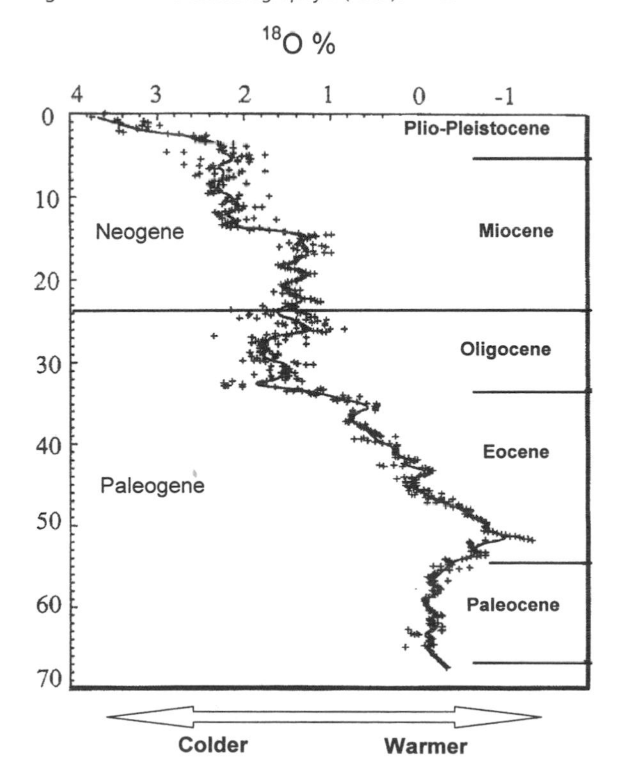

DEEP SEA CORES

Deep sea cores are long cylinders of sediment taken from beneath the ocean floor. The cores contain microscopic fossils of marine animals, sands blown into the ocean from land, and other materials such as volcanic glass, cosmic material (microtektites), or manganese nodules. Deep sea cores have been particularly important for climate research. By analyzing the ratio of ^{18}O to ^{16}O in the shells of microscopic foraminifera that absorb oxygen isotopes in proportion to the presence of those isotopes the surrounding seawater, it is possible to determine glacial ice volume and temperature. The stable isotope of oxygen, ^{18}O, is heavier than the lighter isotope, ^{16}O. During glacial periods, the oceans are rich in ^{18}O because the lighter ^{16}O evaporates more easily and becomes locked in the glacial ice. The ratio of ^{18}O to ^{16}O, the oxygen isotope ratio ($\delta^{18}O$), is also dependent on temperature. The colder the temperature, the more ^{18}O is transferred from the ocean to the foram shells and the higher is $\delta^{16}O$. Planktonic foraminifera provide a record of sea surface temperature, while deep-sea (benthic) species give the best record of ice volume and ocean temperature history.

Analysis of windblown sand in the deep sea cores can also provide information on continental seasonality of rainfall and aridity. Analysis of cores taken off West and East Africa has shown that periods of aridity in Africa since about 3 million years ago correspond to the rhythm of glaciation in the Northern Hemisphere. Before this time, aridity in Africa fluctuated with the Asian monsoon cycle.

—LESLIE C. AIELLO

MILANKOVITCH CYCLES AND ORBITAL GEOMETRY

The Earth's orbit around the sun changes through time and this results in differences in the amount of solar radiation reaching the Earth's surface, producing at times warmer summers or cooler winters. This cyclic change results from variations in the Earth's eccentricity, axial tilt and precession. Together these are known as Milankovitch cycles, after the Serbian astronomer Milutin Milankovitch. Eccentricity is the shape of the Earth's orbit around the Sun, which varies from circular to elliptical on a 100,000-year cycle. Axial tilt is the tilt of the Earth's axis in relation to the orbital plane around the sun and varies from 21.5° to 24.5° on a 41,000-year cycle. Axial tilt largely accounts for the seasons; when the tilt is greatest, the summers are warmer and the winters cooler. Precession is the wobble of the Earth as it spins on its axis. This means that the season when the Earth is closest to the Sun (perihelion) varies, and this variation is on about a 23,000-year cycle.

Northern Hemisphere glaciers form when northern summers are cool and winters are warm. In warm winters the atmosphere can hold more moisture and more snow will fall. The cooler the summer, the less likely it is that the snow from the previous summer melts before the next season's snow begins to accumulate. Northern Hemisphere winters are warmest when the axial tilt is minimal, and summers are coolest when the Northern Hemisphere in summer is farthest from the Sun as a result of precession and greatest orbital eccentricity.

—Leslie C. Aiello

One major area of debate is how these global and hemispheric events relate to more local environments. Particularly in East Africa, faulting and uplift had produced sufficient relief by 7 mya to fragment the environment through local climatic effects, including elevation-related decrease in temperature and an increase in local aridity (or moisture) caused by the rain shadow effect. In general, however, less seasonal forested habitats were replaced in the terminal Miocene by open wooded-grassland habitats, and by the Early Pliocene there was an increased seasonality, with faunas showing a savanna-mosaic character. However, stable isotope analysis of paleosols from the hominin sites suggest that C_4 plants were not a major component of hominin habitats until after 1.7 mya. Before this time hominins seem to have preferred woody C_3 habitats, including riparian C_3 woodlands. This interpretation is also supported by analyses of flora and fauna from the hominin sites.

The second major phase of rapid cooling in the Neogene occurred late in the Pliocene and is associated with the beginning of glaciations in the Northern Hemisphere. Three factors are necessary for Northern Hemisphere glaciations. First, as described above, the overall climate must be cold enough for northern latitude winter precipitation to fall as snow rather than as rain. Second, there has to be sufficient moisture at high northern latitudes for significant precipitation. Third, northern summers must be cold enough so winter snow does not melt.

Increased moisture at high northern latitudes was probably a result of the closure of the Isthmus of Panama, which since 4.6 mya has produced an increased Gulf Stream flow. This introduced warm and saline water into high latitudes, and evaporation of this water resulted in increased northern latitude moisture. Cold northern summers are a direct result of orbital forcing (the Milankovitch cycles). There was a progressive change in the Earth's obliquity (the tilt of the planet in respect to the Sun) between 3.1 and 2.5 mya, resulting in cool northern summers. This followed a longer period of minimum obliquity amplitude fluctuation between 4.5 and 3.1 mya and was probably the trigger for the onset of the Northern Hemisphere glaciations.

Before the onset of northern latitude glaciations, sub-Saharan African climate was tied to monsoon intensity, which was driven by the 19,000–23,000-year Milankovitch precessional cycle. After about 2.8 mya, sub-Saharan African climate was governed primarily by the northern latitude glaciations. During the glaciations, sub-Saharan Africa was more arid; deserts expanded, while forest and savanna habitats were at times severely constricted. Initially, the periods of glaciation and aridity reflected the 41,000-year Milankovitch obliquity cycle. At about 0.9 mya, a shift to a 100,000-year cycle occurred, and both the glaciations and arid periods became more intense. Frequency and intensity of the periods seem to be nonlinearly dependent on orbital forcing against the background ocean topographic changes and decreasing atmospheric CO_2.

These changes had profound effects on human evolution in Africa, migration out of Africa, and subsequent life in northern latitudes. They are related to speciation events and adaptive radiation reflecting environmental fragmentation and change. In the Plio-Pleistocene, the paranthropines may have become dietary specialists while *Homo erectus* and its ancestors became generalists who increasingly relied on cognitive and cultural abilities to adapt to fluctuating environmental circumstances.

The increasingly severe periods of aridity would have resulted in population contraction and local extinctions. Periodic expansion of the Sahara would also have limited southward movement of Eurasian fauna into Africa during the glacial advances, while allowing northward movement of fauna out of Africa during interglacials.

There is also evidence of extreme climate fluctuations in last glacial cycle (c.70,000–11,000 years ago) and earlier. These took place on a human time scale, over a century or even a decade, and had a duration of between 1,000 and 3,000 years. These events would undoubtedly have had a profound effect on hominins in the northern, and possibly also southern, latitudes and must have been implicated in the disappearance of the Neanderthals from Europe, who were simultaneously facing competition from incoming modern humans.

The last glacial maximum occurred between 25,000 and 18,000 years ago, and deglaciation took about another 9,000 years to complete. The period between about 15,000 and 13,000 years ago was very warm, but arctic conditions returned between 13,000 and 11,5000 years ago (the Younger Dryas period in Europe). The effects of this cold period varied regionally in both the northern and southern latitudes. However, by the early Holocene (11,500–5,000 years ago), the Pleistocene megafauna had largely disappeared and there was a marked spread of forest and wetlands. The European forests contained animals that were much more dispersed and less visible than the late glacial fauna, but there were many more accessible plant resources. This provided an entirely different subsistence base for the Mesolithic inhabitants of Europe than was experienced by their Late Pleistocene antecedents.

At the beginning of the Holocene (c.11,500 years ago), the precession of the equinoxes lay at the opposite point on the cycle from the present day, and the Northern Hemisphere received nearly 8 percent more solar radiation during the summer months than at present. This strengthened the monsoons, increasing precipitation in northern latitudes. At the same time, the Sahara Desert was a savanna/lake habitat, and conditions different from those of today existed in many parts of the world. Following this Early Holocene period of thermal maximum, there was progressive change with significant effects on world climate and ecology, including the formation of the modern Sahara, Arabian, and Thar deserts. There is no doubt that the forces operating on climate in the distant past were operating more recently and continue to operate today. Resulting changes in climate, including those induced by human activity, will continue to be of importance to human evolution and cultural adaptation.

[*See also* Agriculture; Hominid Evolution: An Overview.]

BIBLIOGRAPHY

Bromage, T. G., and F. Schrenk, eds. *African Biogeography, Climate Change, and Human Evolution.* Oxford, 1999. A good series of articles focusing on climate, geology, ecology, biogeography, fossil faunas, and hominin evolution.

Cane, M. A., and P. Molnar. "Closing of the Indonesian Seaway As a Precursor to East African Aridification around 3–4 Million Years Ago." *Nature* 411 (2001): 157–162. Presents an interesting alternative idea for regional aridification in Africa and general global cooling in the Pliocene.

Cerling, T. E., J. M. Harris, B. J. MacFadden, M. G. Leakey, J. Quade, V. Eisenmann, and J. R. Ehleringer. "Global Vegetation Change through the Miocene/Pliocene Boundary." *Nature* 389 (1997): 153–158. Links the uplift of the Tibetan Plateau with global reduction of CO_2 and spread of C_4 vegetation in the Late Miocene.

deMenocal, P. B. "Plio-Pleistocene African Climate." *Science* 270 (1995): 53–59. A good discussion of the relationship between northern latitude glaciation and aridification in Africa in the Plio-Pleistocene.

Haug, G. H. and R. Tiedeman. "Effect of the Formation of the Isthmus of Panama on Atlantic Ocean Thermohaline Circulation." *Nature* 393 (1998): 673–676. Links the closure of the Isthmus of Panama with the onset of the northern latitude glaciations in the later Pliocene. This is a very clear discussion of the cause and effect relationships.

Lahr, M. M., and R. A. Foley. "Towards a Theory of Modern Human Origins: Geography, Demography and Diversity in Recent Human Evolution." *Yearbook of Physical Anthropology* 41 (1998): 137–176. An interesting if speculative discussion of the relationship between environmental change in the Middle and Late Pleistocene and human migration out of Africa.

Paillard, D. "The Timing of Pleistocene Glaciations from a Simple Multiple-State Climate Model." *Nature* 391 (1998): 378–381.

Potts, R. "Environmental Hypotheses of Hominin Evolution." *Yearbook of Physical Anthropology* 41 (1998): 93–136. A good review of environmental hypotheses of hominin evolution.

Raymo, M. E., and W. F. Ruddiman. "Tectonic Forcing of Late Cenozoic Climate." *Nature* 359 (1992): 117–122. A good discussion of the relationship between the uplift of the Tibetan plateau and global cooling in the Cenozoic.

Raymo, M. E., K. Ganley, S. Carter, D. W. Oppo, and J. McManus. "Millennial-Scale Climate Instability during the Early Pleistocene Epoch." *Nature* 392 (1998): 699–702. Documents millennial-scale instability in ice age climate and clearly explains its causes.

Reed, K. E. "Early Hominid Evolution and Ecological Change through the African Plio-Pleistocene." *Journal of Human Evolution* 32 (1997): 289–322. One of the best summaries of the inferred ecology of the hominin sites in African during the Plio-Pleistocene.

Roberts, N. *The Holocene: An Environmental History.* 2d ed. Oxford, 1998. An assessable discussion of climate change and human ecology in the Holocene.

Vrba, E. S., G. H. Denton, T. C. Partridge, and L. H. Burckle, eds. *Paleoclimate and Evolution with Emphasis on Human Origins.* New Haven and London, 1995. An extensive compilation of papers by some of the most important researchers in the field of climate and human evolution in the Neogene. This includes a recent discussion of Vrba's Turnover Pulse hypothesis in an article entitled "On the Connections Between Paleoclimate and Evolution" (pp. 24–45).

Wright, J. D., and K. G. Miller. "Control of North Atlantic Deep Water Circulation by the Greenland-Scotland Ridge." *Paleoceanography* 11 (1996): 157–170. Clearly discusses the relationship between North Atlantic deep water circulation and the major periods of climate deterioration in the Miocene and Pliocene.

— Leslie C. Aiello

NEOTENY

In general, the term *neoteny* has been used to describe the retention of larval, juvenile, or early developmental characteristics in organisms that have reached sexual maturity. Julius Kollmann (1885) coined the term to describe the retention of larval features in the Mexican axolotl (*Ambystoma mexicanum*) and other permanently gilled, nonmetamorphosing species of salamanders such as the mud puppy (*Necturus maculosus*) and the olm (*Proteus*). Because these species have lost the ability to metamorphose yet mature normally, they reproduce in what otherwise appears to be a larval body form. Kollmann's original definition describes developmental changes in species (relative to their ancestors) that involves a truncation of somatic development relative to the onset of sexual maturity. Neoteny is therefore somewhat similar to paedomorphosis, a general term described by Walter Garstang (1922) for the truncation of development relative to ancestors. However, neoteny relates somatic developmental timing to sexual maturity, whereas paedomorphosis relates the timing of any trait or portion of development to ontogenetic age. Thus, neoteny is a specific form of heterochrony relating development to the timing of sexual maturity and should not be used as a synonym for paedomorphosis. [*See* Paedomorphosis.]

Although the adult axolotl retains many larval traits and essentially looks completely larval, its reproductive system develops normally, and other aspects of the animal's physiology metamorphose normally. Thus, neoteny can involve many traits (in the case of salamanders, an entire suite of traits controlled by the hormone thyroxin that usually causes metamorphosis) or a single observable morphological or behavioral trait, such as the retention of ancestral juvenile behavior into adulthood.

Over the years the term *neoteny* has been given two more restrictive definitions, producing additional confusion in the literature. To Gavin de Beer, an influential comparative anatomist and embryologist active in the early part of the twentieth century, neoteny required the retardation of somatic development to be linked with acceleration of reproductive development (de Beer, 1930), something that is rarely observed except in insects. Four decades later, Stephen J. Gould (1977), in his review of the relationship between ontogeny and phylogeny, returned the definition of neoteny to Kollmann's original meaning. At about the same time, P. Alberch and colleagues (1979), in their classic work modeling the three simple developmental processes that could produce paedomorphosis (a slower rate, a later onset, or an earlier offset of development), applied the term to the specific case of paedomorphosis produced by a decrease in the developmental rate. This new definition of neoteny, though widely accepted, caused serious semantic confusion because it limited developmental change to only one of several possible ways that truncated development can be produced (rate deceleration) and it changed the developmental timing axis from that of Kollmann (sexual maturity) to ontogenetic age. The use of *neoteny* by Alberch et al. as a descriptor of paedomorphosis, by slowing the rate of development, has subsequently been changed to the term *deceleration* (Reilly et al., 1997).

Today, neoteny is most often used in its original interspecific meaning to describe somatic traits whose development is truncated in otherwise sexually mature individuals. However, additional confusion arises when neoteny is used to describe intraspecific patterns of developmental variation. Neoteny has been used in this intraspecific sense to describe variant individuals within populations that delay somatic or behavioral development beyond the time normally observed in the species. For example, most pond salamander species metamorphose in their first year, but some individuals within some populations may delay metamorphosis for a year or more but still attain sexual maturity at the normal time. Thus, breeding ponds will have some large sexually mature "larvae" breeding with their metamorphosed cohorts that have returned to the ponds. These are often termed neotenic larvae, but they can transform and will metamorphose if the pond dries up. Reilly et al. (1979) discussed the problem of confusing interspecific and intraspecific concepts and terminology and proposed the term *paedotype* for intraspecific variants that temporarily delay development; they also resurrected the term *paedogenesis* for the process that leads to this pattern of phenotypic plasticity within species.

The various definitions of neoteny and the major conceptual differences between them have led to considerable confusion in the literature, making it virtually impossible today to use the term unambiguously. Thus, given that neoteny describes a specific form of heterochrony, it should not be used as a synonym for paedomorphosis or to describe intraspecific patterns of heterochrony; both of these phenomena have formalized terminologies in place (see Reilly et al., 1997).

[*See also* Heterochrony.]

BIBLIOGRAPHY

Alberch, P., S. J. Gould, G. F. Oster, and D. B. Wake. "Size and Shape in Ontogeny and Phylogeny." *Paleobiology* 5 (1979): 296–317. Formal model of simple processes producing paedomorphosis. Incorrectly redefines neoteny.

de Beer, G. R. *Embryology and Evolution.* Oxford, 1930. Confuses the meaning of neoteny by linking truncated somatic development with accelerated reproductive development.

Garstang, W. "The Theory of Recapitulation: A Critical Restatement of the Biogenetic Law." *Zoological Journal of the Linnean Society* 35 (1922): 81–101. Presents the original definition of paedomorphosis.

Gould, S. J. *Ontogeny and Phylogeny.* Cambridge, Mass., 1977. Reviews the relationship of ontogeny and phylogeny. Returns de Beers' definition of neoteny to the original meaning of Kollmann.

Kollmann, J. Das Ueberwintern von europäischen Frosch- und Tritonlarven und die Umwandlung des mexikanischen Axolotl. *Verhandlungen der Naturforschenden Gesellschaft in Basel* 7 (1885): 387–398. Presents the original definition of neoteny.

Reilly, S. M., E. O. Wiley, and D. J. Meinhardt. "An Integrative Approach to Heterochrony: Distinguishing Intraspecific and Interspecific Phenomena." *Biological Journal of the Linnean Society.* 60 (1997): 119–143. Reviews semantic and conceptual bases and history of heterochronic terminology and presents new terms for intraspecific patterns.

— STEPHEN M. REILLY

NEUTRAL THEORY

Until the late 1960s, all differences between species or among members of the same species were understood in terms of their contribution to the fitness of the organism or species bearing them. This view, known as the synthetic theory or neo-Darwinism, claimed that even the most minute features of organisms contribute to their adaptation. It seemed reasonable that the synthetic theory could easily be extended to the molecular level. In other words, if observable phenotypic differences were due to adaptation, differences at the molecular level should be as well. As a result of new developments in biochemistry, it became possible to know the primary structure of proteins as well as to characterize the intraspecific variability of proteins and to check these points.

The first of these technical advances allowed the direct comparison of the amino acid sequences for the same protein among different species. For Motoo Kimura, the great Japanese population geneticist, the amount of sequence divergence seemed to be too high to be explained by positive adaptive selection (Kimura, 1968). Computing the number of differences among vertebrates for globins and cytochrome c, he estimated that, on the average, these proteins undergo 1 percent of amino acid divergence every 28 million years (or one change per nucleotide every 8.4 billion years). Reasoning that the mammalian genome contains about 4×10^9 nucleotide pairs, and assuming that the whole genome diverges at roughly the same rate, it follows that the genome will undergo a nucleotide substitution every two years (at that time it was thought that the whole genome was involved in coding functions, when in fact only about 2 percent encodes for proteins). If we take one year as a representative figure for the generation span in mammals, then mammalian species are undergoing one substitution every two generations. This rate of mutant substitution appeared to be incompatible with the figures obtained by Haldane many years before, based on what he called "the cost of natural selection" (Haldane, 1957). We will explain this key concept in the development of neutral theory. When an allele is substituted by another one through natural selection, the individuals carrying the inferior allele must die at a higher rate or have lower fertility than the individuals carrying the superior allele. If the population size is to be kept constant from generation to generation it is clear that the survivors must produce an excess of offspring. The cost of a gene substitution is the quantity of selective death summed over all generations required for the substitution to take place. The complication in explaining such a high rate of substitution by selection is the enormous amount of selective deaths per generation it would involve, which according to Kimura's view would be much beyond the reproductive capacity of any mammalian species.

Several possible explanations have been proposed in order to account for the incompatibility between substitution rates and selection. Such large numbers of selective deaths would occur only if selection acted independently on each locus. This is a highly unrealistic assumption. Note that when an individual with more than one less-adapted alleles dies, this very often removes several alleles at once, thus reducing the total cost. Other alternatives have been proposed, such as "truncation selection," which proposes that only individuals below a fitness threshold die (Maynard Smith, 1968).

The neutral theory was born with the aim to explain not only the high substitution rates, but also the considerable amount of intraspecific genetic variation observed in most species. During the 1960s, technical advances in the field of biochemistry made it possible to address how much protein variation was present in natural populations. The application of gel electrophoresis revealed a high incidence of protein polymorphisms. Lewontin and Hubby (1966) estimated that the proportion of polymorphic loci in *Drosophila* is about 30 percent, with an average heterozygosity of 12 percent. It should be taken into account that these are conservative estimations, given that gel electrophoresis detects only those variants that produce a change in net charge of proteins. Such a large amount of genetic variation raised a question concerning the evolutionary forces responsible for maintaining these polymorphisms. Overdominance, or heterozygote advantage, has been the chief mechanism invoked by neo-Darwinians to explain protein variability. The best-known example of protein polymorphism maintained by overdominance is the case of the sickle-cell anemia locus (hemoglobin locus). A single amino acid change (E → V at position 6) causes this anemia. Homozygotes VV are almost lethal, while homozygotes EE have normal hemoglobin. The heterozygote (E/V) is, however, more resistant to malaria than normal homozyogotes, and thus fitter than both homozygotes in geographic regions where malaria is endemic (Allison, 1955).

Because under this model heterozygous individuals are favored, this leads to stable polymorphisms, since the matings between heterozygotes always create the three types of genotypes. The main criticism raised against this model is that, for a large number of segregating loci, it generates an unsustainable segregational load. The genetic load (L) measures the extent to which the average fitness of the population is inferior to the best possible individual in that population. With overdominance it is impossible to avoid the genetic load, because populations consisting purely of the optimally adapted heterozygotes are not possible. Following the Mendelian laws, the population inevitably produces some less-adapted homozygotes in each generation, thus reducing the average fitness of that population.

Assuming that natural selection operates independently on each locus, the average fitness of the population for all loci considered simultaneously is found by multiplying the average fitness of each polymorphic locus. As mentioned before, about 30 percent of all loci are polymorphic in *Drosophila*, thus yielding a total of about 3,000 variable loci out of the 10,000 estimated loci. If selection is relatively soft in all loci, say $s = 0.01$, the average fitness for all loci considered together is $\bar{W} = (1 - 0.01/2)^{3000} = 2.94 \times 10^{-7}$, which means that an individual that is heterozygote for the 3,000 loci is about 3 million times fitter than an average individual from the population (that is, heterozygote for 1,500 loci). This result also means that the segregational load is enormous, and hence each reproductive adult must produce several million offspring for the population size to remain constant.

Among many alternative counter-arguments, we will mention only one. In the calculations mentioned above, the segregational load is so large because the average individual is compared with an individual that is heterozygotic for all of the 3,000 variable loci. Note, however, that this individual has an extremely low probability of being produced. For instance, a mating between two full heterozygotes will give a full heterozygote with a probability of 10^{-903}. Therefore, instead of contrasting the mean fitness against this possible yet extremely unlikely genotype, we should contrast it against the fittest genotype that is likely to arise. This is simply because selection will not take place between real and theoretical super-fit individuals (Gillespie, 1991). Similar criticisms were raised by Sewall Wright (1977) against Kimura's way of calculating selective deaths when the populations are undergoing substitutions (i.e., the Haldane cost of evolution).

Proposal of the Neutral Theory. The neutral theory asserts that the vast majority of substitutions at the molecular level are selectively neutral; that is, the substitutions are neither better nor worse, but instead equivalent (neutral) to the allele they replace in a given evolutionary lineage. Since variants are equivalent from the functional standpoint, then natural selection cannot be the driving force in the evolutionary process (Kimura, 1983).

Before going into details, we should distinguish between *mutations* and *substitutions*. A mutation is any change in the genetic material. When a mutation arises, it is present in only one individual in the population, the one that received the mutant gamete. For a mutation to become a substitution, it must replace the pre-existing variant (wild-type allele) in the successive generations and reach a frequency of 1 (i.e., all individuals must have the new mutation). Therefore, a substitution is a fixed difference between species or populations.

According to the neutral theory, the majority of these substitutions are between functionally equivalent alleles, so it is clear that natural selection cannot be the force driving the new allele to fixation, because, by definition, natural selection cannot favor either of two variants that are equivalent from the adaptive point of view. The neutral theory proposes instead that mutation and random genetic drift (random fluctuation in allele frequencies from generation to generation as a result of sampling deviations) are the main forces in molecular evolution.

Pattern of mutations and substitutions under the neutral theory. We can classify mutations according to whether they are beneficial, deleterious, or neutral. It is important to bear in mind that only neutral and advantageous mutations can contribute to evolution, since they are the only two kinds of mutations that can reach fixation in a population. In the case of advantageous mutations, it is clear that the individuals having this kind of mutation have a higher probability of survival (or higher fertility), and thus the most likely fate of these favorable alleles is to increase from generation to generation until they eventually reach fixation. This is called a substitution of a favorable variant driven by *positive* (or Darwinian) selection. If the mutation is neutral, it can reach fixation as well, but the most likely fate of a neutral allele is that it will disappear in the next generation or in the next few generations, and its destiny will depend on chance. Some neutral mutations that are lucky enough can attain considerable frequencies in the population, and some of them can even get fixed in the population. The last category is that of deleterious mutation. It is obvious that individuals having this type of mutation will have less chance of surviving or will be less fertile, and thus it is very likely that the deleterious mutation will disappear quickly from the population. The process by which a deleterious mutation is eliminated from the population because of its detrimental effect is called *negative* (or purifying) selection.

Both the neutral theory and the selectionist view

agree that the majority of new mutations that are introduced in a population are deleterious and thus are eliminated by negative selection. The two theories disagree, however, in the relative frequencies of advantageous and neutral mutations. The neutral theory claims that favorable mutations are far less frequent than neutral mutations. Given this extreme scarcity, they will represent only a minor fraction of all substitutions. Since among the two types of mutations that can reach fixation only the neutral ones exhibit reasonable frequencies, it follows that the majority of mutations that reach fixation (and hence become substitutions) are neutral mutations. The suggestion that favorable mutations are very rare is based on the assumption that the functions of products encoded by the genes were established a long time ago, and thus it is very difficult that a new variant can improve the function of something that is already working well. Favorable mutations can rise to relatively high frequencies either when the function is being established or when it is changing. The selectionist view, in contrast, asserts that favorable mutations, although not very frequent, are much more frequent than neutralists claim, and also, given that they have a higher probability of reaching fixation (since they are positively selected), that they represent a substantial proportion of all possible substitutions (see Figure 1).

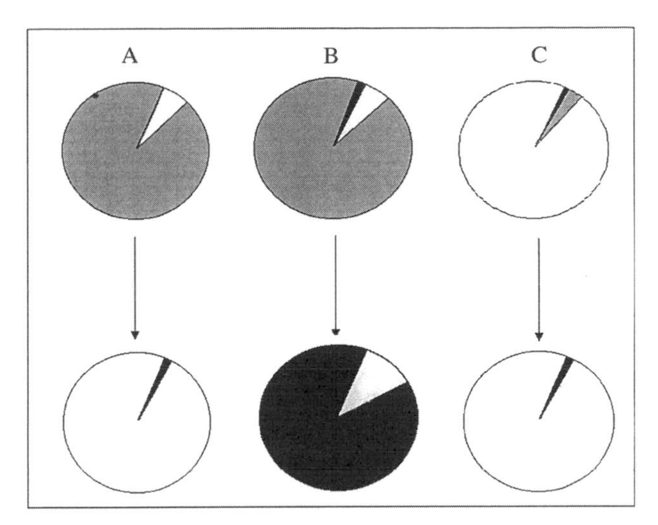

FIGURE 1. Pie Slice Diagrams Representing the Proportion of Different Types of Mutations (Top Panel) and Substitutions (Lower Panel) According to Three Evolutionary Theories. Where (A) corresponds to the neutral theory, (B) to selectionist point of view, and (C) panneutralism. Gray color represents deleterious mutations; white: neutral mutations; black: advantageous mutations. The proportions only represent approximations. Courtesy of author.

The Molecular Clock. Zuckerkandl and Pauling (1965) made one of the most outstanding discoveries in the field of molecular evolution: they found that the degree of amino acid divergence increases almost linearly with time. By plotting the amount of amino acid differences against time (which can obtained from the fossil record), it is possible to see that the divergence increases in a relatively uniform way as the time of divergence increases, which means that molecular evolution is rather constant throughout time. Different proteins differ widely in their rates of divergence, yet each one has an approximately constant rate. This constancy in the divergence rate has been called the *molecular clock*. Although many subsequent analyses have shown dramatic departures from constant rates, the rate of divergence for many genes is constant enough to require an explanation.

The rate of mutant substitutions (K) comprises two terms, the numbers of new mutations that are introduced at each generation in the population, and the proportion of these mutations that happen to attain fixation.

$$K = (\text{number of new mutations}) \times (\text{probability of fixation}).$$

A population consisting of N diploid individuals is formed from $2N$ gametes. If μ is the mutation rate per locus per generation, then $\mu 2N$ new mutations are introduced in the population in each generation. Since the probability of fixation for a neutral mutation is equal to its frequency in the population, and the initial frequency of a newly arisen mutation is $1/2N$, we can substitute in the expression given above to obtain:

$$K = \mu 2N \frac{1}{2N}$$

$= \mu$ for neutral mutations is $1/2N$. The probability of fixation of nonneutral mutations is $2s$, so that $K = \mu 2N$ $2s = 4\mu Ns$, for these mutations.

The remarkable property of the neutral substitution rate is that it depends only on the mutation rate and is independent of the population size. In a small population, the probability of fixation of each new mutation is relatively high (provided $1/2N$ is small), but few mutations are introduced in each generation. In a larger population, each new mutation has a lower probability of fixation but more new mutations are introduced in each generation. The two effects cancel each other, giving a substitution rate that relies on the mutation rate alone. In contrast, selection-driven substitutions depend upon three parameters: the population size (N), the selective advantage (s), and the rate at which favorable mutations are introduced in the population (μ). Neutralists claim that the molecular clock is more compatible with the neutral theory because it is highly improbable that the product of these three parameters remains constant for long periods of time in different evolutionary lineages. By contrast, it seems possible that the rate at which neutral mutations arise can stay relatively constant as long as the function remains the same.

The main criticism against the neutralist explanation

is to ask why the rate is constant per year and not per generation, as the theory predicts. Species with shorter generations are expected to evolve faster at the molecular level, yet they exhibit approximately the same rate. We shall discuss this contradiction again in the section devoted to the nearly neutral theory.

The analyses of substitution rates were not restricted to those cases where the time of divergence is known, because it is feasible to compare relative substitution rates between two species (since their separation from the common ancestor) using as a reference a third species that branched off earlier than the species we are testing (the third is called an outgroup). Applications of this approach, called the *relative rate test*, have shown many departures from constant rates, with some evolutionary lineages exhibiting evolutionary bursts and others dramatic slowdowns. Such variations in the substitution rates could be compatible with the neutral theory if they were related to differences in generation time, but the variability in rates is often much greater than that predicted from the generation-time effect. Gillespie (1991) argued that an episodic clock (characterized by periods of stasis alternating with periods of rapid evolution) is more compatible with a double stochastic process. One way of producing such a double stochastic clock is natural selection in a random varying environment.

Functional constraints, evolutionary rates, and the role of negative selection. The neutral theory predicts that the evolutionary rate depends exclusively on the rate at which neutral mutations are introduced in the population. In turn, the latter rate will depend on the relative proportions of neutral and deleterious mutations. For those genes in which most amino acid positions are important for function, it is to be expected that the majority of mutations will be detrimental, and thus they will be eliminated by negative selection. This is the explanation that the neutral theory gives to account for the existence of different molecular clocks in different genes, given that the rate depends only on the neutral mutation rate, which in turn would be determined by the absolute mutation rate per nucleotide and the fraction of mutations that are neutral. In other words, $K = m(1 - f)$, where m is the absolute mutation rate and f the proportion of mutations that are deleterious. Note that the same argument is valid when applied to the comparisons among different regions or amino acid positions inside a given protein. Those regions of the protein, or those amino acids, that are less important for function are expected to accept mutations without much compromising the function of the whole protein, and thus they will evolve faster. In contrast, selectionism predicts the opposite behavior: what is important for function is expected to be more affected by selection; hence, if selection is the governing force in molecular evolution, it should evolve faster.

This is a strong point in favor of the neutral theory. Many observations have supported the neutralist predictions. A well-known example is the fact that the three different positions of codons evolve at different rates that are inversely proportional to their coding significance.

Nearly neutral, slightly deleterious allele theory. So far we have used a simplistic categorization of alleles and their possible fates according to whether they are neutral or selected. It should, however, be taken into account that in small populations, slightly deleterious alleles can become fixed and slightly advantageous alleles can be lost. Consequently, the distinction between neutrality and selection is less clear when selection is weak. The nearly neutral theory (Ohta, 1973) proposes that the vast majority of mutations fall in two groups: those that are definitively deleterious and are eliminated by purifying selection, and those that are under very weak selection. The latter can be influenced by both random drift and natural selection. An allele behaves as neutral if the product $|S*N| < 1$, where N is the population size and S the selection coefficient. For a given value of S, the mutation behaves as neutral (i.e., is effectively neutral) in small populations and is affected by selection in large populations. Since this hypothesis assumes that most mutations with weak effects are slightly deleterious, the relative proportion of effectively neutral and deleterious mutations increases as the population size decreases. In small populations, more mutations will arise whose fate will be determined by genetic drift, whereas in large populations the proportion of mutations that escape from purifying selection decreases.

This theory has direct implications for the molecular clock. In effect, we expect a negative correlation between substitution rates and population sizes, because even if the absolute mutation rate per generation remains the same, the proportion of these mutations that are "seen" as deleterious by negative selection increases with the population size. In consequence, the proportion of "neutral" mutations decreases as the population size increases. On the other hand, it has been shown that large organisms tend to have long generation times and small populations, while the opposite is true for small organisms. Accordingly, one would expect higher substitution rates per generation in organisms with longer generation times and lower substitution rates per generation in organisms with shorter generation times. The two effects (high rates in long-generation organisms and vice versa) partially cancel each other, resulting in a molecular clock that would be constant per year and not per generation.

Polymorphisms. According to the selectionist point of view, substitutions and polymorphisms are two separate phenomena; for the neutralist, they are two sides of the same coin. A polymorphism is any neutral mutation on its way to fixation or loss. This is a very interesting feature of the neutral theory because it allows us to make several testable predictions.

The simplest prediction is that one would expect higher polymorphism levels as the number of neutral mutations increases. Neutral mutations can increase in response to two different factors, population size and degree of functional constraints. Polymorphisms increase with population size for two reasons: in a larger population, more new mutants are introduced in each generation in the population; and the time required to reach fixation increases as well. Polymorphism levels are expected to increase as functional constrains are relaxed, because in functionally less constrained regions more mutations are expected to fall in the neutral category. Remember that this is exactly the same reasoning that the neutral theory uses to explain differential substitution rates in regions with different degrees of constraints (i.e., less constrained regions evolve faster). Since substitution rates and polymorphism levels are both tightly dependent on functional constraints, it is easy to see that the neutral theory predicts a direct correlation between polymorphisms and substitution rates. This is the basis for several methods developed to test neutrality.

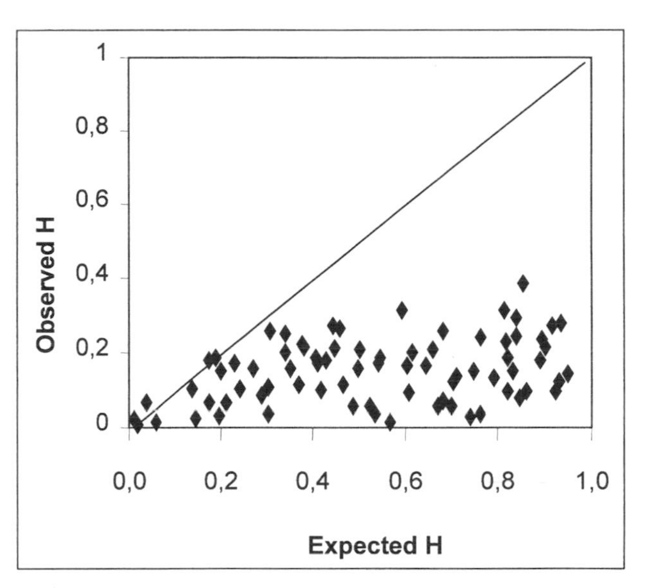

Figure 2. Predicted and Observed Heterozygosity in 77 Species.
After data from Nei and Graur, 1984.

Polymorphism and population sizes. The infinite allele model postulates that, whenever a new mutation occurs, it creates a new, nonpreexisting allele. Therefore, if two given alleles from one population are identical in state, it is because they originated from the same mutational event; that is, they are identical by descent (Kimura and Crow, 1964). This is very useful for addressing the problem of intraspecific variation because it allows us to consider the problem as a decrease or increase in the coefficient of inbreeding F (the probability that in an individual two alleles are identical by descent) in a random-mating population.

In a random-mating population, the increase in the amount of inbreeding (F) can be calculated from the amount of inbreeding in the previous generation and the population size. It should be taken into account that if new mutations are being introduced in the population, any individual that is to receive alleles that are identical by descent from each parent will effectively have alleles that are identical in state, as long as neither of the two gametes mutates. The probability that none of the gametes mutates is $(1 - \mu)^2$, where μ is the mutation rate. Incorporating the probability of nonmutation in the well-known recurrence formula of inbreeding, rearranging the expression and neglecting those terms that include μ^2 (since μ is very small), this gives

$$\hat{H} \approx \frac{4N_e\mu}{4N_e\mu + 1}$$

The equation presented above has been widely used for testing the neutral theory because it makes a very strong prediction concerning the correlation between population size and heterozygosity. Several studies on polymorphisms have tested this prediction. As an example, Ayala (1972) estimated the population size of the fruit fly *D. willistoni* to be much larger than 10^9, and the lower estimate of the mutation rate to be 10^{-7}, thus giving a value of $4N\mu$ (normally called θ) of 401, which in turn would imply that the expected heterozygosity would be 0.9975. However, the actual heterozygosity is only about 0.18, smaller by fivefold. Kimura has attempted to reconcile these negative results with the neutral theory by claiming that the estimates of population sizes, which were based on the actual population size in the stand populations, do not really correspond to the long-term effective population sizes. In the case of *D. willistoni*, he estimated the effective population size to be much smaller because of the reduction of tropical forest regions during the ice ages and also because extinction and recolonization are expected to be quite common in fruit fly species. Nei and Graur (1984) analyzed the relationship between population size and heterozygosity after collecting published data from 77 species. As shown in Figure 2, for most species the observed heterozygosity was much smaller than that expected according to the population size and mutation rate—for example,

$$H = \frac{\theta}{\theta + 1}$$

Although a positive correlation was observed between H and population size, this correlation was due mostly to the extreme values. If one removes from the analysis those species with very small or very large population sizes, the correlation disappears.

Even if the aforementioned results are not in agreement with the neutral predictions, they do not necessarily rule out the neutral theory, since the lower-than-expected heterozygosity could be due to bottleneck effects that reduce the effective population sizes. In summary, the evidence based on polymorphism data can be regarded as inconclusive. On the other hand, it is worth mentioning that these observations fit better with the nearly neutral theory. According to this view, as the population size increases, the proportion of mutations that fall in the "neutral category" decreases, and this can partially cancel the effect of increasing population size.

Polymorphism, substitution rates, and functional constraints. Another strong prediction made by the neutral theory is that the levels of polymorphism within species and levels of divergence between species should be correlated, because they are both affected in the same way by functional constraints. Several tests for neutrality have been developed based on this prediction. One of these tests (HKA), developed by Hudson, Kreitman, and Aguadé (1987), uses polymorphism data from two genes to estimate the parameter θ for each gene, which would be an indirect estimate of $1 - f$, the proportion of putatively neutral sites in each gene. This is used to contrast expected and observed interspecies distances for the same genes. The expected distance for one gene is estimated on the basis of the distance for the other gene and the relative proportions of estimated neutral mutations. McDonald and Kreitman (1991) developed another test that compares the number of synonymous and nonsynonymous polymorphisms with the number of synonymous and nonsynonymous fixed differences between species (substitutions). If polymorphism and substitutions are neutral, then one would expect the ratio of synonymous to nonsynonymous polymorphisms to be the same as the ratio of synonymous to nonsynonymous substitutions. The analysis that these authors conducted using the ADH gene deserves special mention. They found that the nonsynonymous changes were much more frequent in fixed (33 percent) than in polymorphic (4.5 percent) differences, and they interpreted these results as evidence of adaptive amino acid substitutions, because advantageous changes would not remain polymorphic for a long time but go quickly to fixation. Other studies that used these tests have given results that in many cases were compatible with natural selection, but in others with neutral predictions.

Codon Usage: Is Positive Selection So Infrequent at the DNA Level? Since nucleotide changes in the synonymous positions of genes do not alter their coding nature, and hence they are expected to be under few functional constraints, they were proposed to be a paramount example of neutral evolution. However, it is well established that synonymous codon choices can be affected by a diversity of factors, among which translational efficiency and accuracy are perhaps the most important. Efficiency may affect codon bias in highly expressed genes because preferred codons are recognized by the most abundant tRNAs (Ikemura, 1981; Bennetzen and Hall, 1982).

Pressures on codon usage do affect evolutionary rates. Indeed, the synonymous substitution rates are negatively correlated with the strength of codon bias. That is, in genes that are abundantly expressed (and thus more dependent on tRNA concentrations) the synonymous substitution rates are significantly lower than in genes having low expression levels, indicating that the evolutionary rate at silent positions can be lowered by translational efficiency.

Do these results contradict the neutral theory predictions? A first answer to this question is no, because they still support a basic assertion of the neutral theory that higher constraints imply lower substitution rates. But it is valid to ask whether this is enough to support the claim that synonymous positions evolve according to the neutral model. A more specific question should be: Are the majority of synonymous substitutions driven to fixation by genetic drift? Or are positively selected substitutions abundant enough to reject neutrality?

Before answering these questions, we should mention a feature of synonymous changes that makes them an excellent model for testing neutral and selectionist predictions. Suppose one were asked to predict whether a specific type of mutation is favorable, neutral, or advantageous. For the case of an amino-acid-altering mutation, it is not possible to give a general answer to this question. A particular type of change, for example Gly → Arg, could be strongly deleterious in one position of the gene but have no effect in another position. By contrast, the effect of synonymous changes can, in part, be predicted irrespective of the position within the gene where this change occurs, as long as we know which codons are the preferred ones in the species. A change from preferred to nonpreferred is expected to be deleterious in the majority of cases, while a change from nonpreferred to preferred is expected to be advantageous.

This, of course, does not mean that the fate of every synonymous mutation will be determined by selection. Genetic drift can play a determining role if the population size is not big enough, considering the very small selective disadvantage or advantage of synonymous mutations. We have seen that selection on synonymous positions of highly expressed genes can be strong enough to reduce the synonymous rate, something that occurs by decreasing the fixation probabilities of preferred to nonpreferred synonymous mutations (i.e., negative selection). This explanation is accepted by both neutralists and selectionists. Note that the same reasoning can be turned around. For example, if selection is strong enough, then nonpreferred-to-preferred synonymous mutations will have increased probabilities of fixation—that is, positive selection will occur.

Analyses of *Drosophila* genes confirm that advantageous synonymous mutations can attain fixation as a result of positive selection. Akashi (1995) compared the patterns of substitutions and polymorphisms at synonymous positions of five highly expressed genes from *D. simulans* and *D. melanogaster*. He found that the number of segregating polymorphisms that were preferred rather than nonpreferred was much higher than those in the opposite direction. This could be expected because most synonymous sites in these genes are occupied by preferred codons. Nevertheless, in *D. simulans* the number of nonpreferred-to-preferred fixed differences (substitutions) was not significantly different from that of preferred-to-nonpreferred substitutions. The two types of substitutions are expected to appear in equal amounts when codon biases are at equilibrium; otherwise, the genes would increase or decrease their codon biases. It is clear that this pattern of synonymous substitutions cannot be explained by claiming that they correspond to those mutations whose selective differences are so small that they were overwhelmed by genetic drift. If this were the case, one would expect to find the same proportions of preferred:nonpreferred in polymorphisms and in substitutions.

The picture that emerges from these analyses appears to be consistent with the nearly neutral theory; that is, selection and random genetic drift can both participate in molding the pattern of synonymous polymorphism and substitutions, with drift predominating when the populations are small. However, a major difference is that weakly selected mutations are not predominantly slightly deleterious, as the theory claims. Slightly favorable mutations make up a substantial proportion of newly arisen mutations, and among synonymous substitutions in highly expressed genes, they represent half of the total when codon frequencies are at equilibrium.

Conclusions. Since Darwin proposed his theory of natural selection to explain evolution, most evolutionary theories have always been a matter of debate and controversy. The neutral theory was not an exception. The success of the theory was based on its simplicity, which provided a useful conceptual basis for thinking about molecular evolution. Because the neutral theory is quantitative, it is able to make testable predictions. This is the most important contribution of this theory to evolutionary studies. Success and correctness are not synonymous, though, so whether these predictions agree with real data is the very essence of the problem. Nowadays there is agreement that both random drift and selection are important in evolution; there is disagreement, however, on the relative contribution of each force. The debate is still going, and whatever the solution will be, it is clear that is not going to be strictly black or white.

[*See also* Codon Usage Bias; Molecular Clock.]

BIBLIOGRAPHY

Akashi, H. "Inferring Weak Selection from Patterns of Polymorphism and Divergence at 'Silent' Sites in *Drosophila* DNA." *Genetics* 139 (1995): 1067–1076.

Allison, A. C. "Aspect of Polymorphisms in Man." *Cold Spring Harbor Symposia in Quantitative Bioliology* 20 (1955): 239–255.

Ayala, F. "Darwinian versus Non-Darwinian Evolution in Natural Populations of *Drosophila*." In *Proceedings of the 6th Berkeley Symposium on Mathematics and Statistical Problems*, edited by L. M. Le Can, and E. L. Scott, pp. 211–236, Berkeley, 1972.

Bennetzen, J. L., and B. D. Hall. "Codon Selection in Yeast." *Journal of Biological Chemistry* 257 (1982): 3026–3031.

Gillespie, J. H. *The Causes of Molecular Evolution*. Oxford, 1991.

Haldane, J. B. S. "The Cost of Natural Selection." *Journal of Genetics* 55 (1957): 511–524.

Hudson, R. R., M. Kreitman, M. Aguade. "A Test of Neutral Molecular Evolution Based on Nucleotide Data." *Genetics* 116 (1987): 153–159.

Ikemura, T. "Correlation Between the Abundance of *Escherichia coli* Transfer RNAs and the Occurrence of the Respective Codons in Proteins Genes." *Journal of Molecular Biology* 146 (1981): 1–21.

Kimura, M. "Evolutionary Rate at the Molecular Level." *Nature* 217 (1968): 624–626.

Kimura, M. *The Neutral Theory of Molecular Evolution*. Cambridge, 1983.

Kimura, M., and J. C. Crow. "The Number of Alleles That Can Be Maintained in a Finite Population." *Genetics* 49 (1964): 725–738.

Lewontin, R. C., and J. L. Hubby. "A Molecular Approach to the Study of Genic Heterozygosity in Natural Populations. Amount of Variation and Degree of Heterozygosity in Natural Populations of *Drosophila pseudoobscura*." *Genetics* 54 (1966): 595–609.

Maynard Smith, J. "'Haldane's dilemma' and the Rate of Evolution." *Nature* 219 (1968): 1114–1116.

McDonald J. H., and M. Kreitman. "Adaptive Protein Evolution at the Adh Locus in *Drosophila*." *Nature* 351 (1991): 652–654.

Nei, M., and D. Graur. "Extent of Protein Polymorphism and the Neutral Mutation Theory." *Evolutionary Biology* 17 (1984): 73–118.

Ohta, T. "Slightly Deleterious Mutant Substitutions in Evolution." *Nature* 246 (1973): 96–98.

Wright, S. *Evolution and the Genetics of Populations*, vol. 3. *Experimental Results and Evolutionary Deductions*. Chicago, 1977.

Zuckerkandl, E., and L. Pauling. "Evolutionary Divergence and Convergence in Proteins." In *Evolving Genes and Proteins*, edited by V. Bryson and J. Hogel, pp. 97–166. Academic Press, New York, 1965.

— FERNANDO ALVAREZ-VALIN

NICHE CONSTRUCTION

Adaptation typically is regarded as a process by which natural selection molds organisms to fit a preestablished environmental "template." Although the environmental template may be dynamic, the changes that organisms themselves bring about are rarely considered in evolutionary analyses. Yet, to varying degrees, organisms choose their own habitats, choose and consume resources, generate detritus, construct important components of their own environments (e.g., nests, holes, burrows, paths, webs, pupal cases, dams, and chemical environments), destroy other components, and construct environments for their offspring. Thus, organisms not only adapt to environments but in part also construct them. They may do so across a huge range of temporal and spatial scales, stretching from a hole bored in a tree by an insect, to the contribution of photosynthesizing cyanobacteria to the earth's atmosphere. Hence the adaptive fit between organism and environment can come about through two processes: either the organism adapts to suit the environment or the environment is changed to suit the organism. The latter route constitutes niche construction.

Niche construction occurs when an organism modifies the functional relationship between itself and its environment by actively changing one or more of the factors in its environment, either by physically perturbing these factors or by relocating to expose itself to different factors. Niche construction is not the exclusive prerogative of large populations, keystone species, or clever animals. All living organisms take in materials for growth and maintenance and excrete waste products; thus, all organisms must change their local environments to some degree. The evolutionary significance of niche construction stems from the fact that, through their choices, movements, and activities, organisms in part determine the selection pressures to which they and their descendants are exposed.

The niche construction of both past and present generations may influence a population's selective environment. Spiders' webs and caddis fly larval houses modify the selective environments of the constructors themselves. In contrast, birds' nests and female insects' oviposition site choices also modify the environment of the constructors' descendants. The latter cases are examples of ecological inheritance. Ecological inheritance refers to any case in which an organism experiences a modified functional relationship between itself and its environment as a consequence of the niche-constructing activities of its ancestors.

There are numerous examples of organisms choosing or changing their habitats, or of constructing artifacts, leading to an evolutionary response. For instance, many spiders construct webs, which have led to the subsequent evolution of camouflage, defense, and communication behavior on the web. Ants, bees, wasps, and termites construct nests that are themselves the source of selection for many nest regulatory, maintenance, and defense behavior patterns. Many mammals (including badgers, gophers, ground squirrels, hedgehogs, marmots, moles, mole-rats, opossum, prairie dogs, rabbits, and rats) dig burrow systems, some with underground passages, interconnected chambers, and multiple entrances. However, most cases of niche construction do not involve the building of artifacts, but merely the selection or modification of habitats. For example, as a result of the accumulated effects of past generations of earthworm niche construction, present generations of earthworms inhabit radically altered environments where they are exposed to modified selection pressures.

The breadth and scale of niche construction surprise many people. For example, there are more than 34,000 species of spider that construct silken egg sacs, burrows, or webs; more than 9,000 species of birds, the vast majority of which construct nests, and probably as many fish that do the same; and 9,500 known species of ants living in social colonies, with almost all building some kind of nest.

Whereas niche construction is a general process, few species have modified their selective environment to the same extent as humans. It is well recognized that phenotypic plasticity can play an instrumental role in the evolutionary process and that the human capacity for niche construction has been further amplified by culture. Human innovation and technology have had an enormous impact on the environment: they have made many new resources available via both agriculture and industry; they have influenced human population size and structure via hygiene, medicine, and birth control; and they have resulted in the degradation of large areas of the environment. These are all potential sources of modified natural selection pressures. Cultural processes that precipitate niche construction might be expected to have played a critical role in human evolution for many thousands, perhaps millions, of years.

Organisms also alter other species' ecology by modifying sources of natural selection in their environments. In ecosystem ecology, this process is known as ecosystem engineering, and it is known to modulate and partly control the flow of energy and matter through ecosystems. Such modifications can have profound effects on the distribution and abundance of organisms, the control of energy and material flows, residence and return times, and ecosystem resilience.

The evolutionary significance of niche construction hangs primarily on the feedback that it generates. Theoretical analyses have established the evolutionary consequences of niche construction. For example, traits whose fitness depends on sources of selection that are alterable by niche construction coevolve with niche-constructing traits. This results in very different evolutionary dynamics for both traits from that which would occur if each had evolved in isolation. Selection resulting from niche construction may drive populations along alternative evolutionary trajectories, may initiate new evolutionary episodes in an unchanging external environment, and may influence the amount of genetic variation in a population, by affecting the stability of polymorphic equilibria.

Moreover, because of the multigenerational properties of ecological inheritance, niche construction can generate unusual evolutionary dynamics, including inertia effects (where unusually strong selection is required to move a population away from an equilibrium), momentum effects (where populations continue to evolve in a particular direction even if selection changes or reverses), opposite responses to selection, and sudden catastrophic responses to selection. This body of theory supports the view that the complementary nature of organisms and their environment is not just the result of the single process of adaptation by natural selection. Rather, evolution is a two-way process, in which niche construction and natural selection operate in parallel, but also interact.

BIBLIOGRAPHY

Jones, C. G., J. H. Lawton, and M. Shachak. "Organisms as Ecosystem Engineers." *Oikos* 69 (1994): 373–386.

Laland, K. N., F. J. Odling-Smee, and M. W. Feldman. "Evolutionary Consequences of Niche Construction and Their Implications for Ecology." *Proceedings of the National Academy of Sciences USA* 96 (1999): 10242–10247.

Lewontin, R. C. "Gene, Organism, and Environment." In *Evolution from Molecules to Men*, edited by D. S. Bendall. Cambridge, 1983.

Odling-Smee, F. J., K. N. Laland, and M. W. Feldman. "Niche Construction." *The American Naturalist* 147 (4, 1996): 641–648.

KEVIN N. LALAND

NOMENCLATURE

Biological nomenclature concerns the allocation of scientific names to taxa (groups of organisms; singular, taxon). Names of taxa are of two types: vernacular or common names and scientific names. Because the primary role of a name is as a means of communication, so that both parties know what taxon is being referred to, names ideally should be stable over time, universal in their application, and unambiguous in their meaning. Common names typically fail to meet these criteria. Many species have more than one common name (e.g., the cougar has more than 20 other common names, including mountain lion and puma), and different species are sometimes referred to by the same name ("lily" is used for various other plants). Formal biological nomenclature attempts to avoid the communication problems associated with multiple, often regionally based common names through the use of one scientific name for each species and more inclusive taxon, regardless of the language of communication. Scientific names are based on Latin or Greek, the scholarly languages of the eighteenth century. This is a carryover from the work of the Swedish botanist Carolus Linnaeus. Linnaeus employed a system of Latinized "binomial" names, assigning each organism to a genus and species (e.g., *Homo sapiens*). Higher-level rankings were also assigned.

Linnaean Nomenclature. The formation and use of scientific names are regulated through internationally recognized codes of nomenclature that consist of sets of rules and recommendations. Rules are mandatory, whereas recommendations are guidelines that provide preferred options. The current codes have no effect on the circumscription of taxa; instead, once the taxonomist has delimited the taxon, the codes provide an objective means of determining its correct name. The operation of many of the rules and recommendations in the current codes depends on the rank in the Linnaean hierarchy to which a taxon name is assigned. Thus, the current nomenclatural system is tied directly to the Linnaean hierarchy.

There are several nomenclatural codes in current use. The International Code of Zoological Nomenclature concerns the names of organisms considered to be animals, the International Code of Botanical Nomenclature concerns plants and fungi, and the International Code of Nomenclature of Bacteria concerns bacteria, including actinomycetes. There are also codes for cultivated plants and viruses. The nomenclature of different groups of organisms has different starting dates, depending on the date of publication of the systematic work that laid the foundation for the modern nomenclature of that group. For example, the nomenclature of most plants dates from 1753, whereas that of animals dates from 1758. Although the overall goals and objectives of the codes are the same, the codes sometimes differ in detailed format and approach.

Scientific names have traditionally been assigned to ranks in the Linnaean hierarchy. Scientific names of taxa at and above the rank of genus are uninomial and are capitalized (e.g., Felidae—the cat family). Names of species are binomial and consist of the genus to which the species is referred and an uncapitalized species epithet, or trivial name (e.g., *Felis catus*—the domestic cat). Specific and generic names are set in italics. Whereas each species should have a unique binomial, the trivial name may be part of different species names (e.g., *Lontra canadensis*, *Castor canadensis*, *Ovis canadensis*).

Standardized endings are used for names of taxa assigned to some ranks; thus, the rank to which the name is assigned determines its spelling. For example, names of plant families end in *-aceae*, whereas names of animal families end in *-idae*. The family name is usually formed by combining the designated ending with the root of the type genus. Thus, the name of the cat family is Felidae, and the type genus is *Felis*. This association between the spelling of taxon names and their assigned rank effectively determines how these names are defined. In essence, the name Felidae is defined as referring to a taxon at the family level that contains the genus *Felis*. Since the definition of names is the foundation of nomenclature, in the Linnaean system assignment of the taxon to a Linnaean category is often the most fundamental aspect of the meaning of its name.

Taxon names may also have character- or content-based meanings. The codes impose character-based connotations for names through their requirement that names be given character-based descriptions, which are then seen as defining the name. The meaning of names at ranks that are not legislated by a code (zoological names at the rank of order and above) are often strongly character-based. The name Aves is often taken to mean "vertebrates with feathers." However, the reluctance to consider various recently discovered Cretaceous dinosaur fossils with featherlike structures as members of Aves suggests that, for many biologists, the meaning of this taxon name is also based on content (the lineage bracketed phylogenetically by *Archaeopteryx* and modern birds).

A principal goal of the nomenclatural codes is that a taxon have one, and only one, name by which it is referred. This goal is achieved using three operational criteria—publication, typification, and priority. Publication of names ensures the adequate documentation and dating of nomenclatural events. Typification and priority provide an objective basis for determining the correct or valid name of a particular taxon.

Typification is the process whereby the name of a taxon at or below a particular taxonomic category (the level depends on the code) is permanently attached to an element (the name-bearing type). Elements are of two kinds, organisms and taxa. Once a taxonomist has circumscribed a taxon, the name-bearing types that are included in that taxon are used to determine the correct or valid name for that taxon. In other words, a name can apply only to a taxon that contains its name-bearing type.

Under the zoological code, the type of a species is a specimen, the type of a genus is a species, and the type of a family is a genus. In contrast, under the botanical code, the type of the name of any taxon at or below the family level is a specimen, or an illustration of a specimen. Thus, when a taxonomist circumscribes a species, the name of that species must be one whose type specimen is included in that species; similarly, when a zoologist circumscribes a family, its name will be one whose type genus is included in that family. If the taxon does not contain a name-bearing type, a new name is established and a name-bearing type for that name must be designated.

If a taxon contains more than one name-bearing type, the correct or valid name is determined based on the principle of priority. The principle of priority states that the appropriate name is normally the one with the oldest date of publication. Priority applies at a particular rank (botanical code) or group of ranks (e.g., family-group names in the zoological code); therefore, decisions as to the appropriate name depend on the Linnaean rank to which the taxon is assigned. For example, if an animal taxon at the family level has competing names (i.e., the taxon includes more than one genus that is a name-bearing type for a family-level name), the oldest name is usually used. Two or more names that apply to the same taxon are called synonyms. Application of priority is limited by the starting dates of the codes, and by the conservation or rejection of names by nomenclatural committees if the strict use of priority would lead to confusion (e.g., rediscovery of an older name that has not been used for a considerable period of time).

The use of different codes of nomenclature for different groups of organisms is problematic. The fact that the zoological and botanical codes differ in their terminology, the conventions used in citing names, the spelling of names at the same rank, and how a correct name is chosen using priority is confusing. A plant and an animal can have the same scientific name because they fall under the jurisdiction of different codes. More serious problems occur when there is disagreement as to which code should apply to a particular organism. For example, certain organisms, particularly unicellular protists, are considered animals by some workers and plants by others. As a result, a different name might be ascribed to a species depending on which code is used. Although these problems would be avoided through the use of a single code for all organisms, the marked differences between the botanical and zoological codes are seen by many as an insurmountable obstacle to a unified code for names published in the past. Nonetheless, a draft BioCode has been proposed that could provide a unified, consistent nomenclatural system for names published in the future.

Despite the goals of the established codes to provide scientific names that are stable over time, universal in usage, and unambiguous in meaning, Linnaean nomenclature often falls well short of these goals. The shortcomings relate to the pre-evolutionary conceptual origin of the Linnaean hierarchy and are especially apparent given a strictly phylogenetic perspective in which named taxa are restricted to species and clades (monophyletic groups). Because species names include the name of the genus to which the species is referred, the name of the species may change if its inferred relationships change (the species is moved from one genus to another). This change in name occurs because the Linnaean binomial serves two functions: it provides not only the name of

TABLE 1. Possible Phylogenetic Definitions of Aves

Definition	*Content of Aves*
The least inclusive clade that includes *Archaeopteryx lithographica* and *Passer domesticus* (House Sparrow) (node-based definition).	Aves would include the Late Jurassic fossil *Archaeopteryx*, which is considered by most as the earliest bird, modern birds, and all other fossils that share the same most recent common ancestry.
The most inclusive clade that includes *Passer domesticus* but not *Deinonychus antirrhopus* (stem-based definition).	Aves would include all taxa that share a more recent common ancestry to the House Sparrow than to the theropod dinosaur *Deinonychus*. Using this definition some taxa traditionally considered dinosaurs would be included in Aves if they are phylogenetically closer to the House Sparrow than to *Deinonychus*.
The clade stemming from the first species bearing feathers that are synapomorphic with those in *Passer domesticus* (apomorphy-based definition).	Ties the origin of Aves to the origin of feathers that are homologous to those in modern birds. Aves would include various feathered dinosaurs if these structures are considered homologous with the feathers of the House Sparrow.

NOTE: Inferences regarding the content of Aves are based on the assumption that birds originated within the theropod dinosaur clade.

the species, but also information regarding relationships. Because many names have rank-specific endings, if a taxonomist changes the rank associated with a particular taxon, its name will also change. In other words, if taxonomists differ in the categorical assignment of clades, different authors use the same names for different clades and different names for the same clade. Similarly, if a taxonomist chooses to recognize a new taxon at the family level, its insertion into the Linnaean hierarchy may cause a cascade of name changes for other taxa, which have to move to a higher or lower rank in the hierarchy. This last aspect of traditional nomenclature discourages taxonomists from naming new clades at a time when they are being discovered at an unprecedented rate.

Phylogenetic Nomenclature. The inadequacies of the Linnaean rank-based codes of nomenclature have prompted some systematists to explore an alternative approach to biological nomenclature in which the names of taxa are defined not by reference to ranked categories in the Linnaean hierarchy, but by explicit reference to common descent. A draft code of phylogenetic nomenclature (the PhyloCode) was posted on the Internet in 2000. The draft PhyloCode is a proposed alternative set of rules and recommendations for the formation and use of the names of clades; rules for species names will be added in the future. The PhyloCode will be implemented within the next few years after adequate discussion within the systematic community.

Phylogenetic nomenclature provides explicit, stable phylogenetic meanings to the names of clades and species and thereby extends "tree thinking" to nomenclature. As a result, nomenclature would become consistent with the increasing role of tree thinking in taxonomy, whereby recognized taxa are restricted to historical entities, species, and clades. Unlike the traditional Linnaean system, the rules in the PhyloCode are explicitly designed for naming clades.

In phylogenetic nomenclature, names are associated explicitly with clades through the use of phylogenetic definitions. There are three general types of phylogenetic definitions of taxon names. Names of taxa can be defined by referring to (1) the most recent common ancestor of two or more species or specimens (node-based definition), (2) those organisms more closely related to one species or specimen than to another species or specimen (stem-based definition), or (3) the origin of a particular character that is synapomorphic with one in a particular species or specimen (apomorphy-based definition; see Table 1 and Figure 1 for examples). Through these definitions, the association between the name and the clade to which it refers becomes the fundamental aspect of the meaning of the name. Because the definition makes explicit reference to phylogenetic pattern, the content, synapomorphies, and other characteristics of the taxon in question are inferred within the context of a phylogenetic hypothesis (Figure 2).

The intent is not necessarily to replace taxon names currently in use, but instead to define them using phylogenetic definitions rather than by reference to Linnaean ranks. The procedures in the PhyloCode attempt to minimize changes to the current usage of taxon names when they are defined phylogenetically. Linnaean binomial species names are inappropriate for phylogenetic nomenclature because the first part of the name necessarily refers to a genus, a rank in the Linnaean hi-

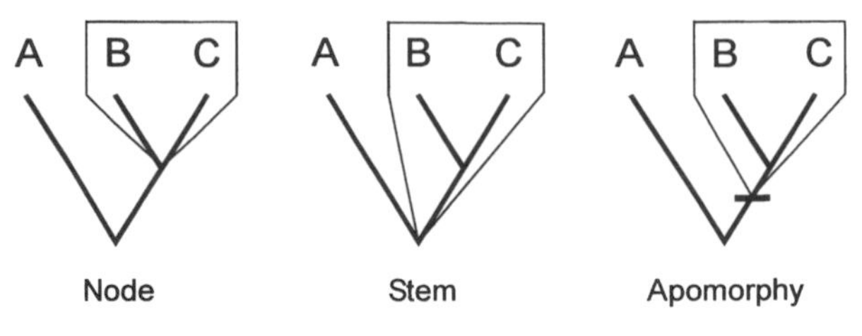

FIGURE 1. The Three Principal Types of Phylogenetic Definitions.
Possible definitions for the node-based clade are "the clade consisting of the most recent common ancestor of B and C and all of its descendants" or "the least inclusive clade that includes B and C." Possible definitions for the stem-based clade are "the clade consisting of all organisms more closely related to B than to A" or "the most inclusive clade that includes B but not A." A possible definition of the apomorphy-based clade is "the clade stemming from the first organism to possess character X synapomorphic with that in B." The bar marks the origin of character X. The species names and characters referred to in phylogenetic definitions are called specifiers. Courtesy of Harold N. Bryant.

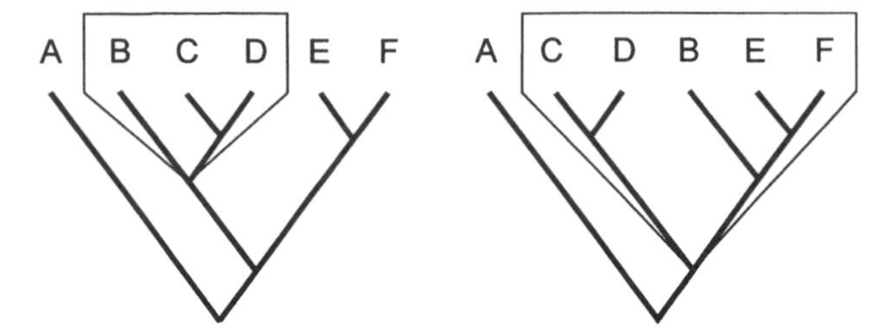

FIGURE 2. The Effect of Changes in Phylogenetic Relationships on the Use of Phylogenetic Definitions.
Given a taxon name with the node-based phylogenetic definition "the least inclusive clade that includes B and D," this name refers to the clade consisting of B, C, and D on the left cladogram, but refers to the clade that also includes E and F on the right cladogram. If the use of this name for the larger clade is considered undesirable, the definition could be worded "the least inclusive clade that includes B and D, but does not include E and F." Given this definition, the name would not apply to any clade on the right cladogram. Courtesy of Harold N. Bryant.

erarchy, whereas phylogenetic nomenclature is not based on ranks. The exact form of species names in phylogenetic nomenclature remains undetermined; it may be either uninomial or binomial. In either instance, unlike Linnaean nomenclature in which species names change with the reassignment of a species to a different genus, species names in phylogenetic nomenclature will remain stable.

In phylogenetic nomenclature, names of clades do not have name-bearing types, at least as this term is used in the traditional codes. Instead, names are associated with clades through the reference in their phylogenetic definitions to specifiers; specifiers can be either specimens, species names, or apomorphies, depending on the type of definition that is used (Figure 1). Specifiers are similar to types in that they provide a reference for the application of the name.

In phylogenetic nomenclature, synonyms are names that, given a particular phylogenetic hypothesis, have definitions that refer to the same clade. As with the current codes, priority (precedence of date of publication of an acceptable phylogenetic definition) will normally determine the correct name for that clade. Unlike the traditional system, disagreements regarding synonymy reflect only objective disagreements regarding phylogeny, rather than subjective disagreements regarding assignments to rank. Priority would also apply to the phylogenetic definitions of names. Because all names will be registered, the frequency of accidental homonyms (use of names with the same spelling for different taxa) will be greatly reduced.

Its proponents argue that phylogenetic nomenclature provides increased stability and universality in the names given to taxa. For example, it eliminates changes to names resulting from changes in rank (e.g., change in assignment of a taxon from family to subfamily level) and changes in position (e.g., change to a species name because of its reassignment to a different genus). New clades can be named without affecting the nomenclature of existing names and clades. In contrast, with the traditional nomenclatural system names have to be assigned to ranks; therefore, the assignment of a name to a particular rank may require a cascade of changes in the rank of other names. Because the spelling of names is linked to their rank, these changes in rank can result in changes to the spelling of the names of several taxa. Thus, using phylogenetic nomenclature would make it easier to name clades one at a time as they are discovered. Thus, contrary to frequent claims, the use of phylogenetic nomenclature does not necessitate a complete knowledge of phylogeny.

Phylogenetic nomenclature allows, but does not require, the abolition of ranks. If ranks are retained, however, they have no bearing on the names that are assigned to taxa. The elimination of Linnaean categories need not result in changes to the spelling of taxon names, thus maintaining access to the taxonomic literature. In contrast, traditional nomenclature sometimes impedes access to the literature because the names of taxa change with changes in their rank. These changes can occur with increases in knowledge about phylogenetic relationships, but also simply because of subjective decisions regarding lumping and splitting of taxa or the assignments of taxa to ranks.

Although phylogenetic nomenclature is more consistent with a phylogenetic approach to the circumscription of taxa than is traditional Linnaean nomenclature, many proponents of phylogenetic classification are resistant to this new approach. Most criticisms of phylogenetic nomenclature are related to the impression that phylogenetic nomenclature will result in greater instability in the content of the taxa to which names apply and a loss of taxonomic freedom in the delineation of taxa. Marked changes in the content of named taxa can occur in phylogenetic nomenclature because of the direct link between the meaning of taxon names and phylogenetic pattern. As phylogenetic hypotheses change, the membership of taxa with particular names may also change (Figure 2). However, if named taxa are restricted to species and clades, changes to the content of taxa must occur with revisions to phylogeny, using either Linnaean or phylogenetic nomenclature. This instability can be confusing, but it is desirable if the content of taxa is to reflect revisions in our understanding of phylogenetic relationships. Qualifying clauses can be added to phylogenetic definitions to reduce changes in content by making the name inapplicable if the membership associated with the name is altered in specific ways (Figure 2). It is too early to tell whether phylogenetic nomenclature would lead to increased overall instability in the content of taxa, as compared to the situation under the traditional system.

Critics of phylogenetic nomenclature are also concerned that taxonomists will lose their ability to modify the content of taxa with changes to the understanding of phylogeny. Once a taxon name is defined, the content associated with that name is determined by the combination of the phylogenetic definition and a phylogenetic hypothesis, not by the taxonomist. Because the contents of taxa cannot be freely altered, phylogenetic nomenclature is perceived as limiting taxonomic freedom. However, proponents of phylogenetic nomenclature argue that this type of taxonomic freedom is inappropriate in phylogenetic taxonomy, in which taxa are considered historical entities discovered through phylogenetic analysis, rather than human constructs.

[*See also* Classification.]

BIBLIOGRAPHY

Cantino, P. D., H. N. Bryant, K. de Queiroz, M. J. Donoghue, T. Eriksson, D. M. Hillis, and M. S. Y. Lee. "Species Names in Phylogenetic Nomenclature." *Systematic Biology* 48 (1999): 790–807. A discussion of the formation of species names in phylogenetic nomenclature.

de Queiroz, K., and J. Gauthier. "Phylogenetic Taxonomy." *Annual Review of Ecology and Systematics* 23 (1992): 449–480. An early and seminal work on phylogenetic nomenclature.

de Queiroz, K., and J. Gauthier. "Toward a Phylogenetic System of Biological Nomenclature." *Trends in Ecology and Evolution* 9 (1994): 27–31. Brief overview of the fundamental aspects of phylogenetic nomenclature.

Greuter, W., F. R. Barrie, H. M. Burdet, W. G. Chaloner, V. Demoulin, D. L. Hawksworth, P. M. Jorgensen, D. H. Nicolson, P. C. Silva, P. Trehane, and J. McNeill, eds. *International Code of Botanical Nomenclature (Tokyo Code)*. Regnum Vegetabile 131, Koeltz Scientific Books, Konigstein, Germany, 1994.

Greuter, W., D. L. Hawksworth, J. McNeill, M. A. Mayo, A. Minelli, P. H. A. Sneath, B. J. Tindall, P. Trehane, and P. Tubbs, eds. "Draft BioCode: The Prospective International Rules for the Scientific Names of Organisms." *Taxon* 45 (1996): 349–372.

International Commission on Zoological Nomenclature 1999. International Code of Zoological Nomenclature, 4th ed. International Trust for Zoological Nomenclature.

Jeffrey, C. *Biological Nomenclature*, 3d ed. Edward Arnold, London, 1989. A concise (89 pp.) introduction to the theory and practice of biological nomenclature as outlined in the Linnaean codes; includes a glossary.

Lapage, S. P., P. H. A. Sneath, E. F. Lessel, V. B. D. Skerman, H. P. R. Seeliger, and W. A. Clark, eds. *International Code of Nomenclature of Bacteria, 1990 Revision*. American Society for Microbiology, Washington, D.C., 1992.

Linnaeus, C. *Species plantarum*. 1753. The foundation of Linnaean nomenclature of plants.

Linnaeus, C. *Systema naturae per regna tria naturae, secundum classes, ordines, genera, species cum characteribus, differentiis, synoymis, locis*. 10th ed. 1758. The foundation of Linnaean nomenclature of animals.

Trehane, P., C. D. Brickell, B. R. Baum, W. L. A. Hetterscheid, A. C. Leslie, J. McNeill, S. A. Spongberg, and F. Vrugtman, eds. *International Code of Nomenclature for Cultivated Plants*. Regnum Vegetabile 133, Wimborne, UK, 1995.

van Regenmortel, M. H. V., C. M. Fauquet, D. H. L. Bishop, E. B. Carstens, M. K. Estes, S. M. Lemon, J. Maniloff, M. A. Mayo, D. J. McGeoch, C. R. Pringle, and R. B. Wickner. "Virus Taxonomy: The Classification and Nomenclature of Viruses." *The Seventh Report of the International Committee on Taxonomy of Viruses*. San Diego, 2000.

ONLINE RESOURCES

"PhyloCode." http://www.ohiou.edu/phylocode/. Draft version of the PhyloCode, which was posted to the internet in early 2000. Authored by P. D. Cantino (Ohio University) and K. deQueiroz (Smithsonion Institution), with the assistance of an advisory board.

— HAROLD N. BRYANT

NOVELTY AND KEY INNOVATIONS

Novelty and key innovation are related but distinct phenomena of evolutionary innovation. Novelty denotes the origination of new phenotypic characters in organismal evolution, while key innovation is the adaptive exploration of new ecological niches that become possible following a specific innovation. Conceptually, the former addresses the question of the causal factors in organismal evolution, and the latter, the macroevolutionary consequences of innovations associated with the speciation process. These two problems of the macroscopic scale of evolution are distinct from the ongoing, microevolutionary innovations at the molecular level. A certain confusion arises from the fact that the general term *innovation*, used indiscriminately for changes that represent departures from earlier states at all levels of the evolutionary process, encompasses both novelty and key innovation.

Novelty-type innovations designate instances of evolutionary change that are distinct from normal variation because they introduce new entities, units, or elements into a certain level of phenotypic organization—physiological, developmental, morphological, functional, or behavioral. A variety of definitions have been proposed, but to be useful, a definition of novelty must be firmly rooted in a character concept that is adequate for the relevant level of organization. Because this requirement has not been systematically applied, in most cases a morphological definition of novelty is used which, with proper adaptations, can be applied to other levels of organization. In morphology, the definition can be based on the homology concept. Accordingly, a morphological novelty is a new constructional element in a body plan that has a homologous counterpart neither in the ancestral species nor in the same organism (serial homologue). This excludes characters that lie within the normal range of variation or that deviate only quantitatively from the primitive morphological condition; however, it includes those cases in which a new homologue has arisen through qualitative individualization of a preexisting, serial element. Novelty always represents a qualitative departure from the ancestral condition, not merely a quantitative one.

Examples of novelty that satisfy this definition exist in plant and animal evolution; they include all tissues and organ systems, such as the nervous system (e.g., the corpus callosum connecting the two brain hemispheres), the feeding system (e.g., jaws and teeth), the locomotor system (e.g., the giant panda's additional "fingers"), skin appendages (e.g., feathers, hair, glands), and many more. Because of its capacity to fossilize, the skeletal system provides a particularly rich field of examples. New cartilaginous and osseous elements were added at all stages of vertebrate evolution. In the beginning, the endoskeleton in itself constituted a novelty, first forming axial support structures and later adding elements of the cranium and of the postaxial skeleton. In the limbs, for instance, the proximal endoskeletal structures arose in the fins of Silurian fish. Later, during the tetrapod transition in the Devonian, the elements of the autopodium (hand and foot) were added, and new support elements forming the pelvic and shoulder girdles arose. Each of these additions, some composed of several individual elements, constitutes a structural novelty and is accompanied by equally important innovations of joints, muscles, and the nervous system.

Mechanistically, morphological novelties can be based on innovations at the molecular or cellular level, or on new developmental modes and interactions. However, they can equally arise from the redeployment of any of these components at a new location, through subdivision or combination of earlier structures, or through developmental individualization of a serial structure. New skeletal elements, for example, arise from the same set of mechanisms on which other skeletal elements are based, but they are elicited at new sites through modified epigenetic interactions. New muscles arise primarily by additional subdivision of existing muscle blastemata, but they can also form anew from the migration of myoblasts to new locations. Complex new body segments arise through integration of formerly separated elements, as in the thorax of insects that unites several units of the serially segmented insect ancestors. In other cases, such as feathers, the novelty seems based primarily on new proteins that arose through molecular evolution. New cell properties can also be at the base of a novelty—for example, in the origination of a migratory cell population, the neural crest, that gave rise to many new characters in the head and nervous system of vertebrates. Developmental individualization of serial elements, based on a variety of molecular and cellular mechanisms, underlies the formation of specialized teeth (canines, molars, tusks), vertebrae (cervical, thoracic, lumbar, sacral), or hair (whiskers, spines).

These proximate mechanisms should not be confused with the evolutionary driving forces that elicit novelty by acting on developmental systems. Novelty, as a deviation from quantitative variation, is not easily explained by the conventional mechanisms of evolution. Neo-Darwinism is concerned primarily with the variation, fixation, and inheritance of existing characters, and less with the mechanisms responsible for generating new ones. It is difficult to implicate selection as the direct cause of novelty, because selection can act only on what already exists. Mutations of the "hopeful monster" kind, which could bring about a complex new character, have been rejected on sound grounds. Three kinds of evolutionary mechanisms through which phenotypic novelties arise can be postulated: (1) rare mutations that generate immediate structural effects; (2) symbiotic unification of previously separate genetic/developmental systems or components thereof; and (3) epigenetic byproducts caused by selection acting on other properties of the organism.

The first scenario is based on the notion that mutation (genetic change) provides the raw material not only for gradual variation but also for kernels of new structural elements—novelty. This is unlikely to be a widespread mode of phenotypic character innovation, but it could underlie such cases as the origin of feathers, which seem to be based on a new macroscopic property of scales introduced by a new keratin resulting from gene duplication and deletion events. Symbiosis, the second scenario, is a radically different mechanism to generate novelty. It assumes the successful combination of genetic material from distantly related organisms. The occurrence of symbiosis has been extensively documented and is recognized as a source of innovation in the evolution of bacteria, fungi, algae, and eukaryote cells, but also in certain evolutionary instances of more complex organisms, although its importance seems to have receded in advanced forms of plants and animals. Symbiosis and other horizontal transfer might not only lead to the combination of pieces of DNA from different sources but could also include the transmission of entire genetic/developmental modules between species.

The third scenario, which seems more common, proposes that novelties are largely epigenetic in origin. Selection acting on other features than the novel character itself—such as organismal proportion, function, or behavior—is thought to elicit epigenetic byproducts that arise from generic properties of the modified developmental systems. For instance, based on the specific properties of cells, tissues, and their interactions, a continuous directional selection affecting parameters such as cell division rates, cell and tissue interaction, position of inductive tissues, timing of processes, and so on can bring the affected developmental system to a threshold point at which incipient new structures arise automatically. Examples are novel skeletal elements that form at new locations based on changes of cell number, blastema size, and mechanical load that accompany proportional modification of a body part. The concept advocating this mode of novelty generation has been called the *side-effect hypothesis*. In all three modes, however, the *incipient* novel structure is nonadaptive in origin. Mechanisms that are able to overcome existing constraints must be postulated, and genetic fixation and developmental routinization of a novelty may arise as secondary consequences through natural selection.

The concept of key innovation provides a connection between structural innovations and the environment. The prediction is that specific innovations, whether based on novelty characters or on physiological or functional innovations, open up possibilities for the exploitation of new energy sources and permit new adaptive radiation of the innovation-bearing group. Key innovation events trigger diversity by cascade effects in the involved structure-function systems.

Examples of the role of key innovations are provided by the diversity of cichlid fish in East African lakes, which exploit a wide range of feeding niches based on the innovative structural versatility of their pharyngeal jaw apparatus. The species diversity of this group is much greater than that of sister groups that do not possess the innovation. The same effect was shown for other groups of fish that bear similar innovations. Another example is the successful evolution and radiation of snakes; a key innovation in the feeding apparatus involves the anterior separation of the left and right lower jawbones and the establishment of various kinds of joints. This permitted a tremendous new versatility of movement, enabling the capture and swallowing of very large prey, which in turn triggered a cascade of other specialized prey capture mechanisms in the mouth area, such as mobile upper jaws, fangs, venom glands, and possibly even prey-following sensors. The ophidian jaw innovation is thus thought to represent the key event that facilitated the utilization of new food resources and was at the origin of the success and wide diversification of snakes.

An example of a key innovation of the physiological kind is endothermal homeothermy, the organism-regulated maintenance of constant body temperature in mammals and birds. It facilitates independence from ambient temperature and a more efficient metabolism, permitting the exploitation of colder climatic regions, nocturnal activity, sustained rapid movement, avian flight, and so on. Endothermy thus was a key event in the explosive radiation of mammals and birds, itself possibly triggered by environmental changes. This underlines an important feature of key innovation: the innovation itself is a consequence of events that occur in the ecological domain, such as changing climatic conditions. Therefore, key innovation provides an important reciprocal link between external environmental (energy-related) factors, internal constructional (genetic and developmental) factors, and population (reproduction and fitness) factors of evolution. At the same time, key innovation supports theories advocating punctuated events in phylogenetic diversification.

Innovations in general, and structural novelties and key innovations in particular, are well recognized problems of evolutionary biology, but they are not yet fully incorporated into the main body of evolutionary theory. The further study of these modes of evolution requires a proper distinction between origination and diversification of form, or between innovation and variation, also with regard to the causal mechanisms responsible for their occurrence. Since innovation is a much less frequent phenomenon, it has been given less attention than variation, but innovations and novelties are found at all levels of organization and are fundamental for the process of phenotypic evolution. The issue of innovation gains importance as the incongruences between molecular evolution and higher levels of organismal organization become more evident, indicating that the origin of innovations might be based on different mechanisms than the steady, quantitative changes underlying variation.

[*See also* Adaptation; Homology; Natural Selection, *article on* Natural Selection in Contemporary Human Society; Symbiosis; Transitional Forms.]

BIBLIOGRAPHY

Galis, F., and E. G. Drucker. "Pharyngeal Biting Mechanics in Centrarchid and Cichlid Fishes: Insights into a Key Evolutionary innovation." *Journal of Evolutionary Biology* 9 (1996): 641–670. A detailed study of a key innovation in fish evolution.

Margulis, L., and R. Fester, eds. *Symbiosis as a Source of Evolu-*

tionary Innovation. Cambridge, Mass., 1991. A comprehensive collection and discussion of the role of symbiosis in the generation of innovation and novelty.

Müller, G. B., and S. A. Newman, eds. *Origination of Organismal Form*. Cambridge, Mass., 2002. A volume devoted to the distinction between generative and variational factors in evolution.

Müller, G. B., and G. P. Wagner. "Novelty in Evolution: Restructuring the Concept." *Annual Review of Ecology and Systematics* 22 (1991): 229–256. A review and definition of morphological novelty.

Nitecki, M. H., ed. *Evolutionary Innovations*. Chicago, 1990. A collection of ideas and concepts of evolutionary innovation, including key innovation and the side-effect hypothesis.

Wagner, G. P., ed. *The Character Concept in Evolutionary Biology*. San Diego, 2001. An encompassing discussion of character concepts with a section specifically devoted to innovation, including the epigenetic origination concept.

— GERD B. MÜLLER

NUCLEUS. *See* Cell Evolution.

NUTRITION AND DISEASE

From an evolutionary perspective, the adage "We are what we eat" may be better translated as "We are what we ate." The search for food has shaped hominid evolution, transformed our species, and molded our bodies and our social system. The development of a bipedal gait, the increase in the size and complexity of the brain, and the cultural institutionalization of food sharing accompanied the dietary adaptations of early hominids. An adaptive (biocultural) model that considers the interaction of biological and cultural factors provides the means for examining food procurement system of early humans and tools for evaluating its ability to deliver the essential nutritional requirements necessary for survival.

An adaptive biocultural perspective predicts that nutritional practices that provide the essential dietary requirements will be favored both culturally and by natural selection. Even if particular food combinations developed fortuitously, the fact that they provide the essential nutrients will ensure the survival of those who adopt them and the food practice will survive culturally. Among many Native American groups, for example, maize, beans, and squash are used as central food items. This combination is effective because maize is deficient in the amino acid lysine and high in the amino acid methionine, whereas beans are low in methionine and high in lysine. The use of maize and beans in the same meal will provide a complementary source of these essential amino acids.

Biology of Food Choice. The basic tastes such as sweet, salty, sour, bitter, and umami (taste for glutamate found in protein) have a neurological and probably an adaptive basis. Humans and other mammals have a preference for sweet substances. When a sweet solution is introduced into the amniotic fluid, the fetus attempts to suckle, demonstrating that taste nerves function even before birth in humans. Sweetness is a predictor of high-energy sources and hence of value in an environment in which food is limited or difficult to acquire.

The distaste of bitter things is a common trait in humans. The human tongue can detect bitterness at dilutions of one part in two million (sweetness can be tasted at a dilution of one part in 200). The ability to taste bitter substances such as PTC (phenylthiocarbamide) or PROP (6-propythiouracil) at low concentration would be advantageous to hunter-gatherers because bitter substances found in plants are often toxic. If the plants are to be consumed, it will require some processing (e.g., cooking) to remove the toxins. In the modern context, the ability to taste PTC or PROP plays a role in food likes and dislikes.

The fact that humans have cravings for sweet substances and an abhorrence of bitter ones is not sufficient to explain our modern pattern of consumption. Our "sweet tooth" may explain the near-universal use of sugars, but it is a socioeconomic system that transforms these from a curiosity and a luxury into a perceived necessity. Conversely, bitter substances such as quinine can be manipulated so that they become a desirable food item, as evident to any drinker of a gin and tonic.

The Primate Legacy. Adaptive inertia refers to the idea that there may be a lag between what is adaptive in contemporary society and what was adaptive in the past. Basic aspects of human diet may be a reflection of our primate legacy. Omnivorous diets in humans can surely be traced back to our primate ancestors. The use of vegetative parts of plants, fruits, seeds, insects, and animals as food sources is a pattern found in primates.

Alternatively, humans may retain these ancient primate preferenced because they are adaptive, or, in other cases, humans have evolved new adaptations. For example, primates evolved changes in the gut to make the processing of secondary compounds and fibrous plant materials possible. It has been shown that gut transport time affects the absorption of plant protein. Humans transport food through the gut more rapidly than do the chimpanzees and orangutans. In humans, the small intestine is larger in volume and the colon smaller than those found in the great apes. This difference suggests that the human gut is well adapted to process high-quality dietary items that are concentrated and can be readily digested, perhaps because the ancestral lines leading to modern *Homo sapiens* adapted to high-density foods. Furthermore, hominids may have developed more efficient food search techniques to reduce the costs of finding dispersed high-quality foods. The time saved by consuming high-density foods increases the occasions for social interaction.

FIGURE 1. Nutrition and Disease.
Many of the diseases that plague modern humans—coronary heart disease, hypertension, diabetes, and many types of cancer—can be traced to the shifts in diet that have occurred over the last century. © Leonard Lessin/Peter Arnold, Inc.

Primates also engage in active search for animal protein sources: many species stalk insects, and the termite- and ant-catching abilities of chimpanzees are well known. Meat eating by apes and baboons has been reported. Reports describe cooperative hunting, and there appear to be sexual differences in some nonhuman primate subsistence activities. Male chimpanzees are more frequently the hunters, and females are more active in the search for termites.

This omnivorous (or more accurately, generalist) feeding behavior of primates and other mammals creates a dilemma for the animal. The omnivore's dilemma is the conflict between wanting new food sources and fearing that they may be toxic. The omnivore's "neophobia" has led to behavioral patterns in which the omnivore will taste small amounts of a novel item and develop an aversion to any that causes illness. Generalist feeding behavior expands primates' range and can accommodate seasonal variation from area to area. Reliance on an omnivorous diet, a pattern of collecting with some predation, and the evidence of at least rudimentary sharing may be features of the primate legacy that may have affected the dietary patterns of our hominid ancestors. Alternatively, these features of primate diets may have been retained in modern humans because they were adaptive.

The development of "cuisine" is a cultural system that helps to mediate the omnivore's dilemma. The cuisine is defined by the foods that are selected from the environment, the manner in which they are prepared, the principles used in flavoring, and the rules about eating.

The Heritage of the Early Hominids. Much early speculation on hominid dietary behavior was based on the "man the hunter" model. This model has been criticized for overemphasizing limited aspects of early hominid behavior and subsistence patterns and for its focus on "hunting" as the major activity shaping human behavior. Following these criticisms, the role of gathering and its influence on behavior have been incorporated in more recent accounts. Studies show that among contemporary hunter-gatherers, women make a substantial contribution to the group's diet, collecting at least 60 percent of the dietary resources.

There is a suite of behaviors that allowed early hominids to function in the mosaic environment of the grassy woodlands. Generalist-omnivorous foraging strategies, tool use, food sharing, the ability to obtain unseen food, cognitive mapping behavior, and female mobility were adaptive features of early hominid subsistence. The dietary shift that characterized early hominid adaptation, involved a decreased reliance on fruits and fruit-bearing trees and greater reliance on plants protected by hard coverings and with their nutrients stored underground. In this scenario, bipedal locomotion would have aided the transport of food, tools, and water between their sources and sites of preparation and consumption.

Lessons from Contemporary Hunter-Gatherers. The analysis of the dietary resources of the !Kung from Botswana provides a model for the behavior of early hunter-gatherers. It has been reported that in two hours of subsistence activity, !Kung women can collect 30 to 50 pounds of food, whereas men acquire 15 to 25 pounds. The women gathered food equivalent to 23,000 calories, enough to feed a person for ten days, whereas men gathered enough food for a five day period (12,000 calories).

The !Kung have a variety of plants and animals from which to choose, recognizing 105 species of plants and 260 species of animals as edible. The 365 food items are selected from over 500 plants and animals identified in the environment by contemporary scientists. One source (the mogongo nut) is considered a primary food source: thirteen plants are considered major food sources (they are widely available), and nineteen species are considered minor foods (locally and seasonally available). The fourteen items in the primary and major food categories make up 75 percent of the foods consumed by the !Kung. A digging stick is required to recover about a quarter of the plants eaten.

The !Kung are obviously selective in their food choices. They maximize their efforts by collecting the fewest species of plants and animals that can satisfy their biological needs. Although meat is the most desired, it is also the most scarce food item. Even with other desired plant foods, there may be variation with respect to individuals' efforts to recover these items.

During periods of marginal caloric intake, hunters

SPICES AS ANTIMICROBIALS

Spices are aromatic plant materials that are used in cooking. According to conventional wisdom, spices are used because they enhance food palatability. This is true, but it is only a proximate explanation. It does not address the ultimate (evolutionary) questions of why people enjoy certain plant chemicals and why flavor preferences vary from place to place.

Sherman and Billing (1999) addressed these questions. They hypothesized that spice use yields a health benefit: reducing food poisoning and food borne illnesses. In support, most spices have antimicrobial properties, which derive from chemicals that evolved to protect the plants against parasities, pathogens, and predators. Analyses of more than 4,500 meat-based recipes in the traditional cookbooks of 36 countries worldwide revealed that as mean annual temperatures increased (an indicator of relative spoilage rates of unrefrigerated foods), proportions of recipes containing spices, numbers of spices per recipe, total numbers of spices used, and use of the most potent antibacterial spices all increased.

A critical prediction of the antimicrobial hypothesis is that spices should be used less in preparing vegetables than meat dishes. This is because cells of dead plants are better protected physically and chemically against bacteria and fungi than cells of dead animals (whose immune system ceased functioning at death), so fewer spices are necessary to make vegetables safe for consumption. Sherman and Hash (2001) tested this corollary by compiling information on more than 2,000 traditional vegetable-only recipes in the 36 countries. By every measure (e.g., proportions of recipes containing spices, etc.), vegetable-based recipes were significantly less spicy than meat-based recipes. Within-country analyses control for possible differences in spice plant availability and degrees of cultural independence, and thus offer strong support for the antimicrobial hypothesis.

When we cook with spices we are borrowing the plants' chemical "recipes for survival" and using them against our own biotic enemies. Cleansing foods of parasites and pathogens contributes to the health and survival of spice users. Observation and imitation of food-preparation habits of healthy families by relatives and friends spreads recipes rapidly within and among societies. Probably spices taste so good because they are good for us.

Sherman, P. W., and J. Billing. Darwinian gastronomy: why we use spices. *BioScience* 49 (1999): 453–463.

Sherman, P. W., and G. A. Hash. Why vegetable recipes are not very spicy. *Evolution and Human Behavior* 22 (2001): 147–163.

—PAUL W. SHERMAN

and gatherers are more likely to hunt large ungulates. Because these animals are sometimes also subsisting on declining resources, their reserves of body fat become depleted. Lean meat becomes a principal source of energy, even though its consumption requires elevated metabolic rates, and higher caloric requirements overall. In addition, deficiencies in essential fatty acids may result.

Gatherers and hunters often develop subsistence strategies that increase carbohydrate and/or fat intake. There are three strategies that can achieve these results: (1) an increase in the selection and procurement of smaller animals with high fat content, (2) storage of fat- and carbohydrate-rich foods, and (3) an exchange of these items with other groups. In most instances, the collecting strategy may be the most reasonable. Given the need for fat, it seems that we will have to add a "fat" tooth to our lexicon.

It has been claimed that many of the diseases that plague modern humans (coronary heart disease, hypertension, diabetes, and many types of cancer) can be traced to the shifts in diet that have occurred over the last hundred years. However, we should not overlook the fact that contemporary food systems have contributed to a life expectancy of over seventy years, far beyond the Paleolithic average.

The omnivore's—or generalist's—dilemma has implications for the food patterns of contemporary populations. In particular, this dilemma may be manifest in humans as a preference for familiar foods combined with a desire for variety. This search for variety can act as an aid to ensuring a nutritionally balanced diet and may even suggest that there are "built-in" mechanisms to ensure that we pursue some variety in our food search. Eating one particular food for an extended period often results in a loss of taste for it. However, the desire to eat other foods is not affected; we maintain palatability at a high level by eating a variety of foods.

The Neolithic Paradox. Following the Neolithic revolution, there was a dramatic increase in population size and density as people adapted agriculture. It was thought that the Neolithic economy generated food surpluses that led to a better-nourished and healthier population with a reduced rate of mortality. The empirical evidence suggests an alternative scenario in the shift from gathering and hunting to agriculture. The picture that has emerged suggests a much bleaker picture of health. Instead of experiencing improved health, there is evidence of a substantial increase in infectious and nutritional disease. The intensification of agriculture reduces the dietary niche, creating new nutritional deficiencies. The development of superfood (maize in many New World cuisines), the expansion of trade, and some social inequalities are consequences of agricultural intensification. Presently, only 103 plant species contribute 90 percent of the food supply of 146 countries.

The legacy of our evolutionary food odyssey has left its mark on the contemporary diet. In the United States, only 14 percent of adults consume the recommended dietary allowances, and only about one-third consume less than 30 percent of their calories from dietary fats. Only 2 percent of Americans are in compliance with both recommendations. Our historical adaptation to prefer high-density foods, coupled with an industrial food system that can deliver vast amounts of foods laced with fats and sugars, contributes to these problems with the contemporary diet. Evidently, we really are "what we ate."

[*See also* Overview Essay *on* Darwinian Medicine; Coronary Artery Disease; *articles on* Human Foraging Strategies.]

BIBLIOGRAPHY

Cohen, Mark N., and George J. Armelagos, eds. *Paleopathology at the Origins of Agriculture*. Orlando, Fla., 1984.

Harris, M., and E. B. Ross, eds. *Food and Evolution: Toward a Theory of Human Food Habits*. Philadelphia, 1987.

Kipple, K. K., and K. C. Ornelas, eds. *The Cambridge World History of Food*. Cambridge and New York, 2000.

Rozin, Elizabeth. "The Structure of Cuisine." In *The Psychobiology of Human Food Selection*, edited by L. M. Barker, pp. 189–203. 1982.

— George Armelagos

ONCOGENES. *See* Cancer.

ONTOGENY RECAPITULATES PHYLOGENY. *See* Recapitulation.

OPTIMALITY THEORY

[*This entry comprises two articles. The overview provides an account of how the assumption of optimality can be used to make predictions about the end-points of natural selection; the companion article discusses the types of decisions animals and plants face in finding, handling, and consuming prey. For a related discussion, see* Human Foraging Strategies.]

An Overview

Charles Darwin proposed evolution by natural selection as a mechanism that could produce organisms that appear to be well designed. Natural selection is based on the idea that organisms differ in their ability to survive and reproduce, and that these differences are heritable. If the environment is reasonably constant and there is sufficient genetic variation, then over evolutionary time the organisms that are good at surviving and reproducing in this environment will spread through the population. Darwin's argument was a verbal one. It can be made formal by developing explicit models that seek to define the optimal form or behavior of an organism. Optimality theory is the field of study directed at specifying and testing such models. There are various advantages associated with developing formal models. They make the underlying assumptions explicit, and also provide a way of checking the logic of a proposed explanation.

In "The Genetical Theory of Natural Selection," R. A. Fisher (1930) introduced many important ideas. His argument about how natural selection should shape the allocation of resources to sons and daughters introduced the concept of an evolutionarily stable strategy [*see* Game Theory] and formed the basis of the theory of sex allocation. His discussion of how an individual's future reproduction depends on its age forms the basis of life history theory. Fisher's pioneering work did not result in an immediate growth in optimality and game theoretic models. These areas only really developed in the 1960s and 1970s with the work of (among others) Eric Charnov, William Hamilton, Robert MacArthur, John Maynard Smith, Geoff Parker, and George Williams. The 1970s also saw the emergence of behavioral ecology as a discipline that uses these models to understand animal behavior. Although optimality models are a fundamental part of behavioral ecology, their use has been the subject of controversy. This article concentrates on the structure of optimality models and the way in which they are used (many of the issues are also relevant to models based on game theory). (For information on the controversy, *see* Abrams, 2001; Krebs and Kacelnik, 1991; Seger and Stubblefield, 1996.)

Although there are various ways to characterize the components of models based on the optimality approach, the basic idea involves the following:

1. A set of strategies, together with their consequences. A strategy is a possible form or way of behaving, that is, a possible trait value that specifies a phenotype.
2. A currency for evaluating the strategies.

From the consequences of a strategy together with the currency, the payoff associated with a strategy can be found. The payoff evaluates a strategy in terms of the currency. An optimization procedure is then used to find the strategy with the highest payoff.

Strategies and Their Consequences. It may not be obvious how to define the strategies. The choice may be based on observed variation in the trait under consideration. In the context of behavior, the strategies for a breeding female might be the number of offspring produced in a particular breeding attempt. The consequences might be the number of young that survive to independence and the probability that the female survives to breed again. A foraging animal might have a choice between how much time to devote to looking for food as opposed to checking for the presence of predators. In this case, we might assume that any proportion (from none to all of the time) of time spent foraging is possible. The proportion adopted will have consequences for the animal's rate of energetic gain and risk of predation. If we are concerned with how fast a bird should fly (see Vignette), the strategies would be the bird's possible flight speeds and the consequence of a speed is the rate of energy expenditure. In dynamic models involving state variables a strategy specifies how behavior depends on state and time, and the consequences of a strategy specify how an organism's state changes (see below).

The possible strategies will be limited by various constraints that stem from the biology of the system under study. (Sometimes the constraints are taken to be a

HOW FAST SHOULD A BIRD FLY?

Any possible flight speed is a strategy. The consequence of flying at speed v is that the bird's rate of energy expenditure (i.e., power) during flight is $P(v)$. This rate is believed to first decrease and then increase as speed increases (see Figure 1). Thus, there is a speed known as the minimum power speed at which the rate of energy expenditure is minimized.

The currency depends on circumstances. Hedenström and Alerstam (1996) considered two cases. In the first case, a bird displays by singing while in flight. Reproductive success might increase with time spent in the air singing. In this case, all else being equal, fitness would increase with time flying, so an appropriate currency might be the time flying for the given amount of energy spent. The payoff to a strategy is this time. If speed is v, then the time spent flying is proportional to $1/P(v)$. Thus, the time is maximized by minimizing the rate of energy expenditure during flight, that is, by minimizing $P(v)$. It follows that the bird should fly at the minimum power speed.

In the second case, a bird migrates toward its breeding area. The bird alternates between flying and foraging to replace the energy spent. Its reproductive success might depend on the date at which it reaches the breeding area, with early arrival resulting in higher success than late arrival. In this case, all else being equal, fitness would increase as the time taken for migration decreases, so an appropriate currency would be the time taken for migration. The speed with the shortest time taken will have the highest fitness, so the optimal solution minimizes migration time. The migration time takes into account not only the time spent flying but also the time spent replacing the energy that is used in flight. If the bird flies as fast as possible, it will minimize the time in flight but it will require a long foraging time to replace the energy used. Because of this trade-off between flight time and foraging time, the overall time may be less if a slower flight speed is adopted.

Let

$$D = \text{distance flown}$$

Then

$$\text{time taken for migration} = \text{flight time} + \text{time to replace energy used in flight}$$

$$\text{Flight time} = \frac{D}{v}$$

$$\text{Energy used in flight} = \frac{DP(v)}{v}$$

$$\text{Time to replace energy used in flight} = \frac{DP(v)}{vg}$$

where g = net rate of energetic gain

Thus, the time taken is

$$\frac{D}{v} + \frac{DP(v)}{vg},$$

which can be written as

$$\frac{D}{g}\left[\frac{g + P(v)}{v}\right]$$

Only the expression in the square brackets depends on the flight speed. It follows that the speed that minimizes the time taken for migration is the speed that minimizes this expression. Following Hedenström and Alerstam, this is referred to as the minimum time speed v_{mt}.

A graphical method for finding this speed is illustrated in Figure 1. It can be seen that the minimum time speed is higher than the minimum power speed. The minimum time speed increases as the rate g at which energy can be replaced.

—ALASDAIR I. HOUSTON AND JOHN M. MCNAMARA

separate component of the model.) In the case of a female producing young, there will be a limit on the maximum number of young. An animal that is foraging may be constrained by the maximum rate at which it can spend energy. [*See* Optimality Theory, *article on* Optimal Foraging.]

Currency. The various strategies are assessed by a currency. Under certain conditions, natural selection will produce organisms that are good at surviving and reproducing. In technical terms, natural selection maximizes fitness, which is a measure of performance based on the number of descendants left by an individual using that strategy. The precise definition of fitness depends on the biological circumstances. In some contexts, this measure can be based on the average number of offspring produced over an organism's lifetime. [*See* Fit-

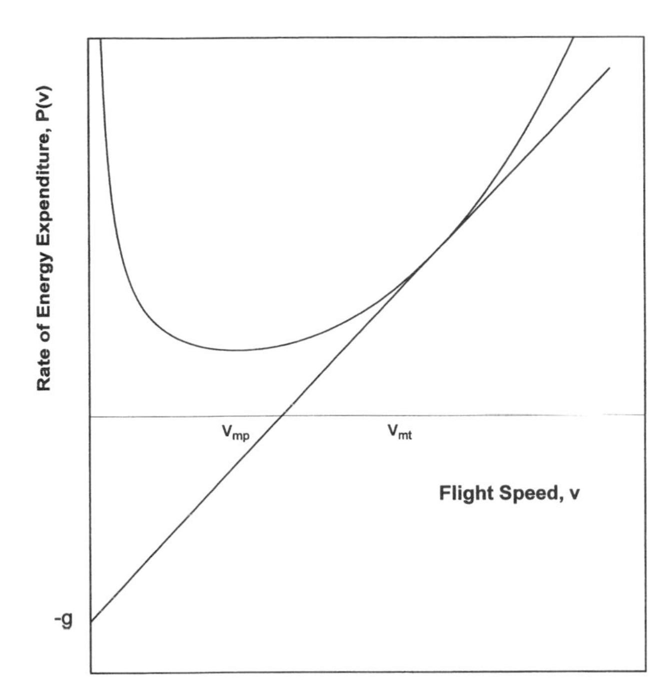

FIGURE 1. How Fast Should a Bird Fly?
The rate of energy expenditure (or power) $P(v)$ during flight as a function of flight speed, v. The minimum power speed v_{mp} is the speed at which $P(v)$ is minimized. The minimum time speed v_{mt} minimizes migration time by minimizing $g + P(v)/v$. This speed can be found by constructing a tangent to the curve $P(v)$ from $-g$ on the y axis. To see why this is so, note that any line to the curve $P(v)$ from $-g$ on the y axis has slope $g + P(v)/v$. The line that is tangent has the lowest possible value of this slope. After Hedenstrom and Alerstam.

ness; Inclusive Fitness.] Models of optimal life histories incorporate explicit assumptions about survival and reproduction and so can work with an appropriate fitness measure. It is, however, not always easy to relate an organism's form or behavior to reproductive success. A currency provides a convenient measure of performance. It is assumed that maximizing (or minimizing) the currency will maximize fitness. In the case of foraging, the rate of energetic gain is often used as currency.

Payoffs and Trade-offs. From the consequences of a strategy and the currency, the payoff associated with the strategy can be determined. This payoff enables us to evaluate the strategy in terms of the currency. The payoff will typically depend on the whole phenotype (not just the trait under consideration) and the environment. For example, in the case of flight speed, the speed v determines (for a bird of a given size and shape) the rate of energy expenditure $P(v)$. If total migration time is to be minimized (second case in Vignette), then the time taken for migration can be calculated from the speed v and the resulting rate of energy expenditure $P(v)$, together with the rate at which the bird can gain energy. This rate will depend on the bird's environment and physiology.

An important general point about payoffs is that there are likely to be trade-offs between various aspects of morphology or behavior, such that a change that increases some aspect of performance will decrease some other aspect of performance. For a reproducing animal, it is likely that it can only increase its success in the current attempt by reducing its performance in future attempts or its chances of surviving to breed again. This means that there is a trade-off between current and future reproductive success. For an animal that is feeding, it may only be possible to increase its rate of energetic gain by also increasing its risk of being killed by a predator, so that there is a trade-off between energy and predation. In this case, the currency needs to take account of both the benefit of gaining energy and the cost of being killed.

Given these components, the strategy that maximizes (or minimizes) the currency is found. This is the optimal solution. In some circumstances, natural selection will result in individuals that adopt this solution (or adopt a simple rule that approximates the solution, see below) spreading through the population; that is, the solution is the end point of evolution by natural selection.

It must be emphasized that it is not assumed that the organism needs to make the same calculations that we have to make in order to find the optimal solution. Just as we do not expect a bird to know about aerodynamic equations in order to be able to fly, so we do not expect animals to make explicit calculations in order to find optimal solutions. Instead, it is assumed that animals follow simple rules that perform well under the conditions in which they evolved. A consequence of this is that the rules might perform badly in environments that are very different from those in which they evolved. This can create problems for attempts to test optimality models in laboratory studies. The animal in the lab may follow rules that take account of the risks of starvation and predation, even though these are not real threats in the lab. It is, of course, possible to study rules without considering natural selection and evolution, but it can be argued that an evolutionary perspective may help us to analyze rules and that a complete account should consider both the rules and their evolution (e.g., Krebs and Kacelnik, 1991). The rules provide a causal explanation, whereas an account based on natural selection provides a functional explanation.

The use of optimality models does not make the assumption that all organisms are optimal. There are many reasons (e.g., genetic drift, gene flow from other populations, lack of genetic variation, and changing environmental conditions) why an organism might not be close to an optimal solution. The aim of constructing an optimality model is to investigate whether the action of selection can explain the form or behavior of a particular species. If the optimal solution for the model agrees with the actual state of affairs, we have not established that the animal under consideration is optimal. All that has been shown is that a particular aspect of the animal can be understood in terms of a particular set of assumptions about natural selection. (The model may, of course, be right for the wrong reasons. In addition to checking a model's predictions, it is necessary to check its assumptions.) If the solution of the model makes the wrong prediction, there are a variety of possible explanations, including the following.

1. Incorrect assumptions have been made about the strategies and constraints. For example, some of the strategies may not be possible.
2. The currency may not be correct. For example, net rate of energetic gain may not be an adequate currency for foraging. [*See* Optimality Theory, *article on* Optimal Foraging.]
3. Natural selection has not yet reached an end point.
4. The aspect of the animal that we are considering has not been shaped by natural selection. For example, it might be a side effect of some other aspects of the organism.

If the model makes a prediction that differs significantly from the data, it is usual to modify the assumptions of the model about the strategies or the currency. Making modifications so as to obtain correct predictions has been criticized, but defenders of optimality argue that modifying assumptions in the light of evidence is a standard feature of science. What is important in this context is that arbitrary changes should not be introduced just to make the model work. (A currency with no relation to fitness would not be acceptable.) The changes that are introduced should lead to new insights and predictions. Opponents of optimality argue that this process of modification is never ending, so that models are not really testable. Defenders reply that the assumption that a particular feature has been shaped by natural selection can be abandoned, but that it is not fruitful to abandon this assumption prematurely. There is, however, no well-defined specification of when to stop, but it can be argued that it is not reasonable to expect such a specification to exist.

Static and Dynamic Models. Optimality models can be divided into two classes, static and dynamic. Static models involve a situation that does not change over time, whereas dynamic models consider changes over time. The models of flight speed presented in the Vignette are static. At first glance, the model based on migration (Vignette, second case) might seem to be dynamic, in that the animal's behavior changes over time. In this case, however, the animal repeats the same cycle over and over again, without any change in the consequences of its behavior. In many circumstances the consequences of an organism's behavior depend on its state, and state is changed by an animal's behavior. For example, a small bird's ability to escape from a hawk may depend on how much fat it has stored on its body. We can take the bird's state to be its level of fat reserves. A high level may make it harder for the bird to maneuver and thus increases its risk of predation. The advantage of fat reserves is that they improve the bird's chances of avoiding starvation if there is an interruption in the food supply. Night is an obvious interruption, because most birds cannot forage at night. Bad weather may also interrupt foraging. The bird can change its fat reserves by choosing how much time to devote to foraging. An increase in reserves will decrease starvation but increase predation. Given this biological background, a dynamic model might focus on winter and assume that natural selection favors the minimization of mortality (starvation plus predation) over this period. The model could consider a sequence of foraging decisions over a day. In this context, a strategy is a rule that specifies the behavior that is performed in each state at each time of day. Even if the bird has just the choice between foraging and resting, there can be a very large number of strategies. Rather than attempting to evaluate the performance of every possible strategy, the optimal strategy can be found by a powerful computational procedure called dynamic programming (Clark and Mangel, 2000; Houston and McNamara, 1999). Using this technique, predictions can be made about how both foraging and the level of fat reserves depend on time of day (e.g., McNamara et al., 1994). Not enough is known about the details of foraging in any species for precise quantitative predictions to be made. What can be done is to establish broad qualitative relationships between, say, fat levels and environmental parameters.

Dynamic models that include the organism's state make it possible to construct models that are relatively realistic. The extra realism of dynamic models comes at a cost. More information needs to be known in order to construct such models, and the optimal solution usually has to be found numerically. This makes the models both harder to analyze and to understand.

Qualitative and Quantitative Predictions. A common occurrence is that a model is able to make correct predictions about qualitative trends, but its quantitative predictions are not accurate. This failure is not surprising. It is likely that in the region of the optimal solution, there is only a small drop in fitness associated with not being at the exact optimum. As a result, selection against deviations is small. (When analyzing models, it may be instructive both to determine the optimum and to investigate the costs of deviations from it.) Furthermore, it is not reasonable to expect that animals will be precisely optimal in terms of any simple model. For one thing, simple models ignore many features of an animal's environment. For example, models of foraging based on rate of energetic gain ignore predation. Many models ignore the need to learn about the environment and the fact that the environment might change. Even if a model provides a good characterization of the selective pressure acting on the animal, crucial parameters or functions might not be known. Without this knowledge, quantitative predictions are not possible. For example, optimality models can predict how fast a bird should fly in a variety of contexts, but these models can only make precise predictions if the rate of energetic expenditure at each possible flight speed is known. This knowledge is difficult to obtain. In the face of these difficulties, it may be better to use qualitative predictions (see Abrams, 2001, for a discussion). In the case of avian flight speed, it is possible to predict how the optimal flight speed should depend on circumstances (see Vignette). Hedenström and Alerstam (1996) show that, as predicted, skylarks fly faster on migration than during song flight. The use of dynamic models may make it possible to obtain relatively rich qualitative predictions.

A general reason for constructing optimality models is that they improve our grasp of the likely outcomes of evolution by natural selection. In this context, Parker and Maynard Smith (1990) distinguish between specific and general models. Specific models are based on a particular case and make quantitative predictions. General models are simple and provide insight, even though they may not be directly applicable to any particular case. (The simplicity of a general model may encourage its use in the wrong context—this can involve distorting or ignoring some features of the species under consideration.) An example of an instructive use of a general model is provided by Abrams' work on the sensitivity of an optimum to various costs. It is sometimes argued that if the cost or benefit associated with an activity is very high, then there will be strong selection acting on the activity and hence we might expect behavior to be close to an optimum. Abrams (2001) shows that this view is not correct. He considers a general model in which the value of a trait determines more than one cost e.g., foraging behavior might determine energetic expenditure and predation risk. He finds that the magnitude of a cost does not necessarily indicate its importance in determining optimal behavior.

Optimality theory and game theory have thrown light on a wide range of topics, including life histories (patterns of growth and reproduction, aging), parental care, foraging, signaling, aggression, and cooperation. They also sharpen our intuition about the consequences of natural selection and suggest directions for empirical research.

BIBLIOGRAPHY

Abrams, P. A. "Optimal Traits When There Are Several Costs: The Interaction of Mortality and Energy Costs in Determining Behavior." *Behavioral Ecology* 4 (1993): 246–253.

Abrams, P. A. "Adaptationism, Optimality Models and Tests of Adaptive Scenarios." In *Adaptationism and Optimality*, edited by S. H. Orzack and E. Sober, pp. 273–302. Cambridge, 2001.

Alexander, R. McN. *Optima for Animals*. London, 1982. A clear account of optimality models of morphology and behavior.

Clark, C. W., and M. Mangel. *Dynamic State Variable Models in Ecology*. Oxford, 2000.

Charnov, E. L. *Life History Invariants*. Oxford, 1993.

Fisher, R. A. *The Genetical Theory of Natural Selection*. Oxford, 1930.

Gould, S. J., and R. C. Lewontin. "The Spandrels of San Marco and the Panglossian Paradigm: A Critique of the Adaptationist Programme." *Proceedings of the Royal Society of London B* 205 (1979): 581–598. A frequently cited and influential attack on explanations based on natural selection. See Maynard Smith (1978) and Parker and Maynard Smith (1990) for a defense.

Haccouu, P., and W. J. Van der Steen. "Methodological Problems in Evolutionary Biology IX: The Testability of Optimal Foraging Theory." *Acta Biotheoretica* 40 (1992): 285–295.

Hedenström, A., and T. Alerstam. "Skylark Optimal Flight Speeds for Flying Nowhere and Somewhere." *Behavioral Ecology* 7 (1996): 121–126.

Houston, A. I., and J. M. McNamara. *Models of Adaptive Behaviour*. Cambridge, 1999.

Krebs, J. R., and A. Kacelnik. "Decision-making." In *Behavioural Ecology*, edited by J. R. Krebs and N. B. Davies, 3d ed., pp. 105–136. Oxford, 1991.

Maynard Smith, J. "Optimisation Theory in Evolution." *Annual Review of Ecology and Systematics* 9 (1978): 31–56.

McNamara, J. M., A. I. Houston, and S. L. Lima. "Foraging Routines of Small Birds in Winter: A Theoretical Investigation." *Journal of Avian Biology* 25 (1994): 287–302.

Parker, G. A., and J. Maynard Smith. "Optimality Theory in Evolutionary Biology." *Nature* 348 (1990): 27–33.

Seger, J., and J. W. Stubblefield. "Optimisation and Adaptation." In *Adaptation*, edited by M. R. Rose and G. V. Lauder, pp. 93–123. San Diego, Calif., 1996.

— Alasdair I. Houston and John M. McNamara

Optimal Foraging

How long should a male dungfly spend copulating with a female? An answer to this question can be given using a simple model from optimal foraging theory. Optimal foraging theory is concerned with how organisms acquire resources. It is a branch of optimality theory, and it makes the same broad assumption that natural selection favors organisms that perform well. Although many of the models are based on animals that are feeding, the basic principles can be applied in other contexts, such as the case of the dungfly.

We start with the assumption that different strategies (ways of behaving) will differ in terms of their results for survival and reproduction. The ability of a strategy to help its practitioner produce descendants provides a measure of the strategy's performance. This measure is referred to as fitness. Thus, in principle, strategies can be compared in terms of fitness; in practice, however, patterns of foraging may be hard to compare in this way. Animals may not be producing offspring during the period in which foraging is being studied, so the link between foraging performance and reproduction may not be clear. Initial work on models of foraging made some plausible simplifying assumptions about how foraging contributes to reproductive success. More recent models have used an animal's state to link foraging and reproduction, so that foraging behavior essentially can be thought of as part of life history theory. [*See articles on* Life History Theory.]

The basic components of a model are a set of possible strategies and their consequences, together with a currency to be maximized (or minimized). The currency serves as a convenient substitute for fitness, so that maximizing (or minimizing) the currency will maximize fitness. The strategy that maximizes the currency is the one that should spread through the population under the action of natural selection. It is important to note that it is not assumed that animals perform the calculations that a human would make in order to find the optimum; instead, it is assumed that animals follow relatively simple rules that perform well.

Optimal foraging theory is often identified with the assumption that animals should maximize their net rate of energetic gain, but this is not the only possible currency. An animal that maximizes this rate will maximize the energy obtained from a given time spent foraging and will minimize the time required to obtain a given amount of energy. There are circumstances in which either of these alternatives would maximize fitness, but it is clear that, in some cases, a different currency will be required. Before looking at this issue in more detail, I will describe various models based on rate maximization, starting with a brief account of two standard cases: patch use and prey choice.

Patch Use. An animal travels through the environment, encountering patches of food. The travel time τ is the (mean) time required to travel from one patch to another. Once in a patch, the animal gains energy at a rate that decreases with time in the patch, perhaps because the animal is using up the resources there. This means that each additional unit of time spent in the patch results in a smaller gain in energy; that is, gains are subject to diminishing returns. A strategy specifies how long to spend in each patch encountered. The optimal time spent in a patch maximizes the rate of energetic gain. Call the resulting rate of energetic gain the maximum rate of gain. When the animal starts foraging in a patch, its rate of gain will be above this maximum rate. As time in the patch increases, the rate of gain decreases, and eventually it becomes worthwhile for the animal to travel to a new patch. The optimal time to leave is when the rate of gain on a patch equals the maximum rate of gain. This is the Marginal Value Theorem (MVT). To see why the MVT gives the optimal solution, note that if the animal stays longer than the optimal time, it is getting a rate on the patch that is lower than the rate it could achieve by leaving the patch and following the optimal strategy. A robust qualitative prediction from the MVT is that the time spent in a patch type increases as the travel time increases.

Models of optimal patch use have been used to predict the foraging behavior of birds and bees. They have also been used by Parker (1978) to predict how long a male dungfly should spend copulating with a female. In this case, the percentage of a female's eggs that a male fertilizes increases with the time that the male spends copulating, but the rate of increase decreases with copulation time—again, diminishing returns.

Prey or Diet Choice. A foraging animal encounters various types of prey item. Each type is characterized by its energy content e and the time h required to "handle" it (the time needed to capture and consume it). Each prey type is encountered at a rate λ. The behavioral options are whether to accept (capture and consume) or to reject an item of a given type when it is encountered. A constraint is that the animal cannot encounter other items while it is handling an item. If the currency is rate of energetic gain, then it can be shown that items of a given type are either always accepted or always rejected—that is, the choice is all-or-none.

Assume that there are two sorts of item, Type 1 and Type 2. Let the profitability e/h be higher for Type 1—that is, $e_1/h_1 > e_2/h_2$. Then Type 1 items should always be accepted, and Type 2 items should be accepted if and only if e_2/h_2 is greater than the rate of gain when only Type 1 items are taken. This rate depends only on the parameters for Type 1 items, so whether Type 2 items should be accepted does not depend on their encounter rate. The optimal policy can be understood from the following intuitive argument. The item with the higher e/h provides the best return from time spent handling, and so it should always be taken. Whether the item with the lower e/h should be taken depends on the energy that it yields compared to the energy that could have been obtained if the handling time had been devoted to searching for and eating more profitable items. (For a general formulation of the optimal policy in terms of the maximum possible rate of energetic gain, *see* Houston and McNamara, 1999, chapter 4.)

It has not proved easy to test the prey choice model with data from animals in the wild, largely because of the difficulty of establishing encounter rates. In the laboratory, Krebs and his colleagues controlled the encounter rate by presenting birds, great tits, with food items on a conveyor belt; this experiment showed that the prey choice model could predict behavior (see Krebs and Davies, 1993, chapter 3). Sih and Christensen (2001) review the success of prey choice models.

Depositing Fuel for a Migratory Journey. The more fuel that a bird carries, the farther it can fly, but the rate of energy expenditure during flight increases with a bird's mass. This means that each extra unit of fuel carried makes a smaller contribution to the distance that the bird can fly before it runs out of fuel; that is, the advantage of depositing fuel is subject to diminishing returns. The resulting curve of distance flown as a function of fuel load can be used to predict the fuel load of birds that can stop to refuel at any point along their route (Lindström and Alerstam, 1992). A critical assumption is that when the bird stops to refuel, a certain time must elapse before the bird becomes established at the refueling site and starts to build up its fuel reserves. This establishment time acts like the travel time in models of optimal patch use. The bird's strategy is the time that it spends refueling, or, equivalently, its level of fuel when it resumes its flight. Lindström and Alerstam argue that an appropriate currency is the overall speed of migration—distance flown divided by time in flight plus time spent refueling. This approach has been used to predict departure fuel loads as a function of the rate at which fuel reserves can be deposited. Initial work on bluethroats predicted the right qualitative trend but did not predict the actual fuel loads.

Central Place Foraging. Some animals do not eat the food that they find; instead, they deliver it to dependent offspring or to a storage site. This is known as central place foraging. Examples include a parent bird bringing food to its young and bees delivering nectar to their colony. The work of Alex Kacelnik on starlings is a classic example of central place foraging (see Krebs and Davies, 1993, chapter 3; Krebs and Kacelnik, 1991, pp. 105–136). Kacelnik trained starlings that were feeding young to come to a hide to collect mealworms. These were delivered to the starling one at a time, and the interval between mealworms increased as the number that had been delivered increased; that is, the gain curve showed diminishing returns. By moving the hide farther from the bird's nest, experiments could increase the travel time. In line with the prediction from foraging theory, the number of mealworms collected increased as the travel time increased. Kacelnik also showed that, although the behavior of the birds was not in perfect agreement with the predictions, the birds tended to avoid costly mistakes.

Factors Beyond Rate of Energetic Gain. The argument behind using net rate as a currency assumes that an animal can gain or spend any amount of energy in a particular time. However, studies of rates of energy expenditure in a variety of animals suggest that there is an upper limit to this rate. The currency that should be maximized in order to maximize the total energy, given this constraint, is a modified form of efficiency (ratio of energy gained to energy spent). Evidence suggests that the speed at which birds fly when bringing food to their young maximizes efficiency.

Using net rate as a currency to evaluate foraging behavior captures certain aspects of foraging that are likely to be important. It also has the advantage of simplicity. There are, however, several aspects of foraging that it ignores. If only the rate of gain of energy is important, animals should not be sensitive to variability about a given mean rate of gain. Animals that show this sensitivity are said to be "risk-sensitive" foragers (this does not refer to predation risk). Explanations of risk sensitivity based on optimal foraging with a currency other than rate of gain have been proposed, but it has being argued that risk sensitivity is really a side effect of fundamental mechanisms that are responsible for learning and memory. Although the importance of risk-sensitive foraging is unclear, there are two important aspects of foraging that are not captured by the rate maximization approach: animals may forage for more than one resource, and they may run the risk of being killed by a predator while they are foraging.

Obviously, if an animal is foraging for a resource other than energy, it might still be possible to use the rate of gain of this resource as a currency (see the discussion of copulation duration in dungflies). The real problem arises when an animal is foraging for more than one resource at the same time. To evaluate behavioral options, we need to be able to compare, in terms of reproductive success, the consequences of obtaining various amounts of each resource.

One way to simplify this problem is to assume that the animal requires a certain minimum level of one nutrient, but that there is no benefit from increasing the intake once the minimum level has been reached. If fitness always increases with the amount of the other resource that is consumed, then the optimal behavior is to maximize the intake of the latter resource, subject to the constraint imposed by obtaining the minimum level of the former. This sort of approach has been applied to the diet of herbivores by Belovsky. For example, the diet of moose has been predicted on the basis of maximizing energy gain subject to obtaining a certain amount of sodium. Although this approach has made successful predictions, the details of the models have been challenged (see Krebs and Davies, 1993, chapter 3).

Another approach is to regard the animal as having a target in terms of the best intake of various resources. The target can be thought of as a point in a space that has an axis corresponding to the intake of each nutrient. Simpson and Raubenheimer (2000) used this approach to study the behavior of locusts that could feed on diets that differed in terms of their protein and carbohydrate contents. Intake could be predicted on the assumption that the locusts minimized the distance between their actual intake of protein and carbohydrate and the nutrient target. Building on their work on locusts, Simpson and Raubenheimer have shown how their approach can

GILLIAM'S CRITERION AND HABITAT CHOICE IN BLUEGILL SUNFISH

In some species of fish, predation risk decreases as size increases. Jim Gilliam developed a model that would predict changes in habitat use as size, and hence predation risk changes. A basic assumption is that an animal cannot reproduce until it grows to a critical size, but that its reproductive success is independent of the time at which it reaches this size. The animal's growth rate depends on its rate of energetic gain. Both this rate and predation risk may depend on the animal's size and its habitat. Gilliam showed that the animal's probability of reaching the critical size, and hence its expected reproductive success, are maximized by choosing at every size the habitat in which the ratio of predation risk to growth rate is minimized (see Stephens and Krebs, chapter 7; Houston and McNamara, chapter 6). Intuitively, the criterion minimizes the mortality per unit of growth. This simple criterion has been used to understand the changes in preferred habitat that are observed in growing bluegill sunfish (Werner and Hall, 1988).

—Alasdair I. Houston

be applied not only to other insects but also to birds and mammals.

The danger of being attacked by a predator is widely recognized to exert a strong influence on foraging behavior. In the face of the threat of predation, a foraging animal may devote time to scanning the environment, at the cost of a reduction in time devoted to foraging. Another response to predation may be to forage in a relatively safe habitat, even if food is not as abundant there as it is in more dangerous habitats. (For the optimal strategy in a special case, see Vignette.)

In order to predict the best foraging behavior in the face of predation risk, we need to understand the basic trade-off that is involved. By foraging, the animal obtains a resource, such as energy. This intake changes the animal's level of energy and hence improves its chances of survival and reproduction. We can represent this by saying that by foraging, the animal can increase its reproductive value. If the animal is killed, it loses the chances of future reproductive success associated with its current condition. The reproductive value of this current state tells us how much the animal's life is worth, and, hence, what it loses if it is killed. Thus, reproductive value enables us to compare the benefit that results from gaining energy with the loss that results if the animal is killed. If the reproductive value of the animal's life is large and the gain in reproductive value from foraging is small, then the animal should be reluctant to risk its life.

The dependence of reproductive value on state provides a basis for understanding foraging decisions. The exact form of the dependence at any time in an animal's life depends on the conditions that it will encounter in the future, together with the behavior that the animal will adopt in these conditions. Therefore, it is difficult to determine how reproductive value depends on state in a particular species. In the face of this difficulty, analysis of foraging decisions can proceed by considering various cases and establishing general qualitative conclusions. Abrams (1992) used this approach to good effect, not just to look at optimal foraging decisions but also to explore their ecological consequences.

An alternative approach is to make an explicit model of the foraging opportunities and dangers of predation, specifying the ways in which an animal's state influences the consequences of its behavior. In these dynamic models, a foraging animal makes a series of foraging decisions. The behavior that is adopted at any time may depend on the animal's state, and it typically will also change that state. The optimal decision at any time depends on the reproductive value at that time, which depends on future behavior. For this reason, it is logical (as well as computationally efficient) to determine reproductive value and optimal behavior by working back from a time at which reproductive value is known or can be estimated. This time might be the end of the animal's life (when reproductive value is zero) or the end of a particular period (such as winter). The procedure of working back to find the optimal strategy is known as dynamic programming (see Houston and McNamara, 1999; Clark and Mangel, 2000).

Optimization or Game Theory? When the consequences of an animal's behavior depend on the behavior of other animals in the population, then optimization by itself does not provide an adequate framework for understanding how natural selection will shape the behavior of members of the population. Instead, it is necessary to use evolutionary game theory. [*See* Game Theory.] The reason for this can be understood by considering the optimal decision (or best response) for a single animal. Because its gains and losses depend on what other animals are doing, the optimal decision depends on the behavior of other members of the population. If the best response for a single animal is different from that of other members of the population and has a higher payoff, then this best response should increase in frequency as a result of natural selection. What we must now seek is a stable outcome at which the best response to the behavior adopted by the population is to adopt the same behavior—that is, we seek a strategy that is a best response to itself. This is known as an evolutionarily stable strategy, or ESS. The use of this concept, developed by John Maynard Smith, has had a profound impact on our understanding of many aspects of animal behavior.

Obviously, there are many contexts in which the consequences of foraging depend on the behavior of other animals and in which an analysis based on the ESS concept is required. A fundamental case concerns how animals in a given environment compete for food. Consider two habitats, one of which contains more food than the other. The first animal to arrive in the area should choose the better habitat, but as subsequent animals arrive and choose this habitat, the intake rate of each animal in the habitat may decrease. This may happen because the animals have to share a limited amount of food, or because the animals waste time interacting with each other. Eventually, it becomes better for an animal to choose the habitat that was initially worse rather than to suffer the adverse effects of a high density of animals in the initially better habitat. When all the animals have chosen between the habitats, the distribution is stable if no animal can do better by changing its location. The resulting distribution is an example of an Ideal Free Distribution.

[*See also* Adaptation; Marginal Values; Mathematical Models.]

BIBLIOGRAPHY

Abrams, P. A. "Adaptive Foraging by Predators as a Cause of Predator-Prey Cycles." *Evolutionary Ecology* 6 (1992): 56–72.

Clark, C. W., and M. Mangel. *Dynamic State Variable Models in Ecology*. Oxford, 2000.

Lindström, Å., and T. Alerstam. "Optimal Fat Loads in Birds: Test of the Time Minimization Hypothesis." *American Naturalist* 140 (1992): 477–491.

Giraldeau, L-A., and T. Caraco. *Social Foraging Theory*. Princeton, 2000. Reviews models of foraging involving game theory.

Houston, A. I., and J. M. McNamara. *Models of Adaptive Behaviour*. Cambridge, 1999. Chapters 4–6 give information on maximizing rate of gain, risk-sensitive foraging, and the trade-off between gaining energy and avoiding predators.

Krebs, J. R., and N. B. Davies. *An Introduction to Behavioural Ecology, 3rd edn*. Oxford, 1993. Standard text gives a very clear account of behavioral ecology; chapter 3 deals with optimal foraging.

Krebs, J. R., and A. Kacelnik. "Decision-making." In *Behavioural Ecology*, edited by J. R. Krebs and N. B. Davies, 3d ed., pp. 105–136. Oxford, 1991.

Milinski, M., and G. A. Parker. "Competition for Resources." In *Behavioural Ecology*, edited by J. R. Krebs and N. B. Davies 3d ed., pp. 137–168. Oxford, 1991. Reviews ideal free distributions.

Parker, G. A. "Searching for Mates." In *Behavioural Ecology*, edited by J. R. Krebs and N. B. Davies, pp. 214–244. Oxford, 1978.

Sih, A., and B. Christensen. "Optimal Diet Theory: When Does It Work, and When and Why Does It Fail?" *Animal Behaviour* 61 (2001): 379–390.

Simpson, S. J. and D. Raubenheimer. "The Hungry Locust." *Advances in the Study of Behaviour* 29 (2000): 1–44.

Stephens, D. W., and J. R. Krebs. *Foraging Theory*. Princeton, 1986. A detailed review of optimal foraging theory, but written before the widespread use of models based on dynamic programming. Chapter 10 discusses problems with models based on optimization.

Werner, E. E., and D. J. Hall. "Ontogenetic Habitat Shifts in Bluegill: The Foraging Rate–Predation Risk Trade-off." *Ecology* 69 (1988): 1352–1366.

— ALASDAIR I. HOUSTON

ORGANELLES. *See* Cellular Organelles.

ORIGIN OF LIFE

[*This entry comprises two articles. The first article discusses the Precambrian fossils and the early metazoan faunas; the second article provides a broad-ranging discussion of the origin of life's building blocks, and how they formed into molecules that stored genetic information and replicated themselves. For related discussions, see the* Overview Essay *on* The New Replicators *at the beginning of Volume 1*; DNA and RNA; *and* Genetic Code.]

The First Fossils

The first fossils have been used to place life's emergence within the context of the four-billion-year record of geologic time preserved on Earth (Orgel, 1998). The central challenge facing students of the origin and evolution of life has been to unravel clues about the nature of the first biota from the ancient (and generally uncooperative) rock record, a record itself modified by processes capable of destroying biological information.

It has been the longstanding tradition of geologists to name historical intervals in the rock record according to recognizable features in the rocks, especially fossils. Classical stratigraphy is related to the presence or absence of particular fossils, such as the disappearance of all dinosaur lineages at the end of the Cretaceous period 65 million years ago. Defining stratigraphy by indexing fossils can be accomplished only under the condition that forms interpreted to be fossil shapes are demonstrably biogenic, and not an abiological artifact of the geology that mimics biological shapes and thereby tricks the eye of the unwary. A problem arises when we travel back into deep time several billion years; it becomes increasingly difficult, even impossible, to rely on fossils to learn about early life and its environment. Thus far, all of the important data about microfossils from the earliest times come from just two localities: Western Australia (Warrawoona Group) and South Africa (Fig Tree Group).

Reports of early Archean (pre-3 billion-year-old) microfossils first appeared in the 1960s from the Onverwacht and Fig Tree rocks of South Africa (Pflug, 1966; Barghoorn and Schopf, 1966). As has often been the case in studies of the early record of life on Earth, these pur-

TABLE 1. Reported Microfossil Assemblages Reported from Archean (pre-2.5 billion year old) Sediments

	Age (byr)	References
Early Archean (3.5–3.0 byr)		
Warrawoona Group	3.5–3.4	Schopf and Packer (1987); Schopf (1992; 1993).
Onverwacht and Fig Tree Group	3.5–3.4	Knoll and Barghoorn (1977); Walsh and Lowe (1985); Walsh (1992)
Late Archean (3.0–2.5 byr)		
Fortescue Group	2.8	Schopf and Walter (1983)
Transvaal Supergroup	2.5	Lanier (1986); Klein et al. (1987)

ported fossil bacteria are now regarded as *pseudofossils*, or even modern contaminants (Schopf and Walter, 1983). Following the initial reports of microfossils from the Early Archean, subsequent research on the rocks of the Onverwacht and Fig Tree cherts revealed structures of clearer biological affinity. Those reported by Knoll and Barghoorn (1977) have a well-defined single mode in size distribution about a mean size (2.5 μm) typical of most bacteria; a significant fraction of the structures appear to preserve cells in the process of binary fission; distinct cell walls in collapsed internal structures were described as seen in younger microfossils; and the structures appear to follow the original sedimentary layering of their host rocks, probably as a consequence of codeposition with the sediments (Table 1). These fossils are dated to 3.4 to 3.5 billion years old.

In the past two decades, new reports have been made of colonial cell structures and filamentlike forms in sedimentary rocks as old as 3.5 billion years. It is no surprise that most of these reports have been met with great criticism. As an example, trains of spheroids and sheathed structures reported by Schopf (1992, 1993) resemble abiotic structures within silicified tuffs from the same sample locality (Buick, 1991; Brasier, 2001). Both the Australian and South African rocks in question contain cherts rich in carbon that occur as minor components within thick and extensive packages of slightly altered volcanic rocks. The capacity of silica (as chert) to preserve the delicate shapes of microbial fossils has been well documented. Younger rocks from the Proterozoic Eon (e.g., Gunflint Chert from 2.1 billion years ago) preserve microbial fossils to a truly remarkable degree, such that an abundant and diverse assemblage of microbial types has been recognized, some of which resemble extant forms. Yet, many (if not all) of the cherts in the two Early Archean geological successions mentioned above were deposited in association with hydrothermal activity and not by sedimentary processes on the ancient sea floor; this observation makes it unlikely that fragile microfossil shapes at these localities would become preserved (Buick, 1984). Paleontological investigation of Archean rocks is not a straightforward enterprise. Even the rocks themselves that contain the putative microfossil shapes (Awramik et al., 1983; Schopf, 1992, 1993) are of arguable age, lacking the kind of detailed mapping necessary to establish their stratigraphic position (Buick, 1984, 1991).

Until recently, it was thought that the nearly 40-year-long search for bona fide microfossils in the Early Archean came to an end with general acceptance of the reports of Schopf (1992, 1993) on poorly preserved but convincing filamentous fossils around 3.46 billion years old, yet even these have been cast out as candidates for the oldest microfossils. Brasier et al. (2001) concluded that the so-called eleven taxa of microfossils of Schopf (1992, 1993) are, regrettably, mineral reaction rims that formed during the later stages of diagenesis within brecciated clasts in a chert dyke. The dyke itself is probably of hydrothermal origin. The forms are abiogenic in that they exist as end members of a morphological continuum, ranging from arcuate to branched, into spherulitic chalcedony. The septation is illusory and results from mineral intergrowths.

Now that the question of life's antiquity appears unsettled through the study of early Archean microfossils, uncertainties about the kind of life present during the emergence of the biosphere become even more acute. Little if anything can be deduced about the physiology of fossil microorganisms merely by looking at them, other than that they probably represent the remains of bacterial lineages. Taken as a whole, what can be interpreted from the Early Archean microfossil record is that, at best, a handful of morphological types is represented. None of these morphologies is informative in regard to physiology or metabolic style of the microorganisms. The apparently low diversity of Early Archean fossil forms is probably a consequence of diagenesis and metamorphism rather than something related to the true diversity of the ecosystem at time of deposition. Often, it is impossible unequivocally to assign a biological origin to small, fragile microfossil shapes. The origin of numerous spheroids mentioned in the literature remains unclear because their morphological simplicity neither supports nor refutes a biological origin. To overcome this, a combination of isotopic, petrographic, molecular, and taphonomic techniques must be employed; rarely, if ever have these four techniques been used on the same sample of ancient rock from the bottom of Earth's sedimentary pile.

BIBLIOGRAPHY

Awramik, S. M., J. W. Schopf, and M. R. Walter. "Filamentous Fossil Bacteria from the Archean of Western Australia." *Precambrian Research* 20 (1983): 357–374.

Barghoorn, E. S., and J. W. Schopf. "Microorganisms Three Billion Years Old from the Precambrian of South Africa." *Science* 152 (1966): 758–763.

Brasier, M. D., O. R. Green, A. Steele, and J. F. Lindsay. "How Old Is Aerobic Photosynthesis? A Fresh Look at the Fossil Evidence." In *Proceedings of the Earth System Processes*, Edinburgh, June 26, 2001.

Buick, R. "Carbonaceous Filaments from North Pole, Western Australia: Are They Fossil Bacteria in Archean Stromatolites?" *Precambrian Research* 24 (1984): 157–172.

Buick, R. "Microfossil Recognition in Archean Rocks: An Appraisal of Spheroids and Filaments from a 3500 M.Y. Old Chert-Barite Unit at North Pole, Western Australia." *Palaios* 5 (1991): 441–459.

Klein, C., N. J. Beukes, and J. W. Schopf. "Filamentous Microfossils in the Early Proterozoic Transvaal Supergroup: Their Morphology, Significance, and Paleoenvironmental Setting." *Precambrian Research* 36 (1987): 81–94.

Knoll, A. H. "Archean and Proterozoic Paleontology." In *Palynology: Principles and Applications*, edited by J. Jansonius and D. C. McGregor, vol. 1 pp. 51–80. American Association of Stratigraphic Palynologists Foundation, 1996.

Knoll, A. H., and E. S. Barghoorn. "Archean Microfossils Showing Cell Division from the Swaziland System of South Africa." *Science* 198 (1977): 396–398.

Lanier, W. P. "Approximate Growth Rates of Early Proterozoic Microstromatolites as Deduced by Biomass Productivity." *Palaios* 1 (1986): 525–542.

Orgel, L. E. "The Origin of Life; How Long Did It Take?" *Origins of Life and Evolution of the Biosphere* 28 (1998): 91–96.

Pflug, H. D. "Structural Organic Remains from the Fig Tree Series of the Barberton Mountain Land." *University of Witwatersrand, Economic Geology Research Unit, Information Circular* 29 (1966): 1–14.

Schopf, J. W., ed. "Geology and Paleobiology of Archean Earth." In *The Proterozoic Biosphere, a Multidisciplinary Study*. Cambridge, 1992.

Schopf, J. W., and B. M. Packer. "Early Archean (3.3-Billion to 3.5-Billion) Microfossils from Warrawoona Group, Australia." *Science* 237 (1987): 70–73.

Schopf, J. W., and M. R. Walter. "Archean Microfossils, New Evidence of Ancient Microbes." In *Earth's Earliest Biosphere: Its Origin and Evolution*, edited by J. W. Schopf. Princeton, 1983.

Walsh, M. M., and D. R. Lowe. "Filamentous Microfossils from the 3,500-Myr-Old Onverwacht Group, Barberton Mountain Land, South Africa." *Nature* 314 (1985): 530–532.

— Stephen J. Mojzsis

Origin of Replicators

The origin of life remains one of the most vexing issues in biology and philosophy. The ancient Greeks were the first to address how life originated. Philosophers of the Ionian school set the stage for Aristotle (384–322 BCE), who advanced the theory of spontaneous generation, according to which living creatures arose from lifeless matter, and animals originated from similar animals. These teachings laid the foundations for the medieval scientific culture, which for nearly fifteen hundred years dominated the thinking of mankind in the Western world.

Spontaneous generation was finally disproved in 1862, when Louis Pasteur responded to a challenge issued by the French Academy of Sciences. The academy offered a prize for convincing experiments that would throw light on the spontaneous generation of living things. Pasteur showed that microbial contamination was responsible for all previous observations of spontaneous generation. Although he concluded that his experiments did not rule out the possibility of spontaneous generation under all conditions, his contemporaries put a broader interpretation on his data. They considered the experiments as an absolute proof of the impossibility of a transition from dead matter to living organisms, and therefore that life could not possibly have arisen on earth.

To address the question raised by this conclusion, in 1865 a German physician, H. E. Richter, advanced the theory of panspermia, in which life is understood to be transported through space and spread from planet to planet, solar system to solar system. At the turn of the twentieth century, the Swedish chemist Svante Arrhenius suggested that microorganisms could be lofted into space on the air currents generated by volcanic eruptions and spread throughout space by the radiation pressure from the sun. But panspermia did not address the question of where life originated. Modern theories of the origin of life for the most part reject the notion of panspermia and seek to explain how life originated and evolved on earth.

It is almost certain that no steps toward the origin of life could have occurred on the earth's surface until about 4.2 billion years ago, after the earth's crust formed and the frequency of meteorite impacts had subsided so that the environment was no longer in constant upheaval. On the other hand, the fossil record strongly supports the existence of microbial cellular life 3.6 billion years ago, and indirect evidence exists for biological activity as early as 3.8 to 4 billion years ago. The time window for life to get started is therefore a surprisingly short 0.4 billion years. While molecular genetics and the fossil record have helped chart the evolution of modern-day life, locating geological remains of the origin of life is probably impossible, due to the reworking of the earth's surface through plate tectonics and erosion.

The oldest known remains of life, from the fossil record, appear surprisingly complex. Presumably, DNA, as is the case for nearly all modern-day organisms, was used to store genetic information. Also, these microfossils appear to be surrounded by some sort of membrane, therefore everything inside them was enclosed. More importantly, the existence of many similar "bacteria" joined together implies that the bacteria were dividing (replicating) and therefore making daughter bacteria. The key to the origin and evolution of life also appears to have been the availability of liquid water. This raises the following questions:

1. How has the earth remained continuously habitable (i.e., retained liquid water) for the past 4 billion years?
2. How did the building blocks of life, such as amino acids and nucleotides, arise from the primitive earth environment?
3. How did these compounds form into molecules that stored genetic information?
4. How did this information make copies of itself (i.e., develop the ability to replicate)?
5. How did membranes evolve to encapsulate this information, and at what stage did membrane-bound life arise?

Habitable Zone Theory. There is a close coupling between life and geological and atmospheric conditions. A planet such as the earth must satisfy a number of conditions to support the evolution of life, based on terrestrial comparisons. It must have liquid water over a biologically significant period of time and other compounds, including the so-called CHNOPS (carbon, hydrogen, nitrogen, oxygen, phosphorus, and sulphur) elements. Liquid water not only acts as a medium for chemical reactions but is also an integral part of biological systems, both at the structural and molecular level. Therefore, habitable planets (and moons), if they are to support surface liquid water, must lie within an orbital zone that is thermally compatible with life—where the temperature lies within the range of water's freezing and boiling points. In the case of the solar system, the orbital zone compatible with surface liquid water (the habitable zone, or HZ) extends from just within the orbit of the earth to just outside the orbit of Mars. It is this perspective that controls our search for other planets outside our solar system that may also harbor life.

The Oparin/Haldane Breakthrough. The first modern theory of the origin of life was independently advanced by the English geneticist and polymath J. B. S. Haldane and the Russian academic Alexander Oparin in his book *The Origin of Life on Earth* (1936). They suggested that the seeds of life arose in space and the atmosphere in the form of various combinations of the CHNOPS elements, under the influence of electrical discharges, radiation, and other sources of energy. According to Haldane (reported in Wells et al., 1934), this material accumulated in the seas until "the primitive oceans reached the consistency of hot dilute soup." In rapidly evaporating inland lakes and lagoons, the soup thickened. In some areas, it seeped deep below ground, and emerged back in hot geysers. All these exposures and chumings induced many chemical modifications and interactions in the original material that was formed in the atmosphere, to produce more complex material that eventually formed life. Haldane's hypothesis was based upon the work of Baly, who noted that under the influence of light, small quantities of sugars and nitrogen-containing compounds were generated from water, carbon dioxide, and ammonia. Similarly, Oparin suggested that the order of events in the origin of life was cells first, proteins second, and genetics third, because he observed that when a suitably oily liquid—perhaps similar to what was formed on the primitive earth—was mixed with water, sometimes the oily liquid dispersed into small droplets that remained suspended in the water, resembling the structures of living cells.

Recreating the primordial soup. The Oparin/Haldane hypothesis requires that the building blocks of life, such as amino acids and nucleotides, were synthesized from components found on early earth. Can we re-create this synthesis? Unfortunately, the composition of the atmosphere at the time life originated is unknown, although several educated guesses have been made. In 1952 the American chemist Harold Urey argued that the primitive atmosphere was hydrogen rich (chemically reducing) and composed mainly of carbon dioxide, water vapor, hydrogen, ammonia, and methane. Urey and Stanley Miller, in a series of pioneering experiments, showed that when such gases were mixed in a pressure vessel and electricity sparked into the vessel (to simulate lightning), many organic compounds formed, including amino acids (which make up proteins). Miller has continued to pioneer such experiments.

In similar experiments, Carl Sagan and his colleagues showed that when gases thought to compose the atmosphere of primitive earth were assembled in a vessel and "sparked," about ten minutes a strange brown pigment formed. Gradually a thick brown tar-like substance that these researchers called tholin covered the interior of the vessel. Analysis of tholin showed that it consisted of a rich collection of organic molecules, including amino acids and nucleotides. Sagan used this apparatus to demonstrate that tholins may also be present in comets and the atmospheres of Jupiter, Saturn, and Titan.

The Ancient Atmosphere. The Oparin/Haldane and Miller/Urey theories for the origin of life require that the atmosphere of the early earth was rich in hydrogen (reducing). However, laboratory experiments and computer models of the ancient atmosphere suggest that UV radiation from the sun, which is today blocked by ozone, would have destroyed hydrogen-based molecules in the atmosphere such as ammonia and methane. The hydrogen released would have been lost to space because it is light (sixteen times lighter than molecular oxygen). In this case, the major component of the atmosphere would have been carbon dioxide and nitrogen that would have been disgorged during volcanic eruptions. Also, the high concentration of carbon dioxide coupled with water vapor would have led to a large greenhouse effect, causing the surface of the earth to be near the boiling point. However, the reducing atmosphere may still have been present, because smoke and clouds from the volcanic eruptions may have shielded chemicals like methane and ammonia from the UV radiation.

Hydrothermal vents as origin-of-life centers. One way to circumvent the chemical state of the early atmosphere and the synthesis of organic material is to forgo the atmosphere completely. An alternative energy source to surface-based systems is provided in the form of geothermal energy at hydrothermal vents and/or hot springs or seepages. Hydrothermal vents can be found at the bottom of the ocean, where magma (liquid rock) spills through the earth's crust and reacts with seawater. These geological features are contenders as sites for the origin of life because thermal, chemical, and electrochemical energy is continuously focused there. Although the idea of hydrothermal vents as origin-of-life centers has been challenged, theoretical modeling and laboratory simulations have shown that amino acids and nucleotides can be synthesized under "vent" conditions. Holm and Charlou have made a study of the Rainbow hydrothermal field on the mid-Atlantic Ridge. They found that water in contact with the mineral olivine is oxidized and reduced to molecular hydrogen, and that complex organic compounds and fatty acids are synthesized abiotically through Fischer-Tropsch type (FTT) synthesis. Mike Russell and colleagues have proposed that life could have emerged via iron sulfide membranes produced at such hydrothermal vents. These membranes would provide a number of features that are beneficial to life, including catalytic sites and the ability to concentrate reactants and protect any burgeoning genetic template.

Exogenous delivery and impact synthesis of organic material. Another alternative source of prebiotic reactants to primordial earth is delivery by the impact of comets and or asteroids containing organic material or synthesis during the impact event. Amino acids have been found in meteorites known as carbonaceous chondrites, which are thought to account for about 5 percent of the meteorites that impact into the earth. Chondrites also contain hydrocarbons, fatty chemicals, and alcohols that may have provided some of the chemicals that formed the membranes of primitive cells. Observations and samplings of Halley's comet suggested that comets might be even richer in organic compounds than meteorites.

From the Synthesis of Prebiotic Reactants to the Origin of Replication. None of the theories provide a completely satisfactory explanation as to how the building blocks of life, such as amino acids, were formed. The difficulty lies in the fact that the composition of the early atmosphere is unknown. It is likely that prebiotic materials (and membranes) were synthesized by more than one route. What these theories lack is a mechanism to account for how replication arose. Amino acids can form proteins, but proteins cannot then make copies of themselves from amino acids. What are required are molecules that can store genetic information and undergo replication.

An RNA world? Almost all life today uses DNA to store genetic information and complex proteins to replicate this molecule. These proteins (and many thousands of others) are encoded by the DNA and made via an intermediary process of transcription (mRNA synthesis) and translation (protein synthesis). How did such complex replication arise from the primordial soup?

Similar to protein molecules whose three-dimensional structure is determined by amino acid composition, RNA molecules can have a three-dimensional structure determined by their base sequence. In 1967 this led Carl Woese, Francis Crick, and Leslie Orgel to suggest that RNA molecules, like proteins, could act as enzymes, capable of catalyzing their own replication, thus negating the need for DNA/protein replication. The term *RNA world*, coined by Walter Gilbert, a biologist at Harvard University, has been used to refer to such a hypothetical time in the evolution of earthly life. This idea led Manfred Eigen to turn the Oparin/Haldane theory upside down. The Eigen theory reverses the order of events in the origin of life; a self-replicating RNA molecule at the very beginning of life, proteins appearing soon afterward to build with the RNA a primitive form of the modern genetic apparatus, and cells appearing later to allow the whole process to occur in a defined volume.

Clay and Minerals as Templates for the Origin of Life. Another information-based system has been advanced by Caims-Smith, who proposed that self-perpetuating structures formed under some conditions by clays, which may have acted as primitive information templates. Caims-Smith argued that the simplicity and abundance of mineral structures makes such a relationship likely. The Caims-Smith theory has the beginning of life as clay first, proteins second, cells third, and genes fourth. Clay crystals directed the synthesis of protein molecules adsorbed to its surface. Later, the clay and the proteins learned to make cell membranes and thus became encapsulated in cells formed from such membranes. Thus the cells contained bits of clay that performed the crude functions of the DNA in modern-day cells. Then a clay-containing cell yielded the discovery that RNA was better genetic material than clay. As soon as RNA was invented, the cells using RNA had an immense advantage in precision during replication over cells using just the clay. Clay-based life then became overrun by RNA-based life.

Laboratory experiments have shown that clay-catalyzed glycine and diglycine oligomerizations are possible. James Ferris and colleagues have shown that prebiotic oligomers (short strings of nucleotides) can form mineral surfaces. One group led by N. G. Holm has investigated the interaction of purine and pyrimidine bases dissolved in water on the surface of graphite, and found that bases have different adsorption characteristics, with guanine being the best and uracil the worst. These researchers suggest that this may have influenced the composition of early genetic templates, and because of this one can easily envisage certain bases such as guanine and adenine being preferred over cytosine and uracil in a primitive genetic system.

RNA, the primordial soup, and hydrothermal vents. There is some difficulty reconciling the RNA world with the idea of RNA arising from a primordial soup. Some of the building blocks of RNA are difficult to synthesize under prebiotic conditions. While purines (guanine and adenine) are relatively easy to synthesize under prebiotic conditions, the synthesis of pyrimidines (cytosine, uracil, and thymine) from simple precursors gives very low yields. Another problem with pyrimidines is the difficulty in attaching them to a sugar molecule to form the basic building blocks of the nucleic acid. This implies that RNA itself was not the first genetic material, but arose later on from some other precursor. One alternative that has been proposed is a peptide nucleic acid (PNA), which consists of a peptide backbone (i.e., amino acids) to which nucleobases are attached. PNA will bind to deoxyribonucleotides according to Watson-Crick base pairing (the pairing of adenine with thymine and guanine with cytosine found in all DNA and RNA). Chemically, PNA bridges the gap between protein and nucleic acids. PNAs can carry genetic information, but the backbone is protein.

D. Penny and colleagues have suggested that hydrothermal vents may have been too hot to allow the use of RNA as a genetic template. Although we traditionally tend to think of genetic templates in terms of the DNA double helix linked by Watson-Crick base pairing, single-stranded RNAs are also likely to form secondary and tertiary structures. Penny has shown that hot conditions found at hydrothermal vents may preclude the correct folding of RNA into structures that may be important for replication. This work suggests that while the precursors of life can be generated at the vent systems, the actual formation of self-replicating templates may have occurred in the cooler regions.

RNA—both an information and an enzyme molecule? One of the primary tests of the RNA world idea is to show that RNA can not only carry information but can act as an enzyme to replicate itself. In the early 1980s Thomas R. Cech and Sydney Altman found certain RNA molecules that have enzymatic activity of one sort or another. These molecules can cut RNA in specified places and are called ribozymes. The discovery of these molecules has removed one of the main objectives to the RNA world hypothesis—that RNA cannot itself act like an enzyme. Biologists have had some success at directing RNA to replicate itself in the laboratory. Jack W. Szostak and colleagues have built customized RNA molecules that can act as enzymes, cutting and pasting together molecules including themselves. In the laboratory, under prebiotic conditions, short RNA sequences can be formed, roughly two to six bases in length. Unfortunately, synthesizing something bigger through "prebiotic" chemistry has not been achieved; this must be done if the idea of an RNA world is to succeed. Manfred Eigen has shown that RNA (with the encouragement of other inducements supplied by the experimenter) can adapt and evolve.

The lipid world and origin of membranes. Lipids can undergo spontaneous self-organization into larger structures such as micelles and bilayers. The formation of membranes is crucial to the development of cellular-based life. Membranes are not inert structures. Mike Russell suggests that primitive membranes contained catalytic sites, and that therefore metabolism may have been possible. These protocells would have had a competitive edge over protocells that relied on diffusion of reactants from the external environment. Experiments have shown that membranes can select different types of amino acids (hydrophobic) and peptides and perhaps permit the formation of oligopeptides that are not possible in an aqueous environment. In addition, membranes can provide compartmentalization and separation of different functions. Certainly, experiments in the laboratory that try to re-create or model a lipid world have shown that a polymerase enclosed in phospholipid vesicles with suitable nucleotides can synthesize RNA.

The origin of protein synthesis. Whether RNA was the genetic material when protein synthesis evolved is unknown. It is possible that RNA and protein coevolved and formed stable complexes from the very first origin-of-life events. John Maynard Smith and Eörs Szathmáry suggested that this first happened not as part of protein synthesis apparatus but as a means of improving the range of efficiency of ribozymes. They suggested that ribozymes acquired amino acids as cofactors, and those that made the ribozymes more efficient catalysts were selected. Maynard Smith and Szathmáry propose that the amino acid was linked to an oligopeptide, which was in turn linked by complementary base pairing to the ribozymes. The ribozymes may then have acquired two or more amino acids that were linked by peptide bonds. It is likely that short oligopeptides were able to direct the synthesis of new RNA daughter molecules from parental templates. If the RNA could actually direct the synthesis of its own "replicase" then it would have a competitive edge over those molecules that could not.

The first primitive sequences capable of protein-catalyzed replication may have been only thirty to sixty bases long and the oligopeptides they encoded only ten to twenty amino acids long. M. Nashimoto has suggested that the primitive translation machinery could have worked in both directions (i.e., RNA to protein and protein to RNA), and thus genetic information could have been obtained from existing proteins via reverse translation. He suggested that one advantage of this is that reverse translation would have ensured biological continuity, with functional proteins encoding RNAs encoding functional proteins.

Toward a grand unified theory—a double origin-of-life event? The Cairns-Smith theory requires a double origin-of-life event, the evolution of a cell-like structure followed by the incorporation of nucleic acids, whereas both the Oparin/Haldane and the Eigen theories were presented as single origin-of-life events. Oparin/Haldane places primary emphasis on metabolism and barely discusses replication. Eigen places primary emphasis on replication and foresees metabolism falling into place rapidly as soon as replication is established. Freeman Dyson suggested that, "the Oparin [and Haldane] and Eigen theories make more sense if they are put together and interpreted as the two halves of a double-origin theory." In essence, Oparin/Haldane described the first origin-of-life event and Eigen the second. Dyson sums up this idea with his usual clarity: "Roughly speaking, Cairns-Smith equals Oparin plus Eigen plus a little bit of clay." All three theories may turn out to contain essential elements of the truth.

The Solar System as an Origin-of-Life Laboratory. One of the main problems with investigating the origin of life on earth is the continuous reworking of the earth's surface, which has destroyed any geological traces of an origin-of-life event. In addition, there is no direct record of the composition of the primordial atmosphere. Possible clues to the origin of life might not be found on the earth at all, and the effort to solve this problem has turned to the new science of astrobiology. There are at least three places in the solar system where possible clues to how life originated might be found—Mars, Europa, and Titan.

Mars. Detailed studies of the images of the surface of Mars taken from orbit by *Mariner 9*, *Viking Orbiter 1* and *2*, and *Mars Global Survey* have provided plausible evidence that liquid water was once present on Mars. Ancient Mars was certainly warmer and wetter and its atmospheric pressure considerably higher than at present. From a biological perspective, there were many similarities between early earth and early Mars. This, combined with the prospect that the clement conditions on primordial Mars could have existed over timescales comparable to the origin of life on earth, has led to the idea that life may also have arisen on Mars. Perhaps the strongest evidence that liquid water once flowed freely across the surface is the existence of dry river channels. Ancient Mars may also have had large, ocean-sized bodies of water. For liquid water to have been stable on Mars, the atmosphere must have been denser in order to provide sufficient greenhouse warming. However, whether ancient Mars was warm and wet or cold and icy is unknown. If liquid water was once stable on the surface of Mars and the martian atmosphere was similar to early earth's, then synthesis of organic molecules via an organic soup model may have been possible. Alternatively, prebiotic synthesis may have occurred around hydrothermal vents, as evidence suggests that Mars was once, and may still be, volcanically active.

Europa. One of the moons of Jupiter, Europa, might have conditions suitable for microbial life. Europa is approximately the size of the earth's moon and is covered largely with smooth white and brownish-tinted ice. Europa is thought to have a thin crust of water-ice less than 30 km thick, perhaps floating on a 50-km-deep ocean, and may be internally active due to tidal heating caused by Jupiter's large gravity field and the decay of radioactive elements. The putative mass of water on Europa exceeds that in the earth's oceans and ice caps. Unlike earth and Mars, Europa cannot be considered to lie in the HZ. However, if Europa contained a liquid water interior for geologically significant periods of time, then conditions may have been suitable for an origin-of-life event. Certainly there could be regions on Europa where conditions lie within the range of adaptation of Antarctic microorganisms. Field exploration in the Antarctic dry deserts—the coldest, driest places on earth—has shown that perennially ice-covered lakes, which have been proposed as a model system for Europa, contain microorganisms. The abundance of the CHNOPS elements on Europa is unknown, although there are strong indications that at least silicates and water are present, as well as hydrated salt minerals such as sodium carbonates and magnesium sulfates. Sulfur dioxide has been identified on Europa and may have been formed when sulfur ions in the Jovian magnetosphere were injected into Europa's water/ice surface. Organic material, such as tholins, may have accumulated on the surface of Europa. Experiments by Jeffrey Bada and colleagues have shown that adenine and guanine as well as amino acids dominated by glycine could form under conditions thought to be prevalent on Europa.

Titan. We have no clear understanding of the composition of the atmosphere on early earth. Ideally, to study this we require a world that retained some of those hydrogen-rich gases, and where the organic building blocks of life are being synthesized in our own era, suitable for study. Titan, a satellite of Saturn, contains such an atmosphere; it is the only satellite in the solar system with a significant atmosphere, approximately 1.5 times the pressure found on the earth, even though the moon is only half the size. However, being so far from the sun, the temperature at the surface is approximately 93 K. Data from the Voyager space probes, the *International Ultraviolet Explorer* laboratory (situated in earth orbit), and laboratory experiments, have indicated that the atmosphere of Titan is composed (in part) of tholins, and therefore may contain the essential building blocks of life. In 1997 a joint NASA/ESA spacecraft called *Cassini* was launched, expected to reach Saturn after a seven-year voyage. Each time the spacecraft passes Titan, the moon will be examined by an array of instruments. *Cassini* will also release a probe called *Huygens* into the atmosphere, which should provide information as to its precise composition, and by inference, what compounds may have been present on the early earth.

Summary. Laboratory experiments have shown that the building blocks of life can be synthesized from compounds thought to be present on early earth, whether in surface lagoons or around hydrothermal vents. In addition, comparisons between different organisms, and the discovery microfossils billions of years old, have provided clues as to how life originated on the earth and how life evolved from a self-replicating molecule to a cell.

[*See also* Genetic Code.]

BIBLIOGRAPHY

Bock, G. R., and J. A. Goode, eds. *Evolution of Hydrothermal Ecosystems on Earth (and Mars)*. Chichester, England, 1996.

Chela-Flores, J., and F. Raulin, eds. *Chemical Evolution: Physics of the Origin and Evolution of Life*. Dordrecht, Netherlands, 1996.

De Duve, C. *Vital Dust: Life as a Cosmic Imperative*. New York, 1995.

Doyle, L. R., ed. *Circumstellar Habitable Zones*. Menlo Park, Calif., 1996. A conference proceedings that includes key papers on habitable zone theory and on the chemical conditions required for origin-of-life events.

Dyson, F. *Origins of Life*. 2d ed. Cambridge, 1999. A unique and scholarly work that attempts to piece together many competing origin-of-life theories.

Fox, S. W., ed. *The Origins of Prebiological Systems and of Their Molecular Matrices*. New York and London, 1965. Perhaps the first conference proceedings on the origin of life; contains some of the thoughts of Haldane.

Jakosky, B. *The Search for Life on Other Planets*. Cambridge, 1998. An excellent account linking the search for life's origins on the earth to the search for life on other planets.

Lahav, N. *Biogenesis: Theories of Life's Origins*. New York, 1999.

Mason, S. F. *Chemical Evolution*. New York, 1995.

Maynard Smith, J., and E. Szathmáry. *The Origins of Life: From the Birth of Life to the Origin of Language*. Oxford, 1999.

Oparin, A. I. *The Origin of Life*. New York, 1938. The break through work described by Oparin.

Segre, D., D. Ben-Eli, D. W. Deamer, and D. Lancet. "The Lipid World." *Origins of Life and Evolution of the Biosphere* 31 (2001):119–145.

Ward, P. D., and D. Brownlee. *Rare Earth: Why Complex Life Is Uncommon in the Universe*. New York, 2000.

Wells, H. G., J. Huxley, and G. P. Wells. *The Science of Life: A Summary of Contemporary Knowledge about Life and Its Possibilities*. London, 1929–1930. A prescient account of the origin of life that sets the scene for the later work of Oparin, Miller, Urey, and one of the first texts to suggest that a comparison of modern life can be used to predict what "primitive" cells might have been like.

Whittet, D. C. B., Ed. *Planetary and Interstellar Processes Relevant to the Origins of Life*. Dordrecht, Netherlands, 1997.

Wills, C., and J. Bada. *The Spark of Life: Darwin and the Primeval Soup*. New York, 2001.

— JULIAN HISCOX

OVIPARITY. *See* Viviparity and Oviparity.

OWEN, RICHARD

The English comparative anatomist and paleontologist Sir Richard Owen (1804–1892) is best known for his pioneering work in the field of paleontology and his role as the founder and director of the British Museum (Natural History).

Owen was born in Lancaster in 1804. His formal university education was restricted to half a year at Edinburgh University, where he matriculated in 1824, moving to London the following year. In 1827, Owen started out on his lifelong career as a museum curator, becoming assistant curator to William Clift (1775–1849) at the Hunterian Museum of the Royal College of Surgeons, then taking over from Clift as curator in 1842. From the outset, Owen was concerned with the expansion of both museum collections and buildings. His goal was to turn the Hunterian Museum into a national museum of natural history. This proved impossible, and in 1856 he resigned to become the first superintendent of the natural history collections at the British Museum (1856–1883). From that time on, Owen fought hard to have a separate museum for his collections constructed, developing his views on museum organization in a rare treatise of its kind, *On the Extent and Aims of a National Museum of Natural History* (1862). After a series of major Parliamentary debates, his plans were realized, and the Natural History Museum in South Kensington was erected (1873–1881).

Early on in his career, Owen was crucially dependent on the patronage of Anglicans who followed the teachings of William Paley. Reverend Paley is best known for his argument that the design of something as complex as the eye compells us to believe in a Creator. As a result, Owen perfected the functionalist approach in comparative anatomy, greatly adding to the argument from design in British natural theology. Exemplications of this functionalist work are his *Memoir on the Pearly Nautilus* (1832), as well as his work of a decade later on the extinct moas from New Zealand. Throughout the 1830s, Owen was an advocate of the doctrine of the creation of species. Through the 1840s, however, he began to diverge from his early, Cuvierian functionalism. Borrowing from German idealist morphology, he wrote his classic *On the Archetype and Homologies of the Vertebrate Skeleton* (1848), in which he put forward the definition and illustration of a vertebrate archetype. This represented the generalized and simplified skeleton of all backboned animals, to which the many parts of real skeletons of fishes, amphibians, reptiles, birds, and mammals could be reduced on the basis of their homological relations.

By changing the emphasis from functional to idealist (or transcendental) anatomy, Owen placed the argument from design on a new footing, shifting the evidence for the existence of a Supreme Designer from concrete adaptations to an abstract plan—from special to general teleology. Divine contrivance was to be recognized not so much by the adaptations to external conditions of individual species as by their common body plan. Thus, in Owen's transcendentalist work, the argument from design did not require a belief in the miraculous creation of species. Owen cautiously began to formulate a theory of theistic evolution, arguing that individual species had come into existence by a preordained process of natural laws, for example, in his *On the Nature of Limbs* (1849). This view met with strong criticism from the Anglican establishment, and for much of the 1850s Owen remained a closet evolutionist. In the wake of Charles Darwin's *Origin of Species* (1859), however, he went public with his so-called derivative hypothesis, first in a *Monograph on the Aye-aye* (1863) and later in such other publications as the *Anatomy of Vertebrates* (3 vols., 1866–1868). Owen's theory differed from Darwin's in several essential respects. Owen explicitly subscribed to the spontaneous generation of life from lifeless matter. He rejected natural selection as a mechanism for producing new species—it could explain only species extinction, he argued. New species do not emerge slowly by an accumulation of small, random changes, but suddenly, in the form of major, directional jumps. Owen visualized the evolutionary process as a saltational one, driven by an organism's innate capacity of change, analogous to the cycle of metagenesis, as in fluke worms, with their remarkably different developmental stages (from egg to mature fluke worm there is an infusorial, a wormlike, and a tadpolelike stage). Saltation allowed Owen to accept major morphological breaks between related species, in particular between humans and the anthropoid apes, which he illustrated by citing particular cerebral features. This led to Owen's infamous hippocampus controversy with T. H. Huxley, who denied the existence of such breaks.

It should be noted that for Owen the question of the origin of species never assumed the significance of a leading research theme, as it did for Darwin. Owen was a museum man whose principal concern was the establishment of a national museum of natural history, and a great deal of his comparative anatomical work was directed toward this end. Darwin and several of his adherents tried to obstruct Owen's museum ideals, and part of the infamous Darwin–Owen conflict must be seen in the light of institutional politics. These tensions were reflected in Owen's notoriously hostile essay for the *Edinburgh Review* (1860) on "*Darwin on the Origin of Species*," but an excessive preoccupation with this review by Darwinians ever since has hindered a balanced appreciation of Owen's considerable accomplishments.

Owen died in London in 1892.

[*See also* Classification; Paleontology.]

BIBLIOGRAPHY

Desmond, A. *The Politics of Evolution: Morphology, Medicine, and Reform in Radical London.* Chicago and London, 1989. Argues in fascinating detail that Owen's idealist morphology served to counter the threat posed by a Lamarckian-radical underground of comparative anatomy teachers in London during the 1830s and 1840s.

Gruber, J. W., and J. C. Thackray. *Richard Owen Commemoration.* London, 1992. On Owen's large network of correspondents.

Owen, R. *The Life of Richard Owen.* 2 vols. London, 1894. The standard Victorian "Life and Letters."

Richards, E. "A Question of Property Rights: Richard Owen's Evolutionism Reassessed." *British Journal for the History of Science* 20 (1987): 129–171. A detailed account of Owen's evolutionism.

Rupke, N. A. "Richard Owen's Vertebrate Archetype." *Isis* 84 (1993): 231–251. Discusses the scientific roots as well as the philosophical and political meaning of Owen's most famous contribution to the life sciences.

Rupke, N. A. *Richard Owen, Victorian Naturalist.* New Haven and London, 1994. A comprehensive account of Owen's work, situating it in the context of the Victorian museum movement.

Sloan, P. R., ed. *Richard Owen: The Hunterian Lectures in Comparative Anatomy, May–June 1837.* Chicago, 1992. Identifies the early influences, at the Royal College of Surgeons, that turned Owen toward idealist morphology.

— Nicolaas A. Rupke

P

PAEDOMORPHOSIS

Paedomorphosis is evolutionary change that results in truncated development in descendant species compared to ancestral species. As a result, adults of the descendant species resemble juveniles of the ancestral species. Paedomorphosis was originally defined by Walter Garstang (1922) to mean "shaped like a child." The classic examples of paedomorphosis are drawn from numerous salamander species, such as the mud puppy (*Necturus maculosus*), that have lost the ability to metamorphose and spend their lives (and reproduce) in the aquatic body form. The development of their bodies is truncated relative to their metamorphosing ancestors via an evolutionary change in the hormonal control of metamorphosis. Paedomorphosis is widespread in evolution because the truncation of previously evolved ontogenies in new lineages is a convenient mechanism to create new traits and forms. It has been used to explain the evolution of taxa from species to phyla. For example, Garstang hypothesized that the chordate body plan arose through a paedomorphic truncation of development of the tadpole stage in the life cycle of urochordates.

Paedomorphosis is identified through the ontogenetic trajectories of traits under study by plotting their patterns of development against age (an ontogenetic trajectory). Ontogenetic trajectories of descendant species traits are compared to hypothesized ancestral trajectories and examined for evolutionary shifts in the timing (onset, offset) and rate (slope) of development. Paedomorphosis is produced by an earlier offset, termed *hypomorphosis* (Shea, 1983), a decrease in rate, termed *deceleration* (Reilly et al., 1997), a later onset, termed *postdisplacement* (Alberch, et al., 1979), or certain combinations of these three simple developmental shifts in the descendant species (see Reilly et al., 1997).

There are several basic assumptions in quantifying paedomorphosis. First, an accurate measurement or descriptor of the ontogeny of the trait or shape under study is needed. Second, heterochronic patterns must be classified with information or convincing inferences on the actual timing (age) of developmental events in the ancestral and descendant ontogenies. One cannot describe relative onsets, offsets, or rates of development without estimates of the time axis. Size can be used as a proxy for age but only when size can be empirically related to age. Third, a phylogenetic hypothesis of the ancestral ontogeny is essential in order to identify the direction of heterochronic change. Fourth, heterochrony involves the development of traits and rarely if ever pertains to whole organisms. Terming whole organisms paedomorphic, as a linguistic shortcut, may obfuscate the real simplicity or complexity of morphogenetic processes. Fifth, heterochrony in somatic traits must be considered independently of heterochrony in reproductive traits. If the timing of sexual maturation is the same in the ancestor and the descendant, only somatic shifts are possible. If the timing of maturation and somatic traits are shifted in the descendant, then two independent heterochronic phenomena have occurred. Many previous definitions of paedomorphosis (e.g., Alberch et al., 1979; de Beer, 1930; Gould, 1977; McNamara, 1986) have unnecessarily linked reproductive and somatic development, confusing the term *paedomorphosis* with the term *neoteny*. [*See* Neoteny.] Finally, heterochrony can be described in terms of "global" and "local" levels on both the shape (whole organism shape vs. coupled or single traits) and time (all or part of the ontogeny) axes. Comparative biologists usually study entire ontogenetic trajectories where the onset point is the beginning of ontogeny and the offset point the final shape. Here, the focus is on how the terminal features of species have come to differ. Developmental biologists, however, generally focus on early segments of ontogeny where heterochrony produces embryological or early ontogenetic differences. On this level, the genetic and mechanistic bases of development are the focus of the analysis and the early perturbations may or may not affect the subsequent, or final, shape of the traits under study. As long as traits are explicitly defined and ontogenetic trajectories are used to define the onset and offset points, any portion of shape ontogeny can be clearly delineated for study (Raff and Wray, 1989).

Since its first use, paedomorphosis has been explicitly used to describe phylogenetic (interspecific) differences in the timing of development (Hall, 1984). Thus, it is invoked as a mechanism to explain how past phenomena have produced the ontogenetic differences observed among species. However, truncated development is occasionally observed in individuals within populations, and the term *paedomorphosis* is often incorrectly applied to this form of heterochrony. Reilly and colleagues (1997) discuss this problem in detail and provide a terminology to be used for intraspecific patterns of heterochrony; variant individuals with truncated development are termed paedotypes and the process producing this paedogenesis.

[*See also* Heterochrony.]

BIBLIOGRAPHY

Alberch, P., S. J. Gould, G. F. Oster, and D. B. Wake. "Size and Shape in Ontogeny and Phylogeny." *Paleobiology* 5 (1979): 296–317. Formal model of simple processes producing paedomorphosis. Incorrectly redefines neoteny and progenesis and confuses intra- and interspecific heterochrony.

de Beer, G. R. *Embryology and Evolution.* Oxford, 1930. Confuses the meaning of neoteny by linking truncated somatic development with accelerated reproductive development.

Garstang, W. "The Theory of Recapitulation: A Critical Restatement of the Biogenetic Law." *Zoological Journal of the Linnean Society* 35 (1922): 81–101. Presents the original definition of paedomorphosis.

Gould, S. J. *Ontogeny and Phylogeny.* Cambridge, Mass., 1977. Reviews heterochronic patterns but confuses intra- and interspecific heterochronic patterns.

Hall, B. K. "Developmental Processes Underlying Heterochrony as an Evolutionary Mechanism." *Canadian Journal of Zoology* 62 (1984): 1–7. Reviews heterochrony from a developmental point of view, showing that formal heterochronic patterns are not simply explained by developmental mechanisms.

McNamara, K. J. "A Guide to the Nomenclature of Heterochrony." *Journal of Paleontology* 60 (1986): 4–13. Heterochrony reviewed from a paleontological point of view. Ignores the problem of relating shape to age to quantify to heterochronic patterns.

Raff, R. A., and G. A. Wray. "Heterochrony: Developmental Mechanisms and Evolutionary Results." *Journal of Evolutionary Biology* 2 (1989): 409–434. Illustrates how developmental patterns often stray from ancestral ontogenies and that any portion of ontogeny can be studied.

Reilly, S. M., E. O. Wiley, and D. J. Meinhardt. "An Integrative Approach to Heterochrony: Distinguishing Intraspecific and Interspecific Phenomena." *Biological Journal of the Linnean Society* 60 (1997): 119–143. Reviews semantic and conceptual bases and history of heterochronic terminology and presents new terms for intraspecific patterns.

Shea, B. T. "Allometry and Heterochrony in the African Apes." *American Journal of Physical Anthropology* 62 (1983): 275–289. First use of the term hypomorphosis (to describe an earlier offset) for the opposite of hypermorphosis.

— STEPHEN M. REILLY

PALEOLITHIC TECHNOLOGY

Paleolithic artifacts are an important and abundant source of information about how human ancestors behaved and how they thought. Even where fossil remains are not preserved, the presence of distinctive stone tools helps to fix the ages of archaeological sites and track the movements of human populations.

Like the earliest hominid fossils, the earliest known artifacts come from the Great Rift Valley in East Africa. Simple stone artifacts from localities such as Hadar and Gona, Ethiopia, and Kanjera, Kenya, have been dated to at least two and a half million years ago. Most researchers believe that early *Homo* (*H. habilis*) produced the first stone tools, but we cannot exclude the possibility that other contemporary hominids such as "robust" australopithecines (*Australopithecus boiseii*) also made and used such implements. The hominid line first diverged more than two million years before this date, so either these earlier hominids were not toolmakers, or they did not use tools of durable materials such as stone.

The earliest stone tool technologies of the Lower Paleolithic period are called Oldowan or mode 1. Although quite simple, they show a good understanding of how stone fractures. Typical artifacts include flakes, sharp-edged chips of stone struck from a larger piece (the core) using a hard stone hammer. Sometimes these were further shaped into flake tools by striking smaller chips from their edges. Another common artifact is the chopper, a pebble with a few flakes removed from one edge, creating a sharp margin. Although the name implies that they were implements, choppers may have simply been cores for the production of smaller flakes.

Stone tools first appear at a time when hominids were becoming more carnivorous, hunting or scavenging the carcasses of large herbivores. Oldowan stone tools are not weapons, but they would have helped hominids to cut through animal hides and crack open bones for marrow, giving them access to foods they could not otherwise obtain. Although carnivory was an important evolutionary development, tool use was more opportunistic than obligatory at this time. Little time or energy was invested in making Oldowan tools, most of which required just a few quick blows with a stone hammer. Although data show that hominids sometimes carried choppers and flake tools away from where they were made, Oldowan tools were seldom moved more than a few kilometers, less than a day's walk for a foraging hominid.

Simple choppers and flake tools served human ancestors for nearly one million years. Around 1.6 million years ago, the basic Lower Paleolithic inventory of flake tools and choppers was augmented by new tool forms and new techniques of manufacture. Acheulean (or mode 2) technology is characterized by bifacial hand axes and cleavers. The term *bifacial* refers to the shaping of these artifacts by removing of flakes from all or part of both sides. Hand axes are pointed or ovate forms, and cleavers have a straight bit. Like Oldowan choppers, hand axes would have been good for heavy-duty tasks, and could have served as cores as well. Studies of microscopic fractures and polishes on some well-preserved examples suggest that they were multipurpose tools. Acheulean bifacial technology did require a more complex and coordinated sequence of actions than the production of Oldowan tools. Along with the symmetry of many specimens, this fact has led some scholars to argue that Acheulean bifaces mark important evolutionary advancements in cognition and technological abilities.

By 1.8 million years ago, perhaps earlier, human ancestors had begun to colonize territories beyond Africa.

Because they preserve much more readily than bones, stone tools are often the best evidence for where human ancestors lived at particular times. Bifacial hand axes are found over much of Europe, Africa, the Middle East, and the Indian subcontinent, but in East Asia only a few sites with bifacial tools have been identified. Instead, most East Asian hominids continued to use simple flake and core (mode 1) technologies up until perhaps 20,000 years ago. The reasons for the discrepancy between East and West have long puzzled archaeologists. Information recently obtained from sites in Java may indicate that human ancestors first reached East Asia by 1.8 million years ago, before the Acheulean had even developed if Africa. Thereafter, technology seems to have developed along diverging lines through much of the Paleolithic period.

The contrasts between East and West do make it clear that the "advanced" Acheulean technology was not integral to geographic expansion of hominids. However, control of fire was a key technological development in hominid colonization of temperate environments (environments with pronounced cold seasons). Because fires can start without human intervention, it is sometimes difficult to know whether evidence for burning at an archaeological site is the result of human actions. The earliest undisputed evidence for controlled use of fire comes from sites in central Europe and Germany dating to between 400,000 and 500,000 years ago, but this date may be pushed back by future discoveries.

Like the Oldowan, Acheulean technology was remarkably stable. Hand axes may have become more refined and symmetrical over time, but the same basic artifact forms persisted for nearly 1.5 million years. This is especially remarkable in light of the fact that at least two hominid species produced Acheulean assemblages. The earliest African Acheulean was the product of *H. ergaster*, but later Acheulean assemblages in Africa, the Middle East, and Europe were made by *H. heidelbergensis* and perhaps early Neanderthals.

Around 250,000 years ago, new kinds of stone tool assemblages appear alongside the late Acheulean. These are called Middle Paleolithic or Mousterian in Europe and Asia, and Middle Stone Age (MSA) in Africa: a more generic term is mode 3. Large "core tools" such as hand axes became more scarce, and there was a corresponding diversification and elaboration of smaller flake tools. Archaeologists have long debated the significance of variation in the forms of Middle Paleolithic/MSA flake tools such as scrapers, points, and denticulates: style, function, and, most recently, progressive reworking of tools have all been suggested as explanations for the diversity of artifact forms. Middle Paleolithic hominids also developed a variety of techniques for producing flakes and blades (long, narrow flakes). Typical of the Middle Paleolithic is the Levallois method, a way of preparing a stone core so that a large, flat flake of predetermined size and shape could be detached.

Hafting is one new technological development in the Middle Paleolithic/MSA. Microscopic damage on tool edges and traces of mastic (adhesive) indicate that some Middle Paleolithic stone tools were affixed to a shaft or handle. Characteristic types of damage on implements, along with the discovery of sharp stone slivers embedded in animal bones from archaeological sites, show that some of these artifacts were spear tips.

As with the Acheulean, more than one variety of hominid produced Middle Paleolithic/MSA artifacts. In Eurasia, the Middle Paleolithic was mainly the product of Neanderthals (*H. neandertalensis*), but in the Middle East some Middle Paleolithic assemblages were apparently made by early anatomically modern humans (*H. sapiens*). Hominid fossil associations for the African Middle Stone Age include both anatomically modern and archaic forms.

By the late Middle Paleolithic period, about 100,000 years ago, hominids were established as far north as Siberia and the Russian Plain. To survive in these environments, they relied on technology more consistently than their predecessors. Middle Paleolithic toolmakers engaged in sophisticated strategies for procuring and conserving supplies of raw material. Although little evidence survives other than polishes on stone scrapers from the working of hides, at least rudimentary clothing technology must have been present. By 60,000 years ago, perhaps earlier, modern humans had colonized the continent of Australia, a feat that some argue required at least simple boats and navigation skills.

As with the Oldowan and Acheulean, little survives of the Mousterian/MSA besides stone tools. However, a few Middle Stone Age sites in southern and eastern Africa furnish evidence of highly precocious forms of technology. Among the most spectacular are elaborate harpoons made of bone from the site of Katanda in Zaire. Dated to about 90,000 years ago, the harpoons look very much like implements produced tens of thousands of years later in other regions. Some researchers would link these finds to the early emergence of anatomically modern humans in sub-Saharan Africa.

Between 40,000 and 45,000 years ago came the last major "revolution" in Paleolithic technology, the Upper Paleolithic (Late Stone Age [LSA] in Africa). A whole range of novel manufacture techniques, artifact forms, and functions appeared at this time. New techniques for producing stone blades made for both more efficient use of raw materials and more regular, standardized tools. Elaborate artifacts of bone, antler, and ivory first proliferated during the Upper Paleolithic/LSA. These implements were more durable than their brittle stone counterparts, but they required a great deal more work to produce. Bone needles and awls testify to the presence

PALEOLITHIC ORNAMENTS AND ART

Most Paleolithic artifacts functioned in interactions between humans and their surroundings. Fire made food more digestible and provided warmth; stone scrapers were used to work hides into clothing, containers, and shelters. During the Upper Paleolithic period, technology also began to play a role in interactions between humans. New forms of material culture, including ornaments, decorated artifacts, and the well-known Paleolithic art served as media for communication. The ornaments that people wore informed others of their group affiliations, statuses, and life histories. Some particularly elaborate artifacts may have been designed to "advertise" the skills of their makers. There are many views about the meaning of Paleolithic art, ranging from portrayals of shamanistic visions to boundary markers. Even if we are not sure about the content of the messages, however, it is evident from the care and effort that went into its production that the art carried important information. Some researchers argue that the first appearance of ornaments and art coincides with the evolution of modern capacities for language. Others hold that they are simply one material expression of long-established capacities and habits.

—STEVEN L. KUHN

of elaborate tailored clothing. Perhaps the most novel feature of the Upper Paleolithic and LSA is the use of material objects as symbols (see Vignette). Ornaments such as beads and pendants appeared first at around 40,000 years ago. By 30,000 years ago, both portable and "parietal" (literally, on the walls) art is found in sites in Europe. For the most part, the Upper Paleolithic and LSA are the work of anatomically modern humans. In southwestern France, however, one early Upper Paleolithic "culture," the Chattelperonian, is now known to have been produced by late Neanderthals.

Archaeologists remain divided as to how rapidly the Upper Paleolithic "revolution" unfolded. Some argue that art, finely made stone blades, bone tools, and ornaments appear as a package around 40,000 years ago. Others see this as a more gradual process, culminating after 18,000 years ago with the late Upper Paleolithic or Epipaleolithic. In either case, it is now clear that the Upper Paleolithic and LSA developed long after anatomically modern humans first evolved. Moreover, the appearance of Upper Paleolithic lifeways did not spell an immediate end to the Middle Paleolithic. In parts of Spain and Portugal, Neanderthals continued to produce typical Middle Paleolithic artifact assemblages until as late as 28,000 years ago.

Upper Paleolithic and LSA artifacts are much more varied and diverse than in earlier time periods. In every region, archaeologists recognize a number of Upper Paleolithic archaeological "cultures," each with its distinctive array of tool forms, ornaments, and sometimes art styles. In western Europe, these include the Aurignacian, Gravettian, Solutrean, and Magdalenian (in chronological order). Quite different sequences of Upper Paleolithic/LSA cultures are found in eastern Europe, the Middle East, East Asia, and Africa. Much of the variation in Upper Paleolithic tool forms reflects the ranges of activities conducted by people living in differing environments. However, divergences in material culture along geographic lines may also represent the first emergence of ethnic divisions among people similar to those known today.

By the later Upper Paleolithic/LSA, about 16,000 years ago, human populations were established in virtually every part of Africa and Eurasia not covered by either high desert or glaciers. From late Upper Paleolithic settlements in northeastern Siberia it was just a short distance across the Bearing Strait to North America. Most researchers agree that the Americas were colonized by peoples from northeastern Asia sometime after 16,000 years ago, a position supported by both linguistic and genetic evidence. However, a small minority argues either for much earlier colonization or for a western European origin for the first North Americans.

[*See also articles on* Hominid Evolution; Modern *Homo sapiens*, *article on* Neanderthal–Modern Human Divergence.]

BIBLIOGRAPHY

Binford, L. *Bones: Ancient Men and Modern Myths*. New York, 1981. A pathbreaking reevaluation of data on Plio-Pleistocene hominid carnivory, this book spurred much discussion and critique but remains highly influential.

Clark, G. *World Prehistory: A New Outline*. London, 1969. This is a classic synthetic work. Although somewhat dated, the system for classification of Paleolithic technological systems into "modes" is still widely used.

Gamble, C. *The Palaeolithic Societies of Europe*. Cambridge, 1999. An innovative overview of technological, economic, and social evolution in Europe during the Middle and Upper Pleistocene.

Isaac, G. L. *Olorgesailie: Archeological Studies of a Middle Pleistocene Lake Basin in Kenya*. Chicago, 1977. This book remains an exemplary study of artifacts from an early African Acheulean site.

Klein, R. G. *The Human Career: Human Biological and Cultural Origins*. 2d ed. Chicago, 1999. The most comprehensive, up-to-date synthesis of both human fossil data and archaeological evidence.

Leakey, M. D. *Olduvai Gorge, vol. 3, Excavations in Beds I and II, 1960–1963*. Cambridge, 1971. A classic, this is the first in a series of volumes presenting results from the research project that established the field of paleoanthropology.

Mellars, Paul. *The Neandertal Legacy: An Archaeological Perspective from Western Europe*. Princeton, 1976. The best synthesis of Middle Paleolithic archaeology and the transition to the Upper Paleolithic in Western Europe.

Schick, K. D., and N. W. Toth. *Making Silent Stones Speak: Human Evolution and the Dawn of Technology.* New York, 1993. A popular account of Paleolithic archaeology, focusing on the earlier time ranges (Lower and Middle Paleolithic); the presentation is lively and well illustrated.

Wynn, T. "Handaxe enigmas." *World Archaeology* 27 (1995): 10–24. This article summarizes much of what is known, and what remains to be learned, about the significance of the Acheulean in human technological and cognitive evolution.

Yellen, J. E., A. S. Brooks, E. Cornelissen, R. G. Klein, M. Mehlman, and K. Stewart. "A Middle Stone Age Worked Bone Industry from Katanda, Upper Semliki River (Kivu) Zaire." *Science* 268 (1995): 553–556. Reports some exciting and potentially revolutionary finds of "advanced" artifacts from the Middle Stone Age in sub-Saharan Africa.

— STEVEN L. KUHN

PALEONTOLOGY

Paleontology is the study of prehistoric life through fossils, the remains of once living organisms that are preserved in the sediments of the earth's crust. Although evolutionary biology has made enormous strides studying living organisms, these studies see evolution only in the thin slice of time known as "the recent." Fossils provide the only direct evidence of 3.5 or more billion years of the history of life, and in many cases they suggest processes that might not be explained by what is known from living organisms. Fossils provide a fourth dimension (time) to the biology of many living organisms. Although indirect techniques such as morphological and molecular reconstruction of phylogenies can be used to infer the past history of life on earth, many groups of organisms, such as conodonts and graptolites, are extinct and would be unknown were it not for the fossil record.

Since paleontology is the study of biological objects (fossils) in a geological context (the sediments in which they were preserved), it is on the border between, and draws from, two major branches of science. Paleontologists must be interdisciplinary in their interests and competent in both geology and biology. Most paleontologists specialize in a particular subdiscipline. Sometimes these subdisciplines are defined by the group of organisms studied. Vertebrate paleontologists study the remains of extinct backboned animals, including fish, amphibians, reptiles, birds, and mammals. Invertebrate paleontologists study the remains of animals without backbones (such as arthropods, molluscs, echinoderms, brachiopods, bryozoans, corals, and sponges). Paleobotanists study the remains of ancient plants. Micropaleontologists specialize in a wide variety of microscopic organisms. These include plants (pollen, diatoms, calcareous algae), animals (ostracods, conodonts, pteropods), and single-celled organisms (foraminifers, radiolarians, and many other Protists). Micropaleontology is useful in locating fossil fuels, and many micropaleontologists are employed by the petroleum industry. Paleontologists who study "megascopic" fossils, on the other hand, tend to be employed by colleges and universities, since their research is more academic in scope.

Paleontologists also define themselves by areas of theoretical interest. Paleobiology describes any application of biological principles to the fossil record; paleobiologists study biological phenomena across many taxonomic groups. Paleobiogeography is the study of the past distribution of organisms in an attempt to understand their origin and dispersal around the world, and sometimes to decipher the motions of continents and land bridges. Paleogeography seeks to determine, by means of the fossil record, ancient continental positions and connections. Some of the earliest evidence for continental drift came from the similarities of fossils on different continents.

Paleoecology is the study of ecological principles as they apply to the fossil record. Paleoecologists try to reconstruct ancient environments and the ecology of extinct organisms. Paleoclimatology is the study and reconstruction of ancient climates; the discipline uses fossils as indicators of past environments, as well as information from geochemistry, climatic modeling, and many other fields.

Fossils are often the only practical means of telling time in geology. Radioisotopic decay methods, such as potassium-argon and uranium/lead dating, work only in rocks that have cooled down from a very hot state, such as igneous or metamorphic rocks. Most of geological history is contained in sedimentary rocks, which cannot be dated by radioisotopes. Biostratigraphy uses the distribution of fossils in stratified sedimentary rocks to correlate and date those rocks. Most paleontologists who are employed by oil and coal companies as economic paleontologists use their knowledge of the fossil record (especially biostratigraphy) to predict the location, quality, and quantity of oil and coal resources.

History of Paleontology. The ancient Greeks interpreted the giant bones of mammoths as the remains of mythical giants, but were puzzled by seashells found hundreds of feet above sea level and miles inland. Had the sea once covered the land, or had these objects grown within the rocks like crystals do? In the sixth century B.C., Xenophanes of Colophon saw the seashells high in a cliff on the island of Malta and suggested that the land had once been covered by the sea.

During the Middle Ages and Renaissance, learned men began to speculate on the meaning of fossils, producing a wide range of interpretations. Originally, the word *fossil* (from the Latin *fossilis*, "dug up") applied to any strange object found within a rock. These in-

PROCESSES OF FOSSILIZATION

The fossil record preserves only a small fraction of the organisms that have existed in the past, and does so in a very selective manner. Some groups of organisms with hard parts (such as shells, skeletons, wood) tend to fossilize readily and much is known about their past. Many others are soft-bodied and rarely if ever fossilize, and paleontology has little to say about their history. The study of how living organisms become fossilized is known as *taphonomy* (Greek for "laws of burial").

From the moment an organism dies, there is a tremendous loss of information as it decays and is trampled, tumbled, broken, and buried. The more of that lost information that can be reconstructed, the more reliable scientific hypotheses are likely to be. In this sense, every paleontologist must act as a forensic pathologist, and determining what killed the victim and trying to reconstruct the events at the "scene of the crime."

The first step is to determine just what type of fossilization has taken place. Most fossils have been dramatically altered from the original composition of the specimen; it is often difficult to determine their original shape and texture, unless one has some idea of the circumstances leading to fossilization.

In a few exceptional cases, organisms are preserved with most of their original tissues intact. Ice Age woolly mammoths have been found thawing out of the Siberian tundra with all their soft tissues essentially freeze-dried and their last meals still in their digestive tracts. Some were so fresh that humans and animals could eat the 30,000-year-old meat with no ill effects. An Ice Age woolly rhinoceros was found intact in a Polish oil seep; the petroleum pickled the specimen and prevented decay. These examples are extremely rare, but when they occur, they give us insight into color, diet, muscles, hair texture, and other anatomical features that paleontologists seldom see.

Some organisms are fossilized when tree resin oozes downward and entraps insects, spiders, and even frogs and lizards. The resin then hardens and forms a tight seal of amber around the organism. Most specimens are only carbonized films, but some are so well preserved that some of their original biomolecules are still intact.

Another mode of fossilization is called permineralization. Many biological tissues contain pores and canals. The bones of animals are highly porous, especially in their marrow cavity, and most wood is full of canals and pores. After the soft parts decay, these hard parts are buried and then permeated with groundwater that flows through them. In the groundwater are dissolved calcium carbonate or silica, which precipitate out and fill up the pores, completely cementing the bone or wood into a solid rock. Although new material comes in, none of the original material is removed. Permineralization can be so complete that even the details of the cell structure are preserved.

Yet another process is called dissolution and replacement. As water seeps through sediments filled with shells or bone, there is a tendency for the original material to dissolve. If the fossil dissolves and leaves a void, then the shape of the fossil is preserved in the surrounding sediments. The internal filling of this specimen is known as an internal mold; the external mold of the specimen is often also preserved. In other cases, the void is filled with sediment and a natural cast of the fossil is formed, mimicking the original in surprising detail. Original bone or shell material can also be replaced without leaving a void. In these cases, the original mineral is dissolved away, and another mineral precipitates almost immediately in its place. This is easiest to detect when a fossil is made of some mineral that is clearly not original.

Finally, remains are preserved through carbonization. Many fossils are preserved as thin films of carbon on the bedding planes of sandstones and shales. When the organism dies, most of the volatile organic materials disperse and leave a residue of coal-like carbon, in the form of a black film that preserves the outline and sometimes the detailed structures of an organism. This kind of preservation is typical of most plant fossils; indeed, coal is the accumulated carbonized films of countless plants.

—DONALD R. PROTHERO

cluded not only the organic remains that we call fossils, but also crystals and concretions and many other structures that were not organic in origin. Most scholars thought that fossils had formed spontaneously within the rock; those that resembled living organisms were thought to have crept or fallen into cracks and then been converted to stone. Others thought that they were grown in rocks from seeds, or were grown from fish spawn washed into cracks during Noah's Flood. Many scholars thought that they were supernatural, "pranks of nature" (*lusus naturae*), or "figured stones" produced by mysterious "plastic forces." Still others considered them to be works of the devil, placed in the rocks to shake religious faith. As quaint and comical as these ideas seem to us today, in their own time they were perfectly rational for people who believed in a literal interpretation of Genesis, and thought that the earth had been created 6,000 years previous and undergone little or no change except for decay and degradation due to Adam's sin.

Essentially modern concepts about fossils were first proposed by Nicholaus Steno. Steno was the court physician to the Grand Duke of Tuscany, so he had ample opportunity to see the shells in the rocks of the Apennine Mountains above Florence, Italy. In 1666 he dissected a large shark caught near the port town of Leghorn. A close look at the mouth of the shark showed that its teeth closely resembled fossils known as "tongue stones," which had been considered the petrified tongues of snakes or dragons. Steno realized that tongue stones were actually ancient shark teeth, and that fossil shells were produced by once living organisms. In 1669 Steno published *De solido intra solidum naturaliter contents dissertationis prodromus* (Forerunner to a dissertation on a solid naturally contained within a solid). The title may seem peculiar at first until the central problem that Steno faced is appreciated: how did these solid objects get inside solid rock? Steno realized that the enclosing material must have once been loose sand, later petrified into sandstone. With this idea, he overturned the long-standing assumption that rocks were permanently formed during the first days of Creation. Steno extended this insight into a general understanding of the relative age of geological features. Fossils that were enclosed in rock that had been molded around them must be older than the rock in which they were contained. On the other hand, crystals that clearly cut across the preexisting fabric of a rock must have grown within the rock after it formed. From this, Steno generalized the principles of superposition, original horizontality, and original continuity that are the fundamental principles of historical geology and stratigraphy.

Although supernatural concepts of the origin of fossils persisted for another century, by the mid-1700s naturalistic concepts of fossils began to prevail. When the Swedish botanist Linnaeus published his landmark classification of all life, *Systema Naturae*, in 1735, fossils were treated and named as if they were living animals. By the time of the publication of Darwin's *On the Origin of Species* in 1859, the realization of the complexity of the fossil record had reached the point where few scholars took Noah's Flood literally.

The Paleontological Perspective. The fossil record provides a unique perspective on life. Without the fossil record, who would have imagined that the world was once ruled by such immense creatures as the dinosaurs, and that the seas were home to equally impressive marine reptiles? Who would have dreamed of some of the bizarre creatures that are now extinct, from the trilobites and ammonites that once dominated the seas, to the incredible plants and animals of the land and air that once existed? Without the fossil record, who would have guessed that through 3 billion years (85 percent of life's history) there were no organisms on earth more sophisticated than bacteria, and no organic structures larger than algal mats? Through most of life's history, much simpler ecological patterns than are seen today prevailed. At one time, the land was not dominated by flowering plants, insects, mammals, and birds (which are all relatively late arrivals on this planet), but by simple plants and (if there were land animals at all) millipedes and spiders and scorpions, and eventually by amphibians and reptiles. Today the sea is the realm of fish and clams and snails and crustaceans, but in the past it was dominated by groups that are either extinct or alive but relatively rare in the modern ocean: trilobites, nautiloids, brachiopods, bryozoans, and crinoids. The air was inhabited by flying insects hundreds of millions of years before the first birds or bats, and even flying reptiles preceded the first birds.

[*See also* Origin of Life, *article on* The First Fossils.]

BIBLIOGRAPHY

Benton, M., and D. Harper. *Basic Paleontology*. Essex, England, 2000. Lower-level college text covering plants, invertebrate, and vertebrate fossils.

Boardman, R. S., A. H. Cheetham, and A. J. Rowell, eds. *Fossil Invertebrates*. Palo Alto, Calif., 1987. Highly detailed, well-illustrated textbook on fossil invertebrates.

Clarkson, E. N. K. *Invertebrate Palaeontology and Evolution*. 4th ed. Oxford, 1998. Highly detailed textbook on fossil invertebrates.

Colbert, E. H., and M. Morales. *Evolution of the Vertebrates*. 4th ed. New York, 1991. General account (although outdated) of vertebrate evolution.

Cowen, R. 2000. *History of Life*. 3d ed. Oxford, 2000. Quirky book (complete with the author's own limericks) that covers basic ideas in the evolution of life for the general market, but does not systematically review the major groups of fossils.

Doyle, P. *Understanding Fossils*. New York, 1996. Basic book on paleontology for the introductory college market.

Lipps, J., ed. *Fossil Prokaryotes and Protists*. Cambridge, Mass., 1993. The most up-to-date and well-illustrated book on microfossils.

Nield, E. W., and V. C. T. Tucker. *Palaeontology: An Introduction*. Oxford, 1985. Well-illustrated textbook covering plants, invertebrates, and vertebrates; more like a lab manual than a theoretical textbook.

Prothero, D. R. *Bringing Fossils to Life: An Introduction to Paleobiology*, 2d ed. New York, 2002. Modern college-level textbook that combines both the principles of theoretical paleobiology with chapters suitable for a systematic lab coverage of invertebrates and vertebrates (as well as plants).

Raup, D. M., and S. M. Stanley. *Principles of Paleontology.* 2d ed. New York, 1978. Revolutionary textbook on theoretical paleobiology, which changed the profession in the early 1970s. However, it does not have systematic coverage of the major groups of fossils.

Stearn, C. W., and R. L. Carroll. *Paleontology: The Record of Life.* New York, 1989. General college textbook on the major groups of fossils, but thin on theoretical principles, and very outdated.

— DONALD R. PROTHERO

PARALLEL EVOLUTION. *See* Convergent and Parallel Evolution.

PARASITES AND PATHOGENS. *See* Disease, *article on* Infectious Disease.

PARENTAL CARE

Parental care typically includes parental behaviors that increase the fitness (i.e., survival and reproductive success) of offspring. The value of such behavior may seem obvious to us, as we spend considerable portions of our lives giving or receiving parental care. However, the nature and scope of parental care vary greatly among species, and may include a variety of forms of investment other than parenting behavior. Here parental care is regarded as one aspect of parental investment, which includes all energy, materials, and time used to support offspring production. Parental investment is in turn one component of reproductive investment, which also includes mating investment. Both parental and mating investments may entail physiological costs (diversion of resources from somatic functions) and ecological costs (increased risk of mortality or reduced future reproductive success resulting from injury, disease, predation, or other harmful ecological factors). These costs are often included in definitions of reproductive investment. Indeed, the degree to which parental care is developed in a species depends on fitness costs to the parent, relative to fitness benefits for offspring.

Parental Care in Relation to Other Aspects of Reproductive Investment. Parental care has been defined in numerous ways and is often difficult to distinguish from other forms of reproductive investment. For some biologists, parental care includes only parental behaviors directed toward the young. At the other extreme, parental care may be regarded as encompassing virtually all forms of parental investment. Moreover, it is sometimes difficult to separate parental care/investment from mating investment. For example, territorial behavior may facilitate mating success, as well as ensure sufficient resources to support the care of offspring. Gamete (sex cell) formation may be considered a component of mating investment, as well as of parental investment (especially with respect to males). And mate choice may determine the magnitude of investment in offspring.

Parental care will be used in this article to mean postzygotic reproductive parental investment, that is, parental investment given to offspring after fertilization or the initiation of embryonic development. However, parental care, in this sense, may overlap extensively with the other two major aspects of parental investment, that is, somatic parental investment (investments by nonreproductive body structures and processes contributing to offspring fitness, at least indirectly) and prezygotic reproductive parental investment (investments by reproductive organs that occur before fertilization or embryonic development of offspring, including the manufacture of gametes and the synthesis and deposition of nutrient reserves in eggs). Examples of overlap between pre- and postzygotic reproductive parental investments include prezygotic yolking of eggs that helps to support the postzygotic growth and development of offspring, the adding of protective coverings and energy reserves to eggs after fertilization (e.g., jelly coats of some amphibian eggs, and albumen, shell membranes, and shells of bird eggs), and the building of nests or acquisition of other beneficial resources for offspring before they are conceived.

Nevertheless, pre- and postzygotic reproductive parental investments are often inversely related. For example, mammalian eggs are tiny (exhibiting negligible prezygotic parental investment) because most parental investment in mammals occurs after zygote formation during gestation and lactation. In contrast, turtle eggs are relatively large because most parental investment in turtles is prezygotic rather than postzygotic.

Parental care and somatic parental investment are often intimately related, as well. For example, feeding of the young cannot occur without appropriate parental feeding/locomotor structures and behaviors. In many birds and mammals, parenting requires an enlarged digestive tract and increased rates of feeding and body metabolism. Evolution of parental care in these animals appears to have favored, or been favored by, the evolution of endothermy (a high body temperature and metabolic rate sustained by internal heat production). This association between endothermy and parental care exemplifies a more general trend in the animal kingdom; that is, the level of parental care is positively correlated with the degree of development of nongonadal reproductive organs (e.g., brood pouches and nursing glands) and other somatic structures/processes supporting reproduction. As a result, somatic and reproductive functions overlap extensively in species with highly developed parental care. This phenomenon should not be surprising, as bodies may be viewed as devices "designed" by natural selection for producing more bodies.

Taxonomic Scope of Parental Care. Parental care can be found in most of the major animal phyla, as well as in land plants (embryophytes or "embryo-bearing plants"). However, it is most developed in some species of insects, crustaceans, and vertebrates.

Kinds of Parental Care and Their Benefits to Offspring. Animals exhibit many kinds of parental care, including supplying offspring with nutrients, warmth, protection, a clean or otherwise favorable nest environment, and opportunities to learn survival or reproductive skills (Table 1). It may seem obvious that these actions benefit offspring, but direct evidence that parental care improves the survival or future reproductive success of offspring is still scanty in many animal taxa. Often this evidence involves preventing or manipulating the level of parental care in question and observing the effects of this manipulation on offspring growth, survival, and reproductive success. For example, unguarded nymphs of the lace bug (*Gargaphia solani*) suffer more predation than nymphs guarded by their mother.

Costs of Parental Care. Parental care is favored by natural selection as long as the benefits to parents and offspring exceed the costs. Limits on parental care include decreased parental survival and breeding opportunities, as well as decreased offspring survival. Parental care is often risky and may require extra energy and nutrients over and above that used for parental maintenance. Resource costs may be met by increased foraging activity (resource acquisition) or by channeling resources away from other life functions (resource allocation). Both of these strategies may entail mortality risks for the parent(s) and offspring. For example, in birds, increased parental activity is positively correlated with rates of nest predation. Draining of body resources for offspring care may also make parents more susceptible to disease, predation, or other mortality risks.

Detecting costs of parental care can be difficult because resource acquisition can affect resource allocation and vice versa. For example, resource acquisition abilities may vary among individuals in a population, thus obscuring costs resulting from resource allocation. This is because "fat" individuals who have acquired large amounts of resources can allocate more resources to both reproduction and survival than "thin" individuals with fewer total resources. As a result, reproductive and somatic (body storage) investments are often positively correlated among breeding females. Generally, parents with good body condition can expend more effort in offspring care, producing more or higher quality progeny than parents in poor condition.

Provisioning of offspring may occur through depletion of somatic energy reserves acquired before offspring development (capital breeding) or by continual acquisition of needed resources during offspring development (income breeding). Oviparous (egg-laying) animals usually make capital investments, whereas viviparous (live-bearing) animals chiefly rely on income investments. However, some viviparous mammals use both capital and income investments to "spread out" the resource cost of parental care over an extended time interval. For example, maternal energy reserves accumulated during pregnancy may be used to help support the large energy cost of lactation (milk production), which is largely met by increased food intake.

Another distinction can be made between variable and fixed costs of parental care. Variable costs depend on progeny number (e.g., parental feeding activity), whereas fixed costs do not (e.g., nest construction).

Despite its net benefits, parental care may in some ways jeopardize offspring survival. For example, shepherding offspring may make them more conspicuous to predators/parasites, or more susceptible to sibling competition/cannibalism. In some insects, fish, and birds, brood parasites fool parents of the same or different species into raising their parasitic young, thus diverting resources away from the host's young, who suffer reduced growth and survival rates.

Despite a growing understanding of specific fitness costs and benefits of parental care, a complete quantitative cost benefit analysis has yet to be carried out in any species.

Allocation of Parental Care. Animals vary greatly in the amount of parental care given by mothers (maternal care) versus fathers (paternal care), and in how this care is divided among offspring.

Sex Differences in Parental Care. In some species, only one parent (usually the mother) provides parental care, whereas in others, both parents participate in the rearing of young. The greater incidence and magnitude of maternal versus paternal care in the animal kingdom may be explained by a simple fact: females produce eggs, and males produce sperm. Because tiny sperm require smaller resource investments than relatively large yolky eggs, males often expend more effort seeking mates than caring for offspring, whereas the opposite is frequently true for females.

Uniparental care may be favored whenever deserting a mate enables successful remating without greatly affecting the care and fitness of a current batch of offspring. Which parent deserts and which cares for the offspring may depend on the timing of fertilization and egg laying. When fertilization is internal, the male can leave before the eggs are laid, thus requiring the female to bear the burden of parental care. However, when fer-

Table 1. Parental Care

Some kinds of parental care exhibited by various animal taxa.

Major type	*Method*	*Representative taxa*
Supplying chemical energy, nutrients, and/or water	Externally as prey or free water	Some arthropods and fish, many birds and mammals
	Externally as skin/oral/anal/esophageal/glandular secretions	Some arthropods, fish, and birds, and all mammals
	Internally as secretions of the ovary, reproductive tract, or special nurse cells	Sponges and some arthropods, fishes, amphibians and reptiles
	Placental or placentalike connections to the parental blood system	Some arthropods, fishes, amphibians and reptiles, and all therian mammals
	Directly from maternal blood system	Some insects
	Offspring consumption of maternal/sibling tissues	Some mollusks, arthropods, fishes and amphibians
Supplying heat to facilitate growth, development, and survival of offspring	From parent's body	Some reptiles and many birds and mammals
	From nests of decaying vegetation constructed by a parent	Some crocodilians and birds
Protection of offspring from enemies and the elements	Carrying offspring externally on parent's body	Some leeches, rotifers, arthropods, echinoderms, fishes, amphibians, birds and mammals
	Carrying offspring inside brood pouches, ovaries, reproductive or digestive tracts, or other body cavities of a parent	Many invertebrates and vertebrates
	Depositing offspring in hiding places or constructing protective shelters or nursery areas, such as nests, cases, holes, and burrows	Some polychaete worms, octopuses, and arthropods, and many vertebrates
	Guarding of offspring	Some octopuses, arthropods, and echinoderms, and many vertebrates
	Retrieval of lost young	Some arthropods, fish, and reptiles, and many birds and mammals
Nest/nursery care	Ventilating brood to facilitate thermoregulation, respiratory gas exchange, and removal of waste products and/or to reduce silting and fungal infection	Some leeches, insects, octopuses, and fish
	Grooming, movement, and/or fungicidal protection of offspring and/or keeping nest/nursery clean to minimize exposure to mold, disease, parasites, and harmful physical conditions	Some arthropods, and many birds and mammals
Learning of survival and reproductive skills	By imitating parents or by receiving instruction from parents	Many birds and mammals

tilization is external, egg laying occurs before fertilization, thus allowing the female to desert the male, who then becomes the primary caregiver, as seen in many fish (see Vignette). However, there are many other possible explanations for the distribution of maternal versus paternal care, including sex differences in parental certainty, in remating opportunities, and in costs and benefits of parental care. Associations between internal/external fertilization and maternal/paternal care may also depend on which evolved first.

Biparental care occurs whenever offspring require more care than can be given by one parent, or when remating is momentarily not possible for either parent. For example, biparental care is common in birds probably because one parent cannot both incubate eggs and forage for food at the same time. Also, birds more often show biparental care than mammals possibly because their nests are more often exposed, thus requiring avian parents to both feed their young and protect them from predation, which again cannot be done at the same time.

Whether parental care is maternal, paternal, or both may also depend on the mating (breeding) system and pattern of sexual selection exhibited by a species.

Allocation of Parental Care among Offspring. Parental fitness is affected not only by how much total resource is devoted to offspring care, but also by how this total resource is divided among offspring. An animal must make two important "decisions": how much resource should be devoted to individual offspring, and should different amounts of resource be given to male versus female offspring? Because of resource and body-space constraints, there tends to be a trade-off between size and number of progeny. Some animals produce many small offspring, whereas others produce few large ones. Limits on milk production in mammals cause nursing young to be smaller in large versus small litters. On the other hand, limits on brood pouch capacity appear to cause stronger negative correlations between egg size and number among small versus large amphipod mothers.

In general, species with parental care tend to produce fewer higher quality offspring than those without such care, a pattern seen in many animal taxa. Various ecological and biological factors may determine the balance between quality and quantity of offspring, and thus the amount of parental care given per offspring. In particular, ecological factors that increase juvenile mortality relative to adult mortality (e.g., size-specific predation, intense competition, harsh physical conditions, and scarce or relatively inaccessible resources) will likely favor enhanced parental care.

Resource allocation to male versus female offspring also varies among animal species. Theory predicts that, in species where male reproductive success is more dependent on body size/condition than female reproductive success, more resources should be given to male than female offspring, especially when mothers have access to abundant resources. In red deer and several other mammals, mothers with high social status or good body condition produce more males than less advantaged mothers, as expected. Other factors such as local mate/resource competition may also affect relative parental investment in male versus female offspring. For example, many wasps parasitize other insects, one female wasp per host. Because hosts and mates are often difficult to find, female offspring usually mate with their brothers on their natal host before seeking a new host. Consequently, many parasitoid wasps have evolved female-biased sex ratios because, from the mother's standpoint, only one son is needed for mating.

Timing of Parental Care Activities. Not only the magnitude but also the timing of parental care can be critical for offspring growth and survival. This timing may involve seasonally restricted breeding or the adaptive scheduling of breeding efforts during a lifetime.

Seasonal timing. To minimize resource costs of parental care, many animals engage in parental activities during times of the year when favorable environmental conditions prevail, including abundant resources and mild weather. A mismatch between the timing of breeding and peak food supply can make parental care excessively demanding, causing increased parental mortality. Such mismatches have been observed in two European populations of blue tits (*Parus caeruleus*) whose breeding schedules are now out of phase with the spring food flush, recently hastened by climatic warming.

Life history timing. Timing of parental care within a lifetime (including ages at first and last reproduction, number and frequency of breeding efforts, and duration of specific phases of parental care, such as incubation, gestation, and lactation) varies considerably among ani-

PATERNAL CARE IN PIPEFISH

Paternal care is common in fish and has apparently evolved to protect eggs and young against predators, anoxia, and pathogens. In syngnathid fish (sea horses and pipefish), males even become "pregnant," caring for broods in a brood pouch or on their ventral surface. Embryos are provided with nutrients and oxygen via placentalike structures. After about one month, juveniles emerge from the brood pouch as independent free-swimming "miniature adults."

In pipefish, this sex role reversal extends to mate competition. Females compete more vigorously for mates than males, contrary to what is observed in most animals. Larger females are more successful in acquiring mates than smaller females. This female–female competition may affect the size of embryos carried by males. Egg/embryo size depends on female size, and large embryos outcompete small embryos for nutrients within the male's brood pouch.

Because of their sex role reversal, pipefish are useful for testing theories of parental care and sexual selection. For example A. J. Bateman (1948) first predicted that the competing sex should exhibit a stronger correlation between number of mates and number of progeny than the choosy sex. As predicted, female deep-snouted pipefish (*Syngnathus typhle*) exhibit a stronger correlation between mating success and fertility than do males (Jones et al., 2000).

DOUGLAS S. GLAZIER

mal species. Generally, large animals have longer periods of parental care and take longer to mature than related smaller animals, probably, at least in part, because of metabolic and developmental constraints. However, even animals of the same size may have different durations of parental care, the causes of which are still not well understood. Generally, birds and mammals with long life expectancies have relatively long periods of parental care. For example, the duration of incubation in birds tends to increase as the risk of predation on offspring and parents decreases. Also, in both birds and mammals, precocial species, whose young are well developed at birth/hatching, tend to have longer incubation/gestation periods than that of altricial species, whose underdeveloped young require extended postnatal/posthatching care.

In sexually reproducing species, different genetic interests of parents and offspring may affect the duration of parental care. Parents enhance their fitness by terminating care of each of their broods as soon as possible, thus maximizing the number of broods produced in a lifetime. In contrast, offspring enhance their fitness by soliciting parental care as long as possible. This parent–offspring conflict not only may cause the duration of parental care to be a highly disputed affair but also may affect the momentary magnitude of parental investment, as appears to be the case for fetal nutrition in mammals and seed nutrition in flowering plants.

In some animals, the magnitude of parental care increases with parental age. This may be because younger parents have less experience than older parents, or because the genetic "tug of war" between parents and offspring over the amount of parental care is tipped toward the parents when they are young and still have many opportunities for further reproduction, whereas it is tipped toward the offspring when the parents are older and have a low expectancy of further reproduction.

Nonparental Care Giving. In some animals, nonparents, such as siblings, aunts, uncles, and grandparents, may care for offspring. Such cooperative breeding, which is especially well developed in social insects and vertebrates, may be favored by kin selection, reciprocal altruism, or limited breeding opportunities.

Parental Care in the Context of the Entire Phenotype. The evolution of parental care affects and is affected by the morphology, physiology, behavior, ecology, and reproductive mode of a species. For example, brooding is more common in small-bodied versus large-bodied marine invertebrates; asexual reproduction has evolved more frequently in brooding versus nonbrooding invertebrates; parental effort in vertebrates may impair the ability of the immune system to fight off disease; parental care and endothermy in birds and mammals appear to have coevolved (see above); and parent–offspring conflict in mammals may affect offspring dispersal patterns, as well as hormonal interactions between the mother and her fetuses.

Parental care not only is a product of evolution, but also may affect the path(s) of evolution. For example, parental care may promote the evolution of animal societies and the survival and diversification of specific taxa through geological time (as has been observed in brooding corals and sea urchins). The maternal environment, in particular, may have profound effects (maternal effects) on offspring phenotypes and thereby evolutionary trajectories.

Future research should examine not only ultimate causes (adaptive values) but also proximate causes (functional mechanisms) for specific parental activities in the context of the entire phenotype. Mechanistic studies can enhance our knowledge of the evolution of parental care by identifying key stimuli and constraints (both biological and environmental), and by providing insights into the costs and benefits of parental care.

Human Parental Care. The evolution of parental care is worthwhile to study in its own right, and because it may provide insight into our own behavior. For example, several studies suggest that human biological parents tend to give better parental care than stepparents. Kin selection theory predicts this result because biological parents are genetically more related to their offspring than are stepparents.

[*See also* Group Living; Kin Selection; Life History Theory, *article on* Human Life Histories; Mate Choice, *article on* Human Mate Choice; Parent–Offspring Conflict; Social Evolution.]

BIBLIOGRAPHY

Allport, S. *A Natural History of Parenting*. New York, 1997. A popular account of parental care in a variety of animals.

Bateman, A. J. "Intra-sexual Selection in *Drosophila*." *Heredity* 2 (1948): 349–368.

Clutton-Brock, T. H. *The Evolution of Parental Care*. Princeton, 1991. A broad-ranging review of the fitness costs and benefits of parental care in animals.

Daly, M., and M. Wilson. "Stepparenthood and the Evolved Psychology of Discriminative Parental Solicitude." In *Infanticide and Parental Care*, edited by S. Parmigiani and F. S. vom Saal. London, 1994. Provocative discussion of the evolution of parental care in humans.

Glazier, D. S. "Trade-offs between Reproductive and Somatic (Storage) Investments in Animals: A Comparative Test of the Van Noordwijk and De Jong Model." *Evolutionary Ecology* 13 (1999): 539–555. Discusses the effects of resource acquisition and body condition on parental investment.

Glazier, D. S. "Smaller Amphipod Mothers Show Stronger Trade-offs between Offspring Size and Number." *Ecology Letters* 3 (2000): 142–149. Example of how body-space limits may cause a trade-off between offspring size and number.

Ito, Y. *Comparative Ecology*. Cambridge, 1980. Discusses effects of parental care on offspring number.

Jameson, E. W. *Vertebrate Reproduction*. New York, 1988. A good overview of the kinds of parental care exhibited by various vertebrates.

Jones, A. G., et al. "The Bateman Gradient and the Cause of Sexual Selection in a Sex-Role-Reversed Pipefish." *Proceedings of the Royal Society of London B* 267 (2000): 677–680.

Jonsson, K. I. "Life History Consequences of Fixed Costs of Reproduction." *Ecoscience* 7 (2000): 423–427. Comparison of fixed and variable costs of reproduction, including parental care.

Koteja, P. "Energy Assimilation, Parental Care and the Evolution of Endothermy." *Proceedings of the Royal Society of London B* 267 (2000): 479–484. Discusses coevolution between somatic and reproductive parental investments in birds and mammals.

Martin, T. E., J. Scott, and C. Menge. "Nest Predation Increases with Parental Activity: Separating Nest Site and Parental Activity Effects." *Proceedings of the Royal Society of London B* 267 (2000): 2287–2293. Excellent study of an ecological cost of parental care in birds.

Mousseau, T. A., and C. W. Fox. *Maternal Effects as Adaptations*. New York, 1998. Broad overview of how the maternal environment affects, and is affected by, evolution.

Poulin, E., and J.-P. Feral. "Why Are There So Many Species of Brooding Antarctic Echinoids?" *Evolution* 50 (1996): 820–830. An example of how parental care can affect evolutionary patterns.

Preston-Mafham, R., and K. Preston-Mafham. *The Encyclopedia of Land Invertebrate Behaviour*. Cambridge, Mass., 1993. Outstanding chapter on the parental behavior of land arthropods, including many illustrations and fascinating details.

Rosenblatt, J. S., and C. T. Snowdon. *Parental Care: Evolution, Mechanisms, and Adaptive Significance*. San Diego, Calif., 1996. Excellent reference on the physiological mechanisms and adaptive significance of parental care in both invertebrate and vertebrate animals.

Sheldon, B. C. "Differential Allocation: Tests, Mechanisms and Implications." *Trends in Ecology and Evolution* 15 (2000): 397–402. Review of evidence showing that mate choice can affect parental investment.

Thomas, D. W., J. Blondel, P. Perret, M. M. Lambrechts, and J. R. Speakman. "Energetic and Fitness Costs of Mismatching Resource Supply and Demand in Seasonally Breeding Birds." *Science* 291 (2001): 2598–2600. Remarkable study of the adaptive significance of seasonally timed parental care.

Trivers, R. L. *Social Evolution*. Menlo Park, Calif., 1985. Insightful, though dated, review of the evolution of parental care and other social behaviors, with especially good treatments of male–female and parent–offspring conflicts.

Willson, M. F. *Plant Reproductive Ecology*. New York, 1983. Includes useful discussion of patterns of parental investment in plants.

— Douglas S. Glazier

PARENT–OFFSPRING CONFLICT

In most sexually reproducing species, offspring carry only one-half of each parent's genes. Because each offspring receives a different 50 percent from each parent, relatedness among siblings is, like that between parents and offspring, also 50 percent; offspring thus have fitness interests likely to be incongruent with both siblings and parents. If there is valuable pre- and post-natal parental investment at stake (e.g., among contemporaneous nestlings), significant and even lethal competitions among sibs may ensue, despite the close genetic relatedness among the players. The theory of parent–offspring conflict (Trivers, 1974) is a formal adaptation of Hamilton's rule, $rb - c > 0$ (where r is the coefficient of relatedness between two genetic kin, b is the fitness gain to the recipient of a social act, and c is the fitness cost to the performer of a social act). The theory shows that each individual offspring should generally favor a skew in parental investment toward itself, while parents should prefer an equal allocation of investment across all offspring, both current and future. To say the same thing more quantitatively and from a gene's perspective, the body a gene inhabits should be impelled to consume the next unit of parental investment up until the point at which that investment becomes twice as valuable to a full sibling, because the 50 percent chance that the sibling carries an identical copy of the gene then makes it a better bet. Getting back to whole organisms and thus to behavior, parent–offspring conflict has been characterized as that which results when natural selection confers greater selfishness in offspring than is optimal for the parent, or, more concisely, mismatched optima for parental investment (Parker, 1985). The idea is important because "interactions between parents and young are among the most widespread and basic social behaviors exhibited by animals (and even by plants)," with obvious implications for humans (Godfray, 1995).

The heart of Trivers's perspective was that offspring must be viewed as active participants in their own care, not mere "passive vessels" into which investment is poured. Offspring traits are molded by natural selection for their efficiency at selfishly extracting additional investment from parents; and parents, being evenly related to all their offspring (0.50), should coevolve traits for resisting selfish extraction. There is, of course, a second category of asymmetry between parents and dependent offspring: parents tend to be larger and stronger than their neonatal opponents. In Trivers's colorful words, "the caribou calf cannot fling its mother to the ground and suckle at will." His solution to this problem was twofold: (1) in some highly social groups (e.g., many bees, ants, and wasps) the primary care-givers for neonates are older siblings, so parentlike control is often abdicated to the next generation; but (2), in most species, the physically inferior offspring must resort to "psychological weaponry" to get what they want. This weaponry was posited as consisting mainly of deceitful misinformation regarding offspring need: if a baby can exaggerate the degree to which it requires extra investment, beyond the parent's power to detect the sham, it stands to gain at the parent's expense. This argument has great appeal. An offspring has direct physiological information about its own strength, condition, and need, but parental information on the same subjects is at least partially filtered through the offspring's own signals. If natural selection inflates those signals and parents cannot risk dismissing them summarily (lest they be accurate), the adults' overall "sales resistance" may be inadequate defense for parental interests. A great many behavioral features of familiar parent–offspring social interactions have been interpreted within that framework, including the timing of weaning in mammals and tantrum displays by human children. In addition, parent–offspring conflict provides insights into begging, an energetically costly behavior in which offspring compete to get more of the parent's resources. Among the many examples of this type of parent–offspring conflict are the brightly colored gapes of canary nestlings, which intensify with increased hunger (Kilner, 1999), variation in begging behavior as an indicator of hunger in ring-billed gulls (Iacovides and Evans, 1998), and the increased provisioning of chicks in response to begging by the pied flycatcher.

The early response to Trivers's idea was decidedly mixed. On the positive side, the whole notion of evolutionary conflict between members of a previously sacrosanct social unit was deliciously counterintuitive. The growing realization that natural selection's potency is concentrated overwhelmingly at the level of individual organisms and even of the gene (e.g., Dawkins, 1989) accommodated the notion of parent–offspring conflict very nicely. On the other hand, a rapid protest was sounded with the argument that excessive selfishness by an offspring could not be favored by natural selection because the supposed beneficiary would lose its short-term gains when it had matured and was parenting its own extra-selfish offspring (Alexander, 1974). The controversy that followed, sometimes called the "battleground" issue (Godfray, 1995), inspired a handful of formal mathematical modeling exercises (e.g., Parker and Macnair, 1978; Stamps et al., 1978; Stamps and Metcalf, 1980), all of which concluded that true evolutionary conflict can exist between the interests of parents and offspring. Specifically, the models showed that the theoretical logic is sound and that traits can evolve by natural selection wherein offspring extract more investment than parents should favor (for an excellent review of the models, see Godfray, 1995). Parent–offspring conflict has thus been established on a conceptual level, and some authors have concluded that its importance is thereby assured. In the 1989 "new edition" of *The Selfish Gene*, for example, Richard Dawkins appended (p. 298), "There is indeed rather little to add to his [Trivers's] paper of 1974, apart from some new factual examples. The theory has stood the test of time."

Problems in Testing Parent–Offspring Conflict. Other observers are considerably less satisfied with the status quo. While conceding that the theoretical heart of the argument is sound, one may still question whether the parent–offspring conflict insight has actually provided a useful explanation for diverse behavioral traits. We may see all kinds of situations in which the behavioral wishes of parents and offspring appear to differ, perhaps even sharply, but observing a petty squabble is not the same as demonstrating that significant fitness costs and benefits are at stake. To take a facetious example, five minutes spent in any grocery store checkout line will expose the observer to a squabble over candy bars or chewing gum, possibly escalating to a full tantrum, but the connection between such an incident and true fitness consequences is less than tenuous (Mock and Forbes, 1992; Mock and Parker, 1997).

A rigorous and compelling demonstration of parent–offspring conflict would offer two key features. First, after a system has been identified in which the behavioral actions of parents and offspring appear to be at loggerheads (the "squabble"), some measurable effect on the fitness interests of both parties must be demonstrated. The most obvious way to approach this would be to manipulate the behavioral squabble experimentally, allowing the parent to prevail in one treatment group (in which case offspring fitness should be depressed) while allowing the offspring to prevail in a second group (with corresponding negative impact on parental fitness). This would establish that a true evolutionary conflict underlies the superficial squabble.

The second feature desired in a demonstration of parent–offspring conflict is simply that the offspring should "win" to some degree. Of course, this is not to say that offspring necessarily do get their way in nature, but it is the tantalizing possibility that Trivers's theory offered by asserting that the offspring is an active and effective player. It is much too easy to think of both hypothetical and real cases where parents simply do things their own way. For example, nestling birds have often been demonstrated to enjoy improved survival and recruitment into the breeding population if they begin life inside a large and nutritious egg and/or in a relatively small brood (less competition for post-hatching food and other resources). So the fitness insult of *not* beginning life in a big egg or small family is clear. But, of course, a zygote has no say in the size of the egg it will inhabit, nor in the number of eggs going into the clutch. Asymmetric relatedness is not always needed to understand the fact that parents do many things that maximize their own fitness and impose the situation on their powerless offspring.

Trivers placed emphasis on the use of psychological weapons by offspring. The most promising behavioral systems that have been studied under field conditions involve species whose offspring appear to wield greater control over how post-zygotic investment is allocated. Here is precisely where one would expect to find observable manifestations of offspring imposing their will over parental interests.

What Parent Birds Really Want. The simplest prediction from an "offspring-wins" scenario is that parental investment will not be allocated evenly across offspring. The original concept held that the individual offspring views itself as a full glass and its sibling(s) as half empty (if they have the same mother and father), so it should try to skew investment toward itself; conversely, the parent, being symmetrically related to all offspring, should avoid any such skew. By this logic, any family system in nature where one finds that parental investment is not fairly distributed might be interpreted as evidence of parent–offspring conflict won by the offspring that got more. Such family systems are exceedingly common in nature, scattered across highly diverse animal and plant taxa. For one well-studied example, consider a small North American songbird called the red-winged blackbird. Typically, four eggs are produced at daily intervals, and incubation commences after number three is in the nest—a widespread parental habit that puts the last egg's embryo at a slight temporal disadvantage. With the first three empowered to begin development a day earlier (parental warmth being essential for that process), the last one hatches one full day later than the nearly synchronous emergence of the other three. This handicaps the runt in terms of its ability to compete (by begging and stretching its neck and gape toward a food-delivering parent), so it grows more slowly and is much more likely to die of starvation if food proves insufficient (Forbes et al., 1997). In another avian example, the three eggs are laid in great egret nests at 1–2-day intervals, and incubation begins even earlier, so the second chick hatches a day behind the first, but one or two days ahead of the third. In this species, the natural size advantages enjoyed by the senior siblings are reinforced by overt aggression (vigorous pecking with a long bill) that quickly establishes a linear dominance hierarchy down the age/size order. Food (boluses of tiny fish regurgitated by the parents) is distributed on average as a 40-40-20 split, with the immediate cause of this unevenness clearly being social intimidation of the youngest chick.

These are two cases (many more exist; see Mock and Parker, 1997) in which parental resources are disproportionately consumed by certain offspring to the possibly fatal detriment of other offspring, with behavioral evidence pointing strongly to the actions of the stronger siblings as creating the disparities. The key question concerning parent–offspring conflict, though, is whether jeopardizing or killing the youngest nestling is necessarily contrary to the parents' fitness interests. The most straightforward interpretation concludes that it must be, but that reasoning is rooted solely in the genetic relatedness asymmetry. On the other hand, there are other aspects, involving either of the other two variables in Hamilton's rule (b, fitness benefit to the recipient, or c, fitness cost to the actor), that are capable of reversing the parental view, leading to the view that avian brood reduction may not represent evolutionary conflict between the parents and the offspring. In particular, under many realistic circumstances, parents may actually gain fitness from losing one or more offspring! This may seem paradoxical, since it seems that parents whose fitness is best served by having, say, two offspring should create that number and no more, in which case no secondary trimming of family size would be required. There are, however, several incentives for parents to "aim high" initially, a life-history strategy that may well lead to a parentally-condoned family size abridgement later in the cycle.

The easiest way to appreciate that parents may actually favor brood reduction is to consider species in which the adults perform the selective executions themselves (see references in Elgar and Crespi, 1992; Mock and Parker, 1997). For example, maternal cannibalism is quite common in rodents, and the selective abortion of fertilized zygotes is widespread in plants and can be profligate. A navel orange tree literally drops tens of thousands of tiny oranges (while they are still inexpensive) each season before giving full investment to a few hundred ripe fruits. Sand tiger shark embryos hatch inside the womb and quickly develop teeth and lateral line organs; both structures are used to hunt and consume sibling embryos and eggs. The mother's complicity in this arrangement can be seen in the fact that she continuous to generate thousands of additional eggs, most of them fertilized, that go into feeding the single, gigantic embryo through its 9–12-month gestation period. Turning back to birds, there are species of moorhens, coots, and gulls where parents seize and kill one or two of their own chicks. In the Magellanic penguin, two chicks hatch several days apart and parents selectively respond only to the begging signals of the larger chick, thus causing its smaller sibling to starve.

In most brood-reducing birds, however, parental favoritism takes much subtler forms. The common habit of hatching eggs asynchronously, for example, is not universal (waterfowl, gallinaceous birds, and other precocial species typically complete laying before starting incubation, with the result that the chicks emerge within a few hours of one another) and is a clear byproduct of parental activity (early incubation). It was proposed more than 50 years ago that this could be a parental means of dealing with ecological unpredictability (Lack, 1947). That is, because parent birds must commit to a certain number of eggs several weeks before those hatch, it may be difficult in most habitats to ascertain precisely how many can be supported when brood demands peak several weeks into the future. For many species, breeding is seasonal, so the option of adding more eggs later is impractical, but the alternative of creating one or two extra eggs at the outset cushions parents against the error of raising fewer offspring than they can afford. If food conditions, for example, turn out to be especially favorable, the extra offspring can be raised well and parental fitness is enhanced. On the other hand, if food conditions are poor (or even average), it may be necessary to trim family size. In light of that possibility, natural selection might favor any parental traits that make the final downward adjustment as efficient (e.g., as inexpensive) as possible. So it was proposed that the parental habit of early incubation might well represent a manipulation of relative competitiveness that could facilitate brood reduction, should that prove necessary (Lack, 1947). This general incentive for parental overproduction, called "resource tracking" (Temme and Charnov, 1987), has been demonstrated experimentally under field conditions for several bird species. When extra food is made available artificially, parents can and do raise more healthy offspring (Magrath, 1989). Of special interest, if parent American kestrels (small falcons) are given supplemental food during the laying period, they adjust their incubation schedule so as to reduce the degree of hatching asynchrony, showing that this parental strategy can be fine-tuned in response to favorable ecological conditions (Wiebe and Bortolotti, 1994).

Two other general incentives have been identified for parents to produce an oversized family initially. By creating one extra offspring in a given cycle, the parents gain a cushion against a host of early mishaps (failure of an egg to hatch, accidental death of a hatchling, loss to a grab-and-run predator, etc.) that might befall a member of the core brood: the extra one can serve as a replacement. At the simplest level, then, modest overproduction provides a form of insurance against parents' having to squander a whole breeding cycle on an undersized family. In an extension of the replacement-offspring argument, it has also been argued that parents might create additional progeny, then choose the most promising specimens from their own array, thus screening for genetic quality in much the same way that sexual partners evaluate each other. Such screening is generally better suited to plants than to animals (Forbes and Mock, 1998)—by virtue of larger "families" and lower control over which male gametes (pollen) will arrive. For example, in a population of *Mimulus guttatus* growing in copper-stressed soil, selection has favored genes for copper-tolerance. When female flowers were experimentally provided with pollen either from local donors (carrying those genes) or from males in normal soil, they selectively aborted offspring in the second group and retained those from the first (Searcy and Macnair, 1993). Such mechanisms can function as the equivalent of animal mate-choice processes, albeit operating relatively late in the cycle.

Progeny-choice parallels have been proposed for animals, but seldom documented satisfactorily. An interesting exception concerns American coots, in which the post-hatching coloration of chicks is unusually garish (blue crown skin, orange head down). The two parents tend to split the large brood of precocial young in half and show conspicuous preferences for the more brightly colored individuals in each subgroup (Lyon et al., 1994). It is not known whether chicks that are especially healthy or otherwise of highest quality express the greatest degree of brightness.

Finally, parental fitness may gain from some over-production in cases where the extra offspring contribute positively to the success of the core brood members. This is most conspicuous in cases of sibling cannibalism (the extra young supply a meal), which is common in small aquatic nurseries that feature very limited alternative foods and/or tend to evaporate (e.g., tadpoles in puddles, odonate larvae in tree-holes) and in species with cooperative breeding (where nonbreeding adults may assist their reproductive siblings), but it might also have value in subtler ways (e.g., by reducing surface-to-volume ratios in cool habitats, thereby providing thermal benefits). The important thing is that these three categories of overproduction incentives (resource-tracking, replacement offspring, and sibling assistance), which may even operate additively, may routinely push parents toward creating more offspring than is in the *parents' own interest* to support fully.

This broader life-history perspective on parental reproduction may explain observed cases of parental favoritism that previously seemed puzzling. In almost every siblicidal bird studied closely, the elder nestlings peck their younger siblings to death right in front of their attending parents and the latter tend not to intervene (e.g., Mock et al., 1990). It is obvious that parents are dozens of times larger than the combatants, ostensibly quite capable of ending any fight by simply brooding their young or by dragging them apart. Parental passivity in this context could be interpreted either as a complete victory for the attacking offspring (such that parents no longer even bother trying to prevent the killing) or, more parsimoniously, as a matter of no fitness consequences for parents (a side-product of overproduction). The only claimed exceptions among siblicidal birds are the South Polar skua and masked booby. In the former, a few accounts describe active separation of fighting nestlings by their parents, though the frequency and effectiveness of such actions remain undetermined (Young, 1963). In the latter, parents have not actually been seen to perform active interference, but when nestlings of this "obligate" siblicide species are reared by parents of a less aggressive congeneric species, mortality is reduced. This has been interpreted primarily in terms of subtle things the foster parents may have done that render sibling aggression less deadly (Lougheed and Anderson, 1999).

Worker-Queen Conflict in Social Insects. The clearest behavioral demonstration of parent–offspring conflict follows from a prediction Trivers made shortly after inventing the theory (Trivers and Hare, 1976). The genetics of the insect order Hymenoptera (ants, bees, and wasps) differ from most other animals by virtue of females being diploid (i.e., having both members of each pair of homologous chromosomes), while males are haploid (having only one member of each pair). This occurs because eggs destined to become sons are left unfertilized (the sperm storage organ inside the mother being held shut by special musculature), while those destined to become daughters receive paternal genes that combine with maternal ones. This difference between the sexes, called haplodiploidy, has several interesting consequences for degrees of relatedness within family units. Briefly, because the sperm donor's gametes are all identical copies, the haploid male having no paired chromosomes to shuffle, full sisters are related by a lofty 0.75, as opposed to the usual 0.50 found in other genetic systems. By contrast, a sister's degree of relatedness to her brother is only 0.25, since he cannot match any of the genes she received from her father. So sisters have triple the interest in accommodating reproductive sisters as they have with reproductive brothers. In a related issue, haplodiploid species are more likely to evolve highly cooperative societies wherein roles become highly specialized. At the pinnacle, "eusocial" species have only one reproductive female, the queen, and many nonbreeding workers, all female. Of relevance to the theory of parent–offspring conflict, the workers typically provide all the post-zygotic investment to eggs and larvae, so they are empowered to an unusual degree. Much of the time new eggs will become nonbreeding workers, so no direct fitness is at stake, but broods of reproductive offspring are produced from time to time, and then workers are predicted to maximize their own interests.

If the workers caring for a mixed-sex brood of future reproductives are all full siblings, their optimum for the sex ratio should be approximately 3:1 in favor of sisters. But the queen's optimum is expected to be close to parity, because she is equally related to her own sons and daughters. When Trivers and Hare (1976) surveyed many species of ants, they found a strong female bias in reproductive broods, which they interpreted as reflecting a parent–offspring conflict that was being won by the workers. More recent comparative studies have also tended to find a similar female bias in reproductive broods (e.g., Seger, 1991). If this is being driven by the relatedness asymmetry between how hymenopteran workers view sisters vs. brothers, then the whole effect would be expected to disappear temporarily whenever the old queen dies. At such a moment, one of the daughters becomes the new queen and all her sisters suddenly agree with the queen's optimal sex ratio. This happens because their relatedness to the next brood of reproductives is channeled directly through the new queen: they are evaluating nieces and nephews, not sisters and brothers. So, for one generation, the colony sex ratio ought to swing back toward parity.

Observations on a population of sweat bees showed a clear swing in the direction of male reproductives after natural queen deaths (Yanega, 1988). This prediction has also been tested experimentally with a different bee species by the simple manipulation of removing old queens. As predicted, the sex ratio of the next reproductive brood swung closer to parity for one generation before returning to the typical female-bias (Mueller, 1991).

Prognosis. Parent–offspring conflict remains an elegant theory fulfilling much of its early promise, and yet having to accommodate new views. It has been applied eagerly to a profusion of natural phenomena, but seldom tested critically. There is no a priori reason to expect this situation to continue: the theory is robust. Furthermore, considerable data exist to show that animal nervous systems can be manipulated "psychologically," so parents are presumably vulnerable in this respect. For example, parent reed warblers have been shown to extract information from two elements of their brood's begging display, the total calling rate and the total gape areas of the exposed mouths, which they integrate and use as the basis for their decisions on how much food to deliver. Unfortunately for them, the European cuckoo, which evicts all the warbler nestlings and then is raised by the host parents, has evolved a display that fits the sensitivity of warbler adults. Remarkably, the cuckoo uses a supernormal vocal signal (it begs at the rate of *eight* warbler chicks) to compensate for its weaker visual cue (having a small mouth for its size), and the parents respond with enough food to have raised four warbler chicks (Kilner et al., 1999). Of course, this interspecific deception is not true parent–offspring conflict in Trivers's sense, but it suggests tantalizing possibilities of within-species parallels.

[*See also articles on* Eusociality; Parental Care; Social Evolution.]

BIBLIOGRAPHY

Alexander, R. D. "The Evolution of Social Behavior." *Annual Review of Ecology and Systematics* 5 (1974): 325–383.

Dawkins, R. *The Selfish Gene*. New edition. Oxford, 1989.

Elgar, M. A., and B. J. Crespi, ed. *Cannibalism: Ecology and Evolution among Diverse Taxa*. Oxford, 1992.

Forbes, L. S., and D. W. Mock. "Parental Optimism and Progeny Choice: When Is Screening for Offspring Quality Affordable?" *Journal of Theoretical Biology* 192 (1998): 3–14.
Godfray, H. C. J. "Evolutionary Theory of Parent–Offspring Conflict." *Nature* 376 (1995): 133–138.
Iacovides, S., and R. M. Evans. "Begging as Graded Signals of Need for Food in Young Ring-Billed Gulls." *Animal Behaviour* 56 (1998): 79–85.
Kilner, R. M. "Family Conflicts and the Evolution of Nestling Mouth Colour." *Behaviour* 136 (1999): 779–804.
Kilner, R. M., D. G. Noble, and N. B. Davies. "Signals of Need in Parent–Offspring Communication and Their Exploitation by the Common Cuckoo." *Nature* 397 (1999): 667–672.
Lack, D. "Clutch Size in Birds." *Ibis* 89 (1947): 302–352.
Lougheed, L. W., and D. J. Anderson. "Parent Blue-Footed Boobies Suppress Siblicidal Behavior on Offspring." *Behavioral Ecology and Sociobiology* 45 (1999): 11–18.
Lyon, B. E., J. M. Eadie, and L. D. Hamilton. "Parental Choice Selects for Ornamental Plumage in American Coot Chicks." *Nature* 371 (1994): 240–243.
Magrath, R. "Hatch Asynchrony and Reproductive Success in the Blackbird: A Field Experiment." *Nature* 339 (1989): 536–538.
Mock, D. W., and L. S. Forbes. "Parent–Offspring Conflict: A Case of Arrested Development?" *Trends in Ecological Evolution* 7 (1992): 409–413.
Mock, D. W., and L. S. Forbes. "The Evolution of Parental Optimism." *Trends in Ecological Evolution* 10 (1995): 130–134.
Mock, D. W., and G. A. Parker. *The Evolution of Sibling Rivalry.* Oxford, 1997.
Mock, D. W., H. Drummond, and C. H. Stinson. "Avian Siblicide." *American Scientist* 78 (1990): 438–449.
Mueller, U. "Haplodiploidy and the Evolution of Facultative Sex Ratios in a Primitively Eusocial Bird." *Science* 254 (1991): 442–444.
O'Connor, R. J. "Brood Reduction in Birds: Selection for Infanticide, Fratricide, and Suicide?" *Animal Behaviour* 26 (1978): 79–96.
Ottosson, U., J. Backman, and H. G. Smith. "Begging Affects Parental Effort in the Pied Flycatcher, *Ficedula hypoleuca*." *Behavioral Ecology and Sociobiology* 41 (1997): 381–384.
Parker, G. A. "Models of Parent–Offspring Conflict. V. Effects of the Behaviour of the Two Parents." *Animal Behaviour* 33 (1985): 519–533.
Parker, G. A., and M. R. Macnair. "Models of Parent–Offspring Conflict. I. Monogamy." *Animal Behaviour* 26 (1978): 97–111.
Searcy, K. B., and M. R. Macnair. "Developmental Selection in Response to Environmental Conditions of the Maternal Parent in *Mimulus guttatus*." *Evolution* 47 (1993): 13–24.
Seger, J. "Cooperation and Conflict in Social Insects." In *Behavioural Ecology: An Evolutionary Approach*, 3d ed., edited by J. R. Krebs and N. B. Davies, pp. 338–373. Oxford, 1991.
Stamps, J., and R. A. Metcalf. "Parent–Offspring Conflict." In *Sociobiology: Beyond Nature-Nurture*, edited by G. Barlow and J. Silverberg, pp. 598–618. Boulder, 1980.
Stamps, J., R. A. Metcalf, and V. V. Krishnan. "A Genetic Analysis of Parent–Offspring Conflict." *Behavioral Ecology and Sociobiology* 3 (1978): 369–392.
Temme, D. H., and E. L. Charnov. "Brood Size Adjustment in Birds: Economical Tracking in a Temporally Varying Environment." *Journal of Theoretical Biology* 126 (1987): 137–147.
Trivers, R. L. "Parent–Offspring Conflict." *American Zoologist* 14 (1974): 249–263.
Trivers, R. L., and H. Hare. "Haplodiploidy and the Evolution of Social Insects." *Science* 191 (1976): 249.
Wiebe, K. L., and G. D. Bortolotti. "Food Supply and Hatching Spans of Birds: Energy Constraints or Facultative Manipulations?" *Ecology* 75 (1994): 813–823.
Yanega, D. "Social Plasticity and Early-Diapausing Females in a Primitively Social Bee." *Proceedings of the National Academy of Sciences USA* 85 (1988): 4374–4377.
Young, E. C. "The Breeding Behaviour of the South Polar Skua, *Catharacta maccormicki*." *Ibis* 1 (1963): 203–233.

— Douglas W. Mock

PARSIMONY. *See* Phylogenetic Inference.

PATTERN FORMATION

The order and diversity of life have fascinated naturalists and evolutionary biologists for centuries. Biological species are discrete entities, but why are there some forms in nature and not others? Which is the main source of order in the discreteness of natural diversity? The traditional approach would rely on natural selection and genetic variation. This approach emphasizes an external source of order and diversity.

The alternative approach is looking at the internal constraints—that is, looking at the developmental process as a dynamic system with organizational principles. As the Spanish developmental biologist Pere Alberch pointed out, natural selection determines the winner of a game, but developmental constraints determine the players. We need to understand the process of development from a physical perspective. We need to have a mechanistic model.

Understanding development from a physical perspective was a difficult task for many decades. The nineteenth century was characterized by two big scientific theories emphasizing the irreversibility of time, but both seemed to point toward opposite directions. On one hand, the Second Principle of Thermodynamics explains that an isolated system will evolve toward maximum disorder. On the other hand, the theory of natural selection provides a mechanism for evolving diversity and complexity. For many decades there was a big gap between biology and physics. In order to bridge this gap, one has to realize that biological systems are open and far from equilibrium. There is no longer a contradiction between physics and biology.

The great British mathematician Alan Turing was the first to think about development from a mechanistic, physical point of view. His paper published in 1952 is considered one of the most influential works in theoretical biology. Turing thought about a simple model

REACTION-DIFFUSION (TURING) MODELS

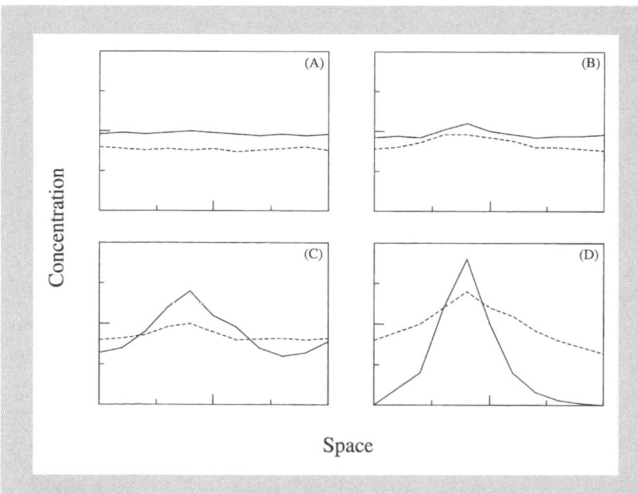

Turing's model can be described as follows:

$$\frac{\partial A}{\partial t} = F(A,I) + D_A\nabla^2 A,$$

$$\frac{\partial I}{\partial t} = G(A,I) + D_I\nabla^2 I,$$

where *A* is the activator concentration, *I* is the inhibitor concentration, and D_A and D_I are the diffusion rates of activator and inhibitor, respectively. $\nabla^2 A$ and $\nabla^2 I$ are the diffusion (Laplacian) operators. *F* and *G* are functions of both morphogens. These nonlinear functions assume that (1) *A* is produced at a rate proportional to its concentration, (2) *I* inhibits production of *A*, and (3) *I* is produced at a rate proportional to *A*. These partial differential equations describe rates of change in both time and space. The figure shows how the spatial distribution of both morphogens changes through time and space. The *x* axis represents a spatial dimension and the *y* axis represents morphogen concentration. Figures (A–D) correspond to successive temporal stages. At the beginning (a) we start from a homogeneous distribution of both morphogens. However, no matter how homogeneous this distribution is, there are continuous fluctuations in the abundance of the activator. Now, imagine that the concentration of the activator is slightly above the average in a spatial location. Since the inhibitor is produced in proportion to the activator, the inhibitor concentration will also be higher at this point. The inhibitor diffuses faster than the activator, so it will diffuse to the neighborhood of the focal point. In the focal point we have a relatively high ratio of activator to inhibitor, and thus a high rate of activator production. The activator will start increasing at this point until reaching a peak. In the neighborhood, the inhibitor will deplete the activator, so the concentration of the latter there will be low (B–C). A final equilibrium distribution will be reached since the growth of the activator at the peak will be compensated by its loss at the neighborhood (D). Thus, we have evolved from a symmetric, homogeneous distribution of morphogens to a patchy distribution of both the activator and the inhibitor.

—JORDI BASCOMPTE

with two morphogens (chemical species): an activator and an inhibitor. The activator is produced at a rate proportional to its concentration; that is, the product of the reaction catalyses its own production. The inhibitor inhibits the activator. Next, imagine that both morphogens diffuse in space, but at different rates. The inhibitor diffuses faster than the activator. Turing wrote a mathematical model based on a couple of partial differential equations encapsulating these elements. Such a model is known as a reaction-diffusion model (See Vignette). The result is compelling. The combination of local autocatalysis and long-range inhibition is able to create spatial pattern, to transform a homogeneous situation to a structured one. This result comes as a surprise because passive diffusion is a force that increases homogeneity (think about a drop of ink in a glass of water).

The activator-inhibitor mechanism can explain pattern formation in developmental systems. Various scientists since Turing have explored the possibilities of this framework to explain particular situations. For example, the mathematical biologist James Murray has used a Turing-like mechanism to explain pattern formation in the pigmentation of mammal coats, such as leopard spots and zebra stripes. Melanin, the pigment producing hair color, is located within specialized cells called melanocytes. These melanocytes are derived from melanoblasts, one kind of precursor cell that can be activated by a morphogen. Murray proposed that the diffusion-driven instability between an activator and an inhibitor could explain the pattern of melanoblast activation. If this were the mechanism at work, the final pattern would depend on the size of the coat. If the size is too small, no differentiation takes place because there is not enough distance between local autocatalysis and the long-range inhibition. The distribution of pigment will thus be homogeneous. This is the case of small ani-

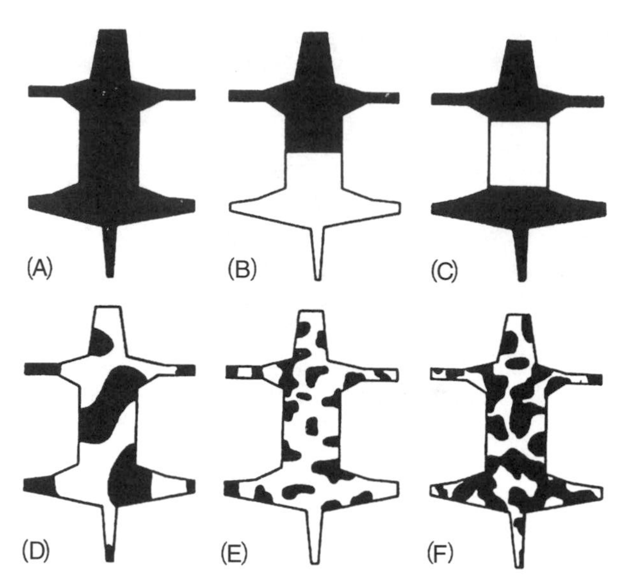

FIGURE 1. Mammal Coat Patterns as Predicted by a Reaction-Diffusion Model.
From (A) to (F) the size of the simulated coat is increased, the pattern becoming more complex. This is in agreement with real coat patterns. J. D. Murray. *Mathematical Biology.* © 1989 Springer-Verlag.

mals like mouse. A first bifurcation in differentiated pattern arises when body size reaches a threshold, creating a half-white, half-black animal such as the honey badger (*Mellivora capensis*) or the Valais goat (*Capra aegagrus hircus*). One can further increase the size of the coat and a new bifurcation takes place, generating an animal with a black-white-black pattern. An example would be the anteater *Tamandua tetradactyla*. Larger animals like the leopard can have a complex coat pattern (Figure 1). Murray's model can even predict small details, such as the pattern on tails and in zebra stripes where the body intersects the extremities. Similar mechanisms have been adduced to operate in the pattern formation of seashells (Figure 2), butterfly wings, and the skin of some fishes (Kondo and Asai, 1995). Murray (1989) presents a detailed review.

FIGURE 2. The Pigment Patterns of Seashells.
Two real examples (top) and the respective simulated patterns (bottom) as predicted by a reaction-diffusion model. Left corresponds to *Conus marmoreus,* and right corresponds to *Conus nobilis marchionatus.* H. Meinhardt. *The Algorithmic Beauty of Sea Shells.* © 1995 Springer-Verlag.

Pere Alberch, working with Murray, Emily Gale, George Oster, and others, extended the reaction-diffusion models to explain developmental sequences in amphibians. Alberch compared reaction-diffusion models with laboratory experiments by modifying the size of the developing limb with colchicine, an inhibitor of cell division. He found that as the size of the forming extremity is reduced, some bifurcations in the process of bone formation disappear. This causes the loss of a phalange or even a digit. Interestingly, the loss of elements is not random but follows a well-defined sequence, which reflects the dynamic process of bone formation. Alberch and colleagues showed that the experimental variation in limb morphology matched natural variation (Figure 3). By modifying the size of the forming limb, a species generates from the bifurcation variability of the dynamic processes that control pattern.

The British developmental biologist Brian Goodwin used a similar approach to understand the formation of the whorls of the giant unicellular green alga *Acetabularia acetabulum.* Such structures apparently have no function, and Goodwin, using similar morphogenetic models, suggested that the dynamic process of development necessarily produces such structures. That is, whorls should not be understood from the point of view of function, but as the outcome of the dynamic processes that control development.

The work of people such as Murray, Goodwin, Meinhart, and Alberch complements traditional views of development and evolution. Genes are certainly important—they determine the production of morphogens and their diffusion rates—but genes only set the parameter for the dynamic processes that shape pattern.

One problem with Turing's theory has been the difficulty in identifying the morphogens. In the case of *Acetabularia*, calcium has been suggested as one of the morphogens involved in the formation of whorls. Also, there are alternative dynamic theories to explain pattern formation in development. One of those is epigenetic selection, the differential survival of cellular lineages competing for limited developmental signals (Sachs, 1991). Other theories are based on cellular automata models considering cell–cell interaction and energetics (Cocho et al., 1987).

[*See also* Cell Lineage; Development, *article on* Developmental Stages, Processes, and Stability; De-

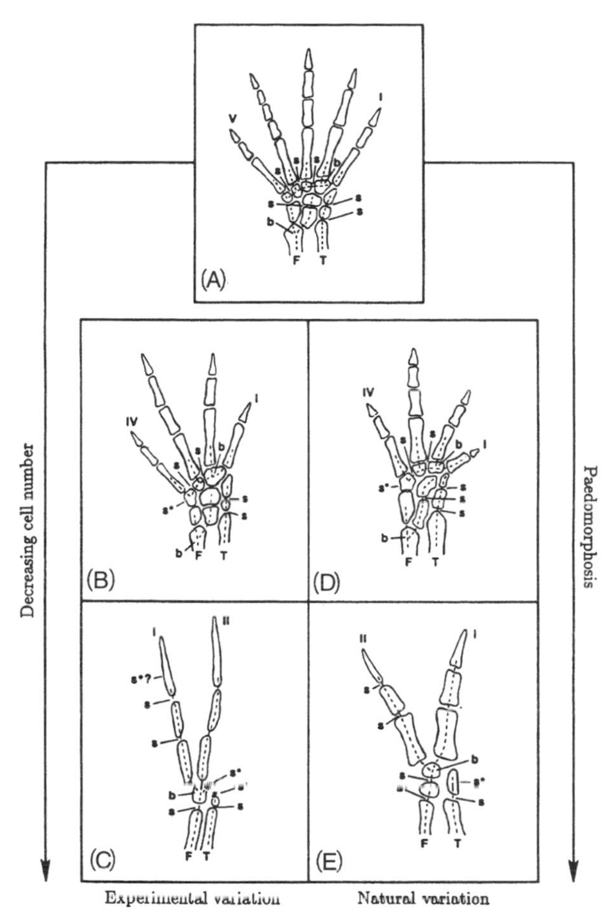

FIGURE 3. Sequential Loss of Skeletal Elements in the Limb of Amphibians as the Size of the Forming Limb is Reduced. Experimental variation (using the cell division inhibitor colchicina) matches evolutionary variation. (A) Normal foot morphology of the salamander *Ambystoma mexicanum.* (B) and (C) are four-toed and two-toed limbs experimentally obtained after treating *Ambystoma* with colchicine. (D) and (E) correspond to the four-toed *Hemidactylium scutatum* and the two-toed *Proteus anguinus* respectively. After G. F. Oster et al., 1998.

velopmental Selection; Multicellularity and Specialization.]

BIBLIOGRAPHY

Alberch, P., and E. Gale. "A Developmental Analysis of an Evolutionary Trend: Digital Reduction in Amphibians." *Evolution* 39 (1985): 8–23.

Cocho, G., R. Pérez-Pascual, J. L. Rius, and F. Soto. "Discrete Systems, Cell-Cell Interactions and Color Pattern of Animals. I. Conflicting Dynamics and Pattern Formation. II. Clonal Theory and Cellular Automata." *Journal of Theoretical Biology* 125 (1987): 419–447.

Goodwin, B. C. *How the Leopard Changed its Spots: The Evolution of Complexity.* New York, 1994. A detailed account of the role of pattern formation in development, with emphasis in structuralism.

Kondo, S., and R. Asai. "A Reaction-Diffusion Wave on the Skin of the Marine Angelfish *Pomacanthus.*" *Nature* 376 (1995): 765–768. One of the best examples of pattern formation through a reaction-diffusion process, providing elegant experimental evidence of Turing's great idea.

Meinhardt, H. *The Algorithmic Beauty of Sea Shells.* Berlin, 1995. Compelling evidence for pattern formation in sea shells. The book is both extremely clear and beautifully illustrated.

Murray, J. D. *Mathematical Biology.* Berlin, 1989. One of the most complete overviews on pattern formation and development.

Oster, G. F., N. Shubin, J. D. Murray, and P. Alberch. "Evolution and Morphogenetic Rules: The Shape of the Vertebrate Limb in Ontogeny and Phylogeny." *Evolution* 42 (1988): 862–884. The dynamics of limb skeletal formation from the perspective of pattern formation. Experimental manipulations are compared with natural variation.

Sachs, T. *Pattern Formation in Plant Tissues.* Cambridge, 1991. An overview of developmental selection as an alternative mechanism for pattern formation.

Thompson, D. W. *On Growth and Form.* Dover Publications, 1992. First published in 1917, this book presented the first comprehensive treatment of biological pattern arising from simple physical processes. One of the great books in the history of biology.

Turing, A. M. "The Chemical Basis of Morphogenesis." *Philosophical Transactions of the Royal Society of London Series B* 237 (1952): 37–72. The first paper dealing with a mechanistic model to understand pattern formation in development. It was the starting point for later work. A classic.

— JORDI BASCOMPTE

PEDIGREE ANALYSIS

In 1865 Gregor Mendel published the laws of genetic inheritance, which he derived from studies of cross-breeding among pea plants with distinct sizes, shapes, and colors. Upon rediscovery of Mendel's work at the beginning of the twentieth century, ample confirmation of Mendelian principles was rapidly obtained in a variety of experimental organisms, particularly the fruit fly (*Drosophila melanogaster*) and the orange filamentous fungus (*Neurospora crassa*). It took several more decades, however, to demonstrate that these laws of inheritance operated in humans as well. This was because Mendel's laws apply to traits determined by a single gene, whereas the great majority of observable traits in humans, such as physical body measurements, are specified by a number of genes plus the effects of environment and other confounding factors. With time, human traits showing simple Mendelian transmission patterns were discovered, including color blindness, ABO blood types, and a number of inherited diseases. Studies of these traits provided convincing demonstrations that Mendel's laws of inheritance were valid in humans.

Since the 1930s, thousands of examples of human Mendelian traits have been described (see McKusick, 1999; "Online Mendelian Inheritance in Man"). The great majority of these traits are inherited diseases that, because of the clinical scrutiny of affected individuals and

RECESSIVE DISORDERS IN "ISOLATED" POPULATIONS

Populations can become genetically isolated because of geographical barriers or cultural choices. In such populations, marriages are often between members of the same group. This inbreeding, coupled with growth of these isolated populations, greatly raises the frequency that carriers of a deleterious recessive mutations will give rise to offspring with the disease phenotype. Two examples can be described in which a deleterious mutation has become frequent in a particular population, and subsequently has resulted in the presence of a genetic disease nearly unique to this population.

Recessively inherited, profound, congenital deafness at the DFNB3 locus (OMIM 600316) is segregating in Bengkala, a Balinese village. The DFNB3 locus was mapped to chromosome 17 and the gene was identified. In Bengkala there are 48 deaf and 2,200 hearing people (Friedman et al. 1995; Wang et al. 1998). The deaf communicate with each other and with the hearing villagers by using a sign language unique to Bengkala suggesting that deafness has been present in this community for many generations. Among the hearing members of this village, one in four carry the recessive DFNB3 mutation. This means that in Bengkala one in sixteen marriages will be between two hearing carriers of the DFNB3 mutation. One-quarter of their children will be born deaf while three-quarters of their children will be hearing of which two-thirds will be carriers like their parents.

A different example in a large population is Werner syndrome (WRN, OMIM 277700), a rare autosomal recessive disorder, which produces symptoms of premature aging first recognizable in young adulthood. Although WRN exists worldwide, roughly 90 percent of all reported cases have been in Japan. The history of the Japanese population has been characterized by isolation from non-Japanese and by extensive inbreeding within the past 1500 years. Studies of the causative gene in WRN showed that about 60 percent of all mutation-bearing chromosomes in Japan represent the same mutation, while another 17 percent carry a second common mutation. Carriers of the first mutation appear to be present at about 1 per 500 individuals in the current Japanese population, which agrees with the expected frequency at which the disease arises, about 1 per million in that country.

Friedman T. B., Y. Liang, J. L. Weber, J. T. Hinnant, T. Barber, S. Winata, I. N. Aryha, and J. H. Asher Jr., "A Gene for Congenital, Recessive Deafness *DFNB3* Maps to the Pericentromeric Region of Chromosome 17." *Nature Genetics* 9 (1995): 86–91.

OMIM http://www.ncbi.nlm.nih.gov/Omim/

Wang A., Y. Liang, R. A. Fridell, F. J. Probst, E. R. Wilcox, J. W. Touchman, C. C. Morton, R. J. Morell, K. Noben-Trauth, S. A. Camper, and T. B. Friedman. "Association of Unconventional Myosin MYO15 Mutations with Human Nonsyndromic Deafness *DFNB3*." *Science* 280 (1988): 447–1451.

Yu C. E., J. Oshima, Y. H. Fu, E. M. Wijsman, F. Hisama, R. Alisch, S. Matthews, J. Nakura, T. Miki, S. Ouais, G. M. Martin, J. Mulligan, and G. D. Schellenberg. "Positional Cloning of the Werner's Syndrome Gene." *Science* 272 (5259, 1996): 258–62.

—THOMAS FRIEDMAN AND DENNIS DRAYNA

their families, have provided us with an understanding of many modes of inheritance, as well as a number of factors that influence patterns of inheritance (Scriver et al., 2001). This knowledge is supported by a vast literature on genetic traits in nonhuman higher organisms, particularly the mouse (*Mus musculus*).

How Inheritance of Traits Is Analyzed. All higher organisms carry two copies of each chromosome; that is, they are diploid. These paired chromosomes are known as autosomes, and traits caused by genes on these chromosomes display so-called autosomal patterns of inheritance. Inheritance patterns of most traits are determined by the relative contribution of the two alleles (alternative copies of the same gene) on the chromosomes. Higher organisms also posses one or two unique sex-determining chromosomes. Traits resulting from genes on these two chromosomes display a different transmission pattern, which is known as sex-linked inheritance. The first step in the analysis of a genetic trait involves performing controlled matings to ensure the causative gene is present in pure form and is inherited in a predictable way, sometimes called "breeding true." In humans, experimental controlled matings are not possible, and thus inferences must be made from the pattern of inheritance of the trait in existing families.

Autosomal dominant inheritance. Dominant inheritance occurs when an organism carrying one copy of a mutant allele (i.e., an alternative to the so-called normal allele) displays the visible trait, known as the phenotype. An example of this type of inheritance is shown in Figure 1. Figure 1 represents the pedigree of a higher organism, such as humans in the examples used here. By convention, the circles represent females and the squares males, and individuals who have the phenotype of interest are shown as filled symbols. Those without the phenotype are represented by open symbols. Lines between them indicate relationships, with parents above giving rise to children below on the pedigree diagram. For example, in Figure 1, the two filled symbols on the top line represent a brother and a sister, both of whom display the trait. The sister married an unaffected individual and together they had six children, one affected daughter and two affected sons, plus one unaffected daughter and two unaffected sons.

As can be seen in Figure 1, autosomal dominant inheritance is characterized by several features. Statistically, males and females are equally affected, and half of all children of an affected parent are affected and half of the children are unaffected. In addition, affected individuals occur in every generation in the pedigree, and the trait is only passed through these affected individuals.

However, in real families there can be exceptions to these rules. Note the individual marked by an asterisk on the left portion of the pedigree. This unaffected male produced an affected son, which is a violation of the rule that a dominant trait only passes through affected individuals. If this trait is rare in the population, it is highly unlikely that his affected son received the gene from anyone other than his father (i.e., paternity by another male). In this case, his father is referred to as a nonpenetrant individual, which means that he carries the variant gene for the dominant trait but for various reasons does not display the trait. Such nonpenetrant individuals are sometimes observed in human families carrying dominant genetic diseases.

Examples of autosomal dominant diseases in humans include amyotrophic lateral sclerosis (also known as Lou Gehrig's disease), Huntington disease, and neurofibromatosis (sometimes referred to as Elephant Man disease).

Autosomal recessive inheritance. Recessive inheritance occurs when an affected individual must have

FIGURE 1. Typical Presentation of Autosomal Dominant Inheritance. Drawing by Dennis Drayna.

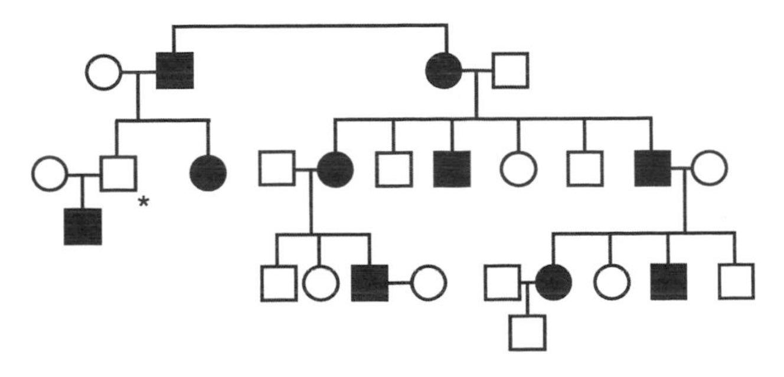

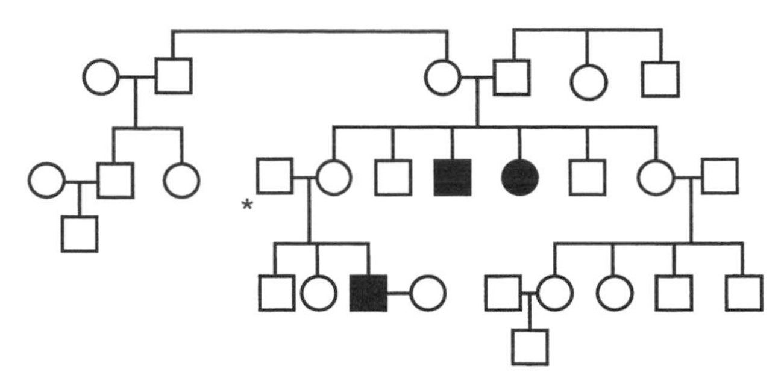

FIGURE 2. Typical Presentation of Autosomal Recessive Inheritance. Drawing by Dennis Drayna.

two copies of the causative variant gene to produce the trait in that individual. Individuals carrying one copy do not have the trait. These individuals are called carriers, and they can pass the variant gene to their offspring who can become affected if the other parent also contributes the variant gene to that offspring. An example of autosomal recessive inheritance is shown in Figure 2.

Autosomal recessive inheritance shows several characteristics. Typically, unaffected parents give rise to both affected and unaffected children. Statistically, one quarter of all the children in such a family will be affected. Most of the mutant genes that produce recessive traits are rare in the general population and follow the pattern shown in Figure 2. Occasionally, however, such variant genes are more common. This can have several different effects. For example, human families can occur in which an affected individual, who carries two copies of the gene, mates with a carrier, who carries one copy of the gene. In this case, statistically half, rather than one quarter, of their children will be affected, and all of the unaffected children will be carriers.

Note the individual marked with an asterisk in the pedigree in Figure 2. This unaffected person is the father of an affected son; both he and his wife are normal parents of an affected offspring, in concordance with the typical presentation of a recessive trait. For this to happen, both parents must be carriers. It is not surprising that his wife is a carrier, because she has affected siblings. Her husband is probably a carrier, because this is the only likely way they produced a son who is affected. A high frequency of the variant gene in the population makes this situation more common.

An example of a rare recessive disorder in humans is adrenyleukodystrophy (sometimes called Lorenzo's Oil disease). Common recessive disorders in humans include sickle-cell anemia, cystic fibrosis, which affects the lungs and other organs, and the iron overload disorder hemochromatosis. The recessive mutant genes for all these common inherited diseases can exist at relatively high frequencies in certain populations, with carriers representing 5–10 percent of the population.

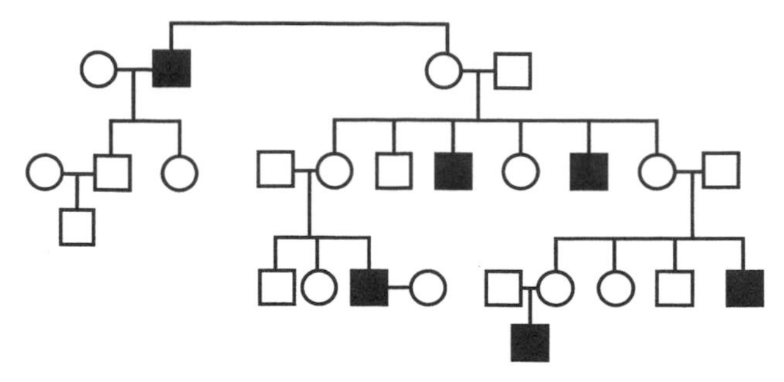

FIGURE 3. Typical Presentation of X-linked Recessive Inheritance. Drawing by Dennis Drayna.

Sex-linked disorders. In humans and many other species, sex is determined by inheritance of a specialized set of chromosomes, the sex chromosomes. In humans, these chromosomes are designated the X chromosome and the Y chromosome, respectively. Males carry one X and one Y, and females carry two X chromosomes. Genes that reside on these two chromosomes specify traits that show marked differences in rates of occurrence between the two sexes.

X-linked recessive. The most common sex-linked inheritance observed is X-linked recessive, which is illustrated in Figure 3. The hallmark of X-linked recessive inheritance is that typically only males are affected. The disorder is usually transmitted through unaffected females. This arises from the fact that human males have only a single X chromosome, thus, mutations in genes on this chromosome are not typically accompanied (or compensated for) by a normal copy of the gene, as for autosomes. Instead, males have a Y chromosome, which is very small and contains few genes. Females in families with X-linked disorders are usually unaffected, because they always posses a second X chromosome, which typically lacks the recessive mutant allele. Among the offspring of a carrier female, statistically half of the males are affected.

One of the most famous example of an X-linked recessive disorder in humans is hemophilia, a disorder characterized by failure of blood to clot properly, causing excessive bleeding. Other sex-linked traits include red-green color blindness and Duchenne muscular dystrophy. The gene encoding red-green color blindness is sufficiently frequent in the population that affected females are occasionally observed. These individuals carry two mutant genes for red-green color blindness, one on each X chromosome, and thus have no functional version of this gene.

Y-linked inheritance. The Y chromosome is small and contains only a few genes. Aside from its major role in determining the sex of the organism, it contains only a small portion of the genetic repertoire in humans. The few known traits caused by variant genes on this chromosome display a unique pattern of inheritance, as shown in Figure 4. Traits encoded by variant genes on the Y chromosome occur only in males. In addition, all the male offspring of an affected male are affected as well.

Mitochondrial inheritance. Mitochondria are distinct substructures (organelles), within eukaryotic cells responsible for energy production. Much evidence suggests that mitochondria evolved from free-living bacteria that developed close ecological relationships with the living cells of higher organisms, ultimately becoming an integral part of the cells of their hosts. Mitochondria are capable of self-replication within the cell, and they contain distinct small circular DNA molecules, which encode a unique set of genes. Mitochondria are maternally inherited. They are passed only through oocytes, not sperm, and thus represent a maternal lineage in higher organisms. Mutations in mitochondrial genes cause a number of diseases in humans and show a distinct pattern of inheritance, as illustrated in Figure 5.

Variants encoded in the mitochondria are transmitted through affected females. Affected males give rise to unaffected offspring (unless the male has children with an affected female) because sperm do not contribute mitochondria to the zygote. It is not uncommon for mitochondrially inherited disorders to show variation in their expression. As an example, note the individual marked by an asterisk in Figure 5. This unaffected person is the offspring of an affected female, and thus violates the expectation that all children of an unaffected female are affected. Examples of mitochondrial inheritance in humans include an inherited blindness (Leber hereditary optic neuropath) and a type of deafness.

The Effect of Inbreeding. Inbreeding is nonrandom mating involving close relatives. Inbreeding is a standard part of experimental methods in genetic studies of model organisms. Some human populations also display inbreeding, either historically or currently. The greatest effect of inbreeding is in the appearance of recessive traits. Inbreeding can bring rare recessive mutated genes together in two copies that previously existed only in

FIGURE 4. Typical Presentation of Y-linked Inheritance. Drawing by Dennis Drayna.

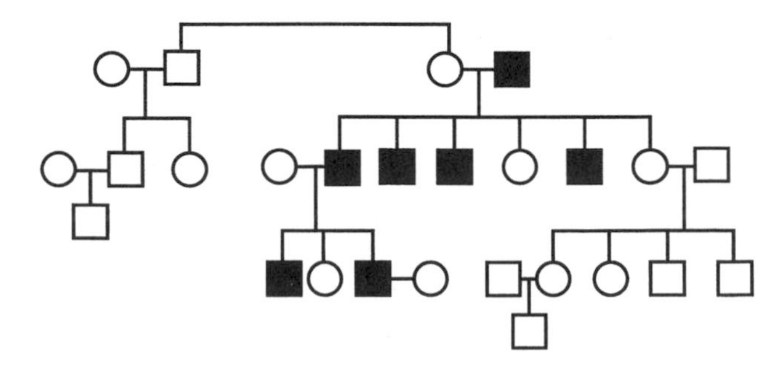

one copy in ancestral individuals in the population. Because recessive mutations show no effect when present in one copy, the existence of the mutant gene is unsuspected until it appears in the offspring of closely related individuals, indicated by the individuals in Figure 6 connected by a double line. Such matings are referred to as consanguineous, and are common in a number of populations in the world today.

The hallmark effect of inbreeding in a population is a greatly increased frequency of rare traits caused by rare recessive mutations. In humans, numerous Mendelian recessive traits, typically genetic diseases, have been described to date only in the offspring of consanguineous matings. These include a number of forms of deafness, several metabolic diseases caused by inherited enzyme deficiencies, and certain skeletal disorders.

Assignment of Inheritance in Humans. Numerous traits, notably common medical disorders, cluster strongly in families but do not readily present any clear mode of inheritance. This requires investigators to determine the degree that these traits arise from inheritance of mutant genes as opposed to the environment. Several methods are used to do this. A classic method is the twin study. In a twin study, the occurrence of a trait in identical twins is compared to its occurrence in fraternal twins. Identical twins are genetic clones of each other, and share all their genes, whereas fraternal twins share on average half their genes, like any two siblings. If, for example, a trait is entirely attributable to genetic factors, identical twins should be completely concordant; that is, one twin will always be the same as the other, whereas fraternal twins will frequently be different from each other.

Another method is an adoption study, in which the occurrence of the trait in adopted children is compared to its occurrence in their adoptive parents (to whom they are genetically unrelated). This process is then repeated in a series of children and their biological parents (who are very closely related genetically). For a trait that is completely attributable to genetic factors, biological children will much more strongly resemble their biological parents, whereas adopted children will resemble their adoptive parents much less or not at all.

FIGURE 5. Typical Presentation of Mitochondrial Inheritance. Drawing by Dennis Drayna.

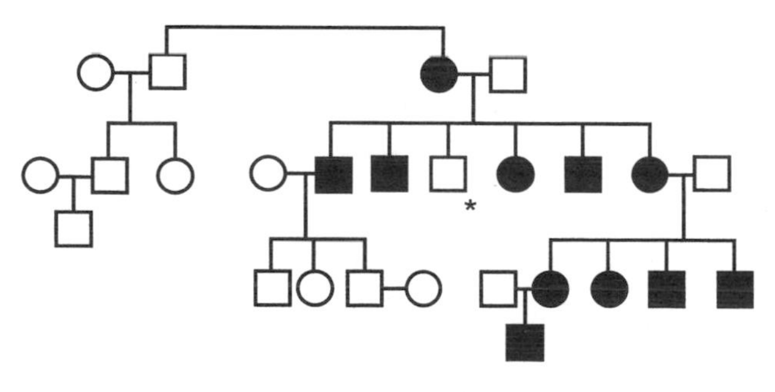

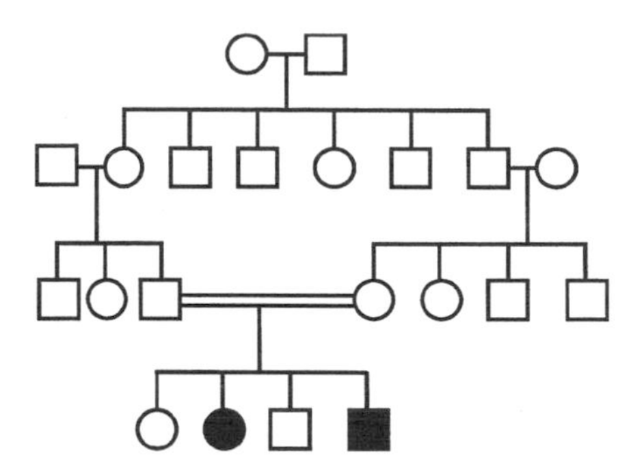

FIGURE 6. Effects of Inbreeding—Recessive Inheritance. Drawing by Dennis Drayna.

Once it is clear that a trait is caused by genetic factors, various confounding factors can still make it difficult to assign a mode of inheritance. In such situations, a statistical method called segregation analysis is used to make a best estimate of the mode of inheritance. In segregation analysis, the occurrence of the trait in families is carefully studied, and the pattern of occurrence is then compared to various models to identify the model that best explains the observed pattern.

Genetic Linkage Studies. Genetic linkage studies seek to identify the chromosomal location of genes that specify traits observed in the intact organism. Linkage studies rely on the observation of coinheritance of two observable traits in an individual as they are passed from parent to offspring. Such coinheritance, called linkage, occurs when the genes encoding the two traits reside closely adjacent to each other on the same chromosome. When this is the case, these genes are rarely or never separated by the natural mixing processes that homologous chromosomes undergo during meiosis (gamete formation). Modern genetic research has provided a useful set of observable markers in the form of readily identifiable, variable DNA sequences, each located at a single known position in the genome. Variations in these markers are inherited in Mendelian fashion. In practice, these markers are systematically tested in families, which carry an identifiable inherited trait, such as a genetic disease, until linkage is observed with one or more of these markers. Because we know the location of our markers at the outset, the observance of linkage tells us that the variant gene responsible for the trait lies in the immediate vicinity of this marker.

Gene identification. Observation of genetic linkage is the first step in a process known as positional cloning, which is defined as the isolation (or cloning) of a gene based solely on its position in the genome. This is a particularly powerful method for identifying genes that cause various human medical disorders, as it does not require additional information about the disease-causing gene and its gene product. In practice, the causative genes are identified by a close examination of all the genes in the region of the chromosome identified by linkage studies. This involves systematic DNA sequencing of all the candidate genes in the region and comparison of the sequence of these genes in affected individuals with the sequence in normal individuals. Identification of causative mutations is typically followed by other confirmatory studies.

Despite the numerous complications in identifying the mode of inheritance of traits, it has proven quite feasible to do this for a large number of inherited traits in higher organisms (Strachan and Read, 1999). Many of these traits in humans are important medical disorders. Information on the mode of inheritance allows linkage studies to be performed, which can ultimately result in the identification of the underlying mutant genes that cause these traits.

[*See also* Cellular Organelles; Chromosomes; *articles on* Disease; Meiosis; Mendelian Genetics.]

BIBLIOGRAPHY

McKusick, V. *Mendelian Inheritance in Man.* 7th ed. Baltimore, 1999.

Scriver, C., A., Beaudet, D., Valle, and W. Sly. *The Metabolic Basis of Inherited Disease.* 8th ed. Baltimore, 2001.

Strachan, T., and A. Read, *Human Molecular Genetics.* 2d ed. New York, 1999.

ONLINE RESOURCES

"Online Mendelian Inheritance in Man." http://www.ncbi.nlm.nih.gov/Omim/. This database is a catalog of human genes and genetic disorders authored and edited by Dr. Victor A. McKusick and his colleagues at Johns Hopkins University and elsewhere, and developed for the World Wide Web by the National Center for Biotechnology Information (NCBI). The database contains textual information, pictures, and reference information. It also contains copious links to NCBI's Entrez database of MEDLINE articles and sequence information.

— Dennis Drayna and Thomas B. Friedman

PEPPERED MOTH

Industrial melanism in the peppered moth, *Biston betularia*, is probably the most widely quoted example of evolution by natural selection, the central mechanism of Darwin's theory of evolution. The reasons for this are easy to appreciate. First, the changes in some populations of this moth were rapid and visually dramatic: a species of white and black moth turned black over a span of about fifty years in the second half of the nineteenth century. Second, the change was associated with a specific environmental factor, the darkening of tree bark owing to air pollution. Third, a single, easily understood selective factor—differing levels of predation of the two forms by birds—appeared to be sufficient to explain the evolutionary change. Finally, the change occurred recently, within the living memory of those who unraveled its causes, rather than as an evolutionary event in the distant past. In many parts of Europe and North America, changes are still happening. As the renowned evolutionary geneticist Sewall Wright wrote, the peppered moth story "is the clearest case in which a conspicuous evolutionary process has been actually observed."

The story is simple. Sometime in the first half of the nineteenth century, there arose a predominantly black mutant form of the peppered moth, named *carbonaria.* A single gene controls the form, with the *carbonaria* allele dominant to the typical allele. *Carbonaria* was first captured in Manchester in 1848. Thereafter, this form increased rapidly in the industrial northwest of England and spread to many other parts of Britain. In many industrial areas, the frequency of *carbonaria* reached over 95 percent, but in areas little affected by air pollution, the original typical form remained at high frequency. Differences in the frequency of the two forms between urban and industrial regions became established and stabilized during the first half of the twentieth century; however, following anti-pollution legislation in the 1950s and 1960s, *carbonaria* began to decline.

A hypothesis to explain the initial rise of *carbonaria* on the basis of camouflage and bird predation was first proposed by J. W. Tutt in 1896 (see Vignette). However, evidence to support Tutt's hypothesis was not obtained for another fifty years. In the early 1950s, Bernard Kettlewell performed a series of experiments to test whether bird predation could account for the different frequencies of the forms in different environments. From a hide, he

Figure 1. Typical and *Carbonaria* Forms of the Peppered Moth, at Rest on a Lateral Tree Branch in Surrey, England.

directly observed wild birds taking live moths from tree trunks in both polluted and unpolluted woodlands. For example, when equal numbers of the two forms of moths were released onto trees in the polluted woodland, 43 typical moths were taken while only 15 *carbonaria* were consumed. Conversely, in an unpolluted wood, the ratio of predation observed was 164 *carbonaria* to 26 typical. The *carbonaria* form appears to have a substantial advantage in polluted woodlands, but to be at a strong disadvantage in unpolluted woodlands, in line with Tutt's hypothesis. Corroborative evidence came from mark-release-recapture experiments in polluted and unpolluted oak woodlands. Large numbers of marked moths of both forms were released and recaptured by means of light and pheromonal assembly traps. The recapture rate for *carbonaria* was approximately double that of the typical form in the polluted wood, but only half in the unpolluted woodland (see Kettlewell, 1973, for full account). It was the reciprocal nature of Kettlewell's results in the two types of woodland that made his evidence so convincing.

The basic components of the rise of industrial melanism in the peppered moth are as follows:

1. The peppered moth has two distinct forms: the typical form is white with black speckling; the other form, *carbonaria*, is almost completely black.
2. A single gene controls the inheritance of these two forms, with the *carbonaria* allele completely dominant to the typical allele.
3. Peppered moths fly at night and rest on trees by day, relying on camouflage as defense against predators that find prey by sight.
4. Birds find resting peppered moths during the day and eat them.
5. The ease with which birds find resting peppered moths depends on how well the peppered moths are camouflaged.
6. Typical peppered moths are better camouflaged than melanics on lichen-covered tree trunks in unpolluted regions, while melanics are better camouflaged in industrial regions, where tree trunks have been denuded of lichens and blackened by atmospheric pollution.
7. The frequencies of melanic and typical peppered moths in a particular area are a result of two factors: the relative levels of bird predation on the two forms in that area, and migration into the area of peppered moths from regions in which the form frequencies are different.

J. W. TUTT'S ACCOUNT OF THE PEPPERED MOTH

In 1896, the great Victorian lepidopterist, J. W. Tutt gave a clear indication that he believed industrial melanism was the result of melanics having increased crypsis against soot-blackened surfaces in industrial regions. Singling out the peppered moth, he wrote:

> The speckled peppered moth as it rests on trunks in our southern woods is not at all conspicuous and looks like a . . . piece of lichen and this is its usual appearance and manner of protecting itself. But near our large towns where there are factories and where vast quantities of soot are day by day poured out from countless chimneys, fouling and polluting the atmosphere with noxious vapours and gasses, this peppered moth has during the last fifty years undergone a remarkable change. The white has entirely disappeared, and the wings have become totally black. As the manufacturing centres have spread more and more, so the 'negro' form of the peppered moth has spread at the same time and in the same districts. Let us see whether we can understand how this has been brought about! Do you live near a large town? Have you a greenhouse which you have tried to keep clean and beautiful with white paint? If so what is the result?, the paint is put on, all is beautifully white, but a little shower comes and the beauty is marred for ever. But in country places . . . it is not spoilt . . . No! Near large towns, when rain falls it brings down with it impurities, the smoke and dirt, hanging in the air. . . . And so we find fences, trees, walls and so on getting black with the continual deposit on them.
>
> Now let us go back to the peppered moth. In our woods in the south the trunks are pale and the moth has a fair chance of escape, but put the peppered moth with its white ground colour on a black tree trunk and what would happen? It would . . . be very conspicuous and would fall prey to the first bird that spied it out. But some of these peppered moths have more black about them than others, and you can easily understand that the blacker they are the nearer they will be to the colour of the tree trunk, and the greater will become the difficulty of detecting them. So it really is the paler ones the birds eat, the darker ones escape. But then if the parents are the darkest of their race, the children will tend to be like them, but inasmuch as the search by birds becomes keener, only the very blackest will be likely to escape. Year after year this has gone on, and selection has been carried to such an extent by nature that no real black and white peppered moths are found in these districts but only the black kind. This blackening we call melanism.

—Michael E. N. Majerus

Since Kettlewell's pioneering experiments, evidence has continued to accumulate in support of this interpretation, and subsequent work has broadened our understanding of the details of the case (reviewed by Majerus, 1998). For example, it is now recognized that peppered moths rarely rest by day in exposed positions on tree trunks, preferring sites on the underside of lateral twigs or branches higher in the canopy (Figure 1). However, the general thrust of the story remains unchanged: the *carbonaria* form increased in frequency where trees were denuded of lichens by sulfur dioxide and black-

TABLE 1. The Decline in Melanism in the Peppered Moth since Anti-pollution Legislation

Figures given are the percentage *carbonaria* in samples at two English sites and one American site. Data are for each fifth year, or nearest equivalent (indicated by asterisk). More detailed data are given in Majerus 1998, Grant et al. 1996.

Year	*Caldy Common, West Kirby, northwest England*	*Cambridge, England*	*George Reserve, Michigan, USA*
1960/1961*	94.2	94.8	91.7*
1965	90.2	—	—
1970	90.8	75.0	—
1975	86.6	64.7	—
1980/1981*	76.9	45.9*	—
1985	53.5	39.5	—
1990	33.1	22.2	—
1995	17.6	19.2	20.0
1998*/2000	11.5*	15.1	

ened by soot fallout, because it was less heavily preyed on by birds.

The decrease in air pollution from the early 1960s led to a gradual decline in *carbonaria* throughout industrial Britain. Theoretical models of this decline are broadly in line with observation. Interestingly, data similar to those obtained in Britain have been collected in the United States, where a melanic form of the same species, *swettaria* (indistinguishable from *carbonaria*) increased and decreased in frequency during the twentieth century (Table 1); the frequency changes were again correlated with pollution levels. These recent changes in Britain and America are effectively natural replicate experiments with consistent results. Furthermore, they show that evolution is not a one-way process. As the environmental factors that led to the initial rise of melanic forms were reversed, the evolutionary change that they led to also went into reverse.

Industrial melanism has evolved in many other species, although the details vary. In Britain, more than 100 species of moths show industrial melanism of one type or another, and cases involving spiders, beetles, and true bugs (Hemiptera) are also known. The case of the peppered moth has been very influential in evolutionary genetics; it demonstrates that evolution may occur as a result of differences in the fitness of different genetic forms of a species. Although this moth has received much academic and public attention, it is only one of many similar cases in which the frequencies of melanic forms have increased and then decreased as industrial pollution has waxed and waned. Each of these cases may be taken as evidence supporting the role of natural selection in biological evolution.

[*See also* Adaptation; Natural Selection; Pesticide Resistance.]

BIBLIOGRAPHY

Brakefield, P. M. "Industrial Melanism: Do We Have the Answers?" *Trends in Ecology and Evolution* 2 (1987): 117–122.

Clarke, C. A., G. S. Mani, and G. Wynne. "Evolution in Reverse: Clean Air and the Peppered Moth." *Biological Journal of the Linnaean Society* 26 (1985): 189–199.

Cook, L. M., K. D. Rigby, and M. R. D. Seaward. "Melanic Moths and Changes in Epiphytic Vegetation in North-west England and North Wales." *Biological Journal of the Linnaean Society* 39 (1990): 343–354.

Grant, B. S., D. F., Owen, and C. A. Clarke. "Parallel Rise and Fall of Melanic Peppered Moths in America and Britain." *Journal of Heredity* 87 (1996): 351–357.

Howlett, R. J., and M. E. N. Majerus. "The Understanding of Industrial Melanism in the Peppered Moth (*Biston betularia*) (Lepidoptera: Geometridae)." *Biological Journal of the Linnaean Society* 30 (1987): 31–44.

Kettlewell, H. B. D. *The Evolution of Melanism.* Oxford, 1973. A comprehensive review of the author's work on the peppered moth and other species exhibiting melanism.

Lees, D. R. "Industrial Melanism: Genetic Adaptation of Animals to Air Pollution." In *Genetic Consequences of Man Made Change.* edited by J. A. Bishop and L. M. Cook, pp. 129–176. London, 1981.

Majerus, M. E. N. 1989. "Melanic Polymorphism in the Peppered Moth *Biston betularia* and other Lepidoptera." *Journal of Biological Education* 23 (1989): 267–284.

Majerus, M. E. N. *Melanism: Evolution in Action.* Oxford, 1998.

Tutt, J. W. *British Moths.* London, 1896.

Wright, Sewall. *Evolution and Genetics of Populations.* Vol. 4. Chicago, 1978.

— MICHAEL E. N. MAJERUS

PERMO-TRIASSIC EXTINCTION. *See* Mass Extinctions.

PESTICIDE RESISTANCE

In her landmark book *Silent Spring* (1962), Rachel Carson observed, "If Darwin were alive today the insect world would delight and astound him with its impressive verification of his theories of survival of the fittest. Under the stress of intensive chemical spraying the weaker members of the insect populations are being weeded out." Not only does evolution of resistance to pesticides in insects and other organisms confirm Darwin's theories, it threatens agriculture and human health worldwide. Understanding pesticide resistance thus enhances fundamental knowledge about evolution and provides insights required for combating an urgent problem.

Pesticide resistance is an increase in the proportion of individuals in a population with the genetically based ability to survive exposure to one or more pesticides. It occurs when there is genetic variation in resistance and repeated pesticide treatments kill susceptible individuals, reducing their proportion in the population. Treated survivors pass genes conferring resistance to their progeny. Because this process entails gene frequency changes in a population, it exemplifies evolution. Indeed, the abundant and diverse cases of pesticide resistance may constitute one of the most compelling sets of evidence for evolution by natural selection.

According to George P. Georghiou (1986), pesticide resistance occurs in at least 100 species of plant pathogens, 55 species of weeds, 5 species of rodents, and 2 species of nematodes. This article focuses on resistance to insecticides in more than 500 species of insects and mites.

History and Extent of Insecticide Resistance. In 1914 A. L. Melander reported the first case of insecticide resistance. He studied the effectiveness of lime sulphur, an inorganic insecticide, against an orchard pest, the San Jose scale (*Quadraspidiotus perniciousus*) in the state of Washington. A treatment with lime sulphur killed all scales in one week in typical orchards, but 90 percent survived after two weeks in an orchard with resistant scales. Although few cases of insecticide resistance were recorded before 1940, the number grew exponentially following widespread use of DDT and other synthetic organic insecticides.

Insects have evolved resistance to all types of insecticides including inorganics, DDT, cyclodienes, organophosphates, carbamates, pyrethroids, juvenile hormone analogs, chitin synthesis inhibitors, avermectins, neonicotinoids, and microbials. Resistance to insecticides derived from the common soil bacterium *Bacillus thuringiensis* (Bt) has special significance because some crop plants have been genetically engineered to produce Bt toxins (see details below).

Resistance occurs in thirteen orders of insects, yet more than 90 percent of the arthropod species with resistant populations are either Diptera (35 percent), Lepidoptera (15 percent), Coleoptera (14 percent), Hemiptera (in the broad sense, 14 percent), or mites (14 percent). The disproportionately high number of resistant Diptera reflects intense use of insecticides against mosquitoes that transmit disease. Agricultural pests account for 59 percent of harmful resistant species while medical and veterinary pests account for 41 percent. Many species have numerous resistant populations, each of which resists many insecticides. Statistical analyses suggest that for crop pests, resistance evolves most readily in those with an intermediate number of generations (four to ten) per year that feed either by chewing or by sucking on plant cell contents.

Unfortunately, resistant pest species outnumber resistant beneficial species such as predators and parasitoids by more than twenty to one. This pattern probably reflects limited attention devoted to resistance in beneficials as well as biological differences between beneficials and pests. Available evidence contradicts the hypothesis that natural enemies evolve resistance less readily because intrinsic levels of detoxification enzymes are lower in predators and parasitoids than in pests. An alternative hypothesis with more support is that natural enemies evolve resistance less readily because they suffer from food limitation following insecticide sprays that severely reduce adundance of their prey or hosts.

Genetics and Biochemistry of Resistance. Evolution of resistance is most often based on one or a few genes with major effect. Before a susceptible population is exposed to an insecticide, resistance genes are usually rare because they typically reduce fitness in the absence of the insecticide. When an insecticide is used repeatedly, strong selection for resistance overcomes the normally relatively minor fitness costs associated with resistance when the population is not exposed to insecticide.

The primary mechanisms of resistance are decreased target site sensitivity and increased detoxification through metabolism or sequestration. Target sites are the molecules in insects that are attacked by insecticides. Decreased target site sensitivity is caused by changes in target sites that reduce binding of insecticides, or that lessen the damage done should binding occur. Metabolism involves enzymes that rapidly bind and convert insecticides to nontoxic compounds. Sequestration is rapid binding by enzymes or other substances with very slow or no processing. Reduced insecticide penetration through the cuticle, and behavioral changes that reduce exposure to insecticide are also mechanisms of resistance. Different mechanisms can occur within an individual insect, sometimes interacting to provide extremely high levels of resistance.

Much progress has occurred in the past decade in understanding the genetic basis of resistance mechanisms. Point mutations that alter the structure of target site proteins and thereby reduce their sensitivity to insecticides can confer high levels of resistance. Most conventional insecticides are nerve poisons that assault one of three target site proteins. Organophosphorous (OP) and carbamate insecticides attack acetylcholinesterase, DDT and pyrethroids attack the sodium ion channel, and cyclodienes attack the γ-aminobutyric acid (GABA) receptor. Resistance-associated point mutations in the genes encoding each of these target sites are similar among *Drosophila melanogaster* and several insect pests. In the most striking case, replacement of a single amino acid (alanine 302) in the GABA receptor is associated with resistance to cyclodienes in at least nine species, including several flies, two beetles, a whitefly, an aphid, and a cockroach. Such similarities suggest that mutations conferring target-site resistance are constrained because target proteins must have reduced sensitivity to insecticide as well as the ability to perform their normal role in the insect.

Mutations that increase the quantity or activity of enzymes that metabolize or sequester insecticides also confer resistance. The three major enzyme groups involved are esterases, cytochrome P450 monooxygenases (P450s), and glutathione S-transferases (GSTs). Gene amplification, increased gene expression, or both can boost levels of these enzymes. For example, *Culex pipiens quinquefasciatus* mosquitoes have up to 250 copies of a gene encoding an esterase that sequesters and confers resistance to OP insecticides. In this disease vector and in the aphid *Myzus persicae*, a limited number of gene amplification events has occurred. Rapid spread of individuals carrying amplifications has been aided by human transportation including aircraft. Regulatory changes that increase expression of genes encoding P450s and GSTs are associated with resistance in mosquitoes and in the housefly, *Musca domestica*.

Point mutations that enhance enzyme activity can also confer resistance. In an esterase gene of the sheep blow fly, *Lucilia cuprina*, each of two single amino acid substitutions confers resistance to OPs. One mutation confers resistance to diazinon and other OPs but not to malathion, while another confers 600-fold resistance to malathion. The amino acid substitution that confers resistance to diazinon in the sheep blow fly also confers resistance to diazinon in the house fly.

Cross-resistance occurs when selection with one insecticide causes resistance to other insecticides. In most cases, cross-resistance involves structurally similar insecticides. However, cross-resistance also occurs between structurally distinct insecticides that attack the same target site, such as DDT and pyrethroids.

Delaying Evolution of Resistance. The goal of resistance management is to delay evolution of resistance in pests. The best way to achieve this is to minimize insecticide use. Thus, resistance management is a component of integrated pest management, which combines chemical and nonchemical controls to seek safe, economical, and sustainable suppression of pest populations. Alternatives to insecticides include biological control by predators, parasitoids, and pathogens. Also valuable are cultural controls (crop rotation, manipulation of planting dates to limit exposure to pests, and use of cultivars that tolerate pest damage) and mechanical controls (exclusion by barriers and trapping).

Because large-scale resistance experiments are expensive, time consuming, and might worsen resistance problems, modeling has played a prominent role in devising tactics for resistance management. Although models have identified various strategies with the potential to delay resistance, practical successes in resistance management have relied primarily on reducing the number of insecticide treatments and diversifying the types of insecticide used. For example, programs in Australia, Israel, and the United States have limited the number of times and periods during which any particular insecticide is used against cotton pests.

Perhaps the greatest challenge facing resistance management today is to avoid rapid evolution of resistance to transgenic crop plants that produce insecticidal proteins from the bacterium *Bacillus thuringiensis* (Bt). Unlike conventional neurotoxic insecticides, Bt toxins kill by binding to and disrupting insect midgut membranes. Transgenic Bt corn, cotton, and potato plants can select intensively for resistance because they expose pests to Bt toxins throughout the growing season. Field-evolved resistance to Bt crops has not been documented yet, but natural populations of the diamondback moth, *Plutella xylostella*, have evolved resistance to Bt sprays. Also, strains of more than ten pest species have evolved resistance to Bt toxins in the laboratory. In most cases, resistance is associated with reduced binding of Bt toxin to midgut membrane target sites.

The refuge strategy has been adopted widely to delay evolution of resistance to Bt crops. By planting refuges of host plants that do not produce Bt toxin, growers enable survival of susceptible pests. For example, refuges of non-Bt cotton are grown near Bt cotton. Ideally, the rare resistant adults emerging from a Bt crop mate with susceptible adults from refuges, and their hybrid progeny are susceptible (i.e., resistance is inherited as a recessive trait). Modeling results show that under these conditions, resistance can be delayed substantially.

The optimal size and spatial distribution of refuges are hotly debated. Another key issue is whether the concentration of toxin in Bt crops is sufficient to kill pests heterozygous for resistance. Resistance to Bt toxins is often partially to completely recessive, but variation in dominance of resistance, inherent pest susceptibility, or toxin concentration could increase survival of heterozygotes and accelerate resistance evolution.

If mutations conferring resistance to two toxins are independent and rare, plants that produce both toxins in sufficient quantity to kill heterozygotes used in conjunction with refuges might greatly delay resistance. In the Bt crops grown commercially now, each individual plant produces only one type of Bt toxin, but Bt plants that produce two or more toxins are being developed to stall resistance.

Whether insecticides are applied externally or produced internally via manipulation of plant genomes, evolution of resistance by pests is inevitable. Resistance to insecticides provides testimony to the power of adaptation and offers abundant opportunities to see evolution in action.

BIBLIOGRAPHY

GENERAL SOURCES

Denholm, I., J. A. Pickett, and A. L. Devonshire, eds. *Insecticide Resistance: From Mechanisms to Management.* London, 1999.

Georghiou, G. P. "The Magnitude of the Resistance Problem." In *Pesticide Resistance: Strategies and Tactics for Management,* pp. 14–43. Washington, D.C., 1986.

Georghiou, G. P., and A. Lagunes-Tejeda. *The Occurrence of Resistance to Pesticides in Arthropods.* Rome, 1991.

McKenzie, J. A. *Ecological and Evolutionary Aspects of Insecticide Resistance.* Austin, Tex., 1996.

Roush, R. T., and B. E. Tabashnik, eds. *Pesticide Resistance in Arthropods.* New York, 1990.

ARTICLES FROM *THE ANNUAL REVIEW OF ENTOMOLOGY*

Denholm, I., and M. W. Rowland. "Tactics for Managing Pesticide Resistance in Arthropods: Theory and Practice." *Annual Review of Entomology* 37 (1992): 91–112.

French-Constant, R. H., N. Anthony, K. Aronstein, T. Rocheleau, and G. Stilwell. "Cyclodiene Insecticide Resistance: From Molecular to Population Genetics." *Annual Review of Entomology* 48 (2000): 449–466.

Gould, F. "Sustainability of Transgenic Insecticidal Cultivars: Integrating Pest Genetics and Ecology." *Annual Review of Entomology* 43 (1998): 701–726.

Hemingway, J., and H. Ranson. "Insecticide Resistance in Insect Vectors of Human Disease." *Annual Review of Entomology* 45 (2000): 371–391.

Roush, R. T., and J. A. McKenzie. "Ecological Genetics of Insecticide and Acaricide Resistance." *Annual Review of Entomology* 32 (1987): 361–380.

Tabashnik, B. E. "Evolution of Resistance to *Bacillus thuringiensis.*" *Annual Review of Entomology* 39 (1994). 47–79.

Wilson, T. G. "Resistance of *Drosophila* to Toxins." *Annual Review of Entomology* 46 (2001): 545–571.

ADDITIONAL SOURCES WITH AN EVOLUTIONARY PERSPECTIVE

Claudianos, C., R. J. Russell, and J. G. Oakeshott. "The Same Amino Acid Substitution in Orthologous Esterases Confers Organophosphate Resistance on the House Fly and a Blowfly." *Insect Biochemistry and Molecular Biology* 29 (1999): 675–686.

Gahan, L. J., F. Gould, and D. G. Heckel. "Identification of a Gene Associated with Bt Resistance in *Heliothis virescens.*" *Science* 293 (2001): 857–860.

Groeters, F. G., and B. E. Tabashnik. "Roles of Selection Intensity, Major Genes, and Minor Genes in Evolution of Insecticide Resistance." *Journal of Economic Entomology* 93 (2000): 1580–1587. Reviews empirical estimates of selection intensity and simulates various genetic models to test theories about resistance.

Rosenheim, J. A., M. W. Johnson, R. F. L. Mau, S. C. Welter, and B. E. Tabashnik. "Biochemical Preadaptations, Founder Events, and the Evolution of Resistance in Arthropods." *Journal of Economic Entomology* 89 (1996): 263–273. Tests evolutionary hypotheses about resistance using an extensive database on crop pests.

Tabashnik, B. E., and M. W. Johnson. "Evolution of Pesticide Resistance in Natural Enemies." In *Handbook of Biological Control,* edited by T. Fisher, T. S. Bellows, L. E. Caltagirone, D. L. Dahlsten, C. Huffaker, and G. Gordh, pp. 673–689. San Diego, 1999. Analyzes evolutionary hypotheses and evidence about the scarcity of resistance in arthropod predators and parasitoids.

— BRUCE E. TABASHNIK

PHENOTYPIC PLASTICITY

Organisms have long been known to be able to change their appearance in response to environmental conditions. For example, arctic foxes and hares in winter change coat colors to white, various aquatic plants produce radically different emergent and submerged leaves, and light-skinned humans experience melanization after exposure to the sun. Such environmentally induced changes in the phenotype of an organism are referred to as phenotypic plasticity. These changes are the result of alterations of developmental programs. Not all plastic responses are necessarily adaptive, however; the extreme phenotypic response of sunburn, for example, may be a precursor to skin cancer.

The spectrum of changes encompassed by plasticity is considerable; practically any type of biotic or abiotic environmental variation may elicit phenotypic plasticity. Physiological alterations range from those as transient as a blush to long-term acclimatization to heat or altitude. Behavioral responses to environmental stimuli also represent examples of plasticity. For example, octopuses can shift their color patterns an average of three times per minute as they forage across divergent substrates; the patterns they generate include mimicry of local fish, cryptic blending with the background, and boldly conspicuous coloration. More lasting changes are produced in morphological structures, such as temperature-induced shifts in the color pattern of the wings of butterflies or helmet and spine formation by the crustacean *Daphnia* in response to predators.

Even single-celled organisms can exhibit substantial plasticity—many bacteria and algae, for example, exhibit distinctive characteristics when conditions change. For multicellular organisms, plasticity can be manifested at any stage of the life cycle, from larval to juvenile and adult stages for animals, and during both vegetative growth and reproduction of plants and fungi. Plastic responses early in the organism's life may resonate throughout the life cycle, changing probability of survival and altering the amount of reproductive output. In other cases, for example morphological responses during a larval stage, there may be no visible effects on the morphology of the adult phase.

Many phenotypically plastic responses are continuous—the characteristics of the organism change gradually with the level of the environmental factor. Examples include the response of plants to changes in levels of water and nutrients and the growth rate of many animals as food supply changes. Other responses, however, are discontinuous, apparently triggered by a certain level of an environmental factor, such as the switch from a wingless to a winged form in various insects or the shift from vegetative growth to flowering in plants as photoperiod changes.

Polyphenism is a term used to describe the production of distinct morphological forms by a single species. Most polyphenisms, such as the wing polymorphisms mentioned above and the production of horns by dung beetles, are the result of plastic responses. A spectacular and often overlooked example of plasticity and polyphenism is the system of morphologically and physiologically diversified castes in social insects. For example, in the insect order Hymenoptera, genetically identical individuals develop into the phenotypically and functionally distinctive soldiers, workers, and reproductives. These differences result from responses to varying conditions within the nest, for example, the amounts of juvenile hormone. A curious category of polyphenism is cyclomorphosis, seen in some arthropods and algae, with a succession of different phenotypes produced in response to seasonal changes. The green alga *Scenedesmus* exhibits such changes, ranging from unicellular to four-cell colonies, and cells with varying shapes, sizes, and length and number of spines.

Many of these more dramatic alterations in phenotype are produced by changes in numerous subcomponents of the phenotype. Take, for example, high-altitude acclimatization. The increased oxygen uptake ability that we recognize as an adaptation to high altitudes is produced by a variety of changes: kidneys secrete the hormone erythropoietin, increasing red blood cell production; more capillaries are formed; heart and lung volumes are increased; and the production of the molecule 2,3-diphosphoglycerate is enhanced, allowing increased oxygen uptake at lower partial oxygen pressures.

A closely related concept to phenotypic plasticity is the reaction norm, which refers to the set of phenotypes that can be produced by an individual genotype that is exposed to different environmental conditions. Reaction norms of a given trait may be plastic or nonplastic (i.e., with no response of a trait to the environment). The German biologist R. Woltereck proposed the term in 1909, describing differences in the morphology of individuals of a clonal form of the crustacean *Daphnia* in response to different diets. The evolutionists T. Dobzhansky and R. Goldschmidt also referred to reaction norms in their works of the 1930s and 1940s, but the concept received its most thorough treatment in the work of the Russian biologist I. I. Schmalhausen (see the following discussion).

Plasticity and reaction norms are ideally studied using individuals of identical genotype raised in different environments. A classic example is the work of J. Clausen, D. D. Keck, and W. M. Hiesey (1940), who used cloned individuals of *Potentilla* in a series of reciprocal transplants from sea level to 3,000 meters (9,842 feet) in the Sierra Nevada of California and demonstrated that individuals originating at different altitudes were genetically distinct, yet locally adapted. Cloning, however, is not possible for many species, so alternatives such as the use of full- or half-sibling individuals are employed. [*See* Quantitative Genetics.] For example, E. Greene (1989) grew full-sibling *Nemoria arizonaria* caterpillars on two diets, oak catkins (inflorescences) and leaves. Catkin-fed caterpillars developed into a highly ornamented form that looks very much like the catkins themselves, whereas those fed on leaves developed into twiglike individuals.

Evolution and Genetics of Plasticity. If the phenotype that can be produced through plasticity is favorable, why not just produce it all the time? This question has been addressed by theoreticians, whose models show that plasticity is selected for any time that environmental conditions change (especially within an individual's lifetime) and a different phenotype is favored (i.e., has higher fitness). Thus, the ability to respond appropriately, whether physiologically, behaviorally, or morphologically, can be highly adaptive.

Because all environments are variable to some extent, these findings raise an alternative question: Why are all organisms not optimally plastic? There are various reasons for this. Any lag time between the environmental stimulus and the plastic response may result in a mismatch between the phenotype and current conditions. Further limitations arise if the benefits of the new phenotype do not exceed the costs of producing it. Another potential cost to plasticity arises from building the genetic and metabolic machinery that enables a response to an environmental factor that rarely occurs. Adaptive plastic responses are most often seen in response to environmental variability that is frequent or predictable; for example, all flowering plants appear to have the ability to sense shading.

The genetic control of plastic responses has been less well studied. Two general categories are proposed: allelic sensitivity, in which a particular gene locus responds to a change in conditions, resulting in a change in the amount or activity of gene product, and gene regulation, in which a regulatory switch turns various genes on or off in an environment-dependent context. The relative importance of these different modes has been vigorously debated, but regulatory plasticity, with the potential for controlling multiple trait responses, has been suggested as the likely mechanism for adaptive plasticity.

Recent work has examined patterns of gene expression in response to different stresses, and assays of organisms ranging from bacteria and yeast to fruit flies and plants indicate substantial changes. Following infection by a fungus, the plant *Arabidopsis* had 169 genes with significantly increased expression and 39 genes with significantly decreased expression. Many genes had enhanced expression in response to several different stimuli, suggesting that defense responses are to some extent coordinated with each other. Likewise, several workers have now shown that insect castes have sets of genes turned on in a caste-specific manner.

Schmalhausen (1949) discussed the evolution of a phenotype as it is produced as part of a reaction norm and concluded that any variation among individuals in a reaction norm that produced a particular phenotypic response would be subject to natural selection. This selection would reduce variation around the mean response, a process he termed *stabilizing selection.* Ultimately, this would lead to a reduction (or canalization) of the phenotypic variation, due to selection against plastic responses to within-environment variability. [*See* Canalization.] This represents one of the intriguing features of the evolution of plastic reaction norms—different levels of selection can simultaneously favor both phenotypic plasticity and phenotypic stability of the reaction norm.

In addition to the primary role of phenotypic plasticity as a means of dealing with fluctuating conditions, Schmalhausen and C. H. Waddington proposed a mechanism, called genetic assimilation, by which plasticity could also serve as an intermediary in the process of phenotypic evolution. They realized that changes in the environment could expose a previously unexpressed portion of the reaction norm. This plasticity, if even slightly in the appropriate direction, could allow the organism to persist. Subsequent selection would favor any mutations that would further improve the reaction norm in the direction of the environmental change. Finally, stabilizing selection would canalize the new reaction norm. Eventually, the organism may lose the ability to respond plastically altogether. Schmalhausen postulated that this sequence of events implies a shift from plasticity due to differential allelic sensitivity, to regulatory plasticity, and such a system could be capable not only of responding to but of anticipating environmental change. Such a system of anticipatory plasticity is apparent in the loss of leaves by winter deciduous trees: this adaptation is induced by changes in photoperiod, a cue that usually allows the freeze or drought itself to be safely avoided.

Case Studies. Many organisms, including animals, fungi, and plants, are known to produce chemical or structural defenses in response to the presence of predators. The wound-induced hormone jasmonic acid is important as a stimulus for the production of nicotine by wild relatives of tobacco. Nicotine is produced by the roots and transported to the leaves, the youngest leaves receive the highest concentrations. These plants can be induced to produce nicotine in the absence of wounding by applying methyl jasmonate. I. T. Baldwin (1998) compared four categories of plants: those that had nicotine production induced, those without induced nicotine, and individuals of each of these types that were or were not attacked by herbivores. He found that attacked induced plants produced more seeds than attacked noninduced plants, establishing a fitness advantage to the production of nicotine. However, plants that were induced but not attacked produced fewer seeds than noninduced and unattacked plants, demonstrating a cost to the production of nicotine in the absence of herbivory.

A classic set of studies by P. M. Brakefield and colleagues has been made on the butterfly *Bicyclus anynana.* This polyphenic species plastically produces two forms, an active warm/wet season form with large eyespots, which flies in search of mates and oviposition sites, and a more sedentary cool/dry season form with very small eyespots, which rests cryptically on dried grass or leaf litter. The genetic, hormonal, and developmental bases of phenotypic differences resulting from this plasticity have been examined. Molecular genetic techniques have revealed that the presence of the product of the gene Distal-less can induce eyespot formation on the wings, and that the amount and timing of the Distal-less gene product is affected by temperature. Differences in Distal-less reproduce the seasonal differences in eyespots seen in the field—the cooler dry season elicits slower development; thus, less gene product is produced. Interestingly, northern populations of *Bicyclus,* from regions where higher rainfall is associated with cooler temperatures, no longer produce larger eyespots at higher temperatures. This suggests that there has been a change in the inducing stimulus for eyespot development.

Current work on phenotypic plasticity continues to investigate the genetic basis of variation in responsiveness and to understand the mechanism and extent of changes in gene expression mediated by environmental stimuli. Other areas of interest include how the relationships among different traits are altered by the environment, how parents may "prepare" their offspring for future environments, and how fast plasticity may evolve in response to the novel environments produced by anthropomorphic changes.

[*See also* Life History Stages.]

BIBLIOGRAPHY

BOOKS AND REVIEWS

Brakefield, P. M., and V. French. "Butterfly Wings: The Evolution of Development of Colour Patterns." *BioEssays* 21 (1999): 391–401. A summation of how developmental mechanisms may be modified during the evolution of adaptive traits. Discusses how seasonal polyphenism can be produced through environmental sensitivity mediated by ecdysteroid hormones and how artificial selection and single-gene mutants can be used to demonstrate the ways genetic variation influences the number, shape, size, position, and color composition of eyespots.

Clausen, J., D. D. Keck, and W. M. Hiesey. *Experimental Studies on the Nature of Species*, vol. 1, *Effect of Varied Environment on Western North American Plants.* Washington, D. C., 1940.

Emlen, D. J., and H. F. Nijhout. "The Development and Evolution of Exaggerated Morphologies in Insects." *Annual Review of Entomology* 45 (2000): 661–708. An examination of the evolution of some spectacular insect morphologies, such as bizarre sexually selected traits and soldier castes. The authors take a reaction norm approach to understanding how allometric relationships between traits and body size might be altered during their evolution.

Gotthard, K., and S. Nylin. "Adaptive Plasticity and Plasticity as an Adaptation: A Selective Review of Plasticity in Animal Morphology and Life History." *Oikos* 74 (1995): 3–17.

Nijhout, H. F. "Control Mechanisms of Polyphenic Development in Insects." *BioScience* 49 (1999): 181-192. A masterful review of mechanisms of genetic, developmental, and environmental control of polyphenisms.

Scheiner, S. M. "Genetics and Evolution of Phenotypic Plasticity." *Annual Review of Ecology and Systematics* 24 (1993): 35–68. Covers the genetic basis of plasticity, how plasticity evolves, and models of plasticity evolution.

Schlichting, C. D., and M. Pigliucci. *Phenotypic Evolution: A Reaction Norm Perspective.* Sunderland, Mass., 1998. Covers the evolution and genetics of phenotypic plasticity and reaction norms. The authors propose the concept of the developmental reaction norm to unite ontogeny, plasticity, and the intercorrelations among traits and argue that reaction norms can be the direct objects of natural selection. They propose that the developmental process itself can be examined as a reaction norm, where cells' fates are determined by their responses to internal environmental conditions.

Schmalhausen, I. I. *Factors of Evolution.* Philadelphia, 1949; reprint, Chicago, 1986. An early and surprisingly modern book on the evolution of the phenotype from a developmental perspective. Uses the concept of reaction norm extensively as a means of understanding how phenotypic characteristics evolve. Originated the concept of stabilizing selection.

Sultan, S. E. "Evolutionary Implications of Phenotypic Plasticity in Plants." *Evolutionary Biology* 21 (1987): 127–178.

Via, S., R. Gomulkiewicz, G. de Jong, S. M. Scheiner, C. D. Schlichting, and P. H. van Tienderen. "Adaptive Phenotypic Plasticity: Consensus and Controversy." *Trends in Ecology and Evolution* 10 (1995): 212–216. A discussion primarily of models of the evolution of plasticity, contrasting selection on character states versus selection directly on plasticity.

PRIMARY LITERATURE

Baldwin, I. T. "Jasmonate-induced Responses Are Costly But Benefit Plants under Attack in Native Populations." *Proceedings of the National Academy of Sciences USA* 95 (1998): 8113–8118.

Evans, J. D., and D. E. Wheeler. "Differential Gene Expression between Developing Queens and Workers in the Honeybee, *Apis mellifera*." *Proceedings of the National Academy of Sciences USA* 96 (1999): 5575–5580. The authors demonstrate that there are many genes differentially expressed between the queen and worker castes of honeybees (*Apis mellifera*). Their results suggest that not only are there genes for the queen "program" turned off in workers, but that there are also genes turned on that appear to be worker-specific.

Greene, E. "A Diet-induced Developmental Polymorphism in a Caterpillar." *Science* 243 (1989): 643–646. Spectacular example (and cover photo) of the production of two distinctive phenotypic forms, which each mimic a different feature of the host plant. Greene's results indicate that tannins in the diet lead to the twig-morph caterpillar, whereas their absence allows the production of the catkin morph.

Hanlon, R. T., J. W. Forsythe, and D. E. Joneschild. "Crypsis, Conspicuousness, Mimicry and Polyphenism as Antipredator Defences of Foraging Octopuses on Indo-Pacific Coral Reefs, with a Method of Quantifying Crypsis from Video Tapes." *Biological Journal of the Linnean Society* 66 (1999): 1–22.

Moczek, A. P., and D. J. Emlen. "Proximate Determination of Male Horn Dimorphism in the Beetle *Onthophagus taurus* (Coleoptera: Scarabaeidae)." *Journal of Evolutionary Biology* 12 (1999): 27–37. Males of this dung beetle come in two forms, small and hornless and larger with a pair of long, curved horns. Natural variation in the quantity and quality of food available to larvae is responsible for producing these differences.

Schenk, P. M., K. Kazan, I. Wilson, J. P. Anderson, T. Richmond, S. C. Somerville, and J. M. Manners. "Coordinated Plant Defense Responses in *Arabidopsis* Revealed by Microarray Analysis." *Proceedings of the National Academy of Sciences* 97 (2000): 11655–11660. Changes in gene expression analyzed following fungal infection and application of defense-signal molecules.

Schmitt, J., A. C. McCormac, and H. Smith. "A Test of the Adaptive Plasticity Hypothesis Using Transgenic and Mutant Plants Disabled in Phytochrome-mediated Elongation Responses to Neighbors." *American Naturalist* 146 (1995): 937–953. A clever set of experiments in which plants with "inappropriate" phenotypes (due to a lack of plasticity) were shown to be outperformed by plants that responded normally to shading or competition.

Van Buskirk, J., S. A. McCollum, and E. E. Werner. "Natural Selection for Environmentally Induced Phenotypes in Tadpoles." *Evolution* 51 (1997): 1983–1997. A thorough examination of the strength of natural selection by predators on the traits of tadpoles. In the absence of the dragonfly larvae predators, tadpoles with narrow tail fins and tail muscles and deep bodies grew faster; however, when predators were present, those with deep tail fins and wide muscles and narrow body survived best. Plastic responses of the tail shapes of tadpoles mirrored these selection differences—when exposed to dragonfly larvae, tail fins became deeper with wider muscles.

— CARL D. SCHLICHTING

PHENOTYPIC STABILITY

The idea that organisms exhibit stable states and resist the effects of perturbations has deep roots in biology. Although Lerner (1954) traces this idea to Hippocrates, the modern concept of physiologic homeostasis was developed by Cannon (1932). In evolutionary biology, the concept of homeostasis finds its most direct expression in the works of Schmalhausen and Waddington in the 1940s and 1950s. These scholars independently developed the idea that mechanisms have evolved that buffer developmental systems against both environmental and genetic perturbations. This is the concept captured by the term *phenotypic stability.* Largely ignored for several decades, the emergence of evolutionary developmental biology has led to the recognition of phenotypic stability as one of the core concepts for understanding how developmental systems are structured and how this structure affects the mechanisms by which they evolve.

Components of Phenotypic Stability.

Canalization. Canalization is the idea that developmental processes follow pathways that are buffered from influences such as environmental perturbations or mutations. Waddington (1957) was led to this idea by the observation that the components of organisms, such as cells or organs, are discrete types and do not present a gradation of possible forms. Like a ball rolling down a slope with grooves, developmental pathways find their way to discrete end points. Moreover, he observed, when a developing organism is subjected to some insult, such as removing a block of growing tissue, the affected structure very often recovers to produce a normal adult. Processes thus exist that allow development to compensate for things that go wrong.

In a remarkable experiment, Waddington gave his theory of canalization an empirical grounding. When *Drosophila* eggs are exposed to ether, some develop a phenotype characterized by a duplication of the mesothorax. For twenty-nine generations, Waddington mated the adults that developed the bithorax phenotype and divided their eggs into a group exposed to ether and a group that was not exposed. By the eighth generation, individuals with a weak bithorax phenotype began to appear in the unexposed group, and subsequent generations showed an increasing number of such individuals. Waddington referred to this apparent evolution of an acquired character as *genetic assimilation.* His explanation is that selection acted to stabilize the expression of the bithorax trait through selection on modifier loci. Selection thus acted to canalize an environmentally induced phenotype to the extent that the phenotype occurred without the environmental stimulus.

Schmalhausen (1949) arrived at similar ideas through a different process. He argued that how the phenotype reacts to the environment is also subject to stabilizing selection. He proposed that when the external stimuli are sufficiently frequent, they are gradually replaced by internal developmental factors that lead to the development of the "adaptive response" in the absence of the original environmental cue. Schmalhausen thus independently arrived at Waddington's idea of genetic assimilation.

Recently, Günter Wagner and others (1997) have added a quantitative genetic foundation to the ideas proposed by Waddington and Schmalhausen. They developed a mathematically explicit model in which canalization is defined as a reduction in the phenotypic effect of a mutation or environmental change. Their model predicts that stabilizing selection affects the canalization of environmental and genetic changes differently. Whereas stabilizing selection always favors variants that reduce environmental variability, a reduction in the effect of mutations is predicted only for traits that exhibit a high genetic variance. This results in the counterintuitive expectation that the traits most closely related to fitness, and hence with the lowest genetic variance, will be subjected to the weakest canalizing selection.

Developmental stability. Waddington wrote: "It can hardly be expected that any epigenetic mechanisms can operate with complete precision" (1957 p. 39). In his view, developmental noise is distinct from the variation minimized by canalization. Although canalization minimizes the effects of external perturbations to a developmental pathway, Waddington regarded developmental noise as the imprecision inherent in the molecular level processes that compose the pathway. Using his epigenetic landscape metaphor, Waddington described developmental noise as "the imperfection of the sphericalness of the ball which rolls down the valley" (1957, p. 40). This theoretical distinction between canalization and developmental stability is most clearly stated in a recent article by Geoffrey Clarke:

> In other words, canalization enchances phenotypic constancy regardless of the underlying genotype or environment whereas developmental stability enhances constancy for a given genotype and environment. In essence, canalization acts to reduce phenotypic variation among individuals whereas developmental stability reduces variation within individuals. (1998, p. 562)

In this view, developmental noise is variation that is clearly not environmental in origin. In much of the recent literature, however, this distinction between canalization and developmental stability is blurred. In an influential paper, Palmer and Strobeck (1986, p. 391) define developmental noise as "minor environmentally induced departures from some ideal developmental program." Similarly, recent treatments of environmental canalization (Gavrilets and Hastings, 1994; Wagner et al., 1997) clearly equate it with reduction in developmental noise.

A clear conception of what is meant by developmental noise or instability is necessary before we can devise means to measure and explain it. Composed of imperfect materials and constructed with imperfect mechanisms within an environment for which variability can be broken down into finer and finer elements ad infinitum, organisms are subject to noiselike effects at many levels. There is thus a continuum of noiselike effects from stochastic behavior at the molecular level to broader aspects of environmental variability. The distinction made by Waddington and Clarke between developmental noise and environmental canalization represents an arbitrary distinction imposed upon a continuous range of phenomena. This view is implicit in much of the recent work on the topic, in which researchers have devised means to capture limited components of this noise-related variation by measuring aspects of phenotypic variation. Much of this work focuses on analyses of variation of minor departures from symmetry in otherwise symmetrical organisms. Such deviations are thought to derive from microenvironmental variation, minor developmental accidents, and molecular-level noise. Van Valen (1962) distinguished between different types of asymmetry (see Figure 1) and argued that asymmetries that are random with respect to side and normally distributed in magnitude are the appropriate measure of developmental instability. He referred to a distribution of asymmetries of this kind among individuals as fluctuating asymmetry (FA).

Studies of FA have focused on its associations with environmental stress, pathologies and teratologies, phenotypic extremeness, heterozygosity, selection intensity, and sexual selection. Despite the resulting enthusiasm for the use of FA to infer stress, selection intensity, and patterns of sexual selection in natural populations, the causes of variation in FA among populations and species remain largely unknown. The strengths of the various correlations that are claimed for it are also hotly disputed. Recently, Klingenberg et al. (2001) proposed the use of patterns of correlations among FA in different traits to "dissect out" modularity in developmental systems. This exciting avenue of research promises to renew interest in developmental instability from a more explicitly mechanistic developmental perspective.

Phenotypic Stability Versus Plasticity. The reverse of phenotypic stability is phenotypic plasticity. Although phenotypic plasticity is sometimes used to refer to any type of environmentally induced variation, it typically refers to the range of relatively adaptive phenotypes produced by a genotype over a typical range of environmental conditions. This relationship between the distribution of phenotypes and a determining environmental factor is known as a reaction norm. First proposed by Woltereck (1909, cited in Stearns, 1989), the modern concept of the norm of reaction and its evolutionary significance was first fully articulated by Schmalhausen (1949). Schmalhausen argued that the norm of reaction is continually molded by stabilizing selection to produce adaptive phenotypic responses to commonly encountered environmental conditions. Examples include the alteration of jaw morphology with diet in Cichlid fishes (Meyer, 1987) and the increase in strength and muscle mass with exercise in humans. Adaptive phenotypic plasticity is now recognized as a central aspect of any species' adaptive regime.

Conceivably, phenotypes are stabilized by canalizing selection, not around fixed developmental trajectories, but rather around a complex set of norms of reaction that describe the adaptive molding of the epigenetic landscape over the range of expected environmental conditions. This view partitions the domain of variability at the phenotypic level into an adaptive component shaped by selection and a nonadaptive component reduced by stabilizing selection. Yet the relationship between plasticity and stability is almost certainly not that tidy. Simons and Johnston (1997) argue, for instance, that in some plants, developmental instability may be adaptive, functioning as a bet-hedging strategy in variable environments. Several researchers have suggested that if the mechanisms that reduce developmental instability also impede phenotypic plasticity, the two components of variability should be positively correlated. Empirical tests have not provided convincing evidence for this claim (Scheiner et al., 1991).

The Evolution of Phenotypic Stability. Schmalhausen (1949) argued that the "autonomization" of development is one of the central trends in evolutionary history. By this, he meant the evolution of mechanisms that isolate development from environmental influences. Indeed, it can be argued that canalization and developmental stability are adaptive for several reasons. The evolution of increasing developmental complexity, for example, will be accompanied by selection for increased fidelity of development. In an organism in which component parts have complex developmental and functional relationships, the parts must develop in highly predictable ways. Thus, to the extent that complexity actually has increased (McShea, 1996), the evolution of increasing developmental stability might be an important large-scale trend in evolution. Similarly, adaptive responses to environmental change that are integrated across many developmental processes require fidelity of development. For structures to respond to environmental change in an integrated manner, they must develop in highly predictable relations to one another. By this argument, the reduction of developmental instability would actually facilitate adaptive phenotypic plasticity. Finally, as envisioned by Schmalhausen and shown by the work of Wagner et al. (1997) and Gavrilets and Hastings (1994), stabilizing selection will always act to increase developmental stability and environmental canalization. If stabilizing selection is as ubiquitous as is commonly believed, phenotypic stability should be strongly selected in most situations.

Once established, the mechanisms that promote phenotypic stability may well influence the course of evolution. Schmalhausen (1949) and Waddington (1942)

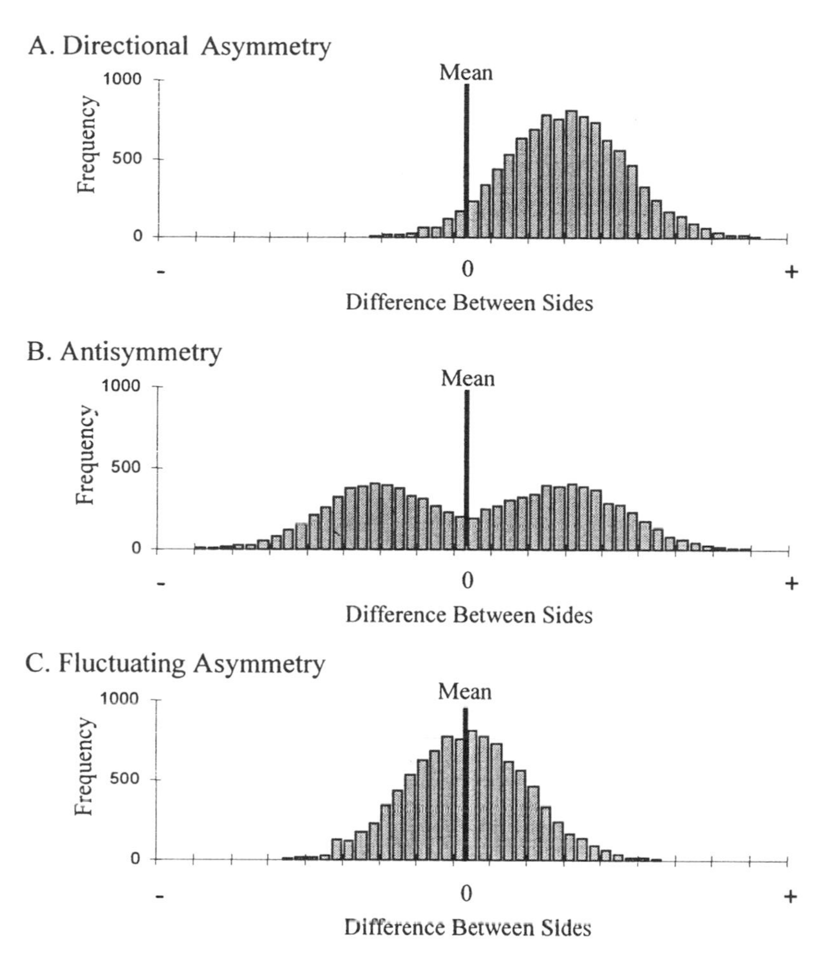

FIGURE 1. Types of Deviations from Bilateral Symmetry.
Directional asymmetry (A) refers to distributions where one side is consistently larger than the other. The left and right ventricles of the human heart are examples of structures that exhibit directional asymmetry. Antisymmetry (B) refers to the tendency to deviate to either side—in effect, a negative interaction between the sides. An example of this would be the claws of fiddler crabs in which one claw is always larger, but the choice of which one becoming larger appears to be random. Fluctuating asymmetry (C) refers to distributions in which the degree of asymmetry is normally distributed and unbiased with respect to side. This type of distribution is consistent with the effects of many small perturbations or stochastic events during development and is therefore used as a measure of developmental instability. Drawing by Benedikt Hallgrímsson.

both argued that by suppressing the effect of mutations, canalization could hide genetic variation from selection. This variation could accumulate until a threshold mutation or extreme environmental effect disrupts the mechanisms of canalization and exposes a reservoir of genetic variation to selection, resulting in a rapid period of evolution at the phenotypic level. Mechanisms that promote canalization and developmental stability can thus be seen to contribute to the dissociation of evolution at the molecular and morphological levels and predispose an evolutionary pattern of stasis interrupted by rapid change. In that they limit the expression of phenotypic variability, the mechanisms that underlie canalization and developmental stability fall into the category of developmental constraints, as broadly defined by Maynard Smith et al. (1985). The constraining effects of these mechanisms, however, are more likely to relate to the rate than the direction of evolutionary change.

Mechanisms of Phenotypic Stability. Although evidence is beginning to accumulate from studies of quantitative trait loci and of the effects of specific mutations as to which genes affect canalization and developmental stability, no obvious mechanisms have emerged as yet. One interesting suggestion made by Klingenberg (2002) is that heat shock proteins may play an important role in buffering developmental processes against some types of environmental variation. These proteins bind to other proteins that have been denatured (and have thus lost their function) and thereby help maintain the functionality of physiologic pathways. As evidence for his claim, Klingenberg notes that *Drosophila* heterozygous for a mutation in one of these proteins, Hsp90, exhibit an elevated frequency of malformations (Rutherford and Lindquist, 1998).

It is also possible that phenotypic stability is affected not by specific genes, but by the genetic architecture of development. Andreas Wagner (1996) has shown that nonlinearity in the epigenetic interactions of transcriptional regulators produces variation in genetic canalization. Another promising suggestion by Wilkins (1997) is that genetic redundancy created by gene duplication has played an important role in the evolution of stabilizing mechanisms. This hypothesis is appealing because of the abundant evidence for compensation of function among paralogous genes—a common source of frustration for gene knockout studies.

Studies of the developmental mechanisms that underlie variation in canalization and developmental stability are in their infancy. However, the increasing understanding of developmental mechanisms promises to create rapid advances in this area. Indeed, Whittle (1998) has suggested that explanations for variation in developmental stability may emerge from fuller understanding of the patterns of epigenetic interactions among genes. A common obstacle may be that so many mechanisms affect phenotypic stability that the effect of experimental perturbations in each one will be too small to be detected. Whether determined by "modifier loci" or emergent properties of epigenetic interactions, unraveling the mechanisms that underlie phenotypic stability will represent a significant leap forward toward understanding how developmental systems interplay with the evolutionary process.

[*See also* Life History Stages.]

BIBLIOGRAPHY

Cannon, W. B. *The Wisdom of the Body.* New york, 1932. Develops the concept of physiologic homeostasis.

Clarke, G. M. "The Genetic Basis of Developmental Stability: 4. Inter- and Intra-individual Character Variation." *Heredity* 80 (1998): 562–567.

Gavrilets S., and A. Hastings. "A Quantitative-Genetic Model for Selection on Developmental Noise." *Evolution* 48 (1994): 1478–1486. A model for how stabilizing selection affects "developmental noise." The concept of developmental noise includes both the variation minimized by developmental stability and environmental canalization.

Klingenberg, C. "A Developmental Perspective on Developmental Instability: Theory, Models, and Mechanisms." In *Developmental Instability (DI): Causes and Consequences*, edited by M. Polak. Oxford, 2002.

Klingenberg, C. P., A. Badyaev, S. M. Sawry, and N. J. Beckwith. "Inferring Developmental Modularity from Morphological Integration: Analysis of Individual Variation and Asymmetry in Bumblebee Wings." *American Naturalist* 157 (2001): 11–23. Uses fluctuating asymmetry as a tool to detect patterns of modularity in developmental systems.

Lerner, I. M. *Genetic Homeostasis.* New York, 1954. In this classic work, Lerner argues for the importance of heterozygosity as a determinant of fitness. This work had significant influence on the livestock industry.

Maynard Smith, J., R. Burian, S. Kauffman, P. Alberch, J. Campbell, B. Goodwin, R. Lande, D. Raup, and L. Wolpert. "Developmental Constraints and Evolution." *Quarterly Review of Biology* 60 (1985): 265–287. This article represents a consensus view of the role of developmental constraints in evolution by several prominent evolutionary biologists.

McShea, D. W. "Metazoan Complexity and Evolution: Is There a Trend?" *Evolution* (1996): 477–492. Having developed morphometric means to quantify aspects of organismal complexity, McShea shows that, although complexity generally has not increased, the upper bound of realized complexity has increased throughout evolutionary history.

Meyer, A. "Phenotypic Plasticity and Heterochrony in *Cichlasoma managuense* (Pisces, Chiclidae) and Their Implications for Speciation in Cichlid Fishes." *Evolution* 41 (1987): 1357–1359.

Palmer, A. R., and C. Strobeck. "Fluctuating Asymmetry: Measurement, Analysis, Patterns." *Annual Review of Ecology and Systematics* 17 (1986): 391–421. Outlines the methodology that is now most accepted for the study of developmental instability using deviations from symmetry.

Rutherford, S. L., and S. Lindquist. "*Hsp90* as a Capacitor for Morphological Evolution." *Nature* 396 (1998): 336–342.

Scheiner, S. M., R. L. Caplan, and R. F. Lyman. "The Genetics of Phenotypic Plasticity: 3. Genetic Correlations and Fluctuating Asymmetries." *Journal of Evolutionary Biology* 4 (1991): 51–68.

Schmalhausen, I. I. *Factors of Evolution.* Chicago, 1949. Develops the concept of reaction norms and the role of stabilizing selection in evolution.

Simons, A. M. and M. O. Johnston. "Developmental Instability as a Bet-hedging Strategy." *OIKOS* 80 (1997): 401–406.

Stearns, S. C. "The Evolutionary Significance of Phenotypic Plasticity." *BioScience* 39 (1989): 436–445. An excellent review of phenotypic plasticity that, though dated, forms a useful point of entry into the more recent literature.

Van Valen, L. M. "A Study of Fluctuating Asymmetry." *Evolution* 16 (1962): 125–142. This classic paper defined most of the central issues dealt with by the later literature on developmental instability.

Waddington, C. H. "The Canalisation of Development and the Inheritance of Acquired Characters." *Nature* 150 (1942): 563. Presents the results of the first genetic assimilation experiment.

Waddington, C. H. *"The Strategy of the Genes".* New York, 1957. Presents Waddington's ideas on genetic assimilation and canalization and their role in evolution. Lays an important part of the foundation for evolutionary developmental biology.

Wagner, A. "Does Evolutionary Plasticity Evolve?" *Evolution* 50 (1996): 1008–1023.

Wagner, G. P., G. Booth, and H. Bagheri-Chaichian. "A Population Genetic Theory of Canalization." *Evolution* 51 (1997): 329–347. Provides a quantitative genetic basis for the evolution of canalization.

Whittle, J. R. "How Is Developmental Stability Sustained in the Face of Genetic Variation?" *International Journal of Developmental Biology* 42 (1998): 495–499.

Wilkins, A. S. "Canalization: A Molecular Genetic Perspective." *BioEssays* 19 (1997): 257–262.

Woltereck, R. "Weitere experimentelle Untersuchungen über Arteveränderung, speziell über das Wesen quantitativer Artunterschiede bei Daphniden." *Verh. D. Tsch. Zool. Ges.* (1909): 110–172.

— BENEDIKT HALLGRÍMSSON

PHILOSOPHY OF EVOLUTION. *See* Overview Essay *on* History of Evolutionary Thought *at the beginning of Volume 1*; Evolutionary Epistemology; Natural Selection, *overview article.*

PHYLA. *See* Classification.

PHYLOGENETIC INFERENCE

[*This entry comprises two articles. The first article discusses the importance of the study of phylogenetics and reveals some of the unusual new applications of phylogenies; the second article provides an outline of the major methods for phylogenetic inference and discusses both optimality criteria as well as algorithms and heuristics for searching for optimal trees. For related discussions, see* Molecular Evolution; *and* Molecular Systematics.]

An Overview

A fundamental concept of the theory of evolution, independently developed by Charles Robert Darwin and Alfred Russel Wallace and published jointly in a letter of 1858, is that species share a common origin and have subsequently diverged through time. Interestingly, both men came to use the metaphor of a great tree to illustrate this notion of descent with modification; ever since, biologists have been using treelike diagrams to describe the pattern and timing of events that gave rise to the earth's biodiversity. The branching pattern of the tree represents the splitting of biological lineages, and the lengths of the branches can be used to signify the age of those events. Today, biologists call these treelike diagrams *phylogenies*.

The biological discipline dedicated to reconstructing organismal phylogenies is called *phylogenetics*. Parallel advances in a number of fields led to a tremendous growth in phylogenetics over the past forty years. First, beginning in the 1960s, sophisticated techniques were developed and refined for the purpose of reconstructing phylogenies from the actual features, or characters, of organisms. Second, phylogenetics grew beyond its traditional application to the classification of living organisms. Recognition that phylogenies can provide an evolutionary framework for studying a variety of problems led to their application in almost every other subdiscipline of biology. Third, rapid increases in the computational power of computers meant that programs implementing phylogeny reconstruction algorithms could accommodate very large amounts of data. Last, the revolution in molecular biotechnology opened up a vast new source of characters—gene and protein sequences—to phylogenetic analysis.

Before discussing the wide-ranging applications of phylogenies, it is necessary to define some essential terminology. An imaginary species phylogeny is presented in Figure 1A as a guide. The lines of the phylogeny, called *branches*, represent species, and the bifurcation points, called *nodes*, represent speciation events. The tips of the terminal branches are present-day species, and each node represents a species that is the common ancestor of all its descendants, or daughter species. For example, in Figure 1A the species at node B is the most recent common ancestor of present-day species 1, 2, and 3, and is not an ancestor of species 4 or 5. Furthermore, the group composed of ancestor B and all its descendants

FIGURE 1. Imaginary Species Phylogeny.
Courtesy of Joseph P. Bielawski.

A

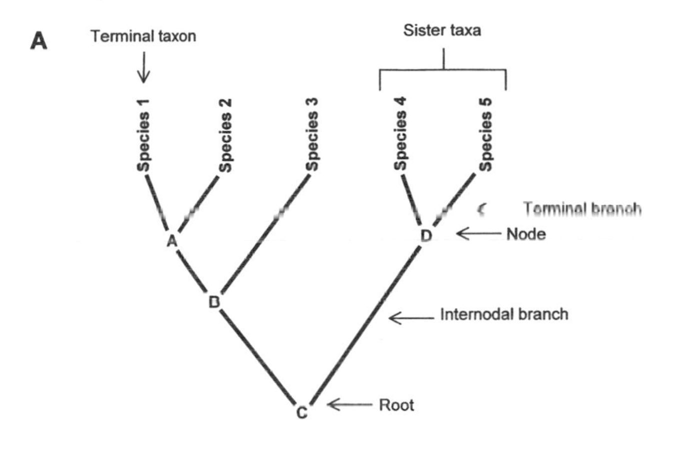

B

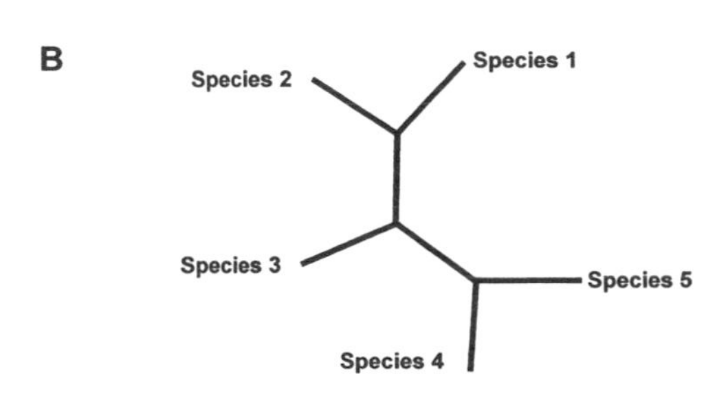

C

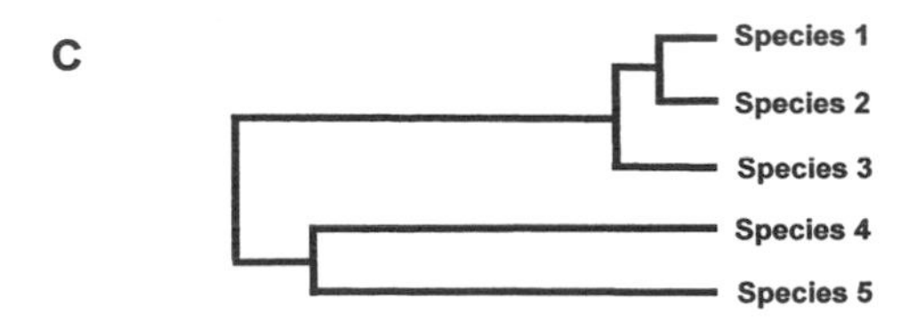

(species 1, 2, 3, and A) is called a *clade*, or a *monophyletic group*. Smaller clades are comprised of A and all its descendants, and D and all its descendants. It must be noted that phylogenetics is not restricted to species. Phylogenetic methods can be used to depict kinship of individuals within a local group or population, relationships among populations or subspecies, relationships among taxonomic lineages above species (e.g., supraspecific categories such as genera, families, etc.), relationships among genes within populations, or relationships among different genes within a gene family.

The phylogeny in Figure 1A is "rooted" at node C, allowing us to infer which ancestral species gave rise to which present-day species. Without a root, a phylogeny looks very different; compare Figure 1A with 1B, which differs only by the placement of a root. The importance of placing a root on a phylogeny should now be clear: without a root, biologists cannot distinguish between what is ancestral and what is derived (descendant). This applies to the biological attributes of species as well as to the species themselves. Rooted phylogenies allow biologists to distinguish similar characteristics owing to common descent (homology) from similar characteristics arising from convergence from different ancestors (analogy) (Figure 2). However, most methods of phylogenetic inference produce unrooted trees, and the location of the root also must be inferred. Today, the overwhelming majority of biologists use the *outgroup* method to root phylogenies (Nixon and Carpenter, 1993).

In the former examples, branch lengths were not intended to convey any information (Figures 1A and 1B). The phylogeny in Figure 1C illustrates how branch lengths can show how much change has occurred along a branch. When the phylogenetic branch lengths are derived from gene sequences, if the rate of evolution is constant over time (the so-called molecular clock), the branches will show the relative divergence times of the lineages. For example, Figure 1C indicates that the divergence of species 1 and 2 was much more recent than the divergence of species 4 and 5. Moreover, if the divergence dates of some points in the phylogeny are known from the fossil record (calibration points), and the characters are evolving in a clocklike fashion, the phylogeny can be used to predict divergences absent from the fossil record.

Data: two characters in five species

Character	Species				
	1	2	3	4	5
1	■	□	□	□	■
2	●	●	○	○	○

Character 1: similarity through convergence (analogy)

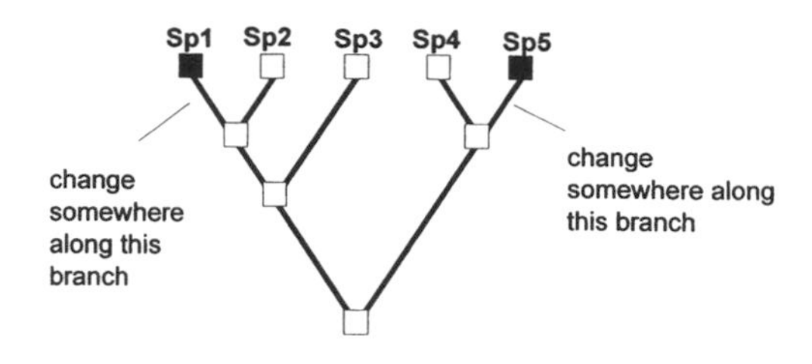

Character 2: similarity through common ancestry (homology)

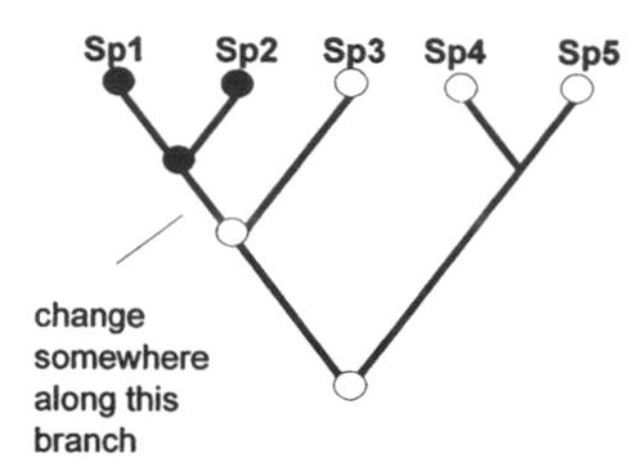

FIGURE 2. Mapping Characters on a Known Phylogeny. Courtesy of Joseph P. Bielawski.

Applications of Phylogenetics. Phylogenies can have practical value in almost every branch of biology, a fact that has become widely recognized only in the 1990s. This expansion, however, makes it impossible to review all the applications of phylogenies; instead, some examples are presented that include both classic and novel applications.

Systematics, classification, and taxonomy. Perhaps the most traditional application of phylogenetics is classification and systematics. Biological classifications are systems that organize the diversity of life, and systematics is the study of that diversity relative to some kind of specified relationship. Biologists generally agree that classification and systematics of species and supraspecific taxa should reflect the natural organization of biological diversity. The discipline devoted to producing a classification that portrays the evolutionary relationships of species and supraspecific lineages is called *phylogenetic systematics*. Narrowly defined, phylogenetic systematics has two basic components: phylogenetic inference, and production of a hierarchal classification system that exactly reflects the phylogenetic relationships. However, this definition has been broadened by some biologists to include many aspects of comparative evolutionary biology.

If a classification system is to be phylogenetic, the naming of species and supraspecific taxa (taxonomy) must reflect their phylogenetic relationships. For this reason, named taxa must comprise monophyletic groups; that is, a named taxon must represent a group descended from a single ancestral species, and all descendants of that ancestor must be included in the named taxon. This means that if a named taxon includes the common ancestor and only some of its descendants (a condition known as *paraphyly*), or does not include the most recent common ancestor (*polyphyly*), it is not acceptable in a phylogenetic classification. Take the traditional Class Reptilia as an example. The traditional Reptilia included the crocodylomorphs (alligators and crocodiles), the lepidosauromorphs (lizards, snakes, and relatives), and the anaspids (turtles and relatives). Phylogenetic analyses, however, indicated that the common ancestor of reptiles also was the ancestor of birds and mammals, which had been placed in different classes. Therefore, the traditional taxonomic grouping called Reptilia was paraphyletic. Practitioners of phylogenetic systematics point out that by using the traditional classification one neglects to recognize a phylogenetic relationship between birds and crocodylomorphs, and between mammals and extinct synapsid reptiles.

The ultimate goal of phylogenetic systematics is a phylogenetic history of all life on earth, the proverbial Tree of Life. A multi-authored Internet project is dedicated to achieving this goal. Individual parts of the Tree of Life are authored by biologists around the world, each working on a specific group of organisms, and are published electronically on the World Wide Web. When completed, this effort will provide a phylogenetic history for all life on earth, a unified taxonomy, and a means of searching and retrieving information about the characteristics of organisms. You can check the progress of this project by visiting the Tree of Life website.

Comparative biology. Evolutionary biologists use the comparative method to discover common evolutionary patterns and to understand the causes of those patterns. The key to this approach is discovering correlated patterns of evolution between different characters of organisms, or between characters of organisms and aspects of the environment that they inhabit. Most comparative studies attempt to address the adaptive significance of biological variation, although many patterns ultimately require nonadaptive explanations.

Since Darwin's time, the comparative method has remained one of the most important analytical tools of evolutionary biologists. However, comparative biology has recently undergone a major transformation: biologists realized that the characteristics of species could be correlated owing to shared ancestry. This means that to distinguish homology from independent evolutionary change, evolutionary biologists had to examine comparative trends together with phylogenetic relatedness. Phylogenetically related species will be more similar in both phenotype and lifestyle than distantly related species, and modern comparative methods must attempt to distinguish between similarities resulting from similar adaptive pressures and similarities attributable to descent from common ancestors. Felsenstein (1985), in the paper that laid the foundation for the modern transformation of comparative biology, wrote "phylogenies are fundamental to comparative biology; there is no doing it without taking them into account."

Biogeography. Biogeography is the study of the distribution of biological diversity in space and time. The subdiscipline devoted to understanding the underlying historical factors that have influenced biogeographic diversity is called historical biogeography. By considering the relationships of taxa, their geographic distributions, and the geological history of the regions they occupy, biogeographers can sometimes infer the historical importance of dispersals and geographic isolation and make inferences about modes of speciation. The methods of historical biogeography also can be applied to uncover geographic patterns of genetic variation within species (a pursuit called *phylogeography*). Phylogeographers use molecular data to infer an intraspecific gene phylogeny that is then mapped onto the geographic distribution of the species.

Health sciences. With recent advances in DNA sequencing technology, phylogenetic analysis of genes has developed into an important tool for tracking the evolution and spread of infectious diseases. Many epidemiological questions that can be addressed by phylogenetic analysis of DNA sequences: What was the origin of an emerging disease? Was there a single origin, or has a disease entered a population in different locations or at different times? How was the infectious disease spread? What was the source of a particular transmission event? How does the disease organism evolve resistance to its host? How does the host immune system evolve resistance to the disease? Are there species closely related to the known pathogens that might be able to cause disease in humans?

The case of HIV (human immunodeficiency virus) illustrates the utility of phylogenetics in epidemiology. Phylogenetic analysis indicated that HIV consists of two main types (HIV-1 and HIV-2) and numerous subtypes. Furthermore, it showed that HIV-1 and HIV-2 entered the human population from different sources: HIV-1 is more closely related to chimpanzee SIVs (simian immunodeficiency virus), and HIV-2 is more closely related to mangabey monkey SIVs. Because different subtypes within HIV-1 are related to different lineages of chimpanzee SIV, and different subtypes of HIV-2 are related to different lineages of mangabey SIV, it seems likely that both HIV-1 and HIV-2 jumped from other primates to humans multiple times. Different subtypes also are prevalent in different human populations or geographic regions, indicating that HIV spread through the human population through different routes and at different times. These phylogenetic analyses illustrate that differences between humans and other primates provide only a weak barrier to transmission of this virus, suggesting the disturbing possibility that new subtypes could enter the human population in the future.

Agriculture. Applications of phylogenetics to agriculture are similar to epidemiology, but the questions are about the origin and spread of pest species rather than of infectious diseases. For example, what was the origin of a pest? How did the pest spread though agriculture? How did some pest organisms evolve resistance to pesticides? Are there species closely related to known pests that might also cause agricultural problems?

Conservation. Tragically, while biologists work to assess and study the diversity of life, the activities of man are causing a loss of biodiversity at a rate unmatched in evolutionary history. Conservation biology is the discipline dedicated to preserving biodiversity. Phylogenetic systematics and taxonomy play a fundamental role in this effort; how can we conserve biological diversity if we do not have a natural system to organize and study it? However, there also are more direct applications of phylogenetics, including identification of genetically distinct breeding populations that require separate protection and management; assessing kinship of individuals to populations so that appropriate breeding stock can be identified for captive breeding programs; assessing kinship of dead or captive individuals for the purpose of conservation law enforcement, known as molecular forensics; and guiding the collection and organization of long-term storage of germ plasm in seed banks. Note that when working with evolutionary divergences below the species level, the discipline of phylogenetics is broadly overlapped by the discipline of population genetics, where sophisticated methods based on gene genealogies are widely used.

Phylogenetic Uncertainty. Because no person was present to observe directly the evolution of a group of organisms, biologists must infer phylogenies from the characters of living and fossil taxa. These inferences will always be subject to at least two sources of error. The first, called *random error*, occurs because a finite set of characters must be used to estimate the phylogeny. The second, called *systematic error*, occurs when the underlying assumptions of the phylogenetic method are incorrect in some important way. Because most of the applications described here rely on the assumption that a phylogeny is known without error, there is always the possibility that some conclusions might be overturned if a subsequent phylogenetic study suggests a different tree topology.

One of the most common methods of minimizing dependence on a single estimate of a phylogeny is to conduct phylogenetic analyses using several methods or several different datasets. The idea is that those parts of the phylogeny that are robust to method or dataset are likely to be good estimates of the true phylogeny. The different topologies can be combined to form a strict consensus tree, which is a single tree that maintains only the monophyletic groups found in all the individual estimates of the phylogeny. Phylogeny-based analyses are then conducted using only the information of evolutionary relationships contained in the consensus tree.

The recent application of Bayesian methods to phylogenetic inference provided biologists with a much more sophisticated tool to account for phylogenetic uncertainty. Here, Markov Chain Monte Carlo (MCMC) methods are used to approximate the Bayesian posterior probabilities of different tree topologies. These probabilities are used to identify the set of trees that have a 95 percent probability of including the true tree. Now it is simply a matter of analyzing each of the tree topologies in this set and weighting each result by the probability of the tree on which the analysis was performed. Although application of Bayesian methods to phylogenetics is still in its infancy, this analytical framework should greatly reduce the sensitivity of future evolutionary studies to the assumption that a single estimate of a phylogeny is known without error.

[*See also* Molecular Evolution.]

BIBLIOGRAPHY

Darwin, C. R., and A. R. Wallace. "On the Tendency of Species to form Varieties; and on the Perpetuation of Varieties and Species by Natural Means of Selection." *Journal of the Proceedings of the Linnean Society, Zoology* 3 (1858): 45–62. The classic letter that first introduced the theory of evolution.

Felsenstein, J. "Phylogenies and the Comparative Method." *American Naturalist* 125 (1985): 1–15. This classic paper introduced the notion that phylogenies are fundamental to comparative biology.

Harvey, P. H., and M. D. Pagel. *The Comparative Method in Evolutionary Biology.* Oxford, 1991. An excellent introduction to comparative evolutionary analysis.

Harvey, P. H., A. J. Leigh Brown, J. Maynard Smith, and S. Nee. *New Uses for New Phylogenies.* Oxford, 1996. An overview of evolutionary questions that can be investigated using phylogenetics.

Huelsenbeck, J. P., and B. Rannala. "Phylogenetic Methods Come of Age: Testing Hypotheses in an Evolutionary Context." *Science* 276 (1997): 227–232. A review of how a statistical approach called maximum likelihood enabled biologists to address evolutionary questions that had been difficult to resolve in the past.

Huelsenbeck, J. P., B. Rannala, and J. P. Masly. "Accommodating Phylogenetic Uncertainty in Evolutionary Studies." *Science* 288 (2000): 2349–2350. This paper discusses how the Bayesian method is used to accommodate phylogenetic uncertainty in evolutionary studies. The authors illustrate the approach with example datasets.

Lutzoni, F., M. Pagel, and V. Reeb. "Major Fungal Lineages Derived from Lichen-Symbiotic Ancestors." *Nature* 41:937–940. This paper combines Bayesian Markov Chain Monte Carlo methods with comparative method, to study the evolution of lichen formation.

Nixon, K. C., and J. M. Carpenter. "On Outgroups." *Cladistics* 9 (1993): 413–426. An excellent review of the outgroup method. The authors also dispel some common misconceptions about rooting trees, outgroups, and polarity.

Ou, C. Y., C. A. Ciesielski, G. Myers, C. I. Bandea, C. C. Luo, B. T. Korber, J. I. Mullins, G. Schochetman, R. L. Berkelman, A. N. Economou, et al. "Molecular Epidemiology of HIV Transmission in a Dental Practice." *Science* 256 (1992): 1165–1171. The study that used phylogenetic methods to determine that a Florida dentist had infected a group of his patients with HIV.

Pagel, M. "Inferring the Historical Patterns of Biological Evolution." *Nature* 401 (1999): 877–884. Case studies are used to illustrate of how statistical phylogenetic analysis combined with DNA sequence data influenced the field of evolutionary biology.

Page, D. M., and E. C. Holmes. *Molecular Evolution: A Phylogenetic Approach*. Oxford, 1998. An excellent introduction to the field of molecular evolution and the inference of phylogenies from molecular data.

Wiley, E. O. *Phylogenetics: The Theory and Practice of Phylogenetic Systematics*. New York, 1981. Although the methods of phylogenetic inference have progressed greatly since its publication, this work contains a very good presentation of the basic concepts of phylogenetic systematics, as well as chapters on related topics such as modes and patterns of speciation, specimens and curation, and publication and the rules of nomenclature.

ONLINE RESOURCES

"The Tree of Life" http://phylogeny.arizona.edu/tree/phylogeny.html. Maintained at the University of Arizona at Tucson; created and edited by Wayne P. Maddison. Provides information on the multi-authored project to electronically publish information on the World Wide Web about the evolutionary history and characteristics of all groups of organisms.

— JOSEPH P. BIELAWSKI

Methods

A phylogeny is a diagram that charts the genealogical relationships and patterns of descent from ancestors to descendants among a group of species, or among individuals within a species. Researchers in all fields of biology use phylogenies to investigate evolution and adaptation, and many of these applications have been spelled out in the companion articles, Phylogenetic Inference Overview and Molecular Systematics. Readers will also find phylogenies appearing in many of the other articles in the *Encyclopedia of Evolution*. Increasingly, anthropologists and linguists are also using phylogenies to investigate issues of human evolution.

This article provides a general and nontechnical overview of the chief methods and ideas of phylogenetic inference. We shall begin by describing what might be called schools of phylogenetic inference, and then move on to describe methods of phylogenetic inference. The distinction is somewhat artificial, but we will use it here as a way of separating ideas about how to use data to reflect phylogenetic history from the methods that have been used to do so.

Schools of Phylogenetic Inference. The phylogenetic relationships among a group of organisms are seldom known and so must be inferred from information collected on the species. The starting point of most phylogenetic studies, then, is to collect the same information on each of a range of species or individuals, and then to use that information to infer their probable relationships. That is, one wishes to identify which species are most closely related, and which are likely to be ancestral to others. Figure 1 provides a very simple example of inferring a phylogenetic tree. The information there can be used to make a guess about the relationships among taxa (species or groups of species), provided that one has in mind some view about how similarities and differences among organisms reflect their phylogenetic history; such a view is a "school" of inference. The tree requiring three events of evolution minimizes the number of times that the particular traits have evolved, implicitly making use of a criterion called *parsimony*. The tree requiring four events of evolution is less parsimonious. We shall return to this criterion when we discuss methods.

An early school of phylogenetic inference was known as *evolutionary taxonomy*. Evolutionary taxonomists recognized that similarity among species could arise either because species were closely related (similarity by descent) or because of convergent or parallel evolution. Convergent evolution arises when two initially dissimilar species evolve to be more similar: seals and penguins, for example, have evolved quite similar body shapes. Parallel evolution occurs when two similar species independently evolve to the same derived state. Evolutionary taxonomists studied groups in great detail and then produced trees based upon their understanding of the evolution of the group, attempting to group as much as possible on the basis of similarity by descent. Many of the trees produced this way were probably accurate, but the method suffered from not having an explicit and objective methodology.

In the early 1960s, a group of statisticians and biologists introduced a new approach, known as *numerical taxonomy* or *phenetics*. The numerical taxonomists argued that there is no sound way to infer the pattern of common ancestry of a group from character-state data. They asserted that organisms should be grouped on the basis of overall (observed) similarity, independently of whether these groupings represented phylogeny. Numerical taxonomists proceeded first by examining as many traits as they could manage, and then they used statis-

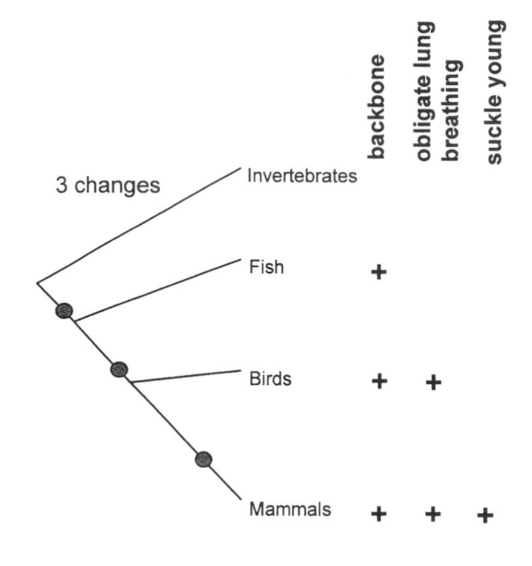

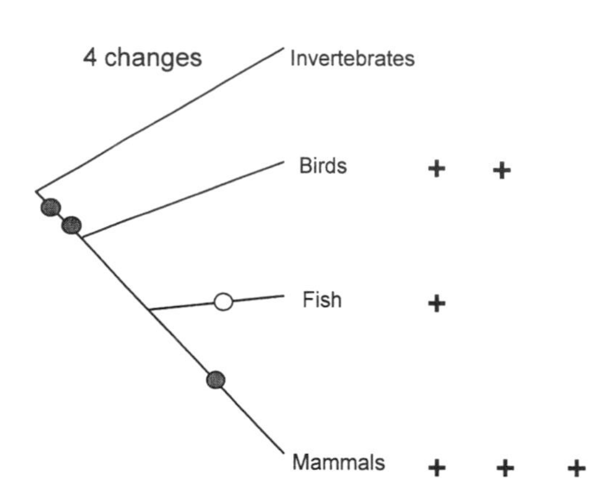

FIGURE 1. Two Phylogenetic Trees Classifying Invertebrates, Fish, Birds, and Mammals.
The upper tree (correct) implies that there have been three events of evolutionary change in the three traits that are used to infer the tree (filled circles indicate likely position of each event): the ancestor to fish, birds, and mammals acquired a backbone, then the ancestor to birds and mammals acquired obligate lung breathing, and finally mammals acquired milk production. The lower tree (incorrect) shows backbones and lung breathing being acquired in the ancestor to birds, fish, and mammals, and then the loss (open circle) of obligate lung breathing in fish, requiring four changes. Alternatively, birds and mammals could be represented as independently acquiring obligate lung breathing, and then no loss of it would have to be implied in fish. This representation also requires four events. Here the more parsimonious tree is also the correct tree, although the two cannot be assumed to be synonymous.
Courtesy of Mark Pagel.

tical clustering methods to arrange taxa into a hierarchies based upon overall similarity. The appeal of the methods was their apparent objectivity, and numerical taxonomy was once popular; however, its use has declined as it became apparent that different clustering statistics could give quite different results, and that clusters did not necessarily reflect true phylogenetic relationships.

In 1966, Willi Hennig's book *Phylogenetic Systematics* was translated into English, giving rise to the highly influential school of systematics known as *cladistics*. Cladists explicitly attempt to reconstruct the hierarchical branching pattern of species. Hennig's insight was that attributes (characters) that are derived (as opposed to ancestral) and shared exclusively by a set of taxa provide evidence for exclusive common ancestry. For example, relative to nonmammals, mammals share the derived trait of milk production, and this character provides evidence of their common ancestry, to the exclusion of, for example, birds and reptiles. Possession of true feathers distinguishes birds, and possession of a placenta distinguishes placental from marsupial mammals. Bat's wings provide evidence that they share common ancestry to the exclusion of other mammals. How to identify derived traits and how to distinguish shared derived traits from convergent evolution play a central role in cladistic inference.

Sets of shared derived traits provide strong evidence that a group is "monophyletic." This is one of the most important principles of phylogenetic inference. A monophyletic group is one that contains all of the species that have descended from a common ancestor, and only those species. By comparison, sets of ancestral traits (ones that are not derived) are less helpful in identifying common ancestry. Milk production being ancestral to mammals is of no use in sorting out relationships among mammals. Most mammals and many reptiles have five toes, a condition ancestral to both groups, but this does not suggest that lions are more closely related to crocodiles than they are to cows (which possess two toes). Using the logic of ancestral and derived traits, cladists construct sets of monophyletic groups to construct phylogenetic trees thought to identify true ancestral–descendant relationships.

By the 1980s, *statistical* and *model-based* methods for inferring phylogenetic trees began to provide an alternative perspective to cladism. Luca Cavalli-Sforza and Anthony Edwards, working in the 1960s, had pioneered their use, and Elizabeth Thompson and Joe Felsenstein offered important refinements in the 1970s and 1980s. Together these researchers developed the techniques for applying the statistical method of maximum likelihood (discovered by the great English statistician and geneticist Sir Ronald Fisher) to the problem of phylogenetic inference. Maximum likelihood methods in a phylogenetics context seek the phylogenetic tree that, along with an explicit model of character evolution, makes the data most likely. A model of evolution describes how characters change from one state to another. The logic of maximum likelihood is that the data (e.g., gene sequences) are presumed to have arisen according to the model of evolution and on a particular phylogenetic tree. Thus, one seeks the combination of tree and parameters of the model of evolution that would have been most likely to have given rise to the observed data. Statistical criteria are used to judge the goodness of fit, and to compare the fit of different possible phylogenies for a given set of data.

It is no coincidence that the model-based methods arose in tandem with the increasing availability of protein-sequence and gene-sequence data. Both are examples of *discrete* data—that is, data for which there exist only a finite number of different states that the character can take (contrast this with, for example, a trait like body size that can adopt an effectively infinite number of different states). Proteins can adopt one of only 20 different amino acids at each site, and gene sequences can have one of only four different bases (A, C, G, or T) at each site in the sequence. The attraction of this discrete sequence data was that it was possible to write down simple and mathematically tractable models of how a given site in the sequence could change to any one of the other possible states.

Do model-based methods identify shared-derived traits and use them to form monophyletic groups? In most circumstances, model-based methods will prefer trees that group together species with similar derived states, because to group them otherwise would cause the model of evolution to fit the observed data poorly. However, we shall see that one of the additional strengths of the model-based methods is that they can accommodate convergent or parallel evolution in ways that can cause trouble to other approaches.

Methods of Phylogenetic Inference. The three principal methods of phylogenetic inference—parsimony, distance methods, and maximum likelihood methods—all draw upon ideas that emerged from the debates among the schools of phylogenetic inference. Parsimony methods are closely linked to cladism, and the maximum likelihood methods arose from the statistical school. Distance-based methods of inference share a number of features with the phenetic school.

Parsimony methods. Parsimony methods seek, out of all the evolutionary trees that could possibly describe the relationships among a group of organisms, the tree that implies the fewest evolutionary changes in the characters being examined. This most parsimonious tree is taken to be the best estimate of the unknown true tree. The principle of parsimony traces its roots to William of Occam, a fourteenth-century philosopher. Occam advocated the view that when alternative explanations for an observed phenomenon exist, the simplest (or most parsimonious) explanation is to be preferred. This principle came to be known as Occam's Razor.

The tree in the upper panel of Figure 1 is the most parsimonious tree. There, four taxa are grouped according to three different pieces of information. The presence of a backbone is a derived feature shared among the fishes, birds, and mammals. This character identifies (within the bounds of this example) these three taxa as a monophyletic group. Next, obligate lung breathing is a derived trait in birds and mammals, identifying them as a monophyletic subgroup of the animals with backbones. Finally, milk production is a derived trait that uniquely distinguishes the mammals. In the language of parsimony methods, this tree is described as having a length of three, denoting the number of character changes or "steps" required to reconstruct it. It has been suggested that, in general, the most parsimonious tree is also that tree on which groups are defined by shared derived characters to the greatest extent possible. Thus, cladists view maximum parsimony as the preferred method for distinguishing among possible trees, and they apply the principles of cladism—notably the identification of monophyly via shared derived characters—to construct trees.

The tree in the lower panel of Figure 1 makes mammals more closely related to fishes than to birds. This tree requires four events of evolution in the three characters. As shown, two derived features are acquired in the branch leading to birds, and one of these (obligate lung breathing) must be secondarily lost in the fishes. Alternatively, birds and mammals could be portrayed on this tree as having independently acquired lung breathing (two events), also requiring a total of four events on the tree. Either way, this tree is less parsimonious than the true tree.

Real phylogenetic problems are far more complicated and seldom produce such clean results. The shortest tree need not be the true tree, and the data typically will not fit the tree perfectly. In the example given, each character evolves to the derived state once. In real data sets, there may be characters that, like lung breathing in the lower tree of Figure 1, are forced to evolve to the derived state more than once (a phenomenon known as *homoplasy*—the same derived state arising more than once from a common ancestral state). Parallel evolution is the chief cause of homoplasy, and convergence can appear as homoplasy. Alternatively, having evolved, the derived state may be lost in some other taxon. The upshot is that in real applications, not all characters will be perfectly compatible with a given tree.

A number of methods for managing these incompatibilities have been proposed. In essence, they are compromises that attempt, as far as possible, to retain the principles of cladism when confronted with data that violate those principles. *Fitch parsimony* is the least restrictive method, allowing characters to be gained and lost more than once on a tree. *Dollo parsimony*, rejects trees that require homoplasies. In place of a derived feature evolving more than once from the same ancestral state, Dollo parsimony solutions find trees that force the derived character to be lost and to return to a more primitive ancestral state. Then, the additional gains of the derived state are treated as a new derived character (because it evolved from a different ancestral state). The most parsimonious tree with these characteristics is preferred.

Sometimes certain characters may be deemed to be more informative than others. The idea is that characters that have evolved slowly might be more likely to be reliable markers of phylogenetic relationships than are characters that evolve rapidly. Rapidly evolving characters typically show homoplasies. Another possibility is that some kinds of characters may be likely to show instances of convergent or parallel evolution. Parsimony methods attempt to accommodate all the situations by applying weightings to characters, with greater weight given to more reliable characters.

An obvious difficulty with weighting according to the amount of change is that the true relationships must be known to identify which characters are reliable markers and therefore deserve more weight. Another problem is to decide what the weights ought to be. Because there is no objective way to answer either question, researchers frequently apply several weighting schemes and assess the influence these have on the resulting phylogeny.

Minimum evolution and distance methods. Minimum evolution (ME) methods find phylogenetic trees whose branch lengths most closely reflect the actual "distances" that are observed among all possible pairs of species. The lengths of the branches of a phylogenetic tree measure the amount of evolutionary change implied to have occurred between an ancestor and a descendant. Tracing the distance between two species via their common ancestor gives the implied distance between those two species. Actual distance can be measured in a variety of ways, but the simplest is merely to ask whether the state of a character is the same or different between two species, assigning these two outcomes a 0 or a 1 respectively. A distance matrix is formed that is a composite of the distances of all the individual characters. (In contrast, maximum parsimony treats each character examined as a discrete piece of evidence, the minimum evolution criterion acts on the aggregate distance score; see Table 1.)

Table 1. Distances Between the Pairs of Taxa in Figure 1
Values below the diagonal are the same as those above.

	Invertebrates	*Fish*	*Birds*	*Mammals*
Invertebrates	—	1	2	3
Fish		—	1	2
Birds			—	1
Mammals				—

Other more complex methods for finding distances are routinely applied when the data are gene or protein sequences. These methods attempt to correct for the possibility that the greater the degree of separation that exists between two species, the more likely it is that the character has changed more than once. For example, during the millions of years of evolution separating invertebrates and fishes on the contrived tree of Figure 1, various genes may have evolved several times at a given position; however, all that is observed is whether the two species are the same or different on that character. Methods for correcting for "double hits" attempt to take into account the amount of time separating two species when calculating distances.

Once a distance matrix has been formed, distance methods search for the phylogenetic tree that minimizes some function of the difference between the observed distances and those implied by the tree. For example, in the lower panel of Figure 1, the branch with two changes along it might be reconstructed to have a length of two, and those with one change might be given a length of one. For this simple example, the upper tree of Figure 1 is the minimum distance tree. Distance methods would reconstruct its branch lengths to reflect the differences implied in the data matrix. For these data, the branches in which a change is implied to have occurred would each have length 1.0, and the remaining branches would have length 0.0. A length of zero implies no difference between the species and its implied ancestral state (which for these contrived data is accurate).

Seldom can a tree be found that perfectly reflects the distances implied in the matrix. Accordingly, a number of different distance matrix methods can be defined depending upon how they minimize the difference between the tree and the observed data. One simple and effective way is to minimize the sum of the squared differences between the distances implied by the tree and those in the distance matrix, a technique known as *squared-change parsimony*. Other distance methods include UPGMA (unweighted pair group method using arithmetic averages) and neighbor-joining. The UPGMA method forces the tree to be *ultrametric*; that is, all of the distances from the root or base of the tree up to the tips must be the same. Parsimony and simple distance methods (and, as we shall see, maximum likelihood methods) do not make this assumption. The neighbor-joining method is not ultrametric and finds an approximate minimum evolution tree by sequentially adding the nearest neighbor that minimizes the total length of the tree.

Distance methods work with overall measures of similarity or difference, without regard to how the patterns arise. Some critics assert that this makes the methods phenetic, and that the distinction of shared versus derived traits is lost. The resolution of this problem depends upon how common convergent evolution is. If convergent and parallel changes are rare, then the observed distance between any two species will reflect evolutionary events that have occurred since the two species separated from their common ancestor. By definition (and assuming convergence is rare), these differences will be derived. If, however, convergence is common, the measured distance between two species will underestimate the true amount of evolution separating them. Whether one believes that distance methods can identify monophyletic groups is largely a question of how common parallel and convergent evolution are. No one knows the answer to this, but it is probably fair to say that parallel and convergent evolution in proteins and gene sequences are far more common than has previously been believed.

Maximum likelihood. Maximum likelihood (ML) methods in phylogenetics are a mathematically consistent set of techniques, derived from the general statistical approach of the same name developed by Fisher. Maximum likelihood methods have elements in common with both parsimony and distance methods. Like parsimony methods, ML methods are character-based in the sense of developing a score separately for each character. Like distance methods, however, ML methods use statistical criteria to arrive at the preferred tree. But the similarities are mostly superficial, because ML methods depart in both logic and methods from all others.

On the simplest view, ML methods estimate the probability of observing the pattern of states that each character displays on the phylogenetic tree. For the example of Figure 1, this would involve calculating a probability of observing backbones in three taxa, but not in invertebrates, and so on for the other two characters. (We assume here that the ancestral states are known. ML methods do not require this assumption but we make it here for convenience.) They then combine these probabilities into a single overall probability. The tree that makes this overall probability—called the "likelihood"—largest is the maximum likelihood tree. To calculate the probability of observing a character requires an explicit model of how the characters evolve. The Vignette illustrates how a model could be constructed for the data of Figure 1 and shows all the steps involved in calculating a likelihood and finding the maximum likelihood value.

The greatest application of ML methods has not been to morphological data, however, but to molecular data—gene sequences and proteins. For these kinds of data, it is relatively easy to develop models of sequence evolution that capture important elements of the process. Consider that for gene-sequence data one might have a model specifying the rates at which each of the four nucleotide bases at a given site of a sequence (A, C, G, or T) evolves over time to another base (Figure 2, upper). Each of the elements of the matrix denoted Q in Figure 2 specifies the rate at which a given nucleotide in a gene sequence changes to one of the three other possible nucleotides. This model requires twelve parameters, and they are assumed to apply to every site in the nucleotide sequence (Figure 2, lower). Each site (like a character in Figure 1) of the gene sequence records the particular nucleotide observed at that site in each of the species.

The model recorded in Figure 2 is the most general model, according a specific rate to each kind of change. For technical reasons, simpler models are conventionally used. Most of these presume that the rates of change from one nucleotide to another are the same in both directions. Such models are known as *time-reversible* (superficially because they look the same going forward or backward in time). This simplifies calculations and reduces to six the number of parameters to be estimated. At the other extreme, one could set all the parameters to the same value, an approach known as the *Jukes-Cantor* model. Alternatively, one could set all of a particular class of changes to be equal to each other. Nucleotide changes called "transitions" ($A \leftrightarrow G$ or $C \leftrightarrow T$)

FIGURE 2. An Example of a Model of Gene-Sequence Evolution Showing the Implied Rates of Transition Among the Four Nucleotides (Upper).
A, C, G, and T are the possible bases of a gene sequence, and the values in the Q matrix correspond to the rates of change from one nucleotide to another. The lower panel displays an example of gene sequence data on N species. Courtesy of Mark Pagel.

$$
\begin{array}{cc} & \begin{array}{cccc} A & C & G & T \end{array} \\ Q= \begin{array}{c} A \\ C \\ G \\ T \end{array} & \begin{bmatrix} - & q_{ac} & q_{ag} & q_{at} \\ q_{ca} & - & q_{cg} & q_{ct} \\ q_{ga} & q_{gc} & - & q_{gt} \\ q_{ta} & q_{tc} & q_{tg} & - \end{bmatrix} \end{array}
$$

```
Species    gene-sequence
1          ACCGTT...C
2          ACGCTG...T
3          GCCGCT...C
.          .........
N          ATGCTT...G
```

HOW TO ESTIMATE A MAXIMUM LIKELIHOOD PHYLOGENETIC TREE

Assume that we have data for the three characters and four taxa of Figure 1. We wish to write down a model of evolution and then use it to find the maximum likelihood tree. We shall write down the simplest of models, but it will illustrate the important principles of ML inference.

Let each of the characters change at a rate given by a parameter we will call λ. We wish to calculate the probability of observing each character on the tree, then combine those probabilities. Let the probability that a character does not change in a branch of length *t* ("time") be given by the equation

$$P_s(t) = Exp[-\lambda * t],$$

where $P_s(t)$ denotes the probability of stasis in a branch of length *t*, and *Exp* is the base of the natural logarithms, e~2.71828. . . . The function $Exp[-\lambda * t]$ takes the value of 1.0 when $t = 0$, and then steadily declines to 0 as *t* grows larger. This accords with our intuitions: the probability of something remaining unchanged is high over short periods of time but steadily declines over time. In the limit, as time grows long, we become increasingly certain that the character will change (if $\lambda > 0.0$).

The probability of a character changing in a branch should increase over time. In fact, the probability of changing is given by:

$$P_c(t) = 1 - P_s(t) = 1 - Exp[-\lambda * t].$$

We now assign the appropriate term to each branch. The branch leading from the root of Figure 1 (upper panel) to invertebrates records no changes in character 1. It therefore is assigned a $P_s(t)$ term. The branch leading from the root to the ancestor to fishes records a change in character 1 (the acquisition of a backbone). It is assigned a $P_c(t)$ term. The appropriate term can be assigned in this way to each branch. Branches of the tree are treated as independent outcomes of evolution. Therefore, the probability of witnessing the set of $P_s(t)$'s and $P_c(t)$'s is given by their product. We repeat this for the other two characters. Then, their product gives the probability of witnessing the events for all three characters. This product over branches and characters we designate the likelihood, L:

$$L = P_s(t_1)P_c(t_2)P_s(t_3)P_s(t_4)P_s(t_5)P_s(t_6) * P_s(t_1)P_s(t_2)P_s(t_3)P_c(t_4)P_s(t_5)P_s(t_6) * P_s(t_1)P_s(t_2)P_s(t_3)P_s(t_4)P_s(t_5)P_c(t_6)$$

The numbers associated with each length *t* correspond to the branches of the phylogeny, with t_1 going to invertebrates, t_3 leading to fishes, t_5 leading to birds, t_6 leading to mammals, and t_2 and t_4 being the internodal branches. The rate parameter and branch lengths are now found by an iterative numerical search procedure in such a way as to maximize *L*. This reveals the following values:

$L = 0.00325153$ $\quad t_1 = 0.0000020$
$t_3 = 0.00000018$
$t_5 = 0.00000057$
$t_2 = 1.06 \quad t_4 = 1.0$
$t_6 = 1.06 \quad \lambda = 0.38$

The estimated branch lengths are pleasing in that all three branches in which we infer a change to have occurred are estimated to have a length of approximately 1.0, and all the others are reconstructed to have a length of zero. The value of the rate parameter suggests a 0.5 probability of a change in a branch of length 0.548. It is easy to verify the overall likelihood and value of λ by substituting the values of the estimated branch lengths into the equation for *L*, and then plotting its value. It will reach its maximum of 0.003 at $\lambda = 0.38$.

Going through the same steps for the lower tree of Figure 1 yields a negligibly lower (worse) likelihood of 0.00326, but a faster rate parameter of 0.60. This accords with our intuition that the lower tree is a worse description of the data, requiring more changes on the tree. It also demonstrates how likelihood can accommodate homoplasy (see text).

Technical notes: more than one solution exists for these data, corresponding to higher or lower values of λ and correspondingly shorter or longer branches, but no conclusions are altered. Calculations have assumed ancestral states are known (see text).

—Mark Pagel

are observed to occur more frequently than a kind of change called a "transversion" (A ↔ C, A ↔ T, C ↔ G, & G ↔ T). Setting all transitions to the same rate and all transversions to another yields a two-parameter model known as the *Kimura* model. Many modifications to this basic structure are possible.

Once a model is specified, it is possible to derive formulae to calculate the probability (called the likelihood) of observing the data at each site of the gene sequence. These probabilities are then combined across all sites into the likelihood. They are combined by finding the product across all sites of the individual probabilities at each site. The product reflects the fact that each character is treated as an independent observation. The probability of observing a set of independent events is the product of their individual probabilities (thus, the probability of witnessing two rolls of a die both come up six is 1/6 × 1/6, or 1/36). The calculations assume that the rate parameters are known; this will seldom be the case, and so they are estimated from the data (see Vignette).

A long nucleotide sequence provides a large amount of information for estimating these parameters. Parameters are estimated by the principle of maximum likelihood. In a phylogenetic context, one seeks the values of the parameters of the model (here the rate parameters of the model of DNA sequence evolution) and the phylogenetic tree (including its branch lengths) that make the observed data (the gene sequences) most probable. This process yields a tree with branch lengths that measure the expected number of substitutions (changes) per site in the sequence data, as well as estimates of the parameters of the model of evolution. These parameters convey important information regarding the nature of the evolutionary process (for example, whether transitions proceed at a higher rate than transversions), and hypotheses about them can be tested statistically.

Maximum likelihood trees may also be the most parsimonious or the minimum evolution trees, but there is nothing in the logic or mathematics of ML estimation that requires this, or even deems it desirable. The reason is that ML methods seek the most *probable* solution. If an ML method detects that characters are changing at a high rate on the tree, and thereby potentially causing homoplasy, it may actually prefer a tree with a relatively large number of changes on it (reflecting the high rates of change). That is, there is nothing in the logic of ML to force it toward a more parsimonious solution. As a result, issues of homoplasy, convergence, and parallel change do not pose logical problems for ML approaches. For the same reasons, ML methods need not resort to the ad hoc weighting schemes of parsimony methods. A site that shows a large amount of change will typically have a lower probability under the model of evolution and will therefore contribute less to the overall likelihood.

This ability of ML methods to deal effectively with large amounts of change corrects a specific weakness of parsimony methods. In addition to having to make use of unsatisfactory ad hoc weighting schemes, parsimony methods can seriously underestimate the true amount of change when rates of change are high. This introduces a systematic source of bias into parsimony trees. ML trees, by virtue of estimating rates of change, tend to be less influenced by this bias.

TABLE 2. Phylogenetic Inference

Shown are the number of different possible phylogenetic trees for a given number of species. The calculations assume that the trees are "rooted" like that in Figure 1.

Number species	*Number of different trees*
3	3
5	105
8	135,135
10	34,459,425
20	8.2×10^{21}
50	2.8×10^{76}
100	3.5×10^{184}
500	1.0×10^{1280}

Finding the Best Tree. Regardless of the method used to find a phylogenetic tree, some procedure for searching the space of possible trees is required. When only a few species are analyzed, it may be possible to search all possible trees to find the best one. However, as Table 2 shows, even for moderate numbers of species the number of possible trees can be large. For only 10 species there are already more than 34 million different possible phylogenetic trees, and for 500 species the number is almost incomprehensible. This makes it impractical to examine every tree for a large number of species, and so a number of approximate methods exist.

Algorithmic methods. Algorithmic methods include stepwise addition, star decomposition, and quartet puzzling. In stepwise addition procedures, the tree is built by adding taxa sequentially. These methods start with three taxa, choose a fourth to add based on one of several criteria, and attach it to the growing tree at the point that incurs the lowest cost under whatever optimality criterion has been chosen. This procedure continues until all the taxa have been added.

Star decomposition approaches begin with all the taxa radiating independently from a single internal point. This is the initial star tree. Taxa are clustered one pair at a time (i.e., the star tree is decomposed), based on which of all possible pairings maximally improves the fit between the tree and the data, until the tree is entirely bifurcating. The neighbor-joining algorithm is a star decomposition method.

Quartet puzzling examines all possible four-taxon trees (quartets), evaluates the best arrangement for each quartet of taxa, and combines those quartets together in the way that minimizes the cost (e.g., tree length).

Exact searches. None of the algorithmic methods is assured of finding the best tree. Exact searches are guaranteed to find the best tree but are limited to trees of relatively small numbers of species. The most obvious

TABLE 3. A Summary of the Principal Features of Methods of Phylogenetic Inference

Method	*Approach*	*Aims*
Parsimony	character-based	minimize "steps" or changes
Distance	phenetic	minimize distance
Maximum Likelihood	statistical (character based)	maximize likelihood
Bayesian (MCMC)	statistical (character based)	estimate features of the phylogeny

exact method is to examine all possible bifurcating trees. A second exact approach is a branch-and-bound search. This approach will not examine all possible topologies, and it can therefore be applied to larger data sets than exhaustive searches, but it still is not possible to apply a branch-and-bound search to more than 18–20 taxa.

Heuristic searches. Most studies include too many species to permit the use of exact searches, so heuristic searches must be employed. The most common approach is to apply one of the algorithmic methods described above (usually stepwise addition) to derive a starting tree and try to improve on this by making small rearrangements called *branch swapping*. There are a number of options for branch swapping, including nearest-neighbor interchanges, subtree pruning/regrafting, and tree bisection/reconnection (arranged from least to most extensive). These methods often find a local optimum, but they may or may not find the globally optimum tree.

Developments in Phylogenetic Inference

Morphology and linguistics. A recent development in phylogenetic inference is the application of maximum likelihood estimation to morphological data. ML models are especially applicable to morphological variables that adopt a discrete number of states. All that is required is to specify a transition rate matrix, such as that in Figure 2, but in which the transitions are not among nucleotides but among character states. This approach has been used for some time to analyze discrete morphological variables on a given phylogenetic tree, but it has only recently been used to infer the tree itself. Similarly, maximum likelihood methods have been extended to linguistic data to infer phylogenies of human cultural groups on the basis of their language similarity and differences.

Bayesian analysis. An important recent advance in phylogenetic inference is the application of Bayesian Markov Chain Monte Carlo (MCMC) methods. These methods generate a long series of phylogenetic trees, each one of which differs from the previous tree in the chain by some transformation. For example, a branch might have its length changed, or a branch might be moved to some new place on the tree. New trees are generated such that the probability of moving to the new tree depends only upon the state of the previous tree in the chain (the no-memory property). Successive trees are then accepted or rejected according to the Metropolis-Hastings algorithm. The Metropolis-Hastings algorithm always accepts a new tree if it improves upon the previous tree in the chain; otherwise, the new tree is accepted in direct proportion to how much worse it is. If it is 90 percent as good as the previous tree, it is accepted 90 percent of the time. Trees are evaluated by their likelihood given a model of gene sequence or protein sequence evolution.

It can be shown that this combination of producing new trees according to the no-memory property and then sampling them with the Metropolis-Hastings algorithm will (if the chain is allowed to run long enough) yield a succession of trees that randomly samples the universe of possible trees according to their probabilities of occurrence under the model of evolution. This is a desirable property because it means that investigators can search the space of trees that are most likely and can ask of the set of trees how reliable certain features are. Rather than searching for the single best tree, one might ask how often in the sampled set of trees a particular monophyletic group or clade appears. The statistical theory that underpins Markov Chain Monte Carlo states that the number of times a clade (or some other feature) appears in a set of sampled trees estimates the posterior probability that the feature exists. Thus, for example, one might sample 50,000 trees from a Markov Chain and then ask in what proportion of the trees some monophyletic group appears. This proportion estimates the probability that that monophyletic group exists, given the model and the data.

MCMC methods therefore provide a straightforward and attractive approach to hypothesis testing in phylogenetic inference. It can also be shown that MCMC methods are more efficient than conventional ML methods in which one searches for the single best tree. MCMC methods using maximum likelihood models look set to become a widely used and important set of techniques in phylogenetic inference.

Concluding Issues. We have discussed at an introductory level the logic and methods of phylogenetic inference (summarized in Table 3). The field is a highly technical one, however, and so we have conveniently ignored several issues and have left others to be discussed at the end.

Our discussion has assumed throughout that our

trees were rooted—that is, that we knew where they began in time. In fact, most methods of phylogenetic inference produce unrooted trees, and then it is up to the investigator to specify the root. Rooting trees can be difficult unless independent information exists as to which of the species or group of species might represent what is known as an *outgroup* to the remaining species. An outgroup is a species or group of species whose ancestors diverged from the remaining ingroup species before any of the ingroup species diverged. By convention, roots are often placed halfway between the in- and outgroups.

The notions of ancestral and derived states play an important role in our discussion, yet we have not explained how one knows which is which. This is called a character's *polarity*, and a number of methods exist to establish ancestral states. One is to assume that the state in the outgroup is the ancestral state. Fossils and features of embryological development may also be informative.

We have implicitly assumed throughout that most parsimonious, minimum evolution, or maximum likelihood trees are the most preferable trees. However, there is no particular reason to believe that the trees that optimize our criteria are the true trees. Nature may not be parsimonious (true homoplasy may exist), and nature is almost certainly more complicated than even our most complex models. Thus we should treat all phylogenetic trees as hypotheses rather than as real objects. To circumvent this problem, some investigators build consensus trees of large numbers of, say, equally parsimonious trees (often a tree search algorithm will turn up hundreds or even thousands of different trees of the same length), or trees of similar likelihood. The assumption is that somehow the consensus is a better representation of the truth than any single tree. The trouble with a consensus tree is that we know it is neither the best tree nor the true tree!

If consensus trees are not the best, and best trees are not necessarily true, then which way shall we turn? Here MCMC methods loom large. These methods, by virtue of collecting a *random* sample of trees from the universe of possible trees, allow one to estimate aspects of the phylogeny. Thus, the proportion of times a particular monophyletic group appears in the random sample of phylogenies estimates the probability that that node really exists, given the data and the model of evolution. The random sample also nudges investigators away from "parsimony," "best," and "shortest" thinking, reorienting their view toward the sample of trees. The sample often contains a few very "good" trees (short, high likelihood), some bad trees, and a large number of middling trees. If we believe our models of evolution, it is in this large number of middling trees that the true tree may lie. This reflects the view that nature may not be parsimonious, and that the most common trees in nature include the messiness of real life.

[*See also* Comparative Method; Convergent and Parallel Evolution; Molecular Clock; Molecular Evolution; Molecular Systematics.]

BIBLIOGRAPHY

Cavalli-Sforza, L. L., and A. W. F. Edwards. "Analysis of Human Evoluton under Random Genetic Drift." *Cold Spring Harbor Symposia on Quantitative Biology* 29 (1964): 9–20. The first paper to attempt to apply maximum likelihood methods to phylogenetic inference.

Cavalli-Sforza, L. L., and A. W. F. Edwards. "Phylogenetic Analysis: Models and Estimation Procedures." *Evolution* 32 (1967): 550–570. Among the first applications of maximum likelihood estimation to phylogeny.

Felsenstein, J. "Statistical Inference of Phylogenies (with Discussion)." *Journal of the Royal Statistical Society A* 146 (1983): 246–272. This is the key paper introducing maximum likelihood methods for the inference of phylogenies from gene sequence data.

Harvey, P. H., and M. Pagel. *The Comparative Method in Evolutionary Biology.* Oxford, 1991. The chapter on phylogenetic inference provides a useful introduction to some of the issues discussed here.

Hennig, W. *Phylogenetic Systematics.* Translated by D. D. Davis and R. Zangerl. Urbana, 1966. The English translation of Hennig's 1950 book in German. This translation gave rise to a debate regarding the manner in which systematics should be conducted.

Hull, D. *Science as a Process.* Chicago, 1988. Reviews, among other issues, many of the debates surrounding phylogenetics in the 1970s and 1980s.

Kitching, I. J., et al. *Cladistics: The Theory and Practice of Parsimony Analysis.* 2d ed. Oxford, 1998. A detailed introduction to cladistical methods.

Larget, B., and D. L. Simon. "Markov Chain Monte Carlo Algorithms for the Bayesian Analysis of Phylogenetic Trees." *Molecular Biology and Evolution* 16 (1999): 750–759. A technical introduction to Bayesian inference of phylogenetic trees.

Lewis, P. O. "Phylogenetic Systematics Turns Over a New Leaf." *Trends in Ecology and Evolution* 16 (2001): 30–37. Clear review of maximum-likelihood estimation of phylogeny and an account of Bayesian estimation methods.

Lutzoni, F., M. Pagel, and V. Reeb. "Major Fungal Lineages Derived from Lichen-Symbiotic Ancestors." *Nature* 411 (2001): 937–940. One of the first articles to use Bayesian MCMC methods for inferring phylogeny. Combines MCMC phylogenetic inference with the inference of ancestral states of lichen formation.

Pagel, M. "Maximum Likelihood Models for Glottochronology and for Reconstructing Linguistic Phylogenies." In *Time-Depth in Historical Linguistics*, edited by C. Renfrew, A. MacMahon, L. Trask, pp. 189–207. Cambridge, 2000. Describes an application of maximum likelihood inference of a linguistic phylogeny.

Pagel, M. "Detecting Correlated Evolution on Phylogenies: A General Method for the Comparative Analysis of Discrete Characters." *Proceedings of the Royal Society* (B) 255 (1994): 37–45. Describes a maximum likelihood model suitable for analyzing morphological data.

Ridley, M. *Evolution and Classification: The Reformation of Cladism.* Harlow, England, 1986. Informative account of the cladistic school of classification.

Sokal, R. R., and P. H. A. Sneath. *Principles of Numerical Taxonomy*. San Francisco, 1963. The central work on phenetic approaches to classification.

Swofford, D. L., G. J. Olsen, P. J. Waddell, and D. M. Hillis. "Phylogenetic Inference." In *Molecular Systematics*, 2d ed., edited by D. M. Hillis, C. Moritz, and B. K. Mable. Sunderland, Mass., 1996. The most comprehensive treatment of phylogenetic methods available today. This chapter describes tree building procedures and gives a clear account of how likelihoods are calculated.

— MARK PAGEL

PHYLOTYPIC STAGES

Stages common to the embryonic development of all animals—the zygote, blastula, and gastrula—have been recognized for at least 150 years. This conservation of embryonic stages across the animal kingdom suggests they are ancient, and indeed such stages have been identified in 500-million-year-old fossils. Given the antiquity of common developmental stages, it was inevitable that scientists would seek stages that typify groups of organisms. Phyla are the highest level of taxonomic organization in the animal kingdom, and so it is natural to look for embryonic stages that typify phyla. Such stages have been called form-building, phyletic, or phylotypic. The phylotypic stage is the point in development when the basic body plan characteristic of animals in that phylum is laid down. As a stage, it is conserved both in embryonic development (ontogeny) and in evolution (phylogeny). Indeed, it has been argued that "the phylotypic stage is the physical embodiment of the link between ontogeny and phylogeny" (Hall, 1997, p. 461).

In insects, the germ-band stage has been identified as the phylotypic stage—the stage when the basic segmentation characteristic of insects (segmented head, thorax, and abdomen) is established. Mechanisms of formation of segments vary among insects. They can be formed simultaneously or in a gradient over a substantial portion of embryonic development. Despite such different developmental mechanisms, the germ-band stage can still be recognized; different developmental mechanisms can produce the same conserved stage.

In vertebrates, a stage called the pharyngula has been identified as phylotypic (Ballard, 1976). The pharyngula is the earliest stage in embryonic development possessing the three structures that characterize vertebrates, these three being the notochord, the dorsal nervous system, and gill slits. The pharyngula also has the rudiments of other vertebrate organs, but their presence does not define the pharyngula.

> The pharyngula exhibits the basic anatomical pattern of all vertebrates in its simplest form: a set of similar organs similarly arranged with respect to a bilaterally symmetrical body axis, possessing chiefly the characters that are common to all the vertebrate classes. . . . One sees in them [the pharyngulas of vertebrates] epidermis but no scales, hair or feathers; kidney tubules and longitudinal kidney ducts are there, but no metanephros; all the little hearts have the same four chambers and there is at least a transient cloaca; there are no middle ears, no gills on the pharynx segments, no tongue, penis, uterus, etc. Basically just vertebrate anatomy, unobscured by the vast array of characters that appear later in development to distinguish the various classes, orders and families. (Ballard, 1981, p. 392).

As seen in both germ-band and pharyngula stages, the phylotypic stage is not necessarily the earliest stage in development; the old notion of the earliest embryonic stage as the most conserved no longer holds. Much basic development takes place before the phylotypic stage, much of it representing basic developmental processes such as cell division and morphogenesis shared by all animal embryos. The essence of the concept of a phylotypic stage is that development before and after this stage will be much more variable than is the phylotypic stage. As P. B. Medawar (1954) characterized embryonic development, there is convergence toward the phylotypic stage and divergence away from it, a metaphor that is often depicted as an hourglass, with the constriction in the hourglass being the phylotypic stage. Thus, in insects that display polyembryony (in extreme cases producing 2,000 embryos from a single egg), which is an extreme modification of insect embryogenesis, early development is very variable, but the phylotypic stage is conserved.

Because timing of development can vary from group to group and because alteration of timing between an ancestor and a descendant (heterochrony) is a mechanism generating morphological evolution, the phylotypic stage may appear at different times in the ontogeny of the members of a phylum, and not all elements of the phylotypic stages may develop synchronously. This is not surprising, because development itself evolves. The unlinking or uncoupling of the fundamental processes of development is an important, if often overlooked, property of developing embryos, the origins of which can be traced back to Joseph Needham, who gave the term *dissociability* to this phenomenon.

Because of such temporal differences, one group of workers have advocated abandoning the concept of the phylotypic stage or replacing it with the concept of a phylotypic period during which phylum-level features arise. Such a suggestion arises in part from defining the vertebrate phylotypic stage more broadly than W. W. Ballard did. This broader definition includes such features as size, presence or absence of limb buds, and variations in the developmental mechanisms that produce structures such as the neural tube. These features and mechanisms, however, are not defining characteristics of members of the phylum Vertebrata and, as already indicated, different developmental mechanisms can give rise to similar conserved stages. Variability in the mechanisms producing features of the phylotypic stage does not negate the concept of a phylotypic stage. Nevertheless, the concept of a phylotypic period does highlight the temporospatial variability seen during embryogenesis, variability that is essential if developmental processes are to be responsive to natural selection. Variability in the time of appearance of a phylotypic stage need not be a sufficient reason to abandon the concept of a phylotypic stage. Indeed, such variability focuses our attention on a search for the developmental mechanisms underlying such a stage.

A focus on shared and conserved developmental mechanisms led to the proposal that a conserved pattern of homeotic genes represents the minimal definition of an animal, the zootype, and to subsequent proposals that phylotypes for different groups could be identified on the basis of shared patterns of gene activity or shared timing of gene expression. Paradoxically, conservation and constraint are important elements of animal evolution. Concepts such as the phylotypic stage, zootype, and phylotype demonstrate the importance of uncovering the different mechanisms by which genes, developmental processes, and embryonic stages have been conserved or enabled to vary over the 500 million years of animal evolution.

[*See also* Body Plans; Constraint; Heterochrony; Heterotopy; Homeobox; Zootypes.]

BIBLIOGRAPHY

Ballard, W. W. "Problems of Gastrulation: Real and Verbal." *BioScience* 26 (1976): 36–39.

Brusca, R. C., and G. J. Brusca. *Invertebrates*. Sunderland, Mass., 1990.

Gilbert, S. F., and A. M. Raunio. *Embryology: Constructing the Organism*. Sunderland, Mass., 1997.

Grbic, M., L. M. Nagy, and M. R. Strand. "Development of Polyembryonic Insects: A Major Departure from Typical Insect Embryogenesis." *Development, Genes, and Evolution* 208 (1998): 69–81.

Hall, B. K. "Phylotypic Stage or Phantom: Is There a Highly Conserved Embryonic Stage in Vertebrates?" *Trends in Ecology and Evolution* 12 (1997): 461–463.

Hall, B. K. *Evolutionary Developmental Biology*. 2d ed. Dordrecht, Netherlands, 1999.

Kranenberg, S. "The Phylotypic Egg Timer." *Netherlands Journal of Zoology* 50 (2000): 289–294.

Medawar, P. B. "The Significance of Inductive Relationships in the Development of Vertebrates." *Journal of Embryology and Experimental Morphology* 2 (1954): 172–174.

Minelli, A., and F. R. Schram. "Owen Revisited: A Reappraisal of Morphology in Evolutionary Biology." *Bijdragen tot de Dierkunde* 64 (1994): 65–74.

Needham, J. "On the Dissociability of the Fundamental Processes in Ontogenesis." *Biological Reviews of the Cambridge Philosophical Society* 8 (1933): 180–223.

Richardson, M. K. "Heterochrony and the Phylotypic Period." *Developmental Biology* 172 (1995): 412–421.

Richardson, M. K., J. Hanken, M. L. Gooneratne, C. Pieau, A. Raynaud, L. Selwood, and G. M. Wright. "There Is No Highly Conserved Embryonic Stage in Vertebrates: Implications for Current Theories of Evolution and Development." *Anatomy and Embryology* 196 (1997): 91–106.

Sander, K. "The Evolution of Patterning Mechanisms: Gleanings from Insect Embryogenesis and Spermatogenesis." In *Development and Evolution*, edited by B. C. Goodwin, N. Holder, and C. C. Wylie, pp.137–160. Cambridge, 1983.

Slack, J. M. W., P. W. H. Holland, and C. F. Graham. "The Zootype and the Phylotypic Stage." *Nature* 361 (1993): 490–492.

Steinberg, M. S. "Goal-directedness in Embryonic Development." *Integrative Biology* 1 (1998): 49–59.

— Brian K. Hall

PLACENTATION

A placenta is an organ that effects nutrient, respiratory, and often waste product exchange between a parent, normally the mother, and her developing young while the embryos are housed in her body. The organ develops from maternal tissue, embryonic tissue, or both, depending on the species. Placentation is the process by which the placenta develops and establishes function. The placentas of many mammals, including humans, are relatively well studied. Such placentas develop as associations of maternal and embryonic cells. However, the mammalian placenta is only one of many kinds—sharks and some rays, a few bony fishes, perhaps an amphibian lineage, several squamate reptiles among vertebrates, and even some invertebrates develop structures for transport between mother and young. Some develop a new exchange structure that involves the vascular supplies; others use or expand parts of the body already present. The latter are often termed pseudoplacentas. Mechanisms for exchange are concomitants of the evolution of live-bearing modes of reproduction. [*See* Viviparity and Oviparity.]

The ancestral reproductive mode is generally assumed to be egg laying, whether before or after the eggs are fertilized, so that the developing young are dependent on the yolk supply of the egg for their nutrition. Retention of developing embryos by a parent is an evolutionarily derived condition, and providing them with nutrients reflects further evolution. Many lineages illustrate such evolutionary trends. Among elasmobranchs, all skates and some sharks lay eggs. However, some sharks retain their embryos in their oviducts, and several develop means of providing nutrition to the young after the yolk is exhausted. They have several modes of placentation. The simplest kind of maternal nutrition is the development of secretory tufts from the uterine epithelium, called trophonemata. A uterine milk is secreted and ingested by the young. Also, the capillary beds of the trophonemata enlarge and project to the surface, forming a membrane for respiratory exchange. Other sharks transform the yolk sac into a placenta. The yolk sac lengthens to form a type of "umbilical cord," and the sac contacts the maternal uterine epithelium, so that the embryonic and maternal circulations are in approximation and exchange can occur.

Among osteichthyans (fish with bony skeletons), only teleosts develop live-bearing modes of reproduction, but infrequently. However, they exemplify several different ways of providing exchange with embryos. Teleosts are unique among vertebrates in lacking true oviducts; hence, embryos are maintained in the lumen of the ovary or the follicle of the ovum. The lining of the ovary becomes highly vascular, and nutrients are either secreted or derived from maternal serum proteins. In follicular development, the follicle cells undergo modifications, including developing microvilli and extensive capillaries. Embryos develop diverse modifications for exchange, depending on the species. Some alter the gut to absorb nutrient material; others develop extraembryonic extensions of the anal region that are involved in intraluminal uptake. Particularly in intrafollicular developers, the embryos modify the yolk sac, coelomic sacs, or even the pericardial sac as their contribution to a pseudoplacenta that includes the modified follicular wall. Respiratory exchange and possibly nutrition are thus facilitated.

Although live-bearing modes have evolved several times among amphibians, nutrient provision and gaseous exchange apparently rarely, if ever, involve placentas. The expanded gills of one group of frogs and one of caecilians function in gaseous exchange, and perhaps nutritional acquisition in the latter, so they qualify as pseudoplacentas.

In contrast to amphibians, live-bearing reproduction and even placentas have evolved many times in reptiles, but only among snakes, lizards, and amphisbaenians. Placental evolution in squamate reptiles and in mammals has many similarities. Both are amniote groups; they have evolved extraembryonic membranes (amnion, allantois, chorion) in addition to the yolk sac. The membranes facilitate terrestrial modes of reproduction. Several squamates have a yolk sac placenta, as described for sharks. In others, the yolk sac placenta is succeeded by a chorionic placenta, in which chorionic modifications constitute the embryonic contribution to a placenta; modifications of the endometrium (epithelial lining of the oviduct) are the maternal contribution. The trends in complexity of modification of the chorion, then its reduction, as in the endometrium, where the circulatory systems of the embryo and the mother approximate so that exchange can be effected, resemble those of mammals. Reduction of the yolk supply is also concomitant to the investment in nutrient production after the yolk is resorbed.

Placentation in mammals is well described. Marsupials have placentas and may first utilize a yolk sac placenta as described for sharks and squamates, then develop a chorioallantoic placenta, as in some squamates and eutherian mammals.

In eutherian mammals, placentation commences as the dividing egg reaches the uterus and its cells differentiate into two types: those forming the embryo and those forming the trophoblast that surrounds: The epithelial lining of the uterus is prepared for implantation by hormones, especially progesterone. The endometrium is highly vascularized and develops secretory glands. The trophoblast layer usually contacts the endometrium in a region specialized for implantation. The outer layer of the trophoblast differentiates into a membranous chorion. Implantation may be superficial, or deep ("invasive"), with the embryonic mass completely embedded in the endometrium. Placentas that result from superficial implantation typically include several cellular layers from both the embryo and the uterus; those that result from deep implantation have fewer layers, so that, in derived placentas, the fetal and maternal capillary systems are in contact, or, in the most extreme condition, the fetal capillaries are embedded in the maternal circulation. The placenta transports nutrients via the mother's bloodstream; the nutrients diffuse into the capillaries of the embryo. Concomitantly, the embryo's waste products diffuse into the maternal circulation. Such exchange depends on the area available for diffusion and the thickness of the cellular barrier between maternal and fetal circulations.

In mammals, the kinds of implantation and the shapes and functions of the placentas that result evolve differently in different lineages. The ancestral condition for placentation has most of the outer surface of the chorion of the embryo covered by folds and extensions (villi) that contact villi or folds, respectively, in the endometrium, forming a diffuse placenta (pigs, horses). In artiodactyls (cows, deer, giraffes), the villi are more restricted, and the wall of the uterus forms pockets (caruncles) into which tufts (villi) of the chorion form extensions (cotyledons) that insert into the caruncles. Such a cotyledonary placenta varies in numbers of cotyledons and caruncles among species. The contact between the maternal and fetal vascular supplies is limited. The evolution of mammalian placentas involves a restriction of the coverage of the embryo's chorionic layer and an increase in contact of maternal and fetal circulations. Carnivores (cats, raccoons, bears) have zonary placentas: the placenta forms a band around the embryo. Most primates and bats, as well as some rodents, have roundish discoid placentas. A reduction of layers between the maternal and fetal circulations correlates with a concentration of placental sites and with shapes, so that the discoid placenta characteristically has the fewest layers and fetal capillaries intertwine the maternal circulation.

The placenta in mammals is an endocrine organ. It secretes hormones that maintain pregnancy. Also, embryos nearly always differ genetically from their mothers, so the absence of immunological rejection of the embryo is of interest. These and other questions of the evolution of placentas in animals of diverse lineages that have live-bearing reproductive modes are current topics of research.

[*See also* Maternal-Fetal Conflict.]

BIBLIOGRAPHY

Amoroso, E. C. "Placentation." In *Marshall's Physiology of Reproduction*, vol 2, edited by A. S. Parkes, pp. 127–311. London, 1956. The classic discussion of placentation in vertebrates.

Blackburn, D. G. "Chorioallantoic Placentation in Squamate Reptiles: Structure, Function, Development, and Evolution." *Journal of Experimental Zoology* 266 (1993): 414–430. An up-to-date summary of information about placentation in vertebrates.

Blüm, V. *Vertebrate Reproduction: A Textbook*. Berlin, 1986. A useful textbook treatment that includes information on placentation.

Hamlett, W. C., A. M. Eluitt, R. L. Jarrell, and M. A. Kelly. "Uterogestation and Placentation in Elasmobranchs." *Journal of Experimental Zoology* 266 (1993): 347–367.

Harder, J. D., M. J. Stonerook, and J. Pondy. "Gestation and Placentation in Two New World opossums: *Didelphis virginiana and Monodelphis domestica*." *Journal of Experimental Zoology* 266 (1993): 463–479.

Luckett, W. P. "Comparative Development and Evolution of the Placenta in Primates." *Contributions in Primatology* 3 (1974): 142–234.

Mossman, H. W. *Vertebrate Fetal Membranes*. New Brunswick, N.J., 1987. The classic discussion of vertebrate, primarily mammalian, placentas.

Schindler, J. F., and W. C. Hamlett. "Maternalembryonic Relations in Viviparous Teleosts." *Journal of Experimental Zoology* 266 (1993): 378–393.

Stewart, J. R. "Yolk Sac Placentation in Reptiles: Structural Innovation in a Fundamental Vertebrate Fetal Nutritional System." *Journal of Experimental Zoology* 266 (1993): 431–449.

Wake, M. H. "Evolution of Oviductal Gestation in Amphibians." *Journal of Experimental Zoology* 266 (1993): 394–413.

— MARVALEE H. WAKE

PLAGUES AND EPIDEMICS

Throughout human history and prehistory, plagues and epidemics have been among the most powerful selective forces acting on our species. The Black Death has had such an impact on Europeans' collective perception of history that it is an instantly recognizable benchmark against which the impact of any disease outbreak is measured.

Plagues and epidemics are usually bacterial or viral in origin, although fungal and protozoan pathogens may also have severe impacts. They occur intermittently or periodically. Plagues such as bubonic plague are marked by severe mortality, but epidemics include milder diseases that may become common but may not cause widespread mortality. Epidemic diseases that affect domesticated animals and important crop plants have also caused severe mortality in human populations that are dependent on them. One of these was the Irish potato famine of 1845–1850, in which, by some estimates, 1.5 million Irish peasants starved because the potatoes on which they depended had been killed by the fungus *Phytophthora infestans*. (The fungus did not act alone—the British rulers of the island made no effort to alleviate the suffering by diverting other sources of food to the starving people of the countryside.)

Causes and Effects of Plagues in Historical Times. Plagues must have been important causes of human mortality before written histories, but we have no firm information about these outbreaks. Table 1 lists the properties of some of the most important plagues of historical times, including their dates and locations, most probable causes, and the numbers of deaths that resulted.

Plagues and epidemics can be contrasted to endemic diseases, which are always present, although they may flare up in local outbreaks. Two of these endemic diseases, which have now largely been controlled, are smallpox and polio. Smallpox, caused by a DNA virus, has shaped human history—it was, for example, the chief cause of the destruction of the Aztec empire of Central America after it was introduced to the Americas by European explorers. Smallpox has now been eradicated worldwide—the first human disease to be conquered completely. Samples of virus remain at the U.S. Centers for Disease Control and Russia's Research Institute for Viral Preparations, stored against the possibility that terrorists will begin to spread smallpox or an allied disease.

Polio, which is caused by a different DNA virus, has apparently been eradicated by immunization in the New World and remains only in parts of sub-Saharan Africa and South Asia. The virus causes mild symptoms in young children but severe symptoms such as paralysis in adults. It was not a serious problem until the twentieth century, when improved public health measures prevented most people in industrialized countries from coming in contact with the virus until adulthood.

The Control and Future of Plagues. Bubonic plague, cholera, influenza, smallpox, polio, and other diseases that can spread as plagues or as severe local outbreaks have in recent decades largely been brought under control through immunization and public health programs. A pneumonic plague outbreak that may have killed as many as 855 people in the industrial city of Surat in northwestern India in 1994 was quickly brought

TABLE 1. Plagues and Epidemics

Plague	*Places and times*	*Symptoms*	*Most probable cause*	*Means of spread*	*Mortality*
Great Plague of Athens	Athens, 430–425 B.C.	Fever, retching, bleeding in mouth and throat	Unknown, perhaps a filovirus allied to Ebola	Unknown	Not known, but several thousand in Athens itself.
Plague of Justinian	Egypt and Europe, A.D. 558–590	High fever, blackened swellings of lymph nodes called buboes in armpits and groin	The bacterium *Yersinia pestis*	Rats and rat fleas (bubonic form) Direct spread through the air from infected lung tissue (pneumonic form).	Perhaps as many as 20 million people throughout Europe, unknown numbers in Africa and the Middle East.
Black Death	China in 1330s, Middle East and Europe, 1347–1350	"	*Y. pestis*. Other possible causes, such as an outbreak of anthrax, are much less likely.	"	25 million in Europe, 1/3 of the population at the time. Unknown numbers in Asia and Middle East.
Later outbreaks of bubonic plague	Over a dozen outbreaks in Europe during the sixteenth and early seventeenth centuries. Outbreaks in China and India at the turn of the twentieth century.	"	*Y. pestis*	"	Major cities such as London and Paris were repeatedly devastated. At least 100,000 deaths in London in 1665. Millions died more recently in India and China.

"Sweating sickness"	Five outbreaks between 1485–1551 in England, also spread in the rest of Europe.	Headache, high fever, severe sweating	Unknown, probably viral. No similar outbreaks since sixteenth century.	Unknown	40,000 died in London in 1528. Mortality widespread throughout England.
Cholera	Seven worldwide pandemics from 1817 to 1960. Most originated in Bengal in northeastern India.	Severe diarrhea, vomiting, and dehydration	The bacterium *Vibrio cholerae*	Contaminated water supply, particularly if water is slightly saline	Uncounted millions worldwide. Mortality drops from 30 to 1 percent of cases if rehydration therapy is given.
Influenza	Severe pandemics occur every 10 to 40 years. The 1918 pandemic, the worst on record, was aided by travel and dislocation of populations following the first World War.	High fever, lungs fill with fluid	A virus with RNA as its genetic material	Droplets in air. Most epidemics originate from 'flu viruses found in pigs, but some may have come from bird viruses.	Twenty million worldwide in the 1918 pandemic. Current high mortality among the young and elderly, especially in underdeveloped countries.
Smallpox	Endemic, but many local epidemics, particularly among indigenous tribes contacted for the first time.	High fever, pustules on skin, inflammation of internal organs	A virus with DNA as its genetic material	Droplets in air	Uncounted millions. Chief cause of destruction of Aztec empire, with deaths in millions.
AIDS	Worldwide epidemic perhaps starting in the 1930s.	Destruction of cellular immune system. Death through opportunistic infections by a wide variety of pathogens.	An RNA retrovirus, HIV-1, is chiefly responsible. Some cases, mostly limited to Africa, caused by a related virus, HIV-2	Sexual contact, contaminated blood and needles	Over 36 million AIDS cases in 2001, most of which will be fatal. 20 million deaths so far.

THE GREAT PLAGUE OF 1665 IN LONDON

Even at the height of the plague, life in the city of London went on. On September 14, 1665, after Samuel Pepys returned to London from Greenwich where he had fled at the height of the plague, he wrote in his diary:

"I spent some thoughts upon the occurrences of this day, giving matter for as much content on one hand and melancholy on another, as any day in all my life. For the first; the finding of my money and plate, and all safe at London, and speeding in my business of money this day . . . Then, on the other side . . . [m]y meeting dead corpses of the plague, carried to be buried close to me at noon-day through the City in Fanchurch-street. To see a person sick of the sores, carried close by me by Gracechurch in a hackney-coach. My finding the Angel tavern, at the lower end of Tower-hill, shut up, and more than that, the alehouse at the Tower-stairs, and more than that, that the person was then dying of the plague when I was last there, a little while ago, at night. To hear that poor Payne, my waiter, had buried a child, and is dying himself. To hear that a labourer I sent but the other day to Dagenhams, to know how they did there, is dead of the plague; and that one of my own watermen, that carried me daily, fell sick as soon as he had landed me on Friday morning last, when I had been all night upon the water, (and I believe he did get his infection that day at Brainford) and is now dead of the plague. . . ."

—CHRISTOPHER J. WILLS

under control by the extensive use of antibiotics. The same approach has been fairly successful in controlling repeated outbreaks of typhus and cholera in refugee camps throughout the world.

Immunization programs and public health measures have been extremely effective in controlling these diseases, but these programs must be pursued vigilantly. Immunization against smallpox has been discontinued through much of the world, and a new generation of unimmunized people are now susceptible to a possible reintroduction of this disease by terrorists. Immunization against the endemic disease measles, which killed 1.1 million children in 1990, has slackened in many parts of the developed world, and measles outbreaks have recently taken place in the United States, Europe, and Japan. Immunization against influenza has increased greatly in effectiveness in recent years, but it is hampered because the virus mutates rapidly and not enough doses of modified vaccine can be manufactured to protect all those at risk with each new outbreak.

Public health measures must also be pursued rigorously. Malaria was almost eradicated in Sri Lanka by the mid-1960s through spraying with DDT and other measures. There were only 17 new cases in 1964. Then lack of funds and of political will caused the spraying program to be halted, and the number of cases rose again to half a million in 1969. Malaria has remained an important public health problem in Sri Lanka since that time.

The HIV Epidemic. The HIV epidemic, primarily transmitted sexually, has spread rapidly throughout the world from central and western Africa, beginning in the 1970s. It appears to be a zoonosis (a disease originating in another species), in which the original transfer perhaps took place in the 1930s. The current chief suspect for an animal source is the chimpanzee, which carries similar but not identical viruses.

To control plagues and epidemics, it is important to be able to detect signs of the disease quickly so that local immunizations and public health measures can be carried out. This has not been possible with AIDS, which is virtually asymptomatic at first and may take a decade to destroy the victim's immune system. In addition, no effective vaccine for AIDS has yet been developed. Drugs that keep the HIV virus multiplication in check have had success in reducing mortality in developed countries. They are, however, expensive, and drug-resistant virus strains are appearing rapidly. The AIDS epidemic is becoming the most devastating, in terms of numbers of deaths, of historical epidemics.

Molecular Evolution of Plague Organisms. As information about the genomes of plague organisms has grown, it has become possible to learn about their history and the sources of their virulence. The history of bubonic plague, caused by the bacterium *Yersinia pestis*, is reflected in the present-day structure of *Y. pestis* populations. Three closely related types of *Y. pestis* have been distinguished on the basis of sequencing and related approaches; the ancestors of these types appear to have been responsible for the three major outbreaks listed in Table 1. The *antiqua* strain from Africa may be descended from the bacteria that caused the Plague of Justinian. The *medievalis* strain from Central Asia may descend from the bacilli of the Black Death, and the *orientalis* strain from eastern Asia is clearly the source of the recent outbreaks in India and China. Although the data are limited, it appears that these strains may have begun to diverge between 1,500 and 20,000 years ago from *Y. pseudotuberculosis*, a closely related bacterium that lives in the soil and infects domesticated animals. It appears that *Y. pestis* is of very recent origin.

The molecular basis of the great influenza epidemic of 1918 is also beginning to be understood. Parts of the genomes of viruses responsible for that epidemic have been recovered from flu victims who were buried in graves in the Alaskan permafrost. It is not yet clear why this strain of the virus was so virulent, but there is evidence that the hemagglutinin gene of the 1918 virus had undergone recombination between human and pig virus strains. The high virulence of a more recent outbreak of avian flu in Hong Kong has been traced to one mutational change in the PB2 protein. Luckily, the latter virus can be transmitted from chickens to humans, but not subsequently from one human to another.

Impact of Plagues on the Human Gene Pool. It is not yet clear what evolutionary impact plagues have had on our species, but evidence is growing that alleles of many genes in the human population have increased in frequency because they confer resistance to endemic and epidemic diseases. One of the most interesting of these mutations is found in a chemokine coreceptor gene, CCR5, that has been shown to code for one of the proteins by which the AIDS virus HIV-1 enters its host cells. Individuals homozygous for this 32-base-pair deletion are resistant to HIV-1 infection. The mutant allele is found at 10 percent frequency in northern Europeans, and its origin has been tentatively dated at between 300 and 1,900 years ago. It seems unlikely that its high frequency is connected to an early epidemic of HIV or an HIV-like virus. One possibility is that the allele was driven to such a high frequency because it conferred resistance to another endemic or epidemic disease, with bubonic plague being a prime suspect. However, no evidence has yet been found that *Y. pestis* interacts with the CCR5 protein on the surface of cells of its hosts.

A less tenuous connection has been found between the CFTR protein, which regulates transport of chloride ions across cell membranes, and the pathogenic bacterium *Salmonella typhi*. This bacterium is responsible for endemic typhoid fever, which often reaches epidemic levels during periods of warfare. When homozygous, mutant cftr alleles are responsible for the severe disease cystic fibrosis. In spite of their severity, high frequencies of these mutant alleles are found in European populations. One reason may be an interaction between the CFTR protein and *S. typhi*. Transgenic mice that carry a defective human cftr gene are less susceptible to *S. typhi* infection than those that do not, apparently because the bacteria cannot enter the mutant mouse cells as easily in the absence of a CFTR protein to adhere to. Thus, being heterozygous for the mutant cftr allele might have protected against typhoid fever.

It seems likely that much more information about the coevolution of our species and the plague organisms that have preyed on us will soon be forthcoming, particularly since the complete genomes of humans and of most of the pathogens discussed in this article have now become available.

[*See also* Antibiotic Resistance; Disease, *article on* Infectious Disease; Emerging and Re-Emerging Diseases; Immune System, *article on* Microbial Countermeasures to Evade the Immune System; Influenza; Malaria; Vaccination; Virulence.]

BIBLIOGRAPHY

Achtman, M., K. Zurth, G. Morelli, G. Torrea, A. Guiyoule and E. Carniel. "*Yersinia pestis*, the Cause of Plague, Is a Recently Emerged Clone of *Yersinia pseudotuberculosis*." *Proceedings of the National Academy of Sciences USA* 96 (1999): 14,043–14,048. Evolutionary analysis of the possible origin of the bubonic plague bacillus.

Holmes, E. C. "On the Origin and Evolution of the Human Immunodeficiency Virus (HIV)." *Biological Reviews of the Cambridge Philosophical Society* 76 (2001): 239–254. Reviews evidence on origins of the virus that causes AIDS.

Korber, B., M. Muldoon, J. Theiler, F. Gao, R. Gupta, A. Lapedes, B. H. Hahn, S. Wolinsky, and T. Bhattacharya. "Timing the Ancestor of the HIV-1 Pandemic Strains." *Science* 288 (2000): 1789–1796. Evidence that the virus causing AIDS entered the human population sometime around 1930.

Pier, G. B., M. Grout, T. Zaidi, G. Meluleni, S. S. Mueschenborn, G. Banting, R. Ratcliff, M. J. Evans, and W. H. Colledge. "*Salmonella typhi* Uses CFTR to Enter Intestinal Epithelial Cells." *Nature* 393 (1998): 79–82. Analysis of a mutation that protects against typhoid fever.

Raoult, D., G. Aboudharam, E. Crubezy, G. Larrouy, B. Ludes, and M. Drancourt. "Molecular Identification by 'Suicide PCR' of *Yersinia pestis* as the Agent of Medieval Black Death." *Proceedings of the National Academy of Sciences USA* 97 (2000): 12800–12803. Demonstration that a medieval plague victim in southern Europe was indeed infected with the bubonic plague bacillus.

Stephens, J. C., D. E. Reich, D. B. Goldstein, H. D. Shin, M. W. Smith, M. Carrington, C. Winkler, G. A. Huttley, R. Allikmets, L. Schriml, et al. "Dating the Origin of the CCR5-Delta32 AIDS-Resistance Allele by the Coalescence of Haplotypes." *American Journal of Human Genetics* 62 (1998): 1507–1515. Examines the time of origin of a mutation that confers some resistance to the AIDS virus HIV.

Wills, C. *Yellow Fever, Black Goddess: The Coevolution of People and Plagues*. Reading, Mass., 1996. An account of the history of plagues and epidemics, from an evolutionary perspective.

— CHRISTOPHER J. WILLS

PLANTS

The word *plant* commonly refers to all autotrophic—literally, self-feeding—eukaryotic organisms. In the process of photosynthesis, these organisms use light energy to produce carbohydrates from carbon dioxide and water in the presence of chlorophyll, inside organelles called *chloroplasts*. Sometimes the term *plant* is extended to include autotrophic *prokaryotic* forms, especially the eubacterial lineage known as the cyanobacteria (blue-green algae). Traditional botany textbooks often also include fungi, which are heterotrophic eukaryotes that degrade living or dead organic material using a battery of enzymes and then absorb the simpler products into their bodies. Fungi, however, appear to be more closely related to animals.

In this essay, I begin by briefly considering the events giving rise to the variety of distantly related lineages referred to as algae, but then concentrate on the green plant lineage, including green algae and land plants. Much of the discussion revolves around Figure 1, which shows progress in resolving relationships among major lineages. In general, our understanding of green plant evolution has been fueled by the discovery that several traditionally recognized taxa are not monophyletic.

Endosymbiotic Events. Chloroplasts are endosymbiotic organelles derived ultimately from cyanobacteria. This well-established view is based on structural evidence (e.g., the form and number of membranes) and especially on evidence that plastid DNA is more closely related to free-living cyanobacterial DNA than to plant nuclear DNA. Endosymbiosis has entailed massive reduction in the size and gene content of the plastid genome relative to free-living cyanobacteria, including the transfer of some genes from the chloroplast to the nucleus.

Recent phylogenetic evidence is consistent with a single primary endosymbiotic event in the common ancestor of a clade containing red algae and green plants. Significant differences between red algal and green plant chloroplasts help us identify instances of secondary endosymbiosis, in which plastids have been acquired by permanent incorporation of either red or green eukaryotes. It appears, for example, that red algal chloroplasts were acquired via secondary endosymbiosis in the heterokont algae, including brown algae and diatoms, and possibly also in apicomplexans, including *Plasmodium*, the malaria parasite. Euglenophytes exemplify the uptake of green algal eukaryotes, while dinoflagellates appear to include a mixture of different types. Clearly, the term *algae* is applied to a wide variety of aquatic photosynthetic organisms that are not all directly related to one another.

Miscellaneous "Algae." There are about 6,000 species of red algae (Rhodophyta), which are found mostly in marine environments, especially in tropical waters. A few are unicellular, but most are filamentous and attach to rocks or other algae (some are even parasitic). They have no motile cells at any stage and often have complex life cycles in which there may be two distinct diploid phases.

Brown algae (Phaeophyta) include about 2,000 species of mostly marine organisms, which are especially conspicuous in cooler regions. Many are filamentous, but some kelplike forms are very large and show differentiation of the body into a holdfast, a stipe, a float, and one or more flat blades, as well as considerable internal differentiation of tissues. Brown algal life cycles run the gamut from alternation between similar-looking diploid and haploid phases ("isomorphic alternation of generations") to the complete elimination of the haploid phase.

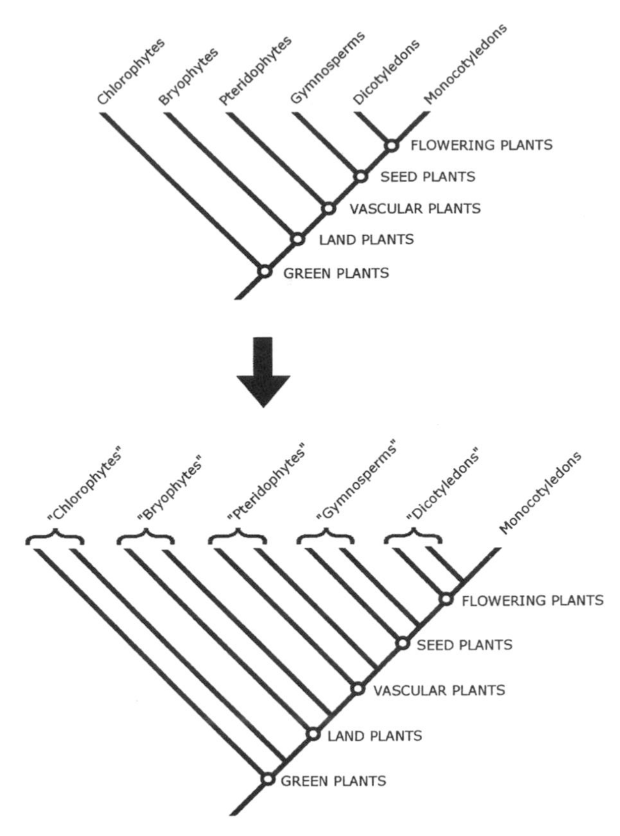

FIGURE 1. Progress in Understanding the Phylogeny of Green Plants.
The upper tree shows the standard view as of the 1970s; the lower tree depicts current understanding. Note that the monophyly of green plants, land plants, vascular plants, seed plants, and flowering plants has been well supported in recent analyses. In contrast, the traditional chlorophytes, bryophytes, pteridophytes, gymnosperms, and dicotyledons are now seen to be paraphyletic. Specifically, charophytes are more closely related to land plants than they are to other "green algae"; mosses are more closely related to vascular plants than to liverworts and/or hornworts; ferns and horsetails are more closely related to seed plants than to lycophytes; extant naked-seed plants are more closely related to angiosperms than to Paleozoic "seed ferns," and eudicots are more closely related to monocots that to *Amborella*, water-lilies, Illiciales, etc. This new understanding has greatly clarified the evolution of key features. Courtesy of Michael J. Donoghue.

Closely related to brown algae are the diatoms, with about 6,000 living species and many more known only as fossils. Like brown algae, they produce chlorophyll c and carotenoid pigments. Their most distinctive feature is the formation of cell walls made of two silicon valves, often elaborately sculptured.

Dinoflagellates, with about 3,000 species, have two flagellae (which produce a spinning motion) located in grooves between cellulose plates embedded in the cell walls. Many are symbiotic with other organisms, including corals, sponges, squids, and clams. The symbiotic forms, or zoozanthellae, typically lack cellulose plates. Dinoflagellates are of great ecological importance in maintaining coral reefs and periodically causing "red tides" that release highly toxic substances.

Green Plants. The traditional "green algae" (Chlorophyta) are directly related to land plants, and together these constitute a green plant clade of about 300,000 species, or nearly one-fifth of all described extant species on Earth. Molecular evidence strongly supports the monophyly of green plants, as do structural features, including the loss of phycobilin pigments (found in cyanobacteria and red algae), the production of chlorophyll b (in addition to chlorophyll a), storage of carbohydrates as starch granules, and motile cells with a characteristic "stellate" structure at the base of the flagellae.

The first green plants were unicellular (so-called micromonadophytes are basal), and a range of unicells still exists, including the model organism *Chlamydomonas*. Multicellularity evolved more than once within green plants. In one lineage, the Volvocales, cells become aggregated into colonies. Others, like the sea lettuce, (*Ulva*), are much larger in size and show greater differentiation of cell functions. The siphonous green algae are characterized by a coenocytic condition, in which extremely large cells contain many nuclei. Alternation of similar haploid gametophyte and diploid sporophyte generations is common, but in *Codium* the gametes are the only haploid cells.

Beginning in the late 1960s, ultrastructural studies of cell division revealed that the charophycean green algae (e.g., *Chara, Coleochaete*) share a characteristic orientation of spindle microtubules (the phragmoplast) with land plants. This suggests that charophytes are the closest living relatives of land plants, which is well supported by recent analyses (Figure 1). Charophytes live in near-shore freshwater habitats and possess several pre-adaptations to life on land, such as flavonoid chemicals that help absorb damaging UV radiation and a glycolate oxidase system that ameliorates the inhibition of carbon dioxide fixation by oxygen.

Land Plants. Morphological and molecular data strongly support the view that green plant life on land originated once, from a single common ancestor. Land plants have a multicellular sporophyte, multicellular reproductive structures, and spores with characteristic trilete (three-armed) marks. They also produce a resting sporophyte "embryo" stage and are thus sometimes referred to as "embryophytes." Adaptations to reduce desiccation (e.g., a cuticle covering the epidermis) and symbiotic associations with fungi were probably critical to the colonization of land.

Land plants have traditionally been classified as either bryophytes or vascular plants, but it is now clear that the classical bryophytes are paraphyletic (Figure 1). Of the three major bryophytic lineages, mosses are probably the most familiar—and, with about 15,000 species, also the most diverse. There are about 9,000 species of liverworts (thalloid forms, such as *Marchantia*, and "leafy" forms) and only about 100 species of hornworts. Morphological phylogenetic analyses initially supported a basal split between the liverwort lineage and all other land plants. Under this view, stomates (surface openings with specialized guard cells) were considered an innovation linking hornworts, mosses, and vascular plants. However, several recent analyses support an alternative interpretation in which hornworts are the sister group of all other extant land plants.

Realization that both the traditional green algae and bryophytes are paraphyletic has greatly clarified the origin of land plant features, especially the life cycle. Charophytes retain the egg on the body of the haploid plant, and the zygote also remains on the parent plant after fertilization until it gives rise by meiosis to haploid spores. In bryophytic plants, and presumably in the common ancestor of land plants, the haploid *gametophyte* stage (producing sperm and eggs by mitosis in structures called antheridia and archegonia) is the more conspicuous, "dominant," phase; the multicellular diploid *sporophyte* forms a single unbranched stalk terminated by a sporangium, where haploid spores are produced by meiosis. The land plant life cycle probably was derived by delaying meiosis and intercalating a multicellular diploid phase via a series of mitotic divisions.

Green plants may be a billion or more years old, but their wholesale occupation of land probably began in the Ordovician, around 450 million years ago. Fossil spores, resembling those of liverworts, provide the first concrete evidence of the existence of land plants. From the Silurian there are well-preserved vascular plant macrofossils, tiny and simple in structure. The sporophyte was a dichotomously branching stem (lacking leaves and roots) with sporangia at the tips of the branches. Careful analyses of excellent fossils (e.g., *Rhynia* from Scotland) indicate that the first branching sporophytes did not produce true tracheids (elongate, thick-walled cells that are dead at maturity and function in water conduction). Fossil gametophytes have also been discovered, which were initially quite similar to the sporophyte. Subsequently within vascular plants, the gametophyte was reduced and the sporophyte was elaborated. One possible explanation for this is that diploid organisms are buffered against deleterious mutations. An alternative is that competition for light and spore dispersal selected for taller sporophytes, whereas gametophyte stature was constrained by dependence on water for fertilization.

Vascular Plants. The basal split in extant vascular plants is between lycophytes and everything else; that is, the classical pteridophytes (seedless vascular plants) are in fact paraphyletic. Lycophytes are characterized by lateral sporangia that open transversely. Leaves and roots evolved independently within lycophytes and again in their sister group. Lycophytes were especially diverse and abundant during the Carboniferous (creating major coal deposits), and some became large trees. Today there are about 1,000 living species, all small in stature. *Lycopodium* and other "club-mosses" are homosporous; they produce just one kind of spore, each giving rise to a bisexual gametophyte. Other lycophytes are heterosporous; they produce microspores, which give rise to male gametophytes, and megaspores, which give rise to female gametophytes. *Selaginella*, with more than 600 species, is most diverse in the tropics, and *Isoetes* is a living representative of the clade that included the giant Carboniferous lycopods.

A key advance of the lineage leading to ferns, horsetails, and seed plants was the differentiation of a main axis and side branches. According to the "telome theory," large (megaphyllous) leaves were derived from such flattened branch systems. Today there are more than 12,000 fern species, many of which have large, highly dissected leaves that unfold from a spiral "fiddlehead." Sporangia are typically produced in small clusters (sori) on the undersides of the leaves. *Psilotum* ("whisk fern") and *Tmesipteris* have long been considered a distinct major lineage of vascular plants, Psilophyta. Lacking true roots and leaves, they are sometimes considered remnants of the first vascular plants. However, molecular phylogenetic studies have established that they are most closely related to some so-called eusporangiate ferns. Horsetails (Sphenophyta), like lycopods, were more abundant and diverse in the Carboniferous, when some also became trees. Today there are only about 15 living species (all *Equisetum*). These have jointed, hollow stems with distinct ridges that accumulate silica; the leaves are reduced and borne in a whorl at each node.

Seed Plants. Seed plants (Spermatophyta) are by far the most diverse lineage within vascular plants, with approximately 250,000 species, most of which are flowering plants. Morphological evidence for seed plant monophyly includes the seed itself and the capacity to produce secondary xylem (wood) and secondary phloem through the activity of a bifacial meristem called the cambium.

To understand the structure of a seed, it helps to think about its evolution. A critical step was *heterospory*—producing two kinds of spores and two kinds of gametophytes (male and female). Heterospory evolved several times within vascular plants; however, in the seed plant line there was a reduction in the number of functional megaspores to just one. That single megaspore produced a female gametophyte (which ultimately produced one or more eggs) entirely within the megasporangium. Finally, the megasporangium became enveloped by sterile sporophyte tissue known as integument, leaving a little opening at the top called the micropyle. One or more pollen grains enter through the micropyle, where they send out a tubular structure that delivers the sperm to the vicinity of the egg.

The seed is a remarkable invention. The young sporophyte is protected from the elements by the seed coat (derived from integument tissue), which may also be modified for dispersal by wind, water, or animals. The embryo is also surrounded by a nutritive tissue, derived in the first seed plants from the female gametophyte.

Our knowledge of the origin and early evolution of seed plants rests on well-preserved fossils from the late Devonian and early Carboniferous. *Archaeopteris* (a "progymnosperm") had a large trunk with well-developed wood and large frondlike branch systems bearing many small leaves; it was heterosporous but lacked seeds. This establishes that wood and heterospory predated the evolution of the seed. "Seed ferns" of the Carboniferous were remarkable in mostly having large frondlike leaves but bearing true seeds.

Today there are five major lineages of seed plants: cycads, ginkgos, conifers, gnetophytes, and flowering plants. Cycads were abundant and diverse during the Mesozoic. Today there are more than 100 species, mostly with short trunks, compound leaves, huge pollen and seed cones borne on separate plants (dioecy), and large colorful seeds. The ginkgo lineage is represented today by just a single species, *Ginkgo biloba*. Its most characteristic feature is the production of fanlike leaves with dichotomous venation. Like cycads, ginkgos are dioecious. The seed coat differentiates into a fleshy outer layer (which smells like rotten butter) and a hard inner layer enclosing the female gametophyte. Cycads and ginkgos have large, multiflagellate, swimming sperm, whereas in the remaining groups the sperm are delivered directly to the egg by a pollen tube (siphonogamy).

There are about 600 living species of conifers. Some, such as pines, spruces, and firs, are dominant around the Northern Hemisphere. Other northern groups include redwoods, junipers, and yews. Some, especially the podocarps, are found on mountains in the tropics, while the araucarias are distributed around the Southern Hemisphere. Conifers are usually trees with drought-adapted, needlelike leaves borne singly along the stem or, in pines, in tight clusters (often in 3s or 5s) on short shoots. Many are *monoecious*, with separate pollen-producing and seed-producing cones on the same plant. The seed cone is a complex structure in which each cone scale (with seeds situated on top) appears to have been derived from a small branch.

There are about 70 living species of gnetophytes (or Gnetales), in three distinct lineages: *Gnetum* (~35 species around the tropics) has broad leaves; *Ephedra* (~35 species in deserts) has scalelike leaves; and *Welwitschia* (1 species in southwestern Africa) has just two large functional leaves that grow from the base. Although these look very different from one another, they share several unusual features, such as multiple axillary buds, circular openings between adjoining vessel elements, and seeds with a micropylar tube.

Flowering Plants. Flowering plants, or *angiosperms*, account for most of green plant diversity, with more than 235,000 extant species. Strong evidence for angiosperm monophyly comes from molecular studies and from derived morphological characters, including the following: (1) seeds produced within a carpel (a folded structure, probably derived from a leaf); (2) reduced female gametophytes, in most cases just eight nuclei in seven cells; and (3) double fertilization leading to the formation of a polyploid nutritive tissue called *endosperm*. Several vegetative features are also noteworthy. Almost all angiosperms produce vessels in the xylem tissue, and the sieve-tube elements (living enucleate cells that transport nutrients, etc.) in the phloem are accompanied by one or more "companion cells."

The flower is a short shoot that bears the characteristic organs: sepals, petals, stamens (producing pollen), and carpels (bearing ovules, or young seeds). Not all flowers produce all of these parts, and the number of each type varies among genera. Each stamen typically has a stalk (filament) and a tip (anther) bearing the microsporangia. Carpels are often differentiated into a lower portion (ovary) that encloses the ovules, an elongate style, and a surface receptive to pollen (stigma).

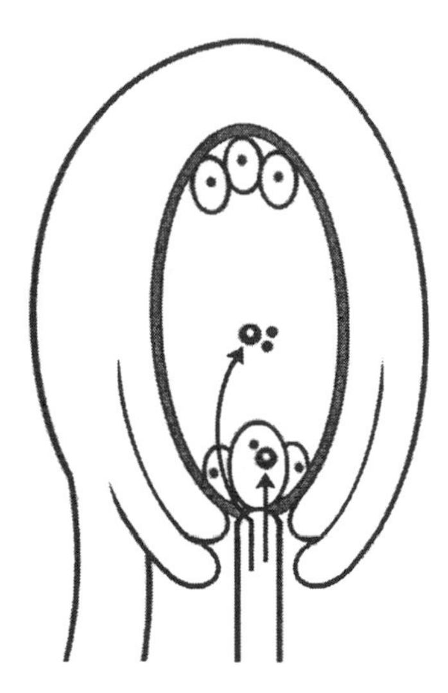

FIGURE 2. An Idealized Angiosperm Ovule at the Time of Double Fertilization.
Note that the ovule (young seed, situated within the carpel) is anatropous with two integuments, and that the pollen tube has entered through the micropyle and has delivered two sperm to the female gametophyte (eight nuclei in seven cells). One sperm will fuse with the egg, giving rise to the diploid zygote; the other will fuse with the two polar nuclei, eventually giving rise to the triploid endosperm that serves as the nutritive tissue for the developing embryo. Courtesy of Michael J. Donoghue.

Inside the microsporangia, meiosis yields haploid microspores that mature into pollen grains (male gametophytes, with two or three cells). A pollen grain that lands on a compatible stigma can send out a pollen tube to deliver the sperm to the female gametophyte inside the ovule. Angiosperm ovules are unusual in several ways (Figure 2). Outside of angiosperms, the seed is orthotropous—borne upright with the micropyle facing away from the seed stalk. In contrast, the angiosperm ovule is typically anatropous: curved over so that the micropyle lies near the stalk. Also, whereas non-angiosperm seeds have one layer of integument tissue, angiosperms typically have two distinct integuments.

In the development of a typical angiosperm female gametophyte, meiosis is followed by the abortion of three products, and the remaining haploid nucleus undergoes a small series of mitotic divisions. Ultimately, the egg is situated toward the micropylar end with two other cells (synergids); there are three cells at the opposite end (antipodals) and two nuclei (polar nuclei) in a large central cell (Figure 2). One of the two sperm nuclei fuses with the egg to give rise to the diploid zygote; the other fuses with the two polar nuclei. This is called "double fertilization." The resulting triploid product undergoes a series of mitotic divisions to form the nutritive tissue called endosperm.

When did flowering plants originate and radiate? The fossil record (pollen and leaves, and more recently, flowers and fruits) documents that angiosperms underwent a major radiation starting in the lower Cretaceous (about 135 million years ago). By the end of the Cretaceous, they were the dominant plants in most terrestrial environments. It is likely, however, that angiosperms actually originated much earlier. This is suggested both by molecular clock estimates and by the fact that all likely close relatives of angiosperms have fossil records extending at least to the Triassic.

Botanists have long puzzled over the relationships of angiosperms to other groups of seed plants. In addition to the four other major extant clades, a number of extinct seed plant lineages may bear directly on the problem. In particular, it has been hypothesized that flowering plants are most closely related to some group of Mesozoic "seed ferns," or perhaps to a group called the Bennettitales (or "cycadeoids"), some of which produced flowerlike structures with complex pollen-producing organs surrounding a central stalk bearing naked seeds.

Morphological phylogenetic studies, including those of fossil taxa, have generally concluded that angiosperms form a clade with Bennettitales and Gnetales—a branch referred to as the *anthophytes*. According to these results, Gnetales are the extant group most closely related to flowering plants. Some molecular datasets suggest that Gnetales are instead related more directly

FIGURE 3. Flowering Plants Show Enormous Variation in Their Reproductive Structures.
Upper left: The lily (Lilium) is a monocot with flower parts in 3's. Upper right: Thunbergia is a eudicot with its five petals fused into a tube. Lower left: Hordeum (which includes barley) is a grass (a large group of monocots), and has very tiny, wind-pollinated flowers packed tightly together. Lower right: The holly (Ilex) has fleshy red fruits that attract bird dispersal agents. Courtesy of Michael J. Donoghue.

to conifers. As of 2001, this question is unresolved, but it is clear that gymnosperms (naked-seed plants) are paraphyletic, because all extant lineages are more closely related to angiosperms than to Paleozoic "seed ferns" (Figure 1).

To interpret the dramatic rise of angiosperms, it is important to know what the first flowering plants were like, and this depends critically on understanding phylogenetic relationships among lineages at the base of the angiosperm tree. Most botanists have held that the first flowering plants were among the so-called Magnoliidae—a paraphyletic group that traditionally includes magnolias, avocados, water-lilies, black peppers, and some other plants. This is not terribly helpful, however, in deriving a picture of the first flowering plants, since magnoliids display an enormous range of morphological forms. Magnolias, for example, have large flowers with many parts (stamens, carpels) arranged spirally on an elongate axis, while black pepper plants have minute flowers with few parts.

Recently, a variety of molecular studies have concluded that the first split within modern angiosperms was between a lineage that now includes just the single species *Amborella trichopoda* (from New Caledonia), and the rest of the 235,000 species. *Amborella* plants are functionally dioecious shrubs with rather small flowers. Unlike almost all other angiosperms, the water-conducting cells in the xylem of *Amborella* are tracheids, implying that the first angiosperms lacked vessels. Also situated near the base of angiosperm phylogeny are branches leading to water-lilies (Nymphaeales), with about 70 modern species living in aquatic habitats, and to a small clade including the star anise and relatives (Illiciales), which are mostly shrubs or woody vines. The remaining angiosperms form a clade within which relationships among major lineages are still not clear.

This picture is at odds with traditional classifications, which imply a basal split into monocotyledons and dicotyledons. Instead, monocots appear to comprise a clade of about 65,000 species, nested within a dicotyledon grade (Figure 1). However, the vast majority of the traditional dicots (~160,000 species) do appear to belong to a single major clade, the eudicots.

Monocots are characterized by a shift from embryos with two cotyledons (seed leaves) to only one. Several other characteristics, though not universally present or restricted to monocots, are nevertheless common. These include scattered vascular bundles in the stem, leaves with major veins more or less parallel, and flowers with parts in 3s. Almost half of the species of monocots are either orchids (Orchidaceae; ~22,000 species) or grasses (Poaceae; ~10,000 species), but also included are palms, bromeliads, bananas, gingers, aroids, lilies, irises, and many other familiar and economically important plants.

Eudicots are characterized by pollen with three germinal furrows (tricolpate grains) or derivative types, as opposed to the monosulcate (single-furrowed) grains found in the basal lineages and in monocots. Many eudicots also have flowers with parts in 4s or 5s. This huge clade includes a number of very speciose lineages, including legumes (beans and their relatives, Fabaceae; ~16,000 species) and composites (daisies and relatives, Asteraceae; ~20,000 species), as well as buttercups, roses, oaks, melons, mustards, cotton and relatives, potatoes and tomatoes, mints, snapdragons, and parsleys, to name only a few familiar groups.

Within angiosperms, a number of changes in flower morphology have occurred repeatedly. These include changes in flower symmetry (e.g., from radial to bilateral symmetry), changes in the number of flower parts (increases and decreases in the number of stamens, carpels, etc.), and the fusion of parts (e.g., the fusion of

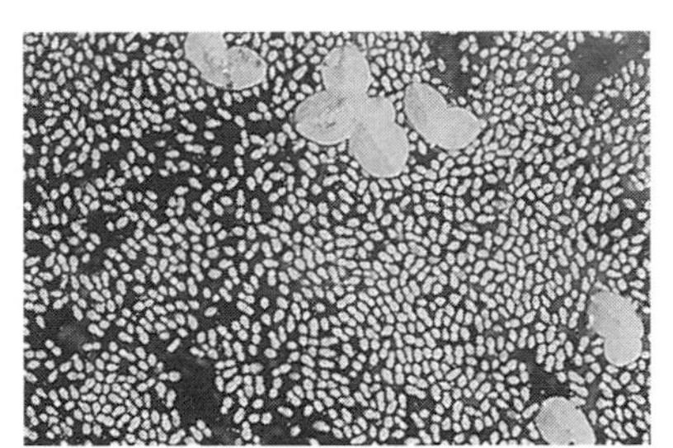

FIGURE 4. Flowering Plants Have Radiated into Many Different Environments, and Show Great Variation in Shape and Size. Upper left: A large cactus (Pachycereus) adapted to life in the desert. Upper right: Tiny plants of two species of "duckweeds" float on the surface of a pond. Lower left: "Pitcher plants" (Sarracenia) have highly modified leaves that capture and digest insects. Lower right: "Dodder" (Cuscuta) plants are parasitic on other flowering plants and have lost chlorophyll in their orange threadlike stems. Courtesy of Michael J. Donoghue.

petals into a tube). In many flowers, the sepals, petals, and stamens arise from below the ovary (which is then a *superior* ovary), whereas in others these parts are borne on the margin of a "floral cup" or arise on top of the (inferior) ovary. Fusion of separate carpels into a compound structure has happened repeatedly, and there are many different arrangements of the ovules within the ovary. The fruits of angiosperms are enormously variable. Some are fleshy, and others dry; some open at maturity to release the seeds, while others remain closed.

Much variation in flower morphology is related to pollination biology. Pollination by various vectors has selected repeatedly for particular suites of floral traits. For example, flowers pollinated by night-flying moths are often white and fragrant, and produce tubular flowers with nectar. In contrast, flowers pollinated by flies often look and smell like rotting meat. Bee-pollinated flowers come in many different forms (generalized radial forms, as in many buttercups and roses, or specialized bilateral forms, as in orchids), but they often have visible "nectar guides" on the petals or produce such patterns in the ultraviolet wavelengths (which bees can see). Vertebrates, especially birds and bats, are also important pollinators. Bird-pollinated plants often have long, red, tubular flowers with copious nectar. Wind pollination has also evolved repeatedly. Wind-pollinated flowers tend to be inconspicuous, lacking in color and odor, and often dangle their stamens and stigmas in the air.

Variation in fruit morphology is related to dispersal agents. Fleshy and colorful fruits tend to be dispersed by vertebrates, especially birds. These tend to have many seeds with hard seed coats, or a single seed surrounded by the hardened inner wall of the ovary (e.g., cherries), which then are protected as they go through the gut of an animal. Many dry fruits are dispersed by wind, including those with conspicuous wings (e.g., maples) or with plumose hairs (e.g., dandelions). In fruits that open at maturity, the seeds are sometimes fleshy, brightly colored, and attractive to animals, or sometimes dry (perhaps winged or hairy) and dispersed by wind.

Angiosperms have repeatedly entered into symbiotic relationships with fungi (especially mycorrhizal ones) and bacteria (especially for nitrogen fixation). Although most angiosperms live on land, they have on numerous ocasions entered freshwater aquatic habitats; however, aside from sea grasses and mangroves, they have seldom moved into marine environments. There have been many shifts from being woody to being herbaceous plants (perennials and, in many lineages, annuals) through the suppression of cambial activity. This shift may correlate with elevated rates of diversification. Although the tropics are spectacularly rich in flowering plants, the radiation of herbaceous lineages in temperate zones and in arid lands has also generated enormous diversity.

[*See also* Algae; Diatoms; Lichens.]

BIBLIOGRAPHY

Crane, P. R., E. M. Friis, and K. R. Pedersen. "The Origin and Early Diversification of Angiosperms." *Nature* 374 (1995): 27–33. Concise synthesis of information on relevant fossil and extant plants, and an overview of ideas on the origin of flowering plants.

Dahlgren, R., H. T. Clifford, and P. Yeo. *The Families of Monocotyledons*. Berlin, 1985. A compendium of information on the biology and major clades of monocots.

Delwiche, C. F. "Tracing the Thread of Plastid Diversity Through the Tapestry of Life." *American Naturalist* 154 (1999): S164–S177. A synthesis of recent ideas and information on endosymbiotic events giving rise to chloroplasts.

Donoghue, M. J., and J. A. Doyle. "Demise of the Anthophyte Hypothesis?" *Current Biology* 10 (2000): R106–R109. An analysis of the controversy surrounding the relationships of gnetophytes to either angiosperms or conifers.

Doyle, J. A., and M. J. Donoghue. "Phylogenies and Angiosperm Diversification." *Paleobiology* 19 (1993): 141–167. Implications of seed plant and angiosperm phylogeny for theories of angiosperm diversification.

Doyle, J. A. "Phylogeny of Vascular Plants." *Annual Review of Ecology and Systematics* 29 (1998): 567–599. An excellent summary of knowledge on vascular plant phylogeny.

Endress, P. K. *Diversity and Evolutionary Biology of Tropical Flowers*. Cambridge, 1994. A wealth of information on flower morphology and pollination biology.

Gifford, E. M., and A. S. Foster. *Morphology and Evolution of Vascular Plants*. 3d ed. New York, 1989. Outstanding reference work on vascular plant morphology; excellent on life cycles.

Graham, L. E. *Origin of Land Plants*. New York, 1993. A synthetic treatment of charophytes and the origin of land plants.

Heywood V. H., ed. *Flowering Plants of the World*. New York, 1978. A thoroughly illustrated compendium of information on angiosperm diversity.

Judd, W. S., C. S. Campbell, E. A. Kellogg, and P. F. Stevens. *Plant Systematics. A Phylogenetic Approach*. Sunderland, Mass., 1999. A phylogenetically oriented systematics textbook emphasizing basic principles and an overview of angiosperm diversity.

Kenrick, P., and P. R. Crane. *The Origin and Early Diversification of Land Plants: A Cladistic Study*. Washington, D.C., 1997. Synthesis of knowledge on the origin and early evolution of land plants; strong on the fossil record and phylogeny.

McCourt, R. M. "Green Algal Phylogeny." *Trends in Ecology & Evolution* 10 (1995): 159–163. A concise summary of evidence on the major lines of "green algae."

Pryer, K. M., H. Schneider, A. R. Smith, R. Cranfill, P. G. Wolf, J. S. Hunt, and S. D. Sipes. "Horsetails and Ferns Are a Monophyletic Group and the Closest Living Relatives to Seed Plants." *Nature* 409 (2001): 618–622. A recent molecular phylogenetic analysis of the major lineages of vascular plants.

Raven, P. H., R. F. Evert, and S. E. Eichorn. *Biology of Plants*. 6th ed. New York, 1999. A standard reference on basic plant biology and diversity.

Stebbins, G. L. *Flowering Plants: Evolution Above the Species Level*. Cambridge, Mass., 1974. An outdated but very stimulating account of the macroevolution of angiosperms.

Stewart, W. N., and G. W. Rothwell. *Paleobotany and the Evolution of Plants*. 2d ed. Cambridge, 1993. A textbook on the morphology and evolutionary implications of plant fossils.

Zimmer, E. A., Y. L. Qui, P. K. Endress, and E. M. Friis. "Current Perspectives on Basal Angiosperms: Introduction." *International Journal of Plant Sciences* 161 (2000): S1–S2. The introduction to a symposium volume devoted to phylogenetic relationships at the base of flowering plants.

— MICHAEL J. DONOGHUE

PLASMIDS

Plasmids are bacterial symbionts. They exist as multicopy extrachromosomal DNA and are found in nearly all bacterial species. They range in size from a few hundred base pairs to several hundred kilobases. The smallest plasmids contain just the basic replicon, a cluster of genes needed to carry out plasmid replication. Larger plasmids house a number of other genes that enhance plasmid spread and maintenance (for example, conjugation and restriction-modification) or code for a variety of host functions (e.g., heavy metal tolerance, UV resistance). In extreme cases, almost half the bacterial genome is present on multiple megaplasmids.

Wide interest in plasmids was spurred by the finding in the 1950s that they were responsible for spreading antibiotic resistance among bacterial species. Plasmids are able to do this because they can propagate themselves by infectious transfer to other bacteria. Some plasmids are restricted to a single bacterial species, but others show far more promiscuous spread and are found across a wide range of unrelated bacteria. The latter create a "horizontal" gene pool, permitting the exchange of genetic information across species boundaries. Plasmids also have a variety of properties that have made them indispensable tools in the biotechnology revolution. For example, plasmids are used as vectors to clone fragments of DNA, as expression systems for foreign genes, and as the source of a number of the restriction enzymes used to cut DNA.

Plasmids present a number of paradoxes to the evolutionary biologist. Little is known about their origins. They are thought to be related to phages (bacterial viruses), but they have dispensed with (or never gained) the ability to form structures that allow an independent existence outside the bacterial cell. This hypothesis is not well founded, however, because plasmids do not contain gene sequences that enable them to be clearly aligned with other phylogenetic groups. Little is known even about how different plasmid groups are related to one another, so the hope of defining a wider phylogenetic context remains remote. As with other symbionts, it is unclear whether plasmids are primarily parasites, or cooperators that are useful to their hosts. In fact, they fulfill both roles, and their success depends on a mix of relationships. They show elements of self-interest by increasing their transmission rate despite the harm this does to the host, but they also have picked up a number of functions that allow hosts to exploit novel environments and flexibly resist stress.

Plasmid Replication. Plasmids are not able to replicate independently. Instead, they rely on the bacterial host's replication machinery. However, they do encode genes that control the initiation and repression of replication. This enables them to turn on replication just after cell division and to turn off replication when copy number rises, leading to an approximate doubling of plasmid number per cell cycle.

The main way this has been investigated is through the isolation of plasmid mutations that abolish the control of replication. Mutants in Co1E1 plasmids (found in *Escherichia coli*) reveal that inhibition is attributable mainly to small antisense RNAs, which are copied from the complementary DNA strand (e.g., for a sense sequence of AUGGCAUGC, the antisense GCAUGCCAU is produced). The antisense RNA binds to the DNA–protein complex at the site of the origin of replication and blocks DNA replication. Antisense RNA is very unstable, so its concentration is very closely related to the number of gene copies per cell. Once the concentration reaches a threshold, it becomes effective at blocking replication and thereby limiting copy number, but below this threshold replication proceeds. Additional, plasmid-encoded inhibitory systems also act at the level of transcription and translation of the replication enzymes. Again, these act in a dose-dependent manner.

For replication to result in stable inheritance, plasmids must not only be replicated but also be distributed to both daughter cells when the cell divides. With random distribution, a cell with n plasmids produces 2^{-n} daughter cells with no plasmids. This places low-copy number plasmids at risk, but even high-copy-number plasmids like pUC8 ($n > 100$) are also notoriously unstable under lab culture. A major cause of instability is plasmid clumping. This increases the asymmetry of plasmids among daughters and makes it more likely that a daughter has an insufficient number to maintain plasmid infection. To counteract this, several plasmids have evolved active partitioning systems that align pairs of plasmids in the plane of cell division, resulting in segregation to both daughter cells. Such systems allow better than random segregation and so contribute to the persistence of plasmids, even with very low copy numbers (i.e., $n = 2$).

Conjugation. Active partitioning helps plasmid maintenance, but it is not as efficient as chromosomal segregation, so copy number variation inevitably arises. Selection tends to favor bacteria with lower copy numbers (or no plasmids) because plasmids impose an energetic burden on their hosts, which increases with plasmid size and copy number. Therefore, unless plasmids encode functions that are beneficial to the bacterial host (see below), selection is likely to drive down copy number and lead inevitably to plasmid-free cells. This has led to the evolution of two further mechanisms, conjugation and post-segregational killing, that contribute to plasmid persistence.

Conjugation allows plasmids to be transferred between bacterial cells. The process of conjugation is entirely controlled by plasmid-borne genes (30+ are known to be involved). The first step is the production of a sex pilus, an approximately 10-nm-diameter hollow tube that projects from the surface of a donor cell. The pilus makes contact with potential recipients and forms a stable cell-to-cell bridge through which plasmid DNA is transferred. The pilus serves to protect DNA from degradation. Conjugation occurs in a unidirectional manner, from the infected donor to recipients that lack plasmids. Bacteria with plasmids actively reject further contact, unless they contain plasmids from different compatibility groups. This avoids the pointless transfer of plasmids to clonal relatives and other bacteria that already contain plasmids.

Some host bacteria (in the Enterococci) actively seek mating partners when they are plasmid-free. They secrete pheromones that trigger donor cell plasmid transfer (i.e., the production of a pilus, etc.). Plasmids control pheromone production by bacterial genes. They produce suppressors that turn off pheromone production by the bacterium. In addition, in the absence of pheromone, plasmids down-regulate genes involved in conjugation. However, if a bacterium loses its plasmids, suppression is released and pheromone production starts. It is not clear whether the pheromone's signaling function is an adaptive trait that benefits the bacterium as well as the plasmid, or just a secreted bacterial byproduct that is exploited by the plasmid.

Some plasmids are transferred only between members of the same bacterial species. Others are found in a broad range of hosts. For example, F-like plasmids found in the human pathogen *Salmonella enterica* are closely related to those found in *E. coli*, even though these two bacteria are thought to have diverged over 140 million years ago. The mechanistic and evolutionary reasons for these differences in host range are not well understood.

Promiscuous conjugation causes bacterial genes to be dispersed across species boundaries. Plasmids pick up host genes through recombination or transposable element movements from the bacterial chromosome. When they cross to another species through conjugation, these genes can be reintegrated into the bacterial chromosome of the new species. Plasmids were creating genetically modified organisms well before biotechnology companies dreamed of them.

Post-segregational Killing. Another mechanism that contributes to plasmid maintenance is post-segregational killing. Daughter cells that fail to inherit plasmids commit apoptosis (suicidal death) after a few further rounds of division. The best-known example of this astonishing behavior is the *hok/sok* system of R1 enterococal plasmids. These are low-copy-number plasmids that suffer high rates of plasmid loss. The *hok* gene produces a lethal toxin and the *sok* gene its antidote. The *sok* gene product is antisense RNA that binds and disables *hok* mRNA. Plasmid-containing cells produce no *hok* toxin and do not commit apoptosis. However, the *sok* antisense RNA is very unstable and is quickly lost from plasmid-free bacteria, leading to the translation of *hok* mRNA, toxin production, and death. The benefits of *hok/sok* killing by R1 plasmids are indirect. The death of bacteria that lose their plasmids reduces local competition and increases the fitness of clonal relatives that retain their R1 plasmids.

It has been proposed by Naito et al. (1995) that plasmid restriction-modification (RM) systems are also examples of post-segregational killing. Like *hok/sok*, RM systems have two component genes. The restriction enzyme (R) recognizes and cuts highly specific DNA sequences. For example, the restriction enzyme from the *Pae*R7 plasmid induces double strand cuts in CTCGAG sequences. The *Pae*R7 plasmid also codes for a modification enzyme (M) that recognizes the same sequence and protects it by methylation of adenine and cytosine residues. Because restriction enzymes have a longer half-life than modification enzymes, plasmid-free bacteria soon fail to methylate their recognition sites, and the bacterial DNA is cleaved by the remaining restriction enzymes. Experimental studies of the population dynamics of RM plasmids show that the elimination of plasmid-free bacteria allows stable maintenance of the RM plasmid.

The *hok/sok* and RM plasmids are good examples of selfish genetic elements. Host killing is advantageous to the plasmid but has no obvious benefit to the bacterial hosts. This view has only recently gained acceptance. Originally, RM systems were thought to have been selected because they defend their host against attack by phage viruses. Phage infects by inserting its DNA (or RNA) into the bacteria and usurping control of the host's cellular functions. RM plasmids attack phage with their restriction enzymes because phage DNA is not appropriately methylated. However, laboratory studies show that defense against phage by RM plasmids is short-lived and has a minor effect compared to that caused by the killing of plasmid-free bacteria.

Genes Encoded by Plasmids. The best-known property of plasmids is their cross-species transfer of antibiotic resistance. Following the introduction of antibiotics in the 1940s, resistance to penicillin and sulphanilamides was quickly reported. This became a major problem from the 1960s onward as the rate of novel antibiotic discovery dried up and multiple resistance developed. It was soon demonstrated that many types of resistance could be transferred between different species of bacteria—for example, from *E. coli* to the dysentery-causing *Shigella*, which are both inhabitants of the gut—and that this spread was caused by plasmids.

Many other bacterial traits are now known to be carried by plasmids. Some, like antibiotic resistance and heavy metal tolerance, have been very widely spread among distantly related bacterial species, others show more limited ranges. However, they all generally show a polymorphic distribution within species: only some bacteria carry trait-encoding plasmids. This reflects the fact that plasmids rarely carry genes that are essential—which the bacteria cannot live. Plasmids tend to encode traits that enable bacteria to thrive under conditions that are restricted in time or space. Bacteria carry trait-encoding plasmids only under these specific conditions; elsewhere plasmids are lost or displaced by other, simpler (non-trait-encoding) plasmids.

These patterns reflect the different levels of selection acting on plasmids and their bacterial hosts. Plasmid retention is clearly favored when the plasmid carries a trait useful to the bacterium. Selection also favors adaptive improvement of the trait, so many plasmid traits are coded by multiple gene complexes. But plasmids impose an energetic load on their hosts, and their loss is favored once selection is relaxed. Because plasmids are easily lost and regained, they represent a flexible route by which bacteria can adapt to transitory or spatially limited stressful conditions. In contrast, plasmids do not code for essential functions because their inheritance is not guaranteed to the same extent as the bacterial chromosome. Frequent transfer of genes between plasmid and bacterial DNA occurs via recombination events and movement of transposable elements, allowing selection to dictate the best location for different genetic functions.

Conclusion. As with all symbiotic relations, there can be cooperation and mutual benefit, or there can be parasitism and exploitation. The interests of bacteria and their plasmids do not coincide exactly, particularly because plasmids spread horizontally through conjugation, whereas bacterial genes generally cannot. These differences have led to the evolution of a number of traits that serve largely to benefit the plasmid, even though they are costly to the bacterial host (e.g., post-segregational killing). To a large extent, however, the interests of plasmids and bacteria coincide, so plasmids have been selected to limit copy number, the rate of conjugation, and their size.

[*See also* Cytoplasmic Genes; Meiotic Distortion; Selfish Gene; Transposable Elements.]

BIBLIOGRAPHY

Boyd, E. F., and D. L. Hartl. "Recent Horizontal Transmission of Plasmids between Natural Populations of *Escherichia coli* and *Salmonella enterica*." *Journal of Bacteriology* 179 (1997): 1622–1627.

Naito, T., K. Kusano, and I. Kobayashi. "Selfish Behavior of Restriction-Modification Systems." *Science* 267 (1995): 897–899.

Pomiankowski, A. "Intragenomic Conflicts." In *Levels of Selection in Evolution*, edited by L. Keller, pp. 121–152. Princeton, 1999. Contains a review of post-segregation killing and related selfish genetic effects.

Summers, D. K. *The Biology of Plasmids*. Oxford, 1996. An introductory review of plasmid biology and evolution.

Thomas, C. M., ed. *The Horizontal Gene Pool*. Amsterdam, 2000. An advanced, multiauthored review of modern ideas about plasmid population biology.

Thisted, T., N. S. Sorensen, and K. Gerdes. "Mechanism of Post-Segregational Killing I." *EMBO Journal* 13 (1994): 1960–1968. A description of the *hok/sok* post-segregation killing mechanism.

— ANDREW POMIANKOWSKI

POPULATION DYNAMICS

Population ecology is the study of the change in the distribution and abundance of species over time and through space. As a quantitative discipline, it relies heavily on mathematics to formalize concepts, test hypotheses, and clarify arguments. This overview provides an introduction to the ideas central to understanding the distribution and abundance of species. We begin with a fundamental description of populations before discussing the ideas of Thomas Malthus and the role of density dependence in population dynamics. More detailed sections examine the role of competition, predation, and multispecies interactions. We conclude with a discussion of spatial dynamics and a brief summary.

Four key processes—birth, death, immigration, and emigration—provide the basis for understanding the patterns of change in the temporal and spatial distribution and abundance of organisms. These changes in numbers of a population can be expressed simply as

$$\text{Changes in population density} = \text{births} - \text{deaths} + \text{immigrants} - \text{emigrants}.$$

This is the *fundamental equation of population ecology.* The dynamics of populations occur as a result of these four processes; although a simple expression, it can mask a wealth of biological detail that is essential to understanding the ecology of a species. Throughout this overview, we will be concerned primarily with the processes of birth and death and how these impinge on the dynamics of populations; in the final section, we will briefly examine the influence of immigration and emigration.

Malthusian Dynamics. In 1798, sixty-one years before Darwin published his ideas of evolution, an English vicar published a treatise entitled *On the Principle of Population.* Thomas Malthus argued that populations increase through a geometric progression (e.g., $2^1 = 2$, $2^2 = 4$, $2^3 = 8$, $2^4 = 16$, $2^5 = 32 \ldots$), whereas the resources that populations use increase in only a linear fashion (2, 4, 6, 8, 10). Malthus suggested that food supply would be insufficient to maintain population growth, and that this geometric increase in populations would lead to mass starvation unless held in check by war, disease, or restraint in reproduction. Malthus's ideas have been interpreted in many ways, but this theory was written to counter the eighteenth-century doctrine of unabated progress by showing that prosperity would have strict limits imposed by the pressure of population growth.

These ideas inspired Darwin's thoughts on evolution by natural selection. Darwin tells us that he was reading "for amusement" Malthus's *Essay on Population* when it became clear that there would be a struggle for existence. Darwin saw this struggle as central to his ideas about natural selection, with the key evolutionary questions being who survived, and why.

For population ecology, Malthus described a key concept: in the presence of unlimited resources, populations will grow exponentially (Figure 1A). However, if

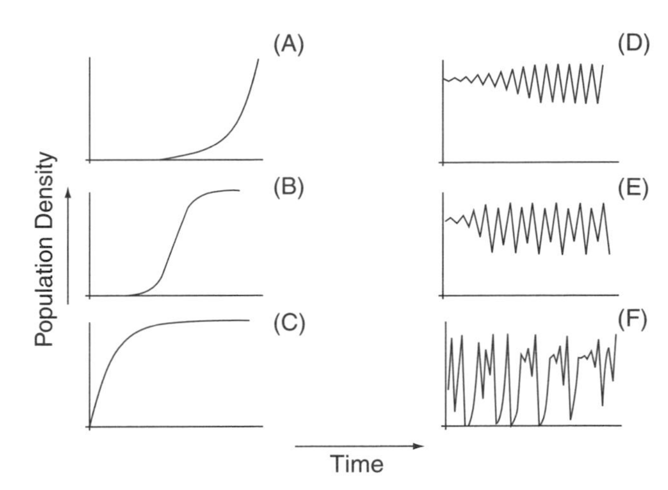

FIGURE 1. Range of Population Dynamics. Populations may exhibit (A) exponential growth, (B) logistic growth to a carrying capacity, (C) stable equilibrium dynamics, (D) 2-point limit cycles, (E) 4-point limit cycles, or (F) chaotic dynamics. Courtesy of Michael Bonsall.

resources are limited, then there will be an upper limit to this growth rate that will impose a ceiling or *carrying capacity* on the population.

Density Dependence. In 1845, the Belgian mathematician Pierre-François Verhulst formalized Malthus's ideas. Verhulst reasoned that during the initial stages of population growth, when resources were abundant, a population would increase exponentially. Eventually, population growth would be limited by resources and the population would reach a carrying capacity. By assuming that the rate of population growth declined linearly with the size of the population, Verhulst was able to obtain a sigmoid curve for the growth of a population: this is known as the *logistic equation* (Figure 1B). Seventy-five years after Verhulst proposed the logistic equation to describe population growth, Raymond Pearl and Lowell Read, unaware of Verhulst's memoirs, introduced the same model for populations. The logistic model is:

$$\frac{dN}{dt} = rN\left(\frac{K - N}{K}\right). \quad (1)$$

This differential equation expresses the notion of *density dependence*, whereby an increase in mortality or decrease in fertility as population size increases can act to restrict the growth of a population. The rate of change of the population (dN/dt), which is r at low density, slows as the size of the population (N) approaches the carrying capacity (K). When $K = N$, the rate of growth drops to zero. This is often referred to as the *equilibrium state* of the population (Figure 1B).

Chaos. In 1974, Robert May showed that very simple models of populations can display a wide range of population dynamic behaviors, from stable population equilibria through limit cycles to chaos (Figure 1C–E). He chose to investigate a simple discrete-time version of the logisitic model,

$$N_{t+1} = N_t exp\left(r - \frac{N_t}{K}\right), \quad (2)$$

where t represents time and "exp" is the base of the natural logarithms (e ~2.718 . . .).

As the growth rate (r) of the population increases, May found that the model displayed a series of different dynamic behaviors of increasing complexity. When $r \leq 2$, the population has a stable equilibrium; as r increases above 2, the population dynamics switch to a 2-point limit cycle oscillating around the equilibrium point. When $r > 2.526$, a further doubling occurs to a 4-point limit cycle. Subsequent doublings, as growth rate increases, lead to 8-point, 16-point, and 32-point limit cycles, and so on until the dynamics of the population appear completely aperiodic, or *chaotic*.

The characteristics of chaotic population dynamics are extreme sensitivity to initial conditions and aperiodic dynamics. The first point highlights that two populations that have slight differences in their initial starting densities and experience chaotic population conditions will eventually show completely different population trajectories through time. The second point is that even though the dynamics appear indistinguishable from random noise, they are described by a completely deterministic process (equation 2) fluctuating between fixed bounds in population size.

Population regulation. Controversies and disagreements have often sharpened our interpretation of the ecology and dynamics of populations. One particular argument centered on the role of density dependence and the persistence of populations. Developing the ideas of Pearl, Reed, and Verhulst, Alexander Nicholson proposed that density dependence was a necessary requisite for population persistence. This engendered a vociferous debate over the role of random fluctuations in birth and death rates versus density-dependent processes. This culminated at a meeting at the Cold Spring Harbor Symposium on Quantitative Biology in 1957. Nicholson and proponents advocated the role of density dependence as the process by which populations are able to persist. Andrewartha, Birch, and Milne contested this point, arguing that density-independent factors such as abiotic processes (e.g., floods, meteor impacts, accidents) were more important in determining the abundance of populations.

In the resolution of this debate, it is critical to distinguish between the determinants of population abundance (i.e., limitation) and population regulation. Regulation is the tendency of a population to increase in size when below a particular level and to decline when above it. This can only occur, by definition, through a density-dependent process. Abundance will be determined by the combined effect of all biotic and abiotic processes. With these definitions in mind, it becomes clear that differences in semantics and understanding fueled this disagreement, rather than any key differences in the underlying ecological processes.

Through the reconciliation, it is now clear that rather than a population being regulated to a particular point (i.e., the equilibrium), a wider notion of regulation is more applicable—a population is regulated *around* an equilibrium. Adopting this definition leads to the idea of a fixed distribution of points around the equilibrium, the so-called *stationary distribution* of population densities. This is clearly the best way to define regulation in any dynamical system such as a population.

Age structure. The simple population models described above make two key assumptions: first, that there is no age structure in the population—all individuals dying or giving birth at the same rate; and second, there are no explicit time lags in the response of the per capita growth rate to changes in population numbers.

The populations of many species are *age-structured*. A simplifying assumption of the models described so far is that the probability of an organism dying or giving birth is constant and independent of age. However, many species have overlapping generations and age-dependent demographic processes. Population ecologists organize data on how births and deaths vary with age in life tables and model these more complex situations using matrix algebra techniques. In the absence of density dependence, populations increase or decrease geometrically, exactly as Malthus predicted, but now the growth rate is a complex function of the different age-dependent birth and death rates. A second important conclusion from these models is that as the population grows exponentially, the relative abundance of the different age classes becomes fixed, a state known as a *stable age distribution*.

Time lags. Time lags or delays in the effects of births and deaths on changes in population numbers constitute *delayed density dependence*. Delays in the key population processes can have a profound influence on populations. Delayed density dependence can lead a wide range of population dynamics, such as limit cycles or chaos. For instance, in intraspecific interactions, delayed density dependence may arise through maternal effects. That is, the fecundity of females and the survival of immatures to adulthood are functions of individual quality. However, individual quality is a function of maternal quality, and these maternal effects introduce delays into populations that can give rise to cyclic dynamics. Although the mechanisms of delayed density dependence may be driven by intraspecific factors, it is more commonly observed through interspecific (between-species) processes such as competition or predation.

Competition. Species do not exist in isolation but are likely to compete for limiting resources such as food and/or territories. A model of the population dynamics of interspecific competition based on the logistic model (equation 1) was originally derived by Alfred Lotka and Vito Volterra. They assumed that the population growth rate of a species is inhibited by both intraspecific and interspecific processes. The Lotka-Volterra equations for the interspecific competitive interaction between two species (N_1 and N_2) are:

$$\frac{dN_1}{dt} = r_1 N_1 \left(\frac{K_1 - [N_1 + \alpha_{12} N_2]}{K_1}\right) \quad (3)$$

$$\frac{dN_2}{dt} = r_2 N_2 \left(\frac{K_2 - [N_2 + \alpha_{21} N_1]}{K_2}\right) \quad (4)$$

α_{ij} is the competition coefficient and denotes the per capita competitive effect on species i by species j. If, for instance, 100 individuals of species 2 have the same competitive effect on species 1 as does a single individual of species 1, then the total competitive effect on species 1 will be equivalent to the effect of $[N_1 + (N_2/100)]$ individuals of species 1. The competition coefficient (α_{12}) is a measure of the interspecific effects of species 2 on species 1. So, if $\alpha_{12} < 1$, species 2 has a lesser effect on species 1 than species 1 has on itself. To understand the dynamics of interspecific competition, we ask under what conditions each species can increase or decrease. When the rate of change of each species is zero $[(dN_i/dt) = 0]$, then equations 3 and 4 reduce to:

$$N_1 = K_1 - \alpha_{12}N_2 \quad (5)$$

$$N_2 = K_2 - \alpha_{21}N_1 \quad (6)$$

These equations can be used to determine the *zero isoclines* for population growth of the two species (Figure 2). Below its isocline, the population of a species increases in size. Above an isocline, the population of a species declines. These isoclines can be plotted in *phase space:* a plot of the density of species 1 versus the density of species 2 (Figure 2A–B).

From the interaction between the species (i.e., how the isoclines cross), we can identify four types of population behavior (Figure 2C–F): in Figure 2C, species 1 outcompetes species 2. The exclusion of one species arises as the density dependence acting to limit the population growth of species 2 occurs at a lower population size than the density dependence acting on species 1. Figure 2D shows the converse of this, where species 2 outcompetes species 1. In Figures 2E–F, the outcome of interspecific competition depends on how the isoclines cross. In Figure 2E, the outcome to the competitive processes is critically dependent on the initial conditions. The point where the isoclines cross is an unstable equilibrium point. Although at this crossover point the two species can coexist, any disturbance away from this point will lead to the exclusion of one of the species. Finally, in Figure 2F the point where the isoclines cross is a stable equilibrium point, and the two competitors coexist indefinitely. Coexistence is favored only when the intraspecific effects are greater than the interspecific effects. That is, a species limits itself more than it limits the growth rate of the other species.

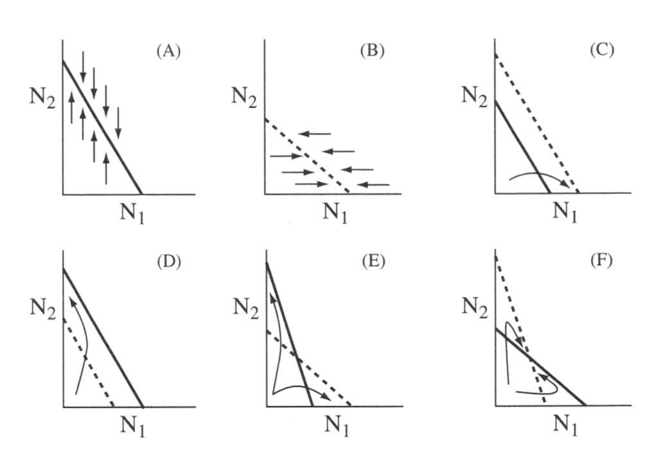

FIGURE 2. Phase-Plane Portraits of Interspecific Competition. Phase-planes are the plot of the density of species 1 versus the density of species 2. Below the isoclines a species increases, and above the isocline a species decreases. (A) The direction of increase and decrease for species 1. (B) The direction of increase and decrease for species 2. (C)–(F) The four outcomes of interspecific competition. Coexistence (F) is favored only when a species limits itself more than its competitors. Courtesy of Michael Bonsall.

In a series of experimental studies on interspecific competition, Georgyi Gause tested the predictions of the Lotka-Volterra competition model. Using the yeasts *Saccharomyces cerevisiae* and *Schizosaccharomyces kephir*, Gause unequivocally demonstrated the importance of interspecific competition in the interactions between species. These yeasts differ in their relative intensities of oxidation and fermentation. *Saccharomyces cerevisiae* is able to reproduce rapidly in the absence of oxygen, while the growth of *Schizosaccharomyces kephir* under anaerobic conditions is exceedingly slow. In the absence of the other species, the growth of each yeast population was shown to follow a pattern described by the logistic equation (equation 1). When in combination, the byproducts of fermentation, together with the utilization of the sugar resources, acted to inhibit the growth rate of each species. The effects of interspecific competition reduced the population growth rate of *Schizosaccharomyces* by two-thirds while the growth rate of *Saccharomyces* was reduced by about half. Over time, it was predicted that *Schizosaccharomyces* would be excluded by *Saccharomyces*.

These experiments by Gause and the Lotka-Volterra competition equations have been central in the development of the *competitive exclusion principle*. This principle states that two competing species cannot coexist in a stable environment if they share the same niche (i.e., the same resource requirements). Coexistence is favored only if there are sufficient differences in the niche. Phrased differently, coexistence is promoted only if the effects of competition within a species are greater than the effects of the competition acting between species. Understanding the ecological processes that alleviate interspecific competition and allow coexistence has been a recurrent theme in population ecology. A wide range of factors has been proposed to explain how the intraspecific effects can outweigh the interspecific effects and promote coexistence. They include the aggregation of competitors into different patches, the availability of exclusive resources, and the effects of life-history trade-offs.

Mechanistic models of interspecific competition predict that the species that persists and excludes all others is the one that requires the least amount of the limiting resource to replace itself. When several species compete for the same limiting resource, then the dynamics and long-term outcome of coexistence or exclusion may be determined by the patterns of resource utilization by the different species, and by the resource supply rate. There have been a number of experimental tests of this theory of resource competition. One clear study examined the competition between two diatoms, *Asterionella formosa* and *Cyclotella meneghiniana*. *Asterionella* needs less phosphate than *Cyclotella*. When they were grown together in a phosphate-limited environment, *Asterionella* displaced *Cyclotella*. Conversely, *Cyclotella* needs less silicate than *Asterionella* and was the dominant competitior in a silicate-limited environment.

Resource-Consumer Dynamics. A second class of interspecific interactions involving two species is predation. Predation occurs when one species consumes another. Four strategies of predation can be identified: herbivory, where the consumer eats a plant; carnivory, where the consumer preys on a herbivore or another carnivore; parasitism, where a consumer attacks or infects a host; and cannibalism, where the consumer preys on individuals of its own species.

Lotka and Volterra also derived two species models to describe the interaction between predators and prey. This model is extremely simple:

$$\frac{dN}{dt} = rN - bN\,P \tag{7}$$

$$\frac{dP}{dt} = cbN\,P - dP. \tag{8}$$

In the absence of predators, prey grow exponentially at a rate r. In the presence of predators, prey die at a rate bP and are converted into new predators with efficiency c. Predators die at some density-independent rate d. The Lotka-Volterra predator–prey model shows that resource–consumer interactions have an inherent propensity to oscillate. When predator numbers are low, prey increase exponentially. As the prey population reaches high numbers, predator numbers increase. Predators overexploit their prey and the prey population collapses. This is followed by a collapse in the predator population, and the cycle starts over again.

One major class of resource–consumer interactions that has attracted considerable research interest is that of host–parasitoid associations. Parasitoids are Hymenoptera, Diptera, or Coleoptera that attack other arthropods by laying eggs in, on, or near their hosts. The juvenile parasitoid develops on or in the host, eventually killing it. The population dynamics of these associations have been shown to be intriguingly complex, with many different aspects of the biology and ecology of arthropod hosts and parasitoids important in the dynamics of the interaction. For instance, the role of host refuges, either temporal or spatial, can lead to the persistence of host–parasitoid interactions. In either case, the population persists because a fraction of the hosts are protected from parasitism in each generation. This breaks the cycle of host overexploitation by the parasitoid that tends to destabilize this type of interaction.

A second class of resource–consumer interaction involves the interaction between hosts and true parasites. It has been shown how the population dynamics of a host can be influenced by both microparasites (e.g., virus, bacteria, or fungi) and macroparasities (e.g., nematodes, helminths). A key concept in host–parasite theory is the *basic reproductive ratio*. This is defined as the number of secondary infections resulting from one primary infection in a pool of susceptibles. If this ratio is greater than one, then an infection will spread. If the ratio is less than one, the infection will not persist and will die out. Understanding how infections are transmitted is fundamental for the population dynamics of this type of predator–prey interaction.

The theory of plant–herbivore interactions has been dominated by the simple Lotka-Volterra models. Unlike predators, however, the principal effect of herbivores is to reduce plant vigor (e.g., number of seeds set, reproductive capacity) rather than to inflict outright mortality. As a result, there is a qualitative difference between plant–herbivore and predator–prey interactions. By adopting a more realistic framework for plant–herbivore interactions, we can make novel predictions about the role of herbivores on the dynamics of plant populations and the strength of interactions between plants and their consumers. For instance, there is experimental evidence for strong, persistent asymmetries in the interaction between plants and herbivores. It appears that plants have a stronger impact on the distribution and abundance of specialist herbivores than the herbivores have on the abundance of the plant.

Cannibalism can also be classed as a form of predation. Predominant in invertebrates, cannibalism can have important implications for population dynamics and has recently been shown to be critically important in the population dynamics of the stored product moth, *Plodia interpunctella*. One reported population pattern in nonseasonal organisms such as *Plodia interpunctella* is that of generation cycles. These cyclic population dynamics involve fluctuations with a period equal to the generation time of the organism. Larval and egg cannibalism are common in this species, and when coupled with asymmetric larval competition (older larvae having a disproportionate effect on smaller larvae), population models predict that aseasonal cyclic dynamics; in particular, generation cycles are highly likely to predominate.

Further, recent empirical evidence has highlighted the influence of predation as the mechanism driving the dynamics of populations. For example, Peter Hudson and colleagues showed how the cycles of red grouse in northen England could be lost if the birds were treated with antihelminthic drugs to remove their intestinal parasites. Populations of treated birds showed reduced tendency to cycles. In a similar study, Peter Turchin and colleagues have shown how the guild of predatory ground beetles are influential in determining the population dynamics of the southern pine beetle (*Dendroctonus frontalis*). After the predatory beetles were removed, the populations of southern pine beetle were less likely to exhibit cyclic population dynamics.

Although competition and predation are key interactions in ecological assemblages, pairwise interactions are rare. Species exist in food webs, and multispecies interactions are common. Extending the ideas of competition and predation to multispecies scenarios is a current focus of population dynamics.

Multispecies Interactions. The organization of ecological assemblages is driven by the biotic processes of a competition and predation. Understanding how these processes can be extended to multispecies interactions is an important goal of contemporary population dynamics.

Darwin assumed that competition was the driving force in evolution. It is competition between species and the structure of the ecological niche that shape assemblages. While competition can have profound effects on the structure and dynamics of ecological communities, the role of predation is likely to be just as important. For example, in 1966, Robert Paine demonstrated that the removal of the predatory starfish *Pisaster ochraceus* from intertidal assemblages along the Pacific Northwest coast led to a dramatic collapse in the invertebrate assemblage. In the presence of the predator, the prey assemblage was made up of fifteen species. After the removal of *Pisaster*, seven species were lost through intense interspecific competition. Predation acted to alleviate competition and favor the coexistence of several species that otherwise would have been excluded by the dominant competitor.

The role of predation in shaping ecological assemblages can be even more subtle. If two species do not compete for resources but share a common predator, then they can still negatively affect one another. Imagine an interaction between a single prey and a single predator, and that a second prey species invades. Although this species feeds on a different resource, it is susceptible to attacks from the same predator. More prey in the diet of the predator leads to an increase in the predator's population. More predators in the environment exert increased predation on both prey species. If this interaction is sustained over time, one of the prey species will eventually be excluded. This, on the face of it, appears similar to what we observe when two species are in competition for resources. However, the species only interact through the shared predator, and the one that is excluded is the species that suffers the higher rate of predation and/or has the lower population growth rate. This ecological process is known as *apparent competition*.

Apparent competition can have a profound effect on the population dynamics of multispecies assemblages. This process may be responsible for the exclusion of red squirrels by gray squirrels through a shared virus in Britain, the replacement of one species of leafhopper by another via shared parasitioids in California vineyards, and the structure of more complicated host–parasitoid assemblages. If apparent competition is a prevalent process in species assemblages, then it is enigmatic how complex interactions persist. A number of mechanisms—such as intraspecific competition, refuges, metapopulations, and predator behaviors—can act to reduce the impact of apparent competition and favor coexistence. As in interspecific competition, it is necessary for the intraspecific effects acting within species to outweigh those acting between species for coexistence to be mediated in these multispecies interactions.

Spatial Population Dynamics. The population dynamics of single species, resource–consumer and multispecies interactions are also shaped by interactions across space. Spatial population dynamics involve the processes of immigration and emigration, and these can influence the distribution and abundance of populations. The simplest way to envisage this is as a *metapopulation*—that is, a population of populations linked by dispersal events. The key processes in metapopulation dynamics are those of colonization and extinction. Four central conditions define a metapopulation. First, populations breed in discrete patches of habitat; second, populations have a high risk of extinction; third, recolonization occurs through a “rescue effect”; and fourth, populations are asynchronous in their dynamics. Rather than describing the dynamics of the individual populations of a species, the dynamics of the metapopulation are often expressed in terms of the changes in the patches of habitat (dh/dt) that are occupied by a species. This model, proposed by Richard Levins in 1969, is very simple:

$$\frac{dh}{dt} = ch(1 - h) - eh. \qquad (9)$$

In this model, the rate of patch colonization (c) by a species increases in proportion to the fraction of patches already occupied (h) and those patches that are unoccupied ($1 - h$). Patches are lost through local extinction (e) of the species. A number of predictions emerge from these metapopulation models. First, the fraction of occupied habitat is expected to increase as the ratio between extinction and colonization declines. Second, the risk of extinction declines as patch area increases. Third, the probability of colonization decreases with distance from the nearest extant patch. Fourth, the fraction of habitat occupied decreases as the average size and density of patches in a network declines.

Although the focus of population ecology has centered around birth and death processes, it is becoming increasingly clear that the dynamics of populations are influenced by spatial events through the processes of emigration and immigration. These ideas about spatial processes have been extended in numerous ways to account for the effects of interspecific competition, predation, and population age structure on the dynamics, distribution, and abundance of populations. Moreover, species interactions across space can lead to interesting and intriguing spatial arrangements or patterns in the abundance of species. Corroborating the predictions of these spatial models with data is a prevailing theme in population dynamics research.

Summary. The principles of population dynamics are founded on four key processes: birth, death, emigration, and immigration. In this overview, we have explored how these different proccess can affect the population dynamics of single species and of resource–consumer and multispecies interactions. This quantitative understanding of population processes provides a framework within which the dynamics, interactions, and coexistence of species can be comprehended.

BIBLIOGRAPHY

Begon, M., J. L. Harper, and C. R. Townsend. *Ecology*. 3d ed. London, 1996. A key introductory text and synthesis of ecology.

Cherrett, J. M. *Ecological Concepts*. London, 1989. An edited symposium volume that gives an in-depth review of the key ecological concepts.

Gurney, W. S. C., and R. M. Nisbet. *Ecological Dynamics*. Oxford, 1998. An introduction to the techniques and applications of population models.

Hanski, I. A., and M. E. Gilpin. *Metapopulation Biology: Ecology, Genetics and Evolution*. London, 1997. An edited volume on spatial population dynamics.

Hassell, M. P. *Spatial and Temporal Dynamics of Host–Parasitoid Interactions*. Oxford, 2000. A detailed monograph on the application of discrete and continuous time population models to predator–prey interactions.

Kingsland, S. E. *Modeling Nature: Episodes in the History of Population Ecology*. Chicago, 1995.

May, R. M. *Stability and Complexity in Model Ecosystems*. Princeton, 1974. A monograph of the extension of competition and predation models to multispecies interactions.

— MICHAEL B. BONSALL

POPULATION GENETICS.

See Frequency-Dependent Selection; Genetic Drift; Genetic Polymorphism; Hardy–Weinberg Equation; Hitchhiking; Inbreeding; Linkage; Mutation; Quantitative Genetics; Shifting Balance.

POPULATION TRENDS

The recovery of ancient population trends, periods of population growth or decline, is central to understanding human evolution. When and why human, or their ancestors’, populations grew, declined, or spread into new environments gives clues to natural selection in the past and to which factors were most important to the evolution of modern humans. Unfortunately, reconstructing ancient population trends has proven a daunting task. Studies of prehistoric demography rely on three primary sources of data: archaeology, archaeologically recovered human skeletons, and the genetic structure of living human populations. Each of these has great potential but presents obstacles that have yet to be surmounted.

Archaeology. Archaeology supplies vital data for reconstructing prehistoric population size, density, and population growth. To make these estimates, archaeologists rely on object counts. They count the number of structures or rooms at a site. They count the square footage of a site, or the number of diagnostic artifacts, especially pottery fragments, at a site. They multiply this count by some estimate of modal population per counted unit to turn it into an estimate of population size.

Some of the key problems inherent to reconstructing ancient demography from archaeological remains are (1) how to account for variation in the size of the material indicator; for example, what to do with room counts when rooms vary in size; (2) whether the indicator being counted varies functionally (e.g., sleeping rooms vs. storage rooms); (3) contemporaneity of units; for example, the number of pottery fragments in a garbage midden is affected by both the number of people throwing trash into it and how long it was used; (4) a lack of demonstrable associations between material items, whether they are hearths, rooms, or pottery fragments, and absolute population counts; (5) how to derive an estimate of modal population per entity; although investigators (e.g., Naroll, 1962) have long attempted to identify cross-cultural regularities in factors such as roofed-over space per person, ratios of individuals per entity are highly culture-bound; and, especially, (6) the assumption that the number of entities in the study universe is complete or that a representative sample of entities can be identified. This is a particularly difficult issue when archaeological sites are poorly preserved or difficult to recover.

Long-term trends, in particular those associated with large-scale changes such as the adoption of agriculture, are particularly difficult to assess reliably. Prehistoric lifestyle changes (as well as the age of sites) may affect archaeological recovery, or they may affect the relationship between population size and indicator frequency. Forager sites, especially in resource-rich environments, are generally more difficult to locate than agriculturalist sites. Foragers tend to live in smaller groups, to be more mobile, and to leave less material remains, whereas agriculturalists tend to be more sedentary and have more possessions. From a purely archaeological-recovery perspective, one would expect to find more evidence of younger agricultural populations than older foraging ones.

Paleodemography. The reconstruction of prehistoric populations from human skeletons is called paleodemography. Although osteologists have been using skeletal samples to make inferences about prehistoric demography for decades, paleodemography is still in its scientific infancy. Paleodemographers are currently struggling to improve their age and sex estimation techniques and to better understand the relationship between a living population and the characteristics of its dead. Two central issues in reconstructing basic demographic trends from skeletons are the accuracy and bias of age estimates themselves and the lack of knowledge of the "population at risk."

As it ages, the human skeleton undergoes a series of changes. Until an individual is about eighteen years old, these changes relate to growth and development. After that, changes relate primarily to degeneration. Osteologists have used patterns of age-correlated change in reference samples of known-age individuals to identify patterns of age-related change in the skeleton; these are then used to estimate the age of other skeletons that display the same age-related changes. There are two problems with this approach. First, individuals of all ages vary in how they show these age markers; for example, some people get their wisdom teeth when they are sixteen years old, some do not see them erupt until they are twenty-eight years old, if ever. Second, osteologists are able to observe more completely the range of variation in markers for ages that are well represented in their reference samples, but not in ages that are less well represented. This means, statistically, that osteologists are more likely to place an archaeologically recovered skeleton in a well-represented age category, and their reconstructions of whole cemetery samples will be biased toward the age pattern of the reference sample. This point was made by Bocquet-Appel and Masset in 1982, but it was not addressed seriously until the mid-1990s. Many osteologists have worked to ameliorate this problem (see *Advances in Paleodemography: Reconstructing Mortality from Skeletal Series*, 2001), but it has proven very difficult to solve.

These problems coalesce in arguments over ancient human (and ancestral) life spans. Several authors, including Trinkaus (1995), have argued that humans have only recently begun to reach ages of sixty years or more. They cite the general lack of older individuals in the archaeological record. However, this may be a result of archaeological conditions and age estimation bias rather than ancient human mortality patterns. Walker and colleagues (1988) have shown that older individuals' bones are less likely to survive archaeological processes. They conclude that this is most likely the result of changes in skeletal mass with increased age. Reference-sample bias (see above) makes the misidentification of older individuals likely.

Even if individual age estimates were perfect, cemetery samples would still present challenges. The distribution of deaths (by age) in a population results from two factors: the rates of death at various ages and the size or proportion of the living population at each given age (the population at risk). Paleodemographers would like to reconstruct those rates of death, but a cemetery sample does not provide the population at risk. Often, to get around this problem, paleodemographers have assumed that the living population was stationary (neither growing nor shrinking), but this can lead to incorrect reconstructions. In some archaeological cases, vital characteristics, such as whether it was growing or shrinking, could be provided by archaeology, but most of the time the living population that produced the cemetery is unknown.

Genetic Studies of Living Populations. Another source of information about prehistoric population trends is the distribution of genes in living peoples. To reconstruct effective population size, the number of breeding males and females a current population is descended from, population geneticists first must determine how many alleles exist for a specific gene. The means of estimating effective populations derives from the coalescence theory (Hudson, 1990). The premise of the coa-

A LATE PLEISTOCENE POPULATION BOTTLENECK

Stanley Ambrose (1998) has proposed an intriguing hypothesis to explain Late Pleistocene population bottlenecks and releases. Approximately 70,000 years ago, world temperature fell dramatically, possibly as a result of the massive eruption of the supervolcano Toba, in Sumatra. The Toba eruption was the largest known eruption of the Quaternary—the second largest known in 454 million years. It displaced 800 km^3 of rock as volcanic ash. By comparison, Tambora 5, the largest known historic eruption, displaced 20 km^3 (Mount St. Helens, in Washington State, displaced 0.2 km^3) in 1816, causing a year "without summer." Toba created a "volcanic winter" that may have lasted six years. Several data sources, including ice cores and prehistoric vegetation movements, indicate the thousand years after the Toba eruption were among the coldest of the Late Quaternary. Ambrose (1998) argues that Toba's volcanic winter could have decimated human populations outside of isolated tropical refugia. The largest tropical refugia would have been found in equatorial Africa. Based on Ambrose's scenario, the high genetic diversity observed in modern African populations would be the result of higher population survival through the Toba bottleneck than found in other parts of the world. At the same time, Ambrose's scenario indicates that contemporary populations grew from a very small founding population of survivors of the Toba nuclear winter just seventy thousand years ago. Ambrose also argues that world populations at the end of Toba's volcanic winter may have been small enough for founder effects, genetic drift, and local adaptation to have caused very rapid population differentiation. Thus, he argues, contemporary human "races" would have differentiated only in the last seventy thousand years.

—RICHARD PAINE

lescence theory is that more diversity between two alleles implies, first, that more time has passed since they split from a common ancestor, and, second, that a larger population was needed to maintain the diversity through time. A typical study includes comparisons of multiple pairs of alleles.

The timing of ancient population events uncovered by genetic coalescence studies has been the subject of considerable debate. Genetic coalescence studies are only as accurate as the investigator's estimate of the mutation rate. Uncertainty in mutation rate estimates means that estimates of time since populations split from a common ancestor and estimates of effective population size must cover wide ranges.

Natural selection itself may have confounding effects on studies of population growth based on genetic distribution (Harpending, and Rogers, 2000). A favored trait may expand quickly, even when the population is not really growing. If researchers focus on carriers of a favored allele, their results will appear as population growth; the two are indistinguishable at a single locus.

Prehistoric Population Growth. Contemporary research is focused on two possible population expansions: one in the later Pleistocene epoch (1.7 million years ago to 10,000 years ago) associated with anatomically modern humans, and the second associated with the adoption of agriculture. Whether the spread of anatomically modern humans corresponds with the spread of more efficient Paleolithic technology is also a central question.

Genetic studies of extant populations suggest that human populations are descended from a small effective population, in the range of ten thousand (Harpending and Rogers, 2000). This could mean that human populations were small (in the range of twenty thousand people) and nearly stationary until very recently in evolutionary time. The point of takeoff is often argued to have been near the adoption of agriculture in the Early Holocene, about ten thousand years ago. Alternatively, human populations may have passed through a bottleneck, a period of small or reduced population size, at some point in the Late Pleistocene, probably sometime between 30,000 and 130,000 years ago (see Vignette for a more specific argument). Estimates of how small the human population got during this bottleneck vary from a low of five hundred to three thousand females (Harpending et al., 1993) to a high of ten thousand females (Takahara et al., 1995). There appears to be widespread consensus that the highest number of reproductive females when populations were at their smallest would be about five thousand. This number is supported by several lines of evidence, including mitochondrial DNA (mtDNA), neutral gene studies, and the genes of the major histocompatibility complex (MHC). Harpending (quoted in Gibbon, 1993) suggests that humans survived a period when they were as endangered as "pygmy chimpanzees or mountain gorillas" are today. Others argue that ten thousand is too many to be considered a bottleneck.

The bottleneck model is supported by analyses of mtDNA, Y chromosome, and short tandem repeat (STR) loci. Other genes seem to suggest a long history of constant population size (Rogers, 2001). The problem is how to differentiate past growth from natural selection (see above). Rogers (2001) points out, interestingly, that the genes that imply an expansion are those assumed to be least effected by natural selection. Rogers also argues that the broad simultaneity of expansions, identified at different loci, tends to fit the hypothesis of population expansion better than that of natural selection.

Presumably, archaeology could provide an independent test by documenting past population expansions, even if it could not provide accurate estimates of overall size. Stiner and colleagues (1999) used evidence of hunting shifts from easily caught species, such as tortoises, to more elusive species, such as partridges and rabbits, as an indicator of Upper Paleolithic population growth in the eastern and northern Mediterranean regions. Their study also suggests populations were exceptionally small during the Middle Paleolithic There is also archaeological evidence of population growth in Europe around forty thousand years ago. However, most of this archaeological evidence is based on the spread of tool types, which, like favored genetic traits, could spread through a population without the overall population actually growing. Archaeologists (e.g., Hassan, 1981) have attempted broad synthetic studies of ancient population trends, but these have been forced to rely heavily on assumptions based on carrying capacity estimates for contemporary foragers. They have not been successful in documenting with confidence Pleistocene population growth patterns.

Changes in Demographic Structure Associated with Settlement and Agriculture. Numerous osteological studies (many summarized in Acsadi and Nemeskeri, 1970; Cohen and Armelagos, 1984) have documented changes following the adoption of agriculture. These changes include lower mean age at death, reduced stature, increases in the frequency of both specific and nonspecific lesions, and increases in frequencies of stress markers.

Currently, the best interpretation of lower mean age at death is that growth rates probably increased. This lends some support to the second population expansion, associated with the adoption of agriculture. Interpreting lesion frequencies is difficult, but the most widespread interpretation of the increase in lesion frequencies (see, e.g., Cohen and Armelagos, 1984) has been that these changes reflect general increases in the level of infection and declines in overall health with the advent of densely settled agricultural communities. However, as Wood and colleagues (1992) have demonstrated, the link between lesion frequencies in skeletal samples and health conditions in living populations is uncertain. Wood and colleagues have shown that higher lesion frequencies may reflect increased survivorship through disease episodes rather than increased rates of disease, indicating improving (vs. declining) health conditions.

Unfortunately, the basic conclusion is that we are currently unable to say very much, with confidence, about population trends before the beginning of historical demographic records. Both archaeology and paleodemography seem to indicate significant growth in recent (within the last five thousand years) history, but that is not much of a surprise. Methodological advances, particularly in genetic techniques and paleodemography, have the potential to change this picture in the relatively near future.

[*See also* Archaeological Inference; Human Sociobiology and Behavior, *article on* Human Sociobiology; Life History Theory, *article on* Human Life Histories.]

BIBLIOGRAPHY

Acsadi, G., and J. Nemeskeri. *History of Human Life Span and Mortality*. Budapest, 1970. Classic study of ancient population trends. Interpretations may be dated, but this book has had enormous impact on our view of ancient populations.

Ambrose, S. "Late Pleistocene Human Population Bottlenecks, Volcanic Winter, and Differentiation of Modern Humans." *Journal of Human Evolution* 34 (1998): 623–651.

Bocquet-Appel, J., and C. Masset. "Farewell to Paleodemography." *Journal of Human Evolution* 11 (1982): 321–333.

Cohen, M. N., and G. R. Armelagos, eds. *Paleopathology at the Origins of Agriculture*. Orlando, Fla., 1984. Large compendium of skeletal studies focused on the shift to agriculture. Appeared just as paleodemography was beginning to look more carefully at the nature of skeletal data. Conclusions reached with confidence in 1984 would be treated with much more caution today.

Gibbons, A. "Pleistocene Population Explosions." *Science* 262 (1993): 27–28.

Gibbons, A. "Mystery of Humanity's Missing Mutations." *Science* 265 (1995): 35–36. Gibbons's descriptions are clear and concise, written for a scientifically educated, but not specialized, audience.

Harpending, H., and A. Rogers. "Genetic Perspectives on Human Origins and Differentiation." *Annual Review of Genomics and Human Genetics* 1 (2000): 361–385. Up-to-date review of genetic evidence for ancient population trends. Written primarily for a professional audience.

Harpending, H. C., S. T. Sherry, A. L. Rogers, and M. Stoneking. "The Genetic Structure of Ancient Human Populations." *Current Anthropology* 34 (1993): 483–496.

Hassan, F. A. *Demographic Archaeology*. New York, 1981. Wide-ranging synthesis of methods and interpretation of prehistoric demography. Somewhat dated.

Haupa, R., and J. Vaupel, eds. *Advances in Paleodemography: Reconstructing Mortality from Skeletal Series*. Cambridge Studies in Biological and Evolutionary Anthropology. London, 2001.

Hudson, R. R. "Gene Genealogies and the Coalescent Process." *Oxford Surveys of Evolutionary Biology* 7 (1990): 1–44.

Naroll, R. "Floor Area and Settlement Population." *American Antiquity* 27 (1962): 587–589.

Paine, R. R., ed. *Integrating Archaeological Demography: Multidisciplinary Approaches to Prehistoric Population*. Occasional Papers No. 24. Carbondale, Ill., 1997. Contains a range of studies of ancient population and trends based on settlement archaeology, paleodemography, and genetic data (some are quite technical). J. E. Buikstra's closing chapter is a particularly useful summary of current issues in paleodemography.

Rogers, A. R. "Order Emerging from Chaos in Evolutionary Genetics." *Proceedings of the National Academy of Sciences USA* 98 (2001): 779–780.

Stiner, M. C., N. D. Munroe, T. A. Surovell, E. Tchernov, and O. Bar-Yosef. "Paleolithic Population Growth Pulses Evidenced by Small Animal Exploitation." *Science* 283 (1999): 190–194.

Takahara, N., Y. Satta, and J. Klein. "Divergence Time and Population Size in the Lineage Leading to Modern Humans." *Theoretical and Population Biology* 48 (1995): 198–221.

Trinkaus, E. "Neanderthal Mortality Patterns." *Journal of Archaeological Science* 22 (1995): 121–142.
Walker, P. L., J. Johnson, and P. Lambert. "Age and Sex Biases in the Preservation of Human Skeletal Remains." *American Journal of Physical Anthropology* 76 (1988): 183–188.
Wood, J. W., G. R. Milner, H. C. Harpending, and K. M. Weiss. "The Osteological Paradox: Problems of Inferring Prehistoric Health from Skeletal Samples." *Current Anthropology* 33 (1992): 343–358. An important critique of paleodemography. Written primarily for a professional audience.

— Richard Paine

PREFORMATION. *See* Epigenesis and Preformationism.

PRICE, GEORGE

George Price, 1922–1975, trained as a chemist, contributed to the Manhattan bomb project during World War II, and worked at various biology and engineering research positions until the mid-1960s. Throughout this period, he read widely, corresponded with many scientists on various topics, and wrote a few papers, including a scientific analysis of paranormal phenomena. Price, born in the United States, left for England in 1967 and took up the study of evolutionary genetics.

Price made three fundamental contributions to population genetics and social behavior. His first contribution was a mathematical expression of natural selection known as the Price equation. This equation shows how a character will evolve over time, depending on the character's association with fitness and the fidelity with which the character is transmitted to offspring. W. D. Hamilton (1996) used the Price equation to develop his theory of biological altruism based on kin selection. [*See* Kin Selection.] Hamilton showed that an individual actor may behave in an altruistic way toward relatives. Such altruism evolves because relatives who benefit from altruism produce offspring with genes and thus characters that are similar to the actor's. Before he met Price, Hamilton had formulated his theory in terms of Wright's correlation coefficient of relatedness between kin. [*See* Wright, Sewall.] Price's insight, embodied in the Price equation, led Hamilton to the correct measure of relatedness for kin selection, the regression coefficient of recipient on altruist. By adopting this new measure of relatedness and by using the Price equation, Hamilton developed a far more powerful expression of his altruism theory.

In his second contribution, Price developed a model for how an animal behaves when in conflict with a neighbor. The puzzle is why animals often settle fights in a ritualized way rather than inflicting serious or deadly wounds. For example, male deer often fight furiously by crashing antlers in head-on battle, but they refrain from attacking when an opponent turns away and exposes an unprotected side. Price solved this problem by recognizing that fighting is a strategic game in which the best strategy of each individual depends on the strategies of neighbors. If everyone else escalates and fights to the death, it would be best to value life over the gains of battle and yield readily in a fight. By contrast, if all opponents yield quickly against escalating aggression, then aggression would lead to victory in all battles with little risk of a serious fight. Populations therefore settle into mixtures of aggressive and yielding interactions, where frequencies of alternative strategies depend on the balance between the danger of escalated fights and the potential gain in each battle. This idea of strategic equilibrium led John Maynard Smith and George Price (1973) to develop the theory of evolutionarily stable strategies, a central theoretical concept of modern evolutionary theory. This concept has been applied to diverse problems, including the sex ratios of populations and the virulence of parasites.

Price's third contribution solved the mystery of R. A. Fisher's fundamental theorem of natural selection, perhaps the most widely quoted theorem in evolutionary genetics. Fisher claimed in 1930 that his theorem held "the supreme position among the biological sciences" and compared it with the second law of thermodynamics. Yet for forty-two years no one could understand what the theorem was about, until Price in 1972 explained the theorem and its peculiar logic. The usual interpretation of the theorem is that the rate of increase in the adaptation (mean fitness) of a population is equal to the genetic variance in fitness of that population. However, this interpretation requires special assumptions about mating, competition, and other factors. Price solved the contradiction between Fisher's claim for generality and the limited scope of the usual interpretation. Price showed that Fisher partitioned the total change in mean fitness into two components. The first is the change caused directly by natural selection, and it is this component that increases at a rate equal to the genetic variance in fitness. The second component accounts for how the changing environment, including changes caused by natural selection, often causes a reduction in fitness. Changing environment includes enhanced adaptation of parasites, predators, and competing individuals from the same species. Price showed how these parts combine into a theorem of universal scope, as Fisher originally claimed.

[*See also* Altruism; Fundamental Theorem of Natural Selection; Game Theory; Hamilton, William D.]

BIBLIOGRAPHY

Frank, S. A. "George Price's Contributions to Evolutionary Genetics." *Journal of Theoretical Biology* 175 (1995): 373–388. Provides additional history and technical details about Price's three contributions to evolutionary theory.
Hamilton, W. D. *Narrow Roads of Gene Land.* Oxford, 1996. In-

cludes biographical information about Price and his key role in the development of Hamilton's own theory of altruism by kin selection.

Maynard Smith, J., and G. R. Price. "The Logic of Animal Conflict." *Nature* 246 (1973): 15–18. Classic article that introduces the strategic theory of conflict resolution and the evolutionarily stable strategy concept.

Price, G. R. "Science and the Supernatural." *Science* 122 (1955): 359–367.

Price, G. R. "Selection and Covariance." *Nature* 227 (1970): 520–521. The original, rather terse article presenting what is now known as the Price equation.

Price, G. R. "Fisher's 'Fundamental Theorem' Made Clear." *Annals of Human Genetics* 36 (1972): 129–140. Clarification of Fisher's theorem, including historical commentary.

Schwartz, J. "Death of an Altruist: Was the Man Who Found the Selfless Gene Too Good for This World?" *Lingua Franca* 10.5 (2000): 51–61. Good biographical summary of Price's life and work. Describes his wide interests and contributions to many different scientific fields.

— Steven A. Frank

PRIMATE COMMUNICATION. *See* Language.

PRIMATES

[*This entry comprises three articles:*

Primate Classification and Phylogeny
Primate Societies and Social Life
Primate Biogeography

The first article discusses the taxonomic definition and defining traits of primates; the companion articles focus on primate development, behavior, and geographic and ecological distribution. For a related discussion, see the Overview Essay *on* Culture in Chimpanzees *at the beginning of Volume 1.*]

Primate Classification and Phylogeny

There are approximately 300 species of living primates placed in 60 genera. Another 500 species and 200 genera are known from the fossil record. Defining primates has long been a difficult task because there is no clear consensus regarding which other order of mammals is the sister group of living primates. Flying lemurs (Dermoptera), tree shrews (Scandentia), and megabats (Megachiroptera) are the usual suspects. In addition, there are several extinct groups of mammals that are either basal primates or primate avatars and hence tend to blur any clear boundaries. However, if we restrict our definition to extant primates and their collateral relatives, the order can be identified reliably on the basis of their common possession of a postorbital bar on the skull, an auditory bulla surrounding the middle ear that is composed of the petrosal bone, and digits that end in nails rather than claws—usually on both hands and feet, but always on the hallux (or big toe).

Within the order, the major division is into two suborders—strepsirhines (lemurs and lorises) and haplorhines (tarsiers and anthropoids). This is a cladistic division and differs from the traditional classification into anthropoids and prosimians (the latter include both strepsirhines and tarsiers). The traditional taxonomy reflects the great difference between anthropoids and both strepsirhines and tarsiers. However, most biologists pre-

THE MOSAIC NATURE OF HUMAN ORIGINS

Humans, like all other organisms, did not appear full-blown in their modern form a few thousand years ago. Rather, different features of our anatomy, physiology, and behavior have evolved in a mosaic pattern over many millions of years. The hierarchical pattern of evolution represented in primate phylogeny and taxonomy reflects both the number of aspects of our anatomy and behavior that we share with other species and the sequence in which these shared features have evolved.

For example, our broad fingers, tipped with nails rather than claws, are a feature we share with (almost) all primates and first appeared in our lineage over 50 million years ago. A reduced olfactory region and increased reliance on vision are features we share with other anthropoids and tarsiers and probably date to about 40 million years ago. Our lack of a tail, mobile upper limb joints, and short lower backs are features we share only with apes and evolved only about 20 million years ago. Our short pelvis is a feature that is not shared with any other primate, but only with other now extinct hominids, and solid fossil evidence dates it to only 3.5 million years ago. Our prominent chin is an even more exclusive feature, shared solely with members of our own species, *Homo sapiens*; it dates from only about 100,000 years ago.

Although they are more difficult to read directly from the fossil record, we can also reconstruct the history of many other aspects of human behavior and physiology based on their distribution among primates and other mammals. Thus, the basic features of live birth and lactation are aspects of behavior we share with all placental mammals and have been in our lineage for over 100 million years. Gregarious sociality, in contrast with solitary foraging, is a feature we share with most primates and has probably been in our lineage for over 50 million years. Trichromatic color vision and monthly ovulatory cycles are features we share only with Old World monkeys and apes, and these have only been in our lineage for 25 million years.

—John Fleagle

AN APE BY ANY OTHER NAME

With the burgeoning genetic information on primate systematics that has become available in the past several decades, the phylogeny of apes is now well established at the generic level. There is overwhelming evidence that the closest living relatives (sister taxa) of humans (*Homo*) are the chimpanzees (*Pan*), followed by gorillas (*Gorilla*), orangutans (*Pongo*), and then gibbons or lesser apes (*Hylobates*). Although there is near universal agreement that this is the correct phylogeny, there is much less agreement on the best taxonomic arrangement for the superfamily Hominoidea, which includes all (living and fossil) apes and humans.

In the traditional taxonomy, which was established before it was widely appreciated that humans are so closely related to chimpanzees, humans (and our numerous extinct humanlike relatives) have been placed in a separate family, Hominidae, colloquially referred to as hominids. All great apes are placed in a common family, Pongidae or pongids. With our current understanding of ape and human phylogeny, this is a gradistic classification that conveys no relationship about which of the great apes is more closely related to humans. Rather, it separates out humans (and our extinct relatives) from all great apes and lumps all great apes in a single family largely on the basis of their lack of human features. Rather than being classified according to their phyletic relationships, great apes are grouped as "not hominids." The main advantages of this scheme are that it is a traditional classification that people are familiar with, and it provides a familiar and appropriate name—Hominidae or hominids—for humans and our extinct relatives. It is also unaffected at a gross level by alternative views on the phylogeny of fossil apes.

However, most biologists today feel that taxonomy should reflect phylogeny as much as possible, and so many authorities have reallocated the nomenclature for great apes and humans accordingly. Under this scheme, the family Hominidae contains all great apes and humans and their fossil relatives. Hominidae is divided into two subfamilies: Ponginae for the orangutan, and Homininae for African apes and humans. The subfamily: Homininae can be further divided into tribes—Gorillini for the gorilla and Hominini for chimps and humans. The advantage of this approach is that it is a phylogenetic classification that reflects the known phylogeny of the living apes such that all descendants of a common ancestor are placed in the same taxonomic group. The disadvantages are that the traditional name for humans and our extinct relatives, Hominidae or hominids, now means something different; there is no commonly accepted name for humans and our extinct relatives; and the taxonomy is very unstable. The genera included in the subfamilies and tribes differ from authority to authority according to their views on the phylogeny of fossil apes, a situation that is much more poorly understood than the phylogeny of living apes, whether or not fossils are included.

—JOHN FLEAGLE

fer cladistic schemes that reflect actual phylogenetic relationships as much as possible. The cladistic dichotomy reflects the phylogenetic relationship between tarsiers and anthropoids. This strepsirhine/haplorine division is evident in the morphology of living taxa, and it receives mixed support from molecular systematics, but it is more muddled when the fossil record in considered. Strepsirhines include the lemurs of Madagascar and the lorises and galagos of Africa and Asia. These groups are united by their common possession of a dental tooth comb and a flaring talus. Although the strepsirhines are currently restricted to these two tropical groups, in the early part of the Cenozoic basal strepsirhines were widespread and abundant in North America, Europe, Africa, and Asia.

The lemuriformes of Madagascar are a diverse radiation and are the largest mammals on the island. There are 16 genera and 45 species of living Malagasy strepsirhines and another 6 genera that recently became extinct, many within the past millennium. Malagasy strepsirhines are divided into five families. The bizarre aye-aye (*Daubentonia*) is placed in its own family and is the sister taxon to all other Malagasy families. The Indriids, characterized by a tooth comb with fewer teeth and a reduced number of premolars, are the most diverse Malagasy family. Although there are only three living genera, there are six recently extinct genera, all substantially larger than any living Malagasy primate. Megaladapidae is an odd family, containing two very different genera, the giant extinct taxon *Megaladapis* and the small, extant *Lepilemur.* Lemurids are the common lemurs, including the speciose brown lemurs (*Eulemur*), the striking ruffed lemurs (*Varecia*), the ringtail lemur (*Lemur*), and the three bamboo lemurs (*Hapalemur*). The smallest and perhaps the most primitive of the Malagasy families are the cheirogaleids. Morphological studies group cheirogaleids with lorises and galagos because of an unusual pattern of cranial blood supply, but molecular studies place them well within the Malagasy group. There are five small, nocturnal genera: the tiny mouse lemurs (*Microcebus*), the dwarf lemurs (*Cheirogaleus*), Coquerel's dwarf lemur (*Mirza*), the fork-marked lemur (*Phaner*), and the poorly known, nearly extinct hairy-eared dwarf lemur (*Allocebus*).

The Lorisiformes is made up of two very distinct families, lorises and the galagos, joined by common features of their skull and reproductive system. They have a fossil record dating back approximately 20 million years. The galagos are characterized by long hindlimbs, large ears, and long tails. They are restricted to Africa. The systematics of this group is in constant flux because new taxa are constantly being described and the generic affinities of many species are debated. The most widely accepted systematics arrangement recognizes four genera: *Otolemur* for the large galagos of eastern and southern Africa, *Galago* for the middle-sized species, *Galagoides* for the small dwarf galagos, and *Euoticus* for the needle-clawed galagos. In contrast, the lorises are characterized by reduced ears, arms and legs of equal length, and a reduced tail. There are two genera of lorises in Africa and two in Asia. The Asian lorises are the slow loris (*Nycticebus*) from Southeast Asia, and the slender loris (*Loris*) from southern India and Sri Lanka. The two African genera are the golden potto (*Arctocebus*) and the potto (*Perodicticus*).

The second major division of primates, and the one that includes humans, is the haplorhines. Haplorhines are characterized by several derived features that reflect a reorganization of the senses, including reduced olfactory region, a posterior septum behind the orbit, a retinal fovea, and loss of the tapetum lucidum in the retina. They are also characterized by hemochorial placentation. There are two groups of living haplorhines: the tiny tarsiers of Southeast Asia, and the anthropoids, or higher primates, found mainly in the Neotropics, Africa, and Asia.

In earlier epochs there were numerous small tarsier-like primates in North America, Europe, and Asia. There is debate over whether the living *Tarsius* is more closely related to these extinct basal haplohines or to the living anthropoids.

The anthropoids, or higher primates, are one of the most distinctive clades among living primates. Gradistic classifications generally distinguish the derived anthropoids from the more primitive prosimians (including both strepsirhines and tarsiers). As a group, living anthropoids are diurnal, visual, manipulative, social animals. They are characterized by a suite of derived dental and cranial characters, including a fused mandibular symphysis, complete postorbital closure, a fused metopic suture, vertical incisors and semimolariform premolars, and an anterior accessory cavity of the middle ear. However, the fossil record shows that this suite of features was acquired in a mosaic, piecemeal fashion.

The major division among anthropoids is between the platyrrhines of the Neotropics and the catarrhines of Africa and Asia. The New World platyrrhines are very difficult to define as a group because many of the features that distinguish them from catarrhines are primitive anthropoid retentions, such as three premolars in each quadrant and absence of an external ear tube or of molar hypoconids. The living fauna of New World monkeys is a diverse radiation of 16 genera and 50 species with a fossil record extending back about 25 million years. The familial taxonomy of New World monkeys has been in flux for several decades, but it now seems relatively well resolved on the basis of very consistent molecular results. The most commonly accepted arrangement divides the 16 genera among five subfamilies grouped in two families. The family Atelidae contains two subfamilies of large monkeys: the prehensile-tailed atelines and the seed-eating pitheciines. The atelines include the howling monkeys (*Alouatta*), the woolly monkeys (*Lagothrix*), the spider monkeys (*Ateles*), and the wooly spider monkey or murique (*Brachyteles*). The pitheciines are a distinctive group of seed-eating monkeys characterized by their robust canine teeth and procumbent incisors—the sakis (*Pithecia*), the bearded sakis (*Chiropotes*), and the uakaries (*Cacajao*). This group also includes the titi monkeys (*Callicebus*).

The second family of platyrrhines is the Cebidae, which contains three subfamilies of smaller monkeys. The most distinctive are the callitrichines, characterized by their small size, reduced dentition, clawed digits, and reproductive twinning. There are five genera: the marmosets (*Callithrix*) and the closely related pygmy marmoset (*Cebuella*); the diverse and speciose tamarins (*Saguinus*); the lion tamarin (*Leontopithecus*); and Goeldi's monkey (*Callimico*). Morphological studies have always placed *Callimico* at the periphery of this group or even in its own family, but molecular studies link it closely with *Callithrix*. The capuchin monkeys (*Cebus*) and the squirrel monkeys (*Saimiri*) make up the Cebinae. Although the night monkeys (*Aotus*) are traditionally grouped with the titi monkeys (*Callicebus*), molecular studies always group *Aotus* with the Cebidae.

The Old World higher primates, the catarrhines, are divided into two very distinct superfamilies: the Old World monkeys (Cercopithecoidea), and the apes and humans (Hominoidea). Both have fossil records going back about 20 million years, but the Old World monkeys are the most derived. There is a single living family

(Cercopithecidae) divided into two subfamilies—cercopithecines and colobines. The cercopithecines are generally frugivorous monkeys with cheek pouches. Most are from Africa. They include the mangabeys (*Cercocebus* and *Lophocebus*), baboons (*Papio*), mandrills and drills (*Mandrillus*), guenons (*Cercopithecus*), vervets (*Chlorocebus*), swamp monkeys (*Allenopithecus*), Patas monkeys (*Erythrocebus*), and talapoins (*Miopithecus*). The only Asian cercopithecine is the widespread and speciose *Macaca*. Colobine monkeys are leaf and seed eaters with three genera in Africa (*Colobus*, *Procolobus*, and *Piliocolobus*) and seven in Asia (*Semnopithecus*, *Trachypithecus*, *Presbytis*, *Nasalis*, *Simias*, *Pygathrix*, and *Rhinopithecus*).

The living hominoids are much less diverse than either the cercopithecoids or the extinct hominoids during most of the past 20 million years. There are only five living genera: the gibbons (*Hylobates*) and orangutans (*Pongo*) from Asia; gorillas (*Gorilla*), chimpanzees and bonobos (*Pan*) from Africa; and the cosmopolitan humans (*Homo*). The phylogenetic relationships among living hominoids is well established. Humans are most closely related to chimpanzees and bonobos, with gorillas, orangutans, and gibbons as successive branches. However, the best taxonomy for this group is less clear, in part because it must also accommodate a large number of extinct human ancestors and collaterals. A strictly phylogenetic framework recognizes two families—hylobatids (gibbons) and hominids (great apes and humans). Within hominids, the orangutans are placed in the Ponginae, with African apes and humans in Homininae. The less rigorous but more common approach recognizes three families within Hominoidea: Hylobatidae (gibbons and siamangs), Hominidae (humans and their extinct ancestors), and a paraphyletic Pongidae, including orangutans, gorillas, chimpanzees.

BIBLIOGRAPHY

Fleagle, G. *Primate Adaptation and Evolution.* New York, 1999. A general source for information of primate evolution and naturalistic behavior of living primates. Organized by phylogeny, with extensive discussions of debated issues.

Goodman, M., C. A. Porter, J. Czelusniak, S. L. Page, H. Schneider, J. Shoshani, G. Gunnell, and C. P. Groves. "Toward a Phylogenetic Classification of Primates Based on DNA Evidence Complemented by Fossil Evidence." *Molecular Phylogenetics and Evolution* 9 (1998): 585–598. A somewhat idiosyncratic attempt at a global classification of primates.

Groves, C. *Primate Taxonomy.* Washington, D.C., 2001. The latest summary of primate taxonomy, with excellent introductory chapters on the goals and methods of taxonomy.

Groves, C. P. "Why Taxonomic Stability Is a Bad Idea." *Evolutionary Anthropology* 10 (2001). A provocative essay emphasizing that taxonomy must always change to reflect our growing knowledge of biology.

Nowak, R. M. *Walker's Primates of the World.* Baltimore, 1999. A review of primate genera with summaries of basic information.

Purvis, A. "A Composite Estimate of Primate Phylogeny." *Philosophical Transactions of the Royal Society, London, B* 328 (1995): 405–421. A summarizing phylogeny of primates, widely used in comparative analyses.

Rowe, N. *The Pictorial Guide to the Living Primates.* East Hampton, 1996. A beautifully illustrated and authoritative guide to primate taxonomy, with photos and summaries of ecology and morphology each species.

Schneider, H. I., I. Sampaio, M. L. Harada, C. M. L. Baroso, M. P. C. Schneider, J. Czelusniak, and M. Goodman. "Molecular Phylogeny of the New World Monkeys (Platyrrhini, Primates) Based on Two Unlinked Nuclear Genes: IRBP Intron1 and Epsilono-Globin Sequences." *American Journal of Physical Anthropology* 100 (1996): 153–179. The broadest treatment of the molecular systematics of New World monkeys.

— JOHN FLEAGLE

Primate Societies and Social Life

Primates are primarily arboreal, relatively large-brained mammals. They cover a wide range of body sizes, but for their size primates have young that are relatively precocial and yet develop slowly and live long lives. The major social distinction within the order is that between infant cachers and infant carriers. Infant cachers are overwhelmingly nocturnal, and all are prosimians. These small primates tend to forage alone, although they clearly live in social networks. Primates that carry their young, prosimian or anthropoid, are usually larger. The nocturnal ones generally live in pairs, whereas diurnal ones predominantly live in larger social groups containing adults and immatures of both sexes. In some of the latter, animals form relatively complex social relationships and show extensive sexual behavior. This set of features is virtually unique among mammals.

Beyond this broad division, the primate order shows an enormous diversity of social systems. Explaining this diversity is a monumental task, and this article will have to paint with a very broad brush. The major organizing framework is the socioecological paradigm developed over the past few decades for birds and mammals, including primates. It is based on a priori notions about fundamental sex differences in mammals, where males, but not females, derive major fitness benefits from gaining access to as many mates as possible. Thus, although variation in male fitness is mainly due to variation in success in competition for access to fertile mates, variation in female fitness is generally linked to success in raising surviving young, and therefore in gaining access to food and shelter and in evading risks in the environment. It follows from this that the associations and social relationships formed by animals of the two sexes should reflect these different priorities. Hence, the socioecological paradigm postulates that female associations and relationships are a response to ecological factors, whereas the social behavior among males largely reflects the distribution in space and time of mating opportunities.

As we shall see, this well established and experimentally validated approach explains much of the variation in social life among primates, but it does not account for the fact that most primates, unlike most other mammals, live in stable groups containing adults of both sexes. These and various other aspects of male–female interactions are best explained by intersexual conflict, which arises from the different mating preferences of males and females. Where males win this conflict, they can force females to mate or even kill their dependent offspring to improve their own mating chances, a phenomenon known as sexual coercion.

Females in Groups. Our first focus is on females and their associations and relationships. Nocturnal primates almost invariably forage alone or in pairs, but may sleep in larger aggregations. They have nonetheless various kinds of social organization. In most, maturing daughters settle near their place of birth, sometimes inheriting part of their mother's home range, so that related females tend to cluster. Diurnal primates almost invariably live in pairs or larger groups. The size of the groups varies, and much work has examined correlates of variation in group size. Group size is smaller among the smallest (<2 kg) and largest (>25 kg) primates; groups of folivores tend to be smaller than those of frugivores of the same body size; forest-living primates traveling on the ground tend to live in larger groups than arboreal ones; and those living in open, treeless habitats tend to form the largest groups. Groups tend to be cohesive, so that all members can be seen together, but some of the largest species form parties of ever-changing composition and live in so-called fission-fusion societies.

For most species of diurnal primates, mean group size can be seen as a compromise between pressures toward larger size, in particular, predation risk and sometimes between-group competition, and pressures toward smaller size, in particular, competition for food among group members. We can observe only species with those combinations of predation risk (often linked to body size) and sensitivity to competition (often linked to dietary requirements) that have produced viable ranges of group sizes in nature.

Several comparative studies have shown that primate group sizes, in terms of the number of females per group, are greater when risk of predation is higher. Risk of predation is measured indirectly through a proxy measure that includes the primate's body size, habitat, and geographic region to estimate the abundance and size of predators in their natural habitats. (Observed rates of predation are not useful for such comparisons because they reflect the risk of predation after the hypothesized adaptive response, here group size, is in place.) A relatively easy way to increase effective group size is to associate with another species whenever that is ecologically possible, especially since the others' vigilance may complement that of the own species. Indeed, forest monkeys often form associations with monkeys of other species in regions where large monkey-eating eagles occur (most of the Neotropics and Africa), but do not do so where the eagles are absent (most of tropical Asia, smaller areas elsewhere). More detailed intraspecific comparisons that attempt to control for habitat quality also show that on islands without large cats, monkeys tend to live in smaller and usually more dispersed groups; they also tend to be more terrestrial. Critical evidence linking survival to group size is still

LIFE HISTORY AND SOCIAL BEHAVIOR

Social systems vary among lineages. For instance, most primate groups show stable male-female association, whereas most ungulate groups do not. Likewise, communal breeding is rare among primates, but common among carnivores. Clearly, ecological factors alone cannot explain such differences.

A major missing ingredient is life history. Life-history characters are tightly inter-correlated across species, creating a gradient from fast (early maturation, large litters, short lives, etc.) to slow life history. Among mammals, primates have slow life histories; and among primates, great apes have the slowest life history.

Slow development allows plenty of scope for learning. It also increases birth spacing, and thus vulnerability to infanticide by males, which may have favored the evolution of various social counter-strategies (e.g., male-female association, female breeding dispersal, small group size, multi-male groups). Potentially long life spans and group living also make cooperative relationships especially important as the same other individuals can be either rivals or allies. Many of these consequences of slow life history may favor cognitive adaptations; indeed, enhanced cognition and slow life history show correlated evolution among mammals.

Not surprisingly, then, primates are among the most cognitively advanced mammals; and among primates, great apes show evidence of cognitive abilities well beyond the reach of most other primates, in both ecology (e.g., complex extractive foraging, routine tool use), and the social realm (opportunistic coalitionary support, triadic reconciliation, imitation, etc.). Humans are great apes: many of our outstanding features are elaborations of traits shared by other hominoids.

—CAREL VAN SCHAIK

missing, although solitaries tend to have dramatically higher mortality.

There are also numerous behavioral indications for the role of grouping in predation avoidance. Spontaneous or experimental encounters with predators tend to make groups more cohesive. Forest primates crossing open terrain between forest patches often do so in remarkably tight groups. Smaller groups of savanna monkeys often do not stray as far from refuges, such as trees or cliffs, as larger groups do; small groups of forest monkeys tend to stay away from the forest floor where cats may lurk, or from large open tree crowns where eagles might strike.

Competition over access to resources will inevitably have both a contest component, in which dominance decides access, and a scramble component, reflecting the impact of group size. Favored food that is found in clear clumps that are small relative to the number of individuals trying to gain access, such as a small fruit tree, is a contestable resource. But even when access to all resources is perfectly contestable, dominants are forced to share somewhat or else they would find themselves alone very soon, losing the benefits of group living. As groups become larger, they usually have to travel more each day, and above a certain size, net food intake will begin to decline, and reproduction suffers. The steepness of the rise in daily energy expenditure is a good index for the cost of group living, and generally populations showing a steeper cost curve live in smaller groups.

Overall, then, the balance between competition and predation explains group size tendencies of different species reasonably well. Some important exceptions occur, however. First, small neotropical primates, mainly belonging to the family Callithrichidae, form small groups in which only one female breeds and the rest help in rearing the young, much as in social carnivores. Other adult females, if present, show reproductive inhibition. There may even be multiple adult males, all of whom help to carry and sometimes to provision the young. In species with such helper systems, infants develop much faster than in all other primates.

Another exception is that folivorous primates often seem to live in smaller groups than expected. Examples include many langurs (*Presbytis*) and some howlers (*Alouatta*); there seems to be little competition within these groups. It is likely that these groups are small (up to four females) because larger groups make more attractive takeover targets by outside males, and such takeovers are often accompanied by infanticide by the incoming male (see below). Female fitness may be greater when they stay in smaller groups, despite somewhat higher loss of immatures to predation.

Animals that live together in stable groups form social relationships, the nature of which depends on the kind of feeding competition among them. In primate groups, if competition is generally weak or mainly through scramble, females form neither dominance relations nor social bonds, but largely coexist peacefully. If competition is strong and has a significant contest component, dominance relationships can be recognized: female dyads show a decided dominant-subordinate relationship, and a linear dominance hierarchy is usually apparent. Although occasional exceptions occur, a female higher in the dominance hierarchy generally has increased lifetime reproductive success, not so much by giving birth to more infants, but rather because her infants survive better and mature more rapidly or she herself enjoys a longer life.

Dominance does not exclude affiliation. Indeed, especially in groups with dominance, some female dyads form strong social bonds characterized by proximity, grooming, and frequent coalitionary support. These females are usually relatives. Mammalian mothers generally protect their offspring, but in these primates, protection extends beyond weaning until adulthood, after which the support becomes mutual between mother and daughter. Sisters likewise support each other. As a result, maternally related females tend to occupy similar ranks in the dominance hierarchy.

Unrelated females also sometimes come to each others' aid, usually to snip challenges from females ranking below both of them, thus helping to maintain the status quo, which is in their mutual interest. In a few species, especially of macaques, females show unusual tolerance to nonrelatives, probably because large female coalitions are required to deal with common enemies, probably coercive males (see below). This example illustrates the possible leverage of subordinates and its effect on the power balance in social relationships.

Females of all species, regardless of the kinds of their social relationships, show great interest in infants. Mothers of young infants, therefore, enjoy temporary leverage as a grooming partner. In some species, mothers routinely allow other females, often young ones, to take their infants and carry them or play with them. Although the distribution of such allomothering is as yet poorly understood, infants in these species tend to develop faster than in species without it.

Philopatric tendencies are related to these patterns in relationships. Females in most high-contest species are philopatric, probably because they need female allies to maintain their dominance, an important influence on their fitness. Those at the low-contest end have more freedom to move, and often disperse between groups. In small groups, as in some of the folivores mentioned above, female philopatry is often facultative, and most of the competition among females may be for group membership.

Because among primates, inbreeding avoidance is

generally a very strong force, female philopatry forces males to disperse, and vice versa. Female philopatry is more common, probably because female benefits are pronounced and males benefit from dispersal whenever females prefer philopatry, but in species with smaller groups, including all pair-living ones, opportunistic dispersal by both sexes is the rule.

Male Sociality. Male primates are expected to map themselves on to the distribution in space and time of mating opportunities, with the spatial distribution of the females being a major determinant. Among the solitary infant-caching species, the larger male home ranges are superimposed on several female ranges, although male–female pairs that share a common range are found in several species. The extent to which male ranges are exclusive is related to their ability to exclude each other from areas of high female density in the mating season. This seems to work in some lorisiforms, but especially in lemurs, many male ranges can overlap, and all of them may simultaneously pursue estrous females.

Where females live in groups, males can join these groups. Their number should be predicted by the "priority of access" model: if there is never more than one fertilizable female, a single dominant male can monopolize all fertilizations. Once multiple females are sexually receptive simultaneously, this male often can no longer monopolize them all, and we expect that others will join the group. This simple model is remarkably powerful. The number of females is a major predictor of the number of males per group, and groups also contain more males where females mate more seasonally or require more mating days for each conception.

Where female groups contain a single male, we often see males that are solitary or live in all-male bands, but both of these are rare when groups contain multiple males. It is not clear exactly how males determine how many would fit in the group, but once groups contain multiple males, the males in the population probably simply distribute themselves over the available groups. Thus, the main distinction is between one and more than one male in the group. Indeed, compared to his counterparts in multimale groups, the male in a single-male group shows much more hostility toward extra-group males, whether would-be immigrants or residents of other groups, and is also far more likely to break up fights among females. In multimale groups, no male will exert himself like that because free-rider males would benefit from his excluding male rivals or the higher reproduction of the females without incurring any costs (an example of the collective action problem).

The more mating competition is through scramble, and thus the weaker the link with male dominance rank, the less we expect males actively to contest the top dominance position. Indeed, single-male groups are often subject to violent takeovers by outsiders, as are multimale groups in which the top-dominant male sires the majority of offspring, but in larger multimale groups, often breeding seasonally, such takeovers are rare and nonviolent succession is more common. In the latter, male turnover is also much lower. Relative testis size of males, a proxy measure for ejaculate size and ejaculation frequency, also increases as the number of males per group increases.

If males can consistently monopolize access to fertile females, we expect that more successful fighters sire more offspring, and we therefore expect males to become bigger and grow larger weapons (in primates, canine teeth) up to the point that the cost of further growth reduces fitness. Indeed, across primates as a whole, those that live in pairs are monomorphic in body and canine size, whereas those that do not are often sexually dimorphic, with males exceeding females. But the fit with mere numbers is rather poor; the fit with paternity monopolization in the dominant male is much better, but still not perfect, especially for body size. There are unknown influences on sexual dimorphism in body size, perhaps related to the fact that dimorphism is a ratio measure, which must treat females as invariant across mating systems. This is hardly likely, but we cannot relax this assumption because we cannot control for body size. For canines, we can examine variation more easily by controlling for body size (canines scale isomorphically with body size across all primates). Among both males and females, relative canine size turns out to be larger where more behavioral contest is observed, but in both sexes we also see an effect of coalitionary aggression: where much of the escalated fights are between coalitions, canine size is reduced, probably because the number of contestants becomes relatively more prominent in determining the outcome of fights and thus selection on canine size is relaxed. Larger male body size in primates is achieved by longer periods of growth; as a result, there are always fewer adult males than adult females per group (except in the monomorphic gregarious lemurs).

Males may find themselves together in a group, but this does not make them friends: paternity cannot be shared the way access to food usually can be. Males sometimes form alliances, either within or between groups. Within-group alliances require large numbers of males per group, which is rare. They are most often seen in species where the effect of dominance rank on mating success is relatively weak: older males may form effective alliances to keep younger immigrants from gaining top dominance, as in large seasonally breeding groups of macaques. In baboons, also a species with large numbers of males per group, aging males may form coalitions against prime mate-guarding males and so displace them from females. Between-group alliances are also not very common. Males may, however, form teams that effectively prevent other males from immigrating. In howlers (*Alouatta*), these teams are more stable when the males are relatives, whereas in chimpanzees males are always philopatric and form strong bonds with each other, although they opportunistically may change allegiance in their within-group power struggles.

The number of males in a group is essentially a byproduct of the number of females and mating seasonality. However, their number also depends on the duration of each female's period of active mating and the degree of mating synchrony beyond mere seasonality. This suggests that natural selection may have modified female mating periods to manipulate the number of males to the females' benefit. Primate males tend to be better at detecting predators and more vigorous in mobbing or even attacking them. There is indeed a tendency for primate groups to contain more males than expected where predation risk is unusually high, but it has so far not been shown that the presence of multiple males produces fitness benefits for the females. Conversely, average rates of infanticide are lower in multimale groups, but it is unknown whether higher a priori infanticide risk is linked to an excess number of males beyond that expected for female group size and mating seasonality.

Males and Females. The females of many primate species are vulnerable to fatal attacks on their dependent infants by males unlikely to have fathered them. Although infanticide by males as a source of infant mortality varies from near zero to over 50 percent, it has been recorded in almost forty species, including most of the well-studied ones. Primate mothers are vulnerable because their infants are dependent on them for a long time, and loss of the infant will advance the female's next reproductive event. Thus, if a male has not sired the female's current dependent offspring, but is likely to sire the next, he may gain reproductively from killing the current offspring.

The great majority of directly observed cases of infanticide in wild primates is found where likely sires had disappeared or were incapacitated. [*See* Infanticide.] This suggests a major role for the likely sire in protecting infants, an inference supported by observations of close association and affiliation between infants and likely sires in various species, including macaques, baboons, and capuchin monkeys. Whenever strange males may come into a position of dominance, females and likely sires would benefit from being in close association.

The threat of infanticide therefore probably explains why virtually all primates that carry their offspring and have long lactation periods show year-round association between males and females, whereas it is absent among those primates caching young (who also have relatively shorter lactation periods). Among other mammals less than one third of species shows year-round male–female association.

Sex in primates is not the brief and furtive affair it is in most mammals. Similarly, there is little evidence for female preferences for mating with males possessing certain morphological characteristics. Instead, many primate females have active tendencies toward polyandrous mating, and show opportunistic sexual proceptivity when encountering new males, even if ovarian activity is impossible, such as during pregnancy. Mating among prosimians approaches that of the estrous cycles of many other mammals, but New World monkeys have longer and more opportunistic mating periods. Females of Old World primates (Catarrhini) have taken this one step further, often mating during half or even all of an ovarian cycle. Their cycles have follicular phases twice as long as those of other primates and are of unusually variable duration, indicating unpredictable ovulation. Many of them also sport exaggerated swellings of the perineal skin around midcycle and give distinct calls toward the end of a mating.

This unusual syndrome of sex-related traits may reflect sexual conflict. Harassment by males of sexually attractive females is most salient among the Old World primates, especially in species in which males are much bigger than females or females range alone at least some of the time. Possessive mate guarding is also most pronounced in this taxon, as is the incidence of males injuring females. In other primates, females seem to have much more control over their mating behavior, including the identity of their mating partners. It is therefore likely that unusual sexuality of catarrhine females evolved in part to reduce the risk of infanticide, although it is also possible that they must enhance their attractiveness as mates more than others by flaunting their viability.

Male mammals cannot unequivocally recognize their infant offspring, and must estimate their chances of paternity based on their mating history with the female, relative to her degree of attractivity. These circumstances allowed selection to manipulate female sexuality so as to dilute the chances of paternity of the dominant male and distribute some to subordinate males, as well as to confuse newly immigrated males as to his paternity chances. Whereas confusion is found in all species in which infanticide by males is common, paternity dilution probably requires more elaborate measures in species where male harassment reduces the females' behavioral freedom to select their mates.

The nonsexual social relationships between males and females vary widely, in ways as yet poorly understood. In some species, conspicuous male–female friendships are found, whereas in others males and females hardly interact. The functional aspects of this variation are poorly understood. When the male friend is the likely sire of the female's current infant, infanticide prevention is implicated, an interpretation supported by playback experiments in baboons. In others, male or mutual agonistic aid may be important. Finally, in bonobos and perhaps some chimpanzees populations, where maturing females emigrate into another community, the friends may be mother and son, with the mothers supporting their sons in their conflicts with other males.

Immatures. The close and long-lasting relationship between primate mothers and their infants provides a safe springboard for the youngsters' exploration of both the social and physical environment, including social learning. The presence of adult males also provides much richer opportunities for male–infant interactions than found in most other mammals. Interestingly, several examples exist where male–infant relationships do not reflect the likelihood of paternity. In several such cases, males use infants as agonistic buffers to regulate relationships with other males. In species with helpers, unrelated males may also carry and provision infants, which may make them more attractive as mates.

Primates grow more slowly than all other mammals. The extended juvenile period provides the opportunity for more learning and social play, which increases with degree of encephalization. Social interactions of immatures tend to foreshadow their adult roles and niches, and sex differences arise much earlier than expected on the basis of immediate function.

Social Complexity. Many primates form social bonds with at least some group members. These bonds reflect long-term investments that need to be defended against possible disruptions, and renegotiated due to changes in the partners' values. Thus, they show anticipatory affiliation in contexts likely to lead to conflict, and tolerance-restoring reconciliation soon after real squabbles. Indeed, reconciliation is rare among animals that do not have a social bond. The power balance among the partners in the bond depends only in part on their dominance relationship, but also on the importance of support (e.g., in fights) and services (e.g., grooming) proffered by the subordinate. The power balance may change when new potential partners become available or old ones disappear (the "market effect"), as reflected in the relative amount of grooming going one way or the relative frequency of support. Dominant male chimpanzees display an awareness of this market effect: they may physically prevent affiliation between other males, who might otherwise establish a closer relationship that could eventually threaten their social position.

Bonds also provide the backdrop for a secondary benefit of sociality: the opportunity to learn novel skills from other group members. Social learning is much enhanced where animals form relaxed social relationships, as demonstrated by the pattern of spread of new inventions. Local traditions (or incipient culture), maintained by social learning, documented in several monkeys and apes, find their most remarkable expression in the tool-use traditions of chimpanzees and orangutans.

Many social phenomena in nonhuman primates show obvious parallels with human behavior, yet we must be careful to distinguish homology from convergence by careful study of function, causation, ontogeny and evolution. The study of functional aspects can reveal the selective benefits of behaviors, but this study requires natural habitats. These habitats are disappearing at alarming rates, and a massive research effort must be mounted before we forever lose the opportunity to study the deeper roots of human behavior.

[*See also* Life History Theory, *article on* Human Life Histories; Mating Systems, *article on* Animals; Sexual Selection, *article on* Bowerbirds; Sperm Competition.]

— CAREL VAN SCHAIK

Primate Biogeography

Researchers investigating primate biogeography seek to explain the geographical distribution of primate species over evolutionary time. The vast majority of living nonhuman primates are found in the tropical and subtropical regions of America, Asia, and Africa, including Madagascar. Only one species, the Barbary macaque (*Macaca sylvanus*), is found in southern Europe (Gibraltar); this species is also the only living primate of North Africa. In Asia, only seven species occur north of the Tropic of Capricorn (in Nepal, southern China, and Japan), and in southern Africa and South America, six and five species, respectively, cross the Tropic of Cancer. The remaining 249 species, distributed across 67 genera from 14 families, inhabit the intermediate tropical forests and savannas. *Homo sapiens* is the only primate species that has successfully colonized all continents and temporal and tropical habitats. It is important to remember, however, that several ancestral and now extinct lineages of today's primates were also widespread in North America, Europe, and North Africa.

The phylogenetic relationships among living primates mirror their biogeographical distribution. The ancestral suborder of prosimian primates is confined to the Old World. For example, Africa is home to all of the at least seventeen species of bush babies (Galagidae) and six species of lorises (Lorisidae), most of which are small, solitary, exclusively nocturnal omnivores that do not come to the ground. The remaining twelve species of lorises occur in Asia, together with five species of tarsiers (Tarsiidae), small specialized carnivores that became secondarily nocturnal and that live either in pairs or in small groups, or lead a solitary life.

The majority of prosimians are found on the island of Madagascar, where the endemic lemurs have radiated into five living families (Cheirogaleidae, Lepilemuridae, Daubentoniidae, Indriidae, and Lemuridae). At least sixteen so-called subfossil species from another three families should be included with the thirty-eight living lemur species because they went extinct only within the last two thousand years, following human colonization of the island. This radiation has produced the smallest known primate, the pygmy mouse lemur (*Microcebus berthae*), weighing 30 grams (1.05 oz) as well as subfossil taxa that exceeded 250 kilograms (about 550 lbs) and thus silverback gorillas, in size. Most lemurs are active at night, but partial (i.e., cathemeral) or exclusive daytime activity has evolved at least twice independently among their ancestors. Many lemurs are omnivores, but several others are specialized folivores or frugivores. All but one species spend most of their time in trees and rarely come to the ground. Lemurs originated from ancestors with a solitary lifestyle but have on several independent occasions evolved to live in pairs or in larger groups.

The other primate suborder, the anthropoid primates, evolved in Africa, where they are today represented by forty-eight species from two families of Old World monkeys and apes (Cercopithecidae and Pongidae). Anthropoids also colonized most of tropical Asia, where there are now thirty-seven species of Cercopithecidae, the Hylobatidae, and one genus of the Pongidae. These Old World primates are all relatively large, exclusively active during the day, and most of them feed on fruits or leaves. Several Old World primates inhabiting savannas or other sparsely forested habitats spend most of their activity periods on the ground. The vast majority of Old World monkeys live in groups composed of several males and females, a few in pair-based family groups, and only the orangutans (genus *Pongo*) lead a solitary life.

Anthropoid monkeys also colonized the New World, where they radiated into eighty-seven species that are now grouped into two large families (Cebidae and Atelidae). Most New World monkeys are small to medium-sized diurnal tree dwellers. Only the night monkeys (genus *Aotus*) became secondarily nocturnal, and no New World monkeys spend much time traveling and foraging on the ground. Their diets are omnivorous or frugivorous, and no species is a specialized leaf eater. These monkeys live either in pairs or in groups with variable composition.

The colonization of Africa, Asia, Madagascar, and the Neotropics by primates can be considered as partly independent evolutionary experiments, resulting in taxa that have evolved in isolation from each other for millions of years and generations. The adaptive radiation of Malagasy lemurs is the result of a single colonization event, presumably by rafting African prosimians in the Eocene epoch (more than fifty million years ago). The radiation of neotropical primates probably dates back to a single colonization in the Oligocene epoch (about thirty million years ago). These two groups represent natural experiments, where either a prosimian or an anthropoid radiation took place in the absence of representatives of the other suborder. The living primates of Africa and Asia are heterogeneous groups of anthropoids and prosimians. Today, however, these two groups have only one genus in common: the Barbary macaque is now the only African representative of what has become an otherwise Asian genus. Even though these four geographically defined groups are members of the same order and two of the faunas are clades, they can be considered as independent groups for the purpose of broad comparisons.

The majority of primates have evolved in climatically and structurally similar environments in tropical and subtropical forests around the world, so that similar diversities in body size, activity, diet, and group size are to be expected. The evolutionary mechanism underlying this expectation is convergence. Convergence is a concept in evolutionary biology that is implicated whenever similar adaptations to comparable environmental factors have arisen independently. Convergences can be recognized on several levels of organization, that is from physiology and morphology to social behavior and community structure, because they are ultimately all shaped by natural and sexual selection on individuals.

Systematic comparisons reveal that primates of Africa, America, Asia, and Madagascar exhibit surprisingly few similarities in basic life history and socioecological variables at the community level. Body size, which is the most fundamental life history trait and which is related to many aspects of an organism's physiology and ecology, is not equally diverse among the four groups. Although the ranges of body sizes exhibited by the members of these four group are broadly similar, the Neotropics lack species with more than about 10 kilograms (22 lbs) and harbor primates that are on average smaller than those in all other regions (if size estimates of subfossil lemurs are included in the comparison). Activity is a fundamental ecological variable because a species' activity pattern (active at day or at night) has important consequences for basic aspects of its ecology and behavior. Across all primates, activity during the day is the most common pattern, but there is significant variation among the four regions in the proportions of diurnal and nocturnal species. This heterogeneity is caused by the large proportion of nocturnal species in Madagascar, which is also home to all but one of the cathemeral primates.

Behavior is a target of natural selection where convergences are expected because behavior mediates the interactions between the organism and its environment. Because feeding animals interact most directly with structural aspects of their habitats, the ecological diversity of a group of species can be indexed by characterizing the feeding behavior of each species. Feeding strategies of primates exhibit a number of dissimilarities among regions. The Asian genus *Tarsius* contains the only exclusively faunivorous primates, whereas primarily gummivorous species are found in all regions but Asia. The most notable qualitative difference concerns the lack of primarily folivorous primates in the New World. It is also striking that Africa lacks one clearly dominating dietary category, whereas folivorous-frugivorous species in Asia, folivorous species in Madagascar, and frugivorous-faunivorous species in South America clearly form the largest guilds in their respective regions.

The behavior of individuals toward conspecifics is also thought to be largely shaped by ecological factors, such as the distribution of resources or the risk of predation. Individuals decide at the behavioral level whether they lead a solitary life or form permanent groups, and which group size is optimal under a given set of ecological conditions. The primates of the four regions differ in their diversity of group sizes. At the qualitative level, Africa is the region with the greatest diversity in group sizes. Asia lacks species that form very large groups, that is, those with more than fifty or so members, even though some species may form large aggregations, consisting of several hundred individuals from several groups. The average size of lemur groups is even more reduced; species with more than fifteen to twenty animals per group are lacking entirely in Madagascar. However, a much larger proportion of lemurs is solitary or pair-living, compared to all other regions. South America lacks solitary species, and species that form very large groups are rare in the New World. Of all group-living primates, lemurs live on average in much smaller groups than primates elsewhere in the world.

Comparisons among local primate communities within and between biogeographical regions provide a second level of analysis for the search of convergences. This approach is instructive because it makes it possible to separate ecological and phylogenetic factors in the evolution of primate communities: ecological adaptations in the presence or absence of particular competitors can be compared among more or less closely related taxa. After characterizing the niche of a primate species with ten ecological variables, comparisons among representative communities from Africa, Madagascar, Asia, and South America revealed more similarities between communities in the same biogeographical regions, even though the number and identity of species were quite variable. This result complements findings from other studies, which found consistent differences among regions in ecological variables, such as rainfall, primate biomass, and alpha-diversity (the number of coexisting species in a single site). In all of these analyses, Asia stands out as the region with the lowest rainfall, primate biomass, and alpha-diversity. Thus, primate social and ecological evolution has been affected by the biogeographic region in which it evolved.

Ecological adaptations are also linked to phylogenetic affinities. Members of a genus, or even of the same family, tend to occupy very similar niches and therefore exhibit great overlap in ecological variables that define niches. For example, all bush babies are small, nocturnal, arboreal frugivorous and insectivorous quadrupeds, whereas all gibbons (Hylobatidae) are large, diurnal, arboreal frugivorous brachiators. Nevertheless, phylogenetic groups also overlap in ecological space. For example, cheirogaleids and lorises share the above mentioned adaptations with bush babies. In all four biogeographic regions, ecological divergence and phylogenetic divergence are positively correlated; that is, more distantly related taxa tend to occupy more different ecological niches. Similarly, the more phylogenetically diverse faunas of Africa, Asia, and Madagascar are ecologically more diverse than the less phylogenetically diverse American primate community. Finally, the correlation between ecological and phylogenetic distance is much lower for lemurs and New World monkeys, which represent the living endpoints of explosive adaptive radiations. Thus, the ecological and morphological diversity of living primates in different regions of the world has been affected by both abiotic characteristics of the respective region and the phylogenetic history of the respective primate community.

Which specific factors determine the structure of a particular community is not yet fully understood. Local factors, such as soil fertility and the resulting food abundance and quality, intensity of competition from other primates and mammals, as well as hunting pressure, have measurable effects on primate densities and diversity. The degree of environmental seasonality, or more specifically the nature and duration of the lean season, is equally important. The current composition of a given primate community is also affected by more regional factors, such as latitude, topography, recent climate change, and previous speciation and extinction rates.

[*See also articles on* Biogeography.]

BIBLIOGRAPHY

Fleagle, J. G. *Primate Adaptation and Evolution.* New York, 1998.

Fleagle, J. G., and K. E. Reed. "Comparing Primate Communities: A Multivariate Approach." *Journal of Human Evolution* 30 489–510.

Fleagle, J. G., C. Janson, and K. E. Reed, eds. *Primate Communities.* Cambridge, 1999.

Kappeler, P. M., and E. W. Heymann. "Nonconvergence in the Evolution of Primate Life History and Socio-ecology." *Biological Journal of the Linnean Society* 59 (1996): 297–326.

Kay, R. F., R. H. Madden, C. P. van Schaik, and D. Higdon. "Primate Species Richness Is Determined by Plant Productivity: Implications for Conservation." *Proceedings of the National Academy of Sciences USA* 94 (1997): 13023–13027.

Martin, R. D. *Primate Origins and Evolution.* London, 1990.

Martin, R. D. "Primate Origins: Plugging the Gaps." *Nature* 363 (1993): 223–234.

Reed, K., and J. Fleagle. "Geographic and Climatic Control of Primate Diversity." *Proceedings of the National Academy of Sciences USA* 92 (1995): 7874–7876.

PETER M. KAPPELER

PRIONS

Prions, or *pro*teinaceous *in*fectious particles, are generally accepted as the agents responsible for a class of fatal neurodegenerative diseases known as transmissible spongiform encephalopthies (TSEs). These include scrapie in sheep, bovine spongiform encephalopathy (BSE or mad cow disease) in cattle, and Creutzfeldt-Jakob disease (CJD), fatal familial insomnia (FFI), Gerstmann-Sträussler-Scheinker disease (GSS), and Kuru in humans (Prusiner, 1998). Prion diseases are classified as acquired (infectious), familial (inherited), or sporadic (spontaneous). In each case the disease-causing agent appears to be a protein that is normally found in the body but for some reason has adopted a different three-dimensional shape or conformation. Astonishingly, these abnormal forms of the protein proliferate by interacting with the normal nonprion forms, causing them to adopt the pathogenic conformation. The ability of prions to pass on their altered three-dimensional shape to the normal forms, that is, to replicate themselves, challenges the long-standing dogma that only nucleic acids can carry genetic information.

For decades, efforts failed to isolate a nucleic acid–based infectious agent responsible for mammalian TSEs, such as a virus. Numerous experiments show that the infectivity of material transmitted from one animal to another is resistant to experimental manipulations that damage DNA or RNA. However, the agent is readily affected by manipulations that denature or destroy proteins. Prion diseases are genetically linked to a gene called *Prn-p*, which encodes a cell surface glycoprotein, PrP. Specific mutations in the *Prn-p* gene result in dominantly inherited prion disease. Expression of hamster *Prn-p* that has been transgenically inserted into mice renders them highly susceptible to hamster prions. Moreover, transgenic mice devoid of *Prn-p* expression are resistant to prion infection. A wealth of data now support the contention that TSEs result exclusively from an altered conformation of the PrP protein, designated as PrP^{Sc}, where the "Sc" refers to "scrapie," named after the TSE disease that is prevalent in sheep. The normal cellular isoform of the protein is denoted PrP^{C}. Unlike other types of protein conformational changes, it is hypothesized that PrP^{Sc} influences PrP^{C} to adopt the same prion conformation. That is, the prion conformation is self-perpetuating and gradually converts more and more PrP^{C} to the pathogenic infectious form, PrP^{Sc}. The function of normal mammalian prion protein is unclear, and its role in evolution remains to be determined. However, the gene is quite conserved across mammals, suggesting that it may play an important role in its normal form.

Stanley Prusiner was awarded the Nobel prize in 1997 for providing substantial evidence to support the "protein-only" hypothesis for prions. The revolutionary concept of prions was also extended to explain some unusual genetic phenomena in yeast and other fungi (Wickner, 1994). There are several prions in the yeast *Saccharomyces cerevisiae*, [*PSI*$^+$], [*URE3*], and [*RNQ*$^+$], and at least one in the filamentous fungus *Podospora anserina* [Het-s]. The yeast genome does not encode a protein related by function or by amino-acid sequence to the mammalian prion protein, PrP. That is, yeast prions are not homologous to their mammalian counterparts. As a result of these discoveries in yeast and other fungi, the word *prion* now refers to any protein that is self-perpetuating: the term is no longer confined to the proteinaceous infectious agents responsible for TSEs. Like the PrP protein, fungal prion proteins can exist in at least two different stable conformations, nonprion or prion. Fungal prions are usually associated with distinct heritable phenotypes. Unlike the mammalian prion, however, fungal prions do not cause diseases but modify the functions of their protein determinants in a self-sustaining way. The yeast prion [*PSI*$^+$], for example, causes ribosomes (they synthesize proteins inside the cell) to read through stop codons inappropriately, thereby producing abnormal, longer proteins. Another, [*URE3*], causes yeast to utilize poor nitrogen sources in the presence of better ones, a process called *nitrogen catabolite repression*. The biological function of yeast [*RNQ*$^+$] is unknown. [Het-s] prevents the fusion of cells of incompatible strains of *P. anserina* by triggering a lytic reaction that kills the fusing cells. These fungal prions and their associated phenotypes are propagated from generation to generation as the prion protein is transferred from mother to daughter cell. This continues the cycle of conformational conversion. Thus, these prions act as heritable protein-based genetic elements that can cause biologically important phenotypic changes without any underlying changes in DNA or RNA.

The wealth of evidence supporting yeast prions is more compelling than those supporting mammalian prions (Caughey, 2000). All three yeast prions show unusual genetic properties. First, they are propagated through the cytoplasm. Other genetic determinants that are not prions are transmitted cytoplasmically, including mitochondrial DNA and viruses, but none of these encode [*PSI*$^+$], [*URE3*], or [*RNQ*$^+$]. Second, transient increases the level of the protein determinant of a prion dramatically increase the frequency at which the prion is formed. Third, yeast can be cured of these prions by growing the cells for several generations in the presence of some nonmutagenic reagents, and the prions can arise again naturally at a low frequency or upon experimental manipulation. Thus, once the prion conformational state is established, it is self perpetuating, and once it is lost, it has a low probability of reforming.

There is direct evidence of the conformational switch for both mammalian and fungal prions. PrP^c, the normal form of the protein, is more susceptible to enzymes that digest proteins (protease) than $PRPS^c$, the prion form. Sup35 and Ure2, the protein determinants of [*PSI*$^+$] and [*URE3*], respectively, are insoluble in their prion states whereas their normal forms are soluble. Moreover, both of these yeast prions are more resistant to protease than the normal forms. The conformational difference between the soluble and prion forms can also be visualized in living yeast cells. Fusions of the prion proteins to green fluorescent protein appear as bright, tight, green fluorescent foci in cells containing the prion. In cells lacking the prion, the fluorescence is diffusely distributed throughout the cell.

Models for conformational conversion provide remarkable insight into the nature of prion formation and propagation. Purified soluble prion proteins slowly but spontaneously form amyloid fibers. Fiber formation can be greatly stimulated by adding preformed amyloid, mimicking the ability of previously converted prion determinants from a mother cell to perpetuate the conversion process when it is passed to the daughter cell. Sup35 prion aggregates are capable of self-seeded propagation in cell-free extracts, and the efficiency of conformational conversion can be monitored quantitatively. It is uncertain whether the prion conformation in yeast is an amyloid, but the conformational changes that produce prions in yeast and amyloid fibers in the test tube are certainly very closely related. For example, mutations that increase or decrease the frequency that prions appear spontaneously alter the rate of amyloid formation in a similar way. Moreover, amyloidlike filaments were detected in cells producing large quantities of the prion protein in cells containing the [*URE3*] prion.

The ability of Sup35, Ure2, and Rnq1 to adopt a prion conformation is conferred by regions of unusual amino-acid composition in the protein. These prion-forming domains (PFDs), which contain a high number of glutamine and/or asparagine residues, are essential for prion propagation. Remarkably, the PFDs are transferable to completely unrelated proteins. A chimeric protein containing the PFD of Sup35 and a transcriptional regulatory protein can exist in two stable but interchangeable states: a normal functional state or an altered, nonfunctional state. Astonishingly, the novel prion affects the transcriptional regulation rather than translational fidelity (Li and Lindquist, 2000). Fusing of the PFDs of Sup35, Ure2, and Rnq1 to the green fluorescent protein also gives rise to fusion proteins with two distinct fluorescence patterns: diffused or aggregated. The fact that the prion domain is modular, transferable, and sufficient for transmission of prion-controlled phenotype provides a convincing piece of supporting evidence for the protein-only prion hypothesis. It also suggests that new prions might be created through an exon-shuffling mechanism by appending prion domains to other functional domains thereby modulating the inheritance of other phenotypes.

One fascinating question regarding prions is whether they perform a needed biological function or are merely an interesting accident of nature. Yeast rapidly lose even mildly deleterious markers. The fact that prion domains of Sup35 and Ure2 are retained among many yeast species suggests that those domains perform some needed function. The presence of similar N-terminal extensions in *SUP35* homologs from other species have led some to propose that the prion domain has an adaptive function, separate from its function in translation. To address this question, the extent of nucleotide polymorphism within different laboratory, commercial, and clinical isolates of *S. cerevisiae* was compared with the extent of sequence divergence of a related species, *S. paradoxus*. The amino-acid sequences for prion domain and the translational termination domain are constrained, presumably by purifying selection against mutations that change amino-acid codons. However, the prion domain is under weaker selection than the translational termination domain. This suggests that the prion domain of Sup35, if not a beneficial adaptation, is at worst only weakly deleterious.

One intriguing possible function for prion domains that would merit conservation is that prions provide yeast a means to cope with changing environmental conditions by producing a heritable, reversible conformational and phenotypic switch without necessitating any alteration in DNA sequence. Indeed, recent studies suggest that [*PSI*$^+$] provides a mechanism for generating genetic variation and phenotypic diversity. True and Lindquist (2000) examined the growth characteristics of seven sets of [*PSI*$^+$] and [*psi*$^-$] strains in more than 150 phenotypic assays. In many conditions they tested, [*PSI*$^+$] strains grew or survived better than [*psi*$^-$] strains. In one genetic background, they also observed prion-dependent changes in colony morphology and flocculance. They also found that different genetic backgrounds displayed unique and diverse constellations of phenotypes in response to different testing conditions. Such diversity may arise through translational read-through caused by [*PSI*$^+$], which presumably appends extra amino acids to the ends of proteins and may activate cryptic or pseudogenes containing translation stop codons. Thus, [*PSI*$^+$] enables the expression of previously silent information under certain environmental conditions. Moreover, since [*PSI*$^+$] is a metastable element that can arise or be cured in yeast populations at a frequency of about one in a million cells, this increases the chance that at least some of a natural population will be able to survive when environmental conditions change. Finally, because the level of Hsp104 affects the induction and curing of [*PSI*$^+$] and is induced by environmental stress, this phenomenon provides an intriguing and plausible mechanism for yeast to preadapt to different environmental niches.

To predict candidates for novel prions, Michelitsch and Weissman (2000) devised an algorithm to search available proteomic sequences for protein regions with high glutamine (Q) and asparagine (N) content (the *proteome* is the name given to the collection of proteins that an organism expresses). They found that a large number of eukaryotic proteins (107–472 per proteome) are Q/N-rich, having regions containing at least 30 Q/N within 80 consecutive residues. Remarkably, there are essentially no Q/N-rich regions in thermophilic bacterial or archaeal proteomes. A very small number (0–4) was found in mesophilic bacterial proteomes. Using a similar but more stringent algorithm, 50 consecutive residues containing at least 45 percent Q/N and 60 percent polar amino acids, about one hundred prion candidates were identified from yeast proteomes. The abundance of Q/N-rich region in eukaryotic proteomes suggests that prionlike regulation of protein function might be a normal regulatory process.

Perhaps the most fascinating phenomenon of prion biology is that multiple prions can exist in a single cell. Recent evidence suggests that one prion can affect the appearance of other prions. Thus even in a single-celled eukaryote like yeast, different combinations of prions within a cell may produce complex combinations of phenotypes, without necessitating any underlying changes in nucleic acids. Glutamine- and asparagines-rich domains that resemble those of other prions are found in many higher eukaryotic proteins and in some cases, number as high as 1 to 3 percent of the total proteome. Elucidating the spectrum of proteins that exhibit similar types of conformational changes to establish self-perpetuating phenotypic states and ascertaining their roles in biology will occupy scientists for years to come.

BIBLIOGRAPHY

Caughey, B. "Transmissible Spongiform Encephalopathies, Amyloidoses and Yeast Prions: Common Threads?" *Nature Medicine* 6 (2000): 751–754.

Li, L., and S. L. Lindquist. "Creating a Protein-Based Element of Inheritance." *Science* 287 (2000): 661–664.

Michelitsch, M. D., and J. S. Weissman. "A Census of Glutamine/Asparagines-Rich Regions: Implications for Their Conserved Function and the Prediction of Novel Prions." *Proceedings of the National Academy of Sciences USA* 97 (2000): 11910–11915.

Prusiner, S. B. "Prions." *Proceedings of the National Academy of Sciences USA* 95 (1998): 13363–13383.

True, H. L., and S. L. Lindquist. "A Yeast Prion Provides a Mechanism for Genetic Variation and Phenotypic Diversity." *Nature* 407 (2000): 477–483.

Wickner, R. B. "[URE3] As an Altered URE2 Protein: Evidence for a Prion Analog in *Saccharomyces cerevisiae*." *Science* 264 (1994): 566–569.

— LIMING LI, SUSAN M. UPTAIN, AND SUSAN LINDQUIST

PRISONER'S DILEMMA GAMES

In his classic monograph titled *Sociobiology* (1975), E. O. Wilson argued that one of the most important challenges facing evolutionary biology is to develop a comprehensive theory for the evolution of altruism. In an attempt to develop such a theory and to pick up the gauntlet thrown down by Wilson and others before him, behavioral ecologists have relied heavily on evolutionary game theory. Evolutionary game theory has borrowed extensively from economic game theory, and a leading workhorse in both these areas is the Prisoner's Dilemma game—a game well suited to analyze the evolution of both altruism and cooperation. [*See* Game Theory.]

Description of Game. The Prisoner's Dilemma game was the 1950s brain child of Mercill Flood and Melvin Dresher of the RAND Corporation (and was popularized by Jon Von Neumann and Oscar Morgenstein). To see how this game works, imagine two suspects of a crime being interrogated separately by the police. Each suspect can either cooperate with the other by revealing nothing to the police or fail to cooperate (defect). To defect in this context is to "squeal" and tell the authorities that the other suspect is guilty.

The possible payoffs in this game are set as follows (Figure 1): The police have enough circumstantial evidence to put away both suspects for one year, even if they receive no additional information from the suspects; that is, both suspects cooperate (call this the R payoff). Should each suspect "rat" on the other (defect), however, both go to jail for three years (P). Finally, should only one suspect snitch (defect) on his (cooperating) partner, such "state's evidence" allows the cheater to walk away a free man (T), but causes his partner to go to jail for five years (S). As can be seen in Figure 1, on any single play of the game, player 1 always obtains a greater payoff for defecting, regardless of what player 2 does, as $T > R$ and $P > S$.

The same logic that leads player 1 to defect, leads player 2 to do so as well. The dilemma in this game is that both suspects would have received a higher payoff if they both cooperated rather than jointly defected, as $R > P$. Yet achieving such mutual cooperation seems impossible, as the temptation to defect looms so heavily over each player. This dilemma exists even if the game is played many times between the same individuals (the iterated Prisoner's Dilemma) so long as it is a fixed num-

		Player 2	
		Cooperate	Defect
Player 1	Cooperate	R 1-year prison term	S 5-year prison term
	Defect	T 0-year prison term	P 3-year prison term

FIGURE 1. Payoffs to Player 1 in the Prisoner's Dilemma Game. Drawing by Lee Alan Dugatkin.

ber of times. If players know when the last move of the game will be, it is always in their interest to defect (this aspect of the game is known as the backward induction paradox).

The classic economic approach outlined above assumes that players in a Prisoner's Dilemma game are rational, and that each individual is consciously attempting to maximize his expected payoff. Evolutionary game theory assumes neither rationality nor even consciousness on the part of players. Instead, strategies that are often (but not always) assumed to be genetically encoded, compete with each other, and payoffs are defined in terms of some appropriate fitness units (nominally a player's reproductive success). Evolutionary models of the iterated Prisoner's Dilemma game also differ from economic versions in that the game itself is assumed to go on for many generations, adding a multigenerational flavor. In this sense, there are two types of iteration in play when the Prisoner's Dilemma is cast in evolutionary terms: between players within a generation and across multiple generations. In each generation, a tally is made of the fitness payoffs accrued by a particular strategy, and strategies are represented in the next generation in proportion to their fitnesses. So, for example, if we consider just two strategies, x and y, and strategy x accrues 100 fitness units within a generation and strategy y 200 units, strategy y would be set at a frequency of 2/3 in the following generation.

Explanation of Altruism and Cooperation. Robert Trivers, in his paper *"The Evolution of Reciprocal Altruism"* (1971), was the first to mention the Prisoner's Dilemma as a possible tool for examining the evolution of altruism and cooperation. A decade later, Robert Axelrod and William D. Hamilton produced their now classic *Science* paper, *"The Evolution of Cooperation"* (1981). Axelrod and Hamilton were interested in the emergence, and subsequent evolution, of cooperation in a world in which any number of strategies vie for evolutionary success. [*See* Reciprocal Altruism.]

To examine the evolution of cooperation, Axelrod and Hamilton used a two-pronged attack. First, Axelrod ran computer tournaments in which participants submitted a strategy in the form of a computer program, and strategies competed against one another. After an initial small computer tournament, a tournament involving sixty-two different strategies competing with each other over the course of numerous generations was undertaken. Strategies were submitted by computer scientists, physicists, economists, mathematicians, sociologists, political scientists, evolutionary biologists, and even a ten-year-old computer hobbyist.

In each generation of the Axelrod computer tournament, the probability that a pair of strategies would meet again in the future was set at a value such that strategies met an average of 289 times during a given generation, but the exact point at which a game would end was not known to players a priori. Each strategy faced itself and all others in a round-robin tournament. Tit for Tat (TFT), a strategy submitted by psychologist Anatol Rapoport, emerged as the winner of this (as well as the earlier) tournament.

Tit for Tat is a relatively simple strategy that instructs a player to cooperate on the initial encounter with a new partner, and subsequently to copy that partner's most recent move. Axelrod (1984) argued that TFT's success is attributable to its three defining characteristics. (1) "Niceness"—TFT never defects before its opponent does. (2) Swift "retaliation"—TFT defects immediately after its opponent defects. (3) "Forgiveness"—Because TFT remembers only one move back, it "forgives" prior defections, if a partner is currently cooperating (i.e., it does not hold grudges). Essentially, TFT succeeds because it pairs cooperative action with cooperative action and defection with defection.

In addition to computer tournaments, Axelrod and Hamilton used simple mathematical models to examine whether TFT was evolutionarily stable, that is, whether it could resist invasion from alternatives if it were at a frequency in the population of close to 1 (i.e., 100 percent). What they found was that if the probability that two players meet again in the future is sufficiently high, TFT was evolutionarily stable. It is also true, however, that the strategy "always defect" was evolutionarily stable, regardless of how long partners were paired up. This poses a problem for TFT, in that while it is stable at high frequency, in the standard Prisoner's Dilemma it cannot spread from low frequency in a world of cheaters. Axelrod and Hamilton, and subsequently many others, however, have shown how small modifications to population structure can facilitate TFT's spread from very low frequencies.

Since Axelrod and Hamilton's original publication, theoreticians have demonstrated that no strategy in the iterated Prisoner's Dilemma can be evolutionarily stable. It turns out that every strategy is invadable by some conceivable mix of alterative strategies. Despite this, most evolutionary models find that TFT or TFT-like strategies such as Tit for Two Tats generally fare well when competing with less cooperative strategies.

The Prisoner's Dilemma game has spurred significant research among both theoretical biologists and empirical behavioral ecologists. On the theory side, modelers have added a suite of other variables to the basic Prisoner's Dilemma game to examine how cooperation fares in a variety of environments. Such factors include group size, population structure, the stochasticity of the environment, the payoff structure of the game, the role of kinship, the mobility of the players, memory limits, encounter probabilities, and the role of mistakes. For example, numerous teams of investigators have examined the effect of adding spatially structured environments. In a spatially structured environment, the strategies are not randomly distributed, but may appear in "clumps." Although the details of how cooperation fares in spatially complex Prisoner's Dilemma environments depend on the exact model, one interesting finding is that such complexity can actually produce a very simple end product. In certain spatial configurations, a strategy of "always cooperate" can actually coexist with pure defectors in an ever changing landscape; that is, if cooperators tend to be surrounded by other cooperators, cooperation is a viable strategy.

The Prisoner's Dilemma has also been used in empirical studies of animal cooperation. These include studies of reciprocal food sharing in vampire bats, nest defense in tree swallows, territoriality in hooded warblers, simultaneous hermaphroditism and "egg swapping" in fish and worms, coalition formation in primates, antipredator behavior in fish, and grooming in impala. As an example, consider the case of the Prisoner's Dilemma and Tit for Tat in the context of predator-inspection behavior in guppies.

In many species of fish, one to a few individuals move away from their school and "inspect" a potential predator to gain information about this danger; this is a form of cooperation. Experimental work and indirect evidence support the notion that the payoffs to predator inspection satisfy the requirements of the Prisoner's Dilemma game, as $T > R > P > S$ (see Dugatkin, 1997, for more on the long-standing and vociferous debate over the payoffs of predator inspection). The best action for a fish to take is to stay back and watch its partner inspect ($T > R$), but if both fish fail to inspect (and receive P), they are likely worse off than had they both gone out toward the predator (and received R).

Overall, the data gathered on inspection behavior in both guppies and sticklebacks (but not mosquito fish) support the notion that inspectors do, in fact, use TFT while inspecting potential predators. As predicted by TFT, inspectors appear to be nice (each starts off inspecting at about the same point in time), retaliatory (inspectors cease inspection if their partner stops), and forgiving (if inspector A's partner has cheated on it in the past but resumes inspection, A then resumes inspection as well). In addition to this direct support for TFT, evidence exists that inspectors remember the identity and behavior of their coinspectors and prefer to associate with cooperators over defectors when in small, but not large, groups.

Although much remains to be done, the Prisoner's Dilemma game, along with all its variants, has proved to be an excellent starting point for examining cooperation and altruism in animals, both from a theoretical and an empirical perspective.

[*See also* Social Evolution.]

BIBLIOGRAPHY

Axelrod, R. *The Evolution of Cooperation.* New York, 1984. An excellent book-length treatment of Prisoner's Dilemma models and their use in political science and evolutionary biology.

Axelrod, R., and W. D. Hamilton. "The Evolution of Cooperation." *Science* 211 (1981): 1390–1396. The single most important paper introducing evolutionary biologists to the Prisoner's Dilemma.

Dawkins, R. *The Selfish Gene.* Oxford, 1976. Introduced the notion of "selfish genes" to general readers. Has many sections on the evolution of cooperation and altruism.

Dugatkin, L. A. *Cooperation among Animals: An Evolutionary Perspective.* New York, 1997. The most thorough review of cooperation in animals available; includes a table on dozens of Prisoner's Dilemma models and hundreds of examples of animal cooperation.

Dugatkin, L. A., and H. K. Reeve, eds. *Game Theory and Animal Behavior.* Oxford, 1998. An edited volume on the multiple uses of game theory, including the Prisoner's Dilemma game, in the field of animal behavior.

Lombardo, M. "Mutual Restraint in Tree Swallows: A Test of the Tit for Tat Model of Reciprocity." *Science* 227 (1985): 1363–1365. The first experimental paper in animal behavior to use the Prisoner's Dilemma as its conceptual backbone.

Michod, R., and M. Sanderson. "Behavioral Structure and the Evolution of Cooperation." In *Evolution—Essays in Honor of John Maynard Smith*, edited by J. Greenwood and M. Slatkin, pp. 95–104. Cambridge, 1985. The first paper to show that Tit for Tat works because it segregates behaviors at the phenotypic level (cooperate with cooperate and defect with defect).

Nowak, M. A., and R. M. May. "Evolutionary Games and Spatial Chaos." *Nature* 359 (1992): 826–829. An excellent paper that examines the use of the Prisoner's Dilemma in spatially segregated populations.

Poundstone, W. *Prisoner's Dilemma: Jon Von Neumann, Game Theory and the Puzzle of the Bomb.* New York, 1992. An entertaining and historic overview of the use of the Prisoner's Dilemma game in many disciplines.

Ridley, M. *The Origins of Virtue.* New York, 1996. A fine trade book on evolution and social behavior, with numerous sections devoted to the Prisoner's Dilemma.

Sigmund, K. *Games of Life.* Oxford, 1993. A short book on the uses of game theory, with emphasis on the Prisoner's Dilemma.

Trivers, R. L. "The Evolution of Reciprocal Altruism." *Quarterly Review of Biology* 46 (1971): 189–226. In this well-cited paper, Trivers first suggests that the Prisoner's Dilemma may be a useful mathematical tool for the study of reciprocity.

Von Neumann, J., and O. Morgenstein. *Theory of Games and Economic Behavior.* Princeton, 1953. The first major book published on the mathematics of game theory and the Prisoner's Dilemma.

Wilson, E. O. *Sociobiology: The New Synthesis.* Cambridge, Mass., 1975. A watershed book marking the start of the modern study of evolution and social behavior.

— LEE ALAN DUGATKIN

PROKARYOTES AND EUKARYOTES

Fossils of early microbial life are found in rocks from western Australia that are nearly 3.5 billion years old. They resemble contemporary filamentous prokaryotes, cells that lack membrane-bounded nuclei and that in older textbooks are often called bacteria. Prokaryotes are incredibly versatile in the ways they can make the energy required for life, and much of their biochemistry can take place in the absence of oxygen under conditions similar to those we think prevailed in early periods. One prokaryote group whose ancestors are well represented in ancient rocks are the cyanobacteria (sometimes misleadingly called blue-green algae). Cyanobacteria developed photosynthesis to harvest the energy of light to make energy, as well as the carbon compounds required for new cells from carbon dioxide and water. Oxygen is a byproduct, and cyanobacteria probably generated the aerobic atmosphere required by many eukaryotes, cells like our own and those of true algae, which have a membrane-bounded nucleus. The earliest plausibly eukaryotic fossils appear in rocks from northern Australia that are believed to be 1.3 billion to 1.5 billion years old. However, lipids (steranes), which are considered typical of eukaryotes, are found in older rocks, approximately 2.7 billion years old, consistent with an earlier origin.

Fossils, biomarkers, and isotopic signatures provide ample evidence of an early start to cellular life, but they cannot be used to reconstruct evolutionary (phylogenetic) relationships. In *On the Origin of Species* (1859), Charles Darwin argued that classifications should portray relationships based on common ancestry, but it was not until the latter part of the twentieth century that the traits were discovered that could be used to unite all of life. In a groundbreaking paper published in 1965, Emil Zuckerkandl and Linus Pauling argued that, by comparing the sequences of nucleic acids and proteins, which they memorably called "documents of evolutionary history," one might determine the evolutionary relationships between species. There are now thousands of DNA, RNA, and protein sequences that can be used in this way. Carl Woese at the University of Illinois did much of the pioneering work. By comparing the sequences of small subunit ribosomal ribonucleic acids (SSU rRNA), a central component of the ribosome, Woese and his colleagues proposed in 1990 that life could be divided into three fundamental domains (Figure 1). Two of these domains, the Bacteria and the Archaea, are prokaryotic; the third, the Eukarya, contains all cells which have a nucleus (Table 1).

Molecular comparisons of single genes such as the one for SSU rRNA cannot identify where the root of the tree of life lies. The position of the root is important because it defines which of the two domains share most recent common ancestry. In principal, the root can be identified by comparing pairs of genes that duplicated from a single parent gene in the last common ancestor of cellular life (Figure 2). The small number of gene pairs (protein elongation factors, ATPases, transfer RNA synthetases) that can be analyzed in this way suggest that the root lies on the branch separating Bacteria from the other two domains, making the Archaea and Eukarya specific, but distant, relatives.

The three-domain classification proposed by Woese and colleagues appears in many textbooks, but it is important to realize that not everyone agrees with it and that it is still only a hypothesis. Inferring the deep past from molecular sequences is not easy, and analyses of single or small numbers of genes cannot be realistically expected to tell the complete story about an organism's past. For example, there are important differences regarding the position of eukaryotes in the two gene trees depicted here. In the SSU rRNA tree (Figure 1), eukaryotes form a separate group that shares a common ancestor with Archaea, whereas in the elongation factor tree (Figure 2), eukaryotes originate from within the Archaea. At present, it is not clear which tree, if either, more accurately reflects eukaryote relationships. Another caveat to bear in mind is that the mathematical models used to make such trees of life are still crude when compared to the known complexities of sequence evolution. For example, there are already cases in the literature where the use of a better model has produced a tree different from one previously published that was produced using the same data.

The Bacteria (sometimes called eubacteria) contain many of the prokaryotes which are familiar as disease-causing, food spoilage, or industrially useful prokaryotes. The Archaea (sometimes called archaebacteria) include some species that flourish under what are considered environmental extremes. For example, the archaeon *Pyrolobus fumarii* holds the record for life at high temperatures, flourishing at 113°C under high pres-

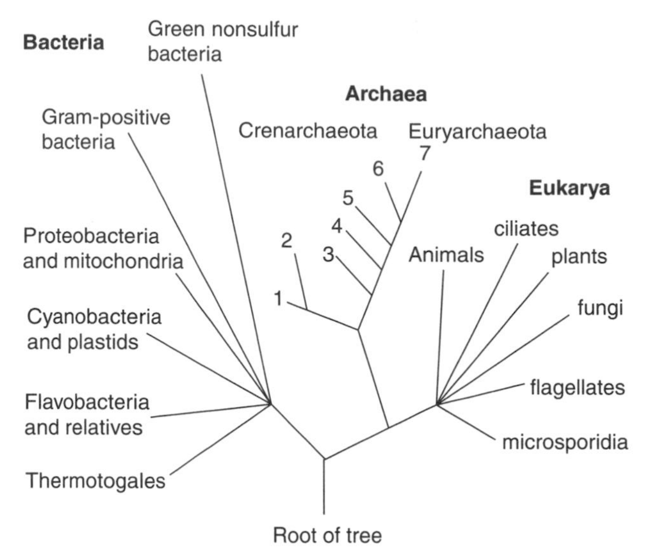

FIGURE 1. Universal Phylogenetic Tree Based on Small Subunit Ribosomal RNA Sequences and Showing the Three Domains. The tree was redrawn from Woese et al. (1990), branch lengths in this figure are meaningless. The position of the root was determined by comparing the few known sequences of paralogous genes that duplicated within the last common ancestor of life, as depicted in Figure 2. The numbers at branch tips correspond to the following archaeal groups: 1, the genus *Pyrodictium*; 2, the genus *Thermoproteus*; 3, the *Thermococcales*; 4, the Methanococcales; 5, the Methanobacteriales; 6, the Methanomicrobiales; 7, the extreme halophiles. Courtesy of author.

sure in "black smoker" hyperthermal vent chimneys situated deep in the oceans. Other Archaea live in hot acid, soda lakes, or saturated salt solutions, still others thrive in the absence of oxygen in the guts of animals or in anaerobic sediments. Archaea are a rich source of useful enzymes for the biotechnology industry. Prokaryotes are the most abundant and ancient species on the planet, and their metabolic diversity is responsible for the recycling of key nutrients in the environment such as carbon, nitrogen, and sulfur.

The Eukarya comprise all cells that have a true nucleus. The nuclear membrane is part of a complex endomembrane system, which divides eukaryotic cells into functional compartments wherein much of the business of eukaryotic cellular life takes place. Eukaryotes also contain an internal protein cytoskeleton that maintains cell shape and provides a means of intracellular transport. Components of this cytoskeleton ensure the accurate segregation of chromosomes in mitosis; other components provide the ability to take in (phagocytose) particles from the surrounding environment. These features, which ultimately underpin sexual reproduction and the ability to ingest food, are absent from prokaryotes. Eukaryotic cells vary in size from tiny free-living protozoa as small as prokaryotes (1–10 μm), to the largest plant cells at around 0.5 millimeter (500 μm). Communities of eukaryotic cells, each with a specialized function, form the animals, fungi, and plants that dominate the visual landscape.

Eukaryotic cells also differ from prokaryotes in having discrete double membrane-bounded compartments that are responsible for energy generation. Mitochondria make energy using an electron transport chain with oxygen as the terminal electron acceptor. Photosynthetic eukaryotes, for example plants and red and green algae, also have plastids that contain the photosynthetic pigments that give leaves their green color. Mitochondria and plastids are the remnants of Bacteria that formed separate endosymbiotic interactions with the ancestors of eukaryotic cells. We know this because each still contains a separate genetic system complete with bacterial-like machinery for replication, transcription, and translation. Moreover, analysis of the genome sequences of mitochondria and plastids clearly demonstrates from which groups of Bacteria the endosymbionts originated (Figure 1). Mitochondria are descended from a member of a group of Bacteria called the alpha-proteobacteria, many of which can still form intimate associations with contemporary eukaryotes. The genome sequences of plastids identify them as descending from a cyanobacterium.

It has long been thought that eukaryotes evolved from prokaryotes, but there are no obvious transitional forms that can bridge the gap between structurally simple prokaryotes and more complex eukaryotes. The two best-articulated hypotheses for eukaryote origins are the serial endosymbiosis theory and the hydrogen hypothesis. The serial endosymbiosis theory has its origins in the early twentieth century based on work by the Russian biologist Constantin Mereschkowsky but was popularized by Lynne Margulis in the late 1960s. It posits, in its most recent version, that an anaerobic Archaea evolved the key eukaryotic traits of a cytoskeleton and membrane system, including a nucleus, by reorganization of existing cellular components (autogenously). The cytoskeleton allowed the now eukaryotic cell to ingest food particles, including Bacteria. Two of these Bacteria were not digested but survived within the host cytoplasm as endosymbionts, eventually giving rise to mitochondria and plastids by reduction and gene transfer to the host nucleus.

The serial endosymbiosis theory posits that the invention of the nucleus occurred before the mitochondrion endosymbiosis. The existence of structurally simple anaerobic eukaryotes, like the microscopic protozoon parasites *Giardia*, microsporidia, and *Trichomonas*, which lack mitochondria, was interpreted as being consistent with this hypothesis. In 1982, these species were called Archezoa by Tom Cavalier Smith to emphasize that they

TABLE 1. Some general features of Bacteria, Archaea, and Eukarya

Property	*Bacteria*	*Archaea*	*Eukarya*
Membrane-enclosed nucleus	No	No	Yes
DNA in circular form	Yes (most species)	Yes	No
Mitosis	No	No	Yes
Operons	Yes	Yes	No (rare cases)
Histone proteins present	No	Yes (some species)	Yes
Initiator transfer RNA	Formylmethionine	Methionine	Methionine
Cytoskeleton	No	No	Yes
Membrane lipids	Ester-linked	Ether-linked	Ester-linked
Energy-generating organelles	No	No	Mitochondria and plastids

might have split from other eukaryotes before the mitochondrion endosymbiosis. However, over the past few years it has become clear that Archezoa contain genes that could have come only from the mitochondrion endosymbiont, the simplest explanation being that they have lost, rather than never had, mitochondria. Loss of mitochondria during adaptation to anaerobic environments is already known for some groups of predominantly aerobic mitochondria-containing protozoa, with ciliates being the best example.

In 1998, the demise of the Archezoa encouraged William Martin and Miklos Müller to propose a radical new hypothesis for the origin of eukaryotes. In the hydrogen hypothesis, a hydrogen-utilizing archaeon and a hydrogen-producing bacterium (the mitochondrion symbiont) came together to forge the common ancestor of eukaryotic cells. The cell formed was still a prokaryote, with the bacterium (the protomitochondrion) living inside the archaeon "host." The endomembrane system, nucleus, and cytoskeleton evolved autogenously only after this event. Metabolic dependencies between different prokaryotes of the kind (interspecies hydrogen transfer) invoked by Martin and Müller are common in nature and are called syntrophic interactions.

We will probably never know which, if either, of these two models (or the many other models in the literature) captures more accurately the origin of eukaryotes. There is the expectation that, if any source of information can illuminate the deep past of cellular life, it will be the sequences of complete genomes. At the time of this writing, there are five complete eukaryotic genomes and around 100 complete or nearly complete prokaryotic genomes. Analyses of the yeast genome have demonstrated that it contains at least two types of genes, consistent with a chimeric origin (an origin from two or more distinct entities) for the eukaryotic genome. Genes encoding components of transcription, translation, and replication generally are more similar to those of Archaea, whereas those encoding enzymes of metabolism often most resemble those of Bacteria. Some of the bacterial genes on eukaryotic genomes can be identified as originating from the mitochondrial and plastid endosymbionts. These data fit both the serial endosymbiosis theory and the hydrogen hypothesis, and the trees in Figures 1 and 2 also predict the presence of Archaea-like genes on eukaryotic genomes. The presence of bacterial genes on the yeast genome, which do not appear to have originated from the mitochondrial and plastid endosymbionts, suggest other, perhaps multiple, contributions to the eukaryotic genome.

The process discussed above is called horizontal gene transfer, whereby genes from one organism are

FIGURE 2. Rooting the Tree of Life Using Ancestrally Duplicated Genes.
The tree shows the results obtained when a tree based upon elongation factor EF-G/2 was used to root a tree based upon EF-Tu/1-alpha. The original analysis was done by Baldauf et al. 1996. Note that in this tree the eukaryotes arise from within the Archaea. Courtesy of author.

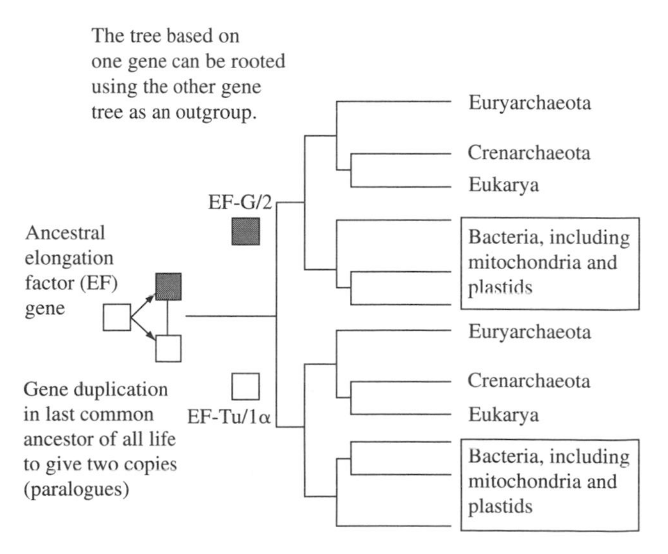

transferred to the genome of another. It had long been known that disease-causing Bacteria can exchange genes that confer drug resistance, but gene transfer from prokaryotes to eukaryotes was thought to be rare. The extent to which horizontal gene transfer has shaped prokaryote and eukaryote genome evolution is one of the most interesting questions facing evolutionary biologists. Put simply, if horizontal gene transfer is very common, then our attempts to reconstruct a single tree of life from molecular data might be doomed to failure, because each gene tree would tell a different story. At present, it appears that many gene trees for eukaryotes are in general agreement with each other, once methodological artifact is factored out, suggesting that horizontal gene transfer has not erased all evidence of eukaryote relationships. The situation for prokaryotes is still unclear. Darwin finished the *Origin of Species* by writing, "From so simple a beginning endless forms most beautiful and most wonderful have been, and are being, evolved." Detailed phylogenetic analyses of prokaryote and eukaryote genomes over the next few years should reveal the extent to which a net of life, rather than a tree of life, is the better explanatory principal for this diversity.

[*See also* Cell Evolution; Cellular Organelles; Cortical Inheritance.]

BIBLIOGRAPHY

Baldauf, S. L., J. D. Palmer, and W. F. Doolittle. "The Root of the Universal Tree and the Origin of Eukaryotes Based upon Elongation Factor Phylogeny." *Proceedings of the National Academy of Sciences USA* 93 (1996): 7749–7754. The analyses upon which Figure 2 in this article is based.

Doolittle, W. F. "Phylogenetic Classification and the Universal Tree." *Science* 284 (1999): 2124–2128. A beautifully written account of the author's view that horizontal gene transfer may make our attempts to produce phylogenies for organisms fundamentally flawed.

Hirt, R. P., J. M. Logsdon, B. Healy, M. W. Dorey, W. F. Doolittle, and T. M. Embley. "Microsporidia Are Related to Fungi: Evidence from the Largest Subunit of RNA Polymerase II and Other Proteins." *Proceedings of the National Academy of Sciences USA* 96 (1999): 580–585. Demonstrates the problems encountered when poor models are used to infer ancient relationships from gene sequences.

Martin, W. "A Briefly Argued Case That Mitochondria and Plastids Are Descendants of Endosymbionts, But That the Nuclear Compartment Is Not." *Proceedings of the Royal Society of London B* 266 (1999): 1387–1395. Critically discusses endosymbiont theories for eukaryote organelles and presents the author's hypothesis for the origin of the nuclear compartment.

Martin, W., and M. Müller. "The Hydrogen Hypothesis for the First Eukaryote." *Nature* 392 (1998): 37–41. The first thoroughly articulated new hypothesis for eukaryotic origins in thirty years. In the same issue of *Nature* (pages 15–16), W. F. Doolittle contributes a lucid and highly readable commentary that compares the new hypothesis to the serial endosymbiosis theory.

Nisbet, E. G., and N. H. Sleep. "The Habitat and Nature of Early Life." *Nature* 409 (2001): 1083–1091. An informative and readable introduction to the literature and speculations regarding early life on earth, including the origins of photosynthesis and of the oxygen atmosphere.

Ochman, H., J. G. Lawrence, and E. A. Groisman. "Lateral Gene Transfer and the Nature of Bacterial Innovation." *Nature* 405 (2000): 299–304. A technical discussion of the evidence for, and role of, horizontal gene transfer as a force in prokaryote evolution.

Pace, N. R. "A Molecular View of Microbial Diversity and the Biosphere." *Science* 276 (1997): 734–470. Gene surveys of natural samples suggest that most prokaryotes have never been isolated and studied in the laboratory. This has important implications for our understanding of life's diversity. A brief and highly readable review by one of the founders of the new field of molecular microbial ecology.

Roger, A. J. "Reconstructing Early Events in Eukaryotic Evolution." *American Naturalist* 154 (1999): 146–163. A thoughtful and clearly written technical summary of what gene trees can tell us about early eukaryotic evolution. Contains a concise and perceptive description of the Archezoa hypothesis and the data that are taken to reject it.

Woese, C. R., O. Kandler, and M. L. Wheelis. "Towards a Natural System of Organisms: Proposal for the Domains Archaea, Bacteria, and Eucarya." *Proceedings of the National Academy of Sciences USA* 87 (1990): 4576–4579. A proposal for a universal tree of life (Figure 1 in this article), representing the culmination of almost twenty years of comparative sequencing of SSU rRNA molecules.

Zuckerkandl, E., and L. Pauling. "Molecules as Documents of Evolutionary History." *Journal of Theoretical Biology* 8 (1965): 357–366. Argues that the sequences of genes and proteins can reveal the evolutionary relationships between species.

— T. Martin Embley

PROTEIN FOLDING

Proteins are chains of hundreds to thousands of amino acids. The protein folds are the basic constructional units of proteins. Each fold consists of a subset of the overall linear chain of amino acids usually between about 80 to 200 amino acids long (its primary structure) folded into a complex three-dimensional arrangement (its native conformation). The native conformation exhibits a hierarchical structure composed of basic secondary structural elements such as the α helical and β strand conformations, as well as turns, and regions that are more or less unstructured. These elements are often arranged into more complex motifs that are combined together to make up the native conformation of the fold (see Figure 1). Some proteins are made up of only one fold, but most are made up of combinations of two or more.

The successive amino acids in the linear chain are linked together by strong or covalent bonds, while the complex spatial arrangement of the folded chain is stabilized by much weaker chemical bonds, that do not involve sharing electrons but arise from electrostatic forces

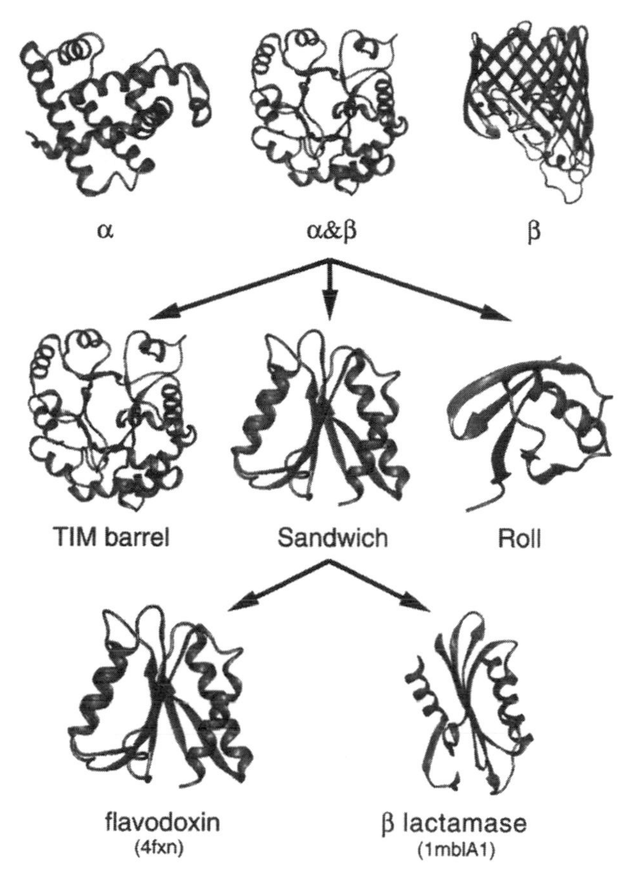

FIGURE 1. Structural Classes of Protein Folds. Showing how the folds can be classified into different structural classes. Top row shows three basic fold classes: α, containing only α helices; α and β, containing α helices and β sheets; and β, containing only β sheets. Middle row shows three different architectural subclasses of the α & β class: TIM barrel, three-layer sandwich and roll. Bottom row shows two different arrangements of the three-layer sandwich. The spiral conformations are the α helices, the broad arrows are the β sheets. Orengo, C. A. et al., "CATh-a Hierarchic Classification of Protein Domain Structures." *Structure* 5 (1997): 1093–1108, with permission from Elsevier Science.

between adjacent atoms; known as "noncovalent bonds" or "weak chemical bonds."

Folding. Immediately after its synthesis at the ribosome, the linear chain of amino acids folds into its native 3-D conformation. Folding occurs in such a way as to bring about the maximum number of favorable electrostatic interactions between the various constituent amino acids. During the process, which generally takes from a tenth of a second to about one minute, negatively charged groups tend to associate with positively charged groups, and hydrophobic (water hating) side chains tend to stack at the center of the molecule away from water while hydrophilic (water loving) side chains tend to arrange themselves on the surface in contact with water.

The result of the folding process—the native stable conformation of the fold—is known as its "minimum energy conformation," because any conformational deviation from this native state, however small—say, moving the relative position of two of its constituent α helices—requires an input of external energy to the system. We can think of the fold in its native conformation to be like a ball at the bottom of a dish. To move the ball in any direction away from its natural position at the bottom of the dish requires an input of energy. During folding, the unstructured amino acid chain is drawn—like the ball at the bottom of the dish—to its minimum energy conformation.

Anfinsen's classic experiments in the late 1950s and early 1960s with the simple protein ribonuclease A, showed that folding can occur in an aqueous medium in vitro—that is, outside the cell—without the assistance of any other molecular species. This finding proved that the amino acid sequence itself contained the necessary information to direct the folding process.

Because of the vast number of possible conformations that a folding amino acid chain could theoretically adopt, there is not time in a fraction of a second for a search of even a tiny fraction of all possible conformational space to find the minimum energy conformation. The process is generally considered to occur in a number of stages involving semistable, partially folded intermediates that lead down an energy gradient or "funnel" to the native minimum energy conformation.

The synthesis of an amino acid chain may take more than one minute, which is sufficient time for the nascent chains to interact with each other and/or other molecular species in their immediate surroundings, forming nonnative aggregates, before proper folding can occur. To avoid such illicit interactions, the cell utilizes a set of special proteins known as chaperones that protect the nascent chains from forming nonnative complexes.

Structure. Although the total possible number of different amino acid sequences composed of the 20 protogenic amino acids is immense (for a chain of 150 amino acids long the number is 20^{150} or approximately 10^{195}), and the number of possible three-dimensional conformations they can theoretically adopt is even greater, the total number of stable three-dimensional structures allowed by physics would appear to be restricted to a tiny finite set of about one thousand unique conformations.

The first line of evidence in support of this conclusion came to light during the 1970s, as the three-dimensional structure of an increasing number of folds was determined by X-ray crystallography, when it became apparent that the folds could be classified into a finite number of distinct structural families that had remained invariant for thousands of millions of years (see Figure 1).

A second line of evidence has come from detailed

studies of the way the various secondary structural elements such as α helices and β sheets are arranged within the folds. Detailed studies by Cyrus Chothia and his colleagues have shown that the topology of many of the folds can be accounted for by what amounts to a set of "constructional rules" governing the way that the various secondary structural motifs are combined and packed into compact three-dimensional structures. These rules are strictly analogous to the laws of crystallography that govern the way molecules are packed into crystals, or the atom-building rules that govern the assembly of subatomic particles into the ninety-two atoms of the periodic table. Consideration of these rules suggests that the total number of permissible folds is bound to be restricted to a very small number of no more than a few thousand folds. Confirmation that this is probably so is provided by a different type of estimate based on the discovery rate of new folds. Using this method, Chothia estimated that the total number of folds (utilized by living organisms on earth) may be no more than one thousand.

Whatever the actual figure, the fact that the total number of folds represents a tiny stable fraction of all possible polypeptide conformations, predetermined by the laws of physics, reinforces further the notion that the folds, like atoms, represent a finite set of built-in physical structures that would recur throughout the cosmos wherever there is carbon-based life utilizing the same twenty amino acids.

Further evidence that the protein folds represent intrinsic natural structures, "givens of physics," are those many cases where protein functions are clearly secondary adaptations of a set of primary immutable natural forms. This is true in the case of some of the more common folds also known as superfolds. In the case of one superfold, the so-called triosephospate isomerase (TIM) barrel, an eight-stranded α/β bundle (see Figure 1), the identical fold has been secondarily modified for many completely unrelated enzymic functions occurring in such diverse enzymes as triosephosphate isomerase, enolase, and glycolate oxidase. Another case where a basic fold has been secondarily modified is seen in the various elegant functional adaptations to oxygen uptake and carriage exhibited by the globin fold in myoglobin and the vertebrate hemoglobins.

That the folds are indeed a finite set of ahistoric physical forms is further supported by the finding that the same fold structure may be specified by many different apparently unrelated amino acid sequences, suggesting multiple separate discoveries of the same intrinsic structure during the course of evolution.

Taken together, the various lines of evidence suggest strongly that the folds represent a finite set of about 1,000 unique natural forms that like atoms or crystals are prescribed by a set of basic physical constructional rules, which restrict stable conformations to a small natural set of forms.

Platonic Forms. Most biologists today view organic forms in Darwinian terms to be basically mutable aggregates of matter—similar to the constructs of a child's erector set—assembled bit by bit under the direction of natural selection. The idea that biological forms might be prescribed by a set of physical laws like atoms or crystals is treated with incredulity. However, in the early nineteenth century among the *Naturphilosophen* (rational and transcendental morphologists), natural law was widely considered to be a major determinant of organic form. Many believed that underlying the adaptive diversity of life was a fixed set of invariable natural kinds like atoms or crystals. These natural kinds were thought to be prescribed by a set of natural laws unique to the vital realm and known as the laws of form. Hence the popularity in the early nineteenth century of the crystal as a metaphor for the order of life. Theodore Schwann, the originator of the cell theory, speculated in his great work *Microscopical Researches*, published in German in 1839, that cells might be a type of crystal. Viewing forms as primary "givens" of physics, functions were considered secondary modifications of primary forms. Such "natural kinds" and immutable laws of form bear a striking resemblance to the philosopher Plato's notion of ideal forms, or Platonic forms as they are known today. It is intriguing that the protein folds, the most thoroughly characterized of all organic forms, seem to be specified by a set of abstract laws of form of precisely the type that the pre-Darwinian biologists were seeking.

[*See also* Proteins.]

BIBLIOGRAPHY

Anfinsen, C. B. "Principles That Govern the Folding of Protein Chains." *Science* 181 (1973): 223–230. A technical account of some of the classic experiments in the field.

Brandon, C., and J. Tooze. *Introduction to Protein Structure*. 2d ed. New York, 1999. One of the best introductory overviews of protein structure, including an excellent chapter on protein folding.

Chothia, C. "One Thousand Families for the Molecular Biologist." *Nature* 357 (1993): 543–544.

Chothia, C., and A. V. Finkelstein. "The Classification and Origins of Protein Folding Patterns." *Annual Review of Biochemistry* 59 (1990): 1007–1039. A technical review of fold structure and classification.

Denton, M. J., and C. J. Marshall. "Laws of Form Revisited." *Nature* 410 (2001): 417.

Mathews, C. K., K. E. van Holde, and K. G. Ahern. *Biochemistry*. 3d ed. San Francisco, 2000. A standard textbook of biochemistry containing good introductory chapters on protein structure and folding.

Orengo, A. C., D. T. Jones, and J. Thornton. "Protein Superfamilies and Domain Superfolds." *Nature* 372 (1994): 631–634. A technical report providing evidence that the same fold may have several different functions.

— MICHAEL J. DENTON AND CRAIG J. MARSHALL

PROTEINS

Proteins are macromolecules constructed of nineteen types of α-aminocarboxylic acids (alanine, arginine, asparagine, aspartic acid, cysteine, glutamic acid, glutamine, glycine, histidine, isoleucine, leucine, lysine, methionine, phenylalanine, serine, threonine, tryptophan, tyrosine, valine) and a cyclic α-iminocarboxylic acid (proline), together comprising the twenty amino acids on which proteins are based. With the exception of glycine, the α carbon atom of amino acids is linked to four different types of substituents, conferring optical activity on these amino acids. The two mirror image forms of these amino acids are called the L-isomer and the D-isomer. In the case of proteins, amino acids are always of the L-isomer.

Proteins are linear strings of amino acids in which the carboxyl group of the amino acyl residue n is joined to the amino group of amino acyl residue $n + 1$ by a chemical bond, called the peptide bond. Information on the primary structure (the sequence of amino acids) of a protein is encoded in the nucleotide sequence of the messenger RNA of the corresponding protein-coding gene. [*See* DNA.] The polypeptide backbone of proteins has a repeated structure that consists of the amide nitrogen, the α carbon, and the carbonyl carbon atoms of the consecutive amino acyl residues. The peptide bonds connecting the amide nitrogen and the carbonyl carbon have a partial double-bonded character; rotation of this bond is therefore restricted. The amide nitrogen, the two α carbons, and the carbonyl carbon are usually in the same plane. In the planar peptide bond, the two α carbons are usually in the *trans* conformation; that is, they are on opposing sides of the polypeptide backbone. Peptide bonds involving the amide nitrogen of prolyl residues, however, may exist in the *cis* conformation; that is, the α carbons are on the same side of the polypeptide chain.

Properties of Amino Acids. The twenty amino acids differ in the chemical nature of their side chains, therefore, they possess distinct physicochemical properties that endow them with unique roles in proteins. Similarities and dissimilarities of amino acid residues affect their replaceability during protein evolution.

Glycine is the smallest amino acid, with only a hydrogen atom side chain. This gives the peptide backbones at glycine residues greater conformational flexibility than at other residues. The side chains of alanine, valine, leucine, isoleucine, and methionine differ in size but are similar in as much as they all have hydrophobic aliphatic side chains that interact more favorably with each other and with other nonpolar residues (e.g., aromatic residues) than with water. Proline also has an aliphatic side chain. However, its most relevant feature is that peptide bonds preceding a prolyl residue are more likely to adopt the *cis* configuration. The large hydrophobic side chains of phenylalanine, tyrosine, and tryptophan residues interact more favorably with each other and with other nonpolar amino acids (aliphatic amino acids) than with water; therefore, these residues frequently form the interior core of protein folds.

Serine and threonine have small aliphatic side chains with a polar hydroxyl group. The carboxyl groups of aspartic acid and glutamic acid side chains are negatively charged and very polar under physiological conditions. The amide side chains of asparagine and glutamine are polar and can serve as hydrogen-bond donors and acceptors. The ε-amino groups of lysyl and the δ-guanidino groups of arginyl residues are usually positively charged and thus are also very polar.

The imidazole side chain of histidine residues frequently participates in enzyme catalysis. Cysteine has a reactive thiol group, and this reactivity of cysteines is frequently exploited by enzymes. In the oxidative milieu of the extracellular space, cysteine residues may form covalent disulfide bonds that can endow extracellular proteins with greater stability.

Structure of Proteins. Thanks to the cooperation of a huge number of noncovalent interactions among backbone and side-chain atoms of the protein molecule (hydrogen bonds, electrostatic interactions, van der Waals interactions, hydrophobic interactions), the amino acid sequence of proteins usually defines a single, stable three-dimensional structure (Figure 1). First of all, noncovalent interactions can stabilize local spatial structures (secondary structure elements) of linear segments of polypeptide chains. A basic type of secondary structure element is the so-called α-helix. A typical α-helical structure usually contains ten to fifteen residues, with the backbone carbonyl oxygen of each residue hydrogen-bonded to the backbone amide of the fourth residue along the chain. A second major type of structural element is the β sheet formed from two or more β strands. β strands usually consist of three to ten consecutive residues; the polypeptide is almost fully extended. In the β sheet, hydrogen bonds are formed between the peptide backbones of adjacent β strands. There is a general tendency that β strands in a sheet are also adjacent in the primary structure. A third type of structural elements of globular proteins are the tight turns or loops that reverse the direction of the polypeptide chain at the surface of the molecules (usually between α helices and β strands) and thus make possible the overall globular structure.

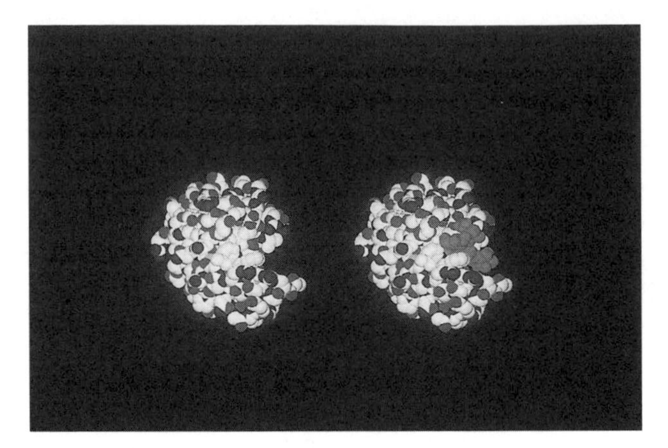

FIGURE 1. Computer Model of a Protein Enzyme.

FIGURE 2. Space Filled Flu Protein Model.

The three-dimensional structure of a protein, the overall spatial topology of its structural elements, is determined by long-range interactions between secondary structure elements and, most importantly, by interactions with water molecules. As a result of the favorable interaction of water molecules with polar and hydrogen bond acceptor and donor groups of proteins, polar side chains tend to be exposed to the solvent on the surface of the protein fold. Conversely, nonpolar side chains do not interact favorably with water. This leads to the exclusion of hydrophobic groups from the aqueous environment. Nonpolar side chains are thus usually buried in the interior of water-soluble proteins and frequently form hydrophobic cores of the protein fold.

The most important general feature of the three-dimensional structure of globular, water-soluble proteins is that they are compact and roughly spherical. Water molecules generally are excluded from protein interiors, and polar groups are on the surface of these proteins. Hydrophobic side chains predominate in the protein interior of globular proteins; the nonpolar residues are primarily involved in packing together the elements of the secondary structure.

Following speciation, proteins fulfilling the same biological function in the diverging species start to accumulate mutations independently. Despite extensive divergence in sequence, orthologous proteins invariably have been found to have very similar three-dimensional structures. The fact that the same protein fold can be defined by very different amino acid sequences indicates that not all positions are equally important for the structural integrity of the protein fold. In general, the sequences defining the main elements of the secondary structure (β sheets, α helices) and the interior hydrophobic residues in the core of the protein fold show the highest conservation. Conversely, surface loops that connect β strands or α helices show the greatest dissimilarities in amino acid sequences. Deletions and insertions occur most frequently at surface loops, usually with little perturbation of the protein fold. Of course, in the case of orthologous proteins with the same function, the most highly conserved residues also include those involved directly in the functional properties of the protein. For example, a comparison of the sequences of orthologous globins from various species has revealed that the most highly conserved residues are those that interact directly with the heme iron and oxygen (and are thus critical for globin function). Similarly, comparison of chymotrypsin sequences of various species has shown that the most highly conserved residues are those that are indispensable for the catalytic activity of these proteases.

Even in regions that show significant sequence variations among orthologous sequences, the choice of amino acid substitutions is not random. A survey of the evolution of many different protein families has identified some strong biases in amino acid substitutions. The most prevalent substitutions occur among amino acids with physicochemically similar side chains: amino acids with aliphatic side chains (isoleucine, valine, leucine, methionine), acidic and amide side chains (aspartic acid, glutamic acid, asparagine, glutamine), basic side chains (lysine, arginine), hydroxylic side chains (serine, threonine), and aromatic side chains (tyrosine, phenylalanine, tryptophan). It is clear from these observations that the biases in amino acid replacements reflect the role of purifying selection: substitutions of physicochemically very similar residues are usually without major effect on the structure and function of proteins and are thus most likely to be accepted. Another noteworthy feature of the pattern of amino acid substitutions is that cysteine residues and large hydrophobic residues (tryptophan, tyrosine, phenylalanine) are replaced least often. These observations also reflect the role of purifying selection in protein evolution. Cysteine residues of extracellular proteins are usually involved in disulphide bonds that are essential for the stability of these proteins; such a cysteine cannot be substituted by any other residue without disrupting that disulphide bond. Similarly, the low mutability of tryptophan, tyrosine, and phenylalanine residues reflects the importance of these residues in forming the hydrophobic core of protein folds.

Evolution of Proteins. Different protein families may evolve at vastly different rates depending on the structural and functional constraints imposed on them by their biological importance for the organism. Proteins under stringent structural and functional constraint evolve at a very low rate because most substitutions would be disruptive and are rejected by natural selection. For example, histones evolve at a very low rate because nucleosome assembly (involving histone/histone and histone/DNA interactions) critically depends on a highly conserved histone structure. Because nucleosomes are the primary determinants of DNA accessibility, their structural integrity affects the vital processes of recombination, replication, mitotic condensation, and transcription. Most residues of histones are critical either for the formation of the histone fold or for their interaction with DNA or with other core histones to form the nucleosomes. Very few substitutions can occur at any site of the histone proteins without impeding some of these vital functions; therefore, histones are among the slowest evolving proteins.

Gene duplication giving rise to copies of a protein coding gene is the most important mechanism whereby new proteins with novel functions can arise. As a result of the divergence of duplicated genes, paralogous proteins with novel and distinct functions may eventually emerge. In some cases, one of the duplicated genes retains its original function, whereas the other paralog accumulates molecular changes that allow it to perform a different task from the genes' common ancestor. For example, one of the lysozyme genes of mouse is expressed in myeloid tissues and the protein fulfills a role in defense against bacterial infections, whereas the other lysozyme gene is expressed in the intestine, where it fulfills a novel, digestive function.

In many cases, both duplicates acquire functions that are different from that of their common ancestor. For example, in the globin superfamily, many duplicates have acquired new functional, regulatory, and expression characteristics (e.g., myoglobin vs. hemoglobins, embryonic vs. fetal vs. adult forms of hemoglobins). Myoglobin has adapted to be the oxygen-storage protein of muscle, whereas hemoglobins became the oxygen carriers in blood. As dictated by their specific roles, myoglobin evolved a higher affinity for oxygen than did hemoglobin, whereas the functions of the various hemoglobins have become much more refined according to their slightly different functional roles. For example, fetal hemoglobin has a higher oxygen affinity than adult hemoglobins and can thus function more efficiently in the low-oxygen environment of the fetus.

Different paralogs may specialize in different subfunctions of their common ancestor. This point may be illustrated by the pancreatic proteases (trypsin, chymotrypsin, elastase) involved in the digestion of proteins present in foodstuffs. These closely related proteases differ markedly in their primary substrate specificity, even though their three-dimensional structures are strikingly similar. Elastase cleaves proteins preferentially in the vicinity of amino acids with small nonpolar side chains; chymotrypsin cleaves primarily at bulky hydrophobic residues, whereas trypsin cleaves almost exclusively at arginyl and lysyl residues.

In summary, studies on paralogous and orthologous members of divergent protein families have revealed that, during evolution, the common protein fold is preserved both in paralogs and orthologs. The structural elements important for the integrity of the protein fold usually accept mutations at very similar rates and patterns in orthologous and paralogous proteins. One general difference between orthologous and paralogous proteins, however, is that orthologous proteins are likely to fulfill very similar functions in different species, whereas paralogous proteins are more likely to have diversified in function. Consequently, in comparing orthologous proteins, residues critical for structure and function are equally likely to be conserved, whereas in comparing paralogous proteins that fulfill very different functions, only residues essential for defining the three-dimensional structure may be conserved.

[*See also* DNA and RNA.]

BIBLIOGRAPHY

Branden, C., and Tooze, J. *Introduction to Protein Structure.* New York, 1991.

Creighton, T. E. *Proteins.* 2d ed. New York, 1993. One of the best textbooks on structure, function, and evolution of proteins.

Dayhoff, M. O., Schwartz, R. M., and Orcutt, B. C. "A Model of Evolutionary Change in Proteins." *Atlas of Protein Sequence and Structure* 5 (Suppl. 3, 1978) 345–352. A classic paper analyzing the pattern of amino acid changes during protein evolution.

Gillespie, J. H. *The Causes of Molecular Evolution.* New York and Oxford, 1991. Discusses many instructive examples of how proteins with novel or modified functions may emerge.

Li, W. H. *Molecular Evolution.* Sunderland, Mass., 1997. One of the best books on molecular evolution, with detailed descriptions of the principles of the evolution of protein-coding genes.

— László Patthy

PROTEOMICS. *See* Overview Essay *on* Genomics and the Dawn of Proteomics *at the beginning of Volume 1.*

PROTISTS

Protists are eukaryotic organisms other than plants, animals, and fungi. They are a "paraphyletic" group, meaning that they do not all share a single common ancestor that is unique only to protists. Although the most accurate term for this diverse assemblage of organisms is "microbial eukaryotes," the term "protist" is still used as a shorthand reference. Understanding the evolution of these lineages requires knowledge of relationships among all eukaryotes, coupled with comparative data on the cellular and molecular biology of protists.

Relationships among Eukaryotes. The evolutionary relationships among protists (also known as protozoa, protoctista, and infusoria) and between protists and other eukaryotes have been subject to much debate and revision. In Ernst Haeckel's 1866 depiction of the "Tree of Life," the "Protista" included such diverse organisms as microbial eukaryotes, fungi, some animals (e.g., sponges), and bacteria. Eukaryotes comprise four of the five kingdoms of life described by Whittaker and Margulis (1978): plants, animals, fungi, and protists. However, from an evolutionary perspective, these divisions are inappropriate because protists are paraphyletic (Figure 1). For example, some protist groups (e.g., choanoflagellates) are more closely related to animals than they are to other protists.

Relationships among the estimated 80–200 eukaryotic groups remain poorly understood. To date, molecular systematic studies have focused on only a small portion of eukaryotic diversity, with a bias toward samplings of plants, animals, fungi, and disease-causing microbial eukaryotes. A working hypothesis for eukaryotic relations, based in part on analysis of multiple molecular markers from a limited number of taxa (Baldauf et al., 2000), is shown in Figure 1. Significantly, data from multiple genes challenge many traditional classification systems. For example, of the three major cell types within microbial eukaryotes—amoebae, flagellates, and ciliates—only one, the ciliates, is monophyletic. Similarly, photosynthetic microbial eukaryotes (algae), including euglenids, red algae, brown algae, and green algae, are scattered across the tree.

Although identifying early-diverging lineages is essential for developing and testing models for the origin of eukaryotes, the basal position of the diplomonads and parabasalids is much debated. The diplomonads include the genus *Giardia*, which is the causative agent of diarrhea associated with "beaver fever." The parabasalids include numerous lineages that live in association with animals as either parasites (e.g., *Trichomonas vaginalis* in humans) or commensals (e.g., the hypermastigotes in the hindguts of termites and wood roaches). Many of the genes sampled from taxa within the diplomonads and parabasalids appear to be evolving rapidly, and there is often little phylogenetic resolution to support the placement of these lineages. Greater taxonomic sampling of potentially early-diverging eukaryotic lineages may help to resolve the base of the eukaryotic tree.

In contrast, relationships among a few lineages of microbial eukaryotes appear robust to both molecular and morphological analyses. As predicted first by ultrastructural data, the sister status of euglenids and kinetoplastids has found support in most molecular genealogies. The euglenids are comprised of autotrophic and heterotrophic flagellates, including the freshwater *Euglena gracilis*. The kinetoplastids, defined by the unusual arrangement of the mitochondrial genome into maxicircles and minicircles, contain a combination of free-living (e.g., *Bodo saltans*) and parasitic flagellates (e.g., *Trypanosoma brucei*, the causative agent of African sleeping sickness, and *Leishmania major*, the causative agent of leishmaniasis). Two other groups that emerge from molecular and recent ultrastructural analyses are the alveolates (ciliates, apicomplexans, dinoflagellates) and heterokonts, or stramenopiles (diatoms, brown algae, golden algae, water molds, slime nets). Alveolates are united by the presence of alveolar sacs underlying the surface of the cells. Heterokonts are flagellates generally characterized by two distinct flagella, one of which is covered with hairlike projections named mastigonemes.

Evolutionary Innovations in Microbial Eukaryotes. Microbial eukaryotes are marked by diverse innovations that can be seen in their life cycles, morphology, genetics, and genome structures. There are many unanswered questions about the evolution of protists, making these organisms particularly fascinating research systems for evolutionary biologists.

Origin of the eukaryotic cell. Eukaryotes are defined by two evolutionary innovations: the nucleus and the cytoskeleton. Although named for the presence of a nucleus (*eu* = true, *karyon* = kernel or seed), it is the cytoskeleton and related proteins that allowed for the dramatic variation in morphology present among eukaryotes. The origin of the eukaryotic cytoskeleton remains a mystery. As part of the serial endosymbiosis theory, Lynn Margulis argues that the eukaryotic cytoskeleton resulted from a motility-driven endosymbiosis between an archaeon and a spirochete (or spirochete-like bacterium; Margulis et al., 2000). However, there are no eukaryotic cytoskeletal proteins that show specific phylogenetic affinity to the spirochetes. Moreover, although homologs of some cytoskeletal proteins have been found in bacteria (e.g., FtsZ and MreB are bacterial homologs of tubulin and actin, respectively), there is considerable evolutionary distance between these prokaryotic and eukaryotic proteins. There are currently no

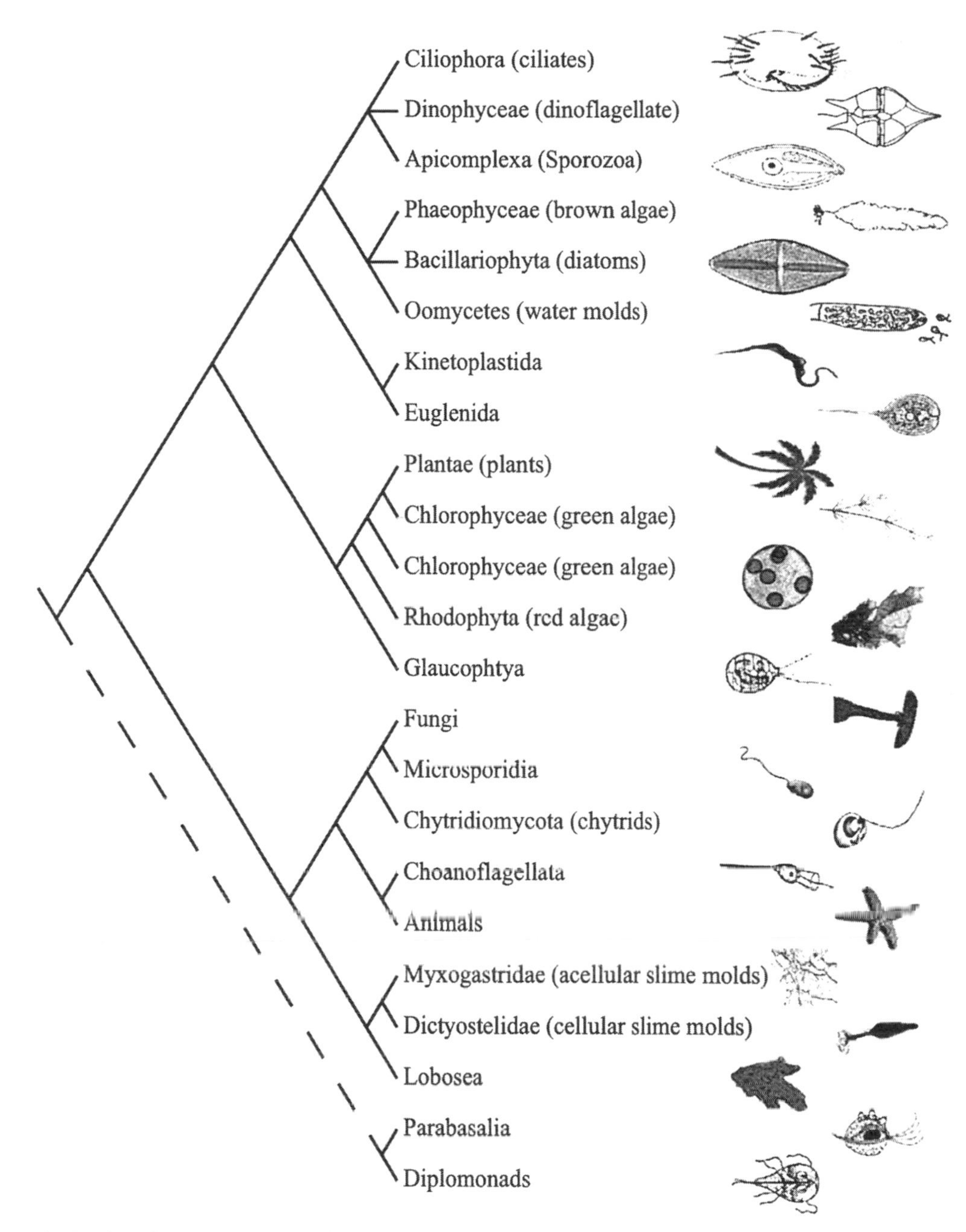

FIGURE 1. Hypothesized Evolutionary Relationships among Eukaryotes.
Phylogeny is redrawn in part from analyses of multiple genes (Baldauf et al., 2000). Dashed line indicates debated placement of basal eukaryotic lineages. Courtesy of Laura A. Katz.

data or convincing models to explain the origin of the eukaryotic nucleus and the associated endoplasmic reticulum.

Origin of mitochondria. One of the few certainties in the origin and diversification of eukaryotes is that mitochondria, found in many but not all eukaryotes, are derived from an α-proteobacterium. Evidence for this symbiotic origin includes the membrane structure of mitochondria, the mode of division, the presence of a circular genome, and the sequence of mitochondrial genomes. Still debated is the exact timing of the acquisition of mitochondria, as is the relationship between mitochondria and hydrogen-producing organelles (hydrogenosomes) found in diverse microbial eukaryotes (e.g., parabasalids, some ciliates, chytrids). Although hydrogenosomes in ciliates and chytrids are derived from mitochondria, the origin of hydrogenosomes in parabasalids is unclear, and the evolutionary history of the enzymes involved in metabolism of all hydrogenosomes is only beginning to be elucidated.

Complex life cycles. Protist life cycles vary in the relative timing of meiosis, mitosis, and cell division. For example, there is a decoupling of karyokinesis (division of the nucleus) and cytokinesis (division of the cell) in many microbial eukaryotes. Karyokinesis occurs without cytokinesis to generate multinucleated forms of acellular slime molds (e.g., *Physarum*), apicomplexans (e.g., *Plasmodium*), green algae, water molds (oomycetes), ciliates, foraminifera, parabasalids, and many other eukaryotic lineages. Multicellularity has also evolved many times and is found in the cellular slime molds, diatoms, ciliates, green algae, brown algae, and red algae.

The life cycle of cellular slime molds includes three distinct phases: single-celled amoebae, multicelled "slugs," and differentiated fruiting bodies (Figure 2). Individual amoebae of the cellular slime mold *Dictyostelium* feed on bacteria in the soil. Appropriate stimuli, including cyclic-AMP, can induce amoebae to aggregate en masse to form a multicellular "slug" (pseudoplasmodium) with differentiated anterior and posterior regions. This "slug" then travels through soil until it transforms into a fruiting body, in which some cells differentiate into stalk while cells at the top become spores capable of transforming into new amoebae.

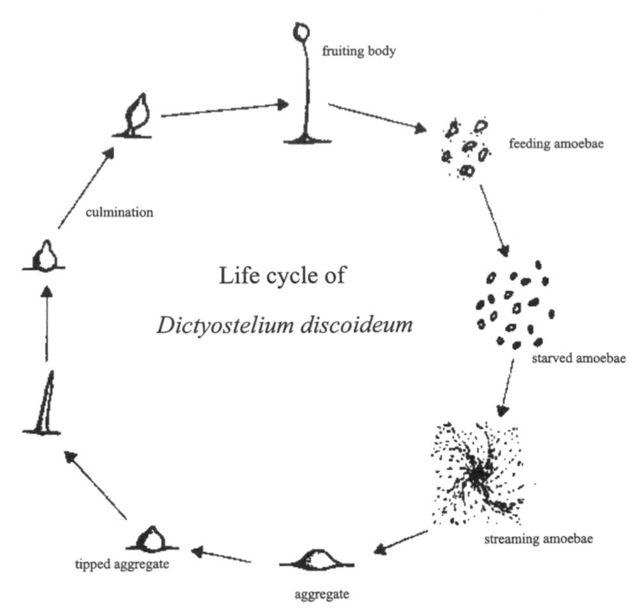

FIGURE 2. Life Cycle of a Cellular Slime Mold. Single-celled amoebae feed on bacteria in the soil. Upon starvation, amoebae stream together to form an aggregate, or "slug." The aggregate then transforms into a differentiated fruiting body and rises above the surface through a process called culmination. Modified from a figure at www.bioch.ox.ac.uk/~pcnlab/cycle.gif

Morphological plasticity of microbial eukaryotes. Some groups of microbial eukaryotes are marked by dramatic morphological plasticity in which individual cells are capable of transforming from one form to another without division. For example, genera such as *Naegleria* and *Psalterimonas* and some myxomycetes and dinoflagellates are all "amoeboflagellates" that alternate between amoeboid and flagellated stages. Likewise, some ciliates (e.g., *Miamiensis*) absorb their oral ciliature and generate new oral structures in response to changing food sources. Perhaps even more striking is the somatic inheritance of induced changes in the cortex of ciliates. For example, microsurgery can induce the formation of a second mouth that is a mirror image of the original oral apparatus in the ciliate *Pleurotricha* (Grimes et al., 1980). The body plan of these altered two-mouthed ciliates is heritable through cell division.

Origin of photosynthetic eukaryotes. Unlike the single evolutionary acquisition of mitochondria, there is considerable evidence that eukaryotes have become photosynthetic numerous times. Lineages thought to be descendants of a primary acquisition of a cyanobacterial endosymbiont include the glaucocystophytes, red algae, and green algae (including plants). Peptidoglycan, a protein common to the cell walls of bacteria, is still found surrounding plastids in glaucocystophytes. All other photosynthetic eukaryotic lineages, including euglenids, dinoflagellates, cryptomonads, diatoms, chlorarachniophytes, brown algae, and coccolithophores, are likely the result of secondary endosymbiosis—that is, endosymbiosis of a photosynthetic eukaryote by a heterotrophic eukaryote. Evidence for secondary (and sometimes even tertiary) endosymbiosis comes from (1) the cell biology of some photosynthetic eukaryotes, and (2) molecular systematics of plastid-encoded genes. The clearest evidence for secondary endosymbiosis is the presence of nucleomorphs, remnant nuclei from symbionts, found associated with plastids in both chlorarachniophytes and cryptomonads (Figure 3). Sequencing of the nucleomorph genes (and genomes) reveals the ancestry of the secondary endosymbionts. For example, sequences of the cryptomonad nucleomorph cluster with red algal sequences, indicating that cryptomonads are the result of a symbiosis between a flagellate and a red alga.

The numerous innovations of dinoflagellates. Dinoflagellates, major components of photosynthesis in coral reefs, can cause red tides and parasitic fish kills in estuaries, and they serve as important components of marine and freshwater food webs. The cell biology of dinoflagellates is marked by numerous unique characteristics. The motile forms of dinoflagellates are often characterized by a distinctive morphology with one flagellum located in a cleft around the body of the cell and a second, trailing flagellum. The life cycle of some di-

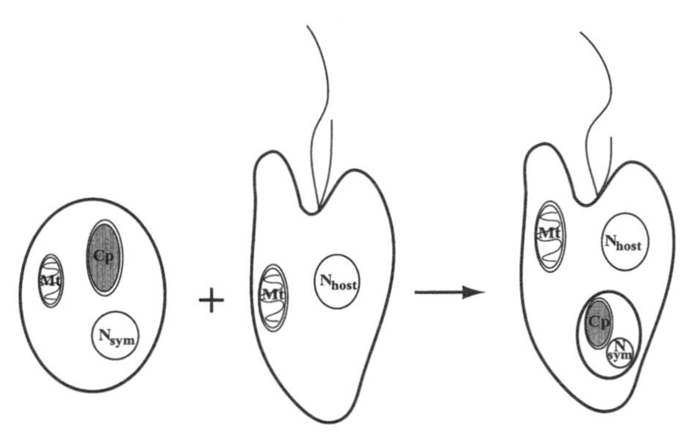

FIGURE 3. Nucleomorphs, Remnant Symbiont Nuclei, Result from Secondary Endosymbiosis.
Secondary endosymbiosis involves a eukaryotic host and a eukaryotic symbiont. Nsym = symbiont nucleus, which becomes the nucleomorph following endosymbiosis; Nhost = host nucleus; Cp = chloroplast; Mt = mitochondria. Courtesy of Laura A. Katz.

noflagellate species can be extremely complex, including numerous (>10) flagellate and amoeboid forms. Some dinoflagellates have also lost histones (found in all other eukaryotes and some archaea) and instead wrap their DNA around novel, basic proteins. Chromatin in dinoflagellates is always condensed; mitosis occurs with the nuclear envelope remaining intact; and, in some taxa, poorly studied "tunnels" form through the nucleus to connect microtubules to chromosomes during mitosis. Finally, analysis of the diversity of chloroplasts in dinoflagellates, interpreted in light of recent molecular phylogenies, suggests that plastids have been acquired (and lost) multiple times within this group through secondary, tertiary, and perhaps even quaternary endosymbioses.

Nuclear dimorphism: ciliates and some foraminifera. At least two groups of microbial eukaryotes, (ciliates and some foraminifera) have differentiated nuclei within single cells. This nuclear dimorphism is best characterized in ciliates, where the two types of nuclei are referred to as micronuclei and macronuclei (Figure 4). Micronuclei are transcriptionally inactive nuclei capable of going through mitosis and meiosis. Macronuclei, the site of virtually all transcription (and hence gene expression) in ciliates, are derived from micronuclei through a series of chromosomal rearrangements that include fragmentation, elimination of internal segments, and amplification of remaining chromosomes. These processes generate macronuclei in some ciliates (e.g., members of the class Spirotrichea) with as many as 25 million "chromosomes," many containing only a single gene, a small amount of untranslated sequence, and telomeres. Comparative data from diverse ciliates suggests that extensive fragmentation of germline chromosomes has arisen at least twice and maybe three times. In an apparently limited group of spirotrichs, these chromosomal processes have generated "scrambled genes" in the micronucleus that must be reordered to create functional genes during the development of the macronucleus (Figure 4). Not surprisingly given their very large number of chromosomes, the macronuclei of ciliates divide through a nontraditional form of division called amitosis.

RNA editing in microbial eukaryotes. RNA editing, on post-transcriptional modification of RNAs, was first described in the mitochondria of trypanosomes and more recently characterized from diverse eukaryotic

FIGURE 4. Dimorphic Nature of Ciliate Nuclei.
The micronucleus (left) is analogous to a germline nucleus and is transcriptionally inactive. Following conjugation, when haploid micronuclei are exchanged between cells, the zygotic nucleus divides by mitosis and one of the resulting daughter nuclei develops into a macronucleus through a series of chromosomal rearrangements. The macronucleus is the site of virtually all transcription in ciliates. Extensive fragmentation and scrambling of chromosomes in ciliates (right). (A) Generation of "gene-sized" macronuclear chromosomes; (B) scrambling of micronuclear chromosomes. MDS = macronuclear destined sequence (in this case, part of a coding region); IES = internally eliminated sequence; telo = telomere. Courtesy of Laura A. Katz.

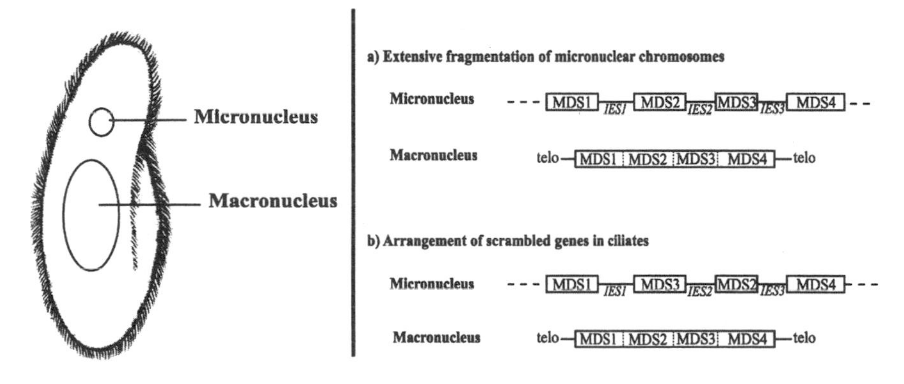

groups including acellular slime molds, *Acanthamoeba*, plants, and animals. In trypanosomes, RNA editing can alter up to 50 percent of the information encoded within mitochondrial genes. For example, the cytochrome oxidase III (coIII) gene, found in mitochondrial genomes of other eukaryotes, exists only as a nonfunctional cryptogene (hidden gene) in *Trypanosoma brucei*. Following transcription, large numbers of uracils are inserted to generate a functional coIII gene. In trypanosomes, the minicircles of the mitochondrial genome serve as guide RNAs to edit the cryptogenes present in the maxicircles. RNA editing in eukaryotes is now known from phylogenetically diverse eukaryotes and encompasses numerous processes, including uracil insertion/deletion, cytosine or dinucleotide insertion, C to U substitutions, and A to I substitutions.

Genome diversity in microbial eukaryotes. Recent sequencing of protist genomes has revealed additional surprises. For example, the microsporidian *Encephalitozoon* not only contains the smallest known eukaryotic genome, but the proteins encoded by this genome are on average 4 percent smaller than homologous proteins in other lineages. This suggests that the selection for small genome size in this lineage has been strong enough to affect diverse proteins. The organization of genes in the genome of the kinetoplastid *Leishmania* is unusual in that transcribed regions are clustered along one strand of the chromosomes or the other, with a low amount of noncoding sequences interspersed between coding sequences. As many as 50 genes can be found in a row on one strand with no genes present on the complementary strand. The structure of the genome of the diplomonad *Giardia lamblia* is highly variable among strains of the same species, such that individual isolates have different karyotypes. As more genomes of microbial eukaryotes are sequenced, even further diversity of genome content and structure will no doubt be discovered.

Biodiversity of Microbial Eukaryotes. Considerable debate exists on the number and distribution of microbial eukaryotes. Are most protist groups cosmopolitan, or does significant endemism exist? Proponents of the "cosmopolitan" viewpoint argue that there are few barriers to gene flow among groups of microbial eukaryotes, and that we have already characterized the majority of species of protists living on the earth (Finlay, 1999). In contrast, others argue that there is considerable endemism among microbial eukaryotes, and that there are many more species of protists yet to be discovered (Foissner, 1999). Current evidence, based largely on morphological assessment of protist diversity, suggests that investigations into novel locations and habitats will reveal additional species of microbial eukaryotes. However, a comprehensive answer to the question of the biodiversity of microbial eukaryotes requires coordinated efforts to examine the correspondence between morphologically defined protist species and the underlying genetics of populations across a diversity of habitats.

[*See also* Clonal Structure, *article on* Population Structure and Clonality of Protozoa; Malaria; Phylogenetic Inference.]

BIBLIOGRAPHY

Baldauf, S. L. "A Search for the Origins of Animals and Fungi: Comparing and Combining Molecular Data." *American Naturalist* 154 (1999): S178–S188.

Baldauf, S. L., A. J. Roger, I. Wenk-Siefert, and W. F. Doolittle. "A Kingdom-Level Phylogeny of Eukaryotes Based on Combined Protein Data." *Science* 290 (2000): 972–977. This paper presents comprehensive, multi-gene genealogies from diverse eukaryotes.

Bonner, J. T. "The Origins of Multicellularity." *Integrative Biology* 1 (1998): 27–36. Contains a comprehensive list of multicelled eukaryotes, focusing on protist groups. Presents evolution of multicellularity in a phylogenetic context.

Cavalier-Smith, T. "Kingdom Protozoa and Its 18 Phyla." *Microbiological Reviews* 57 (1993): 953–994.

Delwiche, C. F. "Tracing the Tread of Plastid Diversity through the Tapestry of Life." *American Naturalist* 154 (1999): S164–S177. Overview of the diverse histories of photosynthetic eukaryotes. Presents models of secondary and tertiary endosymbiosis by combining molecular genealogies with cell biology.

Feagin, J. E., J. M. Abraham, and K. Stuart. "Extensive Editing of the Cytochrome-C Oxidase-III Transcript in *Trypanosoma brucei*." *Cell* 53 (1988): 413–422.

Finlay, B. J., and K. J. Clarke. "Ubiquitous Dispersal of Microbial Species." *Nature* 400 (1999): 828–828. Supports the "cosmopolitan" views of distribution and gene flow among protists.

Foissner, W. "Protist Diversity: Estimates of the Near-Imponderable." *Protist* 150 (1999): 363–368. Argues that many protist species have yet to be discovered.

Grimes, G. W., M. E. McKenna, C. M. Goldsmithspoegler, and E. A. Knaupp. "Patterning and Assembly of Ciliature Are Independent Processes in Hypotrich Ciliates." *Science* 209 (1980): 281–283.

Haeckel, E. *Generelle morphologie der organismen. Allgemeine grundzüge der organischen formen-wissenschaft.* G. Reimer, Berlin, 1866. Contains Haeckel's famous early depiction of the "tree of life."

Hausmann, K., and P. C. Bradbury, eds. *Ciliates: Cells as Organisms.* Gustav Fischer, Stuttgart, 1996. Edited volume containing a variety of articles on ciliates.

Katz, L. A. "The Tangled Web: Gene Genealogies and the Origin of Eukaryotes." *American Naturalist* 154 (1999): S137–S145.

Katz, L. A. "Evolution of Nuclear Dualism in Ciliates: A Reanalysis in Light of Recent Molecular Data." *International Journal of Systematic Evolutionary Microbiology* 51 (2001): 1587–1592. Presents a model for the evolution of the micronuclear/macronuclear structure of ciliates through a combination of molecular systematics, characterizations of ciliate genome structures, and data on ciliate genetics.

Lang, B. F., M. W. Gray, and G. Burger. "Mitochondrial Genome Evolution and the Origin of Eukaryotes." *Annual Review of Genetics* 33 (1999): 351–397. Review of the origin and evolution of mitochondria.

Maier, U. G., S. E. Douglas, and T. Cavalier-Smith. "The Nucleomorph Genomes of Cryptophytes and Chlorarachniophytes."

Protist 151 (2000): 103–109. Description of structural and genealogical analyses of nucleomorphs, remnant symbiont nuclei derived from secondary endosymbiosis in two protist groups.

Margulis, L., M. F. Dolan, and R. Guerrero. "The Chimeric Eukaryote: Origin of the Nucleus from the Karyomastigont in Amitochondriate Protists." *Proceedings of the National Academy of Sciences of the United States of America* 97 (2000): 6954–6959. Most recent presentation of the "serial endosymbiosis theory" of the origin of eukaryotic cells.

Patterson, D. J. "The Diversity of Eukaryotes." *American Naturalist* 154 (1999): S96–S124. Important review of eukaryotic diversity. Presents hypotheses of relationships among eukaryotic groups as well as a long list of taxa of unknown affinities.

Roger, A. J. "Reconstructing Early Events in Eukaryotic Evolution." *American Naturalist* 154 (1999): S146–S163. Reviews models for the origin of eukaryotes in the context of molecular systematics. Includes discussion of the timing of the evolutionary acquisition of mitochondria.

Taylor, F. J. R. "Ultrastructure as a Control for Protistan Molecular Phylogeny." *American Naturalist* 154 (1999): S125–136. Compares relationships based on gene genealogies to those predicted by analyses of morphological characters.

Upcroft, P., and J. A. Upcroft. "Organization and Structure of the Giardia Genome." *Protist* 150 (1999): 17–23. Describes dramatic variation in karyotypes among *Giardia* strains.

Whittaker, R. H., and L. Margulis. "Protist Classification and the Kingdoms of Organisms." *BioSystems* 10 (1978): 3–18.

ONLINE RESOURCES

"Micro*scope" http://www.mbl.edu/microscope. Maintained at the astrobiology institute of the Marine Biological Laboratory (MBL), Woods Hole, Mass. with oversight from David J. Patterson (University of Sydney, Sydney Australia). Micro*scope aims to provide a comprehensive listing of microorganisms with links to images, DNA sequences, education materials, information of cellular biology of organisms, and other relevant web pages. Serves as a starting point to access online information on protists.

"The Leishmania Genome Project." http://www.ebi.ac.uk/parasites/leish.html. Hosted by the European Bioinformatics Institute at the European Molecular Biology Laboratory (EMBL) and chaired by Jennie Blackwell (Cambridge Institute for Medical Research, Cambridge U. K.). Includes data and references from collaborative research effort to completely sequence the genome of the kinetoplastid *Leishmania major*.

— LAURA A. KATZ

PROTOSTOME–DEUTEROSTOME ORIGINS

Within the animal kingdom, the Bilateria form a well-defined, monophyletic group. This is supported not only by the usually very obvious bilaterality with a conspicuous head but also through almost all sequencing analyses of genes, such as 18S ribosomal RNA, and through the presence of a characteristic cluster of *Hox* genes, which are arranged colinearly with the anteroposterior body axis.

The phylogeny of the bilaterians has been an area of much controversy over the last decades, but a consensus now seems to have emerged, with recognition of the two groups Protostomia and Deuterostomia. There is still some disagreement, however, about the position of Phoronida and Brachiopoda (Figure 1).

A classification recognizing three main groups, acoelomates, pseudocoelomates, and eucoelomates, often with the last-mentioned group divided into protostomes and deuterostomes, is frequently presented in papers and textbooks on molecular phylogeny as a background for "modern" phylogenetic results. This old idea is intimately connected with the theory that derives metazoans from ciliates or ciliatelike organisms that by cellularization should give rise to turbellarians. However, ciliates are not closely related to metazoans (the sister group of Metazoa appears to be the Choanoflagellata and further the Fungi), and neither reproduction nor ultrastructure of turbellarians resembles those of ciliates. Most morphologists now regard the three groups as grades rather than as clades, and the "three-level" classification based on body cavities is not supported by modern morphological or molecular studies.

Classification and Characteristics. The division of Bilateria into Protostomia and Deuterostomia was introduced at the end of the nineteenth century and has been based mainly on embryological differences, especially blastopore fate, and morphology of the central nervous system. Parallel pairs of names have been proposed by various authors. The terms *Protostomia* and *Deuterostomia* refer to the fate of the blastopore, which in principle should become the adult mouth (or perhaps mouth plus anus) in the protostomes and the anus in deuterostomes. Unfortunately, there is very considerable variation in blastopore fate within many phyla, so this characteristic cannot alone be used in practical classificatory work.

The architecture of the central nervous system (CNS) is the basis for a parallel pair of names, namely, Gastroneuralia and Notoneuralia. In protostomes, the CNS typically consists of a dorsal brain, a pair of circumpharyngeal commissures, and a paired or fused ventral nerve cord, whereas deuterostomes have a dorsal brain and nerve cord. However, the characteristic deuterostomian CNS is found only in chordates, where new investigations indicate that the brain and neural tube may actually be morphologically ventral.

Almost all recent studies recognize Protostomia and Deuterostomia, but there is not complete agreement about the definition of the two groups. In particular, the lophophorates (i.e., phoronids, brachiopods, and ectoproct bryozoans) appear controversial, and some texts even place them as a group "between" the two. A few studies conclude that the deuterostomes are an in-group of the protostomes.

Many different phylogenies have been proposed; the most common discrepancies between morphology-based

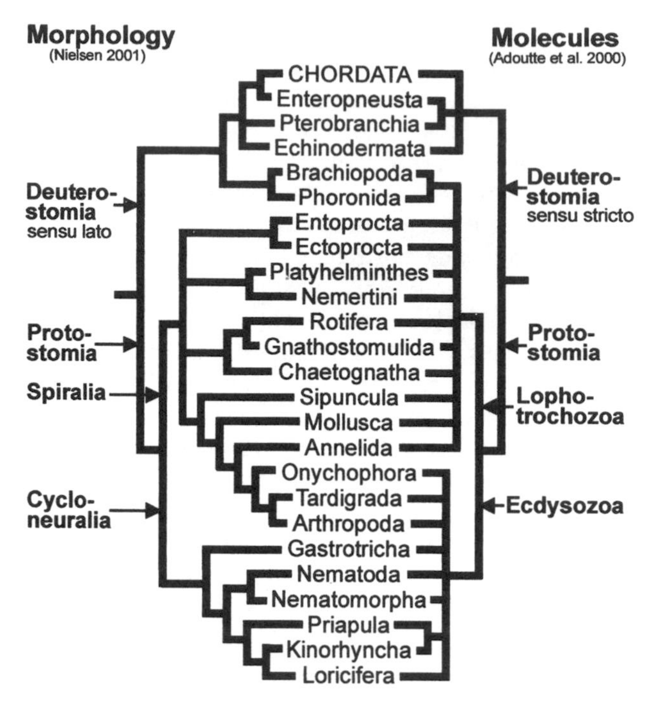

FIGURE 1. Phylogeny of the Bilaterians Based on Morphological and Molecular Characters.
Drawing by Claus Nielsen.

and molecular-based phylogenies are shown in Figure 1. The key problem appears to be the interpretation of the phoronids and brachiopods. Morphologically, it is easy to find synapomorphies characterizing Deuterostomia excluding phoronids and brachiopods, but no synapomorphy can be found characterizing Protostomia including phoronids and brachiopods. This is a strong indication that Protostomia in the wider sense is a paraphyletic group. In contrast, it seems easy to find synapomorphies characterizing both Protostomia in the narrower sense and Deuterostomia in the wider sense, with phoronids and brachiopods forming a sister group of Deuterostomia in the narrower sense.

Some of the more important characteristics separating Protostomia in the narrower sense and Deuterostomia in the wider sense will be mentioned in the following discussion.

Larval distinctions. A life cycle comprising a planktonic larva (probably planktotrophic) and a benthic adult is regarded by many as ancestral to all bilaterians. Morphological studies reveal clear differences both between larval and adult protostomes (including ectoprocts) and deuterostomes (including phoronids and brachiopods). Only a few of the distinguishing characteristics can be mentioned here.

Protostomes have trochophora larvae. Planktotrophic trochophores are known in groups of annelids, molluscs, and entoprocts, and the ciliary bands of larvae of some types of platyhelminths and nemertines and of some adult rotifers may be homologous to proto- and metatroch. The telotroch is lacking in many species. Nonplanktotrophic larvae resembling the planktotrophic types but lacking a metatroch occur in other species of the three first-mentioned phyla and in sipunculans. Some authors interpret the planktotrophic larvae as having evolved independently from nonplanktotrophic forms, but this is based on the assumption that the complicated structures needed for downstream feeding are equally easily gained and lost. Studies of other structures and groups clearly show that larval characteristics can be easily lost, in some cases even through a change in one gene, whereas the evolution of a larval characteristic such as the feeding structures must involve many genes and consequently be a complicated process that is unlikely to have evolved several times with many apparently identical structural details. The planktotrophic larvae have a prototroch and a metatroch of compound cilia on multiciliate cells functioning in downstream-collecting filter feeding (Figure 2).

Deuterostomes have dipleurula larvae. Planktotrophic larvae of this type are known from phoronids, brachiopods, echinoderms, and enteropneusts, and homologous ciliary bands are characteristic of adult phoronids, brachiopods, and pterobranchs. Nonplanktotrophic larvae are found in many groups too, with echinoderms especially demonstrating that planktotrophy has been lost independently in many lineages. Chordates lack larvae with ciliary bands, but many other characteristics indicate their deuterostome affinities. The planktotrophic larvae have a circumoral ciliary band, or neotroch, consisting of single cilia on monociliate cells functioning in upstream-collecting filter feeding.

Central nervous system distinctions. Protostomes generally have a very characteristic central nervous system not found in any deuterostome. It consists of an apical brain (or a brain developing from groups of cells situated just lateral to the apical organ), circumoesophageal commissures, and a ventral nerve cord, which may be paired or fused. During ontogeny, the lateral blastopore lips typically meet and fuse in the midline, dividing the blastopore into the mouth and anus, whereas longitudinal zones of the blastoderm on both sides of the fused blastopore give rise to the ventral nerve cord(s).

Body region differences. Deuterostomes are archimeric; that is, the body of all nonchordate phyla consist of three regions, which in some groups can be recognized externally, but which can always be recognized in the morphology of the mesoderm, both in the adults and during ontogeny. The middle region, or mesosome, carries ciliated tentacles in phoronids, brachiopods, pterobranchs, and echinoderms (where the tentacles are modified to podia). Enteropneusts have a collar-shaped mesosome without tentacles, and the three regions cannot be recognized in the chordates.

Gene differences. Molecular studies, especially sequencing of genes such as 18S ribosomal RNA, almost unanimously demonstrate a monophyletic Deuterostomia (excluding Phoronida and Brachiopoda), but show very little consistency and usually low resolution in the remaining part of the tree. The group Ecdysozoa is very often recognized, but especially the resolution of the Lophotrochozoa is often quite chaotic, for example, with various annelid and mollusc groups mixed together with other phyla. It therefore seems possible that the Lophotrochozoa is a paraphyletic group, as indicated in Figure 1.

The study of the occurrence of *Hox* genes in the bilaterian phyla is still in its infancy, and only a few groups have been investigated. Also here, the existence of the three groups Lophotrochozoa, Ecdysozoa, and Deuterostomia (in the narrow sense) seems indicated, but the interrelationships of the three groups has not been resolved because a *Hox* cluster is not found in the outgroups, so the Lophotrochozoa may well be paraphyletic.

[*See also* Metazoans.]

FIGURE 2. Larval Type and Ciliary Band of Protostome (Trochophora Larva with Downstream-Collecting Band) and Deuterostome (Dipleurula Larva with Upstream-Collecting Band).
Drawing by Claus Nielsen.

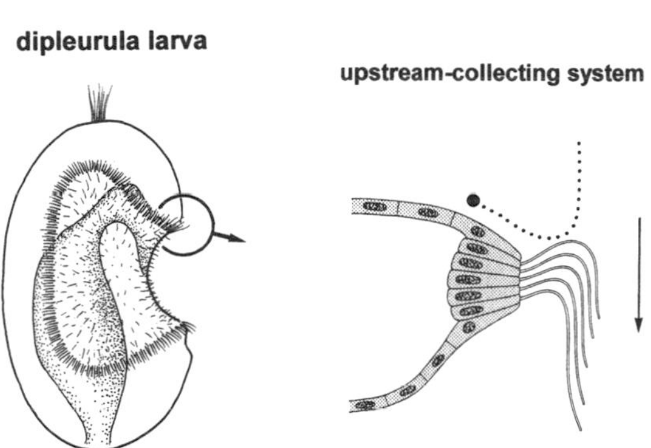

BIBLIOGRAPHY

Adoutte, A., et al. "The New Animal Phylogeny: Reliability and Implications." *Proceedings of the National Academy of Sciences USA* 97 (2000): 4453–4456.

Gilbert, S. F., and A. M. Raunio, eds. *Embryology: Constructing the Organism.* Sunderland, Mass., 1997.

Nielsen, C. *Animal Evolution: Interrelationships of the Living Phyla.* 2d ed. Oxford, 2001.

Raff, R. A. *The Shape of Life.* Chicago, 1996.

Ruppert, E. E., and R. D. Barnes. *Invertebrate Zoology.* 6th ed. Fort Worth, Tex., 1994.

Tudge, C. *The Variety of Life.* Oxford, 2000.

— CLAUS NIELSEN

PUNCTUATED EQUILIBRIUM

Most new ideas in science arise from surprising observations never made before—as when Galileo saw the moons of Jupiter through his telescope and knew that planets could not revolve around the Earth in fixed crystalline spheres, for satellites circling a planet would fracture the sphere in their revolution, and the sphere, therefore, could not exist. Punctuated equilibrium enjoys the odd status of a new interpretation attached to the oldest general observation in the history of our knowledge about the fossil record of species. Cuvier and all pre-Darwinian creationists among paleontologists knew the basic pattern that Darwin himself would also acknowledge but would reinterpret in an evolutionary context: that the great majority of apparent fossil species arise "suddenly" in an unresolvable geological moment represented by a single bedding plane; they die equally suddenly, at a geological boundary of extinction; and, most striking, they remain effectively stable in morphology throughout the strata of a fairly long history (5–10 million years as the average life of a fossil invertebrate species, for example).

For creationists, this pattern merely affirmed the origin of each species by divine fiat, and its stability thereafter in a nonevolutionary world. For Darwin, this pattern could only represent an artifact of a woefully imperfect fossil record; if, say, only 1 percent of actual time lies recorded in strata, then a gradualistic origin of species will generally be expressed as a moment, because all the intermediary stages will lie within the missing 99 percent of the record. Darwin felt so sure about this artifactual explanation that, in a bold claim, he staked his entire theory—the mechanism of natural selection as well as the basic fact of evolution—on his inference:

> The geological record [is] extremely imperfect, and [this fact] will to a large extent explain why we do not find interminable varieties, connecting together all the extinct and existing forms of life by the finest graduated steps. He who rejects these views on the nature of the geological record, will rightly reject my whole theory. (1859, p.342)

Thomas Henry Huxley warned Darwin that he had stuck his neck out too far in making such a bold claim, because geological "suddenness" does not correspond to saltational origin of species in a single generation (which would confute natural selection), and that natural selection does not, in any case, require pure gradualism through every conceivable intermediary degree. Huxley wrote in a famous letter to Darwin, his first after reading the manuscript of the *Origin of Species*: "You load yourself with an unnecessary difficulty in adopting *Natura non facit saltum* so unreservedly" (as Huxley cited the famous Latin aphorism of both Leibniz and Linnaeus: "Nature does not proceed by leaps").

To summarize this situation in the jargon of modern science, all paleontologists agreed (and continue to agree) that most morphospecies of marine invertebrates, by far the largest faunal component of our fossil record, display, in the literal appearance of their remains in strata, a pattern of "sudden" geological origin followed by a long period of stasis (minimal change, usually expressed as effectively random fluctuation around an unaltered mean) before extinction. Cuvier and most pre-Darwinian naturalists interpreted this pattern as the accurate *signal* of a creationist account for the history of life. Darwin, in his radical break with tradition, also affirmed the empirical pattern of stasis and abrupt origin, but he interpreted this literal appearance as the artifact of an imperfect fossil record—that is, as *noise* covering a true signal of gradual change that the intermittant record of strata could not generally preserve.

Darwin's argument remained logically sound, and it did allow paleontologists to assert a gradualistic theory of evolution in the face of an apparently contrary pattern in their primary source of empirical data, but this "argument for imperfection" imposed an enormous price on the emerging profession of evolutionary paleobiology. Darwin's solution implied that the fossil record rarely provides direct evidence for the fundamental phenomenon that any empirically minded scientist would yearn to study—evolution itself, defined (according to theoretical expectations) as a series of transitional fossils linking an ancestral to a descendant species by a stratigraphically ordered series of gradually changing intermediary forms. One can hardly practice science, in any operational sense, when noise degrades signal to such an extent that the primary empirical phenomenon of a subject becomes effectively invisible.

The theory of punctuated equilibrium, as initially proposed by Niles Eldredge and Stephen Jay Gould (the present author) in 1972, made no fundamentally new observation about norms of the fossil record, and it fully accepted the predominance of stasis and abrupt appearance as the literal data of most morphospecies. However, punctuated equilibrium suggested that this literal pattern reflects signal rather than noise after all, and that this signal accurately represents the expected pattern of ordinary evolution as expressed in the fossil record, once we properly scale the mechanisms of Darwinian change into their predicted outcomes in the immensity of geological time. Previous naturalists, Eldredge and I argued, had erred in their expectation that Darwinian gradualism in populations during microevolutionary time would translate into imperceptible intermediacy between ancestral and descendant forms over millions of years in geological strata.

The key argument of punctuated equilibrium includes two assertions. First, most evolutionary change proceeds not by the morphological transformation of ancestral populations into different descendant forms without branching (anagenesis, in technical parlance), but rather by speciation (or cladogenesis), the branching off of a separate and isolated descendant population from an ancestral population that usually persists in its original form. Second, the expression of ordinary speciation in geological time (and thus in the fossil record) should predict a pattern of punctuated equilibrium, not of Darwinian gradualism.

To explain this crucial second point, which initially strikes many people as paradoxical, consider that most of us can easily understand geological immensity in the intellectual sense of knowing how many zeros to place after the 1 when we mean million or billion. But the vastness of geological time proves very hard for most people, even professional scientists, to grasp in the visceral sense of intuitive practice. For example, conventional views of speciation, fully embraced by punctuated equilibrium, envisage a process that is glacially slow at the scale of human lifetimes and observations. Suppose, for example, that an isolated population, in the process of evolving into a distinct species, required 5,000 years to achieve complete reproductive isolation from the parental form. We would view such a process as slow enough to fall effectively below the level of our observation—that is, we could observe the accumulating changes throughout an entire 40-year career and still see less than 1 percent of the total alteration. Thus, we label the process as maximally slow, and we assume that its geological expression should be similarly gradual and recorded as imperceptible transition through substantial sequences of strata.

In the great majority of geological circumstances, however, the results of 5,000 years are compressed onto a single bedding plane, not expressed over meters of strata. Thus, the glacially slow speciation of ordinary Darwinian microevolution will appear as the punctuations of punctuated equilibrium in geological time: the entire microevolutionary transition will be compressed, in nearly all cases, into an unresolvable geological moment. The equilibrium (or stasis) of punctuated equilibrium, then, represents our usual expectation of long and stable persistence for a well-adapted species, securely established in its ecosystem. Moreover, because most species arise allopatrically—by changes within small populations geographically isolated from their ancestors—the chance of finding intermediary forms becomes even smaller. After all, the new species arises within the geological moment of a bedding plane (however slowly by the inappropriate standards of a human lifetime); it also arises in a small population geographically isolated from ancestors. In summary, the new species originates rapidly and in a small area away from the parental form. Most rich collections of fossils sample the successful parental forms in their large area of spread and abundance—and these are the very places where we would least expect to find direct evidence for a descendant caught in the act of speciation.

A Question of Scaling. Many evolutionary ideas and concepts propose various forms and styles of rapidity as challenges to the pervasive gradualism of Darwinian expectations. We must realize, however, that punctuated equilibrium in the strict sense cannot be equated with all such claims at all scales of evolutionary change and time. Punctuated equilibrium is a particular theory about the tempo and mode of speciation, properly scaled to the characteristic timings and geographies of this cardinal evolutionary phenomenon. Thus, at a scale below punctuated equilibrium, the putative origin of new species in a single jump of macromutational change (a popular theory at the origin of the twentieth century, but now largely abandoned)—in other words, speciation by true *saltation* in a single generation—has nothing to do with punctuated equilibrium. The two distinct concepts have often been confused, though, perhaps because Darwinian traditionalists get upset whenever they hear the word "sudden" in any form; the old bugbear of true saltationism rises before them, and they fail to make the proper distinction of scale. In any case, the geologically rapid origin of species under punctuated equilibrium only represents the proper scaling into geological time of ordinary events of speciation that occur very slowly and gradually in microevolutionary time.

Similarly, and at an opposite end of evolutionary scaling, the rapid biotic turnovers of catastrophic mass extinction (with particularly good evidence now available for the triggering of the Cretaceous-Tertiary event by the impact of a large extraterrestrial object) bear no direct relationship to punctuated equilibrium, which is not a theory about catastrophic effects on entire faunas, but rather a description and proposed explanation for the origin and deployment of individual species in geological time. However, the broader theme of "punctuational" rather than gradual change may hold substantial validity, and may at least be worth exploring at several scales, levels, and casual reasons—especially because Western preferences for accumulative and gradual change have so dominated several scientific fields (such as Lyellian geology and Darwinian biology). A punctuational account provides a stimulus for considering plausible alternatives that often are not addressed at all because we fail to articulate our biases and regard them as unquestionable natural facts.

Evidence. Punctuated equilibrium provoked a great deal of controversy and incredulity when Eldredge and I proposed the theory in 1972, expressed much later in book-length refutations by Levinton (1988) and Hoffman (1989), and equally firm and extensive support by Stanley (1979) Gould (2002) and Cope and Skelton (1985), and Somit and Peterson (1992). Admitting my partisan bias as an instigator of the theory, I think that the pattern of punctuated equilibrium has now been confirmed as a common occurrence in the fossil record. Persisting debate now centers on the relative frequency of punctuated equilibrium: Is the pattern rare or dominant in the history of life? and its theoretical meaning: Does the concept suggest alternative ideas for the explanation and interpretation of evolution, or does punctuated equilibrium merely represent the expression of conventional Darwinian processes properly scaled into geological time?

The empirical debate on the existence and importance of punctuated equilibrium has centered on two broad themes. First, how can we know that apparently stable morphological "packages" in the fossil record truly represent biological species, when we cannot in principle apply the accepted criterion of reproductive isolation and potential for interbreeding? I believe that two arguments and sources of data should allay fears on this important point. First, in several studies, ranging from mollusks to bryozoans, living biospecies, genetically and ecologically defined, can also be characterized morphometrically. Each of these living species also has a good fossil record, and the morphometrically defined fossil "species" fully corresponds, in each case, to a single biospecies as recognized among living counterparts. Second, no assertion of punctuated equilibrium can be made when a new species appears in its predicted geological moment, but the ancestral form also dies out at the same time—for then we cannot distinguish the origin of a new species in an event of cladogenetic branching (as punctuated equilibrium requires) from an unresolvable transition within a single unbranched lineage. In practice, proponents of punctuated equilibrium assert their case only when direct evidence exists for survival of the ancestral species *after* the rapid origin of the descendant, because the required event of branching can then be affirmed.

The second broad theme asks if we can affirm that punctuated equilibrium represents a dominant pattern in the history of most clades and not just an unusual event that, although a genuine evolutionary phenomenon, does not contribute in a major way to the causes and patterns of life's history. Punctuated equilibrium may not be a dominant pattern in certain groups, particularly in asexual single-celled protists (an important component of the marine fossil record), where speciation itself cannot be defined in the traditional sense presupposed by punctuated equilibrium. However, overwhelming or effectively exclusive punctuated equilibrium has been rigorously and empirically affirmed for large faunas of many times and regions, and for monophyletic clades, thus reinforcing the argument for punctuated equilibrium as the predominant pattern for the origin and development of species in geological time. For example, punctuated equilibrium applies to all molluscan lineages with extensive fossil records in the Atlantic and Gulf Coast Tertiary and Quaternary strata of the United Stated (Stanley and Yang, 1987), and in all but three of several hundred mammalian species in the classic Big Badlands Oligocene strata of the American West (Prothero and Heaton, 1996). Most impressively, Cheetham's classic study (1986) on the full range of the monophyletic bryozoan clade *Metrarabdotos* affirms an exceptionless pattern of punctuated equilibrium, for abundant fossils in unusually well-resolved strata; each species arises in a geological moment, and stasis is maintained for long periods (up to 10 million years) in each species, based on morphometric measures of all standard parts, and not merely on single characters.

Implications for Evolutionary Theory. But does punctuated equilibrium make a difference to evolutionary theory? Does the establishment of punctuated equilibrium as an important or predominant pattern alter the structure of conventional Darwinian explanation in any major way—especially since the theory has nothing original to say about mechanisms and timings in the microevolutionary origin of species, since punctuated equilibrium emerges as the expected scaling of conventional speciation in geological time? We may at least assert that punctuated equilibrium had strong and salutary benefits for paleontological practice in dispelling the lethargy that had enveloped a field taught to regard its primary observations as an artifact of an imperfect record, hiding an invisible reality of gradualism. In the most immediately practical sense, punctuated equilibrium defined the phenomenon of stasis as an important and genuine phenomenon—and a quite puzzling phenomenon at that, given previous expectations of Darwinian gradualism—and therefore as well worth extensive and rigorous morphometric study. Before punctuated equilibrium, paleontologists tended to treat the well-known phenomenon of stasis as an embarrassment that only illustrated the failure of the record to document evolution, and therefore as not worth active study as an interesting and positive phenomenon in its own right. Eldredge and I insisted that "stasis is data"—and hundreds of documentations for stasis have been published since punctuated equilibrium provided a validating framework for this central empirical phenomenon of the fossil record.

Beyond this operational benefit of reopening basic paleontological data to scientific scrutiny as the expression of a positive signal about evolutionary processes, rather than the artifactual noise of data so inadequate that the signal becomes diluted to invisibility, punctuated equilibrium also makes some novel and even radical suggestions for revisions to evolutionary theory. These proposed revisions arise from both aspects of its name.

Equilibrium and stasis. In the most broadly ontological sense, we have traditionally viewed change as the usual and inherent state of biological entities in an evolutionary world. Change, moreover and in Darwin's canonical conception, is construed as generally slow, accumulative, and gradual—that is, as insensibly incrementing all the time, effectively beneath our notice, and brought to our attention and made effective only by the sheer immensity of geological time. Stability, in this perspective, represents either an impression of the moment (that allows us, for example, to establish discontinuous taxa in our biological classifications), or a transient and temporary pause in the basic workings of nature.

Under punctuated equilibrium, in contrast, stability is the normal and expected state of most entities most of the time; change represents an occasional break in the standard pattern, usually accomplished with relative dispatch, and generally instigated when systems become stressed beyond their usual capacity to respond by absorbing the impact with minimal alteration to their established stability. Evolution remains the basic principle of worldly change, but we must reorient our basic viewpoint by reconceptualizing evolution as rare and exceptional—as a system's last resort rather than its standard mode. Under punctuated equilibrium, stability must be recognized not as the absence of an expectation, but as the usual state of most systems most of the time. The immensity of geological time continues to power the remarkable extent of evolutionary change under both views of life: by granting enough time to add up millions of tiny increments in Darwinian accounts, and in the punctuational alternative, by providing a sufficiently extensive matrix to allow enough rare punctuations to construct the complex pageant and diversity of life.

The concept of stasis as a norm for species also has important implications for the operation of selection as a primary mechanism of evolution. Under Darwin's view, species have no real existence in geological perspective because they can represent only temporary incarnations of the moment for lineages in a constant state of transformation. If populations remain constantly in flux, then species cannot claim any status as "individuals"—that is, as entities or objects that might become active units of evolution in their own right. For this reason (among others), Darwin argued that natural selection can operate only on organisms in their struggle for differential reproductive success and survival of the fittest, and that all large-scale evolution must be explained by accumulation of microevolutionary changes occurring within populations by natural selection at the organismal level.

Punctuation and the speciational reformulation of macroevolution. To continue the argument about stasis from the preceding paragraph, if species are more than arbitrary names for transient stages in a continuum of change—and must be recognized instead as true units with geologically momentary points of birth and death, and with stability for the length (usually millions of years for fossil invertebrates) of their geological duration—then species become Darwinian individuals in their own right, and they may be considered "atoms" of macroevolution, just as organisms serve as agents of natural selection within populations in microevolution. And as organisms prevail in microevolutionary natural selection by differential reproductive success (that is, by living longer in competition with others, and by having more surviving offspring than others), so too may species power macroevolution by differential reproductive success in species selection. That is, some species may live longer than others within their monophyletic clade and may branch off more daughter species than others, thus increasing their representation within the clade, just as organisms increase the representation of their genes within the population by producing more surviving offspring.

This theoretical expansion of natural selection to include not just the Darwinian organism but also selection in an extended hierarchy of other levels both below (the gene) and above the organism (the deme in group selection, the species in species selection) suggests interesting implications for a revised theory of evolutionary mechanisms, and for the explanation of many previously puzzling macroevolutionary phenomena. To give just one theoretical example, the explanation of phyletic trends (such as the increase in hominid brain size over 5 million years of our evolution, or the growing complexity of ammonite suture lines, or loss of toes and establishment of a single-toed hoof in the history of horses)—arguably the central and most important phenomenon of macroevolution—has often eluded us because we have tried to resolve it with only one standard model that may not apply to all cases: accumulating adaptive advantages to organisms as crafted by long-continued natural selection.

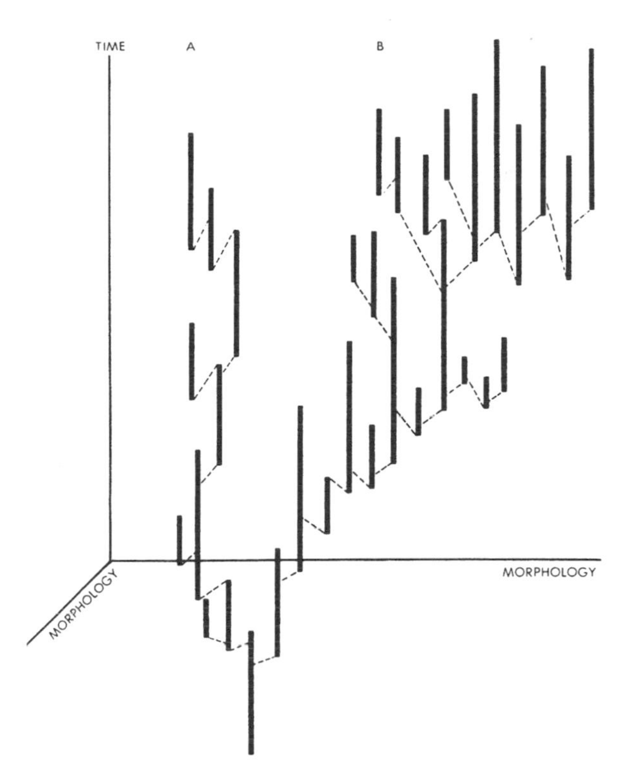

FIGURE 1. Punctuated Equilibrium.
An original diagram of punctuated evolution published by Eldredge and Gould. Courtesy of Stephen Jay Gould.

Under punctuated equilibrium, however, and with the resulting possibility of selection at the species level as an additional explanation of trends, we may need to consider other resolutions with different implications for our understanding of life's history. Consider Figure 1 the hypothetical depiction of punctuated equilibrium drawn by Eldredge and me in our original 1972 article. It shows a monophyletic clade with two subclades, and with evolution proceeding entirely by punctuated equilibrium; that is, each species originates in a geological instant and remains in stasis thereafter. How, one might ask (especially under conventional Darwinian views), could a trend ever occur at all if no changes accumulate *within* any species? But, although we have purposely drawn the left-hand subclade as trendless, note the clear and persistent trend in the right-hand subclade. This substantial trend occurs entirely through differential success of certain kinds of species either in living longer or in branching off more daughters, and not at all by accumulating change within any species. In other words, the trend must be described at the species level as a consequence of the differential reproductive success of some species over others.

Now, the need to describe a trend as differential success of certain species need not imply a radical explanation. After all, the differentially successful species may gain their advantages purely because the organisms that comprise them have won adaptive benefits by the usual working of Darwinian natural selection. However, we must at least allow the possibility (and probably the frequent and important actuality) that some trends among species might be better explained, and not merely described, by direct selection on the species themselves, and not as consequences of ordinary selection on organisms. Suppose, for example, that the trend occurs because some species branch off more daughter species than others. Suppose further that these increased rates of branching occur not because the organisms within these species enjoy any advantages under natural selection, but because the species itself maintains certain traits at the species level (perhaps expressed in terms of geographic range, subdivision of populations, or coherence of groups) that cannot be reduced to the properties of organisms, but that represent "emergent" features of the species itself. In fact, these emergent species-level traits may even counteract the selective benefits of organisms, and the trend could still be powered by species selection. (Perhaps the coherence that binds organisms of a species into tight groups, and therefore promotes isolation and subsequent speciation when subgroups do manage to become separated from the main population, operates to the selective disadvantage of organisms that might gain benefits from higher mobility.)

To close with a metaphor expressing the extent of difference between the standard and stately accumulative view of evolutionary change and the reconceptualization urged by punctuated equilibrium, I cite a quip of my late friend, the geologist Derek Ager, a noted critic of uniformitarianism (the geological analogue to Darwinian evolutionism). He said that the history of the Earth and life may be compared with the experience of a soldier—an existence defined by long periods of boredom interspersed with occasional moments of terror.

[*See also* Overview Essay *on* Macroevolution; Speciation; Species Selection.]

BIBLIOGRAPHY

Cheetham, A. H. "Tempo of Evolution in a Neogene Bryozoan: Rates of Morphologic Change within and across Species Boundaries." *Paleobiology* 12 (1986): 190–202.

Cope, J. C. W., and P. W. Skelton, eds. *Evolutionary Case Histories from the Fossil Record. Special Papers in Palaeontology* 33 (1985): 1–203. A volume of case studies supporting punctuated equilibrium.

Darwin, C. *The Origin of Species*. London, 1859.

Eldredge, N., and S. J. Gould. "Punctuated Equilibria: An Alternative to Phyletic Gradualism." In Schopf, T. J. M. (Ed.) *Models in Paleobiology*, edited by T. J. M. Schopf, 82–115. San Francisco, 1972.

Gould, S. J. *The Structure of Evolutionary Theory*. Cambridge, Mass., 2002.

Hoffman, A. *Arguments on Evolution: A Paleontologist's Perspective*. New York, 1989.

Levinton, J. *Genetics, Paleontology and Macroevolution*. New York, 1988.

Prothero, D. R., and T. H. Heaton. "Faunal Stability during the Early Oligocene Climatic Crash." *Paleaeogeography, Palaeoclimatology, and Palaeoecology* 127 (1996): 257–283.

Somit, A., and S. A. Peterson, eds. *The Dynamics of Evolution: The Punctuated Equilibrium Debate in the Natural and Social Sciences*. Ithaca, N.Y. 1992. A symposium on the theory's influence in biology, and the social and political sciences.

Stanley, S. M. *Macroevolution: Pattern and Process*. San Francisco, 1979.

Stanley, S. M., and X. Yang. "Approximate Evolutionary Stasis for Bivalve Morphology over Millions of Years: A Multivariate, Multilineage Study." *Paleobiology* 13 (1987): 113–139.

– STEPHEN JAY GOULD

Q

QUANTITATIVE GENETICS

Quantitative genetics is the study of the inheritance of phenotypic traits. Evolutionary biologists use quantitative genetics to predict the course of evolution, to interpret history, and to understand the nature of genetic variation and its influence on the phenotype.

Most phenotypes are continuously distributed; that is, they may take on a wide range of values rather than falling into a few discrete classes as expected under single-locus Mendelian inheritance. For example, height and running speed are continuously distributed because they can take on any positive value, whereas sex has a discrete distribution—it can only be male or female. Continuous distributions can arise for two reasons. First, nongenetic effects (such as environmental influences) on phenotypes may be truly continuous in their nature. For example, differences in nutrition during growth can cause a continuum of different sizes, even if all individuals are genetically identical. Second, a trait can be affected by genetic variation at many genes, a common condition known as *polygenic* inheritance. Even a fairly small number of variable loci can produce a continuous distribution of phenotypes.

Quantitative inheritance is detected as resemblances among relatives. Humans have known at least since the origin of agriculture that offspring tend to resemble their parents, and they have used this knowledge to modify domesticated plants and animals into the useful and diverse set of forms we see today. This simple observation can be used to derive one of the most fundamental results of quantitative genetics. Figure 1 shows the parent–offspring relationship and the effects of selection on the phenotype of the offspring. The thin solid line shows the line that best summarizes this relationship (the regression of offspring on parents). The slope of this line is called the *heritability* (h^2). If the average offspring is identical to the mean of its parents, then the line will have a slope of 1. If inheritance plays no role, the slope will be 0.

Heritability determines the proportion of the change in parental phenotype that is attributable to natural or artificial selection and that is realized in the offspring, as shown in Figure 1. The change in average phenotype caused by selection is called the *selection differential*, *S*, and the response to selection is *R*. Thus, from information on the selection differential and the heritability alone, the breeder can predict the response to selection from the *breeder's equation*, $R = h^2S$. If inheritance is perfect, all the selection is transmitted to offspring; in the absence of inheritance, there is no response. Figure 2 shows the relationship between parents and offspring from an actual experiment.

The Relationship between Mendelian and Quantitative Inheritance. To make the connection between the quantitative relationship between parents and offspring, and the Mendelian process of inheritance that we know underlies it, we must examine the causes of the regression slope h^2. The basic quantitative-genetic model assumes that the phenotype of an individual can be represented as the sum of many effects on the phenotype, relative to the phenotype of the average individual, $\bar{z}$. Each of the *n* loci that are variable in a population has some genetic effect on the phenotype, and the effect of the *j*th locus is symbolized as g_j. In addition, the environment also has an effect *e* on the phenotype. The phe-

FIGURE 1. Regression of Mean Offspring Phenotype on Mean Parental Phenotype.
The slope of this line is the heritability, h^2. The effect of a hypothetical selection experiment on this population is also shown. If we apply selection by retaining only the offspring of parents above the thick solid line, the mean of the selected parents changes by an amount $S = \bar{Z}_S - \bar{Z}$ the selection differential. The response to selection $R = \bar{Z}_0 - \bar{Z}$, is the difference between the mean of the offspring of the selected parents and that of the unselected parents. The regression line predicts the mean phenotype of the offspring of the selected parents. The phenotype will change by an amount $R = h^2S$. Courtesy of author.

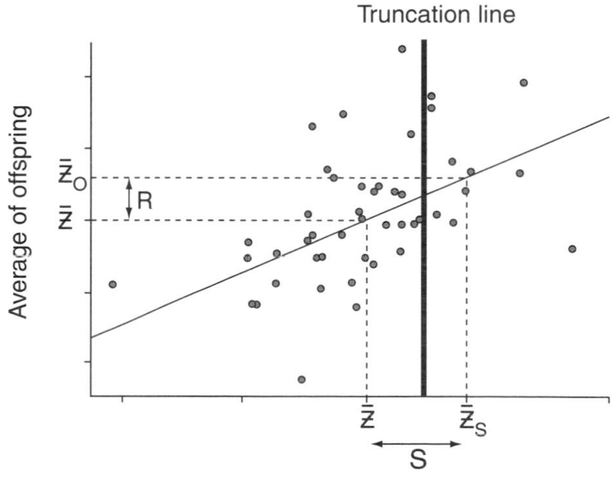

notype of the ith individual is then the sum of these effects,

$$z_i = \bar{z} + \sum_{j=1}^{n} g_{ji} + e_i$$

where Σ is used to denote a sum, and the sum is over the j loci that affect the ith individual's phenotype.

This also makes the variance in the phenotype, V_P, the sum of genetic, V_G, and environmental, V_E, variances,

$$V_P = V_G + V_E.$$

This model involves some very important assumptions. The first is that the genetic impact of each locus and the effects of changes in the environment are independent in their effects. This assumption may be violated if the genotype at one locus affects the impact of genotypes at another locus (a situation known as *epistasis*), or if the effect of an environment depends on the genotype. Such effects are common. The second assumption is that genotypes and environments do not covary; that is, individual offspring are randomly distributed among environments, like seeds scattered at random. This assumption too is frequently violated. For example, in social organisms such as humans, parents usually pass on resources as well as genes to their offspring. A third assumption is that individuals in the population mate randomly, so the allele frequencies are in Hardy-Weinberg proportions before selection. Where these assumptions are known to be violated, a more complex analysis can usually be applied to compensate for the problem.

The ability of selection to affect the phenotype in subsequent generations depends on how much difference the choice of a particular individual parent makes. Although we usually cannot measure the effects of single loci, we can observe how different the offspring of one individual are from those of another. Twice the deviation of the average offspring of a parent from the average offspring in the whole population is known as the *breeding value* of that parent. Parents whose offspring are larger than the average have high breeding values. The variance in breeding values in the population is defined as the additive genetic variance, V_A, and makes up part of the genetic variance V_G. Returning to the parent–offspring regression given above, the slope of the line predicting offspring from the mean of parents is $h^2 = V_A/V_P$. Thus, heritability is the proportion of phenotypic variance that causes parent–offspring resemblance. V_A is what determines the strength of the relationship between parent and offspring and is therefore the most important quantity in quantitative genetics. If all phenotypic differences are heritable, than $V_A = V_P$ and $h^2 = 1.0$. If, on the other hand, none of the phenotypic variation is heritable (for instance, it is all caused by the environment), then $V_A = 0$ and as $h^2 = 0$. The proportion of phenotypic variance that is due to all genetic causes, the broad-sense heritability, $H^2 = V_G/V_P$, can be estimated in studies of genetically identical individuals such as clones or identical twins. H^2 is frequently confused with h^2 in discussions of quantitative genetics in humans.

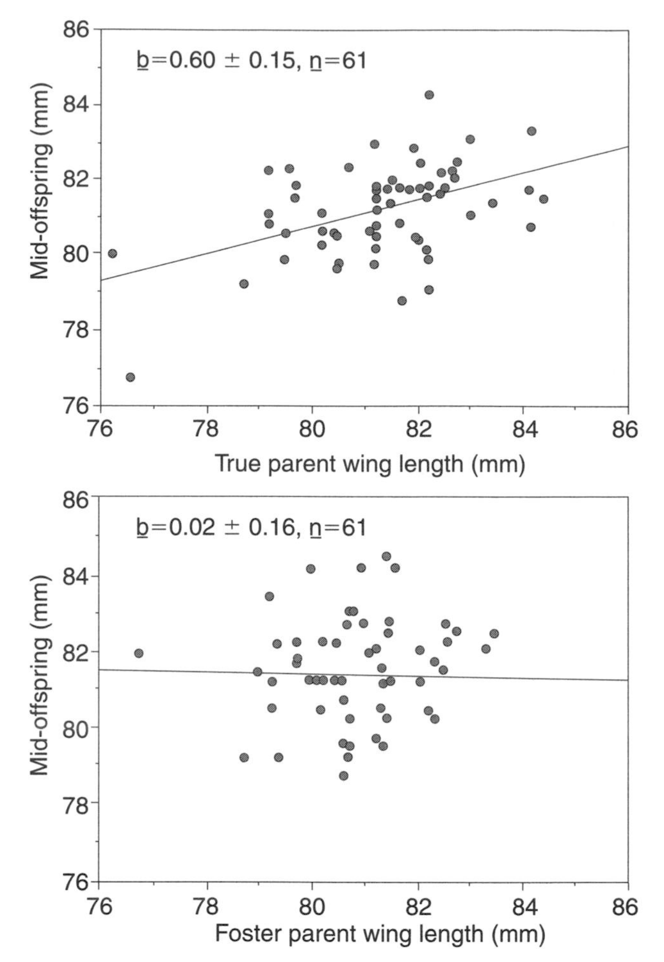

FIGURE 2. Relationship Between Parents and Offspring. Gustafsson and Merilä took eggs from nests of Collard Flycatchers (*Ficedula albicollis*), and placed them in other nests to investigate whether the parents that rear the chick have an influence on the chick's wing length. There was a significant relationship between offspring and their genetic parents despite the transfer, as shown in the upper panel, yielding a heritability of 0.6. There was no significant relationship between the foster parents and the offspring. After L. Gustafsson and J. Merilä. "Foster Parent Experiment Reveals No Genotype-Environment Correlation in the External Morphology of *Ficedula albicollis*, the Collared Flycatcher." *Heredity* 73 (1994): 124–129. Blackwell Science Ltd.

It is important to realize that quantitative genetic parameters are applicable to the chosen population *only* in the environment in which it is measured. V_A depends on allele frequencies, so h^2 may change with selection.

Similarly, if the environment changes, genotypic differences may be magnified or minimized, changing the relative amount of genetic influence.

Environmental influences make it extremely difficult to determine whether phenotypic differences between populations arise from genetic causes. This problem is best illustrated with two examples. Losos and colleagues (1997) transplanted lizards of the genus *Anolis* to an island without lizards. They predicted the evolution of a particular limb morphology in the new habitat, and after a few years they observed that the lizards' morphology had changed in the direction predicted. Subsequent lab studies revealed, however, that a substantial proportion of the observed change was attributable to an environmentally induced change in development and not to evolution (Losos et al., 2000). Similar problems arise in the interpretation of differences between groups of humans in scores on IQ tests. Such differences could have some genetic basis, but environmental effects on IQ are clearly enormous. The best evidence for this is a large increase in unstandardized test scores over the past 50 years, a phenomenon known as the "Flynn effect" (Flynn, 1987). The observed increase is far too large to be the result of evolutionary change.

Estimation of Quantitative Genetic Parameters. Quantitative genetic parameters, such as V_A, are estimated from the degree of resemblance among relatives. In addition, the experimenter must remember that resemblances between individuals also arise from effects of shared environments. Half-siblings are usually used to estimate V_A in experiments, whereas parents and offspring are used in observational studies where matings cannot be controlled, as in studies of humans. Closer relatives are useful for estimating other types of causes of similarity. The effects of parents on their offspring beyond the inheritance of nuclear genes are particularly difficult to deal with experimentally. In mammals, the mother often affects offspring through gestation, nursing, and parental care. These are collectively known as *maternal effects*. Maternal effects include differences attributable to the mother's genotype as well as to nongenetic causes. Paternal effects are also often detected when they are looked for. The magnitude of nongenetic causes of resemblance can be estimated with experimental manipulations, as shown in Figure 2.

More complex experimental designs are particularly important when the simplifying assumptions discussed above may be violated. For example, in humans the assumption of a lack of covariance between genotype and environment is very often violated, particularly when individuals of diverse class or genetic backgrounds are included in the same study. Adoption studies, which can partially eliminate such gene-by-environment covariance, are valuable in estimating how important this effect may be. Even with such information, the interpretations of resemblance still depend on the assumptions of the analysis. For example, the importance of some maternal effects, such as effects of the egg or gestation, cannot be assessed without complex experiments; adoption does not eliminate these maternal effects. Other key assumptions, such as random mating, are often violated in human populations and should be taken into account during analysis.

An illustration of the effects of assumptions on results is the analysis of the inheritance of human IQ scores. Devlin and colleagues (1997) analyzed the results of 212 studies of covariances in IQ scores. When they assumed a simple model in which the only environmental effect was that of childhood environment on children reared together, they estimated the broad-sense heritability as 68 percent and the effect of common environment as only 17 percent of the variance. However, when they fit a more complex model that allowed for an effect of the intrauterine environment as well as rearing, the broad-sense heritability dropped to 47 percent. The intrauterine environment was estimated to account for 27 percent of the variance in IQ scores, whereas the shared childhood environment still accounted for about 17 percent of the variation. Interestingly, h^2 accounted for about 32 percent of the variation in both models.

Selection on Quantitative Traits. Natural selection is caused by differences among individuals in the number of successful offspring produced. The strength of selection can be measured in several different ways besides the selection differential mentioned previously. One alternative measure is the slope of the regression of fitness (e.g., number of offspring produced) on the phenotype, referred to as the *selection gradient* and symbolized β. The gradient is $\beta = S/V_P$, so the breeder's equation can be rewritten $R = V_A S/V_P = V_A/\beta$.

The simple model of selection shown in Figure 1 is unrealistic in a number of ways. The figure represents truncation selection, a special case in which only two fitnesses arise: breeding and not breeding. Truncation is often used in artificial breeding programs, but in nature the fitness of an organism can take any nonnegative value. In general, the mean phenotype after selection is the weighted mean of the phenotypes, where the weights are individuals' fitnesses. The selection differential takes into account only the directional part of selection, which changes the trait mean. Selection may also change the variation in the trait. For example, stabilizing selection favors individuals in the middle of the distribution over those at the extremes. This will reduce the variation among breeding parents, as well as perhaps changing the trait mean.

A second important complication is that each organism has a very large number of traits that are correlated with one another. Large individuals have large heads and long limbs, so selecting the individuals with long limbs

TABLE 1. Median CV_A and h^2 Values and Predicted Relative Rates of Evolution under Three Kinds of Selection for Five Well-Studied Traits in *Drosophila melanogaster*

Under truncation selection, all individuals at one end of the phenotype distribution breed and all other individuals do not; truncation selection is typical of artificial selection. Under linear selection, fitness is proportional to trait value; traits closely related to fitness, such as survival, usually undergo linear selection. Shift optimum indicates a situation in which stabilizing selection favors an intermediate optimum that does not coincide with the trait mean. For each selection regime, the rates are standardized so that the highest rate is 1.

			Relative rate		
Trait	h^2	CV_A	*Truncation*	*Linear*	*Shift optimum*
Sternopleural bristles	0.44	8.39	1.00	0.49	0.02
Wing length	0.36	1.56	0.15	0.01	0.01
Fecundity	0.06	11.90	0.57	1.00	1.00
Longevity	0.11	9.89	0.54	0.69	0.04
Development time	0.28	2.47	0.20	0.04	0.02

SOURCE: Houle, 1992.

will also tend to select for large heads. If a correlation is genetically based, then selection on one trait will result in an indirect response to selection in the other due to *pleiotropy.* This simple effect can readily be appreciated if we relabel the axes in Figure 1 to represent trait 1 in the parents and a different trait, trait 2, in their offspring. The slope of the regression line then depends on the degree to which the phenotype of one trait predicts the phenotype of another in the offspring, a quantity known as the *additive genetic covariance.*

In reality, many traits may be under direct selection, and each may therefore impose indirect selection on others. Evolution of quantitative traits involves a complex balance of forces, resulting in a compromise phenotype rather than one that is optimal in all respects. The balance of selective forces on many traits may cause a particular trait to evolve against the force of selection on that trait alone. To deal with these effects, Lande (1979) generalized the breeder's equation to include any number of selected traits and their indirect effects on each other. This insight also suggested a way to disentangle the direct effects of selection from indirect ones, by means of a statistical approach called multiple regression analysis. Since Lande's proposal, thousands of estimates of selection and inheritance on multiple traits have been made by evolutionary biologists.

Magnitude of Inheritance and Selection. A remarkably consistent finding of quantitative genetic studies is that almost all traits possess significant additive genetic variation and are therefore capable of responding to selection. Traits with genetic variance in humans include, for example, shyness, handedness, and musical ability. A few exceptional traits, such as the asymmetry of body parts in some organisms, do seem to lack additive genetic variance.

Comparisons of the strength of selection and inheritance in different traits or different organisms require measures that are standardized to eliminate the units of measurement. Measures may be standardized either relative to the variation in the population or relative to the mean of the population. We can divide the breeder's equation by the phenotypic standard deviation $\sqrt{V_P}$ to give response in standard deviations $R/\sqrt{V_P} = h^2S/\sqrt{V_P}$. The quantity $i = S/\sqrt{V_P}$ is called the *intensity of selection;* it gives the number of standard deviation units by which the trait is altered by selection within a generation. Estimates of heritability are typically between 30 and 40 percent for morphological traits and are 10 to 20 percent for traits closely connected to fitness, such as survival or fecundity. Intensities of selection from studies of natural populations have a mode near 0 and a median around 0.1 standard deviations.

Alternatively, one can standardize the response to selection by dividing by the mean to give the proportional response to selection $R/\bar{z} = (V_A/\bar{z}^2)\beta\bar{z}$. $V_A/\bar{z}^2 = I_A = CV_A^2$ is the square of the additive coefficient of variation of the trait and gives the amount of variance relative to the mean of the trait. $\beta\bar{z}$ is the "elasticity" and gives the proportional change in fitness for a proportional change in the trait value. CV_A values are generally between 2 and 7 percent for morphological traits and average 12 percent for traits closely connected to fitness.

The traditional form of the breeder's equation suggests that h^2 sums up inheritance, whereas S sums up the effect of selection. This conclusion is misleading, because both h^2 and S depend on the amount of phenotypic variation in the population. Our ability to rewrite the breeder's equation in several equivalent forms argues against relying solely on any one measure of variation, such as h^2, to the exclusion of other ways of summarizing the inheritance of quantitative traits. Table 1 shows that no one standardized measure of genetic var-

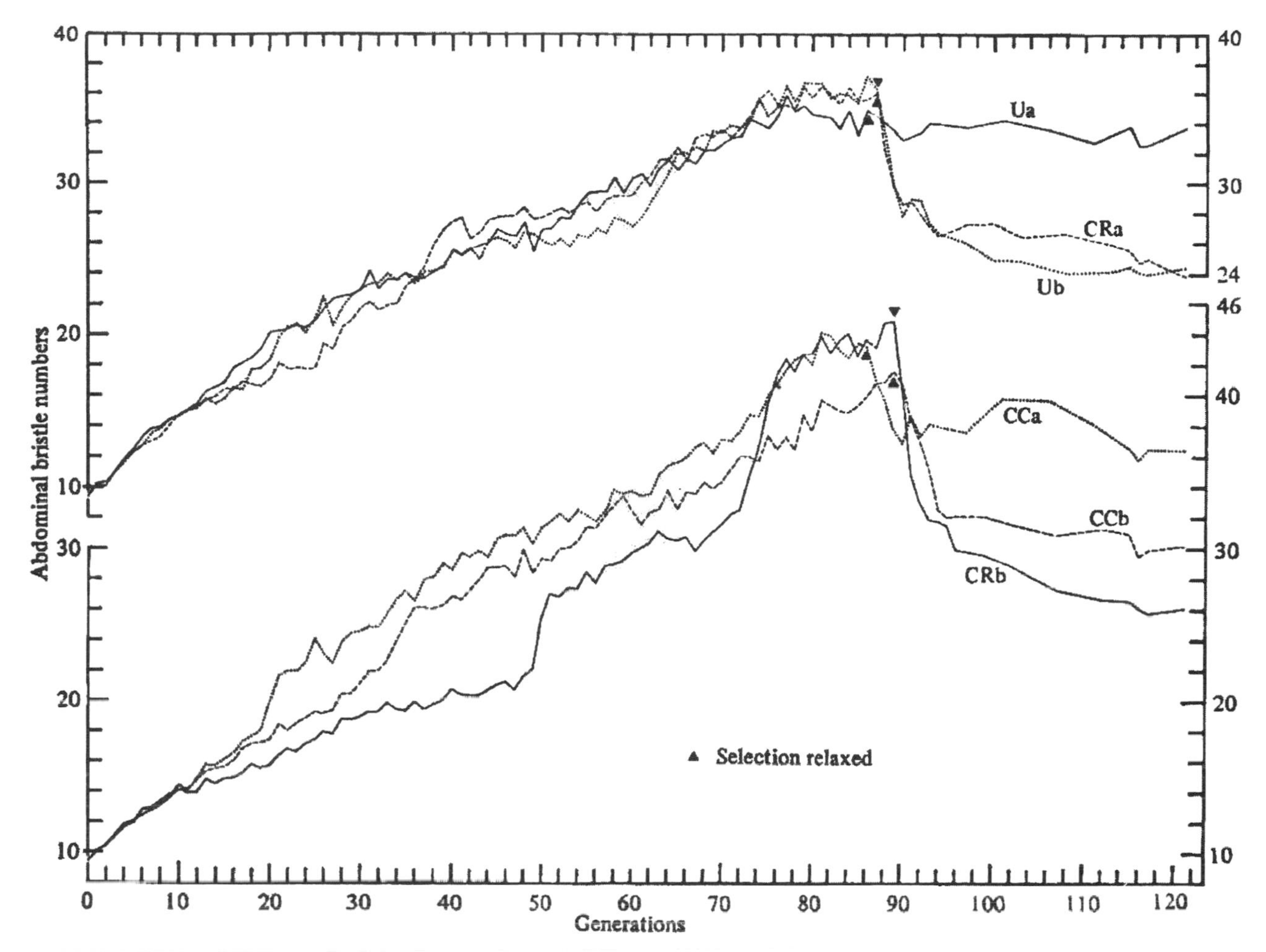

FIGURE 3. Selection for Increased Abdominal Bristle Number in *Drosophila melanogaster*.
B. H. Yoo (1980) selected the top 100 of 500 flies in each population each generation until the generation shown by the triangles. Reprinted with the permission of Cambridge University Press.

iance is the best predictor of the ability of a trait to evolve in all situations. Comparison of h^2 and CV_A values suggests that the low h^2 of fitness traits is best explained by high phenotypic variation, in combination with high V_A.

Predicting the Response to Selection. The usefulness of predictions based on the breeder's equation has been confirmed by many experiments in plant and animal breeding over short time intervals. How well predictions are borne out over longer time periods is less clear. Theory suggests that long-term responses to selection may depart from the initial predictions, for a variety of reasons. The initial response depends on the variation in the base population, summarized by V_A, but V_A may change as allele frequencies change under selection. Ultimately, directional selection may cause one of the original alleles at each locus to become fixed (reach a frequency of 1). As this occurs, the long-term response will come to depend on the rate at which mutation supplies genetic variation, rather than on the initial variation. A second reason for a change in V_A is a change in the environment over time.

Figure 3 shows the result of an unusually long selection experiment that reveals many typical aspects of selection response. Initially, the response was consistent in all replicates, verifying the usefulness of the breeder's equation, but variation in the rate of response increased with time. There were short periods during which some populations responded particularly rapidly, consistent with the occurrence of a new mutation of large effect. Eventually, the population stopped responding, a phenomenon known as a *selection plateau*. When selection is halted, most populations rapidly lose some but not all of the gains from selection, so the selection plateau was not the result of elimination of all variation in the trait. It can be explained by stabilizing natural selection back toward the preselection mean, or by deleterious pleiotropic effects of the selected alleles on other traits.

The Future of Quantitative Genetics. Quantitative geneticists seek a detailed understanding of the relationship between genotypes and phenotypes. New technologies have provided abundant genetic markers for mapping the *quantitative trait loci* (QTLs) that are responsible for the variation within or between populations. A QTL is detected in the form of an association between the phenotype and a variable genetic marker, such as an insertion or single-nucleotide difference in the DNA. Such associations are called *linkage disequilibria.* Disequilibria are strongest when populations differ in phenotype and are distantly related. As a result, QTL mapping has been most successful in domesticated plants and animals—for example, tomatoes or corn—where a strongly selected domestic stock can be crossed with a "wild" strain. It has also been very useful in model organisms like mice and fruit flies. QTLs identified in such species are proving to be informative about the loci that cause quantitative variation in humans.

Direct QTL mapping in humans is much more difficult because we must depend on the naturally occurring linkage disequilibrium, which is usually fairly weak. A useful tool is the identification of families with members who differ in the phenotype of interest, since linkage disequilibrium is large within families. This approach has been used to map and identify the loci responsible for a number of genetic diseases, such as cystic fibrosis.

Genetic mapping only tells us where to look for the correct gene. To identify the actual gene responsible for a particular difference, an array of other genetic tools is needed. Genome sequencing has revealed all of the genes in model organisms, making it possible to infer the function of many loci. Sequences with functions related to the phenotype of interest become *candidate genes* for quantitative variation. This can be combined with QTL mapping to generate hypotheses about the specific locus responsible. Such hypotheses are often incorrect, so they must be tested with additional experiments or observations. Application of these powerful new techniques will allow the marriage of quantitative genetics with population genetics and functional biology to provide an increasingly complete picture of the causes of variation within and among populations.

BIBLIOGRAPHY

Arnold, S. J., and M. J. Wade. "On the Measurement of Natural and Sexual Selection: Theory." *Evolution* 38 (1984): 709–719. This and a companion paper explain how to measure selection using fitness components.

Devlin, B., S. E. Fienberg, D. P. Resnick, and K. Roeder, eds. *Intelligence, Genes and Success: Scientists Respond to* The Bell Curve. New York, 1997. The most reasoned scientific response to naive quantitative genetic arguments about the evolution of human IQ.

Doebley, J., A. Stec, and L. Hubbard. "The Evolution of Apical Dominance in Maize." *Nature* 386 (1997): 485–488. The identification of a key gene in the origin of domesticated maize.

Falconer, D. S., and T. F. C. Mackay. *Introduction to Quantitative Genetics.* 4th ed. Essex, England, 1996. The most accessible introductory text.

Frary, A., T. C. Nesbitt, A. Frary, et al. "A Quantitative Trait Locus Key to the Evolution of Tomato Fruit Size." *Science* 289 (2000): 85–88.

Houle, D. "Comparing Evolvability and Variability of Quantitative Traits." *Genetics* 130 (1992): 195–204. Proposes the use of mean standardized measures of inheritance.

Kimura, M., and J. F. Crow. "Effect of Overall Phenotypic Selection on Genetic Change at Individual Loci." *Proceedings of the National Academy of Sciences USA* 75 (1978): 6168–6171. A technical paper that lays out the relationship between single-locus and quantitative-genetic approaches.

Lande, R., and S. J. Arnold. "The Measurement of Selection on Correlated Characters." *Evolution* 37 (1983): 1210–1226. Proposes the use of multiple regression approach to the estimation of selection.

Lynch, M., and B. Walsh. *Genetics and Analysis of Quantitative Traits.* Sunderland, Mass., 1998. Comprehensive but technical discussion of quantitative inheritance and QTL mapping. A second volume will take up selection.

Mackay, T. F. C. "Quantitative Trait Loci in *Drosophila.*" *Nature Reviews Genetics* 2 (2001): 11–20. QTLs show genotype-by-environment interactions. Regulatory changes are involved in most QTLs.

Mackay, T. F. C., J. D. Fry, R. F. Lyman, and S. V. Nuzhdin. "Polygenic Mutation in *Drosophila melanogaster:* Estimates from Response to Selection of Inbred Strains." *Genetics* 136 (1994): 937–951. Demonstrates that mutation supplies enough new variation to allow a very large response to selection.

Mitchell-Olds, T., and R. G. Shaw. "Regression Analysis of Natural Selection: Statistical Inference and Biological Interpretation." *Evolution* 41 (1987): 1149–1161. Discusses limitations on regression estimates of natural selection.

Mousseau, T. A., and D. A. Roff. "Natural Selection and the Heritability of Fitness Components." *Heredity* 59 (1987): 181–197. This and a companion paper review estimates of heritability.

O'Brien, S. J., G. W. Nelson, C. A. Winkler, and M. W. Smith. "Polygenic and Multifactorial Disease Gene Association in Man: Lessons from AIDS." *Annual Review of Genetics* 34 (2000): 563–591. Exemplifies current efforts to detect QTLs in human populations.

Plomin, R., M. J. Owen, and P. McGuffin. "The Genetic Basis of Complex Human Behaviors." *Science* 264 (1994): 1733–1739. Discusses quantitative genetic evidence for the role of genetic variation in human behavior.

Weber, K. E., and L. T. Diggins. "Increased Selection Response in Larger Populations. II. Selection for Ethanol Vapor Resistance in *Drosophila melanogaster* at Two Population Sizes." *Genetics* 125 (1990): 585–597. Excellent discussion of longer-term responses to selection.

— David Houle

QUASI-SPECIES

Chemists refer to an ensemble of identical molecules as a "species"—for example, the "species" of all H_2O molecules. When the term "species" is used of RNA molecules, however, it denotes not an ensemble of identical molecules, but a wildtype genome accompanied by a distribution of its mutants, hence the term "quasispecies." For biologists, the term is confusing because the biological species is a complicated and pluralistic concept related to gene flow among individuals, whereas the word "quasispecies" is derived from chemistry rather than from evolutionary biology. This term was coined in the 1970s by Manfred Eigen and Peter Schuster during the development of their chemical theory for the origin of life. They described how populations of RNA molecules could reproduce themselves and noted that the spontaneous chemical reproduction of such comparatively simple molecules was subject to a high rate of replication error. As a result, the population of RNA molecules arising from the inaccurate replication process would not be homogeneous, but rather a mixture of RNA molecules with different nucleotide sequences.

Eigen and Schuster were interested primarily in the origin of life. They assumed that RNA was the first biological replicator. In the "primordial soup," the four nucleotides (adenine, guanine, cytosine, and uracil) would spontaneously have formed short chains, so-called polymers. These polymers could have reproduced by base-pairing of each nucleotide to its complementary base (adenine to uracil, and guanine to cytosine). In time, these base pairs of RNA are thought to have been replaced by a more stable, DNA-based double helical structure. Subsequently, the double helix could have split into two single strands that, in turn, could have formed new double helices, thereby imprinting their sequences on new polymers. This process, which is the basis of all life, is now orchestrated by a set of highly sophisticated enzymes that ensure accurate replication. In the absence of these enzymes, the process of replication would have been slow and subject to high error rates.

This primitive genetic replication is a chemical process and can be described by chemical kinetics—that is, by equations specifying how the concentrations of certain molecules change over time. The kinetics of RNA self-replication assumes that molecules have different replication rates according to their sequence. This means that some may produce "offspring" faster than others and hence are "fitter." In addition, the theory takes into account that replication is inaccurate. An offspring sequence need not be identical to its parent; it may differ in certain positions. A substitution of one base for another is called a "point mutation." The equations lead not to a population of identical sequences, but to an ensemble of related but different sequences.

Eigen and Schuster referred to the equilibrium distribution of sequences formed by this mutation and selection process as a "quasispecies." "Mutation" is used in this context because reproduction is subject to errors, and "selection" because sequences have differing fitnesses. The term "equilibrium" here denotes a population in which the relative frequencies of the various distinct RNA molecules remain the same over time. This state arises when the rate of mutation away from a given type is exactly balanced by the rate of mutations that produce that type. Eigen and Schuster went on to argue that the target of natural selection is not the fittest sequence, but the quasispecies as a whole. Natural selection will choose not just the fittest type, but also the fittest ensemble of variants. The fittest sequence may represent only a very small fraction of the quasispecies; indeed, it may not be present at all.

Error Thresholds. An important concept in quasispecies theory is the error threshold. If the mutation rate is too high—that is, if too many mistakes occur in any one replication event—then the population will be unable to maintain its genetic information. The abundance of an individual sequence is dependent on its fitness. When the mutation rate is too high, however, selection is insufficiently strong to maintain fitter sequences. In the long run, the composition of the quasispecies is determined only by random drift in sequence space. Thus, the error rate must be below a critical threshold level for the system to maintain information.

If the error rate is expressed as a per base probability of making a mistake, then the error rate of an RNA sequence increases with its length. So, for any given per-base replication fidelity, the RNA sequence can reach only a certain length. An error threshold can be written as a condition that limits the maximum length of the RNA sequence. Eigen and Schuster estimated that this critical length should be roughly 100 bases for self-replicating RNA molecules in the primordial soup. The error threshold concept can also be expanded to larger organisms. Very roughly, the genome length should not exceed one over the per-base mutation rate. Viral RNA replication, in the absence of any error-correcting mechanisms (proofreading), has a per-base mutation rate of about 10^{-4}. As predicted, the genome length of such viruses is thus about 10^4. The human genome is about 3×10^9 bases long, and the high-quality DNA replication enzymes in human cells ensure an error rate of about 10^{-9} per base. It is important to realize that most organisms are expected to maintain their genomic mutation rates well below the error threshold, because most mutations are deleterious. Within the framework of the genetic quasispecies, all mutations are assumed to be available for selection; however, many mutations are neutral as a consequence of redundancy in the genetic code and will therefore accumulate, producing diversity beyond that estimated by the theory. Thus, there is a phenotypic error threshold above the genetic error threshold.

Because all biological reproduction is error-prone, the quasispecies concept can be applied readily to genetic processes other than RNA self-replication. Small populations of viruses, bacteria, plants, or animals form quasispecies. Their genetic reproduction is, of course, more complicated than a simple copying of the sequence and includes more sophisticated mutational events (such as sexual reproduction and recombination), but the underlying principles remain the same. Any natural biological population is a mix of genomes, and therefore a quasispecies. The distribution of genotypes present in the quasispecies is familiar to evolutionary biologists as the mutation-selection balance.

A quasispecies is a small cloud in sequence space, moving under the influence of mutation and selection. One source of confusion has been the tendency to refer to all variants observed in a population as members of a single quasispecies. In other words, "quasispecies" has become synonymous with any genetic variation observed in a population. Although all sequences are certainly members of some quasispecies, they need not be members of a single quasispecies. There are expected to be many wildtype sequences, each occupying some local optimum in sequence space, surrounded by less fit mutants. When we look at sequence data, we must understand that there may be several quasispecies present, representing positive selection for different traits in different microenvironments. Positive identification of a single quasispecies requires relating putative mutants to their wildtype.

Error Thresholds and the Hypercycle. One immediate consequence of the error threshold is that, for a given error rate, the length of a genome is limited. For low error rates, this does not present a serious restriction. However, during prebiotic evolution, when mutation rates are expected to have been very high, this would have reduced the adaptability of the genome. We therefore arrive at a paradox for the origin of life: in order to have evolved mechanisms of error repair (for example, the capability to code for an error-correcting enzyme), genomes had to reach a critical size; but to reach that size, error repair mechanisms had to be available to overcome the error threshold.

As a potential solution to this evolutionary impasse, Eigen and Schuster suggested the hypercycle. In this theory, the genetic information required to encode necessary proteins is distributed over a population of different molecules, all of which individually remain below the error threshold, but whose summed length crosses the threshold. Each sequence is assumed to be able to influence the replication rate of another sequence in the population. If each of these catalytic steps is arranged in order of catalytic support, they can potentially form a closed, autocatalytic cycle. For example, sequence A catalyses B, which catalyses C, which catalyses A. Such a cycle leads to positive feedback and magnifies the abundance of all members of a hypercycle. Unfortunately, such a cycle is destabilized by parasitic sequences, as first pointed out by John Maynard Smith. If we introduce a sequence X that is catalysed by A but does not itself assist in the replication of another sequence, then the cycle will degenerate, with the X sequence taking over the population. This problem could have been solved by restricting the number of interactions among sequences, for instance through spatial compartmentalization. In this case, the parasitic sequence X might not have the benefit of A and therefore would become extinct locally.

Quasispecies and Sequence Space. The correct metric in which to understand quasispecies is not the usual Euclidian space, but the so-called sequence space. In the sequence space, all possible variants of a given length are arranged so that adjacent neighbors differ by only one base substitution. More generally, the distance between two sequences equals the number of substitutions between them; for example, the sequence AATCG differs from ATCCG in two positions. The dimension of the space is given by the length of the sequence. In each dimension there are four possibilities, corresponding to the four nucleotides A, T, C, and G. The sequence space that contains all sequences of length five has five dimensions and $4^5 = 1{,}024$ points (different sequences).

The three most important features of this sequence space are its high dimensionality; the large number of shortest mutational routes between two distant mutant sequences—for two sequences separated by d point mutations there are $d!$ shortest mutational routes; and the fact that many sequences are confined to a neighborhood close to one another. The diameter of a sequence space that contains 10^{80} points is only 133 "length units," or point mutations. This means that relatively few point mutations can lead from one region in the sequence space to a completely different region, providing that there exists something like a guiding gradient to avoid going in the "wrong direction." In evolution, this gradient is provided by natural selection.

Quasispecies in the Fitness Landscape. Every point in sequence space can also be assigned a fitness value representing the reproduction rate of this sequence. This leads to the concept of a fitness landscape. Fitness landscapes have one more dimension than the corresponding sequence space, because for every point there is a "height." Quasispecies wander over the fitness landscapes searching for peaks, which represent regions of high fitness values. Under the guidance of natural selection, quasispecies climb the mountains in the high-dimensional fitness landscape.

Here, again, we can easily envision how natural selection does not simply choose the fittest sequence, but the fittest quasispecies. Imagine two sequences, A and

B. Assume that A has a higher replication rate than B; thus, it has a higher intrinsic fitness value. Suppose A is surrounded by mutants with very low fitness, while B is surrounded by mutants with high fitness. (Both A and B are local optima, but A is a very sharp peak in the fitness landscape, whereas B is the top of a flat mountain.) In the absence of mutation, A will be selected and B will disappear. With mutation, however, the situation can change, and B could be the winner. In fact, the mathematical equations show a critical mutation rate below which A is the winner, but above which B and its neighbors are favored.

[*See also* Acquired Immune Deficiency Syndrome, *article on* Origins and Phylogeny of HIV; Origin of Life, *article on* Origin of Replicators; Viruses.]

BIBLIOGRAPHY

Eigen, M. "Self-organization of Matter and the Evolution of Biological Macromolecules." *Naturwissenschaften* 58 (1971): 465–523.

Eigen, M., and P. Schuster. "*The Hypercycle: A Principle of Natural Self-organization.*" Berlin, 1979.

Maynard Smith, J. "Hypercycles and the Origin of Life." *Nature* 280 (1979): 445–446.

— David C. Krakauer and Martin A. Nowak

R

RACE

[This entry comprises two articles. The first article discusses population genetics perspectives on race including body form, skin color, and sexual selection hypotheses; the second article discusses the sociological perspectives of race, including ethnic boundaries and identities and the politics of race.]

Population Genetics Perspectives

The term *race* is used colloquially to refer to regional populations that share visible characteristics. It is not technical. Rough synonyms are *breed* and *subspecies*, as in breeds of dog and subspecies of wolf. There are several models or mechanisms that are implicit in most discussions of race differences in humans. They are not mutually exclusive, and any set of differences among people could reflect the action of any or all of them.

The Hierarchical Model. In the nineteenth and early to middle twentieth centuries, anthropologists worked to extend the Linnaean classification of life to human subspecies. Their systematic observations produced classifications, but there was never any general agreement about the number of races that could be enumerated or about the numbers of nested subraces. Implicit in the literature was the idea that races had a branching history, such as species and higher taxonomic units. Mixture between populations was regarded as a nuisance by many researchers, whereas others developed elaborate schemes of the racial origin of certain populations as mixtures of purer races.

In *Origin of Races* (1962), C. S. Coon proposed five major races of humanity: the Capoids, or aboriginal Khoisan-speaking people of southern Africa; the Congoids, or dark-skinned people of the rest of sub-Saharan Africa; the Caucasoids, or people from northern Europe to North Africa, including most of the Indian subcontinent; the Mongoloids, or people of eastern Asia and the New World; and the Australoids, or people from central India, much of the Pacific Rim, and all of Australia. Other authors have come up with other classifications. The model of a branching history of discrete races appears occasionally in the literature of human genetics in which estimates are given of separation times of races.

The underlying assumption of classification is that the units are tips of a tree of descent and that if a collection of units is followed backwards in time, the number of ancestors decreases. The "treeness" of species and of higher taxonomic units is a fundamental postulate of contemporary biology. In contrast, individual humans or other diploid organisms do not have a tree of ancestry: as we follow our own individual genetic material backward in time, for example, the number of ancestors increases; it does not decrease. We each have two parents, four grandparents, eight great-grandparents, and 2^n ancestors n generations ago. This comparison motivates the issue of whether races may be more like individuals, with increasing numbers of ancestors going backwards, than they are like species, with shrinking numbers of ancestors.

The Isolation by Distance Model. This alternative point of view of race differences was stated most clearly by Frank Livingstone in 1962. Instead of viewing races as discrete populations with a branching history, Livingstone proposed that populations are dispersed more or less uniformly over large areas. The forces of natural selection are different in different parts of the range, so that regional adaptations occur. Differences occur at selectively neutral genetic markers because of genetic drift—the accumulation of differences purely by chance over long time periods. The extent of differentiation of populations depends after a long time on the magnitude of local gene exchange or interbreeding: if there is not much gene exchange with neighbors, then regional differences are greater. Livingstone held that human diversity should be understood as a consequence of this population structure, called isolation by distance. In the isolation by distance model, humans would be most similar to their geographic neighbors and similarity would decline smoothly with distance. Different traits or genetic markers should not be correlated with each other: this is the crucial difference between the hierarchical and the isolation by distance models of race. Under Livingstone's scheme, a map of skin color and a map of, for example, the frequency of allele B from the ABO blood locus are in principle independent of each other. We identify races by external appearance, but this is arbitrary and meaningless because these traits just happen to be visible if this model accounts for all human diversity. Were we to make up races from ABO blood group frequencies, we would come up with a completely different and equally meaningless scheme.

The magnitude of differences among populations can be quantified by computing answers to two questions. First, if we randomly pick two copies of a gene from within a population, what fraction of pairs chosen this way are different? Second, if we pick two copies ran-

TAY–SACHS DISEASE

Tay–Sachs disease is an example of a self-destructive defense against an infectious disease. This is a recessive lethal disease of lipid storage in cells, affecting individuals who are homozygous for mutant forms of the enzyme hexosaminidase A. It is particularly prevalent in Ashkenazi Jews of central European ancestry. The gene frequency may be as high as 5 percent among Jews with ancestry in Austria, Hungary, and the former Czechoslovakia, so that as many as 2 per thousand births would be lost to the disease in these populations. In the United States before extensive prenatal testing programs were in place, the incidence in couples of European Jewish ancestry was approximately one in four thousand births.

One of each pair of grandparents of affected individuals must be a carrier of the mutant gene, and these carrier couples have higher reproductive success than controls in retrospective studies. In particular, death rates from tuberculosis are lower in their descendants. Jewish populations in central Europe were more confined to cities than non-Jews, and selection by tuberculosis may have caused the elevated frequency of Tay–Sachs as well as of two other lipid storage disorders, Niemann–Pick disease and Gaucher's disease, in this population.

—HENRY HARPENDING

domly from the world population, what fraction of those pairs are different? Call the first quantity H_0 and the second quantity H_e, then form the statistic

$$F_{st} = (H_e - H_0)/H_e$$

It is easy to show that H_e, overall diversity, is always greater than H_0, diversity within populations. The greater the differentiation of populations, the larger F_{st} will be. F_{st} can be thought of as a measure of the fraction of genetic diversity that is found between populations and $1 - F_{st}$ the fraction of diversity within populations. It is consistently found that F_{st} is 10–15 percent for world scale comparisons among human populations. In other words, 85–90 percent of our neutral genetic diversity is found within populations.

Many argue that these quantities imply that differences between populations are trivial and insignificant and that race differences are therefore ephemeral and without meaning. The problem with this argument is that the race differences that we see may not be neutral. Instead, they may be consequences of natural or sexual selection. If so, then they should not follow the same dynamics as the neutral genome follows. Furthermore, because they are or have been shaped by selection, they may be highly significant markers of our differences. For example, F_{st} estimated from world diversity in skin color is six to nine times as great as F_{st} of neutral markers, suggesting that it has a different evolutionary history and different dynamics.

The Adaptation Model. One of the best-understood genetic markers is the sickle-cell gene, which, in its homozygous state, causes sickle-cell anemia. The sickle-cell gene is found in populations in sub-Saharan Africa north of the Zambezi River, around the Mediterranean, and in India. It is one of a class of self-destructive defenses against an infectious disease. [*See* Vignette for information on Tay–Sachs disease, another self-destructive defense.] Heterozygotes for the sickle-cell gene are protected against many of the effects of the falciparum strain of malaria, whereas homozygotes for the gene have severely reduced viability. The world distribution of the gene has nothing at all to do with race: it has to do with the distribution of malaria. In North America, the homozygous disease, sickle-cell anemia, is a "racial" disorder because most cases occur in the American Black population. Many of the ancestors of the American Black population came from central and western Africa, where the sickle-cell gene is found at high frequency, accounting for the association between race and the disorder in North America. In Johannesburg, South Africa, however, the association is completely different. Sickle-cell anemia there is found in people from the north and is not closely associated with race.

Perhaps visible traits that conventionally define race are also naturally or sexually selected responses to the environment. Skin color in the human species is correlated with incident solar radiation: humans are darker near the equator and lighter toward the poles. Many hypotheses to explain the pattern have been advanced. The most persuasive suggests that light skin facilitates vitamin D synthesis in the skin in climates where exposure to sunlight is limited. Dark skin, in contrast, may protect against sunlight's denaturing circulating vitamins when exposure to sunlight is high. Other plausible hypotheses have been advanced to explain diversity of eye color, the epicanthic fold of the eyelid in people from East Asia, and even hair morphology. Human trunk and limb morphology does follow regularities observed in other species: in colder climates, bodies are less elongated and limbs are relatively shorter. The compact short-limbed morphologies in cold climates dissipate less heat.

The arguments that visible traits have been shaped by adaptation to the environment are shaky and unconvincing. For example there are both very light people and very dark people in cold, cloudy regions (Sweden and Tasmania, respectively). A clear dismissal of this tradition of speculation is given by Jared Diamond (1992), who advocates instead the model that racial traits have been shaped by sexual selection.

Sexual Selection and the Language Model. Charles Darwin believed that visible race differences in the human species were consequences of the process of sexual selection. Sexual selection refers to selection for characteristics that increase mating success. However the process starts, it can maintain itself and lead to the evolution of characteristics with no obvious engineering value to the organism in its confrontation with the environment. The tail on the peacock, for example, evolved because females prefer males with gaudy tails. In turn, the preference is maintained in the females because females lacking the preference for showy tails in their mates have sons with drab tails, and they suffer reduced fitness. This mechanism seems to account for much of the colorful exuberance of nature as well as much speciation and maintenance of separation of species.

Diamond (1992) gives an articulate argument that this process of sexual selection accounts for many of what we call racial traits: skin color, hair morphology, facial anatomy, and breast and genital morphology. With this mechanism, traits that are otherwise bad for the organism can evolve. White skin, for example, may be a biological handicap maintained only by mating preference.

The dynamics of sexually selected traits could resemble the dynamics of language. Imagine a trickle of migration from a continent to an island persisting over many generations. After a long time, the neutral gene frequencies on the island would be close to those on the continent, but the language of the island could easily persist because languages do not blend, although they may exchange words. If the trickle were large enough, the language on the island might change to the language of the continent, and the change would be relatively sudden. Similarly, sexual selection can maintain differences in appearance, in theory, even as neutral gene frequencies are homogenized.

A possible test case is the Khoisan-speaking aboriginal people of southern Africa. They do not look at all like their dark-skinned neighbors who speak Bantu languages: they have a dark yellowish skin, epicanthic folds on their eyelids, and other traits that are supposed to be characteristic of peoples of East Asia. Their hair, however, is like that of their neighbors. Despite the different appearances, the neutral gene frequencies of Khoisan speakers are generic African. In other words, the neutral genes show a homogeneous group, whereas external appearance separates the two groups. This situation is exactly what the sexual selection model predicts.

Conclusion. Our current understanding is that most of the human genome is derived from Africa over the past hundred thousand years, and this is not enough time for differences to have arisen purely by isolation by distance in a large population distributed over the termperate and tropical Old World. The hierarchical model may describe the separation between Africans and non-Africans, but no other such population splits are known. Because skin color and other "racial" traits seem to be more highly differentiated in the human species than neutral traits, these differences must reflect some history of selection. Whether any or all of these are adaptations to the environment is not known. If they are products of sexual selection, it is not known whether packages of traits, gene complexes, or individual traits such as skin color have been the targets of selection.

BIBLIOGRAPHY

Cavalli-Sforza, L. L., et al. "Reconstruction of Human Evolution: Bringing Together Genetic, Archaeological, and Linguistic Data." *Proceedings of the National Academy of Sciences USA* 85 (1988): 6002–6006. An attempt to show congruence between hierarchical reconstructions of linguistic and gene differences.

Cavalli-Sforza, L. L., P. Menozzi, and A. Piazza. *The History and Geography of Human Genes*. San Francisco, 1994. Encyclopedic descriptions of worldwide distribution of classical genetic markers, most but not all of them probably neutral.

Coon, C. S. *Origin of Races*. New York, 1962.

Diamond, J. *The Third Chimpanzee*. New York, 1992.

Garn, S. M. *Human Races*. 2d ed. Springfield, Ill., 1965. A statement of anthropological knowledge about human diversity; essentially the end of a tradition, because the topic became politically unacceptable soon after this book was published.

Harpending, H., and E. Eller. "Human Diversity and Its History." In *The Biology of Biodiversity*, edited by M. Kato. Tokyo, 1999. A description of genetic differences among human populations and an algebraic treatment of the sexual selection model.

Lewontin, R. C. "The Apportionment of Human Diversity." In *Evolutionary Biology*, edited by T. Dobzhansky, vol. 6. New York, 1972. The first clear statement that F_{st} among human populations for neutral genes is 10–15 percent, although a different but related statistic is used.

Livingstone, F. "On the Non-existence of Human Races." In *The Concept of Race*, edited by A. Montagu, pp. 46–60. New York, 1962.

Nei, M., and A. Roychoudhury. "Genic Variation with and between the Three Major Races of Man, Caucasoids, Negroids, and Mongoloids." *American Journal of Human Genetics* 26 (1974): 421–443. A genetic treatment of the hierarchical model of race differences using classical genetic markers.

Rushton, J. P. *Race, Evolution, and Behavior*. New Brunswick, N.J., 1997. A controversial model of race differences that proposes a wide range of adaptive consequences of human evolution in warm and cold climates.

— HENRY HARPENDING

Sociological Perspectives

Social psychologists working in Western societies have noted that race is one of a handful of personal characteristics, along with sex, age, and physical attractiveness, that people normally register automatically and involuntarily at the beginning of any interpersonal encounter. Racial categorization has consequences; it influences expectations that people have of one another, and how they interact. Experiments show that this is true even for individuals who claim that they are free of prejudice and do not to judge others on the basis of their race. On a larger scale, race often serves as a basis for group politics, and race consciousness and racial conflict have been major historical forces.

Race, then, is an important social fact. Race as a social fact depends not just on how populations differ biologically, but also on how people think about biological and other group differences. This article reviews how concepts and evaluations of racial differences overlap with and draw on other sorts of representation, including ideas and values regarding (1) individual differences in appearance and behavior, (2) social groups—including ingroups and outgroups—and political coalitions, and (3) kinship and living kinds. "Race" and "race differences" are used here to refer to biological population differences on a large, geographic scale; "ethnic groups" and "ethnicity" refer to descent, actual or imagined, on a smaller scale. The terms are loose, and different authors use them differently.

Race as an Individual Trait. Populations originating in widely separated geographic areas often show large differences in visible features like skin color, face shape, hair form, and physique, even when overall differences in gene frequencies are modest. These differences in appearance are important for several reasons. They make race more conspicuous and harder to disguise than many other personal characteristics, making it harder to "pass" from one racial category to another. Also, since human beings are very prone to judge one another on the basis of looks, ideas of race often overlap with ideas of physical attractiveness, resulting in *somatic prejudice:* positive or negative responses to members of particular racial or ethnic groups based on their physical appearance. Historian George Mosse (1985:xii, xxvi) notes how important somatic prejudice was in European racism: "Racism was a visual ideology . . . That was one of its main strengths. . . . All racists held to a certain concept of beauty—white and classical." In modern multiracial societies, physical features associated with low status and minority groups are often devalued, even within such groups, with consequences for individuals' bargaining power in personal relationships.

Race is associated not just with appearance but also with behavior. People use information about race in social inference, making educated guesses about how others will behave based partly on *racial stereotypes*—beliefs about differences in behavior associated with racial differences. Researchers disagree about how reliable stereotype-based inferences are. Some argue that stereotypes, usually or even by definition, are either pure fictions or caricatures that invent, exaggerate, overgeneralize, and moralize group differences. Exaggerated stereotypes might result from well-documented perceptual and cognitive processes, like the *contrast effect:* just as a gray figure looks lighter against a dark background than against a light one, so the very act of assigning people to categories and making comparisons across categories might promote exaggerated perceptions of group differences. Other researchers, however, argue that many stereotypes accurately reflect group differences. If race is correlated with behavior, for whatever reason, then it may be rational to use race to predict behavior, albeit not to the exclusion of other sources of information. The actual evidence regarding stereotype accuracy is mixed. In some studies, beliefs about group differences are moderately exaggerated; in others, they are accurate or even understated. It may be that negative exaggerated stereotypes are most significant in situations of intergroup conflict. A further complication is that even accurate stereotypes may involve *self-fulfilling prophecies*, in which the very existence of a stereotype evokes behavior conforming to the stereotype. For example, beliefs that members of a group are unfriendly or unreliable may provoke unfriendly or unreliable behavior.

Other research shows that increased contact between groups, sometimes recommended to reduce stereotypes, often strengthens them. Moreover, children acquire stereotypes early, often by preschool, and often before they can reliably identify members of different groups.

Races as Social Groups. Race matters not just to one-on-one relationships but also on a larger scale, in group formation and intergroup conflict. Conversely, large-scale political and institutional factors play a role in how people define races and draw racial boundaries.

Social psychologists concerned with prejudice and group conflict have demonstrated that human beings are very ready to assume group identities. Individuals assigned to groups on the basis of modest differences or no differences develop positive attitudes and show favoritism toward their own group (the *ingroup*) and negative attitudes toward members of other groups (*outgroups*). Group identity and group solidarity may intensify in the face of resource and status competition between groups, but they can exist even in the absence of such competition. The psychology of ingroups and outgroups is involved in the universal or near-universal phenomenon of *ethnocentrism*, the tendency to regard one's own people and their way of life as superior.

Organized group activity requires more than group sentiments; especially in large groups, it also requires some institutional structure to enforce participation in collective endeavors. Because group sentiments and group institutions may be mutually reinforcing, there is a "chicken-and-egg" problem in understanding which comes first. One theoretical school, *social constructionism*, emphasizes the role of political expediency in group formation: people develop feelings of group identity mostly in the wake of opportunistic coalition building. Applied to race, social constructionism emphasizes how politics affects definitions of race and racial boundaries. From this perspective, the history of race in twentieth-century America is the story of how, as a result of changing political alignments, a division between an Anglo-Saxon (or Nordic) "race" and non-Anglo-Saxon "races" like the Irish, Jews, and Sicilians was eventually eclipsed by the division between whites and nonwhites. The very categories of white and black are at least partly social constructs rather than purely objective biological facts. The definition of black in the United States has mostly followed the "*one drop rule*": a person with even a little black African ancestry is classified as black (or African American). In Brazil, however, the same person would probably be classified as mulatto or white; the difference between the two countries stems from different histories of race relations.

Races as Descent Groups and Living Kinds. Social constructionism is probably only part of the truth about races as social groups. Another theoretical school, *primordialism*, argues that feelings of group identity based on shared language, religion, culture, appearance, or ancestry develop first and then motivate the development of institutions like ethnically based political parties and states. Taking a primordialist approach to racial identity, two ideas seem to be especially important: *kinship* (or *descent*) and *living kinds*.

Kinship and descent are central concepts in anthropology. In tribal societies, large kin groups encompassing kin well beyond the limits of the family are often socially important. In these groups, altruistic sharing, with needy members entitled to call on others for help, is often the ideal—never entirely realized in practice, of course. This contrasts with the ideal of balanced exchange that is more common in relationships outside the kin group. Biologically speaking, races can be regarded as vastly extended kin groups, sharing traits partly by virtue of common descent; in modern societies, races and ethnic groups seem to enlist some of the same feelings that attach to kin groups in tribal societies. The language of kinship—"brother" and "sister"—is often used for fellow ethnics, and ideas of common descent can inspire ethnically and racially based political movements.

If kin groups are one powerful model for thinking about race, living kinds—species and higher-level biological categories—are another. The boundaries between races are often seen as analogous to those between species, especially when races are endogamous. This may be reflected in language: a single word may mean both "race" or "ethnic group" and "species," or one term may refer both to one's own people (in contrast to outsiders) and to human beings (in contrast to nonhuman animals). Taken to an extreme, the combination of ethnocentrism with a model of races as living kinds can lead to doubts about the very humanity of racial outgroups.

Conclusion: Ideas about Race, and Where They Come From. Where do people get the ideas—about physical appearance and behavior, social groups, kinship, and living kinds—that they draw on in thinking about race? Two different research traditions offer very different answers. The dominant tradition in the social sciences follows a *blank slate* theory of the human mind, suggesting that experience, including cultural experience, is the source of all ideas. According to some researchers in this tradition, seemingly natural ideas of kinship, descent, and living kinds are inventions of the modern West, devised to provide an ideological warrant for racially based colonialism, slavery, and expropriation. Societies lacking these ideas may have no basis for constructing concepts of race and ethnicity.

A different intellectual tradition suggests that all human beings share a stock of innate concepts, including some relevant to racial thinking. For example, some anthropologists and psychologists believe that human beings spontaneously take an *essentialist* stance toward some kinds of biological and social classification. This means that people naturally assume that living kinds and human kinds (like ethnic groups) have underlying essences, and that these essences are normally conserved over the course of growth and reproduction, so that even major changes in outward appearance do not change individuals from one species or human kind into another. On this view, children have to learn *which* human kinds are present locally and how to recognize them; however, the idea that there are human kinds sharing enduring inherited commonalities is not learned but part of an innate theory of the social world (Hirschfeld, 1996.)

These claims about innate ideas, like much else in the study of race, are matters of controversy. We still have much to learn about how local cultural traditions, universal principles of human psychology, and the facts of human biological variation combine to produce varying ideas and evaluations of racial differences.

BIBLIOGRAPHY

Gerstle, G. *American Crucible: Race and Nation in the Twentieth Century.* Princeton, 2001. Twentieth-century U.S. history as the interaction between changing versions of racial nationalism and nonracial civic nationalism. Demonstrates how politics influences definitions of race and ethnicity, and sheds light on the rise of racial liberalism and multiculturalism in modern intellectual life.

Hirschfeld, L. *Race in the Making: Cognition, Culture, and the Child's Construction of Human Kinds.* Cambridge, 1996. Evidence from children's concepts of race for an innate theory of "human kinds." Also includes a fair summary of alternative social constructionist theories of race.

Horowitz, D. *Ethnic Groups in Conflict.* Berkeley, 1985. An encyclopedic review and major theoretical synthesis of ethnic conflict since World War II.

Jones, D. "Physical Attractiveness, Race, and Somatic Prejudice in Bahia, Brazil." In *Adaptation and Human Behavior: An Anthropological Perspective*, edited by Lee Cronk, Napoleon

Chagnon, and William Irons. New York, 2000. How definitions of "white" and "black" differ between the United States and Brazil, and the importance of somatic prejudice to race relations in Brazil and elsewhere.

Lee, Y.-T., L. J. Jussim, and C. R. McCauley, eds. *Stereotype Accuracy: Toward Appreciating Group Differences.* Washington, D.C., 1995. Reviews psychological research on stereotypes and stereotype accuracy.

Mosse, G. *Toward the Final Solution: A History of European Racism.* Madison, 1985. Argues that from the eighteenth to the mid-twentieth century racism was a major European ideology, comparable in intellectual and political importance to liberalism and socialism.

Tajfel, H. *Human Groups and Social Categories.* Cambridge, 1981. The social psychology of ingroups and outgroups, by a pioneer in social identity theory.

Van Den Berghe, P. L. *The Ethnic Phenomenon.* New York, 1981. A sociobiological account of how kinship sentiments influence ethnic and racial politics.

— Doug Jones

REACTION–DIFFUSION MODELS. *See* Pattern Formation.

REACTION NORMS. *See* Phenotypic Plasticity.

RECAPITULATION

Recapitulation, or the tendency of an embryo to pass through developmental stages that resemble lower or more primitive forms of life, is a notion that emerged at the beginning of the nineteenth century but whose exact nature changed over the century. Its formative roots were buried in three distinct traditions, two of a philosophical nature and one based on observational evidence.

Beginning with the early Greek philosopher and natural historian Aristotle, naturalists have regarded all life to be organized from simple forms to complex forms along a scale of nature (*scala naturae*), thus emphasizing that the natural world was full, complete, and interconnected. The Swedish botanist Carolus Linnaeus referred to this notion in the eighteenth century as the *Systema naturae.* Like its earlier version, the Linnaean interpretation emphasized the unity of natural beings, a conception of nature accepted by most naturalists at the end of the eighteenth century, including the French naturalist Jean Baptiste de Lamarck, who carried the idea into the nineteenth century.

The scale of nature, also referred to by historians as the "Great Chain of Being," received additional support from a second philosophical source, German natural philosophy, especially the version influenced by the peculiar, complicated, and influential notion of *naturphilosophie.* This philosophy, most closely associated with Immanuel Kant and popularized by the German romantic naturalist Johann Wolfgang von Goethe, explained the unity of nature as the product of ideal forces or archetypes that regulated the formation of all natural objects. Again, this idealistic philosophy noted the organization of life from simple forms to the most complex forms, the human species, toward which all nature strived. In the work of Johann Blumenbach (*Handbuch der Naturgeschichte*, 1802), the notion was complemented by the idea that understanding the past led to understanding the present. In other words, organisms had a historical context.

Regardless of which philosophical position one adopted, therefore, the development of organisms provided the best example to illustrate either the unity of nature or the gradual emergence of the idealized form. After all, embryonic development progressed inexorably from undifferentiated material toward an ultimate form. The German philosopher and naturalist Lorenz Oken claimed in 1802 that one reason for these characteristics was that all forms of life were made of the same building blocks, with more complex forms merely representing more complex *polystocks* (combinations) of primodial parts. Oken did not suggest these parts were cells, but his ideas were suggestive of this later idea. His colleague, Ignaz Döllinger, in 1817 described the developmental process as *Entwicklunsgeschichte*, again emphasizing the historical aspect of development and underscoring its importance in explaining the unity of nature. The French anatomist Geoffroy Saint-Hilaire, writing in his *Philosophie anatomique* at the same time, suggested that all forms of higher animal life were modeled on the structure of the vertebrate, as one could observe by examining embryonic development. Also, in the early 1820s, both Johann Friedrich Meckel and Etienne Serres articulated the notion of recapitulation, suggesting that embryonic structures of highly diversified or advanced organisms repeated the adult stages of the less derived structures within the same animal series. In other words, higher vertebrates, in their developmental history, passed through embryonic stages similar to the adult forms of lower vertebrates.

By this time, the German optical industry had vastly improved microscopes, which were put into service to investigate developmental questions. Meckel and Serres were soon able to offer observational evidence for recapitulation, but their work was quickly attacked by the nineteenth-century's greatest embryologist, the Prussian Karl Ernst von Baer. Combining extremely skillful and thorough observational evidence with a keen argumentative style, von Baer wrote his own work on *Entwicklunsgeschichte* in 1828, but he forcefully repudiated recapitulation. Arguing that there was no unity in the natural world, von Baer noted the existence of four discrete and unique animal plans, the vertebrate, articulate, molluscan, and radiate. Within each of these plans, furthermore, one did not observe recapitulation, but the gradual emergence of diversified individual structures. Development was not a process of the repetition of lower forms of life, but the growing individuality of organisms from an archetypal foundation. All vertebrates diverged from the vertebrate form; there was no recapitulation among the vertebrate group.

Perhaps in part because of the growing dissatisfaction for scientific ideas based primarily on philosophical argument and in part because of von Baer's careful observational work in development, recapitulation ideas soon were discarded. The Swiss paleontologist Louis Agassiz, who learned from Döllinger and was a devotee of von Baer, moved to the United States at mid-century and began to expose his American students to the notion of divergence and shared ancestry along phyletic lines. But his interpretation was faithful to both the idea of limited unity of nature of von Baer and to the notion of divergence in the embryonic history. Then, in 1859 and following his reading of T. H. Huxley's translation of von Baer, Darwin also noted the importance of studying development for evidence of the animal's ancestral characteristics. After all, the divergence von Baer had discussed was remarkably similar to the descent by modification that Charles, Darwin envisioned.

Darwin's epochal book *On the Origin of Species* appeared in 1859. In it, Darwin carefully developed his new interpretation of animal life, emphasizing that forms of life shared an ancestral relationship. He supported his work with an abundance of data from comparative anatomy, paleontology, and what would now be called biogeography, areas in which he was relatively well versed. However, near the end of the book, he made a prescient suggestion that the best evidence for unity of descent would come from embryology, because naturalists had illustrated how important development was for the reconstruction of an organism's ancestral history. Although Darwin had little personal experience to support his assertion, it was in this arena that his new theory received substantial support.

Ernst Haeckel, the self-styled Darwin of Germany, became an immediate adherent to evolution theory when he read the *Origin*, reportedly in one sitting. He immediately set himself to the task of providing the embryological support that Darwin had suggested. Examining a host of different organisms, Haeckel soon noted that all higher forms of life (metazoans) emerged from a similar embryonic stage, the gastrula. This led Haeckel to formulate his *gastraea-theorie*, suggesting that all forms of higher life emerged from a universal primordial gastrula form. From this, Haeckel then argued that, because all forms of life had the same ancestral history, higher forms of life repeated the adult stages of lower forms of life, from which they had diverged. This was a new and revised version of the Meckel–Serres law, now arguing that forms of life were unified in their ancestral descent. Using Darwin's theory for support, organisms now had similar developmental stages because they were related genealogically. In *Morphologie Generelle* (1866), Haeckel framed recapitulation as the biogenetic law, Ontogeny recapitulates phylogeny. In other words, the developmental history of the organism repeated the organism's ancestral (phyletic) history. Futhermore, Haeckel considered the law to be mechanical; that is, phylogeny was the mechanical cause of development.

Biologists at the end of the nineteenth century, like their earlier colleagues, turned to investigate the validity of Haeckel's speculations. Using improved microscopical methods in laboratories throughout Europe and the United States, they soon discovered problems with the universal notion of the *gastraea-theorie* and, subsequently, began to question the applicability of recapitulation. Studies of the exact nature of development (cell lineage), comparisons of the developmental process among different organisms, and questions of the control of development soon pushed aside the more theoretical suggestions of Haeckel. With the sudden emergence of chromosome theory and the newfound appreciation for Gregor Mendel's particulate view of inheritance in the first decade of the twentieth century, recapitulation fell even further from favor in biology. Finally, the English embryologist Gavin de Beer wrote a scathing indictment of recapitulation in *Embryos and Ancestors* (1932), relegating the notion to a long list of historically influential but discredited ideas.

[*See also* Development, *article on* Developmental Stages, Processes, and Stability; Epigenesis and Preformationism; Lamarckism.]

BIBLIOGRAPHY

Coleman, William. *Biology in the Nineteenth Century: Form and Transformation.* New York, 1972.

Gasking, Elizabeth. *Investigations into Generation, 1651–1821.* Baltimore, 1967.

Mayr, Ernst. *The Growth of Biological Thought: Diversity, Evolution and Inheritance.* Cambridge, Mass., 1982.

Oppenheimer, Jane. *Essays in the History of Embryology and Biology.* Cambridge, Mass., 1967. Press.

— KEITH R. BENSON

RECIPROCAL ALTRUISM

In a number of situations, one animal aids another despite obvious incentives to do otherwise. Such *altruistic* behavior is not predicted by Charles Darwin's theory of natural selection. One solution to this problem was provided by Robert Trivers (1971), who proposed that *reciprocal altruism* could account for aid giving among unrelated animals. Trivers pointed out that as long as two individuals take turns helping each other and neither can predict the last exchange, then over time cooperative individuals will survive and reproduce better than noncooperative individuals. For each participant to profit from a reciprocal exchange, the benefit returned to the recipient must exceed the cost of performing the act.

Reciprocity is distinguished from *mutualism*, in which both parties benefit immediately from an act, by the presence of a time lag between when an individual acts as a donor and when it receives a benefit. A time lag provides an opportunity for selfish individuals to cheat either by returning a lesser benefit or nothing at all. In the absence of retaliation against such selfish behavior, cheaters will prosper (and in evolutionary terms, leave more offspring) compared to individuals who reciprocate in kind. Thus, for reciprocity to spread and persist, cooperative individuals must either confine their aid giving only to other cooperative individuals, or retaliate against noncooperative individuals. Trivers suggested that long-lived animals that have stable social groups with many opportunities to exchange aid, such as many primates, would be most likely to engage in reciprocity.

The primary alternative explanation for altruistic behavior was provided by William Hamilton (1964) and has since become known as *kin selection*. Hamilton derived conditions under which reproductive altruism is beneficial owing to the likelihood that donor and recipient share a genetic predisposition for the behavior. Specifically, Hamilton showed that when genetic relatedness between the donor and recipient exceeds the ratio of the costs to the donor of the altruistic behavior, to the benefits to the recipient, then copies of genes carried by a recipient will spread to future generations more rapidly than if the donor acted selfishly by withholding aid. This idea has been used with success to explain the evolution of many different aid-giving behaviors in a variety of animals, as well as other forms of social behavior. If relatedness is low, as for the situation of reciprocal altruism, an altruistic act may still be favorable if the costs to the donor are low enough, or the likelihood of receiving a benefit at a later time is high.

Reciprocal altruism and kin selection are alternative, but not mutually exclusive, evolutionary explanations for altruism. Although Trivers (1971) conceived of reciprocal altruism as a mechanism to explain altruistic behavior among unrelated animals, he noted that related individuals might also exchange aid. Reciprocity can therefore work in concert with kin selection. Trivers also noted that the options available to two individuals involved in a potentially altruistic exchange resemble the *Prisoner's Dilemma*, a well-known game in which two individuals can either cooperate or defect when they encounter one another. In one round of this game, mutual cooperation pays well, but exploiting or taking advantage of a cooperator pays best; refusal by both parties to cooperate pays poorly, whereas cooperating but being exploited by a defector pays worst of all. Thus, in the absence of any information about what the other player may do, the best strategy is to defect (fail to cooperate) because no matter what a partner does, a defecting individual will receive a higher payoff. How then can cooperation evolve?

Robert Axelrod and William Hamilton (1981) showed that cooperation can nevertheless emerge if the same two players interact repeatedly in a Prisoner's Dilemma game. They found that a simple strategy, Tit for Tat (TFT), outperformed all other strategies in a computer tournament. A TFT player always begins by cooperating and on subsequent occasions copies its partner's previous behavior. Thus, TFT requires individual recognition and short-term memory. Although a single individual playing TFT will not increase among a population of defectors, a small group of TFT individuals that preferentially interact among themselves will rapidly outreproduce defectors (here it is assumed that payoffs in the game translate into reproductive success).

Axelrod and Hamilton suggested that families of closely related individuals provide a likely source for TFT groups. Under this scenario, TFT reciprocity would be catalyzed by kin selection and subsequently spread to unrelated individuals. Many additional theoretical studies on the repeated Prisoner's Dilemma game have considered the consequences of alternative assumptions, including errors in partner recognition, sequential rather than simultaneous decisions, using spatial segregation rather than memory to restrict cooperation, and variation in how much altruism to dispense. Among other findings, these studies have revealed that TFT may represent a transitory stage, rather than an end point, for populations evolving toward unconditional cooperation.

The extent to which Trivers' original notion of reciprocal altruism, or some other strategy based on the Prisoner's Dilemma, applies to nonhuman animal behavior is controversial. Skeptics maintain that most putative examples of reciprocal altruism involving nonhuman animals either fail to involve altruism or are better explained as mutualism. One reason for skepticism comes from laboratory studies on pairs of rats and blue jays. When offered food under conditions in accordance with a repeated Prisoner's Dilemma, both species failed to exhibit cooperation. This result is consistent with animal foraging experiments, which show that animals typically assign greater importance to current rather than future rewards. Neither of these species is known to exhibit cooperative behavior in the wild, however, and may not be appropriate models. A second concern is that TFT reciprocity requires relatively complex cognition, that is, memory of the past behavior of individuals, something which remains to be demonstrated for most species. In contrast, proponents argue that reciprocal altruism occurs in a variety of behavioral situations—including predator inspection, sentinel behavior, territorial defense, grooming, gamete trading, food sharing, and coalition formation—and involves animals from fish to primates. Below, these cases are summarized and evaluated particularly with regard to the possibility that the behaviors observed might be better explained as selfish or mutually beneficial.

A particularly controversial example of reciprocity involves predator inspection in fish. Sticklebacks, as well as many other small fishes, often leave a school to approach, rather than flee, from a predator. Several investigators have interpreted this behavior to indicate that approaching fish put themselves at risk to obtain information about a predator's hunger level. Under this interpretation, an approach is dangerous and possibly altruistic, whereas movement away is noncooperative. A number of studies using mirrors, models, and trained fish have shown that individual fish appear to copy the movement of a partner toward or away from a predator. Thus, predator inspection has been argued to be a case of Tit for Tat reciprocity. The problem with this explanation is that some studies show that predator approach reduces risk of predation. Thus, there may be no cost associated with this behavior. Furthermore, an alternative explanation for move copying is that schooling fish tend to move simply to remain close together. Detailed sequential analysis of individual fish movements in the presence and absence of partners and predators do not match TFT predictions. Predator inspection appears, therefore, to result from joint attraction to a conspecific and a predator and is better described as a form of mutualism than as reciprocity.

In a number of group-living birds and mammals, one individual acts as a sentinel or lookout while others forage. Lookouts are often, but not always, related and give alarm calls to alert the group to approaching predators. Reciprocal altruism has been suggested as a possible mechanism for the maintenance of sentinel behavior in some species of jays and mongoose. Recent studies on meerkat, a highly social mongoose found in South Africa, by Tim Clutton-Brock and associates (1999) have confirmed that adult animals in a group alternate sentinel duty. However, individuals do not follow a regular rotation, as would be expected if sentinels reciprocate. Furthermore, sentinel behavior may not be costly. Animals that have recently eaten to satiation are more likely to act as sentinels than are actively foraging individuals, so sentinel behavior does not detract from foraging. Meerkat sentinels also show no evidence of increased predation risk. In contrast, sentinels were often one of the first animals to reach the safety of a burrow. Thus, reciprocity does not appear to be necessary to explain why meerkats exhibit sentinel behavior.

Many species of birds defend territories to exclude competitors from mating or feeding in an area. Although fighting often precedes territory establishment, aggression is much less common among established neighbors, especially at breeding territories. This apparent restraint has been referred to as the "dear enemy" effect, and some investigators have proposed that it corresponds to a Prisoner's Dilemma. Studies on hooded warblers indicate that these birds use a TFT-like strategy when responding to incursions of neighbors. Hooded warblers fit the requirements for reciprocity because they interact with neighbors repeatedly over long periods and can recognize individuals by their song even after migration eight months later. Furthermore, song playbacks that simulate territorial incursions by neighbors elicit a more aggressive response than those by strangers. This is expected if there is retaliation against neighbors for not observing a territorial boundary. In addition, subjects did not respond aggressively to neighbor song prior to an incursion, consistent with initial cooperation. Although this example provides one of the few experimental results consistent with cheater retaliation, no evidence has yet demonstrated that aggressive restraint between a territorial holder and its neighbor represents altruism. Proponents of the dear enemy effect as a case of the Prisoner's Dilemma claim that territory holders pay a cost by respecting their neighbor's borders because they lose opportunities to gain access to food or mates. However, if these individuals determined during territory establishment that subsequent incursions risk injury from escalated fights, then restraint could be mutually beneficial to both parties. If this interpretation is correct, reciprocal altruism need not be invoked to explain this behavior.

More compelling examples of reciprocal altruism involve exchange of valuable commodities or services between individuals. Different systems have been described and involve exchange of grooming, gametes, food, or assistance in acquiring a resource. Although many of these cases involve reciprocal exchange of comparable goods, some involve exchange of different goods. This possibility has led to the development of additional theoretical models based on a marketplace analogy. These models predict that animals compete to participate in reciprocal exchange partnerships. Such competition encourages increasing investment in altruism. Thus, partner choice provides an additional reason besides partner fidelity to continue in reciprocally altruistic situations.

Grooming of another individual (also known as allogrooming) occurs in many different species of birds and mammals and has been carefully studied in several species. For example, impala use specialized comblike teeth to groom the head and neck region of a partner. Grooming is effective at removing disease-causing ticks and

RECIPROCAL FOOD SHARING IN VAMPIRE BATS

Food sharing in vampire bats (*Desmodus rotundus*) illustrates the economic advantages of reciprocity. Female vampire bats, which live in social groups of ten to twenty individuals, will regurgitate blood to roost mates who fail to obtain a blood meal. Feeding failure occurs because their prey, which in most parts of Latin America is now either cattle or horses, sometimes detect the bats and disrupt their feeding. Blood-sharing events often occur between mothers and recently flying young, which are less successful at obtaining a blood meal, but also can involve more distantly related and even unrelated bats. Experiments have shown that the bats not only identify and preferentially feed long-term roost mates, they also are more likely to give blood to an individual that previously fed them. The benefit of sharing food is great because failing to feed on three successive nights results in death by starvation. In contrast, the cost of sharing is low because transfer of a fraction of a blood meal from a well-fed bat to a hungry one can give the recipient much more time until starvation than the donor loses due to differences in their metabolic rates. Consequently, blood exchanges increase the probability of survival of participating bats. In the absence of such reciprocity, no adult should live beyond their third birthday, given how often they fail to get a blood meal. Instead, female vampire bats can survive over fifteen years in the wild.

—GERALD S. WILKINSON

occurs more frequently when ticks are abundant, so the behavior clearly benefits the recipient. Although a cost of grooming has yet to be quantified, grooming takes time away from feeding and reduces the groomer's ability to scan for predators. Each grooming bout is fairly brief but can be exchanged up to forty times in a highly reciprocal manner between animals. Because animals remain in proximity between bouts, there is no need to remember the identity and behavior of others. Impala frequently move between social groups, so it is unlikely that grooming is confined to relatives. Because any one grooming bout is short and a grooming episode can be easily terminated, the advantage to cheat in this system is very low. Impala allogrooming, therefore, appears to meet all of the requirements of reciprocal altruism.

In some species of primates (monkeys and apes), such as baboons and macaques, allogrooming is asymmetrically exchanged with dominant individuals receiving more episodes of grooming than they provide. In other primates, such as chimpanzees, allogrooming is much more symmetrical. One explanation for these differences is that species that form dominance hierarchies may exchange bouts of grooming for access to food or mates. Evidence from baboons consistent with this interpretation shows that grooming reciprocity is predicted by levels of resource competition within a group, with more reciprocal grooming occurring when there is little competition for resources. However, an alternative explanation for variation in reciprocity is that individuals differ in relatedness. Cases of allogrooming that exhibit low levels of reciprocity often involve closely related animals from the same matriline. These cases may be more influenced by kin selection than by reciprocity, however. The degree to which reciprocal allogrooming among primates is mediated by relatedness, rather than resource competition, requires further study.

Egg-trading by sequential hermaphrodites (animals that can alternate between adopting male and female reproductive roles), such as some sea basses and polychaete worms, represents a second example of a commodity exchange system. Sea bass spawn daily by trading eggs and sperm in small parcels, that is, one member of the pair donates eggs, the other sperm, and each partner makes use of the other's contribution. Such parcelling behavior presumably minimizes any benefit of cheating that could occur if only small energetically inexpensive sperm were transferred by both parties, rather than large energetically costly eggs. Given that up to 80 percent of egg-trading episodes are exchanged and the larvae of these species are planktonic—greatly reducing the chance that two adult fish are related—reciprocity provides a better explanation than kin selection for this behavior. Confirmation of this example as a case of reciprocal altruism awaits estimation of the costs and benefits of producing eggs and sperm as well as the consequences of withholding eggs.

Food sharing provides a third example of commodity exchange. Reciprocal food sharing has been reported for vampire bats, capuchin monkeys, and chimpanzees. Observations by Gerald Wilkinson (1984) indicate that vampire bats exhibit reciprocal altruism by sharing blood with roost mates that fail to feed on their own (see Vignette). Although it remains unknown if vampire bats can recognize and exclude cheaters, reciprocal blood sharing need not have arisen from a TFT-like process that requires mental score keeping. As Nowak and May have shown, cooperation can emerge and persist in a spatially heterogeneous world. Consistent with this possibility, female vampire bats segregate into stable social groups.

Capuchin monkeys are highly social New World monkeys that have been observed sharing food in captivity. Experiments by Frans de Waal (2000) show that two capuchins separated by a mesh partition will preferentially sit next to each other and tolerate food being taken by their neighbor. This facilitated food taking is reciprocated between animals over time. In addition, two animals that are forced to work together to obtain food were more inclined to share their food, suggesting that interchange of goods also occurs. De Waal points out that such cooperation does not require mental record keeping, only affiliative tolerance of a conspecific. The importance of relatedness in accounting for capuchin sharing is unknown.

Food sharing has also been observed among chimpanzees. In both captive and field situations, food transfer often occurs without aggression when one animal feeds simultaneously on the same food source, removes a piece of food from the hands of another, or collects food remnants from beneath a feeding individual. In captivity, plant food is either transferred asymmetrically from older dominants to younger subordinates or shared symmetrically among similar-aged adults. Food transfer is also enhanced after recent grooming episodes. In the Tai Forest of Ivory Coast, a successful hunt is accompanied by unique calls that recruit other members of a band. The degree to which a chimp then shares meat depends more on his participation in a hunt than on relative dominance rank or age. John Mitani and David Watts (2001) found that in Kibale National Park meat sharing occurs reciprocally between males that participate in coalitions and is not related to nutritional shortfalls or access to estrous females. Thus, in both captive and field situations, food is sometimes exchanged for other services, such as grooming or participation in a coalition. Further work is needed to determine if kinship also influences food sharing. Kinship could be important since males typically remain in their natal groups.

Reciprocal altruism has also been reported to occur among coalition members in several species of primates and dolphins. Craig Packer (1977) provided the first report of reciprocal alliances in olive baboons. He observed subordinate males soliciting assistance from another male to gain access to an estrous female that was consorting with a dominant male. On some occasions, the soliciting male ended up with the female and the animal that helped was more likely to be assisted later when it solicited assistance. A number of subsequent studies have confirmed that baboon alliances are often long-lasting and increase mating opportunities for participants. However, in some cases the helping male, rather than the soliciting male, gained access to a mate. This has led some to conclude that baboon alliances are better explained as a form of mutualism. Others have argued that primate alliances represent a case of commodity interchange in which participation in an alliance is exchanged for other services, such as grooming. In bonnet, rhesus, and stump-tailed macaques, the number of times male A provided support to male B correlated with the number of times B supported A, as predicted by reciprocal altruism. Males did not, however, withhold support from those that intervened against them in any of these species, suggesting that they do not display TFT-like reciprocity. In contrast, chimpanzees show reciprocity for delayed retaliatiation as well as for support intervention. Kinship plays little role in determining coalition partners in any primate species. Dolphin males also form long-lasting alliances to gain mating access to receptive females. However, the role of kinship, reciprocity, or mutualism remains to be determined for dolphin alliances.

This discussion makes clear that much of the controversy over reciprocal altruism can be attributed to the difficulty in determining if a particular behavior decreases the reproduction or survival of the donor. An additional difficulty is that studies of reciprocity require continuous observations of animal groups—something that is difficult to achieve. Nevertheless, as Trivers (1971) originally suggested, some of the best examples of reciprocity involve nonhuman primates. Reciprocal altruism remains a plausible explanation for exchanges involving grooming, gametes, food, or assistance, but it appears to be unnecessary to explain most other cases of aid-giving behavior among unrelated animals. Good evidence for mental record keeping remains absent, perhaps with chimpanzees being the sole exception. Thus, although the iterated Prisoner's Dilemma has provided a useful framework for thinking about reciprocity, it is questionable if any animal employs Tit for Tat. More likely, asymmetries exist either in the degree to which one individual participates in an exchange or in the relative benefits or costs to each individual, which make more complicated conditional strategies preferable.

[*See also* Cooperation; Kin Selection; Prisoner's Dilemma Games; Social Evolution.]

BIBLIOGRAPHY

Axelrod, R., and W. D. Hamilton. "The evolution of cooperation." *Science* 211 (1981): 1390–1396. Presents the results of a computer tournament, discusses how Tit for Tat is the best evolutionary strategy to adopt in an iterated Prisoner's Dilemma, and argues that this paradigm can account for cooperative behavior throughout the animal kingdom.

Barrett, L., S. P. Henzi, et al. "Market Forces Predict Grooming Reciprocity in Female Baboons." *Proceedings of the Royal Society of London Series B-Biological Sciences* 266 (1999): 665–670. Provide evidence that grooming is a commodity that female baboons trade. In troops where resource competition was high, lower ranking partners contributed more grooming.

Boesch, C. "Cooperative Hunting in Wild Chimpanzees." *Animal Behaviour* 48 (1994): 653–667. Summarizes observations of red colobus captures and subsequent meat exchange by wild chimpanzees in the Tai Forest.

Clements, K. C., and D. W. Stephens. "Testing Models of Nonkin Cooperation—Mutualism and the Prisoner's Dilemma." *Animal Behaviour* 50 (1995): 527–535. Tested captive blue jays in controlled iterated mutualism and Prisoner's Dilemma games and found that, although the jays readily cooperated in the mutualism game, cooperation neither developed nor persisted in a Prisoner's Dilemma.

Clutton-Brock, T. H., M. J. O'Riain, et al. "Selfish Sentinels in Cooperative Mammals." *Science* 284 (1999): 1640–1644. Considers kin selection, reciprocal altruism, and by-product mutualism as possible explanations for sentinel behavior in meerkats.

de Waal, F. B. M. (1989) "Food Sharing and Reciprocal Obligations among Chimpanzees." *Journal of Human Evolution* 18(5, 1989): 433–459. Summarizes observations of captive chimp food sharing of plant material and documents reciprocal exchanges, especially among adult individuals, as well as interchange of food for access to other resources.

de Waal, F. B. M. "Attitudinal Reciprocity in Food Sharing among Brown Capuchin Monkeys." *Animal Behaviour* 60 (2000): 253–261. Provides evidence that capuchin monkeys (*Cebus apella*) share food reciprocally as a consequence of feeding near one another.

Fischer, E. "Simultaneous Hermaphroditism, Tit-for-Tat, and the Evolutionary Stability of Social Systems." *Ethology and Sociobiology* 9 (1988): 119–136. Summarizes the evidence for egg trading in sea bass as a case of Tit for Tat reciprocity.

Frean, M. R. "The Prisoners-Dilemma without Synchrony." *Proceedings of the Royal Society of London Series B-Biological Sciences* 257 (1994): 75–79. Shows that the timing of decisions is critical in determining which strategy is successful in an iterated Prisoner's Dilemma in the long run.

Godard, R. "Tit for Tat among Neighboring Hooded Warblers." *Behavioral Ecology and Sociobiology* 33 (1993): 45–50. Examines song playbacks of neighbor and stranger territorial birds and finds evidence for retaliation against a cheat that does not respect territory ownership.

Hamilton, W. D. "The Genetical Evolution of Social Behavior." *Journal of Theoretical Biology* (London) 7 (1964): 1–51. Covers the theoretical foundation for kin selection as well as how this theory can be applied to the evolution of reproductive altruism, particularly in the social insects.

Harcourt, A. H. and F. B. M. de Waal, eds. *Coalitions and Alliances in Humans and Other Animals*. New York, 1992. A series of excellent review articles on coalition formation in nonhuman primates, dolphins, and humans.

Hart, B. L. and L. A. Hart. "Reciprocal allogrooming in impala, *Aepyceros melampus*."" *Animal Behavior* 44 (1992): 1073–1083. Shows that young and old impala of both sexes routinely groom each other in reciprocated bouts.

Milinski, M. "Tit for Tat in Sticklebacks and the Evolution of Cooperation." *Nature* 325 (1987): 433–435. The first study to propose that small schooling fish adopt a Tit-for-Tat strategy during predator inspection.

Mitani, J. C., and Watts, D. P. Why do Chimpanzees Hunt and Share Meat? *Animal Behaviour* 61 (2001): 915–924. Provides data showing that chimpanzee meat sharing does not occur during nutritional shortfalls or prior to mating. Instead, males reciprocally exchange meat with coalition partners.

Nowak, M. A., and R. M. May. "Evolutionary Games and Spatial Chaos." *Nature* 359 (1992): 826–829. Shows that a spatial version of the Prisoners' Dilemma, with no memories among players and no strategical elaboration, can lead to indefinite persistance of cooperation.

Packer, C. "Reciprocal Altruism in *Papio anubis*." *Nature* 265 (1977): 441–443. The first empirical study on coalition formation in baboons. Those males who solicited help in gaining access to an estrous female were more likely to provide assistance.

Roberts, G. "Competitive Altruism: from Reciprocity to the Handicap Principle." *Proceedings of the Royal Society of London Series B—Biological Sciences* 265 (1998): 427–431. Discusses how introducing differences in individual generosity together with partner choice in models of reciprocity can lead to an escalation in altruistic behavior and to competition for altruism. Considers how reciprocity and competitive altruism are related and how they may be distinguished.

Stephens, D. W., J. P. Anderson, et al. "On the Spurious Occurrence of Tit for Tat in Pairs of Predator-approaching Fish." *Animal Behaviour* 53 (1997): 113–131. Analyzes movements of predator-approaching fish and argues that social cohesion in the absence of a predator, combined with orientation to a predator in the absence of a companion, accounts for apparent Tit for Tat behavior by small predator-approaching fish.

Trivers,R. L. "The Evolution of Reciprocal Altruism." *Quarterly Review of Biology* 46 (1971): 35–57. The original explanation for how unrelated individuals could benefit by reciprocal exchange. Discusses several possible examples, including cleaner fish, alarm calls, alliance formation, and human moralistic aggression.

Wilkinson, G. S. "Reciprocal Food Sharing in Vampire Bats." *Nature* 309 (1984): 181–184. Documents by field observations how vampire bats share blood by regurgitation and provides evidence to show that individuals benefit from frequent reciprocal exchanges.

— GERALD S. WILKINSON

RECOMBINATION. *See* Sex, Evolution of.

RED QUEEN HYPOTHESIS

The Red Queen hypothesis describes situations in which there is a constant rate of evolutionary change in response to a continually changing environment. The phrase was first used by Leigh Van Valen (Van Valen, 1973). Van Valen demonstrated that extinction rates (excluding mass extinctions) were constant within clades for a diverse array of organisms. Such constancy for the rate of extinction was unexpected, as well as difficult to understand if species generally increased their levels of adaptation over time, which might be expected in stable abiotic environments (Lewontin, 1978). However, Van Valen suggested that, for species engaged in complex interspecific interactions, evolutionary "moves" and "countermoves" could have dramatic effects even though the physical environment had not changed. These biotic changes could occasionally result in the extinction of those species that could not adapt to the shifting biotic pressures; and, if so, the probability of extinction should be independent of lineage age. In other words, failure to adapt to the shifting biotic environment could occur at a stochastically constant rate, leading to a constant rate of extinction.

Consider, for example, a prey species that is under intense natural selection owing to interactions with a specialized predator. If the prey evolved a very effective defense in response to selection imposed by this specialized predator, the predator could go extinct. (The risk of extinction would increase for very specialized predators with limited additive genetic variation for resource utilization.) The basic idea reminded Van Valen of Alice's interaction with the Red Queen in Lewis Carroll's book *Through the Looking Glass*, wherein the Red Queen tells Alice one must run as fast as one can to remain in the same place (Carroll, 1872). Van Valen reasoned that species must similarly run (evolve) in order to stay in the same place. He called the idea the Red Queen's hypothesis. (Note that Van Valen used the possessive form "Queen's," but the more common present usage drops the possessive.) Joseph Felsenstein (1971) also used the idea of running as fast as one can as a metaphor for adaptation in changing environments, but he did not explicitly tie the idea to Carroll's Red Queen.

Since its inception, the Red Queen idea has been used by evolutionary biologists to describe two different kinds of selection resulting from interspecific interactions. One type can be thought of as directional selection resulting in "arms races," particularly between predators and prey. For example, garter snakes (*Thamnophis sirtalis*) overlap with and prey on newts (*Taricha granulosa*) over part of their geographic range (Brodie and Brodie, 1999). In some, areas, the newts are highly toxic owing to the presence of tetrodotoxin (TTX) in their skins. This neurotoxin causes death in unspecialized predators, but the snakes that coexist with toxic prey have evolved a resistance to the toxin (Brodie and Brodie, 1990). As such, there seems to have been a coevolutionary arms race between predator and prey. Where the prey has evolved TTX as a defense against predators, the predators have evolved resistance to the toxin. This is the kind of move-countermove envisioned by Van Valen in his original formulation of the Red Queen hypothesis.

The other type of selection invoking the Red Queen idea is fluctuating rather than directional. For example, Graham Bell (1982) borrowed the Red Queen metaphor to describe interactions between hosts and their parasites. He was particularly interested in parasites that imposed selection against the most common genotypes in the host population (frequency-dependent selection). Such selection would drive these common host genotypes down in frequency, and new, previously rare host

THROUGH THE LOOKING GLASS OF COEVOLUTIONARY INTERACTIONS

Van Valen borrowed from Lewis Carroll's story of the world encountered by Alice in a dream about the looking glass house. In this dream world, Alice first finds that objects in the looking glass house appear left-to-right, as if shown in a mirror. She then finds that chess pieces are animated and quite vocal. Alice will later encounter several of these pieces (notably the Red Queen) after she leaves the looking glass house to see the garden.

Alice decides that it would be easier to view the garden if she first climbs a hill, to which there appears to be a very straight path. However, as she follows the path, she finds that it leads her back to the house. And when she tries to speed up, she not only returns to the house, she crashes into it. Hence, forward movement takes Alice back to her starting point (Red Queen dynamics), and rapid movement causes abrupt stops (extinction).

Eventually, Alice finds herself in a patch of vocal and opinionated flowers; the rose is especially vocal. The flowers inform Alice that someone like her (the Red Queen) often passes through, and Alice decides to seek this person, mostly as a way to escape more verbal abuse. When Alice spots the Red Queen, she begins moving toward her, but the Red Queen quickly disappears from sight. Alice finally decides to follow the advice of the rose and go the other way ("I should advise you to walk the other way"). Immediately she comes face to face with the Red Queen (see Lythgoe and Read, 1998).

The Red Queen then leads Alice directly to the top of the hill. Along the way, the Red Queen explains that hills can become valleys, which perplexes Alice. Already, in this world, straight can become curvy, and progress can be made only by going the opposite direction; now, according to the Red Queen, high can become low. At the top of the hill, the Red Queen begins to run faster and faster. Alice runs after the Red Queen but is further perplexed to find that neither one seems to be moving. When they stop running, they are in exactly the same place. Alice remarks on this, to which the Red Queen responds: "Now, *here*, you see, it takes all the running *you* can do to keep in the same place." And so it may be with coevolution: evolutionary change may be required to stay in the same place. Cessation of change may result in extinction.

—C. M. Lively

genotypes would then become common. The parasites would then be under selection to infect these newly common host types. These coupled responses to frequency-dependent selection by both hosts and parasites can easily lead to oscillations in the genotypic frequencies for both species, and they have been observed in computer simulations of host–parasite interactions (e.g., Seger and Hamilton, 1988). Such oscillations are sometimes called "Red Queen dynamics," since both host and parasite seem to be "running" without getting anywhere (Figure 1).

Bell's primary purpose was to illustrate a theory for the evolutionary maintenance of sexual reproduction. The main idea here was that, for hosts engaged in coevolutionary struggles with parasites, sexual reproduction may be favored over asexual reproduction as a way to produce rare genotypes among the offspring, which would be more likely to escape infection. Conversely, parasites may be under selection to reproduce sexually to track changes in sexual host populations. Hence, host–parasite coevolution could select for sexual reproduction in both the host and the parasite, even if sexual reproduction carries costs not associated with asexual reproduction. This idea, now known as the Red Queen hypothesis for sex, has its conceptual roots in the work of W. D. Hamilton (1980) and John Jaenike (1978).

The Red Queen hypothesis for sexual reproduction makes several falsifiable predictions, which have been examined in a freshwater snail (*Potamopyrgus antipodarum*) from New Zealand. The snail is especially well suited for tests of predictions and assumptions because sexual and asexual females are both found, sometimes in the same lakes and streams.

1. Sexual individuals should be favored where the risk of infection is high. Sexual females of the snail were found to be more common in areas where the frequency of infection by trematode worms was high (Jokela and Lively, 1995; Lively, 1992). This result suggests that asexual females have displaced sexual females where the risk of infection is low.

2. Host genotypes should oscillate over time. Changes in the frequencies of clonal host genotypes were found to be correlated with changes in the prevalence of infection in these same clones, exactly as expected under the Red Queen hypothesis (Dybdahl and Lively, 1998).

3. There should be selection against genotypes that were common in the recent past. Laboratory infection experiments showed that clonal genotypes that were common in the recent past were also more susceptible to infection by local populations of parasites (Dybdahl and Lively, 1998; Lively and Dybdahl, 2000). Similar results have been shown for flatworms (Michiels et al., 2001) and grasses (Schmitt and Antonovics, 1986).

4. Parasites should become adapted to infecting local populations of their hosts. If parasites are tracking locally common host genotypes, they should also become better (at least periodically) at infecting hosts from their local population than hosts from remote populations. There is strong support for this idea from experimental studies of *P. antipodarum* involving seven different populations (Lively, 1989; Lively and Dybdahl, 2000). Local adaptation has also been shown in the parasites of waterfleas (Ebert, 1994) and in the trematode parasites of fish (Ballabeni and Ward, 1993). However, local parasite adaptation is not a universal feature of host–parasite interactions in structured populations (Kraaijeveld and Godfray, 2001).

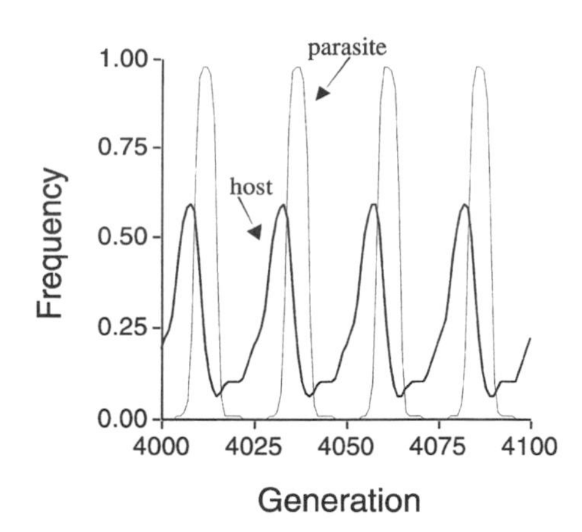

FIGURE 1. Red Queen Dynamics.
Results of a computer simulation for host–parasite coevolution. The thicker line gives the frequency of one host genotype, and the thinner line gives the frequency of the parasite genotype that can infect it. Note that both genotypes oscillate over time, as if they were "running" in circles. The model assumes that hosts have self/nonself recognition systems, which can detect foreign organisms. The model also assumes that hosts and parasites both reproduce sexually. Courtesy of C. M. Lively.

In summary, the Red Queen hypothesis is used to describe similar ideas based on antagonistic coevolution. The original idea was that coevolution could lead to situations for which the probability of extinction is relatively constant during evolutionary time. The gist of the idea is that, in tightly coevolved interactions, evolutionary change by one species could lead to extinction of the other species, and that the probability of such changes might be independent of species age (in millions of years). Similarly, the Red Queen hypothesis is used to describe coevolutionary arms races (particularly between predator and prey), without reference to macroevolutionary events such as extinction. The gist of this idea is nonetheless identical to Van Valen's original suggestion that adaptation/counteradaptation could lead to constant evolutionary change, even in physically constant environments. The final idea is that coevolution, particularly between hosts and parasites, should lead to sustained oscillations in the genotypes of sexual populations, and it could lead to selection against asexual forms of reproduction in both hosts and their parasites. This idea differs from the arms race situation, in that "defenses" are recycled as a result of frequency-dependent selection.

[*See also* Sex, Evolution of.]

BIBLIOGRAPHY

Ballabeni, P., and P. I. Ward. "Local Adaptation of the Trematode *Diplostomum phoxini* to the European Minnow *Phoxinus phoxinus*, Its Second Intermediate Host." *Functional Ecology* 7 (1993): 84–90.

Bell, G. *The Masterpiece of Nature: The Evolution and Genetics of Sexuality*. Berkeley, 1982.

Brodie, E. D., III, and E. D. Brodie Jr. "Tetrodotoxin Resistance in Garter Snakes: An Evolutionary Response of Predators to Dangerous Prey." *Evolution* 44 (1990): 651–659.

Brodie, E. D., III, and E. D. Brodie Jr. "Predator–Prey Arms Races and Dangerous Prey." *Bioscience* 49 (1999): 557–568.

Dybdahl, M. F., and C. M. Lively. "Host-Parasite Coevolution: Evidence for Rare Advantage and Time-Lagged Selection in a Natural Population." *Evolution* 52 (1998): 1057–1066.

Ebert, D. "Virulence and Local Adaptation of a Horizontally Transmitted Parasite." *Science* 265 (1994): 1084–1086.

Felsenstein, J. "On the Biological Significance of the Cost of Gene Substitution." *American Naturalist* 105 (1971): 1–11.

Hamilton, W. D. "Sex versus Non-Sex Versus Parasite." *Oikos* 35 (1980): 282–290.

Jaenike, J. "An Hypothesis to Account for the Maintenance of Sex within Populations." *Evolutionary Theory* 3 (1978): 191–194.

Jokela, J., and C. M. Lively. "Parasites, Sex, and Early Reproduction in a Mixed Population of Freshwater Snails." *Evolution* 49 (1995): 1268–1271.

Kraaijeveld, A. R., and H. C. J. Godfray. "Is There Local Adaptation in *Drosophila*–Parasitoid Interactions?" *Evolutionary Ecology Research* 3 (2001): 107–116.

Lewontin, R. C. "Adaptation." *Scientific American* 239 (1978): 213–230.

Lively, C. M. "Adaptation by a Parasitic Trematode to Local Populations of Its Snail Host." *Evolution* 43 (1989): 1663–1671.

Lively, C. M. "Parthenogenesis in a Freshwater Snail: Reproductive Assurance versus Parasitic Release." *Evolution* 46 (1992): 907–913.

Lively, C. M., and M. F. Dybdahl. "Parasite Adaptation to Locally Common Host Genotypes." *Nature* 405 (2000): 679–681.

Lythgoe, K. A., and A. F. Read. "Catching the Red Queen? The Advice of the Rose." *Trends in Ecology and Evolution* 13 (1998): 473–474.

Michiels, N. K., L. W. Beukeboom, N. Pongratz, and J. Zeitlinger. "Parthenogenetic Flatworms Have More Symbionts than Their Coexisting, Sexual Conspecifics, But Does This Support the Red Queen?" *Journal of Evolutionary Biology* 14 (2001): 110–119.

Schmitt, J., and J. Antonovics. "Experimental Studies of the Evolutionary Significance of Sexual Reproduction. IV. Effect of Neighbor Relatedness and Aphid Infestation on Seedling Performance." *Evolution* 40 (1986): 830–836.

Seger, J., and W. D. Hamilton. "Parasites and Sex." In *The Evolution of Sex: An Examination of Current Ideas*, edited by B. R. Levin, pp. 176–198. Sunderland, Mass, 1988.

Van Valen, L. 1973. "A New Evolutionary Law." *Evolutionary Theory* 1 (1988): 1–30.

— C. M. Lively

REGULATORY GENES

All cells of an organism normally carry the same genetic information. Yet the cells express different parts of this information, depending on physiological and developmental requirements. This is due to a multitude of regulatory genes, which are organized in interdependent networks that can be compared to software programs that run the physiological reaction machinery, organize development, and eventually generate the phenotype. Regulatory genes are distinct from housekeeping genes that serve to maintain the basic metabolism of the cells.

Transcription Factors. From the completed genome projects, we can estimate that 5–10 percent of genes fall into the class of regulatory genes. However, because of the multiple levels at which regulation can take place, it is difficult to provide an exact figure. Regulation can be at the level of transcription of the genes, splicing, translation, differential transport, and others. In addition, the regulatory networks are triggered by signals coming from neighboring cells and hormones from inside or outside the body. The transcription factors are a key element in the regulatory cascade. These are proteins that bind directly to specific DNA sequence stretches within the regulatory regions of the genes and can in this way promote or inhibit the transcription of the gene. Transcription factors can be grouped into families, according to their DNA binding protein domain, for example, zinc fingers, helix-loop-helix domain (HLH), or homeodomain.

Master Control Genes. The action of specific transcription factors is almost always required when cells undergo differentiation processes. For example, a cell that is destined to become a muscle cell must express the transcription factor *MyoD*, which itself turns on a variety of genes that are specifically required for a muscle. If one brings *MyoD* artificially into a competent cell, then it will switch its program and become a muscle cell (Weintraub et al., 1989). Such transcription factors that control whole developmental programs have been called master control genes. The term was made popular by Walter Gehring in connection with the stunning finding that one could convert the legs or antennae of a fly into eyes by artificial expression of the gene *eyeless* in the anlagen of the respective structures (Halder et al., 1995). Master control genes are often highly conserved. *MyoD* or *eyeless*, for example, can be found throughout the animal kingdom performing the same function. Moreover, the *eyeless* gene of the mouse (called *pax6*) can, for example, functionally substitute for the *eyeless* gene of the fly.

Table 1. Major Groups of Genes Involved in the Developmental Hierarchy
Note that these classifications are idealized and that the same gene can be involved in different processes.

Gene Category	*Classes Included*	*General Functions*	*Mutant Phenotypes*
Bauplan genes	Maternal coordinate genes Segmentation genes Dorsoventral genes Asymmetry genes Neurogenic genes	Defining polarity Subdivision of body regions Defining the regions for the organ anlagen	Cell death and loss of specific body regions
Master control genes	Differentiation genes Homoeotic genes	Determination of cell fates Specifying the identity of body regions and organ anlagen	Changes in cell fate Transformation of body regions
Realizator genes	Effector genes Target genes Housekeeping genes	Providing the material for the morphological structure Maintaining the metabolism	Disturbance of specific aspects of the phenotype and metabolism or unspecific lethality

***Hox* Genes.** A special class of master control genes are the *Hox* genes. Mutations in *Hox* genes cause the transformation of body regions. Flies with legs instead of antennae or with a second pair of wings have become particularly famous. Such mutations have been called homoeotic, because they do not lead to loss of particular structures, but to the respecification of structures. The molecular characterization of *Hox* genes has shown that they occur in clusters and code for transcription factors with a specific DNA binding domain, which was called the homeobox. Intriguingly, the function of a particular *Hox* gene along the body axis correlates with the relative position on the chromosome, a feature that seems to be conserved throughout the animal kingdom.

Many other genes also contain homeoboxes but are not *Hox* genes; that is, they function in other kinds of processes. Also, the homeobox is just one of many different DNA binding domains found in transcription factors. The term *Hox genes* therefore refers specifically to genes in the homoeotic cluster. It should be noted, however, that clustering is not a prerequisite for function in common pathways. Most genes driving developmental pathways are not clustered. Also, the functional equivalent of *Hox* genes in plants, namely, the MADS-box genes that control the differentiation of the building blocks of the flower, are not clustered (Theissen et al., 2000).

Regulatory Hierarchies. Development is controlled by hierarchical cascades of interdependent regulatory genes. This has been particularly well studied in *Drosophila*, where systematic mutagenesis screens for identifying the genes involved in early development have yielded an almost complete picture of the processes that convert the fertilized egg into the complex three-dimensional structure of the embryo (Lawrence, 1992). Polarity information is already generated during oogenesis by a cascade of regulatory interactions involving cell–cell signaling. The first steps of embryogenesis are then largely controlled by transcription factors that form concentration gradients within the embryo. In the anteroposterior direction, this results in setting up the segments. In the dorsoventral direction, this results in the differentiation of the mesoderm and the neurogenic region. The position of the appendages is determined by intersection points between the dorsoventral and anteroposterior system. The genes controlling these basic building blocks of the insect body plan can be called bauplan (body plan) genes (Tautz and Schmid, 1998). Further differentiation of the blocks is then achieved by the master control genes that convert a segment, for example, into the thoracic segment bearing the wings, or that trigger the program for eye development in the appropriate tissue. The master control genes need realizator genes (Garcia-Bellido, 1975) to achieve this. Thus, there is a hierarchy of gene action, in which the body plan genes are the ones that determine which tissues become competent to respond to the action of the master control genes, which are required to switch on the corresponding realizator genes (Table 1).

Recruitment of Genes and Modules. An intriguing feature of regulatory genes is that they can be involved in multiple processes throughout development. For example, many of the transcription factors acting as bauplan genes in *Drosophila* are also involved in the specification of particular cells of the nervous system. It was shown that the *Hox* gene *Ultrabithorax (Ubx)* from *Drosophila* has an additional very specific function in determining the extent of bristle patterning on the legs of adult flies, a feature that evolves very quickly between closely related species (Stern, 1998). The function performed by these genes may be very different in different tissues, and even their interaction partners might differ substantially. However, it seems conceivable that there are modules of coregulated genes that can be recruited as a whole to perform their function in another regulatory context. One could speculate that the recruitment of a regulatory module to a new process might lead to a major shift in phenotype, that is, to an evolutionary innovation. Alternatively, identical regulatory modules might become independently recruited to perform similar functions. It is assumed, for example, that the involvement of the gene *Distalless* in many kinds of body outgrowths could be due to such a convergence effect (Panganiban et al., 1997).

Enhancer Networks. The regulatory regions of the genes that respond to transcription factors are called enhancers. Such regions are typically a few hundred bp long and reside in the noncoding DNA, usually in the vicinity of the gene that is regulated, but also can occur at some distance (up to 100 kb away). Several enhancers can independently regulate a gene, allowing it to be expressed in several different contexts. Each enhancer contains multiple binding sites for a number of different transcription factors. It is assumed that this serves to integrate independent input signals into a single output signal, which conveys a precise temporal and spatial information. This system can work in a combinatorial manner that can be simulated in quantitative computational models (Yuh et al., 1998).

Evolution of Regulatory Genes. It has long been speculated that most morphological evolutionary changes are caused by changes in regulatory genes and their interactions. The original argument was made by Allan Wilson on the basis of the finding that the rate of molecular evolution is lower in mammals than in frogs, although the rate of morphological evolution is similar (Wilson et al., 1974). A similar argument can be made with respect to the stunning conservation of the function of master control genes, suggesting that differences among animals are mainly brought about by differences in regulatory interactions (Carroll, 1995).

However, one of the first tests for conservation of regulatory elements yielded a surprising conservation rather than divergence. S. A. Mitsialis and F. C. Kafatos (1985) found that, when they used the regulatory region of moth chorion genes to drive a reporter gene in *Drosophila*, they got faithful expression of the gene in the follicle cells, where the endogenous chorion genes of *Drosophila* are expressed. This suggested that a regulatory circuit could be conserved over long periods of time (250 Myr) even for tissues where one would have assumed that fast evolution should occur easily. There are now several more examples (Tautz, 2000). Intriguingly, these show that regulatory function can be conserved despite sequence divergence in the regulatory regions. M. Z. Ludwig and colleagues (2000) have suggested that this could be due to coevolution among binding sites. Enhancers often contain multiple binding sites for the same transcription factor. It seems conceivable that a mutation in a single binding site might have only a small effect. Because weakly disadvantageous mutations can spread through a population by genetic drift, it is possible that at some point, when a higher allele frequency has built up, a second mutation happens at another site in the enhancer and compensates for the first one. This could restore the original function, although the sequence itself has changed.

The domestication of maize from its wild relative, teosinte, is an example of a regulatory change causing a major effect in phenotype. Here it was shown that the gene promoting apical dominance (i.e., a single strong shoot instead of multiple weak ones) was specifically changed in its expression level during the domestication process. The coding region of the gene, however, did not show signs of specific change (Wang et al., 1999).

Although hard data are still scarce, it seems that changes in regulatory genes have contributed to evolutionary change (Tautz, 2000). This may have included drastic changes in master control genes or shifts of regulatory modules. Such major changes may be only an occasional evolutionary event. Most continuing evolution is still likely to occur in small steps, in both coding and regulatory regions.

[*See also* Canalization; Cell Lineage; Constraint; Phylotypic Stages.]

BIBLIOGRAPHY

Carroll, S. B. "Homeotic Genes and the Evolution of Arthropods and Chordates." *Nature* 376 (1995): 479–485. Review discussing the idea that changes in the regulatory interactions of *Hox* genes have driven major innovations in animal evolution.

Garcia-Bellido, A. "Genetic Control of Wing Disc Development in *Drosophila*." In *Cell Patterning*, edited by R. Porter and K. Elliott, 161–178. Amsterdam, 1975. First discussion of the term *realizator genes*.

Halder G., P. Callaerts, and W. J. Gehring. "Induction of Ectopic Eyes by Targeted Expression of the *eyeless* gene in *Drosophila*." *Science* 267 (1995): 1788–1792. The experiment that triggered the general discussion about master control genes, as well as the evolutionary origin of eyes.

Lawrence, P. A. *The Making of a Fly: The Genetics of Animal Design*. London, 1992. Lucid account of the genetic experiments and their outcomes that led to the understanding of early development of *Drosophila*.

Ludwig, M. Z., C. Bergman, N. H. Patel, and M. Kreitman. "Evidence for Stabilizing Selection in a Eukaryotic Enhancer Element." *Nature* 403 (2000): 564–567. Convincing demonstration of coevolutionary effects that can explain the functional conservation of enhancer sequences despite sequence divergence, using the comparative functional analysis of an enhancer element of the *Drosophila even skipped* gene from different *Drosophila* species.

Mitsialis, S. A., and F. C. Kafatos. "Regulatory Elements Controlling Chorion Gene Expression Are Conserved between Flies and Moths." *Nature* 317 (1985): 453–456. First paper that used a functional assay in transgenic flies to show that regulatory regions can be conserved over large evolutionary distances.

Panganiban G., S. M. Irvine, C. Lowe, H. Roehl, L. S. Corley, B. Sherbon, J. K. Grenier, J. F. Fallon, J. Kimble, M. Walker, G. A. Wray, B. J. Swalla, M. Q. Martindale, and S. B. Carroll. "The Origin and Evolution of Animal Appendages." *Proceedings of the National Academy of Sciences USA* 94 (1997): 5162–5166. Systematic survey of the expression of the *Distalless* genes in animals from different phyla. Concludes that the regulatory module including *Distalless* has evolved only once, but has been recruited independently for different types of body outgrowths.

Stern, D. L. "A role of *Ultrabithorax* in Morphological Differences between *Drosophila* Species." *Nature* 396 (1998): 463–466. Demonstration of a functional role of *Ubx* in the diversification of small morphological differences between closely related *Drosophila* species.

Tautz, D. "Evolution of Transcriptional Regulation." *Current Opinion in Genetics and Development* 10 (2000): 575–579. Review discussing the current experimental evidence for the occurrence of evolution of regulatory genes and enhancers.

Tautz, D. and K. J. Schmid. "From Genes to Individuals: Developmental Genes and the Generation of the Phenotype." *Philosophical Transactions of the Royal Society of London Series B* 353 (1998): 231–240. Review of the question of the differential contribution of the various classes of genes to adaptations and the adult phenotype; introduces the term *bauplan genes.*

Theissen, G., A. Becker, A. Di Rosa, A. Kanno, J. T. Kim, T. Munster, K. U. Winter, and H. Saedler. "A Short History of MADS-Box Genes in Plants." *Plant Molecular Biology,* 42 (2000): 115–149. Most recent review of the distribution, function, and possible evolutionary effects of the MADS-box genes in plants.

Wang, R. L., A. Stec, J. Hey, L. Lukens, and J. Doebley. "The Limits of Selection during Maize Domestication." *Nature* 398 (1999): 236–239. Microevolutionary approach to study the effects of selection during domestication on the *tb1* locus in maize. Demonstrates that strong selection acted on the regulatory region of the gene, but not on the coding region.

Weintraub, H., S. J. Tapscott, R. L. Davis, M. J. Thayer, M. A. Adam, A. B. Lassar, and A. D. Miller. "Activation of Muscle-specific Genes in Pigment, Nerve, Fat, Liver, and Fibroblast Cell Lines by Forced Expression of *MyoD.*" *Proceedings of the National Academy of Sciences USA* 86 (1989): 5434–5438. Classical experiment for demonstrating that *MyoD* is a master regulatory gene for myogenesis.

Wilson, A. C., L. R. Maxson, and V. M. Sarich. "Two Types of Molecular Evolution: Evidence from Studies of Interspecific Hybridization." *Proceedings of the National Academy of Sciences USA* 71 (1974): 2843–2847. Classic paper raising the question of whether the evolution of regulatory interactions among genes has caused the rapid morphological divergence of mammals.

Yuh, C. H., H., Bolouri, and F. H. Davidson. "Genomic *cis*-Regulatory Logic: Experimental and Computational Analysis of a Sea Urchin Gene." *Science* 279 (1998): 1896–1902. Computational model for a regulatory element of a sea urchin gene, based on detailed experimental evidence for the function of the different binding sites in the element.

— DIETHARD TAUTZ

REPETITIVE DNA

An onion has a genome over four times the size of the human genome; the amoeba (*Amoeba dubia*) has a genome size forty-six times that of an onion. This phenomenon can largely be explained by the volume of repetitive DNA (nucleotide sequences that occur several times) contained within each species' genome. The repeats can range in size from just a few nucleotides to thousands, and can vary in both composition and the number of times the sequence is repeated. The genetic code of all organisms contains some repetitive DNA, but the extent to which it occurs varies between species. In humans, although only approximately 2 percent of the genome is thought to be made up of coding sequences, at least 50 percent is made up of repetitive DNA.

Repetitive DNA can be classified in a number of ways, one being to focus on the distribution of these fragments within the genome. Repeat sequences are either found in tandem as a continuous sequence of repeats, or alternatively, a single sequence can be repeated in different locations in the genome. Whether the repeat sequences are in tandem or are dispersed throughout the genome reflects the mechanism by which the repeat has arisen. Interspersed genome-wide repeats tend to originate from elements capable of autonomous or semiautonomous replication. Tandemly repeated elements take no active role in their own replication and are thought to arise by the expansion of an initial sequence, either through slippage during DNA replication or during the DNA recombination processes.

Tandemly repeated DNA elements, though infrequently found in prokaryotes, are very common in eukaryotes. In the kangaroo rat (*Dipodomys ordii*), for example, it has been reported that three tandemly repeated elements (AAG, TTAGGG, and ACACAGCGGG) make up 50 percent of the genome, each sequence occurring at least 1.2×10^9 times. The kangaroo rat has a relatively large genome, but even many species with very small genomes such as the fruit fly (*Drosophila virilis*) and the pin mould (*Absidia glauca*) contain a high proportion of repetitive DNA (40 percent and 35 percent respectively). There are three main types of tandemly repeated DNA elements: satellites, minisatellites, and microsatellites. Satellite DNA is made up of long series of tandem repeats, usually thousands of repeats of between 5–200 base pairs (bp) each. In eukaryotes this DNA is mainly found in the centromeres and has been hypothesised to play a structural role there. Alternatively, it has been suggested that the repeat sequence reflects the centromeres' role as the last region of the chromosome to replicate: in order for the prevention of early replication to be ensured, the centromere must not contain any origin of replication; hence, it is argued, the need for noncoding DNA. The second type of tandemly repeated DNA elements, minisatellites, usually contain between 10 and 100 repeats of 9–100 bp units. They too are found in specific sites on the chromosome (noteably the telomeres) but functions for most minisatellite clusters are unknown. The function of microsatellites (typically 10–100 repeats of four or fewer bp units) is also unclear. However, despite uncertainty as to function,

TABLE 1. Table Showing the Genome Size and Percentage of the Genome Made up of Genome-Wide Dispersed Repetitive Elements

Organism	*Genome size (Mbp)*	*Percentage of genome made up of genome-wide dispersed repetitive elements*
Human	3220	>45%
Maize	2400	>50%
Pufferfish	400	0%
Fruit fly	180	15%
Thale cress	80	2%
Nematode	80	1.8%
Yeast	12	3.1%

both micro- and minisatellites have been invaluable in genetic studies. Owing to the high mutation rate of these elements (that is, the rate at which repeat units are lost or gained), there is wide variation in satellite DNA in the human population. Because of this, micro- and minisatellites are widely used as genetic markers of disease causing agents, in paternity testing, for investigating population structure, and in forensic applications (so-called DNA fingerprinting).

Genome-wide dispersed repeat sequences, like tandem repeats, make up a variable percentage of the genome, depending on the species (see Table 1). There are four main groups of dispersed repetitive elements: long interspersed nuclear elements (LINEs), short interspersed nuclear elements (SINEs), elements with long terminal repeats (LTRs; named because of the section of repetitive DNA found at either end of the element), and DNA transposons. A summary of the characteristics of these four main types can be found in Table 2. Three types (LINEs, LTRs, and DNA transposons) are capable of autonomous replication, while the fourth (SINEs) are only able to replicate with the help of another element, usually a LINE. SINEs are very successful, having left behind 1.5 million copies, making them the single most numerous class of interspersed repeat. The least common form of dispersed repetitive DNA in eukaryotes is the DNA transposons. They make up just 3 percent of the human genome compared to the 21 percent, 13 percent, and 8 percent contributed by LINEs, SINEs, and LTRs respectively. Unlike the other three types of dispersed repeats, which use an RNA intermediate during replication, DNA transposons, as the name implies, do not require an intermediate, transposing using only DNA.

The most common of the interspersed repeats, LINEs and SINEs, are highly repetitive sequences distinct from each other as LINEs are longer (around 6 kilobases compared to SINEs of 100–400 bp). LINEs are able to code for the proteins needed for replication and reinsertion into the genome. SINEs do not encode these proteins and are thought to use the machinery of LINE elements to move around the genome. LTRs, the fourth kind of interspersed repeat, contain in the long terminal repeats regions the sequences coding for all the necessary elements used in the replication of the sequence. It is thought LTR's may have given rise to retroviruses—elements important as causes of disease, for example, HIV.

Semiautonomous and autonomous repetitive DNA elements are frequently truncated, often severely so, on insertion into the genome. These truncated elements consequently lose the ability to self replicate. The majority of LINEs and SINEs, within the human genome at least, are inactive and as such evolve as neutral DNA. For example, of the 500,000–1,000,000 LINE fragments in the human genome, perhaps as few as 20–40 are active.

Repetitive elements are often dismissed as having no function or role, being present in the genome simply because there is no selective pressure on a cell to remove it. However, with respect to evolutionary biology, the study of repetitive elements has much to offer. In particular, it has recently been recognized that sequences that are able to jump around the genome (identified as those with a dispersed distribution) have the potential to make significant changes within the genome. It had been thought that repetitive elements played a minor role in the evolution of species. As many repetitive elements are inactive and/or found in noncoding regions of the genome, they were hypothesized to be subject only to neutral evolution. Any insertions into coding regions of the genome were considered to be rare, and when they did happen, they were assumed to cause deleteri-

TABLE 2. Summary of the Four Main Classes of Interspersed Repeat

LINE = Long interspersed repeat, SINE = short interspersed repeat, and LTR = element with long terminal repeats.

Type	*Replication*	*Length*	*Estimated number of copies in the human genome*	*Fraction of human genome consisting of repeat type*
LINEs	Autonomous	6–8 kilobases	868,000	21%
SINEs	Semiautonomous	100–300 base pairs	1,558,000	13%
LTRs	Autonomous	6–11 kilobases	443,000	8%
DNA transposons	Autonomous	2–3 kilobases	294,000	3%

ous mutation and to be subject to negative selection pressure. While these ideas hold generally, there are important exceptions—instances in which repeat element insertions are not deleterious. It is now acknowledged that repetitive elements may play a fundamental role in the evolution of mammals and are likely to have been influencing genome structure since the time of the first eukaryotes.

There are two main ways in which interspersed genome-wide repeats may promote positive selection on a host organism. First, repetitive elements may have the ability to alter regulatory pathways within a host organism. The elements often carry their own regulatory signals, so giving them the potential to alter gene expression in the area of the genome into which they insert. In the human genome, for example, there are at least a few hundred genes that rely on regulatory sequences thought to have been donated by repetitive elements, one such case being the apolipoprotein (a) gene.

The second way in which genome-wide repeats may be actively involved in the process of evolution is as a potent genomic reorganizer. Although genes can be altered and refined through simple base changes, the permutations are limited. Insertion of new sequences throughout the genome increases the potential of new beneficial genes being created, and the sequences inserted are not always limited to the coding normally carried by the repeat sequences. In LINEs, for example, the machinery that replicates the element can overshoot, replicating parts of adjoining genes. There is then the potential for these sequences also to be inserted as the LINE moves around the genome (a process called *exon shuffling*). Insertion may be into protein-coding regions of the genome leading the production of novel proteins. In more extreme cases, whole genes can be created as repetitive elements insert; at least twenty human genes are considered to be derived from such elements. Wherever the insertion occurs, alteration of the genome is immediate and far more radical than could occur by gradual base change. That autonomously or semiautonomously replicating elements can alter genomes dramatically without causing fatal mutation in an organism is demonstrated in hybrids of wallaby species (Macropodidae). Such first-generation hybrids show massive amplification of repetitive DNA elements (a situation mirrored in other mammalian hybrids), while the normal function of somatic cells is not disrupted. Although the wallaby hybrids are sterile, other hybrids are not, leading to the interesting possibility that fertile female hybrids could give birth to new species. Whether or not this is the case, the potential for repetitive elements to reorganize the genome without fatality in a less extreme situation than a hybrid cross is demonstrated.

Although repetitive elements have been studied for over thirty years, the near completion of the Human Genome Sequencing Project has provided the means to examine repetitive DNA in the context of a complete genome sequence for the first time. Using data obtained from the human genome, the case for repetitive elements increasing genomic diversity, with a possible subsequent role in speciation, has been advanced. The activity of LINE elements, for example, can be linked with bursts of speciation in mammalian evolutionary history. Once LINE elements enter the genome, they are unlikely to be lost. Though frequently inactivated, they remain within the genome, with the result that the human genome contains a nearly complete fossil record of LINE activity, stretching back through the apes and other primates, to at least the origin of the mammals. Assuming that under neutral evolution, change occurs at a constant rate, the approximate age of each fossil can be estimated by counting how many mutations have accumulated in the DNA sequence. In this way, the pattern of LINE activity over time has been retrieved from the fossil record hidden within the human genetic code. LINE activity has fluctuated, and intriguingly, the estimated dates of bursts in LINE activity are coincident with periods in the fossil record where mammalian evolutionary explosions are thought to have occurred.

Repetitive DNA can no longer be sidelined as a neutral influence in the genome. These elements may be implicated in the creation of novel proteins, in the alteration of gene expression and in the increase of genetic diversity. They may even play a part in one of the most fundamental processes of evolution—speciation. Further research into these enigmatic sequences is likely to reveal an importance that has been unrecognized in the past, and, with the daily increase in the amount of genomic DNA for which sequence data is available, the information with which to examine questions about repetitive DNA is becoming increasingly accessible.

[*See also* Forensic DNA; Genetic Markers.]

BIBLIOGRAPHY

Brown, T. A. *Genomes*. Oxford, 1999. An general overview of the genome and the elements found within it.

Graur, D., and W.-H. Li. *Fundamentals of Molecular Evolution*. Sunderland, Mass., 2000. Good summary of information about repetitive DNA elements.

Kidwell, M. G., and D. R. Lisch. "Transposable Elements and Host Genome Evolution." *Trends in Ecology and Evolution* 15 (2000): 95–99. More detailed look at how dispersed DNA can drive evolution.

Lander, E. S., L. M., Linton, Bruce Birren et al. "Initial Sequencing and Analysis of the Human Genome." *Nature* 409 (2001): 860–921. In-depth examination of repetitive elements in the context of the human genome.

Li, W. H. *Molecular Evolution*. Sunderland, Mass. 1997. In-depth information about the repetitive elements found in the genome.

Li, W. H., Z. L. Gu, H. D. Wang, and A. Nekrutenko. "Evolutionary

Analyses of the Human Genome." *Nature* 409 (2001): 847–849. Summary of initial evolutionary analyses using data from the Human Genome Sequencing Project.

Venter, J. C., M. D. Adams, E. W. Myers, et al. "The Sequence of the Human Genome." *Science* 291 (2001): 1304–+. In-depth look at initial findings of the Human Genome Sequencing Project.

— ANDREA J. WEBSTER

REPLICATORS. *See* Overview Essay *on* The New Replicators *at the beginning of Volume 1*; Group Selection; Levels of Selection; Meme; Origin of Life, *article on* Origin of Replicators; Selfish Gene.

REPRODUCTIVE PHYSIOLOGY, HUMAN

The recent application to the reproductive sciences of data and theory from evolutionary biology has provided new insight into human reproductive physiology. A significant finding is that the reproductive biology of women is adapted to spending most of the reproductive years in lactational amenorrhea, a time when breast feeding shuts down ovarian steroid hormone production, and both ovulation and menstruation are absent. Humans, like other primates, have relatively dilute milk adapted to a suckling pattern in which infants nurse for a few minutes several times per hour over a period of many months. Although the duration of ovarian quiescence is variable, in many noncontracepting populations, women have a median of nearly two years of amenorrhea following the birth of a child who survived to the age of weaning. Following the death of a nursing infant, the mother typically resumes menstruation within two months. On a global scale, breast feeding prevents more births than all other methods of family planning combined and has significant health benefits for both mother and infant.

Abnormality of Regular Menstrual Cycles. From an evolutionary standpoint, regular, recurrent menstrual cycles (as are common in many Western populations) are the abnormal consequence of the curtailment of pregnancy and lactational amenorrhea and, to a lesser extent, the early onset of menarche (age 12.5 years, on average, in North Americans vs. 16.0 years in hunter-gatherers). North American women have about 350 to 400 menses in their lifetimes, which is three or four times more menses than are found in populations that have not experienced the demographic transition to low fertility. An example of such a population is the Dogon of Mali, who have a median of about 110 menses, twenty months of postpartum amenorrhea, nine livebirths, and reach menarche at age sixteen (all values for women in the fiftieth percentile).

Epidemiological evidence suggests that the increase in the proportion of the life span spent in menstrual cycling is a major contributor to women's reproductive cancers. The ovarian steroids estrogen and progesterone stimulate cell divisions in breast epithelia, enhancing the risk for DNA copying errors that can lead to malignancy. Another major contributor is postponed childbirth, which delays the maturation of the lobulo-alveolar duct system in the breasts. The evidence that incessant menses are less normal than amenorrhea has called into question the design of combination oral contraceptives, which mimic the menstrual cycle through scheduled monthly bleeding episodes, and has stimulated the development of new contraceptives with altered dosing regimens. The estrogen in these new contraceptives needs to be high enough to protect against osteoporosis and other conditions associated with low estrogen, but low enough to mitigate the risk of reproductive cancer. It is unlikely that menstruation will be entirely eliminated, but it can safely be reduced to three or four times per year.

Evolution of Menstruation. Discussions about whether menstruation is medically necessary dovetail with a recent debate about the adaptive significance of menstruation. This dialogue focuses on the question why did menstruation evolve in the first place? Most mammals advertise ovulation via estrus and do not undergo periodic bleeding; especially rare is the copious blood loss found in humans. In 1993, Margie Profet published the provocative hypothesis that menstruation is an adaptation to cleanse the uterus of pathogens transmitted by sperm. She reasoned that if menstruation is a defense against infection, then contraceptives that suppress the menses will undermine the body's natural protective mechanisms.

A subsequent analysis, however, showed that major predictions of the pathogen defense hypothesis were not supported. One of these predictions is that uterine pathogens should be more prevalent before than after menses. Contrary to this prediction, menstruation actually exacerbates infection, presumably because blood is an excellent culture medium for bacteria. A second prediction is that menstruation tracks pathogen burden. In preindustrial societies, however, sexual activity often occurs in the absence of menstruation, for example, during pregnancy, lactational amenorrhea, and the postmenopausal years. In populations that do not use modern contraceptives, menstruation is a rare event among fecund women of reproductive age. For example, Dogon women between the ages of twenty and thirty-five years had a median of only one period per year. In this age group, the frequency of menstruation was inversely related to the probability of conception, and only sterile women experienced regular menses. Young women who were still gynecologically immature and older women approaching menopause had most of the menses. Assuming that menstruation was also a rare event among fecund women in ancestral populations, it is doubtful that it evolved as a defense against sexually transmitted pathogens.

An alternative hypothesis is that there are two distinct phenomena to be explained: (1) the cyclicity of the endometrium (uterine lining), which is found throughout the mammals, and (2) bleeding through the vaginal opening, which is found in Old World primates and shrews. Strassmann (1996) proposes that endometrial cyclicity spares energy whereas vaginal bleeding is a mere side effect that occurs when regression of the endometrium entails too much blood and other tissue for efficient reabsorption. In support of her argument, she cites data showing that keeping the endometrium primed for implantation requires nearly sevenfold more oxygen consumption per milligram of protein per hour than is needed in the regressed state. Moreover, if a woman forgoes the cost of the luteal phase of the menstrual cycle for twelve months during amenorrhea, she spares an estimated 130 megajoules, or her food supply for half a month. The energy-sparing reductions in tissue mass and metabolic rate that occur during menstrual cycling have many parallels elsewhere in the mammalian body and in other vertebrates.

Myth of Menstrual Synchrony. No area of human reproductive physiology has been subject to more debate, and indeed, mythology, than the notion of menstrual synchrony. The upshot of thirty years of research, however, is that there is no statistically sound evidence for this phenomenon. Martha McClintock's (1971) original paper claimed that women's cycles got two days closer together over four to six months, a far cry from the popular notion of concordant menses. Even this slight effect could not be replicated in studies that controlled for the fact that mere differences in cycle length between women (e.g., thirty days vs. twenty-eight days) are sufficient to bring about synchronization over time, despite the mutual independence of the onsets of different women. Women's subjective impressions of synchrony are often guided by incorrect assumptions about how disparate women's onsets should be by chance. If two women both have a twenty-eight-day menstrual cycle, the maximum that they can be out of phase is fourteen days (not twenty-eight days). On average, they will be seven days apart, and half the time they should be less than seven days apart by mere chance. Given that menstruation lasts about five days, concordant menses should be very common and are enthusiastically noticed, especially among friends and roommates.

Paul Turke (1983) proposed that menstrual synchrony evolved as a female tactic to promote paternal investment in offspring. But his hypothesis assumes that menstrual synchrony implies ovulatory synchrony, which has yet to be observed among female interactants. Moreover, if menstrual synchrony is an evolved feature of human reproductive biology, then it should be found in noncontracepting populations in which menstruation is a rare event. In such populations, the subfecund and sterile women experience most of the menses. Outside these two groups, it is uncommon for two women to both cycle concurrently (as one is likely to be pregnant or in amenorrhea). An explicit test for menstrual synchrony among the Dogon, using 553 menses for 67 women over two years, found that women's onsets were independent of each other and had no relation to lunar phase—despite the absence of electric lighting.

The myth of menstrual synchrony got a recent boost from a report that announced definitive evidence of human pheromones and confirmation of the mechanism underlying menstrual synchrony. Critics, however, cast doubt on the conclusion that the change in cycle length of the subjects was caused by a pheromone, rather than by the known variation in cycle length in any given woman over time. They also raised several other methodological issues. Above all, the proposed pheromone has not been chemically isolated.

Debate Regarding Concealment of Ovulation. Another equally vigorous debate has focused on whether ovulation in women is concealed, why it is concealed, and whether or not this characteristic is unique to humans. Our closest relatives, the chimpanzees, are unique in the hominoids for having exaggerated sexual swellings that peak around the time of ovulation. Gorillas have scarcely detectable swelling of the vulva or labia, but these visual cues are usually evident to humans only in captive animals. Orangutans lack sexual swellings altogether, and in gibbons and siamangs the vulva may redden, but the relation to the cycle is not known. Across the order Primates, visual signs of estrus have been lost at least eight times, but most species have retained visual cues of some kind. Exaggerated swellings, such as those of chimps, have evolved on multiple occasions, with different sources disagreeing on the precise number. The ancestral character state for any clade (or phylogenetic grouping) can be inferred from the closest outgroups, which clearly indicates that the ancestor of the anthropoids (Old World monkeys, apes, and humans) had slight visual signs, as was probably also true for the ancestor of the hominids.

The absence of sexual swellings in species other than humans should not be mistaken for evidence that ovulation is concealed. In all nonhuman primates, the frequency of copulation varies with cycle phase in a fashion that is more pronounced than in humans, with the highest frequency occurring in the ovulatory phase. Moreover, in all species other than humans, olfactory cues are probably present, although these may not be as tightly linked to ovulation as in rodents. Some women experience ovarian pangs (caused by the leaking of follicular fluid) that provide clues to the timing of ovulation, but making the connection between this mid-cycle pain and ovulation has required modern medical knowledge. If we accept the argument that concealed ovulation is unique to humans, then it is clear why it has been so difficult to figure out its evolutionary origins. All adaptive hypotheses (of which there is no shortage) must confront a host of confounding variables with a sample size of one.

Other Peculiarities of Human Reproductive Physiology. Another anomaly of human reproductive physiology is the long postmenopausal life span of women. Other apes have a maximum life span of about fifty years, and fertility fails at the same time as other physiological systems. In humans, by contrast, maximum life span is about a hundred years, but female fertility ends already by age fifty while other physiological systems keep running. Thus, postmenopausal longevity, not early termination of fertility, may be the derived trait in humans. The selective background for the long postmenopausal component of female life histories has been a topic of considerable interest. One of the more prominent proposals is that of Kristen Hawkes and her colleagues, who suggested that our ancestors evolved a slower rate of aging when senior females whose fertility was ending could increase the number of their descendants by provisioning grandchildren. Other researchers, such as Peter Ellison, have focused on why females have not evolved the male pattern of continuous gamete production in concert with the evolution of long life spans. Ellison proposes that that this would require either too large an ovary or a sacrifice in gamete quality.

At present, there are no comparative studies of ovarian size in primates, but as first noticed by Roger Short, a striking pattern emerges with respect to the size of the male gonad. Chimps and other species that live in multimale promiscuous groups have larger testes relative to body mass, whereas gorillas and other species that live in unimale units and enjoy a sexual monopoly have smaller testes to body mass ratios. Larger testes produce more sperm and can swamp the competition, which is advantageous when females mate with more than one male around the time of ovulation. The relationship between testes mass to body mass, on the one hand, and female promiscuity, on the other, is so reliable that if we know one of these two variables, we can predict the other. By this standard, over evolutionary history human females were only mildly promiscuous.

During orgasm, spasmodic contractions of the uterine musculature may produce negative pressure and an up-suck effect that draws semen into the uterus. In a study of women in the United Kingdom, Robin Baker and Mark Bellis reported that unfaithful women experienced orgasms whose occurrence, pattern, and timing suggested high sperm retention during 70 percent of copulations with their lovers, but during only 40 percent of copulations with their established partners. In faithful women, 55 percent of the orgasms were of the high-sperm-retention type. Moreover, sex with the lovers was supposedly more likely to occur during the fertile period of the menstrual cycle. Baker and Bellis interpret these two results as evidence for cryptic female choice: women keep their investing husbands on hand while subconsciously shopping for good genes outside the pair-bond. Like the menstrual synchrony research, however, Baker and Bellis's work raises pressing questions of methodology.

Evolutionary approaches to reproductive biology are new and exciting, but in a young field such as this, we can expect that many ideas will not withstand the scientific process of replication and revision. Hopefully, those ideas that prove wrong will help to clarify the issues and then be swiftly discarded. Reproductive physiology is so undeniably linked to fitness effects that an evolutionary approach is both long overdue and highly promising.

[*See also* Cryptic Female Choice; *articles on* Hominid Evolution; Human Families and Kin Groups; Life History Theory, *article on* Human Life Histories; Maternal-Fetal Conflict; Primates, *article on* Primate Societies and Social Life; Sperm Competition.]

BIBLIOGRAPHY

Alexander, R. D., and K. M. Noonan. "Concealment of Ovulation, Parental Care, and Human Social Evolution." In *Evolutionary Biology and Human Social Behavior: An Anthropological Perspective*, edited by N. A. Chagnon and W. G. Irons, pp. 436–453. North Scituate, Mass., 1979.

Ellison, P. T. *On Fertile Ground: A Natural History of Human Reproduction*. Cambridge, Mass., 2001. A provocative, new introduction to human reproductive physiology from the perspective of natural selection.

Gladwell, M. "John Rock's Error." *New Yorker*, 13 March 2000, pp. 52–63. Fascinating popular article on menstruation, the Pill, and reproductive cancers.

Harcourt, A. H. "Sperm Competition and the Evolution of Nonfertilizing Sperm in Mammals." *Evolution* 45 (1991): 314–315.

Harcourt, A. H., P. H. Harvey, S. G. Larson, and R. V. Short. "Testis Weight, Body Weight and Breeding System in Primates." *Nature* 293 (1981): 55–57.

Hawkes, K., J. F. O'Connell, N. G. Blurton Jones, H. Alvarez, and E. L. Charnov. *Grandmothering, menopause, and the evolution of human life histories. Proceedings of the National Academy of Sciences USA* 95 (1998): 1336–1339.

Hrdy, Sarah Blaffer. *Mother Nature: A History of Mothers, Infants, and Natural Selection*. New York, 2000. The best book on the physiology of motherhood. Emphasizes the trade-offs that go with reproduction and comparisons with other primates; highly literary and readable.

McClintock, M. K. "Menstrual Synchrony and Suppression." *Nature* 229 (1971): 244–245.

Nunn, C. L. "The Evolution of Exaggerated Sexual Swellings in Primates and the Graded-Signal Hypothesis." *Animal Behavior* 58 (1999): 229–246.

Rogers, A. R. "Order Emerging from Chaos in Evolutionary Genetics." *Proceedings of the National Academy of Sciences USA* 98 (2001): 779–780.

Stiner, M. C., N. D. Munroe, T. A. Surovell, E. Tchernov, and O. Bar-Yosef. "Paleolithic Population Growth Pulses Evidenced by Small Animal Exploitation." *Science* 283 (1999): 190–194.

Takahara, N., Y. Satta, and J. Klein. "Divergence Time and Population Size in the Lineage Leading to Modern Humans." *Theoretical and Population Biology* 48 (1995): 198–221.

Turke, P. W. "Effects of Ovulatory Concealment and Synchrony on Protohominid Mating Systems and Parental Roles." *Ethnology and Sociobiology* 4 (1983): 157–168.

— BEVERLY STRASSMANN

REPRODUCTIVE VALUE

In 1930, R. A. Fisher introduced the concept of reproductive value for an age-structured population, as a measure of the relative contributions of individuals of different ages to the ancestry of future generations. The reproductive value of individuals of a given age takes into account their prospects of survival and rate of reproduction over all future ages.

Intuitively, one might expect that an individual's reproductive value during infancy would have an intermediate value (owing to infant mortality) that rises until about the time of sexual maturity. After this it declines, reflecting the declining future reproductive contributions as one gets older.

Fisher intended reproductive value to provide a quantitative measure of the intensity of selection acting at a given age. Although this has proved not to be generally true (Hamilton, 1966), reproductive value appears in models of life history evolution as a weighting factor for the effect on net fitness of a change in survival at a given age. Under many circumstances, natural selection can be thought of as maximizing reproductive value at each age, subject to a specified set of trade-offs among different life history traits (Charlesworth, 1994, p. 238; Schaffer, 1974). Recent work has shown that the concept of reproductive value can be extended to populations divided into more general categories than different ages, such as size and sex, and plays an important role in calculating the expected effect of selection on the characteristics of such populations (Caswell, 1989; Frank, 1998; Pen and Weissing, 2001; Taylor, 1996).

The Mathematics of An Age-Structured Population.

Basic definitions and concepts. To obtain a deeper understanding of reproductive value and its application, some mathematical concepts need to be used. With sexual reproduction, for purely demographic purposes it is most convenient to study the female population alone (Charlesworth, 1994, p. 4). It is also easier to study a discrete time population process, for example, an annually breeding bird or mammal species, where the population is censused each breeding season. In this case, the state of the population at a given time t can be represented by a column vector $n\ (t)$, whose components $n_1(t)$, $n_2(t)$, and $n_d(t)$ are the numbers of females age one, two, and so on, alive at time t, where d is the last age to which fertile individuals survive. Fisher originally defined reproductive value for a population reproducing in continuous time, such as humans. This case can be approximated using discrete time, by increasing the numbers of age classes indefinitely.

If P_x is the probability of survival from age x to x + 1, and m_x is the number of daughters produced by a female age x, the expected number of daughters contributed to age class 1 at time $t + 1$ by a female aged x at time t is $f_x = P_0\ m_x$, where P_0 is the probability of survival from birth to age 1. The population of females then obeys the matrix equation

$$n(t + 1) = L\ n(t), \qquad (1)$$

where the only nonzero elements of the "Leslie matrix" L (Leslie, 1945) are the first row, consisting of the elements $f_1, f_2, \ldots f_d$, and the set of subdiagonal elements $P_1, P_2, \ldots P_{d-1}$.

Matrix theory (Caswell, 1989; Charlesworth, 1994, chap. 1) tells us that, under "reasonable" conditions on the components of the Leslie matrix, the population asymptotically approaches a state in which the numbers of individuals in each age class increase by the same factor of λ_0 each time interval; that is, $n_x(t + 1) = \lambda_0\ n_x(t) =$ for each x. λ_0 is called the "leading eigenvalue" of L.

The population thus approaches a state with constant frequencies of individuals in each age class, with a rate of growth in the logarithm of population size $r = \ln \lambda_0$. r is often referred to as the "intrinsic rate of increase" of the population. Write $l_x = P_0\ P_1\ P_2, \ldots P_{x-1}$, for the probability of survival of an individual from birth to age x. λ_0 can then be shown (Charlesworth, 1994, p. 23; Leslie, 1945) to satisfy the Euler–Lotka equation:

$$\sum_{x=1}^{d} \lambda_0^{-x}\ l_x m_x = 1 \qquad (2)$$

At this asymptotic state, the numbers of individuals at each age, n_x, is proportional to the components of the vector u_0, the "right eigenvector" of the Leslies matrix, which satisfies the equation

$$L\ u_0 = \lambda_0\ u_0 \qquad (3a)$$

A "left eigenvector" of L, v_0, can similarly be defined by the relation

$$v_0 \mathrm{L} = \lambda_0\ v_0 \qquad (3b)$$

One biological interpretation of v_0 follows from the fact that the expression for $n(t)$ in terms of the initial value $n(0)$, and the expansion of L^t in terms of its eigenvalues and eigenvectors (Charlesworth, 1994, p. 22), implies that females initially in a given age class x make a contribution to the population size at time t that is proportional to the corresponding component of v_0 (Charlesworth 1994, p. 37). Another interpretation is that the components of v_0 are proportional to the reproductive values of the corresponding ages, as is shown next.

The definition of reproductive value. Fisher (1999, chap. 2) defined the reproductive value, v_x, of a female currently age x as her contribution to the ancestry of future generations in a population growing at rate r, discounting the contribution to a given time in the future by the growth in population size from the present. Normalizing so that a newborn female has a reproductive value of 1, we have

$$v_x = \frac{\lambda_0^{-x}}{l_x} \sum_{y=x}^{d} \lambda_0^{-y} l_y m_y \tag{4}$$

It turns out that v_0 has components proportional to v_1, v_2, and so on, so that the reproductive values at each age can be identified with the components of this eigenvector (Charlesworth, 1994, p. 37). In particular, this implies that the total reproductive value of a population always grows at rate r, as noted by Fisher.

The Uses of Reproductive Value.

Reproductive value and life history evolution. Fisher (1999) stated that "the direct action of Natural Selection must be proportional to [reproductive value]" (p. 27). This was later taken up by Medawar (1952), who proposed that the intensity of selection acting on a gene affecting survival at a given age x is proportional to v_x. Plots of reproductive value against age show that it generally increases during the juvenile period, reaches a maximum just after the age at maturity, then declines with age. But Hamilton (1966) showed that the effect on r of a small change in the natural logarithm of P_x is proportional to

$$v_{x+1}\, l_{x+1}\, \lambda_0^{(x+1)} = \sum_{y=x+1}^{d} \lambda_0^{-y}\, l_y\, m_y \tag{5}$$

This is independent of age during the juvenile period, then decreases monotonically with age during the rest of life, in contrast to reproductive value.

Further analysis shows, however, that reproductive value is useful in the context of models of life history evolution, when selection is assumed to maximize λ_0 or r under a specified set of constraints among different life history variables (see Charlesworth, 1994, chap. 5 for a discussion of this assumption). The simplest case is the "reproductive effort" model, in which age-specific survival, P_x, is assumed to be a decreasing function of age-specific fecundity, m_x (Charlesworth, 1994, p. 215; Schaffer, 1974). In this case, the optimal life history satisfies

$$1 + \frac{\partial P_x}{\partial m_x} \lambda_0\, v_{x+1} = 0 \tag{6}$$

so that v_{x+1} can be regarded as a weighting factor for the effect on survival at age x of a change in fecundity, in determining the solution for the optimal life history.

More generally, for any optimal life history, the maximization of λ_0 or r can be shown to be equivalent to the maximization of v_x with respect to perturbations at ages x, $x + 1, \ldots d$, with certain technical restrictions (Charlesworth, 1994, p. 238; Schaffer, 1974). Calculations of optimal life history solutions can therefore often be reduced to maximizations of reproductive value at each age.

Generalizing reproductive value. The concept of reproductive value as a left eigenvector of a matrix describing the change in state of a population can usefully be extended to any population divided into a set of discrete categories, whenever a matrix representation of its transition from one time to the next is appropriate (Caswell, 1989; Taylor, 1996). For any matrix A with nonnegative components that satisfies "reasonable" conditions, there will be a nonnegative leading eigenvalue λ_0, with associated left and right eigenvectors v_0 and u_0, where v_0 is a generalization of reproductive value, and the components of the column vector u_0 are proportional to the numbers of individuals in the different classes in a population that has reached its asymptotic growth rate λ_0. The derivative of λ_0 with respect to a given component a_{ij} of A is given (Caswell, 1989, p. 120; Charlesworth, 1994, p. 39) by

$$\frac{\partial \lambda_0}{\partial a_{ij}} = \frac{v_i u_j}{(v_0 u_0)} \tag{7}$$

This can be used for determining the evolutionarily stable strategy (ESS) (Frank, 1998; Maynard Smith, 1982) of a structured population with respect to a trait of interest for example, sex ratio in a system when sex ratio is dependent on the age of the mother (Charlesworth, 1994, pp. 231–236; Pen and Weissing, 2001). If the population is at an ESS with respect to the trait, any invading mutant must have a lower rate of asymptotic growth in numbers than the population as a whole. λ_0 for the mutant must thus be at a maximum when it adopts the trait value in question. Hence, it is only necessary to replace the matrix A that describes the population as a whole with a matrix B that describes the numbers of mutant individuals in each category, and apply equation (7) to the standard conditions for a maximum. This provides a unified approach to solving many important problems in evolutionary ecology.

[*See also* Demography; Life History Theory; Senescence.]

BIBLIOGRAPHY

Caswell, H. *Matrix Population Models.* Sunderland, Mass., 1989.

Charlesworth, B. *Evolution in Age-structured Populations.* 2d ed. Cambridge, 1994.

Fisher, R. A. *The Genetical Theory of Natural Selection.* Oxford, 1930. Variorum edition, J. H. Bennett, ed., Oxford, 1999.

Frank, S. A. *Foundations of Social Evolution.* Princeton, 1998.

Hamilton, W. D. "The Moulding of Senescence by Natural Selection." *Journal of Theoretical Biology* 12 (1966): 12–45.

Leslie, P. H. "On the Use of Matrices in Certain Population Mathematics." *Biometrika* 33 (1995): 183–212.

Maynard Smith, J. *Evolution and the Theory of Games.* Cambridge, 1982.

Medawar, P. B. *An Unsolved Problem of Biology.* London, 1952.

Pen, I., and F. J. Weissing. "Optimal Sex Allocation: Steps towards a Mechanistic Theory." In *The Sex Ratio Handbook*, edited by I. Hardy. Cambridge, 2001.

Schaffer, W. M. "Selection for Optimal Life Histories: The Effect of Age Structure." *Ecology* 55 1974: 291–303.

Taylor, P. D. "Inclusive Fitness Arguments in Genetic Models of Behaviour." *Journal of Mathematical Biology* 34 (1996): 656–674.

— BRIAN CHARLESWORTH

REPTILES

Reptiles, the Reptilia, are a group of amniotes, four-legged vertebrates that differ from amphibians in possessing adaptations permitting full terrestriality, such as a waterproof epidermis, internal fertilization, and shelled eggs with large yolk stores and protective membranes, the innermost being termed the amnion. Historically, the taxon Reptilia usually referred to a paraphyletic group (a group that includes some but not all descendants of the common ancestor) consisting of ectothermic or "cold-blooded" amniotes; thus, it excluded birds but included primitive relatives of mammals (the "mammal-like reptiles"; pelycosaurs and therapsids). More recently, the term Reptilia has been redefined by biologists to refer to a true lineage or monophyletic group (a group containing all the descendants of their common ancestor) of amniotes, by including birds but excluding pelycosaurs and therapsids, which have now been transferred to the Synapsida (Figure 1).

This newer definition is increasingly being adopted by the general scientific community, partly in response to the recent evidence that birds are directly descended from dinosaurian reptiles, and is the one adopted here. As thus understood, reptiles consist of three major lineages (Figure 1): lepidosaurs, represented today by lizards, snakes, and the tuatara; archosaurs, represented today by crocodilians and birds; and parareptiles, represented today perhaps by turtles, though their inclusion within parareptiles is contentious. In addition to the traits found in all amniotes, reptiles possess highly keratinized skin, the ability to conserve water by excreting uric acid, and novel eye structures.

Parareptiles and Other Primitive Reptiles. Most early reptiles possessed *anapsid* skulls—that is, skulls with a solid temporal (cheek) region, the primitive condition inherited from their amphibian ancestors. Most, but not all, of these anapsid-skulled reptiles belong to a lineage termed the Parareptilia. Examples include mesosaurs, procolophonids, and pareiasaurs. Mesosaurs were small, aquatic forms with long necks, webbed feet, and narrow snouts bearing needlelike teeth. They were weak swimmers incapable of transoceanic crossings, and the discovery of closely related species on opposite sides of the present Atlantic Ocean was early evidence for continental drift. Procolophonids superficially resembled stout lizards with spiny skulls and possessed molarlike teeth for crushing hard invertebrates. Pareiasaurs were large (up to 3 meters), slow-moving herbivores with leaf-shaped teeth, heavy and highly ornamented skulls, and armor plating over their back and sides.

A few early, anapsid-skulled reptiles did not belong within the parareptile lineage. Protorothyridids were tiny, slender, long-legged insectivores, while captorhinids were similar but larger and more robust. Protorothyridids are the earliest reptiles known, being found inside petrified tree hollows that are over 300 million years old, and partly on this basis they were long assumed to be ancestral to all other reptiles. However, rigorous analyses suggested that protorothyridids are not ancestral (basal) to all other reptiles, but like captorhinids are relatives of the diapsid radiation (lepidosaurs and archosaurs).

Turtles (Testudines). Turtles (300 living species) include such forms as amphibious terrapins, terrestrial tortoises, and marine turtles. They are among the most distinct vertebrates, exhibiting striking morphological specializations that involve not just the shell but also associated modifications of the vertebrae, limbs, and skull. The turtle shell is a boxlike structure consisting of a dorsal carapace and a ventral plastron, joined laterally by the "bridge." It is open anteriorly for the head and forelimbs, and posteriorly for the tail and hindlimbs. The shell is unique among tetrapods in incorporating both dermal armor and internal skeletal elements. It is secondarily reduced in certain forms, especially aquatic taxa such as sea turtles and soft-shell turtles. The dorsal vertebrae and ribs of turtles are immobile, completely fused to the inside of the carapace. The limb girdles of turtles lie within (rather than outside) the rib cage, inside the protective shell. The limbs project horizontally through the anterior and posterior shell openings, resulting in a low sprawling stance and broad trackway. Except in sea turtles, the limbs can be retracted into the shell.

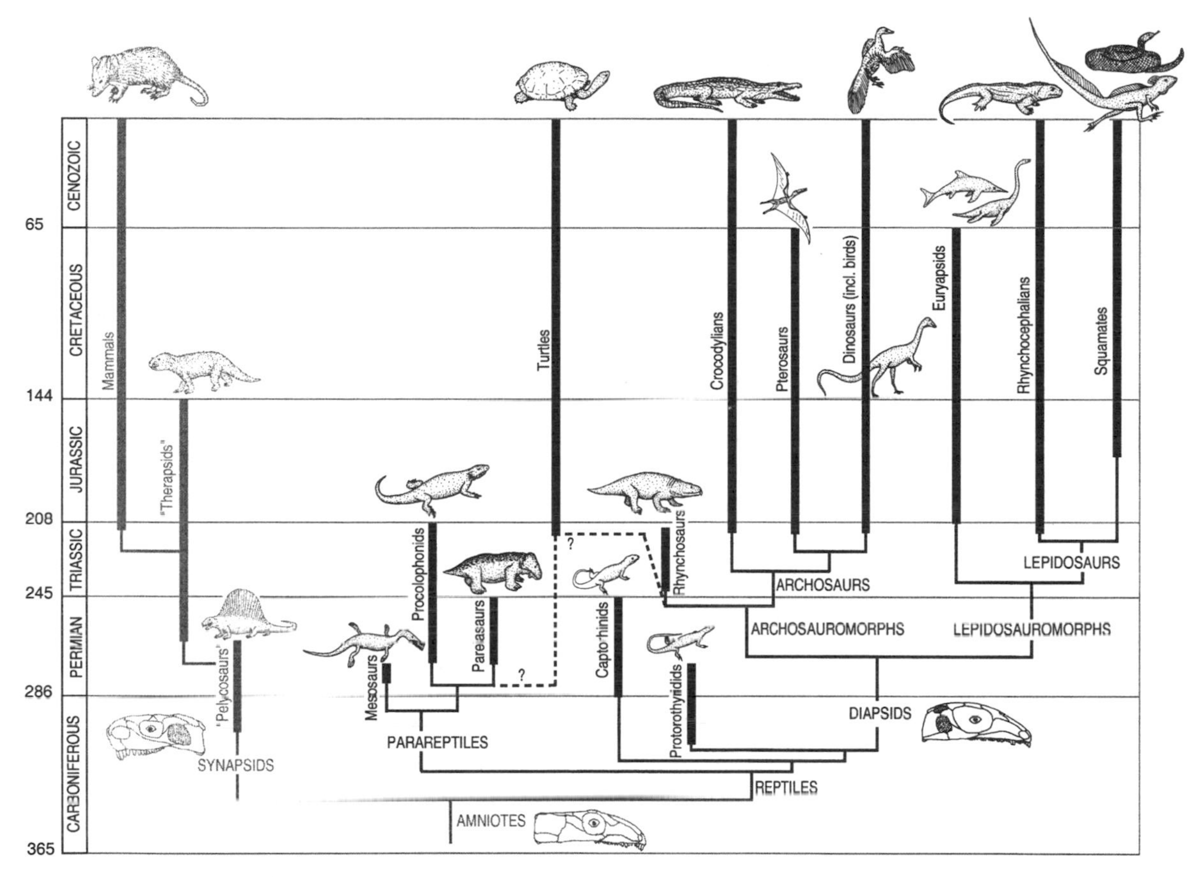

FIGURE 1. The Major Groups of Amniote Vertebrates and Their Interrelationships.
Reptiles are shown in black, synapsids are shown in gray. The thick lines depict the known fossil duration for each group, excluding controversial finds (such as the Triassic "bird" *Protoavis* and the Cenozoic "therapsid" *Chronoperates*). Examples from each lineage are illustrated. Living forms are drawn above the geological time scale, extinct forms drawn within the scale. The skull diagrams show the three major skull types found in amniotes: synapsid (found in synapsids), diapsid (found in diapsid reptiles), and anapsid (found in turtles, parareptiles, captorhinids and protorothyridids). Note that synapsid and diapsid skulls each characterize discrete lineages, but the anapsid skull does not. Drawing by Michael S. Y. Lee.

The skull of turtles is also highly modified, and very different from that of other anapsid-skulled reptiles. All teeth on the jaw margins are lost, replaced by keratinous beaks (rhamphothecae). The orbits are positioned anteriorly, resulting in a short facial region and a long cheek region. In all primitive turtles, the cheeks are solid walls. However, extensive emarginations along the posterior and ventral cheek margins have evolved in more derived turtles. The neck is lengthened, resulting in greater mobility of the head, while the body and tail are shortened to fit within the confines of the shell. The most primitive turtles could not fold their necks to retract their heads into the shell; however, this useful ability has evolved twice, once in pleurodires (side-necked turtles), which bend their neck laterally, and once in cryptodires (hidden-necked turtles), which bend their neck in a vertical plane and thus tuck it behind the skull during cranial retraction. Living pleurodires are all "terrapinlike" in morphology and are restricted to the Southern continents (Africa, South America, and Australasia). Living cryptodires are more diverse and cosmopolitan, including such forms as snapping turtles, soft-shelled turtles, sea turtles, and box tortoises.

Morphological studies place turtles within the parareptile radiation, as relatives of either pareiasaurs or procolophonids, based on such traits as the anapsid skull of turtles and the presence of dermal armor in pareiasaurs and turtles. Recent analyses of mitochondrial and nuclear genes contradict this view and instead consistently suggest that turtles are closely related to archosaurs, and thus members of the diapsid lineage. If so, the apparently primitive skull structure of turtles would represent an evolutionary reversal. Disturbingly, reexaminations of turtle anatomy by several workers in light of the molecular results still failed to find any trace of an archosaurian ancestry. Nevertheless, if analyses are performed so that turtles are "forced" to cluster with archosaurs according to the molecular data, they then group with rhynchosaurs (see below), based on shared features such as toothless, beaklike jaws and fat, squat bodies. The conflict between the morphological and molecular evidence for turtle relationships remains an intriguing dilemma (Figure 1).

Diapsids (Archosaurs and Lepidosaurs). Lepidosaurs, archosaurs, and related forms all have skulls with two large fenestrae (holes) in the cheek region, a condition termed *diapsid.* These fenestrae lighten the skull, and their rims provide insertion areas for the jaw-closing muscles. In addition, these forms possess a third fenestra in the roof of the mouth. These novel cranial features, and some other traits, unite all these reptiles as a distinct lineage, the Diapsida, to the exclusion of all other reptiles (except perhaps turtles). The Diapsida split very early in its history into two great lineages, one (the lepidosauromorphs) leading to living lepidosaurs, and the other (the archosauromorphs) leading to living archosaurs. Most diapsid reptiles, except some early primitive forms, can be assigned confidently to one of these clades.

Lepidosaurs (Tuataras, Lizards, and Snakes). Extant lepidosaurs include *Sphenodon* (the tuatara) and the Squamata (lizards and snakes). They all share a distinctive skin that contains a unique type of keratin and is shed in large pieces. The tuatara (*Sphenodon*) is a famous reptilian "living fossil" and consists of two very similar species (distinguishable genetically) that are restricted to small, rat-free islands off New Zealand. They superficially resemble slow-moving, stout iguanas and have an unusually slow metabolisms and life cycles, perhaps adaptations for their harsh, cold habitat.

Squamates (lizards, amphisbaenians, and snakes) are a diverse and successful radiation. All squamates share numerous distinctive evolutionary novelties, such as a mobility (streptostyly) of the quadrate bone which suspends the lower jaw, paired eversible male copulatory organs (hemipenes), and a distinct type of vertebral joint (procoely).

Iguanian lizards (1,000 living species), the most basal or "primitive" living squamates, consist of iguanids (iguanas), agamids (dragon-lizards), and chamaeleonids (chameleons). They are generally large forms with elaborate crests and ornaments over their skulls and bodies, and the ability for rapid and profound color change (metachromism). These traits (exaggerated in chamaeleons) appear to be linked to male territoriality and visual displays, which are more highly developed in iguanians than in other lizards. Most of the (relatively few) herbivorous lizards are iguanians. Chameleons are among the most famous and bizarre lizards, and many of their unusual features are related to their sit-and-wait predation strategy: rapid and extensive color changes (camouflage), grasping digits and prehensile tail (facilitating a permanent tight grip on branches), independently movable eyes on turrets, and long projectile tongue (enabling visual sweeps and prey capture without head movement).

Gekkotan lizards (geckos and flap-footed lizards, 1,050 living species) are another basal ("primitive") group of squamates. They are usually nocturnal and accordingly have large and distinctive eyes. These eyes have vertical pupils, and the eyelids are fused into a transparent "spectacle," which is cleaned by licks from the fleshy tongue (rather than blinking). Most members have enlarged toe-pads that enable them to scale smooth vertical surfaces. Gekkotans also lack many skull bones found in other squamates, probably because of early cessation of ossification (paedomorphosis). Vocal communication is highly developed, with some members having elaborate reportoires similar to those of many frogs. Accordingly, gekkotans have well-developed larynxes ("voiceboxes") and highly sensitive auditory structures. One lineage of gekkotans has become very snakelike and is often separated off as the Pygopodidae (flap-footed lizards).

Scincomorph lizards (e.g., skinks, cordylids, lacertids, teiids—2,000 living species) are the most diverse and "typical" group of lizards, consisting mainly of small-bodied, generalized forms. Accordingly, most of the features they share are widespread in other lizards as well, and the evidence that scincomorphs form a true lineage (clade) is thus weak. Most scincomorphs are agile, secretive, smallish animals that shelter beneath leaf litter or loose rocks. This intimate association with the substrate has probably facilitated the frequent (>30 times) evolution of burrowing habits and thus of limb reduction and body elongation within this group.

Anguimorph lizards (180 living species, not counting snakes) are generally medium to large predators, and include the two species of Gila monsters (the only venomous lizards), as well as the Komodo dragon (a large monitor) and mosasaurs (gigantic, extinct marine predators). All anguimorphs possess a distinctive pattern of tooth replacement and a retractile, deeply forked tongue that is used to detect prey odors. Many also have highly flexible lower jaws and sharp recurved teeth. All these traits are related to feeding on large prey and are also found in snakes, which might be related to anguimorphs.

Snakes and amphisbaenians are two of the many lineages of squamates that have undergone body elongation and limb reduction. Because of their diversity and anatomical distinctiveness, they have often been removed from the remaining squamates and the leftovers formally recognized as the taxon Sauria ("lizards"). However, the latter taxon is a grade rather than a true lineage and thus not formally recognized here.

Snakes (Serpentes, 2,900 living species) are one of the most rapidly diversifying groups of vertebrates. They range from tiny forms such as the primitive wormlike blindsnakes, to giant constrictors such as boas and pythons, and deadly mambas, cobras, and sea snakes. Characteristic external features include eyelids fused into a transparent "spectacle," absence of the external eardrum, retractile forked tongue and long, limb-reduced bodies. Each of these traits, however, has evolved independently in some "lizards," and the only reliable features of snakes are internal. Snakes are characterized by extremely loose skulls with highly flexible upper and lower jaws loosely suspended from a central bony braincase. In many snakes, including most advanced forms, the left and right lower jaws are connected anteriorly by elastic ligaments and thus can separate to engulf proportionately huge prey. There are usually between 140 and 600 trunk vertebrae (more than in even the most elongate lizards), and the forelimb and pectoral girdle are totally lost (vestiges remain in all limb-reduced lizards).

Even the earliest snakes had extensive adaptations for predation, and this constraint appears to have prevented snakes from evolving into omnivores or herbivores. *Aglyphous* snakes lack fangs and venom glands; this category includes all primitive snakes and indeed, most snakes (80 percent). Aglyphous snakes that take larger prey kill by constriction and continuous bites. However, several groups of snakes have independently evolved venom fangs (enlarged teeth with grooves or canals for injecting venom) and venom glands (modified salivary glands). These venomous forms often do not constrict, but adopt a strike-and-release strategy to avoid being injured by large struggling prey. *Opisthoglyphous* snakes have fixed fangs at the back of the jaws. Some colubrids (e.g., boomslangs) have this arrangement. *Proteroglyphous* snakes have fixed fangs at the front of the jaws. This characterizes elapid snakes (e.g., cobras, sea snakes). *Solenoglyphous* snakes have mobile fangs that are erected only while striking. Because the fangs can be folded away when not in use, they can be very large. Vipers (e.g., rattlesnakes and adders) have this arrangement.

Though there is widespread agreement that snakes evolved from lizards, the more precise details are debated. It has long been assumed that snakes evolved from a lineage of burrowing lizards, based on the close association of burrowing habits with limb reduction in living lizards, and highly divergent eye structure, which suggested that the eyes of snakes became reduced and then re-elaborated. However, an even older idea is that they evolved in a marine habitat, for eellike swimming. This hypothesis has been revived recently by studies linking snakes with marine lizards (mosasaurs), and the discovery of well-preserved early marine "proto-snakes" such as *Pachyrhachis* and *Haasiophis*. These long-bodied squamates appear to be true snakes except for the retention of hindlimbs, and they are most plausibly interpreted as transitional forms (missing links) between lizards and snakes; however, it has also been proposed that they are true (advanced) snakes that have re-evolved legs. A re-evaluation of eye anatomy has also revealed close similarities between the eyes of snakes and those of marine vertebrates (e.g., fish, whales), again consistent with marine origins.

Amphisbaenians (160 living species) are a highly aberrant group of long-bodied, limb-reduced squamates that superficially resemble large, fat earthworms. They are highly specialized and efficient burrowers, with extremely solid skulls for ramming their way through the substrate, and scales and muscles arranged in rings around the body for gripping the sides of burrows. Their eyes are among the most degenerate in vertebrates, and they rely largely on chemical and vibrational cues to locate prey. Their precise position within Squamata remains unclear.

Large extinct marine reptiles, such as the long-necked plesiosaurs, the short-necked pliosaurs, and the fishlike ichthyosaurs, are all members of a single radiation termed the Euryapsida. They are characterized by a diapsid skull with an extremely wide cheek region lacking the lower strut of bone, a condition termed *euryapsid*. This diverse group appears to be related to lepidosaurs; it suffered a mass extinction at the end of the Cretaceous, along with the (nonavian) dinosaurs and many other organisms.

Archosaurs (Crocodiles, Dinosaurs, and Birds). The Archosauria include some of the most spectacular reptiles, such as crocodiles, pterosaurs, dinosaurs, and birds. They are characterized by anatomical traits such as a fully divided heart venticle, a new pair of bones (the laterosphenoids) forming the front of the braincase, a system of air sacs within the skull, and fenestrae in the snout and lower jaw (the snout fenestra has, however, been secondarily closed in living crocodilians). Living archosaurs (crocodilians and birds) share behavioral traits such as use of stomach stones (gastroliths), nest-building, parental care, and vocalizations (chirping) by nestlings. These habits are difficult to confirm in fossil archosaurs, but smoothly worn stomach stones have been found within complete dinosaur skeletons, and dinosaurs have recently been found fossilized while incubating nests of eggs. Molecular studies reveal that the DNA of crocodiles and birds is very similar. The large number of advanced morphological, behavioral, and genetic features shared by birds, crocodilians, and (where known) fossil archosaurs reflect their close evolutionary relationship and justify the current practice of classifying birds with archosaurian reptiles, rather than the older approach of separating birds off from all reptiles into their own group. The latter approach is further complicated by recent discoveries of several feathered, birdlike dinosaurs that blur the distinction between birds and dinosaurian reptiles.

Living crocodilians (crocodiles and alligators, 23 living species) are all large, semiaquatic predators. They are all morphologically quite uniform, with long snouts, conical piercing teeth, longish bodies, short but robust limbs, laterally compressed tails, and leathery skin containing bony plates. There are two major lineages, the alligators and the crocodiles (including false gavials). The relationships of the true gavial are unclear; anatomical evidence suggests that it represents a third lineage lying outside of both alligators and crocodiles, but analyses of DNA sequences tend to place it within crocodiles. All living crocodilians are ambush predators that (as adults) take sizable vertebrate prey such as fish, amphibians, birds, and mammals, captured either underwater or at the water's edge. Although living crocodilians are neither taxonomically or anatomically diverse, the group radiated widely in the past. There were cursorial, long-legged, fully terrestrial forms such as *Sphenosuchus*, which actively chased prey and perhaps climbed trees; giants such as *Deinosuchus*, larger than the largest carnivorous dinosaurs; ocean-going forms such as teleosaurs, with flippers and caudal fins; and small, heavily armored herbivorous forms.

Rhynchosaurs were primitive relatives of archosaurs, though they were formerly grouped incorrectly with rhynchocephalians (which are lepidosaurs). The dominant herbivores during the Triassic, they had stout bodies, wide, short skulls, and crushing beaks instead of toothed jaws. If turtles are related to archosaurs, as has been suggested by recent molecular analyses, then they most likely have affinities with rhynchosaurs.

Pterosaurs were an extinct group of archosaurs closely related to dinosaurs and were the first vertebrates to evolve powered flight. Their bones were extremely hollow and light (like bird bones), and their membranous wings were suspended by a greatly elongated fourth finger and stiff internal fibers. The shape of their wings has long been debated, but fossils preserving soft tissue have revealed that (at least in some taxa) the wing membrane was wide and stretched between the forelimbs and hindlimbs, resulting in batlike morphology and sprawling, clumsy gait. These fossils have also revealed that pterosaurs were covered in fine, hairlike structures and thus might have evolved endothermy ("warm-bloodedness") in response to the high metabolic demands of flapping flight.

[*See also* Birds; Dinosaurs; Vertebrates.]

BIBLIOGRAPHY

Benton, M. J. *Vertebrate Palaeontology: Biology and Evolution. 2d ed.* London, 1997. Concise summary of the fossil record of reptiles.

Brochu, C. A. "Morphology, Fossils, Divergence Timing, and the Phylogenetic Relationships of *Gavialis*." *Systematic Biology* 46 (1997): 479–522. Evaluates the morphological and molecular evidence for evolutionary relationships within crocodilians.

Coates, M., and M. Ruta. "Nice Snake, Shame about the Legs." *Trends in Ecology and Evolution* 15 (2000): 503–507. Concise overview of the current debate on snake origins.

Eernisse, D. J., and A. G. Kluge. "Taxonomic Congruence versus Total Evidence, and Amniote Phylogeny Inferred from Fossils, Molecules and Morphology." *Molecular Biology and Evolution* 10 (1993): 1170–1195. An important early morphological and molecular analysis of relationships between the major groups of reptiles and other amniotes.

Estes, R., and G. Pregill, eds. *Phylogenetic Relationships of the Lizard Families*. Stanford, 1988. Contains numerous landmark papers covering evolutionary relationships between all lepidosaurs (not just lizards).

Gans, C., et al., eds. 1969–1998. *Biology of the Reptilia. 19 vols.* New York: Academic Press (vols. 1–18). St Louis: Society for the Study of Amphibians and Reptiles (vol. 19 and future volumes). Comprehensive ongoing multivolume work with chapters each reviewing specific aspects of reptile biology and evolution; some early pieces are dated but many later pieces are cutting-edge.

Gauthier, J., A. G. Kluge, and T. Rowe. "Amniote Phylogeny and the Importance of Fossils." *Cladistics* 4 (1988): 105–209. Classic paper containing the first comprehensive analysis of the morphological (including fossil) evidence regarding relationships between reptiles and other amniotes.

Greene, H. W. *Snakes—The Evolution of Mystery in Nature*. Berkeley, 1997. Highly readable and well illustrated, yet scientifically rigorous, review of snake biology and evolution by a leading authority.

Lee, M. S. Y. 2001. "Molecules, Morphology and the Monophyly of Diapsid Reptiles". *Contributions to Zoology:* in press. Analyzes the morphological and molecular evidence for turtle–diapsid relationships.

Pough, F. H., R. M. Andrews, J. E. Cadle, M. L. Crump, A. H. Savitzky, and R. D. Wells. *Herpetology*. Englewood Cliffs, N.J., 2000. General overview of the biology of (mainly living) reptiles.

Pritchard, P. C. H. *Encyclopedia of Turtles*. Neptune, N.J., 1979. Exhaustive illustrated compendium of all turtle species.

Ross, C. A., ed. *Crocodiles and Alligators*. Sydney, 1989. Readable review papers on crocodilian biology and evolution.

Sereno, P. C. "Basal Archosaurs: Phylogenetic Relationships and Functional Implications." *Society of Vertebrate Paleontology Memoir* 2 (1991): 1–53. Analysis of relationships between fossil and living archosaurs.

Shaffer, H. B., P. Meylan, and M. L. McKnight. "Tests of Turtle Phylogeny: Molecular, Morphological, and Paleontological Approaches." *Systematic Biology* 46 (1997): 235–268. Analysis of evolutionary relationships within turtles based on morphology and molecules.

— MICHAEL S. Y. LEE

RESISTANCE, COST OF

Many organisms have evolved resistance to agents in their environment that cause morbidity or mortality, including chemicals and diseases. Some examples of evolved resistance to chemical agents include HIV strains that are resistant to protease inhibitors used to treat patients with AIDS; pathogenic bacteria, such as *Staphylococcus aureus*, that have become resistant to multiple antibiotics used to treat infections; resistance by certain plants to heavy metals present in mine tailings and other contaminated soils; and resistance by insects to chemicals, such as DDT, that have been used to control agricultural pests. Examples of evolved resistance to infectious diseases range from bacteria that become resistant to viruses that infect them, all the way to humans, who have genotypes that make them less susceptible to malaria infections in regions of the world where malaria is common.

In all these examples, there is a strong selective advantage to the resistant genotype in the environments where the relevant agent is present. But resistance sometimes has a deleterious side effect, which is often called the cost of resistance. This cost is one example of an evolutionary tradeoff. The simplest genetic mechanism for producing a tradeoff of this sort is that the same mutation that confers resistance interferes with some other aspect of the organism's performance. Such multiple effects of the same mutation are termed pleiotropy in genetics.

A cost of resistance has been demonstrated in several of the examples of evolved resistance described above. The mutations that make humans resistant to malaria, for example, do so by causing changes in the red blood cells that the parasite infects. When an individual is homozygous for (carries two copies of) this mutation, the result is a devastating illness called sickle-cell anemia. Thus, although there is selection to maintain this allele in the human population in geographical regions where malaria is prevalent, there is selection against this same allele in the absence of the parasite. The sheep blowfly (*Lucilia cuprina*) is a pest in Australia, which for some time was controlled using the insecticide diazinon, although mutations eventually appeared that made the flies resistant to this compound. The resistant flies, however, were noticeably inferior to their sensitive counterparts in certain other respects, such as requiring a longer time to develop from eggs into adults in the absence of the insecticide. Bacteria can evolve resistance to the viruses that infect them, and the bacteria often become resistant by mutations in the genes that encode surface receptors to which the viruses attach to initiate the infection. However, some of the same receptors are used by the bacteria to transport nutrients into the cell or to maintain the structural integrity of their cell wall. Consequently, mutations that make the bacteria resistant to viral infection are often deleterious when viruses are absent from the environment.

The evolutionary process may not stop with the emergence of a resistant population. In the case of diseases, resistance in the host population may often promote the evolution of the parasite or pathogen to overcome the host defenses. For example, "host-range" mutations sometimes occur that allow viruses to infect bacteria that have evolved resistance to the original virus, and some herbivorous insects have evolved mechanisms to detoxify, sequester, or avoid toxic compounds produced by their host plants. The reciprocal evolution of defenses and counterdefenses is a compelling example of coevolution.

Even without the coevolution of another species, the emergence of resistance may promote further evolution in the resistant population, owing to the cost of resistance. That is, selection should favor a mutation that allowed an organism to minimize the cost of resistance by ameliorating a deleterious side effect of the mutation that conferred resistance. Compensatory mutations that reduce or eliminate the cost of resistance have been demonstrated in several cases. The Australian sheep blowfly that evolved resistance to diazinon and suffered from slower development as a side effect later acquired a second mutation that restored normal development without diminishing its resistance. Similarly, bacteria that have become resistant to viruses and antibiotics have been seen to undergo further evolution that reduces the initial cost of resistance. Also, many organisms with genotypes that confer resistance to a particular agent actually express the resistance phenotype only when that agent is present. Thus, the organisms can avoid the cost when being resistant is unnecessary. Such inducible expression occurs in many cases of bacterial resistance to antibiotics. For example, the gene that encodes a chemical pump that exports tetracycline from the cell is induced when that compound is present in the cell, but that gene is otherwise repressed. Some aquatic invertebrates produce protective armor only when they are raised in the presence of their predators, indicating that they can sense some chemical that is produced by their attackers. Thus, just as organisms can adapt genetically to changes in their external environment, so too can they adapt to changes in their internal genetic state by evolving ways to minimize the deleterious side effects of other mutations.

Whether there is a cost of resistance in a given case may have important consequences. As with other tradeoffs, the cost of resistance may contribute to the maintenance of genetic diversity. In the simplest case, resistant genotypes or species will dominate in geographical areas where the corresponding chemical agent or disease is present, whereas sensitive forms will dominate elsewhere. In some cases, sensitive and resistant forms may even coexist in the same local area as a result of the balance between the costs and benefits of the mutations that confer resistance. For example, in regions with malaria, there is balancing selection that maintains two alleles because the heterozygote is most fit, with one homozygote suffering from a higher incidence of malaria and the other homozygote suffering severe anemia. In the case of bacteria, which are haploid, both sensitive and resistant genotypes can often coexist in the presence of the virus because the former type is superior at competition for limiting resources, whereas the latter avoids infection.

From a practical standpoint, a cost of resistance can be helpful for managing a pest or pathogen population by the prudent usage of the pesticide or antibiotic. The cost of resistance implies that the sensitive genotype will tend to become dominant when and where the chemical agent is not used. In contrast, indiscriminate use of a pesticide or antibiotic favors the evolution and spread of resistant forms, making it more difficult to control a more serious outbreak or infection. Thus, understanding the evolution of resistance, including its physiological cost and the potential for adaptation to ameliorate the cost, is an important aspect of pest management and public health.

[*See also* Antibiotic Resistance; Disease, *article on* Infectious Disease; Heterozygote Advantage, *article on* Sickle-Cell Anemia and Thalassemia; Pesticide Resistance.]

BIBLIOGRAPHY

Bohannan, B. J. M., and R. E. Lenski. "Linking Genetic Change to Community Evolution: Insights from Studies of Bacteria and Bacteriophage." *Ecology Letters* 3 (2000): 362–377. Review of interactions between bacteria and viruses, focusing on the ecological consequences of resistance and its cost.

McKenzie, J. A., M. J. Whitten, and M. A. Adena. "The Effect of Genetic Background on the Fitness of Diazinon Resistance Genotypes of the Australian Sheep Blowfly, *Lucilia cuprina*." *Heredity* 49 (1982): 1–9. A case study from nature that shows the cost of resistance for an insect to an insecticide and demonstrates subsequent evolution to reduce that cost.

Schrag, S. J., V. Perrot, and B. R. Levin. "Adaptation to the Fitness Costs of Antibiotic Resistance in *Escherichia coli*." *Proceedings of the Royal Society, London, B* 264 (1997): 1287–1291. A laboratory experiment demonstrating the cost of resistance for a bacterium to an antibiotic and amelioration of that cost by subsequent evolution.

Tollrian, R., and C. D. Harvell, eds. *The Ecology and Evolution of Inducible Defenses*. Princeton, 1998. This multiauthor volume examines resistance in many ecological contexts, with an emphasis on how organisms have evolved to balance the costs and benefits of resistance.

— Richard E. Lenski

RETICULATE EVOLUTION

The use of phylogenies to study biological processes in a comparative, evolutionary framework has grown in application. Phylogenies have been used traditionally to classify organisms into species and higher taxonomic groups. However, phylogenies have also been used to document the growth rate and spread of diseases (even to predict future outbreaks), as evidence in court for epidemiological linkage, and to study ecological and physiological changes in a comparative framework. In large part, this increase in the use of phylogenies arises from the increase in gene-sequence data used to estimate these evolutionary relationships, coupled with sophisticated analytical approaches and increased computational power.

Central to phylogeny is the assumption of a bifurcating tree to represent relationships among organisms or their parts. Such a representation becomes problematic at the population genetic level of individualy within a species because relationships are no longer necessarily bifurcating. Genealogical relationships are often said to be reticulating (multiple interconnections between ancestors and descendants) or tokogenetic instead of bifurcating (one ancestor gives rise to two descendants) or phylogenetic (Figure 1) and are more appropriately represented by a network (a tree with cycles) rather than an acyclic tree.

Causes of Reticulate Evolution. Reticulate evolutionary relationships can be classified into two distinct categories. First, there are the apparent reticulate relationships, that is, those sequences that appear more closely related to one another for reasons other than identity by descent. Second, there are true reticulate relationships; that is, some biological force has brought together previously diverging lineages. Both types of reticulate relationships are often ignored in population genetic studies. Instead, bifurcating trees are often applied to population genetic data under the assumption that a true bifurcating tree underlies the evolutionary history of the nucleotide sequences being analyzed. Even if this were true, the true bifurcating tree would not be marked by unique mutational events at each branching point, as is typically true of trees of species. Therefore, the bifurcating relationships may not accurately reflect the underlying genealogy of the gene sequences at the population level.

To avoid this lack of correspondence between unique individuals and unique gene sequences, researchers attempt to estimate haplotype or allele trees, where haplotypes or alleles are defined simply as unique sequences in a sample of sequences from a population. The representation of the relationships among these sequences as a bifurcating hierarchy can fail at the population level owing to the fact that mutational differences are not all unique. Some sites in a sequence may have mutated more than once, causing "multiple hits" on this site (sometimes even back mutations to the original state). Mutations can also occur at the same site and in the same direction but in different lineages. These changes can be either parallel (changes along independent lineages from the same ancestral state to an identical derived state) or convergent (changes along independent lineages from different ancestral states to the identical derived state) if associated with adaptive change. They can also simply occur by chance. Such changes cause homoplasy—characters that are identical in state but not identical by descent. Homoplasies create loops of evolutionary relationships (i.e., apparent reticulations).

Figure 2 offers a true bifurcating genealogy (the "gene tree" that illustrates many of these points). The mutational changes are mapped along the tree. Note that some members of the genealogy have gone extinct (haplotype D). Not all terminals or tips of the tree are identified by unique mutational events (e.g., A_1 and A_2 are indistinguishable owing to the lack of a mutational event). The haplotype tree is shown as well to the right of the gene tree. Note that there are identical changes along different branches (terminals I_1 and I_2 are identical in state but not identical by descent).

In addition to mutations that may fail to identify individuals uniquely, other biological phenomena, which occur most frequently at the population level, can also cause reticulate relationships. The first of these is hybridization. Hybridization results in the coming together of two distinct populations that previously were on their own evolutionary trajectories. This mixing of previously distinct genetic material causes reticulate evolutionary relationships. Likewise, recombination causes reticulate relationships via the same mechanism, that is, the mixing together of genes from previously distinct evolutionary lineages. Thus, aspects of the mutational process itself and other biological phenomena cause reticulate relationships, either true or apparent. Either way, the assumption of a bifurcating tree of hierarchical relationships is inappropriate at the population level. Instead, we look to methods that can depict these reticulating relationships in a network.

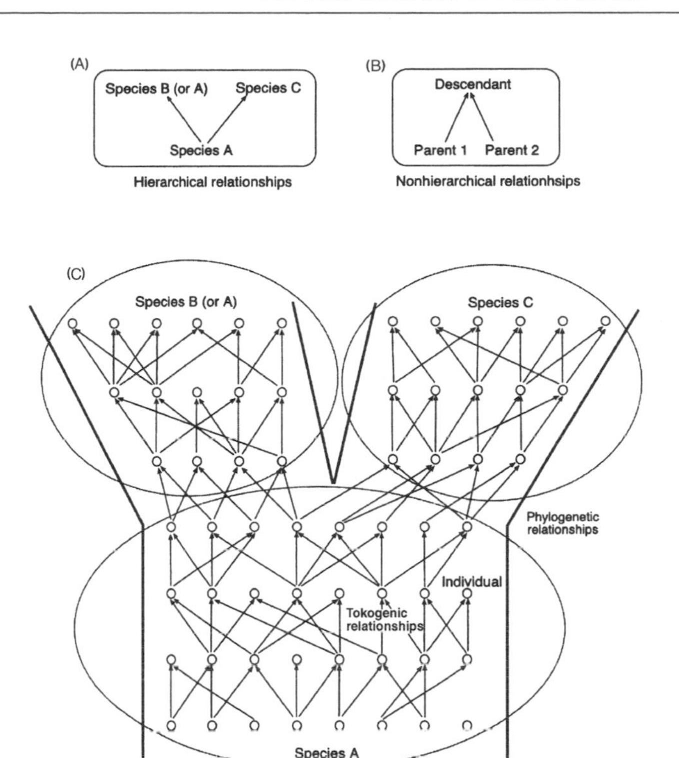

Figure 1. Tokogeny (Reticulate Evolution) Versus Phylogeny (Bifurcating Evolution).
(A) Processes occurring among sexual species (phylogenetic or bifurcating processes) are hierarchical. That is, an ancestral species gives rise to two descendant species. (B) Processes occurring within sexual species (tokogenetic or reticulate processes) are nonhierarchical. That is, two parentals combine their genes to give rise to the offspring. (C) The split of two species defines a phylogenetic relationship among species (thick lines) but, at the same time, relationships among individuals within the ancestral species (species 1) and within the descendant species (species 2 and 3) are tokogenetic (arrows). Drawing by David Posada and Keith A. Crandall, "Intraspecific Gene Genealogies: Trees Grafting into Networks." *Trends in Ecology and Evolution* 16 (2001): 37–45.

Methods for Representing Reticulate Evolution. In addition to reticulate evolution, other population genetic phenomena occur that are ignored by traditional methods of phylogeny reconstruction, such as ancestral taxa remaining in the population, ancestral taxa giving rise to multiple mutational offspring, relatively low levels of divergence, and large sample sizes. Network approaches for representing genealogical relationships account for these population genetic realities, including reticulate relationships. Methods for reconstructing networks of genealogical relationships are generally genetic distance–based methods, including split decomposition, pyramids, statistical geometry, median networks, statistical parsimony, molecular variance parsimony, and netting. There is also a recent approach using maximum likelihood called likelihood network, as well as two least-squares methods called reticulogram and reticulate phylogeny. Descriptions of these software packages and information on obtaining these packages are reviewed by Posada and Crandall (2001). Applying a handful of these methods [as well as traditional bifurcating methods such as UPGMA (unweighted pair group method using arithmetic averages) and maximum parsimony] to the data resulting from the true bifurcating genealogy in Figure 2, we see the different representations of the haplotype relationships from these different networking methods (Figure 3).

Applications of Network Approaches. The visualization of gene genealogies as networks has aided research in a number of areas. For example, the network approach was used by Crandall (1995) to investigate the transmission links between a dentist in Florida infected with the human immunodeficiency virus (HIV) and his HIV-infected patients. In this case, it was argued that the dentist's patients received the virus from other sources, whereas at least one patient claimed to receive the virus

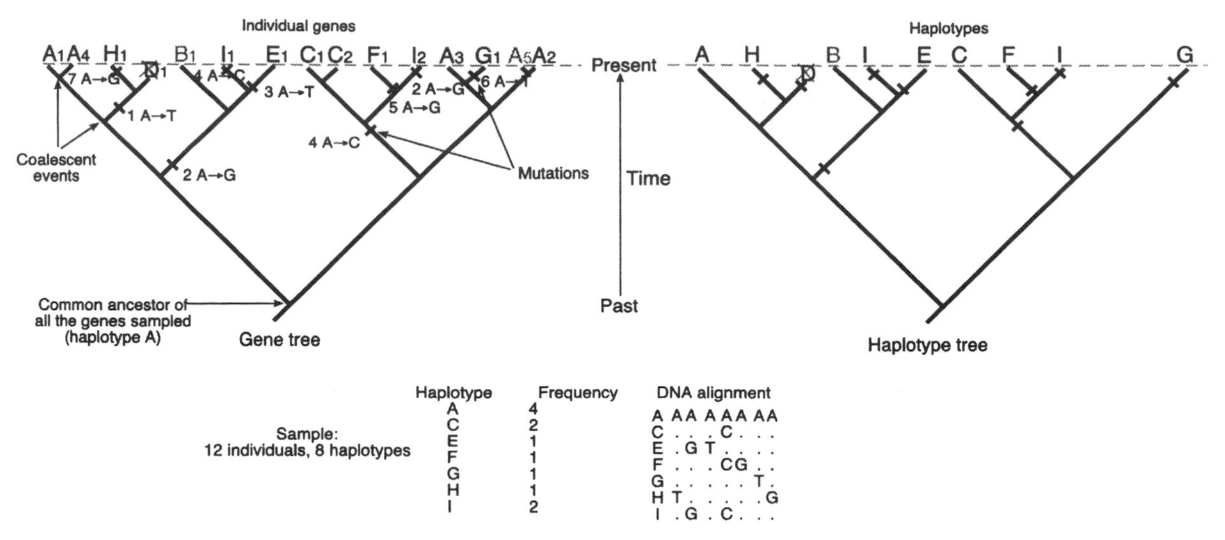

FIGURE 2. A True Bifurcating Gene Genealogy Compared to a Sampled Haplotype Tree.
Note the parallel changes in red creating identical haplotypes. Likewise, there are bifurcation events (C1 and C2) that are not marked by mutation and are therefore indistinguishable in a haplotype tree C). In addition, there are unsampled haplotypes B) and extinct haplotypes D). Drawing by David Posada and Keith A. Crandall, "Intraspecific Gene Genealogies: Trees Grafting into Networks." *Trends in Ecology and Evolution* 16 (2001): 37–45.

from the dentist. Nucleotide sequences were collected from virus present in the dentist, HIV-positive patients, and HIV-positive members of the local community (controls). This genealogical analysis incorporated all the available sequence data and allowed for the biological reality of recombination among sequences. Analyses performed using the bifurcating tree approach could not obtain good resolution with all the data because many of the sequences were very closely related and showed reticulate patterns of relationship. The network approach clearly showed a linkage between some patients and the dentist.

Templeton and colleagues (2000) used a network approach to study the influence of recombination and mutation on the genetic diversity of a candidate gene associated with coronary artery disease. They analyzed a 9.7-kilobase region of the human lipoprotein lipase gene from seventy-one individuals over three culturally distinct populations. Using a networking approach, they identified twenty-nine recombination events, one gene conversion event, and mutational events contributing to the genetic diversity of this gene associated with risk for coronary artery disease. They also demonstrated a concentration of recombinational and mutational events in relatively small portions of the gene, thus challenging a basic assumption in population genetic analysis of infinite sites (where every new mutation is assumed to occur at a new site).

Finally, the classic application of network approaches is in the area of phylogeography—examining phylogenetic relationships at the population level relative to geographic location to infer historical events. Even with the traditional mitochondrial (mtDNA) markers used, reticulate evolution is commonplace. This coupled with lower levels of divergence, larger sample sizes, and ancestral haplotypes remaining in the population lead to the use of network approaches. With the ever increasing use of nuclear markers for population genetic inference, it is even more important to incorporate network approaches as the potential for reticulate evolution arising from recombination is real. A recent example of the powerful inference from the network approach comes from a study by Vilá and colleagues (1999), in which they examined phylogeographic relationships among coyotes and wolves. Using the network approach, they inferred historical isolation of the coyote populations and historically panmictic populations of wolves. However, the recent population structure for these two species was shown to be inversed, with the coyotes showing recent panmixia and the wolves showing fragmented populations (both presumably resulting from the influence of humans). One of the many strengths of the network approach is the ability to partition out these historical from current population structures.

The network approaches allow researchers to visualize the true reticulate nature of gene genealogies below the species level (or even at the species level with hybridizing species). As nucleotide sequence data sets become increasingly large and commonplace in population genetic studies, these methods for visualizing re-

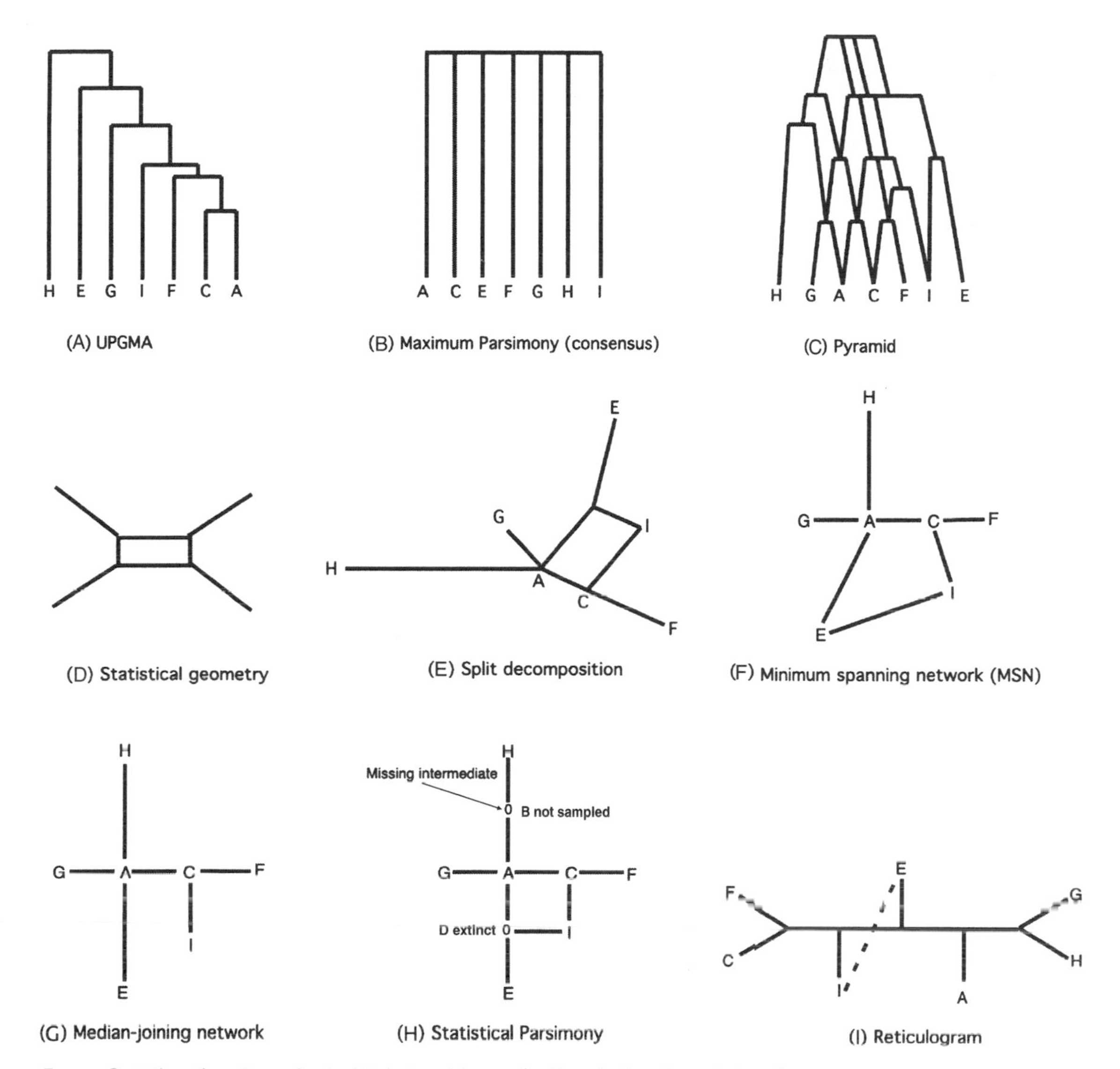

FIGURE 3. Estimating Genealogical Relationships at the Population Genetic Level.
Different genealogy reconstruction techniques, including network techniques, were applied to the data associated with the true bifurcating gene genealogy shown in figure 2. (A) UPGMA, (B) maximum parsimony, (C) pyramid, (D) statistical geometry, (E) split decomposition, (F) minimum spanning network, (G) median-joining network, (H) statistical parsimony, and (I) reticulogram. Drawing by David Posada and Keith A. Crandall, "Intraspecific Gene Genealogies: Trees Grafting into Networks." *Trends in Ecology and Evolution* 16 (2001): 37–45.

ticulate evolution will become increasingly important as tools to aid in the genealogical analysis of these data. The bifurcating paradigm of systematics, though useful, does not accommodate many of the unique features of population genetic data, especially reticulate evolutionary relationships. The network approaches allow an explicit incorporation of these population genetic realities and therefore give a more accurate picture of the true underlying relationships at the population genetic level.

[*See also* Molecular Evolution; Molecular Systematics; Phylogenetic Inference, *article on* Methods.]

BIBLIOGRAPHY

Avise, J. C. *Phylogeography.* Cambridge, Mass., 2000. Text on the application of phylogenetic techniques combined with geographic data, including applications of network approaches.

Bandelt, H.-J., P. Forster, B. C. Sykes, and M. B. Richards. "Mitochondrial Portraits of Human Populations Using Median Networks." *Genetics* 141 (1995): 743–753. The median network approach applied to human mtDNA.

Crandall, K. A. "Intraspecific Phylogenetics: Statistical Support for Dental HIV Transmission." *Journal of Virology* 69 (1995): 2351–2356. An example of the use of statistical parsimony network approach in an HIV transmission case.

McDade, L. A. "Hybrids and Phylogenetic Systematics: II. The Impact of Hybrids on Cladistic Analysis." *Evolution* 46 (1992): 1329–1346. An examination of hybridization resulting in reticulate relationships.

Posada, D., and K. A. Crandall. "Intraspecific Gene Genealogies: Trees Grafting into Networks." *Trends in Ecology and Evolution* 16 (2001): 37–45. A through review of network approaches for reconstructing gene genealogies, complete with references to software.

Templeton, A. R., A. G. Clark, K. M. Weiss, D. A. Nickerson, E. Boerwinkle, and C. F. Sing. "Recombinational and Mutational Hotspots within the Human Lipoprotein Lipase Gene." *American Journal of Human Genetics* 66 (2000): 69–83. An application of the study of reticulate evolution caused by recombination and its implication for risk factors associated with coronary artery disease.

Vilá, C., I. R. Amorim, J. A. Leonard, D. Posada, J. Castroviejo, F. Petrucci-Fonseca, K. A. Crandall, H. Ellegren, and R. K. Wayne. "Mitochondrial DNA Phylogeography and Population History of the Grey Wolf *Canis lupus*." *Molecular Ecology* 8 (1999): 2089–2103. An application of networking approach to examine phylogeography of the gray wolf, partitioning historical population structure from current structure.

— Keith A. Crandall

RNA. *See* DNA and RNA.

S

SEGMENTATION

Many multicellular organisms show a repetition of body structures. These repetitions can involve the whole body, such as the segments in annelids and arthropods, but they also can be seen in parts of the body, such as the antennae of insects (see Figures 1 and 2). The correct term for the subdivisions of the whole body is *metamerism*, but *segmentation* is often used synonymously.

The principle of taking an established structure, repeating it, and modifying it for different tasks seems like an excellent design principle for creating evolutionary novelties. If this principle has indeed governed the course of evolution, one should conclude that animals with unmodified segmental repetitions should represent more ancestral forms, whereas animals with increasingly differentiated segments should be more advanced. Furthermore, because this is such a useful principle, it would seem likely that it has evolved independently several times. Indeed, the current view is that segmentation in animals has arisen at least twice: once in the lineage of annelids, which themselves are considered to be the progenitors of arthropods, and independently in the lineage leading to chordates.

Segmentation in annelids is primarily mesodermal, the ectoderm is secondarily segmented along the mesodermal borders. In arthropods, it is the other way around: the ectoderm is segmented first, and the mesoderm follows. Chordates have only mesodermal segments, which are generally called somites. An important task of somites is to form the coelom, a body cavity that is the hallmark of all advanced animals. However, the process of coelom formation differs among the major groups of animals.

Segments and somites are usually produced in a sequential budding process. They are generated in a growth zone at the posterior end of the embryo, which buds off new segments or somites in regular intervals. The number of segments produced is strictly fixed in some animals (mainly in several arthropod groups), but it can be individually variable in others (annelids and chordates). The generation of the segments in the head region appears to be different from that in the trunk region, although the basic structure of including mesodermal subdivisions that form a coelomic cavity is the same.

The molecular basis of segmentation has been best studied in *Drosophila*, which shows, however, a highly derived body plan. Here, all segments are formed simultaneously at an early stage of embryogenesis. Still, many of the genes found to be involved in segmentation in *Drosophila* appear also to be active in insects that show the ancestral mode of sequential generation of segments (Tautz and Sommer, 1995). The most intriguing group in this process are pair-rule genes. These are involved in generating a transient double segmental periodicity, from which the segmental periodicity arises. From a constructional point of view, this seems to be a nonintuitive way of generating segments, namely, to use a molecular double segmental input in a process that buds off the morphological structures one by one. In vertebrates, this process is indeed radically different. Here, it is an oscillator at the most posterior end of the growth zone that generates cycling waves of gene activity, which eventually lead to the formation of new somitic borders. Intriguingly, however, key genes in this process are homologues of the pair-rule gene *hairy* in *Drosophila* (Pourquie, 2000), suggesting a possible link.

Molecular phylogenies have started to question the long established view of a close relationship between annelids and arthropods (Adoutte et al., 1999). Instead, annelids are found to be more closely related to nonsegmented molluscs. This could suggest that the segmentation process in annelids is different from both that of arthropods and of chordates. Alternatively, there might have been a segmented ancestor at the base of the bilateria, with segmentation independently lost in different groups (Davis and Patel, 1999).

Important answers for understanding the evolution of segmentation will come from the analysis of the molecular mechanisms of segmentation in annelids, as well as

Figure 1. Male Mosquito Antennae.

FIGURE 2. Adult Luna Moth Showing Male's Large Antennae. © Dwight R. Kuhn.

from finding out whether there is a relationship between the cycling way of generating somites in vertebrates and the pair-rule mechanism in insects.

As mentioned previously, there are further instances of segmentation within body regions, such as the appendages in arthropods, the generation of fin rays in fish, and the limbs and digits in higher vertebrates. Repetitive reuse of structures is also a design principle in most plants. There are even patterns of segmentation in cnidarians, namely, when new medusae are formed in a process that is called strobilation. It remains a challenge to compare the molecular basis for these design principles, which will eventually provide deep insights into the contingencies of evolutionary pathways.

[*See also* Metazoans; Phylotypic Stages.]

BIBLIOGRAPHY

Adoutte, A., G. Balavoine, N. Lartillot, and R. de Rosa. "Animal Evolution: The End of the Intermediate Taxa?" *Trends in Genetics* 15 (1999): 104–108. Review on the new data from molecular phylogeny, suggesting that a complete revision of the phylogenetic relationships among basal metazoan lineages is necessary.

Davis, G. K., and N. H. Patel. "The Origin and Evolution of Segmentation." *Trends in Cell Biology* 12 (1999): M68–M72. Topical review on the molecular knowledge about segmentation mechanisms in various taxa, including a discussion of different evolutionary hypotheses.

Pourquie, O. "Vertebrate Segmentation: Is Cycling the Rule?" *Current Opinion in Cell Biology* 12 (2000): 747–751. Review on the data for cycling mechanisms of somitogenesis in different veretbrates.

Tautz, D., and R. J. Sommer. "Evolution of Segmentation Genes in Insects." *Trends in Genetics* 11 (1995): 23–27. Discusses possible similarities and differences in segmentation genes in insects, with special focus on pair-rule genes.

— DIETHARD TAUTZ

SELECTION. *See* Artificial Selection; Density-Dependent Selection; Developmental Selection; Frequency-Dependent Selection; Fundamental Theorem of Natural Selection; Group Selection; Kin Selection; Levels of Selection; Natural Selection; Sexual Selection.

SELF-DECEPTION

Findings from psychology support common experience in telling us that self-deception—active misrepresentation of reality to the conscious mind—is an everyday human occurrence. At the level of both individuals and societies, this tendency can help produce major disasters (e.g., in aviation and misguided wars—see Trivers and Newton, 1982; Wrangham, 1999; Feynman, 1988). Since we know that selection *can* fine-tune organisms to assess their circumstances accurately, and since the potential costs of misrepresentation can be so great, the question naturally arises: What evolutionary forces *favor* self-deception?

Benefits for deceiving others may be the key. Thirty years ago, Trivers noted that self-deception can make hiding the truth from others easier and more effective. In our own species, we recognize that shifty eyes, sweaty palms, and croaky voices may indicate the stress that accompanies conscious knowledge of attempted deception. By becoming unconscious of the deception, the deceiver hides these signs from the observer. He or she can lie without the nervousness that accompanies conscious deception.

Self-deception also allows us to be unconscious of ongoing motivation. We may experience a conscious stream of thoughts that act, in part, as rationalizations for what we are doing so that we may more effectively conceal the true motivation from others. When actions are challenged, a convincing alternative explanation is then at once available, complete with an internal scenario: "But I wasn't thinking that at all, I was thinking . . . ."

One major arena of self-deception is self-promotion—exaggeration of our status or abilities on the positive side, denial on the negative. If you ask high school seniors in the United States to rank themselves on leadership ability, fully 80 percent say they have better than average abilities. Professors are even more adept at self-deception—an almost unanimous 94 percent rate themselves as being in the top half of the profession! Some tricks of the trade are biased memory, biased computation, and changing from active to passive voice when changing from describing positive to negative outcomes.

People often deceive themselves about their role in social relationships. Husband and wife, for example, may agree that one party is a long-suffering altruist while the other is hopelessly selfish, but they may disagree over which is which. These may be thought of as biased social theories—theories that, through biased use of facts and logic, give ourselves a preferred position. We have social theories regarding all of our relationships, including employer–employee and the structure of our larger society (for example, is it fair regarding people such as ourselves?).

There is probably an intrinsic benefit in keeping a more optimistic view of the future than facts would seem to justify. It has been known for some time that depressed individuals tend not to go in for the routine kinds of self-inflation that we described earlier. This is sometimes interpreted to mean that we would all be depressed if we viewed reality accurately, but some researchers have suggested that the depressed state may be a time of personal reevaluation, where self-inflation would serve no useful purpose. Life is intrinsically future-oriented, and mental operations that keep a positive future orientation at the forefront appear to result in better future outcomes. The existence of the placebo effect may be another example of this principle.

Because it must be advantageous for the truth to be registered somewhere, mechanisms of self-deception presumably reside side by side with mechanisms for the correct apprehension of reality. This was demonstrated in an elegant experiment by Gur and Sackeim (1979). Their experiment was based on the fact that humans respond physiologically upon hearing a human voice, as measured, for example, by a jump in the galvanic skin response (GSR), but the person responds more intensely upon hearing his or her own voice than the voice of another. A measure of GSR in response to various recorded voices, therefore, can be used as a measure of unconscious self-recognition.

People tested for self-recognition in this way fell into four categories: some made no mistakes; some denied their own voice some of the time; some projected their own voice some of the time; and some people made both kinds of errors. When the skin response was tallied for all these verbal errors, a striking pattern emerged: in almost all cases, the skin knew better. People who verbally claimed that the voice was not their own (deniers) showed the high GSR typical of self-recognition. Those who claimed that another's voice was their own (projectors) showed the smaller jump in GSR typical of hearing another person's voice. Deniers showed the greatest change in GSR of the four groups; presumably, denial of significant features of reality requires hyperarousal, whereas projection of reality is a more relaxed enterprise. Gur and Sackeim also showed that denial and projection were motivated in a logical fashion: people who were made to feel bad about themselves started denying their own voices, while those made to feel better about themselves started projecting their voices.

If the verbal reports of Gur and Sackeim's subjects reflect their conscious beliefs rather than an intent to deceive the experimenter (an assumption that seems likely but cannot be proven), this experiment suggests the following three attributes of self-deception. (1) True and false information is simultaneously stored in a single person. (2) The false information is stored in the conscious mind while the true information is unconscious. (3) Self-deception is often motivated with reference to others.

Although the neurological bases of self-deception are not well understood, we do know enough to question the common impression that information reaching our brain is immediately registered in consciousness, and that signals to initiate activity always originate in the conscious mind. While a nervous signal reaches the brain in only 20 milliseconds, it takes a full 500 milliseconds for the signal to register in consciousness. This is all the time in the world, so to speak, for emendations, changes, deletions, and enhancements to occur. Indeed, neurophysiologists have shown that stimuli can affect the content of an experience at least as late as 100 milliseconds *before* the occurrence reaches consciousness.

Neurobiologists have recently produced striking evidence that at least some processes of denial and rationalization reside in the left hemisphere of the brain. People who have suffered a stroke in the left hemisphere (with paralysis on the right side of the body) recognize the seriousness of their condition, but a fraction of those with a right-hemisphere stroke (left-side paralysis) vehemently deny that there is anything wrong with them, a condition known as anosognosia. When confronted with strong counter-evidence, these patients indulge in a remarkable array of rationalizations to deny the cause of their inability to move (arthritis, general lethargy, etc.).

Ramachandran (1998) has noted that there is a strong similarity between the strategies of these patients and what the psychoanalyst Sigmund Freud referred to as "psychological defense mechanisms" in normal people (e.g., rationalizations, denials, and repression of unpleasant memories). The anosognosics, however, display these mechanisms in exaggerated form. For example, one anosognosic woman with a paralyzed arm had been asked to tie shoelaces. When asked later whether she had done so, she replied, "Oh, yes, I tied it successfully with both my hands." The odd wording of this response (that she tied it with "both my hands") suggested to Ramachandran that the lady "doth protest too much," and may have had an unconscious recognition of the truth of her condition despite the vehemence of her denial. Further experimentation allowed him to show that such patients do indeed have an unconscious awareness of their paralysis.

What could be the benefit of this kind of self-deception? Ramachandran suggests that it allows people to create a coherent belief system that makes stable, consistent behavior possible in the face of a confusing flood of incoming sensations. In order to act consistently (or at all), the brain must put this information together into a story that makes sense. As Ramachandran explains, "When something doesn't quite fit the script . . . you very rarely tear up the entire story and start from scratch. What you do, instead, is to deny or confabulate in order to make the information fit the big picture. Far from being maladaptive, such everyday defense mechanisms keep the brain from being hounded into directionless indecision by the 'combinational explosion' of possible stories that might be written from the material available to the senses." The right hemisphere presumably acts as a kind of "reality check" on this tendency, and when it is disabled by injury, the exaggerated denial of anosognosia results. This explanation is probably not the entire story, however, because it does not explain why the deceptions of the left hemisphere are consistently biased in the direction of self-promotion (individuals with anosognosia insist that they are able to do things they cannot). Perhaps the same things that lead the rest of us to rate ourselves above average operate here.

Ramachandran's arguments are consistent with Gazzaniga's (1992) observations on "split-brain" patients (people whose severe epilepsy has been controlled by surgically severing the connections between the two cerebral hemispheres). In these patients, the left brain is unaware of what the right brain knows, and vice versa. Show a salacious picture to the right hemisphere only, and the patient will respond appropriately (for example, with embarrassed laughter), but the verbal left hemisphere won't know why. That won't stop it from inventing a reason, however. The reason given by the patient will be a fabrication consistent with what the left hemisphere knows; hence, Gazzaniga calls the left hemisphere the brain's "interpreter."

Self-deception, therefore, may have several adaptive functions. Trivers has suggested that self-deception was favored by selection in part to make deception of others more effective. Ramachandran has argued that self-deception enables us to act consistently in spite of a confusing and sometimes conflicting barrage of sensory inputs. In both cases, the truth is perceived somewhere in the brain, but the most adaptive behavior requires the conscious mind to alter or suppress the evidence.

BIBLIOGRAPHY

Feynman, R. *What Do You Care What Other People Think? Further Adventures of a Curious Character*. New York, 1988. The last chapter is a brilliant analysis of institutional self-deception as a cause of the *Challenger* disaster.

Gazzaniga, M. S. *Nature's Mind: The Biological Roots of Thinking, Emotions, Sexuality, Language, and Intelligence*. New York, 1992. Includes discussion of the author's work with split-brain patients.

Greenwald, A. G. "The Totalitarian Ego: Fabrication and Revision of Personal History." *American Psychologist* 35 (1980): 603–618. Excellent review of the human tendency toward self-inflation.

Gur, R., and H. A. Sackeim. "Self-deception: A Concept in Search of a Phenomenon." *Journal of Personality and Social Psychology* 37 (1979): 147–169. The classic paper demonstrating self-deception experimentally.

Libet, B. "Neuronal Time Factors in Conscious and Unconscious Mental Functions." In *Toward a Science of Consciousness: The First Tucson Discussion and Debates*, edited by S. R. Hameroff, A. W. Kaszniak, and A. Scott, pp. 337–346. Cambridge, Mass., 1996. A review of experiments revealing neuronal times in consciousness.

Lockard, J. S., and D. L. Paulhus, eds. *Self-deception: An Adaptive Mechanism?* Englewood Cliffs, N.J., 1988. A collection of articles with both theoretical speculation and empirical findings.

Ramachandran, V. S., and S. Blakeslee. *Phantoms in the Brain: Probing the Mysteries of the Human Mind*. New York, 1998. Chapter 7 of this engaging book describes anosognosia and the self-deception of the left hemisphere.

Trivers, R. *Social Evolution*. Menlo Park, Calif., 1985. See chapter 16, "deceit and self-deception."

Trivers, R. "The Elements of a Scientific Theory of Self-Deception." *Annals of the New York Academy of Sciences* 907 (2000): 114–131. Detailed review of the evolutionary approach to self-deception.

Trivers, R., and H. Newton. "The Crash of Flight 90: Doomed by Self-Deception?" *Science Digest* (November 1982): 66, 67 and 111. Detailed analysis of the possible effect of self-deception in the cockpit on an airplane crash.

Wrangham, R. "Is Military Incompetence Adaptive?" *Evolution and Human Behavior* 20 (1999): 3–17. Processes of self-deception as they have evolved in the context of warfare.

— Elizabeth Cashdan and Robert L. Trivers

SELFISH GENE

A key feature of Mendelian inheritance is that it is usually fair: organisms pass on the two copies of each gene with equal frequency to the next generation. This means that the process of inheritance per se has no directional effect on allele frequencies, and it is only natural selection—the differential survival and reproduction of individuals with different genotypes—that can have such an effect. Thus, most genes persist in populations because they increase the fitness of the organisms in which they reside. However, not all inheritance is fair, and in particular some genes contrive to be inherited at a greater-than-Mendelian rate. This allows the genes to increase in frequency and persist in populations even if they are harmful to the host organism. These are known as selfish genes, or selfish genetic elements. Because they are harmful to the host organism, selection can occur at other loci to suppress their action—that is, selfish genes can engender intragenomic conflict.

As an example, consider homing endonuclease genes. These are optional genes, present in some members of a species but not others, and they have no known host function. Rather, the only thing they do is encode an enzyme that specifically recognizes a particular twenty-to-thirty-base-pair sequence of DNA and cuts it. As it happens, this sequence exists only on chromosomes that do not contain a copy of the endonuclease gene; for chromosomes carrying the gene, the gene interrupts the recognition sequence and so protects those chromosomes from being cut. Therefore, in heterozygous individuals, in which one copy of the chromosome carries the gene and one copy does not, the chromosome without the gene is cut by the enzyme. This turns on the cell's broken-chromosome repair system, which uses the homologous chromosome (which contains the gene) as a template for repair. The consequence is that after repair, both copies of the chromosome contain a copy of the gene, and the heterozygote has been converted into a homozygote. All meiotic progeny will carry a copy of the gene, rather than the Mendelian 50 percent. Such a gene is of no use to the host, and probably even imposes some cost; nevertheless, it can spread rapidly through a population because of the biased inheritance. Homing endonuclease genes are particularly common in fungi and algae.

Homing endonucleases follow a strategy of "overreplication," in which the selfish gene manages to get replicated more frequently than the rest of the genome, which replicates only once per cell cycle. Transposable elements, the most widespread class of selfish genes, also adopt this strategy. Another approach is used by B chromosomes. These are optional, supernumerary chromosomes not obviously homologous to the normal set of A chromosomes. In most well-studied cases, B chromosomes lower the fitness of their host, and persist in populations only because of a super-Mendelian rate of inheritance, in which they are inherited by more than 50 percent of the gametes. This increased rate of transmission usually results from the B chromosome preferentially moving during some cell division toward the germ line and away from the soma. Most frequently, this "gonotaxis" is observed during female meiosis, in which only one of the four meiotic products is viable (the other three being sloughed off as degenerate polar bodies). B chromosomes in many species manage to position themselves at a location on the meiotic spindle such as to increase their probability of being incorporated into the egg and avoiding the polar bodies. B chromosomes in other species (particularly grasses) show gonotaxis in pollen grain mitosis, with the B's preferentially going to the generative nucleus (which gives rise to the sperm nuclei) and away from the tube nucleus (which is responsible for the growth of the pollen tube toward the ovule but is then lost). B chromosomes are thought to occur in 10–15 percent of plant and animal species.

The third main strategy is the most diverse. When diploid organisms undergo meiosis, the two alleles at a locus can be seen as competing for representation in the haploid products, and another way for a gene to increase its rate of inheritance is to sabotage or cripple the competing allele. For example, the well-studied t-haplotype of mice refers to a cytologically detectable region on the seventeenth chromosome with many inversions relative to the normal chromosome. Somehow (the exact mechanism is not yet known) this chromosomal region manages to incapacitate the sperm carrying normal or wild-type seventeenth chromosomes. That is, in males heterozygous for the t-haplotype, the sperm carrying the normal chromosome shows flagellar dysfunction and premature acrosome reactions. Consequently, some 90 percent of that male's progeny carry the t-haplotype, rather than the Mendelian 50 percent. This is not some failure of the wild-type chromosomes, because in males that are homozygous for wild type, all the sperm are fine—rather, the t-haplotype is somehow sabotaging the normal sperm. Were this the only effect of the t-haplotype, it would rapidly go to fixation in mice populations. However, in the homozygous state it also causes male sterility and some lethality, and so the t-haplotype remains instead at some intermediate equilibrium frequency in mouse populations of about 5–10 percent. Comparative DNA sequence analysis suggests the t-haplotype has persisted in mouse populations for about 3 million years.

The X* chromosome of some lemming species takes a more roundabout route to killing its competition. X* is a third type of sex chromosome, in addition to the usual X and Y. It is like a normal X except that it is a dominant feminizer, meaning that X*Y individuals are female instead of male. When such females mate with a normal XY male, one quarter of the zygotes formed will be YY and will die. Therefore, in the surviving progeny, fully two-thirds of them will carry the X* chromosome, instead of the Mendelian 50 percent. The X* chromosome can therefore increase in frequency in the population. As it does so, it will bias the sex ratio toward females. Because of the inherent frequency-dependent selection on sex ratio, the X* chromosome will not go to fixation (which is just as well for the population, as it would then go extinct from lack of males), but rather come to some intermediate equilibrium frequency. Because of the biased sex ratio, other host genes will be selected to suppress the feminizing action of X*, or to bias the sex ratio in the opposite direction, toward males.

In plants, cytoplasmic genes such as those found in mitochondria and chloroplasts can also be selected to feminize individuals, by disrupting pollen production. Such an action is not harmful to the fitness of a cytoplasmic gene because it does not get transmitted to the next generation through pollen anyway, being instead purely maternally inherited. Indeed, abolishing pollen production can be positively selected if it in any way augments ovule production. However, nuclear genes get half their transmission to the next generation through pollen, and so are selected to suppress any mitochondrial male sterility gene. As a consequence, a mitochondrial male sterility gene can arise and be selected, followed by a nuclear restorer of male fertility. Both genes may sweep through the population to fixation, with the result that the population returns to being fully hermaphroditic, with little external indication of what has happened. However, if the population is crossed with another population, then the male sterility gene and the restorer can be separated, and the male sterility phenotype observed. In fact, cytoplasmically inherited male sterility is a common feature of plant hybrids.

Selfish genes are also thought to underlie the evolution of genomic imprinting, and perhaps also mutation and recombination. Conceptually, their main importance is in illustrating how not all the genes in an organism exist for the good of that organism, and not all the adaptations it exhibits are for its own good. Only by understanding the logic of selfish genes can one understand why there might be genes for cutting chromosomes at particular sites, or for incapacitating half of one's sperm, for feminizing Y-bearing individuals, or for abolishing pollen production.

Selfish genes have long been known, with key papers by Gershenson in 1928 on X chromosome drive in *Drosophila;* Lewis in 1941 on mitochondrial male sterility genes in plants; and Östergren in 1945 on the selfish nature of B chromosomes. Most of what we know about them, including most of the diversity of types of selfish genes, has only been discovered since the 1980s, during which time the phrase *selfish gene* was adopted. The phrase has an alternative (and prior) usage in the nontechnical literature, deriving from *The Selfish Gene*, an influential book by Richard Dawkins celebrating the gene-centric view of evolution that slowly emerged after the rediscovery of Mendelian inheritance. The phrase made for a catchy title of a justifiably influential book, and can still be used as shorthand to denote this gene-centric view of evolution. However, it is less useful as a technical term. If all genes are selfish, then the adjective is redundant. And the phrase does not quite seem accurate: after all, the vast majority of genes in an organism cooperate to make it a coherent whole. It is a "win-win" interaction, not the "win-lose" interaction implied. Hence the more restrictive definition of *selfish gene* is used in the technical literature.

[*See also* Cytoplasmic Genes; Genomic Imprinting; Levels of Selection; Transposable Elements.]

BIBLIOGRAPHY

Burt, A., and R. Trivers. "Selfish DNA and Breeding System in Flowering Plants." *Proceedings of the Royal Society of London. Series B: Biological Sciences* 265 (1998): 141–146. Develops the idea that outcrossed species will be more susceptible to some selfish genes than inbred species, focussing in particular on B-chromosomes.

Gershenson, S. "A New Sex-Ratio Abnormality in *Drosophila obscura.*" *Genetics* 13 (1928): 488–507. With Lewis and Östergren, one of three classic papers on selfish genetic elements.

Goddard, M. R., and A. Burt. "Recurrent Invasion and Extinction of a Selfish Gene." *Proceedings of the National Academy of Sciences of the United States of America* 96 (1999): 13880–13885. A study on the evolutionary dynamics of homing endonuclease genes, suggesting that they frequently go extinct within host species, and then reinvade by horizontal transmission from other species.

Hurst, L. D., A. Atlan, and B. O. Bengtsson. "Genetic Conflicts." *Quarterly Review of Biology* 71 (1996): 317–364. A thorough review of selfish genes, the genetic conflicts they engender, and their role in evolution.

Lewis, D. "Male-Sterility in Natural Populations of Hermaphrodite Plants." *New Phytologist* 40 (1941): 56–63. With Gershenson and Östergren, one of three classic papers on genetic elements.

Östergren, G. "Parasitic Nature of Extra Fragment Chromosomes." *Bot. Not.* 2 (1945): 157–163. With Gershenson and Lewis, one of three classic papers on selfish genetic elements.

— Austin Burt

SELF-KNOWLEDGE. *See* Emotions and Self-Knowledge.

SENESCENCE

Senescence is defined as a progressive increase in the probability of dying as an organism ages. Senescence has deleterious effects on the survival and reproductive capacity of organisms, which at first sight makes its evolution hard to explain. The extent of senescent changes varies among species. Some animals, such as *Hydra*, do not show increases in age-specific mortality and are regarded as immortal in the sense that, barring accidental causes of death, there is no apparent limit to individual survival. Most animals, however, exhibit clear age-related increases in intrinsic mortality (that is, mortality not caused by environmental factors such as predation) when reared under protected conditions that allow them to live long enough to grow old. Sometimes the term is applied to describe changes to fertility with age—menopause in human females being a case of "reproductive senescence." In plant biology, the term *senescence* has a variety of meanings, including the annual cycle of leaf fall. The general principles of life history theory apply to plants as well as to animals, but this article will focus principally on understanding the evolution of senescence within animal life histories.

Evolutionary explanations for senescence are either adaptive, in which senescence is suggested to confer some advantage, or nonadaptive, in which senescence is seen as arising either through the failure of natural selection to prevent it or as the byproduct of selection for some other, beneficial trait. Adaptive explanations for evolution of senescence—such as that it provides a mechanism to prevent overcrowding or to accelerate the turnover of generations—mostly rely on group selection (here, individuals sacrificing their own longevity for the good of the group) and are correspondingly hard to sustain. Not only is it difficult to see how the hypothetical advantage to the group can outweigh the obvious fitness costs of senescence for individuals; there also is scant evidence that senescence makes any important contribution to mortality in wild populations. Most wild animals die from hazards such as infection, predation, starvation, or cold long before any senescent effects might occur. There is therefore little opportunity for selection to act directly to produce an intrinsic process of senescence.

It is the fact that natural selection has little opportunity to act strongly on older organisms that lies at the heart of the nonadaptive theories of senescence, which now have the greatest empirical as well as theoretical support. The challenge for the nonadaptive theories is to explain how senescence might evolve in a population that does not already undergo intrinsic aging. Because survivorship (the probability of living to a given age) necessarily declines as a result of extrinsic (environmental) mortality, there must come an age at which the residual force of natural selection on later stages of the life history is negligible. In other words, there comes an age when there begins what we can call a "selection shadow." Within the selection shadow, the statistical chance of remaining alive and contributing further offspring to succeeding generations is so small that selection is effectively powerless because there are so few members of the population left on which it can act. If mutations were to arise that had late-acting deleterious effects affecting viability of the organism only at ages within the selection shadow, such mutations might accumulate within the genome, unchecked by natural selection. This is the basis of the "mutation accumulation" theory of aging, originally proposed in 1952 by Peter Medawar.

The mutation accumulation theory offers an essentially neutral explanation for the evolution of senescence. An alternative suggestion is that senescence has come about as the byproduct of positive selection for other beneficial traits—that is, through tradeoffs. If there are genes that have beneficial effects on the viability and fertility of organisms during the important early years, when the force of selection is relatively intense, these will be favored by natural selection even if they also have deleterious effects at ages within the selection shadow. Such genes are said to have *pleiotropic* effects on the life history and form the basis of the *pleiotropy* (sometimes *antagonistic pleiotropy*) theory of senescence. For example, genes that confer a commitment to rapid growth early in life could conceivably have less beneficial or even harmful effects later in life.

Evolutionary theories answer the ultimate ("Why?") questions about the place of senescence within an organism's life history; however, it is also important to consider proximate ("How?") questions about mechanisms. A direct connection between the ultimate and proximate explanations for the evolution of senescence is provided by the *disposable soma* theory, which is based on optimal allocation of resources between somatic maintenance and reproduction. Maintenance of the soma (those parts of the organism not directly contributing genetic material to the next generation) only needs to be good enough to keep the organism in sound condition for as long as it has a reasonable chance of survival in the wild. Somatic maintenance systems, such as DNA repair or the antioxidant defense, consume metabolic resources. The disposable soma theory therefore predicts that senescence evolved because under pressure of natural selection there did not exist any advantage in investing in better somatic maintenance than was actually required. Instead of investing extra resources in better somatic maintenance, the optimum will be to invest in faster growth or more prolific reproduction. The disposable soma theory specifically predicts that senescence is caused by the gradual accumulation of unrepaired somatic damage, and that longevity (rate of aging) is genetically regulated by the levels of somatic maintenance systems, such as DNA repair.

Tests of the evolutionary theories have thus far concentrated on looking for (1) an age-related increase in the variance of fitness components (survival and fecundity) as predicted by the mutation-accumulation theory, (2) tradeoffs between early- and late-life fitness components as predicted by the pleiotropy theory, (3) tradeoffs between fertility and longevity as predicted by the pleiotropy and disposable soma theories, and (4) correlations between somatic maintenance and species longevity as predicted by the disposable soma theory. A range of experiments using the fruitfly *Drosophila melanogaster* has confirmed the existence of tradeoffs between fertility and longevity in this species; recent analysis of records for British aristocrats extending across twelve centuries suggests that a similar tradeoff exists in humans. Positive correlations have been found between species longevity and cellular DNA repair capacity and resistance to oxidative stress. For example, the activity of the enzyme poly(ADP-ribose) polymerase-1, which mediates the immediate response to DNA damage (e.g., caused by irradiation), has been shown to correlate strongly with species lifespan, as has the cellular capacity to withstand a broad range of chemical stressors (e.g., hydrogen peroxide). Long-lived fruitfly populations produced by artifical selection generally show increased resistance to stress, as do mutant nematode worms carrying mutations that increase the lifespan. In many instances, it appears that genes affecting rate of aging map onto pathways involved in energy regulation (e.g., insulin-receptor signaling).

The findings from a wide range of model systems confirm the predicted links between somatic maintenance and longevity and point to the central importance of energy allocation within the organism's life history. Similar conclusions have also been derived from evolutionary analyses of cases of life-history plasticity, such as the life-extending effects of rodent calorie-restriction and the availability of switching into a long-lived, stress-resistant larval form (the dauer) in the nematode *Caenorhabditis elegans*. The evolutionary theories of senescence are not mutually exclusive, and it is likely that a combination of genetic factors may be involved, some broadly shared across individuals, populations, and species, and others more particular.

[*See also* Life History Theory, *article on* Human Life Histories.]

BIBLIOGRAPHY

Finch, C. E. *Longevity, Senescence and the Genome.* Chicago, 1990. Exceptionally comprehensive review of the general biology of senescence.

Holliday, R. *Understanding Ageing.* Cambridge, 1995. Clear overview of ideas about evolution and mechanisms of senescence.

Kirkwood, T. B. L., and S. N. Austad. "Why Do We Age?" *Nature* 408 (2000): 233–238. General review of the evolutionary theories of senescence and of the tests that have been made.

Kirkwood, T. B. L., and T. Cremer. "Cytogerontology since 1881: A Reappraisal of August Weismann and a Review of Modern Progress." *Human Genetics* 60 (1982): 101–121. Review of early ideas about evolution of senescence, focusing on the pioneering work of Weismann and including a comprehensive, critical discussion of the adaptive theories.

Rose, M. R. *Evolutionary Biology of Aging.* New York, 1991. Clear discussion of the general principles behind the mutation-accumulation and pleiotropy theories.

Westendorp, R., and T. B. L. Kirkwood. "Human Longevity at the Cost of Reproductive Success." *Nature* 396 (1998): 743–746. Analysis of life-history records from British aristocrats showing evidence for a tradeoff between human longevity and fertility.

— Thomas B. L. Kirkwood

SEX, EVOLUTION OF

A very broad definition of sex is that it is the process that forms individual organisms containing genes from more than a single source (or parent). This covers a number of processes occurring in organisms ranging from bacteria to humans. In bacteria, individuals actively take up DNA from the environment and recombine it into their genomes (transformation). Gene transfer can also be caused by parasitic plasmids and phages that move between individual bacteria, carrying genes with them. In most of this article, however, we will be concerned with the more commonly known form of *sexual reproduction* that occurs in eukaryotes. This involves (1) the alternation between a haploid stage (single set of chromosomes) and a diploid (double set of chromosomes) stage; and (2) the exchange of genes between the chromosomes of a diploid pair (*recombination*), which can lead to the production of individuals with novel genotypes.

More specifically, sexual reproduction in eukaryotes involves the alternation of (1) the creation of haploid germ cells or gametes from a diploid cell by meiosis (a special kind of cell division), and (2) fertilization, the fusion of haploid cells (gametes) to form a diploid cell (the zygote). During meiosis, the chromosomes of a diploid pair can exchange genetic material by a process that is termed *crossing-over*. This recombination can break links between genes that were previously linked on the same chromosome, or create links between previously unlinked genes.

The process by which sexual reproduction occurs is remarkably diverse across the eukaryote species. In some species, sex is an obligate part of the life cycle (e.g., humans and other mammals, many insects), while in others it is an optional component of reproduction, induced under special circumstances (e.g., water-fleas, some insects, some parasitic nematodes, most protists and fungi). In some species, fertilization is followed immediately by (zygotic) meiosis, and so most of the life cycle is spent as haploid (e.g., algae or protozoa such as malaria parasites); in others, gametic meiosis is followed by fertilization, and so most of the life cycle is spent as diploid (e.g., humans). Some organisms have clearly differentiated (*anisogamous*) "male" and "female" gametes (e.g., humans), and others are *isogamous* (e.g., some protozoa).

The Problem of Sex. The relatively widespread occurrence of sex is often described as one of the greatest problems for evolutionary biology. Why is this so? Sex is puzzling because it can involve a large cost. First, sex or recombination breaks up favorable gene combinations that have increased in frequency under the action of natural selection. Put simply, it destroys the genotypes that were good enough to win the natural selection lottery to survive and reproduce (this is sometimes termed the *recombination load*, because it results from recombination breaking up genotypes).

Second, in species with anisogamy (separate male and female gametes), sexual reproduction involves a "cost of producing males." Asexual females can potentially produce twice as many daughters as sexual females, because sexual females must "waste" half of their resources on producing males (assuming, as is commonly expected and observed, 50 percent males). Consequently, the ratio of asexual to sexual females should double each generation, resulting in a "twofold cost of sex" (Figure 1). For example, suppose that each female produces two offspring that survive to reproduce. In an asexual species, both these offspring would be females, and so the asexual population size would double each generation. In contrast, with sexuals, one of the offspring would be male and one female, and so the sexual population size would remain constant. In this simple numerical example, it is clear that asexuals could rapidly outcompete sexuals.

Given these costs of sex, we would therefore expect natural selection to favor asexual reproduction in wild populations. However, it often does not: sexual reproduction is widespread throughout the animal and plant kingdoms. The rest of this article will focus on why sex occurs. This actually involves a number of subtle questions: on the evolution of sex, the maintenance of sex, the timing of sex, and the amount of sex. We discuss sex in bacteria and eukaryotes separately, because they have independent origins and are likely to have different explanations.

Sex in Bacteria. Why do bacteria actively take up DNA from the environment and recombine it into their genomes (transformation)? When considering sex in bacteria, it is important to realize that this process is split into two components—*competence*, the active uptake of DNA fragments from the environment, and recombination of the host genome with those fragments. Did competence evolve to facilitate recombination, or did it evolve for another reason?

Three hypotheses have been suggested to explain competence: (1) competence evolved to allow recombination, which provides some fitness advantage (much as we believe sex is an adaptation in eukaryotes, providing a fitness advantage, as described in later sections); (2) competence evolved to provide a template to repair damaged DNA; or (3) competence evolved for the acquisition of nucleotides as a source of food. In recent years, experimental work by Rosemary Redfield has tested these hypotheses. Redfield (2001) found that DNA uptake was not induced by DNA damage, as would be predicted by the DNA repair hypothesis, but it was induced by starving, supporting the acquisition of food hypothesis. Furthermore, the amount of competence induced under starvation can be reduced by supplying specific nucleotides (i.e., an alternate food source).

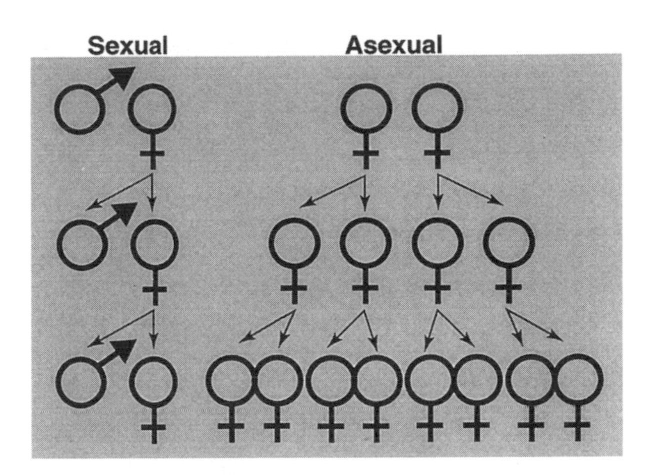

FIGURE 1. The Twofold Cost of Sex.
Asexual females can potentially produce twice as many daughters as sexual females, because sexual females must "waste" half of their resources on producing males (sons). This means that the ratio of asexual to sexual females should double each generation (the twofold cost of sex). Here is an example in which each female produces two offspring that survive to reproduce. In an asexual species, both these offspring would be females, and so the asexual population size would double each generation. In contrast, with sexuals, one of the offspring would be male and one female; and given that it takes two parents to make an offspring, the sexual population size would remain constant. Courtesy of Andrew D. Peters.

If DNA uptake by bacteria is to acquire food, why does recombination occur? One strong possibility is that it occurs by accident. DNA uptake is targeted toward fragments similar to those in the bacteria's own genome. Although this might suggest the importance of DNA repair or some advantage of recombination, most of the DNA is degraded for food before anything else can happen. An alternative (untested) explanation is that similar DNA might be easier to get across the cell membrane. All the genes involved in recombination have other functions in the cell that can explain why they cause recombination (e.g., DNA repair). Selection for these functions dwarfs any benefits that would be obtained from recombination, which generally leads to a decrease in fitness. Consequently, a strong possibility is that DNA uptake occurs to acquire food, and that occasionally an accident then leads to some DNA being recombined into the genome.

This suggests that selection for the ability to recombine has been weak or nonexistent in bacteria, and that transformation did not evolve for genetic exchange. Although recombination is relatively rare in bacteria (not a part of its normal life cycle) and involves only small fragments, this does not downplay its importance; there is abundant evidence for horizontal gene transfer, and this can often have considerable consequences, such as when it confers antibiotic resistance. Nonetheless, this makes sex in bacteria very different from that in eukaryotes, where meiotic sexual reproduction evolved for recombination.

Sex in Eukaryotes. How is sexual reproduction distributed among the eukaryotes? Sex occurs in almost all eukaryote groups. In particular, it is typical among the crown taxa (animals, plants, fungi). However, even among these groups (e.g., lower vertebrates, invertebrates, flowering plants), asexual (parthenogenetic) species occur, showing that asexuality is mechanistically possible and has evolved many times, and that there must be some factor maintaining it. Although the eukaryote phylogeny is not sufficiently resolved to know how many times meiotic sex has evolved, current opinion suggests that sexual reproduction probably arose in a protist, a unicellular eukaryote, as an optional component of a reproductive cycle that was usually asexual (mitotic).

The taxonomic distribution of asexuality in the eukaryotes is not random. Asexual reproduction tends to be found in odd populations, species, or even whole genera, within a larger taxonomic group that reproduces sexually. Consequently, asexuality is a derived state and apparently not long-lived on an evolutionary time scale. There are a couple of exceptions to this, notably a suborder of rotifers (Bdelloidea) and the Darwinulid ostracods.

Before we move on to discuss possible advantages to sex, we note that there is an important difference between explaining the initial evolution and the maintenance of sex. In species that lack a specialization of germ cells into male and female forms (called *isogamous* organisms), the twofold cost of sex (i.e., production of males) does not need to be paid, and so only a slight advantage would be required for the initial evolution of sex to be favored. In anisogamous organisms, females pay all the costs of provisioning the large egg with cytoplasm and nutrients. Males do not provision their tiny sperm. In isogamous organisms, males and females do not exist, although mating types do, often denoted "+" and "−." Here both mating types provision the gametes. So females of anisogamous species pay double the cost of producing gametes. It is only in anisogamous species with more advanced sexual reproduction that explaining the maintenance of sex becomes more problematic. Consequently, the major problem is explaining why so many plants and animals always reproduce sexually.

Theoretical Advantages of Sex. In the nineteenth century, August Weismann suggested that the advantage of sexual recombination was that it provides new genetic variation for selection to act upon. Sex leads to the mixing of genes from different individuals and the production of novel genotypes. The crux of Weismann's idea was that some of these novel genotypes would have a relatively high fitness and some of them would have low fitness. Consequently, by creating new genotypes, sex increases the distribution (variance) of fitness values. Natural selection would then favor the individuals with relatively high fitness. The power of natural selection to increase the mean fitness of a population depends upon the variation in fitness (in the extreme, if there was no variation, and all individuals in a population had equal fitness, then natural selection could not increase mean fitness). Consequently, by increasing the variation in fitness, sex allows the efficiency of natural selection to be increased and hence accelerates the increase in mean fitness (i.e., increases the fitness of grandchildren, etc.). Put simply, sex allows the better genes to be concentrated in a few "extra-fit" individuals, and the worst genes to be lost via particularly low-fitness individuals.

In the latter half of the twentieth century, this idea was dismissed by George Williams as being based upon group (or clade) selection, a force generally accepted not to be important, and a number of models were developed that could provide an advantage to sex at the level of the individual or gene. However, Weismann's idea has recently regained support. In particular, it has been realized that Weismann's mechanism underlies most of the formal genetic models that provide an advantage to sex; that the Weismannian mechanism provides an advantage at both the group and gene level; and that competition between asexuals and sexuals is at the group level in that genes do not flow between the groups by reproduction. Burt (2000) provides a recent and excellent review of this modern Weismannian perspective.

Of course, this does not answer the question of why sexual reproduction should lead to an increase in the variance of offspring fitness. Mechanisms that do this can be broadly classified into two groups: environmental (or ecological) models, and deleterious mutation models. Furthermore, these models can be distinguished depending upon whether they are deterministic and work in large (infinite) panmictic populations, or stochastic and require finite (and usually small) populations or certain population structures. In the rest of this section, we briefly summarize the most popular hypotheses for sex; a thorough review is provided by Kondrashov (1993). Although more then 20 hypotheses have been suggested and new ones are published with regularity, we will not (and could not hope to) cover all of these. In addition, many require extremely restrictive assumptions and so are not likely to be generally important, and it is not unusual for new models to be merely reformulations of or tweaks on old ideas.

Fisher-Muller hypothesis. One of the earliest hypotheses for sex states that sexual reproduction and recombination allow beneficial mutations that initially appear in separate individuals to be combined. In an asexual species, multiple beneficial mutations will only fix if the second (and later) beneficial mutations occur in an individual that already carries the first beneficial mutation. Consequently, sex increases the rate at which adaptive mutations are fixed in a population and hence accelerates evolution. However, the usefulness of this explanation for providing a short-term benefit to sex sufficient to balance the twofold cost has been questioned, because it requires relatively small populations and the frequent occurrence of beneficial mutations with sufficiently large fitness consequences.

Deterministic environmental models. There is a wide range of environmental models that can provide an advantage to sex. These models rely upon epistasis (nonadditive gene interactions, which are not required for the Fisher-Muller hypothesis) and suggest that sex accelerates adaptation to a changing environment by creating new gene combinations. The biological basis of such selection pressures may involve a variety of biotic or abiotic mechanisms.

Early models suggested that sex provides an advantage because different genotypes may do better in different environments, and that spatial or temporal variation in the environment would favor new gene combinations that can utilize the currently abundant habitat. The version of this idea that is reliant upon spatial variation in the environment is termed the "Tangled Bank." It assumes that different genotypes do better in different environments (niches). This provides an advantage to sex, because an asexual clone is restrained by the narrowness of the environments that it can exploit (i.e., the narrowness of its ecological range). Sexual reproduction leads to a diversity of genotypes that are able to exploit different environments. Bell termed this kind of model the Tangled Bank from a phrase in the concluding paragraph of Darwin's *Origin of Species*, which expresses the complexity of natural habitats that must strike anyone who contemplates a clump of moss or a bed of weeds. Another analogy sometimes used to describe this kind of idea, introduced by George Williams, is that of buying lottery tickets. Asexual reproduction is like buying several lottery tickets with the same number, whereas sexual reproduction is like buying fewer tickets, but each with a different number—consequently, the sexual strategy is more likely to provide a winning ticket. Finally, this advantage to sex can also be increased by competition between siblings. The reasoning here is that asexual reproduction will lead to offspring with the same genotype, who will therefore compete for the same habitat, reducing their mean fitness. In contrast, sexual reproduction leads to genetically diverse offspring, who will utilize different environments, decreasing competition among them.

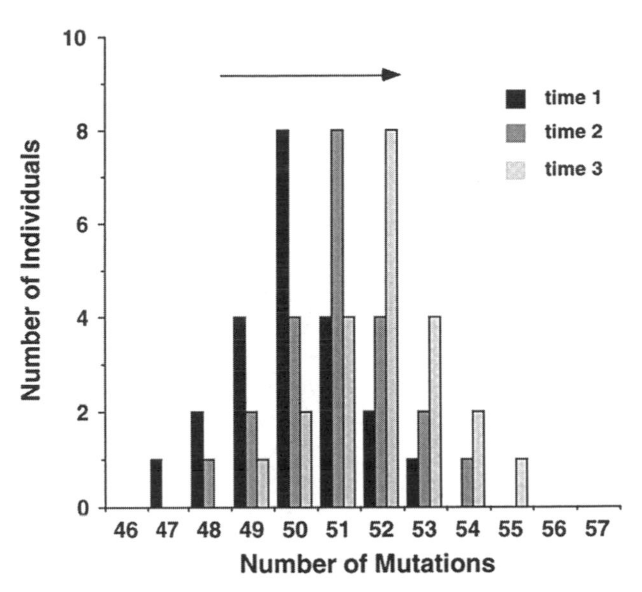

FIGURE 2. Muller's Ratchet.
Shown is the distribution of the number of deleterious mutations per individual in a population, at three time periods. Between each time period, the class of individuals that have the least number of deleterious mutations is lost as a result of chance events (failure to reproduce or mutation). In asexual populations, this class of individuals cannot be re-created, and so the accumulation of deleterious mutations is irreversible and inevitable. Courtesy of Curt M. Lively.

Currently the most popular environmental hypothesis is the "Red Queen." This states that sex provides an advantage in competitive interactions between competing organisms (antagonistic coevolution). Several people, especially Hamilton, have championed the idea that parasites could provide the driving force, favoring sexual reproduction in their hosts. The idea here is that parasites evolve to be able to infect the currently common host genotypes. Sex is then favored in the hosts, because it allows the production of novel genotypes that may be more resistant to the parasites. Over time, this will lead to continuous oscillations in host and parasite gene frequencies (host-parasite coevolution)—a "coevolutionary dance." The title of the Red Queen hypothesis was assigned by Graham Bell from the episode in which Lewis Carroll's Alice learns from the Red Queen that "it takes all the running you can do, to keep in the same place"—that is, you keep having to have sex just to stay ahead of the parasites, but you can never escape them. We have focused here on a scenario in which parasites drive the Red Queen and favor sex in their hosts, but there are other possibilities, such as host immune responses driving sex in their parasites.

Muller's ratchet. Muller's ratchet is the inevitable accumulation of deleterious mutations in asexuals that occurs through chance events (Figure 2). In any population, there will be a distribution of the number of deleterious mutations per individual. Now consider the class of individuals that has the least number of deleterious mutations. Given a finite population size, there is a probability, in each generation, that this class of individuals will be lost. This might occur through the random chance that none of them reproduce or that they all mutate at the same time. Ignoring the possibility of back or compensatory mutations, this accumulation of deleterious mutations is irreversible. In sexual populations, recombination allows this class of individuals with the least number of mutations to be recreated. Consequently, the mean number of deleterious mutations in an asexual lineage can only increase, and the ratchet clicks forward one notch. Although Muller's ratchet clearly provides an advantage to sex, it is a stochastic process and is thought to operate too slowly to compensate for the twofold cost (i.e., asexuals would replace sexuals before the fitness of the asexuals declined below that of sexuals). In addition, it has been shown that only a small amount of sex is required markedly to reduce the accumulation of deleterious mutations by Muller's ratchet.

Mutational deterministic hypothesis. Alexey Kondrashov pointed out that deleterious mutations can provide a short-term advantage to sex that can compensate for its twofold cost if each additional deleterious mutation leads to a greater decrease in fitness than the previous mutation (termed *synergistic epistasis* between deleterious mutations) (Figure 3). When this is the case, sexual reproduction increases the variance in the number of deleterious mutations that will be carried by offspring. The low fitness of individuals carrying above-average numbers of deleterious mutations will then lead to a larger number of deleterious mutations being eliminated. If the deleterious mutation rate is sufficiently high, then this process can compensate fully for the twofold cost of sex. Early theoretical work suggested that the deleterious mutation rate needed to be greater than 1.0 per genome per generation to balance the twofold cost of sex, but more realistic models suggest that several other factors (e.g., finite populations, variation in the extent of epistasis) mean that a mutation rate greater then approximately 2.0 is required.

Some recent models have examined the consequences for the mutational deterministic hypothesis of greater selection against mutations in males. This could occur via mate choice or competition for mates (sexual selection). In these models, a lower deleterious mutation rate can be required to explain sex. Furthermore, the requirement for synergistic epistasis can be removed. Although these models can make it easier to explain sex, they are restrictive in that they do so only for sexual species in which sexual selection is important.

A common feature of most models discussed above (but possibly not some environmental models such as

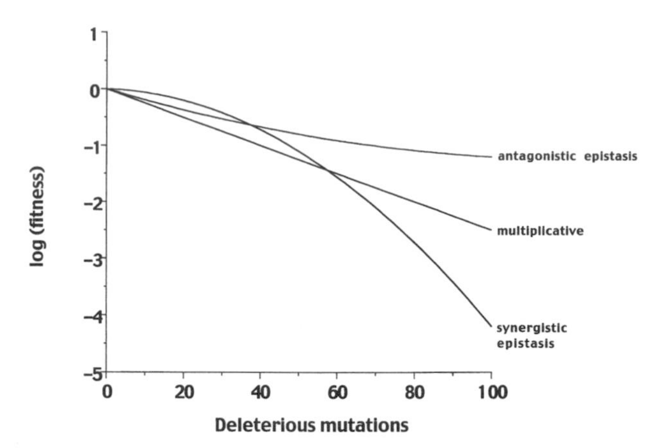

FIGURE 3. Interactions Between Deleterious Mutations. The relationship between (log) fitness and the number of deleterious mutations may be linear (multiplicative); each extra mutation lead to a greater decrease in fitness (synergistic epistasis); or each extra mutation lead to a smaller decrease in fitness (antagonistic epistasis). The mutational deterministic hypothesis requires synergistic evidence. To date, most experimental data suggest, on average, multiplicative interactions. Courtesy of Stuart A. West.

the Red Queen) is that sex provides an advantage because it increases the variation in offspring fitness. Hence, sex increases the response to selection, as suggested by Weissman—despite leading to the production of less fit offspring in the short term, termed the recombination load. This increase in the variance of fitness occurs because sex breaks up nonrandom gene associations (linkage disequilibrium). Consequently, these models work because they provide a mechanism for generating nonrandom gene associations (linkage disequilibrium); the mechanism is either genetic drift in stochastic models with additive interactions, or epistasis in deterministic models. These mechanisms relate to whether a hypothesis can explain the maintenance of sex in the short term by preventing the successful invasion of an asexual clone (e.g., deterministic mechanisms such as the Red Queen or the mutational deterministic hypothesis), or whether it can lead only to the long-term extinction of asexuals (e.g., stochastic processes such as Muller's ratchet).

What Hypothesis Do the Data Support?

Comparative data. The earliest empirical work that attempted to explain sex was correlational and focused on the ecological hypotheses. In these studies, the occurrence of sex or rate of recombination is examined, either across or within species, with respect to key ecological variables. These studies emphasize finding predictions that discriminate among different hypotheses. For example, Curt Lively (1987) examined the snail *Potamopyrgus antipodarum*, in which both sexual and asexual forms can be found living in the same area.

Lively found that, across different populations in New Zealand, the proportion of sexual individuals was positively correlated with the frequency of their parasites, supporting the Red Queen hypothesis (Figure 4). In contrast, the proportion of sexual individuals was not correlated with the temporal or spatial variation in habitat, as predicted by the Tangled Bank hypothesis. Similarly, Austin Burt (2000) and Graham Bell (1982) found that the recombination rate was greater in mammals with longer generation times, which are likely to suffer greater parasite pressure, supporting the Red Queen hypothesis. In contrast, the recombination rate did not correlate with brood size, a likely correlate of the amount of competition between siblings, as predicted by the Tangled Bank hypothesis.

These comparative studies have provided support for the Red Queen hypothesis, and they have played an important role in the rejection of some of the environmental models (such as the Tangled Bank). Furthermore, it is fundamental that any theory of sex must be able to explain the distribution of sex both across and within species. However, because these studies are correlational, they are open to multiple explanations, and *post hoc* scenarios can be developed that allow the results to be explained by environmental or mutation-based models. In particular, with some ingenuity, the mutational deterministic hypothesis can explain most patterns that are predicted by the Red Queen. For example, patterns such as those observed by Lively could be explained by arguing that parasites are the factor that selects against individuals with high mutation loads (although in this case with the snails, this is unlikely because the parasites are sterilizing, and there is no evidence that asexuals, who should have higher mutation loads, are more susceptible to infection per se). Similarly, the pattern observed by Burt and Bell could be explained if species with longer generation times have higher deleterious mutation rates (for which there is much evidence—see Figure 5). Thus, while there is no doubt that the comparative data support the Red Queen hypothesis, they seem incapable of rejecting the mutational deterministic hypothesis. This fact, along with their theoretical plausibility, has made Red Queen and the mutational deterministic hypotheses the major contending explanations for sex.

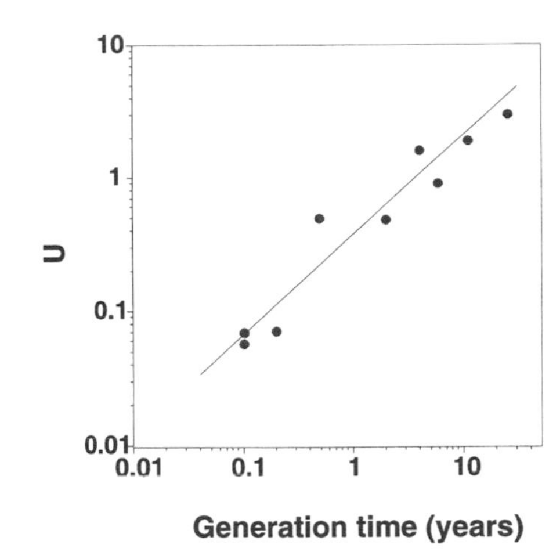

FIGURE 5. Estimates of Deleterious Mutation Rate (*U*) Obtained by Comparing Genomic Data from Different Species, Plotted Against Generation Time.
These data suggest that the deleterious mutation rate increases strongly with generation time but is below 1.0 for a large number of sexual species (e.g., many insects and nematodes). Data from Keightley and Eyre-Walker (2000).

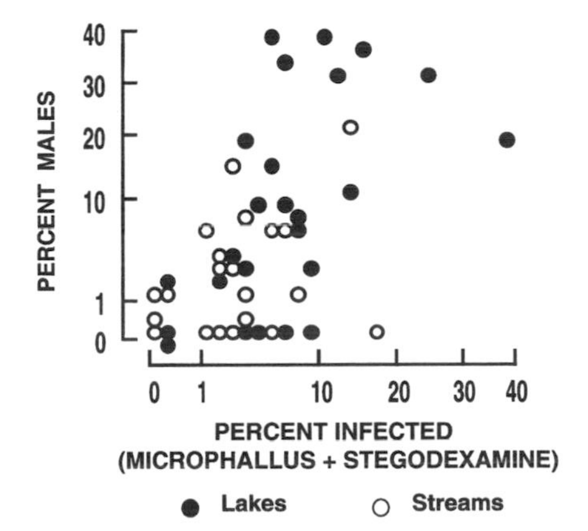

FIGURE 4. The Relative Frequency of Sexual and Asexual Forms of the Snail *Potamopyrgus antipodarum* Across Different Populations in New Zealand.
The proportion of sexual individuals is positively correlated with the frequency of their parasites, supporting the Red Queen hypothesis, and not correlated with the temporal or spatial variation in habitat (lakes versus streams), as predicted by the Tangled Bank hypothesis. Data from Lively (1987).

Experiments and parameter estimation. Given that comparative work cannot decisively distinguish between the different hypotheses, it is necessary to test the assumptions of different models and measure crucial parameters. In recent years there has been a surge in such work.

The mutational deterministic hypothesis was thought to offer excellent opportunities for empirical testing. It clearly requires a deleterious mutation rate greater then 1.0–2.0 and synergistic epistasis between deleterious mutations. However, in practice it is not so easy to test these assumptions. The first attempts to estimate the deleterious mutation rate were by mutation accumulation experiments pioneered by Terumi Mukai in the 1960s on fruit flies (*Drosophila*). In these, selection was relaxed, deleterious mutations were allowed to accumulate for a large number of generations in replicate lines, and then the deleterious mutation rate was estimated by the rate and form of fitness decline over time. Similar experiments have since been carried out on other organisms, including bacteria (*Escherichia coli*) and nematodes (*Caenorhabditis elegans*). Unfortunately, estimates of the deleterious mutation rate have varied widely among studies, and even between analyses of the same datasets. While some argue that the data are consistent with the mutational deterministic hypothesis, others argue that they are not. More studies are required from a range of sexual organisms.

More recently, indirect estimates of the deleterious mutation rate have been obtained by comparing genomic data from different species. Using this approach, Peter Keightley and Adam Eyre-Walker (2000) have estimated the deleterious mutation rate in a range of organisms, including humans, sheep, dogs, mice, and fruit flies (Figure 5). They found that the deleterious mutation rate increased strongly with generation time, providing a possible explanation for variation in the deleterious mutation rate across species. Importantly, their estimate suggests that the deleterious mutation rate would be below 1.0 for a large number of sexual species (e.g., many insects and nematodes). These data suggest strongly that the mutational deterministic hypothesis cannot explain sex. However, it should be noted that their estimates of the deleterious mutation rate are underestimates, and it has been argued that they may even be sufficiently underestimated to save the mutational deterministic hypothesis, and so the debate is certainly not closed.

The mutational deterministic hypothesis also requires another assumption that is in principle easily tested: each deleterious mutation must lead to a greater decrease in fitness then the last (synergistic epistasis) (Figure 3). Early experiments on fruit flies and algae supported the occurrence of synergistic epistasis, but these were either carried out without proper experimental controls or lacked a proper theoretical basis. More recently, experiments testing for epistasis have been carried out by examining how fitness varies with the insertion of a known number of deleterious mutations. Such experiments on bacteria (*E. coli*) and nematodes (*C. elegans*) showed no tendency toward synergistic epistasis, whereas work on fruit flies did (although this result relied on a single data point, consisting of the least fit individuals). Overall, this suggests that a general tendency toward synergistic epistasis is either absent or weak. Although this may suggest doom for the mutational explanation of sex, recent theoretical work suggests that the need for synergistic epistasis can be reduced by sexual selection (as discussed earlier) or by patterns of dispersal leading to populations being structured into subpopulations. In addition, methodological constraints mean that estimates of the importance of epistasis are usually expected to be underestimates.

In contrast, it has been less clear how to carry out such critical tests of the environmental models. To some extent, this is because theory has not been able to make such clear predictions; however, recent work has started to address this problem. For example, Curt Lively and Mark Dybdahl (2000) have shown that parasites are better at infecting locally common genotypes of the snail *P. antipodarum*, as predicted by the Red Queen hypothesis. This suggests that parasites track locally common host genotypes, a necessary condition of the Red Queen hypothesis. Furthermore, they have shown not only that locally common genotypes are disproportionably infected, but also that they subsequently show a drop in frequency, presumably because of their high infection rate. Future work is required to determine if genotypes then increase in frequency again after having been rare for some time, showing the kind of cycle in frequency predicted by the Red Queen. Experimental work has also supported the comparative data in its rejection of the Tangled Bank hypothesis. Experiments on plants and algae have shown that populations consisting of mixed genotypes show carrying capacities not much greater than that of the average of their components, and rarely greater than that of the best component.

A Pluralist Approach. Given the lack of a clear winning hypothesis, what can be said firmly about the explanation for sex? First, the Red Queen and the mutational deterministic hypothesis (including its more complex variants) appear to be the leading contending explanations for sex. Second, the empirical data, both comparative across species, and more detailed work within species are consistent with the Red Queen hypothesis but are not sufficient to prove decisively that it explains sex. Third, although this work is the subject of controversy, the best data that we currently have suggest that the mutational deterministic hypothesis cannot explain sex. In particular, deleterious mutation rates may be appreciable (e.g., >0.1) but not above the necessary 1–2 in a wide range of sexual species, and synergistic epistasis between deleterious mutations appears to be either absent or extremely weak.

This search for one hypothesis to explain sexual reproduction may be misleading, however. Considerable advantages may be gained from a pluralist approach that considers that different mechanisms may work simultaneously. First, acting alone, each of the various theories requires extreme, and possibly unreasonable, assumptions to be able fully to explain the maintenance of sex. However, even if a model is not able fully to explain the twofold cost of sex, it may play an important role. For example, the genomic estimates of the deleterious mutation rate suggest that, across a wide range of sexual species, it is large enough to play an important role but cannot explain sex alone. Perhaps deleterious mutations pay half the cost and coevolving parasites pay the other half.

Second, different mechanisms may interact synergistically. In particular, environmental and mutational mechanisms may complement each other, covering each other's weaknesses. For example, host-parasite coevolution, as envisaged in the Red Queen hypothesis, provides an extremely short-term advantage to sex that slows down the spread of asexual clones. This allows more time for deleterious mutations to accumulate before the clone could potentially replace the sexual population; in addition, it keeps the asexuals at smaller population sizes and hence speeds up the rate at which they accumulate deleterious mutations. As another example, in regard to the fixation of beneficial mutations as described above in the context of the Fisher-Muller hypothesis, deleterious mutations can make it even harder for fixation to occur. The reason for this is that beneficial mutations must arise not only in the same individuals, but also in individuals with a relatively low number of deleterious mutations, or else they will be dragged down by their low-quality genetic background (termed the "ruby in the rubbish" problem by Joel Peck). These and other possible interactions between mechanisms are discussed in detail by Stuart West, Curt Lively, and Andrew Read (1999). Although it is a messy answer, a pluralist approach (especially one involving deleterious mutations and coevolving parasites) seems currently the most plausible explanation for sex to a wide range of evolutionary biologists.

Why So Much Sex? Before concluding, we briefly consider a more subtle and possibly more perplexing problem. Why do animals and plants have so much sex? Specifically, why is obligate sex (in every round of reproduction) so common in animals and plants? Models of most mechanisms that provide an advantage to sex suggest that occasional sex provides approximately the same benefit as obligate sex. In most unicellular eukaryotes and some animals (e.g., water-fleas, aphids, and some wasps), sexual reproduction is an occasional component of the life cycle. Why is this not the case for most plants and animals?

One possible explanation for this, covered in detail by Burt (2000), is that facultative or occasional sex is unstable. Specifically, selection could always favor a little less sex, leading eventually to asexual reproduction, even in a situation where obligate sexuals outcompete asexuals. The crucial point is that asexuals competing against sexuals involves competition at a different level (between groups) than does competition between facultatively sexual individuals that have different amounts of sex. Cases in which facultative sex occurs seem to represent historical serendipity, as suggested by the fact that sex is usually associated with some other life history difference, such as dispersal (aphids) or the production of overwintering eggs (water-fleas). Interestingly, the taxa in which facultative sex is common, such as protists and fungi, are generally isogamous, and so the twofold cost of sex is not paid; thus, it becomes much easier to explain the stability of occasional sex—it has a low cost and a high advantage. It is easy to imagine a situation in which occasional sex initially evolved under conditions in which obligate sex would not be stable; then a little sex opened the door for an increased rate of evolution that led to bigger more complex organisms with longer generation times; and finally, this led to an increase in deleterious mutation rates and parasite loads, until the point at which obligate sex could be favored in anisogamous species. Furthermore, recent theory suggests that sexual selection (mate choice and mate competition) increases the advantage of sex by purging less fit individuals.

This discussion of occasional sex also makes an important point for the highly influential balance argument of George Williams. Williams argued that the existence of facultative or occasional sex provided strong evidence against Weismann's hypothesis, because sex must be providing an immediate short-term fitness advantage, and this could not be done merely by increasing the variance in fitness. The discussion above suggests an alternative explanation for why facultative sex could be maintained.

Conclusions. Explaining sex has been termed "the queen of problems in evolutionary biology" (Bell, 1982). Although there is still no entirely conclusive answer, the increasing attention paid to this problem in the past three decades by theoretical and empirical biologists has at least led us to a better idea of the plausible explanations, and to a strong suspicion that the answer is complex, involving deleterious mutations and antagonistic coevolution, particularly with parasites.

[*See also* Red Queen Hypothesis.]

BIBLIOGRAPHY

Bell, G. *The Masterpiece of Nature: The Evolution and Genetics of Sexuality*. Berkeley, 1982. An exhaustive survey of the distribution of asexual and sexual reproduction in nature.

Burt, A. "Sex, Recombination and the Efficacy of Selection—Was Weismann Right?" *Evolution* 54 (2000): 337–351. A recent and excellent review of the Weismannian perspective on sexual reproduction.

Elena, S. F., and R. E. Lenski. "Test of Synergistic Interactions among Deleterious Mutations in Bacteria." *Nature* 390 (1997): 395–398. Experimental evidence for no general tendency toward synergistic epistasis in deleterious mutations.

Hamilton, W. D., Axelrod, R., and R. Tanese. "Sexual Reproduction as an Adaptation to Resist Parasites." *Proceedings of the National Academy of Sciences USA* 87 (1990): 3566–3573. A highly influential model and review of the Red Queen hypothesis.

Keightley, P. D., and A. Eyre-Walker. "Deleterious Mutations and the Evolution of Sex." *Science* 290 (2000): 331–333. Indirect estimates of the deleterious mutation rate obtained by comparing genomic data from different species, which suggest the rate is not sufficient for the mutational deterministic hypothesis to explain sex.

Lively, C. M. "Evidence from a New Zealand Snail for the Maintenance of Sex by Parasitism." *Nature* 328 (1987): 519–521. Correlational field data supporting the Red Queen hypothesis.

Lively, C. M., and M. F. Dybdahl. "Parasite Adaptation to Locally Common Host Genotypes." *Nature* 405 (2000): 679–681. A recent introduction into Lively's highly influential long-term field study on asexual and sexual snails, which has provided some of the best support for the Red Queen hypothesis.

Lynch, M., J. Blanchard, D. Houle, T. Kibota, S. Schultz, L. Vassilieva, and J. Willis. "Perspective: Spontaneous Deleterious Mutation." *Evolution* 53 (1999): 645–663. A recent review of attempts to estimate the deleterious mutation rate, especially with mutation accumulation experiments.

Peters, A. D., and C. M. Lively. "The Red Queen and Fluctuating Epistasis: A Population Genetic Analysis of Antagonistic Coevolution." *American Naturalist* 154 (1999): 393–405. A recent analysis to determine the population genetic basis (why it works) of the Red Queen hypothesis.

Redfield, R. J. "Do Bacteria Have Sex?" *Nature Genetics* 2 (2001): 634–639. An introduction into Redfield's collection of extremely clear experimental studies on why bacteria have sex.

West, S. A., C. M. Lively, and A. F. Read. "A Pluralist Approach to the Evolution of Sex and Recombination." *Journal of Evolutionary Biology* 12 (1999): 1003–1012. A detailed discussion of the advantages to be gained from considering how different mechanisms that provide an advantage to sex may act simultaneously and interact synergistically.

— STUART A. WEST

SEX CHROMOSOMES

Many species of animals and some species of flowering plants have separate sexes. The differences between the sexes are often controlled by a pair of sex chromosomes, which are genetically and often structurally distinct. Most commonly, there is an X/Y chromosome system, with XX females and XY males, where sex is determined either by a gene or genes on the Y chromosome, as in mammals, or by the ratio of the number of X chromosomes to the number of the other chromosomes (autosomes), as in *Drosophila*. Less often, there is female heterogamety, with ZW females and ZZ males, as in birds, Lepidoptera, and some species of plants and lower vertebrates (Bull, 1983; Westergaard, 1958). The term *heterogametic sex* is used to refer to the sex that is heterozygous for the sex chromosomes, the term *Y chromosome* for systems with both male and female heterogamety.

In some, but not all, species with chromosomal sex determination, the X and Y chromosomes can be recognized as different in appearance by light microscopy. This difference is associated with the accumulation of various types of repetitive DNA sequences in disproportionate abundance on the Y chromosome (Bull, 1983; Charlesworth, 1996), and the Y chromosome is commonly described as being largely composed of heterochromatin (Bull, 1983; Graves, 1995). In common with other heterochromatic regions of the genome, it is often late-replicating in cell division.

Another difference from the autosomes is that the pair of sex chromosomes fail to cross over with each other in the heterogametic sex along all or part of their length. Crossing over occurs only in a segment of the sex chromosomes, called the pairing region or pseudoautosomal region (Bull, 1983; Graves, 1995; Westergaard, 1958).

A striking common feature of sex chromosome systems is the almost complete erosion of genes from the Y chromosome. The best-studied sex chromosome system is that of humans. Only nineteen homologous X–Y pairs of loci have been found after exhaustive searches (Lahn and Page, 1999). This is a much smaller number than would be predicted from the total amount of DNA in the human Y chromosome, whereas the X chromosomes have thousands of active genes. The Y chromosomes of groups such as mammals or *Drosophila* have thus lost most of their genetic activity, leading to the familiar phenomenon of sex-linked inheritance. In some groups, such as many Orthoptera, the Y chromosome has been completely lost, so that males are described as X0 in constitution (Bull, 1983; Charlesworth, 1996).

It is not yet clear whether plant Y chromosomes are as genetically degenerate as those of animals. There is evidence that the Y chromosome of the white campion *Silene latifolia* has lost essential genes, as plants having a Y but no X chromosome are usually inviable (Vagera et al., 1994; Westergaard, 1958), but it does carry functional copies of some genes that are present on the X (Delichère et al., 1999).

FIGURE 1. Diagram of the Sex Chromosomes of Mammals and a Plant. Drawing by Deborah and Brian Charlesworth.

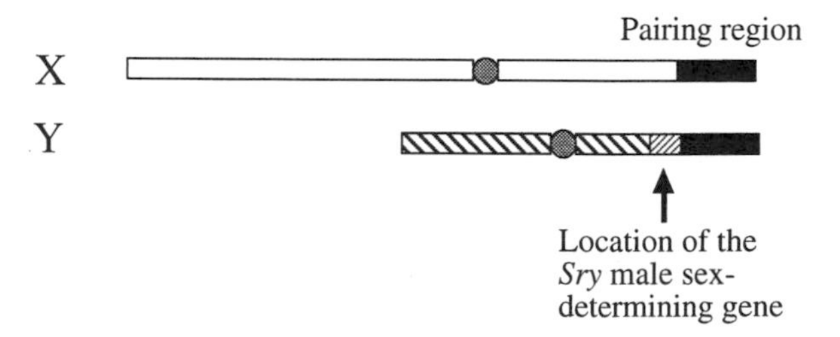

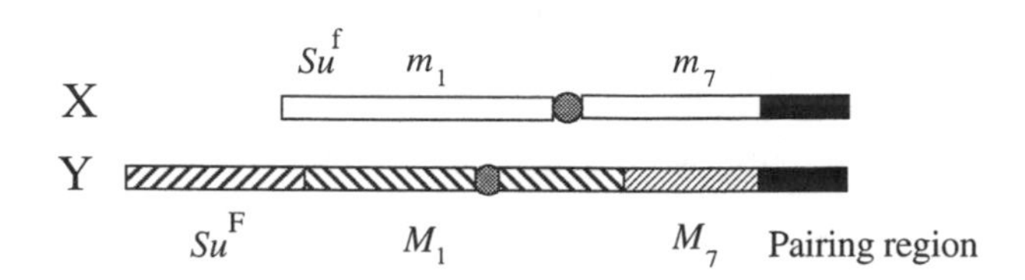

TABLE 1. Effects of the Same Female-Sterility Gene Change ($Su^f \rightarrow Su^F$) in Hermaphrodites and Females

When females are present, this substitution has an advantageous effect on hermaphrodites because it increases male fertility, but is disadvantageous in females because it decreases female fertility.

Second mutation $Su^f \rightarrow Su^F$ (female-sterility locus)	First mutation M → m (male-sterility locus) Genotype at male-sterility locus M/m or M/M	m/m
Su^f/Su^f	Hermaphrodite ↓	Female ↓
Su^F/Su^f or Su^F/Su^F	Male	Neuter

The possession of genetically eroded Y chromosomes is sometimes associated with dosage compensation, such that the activity of most X-linked genes is equalized in males and females by special processes of gene regulation, despite their different dosage in the genomes of the two sexes (Bull, 1983; Charlesworth, 1996). In mammals, this is achieved by inactivating one of the two X chromosomes in females; in *Drosophila*, by doubling the rate of transcription of X-linked genes in males. In Lepidoptera, however, there is no evidence for dosage compensation.

The Origin of Primitive Sex Chromosomes. Sex chromosomes have evolved independently in many groups of animals and plants, and intermediate stages, ranging from minor genetic differences between the karyotypes of the two sexes to fully developed sex chromosomes, have been observed. In plants, separate sexes (dioecy) has a widespread taxonomic distribution in roughly half of all flowering plant families, but it is at low frequency within most families. This strongly suggests recent and independent evolutionary origins of dioecy, and thus of sex chromosomes, in different angiosperm families (Darwin, 1877; Westergaard, 1958).

Reduction in crossing over between X and Y chromosomes. As suggested by H. J. Muller in 1918, the first step in the evolution of sex chromosomes must have involved the establishment of restricted recombination between a pair of proto-X and proto-Y chromosomes. [*See* Muller, Hermann Joseph.] This is a response to the need to prevent recombination among genes with primary sex-determining roles. The evolution of separate sexes from an initially hermaphroditic ancestor, as in plants, gives some insight into how this probably happened. The first step is most likely the invasion of the population of hermaphrodites (or other functionally hermaphrodite cosexual individuals such as monoecious plants, with separate male and female flowers) by a mutation creating females, as a result of loss of male function. Femaleness may be advantageous either because females cannot self-fertilize, and hence produce fitter progeny (not suffering from inbreeding depression), or because they produce more seed than hermaphrodites by diverting resources from male functions (thus compensating for the lost opportunities for reproduction through sperm or pollen). Most likely, both these advantages are involved. The presence of females in the population provides an advantage for the hermaphrodites to evolve toward maleness (Charlesworth and Charlesworth, 1978).

The simplest genetic model, supported by data on the genetics of dioecy in plants (Westergaard, 1958), involves two mutations at distinct loci, one producing females from hermaphrodite (*M* to *m* in Table 1), and another producing males from hermaphrodite (Su^f to Su^F). Unless these mutations are closely linked, the production of neuters and hermaphrodites by recombination between them will prevent the spread of the female-sterile mutation Su^F. If linkage is sufficiently close, and Su^F can spread, genetic modifiers or chromosome rearrangements that further reduce crossing over between the two loci are favored by selection (Charlesworth and Charlesworth, 1978), leading to suppression of crossing over in part of the new Y chromosome.

Subsequently, selection for alleles that are advantageous in males but disadvantageous in females can lead to genetic differentiation between the two sex chromosomes at other loci, as well as to selection for suppression of recombinational exchange over most or all of their length (Rice, 1996). This sets the scene for the further evolution of an incipient Y chromosome. The suppression of recombination may occur gradually over time; for example, data on the molecular evolution of Y-linked genes in humans suggest a succession of several stages in the reduction of crossing over between X and Y (Lahn and Page, 1999).

The Degeneration of Y Chromosomes. This first step, the evolution of restricted recombination between primitive X and Y chromosomes, creates a new and unusual evolutionary situation. Any Y-linked genes with sex-specific functions will presumably be under strong selection to maintain their functions. The other genes in the nonrecombining part of the Y initially will have homologues on the evolving X chromosome, but these are in a very different genetic situation from that of autosomal or X-linked genes. Muller (1918) suggested that Y-chromosome genes lose function because they never become homozygous. Recessive loss-of-function mutations at Y-linked loci are not selected against, because they are covered up by active X-linked copies, and can thus increase in frequency under the pressure of mutation. But this explanation is inadequate when one recalls that dosage compensation has evolved in many systems; this would not be favored if loss of function were fully recessive, and there is independent evidence that this is usually not the case (Charlesworth, 1996).

Our understanding of the degeneration of Y chromosomes is based on the population genetics of nonrecombining genomes and asexual populations. Several possible models have been proposed, but tests of their predictions are only just starting to be carried out, and we are far from being able to decide definitively between them. One possibility is that, when a favorable mutation in a gene on the Y chromosome spreads through the population, it will drag along any deleterious mutations present on the chromosome in which it originated; that is, there is hitchhiking by favorable mutations (Rice, 1987). A less obvious theory, also involving a hitchhiking process that occurs when there is no recombination, is that selection against deleterious mutations effectively reduces the population size for genes on Y chromosomes (much as selection against unsatisfactory individuals in a herd of farm animals reduces the number of breeding individuals). This increases the chance that the random sampling effects of small population size cause weakly selected deleterious mutations to accumulate on the Y chromosome, whereas they would be eliminated from genes on a recombining chromosome (Charlesworth, 1996). Selectively favorable mutations are also less likely to become fixed on the Y chromosome compared with the X (Orr and Kim, 1998). Another process to which nonrecombining chromosomes are subject is "Muller's ratchet," whereby deleterious mutations accumulate in a series of irreversible steps as a result of the random losses of the individuals who carry the smallest number of deleterious mutations (Charlesworth, 1996).

The net result of these processes is the progressive reduction in the mean level of functioning of genes on the Y chromosome. This should favor the evolution of increased expression levels of their more functional X-linked homologues in the heterogametic sex, resulting in the evolution of dosage compensation (Charlesworth, 1996). Lack of gene activity on the Y chromosome means that insertions of transposable elements are less likely to cause harmful mutations, so that their spread into the Y is unopposed by selection. In addition, other population genetic processes (including Muller's ratchet) can lead to the accumulation of transposable elements and other types of repetitive sequence in nonrecombining genomic regions (Charlesworth et al., 1994). The accumulation of such sequences on the Y can thus be viewed as a predictable evolutionary consequence of its lack of crossing over.

Drosophila *neo-Y chromosome systems.* One source of data for testing the evolutionary theories is to study recently evolved Y chromosomes. The sex chromosomes of dioecious flowering plants are one interesting system, but little is known about them at present. Systems of neo-X and neo-Y chromosomes, in which a chromosome fusion between an autosome and a sex chromosome causes the autosome to be transmitted in the same way as the sex chromosomes, are also potentially illuminating. Given the lack of crossing over in male *Drosophila*, an autosome that becomes a neo-Y chromosome as a result of translocation of a chromosome arm to a sex chromosome (either Y or X, see below) will immediately be exposed to any forces that cause degeneration. The neo-Y must experience the same evolutionary forces as a newly evolved Y chromosome, and there is clear evidence in such cases that genetic degeneration occurs (Charlesworth, 1996; Steinemann and Steinemann, 1998). Such situations can provide evidence about how long it may take for Y-linked genes to degenerate genetically and what factors may be important for the degeneration process.

Several relatively recent Y chromosome systems, of varying age, are known in the genus *Drosophila*. The gene contents of *Drosophila* chromosome arms are highly conserved in evolution (Powell, 1997). The opportunity thus exists to use data on the molecular evolution of neo-X and neo-Y chromosomal genes to test the predictions of the evolutionary models. These all predict that the evolution of a completely degenerate neo-Y chromosome, as well as of full dosage compensation of the genes on the neo-X chromosome, should be very slow (Charlesworth, 1996). The comparative evidence from *Drosophila* supports this prediction. The *D. pseudoobscura* neo-X chromosome was formed by a fusion of an autosome to the true X about thirteen million years ago. There is apparently complete dosage compensation of genes on the neo-X and loss of activity of all genes on the neo-Y (Charlesworth, 1996). The related species, *D. miranda*, which diverged about two million years ago, has a neo-Y chromosome formed by the fusion of the *Drosophila* autosome "element C" with the true Y. Although many of the genes on the neo-Y still retain their function, a substantial fraction have become nonfunctional or completely lost, and their homologues on the corresponding neo-X chromosome have become dosage compensated (Steinemann and Steinemann, 1998). Data on genetic variability on the neo-Y suggest that its effective population size is very low, as expected in the models described above (Bachtrog and Charlesworth, 2000).

[*See also* Dosage Compensation; Genomic Imprinting; Hitchhiking; Sex, Evolution of.]

BIBLIOGRAPHY

Bachtrog, D., and B. Charlesworth. "Reduced Levels of Microsatellite Variability on the Neo-Y Chromosome of *Drosophila miranda*." *Current Biology* 10 (2000): 1025–1031.

Bull, J. J. *Evolution of Sex Determining Mechanisms*. Menlo Park, Calif., 1983.

Charlesworth, B. "The Evolution of Chromosomal Sex Determination and Dosage Compensation." *Current Biology* 6 (1996): 149–162.

Charlesworth, B., and D. Charlesworth. "A Model for the Evolution of Dioecy and Gynodioecy." *American Naturalist* 112 (1978): 975–997.

Charlesworth, B., P. Sniegowski, and W. Stephan. "The Evolutionary Dynamics of Repetitive DNA in Eukaryotes." *Nature* 371 (1994): 215–220.

Darwin, C. R. *The Different Forms of Flowers on Plants of the Same Species*. London, 1877.

Delichère, C., J. Veuskens, M. Hernould, N. Barbacar, A. Mouras, I. Negrutiu, and F. Moneger. "*SlY1*, the First Active Gene Cloned from a Plant Y Chromosome, Encodes a WD-Repeat Protein." *EMBO Journal* 18 (1999): 4169–4179.

Graves, J. A. M. "The Origin and Function of the Mammalian Y Chromosome and Y-borne genes—an Evolving Understanding." *BioEssays* 17 (1995): 311–321.

Lahn, B. T., and D. C. Page. "Four Evolutionary Strata on the Human X Chromosome." *Science* 286 (1999): 964–967.

Muller, H. J. "Genetic Variability, Twin Hybrids and Constant Hybrids in a Case of Balanced Lethal Factors." *Genetics* 3 (1918): 422–499.

Orr, H. A., and Y. Kim. "An Adaptive Hypothesis for the Evolution of the Y Chromosome." *Genetics* 150 (1998): 1693–1698.

Powell, J. R. *Progress and Prospects in Evolutionary Biology: The Drosophila Model*. New York, 1997.

Rice, W. R. "Genetic Hitch-hiking and the Evolution of Reduced Genetic Activity of the Y Sex Chromosome." *Genetics* 116 (1987): 161–167.

Rice, W. R. "Evolution of the Y Sex Chromosome in Animals." *Biosciences* 46 (1996): 331–343.

Steinemann, M., and S. Steinemann. "Enigma of Y Chromosome Degeneration: Neo-Y and Neo-X Chromosomes of *Drosophila miranda*, a Model for Sex Chromosome Evolution." *Genetica* 102/103 (1998): 409–420.

Vagera, J., D. Paulikova, and J. Dolezel. "The Development of Male and Female Regenerants by in vitro Androgenesis in Dioecious Plant *Melandrium album*." *Annals of Botany* 73 (1994): 455–459.

Westergaard, M. "The Mechanism of Sex Determination in Dioecious Plants." *Advances in Genetics* 9 (1958): 217–281.

— Deborah Charlesworth

SEX DETERMINATION

The term "sex determination" (SD) denotes the process by which an individual or a reproductive structure becomes male or female. For dioecious species (those with two distinct sexes), it refers to the mechanisms by which an individual embryo is directed to become either male or female. For hermaphroditic species, it denotes the process by which individual sexual organs are assigned to produce either male or female gametes. For both dioecious and hermaphroditic species, SD is a vital biological process: imprecise SD leads to the production of faulty, intersexual individuals or sex organs, and consequently to reproductive impairment. SD is thus an important component of Darwinian fitness, and as such, it must be subject to strong selective pressures. Nevertheless, SD mechanisms can undergo rapid evolutionary change. This evolutionary lability is indicated by the diversity of these systems, not only within major phylogenetic groups such as insects or vertebrates, but occasionally even within species. The evolution of SD mechanisms thus presents a puzzle: the trait itself is under rigorous selective control, yet it can undergo frequent and rapid evolutionary change.

This article explores three facets of SD, focusing on animal systems, for which the data are most extensive. First, we will review the diversity of known animal SD systems. Second, we will look at evidence that sex-determining systems can undergo rapid evolutionary change, and at the nature of the transitions between systems. Third, we will examine the underlying genetic foundations of SD and how these genetic architectures can change. Finally, we will return to the conundrum of rapid evolutionary change in these systems despite strong directional selective pressures for maintaining correct sex determination. The related phenomenon of sex ratio control, though relevant to the evolutionary pressures affecting SD, is covered separately in the entry "Sex Ratio."

Diversity of Sex Determining Systems. Although the number and range of modes of SD is large, they can be divided into three major categories: genetic sex determination (GSD), environmental sex determination (ESD), and cytoplasmic sex determination (CSD) systems (see Table 1). GSD systems, found in both animals and plants, are those in which the development of one sex or the other is triggered by the presence or absence of one or more critical genetic factors. ESD systems are those in which the sex of an individual is determined by the setting of a particular environmental variable along a range of possible values, with males developing at one set of values and females at the other. CSD systems are known to exist only in a few crustacean groups; in these, the presence of a particular cytoplasmic element triggers development of a particular sex, usually female, regardless of other factors. Table 1 shows some of the groups in which the different types of SD have been found and something about the mechanisms involved. The following discussion describes some key features of the three categories.

Most animals have GSD systems, of which there are four distinct subcategories. The great majority of GSD

TABLE 1. Modes of Sex Determination (SD) in Animals

Category	*Found in*	*Mechanism*
Genetic Sex Determination (GSD)		
Male heterogamety XY or XO (male); XX (female)	Mammals; fish; lizards; nematodes; arachids; insects; crustacea; nematodes	Dominant M factor on Y, recessive F factor on X, or double dose of X for feminization
Female heterogamety ZZ (male); ZW (female)	Birds; snakes; fish; insects; crustacea	Dominant F factor on W or recessive M factor on Z
Multifactorial and polyfactorial	Fish; mammals; insects; crustacea	Mix of dominant F or M factors (multifactorial) or additive M or F factors (polyfactorial) from multiple loci
Arrhenotoky haploid males; diploid females	Bees; wasps; scale insects; beetles; ticks; mites; rotifers	"Complementary sex determination" requiring different alleles at single locus to give female development
Environmental Sex Determination (ESD)		
Temperature as key variable	Fish; turtles; crocodiles; insects	Temperature-sensitive production of estrogens
Crowding as key variable	Fish; nematodes	??
Contact as key variable	Echiurid worms	Larva settling on adult female become male; otherwise, female development
Cytoplasmic Sex Determination (CSD)		
Bacterial, microsporidial or protozoan symbionts in germline	Crustacea	Symbiont feminization by over-riding of M factors.

systems have a single segregating pair of chromosomes, termed the sex chromosomes, that determine sex. Among these, most are so-called XX/XY systems, in which males have two different sex chromosomes (XY)—a situation of "male heterogamety"—while females carry two of the same sex chromosome (XX). The other major class of GSD system employing sex chromosomes involves female heterogamety (ZZ/ZW); in this case, females have two different sex chromosomes (ZW), while males have two of the same sex chromosome (ZZ). Within both classes, the degree of visible (karyotypic) differentiation of the two sex chromosomes can vary enormously, from none to dramatic. Where the differences are visible, the chromosomes are said to be "heteromorphic." Furthermore, the chromosome unique to the heterogametic sex (either the Y or the W) can be reduced to a vestige or even eliminated entirely, as in certain XX/XY systems that have become XX/XO. Systems in which the degree of heteromorphy is slight or undetectable are probably recently originated ones, and those with pronounced sex chromosome differences are generally longer entrenched. Segregating single-gene GSD systems can be considered the simplest form and probably represent the earliest stage of chromosomal differentiated GSD systems. In other words, the formation of distinctive sex chromosomes itself reflects an evolutionary process. Another important point is that, even among chromosomal GSD systems of the same nominal class, the mechanisms of the sex-determining trigger can be quite different. Thus, both the fruit fly and the mouse have XX/XY systems, but in the fruit fly, the triggering agent is the ratio of sex chromosomes to non-sex chromosomes (the autosomes)—the so-called X:A ratio—while in the mouse (and other eutherian mammals), sex is determined by a dominant masculinizing gene on the Y chromosome (a gene designated *Sry*).

The two commonest forms of ESD systems are those in which the environmental trigger is provided either by temperature or by degree of crowding. Temperature-dependent setting of sexual phenotype is found in crocodiles, turtles, and certain species of lizards, fish, and insects. Crowding as a sex-determining trigger is rarer, known principally in a few fish and nematode species.

CSD systems typically involve microbial or protozoan infection in the germ line of certain crustacean groups. The mechanisms behind CSD can differ, however. They include selective killing of one sex (a sex ratio effect, rather than a SD effect per se), altered segregation of sex chromosomes, and true feminization of those embryos whose genetic constitution would otherwise dictate male development. Since all known CSD systems originated with an external infection, CSD systems may be regarded as a special case of ESD.

Evolutionary Transitions Among Sex-Determining Systems. Despite the fact that most organisms can be classified as having GSD, ESD or CSD systems, these typologies are not immutable characteristics. There is much evidence for evolutionary transitions between major categories and between different kinds of GSD systems. The inference that SD is a rapidly evolving trait is based on that evidence.

The occurrence of multiple kinds of SD system within one species or a group of closely related species provides particularly strong support for this inference. Polymorphism of this sort has been documented in both vertebrates and arthropods. For instance, both the platyfish (a vertebrate) and certain isopod species (a crustacean group) are known to have both XX/XY and ZZ/ZW populations; among the platyfish, at least, matings between individuals of different systems can take place. In the housefly, *Musca domestica*, three different GSD systems have been found in different populations: one involves cytologically differentiated sex chromosomes with male heterogamety; another involves a dominant M gene residing on a transposable genetic element; and one involves a dominant female-determiner. In principle, conversion of one form of heterogametic system to the other need only involve invasion of a population by a new sex-determining gene that is epistatic to the previously existing one.

Furthermore, transitions can occur not just between different GSD systems, but even between different major categories of SD systems. For instance, the isopod species *Armadillium vulgare* has a ZZ/ZW system of genetic sex determination. When ZZ males, however, are infected by a species of the alpha purple bacteria group, *Wolbachia*, they are converted into females. In some populations, the *Wolbachia* infection has spread so widely that it has replaced the usual ZZ/ZW system, with infection alone determining sex (females infected, males uninfected). There is also some evidence for conversion between a temperature-dependent ESD system and an XX/XY temperature-independent GSD system in the Atlantic silverside fish. Such a system is a hybrid ESD-GSD system.

Apart from comparative studies that strongly imply the lability of and transitions between different GSD systems, there is experimental evidence. The nematode *Caenorhabditis elegans* has a male heterogametic system of the XO/XX form, but it is possible, with appropriate genetic manipulation of the female-determining autosomal gene *tra-1* (see Figure 1), to convert a population consisting entirely of XX animals into a ZW/ZZ system.

The principal exceptions to the general lability of SD systems appear to be those in which the sex chromosomes are highly physically differentiated, and those in which the key SD control gene (see below) also controls dosage compensation. Such systems are probably no longer capable of evolving new SD mechanisms.

Genetic Pathways of Sex Determination and Their Evolution. In recent years, it has become apparent that classifying SD systems in the manner shown in Table 1 is, in a certain sense, highly misleading. First, it seems probable that all SD systems have *some* genetic component; ESD and CSD systems involve an initiating external agent, but genetic machinery is involved in carrying out the "decision." Furthermore, the classification in Table 1 creates a taxonomy based on the *initiating genetic signal*; yet GSD systems consist not just of such "upstream" switch genes but of sequences of genes, each one affecting the activity of the next, such sequences being termed "genetic pathways." It is the setting of the final, or most "downstream," gene activities

FIGURE 1. The Somatic SD Pathway of the Nematode, *C. elegans*.
As described in the text, the sequence of gene activities is a sequence of differential inhibitions. There are two different outcomes (production of males and of hermaphrodites, which are modified females) because the initial X:A ratio difference sets a difference in the first gene of the pathway, that of *xol-1*. The critical difference is the setting of the downstream-most gene, *tra-1*: high activity gives hermaphrodite (female) development and low activity, male development. Downstream of *tra-1*, there are "sub-contracted" regulator genes. One of these, active in the male, is *mab-3*, a member of the *Dmrt* gene family (see text and Raymond et al., 1998). For a detailed description of the pathway, see Hansen and Pilgrim, 1999 or, for a simpler version, Kuwabara, P. and Kimble, J. (1992). Molecular genetics of sex determination in *C. elegans*. *Trends in Genetics* 8, 164–168.

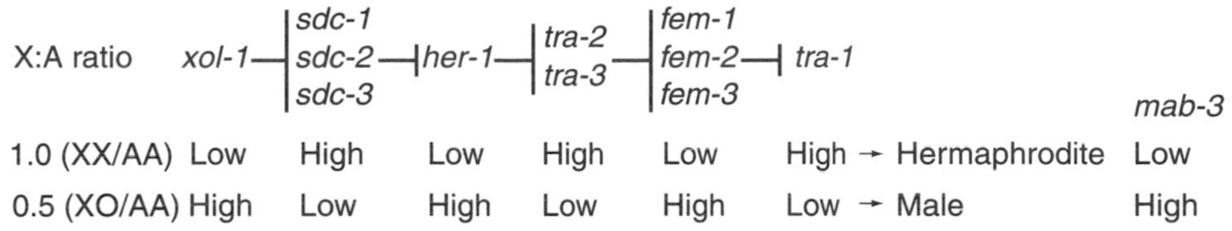

X:A ratio	*xol-1* —	*sdc-1* / *sdc-2* / *sdc-3* —┤	*her-1* —┤	*tra-2* / *tra-3* —┤	*fem-1* / *fem-2* / *fem-3* —┤	*tra-1*		*mab-3*
1.0 (XX/AA)	Low	High	Low	High	Low	High →	Hermaphrodite	Low
0.5 (XO/AA)	High	Low	High	Low	High	Low →	Male	High

in these pathways that actually determines sex. For instance, in the SD pathway of the fruit fly *Drosophila melanogaster*, there are two alternately spliced forms of the downstream gene, *doublesex* (*dsx*); female development follows expression of the dsx^f form, while male development occurs if the dsx^m form is expressed. In the other well-characterized GSD pathway, that of *Caenorhabditis elegans*, the gender of a particular embryo is determined by the on/off state of the downstream gene, *tra-1*. High activity yields hermaphrodite (female) development, and low activity yields male development.

A simplified version of the *C. elegans* SD pathway for the nematode soma is shown in Figure 1. (The pathway for the nematode germ line is somewhat different.) Two features of the main *C. elegans* pathway stand out. First, it consists of a series of inhibitory activities involving diverse molecular mechanisms; this structure ensures that opposite final states of *tra-1* are obtained from the initial setting, produced by differences in the X:A ratio. Second, the pathway is far more complicated than it needs to be simply to produce a difference in state of *tra-1*. Such unwarranted complexity hints at a construction driven by evolutionary vagaries. In particular, the fact that the pathway consists of a series of inhibitory activities suggests that it evolved, step by step, via selection for and "recruitment" of successive inhibitory activities. Such a pattern would produce an expansion of the pathway during evolution from downstream-most to upstream-most elements. Furthermore, it predicts, among related organisms, the conservation of downstream elements with probable divergence among upstream elements. Although this has not been demonstrated for the nematode SD pathways, there is now evidence for preferential functional conservation of downstream elements, relative to upstream genes, in both vertebrate and insect SD pathways. New findings even indicate that there is a small gene family, the *Dmrt* genes (for "*d*oublesex *m*ab-3 *r*elated *t*ranscription factor"), whose members serve as downstream male-determining genes in nematodes, insects, and vertebrates.

Conclusions. In recent years, the investigation of the evolution of sex determination has moved into a new phase. Increasingly, one can view the control of sex determination in terms of sequences of gene action. This holds true even for ESD, where the environmental trigger can be seen as setting in motion a sequence of gene actions that differ in the two sexes to produce two different pathways of sexual differentiation. Much of the lability of SD pathways can now be interpreted as involving recruitment of new upstream genetic factors that modify or modulate the action of pre-existing and shorter SD pathways. Such recruitment both lengthens pathways and makes them more complex than their ancestral forms. In contrast, where SD differs between the soma and germ line within an organism, as shown for both *D. melanogaster* and *C. elegans*, some of the differences almost certainly have arisen via recruitment of more downstream elements and, possibly, displacement of others.

The forces that drive rapid evolutionary change in SD pathways remain obscure, however. Selection for different sex ratios in certain conditions almost certainly plays a part, but such pressures may not be the whole story. It is also possible that selected changes in other genetic pathways and networks impinge on the operation of the SD pathways, indirectly generating selective pressures for change in the latter. The unusually high rates of sequence evolution that have been observed for many SD genes are consistent with such internal "arms races." Since diminution of activity in one setting of the pathway simply causes the production of an individual of the opposite sex, SD pathways may have built-in compensatory responses for the effects of change that other genetic pathways lack. As means of manipulating individual genetic activities in different organisms develop through the techniques of functional genomics, such ideas should be testable. Nevertheless, a classic problem in evolutionary biology, the evolution of sex determination, has already been brought into the realm of molecular genetic and developmental thinking.

[*See also* Dosage Compensation; Sex Chromosomes; Sex Ratios.]

BIBLIOGRAPHY

Baker, B. S. "Sex in Flies. The Splice of Life." *Nature* 340 (1989): 521–524. An excellent review of the somatic sex determination pathway in *Drosophila melanogaster*.

Bull, J. J. *Evolution of Sex Determining Mechanisms*. Menlo Park Calif., 1983. The classic account of the variety of SD systems and their evolution.

Bull, J. J., and E. L. Charnov. "Changes in the Heterogametic Mechanism of Sex Determination." *Heredity* 39 (1977): 1–14. An analysis of how one form of heterogametic GSD can readily evolve into the other form.

Charlesworth, D., and D. S. Guttman. "The Evolution of Dioecy and Plant Sex Chromosome Systems." In *Sex Determination in Plants*, edited by C. C. Ainsworth, pp. 25–49. Oxford 1999. A good review of the evidence that the hermaphroditic state is ancestral in angiosperms and the genetic mechanisms involved in the evolution of dioecy and sex chromosomes in plants.

Cline, T. W. "The *Drosophila* Sex Determination Signal: How Do Flies Count to Two?" *Trends in Genetics* 9 (1993): 385–390. A discussion of how a mere twofold difference in X:A ratio can affect the sex determination decision.

Conover, D. O., D. A. van Voorhees, and A. Ehtisham. "Sex Ratio Selection and the Evolution of Environmental Sex Determination in Laboratory Populations of *Menidia menidia*." *Evolution* 46 (1990) 1722–1730. A laboratory demonstration of the evolution of an ESD system toward GSD.

Hansen, D., and D. Pilgrim. "Sex and the Single Worm: Sex Determination in the Nematode *C. elegans*." *Mech. Dev.* 83 (1999): 3–15. A good review of the SD pathway in the nematode and of the molecular evolution of the genes in SD pathways.

Hodgkin, J. "Two Types of Sex Determination in a Nematode." *Nature* 304 (1983): 267–268. An early experimental conversion of a male heterogametic SD system to a female heterogametic SD system, possibly mimicking some of these conversions in nature.

Pieaud, C. "Temperature Variation and Sex Determination in Reptiles." *BioEssays* 18 (1996): 19–26. A discussion of the biochemistry of one form of ESD, that found in turtles.

Raymond, C. S., et al. "Evidence for Evolutionary Conservation of Sex-determining Genes." *Nature* 391 (1998): 691–694. The first report indicating that the various elucidated SD pathways in bilaterian animals may have a shared link after all, a downstream male-determining gene of the *Dmrt* family.

Rigaud, T., P. Juchault, and J.-P. Mocquard. "The Evolution of Sex Determination in Isopod Crustaceans." *BioEssays* 19 (1997): 409–416. A review of CSD mechanisms in one group exhibiting this form of SD, the isopods.

Tomita, T., and Y. Wada. "Multifactorial Sex Determination in Natural Populations of the Housefly (*Musca domestica*) in Japan." *Japanese Journal of Genetics* 64 (1989): 373–382. A survey of the range of SD systems found within populations of a particular species, the housefly *Musca domestica*, illustrating the evolutionary lability of SD mechanisms.

Wilkins, A. S. "Moving up the Hierarchy: A Hypothesis on the Evolution of a Genetic Sex Determining Pathway." *BioEssays* 17 (1995): 71–77. An explanation of how sex determination pathways might evolve, becoming more complex in structure, during their evolution.

— ADAM S. WILKINS

SEXES AND MATING TYPES

The evolution of sexes is not one but many problems. First, we need to define what a sex is. Typically, we refer to males and females as different sexes. There are many differences between the sexes, for example, in size, coloration, sexual organs, and parental care. Yet none of these are consistent differences across the range of animal and plant species. The only consistent difference between the two sexes is size of the gametes: males produce small gametes (sperm or pollen), whereas females produce large gametes (eggs or more technically oocytes). This difference is also present in hermaphroditic species (many plants, numerous invertebrates, but notably rare in insects), where a single individual produces both male (small) and female (large) gametes.

This definition of the sexes poses a major question: Why do all higher organisms have two forms (sexes) that differ in gamete size? This asymmetry is fundamental to reproduction as only sperm can fertilize eggs. An egg (i.e., large gamete) never fuses with another egg to produce an offspring, nor do two sperm ever fuse with each other. A way of approaching this question is to note that the ancestral state appears to be isogamy, that is, same-sized gametes. This is seen in numerous protists (single-celled organisms) such as yeast and the green algae *Chlamydomonas*. Why, then, did isogamy evolve into anisogamy, that is, unequal-sized gametes?

This question still does not get to the real problem because in isogamous forms there are so-called mating types. In *Chlamydomonas*, for example, these are referred to as + and − types. In yeast, they are known as α and alpha. For the most part, isogamous organisms have two mating types, but there are rare exceptions (e.g., slime molds).

What these mating types have in common with sperm and eggs is the most elemental form of mate choice. Just as sperm fuse only with eggs (and vice versa), so + gametes will fuse only with − gametes in *Chlamydomonas*. The genetics of mating preference in *Chlamydomonas* are well described. The mating type loci segregate as if they were one locus with two alleles (though in fact a large multigene complex). The + and − types both code for proteins on the cell membrane that mediate the preference to fuse with cells of the opposite type. The most fundamental questions, then, are why do isogamous organisms have mating types, and why are there typically just two of them?

The Evolution of Mating Types. That mating types should exist at all is a paradox. Imagine a species of isogamous organisms that have a means to recognize other members of the same species but in which there are no mating types: any cell from the same species is a potential mate. When it comes to mate finding, the probability that a randomly encountered cell from the same species is a potential mate is therefore 1. In contrast, when there are two mating types, this figure goes down to 0.5. Therefore, the move from no mating types to two mating types appears to be one that decreases fitness, in that it is likely to increase the costs associated with mate finding.

The problem does not stop there. Imagine now a world with three mating types. We can call them *a*, *b*, and *c*. Under this scheme, we could imagine numerous rules for mating. Two of them could mate with only one other type in a pair; *a* might mate with *b* but not *c*, *c* might mate with *b* but not *a*. As *a* and *c* both mate with *b*, gene flow is maintained, so that the three types are all members of the same species. Alternatively, mating may not be pairwise: it may be that *a*, *b*, and *c* all must fuse to form a zygote. Neither of these methods is observed in nature. The most likely reason is that both make mate finding even more costly.

A third possibility is that each is prepared to mate with the other two in pairwise interactions. There are a few examples of mating systems like this, demonstrating that its rarity is unlikely to be due to a constraint (i.e., the impossibility of having a third type). The enigma is why this is not more common. In such a system, two-thirds of the population, on average, are potential mates. Furthermore, at the point at which a third sex first arises, all members of the population are potential mates. Therefore, having two sexes is not only worse than having none, it is also worse than having three, which in turn is unstable to invasion by a fourth, and so on. Two stable positions seem to exist: no mating types or an infinite number.

Naturally, the figure infinity is a mathematical curiosity, and the actual upper limit is likely to be dictated by the loss of mating type alleles by drift. Under some mating schemes, it is argued that drift can also take the population back to two sexes. However, this is unlikely when mate finding is costly. The problem can then be stated as why, if mating finding is costly, do we see so many examples of species with two and only two mating types? The resolution of this question must have something to say about the fundamental asymmetry between the sexes.

The Inheritance of Organelles. One answer, which appears to be both theoretically and empirically robust, centers on one of the other roles of the mating type loci in *Chlamydomonas*. As is typical of mating type loci, this gene complex not only controls whom to mate with but also coordinates the inheritance of organelle genomes (i.e., mitochondrial and chloroplast genomes).

It is well known that anisogamous organisms typically have uniparental inheritance of organelles (e.g., humans receive theirs, the mitochondria, predominantly, if not exclusively, from their mothers). Often this is put down to a simple size asymmetry between the gametes: sperm have little room for organelles. However, many facts point to this not being the correct answer. First, in humans, the paternal mitochondria are actively destroyed in the fertilized egg rather than simply diluted out. Second, in numerous plants, the transmitting parent can be the father, and it is the maternal contribution that is destroyed. This is the case in giant redwoods, for example, in which both chloroplast and mitochondria are paternally inherited. In other gymnosperms, inheritance of one organelle may be maternal while the other may be paternal. Third, in mussels, there are two types of mitochondrial genome. One is transmitted from mothers to all progeny, the other by fathers to sons (but not to daughters). Fourth, in the majority of unicellular isogamous organisms, *Chlamydomonas* included, inheritance of organelles is uniparental. This too is the result of active destruction of the organelle genomes of one of the two parents. As with the gymnosperms, it can be the case that the transmitting parent for the mitochondria is not the transmitting parent for the chloroplast genome (this is so in *Chlamydomonas*). This all points to some selection pressure that acts strongly to achieve the uniparental transmission of organelle genomes.

The dominant theory for the evolution of two sexes assumes that originally organelles were inherited from both mating types. Imagine that there was some cost to having biparental inheritance. A mutant that destroyed

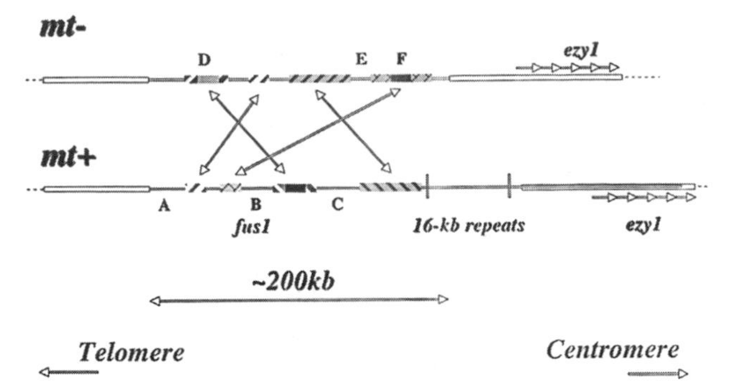

FIGURE 1. The Mating Type Loci of *Chlamydomonas reinhardtii*.
The region contains genes known to be cell surface recognition proteins (e.g., *fus 1*) and others involved in organelle inheritance (the unnamed 16-kilobase [kb] repeats and probably *ezy 1* as well). From the theory outlined in the text it is predicted that (1) *mt*− and *mt*>+ need not code for the same proteins (genes A–F are sequences unique to either + or −) and (2) there should be blocks to recombination between the two clusters. The regions have many features that limit recombination, including rearrangements of genes and gene orders (arrows indicate othologous segments), insertions to limit homology searching, as will inversions. Classical genetics indicates that these are effective blocks to recombination. With permission, from the *Annual Review of Plant Physiology and Plant Molecular Biology*, Volume 46 © 1995 by Annual Reviews www.AnnualReviews.org.

the organelles of the partner thus would be advantageous. However, as the mutant allele spreads through the population, it starts to meet itself, which is likely to be costly—both of the cells attempting to fuse to form a zygote will now attempt to kill off each other's organelles. This frequency-dependent selection ensures that the mutant spreads when rare but cannot go to fixation. This position is now ripe for the evolution of mate choice. The cells with the organelle elimination gene will prefer to mate with those without and vice versa. The genes for mate choice and for organelle elimination need to be closed linked (i.e., nonrecombining); otherwise, offspring will end up with no organelles or the disadvantage of bidirectional inheritance of organelles. This more or less describes the anatomy of the mating type locus of *Chlamydomonas* (Figure 1).

Why is biparental inheritance likely to be costly? Most models concentrate on the problem of "selfish" organelle genomes. Unusually, organelle genomes are free to replicate during the cell cycle. This can be contrasted with nuclear genes that are replicated once per mitosis. Consequently, we can expect a faster replicating organelle genome to have an intracell competitive advantage. Such fast replicators have been observed in yeast and are typically organelle genomes with deletions. These deletions usually compromise function. Despite the reduced function, the fast replicators can still spread in the population so long as they receive biparental inheritance and are not too costly in comparison with their intracell advantage.

The spread of the organelle elimination allele depends on its being associated with the more fit of the two organelle types. That is, if the elimination allele first occurs in an individual without deleterious organelles, then the allele stays associated with the organelle genome that confers high individual fitness. The presence of such organelles can also drive a system with three mating types back to one with two, so long as there is some leakage in the system of uniparental inheritance. This so far is the only model to show how three mating types can evolve to two when mate finding is costly.

There is comparative evidence to support this sort of model. First, the model predicts that if there is no cell fusion of gametes, and hence no mixing of cytoplasm, then there should be no restriction on the number of mating types. In both ciliates and numerous basidiomycete fungi, this is what is seen. Second, in some ciliates, gamete fusion appears to have been reevolved, and with it has come a reduction in the number of mating types down to two. In some populations, both fusion and nonfusion mating is seen. There are very many matings types when the latter is performed but not for the former. Third, the model predicts that patterns of organelle inheritance may vary as a function of the inbreeding level, which is seen. For example, if identical twins were to meet, then if one has the deleterious organelle, the other will as well. Hence selection for the control of organelle inheitance is weakened.

This is by no means the resolution of the issue. Many isogamous organisms have more than one sort of mating type locus. The slime molds, for example, in addition to having a locus that controls both mate choice and organelle inheritance, also have at least one other unlinked locus that controls mating preferences. Additional incompatibility loci have also been reported in plants and fungi. Why they exist is a separate and still unresolved issue.

Whether the fast replicating mitochondria are the reason for uniparental inheritance is unclear. Other models postulate a simple cost to mixing organelle genomes from different sources (rather than a cost associated with one particular type of organelle genome), in which case the nuclear allele eliminating organelles is equally beneficial if it eliminates its own or if it eliminates others. So far, there is little evidence for such costs. The distinction between these subtly different sorts of models is of some importance when considering the evolution of anisogamy.

The Evolution of Anisogamy. The second transition to consider is that from a system with two mating types in an isogamous species to one with different-sized gametes. What forces may have been involved here?

Extending the previous logic, some experts have argued that the evolution of small sperm may have been a general mechanism for forcing uniparental inheritance of all cytoplasmic factors (viruses and bacteria as well as organelles) or for improving the efficiency of uniparental inheritance. In *Chlamydomonas reinhardtii*, for example, there is 10 percent leakage of chloroplast genomes from the "nontransmitting" mating type. However, in the equally isogamous *C. monoica*, uniparental transmission is near perfect.

Although this may appear to be an attractive idea, its feasibility is very much dependent upon the sorts of costs that arise as a result of cytoplasmic fusion. At the crux of the problem is the idea that the evolution of anisogamy requires a mutation that acts in one sex (the putative male) to reduce the size of its own gametes, and in so doing limit the amount of biparental inheritance. If there are costs of cytoplasmic mixing, then this is no problem. However, as noted above, evidence for such costs is weak. If, by contrast, the main cost arises from fast replicating organelles, there is a problem. There is only very weak selection in favor of the allele that forces the elimination of its own organelles, because it does not build up a strong association with the more fit of the two organelle types.

The model therefore has a hard time explaining the evolution of anisogamy, but it makes sense of many details of control of organelle inheritance. Most notably, it predicts that the transmitting sex (typically females) should prevent the partner's organelles from being transmitted or inherited. Evidence in humans and in *Chlamydomonas* indicates that uniparental inheritance is indeed controlled by such alleles. The fact that some sperm such as those of the sea squirt *Ciona* have their cytoplasm removed by passage through a narrow canal prior to entry into the egg also makes sense in this context. The same is seen in some ferns.

This apparently small detail provides evidence that the most discussed model for the evolution of anisogamy is also unlikely to be wholly correct. This model, named for its developers Geoff Parker, Robin Baker, and Vic Smith (PBS), postulates that anisogamy evolves when zygote size and zygote fitness are related in a particular way. The model requires that a unit increment in zygote size causes a more than unit increment in zygote fitness. Why this may be is not obvious. If it is true, then it pays to invest more into a given gamete, that is, to make the gametes larger. The second component in the model supposes that as this goes on, cheats can prosper that make smaller gametes that parasitize larger gametes.

These cheats make many small gametes, thus fertilizing more eggs (the larger gamete) per unit investment. This disruptive selection therefore gives rise to the evolution of anisogamy.

An attractive feature of the model is that it does suppose that isogamy can exist stably. Tests of the PBS model have been limited. As mentioned, the model requires that females attempt to maximize zygote volume. In this light, the activity of stripping off sperm cytoplasm prior to fusion is clearly contrary to expectations. However, this may be derived secondarily.

Only two sorts of tests have been performed. First, it has been questioned whether with increasing organism size the degree of anisogamy goes up. It appears that a prediction of the model is that as organisms get larger, the potential benefits for producing larger zygotes may accelerate, and thus larger organisms should be more anisogamous. The volvocales, relatives of *Chlamydomonas*, have been used to test this prediction. The volvocales are a group of organisms that are simple multicellular balls of relatively few cells. A few tests have been done to see whether the size of the parent that produces the eggs is a predictor of the degree of anisogamy (ratio of egg to sperm volume). The evidence appears to suggest that this is true, although the answer from the comparative analysis is sensitive to the choice of which phylogeny is employed.

Second, the model predicts that within the same group, as the size of the organism goes up, so should the size of the egg. Phylogenetically controlled analyses suggest that this may be true, but, again, the result is sensitive to the choice of phylogeny. Alternative reasons as to why these two predictions may be upheld within the volvocales have also been presented.

The situation remains ambiguous, and other models should be considered. For example, it has been suggested that anisogamy arose through disruptive selection, with one sex favoring staying put and emitting pheromones to attract the other, whereas the other was not a pheromone producer but an active searcher instead.

BIBLIOGRAPHY

Cosmides, L. M., and J. Tooby. "Cytoplasmic Inheritance and Intragenomic Conflict." *Journal of Theoretical Biology*, 89 (1981): 83–129. The first article to propose a link between the evolution of anisogamy and selfish organelles. No mathematical analysis is presented.

Hoekstra, R. F. "Evolution of Uniparental Inheritance of Cytoplasmic DNA." In *Organisational Constraints on the Dynamics of Evolution*, edited by J. Maynard Smith and G. Vida, pp. 269–278. Manchester, 1990. The first article to note the importance of linkage disequilibrium in models of uniparental inheritance invoking selfish organelles.

Hurst, L. D. "Selfish Genetic Elements and Their Role in Evolution: The Evolution of Sex and Some of What That Entails." *Philosophical Transactions of the Royal Society of London Series B—Biological Sciences* 349 (1995): 321–332. A review of all the comparative evidence concerning the selfish cytoplasm model for the evolution of the number of mating types.

Hurst, L. D. "Why Are There Only two Sexes?" *Proceedings of the Royal Society of London Series B—Biological Sciences* 263 (1996): 415–422. A model of the selfish organelle hypothesis for the evolution of two mating types showing that a reduction from three to two mating types is possible even if mate choice is costly.

Hurst, L. D., and W. D. Hamilton. "Cytoplasmic Fusion and the Nature of Sexes." *Proceedings of the Royal Society of London Series B—Biological Sciences* 247 (1992): 189–194. The first article to note that the selfish cytoplasm model could explain much of the variation in the number of mating types.

Iwasa, Y., and A. Sasaki. "Evolution of the Number of Sexes." *Evolution* 41 (1987): 49–65. An investigation into the possible role that drift might have in returning populations from many to two sexes.

Levitan, D. R. "Effects of Gamete Traits on Fertilization in the Sea and the Evolution of Sexual Dimorphism." *Nature* (London) 382 (1996): 153–155. Provides evidence that, under sperm limitation, eggs can be favored to be large so as to be easily found.

Parker, G. A., R. R. Baker, and V. G. F. Smith. "The Origin and Evolution of Gamete Dimorphism and the Male–Female Phenomenon." *Journal of Theoretical Biology* 36 (1972): 181–198. Presents the disruptive selection argument for the evolution of sperm and eggs. A weakness of the article is that the authors assume that the absence of mating types is the ancestral condition.

Randerson, J. P., and L. D. Hurst. "Small Sperm, Uniparental Inheritance and Selfish Cytoplasmic Elements: A Comparison of Two Models" *Journal of Evolutionary Biology* 12 (1999): 1110–1124. Shows that the selfish organelle model for the evolution of anisogamy is unlikely to be a valid explanation if the most likely sort of costs of possession of a selfish organelle are the only costs.

Randerson, J. P., and L. D. Hurst. "A Comparative Test of a Theory of the Evolution of Anisogamy." *Proceedings of the Royal Society of London Series B—Biological Sciences* 268 (2001): 879–884. A comparative test of the PBS model for the evolution of anisogamy. The two predictions of the PBS model are shown to be upheld; this result is sensitive to the choice of phylogeny, and alternative explanations for the patterns, particular to the volvocales, are discussed.

Randerson, J. P., and L. D. Hurst. "The Uncertain Evolution of the Sexes." *Trends in Ecology and Evolution* 16 (2001): 571–579. Argues that the PBS model is unlikely to be correct and that mate attraction models should be given more prominence.

— LAURENCE D. HURST

SEX RATIOS

The sex ratio is the numerical ratio of male to female individuals, usually expressed as the proportion of males within a group (such as a family) or population. Evolutionary studies of sex ratio variation are some of the most successful applications of predictive theoretical biology, and they have provided classic tests of kin selection theory, with great relevance for the debate concerning levels of selection. However, much is still not known about sex ratios—in particular, why the same theory works well in some taxa but less well in others. Nevertheless, in studies of those organisms where the theory works well, there is currently great interest in using measurements of the sex ratio to perform "reverse engineering," and to deduce other aspects of an organism's biology that may be much more difficult to measure directly.

Development of Theory. Most animal populations seem to be composed of approximately equal numbers of males and females. It is sometimes claimed that Darwin avoided trying to explain this observation on the grounds that the problem seemed too difficult, leaving R. A. Fisher to suggest what is now accepted as the correct explanation in 1930 (see Edwards, 1998). In fact, Darwin suggested an explanation in the first edition of *Sexual Selection and the Descent of Man* (1871), closely matching that attributed to Fisher, and it seems that the idea was widely known in the early decades of the twentieth century—perhaps so widely that Fisher felt no need to give any attribution to the explanation. Nevertheless, Fisher expressed the idea very succinctly, and his explanation remains the first clear statement of the principle of frequency-dependent selection, and of evolutionarily stable strategies.

Fisher reasoned that if there were an imbalance of one sex, say fewer females, in a population, then any individuals that had a heritable tendency to produce more of the rarer sex (in this case, females) would produce offspring that had higher reproductive success. Hence, the frequency of genes causing overproduction of females would increase in the population, so that this sex would became commoner. As the population sex ratio approached equality, the strength of selection on the genes producing more of the rarer sex would decline, until at equality they would have no net selective advantage. This argument works no matter which sex is initially rarer, and so it predicts that an equal sex ratio is the only evolutionarily stable outcome. One criticism of Fisher's theory is that if sex is determined by sex chromosomes inherited in a Mendelian fashion, then an equal sex ratio is also expected as an outcome of random Mendelian inheritance. However, the few experimental tests of Fisher's idea that have been performed have supported it.

In cases where the two sexes cost differing amounts for the parents to produce (for example, because of one sex being larger than the other), Fisher's theory still works, but in this case it is the total amount of resources invested in each sex that is predicted to be equal. This will lead to an unequal sex ratio. The quantity that is predicted to be equal is the amount *allocated* to the two sexes; in fact, most sex ratio theories are actually theories about *sex allocation*, although the sex ratio is often assumed to be closely correlated with sex allocation in empirical studies. In this broader guise, simple theory is also able to predict, for example, how simultaneous hermaphrodites should divide their resources between male and female reproductive function. Because the theory predicts what parents should do, and because it is based on an economic concept of expenditure, processes happening after parents cease investing in their offspring may not influence the parents' optimum sex ratio. For example, if sexual dimorphism develops after offspring become independent, or if mortality is sex-biased after independence, biased sex ratios at birth are not expected.

Hamilton's Sex Ratio Theory. Fisher's theory makes many simplifying assumptions (for instance, random mating within an infinite population and parental control of the sex ratio), and it is the cases where these assumptions are violated that have yielded some of the most spectacular predictions and confirmations of modern sex ratio theory. The first major development was that by W. D. Hamilton in a groundbreaking paper published in 1967. Hamilton pointed out that if individuals mate within highly structured populations, then the logic of sex ratio selection will be changed. To take the most extreme example, if one imagines that matings are only between full siblings, then much of the investment in males would be wasted if the sex ratio were equal, because a single male might potentially fertilize all of his sisters. In such a situation, a sex ratio biased very strongly in favor of females would be the evolutionarily stable strategy. Hamilton termed this *local mate competition.*

In the same paper, Hamilton pointed out that the optimal sex ratio for a gene, or a chromosome, might depend on its mode of inheritance. For example, because the cytoplasm present in sperm generally does not become incorporated in the zygote, males represent an evolutionary dead end for any genetic element residing in the cytoplasm of the cell (for example, mitochondrial DNA), and therefore they should "prefer" a sex ratio composed of all females. On the other hand, for a gene on a Y chromosome, female offspring represent an evolutionary dead end (or "no entry" sign), so genes on the Y chromosome should prefer a sex ratio composed of all males. If either cytoplasmic genetic elements or genes on the sex chromosomes influence the sex ratio, then Fisher's assumption of parental control of the sex ratio is violated, and spectacular departures from sex ratio equality may be seen. Hamilton's arguments can be interpreted in terms of relatedness, and they were one of the first examples of explicit gene-level selection thinking; the arguments can also be extended easily to organisms where relatedness differs between different classes of individuals—for example, in many social insects.

The most familiar social insects (ants, bees, and wasps) have haplodiploid sex determination: females are diploid and develop as normal from a fertilized egg, but males are haploid and develop from unfertilized eggs. This creates a *relatedness asymmetry* and affects the reproductive value of different sexes of relatives. For queens, sons and daughters are worth an equal amount. For workers (female offspring of the queen), however, the relative worth of sisters and brothers depends upon how many times their mother has mated. In the simplest scenario, when a queen has mated only once, sisters are worth three times as much brothers. Sisters receive an identical genetic endowment from their fathers and have a 50 percent chance of sharing genes via their mother, so they are related by 0.75; their relatedness to their brothers is only 0.5, however (genes shared via mother only). When these relatedness values are multiplied by the mean reproductive value of each sex of siblings, they yield a 3:1 female:male ratio. Thus, the queen and workers are potentially in conflict over the sex ratio of reproductives produced by the colony: for the queen, an equal sex ratio would be best, but for workers, three females for every male would be best. Data from a wide range of ant, bee, and wasp species suggest that, on the whole, workers win this conflict, since sex ratios of reproductives are strongly female-biased.

Variable Sex Ratios. The theory discussed above has generally been concerned with fixed situations in which there may be a single optimum sex ratio for a population, or for a group of individuals. R. L. Trivers and D. E. Willard first pointed out, in 1973, that the sex ratio that individuals produce might also be subject to selection. Their idea was related to ungulate mammals, but it can be expanded greatly to cover any situation where there are potentially different fitness returns to be obtained from investing in offspring of one sex or the other (see Charnov, 1982, for discussion of many applications of this idea).

As an example, imagine that a female insect lays solitary eggs on patches of food of limited size, and that the food patch represents all of the resources that the developing offspring will have access to before reaching sexual maturity. In this imaginary species, the fitness of female offspring is determined largely by how many eggs they lay, which in turn is determined by their body size at maturity. For male offspring, however, fitness is independent of size. Under these conditions, the fitness of daughters will be higher than sons when a brood is produced from a large food patch, but the reverse will be true in small patches. Selection then favors a male-biased sex ratio in small food patches, and a female-biased sex ratio in large food patches. This simple model can be extended to almost any situation in which an environmental variable has differential effects on the value of male and female offspring—for example, mate attractiveness in birds, or rearing temperature in reptiles. Precisely the same argument can be used to predict at what point a sequential hermaphrodite (e.g., sex changing fish) should switch from being one sex to being the other, or how much a simultaneous hermaphrodite should expend on male and female gametes.

Flexibility in the sex ratio may also be expected when male and female offspring cost their parents different amounts depending on the circumstances in which they are produced. For example, in some vertebrates, one sex of offspring may stay with the parents and assist them with rearing subsequent broods of offspring. In that case, they are sometimes said to "repay" the cost of their production, which leads to selection for overproduction of the helping sex. However, producing more of the helping sex may not always be beneficial: if the family inhabits a limited range, at some point production of the sex of offspring that remain behind to help may cost more than producing a dispersing offspring, because the helping offspring will deplete resources that could be used for further offspring by the parents.

Empirical Evidence—Invertebrates. Evidence in support of the predictions of sex allocation theory has been gathered in many ways. In some organisms (particularly invertebrates with haplodiploid sex determination), the support for a range of predictions is so good that it is probably safe to assume that the theory is correct. In some cases where data do not fit theoretical predictions, it may be that this represents a failure by the researcher to understand the biology of the system, rather than a failure of theory.

Support for the effect of local mate competition on the evolution of sex ratios has come from a wide range of invertebrate species, especially wasps, ants, beetles, mites, and spiders. Some of the most striking of these studies have been carried out on fig wasps. Figs and the tiny wasps that pollinate them have evolved intricate interrelationships throughout the tropics; each species of fig is associated with one or more species of pollinating wasp. Figs produce flowers inside a closed fruit (the synconium) and rely on females of their particular pollinating wasp species to crawl into this fruit and pollinate their flowers. While doing so, the wasps lay their own eggs, which develop within the fruit, feeding upon some of the flowers. The wasps mature just before the fruits ripen. In many cases, the male wasps are nondispersing (they may lack wings altogether) and mate with females that hatch within the same fruit as them; it is the mated females that disperse to other fruits. This life history creates the possibility for strong local mate competition. If only one female has laid eggs within the fruit, then her sons will compete among themselves to fertilize their sisters. In this case, the optimum sex ratio is to lay just enough male eggs to fertilize all of the females;

the clutch should consist of a small number of males, with all the rest females. However, as increasing numbers of females lay eggs within the same fruit, the relatedness among competing males falls, with the result that the optimum sex ratio creeps up toward an equal number of males and females.

Work by Allen Herre in Panama has shown that multiple species of fig wasps obey this rule relating number of ovipositing females to sex ratio, but it has also added an intriguing twist that demonstrates the explanatory power of evolutionary theory. Although the same basic theory governs the optimum female response, different species may not all evolve identical responses. For example, some species almost never encounter a situation in which another female has oviposited in the same fruit, while others may do so frequently. In that case, one may expect the species that frequently encounters a range of situations to demonstrate a closer fit to the predicted relationship, because it is more often exposed to that particular selective regime. Herre showed that this is indeed the case: wasp species that are exposed regularly to a range of degrees of local mate competition show a closer fit to the theoretically expected relationship.

Experimental studies of parasitic wasps (parasitoids) also have confirmed the predictions of local mate competition, and these results have been used frequently to demonstrate that sex ratios are adjusted in response to environmental conditions. The fitness of female offspring is more strongly affected by the size of the host in which they develop than is male fitness, because female fitness is determined largely by fecundity (number of eggs laid), which is in turn determined by body size when the wasp emerges from its host. Consequently, there is selection for females to lay male eggs in small hosts, and female eggs in large hosts. Careful experiments show that it is the relative size of the host, rather than its absolute size, that determines the egg-laying female's sex ratio response. By exposing a newly emerged female to undersized hosts followed by an average-sized host it is possible to shift the sex ratio response relative to that of females that have experienced a normal range of host sizes.

Empirical Evidence—Vertebrates. In contrast to the work on invertebrates such as parasitoid wasps and social insects, which often shows a beautiful match between theory and data, the question of whether vertebrates (particularly birds and mammals) can adjust their sex ratio in an adaptive fashion has long been controversial. One reason for this is that in birds and mammals sex is determined chromosomally, rather than by whether an egg is fertilized or not (as in haplodiploid systems). It thus seems reasonable to assume that the sex of an offspring will ultimately be determined by which sex chromosome it inherits from its heterogametic parent (father in mammals, mother in birds), which will in turn be determined by random Mendelian segregation. However, a large number of studies have reported finding correlations between the sex ratio within families and some environmental factor, and a small but increasing number of experimental studies have demonstrated similar effects on the sex ratio. In many cases, these findings are still compatible with the idea that random Mendelian segregation controls the sex ratio, because it is virtually impossible to measure the sex ratio at fertilization in either birds or mammals. As a consequence, it is possible that mortality may be sex-biased between fertilization and the point at which the sex ratio is measured.

Sex-biased mortality is of particular concern in studies of mammals, where the sex ratio is rarely recorded before birth, and in which it is well established that the larger sex of offspring (usually males) is more likely to die before birth, particularly if environmental conditions are harsh. This association between the environment and sex-biased mortality can make it difficult to determine whether a correlation between the environment and the sex ratio represents an adaptation by the parent or not. For example, in red deer (*Cervus elaphus*), males are larger than females, and sons born to dominant females are disproportionately successful as adults. These conditions create selection for the sex ratio to be adjusted in relation to dominance status, with dominant females giving birth to a higher proportion of sons. However, recent analysis of a long-term dataset from the Scottish island of Rum shows that the effect of female dominance on the sex ratio is environmentally dependent. A correlation was present when the population's density was low, but this correlation disappeared as density increased, perhaps because sex-biased mortality then affected all females equally (Kruuk et al., 1999). It is not known whether the correlation between female dominance and the sex ratio at low density represents maternal control over the sex ratio, or simply the fact that only subordinate females suffered from male-biased offspring mortality.

In birds, where eggs are laid into a nest soon after fertilization, sex-biased mortality before laying seems less likely than in mammals, although it is hard to rule out with certainty. Recent development of genetic markers for the avian sex chromosomes has made possible some studies suggesting remarkable degrees of adaptation in the sex ratio. Foremost among these is a study of cooperatively breeding Seychelles warblers (*Acrocephalus sechellensis*) by Komdeur et al. (1997). In this bird, confined to a few islands in the Indian Ocean, daughters help their parents to rear subsequent offspring and thereby increase their parents' fitness. However, helpers are useful only on high-quality territories where there are plenty of insects to eat; on low-quality territories, the presence of helpers increases competition for food, leaving less for nestlings. Parents show strong, and reproducible, biases in the primary sex ratio of their offspring depending on the quality of territory that they breed on, producing daughters on high-quality territories and sons (which disperse) on low-quality territories.

Despite some successes like those mentioned above, there are numerous studies of sex ratio variation in vertebrates where it appears that sex ratios are determined by chance processes. Vertebrates tend to have more complex life histories, with overlapping generations, than do the haplodiploid insects to which sex allocation theory has been applied so successfully. This may mean not only that it is harder for biologists to understand why (and whether) sex ratios should vary, but also that selection for sex ratio adaptation does not act in a consistent manner.

Applications of Sex Ratios. In cases where sex allocation theory can be shown to work well, it can then be used to deduce other things about an organism. One example of this with considerable applications to biomedical science concerns the use of Hamilton's local mate competition theory to work out the population structure of parasites, such as human malaria as pioneered by Andrew Read. Malaria parasites reproduce clonally inside vertebrate hosts, but they also produce sexual stages (gametocytes) that are taken up in blood meals by blood-feeding insects such as mosquitoes, which act as vectors of the parasite between hosts. Sexual reproduction occurs inside the gut of the vector. Because blood meals are taken so infrequently, mating is likely to take place between the gametocytes from a single blood meal. Consequently, mating will generally take place between the different parasite genotypes (clones) that infect a single host. This sets the stage for potential local mate competition if only a small number of parasite clones infect each host. If one then measures the sex ratio of gametocytes within a host, the simple theory developed by Hamilton can be used to predict how many clones the average host contains. This information can be obtained in other ways—for example, by using genetic markers to investigate the population structure of the parasite—but this is both expensive and time-consuming compared to the "reverse engineering" method using the sex ratio. All this involves is counting gametocytes in blood smears under a microscope. Remarkably, in the few cases where population structure has been estimated in both ways, there is a close agreement between the two estimates. Knowledge of the population structure of parasites is of considerable importance for the development of vaccines and in predicting the spread of drug resistance among parasites.

Why is simple theory sometimes so successful in predicting something so fundamental about an organism as how many male and female offspring it produces? There are probably two contributing factors. First, sex ratios are simple because there are often only two quantities—the number of males and the number of females—which trade off with each other. Second, sex ratios may be under very strong selection. As a consequence, the study of sex ratios represents an excellent model for the study of the limits of adaptive evolution.

[*See also* Cytoplasmic Genes; *articles on* Eusociality; Fisher, Ronald Aylmer; Meiotic Distortion.]

BIBLIOGRAPHY

Bourke, A. F. G., and N. R. Franks. *Social Evolution in Ants.* Princeton, 1995. Contains a superb explanation of the evolutionary reasoning behind sex ratio theory in social ants, with extensive discussion of supporting data.

Charnov, E. L. *The Theory of Sex Allocation.* Princeton, 1982. The major theoretical treatment of sex ratios and sex allocation; written in accessible style.

Edwards, A. W. F. "Natural Selection and the Sex Ratio: Fisher's Sources." *American Naturalist* 151 (1998): 564–569. An interesting account of the history of the idea attributed to Fisher.

Godfray, H. C. J. *Parasitoids: Behavioral and Evolutionary Ecology.* Princeton, 1994. Contains an extremely clear review of one of the most productive and successful areas of sex ratio research.

Hamilton, W. D. "Extraordinary Sex Ratios." *Science* 156 (1967): 477–488. An extraordinary paper, which as well as defining the field of post-Fisherian sex ratios introduced many other key concepts in current evolutionary biology.

Herre, E. A. "Optimality, Plasticity and Selective Regime in Fig Wasp Sex Ratios." *Nature* 329 (1987): 627–628. An example of the close fit between theory and data that can be expected in studies of sex ratio, and how this may be used to investigate the limits of adaptation.

Komdeur, J., S. Daan, J. M. Tinbergen, and A. C. Mateman. "Extreme Adaptive Modification in Sex Ratio of the Seychelles Warbler's Eggs." *Nature* 385 (1997): 522–525. A striking example of adaptive sex ratio variation in a bird.

Kruuk, L. E. B., T. H. Clutton-Brock, S. D. Albon, J. M. Pemberton, and F. E. Guinness. "Population Density Affects Sex Ratio Variation in Red Deer." *Nature* 399 (1999): 459–461. One of the classic studies of sex ratio variation in mammals.

Trivers, R. L., and H. Hare. "Haplodiploidy and the Evolution of the Social Insects." *Science* 191 (1976): 249–263. An exhaustive test of parent-offspring (worker-queen) conflict in social ants, using predictions about the optimum sex allocation for the two warring parties. The paper spurred lively debate in subsequent issues.

Trivers, R. L., and D. E. Willard. "Natural Selection of Parental Ability to Vary the Sex Ratio of Offspring." *Science* 179 (1973): 90–92. The paper which introduced the idea that individuals might be under selection to change their sex ratios.

West, S. A., E. A. Herre, and B. C. Sheldon. "The Benefits of Allocating Sex." *Science* 290 (2000): 288–290. Discusses several areas of recent research, in particular, work pointing to broader applications of sex ratio theory in biology, such as examining constraints on adaptation, and estimating the population structure of malaria parasites.

— BEN C. SHELDON AND STUART A. WEST

SEXUAL CONFLICT. *See* Cryptic Female Choice.

SEXUAL DIMORPHISM

Sexual dimorphism refers to differences between the sexes in physical characteristics and behavior. These differences are sometimes so striking that males and females have initially been classified as different species. In the introduction to his theory of sexual selection, Charles Darwin (1871) considered two categories of sexual dimorphism. Primary sexual traits are directly concerned with and necessary for reproduction: the reproductive organs themselves, structures in which eggs or young develop and intromittent organs such as penises. Secondary sexual characteristics are not absolutely required for reproduction, and Darwin was especially concerned with the subset of these traits that influence an individual's relative mating success. Secondary sexual traits can affect mating success in the competing sex (usually males) in two contexts: (1) in direct struggles for access to mates (usually females) or to resources needed by mates, a process called intrasexual selection, and (2) in the attraction and courtship of members of the opposite sex, a process called intersexual selection. The different roles typically adopted by the two sexes in these contexts arise because males usually invest relatively little in a multitude of small gametes (sperm) and can, in principle, fertilize the eggs of many females. By contrast, the reproductive potential of females is more limited because females invest heavily in a smaller number of larger gametes (eggs) and may also be obligated to provide parental care to their young. Female mammals, for example, must provide milk to ensure survival of their young. The idea that differences in reproductive potential and parental investment underlie sexual differences in reproductive behavior is supported by examples of sex-role reversal. In pipefish and sea horses, for example, males have pouches in which they can brood a limited number of young at any one time, and females compete with one another for male parental care.

Ecological Differentiation of the Sexes. Before discussing how secondary sexual characteristics might be favored and shaped by sexual selection, Darwin (1871) mentioned sexual differences that reflect different habits of life. Among his examples, Darwin considered certain flies, such as species in the family Culicidae, whose mouth parts differ because females feed on blood and males on flowers, some insects in which males but not females have wings, and some birds having sexual differences in beak size and shape. In the huia, an extinct New Zealand bird, the female had a long decurved bill used to probe for insects in soft, decayed wood, whereas the male had a straight woodpecker-like bill used to chisel insects from hard wood. Darwin also recognized that some secondary sexual characters, such as the more highly developed sensory and locomotory organs of males found in many species, are unlikely to fit neatly into one or the other of these categories. Males that are especially well endowed in these respects might not only be better able than females to find and catch some kinds of prey items but might also be competitively superior to other males in detecting and mating with females.

Size Dimorphism. In many social animals, males are significantly larger than females. This is especially likely in species in which males fight for access to females that can be herded or that exist in a cohesive social group. This kind of mating system is called a female-defense or harem-defense polygyny, in which males can have more than one sexual partner. Sea lions, elephant seals, many ungulates (hoofed mammals), lions, and some primates are good examples. Male body weight ranges from 1.6 to 7.0 times that of females, and males usually fight directly for access to receptive females. Male mating success, at least over the short run, is highly skewed, such that a few males will obtain all or most of the matings during the period in which they are dominant. In southern elephant seals, for example, the top-ranking males in dominance hierarchies can account for more than 80 percent of the copulations (McCann, 1981). High-ranking males are often challenged, however, and seldom survive as harem masters for more than a year or two. In ungulates with well-developed weapons, as many as 10 percent of the bulls can die from injuries each year (Clutton-Brock et al., 1982). Sexual size dimorphism is also common in species in which males compete for territories—a kind of mating system that is termed a resource-defense polygyny. Because the resources in territories are required by females, males that do not possess territories have little chance of mating. Indeed, even if females have the chance to choose their mates, they probably assess resources in the territories as well as physical and behavioral attributes of the male defending it.

Sexual dimorphism in size can also favor females. In many insects and lower vertebrates such as fish and amphibians, for example, selection for fecundity (egg production) is strong and positively correlated with body size, and in a few species, males choose among females on this basis. In social insects, females that reproduce (queens) are larger than the males that guard the colony, which, in turn, are larger than the nonreproductive females (workers). The most extreme examples of female-biased size dimorphism are found in deep-sea anglerfish. Females can be nearly ten times larger than males, and in some species, the line between sexually and ecologically selected dimorphism is blurred. That is, in addition to small, free-living males, some males attach to the female, their circulatory systems become fused, and the male receives all of his nutrition from the female (Nelson, 1984).

Weapons. Weapons such as horns and antlers have evolved numerous times in insects (flies, beetles) as well as in many kinds of ungulates in which males compete among themselves for access to females. With rare exceptions (reindeer and caribou), only males develop elaborate antlers, whose shape, location, and orientation are usually more suited for fighting rivals than for defense against predators. Antlers are also longer (relative to male body size) in ungulates that form large breeding parties and shortest in species in which females are widely dispersed (Clutton-Brock et al., 1982). Among primates, males have distinctly longer canine teeth than females in species with intense male–male competition than in species without such competition (Harvey et al., 1978).

Ornaments and Displays. If females have the ability to choose males, even if only among a subset that succeeds in obtaining a territory, then selection favors males that are most attractive to females. Males of many animal species advertise to females with conspicuous visual displays or acoustic signals. The conspicuousness of visual displays is typically enhanced by bright plumage or coloration, long tails, and other appendages, such as the "swords" of swordtail fish, which can be considered ornaments. These structures are often exaggerated in males and usually absent or relatively poorly expressed in females. Exceptions include some shorebirds such as phalaropes, in which large and brightly plumaged females compete for male parental care.

In some species, displays traditionally take place in areas where males gather solely for the purpose of attracting females. These areas are called leks, and the mating system is termed a lek mating system (Andersson, 1994). Even though males may defend small territories within the lek, these territorities do not contain resources needed by females. Moreover, because males do not contribute to parental care, the female receives only sperm. As in some female- and resource-defense systems, male mating success is frequently highly skewed. The most successful males may be especially attractive to most females or may control advantageous places within the lek, where their displays are more conspicuous or through which females are most likely to pass. In some kinds of fishes and birds, some males are especially successful because they are the first to attract a female, and other females copy their choice. Female choice can also be a dominant force in nonlekking species. In bower birds, for example, males not only display but build special structures (bowers) that are decorated with objects, such as feathers, shells, or beetle casts.

Ornaments and displays can also influence female attraction in more subtle ways. For example, even in monogamous species, males with more effective displays or songs often are the first to be chosen by females, and by virtue of this head start, these males may be able to raise more than one set of young in a breeding season. Attractive males might also be more successful in mating with females other than their mate, which is common in many nominally monogamous species of birds.

Secondary Sexual Traits as Indicators of Benefits. Whenever females are choosy enough to incur costs in mate assessment, which can include increased predation risks and delays in reproduction, the following question arises. Given that being choosy incurs some costs, what might females gain from choosing particular males? This question can be especially difficult to answer in lek-breeding species in which females gain no resources from the male or his territory. The answer is complex and controversial because besides their possible genetic contribution (indirect benefits) to the offspring, males might also provide more subtle, direct benefits. Direct benefits can include higher fertilization success, a lowered risk of infecting the female or offspring with a contagious disease or parasites, and even the reduction in assessment costs that could occur if the male's displays make him easier to find. Indirect, genetic benefits of mating with particular males might increase the attractiveness of male offspring (Fisher's runaway process of sexual selection), the viability of offspring of both sexes (good genes or handicap models), or both. In any event, secondary sexual traits and particular aspects of sexual signals are often correlated with the direct and indirect benefits that a female might receive by choosing to mate with particular males.

Traits and signal attributes associated with sexual selection should be honest indicators that cannot be easily assumed or produced by inferior males. For example, the pitch of acoustic signals is often correlated with and constrained by the size of the advertising male. Size, in turn, can affect the ability of the male to fertilize eggs or to provide protection for the female, her young, or both. As emphasized by Zahavi (1975), another way of assessing signal reliability is to focus on the properties of signals that correlate with the cost of displaying or signaling because inferior individuals are unlikely to be able to bear these costs. In male insects, frogs, and birds, for example, mating success is often positively correlated with the signal duration, the rate of signaling, and the proportion of the season in which the male displays or signals (Andersson, 1994). Measurements of metabolic rates during displaying or calling show, in fact, that energetic costs are highly correlated with duration, rate, or both (e.g., Prestwich, 1994). In male frogs, the trunk and laryngeal musculature that functions in signal production is far more massive than that in females. These muscles also have a greater density of the mitochondria and higher concentrations of enzymes needed to supply them with energy needed for many cycles of strong contractions. In species with prolonged breeding seasons, calling or displaying males may lose body weight, fail to grow, or be limited in the number of breeding bouts in which they participate. In some species of frogs, males producing costly signals that are most attractive to females fertilize a higher proportion of eggs or produce offspring that grow faster in comparison with the offspring of males producing less costly, relatively unattractive signals (Welch et al., 1998).

Males that produce reliable signals that are subject to strong female choice also pay considerable costs precisely because of those aspects of their appearance and behavior that increase their mating success. Brightly colored individuals are conspicuous to predators as well as to females, and long tails or other appendages can hinder escape from predators. Selection by predators has been shown to reduce the proportion of brightly colored and spotted male guppies in a matter of several generations (Houde, 1997), and the sex ratios of adult birds with long tails are often female-biased, suggesting that predation on males is especially severe. Acoustically orienting predators (bats) and parasitoids (organisms that lay eggs that develop inside the host and usually kill it) also take a toll on singing frogs and insects (Zuk and Kolluru, 1998). Moreover, the calling properties of some kinds of male crickets and frogs that make them more attractive to females also make them most likely to be attacked by parasitoid flies and frog-eating bats. Thus, even though males that are most subject to female choice seldom risk injury in fights with rivals, their efforts to attract females can be just as costly.

[*See also* Fluctuating Asymmetry; Mate Choice, *article on* Human Mate Choice; Sexual Selection, *article on* Bowerbirds.]

BIBLIOGRAPHY

Andersson, M. *Sexual Selection.* Princeton, 1994. Overview of sexual selection theory and empirical studies.

Clutton-Brock, T. H., F. E. Guinness, and S. D. Albon. *Red Deer: Behavior and Ecology of Two Sexes.* Chicago, 1982. A readable review that discusses sexual competition among red deer and other species of ungulates.

Darwin, C. *The Descent of Man and Selection in Relation to Sex.* London, 1871. Darwin's book is still useful today and contains many examples.

Harvey, P. H., M. Kavanagh, and T. H. Clutton-Brock. "Canine Tooth Size in Female Primates." *Nature (London)* 276 (1978): 817–818.

Houde, A. E. *Sex, Color, and Mate Choice in Guppies.* Princeton, 1997. A review of the experimental studies showing immediate and long-term effects of the opposing forces of sexual selection and predation on male appearance and courtship in guppies.

McCann, T. S. "Aggression and Sexual Activity of Male Southern Elephant Seals, *Mirounga leonina.*" *Journal of Zoology* 195 (1981): 295–310.

Nelson, J. S. *Fishes of the World.* New York, 1984.

Prestwich, K. N. "The Energetics of Acoustic Signaling in Anurans and Insects." *American Zoologist* 34 (1994): 625–643. A review of energetic costs of variation in sexual signals.

Welch, A. M., R. D. Semlitsch, and H. C. Gerhardt. "Call Duration as an Indicator of Genetic Quality in Male Gray Tree Frogs." *Science* 280 (1998): 1928–1930. A robust demonstration that offspring benefit when females pick males who produce costly signals.

Zahavi, A. "Mate Selection—a Selection for a Handicap." *Journal of Theoretical Biology* 53 (1975): 205–214. First verbal models suggesting that costly signals or structures are likely to be the most reliable criteria for female choice.

Zuk, M., and G. R. Kolluru. "Exploitation of Sexual Signals by Predators and Parasitoids." *Quarterly Review of Biology* 73 (1998) 415–438. A comprehensive review of the costs of displays and other signals used to attract members of the opposite sex.

— H. Carl Gerhardt

SEXUAL SELECTION

[This entry comprises two articles. The first article provides an overview of sexual selection as differential mating success; the second article discusses the evolution of the bower building habits of bowerbirds and the transfer of displays to bowers. For related discussions, see Mate Choice *and* Mating Systems.*]*

An Overview

Natural selection, one of the main underpinnings of evolution, occurs when individuals vary in their ability to survive and reproduce. Sometimes, however, individuals with traits that appear detrimental to their survival may still have high reproductive success. Such an apparent paradox arises because of the process of sexual selection, selection based on differential mating success. In addition to proposing the existence of natural selection, Charles Darwin suggested that traits arising through sexual competition could be attributed to the distinct process of sexual selection. A male elephant seal's gargantuan size, a male bird of paradise's showy plumes, and the songs of many insects, frogs, and birds are all thought to be the result of sexual selection. In some cases, traits favored by sexual selection because they increase an individual's likelihood of mating simultaneously decrease the animal's ability to escape predators or forage effectively.

Mechanisms of Differential Mating Success. Individuals will differ in their ability to attract mates, and hence in their reproductive potential, through two major mechanisms of sexual competition, one occurring within the members of a sex (intrasexual) and one occurring between members of opposite sexes (intersexual).

Intrasexual selection generally takes the form of competition between males for access to females, while intersexual selection usually occurs through female choice of particular males. The reason for the asymmetry lies in the factors that limit reproductive success for each sex. Because males often invest relatively little in an individual mating, they leave the most genes in subsequent generations by inseminating as many females as possible; within some constraints, discussed below, the more mates a male has, the more offspring he sires. In contrast, because females are the sex that supplies the nutrient-rich egg, and often the sex that cares for the young, they are limited by the number of offspring they can successfully produce and rear. That number, while variable, is much more confined than the number of offspring a male might hypothetically be able to sire. Thus, with some exceptions, male reproductive success has a higher variance than female reproductive success. Males therefore have much to gain by competing with other males to gain more fertilizations, and females have much to gain by ensuring that the males they do mate with are of high quality and likely to sire high-quality offspring. Sexual selection is therefore often more intense in males than in females.

Intrasexual competition and its consequences. Male competition is often overt, with males fighting during the breeding season. Males may also compete indirectly, perhaps through territorial defense or by establishing a hierarchy of social dominance. Male giraffes, for example, slam their long necks against their rivals, and male antlered flies in the genus *Telostylinus*, named for the long extensions on their heads, rear up on their legs and press their heads together like tiny deer. The antlers of the flies as well as those in many male ungulates such as sheep and deer or antelopes are thought to have evolved because of their function as weapons. Similarly, males tend to be larger than females in species in which such large size can be advantageous in fights. One study of seals and their relatives showed that in polygynous species such as elephant seals, where males fight vigorously for dominance and the most dominant males sire the most pups, males are much larger than females, whereas in monogamous species such as the harbor seal, the two sexes are similar in size. Such a difference in appearance between the sexes is termed sexual dimorphism, and while sexual selection is often the explanation for sexual dimorphism, it is important to note that it is not the only one. Males and females may, for example, forage in different ways and hence have evolved different mouthparts or other aspects of morphology. In addition, males may compete without obvious weaponry, as is the case for many species of butterflies.

Winning fights or attaining a high rank can yield higher reproductive success in a variety of ways. First and most obvious, the victorious individual may be able to sequester mates either directly or through control of a necessary resource. Dominant male elephant seals, for example, herd groups of females onto areas of beach which they defend against other males. Success following intrasexual competition may also come from more indirect means, such as via female preference for winners of fights or for dominant males. In this manner, both modes of sexual selection may interact.

Intersexual selection and its consequences. In contrast to intrasexual competition, intersexual selection or mate choice is often quite subtle. Indeed, although Darwin proposed that female choice was a fundamental component of evolution, few of his contemporaries shared his enthusiasm. [*See* Mate Choice: An Overview.] They found the idea that female animals were capable of such fine-tuned distinctions to be implausible. Largely because of this opposition to the idea of female choice, sexual selection as a theory lay dormant for several decades. Even after genetics became incorporated with Darwin's ideas on evolution to form the new synthesis, the major evolutionary biologists of the early twentieth century were largely uninterested in sexual selection.

In the 1970s, however, sexual selection and female choice rose to the forefront again, partly as a result of a paper by the evolutionary biologist Robert Trivers outlining the way in which females and males inherently differ because of how they put resources and effort into the next generation, which he termed *parental investment.* As mentioned above, females normally have greater parental investment in each individual offspring, and hence they are more likely to be choosy about with whom they mate. And indeed, numerous studies have demonstrated the occurrence of female choice for particular males. The bowerbirds are one example of such preference; others include preference for male guppies with more orange spots, male crickets with more continuous songs, and male frogs with deeper croaks. Unlike male competition, female choice selects for ornaments, also called epigamic characters, rather than for weapons used directly in combat. Of course, some traits may function in both sexual competition and mate choice, so that the songs of many birds serve both to warn other males off and to attract potential mates. And females may use several different ornamental traits during mate choice; in red jungle fowl, the ancestors of domestic chickens, females prefer males with long, red combs, red eyes, and to a lesser extent, more colorful feathers. Manipulation of only one of these traits does not cause the females to shift their preferences, suggesting that multiple cues are used in nature. The different ornaments may function to reinforce one another or may give independent information about aspects of a male.

Finally, it is important to remember that even if individuals have preferences for particular characteristics in a mate, they may not be able to exercise those preferences because of constraints on their choices. For example, a female butterfly may need to lay eggs when she finds an appropriate host plant, and cannot always wait for a preferred male to come along. These constraints may also play a role in helping to distinguish between various models of sexual selection.

Role Reversal and Parental Investment: Some Complications. Although biologists usually consider competition only among males and choice only by females, this dichotomy is by no means absolute. More and more studies are finding intense female competition as well as male choosiness. Below are outlined some of the circumstances that can lead to these situations.

If males invest much time, effort, or resources into their offspring, they are likely to be limited in the number of mates they can fertilize. For example, many male insects produce nutritious "nuptial gifts" at mating along with the sperm used to fertilize a female's eggs. They may also present the female with a prey item that she consumes during copulation. Because the gifts are costly to produce or acquire, males are limited in the number of matings they can achieve, and hence have evolved to be choosy, mating only with females particularly likely to produce large numbers of eggs. In many such cases, larger females have higher fecundity, causing selection for larger females via male mate choice.

Males are also likely to be choosy when a species is monogamous, and both members of a mated pair remain together for an entire breeding episode. Thus, in many songbirds, seabirds, and a few other types of animals, courtship is a process of mutual evaluation by both partners. Sexual selection operates in a similar way on both sexes, and males and females may share secondary sexual characters such as bright plumage in many songbirds. In other monogamous species, males are still the larger or more ornamented sex, and the source of this sexual dimorphism is still not fully understood. Some researchers have suggested that more colorful males are still able to attract more females for extra-pair copulations, while others have proposed that sexual selection operates in monogamous species by favoring earlier breeders.

Along with male mate choice, female competition is also being recognized as more common than previously thought. Females compete for access to mates in polyandrous species, those in which a female is mated to more than one male. Polyandry is seen in several species of water birds, including the jacanas or lilytrotters. In these situations, males generally perform much of the parental care, and females leave a clutch of eggs with each mate. Females also compete in other species in which males provide a substantial contribution to the offspring.

Operation of Sexual Selection. The mechanism by which males evolve weapons used in competition for females is straightforward: males better at combat sire more offspring, selecting for further elaboration of the weapon or even larger body size. Eventually, natural selection is expected to curb the extent of the development of the secondary sexual character as it becomes too costly in terms of survival of its bearer. Although males are potentially selected to exaggerate their fighting prowess, it is expected that most types of signals

SEXUALLY SELECTED ORNAMENTS IN FEMALES

In most species the sexes are distinguishable. Males are often big, bright and beautiful, adorned with impressive ornaments and weapons. In contrast, females are often referred to as "the little brown jobs," the sex that is designed to blend into the background for safety. Darwin questioned why males would develop such pronounced trait differences, particularly if elaborate ornaments are costly to produce and maintain, and might attract predators, thereby reducing the bearer's chance of survival. Darwin's answer was sexual selection: individuals with the exaggerated traits would enjoy higher mating success through direct competition over, or greater attraction of, the opposite sex. Sexual selection operates predominantly on males because their reproductive success is most constrained by access to fertile females. In contrast, female reproductive success is not constrained by partner number, but rather partner quality and resources to raise offspring. Females invest more in offspring production from the start, making females the limiting reproductive resource for males. Males should develop traits to maximize access to reproductive opportunities, and females should choose males according to their quality.

In a few species, sexually selected ornaments can be observed in both sexes. In baboons (*Papio cynocephalus*), for example, males are much bigger and have much larger canine teeth than females, but females too are ornamented. They develop an elaborate, colorful swelling of the perineal skin around the time of ovulation. These sexual swellings are not necessary for the mechanics of copulation, but appear to be solely designed for mate attraction. Indeed, Domb and Pagel ("Sexual Swellings Advertise Female Quality in Wild Baboons." *Nature* 410 [2001]: 204–206) found that male baboons spend more time fighting over females with larger swellings. This appears to be a sound behavioral strategy as it was also found that sexual swelling size reliably advertised a female's reproductive value. Females with larger sexual swellings began reproducing earlier and a higher proportion of their offspring survived. If females are normally the choosy sex, then why would they need to produce sexually selected ornaments? Male mating effort is likely to be particularly costly in species with female ornaments. Costs such as male-male fighting and mate-guarding place a premium on choosing the right mate. Thus, this study of baboons demonstrates an unusual case of a sexually selected ornament in females. The ornament evolves because it pays males to be selective in their choice of a mating partner, and thus for females to advertise their quality.

—Leah Gardner Domb

that evolve will communicate essentially accurate information. [*See* Signalling Theory.]

The evolution of ornamental traits used in mate choice is more complicated. How does selection favor traits that seem to have no function in everyday life and that are not useful in combat? When females gain what are called direct benefits for themselves and their offspring by mating with a particular male, the traits may serve as indicators of a male's ability to bring food or help with parental care. More puzzling are species in which males do not give any contribution to the female or her offspring beyond the sperm to fertilize her eggs, and yet it is in those species where sexual selection appears to have been the most intense. These include the birds of paradise, with their elaborate plumage, as well as the bowerbirds.

Darwin simply thought that females found males with ornaments attractive, but the statistician and scientist R. A. Fisher was probably the first to develop formally a theory to explain the evolution of exaggerated traits. Fisher suggested that female choice in these species is arbitrary. Imagine an ancestral population in which some females have a slight preference for a male with, say, a longer tail than average. Perhaps the length gives the male a slight advantage in flight. The offspring of males with a slightly longer tail and females with a preference for them will inherit both the female preference and the male trait. Thus males will tend to have longer tails and carry genes that in females would lead to expression of preference, while daughters will tend to exhibit preference and carry genes that if present in males would bring about longer tails. The trait and the preference for it thus become genetically correlated, and if the correlation is strong enough, selection will proceed to exaggerate the character until natural selection stops it, say because it makes the male too conspicuous to predators. This snowballing effect is called "runaway sexual selection," or sometimes "Fisherian selection." According to this view, the trait does not "mean" anything other than that it happened to have been preferred a long time ago. In other words, the ornament does not indicate that the male carrying it has greater survival ability or would be a better parent.

In contrast to this idea, the "good genes" or adaptive sexual selection hypothesis suggests that males with elaborate traits are indicating their viability and fitness to females, so that a female mating with a male who has an exceptionally long tail has offspring who not only have long tails themselves but are also more fit than offspring of less ornamented males. The secondary sexual characters are therefore indicators of quality and not just arbitrary signals. This hypothesis differs from the previous one because here the trait actually is linked to higher survival, whereas in runaway sexual selection the trait is neutral or even negative with respect to survival. Under the good genes scenario, secondary sexual characters are costly to produce and depend on the condition of their bearer, so that males cannot "cheat" and produce an ornament when they lack the underlying quality such a trait would indicate.

Evidence has been found in favor of both hypotheses. In fruit flies, peafowl, crickets, and a variety of other animals, the offspring of females allowed to choose their mates have higher fitness, usually measured as the number of surviving juvenile offspring, than the offspring of females mated at random to males. This result supports the good genes hypothesis. In guppies, however, one study found that more attractive males appeared to sire offspring with worse survival rates, not better, which the author concluded was support for a Fisherian process. A meta-analysis, a kind of large-scale analysis of many different tests of the same hypothesis, showed that benefits arising from good genes appear to be present but rather weak in a wide range of taxa. It is difficult to know how to interpret these results, because it can always be argued that the studies in question did not examine the appropriate aspect of offspring fitness.

It has been suggested that runaway sexual selection can operate only so long as mate choice has no costs. In other words, as long as females are constrained by time, social behaviors such as harassment by prospective mates, or the necessity for avoiding predators, mate choice becomes costly, and this cost must be offset by some benefit to females of being choosy. Having offspring of higher fitness, as in the good genes model, would constitute such a benefit, so some scientists have concluded that because mate choice is likely to be costly in many animal species, a purely runaway process probably does not by itself account for sexual selection in nature.

The two processes are not mutually exclusive. It is possible, for example, for a Fisherian process to act to elaborate further a trait that originally evolved in the context of conferring good genes to a female's offspring. Furthermore, some recent models have questioned the dichotomy between runaway sexual selection and the good genes hypothesis. These point out that even under runaway sexual selection, if males with elaborate ornaments have sons achieving higher mating success, females still obtain "good genes" benefits because their offspring have higher fitness, albeit not in the form of higher viability.

Sexual Conflict and Sexual Selection after Mating. Even after copulation, mating success of individual males can vary. As discussed in the entry on Sperm Competition and cryptic female choice, if a female mates with more than one male, the sperm of each of her mates may not be equally successful at inseminating her eggs. Differential fertilization may occur because some males' sperm outcompetes the rest, or because females are able to control which males fertilize their eggs; this latter notion is controversial but has been attracting much recent interest.

Because males and females have different limitations on reproductive success, with males being limited by the number of mates and females by the number of offspring produced, we expect that what benefits males will not necessarily benefit females and vice versa. This conflict may apply at several levels. First, in species with parental care, females are likely to benefit if males are monogamous and devote their attention to a single female's offspring, whereas males increase their reproductive success when they are polygynous. Such conflicts over the mating system have been documented in species as diverse as fish, mammals, and beetles.

Second, females and males may have different costs and benefits during mating itself. In water striders, for example, males jump onto females on the water surface and attempt to mate; females expend much energy in dislodging unwanted suitors or in carrying them around, and may benefit by mating with such harassing males even if they are not preferred.

Third, the nature of male and female biology may be shaped by conflicts over mating. In *Drosophila melanogaster*, male seminal fluid contains toxins that reduce female life span and fecundity and make remating less likely; such toxins may also be present in other animals. Several biologists have suggested that males and females are in a sort of arms race, so that if males evolve a mechanism for preventing females from mating with subsequent males, this mechanism can benefit the males even if females do not live as long or produce as many offspring, just so long as the offspring they do produce are sired by that male. Females, however, would then be expected to evolve some form of resistance to the male trait, which in turn selects for an increase in the male's ability to overcome that resistance. Whether males or females are "winning" in any particular species may depend on the ecological circumstances that govern the ability of either sex to respond to the other, or perhaps on which sex has most recently evolved the latest adaptation or counter adaptation.

[*See also* Assortative Mating; Signalling Theory.]

BIBLIOGRAPHY

Andersson, M. *Sexual Selection*. Princeton, 1994. A good, even-handed treatment of most topics in sexual selection.

Cronin, Helena. *The Ant and the Peacock: Altruism and Sexual Selection from Darwin to Today*. Cambridge, 1991. A popular and well-written discussion of two of the most perplexing ideas in Darwin's writings.

Darwin, C. *The Descent of Man, and Selection in Relation to Sex*. Princeton, 1981. A reprint of the 1871 edition. Fascinating to read even if the English is a bit archaic, as Darwin painstakingly explores the foundation of sexual selection.

Ridley, M. *The Red Queen: Sex and the Evolution of Human Nature*. New York, 1995. A popular book examining both the evolution of sexual reproduction and its repercussions. Well-written and entertaining.

Trivers, R. L. "Parental Investment and Sexual Selection." In *Sexual Selection and the Descent of Man, 1871–1971*, edited by B. Campbell, pp. 136–179. London, 1972. The classic explanation of male and female reproductive strategies.

Williams, G. C. *Sex and Evolution*. Princeton, 1975. Concise, well-written explanation of the basic differences between males and females and how these are manifested in various species. Contains some mathematical treatments requiring an understanding of algebra.

— MARLENE ZUK

Bowerbirds

The evolution of complex male sexual display traits remains one of the most controversial issues in evolutionary biology. Bowerbirds (family Ptilonorhynchidae) provide an outstanding model system for testing hypotheses about the evolution and functional significance of complex display. This diverse group is best known for its polygynous species that build stick bowers on the ground associated with a decorated display; females provide parental care at nests built in trees. Multifaceted male displays involve the decorated display court and the bower that act as a stage for energetic vocal and dancing male display that is observed by the female from inside the bower. Courtship and matings occur in the bower. Because mating and courtship occur at a known location, they can be monitored with video cameras. This has allowed the collection of highly detailed information on courtship and male mating success.

Bower quality, numbers of preferred decorations, and vocal/dancing elements all contribute to male mating success. Male reproduction is skewed; one male may mate with twenty-five different females at his bower in one season. Most females mate with a single male after visiting bowers of multiple males. Females previously mated with high-quality males show reduced mate searching and typically return to mate with these males in successive years.

Bowers likely originated as devices that protect females from forced copulation by courting males. For females to benefit from mate choice, they must be able to visit and observe courtship by different males and reject those who are unsuitable. Female bowerbirds are susceptible to forced copulation when on the ground. The different bower types require the male to move around a barrier to copulate with the female. This allows disinterested females the opportunity to safely escape the display court. Archbold's bowerbird, a species that has lost bower building, shows unique compensatory courtship displays in which males stay low and press their bodies close to the display court. Courting from this low position prevents the male from capturing the female from above as required for a forced copulation. Although bowers reduce male opportunity for forced copulations, males with high-quality displays likely gain from increased visitation of unfertilized females attracted to display courts.

E. T. Gilliard suggested that in bowerbirds bright male plumage traits were transferred over evolutionary time from the bird to the bower, however this is not supported in a phylogenetic analysis. A mitochondrial DNA (mtDNA) phylogeny of the bowerbirds shows that the monogamous catbirds are ancestral followed by evolution of two clades that build distinctive bower types. Within each clade, bower form and color's used in decorating the bower are highly variable. Reversals and convergences of bower types and decorations obscure the effects of shared ancestry and indicate no consistent trend from ancestral to derived species when contrasted with male plumage development. Results from several studies indicate that variation in bower form and decoration, on the one hand, and male plumage, on the other, has arisen for different functional reasons.

Evolutionary changes to bowers allow improved conditions for the presentation of male display. In spotted bowerbirds, females prefer males who give high-intensity displays even though these displays are threatening. Changes in bower structure associated with high-intensity display allow females to stand sideways in the bower and view male courtship displays protected by the see-through bower wall. Protection offered by the modified bower allows males to give higher intensity displays and ultimately more attractive displays. This "threat reduction" may be important in understanding display adaptations in other species. For example, male birds of paradise hang upside down in trees while displaying. This may have evolved as a device for reducing the threat to females from high-intensity displays.

Some male display traits have been co-opted from other functions. Phylogenetic analysis of "Skraa" calls are used across bowerbirds, as aggressive calls suggest that they have been borrowed for use as courtship calls in the *Chlamydera* bower birds. Co-option of displays that signal male quality in aggressive contexts for use in courtship provides a simplified mechanism by which females can select for "good genes." This does not require the difficult process of coevolving male traits and female preferences that is required in most recent models. Bower quality is also used in mate choice in one clade, but this is not a likely cause of bower evolution because incipient bowers would not reliably indicate quality differences among males. Similarly, decorations may have initially functioned as indicators of bower location and then been used in mate assessment. Thus, co-option and additions of secondary functions can have an important role in display trait evolution and can explain complex trait evolution without requiring coevolution of male traits and female preferences.

Males of different species choose colors of decorations in response to different light regimes. In *Amblyornis* bowerbirds, males in species using foggy ridge tops use predominantly black decorations, whereas closely related species displaying on more sunlit slopes use a variety of decorations with a wide array of bright colors. The added cost of collecting bright decorations are justified on brightly lit slopes but may not be justified on poorly lit ridge top display sites. The effectiveness of functional explanations in describing shifts in bower structure and decoration provides an important alternative to runaway sexual selection as a cause of interspecific variation in display trait evolution.

Long-term monitoring of satin bowerbird bowers shows that high male skew in mating success is maintained across multiple years, causing variation in male lifetime reproductive success to be high. It has been suggested that costly male displays are necessary for honest advertisement of male quality, but the tendency for males to hold top reproductive positions for as long as six years suggests that display may not be extremely costly for them. Male bowerbirds do not obtain adult plumage until they are seven years, and before then they frequently engage in male–male courtships, where displays may be learned. Once acquired, male ability to displays can be utilized in successive years at little additional cost. Recent studies suggest that attractiveness of male display comes from the intensity of display and from male ability to modulate display intensity in relation to female signals of comfort. Thus, in bowerbirds, experience and learning ability may ensure quality signals rather than just difference in the costs of male traits that males can support. Honest advertising need not be highly cost dependent.

Speculative discussions of bowerbirds and other species with elaborate display characterize male displays as arbitrary and tend to emphasize single models such as runaway sexual selection. Information from detailed studies of bowerbirds suggests a different view. Male displays evolve by a variety of processes. Rapid, large changes in display can occur, resulting in highly integrated displays meeting female needs, and are tuned to habitats occupied by individual species. Although this adaptive view of mate choice has long been accepted in discussions of mate choice in species with resource-based monogamous mating systems, evidence suggests that it is now appropriate to extend it to species with nonresource-based mating systems that have elaborate male diplays.

[*See also* Leks; Male–Male Competition; Mating Systems, *article on* Animals; Sexual Dimorphism.]

BIBLIOGRAPHY

Borgia, G. "Bowers as Markers of Male Quality: Test of a Hypothesis." *Animal Behaviour* 35 (1985): 266–271. Discusses male bower decorations and shows that decoration reductions reduce male mating success.

Borgia, G. "Why Do Bowerbirds Build Bowers?" *American Scientist* 83 (1995): 542–547. Discusses the origin of bower building and presents and provides evidence for the hypothesis that bowers evolved and are attractive to females because they provide protection from forced copulation.

Borgia, G., and S. Coleman. "Co-option of Aggressive Courtship Signals from Aggressive Displays in Bowerbirds." *Proceedings of the Royal Society of London B* 267 (2000): 1735–1740. Co-option of aggressive male displays for use in courtship has been suggested as potentially important in the evolution of male display, but there were no clear-cut cases showing whether courtship calls were co-opted for aggressive calls or vice versa.

Borgia, G., and D. Presgraves. "Coevolution of Elaborated Male Display Traits in the Spotted Bowerbird: An Experimental Test of the Threat Reduction Hypothesis." *Animal Behavior* 56 (1998): 1121–1128. Presents results of bower manipulation experiments designed to test the hypothesis that bower walls in spotted bowerbirds function as protective screens for females during courtship.

Kusmierski, R., G. Borgia, A. Uy, and R. Crozier. "Molecular Information on Bowerbird Phylogeny and the Evolution of Exaggerated Male Characters." *Proceedings of the Royal Society of London* 264 (1997): 307–313. Shows that bower types are not predicted by phylogenetic relationships within the two major clades of bowerbirds.

Uy, J. A. C., and G. Borgia. "Sexual Selection Drives Rapid Divergence in Bowerbird Display Traits." *Evolution* 54.1 (2000): 273–276. Considers the role of sexual selection as a factor driving speciation.

Uy, A., G. Patricelli, and G. Borgia. "Dynamic Mate Searching Tactic Allows Female Satin Bowerbirds to Reduce Searching." *Proceeding of the Royal Society of London B* 267 (2000): 251–256. Studies of marked females show that females search multiple males before mating, usually returning to mate with the highest quality male they have encountered.

— GERALD BORGIA

SHIFTING BALANCE

The shifting balance theory of evolution is a controversial proposal to describe how a population may evolve from one locally stable state of adaptation to another (Figure 1), even though the intermediate states may be a lower fitness. Sewall Wright (1932, 1977) proposed the idea of an *adaptive landscape* to describe the relationship between the mean fitness of a population and its distribution of genotypes. He suggested that there will commonly be several alternative genetic states that give a better adapted phenotype than all other alternative genotypes. He called these high points in the relationship between genotype and fitness "adaptive peaks" and the low points between "adaptive valleys." Under natural selection alone, Wright reasoned, a species on an adaptive peak would remain there. Deterministic evolution would not move the genotype frequencies of the population away from this peak, regardless of whether this locally stable state was also the most fit combination possible, that is, regardless of whether there were other, higher adaptive peaks. Wright (1932, p. 358) said, "The problem of evolution as I see it is that of a mechanism by which the species may continually find its way from lower to higher peaks. . . ."

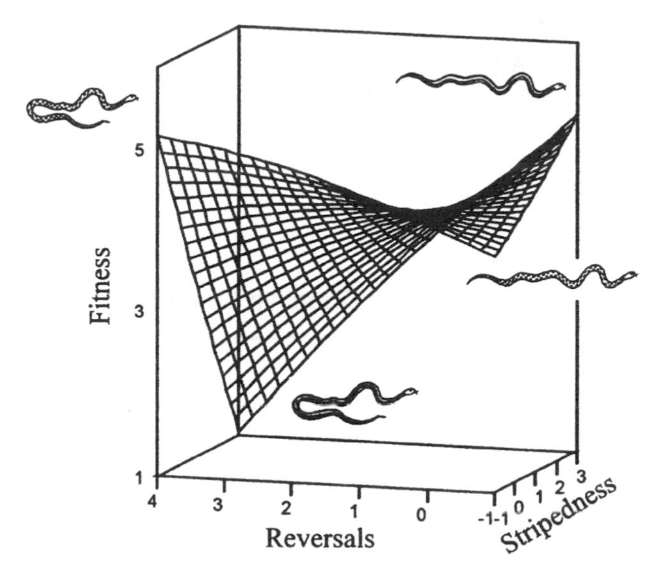

FIGURE 1. The Adaptive Landscape.
An example of an adaptive landscape, showing the fitness associated with various coloration and behavioral patterns in the garter snake *Thamnophis ordinoides.* These snakes show a variety of color patterns, ranging from checkered patterns to complete longitudinal stripes, the contrast between which is represented on one of the axes below as stripedness. The snakes are also polymorphic for the behavior they exhibit when confronted with a potential predator: some snakes go a short distance and reverse direction suddenly, whereas others attempt to escape in a more or less linear path. The striped snakes are easy to see when stationary, but more difficult to catch when moving linearly because the longitudinal stripes give few visual clues as to their speed. Checkered snakes, in contrast, are relatively easy to capture when moving linearly, but they are more likely to escape attention by blending into their background when stationary. Thus, two forms are relatively fit, and other intermediate combinations are less fit. This graph comes from the work of Butch Brodie [1992] and was kindly provided by him.

Explanation of the Theory. Wright's own answer to this question was the shifting balance theory (SBT). SBT proposes that genetic drift is important in allowing a population to diverge from a locally stable state and thereby allow it to evolve to another, higher adaptive peak. The model was divided by Wright into three phases (Figure 2). In phase 1, a population moves away from a local adaptive peak by genetic drift. If the population drifts far enough from the peak that the direction of deterministic evolution is toward another peak, then in phase 2 individual selection can take the population toward that new peak. Finally, if a population has shifted

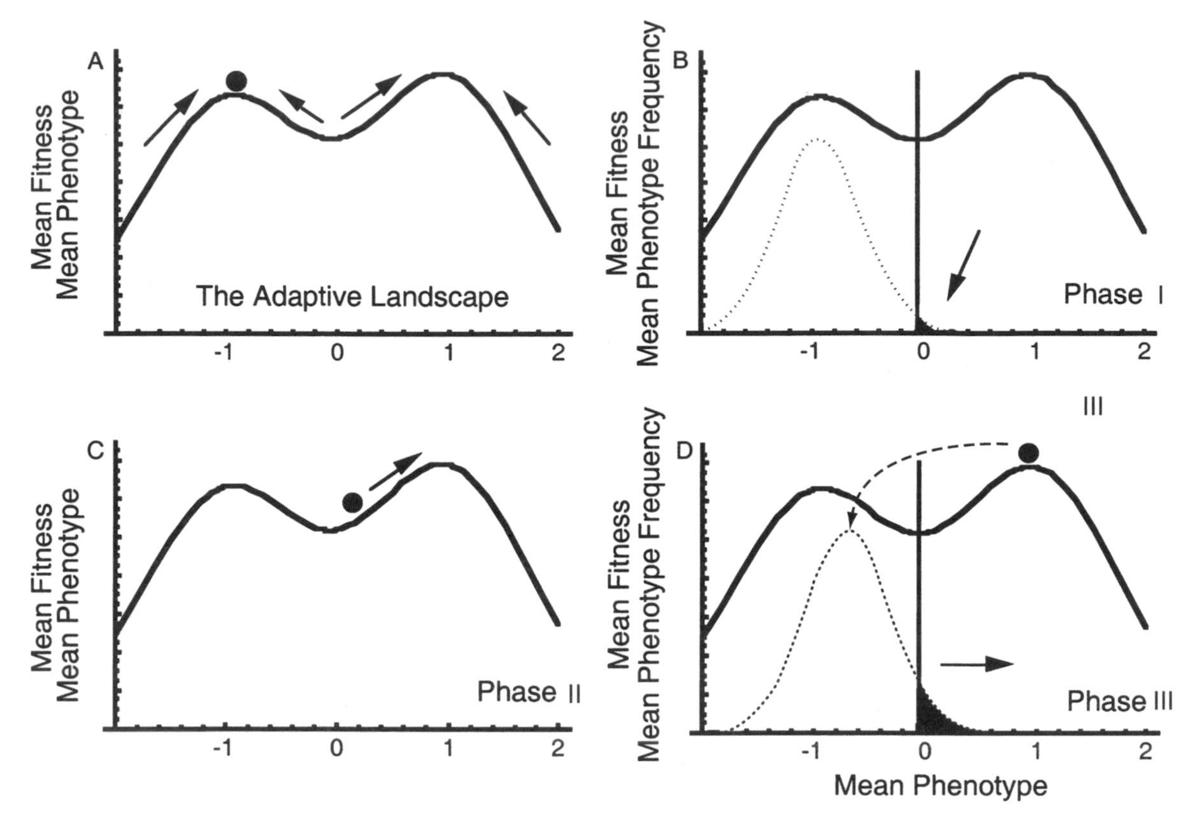

FIGURE 2. A Cartoon Version of the Shifting Balance Theory.
This figure represents a reduced, one-dimensional version of the shifting balance theory, using a quantitative genetic model of phenotypic evolution. The *x* axes in all parts of this graph represent the mean phenotype of a population, with the *y* axes representing the mean fitness of a population (broken line) or the frequency of populations with a particular mean (dotted line). In panel A, there is an adaptive landscape with two peaks. Evolution by selection alone will cause a population to evolve to higher mean fitness; therefore populations are expected to evolve "uphill" on this landscape, as shown by the arrows. A population at the lower peak to the left is "trapped" if the landscape does not change and there are no random effects. The next three panels represent the three phases of shifting balance theory. Panel B shows that with genetic drift, there may be a distribution of local population means; with sufficient drift some populations (indicated by the arrow) may drift so far as to be in the "domain of attraction" of another adaptive peak. In panel C, selection acts to take those populations toward the new peak. In panel D, the populations that have shifted successfully to a new peak may act via migration to affect other populations and move their mean phenotype toward the new peak, thus making further transitions easier. This representation, although similar to Wright's description of the shifting balance, is unlikely to show a realistic possibility, since in practice drift, selection, and migration all will occur simultaneously. Michael C. Whitlock.

from one stable state to another with a higher mean fitness, then that population may have a higher reproductive success than other populations, by being less likely to go extinct or by contributing differentially by migration to other populations. This group selection, which exports the adaptive shift to other populations, has been referred to as phase 3. In practice, these three phases need not be absolutely distinct but may occur simultaneously.

SBT served as a major model of evolution for much of the twentieth century because it tackled a very important problem (release of evolutionary constraint) with a captivating metaphor (the adaptive landscape) and pleasing complexity (encapsulating migration, genetic drift, and mutation as well as the more widely studied selection), as proposed by one of the founders of the field. It therefore attracted much attention.

Objections to the Theory. Wright's theory was controversial from its beginning, however. One of the other founders of population genetics, Sir Ronald Fisher, thought that the shifting balance process was both unnecessary (in the sense that he thought species would never be stuck on adaptive peaks but would always have avenues available for improvement) and unfeasible (because he thought that population sizes in nature were sufficiently large that genetic drift would be ineffective). This controversy continues today, with doubt among many evolutionary biologists about the plausibility and applicability of the strict shifting balance model (Coyne et al., 1997, 2000; but *see* Wade and Goodnight, 1998). Theoretical work (e.g., Barton and Rouhani, 1987) and simulations (e.g., Moore and Tonsor, 1994) have shown that most aspects of SBT are possible. In general, however, this work shows the theory of shifting balance to be so unlikely relative to alternatives that it is not considered an important part of the process of evolution.

The theoretical objections to SBT are as follows. (1) Low migration between populations seems to be required for phases 1 and 2 to get a local population to a new peak, yet high migration seems to be necessary for phase 3 to be efficient. (2) Populations in nature are unlikely to be often small enough for phase 1 to allow enough genetic drift. (3) Group selection is likely to be relatively weak and therefore unable to export adaptive peaks. (4) Perhaps most importantly, alternatives seem to be more likely.

Wright (1932) initially proposed several alternative modes of evolution on complex landscapes. These were an increase in the mutation rate, a decline in the strength of selection, and a qualitative change in the environment, each of which could allow a population to move to another adaptive peak. Another possibility is a new mutation that allows the population to evolve immediately to a higher fitness without deleterious intermediates. It is also possible that a population always has some direction in multidimensional space in which it can evolve to a higher mean fitness, thereby never getting "stuck" (a possibility favored by Fisher). Finally, we should consider that populations in fact may get stuck on adaptive peaks without any possibility of evolving to higher ones (that the constraints to evolution implied by adaptive peaks are more or less permanent). Examples of some of these alternatives are known. For instance, a macromutation has allowed evolution to a new peak in *Pyronestes* finches (Smith, 1990, 1993), and a change in environment significantly alters the adaptive landscape of interactions between alleles of α-glycerol phosphate dehydrogenase and alcohol dehydrogenase in *Drosophila* (Cavener and Clegg, 1981). Furthermore, theoretical analysis has shown that subtle changes in the environment are much more likely to allow peak shifts than SBT (Whitlock, 1997).

Examples. Empirically, SBT has been challenged by the observation that there is not a single evolutionary transition known that can be best explained by SBT. Several phenomena in nature, however, show many of the aspects of SBT. These include the evolution of chromosomal rearrangements, Müllerian mimicry, paired changes in RNA molecules, and selfing in self-incompatible plants. In most of these cases, there is strong circumstantial evidence that genetic drift has allowed populations to move from one stable state to another, although for none is there evidence that the new stable state has a higher mean fitness or that adaptation has been increased.

A closer look at one of these examples shows some of the strengths and weaknesses of the SBT as a model of adaptation. One of the best documented examples of the importance of drift influencing the direction of adaptive evolution is the evolution of selfing in plant species with heterostyly (Barrett, 1993). Heterostyly is the genetic polymorphism for the height of the style (a prolongation of the ovary that bears the pollen-receiving organs) and anther (the pollen-bearing structure). Individuals are unlikely to pollinate other individuals of the same morphology; therefore, they are less likely to mate with themselves. The evolution of selfing in an outcrossing species requires a low level of inbreeding depression or selection for reproductive assurance (i.e., selection caused by a lack of available other mates), which are both more likely to occur in populations that have already been inbreeding or in small populations. In some species, especially well studied in a Brazilian population of *Eichhornia paniculata*, genetic drift in small populations often causes the loss of one of the three style-length morphs, which allows more selfing, causes selection to reduce inbreeding depression, and results in selection for a modification of the other styly morphs to allow even more selfing. Thus, genetic drift changes the selective context of the populations, which then causes selection to take the populations to a new stable state. This sort of evolution is not possible by a strict selectionist model of evolution, and thus this system is made easier to understand with SBT. However, the system is not a perfect match to SBT in at least three ways (Coyne et al., 1997). First, part of the change in the selection regime is not due to genetic drift but to the direct effects of small population size on the difficulty of finding mates. Second, there is no evidence that the mean fitness of the population has increased by the evolution of selfing, whereas Wright (1932, 1977) intended the SBT to be a model of increasing adaptation. Finally, there is no evidence that group selection has acted to export the adaptive peak to other populations; there is no evidence for phase 3. Similar problems exist for using SBT as a model of adaptation in other potential examples (Coyne et al., 1997).

SBT, however, is almost impossible to document. Shifting balance is a model in which rare events (genetic drift to the domain of attraction of a new peak) have large evolutionary impacts. Because they are rare, it would be extremely unlikely to observe the shifts in progress; therefore the lack of observed shifts over a short time scale cannot falsify the model. Furthermore, the theory requires that there is an "adaptive valley" between peaks, yet we can never directly observe the multidimensional landscape. Thus, we can never know if there are "ridges" of increasing mean fitness through other, unstudied, dimensions of the landscape. We therefore can never rule out a strict selectionist model (Whitlock et al., 1995). This means that SBT can never be proven, so its value is as a metaphor and as a goad to further research in complex evolution. SBT has encouraged, among other topics, the development of the theory of genetic drift, of group selection, and of complex adaptive landscapes and epistasis. Whether or not the strict SBT is true, its contribution has been immense.

[*See also* Fisher, Ronald Aylmer; Genetic Drift; Genetic Polymorphism; Wright, Sewall.]

BIBLIOGRAPHY

Barrett, S. C. H. "The Evolutionary Biology of Tristyly." In *Oxford Surveys in Evolutionary Biology*, edited by D. Futuyama and J. Antonovics, vol. 9, pp. 283–326. Oxford, 1993.

Barton, N. H., and S. Rouhani. "The Frequency of Shifts between Alternative Equilibria." *Journal of Theoretical Biology* 125 (1987): 397–418.

Brodie, E. D. "Correlational Selection for Color Pattern and Antipredator Behavior in the Garter Snake *Thamnophis ordinoides*." *Evolution* 46 (1992): 1284–1298.

Cavener, D. R., and M. T. Clegg. "Multigenic Response to ethanol in *Drosophila melanogaster*." *Evolution* 35 (1981): 1–10.

Coyne, J. A., N. H. Barton, and M. Turelli. "Perspective: A Critique of Sewall Wright's Shifting Balance Theory of Evolution." *Evolution* 51 (1997): 643–671.

Coyne, J. A., N. H. Barton, and M. Turelli. "Is Wright's Shifting Balance Process Important in Evolution?" *Evolution* 54 (2000): 306–317.

Moore, F. B. G., and S. J. Tonsor. "A Simulation of Wright's Shifting-balance Process—Migration and the three Phases." *Evolution* 48 (1994): 69–80.

Smith, T. B. "Natural Selection on Bill Characters in the two Bill Morphs of the African Finch *Pyrenestes ostrinus*." *Evolution* 44 (1990): 832–842.

Smith, T. B. "Disruptive Selection and the Genetic Basis of Bill Size Polymorphism in the African Finch *Pyrenestes*." *Nature* 363 (1993): 618–620.

Wade, M. J., and C. J. Goodnight. "Perspective: The Theories of Fisher and Wright in the Context of Metapopulations—When Nature Does Many Small Experiments." *Evolution* 52 (1998): 1537–1553.

Whitlock, M. C. "Founder Effects and Peak Shifts without Genetic Drift: Adaptive Peak Shifts Occur Easily When Environments Fluctuate Slightly." *Evolution* 51 (1997): 1044–1048.

Whitlock, M. C., P. C. Phillips, F. B. G. Moore, and S. Tonsor. "Multiple Fitness Peaks and Epistasis." *Annual Review of Ecology and Systematics* 26 (1995): 601–629.

Wright, S. "The Roles of Mutation, Inbreeding, Crossbreeding, and Selection in Evolution." *Proceedings of the Sixth International Congress of Genetics* 1 (1932): 356–366.

Wright, S. *Evolution and the Genetics of Populations: 3. Experimental Results and Evolutionary Deductions*. Chicago, 1977.

— MICHAEL C. WHITLOCK

SIBLING RIVALRY

Sibling rivalry arises when brothers and sisters demand more resources than their parents or the environment can supply. It is common especially when offspring are confined to "nurseries" such as a womb, a den, or a nest. Such sibling rivalry has been referred to as a "lifeboat dilemma," focusing attention on the fundamental decisions between selfishly consuming resources or allowing a sibling to consume them instead. If resources turn out to be inadequate for all nursery mates to survive, then sibling rivalry can escalate to lethal extremes.

The theoretical importance of sibling rivalry is easy to appreciate when it is remembered that the spread of a gene through a population is affected both by how individuals with a copy of the gene succeed in survival and mating and by how close genetic kin do as well. A nestling bird's full sibling (same mother and father) has identical copies of half, on average, of all the genes in the nestling's own genome. This means that the nestling bird receives a potential 50 percent dividend to its Darwinian fitness from its sibling's future reproductive success. In this sense, a social glue exists to bind genetic relatives together. [*See* Hamilton, William D.] However, embryos and neonates of most species grow rapidly, a process that must be fueled by copious nutrients and other resources. In many species, one or two parents must personally deliver these resources. If and when there is a shortfall, decisions must be made on how the available investment should be distributed. Thus, sibling rivalry can be thought of as a very common model for studying the evolutionary limits of selfishness. The basic prediction is fairly simple: a selfish individual should want to consume the next unit of parental investment until the point is reached beyond which that unit would be twice as valuable to a full sibling, or four times as valuable to a half-sib, as to self (Hamilton, 1964; Parker et al., 1989).

The solutions to sibling competition vary widely, according to the severity and duration of shortages, nursery geometry, brood size, genetic relatedness, and any "weaponry" possessed by the rivals. The simplest behavioral manifestation is begging, by which each offspring communicates its desire to its parents to bias whatever food or other resources are brought toward the individual signaler. Sometimes parents respond selectively to the signals they receive, for example, favoring the smallest nestling in certain birds (Gottlander, 1987; Stamps et al., 1985). In other species, the older of two hatchlings may be the only one that receives any food, so the younger nest mate starves from parental neglect (Boersma, 1991).

In addition to the familiar postures and cries of nestling birds and mammals, offspring often move into more advantageous positions (jockeying). For example, if parents must always enter the nursery from one predictable direction, (e.g., through a burrow), then the spot by the entrance may be sought after. Such jockeying blends into strategic displacement of rivals (jostling), which in

SIBLICIDE BY SPOTTED HYENAS

Only a few mammals have been documented to practice siblicide on a regular basis, namely, domestic swine (Fraser and Thompson, 1990), pronghorn embryos (O'Gara, 1969), Galapagos fur seals (Trillmich, 1986, 1990), and spotted hyenas (Frank et al., 1991; Golla et al., 1999; Hofer and East, 1997). Anecdotal evidence also suggests siblicide in two foxes (Henry, 1985; MacPherson, 1969) and one bat (Leippert et al., 2000). It is not clear whether young mammals are biologically less well suited to gain from sibling aggression than birds. Perhaps because milk is provided through multiple teats, it is less easily monopolized through intimidation (Mock and Parker, 1997).

The mammal most closely studied for its practice of siblicide is the spotted hyena, a species regarded as peculiar in various other respects. Most famously, females possess a hypertrophied clitoris and are both larger and socially dominant over males in the matrilineal clans. They give birth through the clitoris to typically two cubs at the mouth of a subterranean nursery that is too small for the mother to enter, often a vacant aardvark den, returning sporadically to feed the cubs. Intriguingly, cubs have fully erupted teeth at the moment of birth, a trait also shared with domestic swine, that are used to attack the nest mate. In fact, the characteristic "bite-shake" behavior used by hyenas to kill prey has been observed in captivity within the first hour after birth. The first full description of this activity, based mainly on close observations of captive litters, also reported a significant sex ratio effect, with same-sex litters appearing to engage in siblicide more often than mixed-sex litters (Frank et al., 1991). Attempts to provide a functional explanation for this effect, which is unparalleled in siblicidal birds and the reverse of what has been reported for parasitoids, focused initially on the extraordinary masculinization of females. It was proposed that a dominant female cub might do well to eliminate a sister immediately, rather than have to compete with her for social status later. Alternatively, the high levels of circulating testosterone found in female spotted hyenas may simply provide a proximate explanation for heightened aggressiveness.

It now appears that the same-sex effect may not be characteristic of the species at all. With a much larger sample from multiple clans under field conditions, Golla et al. (1999) reported no sex ratio bias in the siblicide process. Indeed, the bulk of the data now point toward spotted hyenas as practicing a familiar "avian" type of facultative siblicide. Single cubs grow more rapidly than paired ones (Golla et al., 1999; Hofer and East, 1993), so when food is short and milk deliveries are increasingly infrequent, the slightly older and socially dominant cub stations itself at the den's mouth and prevents its sibling from nursing until it starves.

—Douglas W. Mock

turn can escalate into the use of overt aggression. Sib fighting, such as pecking by nestling birds and nipping by mammal pups, can intimidate rivals into begging less vigorously. Sibling aggression leading to the victim's death, or siblicide, typically combines the physical injury caused by the attacks and the cumulative effects of food deprivation. In some eagles, pelicans, cranes, and boobies, siblicide occurs in virtually every brood (Simmons, 1988); in other birds, fatalities are conditional on various outside forces, mainly associated with food deliveries. Nestling aggressiveness may itself be adjusted according to food amounts, as shown experimentally for blue-footed boobies (Drummond and Osorno, 198x) and ospreys (Machmer and Ydenberg 199x), or it may remain relatively steady, as in egrets, with the subordinate chick's ability to withstand physical abuse conditional on the amount of food it receives (Mock et al., 1987).

Examples of Siblicide. Although avian siblicide has been known for at least two and a half millennia (Aristotle wrote about it in his *History of Animals*, Book 6), the quantitative study of its many varieties was impractical until about two decades ago, when it was found to be quite common in several species that nest in dense colonies or in nest boxes that researchers could provide and monitor. More recently, it has been studied in amphibians (Pfennig), fish (Wourms, 1981), and mammals (see Vignette). Insects of the order Hymenoptera (ants, bees, and wasps) are also famous practitioners. When a queen honeybee dies, for example, larvae developing in special "royal cells" (which are larger in dimension and whose inhabitants are fed with special food) grow into potential successors. The first of these proto-queens to pupate quickly stings and kills all the others, thereby eliminating reproductive rivals.

One of the most extreme cases brought to light is a parasitoid wasp (*Copidomopsis floridanum*) that para-

FIGURE 1. In species such as the Magellanic penguin, the first of two hatchlings is usually the only one that receives any regurgitated food, so the younger nestmate starves from parental neglect. © Joe McDonald/Visuals Unlimited.

lyzes a caterpillar and then oviposits two eggs, one male and one female, inside the host's body. Here we have a nursery (the host) and a finite supply of food that will not be enlarged by parental activity: the mother departs, never to return. The two eggs quickly undergo a rapid cloning phase called polyembryony, producing roughly 200 twin sons and 1,200 twin daughters. About fifty of the daughters grow more rapidly than the rest and develop huge jaws but no sex organs. These "precocial larvae" were initially thought to be specialized protectors of the whole brood (against the potential threat of a second mother ovipositing a rival brood in the same host), but more recent scrutiny has revealed quite the opposite: they seek and cannibalize their own brothers (Grbíc et al., 1991). The explanation for this extraordinary behavior centers on the fact that brothers fertilize their own sisters before any siblings pupate and disperse. The precocial larvae have no sexual potential of their own, but their reproductive twin sisters carry four times as many identical copies of their genes as do their brothers. Furthermore, a single brother can fertilize numerous sisters. So the optimal brood sex ratio, as viewed by a precocial larva, requires very few males. By killing brothers, precocial larvae reserve more host tissue for their reproductive sisters.

Manifestations of sibling rivalry are likely to be even more diverse in plants, although relatively little study has been made of botanical parallels. Many plants germinate from seeds that fall directly beneath the parent, so siblings are commonly crowded for space, light, and soil nutrients. Tiny rain forest trees may live for decades in a stunted limbo, effectively waiting for an opening in the continuous canopy that will support full growth. That opportunity may not come until the parent falls, at which time many offspring may compete for the space created. Sibling rivalry between seeds may even occur before the offspring disconnect from the parent. In Indian rosewood (*Dalbergia sissoo*), for example, half a dozen ova develop inside a flattened pod that will eventually become a lightweight glider, enabling its contents to drift an uncertain distance from the parent's site. A flower at the pod's tip receives pollen, so the pollen tubes entering the pod fertilize the most distal seed first, then the next most distal, and so on. As soon as it is fertilized, the distal seed begins production of a water-soluble chemical that is borne back toward the stem, stunting and eventually killing all other pod mates. Soon the average number of seeds has fallen to one, which enlarges considerably before the pod is abscised for dispersal. Experimental testing has shown that the killing of victim seeds is definitely caused by the distal sibling's product: extract from it will kill healthy seeds on an agar base, whereas a similar extract from the parent's own pod tissue has no harmful effects (Ganeshaiah and Uma Shaanker, 1988).

[*See also* Parent–Offspring Conflict; Parental Care.]

BIBLIOGRAPHY

Boersma, P. D. "Asynchronous Hatching and Food Allocation in the Magellanic Penguin *Spheniscus magellanicus*." *Acta XX Congressus Internationalis Ornithologici* II (1991): 961–973.

Frank, L. G., S. E. Glickman, and P. Licht. "Fatal Sibling Aggression, Precocial Development, and Androgens in Neonatal Spotted Hyenas." *Science* 252 (1991): 702–704.

Fraser, D., and B. K. Thompson. "Armed Sibling Rivalry among Piglets." *Behav. Ecol. Sociobiol.* 29 (1990): 9–15.

Ganeshaiah, K. N., and R. Uma Shaanker. "Seed Abortion in Wind-dispersed Pods of *Dalbergia sissoo*: Maternal Regulation of Sibling Rivalry?" *Oecologia* 75 (1988): 135–139.

Golla, T., H. Hofer, and M. L. East. "Within-Litter Sibling Aggression in Spotted Hyaenas: Effects of Maternal Nursing, Sex, and Age." *Animal Behaviour* 58 (1999): 715–726.

Gottlander, K. "Parental Feeding Behaviour and Sibling Competition in the Pied Flycatcher *Ficedula hypoleuca*." *Ornis Scand.* 18 (1987): 269–276.

Grbíc, M., P. J. Ode, and M. R. Strand. "Sibling Rivalry and Brood Sex Ratios in Polyembryonic Wasps." *Nature* 360 (1992): 254–256.

Henry, J. D. *Red Fox: The Cat-like Canine*. Washington, D.C., 1985.

Hofer, H., and M. L. East. "Skewed Offspring Sex Ratios and Sex Composition of Twin Litters in Serengeti Spotted Hyaenas (*Crocuta crocuta*) Are a Consequence of Siblicide." *Appl. Anim. Behav. Sci.* 51 (1997): 307–316.

Leippert, D., W. Goymann, and H. Hofer. "Between-Litter Siblicide in Captive Indian False Vampire Bats (*Megaderma lyra*)." *J. Zool. Lond.* 251 (2000): 537–540.

MacPherson, A. H. "The Dynamics of Canadian Arctic Fox Populations." *Can. Wildl. Serv. Rep. Ser.* 8 (1969).

Mock, D. W., and L. S. Forbes. "The Evolution of Parental Optimism." *Trends in Ecol. Evol.* 10 (1995): 130–134.

Mock, D. W., and G. A. Parker. *The Evolution of Sibling Rivalry*. Oxford and New York, 1997.

O'Gara, B. "Unique Aspects of Reproduction in the Female Pronghorn (*Antilocapra americana*)." *Am. J. Anat.* 125 (1969): 217–232.

Parker, G. A., D. W. Mock, and T. C. Lamey. "How Selfish Should Stronger Sibs Be?" *The American Naturalist* 133 (1989): 846–868.

Stamps, J., A. B. Clark, P. Arrowood, and B. Kus. "Parent–Offspring Conflict in Budgerigars." *Behaviour* 94 (1985): 1–40.

Trillmich, F. "Attendance Behavior of Galapagos Fur Seals." In *Fur Seals: Maternal Strategies on Land and at Sea*, edited by R. L. Gentry and G. L. Kooymann, pp. 168–185. Princeton, N.J., 1986.

Trillmich, F. "The Behavioural Ecology of Maternal Effort in Fur Seals and Sea Lions." *Behaviour* 114 (1990): 3–20.

— DOUGLAS W. MOCK

SICKLE-CELL ANEMIA. *See* Heterozygote Advantage, *article on* Sickle-Cell Anemia and Thalassemia.

SIGNALLING THEORY

Animal signals or displays are traits that are specialized for the purpose of communication. This definition encompasses a wide range of different structures and behaviors, from the chirping of crickets and the calling of frogs to the bright plumage colors and ornaments of birds; from the pheromone trails laid by ants to the electrical signals produced by some species of fish. Signals may be used to threaten a rival or to attract a mate, to ward off a predator or to solicit food from a parent, to mark out a territory or to warn others of danger. But however diverse in form and function, all signals share a common purpose: they serve to influence the behavior of other animals.

Signal Design and Selection for Efficient Displays. Because the purpose of a signal is to influence the behavior of receivers, it must be detectable. A display that goes unnoticed or unrecognized by the intended recipient(s) will not be favored by selection. But communication often takes place over long distances, in a noisy environment. By the time it reaches the receiver, a signal may be attenuated, degraded, and overlaid with irrelevant stimuli. Natural selection therefore tends to favor effective displays that are easier to detect and recognize.

At the same time, a distinctive and striking signal may prove costly for the signaller. The loud calls of male frogs and crickets are effective in attracting mates, but they may also attract the unwanted attention of predators and parasites. Furthermore, calling can be both time consuming and energetically demanding. Natural selection therefore tends to favor displays that are less risky and cheaper to produce.

The signals that we see in nature should be those that strike the best possible balance between the opposing selective pressures for greater efficacy and lower cost. In other words, they should exhibit efficient design.

Common design features of animal signals. A number of features, common to many different kinds of display, serve to increase efficiency and to facilitate detection and recognition.

First, signals tend to be conspicuous. That is, they tend to contrast markedly with irrelevant background stimuli. This makes it easier for the receiver to distinguish the signal from "noise."

Second, most signals are highly stereotyped. Compared with other traits and behaviors, displays tend to be relatively invariant in form. This standardization reduces the difficulty of the task facing receivers by minimizing the variation with which they must cope. It is far easier to classify incoming signals into a small number of distinct categories than to deal with a highly variable display.

Third, signals frequently exhibit extensive redundancy. In other words, there are often predictable relationships among different parts of a display. In its simplest form, redundancy may simply entail repetition; alternatively, a signal may feature a predictable sequence of behaviors or sounds, or may combine several different elements (e.g., auditory and visual) in parallel. Redundancy facilitates detection because the receiver need only recognize one part of the signal to identify the whole.

Finally, signals often feature alerting components. These are introductory elements that are particularly easy to detect (perhaps because they are strikingly conspicuous and stereotyped), but which need not convey much information. Alerting components increase the chance that the receiver will detect and recognize the subsequent, information-bearing parts of the display, by specifying the interval of time during which it can expect to receive them.

Selection for efficacy and signal diversity. The design features mentioned above are common to many different kinds of display, and serve very generally to increase efficiency. Beyond these basic properties, however, the design of an efficient signal will reflect the physical environment in which communication occurs, and the sensory capacities and "psychology" of receivers.

A series of studies conducted by John Endler on the coloration of male guppies nicely illustrate such influences. The natural habitats of these fish are mountain streams in the tropical forests of northeastern South America and the Caribbean. Such streams feature a complex mosaic of different light environments: forest shade, in which the light has been transmitted through or reflected from vegetation, and thus is rich in green wavelengths; woodland shade, which is dominated by blue sky light, but not illuminated directly by the sun; small gaps, which are illuminated by the redder wavelengths of direct sunlight; and large gaps, which feature "white" light from the sun and open sky together. Moreover, a fifth, purplish light environment is created when the sun is rising or setting.

Different populations of guppies exhibit an equally complex array of male color patterns, comprising patches of red-orange, yellow, bronze-green, cream-white, blue, silver, brown, and black. These color patterns (which vary among populations) play a role in mate choice, but they may draw the attention of predators as well as of females. Endler showed that the colors characteristic of different populations tended to be those that appeared most conspicuous to guppies while least conspicuous to predators under local conditions. For instance, in a population that suffered heavily from predation by a species of freshwater prawn, the most abundant hue was orange, a color to which this particular predator is relatively insensitive. Where predatory fish, which do not share the same insensitivity, posed a greater threat, the guppies displayed less of this color. Furthermore, guppies appear to time their courtship so as to enhance the efficiency of communication. Male display occurred either early or late in the day, when purplish light from the rising or setting sun dominates the ambient spectrum. Endler found that switching to any of the other light environments would lead to a reduction in conspicuousness to other guppies and/or to an increase in conspicuousness to predators.

Coevolution Between Signaller and Receiver. Both the physical environment and the psychology of receivers exert strong selective pressures on signal design. But while the physical environment is relatively static, the psychology of receivers may itself change in response to the evolution of a signal. The reason for this is that receivers are under selection to make best use of any information that signallers provide. Ultimately, therefore, animal displays are the product of coevolution between signallers and receivers, two parties whose interests do not always coincide.

Natural selection favors signals that influence the behavior of receivers in ways beneficial to the signaller. In some cases, what is advantageous for the signaller may also be beneficial for the receiver. Thus when a worker honeybee returns to the hive with nectar, she performs a "waggle dance" that informs her fellow workers about the direction and distance of the food source. Both signaller and receivers in this case share a common interest; the hive as a whole, including the dancing worker and her sisters, gains by exploiting the rich food source that has been located. In situations like this, we speak of cooperative communication, or a cooperative signalling system.

In many cases, however, conflicts of interest arise between signaller and receiver. The former may often stand to gain by eliciting a response that is not to the benefit of the latter. Consider, for instance, a threat display like the roaring of red deer. The receiver of such a signal stands to gain from an accurate assessment of its opponent's fighting ability or willingness to escalate; at the same time, the signaller stands to gain by misleading the receiver as to its own fighting ability or aggressive motivation, so as to deter opposition. A more extreme example may be seen in signals employed by predators to attract prey. The bolas spider employs a pheromonal "lure" to attract male moths, through mimicry of the sex pheromones released by females of their species. The spider benefits from this response, drawing in prey to capture and consume, but the moths clearly do not.

Fighting and predator-prey interaction feature obvious conflicts of interest. But differences arise even in contexts that at first sight seem harmonious. Sexual signalling is a good example. Both sexes share a common interest in finding and pairing with conspecifics. However, males are usually under stronger selection to acquire more mates, and females to acquire better mates. Females thus stand to gain from an accurate assessment of mate quality, while males stand to gain by misleading potential partners as to their own worth. Similarly, parents share a common interest with their offspring in the latter's survival. But while a parent stands to gain by allocating food in relation to the hunger or need of its young, individual offspring stand to gain by misleading the parent as to their own level of need, because they are selected to acquire more food than the parent is selected to give.

When the interests of signaller and receiver do not coincide, communication can no longer be viewed as a cooperative exchange of information. Rather, it is the focus of an "arms race" between signallers as "manipulators" and receivers as "mind readers." Selection favors signallers that can elicit responses beneficial to themselves. At the same time, it favors receivers that devalue and ignore manipulative signals, and make best use of what information they can obtain from the signaller's behavior.

Arms races and exploitation. The view of communication as an arms race seems to suggest that, on an evolutionary time scale, informative or honest signalling cannot persist. Suppose, for example, a population exists in which signallers reliably advertise their aggressive intentions by means of a threat display. Receivers would do best to make use of this information, and retreat when facing an opponent announcing greater motivation than their own. As a result, it seems that a "deceitful" mutant that always announces a high level of aggression can prosper, because its opponents will always retreat. With the spread of such mutants, the signal will be devalued and it will no longer pay to respond to it.

Signal evolution may thus be characterized by repeated but transient bouts of exploitation, in which a novel signal evolves that elicits responses beneficial for signallers, but ultimately loses its effect, owing to selection on receivers for reduced responsiveness. New opportunities for exploitation will repeatedly occur, because selection cannot influence responsiveness to a particular display until it is actually encountered. Consequently, "hidden preferences" for a signal may arise in a population of receivers by random drift, which signallers can then exploit (though once they begin to do so, selection will favor reduced responsiveness to the display).

Can there be any stable outcome to the arms race between signaller and receiver? Stable exploitation may be possible when signallers take advantage of perceptual sensitivities or biases that serve receivers well in contexts other than communication. For example, male water mites of the species *Neumannia papillator* attract the attention of females by means of "courtship trembling," vibrating the water surface in a manner resembling that of the species' copepod prey. Females, particularly when hungry, are likely to orient to and clutch at a courting male as if he were a prey item, but will subsequently mate rather than consume him. In this situation, selection is unlikely to favor a reduction in female responsiveness to the male display, because sensitivity to vibrations of the water surface is essential for prey detection.

Deceit may also persist over evolutionary time if it occurs at a sufficiently low frequency (and the cost to receivers is small enough). For instance, stomatopod crustaceans of the species *Gonodactylus bredini* employ characteristic threat displays in defense of their burrows, even if they have recently molted and thus are unable to fight effectively because their carapace has not yet hardened. Why does selection not favor individuals that are less responsive to these deceptive displays? Because molting occurs relatively infrequently (and the subsequent period of vulnerability is short), a threat is only occasionally deceptive. Consequently, receivers are unlikely to benefit by ignoring the display altogether.

The maintenance of honesty. Although deceit can persist at low frequency, or through exploitation of a sensory bias, the majority of signals seem to convey at least some reliable information. In other words, stable honesty is the most common outcome of the arms race between signallers and receivers. How can honest signalling persist, despite an evolutionary incentive to deceive?

A signal may prove "unfakeable" for reasons of physical necessity. If there is a material link between the signal and some underlying aspect of the signaller's state, then deceit may simply be impossible. An example is coloration based on carotenoid pigments. Most animals cannot synthesize carotenoids themselves, and have to acquire them as part of their diet. So the intensity of carotenoid coloration is linked to foraging success; an individual that is unable to find much food simply cannot manufacture the carotenoids needed for vivid color.

While some signals seem to be physically tied to the signaller's underlying condition, this is not always the case. Honesty can nevertheless be maintained if the signal is costly to produce. Though possible, cheating may then prove unprofitable. This idea, first proposed by the Israeli biologist Amotz Zahavi, is known as the handicap principle. The principle was first applied to sexual signals (though it is relevant to many forms of communication). Suppose that a male sexual display that serves to attract females is costly to produce, and more so for inferior individuals. Under these circumstances, only superior males may gain a net benefit from signalling. For inferior individuals, the greater cost of the signal outweighs the advantages of attracting additional mates. Females can therefore rely on the signal for information about mate quality.

In the context of sexual display, signallers are assumed to differ in their ability to bear the cost of signalling. In other circumstances, signallers may vary in the benefits that they stand to gain. Suppose, for example, that an offspring display which serves to solicit food from parents is costly to produce. Under these circumstances, only hungry young may gain a net benefit from signalling. For satiated offspring, the benefit of soliciting additional food is lower, and may be outweighed by the cost of the display. Parents can therefore rely on the signal for information about offspring hunger. This alternative possibility has been termed signalling of *need*, as opposed to signalling of *quality* (although in reality, the two situations are not mutually exclusive, because signallers may often vary both in their ability to bear the cost of display and in the benefit they stand to gain thereby).

The "drumming" display of the male jumping spider *Hygrolycosa rubrofasciata* provides a good example of an honest, costly signal. Males of this species attract mates by drumming their abdomen vigorously on the ground. Experimental studies in which males are encouraged to drum by exposure to females have shown that the display is energetically demanding, and that repeated drumming leads to a decline in body weight and to increased mortality. Better-fed males, however, naturally drum more vigorously. Moreover, they are better able to cope with the demands of experimentally induced display. It thus seems that this signal provides females with honest information about male condition, and that the reliability of the display is maintained by its energetic cost, which is greater for males in poorer condition.

Social constraints on deception. Physical cost is not, however, the only factor that can maintain the reliability of a signal. In some cases, cheap yet honest signals may persist because cheats incur social costs. In a number of bird species, for instance, fights are settled by "badges of status," colored plumage patches that vary in size between individuals. Production of a large badge indicative of high status may seem to entail little physical cost, but experimental manipulation of badge size (e.g., in house sparrows, *Passer domesticus*) has shown that individuals with larger plumage patches can suffer increased aggression from other, dominant competitors. This social "punishment" may deter weak or subordinate birds from developing large badges, if they are less able to cope with aggressive encounters.

More sophisticated forms of punishment have also been reported in primates. For example, upon discovering food, free-living rhesus macaques (*Macaca mulatta*) sometimes give distinctive calls that encourage those nearby to approach. Individuals who fail to call but are subsequently caught with food incur greater levels of aggression from the other members of their group. Once again, therefore, reliable signalling seems to be maintained by social punishment. In this case, moreover, receivers seem to be able to recognize deception when it occurs (since individuals who fail to call are punished only if they are caught with food).

BIBLIOGRAPHY

Adams, E. S., and R. L. Caldwell. "Deceptive Communication in Asymmetric Fights of the Stomatopod Crustacean *Gonodactylus bredini.*" *Animal Behaviour* 39 (1995): 706–716. An interesting case study of occasional deceit.

Bradbury, J. W., and S. L. Vehrencamp. *Principles of Animal Communication.* Sunderland, Mass., 1998. A comprehensive survey of the study of animal signals and their evolution.

Endler, J. A. "Variation in the Appearance of Guppy Colour Patterns to Guppies and Their Predators under Different Visual Conditions." *Vision Research* 31 (1991): 587–608. A comprehensive treatment of the effect of the light environment on the conspicuousness of different colour patterns.

Espmark, Y., T. Amundsen, and G. Rosenqvist. *Animal Signals: Signalling and Signal Design in Animal Communication.* Trondheim, Norway, 2000. A recent edited volume illustrating current approaches to the study of animal signalling.

Hauser, M. D., and P. Marler. "Food-Associated Calls in Rhesus Macaques (*Macaca mulatta*). 2. Costs and Benefits of Call Production and Suppression." *Behavioural Ecology* 4 (1993): 206–212. A nice illustration of the social control of deception.

Hill, G. E., and R. Montgomerie. "Plumage Colour Signals Nutritional Condition in the House Finch." *Proceedings of the Royal Society of London, Series B,* 258 (1994): 47–52. Evidence that carotenoid coloration conveys reliable information about condition.

Kilner, R., and R. A. Johnstone, "Begging the Question: Are Offspring Solicitation Behaviours Signals of Need?" *Trends in Ecology and Evolution* 12 (1997): 11–15. A review of evidence that offspring begging signals need for food.

Krebs, J. R., and R. Dawkins, "Animal Signals: Mind-Reading and Manipulation." In *Behavioural Ecology: An Evolutionary Approach,* 3d ed., edited by J. R. Krebs and N. B. Davies. Oxford, 1984. A classic argument for signal evolution as an arms race.

Mappes, J., R. V. Alatalo, J. Kotiaho, and S. Parri. "Viability Costs of Condition-Dependent Sexual Male Display in a Drumming Wolf Spider." *Proceedings of the Royal Society of London, Series B,* 263 (1996): 785–789. An elegant study of an honest, costly signal.

Møller, A. P. "Social Control of Deception among Status Signalling House Sparrows, *Passer domesticus.*" *Behavioural Ecology and Sociobiology* 20 (1987): 307–311. Evidence that deception can incur socially imposed costs.

Proctor, H. C. "Courtship in the Water Mite *Neumannia papillator*: Males Capitalize on Female Adaptions for Predation." *Animal Behaviour* 42 (1991): 589–598. A possible example of male exploitation of a female sensory bias.

Zahavi, A. *The Handicap Principle.* Oxford, 1997. A compelling though tendentious argument for the importance of the handicap principle in signal evolution.

— RUFUS A. JOHNSTONE

SIMPSON, GEORGE GAYLORD

George Gaylord Simpson (1902–1984), paleontologist and evolutionary biologist, was born in Chicago, Illinois, the first son and third child of Helen J. (Kinney) and Joseph A. Simpson. Simpson received a bachelor's degree in geology from Yale University in 1923 and a doctorate in paleontology in 1926. He wrote his doctorate dissertation on Mesozoic mammals, some 200 to 65 million years old, under the supervision of Richard Swann Lull, and continued this research during a postdoctoral year at the British Museum in London. His work on these heretofore little known, but important, Mesozoic fossils quickly earned him an international reputation as a paleomammalogist and led to a position at the American Museum of Natural History in New York City in 1927. His senior colleagues at the museum included H. F. Osborn, W. K. Gregory, and W. D. Matthew, the few paleontologists of that era who consistently addressed biological questions in their study of fossils.

Simpson's continuing research on fossil mammals took him to Patagonia in the 1930s, and his first book, *Attending Marvels* (1934), combines travel journalism with popular science. In the late 1930s and early 1940s, Simpson began to address broader issues in evolutionary theory. In 1939, he published *Quantitative Zoology* with his second wife, Anne Roe, a distinguished clinical psychologist, and by the end of 1942, just before Simpson entered military service, he completed two book-length manuscripts, *Tempo and Mode in Evolution*

(1944) and *Principles of Classification and a Classification of Mammals* (1945).

In *Tempo and Mode in Evolution*, his most significant book, Simpson argued that Darwinian natural selection, operating on inherited variation within species, provided a parsimonious interpretation of the adaptations, specializations, and extinctions observed by paleontologists among fossils. Consequently, there was no need to rely on inherent, metaphysical mechanisms popular with most contemporary paleontologists to explain rates and patterns of evolution. *Tempo and Mode in Evolution* thus joined a half dozen other books that formed the foundation for the "modern evolutionary synthesis"—a synthesis because the new theory of evolution came from a number of disciplines, including genetics, ecology, anatomy, field biology, and botany, as well as paleontology.

After two years of service in U.S. Army Intelligence in North Africa and Sicily, Simpson returned to the American Museum in late 1944, and in the following year he became chairman of the new Department of Geology and Paleontology; he was also appointed professor of vertebrate paleontology at Columbia University. In 1949, Simpson published *The Meaning of Evolution*, a popular and widely read account of modern evolutionary theory that sold a half million copies in a dozen translations. In 1953, he completed another semitechnical volume, *Evolution and Geography*, which argued for the wide dispersal of animals, particularly mammals, across fixed continents over long spans of geologic time. In so doing, Simpson strongly challenged the theory of drifting continents as earlier formulated by the German geophysicist Alfred Wegener. That same year Simpson saw the publication of *Major Features of Evolution*, a much enlarged revision of the earlier *Tempo and Mode in Evolution*.

In 1958, Simpson resigned as chairman of the Department of Geology and Paleontology of the American Museum after a dispute with the director, and a year later he joined the staff of Harvard University's Museum of Comparative Zoology as Alexander Agassiz Professor of vertebrate paleontology. In 1964, Simpson published his "favorite book," *This View of Life*, a collection of essays on various topics, including Charles Darwin and evolution, historical biology, the apparent purpose in animate nature, speculations about evolution beyond the earth, and the human evolutionary future.

In 1967, Simpson left Harvard and joined the faculty of the University of Arizona, where he remained until his death in 1984. Simpson did a little teaching, but mostly he continued to publish technical articles and books—on South American mammals, penguins, Darwin, fossils, and the history of life—as well as collections of essays and an autobiography.

As one of the leading authorities in vertebrate paleontology and evolutionary theory during the middle half of the twentieth century, Simpson received many honors, including membership in the American Philosophical Society (1936), the National Academy of Sciences (1941), and the Royal Society of London (Foreign Member, 1958), as well as election as president of the Society of Vertebrate Paleontology (1942), the Society for the Study of Evolution (1946), the American Society of Mammalogists (1962), the Society for Systematic Zoology (1962), and the American Society of Zoologists (1964).

PALEONTOLOGY OF SOUTH AMERICA

Simpson centered much of his research on the Cenozoic mammals of Argentina, Brazil, and Venezuela that had evolved independently, separated from the rest of the world, for 50 million years before the Pleistocene ice ages. His first book, *Attending Marvels* (1934) recounted his experiences of field work in Patagonia, and his last book, *Discoverers of the Lost World* (1984), told of the scientists who uncovered and described the abundant fossil mammals of South America. In the fifty years between, Simpson himself made lasting contributions to the discovery and description of many important South American fossils as well as clarifying their ages relative to one another and to the global geologic time-scale. Indeed, more than one-third of Simpson's primary publications was based on South American mammals. His book, *Splendid Isolation* (1980), is a popular account of what he termed "the curious history" of these diverse and unique organisms. Simpson's South American research also stimulated the formulation of his principles of historical biogeography that, ironically, led him to oppose Wegenerian continental drift. For Simpson, mobile organisms dispersing through geologic time across stable continents was sufficient to explain the past distributions of terrestrial life, as so well exemplified by North and South American mammals.

Despite his many South American successes, his field expedition to the headwaters of the Amazon near the Brazil-Peru border in 1956 resulted in a near-fatal accident that left him partially crippled and thereafter unable to carry out further serious field work. His prolonged recovery from the accident also created difficulties with the American Museum of Natural History where he was chair of the Department of Geology and Paleontology, so that he soon resigned and took up an appointment as Alexander Agassiz professor at Harvard's Museum of Comparative Zoology.

—Léo F. Laporte

He also received the National Medal of Science from President Lyndon Johnson (1966) and more than a dozen other medals and prizes from international scientific societies and organizations.

[*See also articles on* Biogeography; Paleontology.]

BIBLIOGRAPHY

Laporte, L. F. *George Gaylord Simpson, Paleontologist and Evolutionist.* New York, 2000. Centers on major areas of Simpson's research.

Simpson, G. G. *Concession to the Improbable: An Unconventional Autobiography.* New Haven, 1978. Simpson's idiosyncratic reflections on his long life.

Simpson, G. G. *Simple Curiosity: Letters from George Gaylord Simpson to His Family,* 1921–1970. Edited by L. F. Laporte. Berkeley and Los Angeles, 1987. Personal letters written over a half century, mostly to his older sister Martha, that reveal much more about Simpson than his autobiography that was written for the public.

— Léo F. Laporte

SIMULATIONS OF EVOLUTION

Simulations of evolution can be divided into two kinds: those aimed at creating complex objects using only simple rules, such as descent with modification and subsequent selection, and those in which the nature of the evolution process itself, as well as its influence on the products of evolution, is the object of inquiry.

In general, simulation refers to a technique in which a process and the relevant variables are abstracted mathematically, so that they can be studied, usually, but not necessarily, on a computer. Strictly speaking, simulations of evolution are not really simulations (as opposed to, say, the simulation of a tornado) because the actual rules of evolution are implemented and act on information (e.g., inside the computer) that has a physical, rather than a mathematical, representation. It can be said that the objects or their representations whose evolution is studied are actually evolving and thus real. The environment in which the process takes place, however, is not.

Evolution is often used as an algorithm for optimization because it is known to be an effective tool for searching through large numbers of possible solutions that can vary in a number of dimensions. This use represents the first kind of application of the principles of evolution within computers. Evolution is then used for discovery (as in the use of a genetic algorithm to discover the fastest algorithm for sorting a list of objects) or for optimal adaptation of an object to its environment (as in the design of optimal airplane wing shapes). Such applications are characterized by fixed "fitness functions" (which determine the success of a candidate solution) commensurate with the goal at hand. [*See* Genetic Algorithms.]

Because the rules of evolution can be so easily implemented within a computational environment such as a computer, it is tempting to study how these rules affect actual or simulated populations, as a function of simulated environments. This is an example of a simulation in which the process of evolution is itself part of the inquiry. For example, one might want to check whether, and under what circumstances, simple rules of inheritance with variance, coupled with selection can lead to complex structures. A well-known example is provided by Richard Dawkins's (1976) evolution of "biomorphic" structures within a simple program called Biomorph. Biomorphs are treelike structures whose physical appearance is determined by nine "developmental genes" that determine the biomorph's appearance, such as the number of branchings, angle between branches, elongation of branches, and color. Each biomorph is represented by its genotype, which is the sum of all the genes that determine its structure. These genotypes are subject to mutation, in which a gene is altered slightly (e.g., by changing the angle between branches), and crossover, in which a new genotype is created by combining some of the genes of one biomorph with those of another. Selection is carried out by the user, who selects one particular biomorph from the population to serve as the progenitor for the next generation, which consists of mutated offspring of that progenitor. By these simple rules, biomorphs of stunning complexity can be evolved, depending on the user's whim and fancy. In this case, the criterion that determines success or failure is not a fixed function, as it can change with the user's visual appreciation of the evolved product. This simulation also teaches that the number of possible evolutionary paths is so large that the complex products of evolution are hardly ever the same, and essentially irreproducible in repeated attempts.

Another well-known demonstration of the power of the evolutionary principles is Karl Sims's (1994) evolution of morphology and behavior in simulated environments. The object of this study was to observe the evolution of (simulated) organisms in an environment of high complexity, where the fitness function had elements external to the population (e.g., the simulated laws of gravity, friction, and fluid mechanics) as well as internal ones, when the success or failure of a lineage depended on competition between lineages within the artificial world. An organism's genotype is given by a "nested graph," which determines both the creature's morphology and the sensors, actuators, artificial neurons, and how they connect to the body. These nested graphs can be mutated and crossed over, and a similar mechanism as in Dawkins's biomorphs leads to successive generations. Selection occurs either by an external criterion (such as requiring locomotion in the given environment leading to either "walking" or "swimming" creatures) or by an internal one, as when organisms were selected according to their ability to succeed in a battle with another organism over a food source. In either case, the resulting organisms and the means by which they accomplish their tasks are sometimes familiar (as in the evolution of swimming motion) and sometimes unfamiliar to the point of giving a decidedly "alien" impression.

The lessons learned from this study are twofold. First, it shows that relatively simple ingredients suffice to create not only structure (as in the biomorphs) but also morphology, forms of locomotion, and even behavior, such as schooling and predator evasion. Second, we learn that the products of evolution invariably reflect the environment to which they adapt, so that the same process iterated in simulated water will give rise to swimming organisms, as opposed to the walking ones evolved on a flat surface. Another lesson learned often painfully in simulations of evolution is the prime law "You get what you select for." For example, in early experiments conducted by Sims to evolve locomotion on a flat surface, the physical law of conservation of momentum was not properly implemented. Although this flaw in the simulated world may not have been immediately obvious, it became so when organisms evolved who propelled themselves forward simply by hitting themselves on the back.

The previous examples strongly suggest that the principles of evolution are universal, in the sense that they apply to systems other than terrestrial biochemistry, and that the manner in which the objects undergoing evolution are described (i.e., the manner in which their information is coded) is immaterial to the process itself. This realization has led to the creation of an artificial form of life ("digital life") that does not share any ancestry with biochemical life, but still embodies the general principles of Darwinism.

The idea for digital life is a product of the computer age, spawned by the appearance of computer viruses. Although the latter evolve only as a result of "arms races" between hackers and operating system designers, it is possible to create artificial, simulated worlds within computers in which colonies of self-replicating computer programs live, adapt, and compete for resources. Early evolution experiments with colonies of computer viruses by Steen Rasmussen and colleagues (1990) were abortive because the artificial chemistry, that is, the set of instructions that compose the self-replicating programs, must not be too fragile under mutations. The first form of digital life that was capable of evolution was developed by Tom Ray (1991), a biologist who designed a system with a robust set of instructions and wrote the first self-replicating program in that language to serve as a seed. Such sets are referred to as artificial computer languages or computational chemistries, because the programs are executed on virtual rather than real processors. The number of different instructions in a computational chemistry is usually chosen to be of the same order as the amino acids in the genetic code. We can thus view these self-replicating programs as proteins that catalyze their own replication: they are both information and function. Because there is no limit imposed on the sequence length of these organisms, evolution is essentially open-ended because success or failure depends only on a program's ability to survive in its environment.

Although for digital organisms information is physically encoded in the bits of computer memory and thus "real," and although replication requires the physical copying of information from one place in memory to another, the world in which these programs live is simulated. This is not necessary in principle, but it is done for reasons of security and control. Adapting computer programs that exploit the actual hardware and software for survival are a threat not only to the host computer but also to other hosts connected to it via a network. A simulated world also allows detailed control of environmental conditions, which is especially important if these organisms are used in experimental evolution. In that respect, this artificial world is no different from the petri dishes or vials used in experiments with biochemical organisms. A particular form of digital life, hosted by the "Avida" program and developed at the California Institute of Technology in Pasadena has been used successfully in such a manner. The issues that can be addressed with these systems are the standard fundamental questions of evolutionary biology, such as the evolution of complexity and complex features. Furthermore, such a system can be used to begin a comparative evolutionary biology by repeating experiments within the digital world that were originally carried out with conventional organisms. This program of research should ultimately identify those aspects of evolutionary and adaptive dynamics that are independent of the particular organism used in the study and those that are not.

BIBLIOGRAPHY

Adami, C. *Introduction to Artificial Life.* New York, 1998. Introduces the methods and principles of digital life and discusses experiments that can be carried out with the accompanying Avida software.

Dawkins, R. *The Selfish Gene.* Oxford, 1976. Contains a description of Biomorph software.

Hofacker, I. L., W. Fontana, P. F. Stadler, L. S. Bonhoeffer, M. Tacker, and P. Schuster. "Fast Folding and Comparison of RNA Secondary Structures." *Monatshefte f. Chemie* 125 (1994): 167–188. This software package, which simulates the chemistry of RNA folding, can be used to study the evolution of RNA sequences.

Kauffman, S. A. *The Origins of Order.* Oxford, 1993. Argues that Darwin's theory alone cannot account for the order displayed by living systems and proposes autocatalytic networks of molecules as the driving force behind the ordering process.

Lenski, R. E., C. Ofria, T. C. Collier, and C. Adami. "Genome Complexity, Robustness, and Genetic Interactions in Digital Organisms." *Nature* 400 (1999): 661–664. The first article in comparative evolutionary biology examining the strength and sign of interactions between mutations in digital organisms as compared to *E. coli* bacteria.

Nowak, M. A. "What Is a Quasispecies?" *Trends in Ecology and Evolution* 7 (1992): 118–121. A short overview of the key concepts in Manfred Eigen's theory of molecular evolution, which allows the mathematical simulation of evolving macromolecules.

Rasmussen, S., C. Knudsen, R. Feldberg, and M. Hindsholm. "The Coreworld: Emergence and Evolution of Cooperative Structures in a Computational Chemistry." *Physica* D 42 (1990): 111.

Ray, T. S. "An Approach to the Synthesis of Life." In *Artificial Life II: Proceedings of an Interdisciplinary Workshop on the Synthesis and Simulation of Living Systems* edited by C. G. Langton, C. Taylor, J. D. Farmer, and S. Rasmussen. Redwood City, Calif., 1991.

Sims, K. "Evolving 3D Morphology and Behaviour by Competition." *Artificial Life* 1 (1994): 353.

Wilke, C. O., J. L. Wang, C. Ofria, R. E. Lenski, C. Adami. "Evolution of Digital Organisms at High Mutation Rates Leads to Survival of the Flattest." *Nature* 412 (2001): 331–333. An experimental study using the Caltech digital lifeform that demonstrates that robustness to mutations is a better predictor of fitness than replication rate when mutation rates are high.

— CHRISTOPH ADAMI

SOCIAL DARWINISM

Through the ages, scientists have developed numerous theories not only to explain human actions and social organization but also to justify them. Social Darwinism, popular in the late nineteenth century, belongs on this long list.

As its name implies, social Darwinism attempted to explain human beings' psychological predispositions and the societies that they form in terms of Charles Darwin's theory of natural selection. As present-day students of science look back at social Darwinism, they tend to emphasize how shoddy the science appears by today's standards and how blatant the biases of social Darwinism were—again, by today's standards. But the proper perspective for perceiving the past is in terms of the past, not the present. The contemporaries of social Darwinists objected to the methods that the social Darwinists employed, but it should be remembered that one of the main objections raised to Darwin's theory as such was that it was not sufficiently "scientific." The biases of social Darwinists seem obvious to us today, but they did not invent these biases; they were widely shared by their contemporaries.

First, and foremost, social Darwinists thought of themselves as providing a genuine science of human beings. They thought that much of what passed as psychology and sociology in the nineteenth century was more philosophy than science. They proposed to take a genuinely scientific theory and extend it to cover human beings in a scientifically legitimate fashion. Of course, the "science" they had in mind was the science of their day, not ours. In general, social Darwinists thought that such groups as the Irish were clearly inferior and proposed biological explanations for this inferiority, but then the general public in England at the time held the same view of the Irish and other groups.

Social Darwinists emphasized the competitive nature of the evolutionary process, Darwin's "survival of the fittest." According to Darwin, species evolve primarily through the action of natural selection. Certain organisms are better able to cope with their environments than others. Those organisms that are better at coping with their environments leave more offspring than those that are not. An organism's environment includes such nonliving elements as droughts and increases in temperature, but of equal importance are other organisms, organisms that belong to other species as well as to their own.

The primary purpose of society is to intercede between human beings and their various environments. Isolated individuals cannot cope with harsh winters or marauding carnivores, but people living in groups can. However, the benefits of living in societies can have undesirable consequences. The human species stands in danger of degenerating. Anyone in our distant past who could not see very well was likely to leave fewer offspring than those who had better sight, but with the advent of such "artificial" devices as glasses, the effects of selection have been greatly reduced. The net result is an increase in genetically "inferior" human beings. Hence, if societies were to disappear, we would be in real trouble. But some characteristics were deemed negative even in the presence of societies. For example, indolent members of a society damage that society, and in the view of English landowners, the Irish were seen as extremely indolent.

The view that increased population pressure influences the course of human events can be traced back to at least T. R. Malthus (1766–1834) and his *Essay on the Principle of Population* (1798). Malthus began his famous essay by noting that there is "no bound to the prolific nature of plants or animals," resulting in their "crowding and interfering with each other's means of subsistence." The same, he continues, can be said of human beings. The full title of his book was *An Essay on the Principle of Population; or A View of its past and present Effects on Human Happiness; With an Inquiry into our Prospects respecting the future Removal or*

Mitigation of the Evils which it Occasions. If we human beings do not control our numbers, we are likely to suffer dire consequences. Thus, Malthus can be seen reasoning from plants and animals in general to human beings in particular. Later, both Darwin and Alfred Russel Wallace would complete this circle, returning the argument to plants and animals. Later critics of social Darwinists complain that they read the competitive nature of Victorian society back into nature, but that is where they got it in the first place.

Darwin was aware of the logic of Malthus's argument applied to people, but he did not make it one of his major themes. The English philosopher Herbert Spencer (1820–1903) did. Not only did Spencer coin the phrase "survival of the fittest," but he also popularized social Darwinism. He presented his views on human nature in the same way that he presented his other scientific views—at such a soporific length that even Darwin complained. Although numerous authors urged social Darwinism on the populace at large, they did not form a school or movement in the sense that they held meetings, elected officers, and the like. Individual authors lectured and published on the topic; that is all. With the rediscovery of Gregor Mendel's laws at the turn of the twentieth century, that all changed as eugenics societies sprang up in Europe, Great Britain, and the United States. The goal of these societies was to encourage people with "good genes" to have more children (positive eugenics) and to encourage people with "bad genes" to have few, if any, children (negative eugenics). At times, negative eugenics took the form of forced sterilization.

One feature of social Darwinists is that they made their proposals in the absence of much in the way of scientific knowledge of heredity. People at the time assumed that racial mixture and inbreeding would have detrimental effects, but none of the theories of heredity then prevalent were all that coherent or powerful. With the advent of Mendelian genetics and the evolution of social Darwinism into eugenics, at last inferences about the good of the human species would be made on the basis of sound scientific knowledge. Unfortunately, the sort of knowledge relevant to eugenics was not developed until well into the twentieth century. One message of this knowledge was that the sort of practices being encouraged by eugenicists would have very little impact on the human species as such. In the United States, thirty states had passed sterilization laws by 1940, but the frequency of sterilization did not begin to approach the frequency needed to modify the human species.

From the start, politics seemed to play a larger role in social Darwinism and eugenics than did science. Many of the champions of these programs meant well. They genuinely feared that the human species was degenerating. Inferior people were outbreeding their betters. They believed that the short-term unhappiness caused by such palliative measures as sterilization would be outweighed by contributions to the general good. Unfortunately, the short-term unhappiness was sure and swift, whereas the general good that was to follow from these actions would occur, if at all, only in the distant future. Among the numerous crimes against humanity perpetuated by the Nazis in World War II was the sterilization and even murder of what they took to be "deficient" human beings. Because eugenics became associated in the minds of many with these Nazi atrocities, it gradually waned in popularity. Eugenics laws remained on the books but were applied less and less frequently.

In the 1970s, scientists once again turned to investigating the influence of biological factors on the human species under the banner of sociobiology. Once again, their opponents accused them of reading their own racist, sexist, and genetic determinist views into nature. As the controversy over sociobiology settled down, evolutionary psychologists took over the project. This time, they would really apply genuine scientific theories to human beings. Looking back at the social Darwinists and eugenicists, their science seems incredibly weak and the influence of their social biases more than a little obvious. Whether a genuine science of the influence of biological evolution on the human species is finally developing, only time will tell.

[*See also* Adaptation; Eugenics; Naturalistic Fallacy; Social Evolution.]

BIBLIOGRAPHY

Adams, M., ed. *The Wellborn Science: Eugenics in Germany, France, Brazil, and Russia.* Oxford, 1990. A summary of the different forms that eugenics took in four different countries.

Bannister, R. *Social Darwinism: Science and Myth in Anglo-American Social Thought.* Philadelphia, 1979. Argues that evolutionary theory can be and was used to justify liberal institutions as well as those that are more conservative.

Greene, J. C. *Science, Ideology, and World View: Essays in the History of Evolutionary Ideas.* Berkeley, 1981. Evaluates the relation between evolutionary theory and society.

Kevles, D. J. *In the Name of Eugenics: Genetics and the Uses of Human Heredity.* New York, 1985. A comprehensive review of the interplay between science and society with respect to eugenics.

Moore, J. "Deconstructing Darwinism: The Politics of Evolution in the 1860s." *Journal of the History of Biology* 24 (1991): 353–408. Traces social Darwinism in Darwin's own day.

Sahlins, M. D. *The Use and Abuse of Biology: An Anthropological Critique of Sociobiology.* Ann Arbor, 1976. One of the first critics of sociobiology to view it as latter-day social Darwinism.

— David L. Hull

SOCIAL EVOLUTION

Animal societies range from the loosely structured schools of fish and flocks of birds to the tightly organized colonies of naked mole rats, honeybees, and termites. The interactions among the members of these societies run the gamut from the minimal (simply following or standing near another individual) to the extraordinary (as when hundreds of weaver ants, both adults and larvae, work together to construct an elegant leaf nest).

The attempt to understand the evolution of societies and social behavior in all their diversity has intrigued evolutionary biologists in part because of curiosity about the origins and adaptive value of our own highly social nature. [*See* Human Sociobiology and Behavior *article on* Human Sociobiology.] But the fundamental reason for the exploration of social evolution stems from the recognition that living and interacting with others carries disadvantages as well as advantages for individuals. Members of a school of fish or herd of wildebeest pay a price for assembling in groups, such as the risk of acquiring a communicable disease or of having to compete with so many others for the food available to them. The social foundation for schooling or herding can only spread through a species if the costs of aggregation are outweighed by the benefits social individuals gain, such as improved defense against predators.

The Genetic Costs of Social Living. Every aspect of social life has costs of some sort and therefore provides an evolutionary puzzle worthy of solution. The paramount challenge in this regard comes from the extreme altruism or self-sacrificing behavior shown by some members of some animal groups. A familiar, if painful, example comes from being stung by a honeybee, an unpleasant event for the person being stung but lethal for the bee. The insect's barbed stinger catches in the individual's skin and as the honeybee tries to pull free, it eviscerates itself. This suicidal act leaves the bee's poison gland attached to the stinger, with the result that the individual being stung receives a full dose of toxin, the better to deter it from harming the colony to which the deceased honeybee belonged.

The bee that made that ultimate sacrifice spent its entire adult life sacrificing in less dramatic ways for the welfare of the hive's queen and its colonymates, by helping to rear the brood, guard the hive entrance, and collect pollen and nectar to make honey to feed the bees that remain at home. Yet the worker bee, as is true for the worker caste in many other social insects, is effectively sterile, severely handicapped reproductively by her greatly reduced ovaries. The worker bee's behavior raises the following question: how can the hereditary attributes of sterile, self-sacrificing individuals be retained within a species, given that their carriers fail to reproduce and so cannot pass on the genetic basis of their capacity for helpful behavior to future generations? If the social members of any species engage in acts that reduce their chances of having offspring and passing on their genes, the frequency of these genes should tend to decline over successive generations. If the pattern persists over many generations, behaviors that reproductively handicap individuals will be eliminated and replaced by hereditarily different characteristics that enable individuals to achieve higher fitness or genetic success (as measured by the number of copies of their distinctive genes they contribute to the next generation).

Table 1. The Reproductive Consequences for Individuals Engaging in Different Kinds of Helpful and Unhelpful Social Interactions

Social interaction	*Effect on the actor*	*Effect on the recipient*
Cooperation or mutualism	+	+
Reciprocity or reciprocal altruism	+ (delayed)	+
Parental care	+	+
Altruism	−	+
Selfish behavior*	+	−
Spite	−	−

*It is possible for some kinds of exploitative behaviors to be superficially helpful but to benefit only the "helper."

The logic of this argument can be applied to the analysis of all forms of social interactions (Table 1). Consider spiteful actions in which an individual harms another but at a cost to its own reproductive chances. One possible example is the lethal assault by some adult gulls on the offspring of their neighbors, an action that harms their neighbors' fitness. But the assailants' reproductive success may also be lowered to some extent because the killers must expend valuable time and energy in battering their victims to death. Moreover, potentially everyone else other than the victim benefits from the spite, but without paying the costs. Therefore, if the murderous gulls really do lose fitness from their socially spiteful tendencies, the behavior would seem unlikely to persist for long. Indeed, unmistakable cases of spite are exceedingly rare throughout the animal kingdom, just as we would expect from an analysis of the genetic consequences of spite.

Group Benefit Selection. Until fairly recently, self-sacrificing behaviors of the sort exhibited by honeybees failed to elicit much puzzlement. Instead, many biologists felt that self-sacrifice by some members of a species was good for the preservation of that species. This view received its formal development in 1962 in V. C. Wynne-Edwards' *Animal Dispersion in Relation to Social Behaviour.* Wynne-Edwards attempted to explain almost every aspect of animal social behavior as an adaptation designed to increase the survival chances of the species, no matter what its effect on the genetic success of individuals. Thus, for example, the fact that red grouse unable to acquire a territory do not attempt to breed was said to be a population-regulating device that helped ensure that grouse did not become too numerous, which could destroy the essential resources needed for that species' survival.

The theory of evolution presented by Wynne-Edwards has been called group-benefit selection. The essence of the theory is that unless species (or groups within species) can prevent overpopulation, they will go extinct. Because many species (and groups) have persisted to the present, they must, according to the group-benefit theory, possess the means to avoid excessive, species-threatening reproduction.

In response to Wynne-Edwards and his group-benefit theory, the British ornithologist David Lack and the American biologist George C. Williams wrote rebuttals. They pointed out that Darwinian natural selection occurs when individuals, not groups, differ hereditarily in ways that affect their reproductive success. If some individuals give up reproduction so that others can breed, thereby helping their species persist, the effect of natural selection would be the eventual elimination of those individuals and their reproductively restrained genes, in favor of individuals that did not show such restraint.

Lack and Williams convinced most biologists that Darwinian selection at the level of the individual for traits that promoted success at gene propagation was likely to be far more powerful than selection at the level of the group for self-sacrificing social behavior. Thereafter, most evolutionary biologists developed and tested explanations for social behaviors in terms of their potential contribution to the genetic success of the social individual. For example, the nonterritorial grouse may refrain from breeding not to keep their population at an optimal level for group benefit but because individuals that tried to breed without a suitable territory would usually fail and would harm their chances for future reproduction. Instead, nonterritorial, nonbreeding birds may be able to conserve their energy and survive until the next breeding season, when they may be able to secure a territory and reproduce successfully.

Helpful Social Interactions and Their Genetic Consequences. The same individual-selection approach has been applied to helpful social behaviors (Table 1), which can be categorized in terms of the genetic consequences for interacting individuals. Consider the possibilities when two lionesses join forces to capture an antelope. If both females benefit from their social alliance by improving the odds of making a kill and equally sharing a large meal, then their actions are considered a case of cooperation, or mutualism. Cases in which both parties immediately gain extra calories or some other benefit that translates into enhanced survival and reproductive success (i.e., greater genetic success) are readily explicable in Darwinian terms. [*See* Mutualism.]

What if, however, one lioness has recently eaten and is in no great need of a meal, but joins her hungry companion anyway and, in so doing, provides critical assistance that results in a large and valuable caloric gain for the lioness she helps? If, later, when the helpful lioness is herself in great need of antelope steaks and her companion returns the favor, then we have an example of reciprocity, or reciprocal altruism. In this instance, the lioness that participated in the first hunt did so at some personal cost given the time, energy, and risks involved in helping her companion to a meal. But this modest cost was then more than repaid when her companion assisted the previously helpful lioness when she was in need of food. [*See* Reciprocal Altruism.]

Reciprocal altruism is really a special case of a mutualism in which both cooperators gain benefits that translate on average into more genes transmitted to the next generation by both individuals. As such, these actions pose no special puzzle for evolutionary theorists, although biologists have noted that reciprocity requires individuals to repay at a later time after having received help from a partner. In such a system, partners that failed to return the favor to their helper might often come out ahead, having secured benefits from their initial interaction without then incurring the small costs associated with the payback phase. It is now agreed that systems of reciprocal altruism are more likely to persist if the participants remain in association and are able to enlist each other's aid in a long series of back-and-forth exchanges rather than a single round of reciprocity.

Parent–Offspring Interactions. Consider another kind of social interaction among lionesses, this one between a mother and her relatively inexperienced daughter, with the mother doing the hard work of bringing down an antelope in order to give her offspring more experience with hunting as well as the bulk of the meal they secure for themselves. This parental interaction carries reproductive costs for the mother because she is expending time and energy that could be spent in other ways that might improve her chances of reproducing in the future. Moreover, every hunt carries with it the risk of life-shortening injury. These costs, however, may be more than matched by an increase in the likelihood that the currently existing daughter may reach the age of reproduction and so pass on some of the genes received from her helpful mother.

Note that, once again, an evolutionary understanding of a social interaction requires genetic calculations. If a parental action saves an existing offspring that otherwise would have been certain to die, the parent has in effect saved a carrier of half its genome. (Children normally receive half of their genes from their mother and half from their father.) If the action also had a 20 percent chance of causing the parent to die sooner than it would otherwise and so fail to produce one surviving offspring at a future date, then the genetic cost of the act can be set at 20 percent of one-half of a genome or one-tenth of a genome. Given these genetic consequences, adult females that behaved in this fashion would increase the frequency of the genes underlying the ability to behave parentally in a certain way. Of course, other kinds of parental solicitude toward offspring could lead to reduced genetic success, in which case these other kinds of parental behaviors should become less and less frequent over time, gradually shaping the limits of parental helpfulness.

In fact, parent–offspring interactions are not always cooperative and harmonious. Members of a family share only some of their genes (except for identical twins or clonally produced offspring, such as the progeny of an asexual aphid mother). Because of their genetic differences, a parent and a child may have different adaptive strategies that bring them into conflict in those situations when a parental action would benefit the child but produce a net loss of lifetime reproductive success for a parent. Some persons have interpreted the weaning tantrums that occur in various mammals as the kind of parent–offspring conflict expected when a youngster stands to gain by monopolizing its parent's largesse but when the parent's fitness would be advanced by conserving energy for the next round of reproduction. Under these circumstances, we expect the youngster to demand and the parent to refuse. [*See* Parent–Offspring Conflict.]

The Evolution of Altruism. Let us return to the example of hunting lionesses. Suppose that the one lioness does the bulk of the work in capturing an antelope, and the other feeds on the kill even though she will not repay her helper in the future and is not an offspring. Here we have an example not of mutualism, reciprocity, or parental care, but of altruism; the altruist sacrifices personal reproductive success (to the extent that the work done to help make a kill reduces the helpful female's chances of having more surviving offspring) while raising the reproductive chances of a companion who is neither a son nor a daughter. [*See* Altruism.]

Various kinds of altruism do occur in lion social groups (called prides). In addition to the possibility that some females help other nonfamily members to meals at some cost to themselves, observers have documented that females will guard and nurse cubs of other females. Moreover, male lions also form small groups whose members guard prides even though some males get to mate relatively rarely compared to other males in their band. In these cases, the males low on the mating totem pole provide costly assistance to those who are reproducing at a much higher rate. Altruism of this sort is not limited to lions but also occurs in animal species ranging from social insects, for example, honeybees, to the bird species known for their helpers-at-the-nest, in which nonbreeding individuals help to defend a nesting territory and rear the young produced by a breeding pair.

Charles Darwin offered an explanation for altruism, noting that if the altruists helped family members, they were in effect perpetuating the family lineage and its heritable attributes, which could include the willingness of some to help other family members. However, Darwin did not have access to modern genetic information, which was needed for the next advance in understanding the evolution of altruism. This advance came from William D. Hamilton in 1964, when he proposed that individuals could sometimes increase the frequency of their genes if they directed their altruism toward relatives, which share some genes in common as a result of having the same recent ancestors. [*See* Hamilton, William D.] Siblings, for example, have 50 percent of the same genes on average as a result of having the same parents; nephews and nieces share 25 percent of their genes with their uncles and aunts because their family lines trace back to the same grandparents.

An individual that saves the life of an adult brother or sister in effect gains the genetic equivalent of having a surviving offspring of its own. A lioness whose altruistic actions keep three nieces alive has made a genetic contribution ($3 \times 0.25 = 0.75$ gene sets) to the next generation, which is equivalent to having 1.5 surviving cubs of her own ($1.5 \times 0.5 = 0.75$ gene sets).

It was Hamilton's genius to realize that there are two ways to increase the frequency of one's genes in the next generation: the direct method of personal reproduction and the less obvious, indirect method of helping extra relatives to survive. By thinking about an animal's contributions to the next generation in the same genetic currency, he provided a way to evaluate whether altruism would spread or disappear over evolutionary time. Hamilton pointed out that if an individual's altruism costs less genetically (reduction in personal reproduction) than the indirect genetic benefits that result from helping relatives to reproduce that would otherwise have perished or not reproduced, then the altruism is adaptive in the sense of raising the helpful individual's lifetime genetic success, or inclusive fitness. [*See* Inclusive Fitness.] The term *inclusive fitness* acknowledges that one's total genetic contribution to the next generation is the cumulative total derived from the individual's genes that are passed on to surviving offspring plus copies of that individual's genes present in relatives. One can calculate inclusive fitness by adding up the animal's direct fitness (number of surviving offspring multiplied by the proportion of genes shared by parent and offspring, which is usually 0.5) and the animal's indirect fitness (the number of extra relatives that lived to reproduce thanks to the help they received from the helper multiplied by the proportion of genes shared by the helper and those that received his assistance). So, for example, a lioness who had two surviving offspring of her own and who also helped keep one niece alive would have an inclusive fitness of $(2 \times 0.5) + (1 \times 0.25) = 1.25$.

Predictions Derived from Inclusive Fitness Theory. As noted above, the concept of inclusive fitness can help to predict the circumstances under which altruism can spread through a population. First, altruists should tend to direct their assistance primarily to close relatives, such as parents, siblings, nephews, and nieces, and not to distant relatives or nonrelatives. Only when the self-sacrificing individuals have a substantial likelihood of sharing genes in common can the indirect genetic benefit gained by keeping relatives alive outweigh the direct genetic cost of reduced personal reproduction by the altruist.

In fact, studies of altruism routinely reveal that the behavior is not directed indiscriminately to other unrelated individuals. As Darwin realized, worker ants and termites act on behalf of their families, assisting their mother queen in the production of a new brood of potentially reproducing brothers and sisters. Likewise, lionesses live with and help their mothers, aunts, nieces, and nephews, thereby indirectly propagating their genes. Avian helpers-at-the-nest also sacrifice as they help to rear additional brothers and sisters, who may eventually breed successfully and so spread the genes of the helpers.

A second expectation derived from the inclusive fitness approach to social behavior is that altruism is more likely to evolve when the genetic costs are low. If an individual has little chance of reproducing successfully on its own, which means that the opportunity to gain direct fitness is slight, then altruism is more likely to generate a net gain in inclusive fitness for a potential altruist. Once again, reality matches expectation. The worker bee or ant with inactive ovaries has little prospect of passing on her genes directly, so she gives up little when she forgoes reproduction to help her mother make future queens (her sisters) and brothers that may reproduce. Similarly, a subordinate male in a coalition of lions that helps his more powerful siblings defend a pride against rival coalitions but has to stand by while his brothers do most of the mating is not really paying a large price by letting the others reproduce. For one thing, if he were to try to displace his brothers from the females in the pride, he might well be injured by his more dominant siblings (or he might harm them and so reduce his indirect fitness); if he were to strike out on his own, the male would have almost no chance of ever taking a pride for himself from coalitions of unrelated rivals.

Along these lines, it is relevant that helpers-at-the-nest in white-fronted bee-eaters or Seychelles warblers are typically young males, the sons of the breeding pair they assist. If such a male is to breed successfully on his own, he must acquire a territory but may fail if the area is saturated with experienced territory holders. Under these circumstances, sons may help their parents for a year or two, boosting the output of siblings, some of which may go on to breed, thereby adding to the male's indirect fitness (and thus his inclusive fitness). Revealingly, young birds are less likely to become avian helpers at the nest if one of their genetic parents has been replaced by a stepparent, a circumstance that results in lower genetic relatedness between the helper and his parents' offspring, which become his half siblings with half the number of shared genes compared to full siblings. The unconscious goal of a potential helper is the maximization of its inclusive fitness; helping half siblings is a less effective means to this end than helping full siblings.

Because of Hamilton's intense interest in the genetic consequences of social life, he realized the possible significance of the correlation between the extreme altruism exhibited by certain social bees, ants, and wasps and the method of sex determination of these insects. Bees, ants, and wasps belong to the same order of insects, the Hymenoptera. In this group, the sex of an individual is usually a function of whether the insect has developed from an unfertilized egg or from an egg that has been fertilized by a sperm. The unfertilized egg with just the maternal set of chromosomes goes on to become a haploid male, which as an adult will produce genetically identical haploid sperm. Fertilized eggs go on to become diploid females whose cells contain two copies of each and every chromosome, one maternal set and one paternal set.

When a female mates with only one male, the sperm she receives and stores for later use in fertilizing her eggs are genetically the same. This means that when she produces her daughters, they will have the same set of paternal chromosomes. The other set of chromosomes that make up the daughters' genomes is drawn from the mother's two-copy set. On average, daughters share half of the maternal chromosomes and genes, so that overall they have a genetic relatedness of 0.75 (an identical half-genome derived from their shared father and half of the other half-genome derived from their mother). All of this means that sister hymenopterans are more closely related than is customary in sexually reproducing species with diploid males as well as diploid females. This fact, in turn, means that a worker bee, ant, or wasp that directs her energies toward making more reproducing sisters gains 0.75 genetic unit for each extra surviving sister compared to the 0.5 genetic unit that an altruistic avian helper at the nest can secure for an additional sibling. In other words, all things being equal, extreme altruism should be more likely to evolve in the Hymenoptera than in other groups.

It is true that species with sterile helpers are best represented among bees, ants, and wasps. However, as Hamilton himself noted, a high degree of genetic relatedness among interactors is not essential for the evolution of insect altruism, as demonstrated most obviously by termites. Males of these insects are diploid, and sisters are no more closely related than are sisters of most other sexually reproducing species. Yet termites are intensely social, some with colonies of millions of individuals organized into several castes, each with its own sacrificial role to play in the colony's economy, all working on behalf of their queen mother and king father to produce a new generation of potential colony founders (and gene propagators). The evolutionary basis for extreme altruism in termites is still not fully understood, and questions remain about what is responsible for the great diversity of social systems within the Hymenoptera. In particular, why is it that one or a few queens monopolize reproduction within some colonies of social bees, ants, and wasps, whereas, in others, reproduction is more or less evenly shared among a group of queens? Although not every issue in social evolution has been resolved, the cost-benefit approach used by gene-counting biologists almost certainly holds the key to advances in our understanding of social behavior.

[*See also* Alarm Calls; Eusociality, *article on* Eusociality in Mammals; Group Living; Kin Recognition; Kin Selection.]

BIBLIOGRAPHY

Alcock, J. *Animal Behavior: An Evolutionary Approach.* Sunderland, Mass., 2001. A general textbook treatment of social behavior and its evolution.

Brown, J. L. *Helping and Communal Breeding in Birds: Ecology and Evolution.* Princeton, 1987. A more advanced review of helpers-at-the-nest in birds in relation to the inclusive fitness theory.

Dawkins, R. *The Selfish Gene.* New York, 1977. A brilliant popularized account of evolutionary theory, especially as applied to social behavior.

Hamilton, W. D. "The Genetical Evolution of Social Behaviour." *Journal of Theoretical Biology* 7 (1964): 1–52. The highly technical, but readable, article that started the modern revolution in social behavior analysis by presenting the inclusive fitness theory.

Wilson, E. O. *Sociobiology: The New Synthesis.* Cambridge, Mass., 1975. The classic and sweeping summary of our understanding of social evolution as of 1975. Still useful as an account of the natural history of social behavior across the animal kingdom as well as a summary of the theoretical foundation of evolutionary analyses of sociality.

Wynne-Edwards, V. C. *Animal Dispersion in Relation to Social Behaviour.* Edinburgh, 1962. The book on group-benefit selection that in some sense started the modern analysis of social behavior by stimulating George C. Williams to write *Adaptation and Natural Selection* (Princeton, 1966).

— JOHN ALCOCK

SOCIAL INSECTS. *See* Eusociality, *article on* Eusocial Insects.

SPECIALIZATION. *See* Multicellularity and Specialization.

SPECIATION

Speciation is the process whereby two or more descendant species evolve from an ancestral species. Most evolutionary biologists use "speciation" to mean the evolution of genetically determined barriers to gene exchange between populations. That is, they employ the *biological species concept*, in which a species is considered a population, or group of populations, of actually or potentially interbreeding individuals who are reproductively isolated by biological differences from other such groups. The concept of speciation as the evolution of reproductive isolation (RI), which is adopted in this article, applies only to outbreeding, sexually reproducing organisms (Futuyma, 1998).

Speciation is responsible for every branch in the phylogeny of sexual organisms, and thus is the source of much biological diversity. Without RI, interbreeding among populations would eventually diminish or erase their adaptive differences. Thus, the long-term persistence of adaptive differences and the possibility of cumulative long-term divergence depend on speciation. It is possible, therefore, that morphological changes in fossil lineages could be associated with speciation, as has been posited by proponents of the controversial hypothesis of punctuated equilibria.

Species. Species are reproductively isolated by genetically based differences in characters called *reproductive isolating barriers* (RIBs) or *isolating mechanisms*. Because RI may be incomplete, not all populations can be definitively classified as members of the same versus different species. Two such "semispecies" sometimes have adjacent distributions and form a *hybrid zone* where their ranges meet.

Differences between species. RIBs are usually classified as prezygotic and postzygotic. *Prezygotic barriers* prevent formation of hybrid zygotes. These barriers include *temporal isolation* (e.g., breeding at different times of year), *ecological isolation* (mating in different habitats), and especially *sexual isolation* (failure to mate despite opportunity) and *gametic isolation* (failure of gametes or their nuclei to unite). *Postzygotic isolation* consists of reduced survival or fertility of F_1 or backcross hybrids and thus reduces gene exchange between populations even if hybrid zygotes are formed. The cause of postzygotic isolation may be extrinsic, as when hybrids survive poorly in the environments to which parent types are well adapted, or intrinsic, often because of interactions between incompatible genes. Hybrid sterility is *not* the defining property of species; many closely related species can form fertile hybrids when crossed in the laboratory but are prezygotically isolated and form distinct gene pools in nature.

Sympatric populations are in the same locality and can encounter each other; *parapatric* populations have abutting geographic distributions; *allopatric* populations have nonoverlapping distributions. The existence of two or more sympatric, closely related species is usually detected by indirect evidence that individuals form distinct groups, based on either phenotypic or molecular characters. For example, sympatric downy and hairy woodpeckers (*Picoides pubescens* and *P. villosus*) in eastern North America do not overlap in size or in several features of plumage pattern. In contrast, eastern ("yellow-shafted") and western ("red-shafted") populations of another North American woodpecker, the northern flicker (*Colaptes auratus*), differ in several plumage features; but all possible combinations of these features, and intermediate states of some of them (such as salmon-colored shafts of the wing feathers), characterize intervening populations. These patterns indicate that the flicker populations hybridize freely, acting as a single species, whereas the coherent combination of differences between downy and hairy woodpeckers implies that they do not interbreed. The phenotypic differences between these forms do not *define* species, but they provide provisional *evidence* of RI, since recombination in a single random-mating population generally would break down associations between different characters and would give rise to a range of intermediate phenotypes. Molecular characters often provide similar evidence. Allopatric populations may be suspected of having acquired RI if they are as different as related sympatric species are, but only experimental crosses can reveal whether reproductive barriers have evolved.

Most species differ in ecological properties (ecological niche), and they usually differ accordingly in physiological or morphological adaptations. Populations may evolve RIBs without evolving ecological differences, in which case they may retain parapatric or allopatric distributions and be unable to coexist in sympatry.

Genetics of reproductive isolation. Sister species (those derived from an exclusive common ancestor) are often reproductively isolated by several RIBs, such as sexual and postzygotic isolation. Likewise, gene substitutions capable of producing RI continue to accrue after speciation has occurred. Because we cannot determine which genetic differences caused speciation and which followed it, evidence on the genetic cause of speciation is best sought by crossing populations that are undergoing or have recently completed the speciation process (Coyne and Orr, 1998).

Hybrid sterility may vary in degree and may be caused by differences between the parent populations either in genes or in karyotype (chromosome number and structure). All other RIBs are usually attributable to gene differences. These can sometimes be characterized by crossing closely related species. For example, differences in the form and color of the flowers of two species of monkeyflowers (*Mimulus*) result in one's being pollinated chiefly by insects and the other by hummingbirds, and so contribute to prezygotic isolation. These species were crossed, and the highly variable F_2 hybrids were characterized for flower features and for a large number of mapped molecular markers that distinguish the species (Bradshaw et al., 1998). If hybrids that differ for any one such marker also differ in, for example, flower color, there must be at least one closely linked gene that contributes to the species' difference in this feature. The number of such markers that are associated with variation in the feature then provides a minimal estimate of the number of gene differences that contribute to the species' difference in that trait. Each of the several flower features that distinguish the monkeyflowers turned out to be based on at least two or three loci.

Similar studies of closely related species of *Drosophila* and other organisms have shown that, although characters conferring sexual isolation may differ by only a few genes, hybrid sterility is typically polygenic (due to many gene differences), with genes from the two parent species interacting unfavorably when combined in hybrids.

The often polygenic basis of RI is consistent with other evidence that speciation is usually a gradual process—in other words, that RI increases incrementally over the course of time. For example, partial sexual isolation exists among geographic populations of the dusky salamander, *Desmognathus ochrophaeus*, with more distant populations displaying greater reluctance to mate with each other. Among populations and species of *Drosophila* and of toads (*Bufo*), the level of both sexual and postzygotic isolation is correlated with the degree of molecular difference, which is correlated with the time since they diverged.

Modes of Speciation. Speciation can be classified by (1) the geographic setting for divergence, (2) the process by which RI evolves, and (3) the kind of variation that confers RI. Classified by geographic pattern, speciation might be *allopatric* (divergence in geographically separated populations), *parapatric* (divergence of spatially adjacent populations despite initial gene flow between them), or *sympatric* (divergence of a single random-mating population into two reproductively isolated entities). Allopatric speciation may occur by *vicariance* (divergence of large populations) or *peripatric* speciation (genetic divergence between a population founded by few individuals and a large, more slowly changing "parent" population). The population-genetic processes that might cause speciation are genetic drift, natural selection, or a combination of the two. RI may be based on polyploidy, structural chromosome rearrangements, genes, or cytoplasmic incompatibility.

Models of Speciation. Speciation is unlikely to be caused by a single mutation, because such a mutation is unlikely to increase in frequency: it causes failure to mate (if it causes prezygotic isolation), or it reduces the heterozygote's fertility or survival (if it causes postzygotic isolation). If, on the other hand, a polygenic feature confers RI (e.g., if genotypes $A_1A_1B_1B_1C_1C_1$ . . . and $A_2A_2B_2B_2C_2C_2$ . . . do not exchange genes), the many intermediate genotypes (e.g., $A_1A_2B_1B_1C_2C_2$) may be able to interbreed with both extreme genotypes and form a conduit for gene flow between them. How, then, can two distinct populations, without intermediate genotypes, be formed?

Vicariant allopatric speciation. The most widely accepted model of speciation is allopatric speciation, in which two populations come to differ by two or more allele substitutions. For instance, an ancestral species with genotype $A_1A_1B_1B_1$ becomes divided into two populations, designated one and two, by a topographic or habitat barrier. In population one, allele A_2 replaces A_1 by either selection or genetic drift, and in population two, B_1 is likewise replaced by B_2. A_2 and B_2 interact epistatically, so that genotypes with both these alleles (e.g., the hybrid $A_1A_2B_1B_2$) have low fitness. If the populations expand their range and meet, they will be postzygotically isolated to some extent. The critical feature of this model is that the alleles that in combination confer RI never encounter each other and so do not reduce fitness during the course of substitution.

Incompatibility is likely to evolve at an accelerating rate in the allopatric populations, as further allele substitutions occur. Moreover, ecological or sexual selection could gradually cause divergence between the populations in a character that influences prezygotic isolation. For example, one population of an ancestrally bee-pollinated plant might gradually become attractive to hummingbirds exclusively, as has occurred in several groups of plants.

The physical isolation between the populations in the allopatric model prevents gene flow from eroding or preventing the early stages of genetic divergence produced by selection or genetic drift. When enough genetic divergence has accrued to produce substantial reproductive isolation, the populations can become sympatric with little or no gene exchange. Divergence between parapatric populations can occur, resulting in parapatric speciation, but only if selection is stronger than gene flow. For example, genotypes with different female preference and male display characters could spread by sexual selection from different sites of origin in a broadly distributed species and be sexually isolated where they meet.

The role of selection. Allopatric genetic divergence resulting in RI is virtually inevitable, given enough time, because it can occur by either genetic drift or natural selection. Ecological selection could cause reproductive isolation directly, by acting on an isolating character such as breeding season, or indirectly, if RI is a pleiotropic byproduct of genetic divergence. Pleiotropy is the most likely origin of postzygotic isolation, since natural selection cannot favor reduced survival or fertility.

Prezygotic isolation could also evolve by sexual selection. For example, a male display trait and female preference for such a trait may evolve together, for several reasons. Females may be selected to choose males with highly developed traits that indicate high fitness; the trait and the preference may reinforce each other's evolution in a "runaway" process; or males may evolve to overcome ever-increasing levels of female resistance to their stimuli ("chase-away" selection; see Holland and Rice, 1998). When the course of such evolution differs between populations, different mate preferences, and thus sexual isolation, may evolve (Pomiankowski and Iwasa, 1998).

Can there be selection for reproductive isolation in itself? The evolution of RI in allopatric populations is not affected by the existence of another population, so in this instance there is no selection for RI per se. However, if two divergent populations hybridize and the hybrids have low fitness, individuals that do not hybridize have fitter offspring than individuals that mate indiscriminately. Thus, genes conferring mating discrimination may be favored, and selection will strengthen, or reinforce, incipient *prezygotic* isolation. Recent models have shown that such reinforcement is likely to occur, especially if the populations have already become partly sexually isolated by the time they encounter each other (Kirkpatrick and Servedio, 1999).

Peripatric speciation. Mayr (1954) observed that isolated, localized populations of birds and other animals, such as those on islands, often differ greatly in phenotype, even though their environments do not obviously differ. These observations led him to postulate *peripatric speciation*, or *founder-effect speciation*, in which a *bottleneck* (reduction) in population size initiates genetic change. Mayr proposed that if a local population were founded by few individuals from a "parent" population, it would have reduced genetic diversity and substantially altered allele frequencies at some loci as a result of initial sampling error and subsequent small population size (an instance of genetic drift that he called the "founder effect"). If, as Mayr thought, fitness is affected by strong epistatic interactions among loci, these initial changes would create selection for allele frequency changes at interacting loci. Such compensatory changes might "snowball" and produce a new species. Thus, genetic drift first increases the frequency of certain alleles that would not have increased in response to selection alone, perhaps because they are deleterious in heterozygous condition. Natural selection, stemming from interactions between these alleles and other loci, then brings about further genetic change. Mayr's hypothesis has attracted much attention, partly because it was adopted by Eldredge and Gould in 1972 as the theoretical foundation of their hypothesis of punctuated equilibria. (Indeed, Mayr himself proposed a similar extension to the fossil record in 1954.)

Many population geneticists believe this genetic scenario is unlikely (Turelli et al., 2001). Suppose that the fitness of both homozygotes at a locus (A_1A_1 and A_2A_2) exceeds that of A_1A_2, that the source population is nearly fixed for A_1, and that if the newly founded population became fixed for A_2, selection would favor changes at other, epistatically interacting loci. When A_2 is rare, natural selection cannot increase its frequency, because most A_2 alleles are carried by inferior heterozygotes. Fixation of A_2 would require that its frequency rise to more than about 0.5 by genetic drift—which is exceedingly unlikely if A_2 were initially rare and if selection against A_1A_2 were strong. If selection against A_1A_2 is very weak, the chance that A_2 increases by genetic drift is greater. However, RI would then depend on changes in allele frequencies at other, interacting loci, collectively giving rise to a reproductively isolated genotype. Population geneticists have argued that such highly interactive alleles would seldom be present in the parent population and would not persist through a population bottleneck.

Sympatric speciation. Sympatric speciation has been controversial. For RI to evolve, selection must overcome gene flow, which is maximal within a random-mating population. Most models of sympatric speciation propose disruptive selection on an ecologically important character, such that two distinct phenotypes are adapted to different resources and have higher fitness than intermediates. Two principal bases for reproductive isolation have been proposed. One possibility is that using different resources automatically creates assortative mating. For instance, many host-specific parasites and herbivorous insects mate only while on the host, so genotypes that differ strongly in host preference may form distinct mating pools. Polymorphism for genotypes (e.g., A_1A_1, A_2A_2) that exclusively prefer different hosts will evolve if there is strong disruptive selection at another locus that affects phenology or physiology, such that B_1B_1 is adapted to one host and B_2B_2 to the other. Two subpopulations may evolve, one consisting of genotype $A_1A_1B_1B_1$, which prefers and survives well on host one, and the other of genetype $A_2A_2B_2B_2$, which prefers and survives well on host two. An association ("linkage disequilibrium") between alleles A_1 and B_1 (and between A_2 and B_2) develops because other combinations are less fit (e.g., animals carrying the A_1B_2 combination would tend to choose host one but be better adapted to host two). This association can persist because the preference of genotypes A_1A_1 and A_2A_2 for different host plants results in assortative mating, and hence in few heterozygous offspring ($A_1A_2B_1B_2$) in which recombination might occur.

Alternatively, we might assume that the ecological character is genetically independent of mate preference, and that selection favors assortative mating on the basis of a "marker" such as coloration. For example, suppose a polygenic group of A loci affect a feature such as gape (mouth size), and that an individual's gape determines the size of prey it can best handle. Another set of loci, B, determines coloration (varying, e.g., from blue through violet to red). If only large and small prey, but not intermediate prey, are available, there is disruptive selection on gape size. It would be advantageous for both large-gaped and small-gaped individuals to mate with like individuals, because those who mate at random may have intermediate, poorly adapted offspring. Suppose individuals choose mates that are similar to themselves in color (since there is no reason to think that mate choice will necessarily be based on mouth size). Then as long as these colors are correlated with gape size, individuals with intermediate color will have unfit offspring with intermediate gape, whereas extremely red or extremely blue individuals will have fit offspring with either large or small gapes. Thus, we might expect the population should split into two assortatively mating groups that differ in gape and color.

The problem with this scenario of sympatric speciation is that recombination breaks down the association between color and the ecologically adaptive character (gape size). Lacking such association, individuals that differ in gape may have the same color and so may produce intermediate offspring, so there is no advantage in mating assortatively by color. Only recently have computer models shown that a population may indeed split sympatrically into two ecologically divergent mating groups (Dieckmann and Doebeli, 1999). However, this is likely to occur only if individuals are highly specialized for different resources, for instance based on their gape size. Moreover, sympatric divergence is less likely if assortative mating is based on a "marker" character, such as color, than if it is based on the adaptive ecological character itself.

Evidence on Modes of Speciation. Although evolution in laboratory populations has provided glimpses into processes of speciation, speciation in natural populations generally takes too long to observe directly. Most evidence on the geographical setting and the causes of speciation is therefore inferential. Because it can be difficult to distinguish among alternative hypotheses of past events, many aspects of speciation are still controversial.

Allopatric speciation. It has long been apparent that speciation often occurs allopatrically (Mayr, 1942). Closely related species are often geographically segregated—for example, pairs of marine species separated by the Isthmus of Panama, which formed in the late Pliocene. Sister species that occupy the same area, such as pairs of stickleback (*Gasterosteus*) species in certain northern lakes, have evidently been formed by "double invasion"; different degrees of DNA sequence divergence from their marine "parent" suggest that one invaded the lake earlier, became phenotypically and reproductively distinct, and was then followed by a second invasion from the marine population (Taylor and McPhail, 2000). As noted earlier, geographic populations of widespread species sometimes display varying degrees of reproductive isolation when experimentally tested. Parapatric populations of a species often differ genetically not only in phenotypic characters, but also in their monophyletic "gene trees," interpreted to mean that they diverged allopatrically over a considerable span of time, only recently expanding and making "secondary contact" (Avise, 2000). Finally, Barraclough and Vogler (2000) reasoned that the amount of range overlap should increase over time if sister species originated allopatrically, but should remain steady or decrease if they originated sympatrically. Using a molecular clock to estimate time since divergence, they found patterns conforming to allopatric speciation in five or six of eight clades of birds, fishes, and insects.

The causes of allopatric speciation have been rather neglected, but there is some evidence that selection has played a role. In sticklebacks (*Gasterosteus*), morphologically different open-water and bottom-feeding species have evolved in parallel in several lakes. Females of each species mate preferentially with males of the same ecological type, regardless of lake of origin, indicating that reproductive isolation is based on characters (perhaps body size) that diverged in response to ecological selection (Rundle et al., 2000). Geographic populations of a leaf beetle (*Neochlamisus bebbianae*) are more strongly sexually isolated if they are adapted to different host plants than if they use the same host, as predicted if divergent ecological adaptation pleiotropically produced reproductive isolation (Funk, 1998). That sexual selection contributes to speciation is suggested by the observation that females may mate only in response to species-specific male signals that are known to be sexually selected. For instance, female frogs of the genus *Physalaemus* not only choose males of the same species on the basis of their vocalization, but also use vocal features to choose some conspecific males over others. Species diversity is higher in some clades of birds and insects that are thought to experience stronger sexual selection than their sister clades (Arnqvist et al., 2000).

Evidence of reinforcement of prezygotic isolation is growing, after an earlier period of skepticism (Noor, 1999). Controlled for time since divergence (indexed by molecular difference), sexual isolation between populations and species of *Drosophila* is greater between sympatric than allopatric pairs of taxa, whereas postzygotic isolation showed no such difference. Male pied flycatchers (*Ficedula hypoleuca*) closely resemble *F. albicollis* where they are allopatric but are colored very differently where the species are sympatric. Female pied flycatchers prefer males from their own kind of population. Such instances of *character displacement*—greater difference between sympatric than allopatric populations of two species—are expected if reinforcement has occurred.

Evidence for founder-effect speciation is sparse. Some laboratory populations of *Drosophila* and *Musca* (housefly) that have been passed through bottlenecks have displayed slight sexual or postzygotic RI, as well as changes in courtship behavior. The significance of these results has been debated because the effects were generally slight and occurred in a minority of replicate populations. Speciation in some organisms has been rapid and prolific on islands or in lakes where populations may have experienced bottlenecks. For example, more than 800 species of *Drosophila* have evolved in the Hawaiian archipelago. However, few such species display the reduced molecular variation that is the expected signature of a past bottleneck. Natural selection, perhaps involving adaptation to unused resources, cannot be ruled out as the cause of speciation (Turelli et al., 2001).

Sympatric speciation. Because sympatric speciation requires rather special conditions, it should not be assumed without evidence—which, however, may be hard to obtain. Sympatry of sister species is not, in itself, evidence of sympatric speciation, because the geographic ranges of species have changed dramatically over time. Even sister species that inhabit a confined area such as a lake or island can be formed "microallopatrically" if the area is complex enough to harbor separate populations of organisms with restricted dispersal.

Among the convincing examples of sympatric speciation are a few instances of two or more fish species that form a monophyletic group within a single lake, such as volcanic crater lakes in Cameroon. These cases are convincing because the lakes form simple conical basins, with no opportunity for fish populations to have become subdivided. Barraclough and Vogler (2000) found that the relationship between range overlap and recency of origin of sister taxa in a few clades fits the expectation for sympatric speciation (constant or decreasing overlap with the passage of time). One such taxon is the fly genus *Rhagoletis* (Filchak et al., 2000). Originally, *Rhagoletis pomonella* mated and developed almost exclusively on hawthorn plants. It started attacking apple in the late 1800s. These plants now support incipient species that differ strongly in allele frequencies at several loci. These are linked to a genetic difference in the timing of the life cycle, such that their mating seasons overlap only slightly. The incipient species differ significantly in host-plant preference and so are somewhat ecologically isolated. Several other cases of very closely related, sympatric insects that mate and feed on different host plants are likely to represent sympatric speciation (Via, 2001).

Other modes of speciation. *Hybrid speciation*, whereby a hybrid between two species constitutes or evolves into a third species, is thought to occur by *recombinational speciation* or by *allopolyploidy*. In recombinational speciation, one or more F_2 or backcross hybrid genotypes is at least partly reproductively isolated and evolves into a new species. Molecular phylogenetic analysis has shown that several diploid species of sunflowers (*Helianthus*) originated from hybrid ancestors. Diploid hybrid speciation is probably rare, because a particularly fit hybrid genotype—the possible progenitor of a new species—is likely to interbreed with some other genotypes and thus be broken down by recombination.

More common by far is polyploidy, probably the only mode of speciation that occurs largely in a single genetic step. For example, unreduced gametes of two diploid species may join, yielding an allotetraploid hybrid. The polyploid is partly postzygotically isolated from the diploid parents because backcrosses produce mostly sterile (e.g., triploid) progeny. This fact also poses a problem to the origin of a polyploid species. Most offspring of a newly arisen tetraploid will be sterile triploids if the tetraploid mates with the more abundant diploids. Thus, tetraploids might not become established unless they occupy different habitats from their diploid parents. Nonetheless, as many as 4 percent of speciation events in plants are due to polyploidy, and a large fraction of plant species have had ancient polyploid ancestors. Some polyploid species, such as allopolyploid goatsbeards (*Tragopogon*), have evolved within the past century.

Speciation by *cytoplasmic incompatibility* has been studied in parasitic wasps in the genus *Nasonia*, and it may prove to be widespread. As in other Hymenoptera, female *Nasonia* develop from fertilized eggs, whereas males develop from unfertilized haploid eggs. *Nasonia* wasps carry *Wolbachia*, intracellular bacteria that are transmitted in the egg from mothers to offspring. Several incompatibility types of *Wolbachia* exist. If sperm from a male with a given type joins an egg that lacks that type of *Wolbachia*, the paternal chromosomes degenerate and the egg develops into a male. Two species of *Nasonia* produce only male (nonhybrid) progeny when crossed, but no hybrid (female) progeny. However, they produced hybrid females when they were treated with antibiotics that killed their *Wolbachia*. These species carry different incompatibility types of *Wolbachia*, so both species' sperm are incompatible with the egg of the other species, resulting in paternal chromosomal degeneration—and thus production of nonhybrid males—in initially diploid hybrid zygotes. Although these species also display some sexual isolation, divergence in their symbiotic bacteria has probably played a role in speciation.

The Time Course of Speciation. How long does the process of speciation take? The time since isolation of some populations can be dated by geological evidence. For example, various sister pairs of snapping shrimp have achieved varying degrees of RI since the Isthmus of Panama separated them about 3.5 million years ago, indicating that this timespan is enough for full speciation to be attained in some cases, but not others (Knowlton et al., 1993). More than 300 species of cichlids have evolved in Lake Victoria since this African lake originated less than 200,000 years ago. Geological evidence suggests that the lake basin was dry only 12,000 years ago; if true, speciation in these fishes has been astonishingly rapid.

The duration of speciation can also be estimated from molecular divergence, if this provides an approximate molecular clock. For example, genetic distances among partially reproductively isolated populations of *Drosophila* indicate that speciation may require more than 2 million years in some cases, and a similar average time for speciation seems to characterize birds and mammals (McCune and Lovejoy, 1998).

McCune and Lovejoy (1998) reasoned that sympatric speciation, which by hypothesis is directly caused by disruptive selection, should be faster on average than allopatric speciation, which may often be a byproduct of genetic change in response to selection or genetic drift. They used DNA sequence divergence to estimate the duration of speciation in taxa proposed to have speciated allopatrically and in taxa that might have speciated sympatrically (e.g., sister species of fishes in the same lake). As predicted, the estimated duration was longer for presumed cases of allopatric speciation.

[*See also* Species Concepts; Species Diversity; Species Selection.]

BIBLIOGRAPHY

Arnqvist, G., M. Edvardsson, U. Friberg, and T. Nilsson. "Sexual Conflict Promotes Speciation in Insects." *Proceedings of the National Academy of Sciences USA* 97 (2000): 10,460–10,464.

Avise, J. C. *Phylogeography*. Cambridge, Mass., 2000. An explication of how genealogies of genes within and among species provide information on population history and speciation.

Barraclough, T. G., and A. P. Vogler. "Detecting the Geographic Pattern of Speciation from Species-Level Phylogenies." *American Naturalist* 155 (2000): 419–434. Introduces a new approach for discriminating among geographic models of speciation.

Bradshaw, H. D., Jr., K. G. Otto, B. E. Frewen, J. K. McKay, and D. W. Schemske. "Quantitative Trait Loci Affecting Differences in Floral Morphology between Two Species of Monkeyflowers (*Mimulus*)." *Genetics* 149 (1998): 367–382.

Coyne, J. A., and H. A. Orr. "The Evolutionary Genetics of Speciation." *Philosophical Transactions of the Royal Society of London, B* 353 (1998): 287–305. A comprehensive overview of the genetic basis of speciation.

Dieckmann, U., and M. Doebeli. "On the Origin of Species by Sympatric Speciation." *Nature* 400 (1999): 354–357.

Filchak, K. E., J. R. Roethele, and J. L. Feder. "Natural Selection and Sympatric Divergence in the Apple Maggot *Rhagoletis pomonella*." *Nature* 407 (2000): 739–742. A recent paper on a "poster child" for sympatric speciation.

Funk, D. J. "Isolating a Role for Natural Selection in Speciation: Host Adaptation and Sexual Isolation in *Neochlamisus bebbianae* Leaf Beetles." *Evolution* 52 (1998): 1744–1759.

Futuyma, D. J. *Evolutionary Biology*. 3d ed. Sunderland, Mass., 1998. Includes expanded coverage of speciation, and literature references to studies mentioned but not referenced in this article.

Holland, B., and W. R. Rice. "Chase-Away Selection: Antagonistic Seduction Versus Resistance." *Evolution* 52 (1998): 1–7. An intriguing model of sexual selection with implications for speciation.

Kirkpatrick, M., and M. R. Servedio. "The Reinforcement of Mating Preferences on an Island." *Genetics* 151 (1999): 865–884.

Knowlton, N., L. A. Weigt, L. A. Solorzano, D. K. Mills, and E. Bermingham. "Divergence in Proteins, Mitochondrial DNA, and Reproductive Compatibility Across the Isthmus of Panama." *Science* 260 (1993): 1629–1632.

Mayr, E. *Systematics and the Origin of Species*. New York, 1942. A foundation of studies of species and speciation.

Mayr, E. "Change of Genetic Environment and Evolution." In *Evolution as a Process*, edited by J. Huxley, A. C. Hardy, and E. B. Ford, pp. 157–180. London, 1954. A classic, influential, and still controversial paper.

McCune, A. R., and N. J. Lovejoy. "The Relative Rate of Sympatric and Allopatric Speciation in Fishes: Tests Using DNA Sequence Divergence Between Sister Species and Among Clades." In *Endless Forms: Species and Speciation*, edited by D. J. Howard and S. H. Berlocher, pp. 172–185. New York, 1998.

Noor, M. A. F. "Reinforcement and Other Consequences of Sympatry." *Heredity* 83 (1999): 503–508. A review of this controversial subject.

Otto, S. P., and J. Whitton. "Polyploid Incidence and Evolution." *Annual Review of Genetics* 34 (2000): 401–437.

Pomiankowski, A., and Y. Iwasa. "Runaway Ornament Diversity Caused by Fisherian Sexual Selection." *Proceedings of the National Academy of Sciences USA* 95 (1998): 5106–5111.

Rundle, H. D., L. Nagel, J. W. Boughman, and D. Schluter. "Natural Selection and Parallel Speciation in Sticklebacks." *Science* 287 (2000): 306–308.

Taylor, E., and J. D. McPhail. "Historical Contingency and Ecological Determinism Interact to Prime Speciation in Sticklebacks, *Gasterosteus*." *Proceedings of the Royal Society of London, B* 267 (2000): 2375–2304.

Turelli, M., N. H. Barton, and J. A. Coyne. "Theory and Speciation." *Trends in Ecology and Evolution* 16 (2001): 330–342. A thorough (but nonmathematical) analysis of theoretical aspects of speciation.

Via, S. "Sympatric Speciation in Animals: The Ugly Duckling Grows Up." *Trends in Ecology and Evolution* 16 (2001): 381–390. A review of proposed cases of sympatric speciation.

— DOUGLAS J. FUTUYMA

SPECIES CONCEPTS

Species are fundamental units of biological diversity; beyond general agreement on this simple fact, there exists little consensus among evolutionary biologists and systematists about how the term "species" should be defined. Recent reviews have listed as many as 20 different species concepts, but many of these are subtle variations on a number of common themes. Here, the most frequently cited or discussed concepts and definitions (see Vignette on Seven Species Concepts) are organized according to a set of defining criteria.

A number of possible contrasts can serve as organizing principles for a taxonomy of species concepts. For example, definitions of species may rely entirely on examining patterns of character variation, without considering the evolutionary history or processes that have led to those patterns. Emphasis may be placed on making species concepts clearly operational, or universal, or both, so that they will be useful for practicing taxonomists and will apply to groups of lineages regardless of whether those are sexual or asexual, whether they consist of a single interconnected population or of many disjunct (allopatric) populations, or whether they represent fossil or living taxa. Alternatively, the motivation for choosing a species concept may derive from a particular model of the evolutionary process, or from an understanding of the forces that are responsible for maintaining similarity or difference among extant populations or groups of populations. It may be deemed more important to imbue species concepts with theoretical significance so that they are consistent with a history of descent with modification, or so that they are appropriate units for studying the evolution of reproductive/genetic isolation. According to this view, species may be viewed as products of evolution, or they may be considered to be units of evolutionary change.

With these possibilities in mind, it is useful to distinguish three classes of species definitions: those that are based on observations of character state similarity or difference; those that require that species are consistent with the principles of phylogenetic systematics and reflect the outcome of a history of descent with modification; and those that derive from the notion that species should be cohesive groups, most often connected by gene exchange. These organizing principles are not mutually exclusive; in fact, all species concepts ultimately depend on inference from patterns of variation. Both evolutionary biologists and systematists differ on whether process inferred from observed pattern should also enter into our conceptual framework.

Character-based Definitions. Perhaps the most straightforward approach is to define species on the basis of observed similarity or difference. Species are simply entities (groups of individuals or populations) that

SEVEN SPECIES CONCEPTS OR DEFINITIONS FROM THE LITERATURE OF SYSTEMATIC BIOLOGY AND EVOLUTIONARY BIOLOGY

The following are organized according to whether they are based on (1) character state similarity or difference; (2) assessment of genealogical or evolutionary relationships; or (3) cohesion or isolation.

Character-Based Species Concepts. All species concepts are based ultimately on assessment of character state similarity or difference. However, character-based concepts do not require any further inferences, but rely only on observed differences. The character-based phylogenetic species concept recognizes two populations as different species if all individuals from the two populations are diagnosably distinct, that is, they can be distinguished with certainty as to the population of origin based on examination of one or more traits. The species definition based on genotypic clusters recognizes that some "intermediates" (individuals of mixed ancestry) may occur when individuals from two distinct populations overlap, but defines these populations as distinct species if the intermediates are relatively few and the parental types remain distinct outside the area of overlap.

Phylogenetic species concept (character-based):

"the smallest aggregation of populations (sexual) or lineages (asexual) diagnosable by a unique combination of character states in comparable individuals" (Kevin Nixon and Quentin Wheeler, 1990).

Genotypic cluster species definition:

"distinguishable groups of individuals that have few or no intermediates when in contact . . . clusters are recognized by a deficit of intermediates, both at single loci (heterozygote deficits) and at multiple loci (strong correlations or disequilibria between loci that are divergent between clusters)" (James Mallet, 1995).

Genealogical and Evolutionary Species Concepts. These concepts recognize species as components of a genealogy or phylogeny. In the genealogical species concept, species are groups of individuals all of whom are more closely related to each other than they are to any individual outside the group. The evolutionary species concept defines species as lineages (sets of ancestors and their descendants) over time; these lineages must remain distinct and have their own unique evolutionary trajectory.

Genealogical species concept:

"'exclusive' groups of organisms, where an exclusive group is one whose members are all more closely related to each other than to any organisms outside the group. . . . Basal taxa . . . that is taxa that contain no included taxa" (David Baum and Kerry Shaw, 1995).

Evolutionary species concept:

"an entity composed of organisms that maintains its identity from other such entities through time and over space and that has its own independent evolutionary fate and historical tendencies" (E. O. Wiley and Richard Mayden, 2000).

Isolation and Cohesion Concepts. These concepts focus on processes that promote similarity or allow differentiation. The widely used (and cited) Biological Species Concept is sometimes referred to as the isolation concept because it defines species both in terms of gene exchange within and barriers to gene exchange (reproductive isolation) between groups of individuals. In contrast, the cohesion and recognition species concepts focus only on characteristics shared by individuals, most often on common mechanisms that promote gene exchange (e.g., mate or gamete recognition systems). The isolation and recognition concepts apply only to sexual organisms. The cohesion concept can accommodate asexual lineages by arguing that ecological equivalence of such lineages implies that they belong to the same species (although they are not connected by gene exchange).

Biological species concept (isolation concept):

"groups of actually or potentially interbreeding natural populations which are reproductively isolated from other such groups" (Ernst Mayr, 1963).

"systems of populations, the gene exchange between these systems is limited or prevented in nature by a reproductive isolating mechanism or by a combination of such mechanisms" (Theodosius Dobzhansky, 1970).

Recognition species concept:

"the most inclusive population of individual biparental organisms which share a common fertilization system [specific mate recognition system]" (Hugh Paterson, 1985).

Cohesion species concept:

"the most inclusive population of individuals having the potential for phenotypic cohesion through intrinsic cohesion mechanisms [genetic and/or demographic exchangeability]" (Alan Templeton, 1989).

—RICHARD G. HARRISON

look different from other such groups. Both the classical morphological (or phenetic) species concept and the phenetic species concept advocated by numerical taxonomists are based on identifying (intuitively or numerically in the respective cases) the boundaries between clusters in morphological character space. These character-based concepts do not rely on a particular model of an evolutionary or historical process that gave rise to the observed diversity. However, many character-based concepts implicitly or explicitly argue that species are the end products of evolution, and that populations are connected by genetic exchange.

Some phylogenetic systematists have argued for a phylogenetic species concept (PSC) in which species are the smallest groups of populations (in sexual species) or lineages (in asexual taxa) that are diagnosably distinct from other such groups. Phylogenetic species defined in this way are viewed as the smallest lineages analyzable by cladistic methods. Although it is apparently independent of models of the evolutionary process, this version of the PSC depends on understanding the reproductive relationships of individuals in space and time. Without such an understanding, males and females within sexually dimorphic species, or distinctive morphs in polymorphic species, would be categorized as different species. Indeed, unless reproductive relationships are known, it is not even possible to define a population.

An alternative character-based definition views species as genotypic clusters; species are then distinguishable groups of individuals that have few or no intermediates (i.e., hybrids) when they exist in contact (i.e., are sympatric). This definition is motivated first by the belief that divergence is not all-or-none—that species may not be more "real" than races or subspecies, and second by the observation that diagnosably distinct entities often hybridize when they co-occur, resulting in local populations with some proportion of individuals that are intermediate (of mixed ancestry). In spite of hybridization, the parental types often remain distinct, and the distribution of genotypes or phenotypes is bimodal. The persistence of distinctive genotypic or phenotypic clusters warrants recognition of these groups as distinct species.

Character-based definitions have the great advantage of being operational, because they require only data on observed patterns of phenotypic or genotypic variation. The character-based PSC applies equally to asexual and sexual lineages, and to allopatric and sympatric populations. If the character-based PSC is strictly applied, however, then the observation of single fixed character differences between members of two populations leads to recognition of these populations as distinct species. Critics of this version of the PSC suggest that its application would result in an unwarranted proliferation of species; advocates counter with the argument that diagnosably distinct populations are the fundamental units of evolution—that is, the units that diverge and become locally adapted.

History-based Species Concepts. A second approach to species concepts gives primacy to history and argues that species must be monophyletic or exclusive groups (see the genealogical species concepts in the Vignette). An exclusive group is one in which all members are more closely related to one another than they are to any entities outside the group. It may seem reasonable that all members of a single species would be more closely related to one another than they are to members of other species; however, exclusivity develops gene by gene (or genome region by genome region). Given a polymorphic ancestor, recently diverged populations (even those that exhibit one or more fixed differences) may not be exclusive groups for segments of their genomes (e.g., some humans have MHC alleles more closely related to alleles in chimps than to any other alleles in humans, presumably the result of the maintenance of a polymorphism that predates the lineage splitting event that gave rise to chimps and humans). Moreover, if a widespread species gives rise to one or more daughter populations that then diverge from the parent species, the widespread species may be paraphyletic with respect to the daughter populations. In such cases, some individuals of the parent species may be more closely related to individuals in the daughter species than to some of their own conspecifics.

Proponents of species concepts that rely on monophyly or exclusivity also describe their concepts as "phylogenetic." The contrast between the character-based and history-based PSCs reflects a fundamental division within the community of phylogenetic systematists. The basic disagreement is over whether species definitions should rely simply on characters, or whether species must be defined by using the historical relationships inferred from those characters and, in particular, over whether the only characters that matter in defining species are shared derived characters (synapomorphies). Species concepts that rely on monophyly or exclusivity are retrospective; species are the smallest (least inclusive) monophyletic groups and are the natural outcomes of a history of descent with modification. The evolutionary species concept (see Vignette) is also explicitly phylogenetic. Rather than being the smallest monophyletic groups, species are viewed as the largest tokogenetic (interbreeding) systems—the largest evolving entities. They are lineages or groups of organisms that maintain their distinctness through space and time and have their own evolutionary fate.

Isolation and Cohesion Species Concepts. The third group of species concepts proposes that current isolation and/or cohesion are the basis for delineating species boundaries. These concepts are prospective rather than primarily retrospective; they are based on considerations of ongoing evolutionary interactions that have implications for the future status of populations. The best-known species concept is the biological species concept (BSC). Until recently, it was the default framework for thinking about species and speciation. It was originally articulated by two leading evolutionary biologists of the twentieth century, Ernst Mayr (an avian systematist) and Theodosius Dobzhansky (a *Drosophila* population geneticist). The BSC views species as groups of populations that are held together by gene flow (interbreeding) and that are isolated from other groups of populations by intrinsic differences that serve as barriers to gene exchange (often referred to as "isolating mechanisms"); that is, a species is a group of individuals that can interbreed. Both cohesion (through gene flow) and isolation are important components of the BSC, but much of the emphasis has been on isolation, which necessarily means that a species (group of populations) can be defined only in relation to other groups. This characteristic of the BSC has led some evolutionary biologists to refer to the BSC as the "isolation concept."

In contrast, two other definitions—the recognition and the cohesion species concepts—attempt to be nonrelational, defining species as inclusive groups held together by particular cohesive forces or mechanisms. According to the recognition concept, species are viewed as groups of individuals held together by a common system of fertilization or mate recognition. For example, species-specific acoustic or chemical communication systems are essential for both mate finding and mate recognition in many insects. Thus, rather than emphasizing the barriers that isolate individuals or populations (which may be the incidental byproducts of genetic divergence arising between populations in allopatry, and not "mechanisms" molded by natural selection for the purpose of isolation), the recognition concept emphasizes genetic continuity actively maintained by natural selection. The cohesion species concept expands on this view; it suggests, in addition to genetic cohesion, the possibility of ecological cohesion (or "demographic exchangeability"). Groups of organisms are demographically exchangeable if they are ecologically equivalent, occupying the same niche. Of course, documenting that two organisms occupy the same niche (use the same resources, attract the same set of predators and parasites, etc.) is extremely difficult. Nonetheless, demographic exchangeability as a criterion for recognizing individuals as conspecifics allows the cohesion concept to apply to asexual lineages. Neither the BSC nor the recognition concept, both of which are based on cohesion through genetic exchange, can be applied to lineages that are asexual, nor can they be applied directly to allopatric populations, because we cannot evaluate whether intrinsic barriers to gene exchange would operate. However, Mayr's original definition for species included the term "potentially interbreeding," suggesting that some assessment might be made of the reproductive status of allopatric populations.

Although the BSC is still popular among population geneticists and evolutionary biologists interested in speciation, this concept (and, by implication, the recognition concept) has been heavily criticized because it is difficult to apply. In practice, evolutionary biologists rarely test directly whether gene exchange occurs; they rely instead on phenotypic or genotypic indicators of gene exchange. Furthermore, tests of interbreeding often cannot be performed—for example, in asexual lineages or in allopatric populations. The BSC is also viewed as inconsistent with the principles of phylogenetic systematics. Both common descent and current interbreeding can be viewed as "processes" through which organisms are related; however, interbreeding (or reproductive compatibility) is a shared ancestral trait—a symplesiomorphy—and therefore does not conform to a cladist's criterion for grouping organisms (since groups must be defined by shared derived characters). Furthermore, if reproductive compatibility is used as a criterion for defining species, then some species will be paraphyletic assemblages (groups of organisms that include some but not all descendants of a common ancestor). If monophyly (exclusivity) is viewed as a defining characteristic for species, the potential for paraphyly is a serious flaw in the BSC. Some cladists even suggest that the BSC interferes with our attempts to recover the history of descent with modification. The counter-argument is that paraphyly simply reflects an expected pattern, given that differentiation leading to intrinsic barriers to gene exchange often arises within a single local population. Indeed, the observation of paraphyly is viewed by some as providing valuable information about the history of diversification and speciation.

Subspecies, Variation, and Continua. The origins of fixed character differences or barriers to gene exchange between populations are not instantaneous events, so it should come as no surprise that evolutionary biologists discover groups of populations at various levels of phenotypic (genotypic) divergence, for which barriers to gene exchange exist but are not complete. For example, wolves and coyotes can interbreed but are not considered the same species. Indeed, some evolutionary biologists (beginning with Darwin) have argued that there is simply a continuum from races (distinctive populations occupying a particular geographic region) to species. The evolutionary biology literature is strewn with names of groups that are viewed as possible stages along this path from conspecific population to "good" species. For example, subspecies are named allopatric or parapatric races, often described on the basis of one or a few distinctive morphological characters.

HYBRIDIZING FIELD CRICKETS

Although the species definitions discussed in this entry are conceptually sound, many are difficult to apply in practice. Consider the case of the hybridizing field crickets, *Gryllus firmus* and *Gryllus pennsylvanicus*. These two eastern North American "species" are very similar in morphology, behavior, and ecology. In fact all North American crickets in the genus *Gryllus* were once considered a single widespread and variable species; only in the 1950s were different species distinguished on the basis of life history, habitat,and song characteristics. Careful comparisons between *G. firmus* and *G. pennsylvanicus* reveal differences in ovipositor length (the ovipositor is the structure with which females deposit eggs in the soil). These differences, together with recently discovered diagnostic differences in DNA sequence at several nuclear genes and mitochondrial DNA clearly suggest that the two crickets are phylogenetic species according to the definition of Nixon and Wheeler. However, analyses of genealogical relationships based on DNA sequences do not show these species to be exclusive groups (i.e., they are not [yet] genealogical species). Along an extensive zone where the two crickets overlap, the "species" hybridize, resulting in some individuals of mixed ancestry. But such individuals are relatively rare, disequilibrium remains high, and away from the hybrid zone the parental types remain distinct. Therefore, the two crickets are probably species according to the genotypic cluster species definition. Are they species according to the Biological Species Concept or the recognition/cohesion concept? That depends on how these concepts are interpreted. Clearly the two crickets can potentially interbreed—in fact they actually interbreed to a limited extent in natural populations. They are reproductively isolated, but not completely! Put a different way, their mate recognition systems are different, but must overlap sufficiently such that limited gene exchange between the two "species" does occur. Such examples of entities that interbreed to some extent, but tend to remain distinct even in the face of this interbreeding, are remarkably common, both in plants and in animals. Given that evolutionary change is an ongoing process, it is not surprising that careful observation should reveal lineages at all stages along the path from conspecific populations to good species. Therefore, some groups will always be problematic when attempts are made to apply the many species concepts available.

—RICHARD G. HARRISON

Other relevant terms that are frequently encountered in the evolutionary literature include the following:

- semispecies: generally, parapatric forms between which there are partial but not complete barriers to gene exchange;
- superspecies: a group or complex of semispecies;
- polytypic species: a geographically variable species that is often subdivided into named subspecies or races;
- ring species: a chain of interbreeding subspecies in which the end members are reproductively isolated;
- sibling species: "good" (i.e., reproductively isolated) species that are difficult to distinguish on the basis of morphological characters (sometimes called "cryptic" species);
- syngameon: a term used by botanists to describe an assemblage of semispecies or of sympatric species that exchange genes to some extent (are partially interbreeding).

Because character variation within species is not always concordant, subspecies boundaries may reflect patterns of variation for only a small subset of the genome. Nonetheless, many subspecies, because they are diagnosably distinct, would be considered species according to the character-based PSC. However, if hybridization occurs freely at subspecies boundaries, generating a hybrid swarm, then the subspecies do not conform to the definition of species based on genotypic clusters, in which genetic bimodality must persist for species to be recognized. Indeed, hybridization between parapatric or sympatric forms poses a problem for virtually all species concepts. Barriers to gene exchange between hybridizing taxa are often semipermeable, with considerable variance in the extent of introgression in different parts of the genome. Such observations have led to the suggestion that species may need to be defined gene by gene. Furthermore, even character-based concepts are difficult to apply in the face of hybridization because the unique combinations of character states that characterize phylogenetic species are broken up and recombined (see Vignette on Hybridizing Field Crickets).

[*See also* Speciation.]

BIBLIOGRAPHY

Baum, D. A., and M. J. Donoghue. "Choosing among Alternative 'Phylogenetic' Species Concepts." *Systematic Botany* 20 (1995): 560–573. One view of the disagreements among cladists as to how to define species.

Baum, D. A., and K. L. Shaw. "Genealogical Perspectives on the Species Problem." In *Experimental and Molecular Approaches to Plant Biosystematics*, edited by P. C. Hoch and A. G. Stevenson, pp. 289–303 St. Louis, 1995. A provocative introduction to the notion of species as exclusive groups, couched in terms of the nature of gene genealogies.

Claridge, M. F., et al., eds. *Species: The Units of Biodiversity*. London, 1997. A collective volume on species concepts, with many chapters dealing with species concepts as applied to particular taxonomic groups; the chapter by Hull provides a good overview of problems in defining species.

Cracraft, Joel. "Speciation and Its Ontology: The Empirical Consequences of Alternative Species Concepts for under standing

Patterns and Processes of Differentiation." In Otte and Endler (1989).
de Querioz, K., and M. J. Donoghue. "Phylogenetic Systematics and the Species Problem." *Cladistics* 4 (1988): 317–338. An early discussion of the relationship between species concepts and the principles and practice of phylogenetic systematics.
Dobzhansky, T. *Genetics of the Evolutionary Process*. New York, 1970.
Howard, D. J., and S. H. Berlochler. *Endless Forms: Species and Speciation*. New York, 1998. A recent summary of issues in species and speciation. The chapters by Harrison, Templeton, de Quieroz, and Shaw focus explicitly on a number of aspects of the species problem.
Maddison, W. P. "Phylogenetic Histories within and among Species." In *Experimental and Molecular Approaches to Plant Biosystematics*, edited by P. C. Hoch and A. G. Stevenson, pp. 273–287. St. Louis, 1995. A thoughtful and intuitive introduction to thinking about gene genealogies and their relationship to species boundaries.
Mallet, J. "A Species Definition for the Modern Synthesis." *Trends in Ecology and Evolution* 10 (1995): 294–299. An argument for defining species as genotypic clusters.
Mayr, E. *Animal Species and Evolution*. Cambridge, Mass., 1963.
Mayr, E. *The Growth of Biological Thought*. Cambridge, Mass., 1982. An overview of issues in modern evolutionary biology from the perspective of one of the most influential evolutionary biologists of the twentieth century; chapter 6 deals directly with Mayr's views about species concepts.
Nixon, K. C., and Q. D. Wheeler. "An Amplification of the Phylogenetic Species Concept." *Cladistics* 6 (1990): 211–223. The clearest argument for a character-based phylogenetic species concept.
Otte, D., and J. Endler, eds. *Speciation and Its Consequences*. Sunderland, Mass., 1989. A collective volume with several important chapters that deal with species concepts, especially those by Templeton (on cohesion species concepts), Cracraft (on phylogenetic species), and Endler.
Paterson, H. E. H. "The Recognition Concept of Species." In *Species and Speciation*, edited by E. S. Vrba, 21–29. Pretoria, 1985. The original justification for the recognition concept of species, and an argument against the notion of "isolating mechanisms."
Sokal, R. R., and T. J. Crovello. "The Biological Species Concept: A Critical Evaluation." *American Naturalist* 104 (1970): 127–153. An early critique of the biological species concept.
Templeton, A. "The Meaning of Species and Speciation: A Genetic Perspective." In Otte and Endler (1989)
Wheeler, Q. D., and R. Meier, eds. *Species Concepts and Phylogenetic Theory*. New York, 2000. A collective volume which includes arguments in support of an array of species concepts, together with critiques of those arguments—essentially a roundtable discussion in print.
Wiley, E. O., "The Evolutionary Species Concept Reconsidered." *Systematic Zoology* 27 (1978): 17–26. An early summary of the evolutionary species concept.
Wiley, E. O., and Richard Mayden. "The Evolutionary Species Concept." In Wheeler and Meier (2000).

— Richard G. Harrison

SPECIES DIVERSITY

Biologists use measures of the diversity of species in a given area to investigate questions in ecology and to document patterns important to conservation. Species diversity is typically measured in one of two ways, either as a simple count of the number of species in an area (species richness) or by an index that takes account of the importance of each species. The best known diversity index is E. H. Simpson's (1949) D, which is calculated as the sum of the reciprocals of p_i^2: $1/p_1 + 1/p_2 + \ldots + 1/p_i$, where the p_i are the proportional importance: often represented by a relative abundance of species. They vary between 0 and 1. Thus, an assemblage of ten equally important species would have a diversity of 10, whereas an assemblage of nine very rare plus one abundant species has a much lower diversity, closer to 1.

Ecologists are usually interested in the diversity of species found within a given homogeneous local habitat, or within-habitat diversity. Because communities differ in species composition, diversity over a broader geographic region also includes a between-habitat component of diversity. Regional diversity is correlated with local diversity but exceeds it. Perhaps the most useful measure is point diversity, which consists of the diversity of species that occur together at a given point in space (it is the diversity in the limit as area is diminished toward zero at a given instant).

Typically, the number of species increases with geographic area, among subsets of biotas, among islands in an archipelago, and in samples within continents, according to a simple species-area rule $S = cA^z$, where S is number of species, A is area, and c and z are constants; z is usually less than 1.0, often close to 0.25. Simple plots of number of species versus area are strongly curvilinear because z is typically small. Therefore, log–log plots are much easier to use. Taking logarithms of the above equation gives $\log S = \log c + z\log A$, which is an equation for the straight line relationship between $\log S$ and $\log A$ with intercept $\log c$ (a constant) and slope z (another constant). On such a log–log plot, the intercept measures point diversity, and the slope is a measure of how within-habitat diversity is related to between-habitat diversity. Interestingly, z values are higher, (i.e., diversity increases more rapidly with area) for islands in archipelagos than they are for samples of mainlands, because islands of a given size support fewer species, especially at higher *trophic* levels, than a similar sized piece of mainland.

Species diversity varies greatly from place to place because of differences in species richness and their relative abundance. One system can support more species than another in several basic ways. A greater variety of available resources will support more species than a less diverse resource base. More species can be packed in on the same range of available resources if, on average, species use a narrower range of resources (i.e., they are more specialized and have narrower niches). Resource partitioning among species reduces competition and promotes diversity. More species can be packed in on the same range of available resources if, on average, species share more resources (i.e., they tolerate greater niche overlap). Each of these mechanisms contributes to local diversity. Another way in which two systems can differ is in the degree to which they support as many species as possible, or the degree of saturation with species.

A prominent geographical pattern in species diversity has attracted the attention of biologists for well over a century: temperate zones tend to contain fewer species than tropical regions. For example, no temperate zone forest contains more than a dozen species of trees, but most tropical rain forests support hundreds. Many explanations have been proposed to explain such latitudinal gradients in species diversity. Several of these hypotheses, which are not necessarily mutually exclusive, are described here.

Area Effects. Flat maps misrepresent surface areas at high latitudes. Because of the earth's spherical shape and the way we define latitude and longitude, a 1° × 1°-square grid diminishes in size as one moves toward the pole. Tropical terrestrial habitats, particularly in central Africa and Brazil, cover an extensive area of the earth's surface, extending contiguously both northward and southward away from the equator. Species-area relationships dictate that such an extensive region will contain many species. Very little land surface exists at temperate latitudes in the Southern Hemisphere, which supports only a low diversity of plants and animals. Although land area is extensive in the Northern Hemisphere, it is broken up into several distinct latitudinal belts (tundra, boreal forest, temperate biomes, and subtropical biomes), each with its own characteristic biota. Moreover, geographic ranges of tropical species tend to be restricted, whereas those of temperate zone species are much larger.

Time Theories. Diversity is postulated to increase with a community's age. Speciation rates could be higher in the tropics than in temperate zones. Temperate habitats are impoverished with species because their component species have not had time enough to adapt to or to occupy their environments completely since recent glaciations and other geological disturbances. However, more "mature" tropical communities are more diverse because there has been a longer period without major disturbances for organisms to speciate and diversify within them. This evolutionary time theory does not necessarily imply that temperate communities are unsaturated with individuals; niche expansion may often allow nearly full utilization of available resources even in a habitat that is impoverished with species.

A variant is the ecological time theory, which deals with a shorter, more recent, time span. Of primary concern is time available for dispersal, rather than with time for speciation and evolutionary adaptation. Newly opened or remote areas of suitable habitat, such as an area of forest burned by lightning, an isolated lake, or a patch of sand dunes, may not have their full complement of species because there has been inadequate time for dispersal into these areas. Dispersal powers of most organisms are probably good enough that this mechanism may be of relatively minor importance in most communities.

Climatic Stability. Over a broad zone that includes most of the tropics between latitudes 23°N and 23°S, temperatures fluctuate as much during a daily cycle as average daily temperatures do over the course of an entire year. Large areas of the tropics also have exceedingly reliable precipitation. More solar energy falls on a unit area in the tropics, and incident solar radiation diminishes with increasing latitude. At higher latitudes above 23°N and 23°S, temperatures fall off with increasing latitude (as does annual productivity—see discussion below). A favored hypothesis is climatic stability, which asserts that such stable tropical climates foster specialization (or narrow niches), hence high diversity. In contrast, variable temperate climates require broader niches, which results in fewer species. This mechanism is closely related to competition hypotheses (see description below) and probably operates in concert with such hypotheses.

Spatial Heterogeneity. The spatial heterogeneity hypothesis postulates that structurally complex habitats will support more species than simple habitats because they offer a greater diversity of microhabitats, hence a broader resource base and more resources to partition. Broad empirical support for this mechanism exists.

An interesting variant is the nutrient mosaic hypothesis. The number of ways in which plants can differ is decidedly limited, especially in the wet tropics, where variation in soil moisture is relatively slight. One mechanism that could help to maintain high plant species diversity involves differentiation in the use of various materials, such as nitrogen, phosphorus, potassium, calcium, and various rare earths. According to this argument, each tree species has its own peculiar set of requirements; soil underneath each species becomes depleted of those particular resources, making it unsuitable for seedlings of the same species. (Eventually, after a tree falls and is decomposed, these materials reenter the nutrient pool. That particular species can then grow there again.) Thus, like the seed predation hypothesis (see description below), this hypothesis predicts a "shadow" around a parent tree where seedlings of that species will be rare or nonexistent.

Productivity Hypothesis. The productivity hypothesis (sometimes called the energy hypothesis) is based on average niche breadth. When food is scarce, foraging animals cannot afford to bypass potential prey and diets must be broad (thus, not as many species can be packed in). Conversely, if food is plentiful, a forager can afford to bypass inferior prey items because it can expect to encounter better ones soon. Confining its diet to better prey results in narrow dietary niches, which allows greater resource partitioning and hence allows more species to coexist. At higher latitudes above the tropics, annual productivity falls off with increasing latitude. Not surprisingly, the most productive, and therefore most diverse, natural communities, coral reefs and tropical rain forests, are in the tropics.

Competition Hypotheses. Under one competition hypothesis, species' populations in diverse communities are viewed as being stable, at equilibrium with their resources, resulting in intense competition, both within and between species. Specialists prevail because they have their own zone of competitive superiority. In contrast, in less diverse systems, species' populations fluctuate more, resources are not exploited as fully, and competition is reduced, allowing more generalized broader niches.

Another competition hypothesis involves circular networks. Under this mechanism, species A is envisioned as being competitively superior to species B, species B in turn excludes species C, whereas species C wins in competition with species A. Under such a circular hierarchy of competitive abilities, the identity of the species occurring at a particular spot will be continually changing from C to B to A, then back to C, repeating the cycle. Circular networks with many more species probably exist. Such nontransitive competitive interactions could well help to maintain the high diversity of coral reefs and tropical trees.

Disturbance Hypotheses. These are essentially alternatives to competition hypotheses. Disturbances, such as fires, floods, and ice or wind storms, can reduce populations and interrupt the process of competitive exclusion locally, thus allowing maintenance of high diversity. Note that disturbances can be small or large, and range over vast spatial and temporal scales. Disturbances can generate spatial heterogeneity. Really intense disturbances reduce diversity. Long undisturbed areas also tend to have low diversity because competitively dominant life-forms usurp all resources available. Intermediate levels of disturbance, however, often facilitate diversity.

A biotic variant of the disturbance hypothesis is the predation hypothesis, which operates with predators acting as agents of disturbance; by lowering prey populations and reducing competition, predators can allow local coexistence of species that would be eliminated in the absence of a predator. Such a "keystone" predator can thus promote diversity. Predators can also enhance diversity in another way known as predator switching behavior: many predators feed preferentially on more abundant prey species; when these are depleted, predators switch to other abundant prey, allowing the first

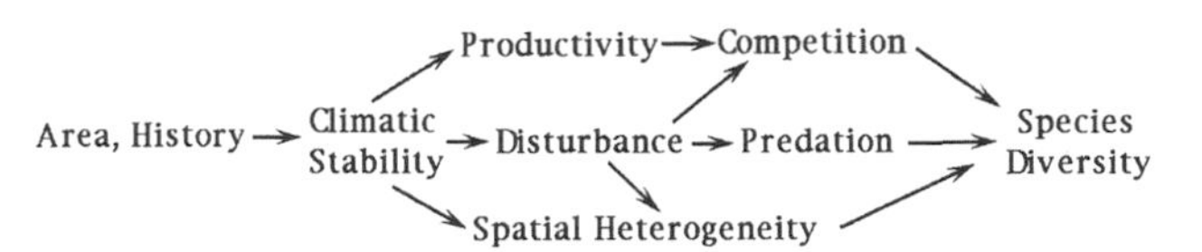

FIGURE 1. Possible Ways in Which Mechanisms Could Interact. Courtesy of author.

prey species to recover. Such frequency-dependent predation can promote prey diversity.

Seed Predation Hypothesis. Seed predation is intense in the tropics, and seedlings cannot establish themselves in the vicinity of parental trees because high seed densities attract many specialized seed predators. This argument predicts that successful recruitment of seeds to seedlings will occur in a ring around (but at some distance from) the parental tree. Inside and outside this ring, other tree species can establish themselves. A wide variety of seed protection tactics (such as toxic matrices) has forced many seed predators to specialize in seeds of particular species. Heavy seed predation coupled with species-specific seed predators holds down densities of various tree species and creates a mosaic of conditions for seedling establishment. Factors other than seed predation also limit the abundance of tropical tree species and prevent single-species dominance.

Still another biotic variety of disturbance is the epiphyte load hypothesis, invoked to explain tropical tree species diversity. Light is a master limiting factor in tropical forests—seedlings endure in the dark understory waiting for a light gap to open up in the canopy above. Epiphytes, plants that grow on other plants such as vines, can act as agents of disturbance. Many more epiphytes occur in the tropics than in the temperate zones, probably because vines must die back during winter. Tree falls due to epiphyte loads are frequent in the tropics, continually opening up patches in the forest and fostering local secondary succession.

Clearly, several or even most of these mechanisms could act together in concert or in series to determine diversity in any given community, and the relative importance of each mechanism doubtless varies widely from community to community. A multitude of ways in which these mechanisms could interact have been suggested, such as the pattern shown in Figure 1.

[*See also* Speciation.]

BIBLIOGRAPHY

Gleason, H. A. "On the Relation between Species and Area." *Ecology* 3 (1922): 158–162.

Janzen, D. H. "Herbivores and the Number of Tree Species in Tropical Forests." *American Naturalist* 104 (1970): 501–528.

MacArthur, R. H. "Patterns of Species Diversity." *Biological Reviews* 40 (1965): 510–533.

MacArthur, R. H. *Geographical Ecology: Patterns in the Distribution of Species.* New York, 1972.
Pianka, E. R. "Latitudinal Gradients in Species Diversity: A Review of Concepts." *American Naturalist* 100 (1966): 33–46.
Pianka, E. R. *Evolutionary Ecology.* 6th ed. San Francisco, 2000.
Rosenzweig, M. L. "Species Diversity Gradients: We Know More and Less Than We Thought." *Journal of Mammalogy* 73 (1992): 715–730.
Rosenzweig, M. L. *Species Diversity in Space and Time.* Cambridge, 1995.
Simpson, E. H. "Measurement of Diversity." *Nature* 163 (1949): 688.
Strong, D. R. "Epiphyte Loads, Tree Falls, and Perennial Forest Disruption: A mechanism for Maintaining Higher Tree Species Richness in the Tropics without Animals." *Journal of Biogeography* 4 (1977): 215–218.

— Eric Pianka

SPECIES SELECTION

In the 1970s, biologists understood that natural selection can act a given level if the entities at that level (e.g., genes, organisms, or groups) vary in heritable traits that influence fitness. With the advent of the theory of punctuated equilibrium, paleontologists hypothesized that species might also be units of selection. (Punctuated equilibrium inspired, but is not logically necessary for, species selection.) Thus arose the idea of species selection: if the species within a higher taxon such as a genus vary in traits that are heritable and that influence the likelihood that the species will survive or reproduce, then the higher taxon will evolve as some kinds of species become more common within that taxon. These might, for example, be specialist species or species with large geographic ranges.

Two important distinctions have improved our understanding of species selection. Early advocates of species selection treated differences in speciation and extinction rates as evidence of species selection. This approach is now widely rejected because evolutionists distinguish sorting from selection. Sorting refers to differential survival and reproduction. Consider a case of sorting at the level of organisms: tall plants have greater reproductive success than shorter forms. Several processes might cause this pattern: (1) tall individuals accidentally occupy more favorable soil conditions, (2) height provides some advantage in the struggle for survival, or (3) the genes for height are linked to genes that offer some selective advantage. Although all three cases involve sorting by height (tall plants have greater reproductive success than short plants), only process 2 counts as natural selection for height per se. Because natural selection is only one possible explanation of sorting, species selection should be distinguished from species sorting.

A second important advance came from recognizing that selection involves two different kinds of entities. In standard examples of selection, organisms compete for scarce resources (e.g., light). Because some traits (e.g., height) help organisms to compete, genes for those useful traits increase in frequency. Thus, natural selection involves interactors (e.g., individuals) whose phenotypes directly interact with the environment and replicators (e.g., genes) that ensure the heritability of phenotypic traits. Selection occurs when the differential success of interactors causes the differential proliferation of replicators. Thus, there are really two distinct questions about whether species are "units of selection": Are species interactors? Are species replicators?

To say that species are replicators means that (1) species reliably pass on some structure or information to their descendants and (2) these inherited structures influence the "phenotype" of species. Species-level gene pools appear to meet these conditions. A gene pool is simply the collection of genes and their frequencies in all of the organisms that comprise a population. During speciation, a single gene pool divides into two distinct (noninterbreeding) gene pools. Although speciation does not perfectly replicate the parental gene pool, descendant gene pools resemble parental gene pools enough for speciation to count as replication. Because gene pools are replicated with fairly high fidelity and influence the phenotype of the species, they can function as replicators.

It has been much harder to determine if species function as interactors. The problem is that even if heritable species-level traits are correlated with species-level fitness, the correlation may result from natural selection acting at lower levels. Consider a case at the level of a group of organisms. Assume that fast-running herds of deer survive longer than slow-running herds. This is evidence that groups are interactors (and thereby units of selection) only if the differences in group survival cannot be fully explained as a result of natural selection at the level of the organisms for fast-running deer. Developing the conceptual criteria and practical methods for determining the level of interaction with the environment—without simply assuming the superiority of either higher- or lower-level explanations—remains the central problem in the units of selection debate.

Recent discussions of species selection have focused on two proposals for determining the level of interaction. The prevailing approach requires emergent traits (i.e., species selection occurs if and only if variation in heritable and emergent species-level traits causes variation in species-level fitness). Population structure (e.g., nonrandom gene flow within or among populations) is an emergent trait because it cannot be attributed to individual organisms per se and is not a simple function of organismic properties. By contrast, aggregate traits are simple functions of lower-level properties. For example, if generalist species are simply collections of generalist organisms, then being a "generalist" is an aggregate species-level property.

Figure 1. Ammonite (*Prionocyclus wyomingensis*) Cretaceous Period, 87 million years ago. © Ken Lucas/Visuals Unlimited, Inc.

The emergent property definition has been criticized for being too restrictive. Advocates of the alternative and subtly different "emergent fitness" definition claim that emergent traits are not necessary for species selection: species function as interactors when aggregate species-level traits contribute to differences in an emergent component of species-level fitness. This technical dispute is important because the emergent fitness approach broadens the definition of species selection, thereby making species selection more common.

One of the most plausible cases of species selection concerns geographic range size among Cretaceous molluscs. Range size has been regarded as an emergent trait because the geographic area over which a species is found is not a simple function of individual properties. Because species with small ranges are more susceptible to extinction, range size increases species-level fitness. Range size appears to be heritable in the sense that species with small ranges tend to give rise to species with small ranges. Although the empirical details are complex, geographic range appears to be an emergent and heritable species-level trait that influences species-level fitness. (See Bibliography for discussion of other possible examples.)

Most biologists believe that species selection is possible but only a minor force in evolution. Three principal arguments support this judgment. First, because only a few species-level traits are both heritable and emergent (candidates include population structure, geographic range, and mating systems), defenders of the emergent trait definition argue that species selection will be rare. (If we adopt the less restrictive emergent fitness criterion, the scope of species selection is somewhat wider.) Second, because organisms have a short generation time relative to species, organismic selection within the lifetime of a species will exert greater influence on trends than species selection (see Gould, 1998, for a response). Third, one might argue that species are not well suited to function as interactors. Interactors are supposed to be cohesive entities that compete for scarce resources in a common environment. But, it is claimed, species are not "cohesive" (the populations of a species may not interact at all). And because species within a clade generally do not share a common environment, it is improper to say that species are being selected for their adaptedness to a common environment (Damuth, 1985, advances this objection; Grantham, 1995, responds).

[*See also* Speciation.]

BIBLIOGRAPHY

Damuth, J. "Selection among 'Species': A Formulation in Terms of Natural Functional Units." *Evolution* 39 (1985): 407–430. An influential presentation of the claim that species are not well suited to function as interactors.

Eldredge, N. *Macroevolutionary Dynamics*. New York, 1989. Advocates a hierarchical expansion of evolutionary theory; summarizes the main objections to species selection, and concludes that species selection is at best a minor force.

Gould, S. J. "Gulliver's Further Travels: The Necessity and Difficulty of a Hierarchical Theory of Selection." *Philosophical Transactions of the Royal Society of London B* 353.1366(1998): 307–314. Responds to a number of criticisms of species selection; claims that entities at various levels display different kinds of individuality and different kinds of adaptations.

Grantham, T. A. "Hierarchical Approaches to Macroevolution: Recent Work on Species Selection and the Effect Hypothesis." *Annual Review of Ecology and Systematics* 26 (1995): 301–326. A moderate defense of species selection; discusses the emergent trait and emergent fitness definitions and responds to Damuth (1985).

Hull, D. L. "Selection and Individuality." *Annual Review of Ecology and Systematics* 11 (1980): 311–332. Introduces the distinction between replicators and interactors.

Jablonski, D. "Heritability at the Species Level: Analysis of Geographic Ranges of Cretaceous Mollusks." *Science* 238 (1980): 360–363. A brief summary of the evidence supporting an influential example of species selection. Presents evidence that species-level traits are heritable and influence species-level fitness.

Lieberman, B. S., and E. S. Vrba. "Hierarchy Theory, Selection, and Sorting: A Phylogenetic Perspective." *Bioscience* 45.6 (1995): 394–399. A nontechnical introduction to Vrba's hierarchical theory of evolution. Argues that explicit attention to phylogeny is crucial and presents phytogenetic data that may undermine Jablonski's defense of species selection.

Lloyd, E. A., and S. J. Gould. "Species Selection on Variability." *Proceedings of the National Academy of Sciences USA* 90 (1993): 595–599. Defends the emergent fitness conception of species selection and provides a possible example of species selection on an aggregate trait.

Vrba, E. S., and S. J. Gould. "The Hierarchical Expansion of Sorting and Selection: Sorting and Selection Cannot Be Equated." *Paleobiology* 10 (1986): 146–171. Distinguishes between sorting and selection and defends the general importance of adopting a hierarchical perspective.

Williams, G. C. *Natural Selection: Domains, Levels, and Challenges*. New York, 1992. Provides an alternative way of framing questions about levels of selection; contains many suggestive, though not fully worked out, examples.

— Todd Grantham

SPERM COMPETITION

Sperm competition is the competition between the ejaculates of different males to fertilize the ova of a particular female. Among species with internal fertilization, sperm competition occurs when females are inseminated by more than one male during a single breeding cycle; among externally fertilizing species, it occurs when more than one male simultaneously releases sperm in the vicinity of a female's ova (Parker, 1998).

Sperm competition, together with cryptic female choice, is part of the process of postcopulatory sexual selection. The evolutionary implications of sperm competition were first recognized in 1970, when it was realized that competition between males could continue beyond copulation and insemination, and that what males really compete for is fertilizations rather than mates. Since 1970, it has become clear that the females of the majority of animal species routinely copulate with more than one male—even in taxa that are generally regarded as monogamous. Researchers now distinguish between social mating systems, such as social monogamy, and genetic mating systems. For example, the European blue tit (*Parus caeruleus*) is socially monogamous (forming a season-long pair bond with one male) but is often genetically polyandrous (engaging in extra-pair copulations with at least one other male), so that broods may contain offspring with the same genetic mother but different genetic fathers.

The term *sperm competition* also refers to adaptations that arise as a result of it and to any of the consequences resulting from females copulating with more than one male.

Sperm competition creates powerful, if subtle, selective pressures, and together with cryptic female choice, it has shaped many life-history characteristics, such as body size and reproductive morphology, physiology, and behavior. Sperm competition generates conflicting selective pressures on males. On the one hand, selection will favor males that manage to fertilize the ova of already-inseminated females, but on the other, it will also favor males that prevent other males from fertilizing the ova of females they themselves have recently inseminated. Initially, it was assumed that females had little to gain from copulating with more than one male, and it was also assumed that they were either passive or reluctant participants in sperm competition. More recently, it has been recognized that not only do females actively seek multiple partners, but that they do so because they may obtain evolutionary benefits from having their ova fertilized by particular males. As a result, females of some species may have evolved the ability to discriminate in their reproductive tract between the sperm of different males (referred to as *cryptic female choice*). Sperm competition may therefore result in the rapid coevolution of male and female reproductive traits.

Adaptations to Sperm Competition. All parts of a male's reproductive anatomy—the testes, sperm storage sites, the penis, accessory glands and the seminal fluids they produce—and the sperm themselves have been shaped by sperm competition (and cryptic female choice).

Testis size. The most ubiquitous adaptation to sperm competition shown by males is relative testis size. Across a range of taxa, including insects, fish, reptiles, birds, and mammals, species that experience more sperm competition have relatively larger testes. This makes sense: larger testes produce more sperm, and the number of sperm a male inseminates is the single most important factor determining his fertilization success in the presence of sperm competition (see below).

Male sperm stores. Ejaculated sperm do not always come directly from the testes. Males of most species have sperm reservoirs on which they draw during copulation. In mammals, the epididymis is the male's store; in insects, it is the seminal vesicle, and in passerine birds, the seminal glomerus. Not surprisingly, those species with more intense sperm competition also tend to have the largest sperm stores.

Accessory glands. The accessory glands produce the seminal fluid in which the sperm are transported to the female and her ova. Among some animals, species with more intense sperm competition tend to have relatively larger accessory glands than those where sperm competition is rare or absent. In certain taxa, the accessory glands produce seminal substances that set hard to form a copulatory plug, a device that can impede further insemination by other males. Plugs occur in insects, crustaceans, reptiles, marsupials, and mammals. In the ghost crab, the male transfers seminal fluid ahead of his sperm, and once inside the female's sperm it sets hard, effectively sealing off any rival's sperm and preventing it from reaching the female's ova. In the fruit fly *Drosophila*, the seminal fluid contains a cocktail of substances that increase male reproductive success. Some seminal products increase the female's oviposition rate, and other products actually poison any previously inseminated sperm—a very effective form of paternity defense. However, the substances in the male fly's seminal fluid that kill rival sperm are also toxic to females, and the more often females copulate, the sooner they die—a clear case of sexual conflict. From the male's perspective, the increased female mortality is not important, as long as she produces some eggs fertilized by his sperm; and because the seminal products increase the female's oviposition rate, this is likely to happen. *Drosophila* has proved a model system for examining the evolution of traits associated with sexual conflict. When males were allowed to evolve (in the laboratory, over several generations) in a situation where sperm competition was intense, their seminal fluid became increasingly toxic. Conversely, when males were maintained with the same female and forced to be sexually monogamous (so it became disadvantageous for males to harm females), their seminal fluid became much less toxic over just a few generations. These elegant experiments (Rice, 1996; Holland and Rice, 1999) demonstrate how quickly different selection pressures can shape traits associated with sperm competition.

Penis. Among primate species, the more complex the male's penis (i.e., whether it contains a baculum, or a stiffening bone, and/or whether it bears spiny appendages), the more likely it is that females routinely copulate with more than one male, and hence that sperm competition is important. Exactly the same pattern exists among insects, renowned for their structurally complex genitalia.

Sperm. Sperm show remarkable diversity in size and structure. They range from the tailless amoebalike sperm of nematode worms, through the classic tadpole-like sperm of most mammals (varying in length from 28 μm to 349 μm), to the corkscrew-shaped sperm of songbirds (30 μm to 300 μm in length), to the giant sperm, 6 cm long, in certain fruit fly species. The function of giant sperm in fruit flies remains obscure. In other taxa, longer or larger sperm tend to be associated with more intense sperm competition, for example in birds (Briskie et al., 1997). However, the way sperm length confers an evolutionary advantage in sperm competition among birds is unclear: there is no evidence that longer sperm swim any faster than shorter ones. In *Caenorhabditis* nematodes, sperm move by means of pseudopodia; larger sperm move faster than small ones and can place themselves in the best position for fertilization.

Mechanisms of Sperm Competition. The rules that determine which males sperm will fertilize a female's ova in a sperm competition scenario have been elucidated for only a few species. The mechanism of sperm competition is best known for some insects, birds, and mammals.

Insects. Sperm competition in most insects has been studied in captivity by allowing two males to inseminate the same female sequentially and then determining the paternity of the subsequent offspring. With this experimental protocol, the second or last male to inseminate the female fertilizes most eggs, a pattern referred to as *last male sperm precedence* (see Simmons and Siva-Jothy, 1998). Understanding the way last male sperm precedence arises provides a way of understanding the mechanism of sperm competition. Given the diversity of insects (c.750,000 species), it is not surprising that the mechanisms of sperm competition are also diverse. *Drosophila's* toxic seminal fluid has already been discussed, and the damselfly's specially modified penis is described in the Vignette. In the rove beetle (*Aleochara curtula*), the male transfers his sperm to a female in a package, or spermatophore. Once inside the female and after the male and female have separated, the spermatophore swells up like a balloon, squeezing out any previously stored sperm from the female's sperm store. At its maximum size, the spermatophore is punctured by "teeth" inside the sperm store, which releases the sperm, which then fertilize most of the female's eggs. In the yellow dungfly (*Scatophaga stercoraria*), last male sperm precedence is achieved in the following way. A male inseminating a virgin female places his sperm in a special bag-like structure, the bursa copulatrix, that is connected to her sperm store (spermatheca) by a long, narrow duct. A pistonlike structure in the female tract pumps the sperm up from the bursa into the spermatheca. If the female then copulates with another male, the internal processes are repeated, but now the piston blows back most of the original sperm before taking up those from the second male (see Vignette on Butterfly Sperm Competition).

Birds. As in insects, last male sperm precedence appears to be the rule when female birds are sequentially inseminated by two males. The processes that generate this pattern have been elucidated for the zebra finch *Taeniopygia guttata* (a passerine) and the domestic fowl *Gallus gallus* (a nonpasserine), and are basically the same. Following insemination, a small proportion of sperm (c.1–2 percent of the ejaculate) enter the female's sperm storage tubules (located at the utero-vaginal junction)—the rest are lost from the female tract when the female defecates. Over the next 10–20 days, sperm are

THE DAMSELFLY PENIS

The penis of the damselfly *Calopteryx maculata* is designed to remove the sperm of rival males from the female's sperm store. Its effectiveness is demonstrated by the fact that when a female copulates with two or more males, the last male typically fertilizes over 90 percent of her eggs. *Calopteryx* copulate for about 90 seconds, during which the male makes distinctive pumping movements. By separating copulating couples after different intervals and examining the tip of the penis, which is covered with tiny hooks and horns, Waage (1979), found that the male's pumping action drags out the sperm from previous males before transferring his own.

—Timothy R. Birkhead

BUTTERFLY SPERM COMPETITION

Butterflies produce two types of sperm: normal sperm with a nucleus (referred to as eupyrene) and non-nucleated sperm (apyrene), which are incapable of fertilization. Why any organism should produce and inseminate nonfertilizing sperm is a mystery, and three hypotheses were proposed. Nonfertilizing sperm may: (1) help to transport their fertilizing counterparts, (2) provide nourishment for fertilizing sperm, or (3) serve a role in sperm competition, either by reducing the effectiveness of sperm from rival males, or by being be a cheap way of filling the female's sperm store and discouraging her from remating, and thereby reducing the likelihood of sperm competition. Experiments on the small white butterfly *Pieris napi* confirmed that the role of nonfertilizing sperm was to fool females into assuming their sperm stores were full and delaying remating (Cook and Wedell, 1998).

—TIMOTHY R. BIRKHEAD

released at a constant rate from the tubules and travel up the oviduct to the infundibulum, where fertilization takes place: each ovum is fertilized separately at 24-hour (or longer) intervals. The number of sperm stored in the female's tubules and available for fertilization therefore declines over time. If a second male inseminates the female with the same number of sperm, his sperm follow the same pattern, but at any point in time those from the second male outnumber those from the first—hence the last male precedence. The longer the interval between two inseminations, the greater is the last male effect. This basic mechanism of sperm competition in birds can be modified by at least three other factors: differences in sperm numbers, differences in sperm quality, and cryptic female choice (see Birkhead et al., 1999; Colegrave et al., 1995; Pizzari and Birkhead, 2000).

Mammals. There are no consistent mating order effects in mammals. If two or more males inseminate the same female (and assuming sperm numbers and sperm quality to be identical), the basic mechanism of sperm competition is an interaction between (1) timing of insemination by the male, (2) timing of ovulation, and (3) the time it takes for inseminated sperm to prepare themselves for fertilization (referred to as to capacitation) (Ginsberg and Huck, 1989). The main difference in the mechanism of sperm competition between mammals and birds and insects is that (with a few exceptions, such as bats) mammals do not usually store sperm for protracted periods. However, just as with birds, this basic mechanism of sperm competition is modified by sperm numbers and sperm quality: animal breeders have known for long time that even when equal numbers of sperm are mixed and inseminated together, some males are disproportionately and consistently more successful than others at fertilizing ova (Dziuk 1996).

[*See also* Cryptic Female Choice; Male-Male Competition; Mate Choice, *article on* Human Mate Choice; Sexual Selection, *article on* Bowerbirds.]

BIBLIOGRAPHY

Birkhead, T. R. *Promiscuity.* London, 2000.

Birkhead, T. R., J. G. Martinez, T. Burke, and D. P. Froman. "Sperm Mobility Determines the Outcome of Sperm Competition in the Domestic Fowl." *Proceedings of the Royal Society of London. Series B: Biological Sciences* 266 (1999): 1759–1764.

Briskie, J. V., R. Montgomerie, and T. R. Birkhead. "The Evolution of Sperm Size in Birds." *Evolution* 51 (1997): 937–945.

Colegrave, N., T. R. Birkhead, and C. M. Lessells. "Sperm Precedence in Zebra Finches Does Not Require Special Mechanisms of Sperm Competition. *Proceedings of the Royal Society of London. Series B: Biological Sciences* 259 (1995): 223–228.

Cook, P. A., and N. Wedell. "Non-fertile Sperm Delay Female Remating." *Nature* 397 (1998): 486.

Dziuk, P. J. "Factors That Influence the Proportion of Offspring Sired by a Male Following Heterospermic Insemination." *Animal Reproduction Science* 43 (1996): 65–88.

Ginsberg, J. R., and U. W. Huck. "Sperm Competition in Mammals." *Trends in Ecology and Evolution* 4 (1989): 74–79.

Holland, B., and W. R. Rice. "Experimental Removal of Sexual Selection Reverses Intersexual Antagonistic Coevolution and Removes Reproductive Load." *Proceedings of the National Academy of Sciences of the United States of America* 96 (1999): 5083–5088.

Parker, G. A. "Sperm Competition and the Evolution of Ejaculates: Towards a Theory Base." In *Sperm Competition and Sexual Selection*, edited by T. R. Birkhead, and A. P. Møller, pp. 3–54. London, 1998.

Pizzari, T., and T. R. Birkhead. "Female Fowl Eject Sperm of Subdominant Males." *Nature* 405 (2000): 787–789.

Rice, W. R. "Sexually Antagonistic Male Adaptation Triggered by Experimental Arrest of Female Evolution." *Nature* 381 (1996): 232–243.

Simmons, L. W., and M. T. Siva-Jothy. "Sperm Competition in Insects: Mechanisms and the Potential for Selection." In *Sperm Competition and Sexual Selection*, edited by T. R. and A. P. Birkhead, pp. 341–434. London, 1998.

Waage, J. K. "Dual Function of the Damselfly Penis: Sperm Removal and Transfer." *Science* 203 (1979): 916–918.

— TIMOTHY R. BIRKHEAD

SPIDERS. *See* Arthropods.

STRATIGRAPHY

The basic principles of understanding earth history and the rock record are known as *stratigraphy*, literally "the study of layered rocks." Although stratigraphy originally developed as a method of describing and interpreting layered sedimentary rocks (rocks such as sandstone and

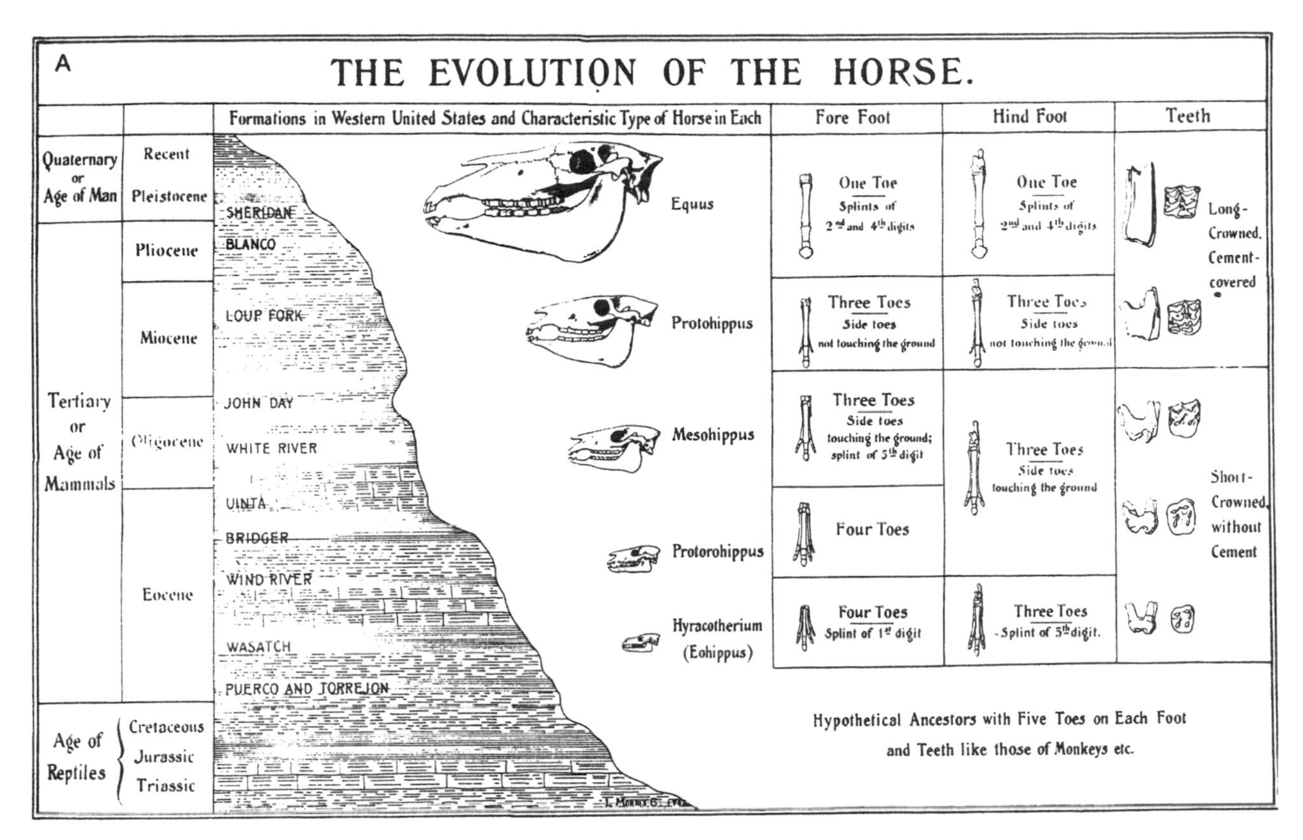

FIGURE 1. Early Representation of Horse Evolution Emphasizing the Linear Transformation in Form Through Time. The old names of the fossil-bearing beds are shown on the left, in stratigraphic order, demonstrating exactly where this sequence of horse fossils can be found. Courtesy the Library, American Museum of Natural History.

limestone formed by or from preexisting deposits), by the late twentieth century its scope had broadened to include layered igneous rocks (rocks cooled from magmas such as lava flows and volcanic ash falls) and even intrusive igneous rocks—any rocks that demonstrate a sequence of geologic events. Based on stratigraphic principles, geologists have been able to reconstruct the past 4.5 billion years of earth history in surprising detail.

There are two primary means of determining the age of events in the geological past. A geologist can determine the relative sequence of events (event A is younger or older than event B), or the numerical age (i.e., this rock is so many millions of years old) of a rock unit, the latter primarily by radioisotopic dating of igneous rocks. (Numerical dating is erroneously called "absolute dating" in older books.)

The basic principles of relative dating were first proposed by Nicholaus Steno in 1669. In Steno's time, most scholars thought of the rocks of the earth's crust as having been created exactly as they then appeared, some 6,000 years previous. They were puzzled by the occurrence of fossils in solid rocks, and thought that the fossils might have grown in the rocks by supernatural forces, or been stranded there at the time of Noah's Flood. Steno realized that the presence of fossils in sedimentary rocks showed that these rocks had not always been solid, but were once composed of loose sedimentary materials (sand, mud, lime) that consolidated around the fossils and later turned into stone. From this insight, Steno derived several principles of relative dating that are the foundation of stratigraphy. The most important of these is the principle of superposition. If sedimentary rocks are deposited as layers of sand or mud, one on top of the other, then the layers on the top of the stack will be younger than those toward the bottom. This is analogous to a stack of papers on a desk that have accumulated for a long time without being shuffled. Those at the bottom of the stack were left there some time ago, while those at the top were placed there most recently.

Dating Methods. The fundamental method of determining the age of geological events is relative dating, using the principle of superposition. This was the way that geologists worked out the sequence of strata in Europe that is the basis of our modern timescale. However, at the same time, they found that many rock units were similar in appearance and hard to distinguish from one another; others change their content (for example, from sandstone to shale) over distance. In the late 1700s the

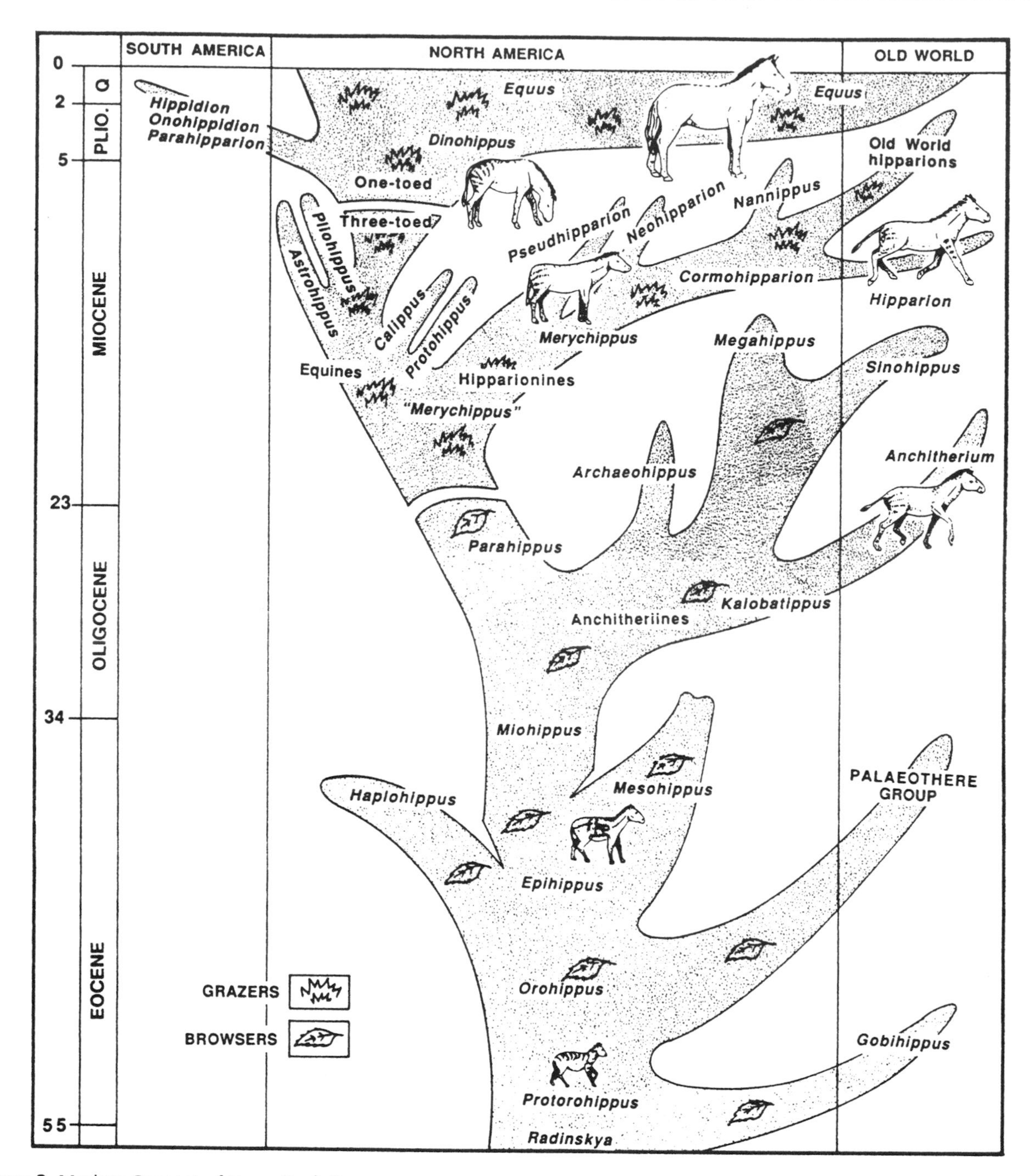

FIGURE 2. Modern Concept of Horse Evolution.
Shown are much more complex, bushy, branching pattern as we discover more horse fossils. Donald Prothero.

English canal engineer William Smith realized that each rock unit in his English succession contained its own distinctive assemblage of fossils, and that each rock unit could also be recognized by its fossil content. The fact that fossils change continuously through time is known as faunal succession and is the basis for biostratigraphy, the scientific study of the distribution of fossils in rock sequences. Biostratigraphy is the only practical means of dating most sedimentary rocks. Rock types may change over distance, or become difficult to distinguish, but distinctive assemblages of fossils are unique to certain segments of geologic time and can be used around the world to recognize that time interval. Most paleontologists employed by oil companies are biostratigraphers, using fossils to date and correlate rocks as precisely as possible.

When the timescale developed in the early 1800s, no one could tell the age and duration of the different geological periods. Some geologists thought that the age of the earth was nearly infinite (Hutton noted that there was "no vestige of a beginning"), but other scientists gave the earth only 20 million years or less for its entire history. The discovery of radioactivity in 1896, however, provided the first method of obtaining a numerical age for geological events. When a radioactive atom, such as uranium or rubidium, spontaneously decays by nuclear reactions, it gives off heat, radioactive particles, and leaves a nonradioactive daughter atom. The rate of this nuclear reaction is well known, and half of the original parent atoms decay to their daughter atoms in a fixed interval of time, known as a half-life. If we can measure the ratio of parent to daughter atoms, we can determine when this atomic decay reaction began and date the material that contains these atoms. However, this system only works for minerals that have cooled down from a very hot state (igneous or high-grade metamorphic minerals), locking in the parent atoms at the time of their crystallization. Thus, it is inapplicable to normal sedimentary rocks, which are formed from preexisting grains eroded from other rocks and not from a molten state. To determine the age of sedimentary rocks, the geologist needs interbedded igneous rocks (such as ash falls or lava flows), which give numerical ages in certain parts of the sedimentary sequence, or cross-cutting igneous dikes, which bracket the age of the rocks through which they cut and which cut across them.

Stratigraphy and Evolution. Although evolutionary biologists have been able to determine much about the process of evolution from studying living organisms, the fossil record is the only direct evidence for how evolution actually occurred. In the early days of stratigraphy, geologists attempted to explain the change in fossils through the rock strata as the result of creatures killed by successive floods (not mentioned in the Bible). But by the 1830s, so many different extinct faunas had been described that it was no longer possible to explain the change in fossils through time by biblical stories.

In 1859 Charles Darwin published his theory of evolution by natural selection, and revolutionized biology. Ironically, Darwin's book had relatively few examples from the fossil record to support his theory of evolution, and his chapters on the subject were largely apologies for the incompleteness of the fossil record. But in 1861 *Archaeopteryx*, the transitional fossil between reptiles and birds, was described, and paleontologists soon began to amass more and more examples of evolutionary transformations in the fossil record. In the 1870s Kowalevsky and Marsh described the evolutionary history of the horse (Figure 1), which remains a classic example of evolution to this day. The oldest horses are primitive, tiny four-toed beasts with low-crowned teeth from rocks 55 million years old. Larger animals with three toes and longer legs are found in 30-million-year-old strata, while modern horses have long one-toed legs and very high-crowned teeth. Although the general outline of horse evolution is still valid, in the twentieth century many more horse fossils were found that made the picture much more complex. Instead of a single lineage getting larger and more advanced, we now know that horse evolution was highly branched, with multiple lineages of horses living at any given time (Figure 2). During the Miocene epoch, horses became even more diverse, with some lineages retaining the old low-crowned teeth for eating leaves, while many others evolved higher-crowned teeth for eating gritty grasses. Even the loss of side toes occurred in several different lineages, and one-toed horses evolved at least twice. About 10 million years ago, there were at least twelve different lineages of horses living at the same time. At the end of the last ice age, many of the different lineages of horses became extinct, so that only a single genus of horse (*Equus*) survives, divided into a number of species of zebras, asses, and wild and domestic horses.

[*See also* Geology; Paleontology.]

BIBLIOGRAPHY

Berry, W. B. N. *Growth of Prehistoric Time Scale.* 2d ed. Palo Alto, Calif., 1987. An excellent account of the historic development of the understanding of geologic time.

Dalrymple, G. B. *The Age of the Earth.* Palo Alto, Calif., 1991. A thorough account of the attempt to date the earth, and how modern methods of radioisotopic dating work.

Eicher, D. L. *Geologic Time.* 2d ed. Englewood Cliffs, N.J., 1976. A good introductory-level paperback text on geologic time and stratigraphic principles.

Prothero, D. R. *Interpreting the Stratigraphic Record.* New York, 1990. A college-level textbook on the principles of stratigraphy.

Prothero, D. R., and R. H. Dott, Jr. *Evolution of the Earth.* 6th ed. New York, 2001. A college textbook in historical geology, with chapters on evolution and stratigraphic principles.

Schoch, R. M. *Stratigraphy: Principles and Methods.* New York, 1989. A more advanced college-level textbook on the principles of stratigraphy.

— DONALD R. PROTHERO

SYMBIOSIS

Symbiosis is often defined as an association from which all participating organisms derive benefit. It is usually assumed that the partners in a symbiosis are members of different species and that the association is persistent, lasting for much or all the life span on the organisms. Fleeting interactions, such as the pollination of flowers, are nonsymbiotic mutualisms.

Although apparently simple, this definition of *symbiosis* can be difficult to apply to many real associations. The principal problems are twofold. First, some organisms benefit from an association only under certain environmental circumstances. This can be illustrated by the symbiosis between plant roots and mycorrhizal fungi, in which the fungal partner enhances the mineral nutrition of the plant and derives photosynthetic carbon from the plant. The nutritional benefit to the plant is evident only in low-nutrient soils, and plant seedlings may even grow more slowly when bearing mycorrhizal fungi because the fungi consume much of the plant's photosynthetic carbon. Should mycorrhizas be described as symbioses, even though the plant may be harmed by the association under certain conditions? The second problem is that it is often not straightforward to demonstrate whether an organism benefits from an association. As an example, lichens are a symbiosis between algae and fungi, but the algae in lichens are rare or unknown in the free-living condition (i.e., apart from the lichenized fungi in nature), indicating that they are adapted to the lifestyle in lichens. However, they derive no known benefit from the association (in particular, nutrient transfer from the lichen fungus to the algal cells is undetectable). Should lichens be considered as symbioses in which the nature of the benefits have not yet been characterized fully, or as parasitic associations on which the algal "victims" are apparently dependent?

As a result of these difficulties, many researchers prefer the meaning of *symbiosis* as originally coined by Anton de Bary in 1879 the living together of differently named organism. This definition makes no reference to the significance (benefit or harm) of the association to the participating organisms, and de Bary explicitly included parasitic and pathogenic associations as examples of symbioses. Although de Bary's definition is widely used in the symbiosis literature today, parasites and pathogens are rarely considered as examples of symbiotic organisms. Following this convention, this article does not address overtly parasitic associations.

Types of Symbioses. Table 1 provides a survey of the best-known symbioses. From a functional perspective, symbioses are characterized by one to all of the participating organisms gaining access to a capability or property of their partner(s). Most capabilities acquired through symbiosis can be categorized as nutrient acquisition, enhanced metabolic function, or protection from predators or pathogens. A few symbioses do not fit neatly into these categories. For example, the flashlight fish (Monocentridae) use the light produced by symbiotic luminescent bacteria in light organs located just below their eyes in intraspecific signaling, including courtship displays; and the trichomonad protist *Mixotricha paradoxa* acquires mobility from the "coat" of motile spirochetes on its surface.

Table 1 illustrates three general points about symbioses. First, one organism may derive multiple benefits from a symbiosis. As examples, the dinoflagellate algae in corals provide their host with photosynthetic carbon (usually in the form of glycerol), recycle nitrogen, and promote calcification; and mycorrhizal fungi may promote both the mineral nutrition and pathogen resistance of plants. Second, the partners in one symbiosis may obtain different types of benefit. For example, the caterpillars of lycaenid butterflies are protected from predators by tending ants, which derive sugars and amino acids from the glandular secretions of the caterpillar. Finally, and as considered above, we are ignorant of the benefit derived by some organisms in symbiosis, for example, algae and cyanobacteria in lichens, luminescent bacteria in fish, and spirochetes associated with *Mixotricha paradoxa*.

Evolutionary Origins of Symbiosis. Mutualistic symbioses can evolve from antagonistic relationships. This is illustrated by research on the association between the protist *Amoeba proteus* and certain bacteria. Initially, the bacteria harmed their host; infected amoebae were smaller and grew more slowly than uninfected amoebae. However, over a period of about two hundred host generations (eighteen months), the detrimental effect of the bacteria on the amoebae declined and, at the end of the period, the amoebae were found to be dependent on the bacteria.

How frequently have symbioses evolved from antagonistic interactions? We have very little direct evidence, but certain symbionts are closely related to, and may have evolved from, pathogens. For example, the endophytic fungi in grasses are so closely allied to the pathogen *Epichloë* that the two taxa can be sexually compatible; the nitrogen-fixing symbiont *Rhizobium* in legumes is closely related to plant pathogenic bacteria of the genus *Agrobacterium*.

Although the notion that symbioses may commonly have evolved from antagonistic relationships is widely accepted, there are indications that many symbioses probably evolved from casual associations. Many of the organisms in symbiosis are not related to pathogens or parasites. For example, neither the bacteria and yeasts absolutely required by various insects (e.g., lice, aphids, timber beetles) nor the algae in animals, protists, and lichenized fungi are allied to pathogens. In summary, many symbioses have probably evolved from casual relationships through selection pressures on the partners to enhance the advantage derived from the association. The evolutionary transition from antagonistic to symbiotic relationships may occur relatively infrequently.

Evolutionary Trends in Symbiosis. Organisms in symbiosis are subject to evolutionary change, commonly in response to the selection pressures associated with the symbiosis. This section addresses three types of evolutionary change that are evident in a range of symbioses: morphological novelty, especially in plants or animals, that enhance the benefit derived from the symbiosis; loss of redundant capabilities; and deterioration

Table 1. A Survey of Symbioses

Association	*Example*	*Nature of benefit*
A. *Plant Hosts*		
Mycorrhizal fungi	Arbuscular mycorrhizal fungi, widespread among all plants. Ectomycorrhizal fungi in many woody perennials. Ericoid mycorrhizal fungi in Ericaceae (e.g., heathers) Orchid mycorrhizal fungi in orchids.	Plants derive nutrients, especially phosphorus, and protection from root pathogens. Fungi derive photosynthetic carbon from plant (except orchid mycorrhizal fungi, which provide C to the plant).
Nitrogen-fixing bacteria	Rhizobia in legumes. *Frankia* (an actinomycete) in various woody dicots. Cyanobacteria in several genera (e.g.,*Gunnera*, water fern *Azolla*).	Plants derive fixed nitrogen from bacteria. Bacteria derive photosynthetic carbon from plant.
Endophytic fungi	Fungi (allied to *Epichloë*) in shoots of various grass species.	Secondary compounds in fungi protect plants from shoot herbivores. Fungi derive photosynthetic carbon from plants.
Ants	Ants associated with various plants (e.g., *Acacia*, *Cecropia*).	Ants protect plants from herbivores. Ants obtain nest site and food (e.g., secretions of extrafloral nectaries, proteinaceous bodies) from plants.
	Ants nesting in epiphytes (e.g., *Myrmecodium*)	Ants obtain protected nest site in hollow plant organ. Plants obtain nutrients absorbed from ant nest site.
B. *Animal Hosts*		
Photosynthetic algae	Dinoflagellate *Symbiodinium* in marine animals (e.g., corals). Chlorella in freshwater animals.	Animals derive photosynthetic carbon and enhanced nitrogen utilization efficiency.
Chemosynthetic bacteria	Bacteria in marine animals, especially deep-sea Pogonophora (e.g., *Riftia*) and some bivalve molluscs.	Animals derive fixed carbon from the bacteria.
Bacteria and yeasts in insects	Various taxa in insects feeding on vertebrate blood, phloem, sap, or wood.	Animals derive essential amino acids, vitamins, and/or sterols from the microorganisms.
Cellulose-degrading microorganisms	Bacteria (e.g., *Ruminococcus*, *Bacteroides* spp.) in most herbivorous mammals. Protists (hypermastigotes and trichomonads) in some termites.	Microorganisms degrade cellulose in ingested plant material to compounds (short chain fatty acids) that can be used by the animal as an energy source.
Luminescent bacteria	*Vibrio/Photobacterium* in some marine teleost fish and squid	Animals use bacterial light for camouflage, as an alarm signal, or for intraspecific signaling
"Tending" ants	Caterpillars of lycaenid butterflies and homopteran insects (e.g., coccids, aphids) with many ant species.	Ants derive food from sugary secretions of lycaenids and egesta of Homoptera, which are protected from natural enemies by the tending ants.
C. *Fungal Hosts*		
Lichens	Mostly ascomycete fungi with algae (e.g., *Trebouxia*) and cyanobacteria (e.g., *Nostoc*).	Lichenized fungi derive photosynthetic carbon from algae and cyanobacteria and fixed nitrogen from some cyanobacteria.
D. *Protist Hosts*		
Methanogenic bacteria	Methanogens associated with anaerobic ciliates (e.g., *Metopus* spp.).	Anaerobic respiration by protests promoted by methanogen-mediated removal of hydrogen. Hydrogen derived by methanogen is substrate of ATP-yielding methanogenesis.
Nitrogen-fixing bacteria	Cyanobacteria (*Richelia intracellularis*) associated with diatoms (e.g., *Hemiaulus*, *Rhizosolenia* spp.).	Diatoms derive fixed nitrogen from cyanobacteria.
Motility symbioses	Spirochaetes and large protists.	Protists derive mobility from spirochetes attached to surface.

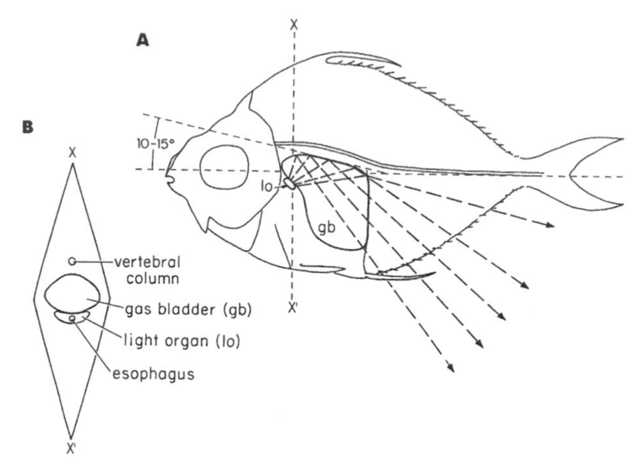

FIGURE 1. The Light Organ in Leiognathid Fish *Leiognathus equulus*.
(A) Diagrammatic vertical section showing the relationship of the light organ (lo) and gas bladder (gb), and the path of light (dashed lines and arrows) emitted from the light organ. (B) Cross section of (A) at X-X′ showing relative position of light organ and gas bladder. McFall-Ngai (1983) *American Zoologist*. © 1983. Reprinted by permission of Wiley-Liss, Inc., a subsidiary of John Wiley & Sons, Inc.

of the symbiosis, potentially resulting in the evolutionary transition of a symbiosis to parasitism.

Morphological novelty. A classic example of morphological novelty is provided by some fish and squid that form symbioses with luminescent bacteria. The bacteria are housed at very high densities in light organs. Under these conditions, they generate light continuously but at rather low intensity. Various anatomical modifications have evolved in the animal hosts, such that the amount and direction of light emitted from the animals is controlled by lenses, mirrors, shutters, and so on. For example, leiognathid fish use luminescence in communication and as camouflage by counterillumination. Counterillumination refers to ventrally directed diffuse illumination that breaks the silhouette of the fish such that a predator lower in the water column cannot see the fish against down-welling light. The emission of light produced continuously from the light organ surrounding the fish oesophagus is controlled by muscles that act as shutters. Counterillumination is achieved by, first, reflection of posteriorly directed light off the dorsal surface of the fish gas bladder lined with highly reflective purine crystals, and, second, passage through the ventral body wall musculature to generate diffuse illumination (Figure 1). In other animal hosts, notably the Hawaiian squid, specific muscles have become highly modified to act as a lens that directs and focuses the emitted light.

Many ant-tended plants display morphological innovations that attract ants to the plant and promote ant "patrolling" over their shoot system, such that the ants are likely to encounter and either kill or deter any herbivores. Shrubs and trees of the genus *Acacia* in Africa and South America provide the ants with dry, dark nest sites (known as domatia)—hollow stems in the African species and swollen, hollow thorns in the American species. The function of ants as bodyguards for the plant is promoted by plant provision of small "parcels" of food at various locations around the shoot system, particularly the vulnerable leaves. Two types of food are provided: sugary secretions from extrafloral nectaries, which are glandular structures anatomically similar to the nectaries in flowers, and Beltian bodies at the tip of individual leaflets, which provide ants with protein and lipid (Figure 2A). Similar morphological innovations—domatia, extrafloral nectaries, and solid nutrient-rich bodies—have evolved in other ant-tended plants. For example, many *Cercropia* species have domatia and nutrient-rich bodies known as Müllerian bodies (Figure 2B), a vivid example of convergent evolution.

Loss of redundant capabilities. Organisms living in symbioses may lack structures or capabilities possessed by related nonsymbiotic taxa. Some features are not required in symbiosis because they are only advantageous to the organism in isolation or because the partner provides the function. The redundant features may have been selected against (e.g., they may consume limiting resources), or they may have been lost neutrally, through the absence of selection pressure for their retention.

The ant-tended *Acacia* species provide an example of loss of redundant function. Many plants contain allelochemicals, that is, secondary compounds whose primary function is to deter natural enemies including herbivores. The shoots of Australian *Acacia* species have a rich array of allelochemicals with antiherbivore function, but the ant-tended *Acacia* species in Africa and South America (see above) are, by comparison, deficient in allelochemicals, presumably linked to the effectiveness of ant-mediated protection.

Loss of function is also evident in symbiotic microorganisms. For example, the symbiotic bacteria *Buchnera* in aphids have a very small genome (0.6–0.7 Mb, about 20 percent of the genome size of the related bacterium *Escherichia coli*). In 2000, Shigenobu and colleagues published the complete genome sequence of *Buchnera* from pea aphids. The gene inventory of this sequence is remarkable for the absence of many genes in free-living bacteria. *Buchnera* has lost the genetic capacity for complete metabolic pathways, including the TCA cycle, glucogenesis, phospholipid synthesis, and lipopolysaccharide production. Other missing genes illustrate the complementarity of *Buchnera* and aphid function. In particular, the amino acid biosynthetic capability of *Buchnera* almost exactly complements the capability

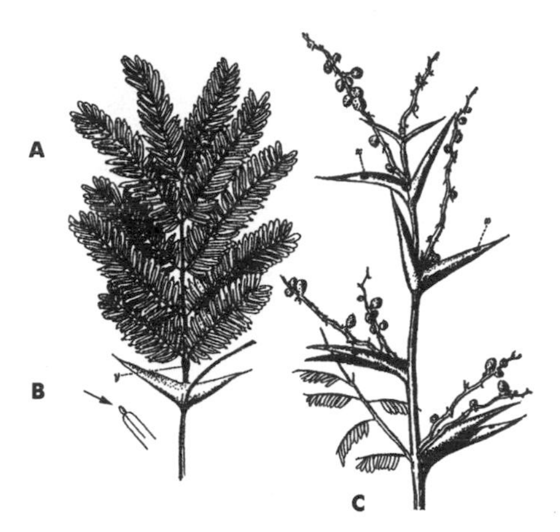

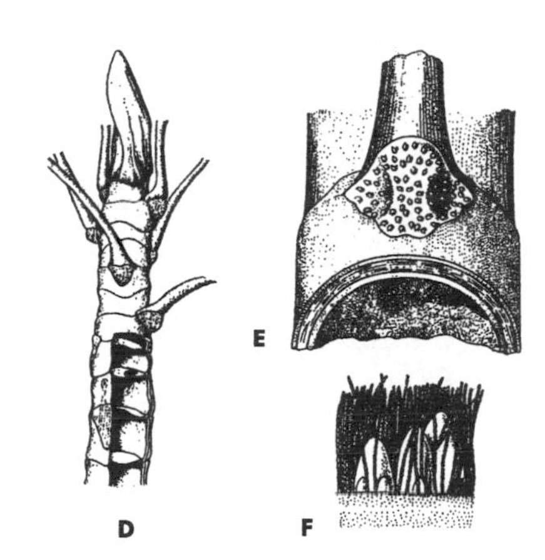

FIGURE 2. Mophological Features of Ant-Tended Plants.
Acacia sphaerocephala (America bull's-horn acacia): (A), Leaf with extrafloral nectary (y). (B) Tip of leaflet enlarged to show Beltian body (arrow). (C) End of a branch, showing pairs of hollow thorns (domatia) occupied by ants with entrance holes at x. *Cecropia adenopus.* (D) The growing tip of a young tree, cut away to show the hollow internodes (domatia) occupied by ants. (E) Müllerian bodies located at the base of the petiole. (F) A magnified view of Müllerian bodies. Reproduced from Hoelldobler and Wilson. *The Ants.* Springer, 1990.

of the aphid, such that the two partners together can synthesize virtually all amino acids that contribute to protein. *Buchnera* retention of the capacity to synthesize just those amino acids that the aphid cannot make is crucial to the capacity of the aphid to utilize its diet of plant phloem sap, which is deficient in amino acids that the aphid cannot synthesize. As a result of the large-scale gene loss, it is inconceivable that *Buchnera* could persist indefinitely in isolation. Furthermore, because *Buchnera* is vertically transmitted, its fitness depends utterly on host fitness. This symbiotic bacterium can be considered as a highly specialized mutualist.

Deterioration of the symbiosis. Organisms vary in their effectiveness in symbiosis. In other words, some symbiotic organisms are of greater selective value than others to their partner. The extreme of this variation in effectiveness is cheats, organisms that can form a symbiosis and derive advantage from their partner(s), but that provide no benefit. For example, the nitrogen-fixing fungal symbionts in the actinomycete genus *Frankia* associated with roots of alder (*Alnus* sp.) trees include strains that colonize the alder host but do not fix nitrogen; and the naturally occurring algae in symbiosis with an intertidal flatworm *Convoluta roscoffensis* include one species of the genus *Prasinocladia* that releases very little photosynthetic carbon to its host worms, which (presumably as a consequence) grow slowly and produce few eggs.

It has been argued that symbioses involving bacteria that are invariably transmitted vertically are particularly likely to deteriorate genetically. Where each host individual invariably derives very small numbers of asexually reproducing symbionts exclusively from its mother, symbionts bearing slightly deleterious mutations may come to dominate, or even be fixed, by chance. Moran (1996) obtained good molecular evidence for such genetic deterioration in the vertically transmitted bacteria including symbionts in insects (Figure 3), but the impact of this process on the symbiotic effectiveness of these bacteria is uncertain.

Although cheats are evident in some types of symbioses, there are very few instances of parasitic taxa that have evolved from organisms in symbiosis. The lycaenid butterflies provide one example. The caterpillars of most lycaenids are ant-tended, deriving protection from predators for the "price" of sugary secretions on which the ants feed. A minority of lycaenids, however, are parasites of their ant tenders. The young caterpillar of the large blue butterfly *Maculinea arion* feeds on thyme plants and is ant tended, but at a certain developmental stage, the caterpillar drops from the plant to the ground and, when contacted by an ant, is taken back to the ant nest. The ants, apparently mistaking the caterpillar for an ant grub, tolerate the caterpillar, which consumes ant grub over many weeks in the nest before pupating and maturing to an adult in the following spring. A single caterpillar can be sufficiently voracious to destroy an entire colony of the ant partner.

It is not well understood why the evolutionary transition from mutualism to parasitism in symbioses appears to occur relatively rarely. Perhaps cheats that arise do not persist over evolutionary time because they are selected against, either by the partner, which may evolve ways to recognize and discriminate against cheats, or at the level of the symbiosis (associations with cheats are less fit than associations with cooperators).

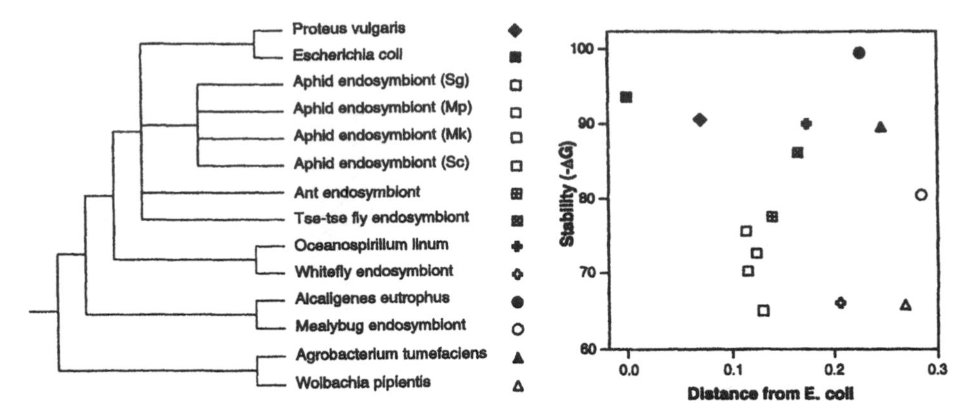

FIGURE 3. The Consequences of Vertical Transmission on the Molecular Evolution of Bacteria.
The graph (right) shows on the y axis the stability of a portion (Domain 1) of the 16S rRNA molecule, as determined from the deduced free energy ($-\Delta G$) of folding. The value of $-\Delta G$ is generally lower for vertically transmitted bacteria (open symbols) than for free-living bacteria (closed symbols). The vertically transmitted bacteria are not more closely related to each other than to free-living forms, as is shown by the phylogeny on the left and the range of distances from *E. coli*, measured as sequence difference from *E. coli*, on x axis of the graph. This indicates that the deleterious consequence of vertical transmission on rRNA stability has evolved independently multiple times. Reproduced from J. D. Lambert and N. A. Moran "Deleterious Mutations Destabilize Ribosomal RNA in Endosymbiotic Bacteria." *Proceedings of the National Academy of Sciences of the United States of America* 95: 4458–4462. © 1988 National Academy of Sciences, U.S.A.

Symbiosis-derived Organelles. It can be argued that without symbiosis, and particularly without symbiosis-derived organelles, the eukaryotes, the group that now contains all organisms with cell nuclei, would have been evolutionarily insignificant, predominantly unicells restricted to anoxic environments. This is because the lineage giving rise to eukaryotes lacked key metabolic capabilities, including aerobic respiration and photosynthesis. Eukaryotes have acquired these capabilities by forming symbioses with bacteria capable of aerobic respiration and photosynthesis, respectively. The bacteria evolved into mitochondria and plastids by large-scale gene loss and gene transfer to the eukaryotic nucleus. We can be confident that these events occurred because these organelles have retained coding DNA with sequence allied to alphaproteobacteria (mitochondria) and cyanobacteria (plastids) and not to any eukaryotic lineage.

Symbiont-derived organelles have been described by David Smith as comparable to the grin of the Cheshire Cat in the children's book *Alice in Wonderland*; through evolutionary time, they may just fade away. The genome of animal mitochondria is certainly very small. For example, the human mitochondrial genome is 17 kilobases and codes for two rRNA and twenty-two tRNA genes, maintained for the synthesis of just twelve protein-coding genes, and the transfer of these genes to the nucleus in a functional state is precluded by deviation of the human mitochondrial genetic code from the universal code used by nuclear genes. The genes coded by the mitochondrial genomes in some eukaryotes and all plastid genomes studied to date conform to the universal code.

Other organelles in eukaryotes, notably peroxisomes and microbodies, have characteristics suggestive of a symbiotic origin. Specifically, they persist by division and are not generated de novo by eukaryotic cells. It is entirely possible that the bacterial ancestors of these organelles have lost all their DNA, and the resultant organelles are just "the ghosts of symbionts past." It can be speculated that such genetic assimilation leaving no molecular evidence of previous independent identity is the ultimate fate of certain symbiotic microorganisms.

[*See also* Mutualism.]

BIBLIOGRAPHY

Corsaro, D., D. Vanditti, M. Pazdula, and M. Valassina. "Intracellular Life." *Critical Reviews in Microbiology* 25 (1999): 39–79. An excellent overview that compares symbiotic and parasitic microorganisms and summarizes current understanding of the evolution of bacterial-derived organelles.

Douglas, A. E. "Nutritional Interactions in Insect-Microbial Symbioses." *Annual Review of Entomology* 43 (1998): 17–37.

Herre, E. A., N. Knowlton, U. G. Mueller, and S. A. Rehner. "The Evolution of Mutualisms: Exploring the Paths between Conflict and Cooperation." *Trends in Ecology and Evolution* 14 (1999): 49–53.

Hoelldobler, B., and E. O. Wilson. *The Ants*. Berlin, 1990. An excellent overview of symbioses involving ants, including relationships with plants and other insects.

Honegger, R. "Functional Aspects of the Lichen Symbiosis." *Annual Reviews of Plant Physiology and Plant Molecular Biology* 42 (1991): 553–578.

Margulis, L. *Origin of Eukaryotic Cells*. New Haven, 1970. A classic book written in very accessible style, in which the author presents the theory that eukaryote cells comprise a symbiosis among different bacterial taxa.

Moran, N. A. "Accelerated Evolution and Muller's Ratchet in Endosymbiotic Bacteria." *Proceedings of the National Academy of Sciences USA* 93 (1996): 2873–2878. This paper, which requires basic knowledge of molecular evolution, addresses the molecular deterioration of vertically transmitted bacteria with small effective population sizes.

Paracer, S., and V. Ahmadjian. *Symbiosis: An Introduction to Biological Associations.* Oxford, 2000. An up-to-date introduction to symbioses that presupposes little biological knowledge.

Shigenobu, S., et al. "Genome Sequence of the Endocellular Bacterial Symbiont of Aphids *Buchnera* sp. APS." *Nature* 407 (2000): 81–86. Describes the gene content of *Buchnera*, the first symbiotic bacterium with a complete genome sequence.

Smith, S. E., and D. J. Read. *Mycorrhizal Symbiosis.* London, 1997.

— ANGELA E. DOUGLAS

SYSTEMATIC COLLECTIONS

Knowledge about the evolutionary history of organisms is obtained by studying variation at different levels of biological organization. Most often the ultimate source of information on variation is that obtained by examining specimens of individual organisms that are preserved in research collections at museums, herbaria, living collections, and frozen tissue ("genetics resources") collections. It is estimated that approximately 2.5 billion specimens are housed in natural history collections around the world (Duckworth et al., 1993). Although this number seems large, it pales in comparison to the number of specimens that would actually represent the earth's species diversity; likely a figure hundreds of times greater would be needed. Nonetheless, systematists and evolutionary biologists use natural history collections to learn about geographic variation, speciation, phylogeny, and biogeography. Systematics collections are also important for studies in other areas of inquiry: "Collections are the permanent record of our natural heritage, and thus contain the materials that support the research of many scientific disciplines, including those working to preserve biodiversity and monitor global change" (*Systematics Agenda 2000*, 2000).

Where and How Are Specimens Obtained? For many types of organisms, collecting specimens requires special permits, usually issued by both local and national authorities. Birds and butterflies are currently among the most regulated types of wildlife. For example, in the United States, permission to collect bird specimens is granted jointly by the U.S. Fish and Wildlife Service and by each state government. An application covers the reasons that specimens are needed and outlines any new knowledge that will result from the research. Only after reviewing an application for a scientific collecting permit is it issued. Thus, a scientist must have a valid reason for collecting specimens. These specimens must be deposited in a public institution so that they are available for study by other scientists (sometimes older specimens can be in private collections). Federal and state authorities also give "salvage" licenses to schools and nature centers, so that birds killed, for example, in accidents with cars or other objects can be preserved for display and educational purposes.

For other types of animals and plants, there are fewer regulations unless a species is known to be threatened or in danger of extinction. Scientists collect only enough specimens required for research projects. These are then permanently housed in research collections, where they can be used by other scientists now and in the future. There are very few cases in which a scientific collector has caused harm to a population or species by overcollection. However, often amateurs or those seeking to sell natural history specimens have damaged natural populations. Populations of some parrots, orchids, and butterflies come to mind. Before attempting any collection, consult with local and national authorities to determine what permits are required. All specimens, if properly labeled, are of value and should be donated to a public repository.

The type of organism under study dictates how the specimen is collected, preserved, and stored, then later studied by systematists. Only a few general examples will be given here. Vertebrates such as birds and mammals are collected in nature using nets, traps, and sometimes guns. Once collected, specimens are skinned (literally turned inside out) and stuffed with cotton. With only the skin and feathers (if a bird) or fur (mammal) surrounding the cotton, the specimen dries and does not decay, assuming it is kept at the correct temperature and humidity. Such specimens were collected as early as the beginning of the nineteenth century. The resultant "scientific study skin" does not look like specimens most museum visitors are familiar with in that they are not prepared in a lifelike pose. Scientific study does not require a lifelike taxidermic mount (such as the duck or deer you might see in a public museum or nature center display). In addition to skins, skeletal parts are often saved (and "cleaned" by using dermestid beetles), and tissue samples are preserved either in freezers or in a DNA preservative (which often does not need refrigeration). Thus, as much material as possible is preserved from each individual to furnish the most information.

Other vertebrates, such as fishes, amphibians, and reptiles, are collected by hand, nets, and traps and preserved in a fixative such as formalin, then transferred to alcohol for permanent storage. These are called fluid-preserved specimens. Although they might lose the colors they had in life, the other anatomical features are preserved. One could take a fluid-preserved snake and later dissect it to learn about its reproductive condition and often what it had eaten.

Insects, especially terrestrial ones, are collected with nets and traps, then killed in a jar of ethyl acetate or cyanide. After death, a special pin is inserted through the specimen along with various labels. Often immature specimens, such as caterpillars, are fluid-preserved because their characteristics are not adequately preserved by drying.

Other nonvertebrates, such as worms and spiders, are fluid-preserved. Preservation of some species, such as snails and clams, involves saving only the hard shell.

Plants are collected by hand, then usually "pressed" between sheets of paper (pressure is applied by wood or cardboard that is placed on either side). This causes the plant to dry, and the resultant specimen sheet is labeled. A collection of plants at an institution is referred to as a herbarium.

Unfortunately, most species are very small and not easily preserved. Examples are bacteria and viruses, forms of special interest because of their relationship to human health. Special methods are required to preserve samples of these types of organisms. The American Type Culture Collection (Manassas, Virginia) is a central collection of microorganisms (http://www.atcc.org). There are few if any permit regulations for these types of organisms.

Paleontological collections, which contain fossils, are one of the most important types of collections. The existence of extinct species, along with intermediates between them and living forms, was and continues to be one of the most forceful proofs of the process of organic evolution. Although fossils are often fragmentary, because of the artifacts of preservation, and are missing altogether for some types of organisms, they provide positive evidence of how species evolve over time.

A relatively new type of collection is that containing tissue samples (Dessauer et al., 1996). With the advent of molecular methods of studying evolution, researchers began to collect new specimens so that they could extract proteins and nucleic acids (DNA, RNA) from fresh tissue. Tissue samples often remained after study, and they are now treated like specimens in traditional museums. They can be loaned and are often separately cataloged. Tissue samples usually are frozen, because freezing, especially at ultracold temperatures (−70°C; termed cryopreservation), preserves proteins and DNA so that they can be recovered structurally intact and physiologically active. Today, for preservation of DNA, tissues are preserved in room-temperature buffers or even ethanol; other uses still require cryopreservation.

The Nature of Specimens. Typically, specimens are labeled to describe where and in what habitat they were obtained, the date of collection, sex, soft part colors (which often fade in preserved or dried specimens), mass, description of reproductive condition, stomach contents, and other information deemed pertinent by the collector. In some cases, each individual in a collection receives a separate number in a catalog. For some types of organisms, specimens are cataloged in lots; for example, a container with many small fish might receive a number. This information, once verified and entered into a database, allows researchers to search collections electronically from remote locations.

Specimens in a scientific collection are like rare books in a library: they provide a permanent record of biodiversity. Photographs or field notes are often inadequate on their own and may not anticipate future questions. With new technologies, specimens in research collections are providing information that the original collections never imagined. For example, it is now routine to obtain DNA from bits of dried skin or from alcohol-preserved specimens (e.g., fish, reptiles, and amphibians). For example, researchers have deduced the nearest living relatives of several recently extinct species such as the quagga (Higuichi et al., 1984) by extracting DNA from dried specimens and comparing the DNA sequences to related forms that are not extinct. Thus, collections are repositories of many different kinds of information that is preserved in specimens.

Information from specimens. Once a specimen is preserved in a research collection, it provides numerous types of evidence bearing on evolutionary topics. The first and most obvious kind of evidence concerns the morphology or anatomy of the organism. The size and shape of external and internal anatomical features can be quantified and summarized for populations or species, thereby allowing quantitative, statistical comparisons to be made among species. For example, a comparison of specimens allows researchers to see the evolutionary changes in the size and shape of closely related species. By comparing specimens of the same species from different geographic regions, researchers can see the effects of natural selection. In many vertebrate species, for example, specimens from populations that live in humid environments are noticeably darker, presumably because of an adaptation for background matching (a phenomenon termed Gloger's rule). A comparison of specimens from different geographic regions also shows where discrete gaps exist in phenotypic characters, which are often associated with the boundaries (limits) of species. Thus, specimens document what species exist and how they differ from closely related species.

A comparison of morphological features preserved in specimens is also used at higher taxonomic levels, to infer the phylogeny or evolutionary history of a group. For example, qualitative features such as the coloration of a particular body region or the appearance of a particular skeletal process, can be coded for different species. The pattern of such features will reveal the hierarchical descent process that systematists illustrate using the metaphor of a tree (which is a genealogy or an evolutionary history). Such a phylogenetic tree helps us understand the distribution of particular features in the study of organisms. For example, the joint possession of a feature (e.g., yellow undersides) in two species will most likely be recognized to have been inherited from a common ancestor with that feature.

Many kinds of molecular information can be obtained from tissue samples, such as DNA sequences. One of the largest research endeavors in evolutionary biology is DNA sequencing, which provides information on phylogeny, species limits, adaptation, and geographic variation. The combination of molecular and morphological information from specimens also helps to detect instances of convergent evolution—that is, unlike the yellow undersides referred to above, the existence of similar features was arrived at via independent evolution in different lineages. For example, it is well known that the cacti of the New World look very much like the euphorbs of the Old World. However, these two groups are not each other's nearest relatives. Instead, a careful inspection of specimens shows that the similarities are due to independently acquired features in response to solving common ecological challenges.

One might also observe parasites on a specimen, which can be collected and preserved. It is known that the pattern of evolution of parasites often mirrors the evolutionary relationships of their hosts (Hafner and Nadler, 1988).

What Are the Major Issues Facing Systematics Collections Today? Several issues are jeopardizing the roles that systematics collections can play. Despite the fact that there are over two billion specimens in collections, only a small fraction of the earth's species diversity is documented by preserved material (*Systematics Agenda* 2000, 2000). Given the current rate of extinction, many species are disappearing before examples can be collected and preserved. There are millions of as yet undescribed species. Fewer and fewer students are being exposed to and trained in methods of specimen collection, preservation, and identification and classification. In fact, there are too few experts for some groups of organisms to conduct research and train students. In many countries that harbor the most undescribed species, there are few adequately funded collections and curators to operate them. Anticollecting attitudes also hinder the collection of some types of organisms, especially birds, butterflies, and some plants such as orchids. Many institutions have drawn funding away from traditional collections because of misguided perceptions that collections have enough specimens and that the research that can be done on specimens is old-fashioned. This perception is false, and we can only hope that scientists preserve enough specimens so that we can learn about creatures that once shared the earth with us.

An enormous problem facing collections concerns electronic databasing. For the specimens that currently exist, relatively little of the information is digitized. Furthermore, thousands of scientific publications based on specimens are not online and linked to databases. There is no electronic catalog of the names of known organisms.

Who Uses Collections? Collections have many uses. Specimens identified by taxonomic experts provide a reference against which new specimens can be identified. Persons writing field guides use collections because specimens document when and where a particular species (and which sexes and ages) occurs. Illustrators use collections so that field guides will be accurate. Scientists describing new species compare their specimens against those already identified in collections. Systematists interested in constructing phylogenies, or determining species limits, will compare specimens from different lineages or geographic areas. Ecologists interested in how species partition "niches" measure specimens to determine differences in size of food-gathering appendages. Behaviorists study specimens to see what features of morphology might be used in interindividual communication. Law enforcement personnel use collections to establish the identification of illegally obtained organisms.

BIBLIOGRAPHY

Dessauer, H. C., C. J. Cole, and M. S. Hafner. "Collection and Storage of Tissues." In *Molecular Systematics*, edited by D. M. Hillis, C. Moritz, and B. K. Mable, 2d ed., 29–47. Sunderland, Mass., 1996.

Duckworth, W. D., H. H. Genoways, and C. L. Rose. *Preserving Natural Science Collections: Chronicle of Our Environmental Heritage.* Washington, D.C., 1993.

Hafner, M. S., and S. A. Nadler. "Phylogenetic Trees Support the Coevolution of Parasites and Their Hosts." *Nature* 332 (1988): 258–259.

Higuichi, R. G., L. A. Wrischnik, E. Oakes, M. George, B. Tong, and A. C. Wilson. "DNA Sequences from the Quagga, an Extinct Member of the Horse Family." *Nature* 312 (1984): 282–284.

Miller, E. H., ed. *Museum Collections: Their Role and Future in Biological Research.* British Columbia Provincial Museum Occasional Papers Series, No. 25. Vancouver, B.C., 1985.

Systematics Agenda 2000: Charting the Biosphere. American Museum of Natural History Technical Report SA2000. New York, 2000.

— Robert Zink

T

TERRITORIALITY

Territoriality is the defense of an area by an individual or group for its exclusive use. This defense may take the form of physical combat or noncontact signals such as scent and song. Territorial boundaries may sometimes be very clearly defined; but when the probability of attack on intruders decreases the farther the distance from the core of the territory, territorial margins are more difficult to identify. In contrast to territories, a home range is an area peacefully shared by several individuals or groups. Territories, like home ranges, typically hold within their boundaries resources, such as food, mates, and shelter, that are often unshareable and are essential to the space owners.

Costs and Benefits of Territoriality. The decision to be territorial depends on the costs and benefits of territoriality. These costs and benefits in turn depend on four main factors: (1) the quality of resources, (2) the spatial and temporal distribution of resources, (3) the density of conspecifics competing for these resources, and (4) individual status or condition.

If resources are of poor quality and are sparsely distributed, an individual will require a large area to satisfy its needs and will probably be unable to defend these resources. By contrast, an equally sparsely distributed resource, but of high quality, may be worth defending. Clumped resources may promote territoriality, because individuals using such resources will need to roam over a small area to fulfill their needs. Distribution in time is also important. In general, if resources are ephemeral, individuals will have to range over a wide area to exploit enough patches (hence, no territoriality), but if resources are predictable in time, then territories can be set up around these good areas.

Spatial and temporal resource distribution will interact with biotic factors to determine whether territoriality is economical. When there is intense competition for limited resources, as may happen if the density of conspecifics is high, the costs of defending a territory from intruders will be increased, making territoriality less likely. Individual condition will also play a role, with the strongest individuals being able to withstand higher costs to be territorial than weaker ones. The latter may opt for alternative strategies such as being nonterritorial floaters or satellites, which results in few opportunities to breed. In some cases, individuals that fail to obtain a territory do not survive.

Because these four factors interact in a complex way, it is difficult to predict the exact conditions under which animals are expected to defend territories; however, animals should be expected to do so only when the benefits of territoriality exceed the costs. Theoretical models based on this assumption, in which the main benefit of territoriality is increased access to food and the main cost is defense from competitors, predict that territoriality should occur at intermediate levels of food availability (Figure 1). At low food availability, the cost of defending a territory is too high (because of competition), and at high food availability, the benefit is too low (because just as much food may be obtained by being nonterritorial), hence in both cases, territory defense is not expected. This prediction is reasonably well supported by a range of observational studies of animals, particularly during nonbreeding periods.

Animals defending territories during nonbreeding periods generally do so on their own or in pairs, and the

FIGURE 1. When to Be Territorial in Relation to Resource Abundance.
The top panel shows how the benefits B (in terms of rate of resources gained) and costs C (in terms of defense) of territoriality change in relation to resource abundance. The net benefit (B-C) of territoriality is shown in the lower panel. Territoriality should occur at levels of resource abundance for which B > C, that is, whenever the net benefits exceed zero (shown by black bar). Degree of resource clumping and population density could be substituted on the *x* axis with similar results. Illustration courtesy of Isabelle M. Côté

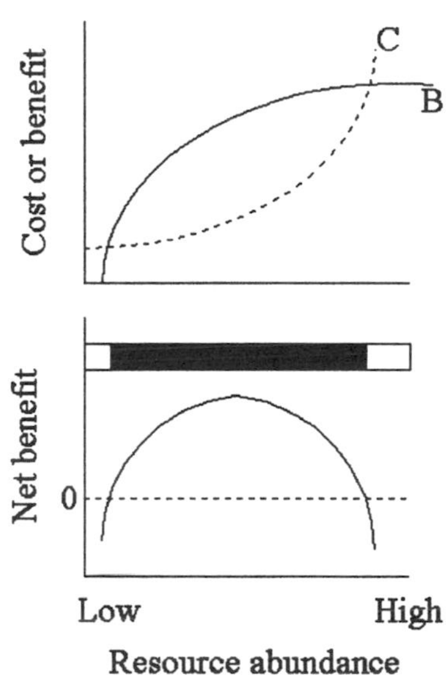

FOOD AVAILABILITY AND TERRITORIALITY: AN EXPERIMENT WITH AUSTRALIAN HONEYEATERS

A fundamental prediction of economic models of territoriality is that animals will not defend territories when food is so abundant that defense does not improve access to food. Manipulating food levels experimentally is the best way to determine the true influence of food on territoriality, but such studies have rarely been carried out. One exception is a field experiment with territorial honeyeaters (*Phylodonyris novaehollandiae* and *P. nigra*) by Doug Armstrong in Australia. Honeyeaters are nectar-feeding birds that display wide seasonal fluctuations in levels of territorial aggression, which seem correlated to nectar availability. But is food level controlling territoriality directly, or is a confounding factor at work?

To answer this question, Armstrong provided feeders with artificial nectar in neutral locations where no birds had territories. These feeders gave territory holders easy access to as much nectar as they wished just outside their territorial boundaries, hence territorial defense could not enhance nectar availability. If food really controls territoriality, the high aggressiveness of birds normally seen in the autumn and spring, when natural nectar levels are low, should be reduced to levels equivalent to the winter, when natural nectar availability is high. Honeyeaters did use the feeders instead of flowers in the autumn and spring when nectar was scarce. However, they continued to defend their territories aggressively at those times. In fact, they showed the same seasonal changes in aggressiveness as birds that did not have access to feeders.

Does this mean that the fundamental prediction made earlier is wrong? Not necessarily, but Armstrong's results highlight the need to consider alternative explanations for correlations between food and territorial aggression. In the case of honeyeaters, the fact that the territories were breeding as well as feeding areas may be important. Territorial ownership is more often challenged in the autumn and spring, hence males must be more aggressive at those times to retain their territories. However, for many other studies of nonbreeding birds in which food appears related to territoriality, there are fewer alternative explanations.

—ISABELLE M. CÔTÉ

resource defended is usually food. However, territories defended during breeding periods usually entail reproductive activity by a breeding pair and the protection of breeding sites or main centers of activity, as well as food resources; models focusing only on increased access to food are of limited usefulness in these cases (see Vignette). Moreover, breeding territories may be occupied not only by a breeding pair but by other individuals that are often related to the breeding pair. Although these additional individuals use up resources within the territory, they can contribute to territorial defense and help to feed or protect the breeding pair's young. Some of the helping young may then inherit their parent's territory or that of neighbors.

Size of the Territory. A cost-benefit analysis can help to predict the likelihood of territoriality as well as optimal territory size, which is the size of territory at which the net benefits of territoriality are maximized (Figure 2). This approach has been used successfully many times, particularly to determine what the territory size of nectar-feeding birds will be in the nonbreeding season, when birds are assumed to be maximizing their rate of food intake.

Food and competitor density have traditionally been considered the major determinants of territory size. When food is abundant or when the density of conspecifics is high, territories should be relatively small. There are, however, several reasons that animals may deviate from this general pattern. For example, territory characteristics, such as vegetation density or water turbidity, that affect the detectability of intruders may change the optimal territory size. Factors intrinsic to the territory holder itself may also be important and may make the identification of an optimal territory size difficult. Males, for example, often have larger territories than females,

FIGURE 2. Effects of Changing Costs and Benefits of Territoriality on Territory Size.
When the cost of territoriality increases from C1 to C2 (top left panel), due for example to an increase in density of competitors, the optimal territory size T (i.e., where the net benefit B-C is greatest) becomes smaller (bottom left panel). Similarly, an increase in the benefits of territoriality from B1 to B2 (top right panel), as would occur if resource quality or quantity increased, also results in a smaller optimal territory size (bottom right panel). Illustration courtesy of Isabelle M. Côté

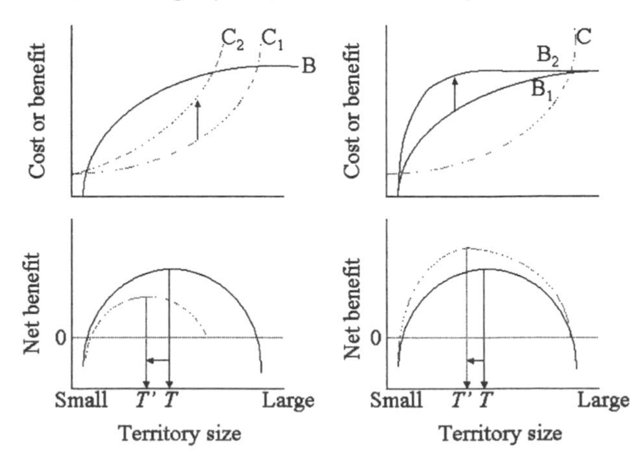

FIGURE 3. Mating Rights.
Bull elk will fight off rivals for the right to mate with up to 60 cows. © Rick Poley/Visuals Unlimited.

and they increase their territory areas during the breeding season. Individuals of the same sex may also be found to be defending territories of very different sizes. Does this mean that only some individuals in a population achieve an optimal territory size, or do different individuals have different optima?

The latter is difficult to show but is likely for several reasons, the most important of which is that different individuals in a population are likely to have different physiological and psychological characteristics that will influence the costs and benefits of defense and hence of individual optimal territory size. One important physiological factor is the level of internal energy reserves. A hungry animal's main priority should be foraging, to avoid starvation, rather than territorial defense, thus resulting in a smaller territory than might be expected for a satiated animal that can spend more time and energy in maintaining a territory. Optimal territory size should therefore vary with the internal state of each individual. In real life, daily changes in territory size, which mirror changing reserve levels, will be limited because territorial boundaries are often constrained by those of neighbors. Territory size may therefore reflect the individual's state at the time of territory establishment but bear little relationship to its current state.

The fact that individuals vary in ways that impinge on the costs and benefits of territorial defense raises the interesting possibility that not all individuals are maximizing the same currency by being territorial. Indeed, in the case of nonbreeding territorial animals, individuals with ample energy resources may try to minimize foraging time in order to maximize the amount of time left for territory defense and other activities. By contrast, individuals with limited reserves may try to maximize their rate of energy intake. For breeding territorial animals, the currency may not be food-related at all. It may be more related to reducing predation risk or increasing reproductive success, but the relationship between these currencies and territory size is sometimes variable. For example, in cases where males are territorial and females choose their mates on the basis of the quantity of resources held within territories or territory size itself, then males should attempt to defend the largest area possible. In such cases, good correlations between territory size and male reproductive success have been shown. Alternatively, females may choose mates on the basis of male physical characteristics. Male reproductive success is then not related to territory size and males are selected to defend the smallest areas possible. The ultimate example of this process is found in lekking species, in which the territories held by males are extremely small, hold no resources, and are used solely as arenas for courtship displays.

Neighbors can have an important effect on territory size. Territory size should theoretically decrease as the density of conspecific neighbors increases if neighbors are responsible for most intrusions onto a territory. But this is not always the case. Once territorial boundaries have been negotiated between neighbors, relatively little time is then spent in recontesting these defined boundaries (i.e., the "dear enemy" phenomenon). Furthermore, neighbors may then collaborate in repelling nonneighbor intruders, which may allow the establishment of larger territories than would be possible in the absence of neighbors.

Territoriality implies monopolization of space by some individuals at the expense of others. Because of this, and because of the impact on breeding and survival of failing to own a territory, territoriality is an important process in the regulation of populations and thus a behavior pattern with profound ecological consequences.

[*See also* Leks; Mate Choice; Mating Strategies, Alternative; Mating Systems, *article on* Animals; Optimality Theory, *article on* Optimal Foraging.]

BIBLIOGRAPHY

Armstrong, D. P. "Correlation between Nectar Supply and Aggression in Territorial Honeyeaters: Causation or Coincidence?" *Behavioral Ecology and Sociobiology* 30 (1992): 95–102. One of the few experimental studies of the effect of food abundance on territoriality.

Brown, J. L. "The Evolution of Diversity in Avian Territorial Systems." *Wilson Bulletin* 76 (1964): 160–169. The first use of a cost-benefit approach to understand territoriality.

Carpenter, F. L. "Introduction to the Symposium: Territoriality: Conceptual Advances in Field and Theoretical Studies." *American Zoologist* 27 (1987): 223–228. The introduction to a series of eleven seminal papers on key aspects of territoriality included in the volume.

Davies, N. B., and A. I. Houston. "Owners and Satellites: The Economics of Territory Defense in the Pied Wagtail, *Motacilla alba*." *Journal of Animal Ecology* 50 (1981): 157–180.

Gill, F. B., and L. L. Wolf. "Economics of Feeding Territoriality in the Golden-Winged Sunbird." *Ecology* 56 (1975): 333–345. A classic study in which costs and benefits of territoriality are measured in the same currency (i.e., calories).

Maher, C. R., and D. F. Lott. "A Review of Ecological Determinants of Territoriality within Vertebrate Species." *American Midland Naturalist* 143 (2000): 1–29. A review of the ecological correlates of territoriality, including a comprehensive list of references.

— ISABELLE M. CÔTÉ

TRADEOFFS. *See* Life History Stages; Life History Theory.

TRAGEDY OF THE COMMONS

When many individuals use a self-renewing resource, such a animal prey or grazing land, users may harvest the resource at a rate that is too high to sustain. Even if they all see the danger of depleting the common resource beyond recovery, those who take more tend to profit more in the short run. Garrett Hardin (1968) called this "the tragedy of the commons." These consequences of conflicting interests among individuals occur throughout the world and have been part of social life since before human ancestors first began to live in traditional hunting and gathering societies. The plight Hardin described in his influential analysis can be represented by the famous game theory payoff structure called a prisoner's dilemma, in which players fail to cooperate because the short-term gains for defection are too large to resist. [*See* Prisoner's Dilemma Games.] Hardin took the term "tragedy of the commons" from English grazing commons, in which the temptation to add sheep degraded the commonly held grazing land.

The tragedy of the commons is similar to the problem of public goods. Amenities such as the signal of a lighthouse are public goods; once supplied, they are consumed by all, even by those who did not contribute to producing them. Because consumers benefit whether or not they pay, individuals do better to allocate their limited resources to things they must pay to get. Thus, the problem of public goods is undersupply. Two variables, exclusion and subtractablity, distinguish public goods and common pool resources from private property. Owners can easily exclude other users from their private property, but not from public goods or commons. Unlike common pool resources that are depleted by additional users, public goods are also nonsubtractable. The use of a public good by one individual does not decrease its availability to other users. As with public radio, more numerous and attentive listeners do not deplete the resource.

Both the tragedy of the commons and the undersupply of public goods are examples of collective action problems (in which actions that best serve the interests of each individual are not the same as the actions that best serve the interest of the collective). Such conflicts of interest tempt free riders (who take benefits but do not contribute to collective welfare). Punishment can reduce incentives for free riding, but monitoring use and enforcing punishment are themselves costly. Because someone must pay the cost of imposing punishment, this only solves one collective action problem by creating another.

History and ethnography enrich our understanding of these social dilemmas. The classic example of a public good is a lighthouse. Despite the temptation for users to free ride, lighthouses were successfully funded by tolls from shipping firms working along the English coast in the eighteenth and nineteenth centuries (Coase, 1974). Also, despite Hardin's argument about the inevitability of commons depletion, common-pool resources sometimes have been and are sustainably managed. Ironically, his example, the English commons, is one such case. Instead of being open-access regimes in which no property rights were defined, these were common-property regimes in which rights of use were specified and enforced. Sustainable common property regimes are more likely when a well-defined set of conditions is met (Ostrom, 1990): small, stable-membership groups (in practice, often kin-based), with the power to make and enforce rules, monitor use, impose graduated sanctions, and exclude outsiders effectively. The most successful groups in remote settings are vigilant in excluding outsiders and in detecting and punishing cheaters within the group; they have relatively stable group membership and use resources for subsistence—it is not possible to "take the money and run." For common-property regimes to be sustainable in close proximity to markets, it is important that their right to exclude and punish is recognized by a larger political entity. These conditions are not easily met, and failure of commons is frequent.

Experimental work on how people solve common-pool problems is leading social scientists to converge with behavioral ecologists on an evolutionary-minded approach (e.g., Ostrom 1990, 2000; Ostrom et al., 1994). Face-to-face, repeated communication lowers the likelihood of overuse. The more people can communicate and agree on rules, the more likely they are to cooperate, all obtaining a better payoff, and thus lowering the need to impose sanctions on cheaters. Successful common-pool resource solutions, across time and space, tend to have emerged when the social benefits for cooperation within a group are relatively high, and a repertoire of effective rules has evolved through experience. Nonetheless, not all actors are identical, and when cooperation does not arise from narrow self-interest, more powerful actors typically pursue their own interests, at the expense of less powerful actors (e.g., Ruttan and Borgerhoff Mulder, 1999). Privileged individuals or groups with more resources under their own control sometimes gain enough from a public good that it is in their interest to supply it.

Rewards other than the consumption value of the goods themselves may also explain some instances of cooperation and the production of common goods. What Thorstein Veblen called "conspicuous giving" is an example of Zahavi's handicap principle. [*See* Signalling Theory.] Displays can honestly signal superior quality because they are too costly to fake. Both signalers and audience then benefit from the reliable information they convey. Costly signaling may remove collective action problems. Instead of contributing the minimum necessary to avoid punishment, the signal function of cooperation may result in showoffs competing to contribute more. Men in foraging societies often devote substantial effort to hunting large game even though little of the meat goes to an individual hunter's own family. Although others are supplied the widely shared meat, the hunter's incentive may lie in status rivalry with other men.

Many modern large-scale ecological dilemmas share characteristics of commons problems: urban sprawl, polluted air and water, the collapse of fisheries and some wildlife populations, acid rain, global warming, and overpopulation. As Hardin (1968) and Lee (1990), argued, human fertility is also a commons problem: as human populations reach carrying capacity worldwide, the children one individual has are children another individual may in consequence not have. These problems are especially difficult, because they exist in evolutionarily novel conditions. They are large-scale problems with high rewards for individual/corporate defection and little incentive for costly cooperation. Even defining the "group" in a way relevant to human evolutionary history is difficult. Modern large-scale commons are not simply small commons writ large (Costanza et al., 2000): they have different, additional constraints. The solutions we seek must encompass these evolutionary novelties. Our evolutionary past offers only clues, not solutions.

[*See also* Human Sociobiology and Behavior, *article on* Human Sociobiology.]

BIBLIOGRAPHY

Coase, R. H. "The Lighthouse in Economics." *Journal of Law and Economics* 17.2 (1974): 357–376.

Costanza, R. B. Low, E. Ostrom, and J. Wilson, eds. *Ecosystems, Institutions, and Sustainability.* 2000.

Hardin, G. "The Tragedy of the Commons." *Science* 162 (1968): 1243–1248.

Hawkes, K. "Why Hunter-Gatherers Work: An Ancient Version of the Problem of Public Goods." *Current Anthropology* 34.4 (1993): 341–361.

Hawkes, K., and R. Bliege Bird. "Showing-off, Handicap Signalling, and the Evolution of Men's Work." *Evolutionary Anthropology* 10 (2001).

Lee, R. D. "Comment: The Second Tragedy of the Commons." *Population and Development Review* 16 (1990): 315–322.

Ostrom, E. *Governing the Commons: The Evolution of Institutions for Collective Action.* Cambridge, Mass., 1990.

Ostrom, E. "Collective Action and the Evolution of Norms." *Journal of Economic Perspectives* 14.3 (2000): 137–158.

Ostrom, E., R. Gardner, and J. Walker. *Rules, Games, and Common-Pool Resources.* Ann Arbor, 2000.

Ruttan, L. M., and M. Borgerhoff Mulder. "Are East African Pastoralists Truly Conservationists?" *Current Anthropology* 40.5 (1999): 621–652.

— Bobbi Low

TRANSITIONAL FORMS

The most revolutionary concept of biological evolution is that all organisms have a common ancestor, from which they have diverged via long sequences of transitional forms. The ultimate proof of evolution is provided by fossils that document progressions of intermediate organisms from a succession of rocks dated from periods of transition.

Nature of Transitions. Charles Darwin's theory of evolution was based on studies of modern populations that showed sufficient variability to indicate their capacity to evolve, via intermediate forms, into distinct species. Darwin assumed that the fossil record would eventually reveal a continuous sequence of transitional forms between all taxonomic groups, but very few had been discovered at the time he wrote *The Origin of Species* (1859). He attributed their rarity to the absence of many rock units resulting from geological processes. However, as knowledge of the fossil record increased over the following 150 years, it became ever more obvious that transitional species were still much less common than those that belonged to well-established groups. As late as 1953, George Gaylord Simpson, a leading paleontologist and evolutionary biologist, admitted that no transitions between major groups were adequately known, even among multicellular plants and animals, for which the fossil record was otherwise very informative. In addition, periods of transition appeared to be much shorter than the duration of ancestral or descendant groups, implying more rapid rates of evolution (Figure 1).

For these reasons, some biologists argued that transitions might be the result of unique evolutionary processes. Darwinian natural selection provides a well-documented explanation for changes within populations and species leading to more effective adaptation to particular environments and ways of life. However, it is difficult to understand how this process can explain shifts between major adaptive zones (e.g., water to land and terrestrial locomotion to flight), the evolution of entirely new structures, or the emergence of different body plans

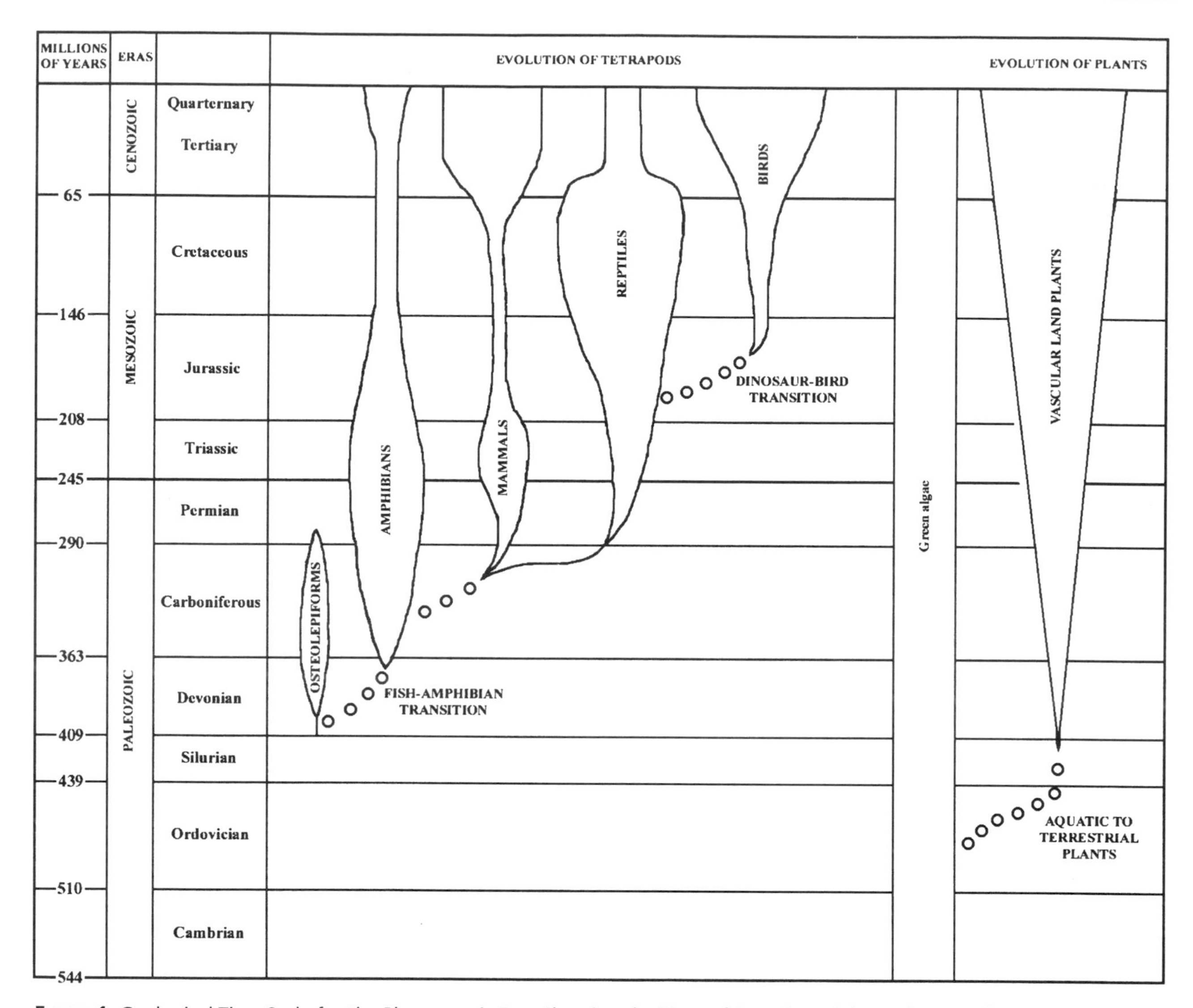

FIGURE 1. Geological Time Scale for the Phanerozoic Eon, Showing the Time of Duration of the Major Vertebrate and Plant Groups Discussed.
The open circles show the relatively short periods of time during which transitions occurred between major groups with basically different ways of life. Prepared by author.

such as those that distinguish the phyla and classes of multicellular animals from one another.

In *The Material Basis of Evolution* (1940), Richard Goldschmidt argued that large-scale changes could be attributed to major reorganization of the genes within chromosomes. Simpson, in contrast, suggested that small, isolated populations in marginal environments might respond very rapidly to selection and lead to what he referred to as "quantum evolution" (an analogy to the gaps between the established orbits of electrons, revealed by changes in the spectra of atoms) but would leave only a very limited fossil record. More recently, in *Macroevolution* (1979), Steven Stanley postulated that major transitions may be driven by rapid successions of speciation events.

The small number of transitional forms may result primarily from the limited extent of adaptive zones that were available for intermediate species between groups characterized by distinct morphology and ways of life. Indeed, relatively few lineages may have actually accomplished major transitions. Despite an increasing rate of discovery and analysis of fossils, gaps remain at all taxonomic levels. However, our understanding of the nature of major transitions has greatly increased over the past twenty years, as our knowledge of the fossil record has increased. In addition, research in molecular biology now provides a basis for understanding the genetic control of the structure and physiology of all organisms and demonstrates how modifications at the genetic level have led to evolutionary change at the morphological level. Molecular biology also provides an effective means for establishing interrelationships between all groups of organisms with living representatives.

Transitions from Water to Land. The most informative transitions are those for which there is a good fossil record of the ancestral and descendant groups, as well as intermediate forms, and for which we have a good knowledge of the genetics of their living representatives. The best examples are provided by vertebrates, arthropods, and multicellular land plants.

The fish–amphibian transition. The fossil history of vertebrates can be studied in detail because their bony skeletons are readily preserved and reflect many features of the soft anatomy and way of life, including locomotion, feeding, nature of major sense organs, and aspects of reproduction, respiration, and circulation. A major event in their history was the emergence of amphibians from fish in the Late Devonian, between 375 million and 363 million years ago (Figure 2). The ancestors of amphibians can be found among a group of fish termed the sarcopterygians, or lobe-finned fish. These include the living lungfish and coelacanths, which differ from most modern fish, the ray-finned or actinopterygian fish, in having thick fleshy fins, with a central axis comparable to the limb axis of terrestrial vertebrates. Their heavy, supporting fins can be associated with life near the bottom, originally in near-shore marine waters.

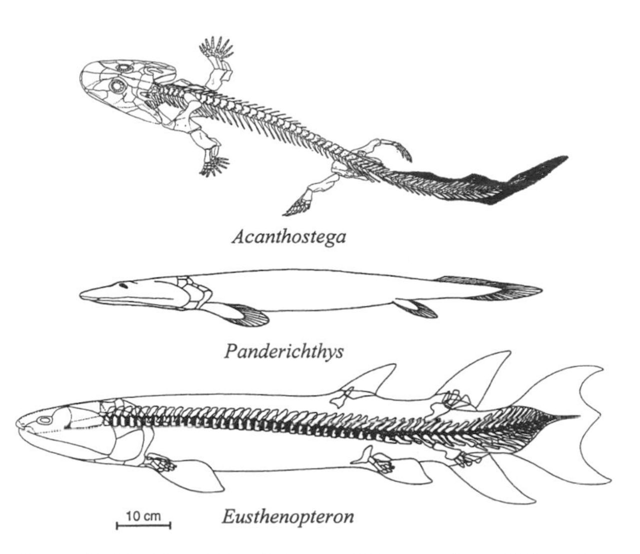

FIGURE 2. The Transition Between Osteolepiform Fish and Primitive Amphibians Represented by *Eusthenopteron* and *Panderichthys* from the Early Upper Devonian and *Acanthostega* from the Late Upper Devonian, a Span of Approximately Fifteen Million Years.
Eusthenopteron from Andrews, S. M., and T. S. Westoll. "The Postcranial Skeleton of *Eusthenopteron foordi* Whiteaves." Reproduced by permission of the Royal Society of Edinburgh from *Transactions of the Royal Society of Edinburgh* 68 (1970): 207–329. *Panderichthys* from Vorobyeva, E. I., and H.-P. Schultze. "Panderichthyid Fishes and Their Relationships to Tetrapods." In *Origins of the Higher Groups,* edited by H.-P. Schultze and L. Trueb, pp. 68–109. New York: Cornell University Press, 1991. *Acanthostega* from M. I. Coates. "The Devonian Tetrapod *Acanthostega gunneri* Jarvik: Postcranial Anatomy, Basal Tetrapod Interrelationships and Patterns of Skeletal Evolution." Reproduced by permission of the Royal Society of Edinburgh from *Transactions of the Royal Society of Edinburgh: Earth Science* 87 (1996): 363–421.

The stout limbs could have been used to push against the bottom, enabling these fish to move into very shallow water, where they may have fed on other fish that were caught between the tides. In common with all early bony fish, sarcopterygians had swim bladders opening into the esophagus that enabled them to achieve neutral buoyancy by gulping air bubbles. The ancestors of land vertebrates could also have used their swim bladders as we do our lungs, for exchange of respiratory gases. This was especially important for life in warm, shallow water that would have had low oxygen content, resulting in reduced efficiency of gas exchange through their gills. The specific lineage leading to land vertebrates, termed the osteolepiforms, were unique in possessing internal nostrils. These openings enabled them to inhale oxygen from the air by expansion of the oral cavity, without opening the mouth.

Although the osteolepiforms may have lived in shallow water and respired like amphibians, they were strictly aquatic, to judge by *Eusthenopteron*, shown in Figure 2, with typically fishy fins. *Panderichthys* (Figure 2) was specialized for life in even more shallow water, to judge by the loss of the dorsal fins and the dorsal extension of the caudal fin. This fish also consolidated the bones of the skull, so that it could be lifted more effectively without the buoyancy of the water. Up to this point, the osteolepiforms were clearly fish, although they had adapted toward life in especially shallow water. However, several anatomical changes that facilitated life in that environment—muscular fins with a massive bony axis, use of the swim bladder as a lung, and evolution of internal nares—made it possible for them to survive out of the water for at least short periods.

The most critical change, toward the end of this transition, was in the structure of the paired fins, leading to the elaboration of digits and effective joints that enabled them to walk on land. Although the changes that were seen earlier in this sequence were primarily modifications of structures already present, the formation of the wrist and ankle joints and the elaboration of distinct digits resulted in new structures. The digits in particular may be attributed to significant changes in the genetic control of development. Specific genes associated with the formation of paired appendages, termed *Hox* genes, are expressed along the posterior margin of the limb in living ray-finned fish. In modern amphibians, the most distal of these genes is expressed for a longer period of time and in a more anterior position, which corresponds with the area where the hand and foot develop. Presumably, the changes in the area and timing of *Hox* gene expression resulted from random mutations, but the effects of these chance events would have been selected for in the lineage of fish that were adapting to life at the interface between water and land.

The oldest fossils of animals with limbs comparable to those of fully terrestrial vertebrates appear at the end of the Devonian period, about 363 million years ago. The best known is *Acanthostega* (Figure 2). Its skull was well consolidated, but the bony operculum, necessary for respiration in fish, was lost, making it possible to move the head freely relative to the trunk. The limbs have clearly distinct digits, but eight were present on each limb, reflecting their origin from even more numerous posterior fin rays in their fish ancestors. The tail retained a strong vestige of its aquatic function, with an extensive caudal fin. Jennifer Clack and Michael Coates (1995), who described *Acanthostega*, argue that it and other Devonian amphibians may have been primarily aquatic despite the structure of the head and limbs.

Although many of the anatomical changes that were necessary for life on land had evolved earlier, the sequence between *Eusthenopteron* and *Acanthostega* may be considered the key period in the transition between obligatorily aquatic and facultatively terrestrial ways of life. However, *Acanthostega* was still primitive in many respects, such as the poorly developed area of articulation between the skull and the neck and the limited integration of the elements of the vertebrae. Other features, including the evolution of reproduction free of the water and the capacity to detect airborne vibrations via the middle ear, did not evolve until tens of millions of years later. Nevertheless, the initial change in habitat occurred over a much shorter period of time (about twelve million years) than the prior history of sarcopterygian fish (as long as fifty million years) or the subsequent history of terrestrial vertebrates.

Panderichthys is the best known, clearly intermediate form, but less complete specimens of several other osteolepiform fish from the Late Devonian also contribute to our understanding of this transition. In contrast with the assumption that transitional forms would show little diversity, several divergent taxa exhibit intermediate characteristics between typical sarcopterygians and amphibians, and some species were represented by large populations. Although changes in specific parts of the skeleton that were critical to life on land occurred relatively rapidly during this transition, the rate of evolution was no more rapid than that for other parts of the skeleton in more recent reptiles and mammals. What makes this and comparable transitions stand out is that natural selection for adaptation to a clearly distinct physical environment may have been more or less unidirectional within particular lineages.

Origin of land plants. The origin of land plants, which were necessary for the evolution of the entire terrestrial ecosystem, shows many parallels with the origin of amphibians. Land plants (tracheophytes) are distinguished from aquatic algae by two specialized tissues: the xylem, which conducts water from the soil to the top of the plant, and the phloem, which distributes the products of photosynthesis from the leaves to the rest of the plant. Xylem also serves for support against the force of gravity. Small, very simple, land plants are known as early as the Late Silurian, some 420 million years ago.

As in the ancestry of amphibians, key changes necessary for the origin of land plants occurred among their aquatic ancestors, whose fossil record can be traced back to the Middle Ordovician, about 470 million years ago. Land plants arose from a specific group of green algae, the charophyceans, whose living descendants exhibit a series of stages leading toward the anatomy of tracheophytes. The most primitive consists of a single biflagellate cell without a cell wall. More advanced species show in progression: the loss of flagella, the assumption of colonial or filamentous aggregations, and the formation of cellulosic cell walls as a result of acquisition of genes for cellulose synthesis from endosymbiotic bacteria. In more advanced species, true multicellularity evolved, with intercellular cytoplasmic connection, the differentiation of numerous cell types, and the concentration of cell division in an apical meristem.

An especially important advance, compared with other green algae, was the retention of the egg within a specialized organ, the archegonium, where it was fertilized. This protects both the ovum and the developing embryo within the body of the adult plant. Without an archegonium, multicellular plants could not have reproduced on land.

Another feature of land plants is the distribution of spores or pollen by the wind. In contrast, most aquatic plants distribute both spores and gametes via the water. Within one charophyte lineage, the embryo developed as a distinct structure, the sporophyte, which extended beyond the body of the parent as a stalked sporangium that raised the spores out of the water so they could be dispersed by the wind. Airborne spores can be transported farther and faster than they could propel themselves in the water. The wind would also have carried them over land, where they could have propagated in small pools.

Fossils of spores resembling those of primitive land plants are known as early as the Ordovician period. They are associated with cuticle, which in land plants covers the surface and limits water loss, but there is no evidence of vascular tissue from this time. This suggests that both airborne spores and cuticle had evolved among primarily aquatic algae. Within the Late Ordovician and Early Silurian, plants comparable to living mosses, the bryophytes, evolved. Mosses were more advanced than charophytes in their apical meristem, which was capable of producing three-dimentional growth.

By the end of the Silurian (410 million years ago), true land plants with xylem, phloem, and stomata (openings in the surface of the stems for passage of respiratory gases) and a shoot meristem capable of branching are observed in a tiny, leafless plant, *Cooksonia*, found in Wales and Australia. All more advanced land plants evolved from genera such as *Cooksonia*. From the appearance of airborne spores in the Ordovician to the formation of vascular tissue in the Late Silurian required approximately fifty million years. Regulatory genes, include the MADS-box genes, analogous to the *Hox* genes in animals, affecting shoot meristems and eventually the formation of flowers, may have been relevant to early land plant evolution.

The Origin of Feathers and Birds. A second transition among vertebrates involving a major change in environment and mode of locomotion was the origin of birds and flight. It was long recognized that birds were closely related to reptiles, but over the past twenty-five years many fossils have been discovered indicating that birds evolved specifically from bipedal, carnivorous dinosaurs. Their ancestors resembled the small but highly intelligent raptors that were featured in the movies *Jurassic Park* and *The Lost World*. The oldest known and most primitive bird is *Archaeopteryx*, from approximately 145 million years ago (Figure 3A). It closely resembles raptors (more technically dromeosaurs) in nearly all features of the skeleton, except for the much larger forearm and hand. More importantly, it had flight feathers of the same structure and pattern as those in modern flying birds. But *Archaeopteryx* was much smaller than dromeosaurs, weighing approximately 0.5 kilogram about the size of a chicken.

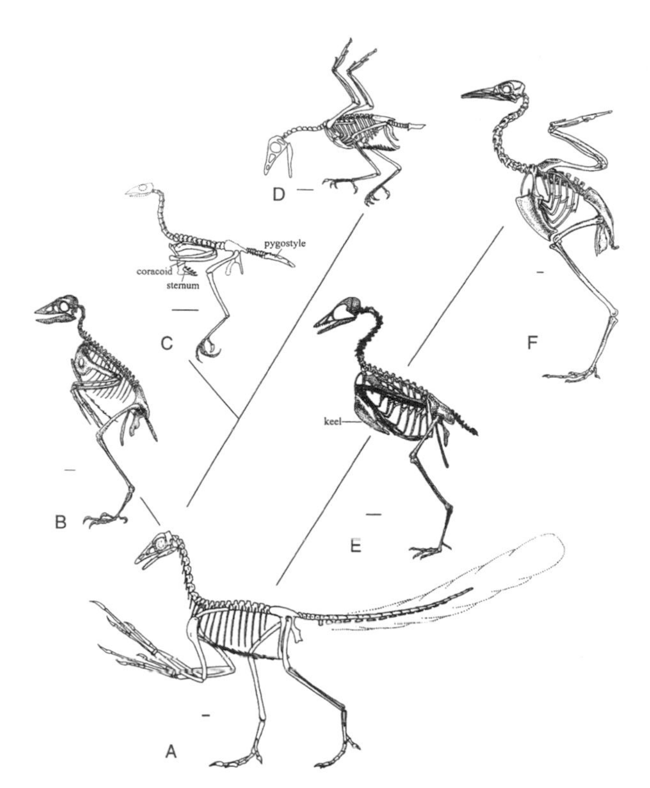

Figure 3. Transitions of the Skeletal Anatomy Between the Most Primitive Known Bird, *Archaeopteryx* from the Late Jurassic, Through a Series of Intermediate Forms from the Early Cretaceous, to the Modern Avian Anatomy Seen in Late Cretaceous Birds.
(A) *Archaeopteryx*. From Ostrom, J. "*Archaeopteryx* and the Origin of Birds." *Biological Journal of the Linnean Society* 8 (1976): 91–182, Academic Press Ltd. (B) *Confuciusornis* from Hou et al. "A Beaked Bird from the Jurassic of China." *Nature* 377 (1995): 616–618, Macmillan Magazines Ltd. (C) *Iberomesornis*. From Sanz and Bonaparte. "A New Order of Birds (Class Aves) from the Lower Cretaceous of Spain." *Natural History Museum of Los Angeles County Science Series* 36 (1992): 39–49. (D) *Sinornis*. Reprinted with permission from *Science* vol. 255, Sereno and Chenggang; copyright 1992, American Association for the Advancement of Science. (E) *Chaoyangia*. Reprinted with permission from *Science* vol. 274, Hou et al.; copyright 1996, American Association for the Advancement of Science. (F) *Lithornia*. Reprinted with permission from *Science* vol. 214, Houde and Olson, copyright 1981, American Association for the Advancement of Science. Scale bars are 1 centimeter in length.

No fossils are known of dinosaurs immediately ancestral to *Archaeopteryx*, but it is assumed that the transition toward birds began with a reduction in size. This is logically associated with a shift in prey from large vertebrates to small insects or other arthropods. The origin of feathers may be attributed to changes in the configuration of scales, which cover the bodies of reptiles. The initial role of feathers in the ancestors of birds may have been for insulation, because such small animals would have lost heat much more rapidly than the large dinosaurs, which probably had a high, constant body temperature.

The manner by which flight evolved is still being debated. Many scientists think that birds were initially capable of climbing trees and that flight evolved via gliding back to earth. John Ostrom (1976), in contrast, hypothesized that birds might have evolved among rapidly running, ground-dwelling dinosaurs that moved their forearms during prey capture in a manner similar to the flight stroke and achieved the capacity for flapping flight from the ground up.

It is not yet possible to establish the time it took for *Archaeopteryx* to evolve from small terrestrial dinosaurs, but there is an informative fossil record of the early stages in the transition from *Archaeopteryx* to the morphology of modern birds. Many specimens found in Spain and China dating from the Early Cretaceous, ranging from about 140 to 120 million years ago, (Figure 3) provide a good record of the progressive modernization of the bones associated with flight. Most conspicuously, many of the Early Cretaceous birds were much smaller than *Archaeopteryx*, comparable in size to a sparrow, and much more maneuverable as a result of a reduction in the length of the tail to a stiff nubbin termed the pygostyle.

Most fossils of *Archaeopteryx* lacked the most important bone for the attachment of flight muscles in modern birds—the sternum, or breast bone. A small, flat sternum is present in one specimen slightly younger than the rest. During the Early Cretaceous, a keel developed along the midline of the sternum, which further increased the surface for muscle attachment. As the sternum evolved, the paired coracoids were elaborated to resist the force of contraction of the flight muscles. Other advances seen among the Early Cretaceous birds were the evolution of a tiny anterior winglet, the alula, which increases lift at slow speed, and the capacity to fold the wings close to the body. The last conspicuous change was the reduction in size and the fusion of the bones of the fingers, which occurred approximately twenty-five million years after the time of *Archaeopteryx*. By then the skeleton of birds was essentially modern, but the radiation of the living orders did not take place until the Late Cretaceous and Early Tertiary.

Most of the changes in the bony skeleton leading to modern birds can be explained by Darwinian selection on preexisting structure. In contrast, feathers, which are key to flight, are unique in both their anatomy and mode of development. Feathers are rarely fossilized, but they are preserved in numerous specimens of *Archaeopteryx* and in several species of birdlike dinosaurs. The configuration of the feathers and the manner of their distribution on the body can be arranged in a series of progressively more advanced patterns, but, unfortunately, this does not correspond with the sequence of their appearance in the fossil record. *Archaeopteryx*, the most primitive true bird, shows fully modern flight feathers, with an asymmetrically placed rachis, capable of generating lift, in the Late Jurassic, about 150 million years ago. More primitive feathers, associated with animals classified as dinosaurs on the basis of skeletal features, are actually known considerably later, from a series of lake deposits in China dating to the Early Cretaceous, between 140 million and 120 million years ago.

On the basis of developmental studies of scales and feathers in modern birds, Cheng-Ming Chuong and colleagues (2000) observed ten levels of increasing structural and genetic complexity in the transformation between simple scales and advanced flight feathers. Among the feathered dinosaurs, *Sinosauropteryx* shows the most primitive condition, with fuzzy fibers (protofeathers) forming a downlike covering over the entire body. This supports the hypothesis that feathers initially evolved for insulation. In *Caudipteryx*, the forelimbs and tail both have long pennaceous feathers. These resemble flight feathers, but are bilaterally symmetrical and restricted to small areas, indicating that they could not have functioned in flight. But they were advanced in another feature. The restriction of these feathers to limited areas of the body indicates the origin of another type of developmental control—the establishment of feather tracts. Instead of being distributed all over the body, as are the scales in most reptiles, the position of each feather in birds is specified so that together they can function as an effective lift-producing surface. However, the presence of feather tracts in a nonflying dinosaur may indicate that pennaceous feathers initially evolved for sexual display, rather than flight.

One of the oldest birds found in China, *Confuciusornis*, shows the fully differentiated pattern of feathers seen in modern flying birds: down and contour feathers on the body and bilaterally asymmetrical flight feathers on the wings. A pair of greatly elongated tail feathers has a dimorphic distribution within populations, suggesting a carryover of the dinosaurian use of feathers for display.

Study of chickens shows that areas of their legs typically covered by scales can, as a result of cellular and molecular changes, produce feathers, indicating that these structures are broadly homologous and that feathers evolved from scales like those in dinosaurs and other reptiles. According to Richard Prum (1999), the evolution of feathers began with the feather follicle that regulates the genetic and mechanical features of feather development, which are very different from those of reptilian scales. Without the appearance of the feather follicle as a developmental novelty among small, flightless dinosaurs, avian flight might never have evolved. As yet, there is no fossil evidence as to how long before *Archaeopteryx* primitive downy feathers first evolved, but the full complement of feather types had certainly appeared before the end of the Jurassic period.

A striking feature of the evolution of birds was its mosaic pattern. Several divergent lineages are recognized. In some, for example, *Confuciusornis*, most of the skeleton remained primitive, but the skull appeared much more advanced in the loss of teeth. Other groups rapidly modernized the flight apparatus, but they retained reptilian teeth. In the lineage leading to modern birds, the sternum increased in length at an early stage, possibly in association with the elaboration of air sacs that characterize advanced birds. It is interesting that early stages in the evolution of feathers, in which they were used for insulation or display but not flight, continued to be exhibited by dinosaurs living thirty million years after modern flight feathers had evolved in *Archaeopteryx*. As in the case of the fish–amphibian transition, most of the skeletal structures crucial to flight evolved in a relatively short period of time, about fifteen million years, presumably as a result of more or less continuous, unidirectional selection.

The origins of amphibians, birds, and land plants were described in detail because all show major changes in the structure, the way of life, and the environment in which they lived, over relatively short periods of time. Other major transitions among vertebrates were the origin of mammals and modern ray-finned fish, the teleosts. Both of these took place gradually over very long periods of time, within a single physical environment, but without short critical periods of major transformation. They involved many structural and physiological changes through several successive radiations. These more complicated transitions are summarized in Robert Carroll's book *Vertebrate Paleontology and Evolution* (1988).

Significance of Transitions. Fossils of transitional forms not only document the evolutionary continuity among organisms but also demonstrate the exceptional nature of transitions. Evolution has not proceeded at a uniform rate over long periods of time, but is strongly episodic, with a few large-scale events occurring at intervals of tens of millions to billions of years. The origins of amphibians and birds show that major morphological and adaptive changes between groups customarily ranked as classes (the rank below phyla) may occur over ten million to fifteen million years, which is less than the duration of many species. The relative rapidity of major transitions can itself account for the limited record of intermediate fossils.

The largest scale transitions are exceptional in other aspects. Among vertebrates, the transition from an aquatic to a terrestrial way of life occurred only once. Powered flight evolved in pterosaurs and bats as well as birds, but the origin of feathers occurred only once. Insects were the only nonvertebrate group to achieve flight, and all multicellular land plants arose from a single ancestral lineage. The similarity of the mitochondria in all eukaryotes points to the perpetuation of a single endosymbiotic union between a particular aerobic eubacterium and an archaeobacterial host.

The origin of limbs in tetrapods, the feather follicle in the ancestors of birds, and the archegonium in the precursors of land plants can be attributed to unique changes in developmental systems. Even larger scale changes in the nature of the genetic control of entire body plans underlay the origin of metazoans and the evolution of the distinctive body plans of the thirty-five phyla of multicellular animals and the origin of vertebrates.

Although evolution at the level of populations and species is largely a reflection of short-term, irregular, and oscillating changes in their physical and biological environment, major evolutionary transitions have occurred in relationships to large-scale and long-lasting changes in the relationship between organisms and their environment, as well as to genetic changes in the capacity of organisms to evolve.

[*See also* Missing Links; Novelty and Key Innovations.]

BIBLIOGRAPHY

Ahlberg, P. E. "*Elginerpeton pancheni* and the Earliest Tetrapod Clade." *Nature* 373 (1995): 420–425. Description of a minor radiation during the transition between osteolepiform fish and amphibians.

Carroll, R. L. *Vertebrate Paleontology and Evolution*. New York, 1988.

Carroll, R. L. *Patterns and Processes of Vertebrate Evolution*. Cambridge, 1997. Includes a chapter on transitions, discussing the origin of whales, secondarily marine reptiles and birds.

Carroll, S. B., J. K. Grenier, and S. D. Weatherbee. *From DNA to Diversity. Molecular Genetics and the Evolution of Animal Design*. Oxford, 2001. Analysis of the genetic basis for evolutionary change among multicellular animals, including the origin of limbs in amphibians and the wings of insects.

Chuong, C.-M., R. Chodankar, R. B. Widelitz, and T.-X. Jiang. "Evo-Devo of Feathers and Scales: Building Complex Epithelial Appendages." *Current Opinions in Genetics and Development* 10 (2000): 449–456.

Clack, J. A. "The Origin of Tetrapods." In *Amphibian Biology Palaeontology: The Evolutionary History of Amphibians*, edited by H. Heatwole and R. L. Carroll, vol. 4, pp. 979–1029. Chipping Norton, Australia, 2000. Detailed description of Devonian amphibians and their origin from sarcopterygian fish.

Clack, J. A., and M. I. Coates. "*Acanthostega gunnari*, a Primitive, Aquatic Tetrapod?" *Bulletin du Muséum National d'Histoire Naturelle* 17 (1995): 359–372.

Fecuccia, A. *The Origin and Evolution of Birds*. New Haven, 1996.

Goldschmidt, R. B. *The Material Basis of Evolution*. New Haven, 1940.

Graham, L. E., M. E. Cook, and J. S. Busse. "The Origin of Plants: Body Plan Changes Contributing to a Major Evolutionary Radiation." *Proceedings of the National Academy of Sciences* 97 (2000): 4535–4540. Excellent, concise review of the transition from unicellular green algae to land plants.

Houde, P., and S. L. Olson. "Paleognathous Carinate Birds from the Early Tertiary of North America." *Science* 214 (1981): 1236–1237.

Hou, L.-H., Z. Zhou, L. D. Martin, and A. Feduccia. "A Beaked Bird from the Jurassic of China." *Nature* 377 (1995): 616–618.

Knoll, A. H., and S. B. Carroll. "Early Animal Evolution: Emerging Views from Comparative Biology and Geology." *Science* 284

(1999): 2129–2137. Genetic transitions leading to advanced multicellular animals.

Maynard Smith, J., and E. Szathmáry. *The Major Transitions in Evolution.* Oxford, (1995). Discusses many transitions that are not adequately documented in the fossil record, including the origin of life and genetic material, the orgin of eukaryotes, sex, animal societies, and language.

Niklas, K. J. *Evolutionary Biology of Plants.* Chicago, 1997. A general review of the major features of the evolution of land plants.

Ostrom, J. 1976. "*Archaeopteryx* and the Origin of Birds." *Biological Journal of the Linnean Society* 8 (1976): 91–182.

Padian, K., and L. M. Chiappe. "The Origin of Birds and Their Flight." *Scientific American* 278 (1998): 38–47. Concise description of the anatomy of transitional forms and stages in the evolution of flight.

Prum, R. O. "Development and Evolutionary Origin of Feathers." *Journal of Experimental Zoology* 285 (1999): 291–306.

Sanz, J. L., and J. F. Bonaparte. "A New Order of Birds (Class Aves) from the Lower Cretaceous of Spain." *Natural History Museum of Los Angeles County Science Series* 36 (1992): 39–49.

Schopf, J. W. *Cradle of Life: The Discovery of Earth's Earliest Fossils.* Princeton, 1999.

Sereno, P. C., and R. Chenggang. "Early Evolution of Avian Flight and Perching: New Evidence from the Lower Cretaceous of China." *Science* 255 (1992): 845–848.

Shubin, N., C. Tabin, and S. Carroll. "Fossils, Genes, and the Evolution of Animal Limbs." *Nature* 388 (1997): 639–648. Genetic basis of origin of limbs in amphibians and wings in insects.

Simpson, G. G. *Major Features of Evolution.* New York, 1953. An outstanding synthesis of natural selection, genetics, and paleontology, based on data available in the mid-twentieth century. A classic work in the field of large-scale evolution.

Stanley, S. M. *Macroevolution.* New York, 1979.

Zimmer, C. "Back to the Sea." *Discover* (1995, January): 82–84. Color reconstructions of transitional animals between primitive land mammals and whales.

— Robert L. Carroll

TRANSMISSION DYNAMICS

Parasites, including infectious transmissible pathogens, depend on a host organism for their reproduction. Because of the limited life span of their hosts, parasites need to find a way to transmit to new hosts to ensure their survival. Consequently, parasites have evolved a large array of modes of transmission. The transmission mode in turn influences other evolved characteristics of the parasite and its host.

Transmission Modes. Transmission modes of parasites can be classified according to several criteria, such as the carrier of the infectious agent, the form of contact between infected and susceptible hosts, and the genetic relation between donor and recipient hosts. On the basis of the carrier, one can distinguish between airborne, waterborne, foodborne, and vectorborne transmission. Typical examples of airborne transmission are human respiratory diseases, such as influenza and tuberculosis, which are transmitted over short distances through droplets produced by a coughing infected individual. Airborne transmission is also common among plant pathogens and can occur over considerable distances through wind. The rust fungus (*Puccinia graminis tritici*) moves every year from the southern parts of the United States, where it overwinters, to Canada. Infections of the gastrointestinal tract, such as cholera and salmonellosis, are frequently transmitted through contaminated water or food, but they also can be acquired through the fecal/oral route of transmission.

Many animal and plant pathogens are transmitted via so-called vectors. Often the vectors are insects, but more generally a vector is defined as a living carrier that actively transports pathogens between hosts. Many important diseases of vertebrates are transmitted by bloodsucking animals. For example, mosquitoes of the genus *Anopheles* are the carriers of the protozoan pathogen *Plasmodium*, the causative agent of malaria. Other examples are *Trypanosoma brucei*, the causative agent of sleeping sickness, which is transmitted by the tsetse fly (*Glossina morsitans morsitans*,) and yellow fever virus, which is transmitted by mosquitoes of the genus *Aedes*. Although for some pathogens the vector acts only as an agent for parasite transmission (yellow fever virus), for others (such as *T. brucei* and *Plasmodium*) the vector organism represents an intermediate host in which the parasite goes through essential developmental stages of its life cycle. In some circumstances, humans may also act as vectors. The Hungarian physician Ignatz Semmelweiss (1818–1865) noted in 1846, well before the germ theory of infectious diseases was established, that hospital personnel acted as vectors for the transmission of childbed fever, at the time a major cause of death of women in childbirth.

In addition to the physical mode of transmission, one can distinguish between direct and indirect transmission. Generally, vector-, food-, and waterborne modes of transmission are indirect. Directly transmitted diseases require close contact between infected and susceptible hosts. Typical examples are pathogens transmitted over short distances by air and by the fecal/oral route of infection. Sexually transmitted pathogens (diseases), such as *Neisseria gonorrhea* (gonorrhea), *Treponema pallidum pallidum* (syphilis), and acquired immune deficiency syndrome (AIDS), fall into the category of directly transmitted diseases. [*See* Acquired Immune Deficiency Syndrom, *article on* Dynamics of HIV Transmission and Infection.]

Occasional transmission of a pathogen from its natural host to novel host species has been responsible for many important diseases. For example, the bubonic plague, which killed about one-third of the population of Europe in the fourteenth century, was caused when *Yersinia pestis*, a bacterium that usually infects rodents, was transmitted by fleas to humans. Such an occasional transmission of an animal disease to humans is referred to as a zoonosis. Often, as is the case for the rabies virus in humans, the infection of the new host species is a biological dead end for the transmission of the parasite. However, cross-species transmission can lead to epidemics in the new host species. For example, the pandemic of human immunodeficiency virus (HIV) was triggered by recent cross-species transmissions of an ancestral virus from Old World monkeys and chimpanzees to humans.

Changes associated with modern-day society have led to new, sometimes major transmission modes for certain pathogens. Infection by contaminated blood products represented an important mode of transmission in the early stages of the HIV and hepatitis C virus epidemics. There is also concern that new developments in medicine such as xenotransplantation (the transplantation of animal organs into humans) could lead to new cross-species transmission of infectious pathogens. New practices in intensive farming have also led to new modes of transmission. For example, the feeding of animal carcasses to cattle in intensive animal farming has likely led to the epidemic outbreak of bovine spongiform encephalopathy (BSE), a new prion disease of cattle.

Evolutionary Consequences. The way in which a parasite is transmitted from one host to the next has profound consequences for the coevolution of host and parasite. This is evident when comparing parasites that are transmitted vertically (from parent to offspring) and horizontally (between any individual). Evolutionary theory predicts that exclusively vertically transmitted parasites should generally evolve to cause no reduction in fitness to their hosts, because their survival is strictly coupled to that of their hosts. In contrast, horizontally transmitted parasites may reduce host fitness, provided the benefits to the parasite of doing so outweigh the costs to the parasite of harming its host.

The host population in which the parasite is usually transmitted is referred to as its natural host population. Parasites differ in their natural host range. Although some are specific to a single host, others can infect many host species. Cross-species transmission of a parasite from its natural reservoir to new hosts puts the parasite into a novel environment to which it is unlikely to be adapted. Often in the new environment the parasite is unable to exploit the host's resources efficiently for reproduction and has little effect on the fitness of the novel host. Sometimes, however, as may be the case for Ebola virus infections in humans, the parasite inflicts substantial harm to the new host without obvious fitness benefits to the parasite.

Parasites frequently modulate the behavior of the infected host in a way that facilitates their own transmission. An example is the trematode *Dicrocoelium dendriticum*, which induces in its intermediate hosts, the ants *Formica rufibarbis* and *F. cunicularis*, a behavioral change that makes the ants climb to the tips of grass, and thereby makes them more vulnerable to predation. In this way the parasite facilitates its transmission to the final host, one of several grazing mammals.

Ecological Factors. Ecological properties of the host population, such as population density, immunity or resistance, host mobility, and population structure, play an important role in the spread of infectious pathogens. Pathogens transmitted via direct contact require a threshold density of the host population to be able to infect and persist in the population. For the measles virus, for example, this threshold density is around a community size of 300,000 individuals. Moreover, it is believed that many of the microparasitic childhood infections only became endemic in the human population after the population density increased with the advent of agriculture. Although for many transmission modes increases in population density increase the severity of an epidemic, this is generally not the case for sexually transmitted diseases, because the likelihood of infection by a sexually transmitted pathogen depends on the relative frequency, not the absolute density, of infected hosts.

The fraction of hosts in a population that are protected against a pathogen by acquired immunity or hereditary resistance is a key determinant of transmission dynamics. This is exemplified by the smallpox epidemic of 1520 in which about half of the population of 2.4 million Aztecs were killed, when the Spanish invaders of Mexico inadvertently introduced the virus into the previously unexposed population. Another example is chestnut blight, which within thirty years following its introduction from Asia largely eliminated the American chestnut population in parts of North America.

Host mobility is another important factor for transmission. Pathogens of immobile hosts, such as plants, typically rely on vectorborne or airborne transmission. Pathogens transmitted by direct contact, however, usually require mobile hosts. Increases in host mobility can accelerate epidemics. Modern means of transportation represent an important factor in the increasingly global spread of agricultural and human pathogens.

Aspects of host population structure, such as temporal fluctuations and locations with high host density, low immunity or resistance, or low genetic diversity, can also have a considerable impact on the spread of epidemics. Seasonal changes in host population density can lead to periodic cycles in endemic infections. The incidences of childhood infections, for example, show clear correlations with the timing of school vacations. The effect of spatial population structure is evidenced, for example, in kindergartens, which from the perspective of the pathogens are breeding grounds with high host density and low immunity. The vulnerability of monocultures to parasitic attack demonstrates the effect of low genetic diversity.

Mathematical Theory. The study of the transmission dynamics and evolution of infectious pathogens has been assisted by the development of the underlying mathematical theory. Perhaps the first attempt to model infectious disease dynamics was undertaken by the Swiss mathematician Daniel Bernoulli (1700–1782) around 1760, who studied the effect of variolation against smallpox well before it had been recognized that the disease is transmitted by infectious "germs." In contrast to statistical analyses of epidemiological data, mathematical models attempt to describe the transmission of infectious pathogens mechanistically in order to develop an understanding of the resulting dynamic behavior.

Central to most mathematical models is the assumption of the so-called mass-action principle, a concept that was originally developed in chemical kinetics around 1870 by the Norwegian scientists Cato Maximilian Guldberg and Peter Waage, but was first applied to infectious disease dynamics by the English scientist W. H. Hamer in 1906. The mass-action principle postulates that the rate of new infection of susceptible hosts is proportional to the product of the densities of the susceptible hosts and the infecting agent, that is, the infected vector, the infectious propagule in the water or air, or the infected host for pathogens transmitted by direct contact. In particular, the mass-action principle assumes random interactions between susceptible hosts and the infecting agent. This is often, but not always, a good assumption. It works well for large, well-mixed populations, but it does not accurately reflect the transmission dynamics if, for example, the spatial distribution of the host or pathogen population is highly nonrandom.

Some basic principles of theoretical epidemiology can be illustrated by a discussion of the population dynamics of a directly transmitted infection, such as the common cold, for which it can be assumed that the encounters of an infected individual are to a good approximation random. According to the mass-action principle, the rate of "gain" of new infected hosts is given by $b\,S\,I$, where S and I are the population densities of susceptible and infected hosts, respectively, and b is a proportionality constant describing the infectivity (per contact and time). The rate of "loss" of infected hosts is proportional to the infected host density and given by $(n + p + r)I$, where the constants n, p, and r describe the rates of natural host mortality, parasite-induced host mortality, and recovery from infection (per time and infected host). The pathogen increases in abundance in the host population if the rate of gain exceeds that of loss, that is, if $R_0 = bS/(n + p + r) > 1$. R_0, the so-called basic reproductive rate, is a key concept of theoretical epidemiology, commonly defined as the number of secondary infections caused when one infected host is placed into a population of otherwise uninfected hosts. R_0 must exceed unity for a parasite to be able to invade and become endemic in a host population. The basic reproductive rate allows one to determine the threshold density of susceptible hosts, $S > (n + p + r)/b$, above which the parasite can persist in the host population. Moreover, the basic reproductive rate determines the minimal fraction of susceptible hosts, $f_c = 1 - 1/R_0$, that need to be immunized to prevent or eradicate an epidemic. This implies that a vaccination program that covers less than 100 percent of the host population may nevertheless eradicate the pathogen in the host population, a concept referred to as herd immunity.

The basic reproductive rate not only determines whether a pathogen can invade a host population but also offers insight into the dynamics of adaptation of the pathogen to its host population. For many host–parasite systems, theory suggests that among a population of pathogen genotypes spreading in a host population, the genotype with the highest basic reproductive rate is the best competitor. In these simple host–parasite systems, the basic reproductive rate is a measure of parasite fitness, thus, natural selection tends to maximize the basic reproductive rate.

[*See also* Basic Reproductive Rate (R_0); Coevolution; Disease, *article on* Infectious Disease; Frequency-Dependent Selection; Immune System; Influenza; Malaria; Mathematical Models; Myxomatosis; Resistance, Cost of; Vaccination; Virulence.]

BIBLIOGRAPHY

Anderson, R. M., and R. M. May. *Infectious Diseases of Humans.* Oxford, 1991. The most comprehensive work on theoretical epidemiology, also highly valuable for extensive compilation of data. Focuses on human diseases.

Bailey, N. T. J. *The Mathematical Theory of Infectious Diseases and Its Applications.* 2d ed. London, 1975. An early, classic textbook on theoretical epidemiology. Comprehensive in the mathematics, but sparse in data.

Cox, F. E. G. *Modern Parasitology.* 2d ed. Oxford, 1993. Parasitology textbook that contains detailed section on epidemiology and ecology.

Levin, B. R., M. Lipsitch, and S. Bonhoeffer. "Population Biology, Evolution, and Infectious Disease: Convergence and Synthesis." *Science* 283 (1999): 806–809. Short review of recent applications of population biological modeling in medicine.

Mims, C., J. Playfair, I. Roitt, D. Wakelin, and R. Williams. *Medical Microbiology.* 2d ed. St. Louis, 1998. Short, but comprehensive, reference work on infectious pathogens, mostly those of humans.

Poulin, R. *Evolutionary Ecology of Parasites.* Dordrecht, Netherlands, 1997.

— SEBASTIAN BONHOEFFER

TRANSPOSABLE ELEMENTS

Transposable elements are fragments of DNA that replicate within a host organism's genome. They are typically 1,000 to 10,000 base pairs long, and encode one or more proteins whose sole function is to copy the element to a new location in the genome. Such reproduction allows a transposable element family to spread through a host gene pool. Though they can be highly invasive when first introduced into a species, transposable elements are susceptible to degenerating over relatively short evolutionary time spans because some defective elements that do not encode functional proteins are still able to replicate (at least) as fast as functional elements. Purifying selection maintaining function is therefore weak. For many transposable elements, persistence over long evolutionary time spans therefore depends upon regular colonization of new gene pools by horizontal transmission. On occasion, particular inserts can be beneficial to the host, and transposable elements may be important sources of genomic variation. Rarely, transposable elements are domesticated by their hosts and become positively useful to them. Normally, though, the net effect of transposable element activity is harmful to the host organism, and so they are examples of selfish genetic elements.

Each of these points is elaborated upon below. The essay only deals with transposable elements of animals, plants, and fungi; those found in bacteria may differ in some important respects, but their evolutionary biology is not well understood.

Structure and Mechanism. There are three main types of transposable elements. The simplest are the DNA transposons, which typically contain a single gene encoding a "transposase" and have inverted repeats at their ends, which may be 10s or 100s of base pairs long. Once made, the transposase protein recognizes the inverted repeats of the element, cuts the element out of the genome, and inserts it elsewhere, often nearby. This transposition reaction occurs during or after the host cell has replicated its DNA, prior to cell division, and can lead to an increase in the number of transposable elements in two ways. First, the element may preferentially move from replicated DNA to unreplicated DNA, in which case it will be replicated twice in a single cell cycle. Second, the chromosomal break left by the excised element can be mended by the cell's recombinational repair system, which uses a template, usually the sister chromatid (which still contains a copy of the element). The result is that the element is reconstituted at the site from which it was excised, as well as being inserted somewhere else in the genome. *Activator (Ac)* elements of maize and *P*-elements of *Drosophila* are among the best studied DNA transposons.

The second group of transposable elements are the LINEs (*L*ong *I*nterspersed *N*uclear *E*lements; also known as non-LTR retrotransposons), which transpose via an RNA intermediate. Elements are transcribed into RNA (leaving the original DNA copy intact), and the transcript is then translated into one or two proteins, which bind to the RNA molecule from which they were synthesized. The protein-RNA complex moves back to the nucleus, and one of the proteins nicks the host DNA, reverse-transcribes the RNA into DNA, and causes the DNA to be ligated into the host genome. Some LINEs are like DNA transposons and can integrate at very many places in the genome, while others are site-specific, targeting particular sequences in multi-copy host genes. For example, *R1* and *R2* are LINEs of insects which insert only into specific places in the 28S rRNA genes. Amusingly, *Tx1L* elements of *Xenopus* frogs specifically target a DNA transposon, *Tx1D*, and *Zepp* elements of *Chlorella* algae have a tendency to insert into pre-existing copies of themselves!

Finally, there are the LTR (*L*ong *T*erminal *R*epeat) retrotransposable elements, which also transpose via an RNA intermediate. These are more complex, encoding two structural proteins and three enzymes, and have at their ends the direct repeats that give them their name. Again, transposition begins with transcription. Some transcripts are translated into proteins, and those proteins work on other transcripts to encapsidate them, reverse transcribe them into DNA, and insert the DNA back into the host genome. These elements are closely related to retroviruses, including HIV, the cause of AIDS.

Population Biology. Transposition rates are generally low. In *Drosophila*, elements transpose about once every 10,000 generations, though there is much variation (at least a factor of 10) among families of transposable elements and among strains of flies. In special "dysgenic" crosses, up to a quarter of *P*-elements can transpose in a single generation. Even the lower rates of transposition are enough to have a dramatic effect over short evolutionary time scales. For example, with a transposition rate of 10^{-4}, the abundance of elements is expected to double in about 7,000 generations (c.600 years for *D. melanogaster*), or increase ten fold in 23,000 generations (c.2,000 years), if unopposed by selection (assuming all new inserts are functional; calculated using $y = (1 + t)^g$, where y is the proportional increase in copy number, t is the transposition rate, and g is the number of generations).

In fact, the spread of a transposable element through a host population will be somewhat slower than these calculations imply, as effects on the host are usually deleterious, and so natural selection on the host population will work against them. Again, the best data are for *Drosophila*. High rates of *P*-element transposition cause chromosome breakages and rearrangements, which in turn cause reduced fertility and increased embryonic lethality. Moreover, some 5–10 percent of insertions cause recessive lethal mutations. The nonlethal inserts also appear to reduce fitness: in one study of *P*-element insertions on the 3rd chromosome, viability in the laboratory was reduced about 5.5 percent per insert when heterozygous, and 12.2 percent when homozygous. In *D. simulans*, flies with just two active *mariner* elements die about four days earlier than those with none (57.6 v. 61.4 days).

The deleterious nature of transposable elements is also apparent in other ways. About 50 percent of visible mutations in *D. melanogaster* are due to transposable element insertions, and about 10 percent in mice. In humans, about 0.2 percent of *de novo* mutations causing genetic disease are due to retrotransposable element insertions. Transposable elements can also be harmful in causing recombination between elements at different locations in the genome. In humans, 0.3 percent of genetic disease is caused by such ectopic recombination between transposable elements. Finally, formation of transpositional intermediates is also likely to put a burden upon the host cell.

Despite these harmful effects, transposable elements can still invade a host gene pool. Indeed, this is possible even if every single insert is harmful to the host organism and reduces its fitness. Deleterious inserts are inevitably driven extinct by natural selection on the host, but as long as an insert transposes to a new location on average more than once before this happens, then it will increase in frequency within the population.

The ability of transposable elements to spread through a population has been dramatically—and fortuitously—demonstrated by *P*-elements. These have established themselves in the *Drosophila melanogaster* gene pool in just a few decades, having been introduced, somehow, from *Drosophila willistoni*, a distant relative. Of Seven American *Drosophila melanogaster* strains collected before 1950, only one has any *P*-elements, and of two French and Soviet strains collected before 1966, none has any *P*-elements. However, strains collected since that time are increasingly likely to carry *P*s, and now they appear to be present in all natural populations.

Further insight into this invasion comes from analysing the frequencies of P-elements at particular sites in the genome. Surveys of natural populations show that at any one site, insertions tend to be rare. This indicates that *P*-elements have not spread because some inserts have been beneficial and increased in frequency by natural selection. Rather, most inserts appear to be deleterious (though neutral or very small beneficial effects cannot be ruled out simply because there has not been enough time for such mutations to increase in frequency, given that *P*-elements only invaded the *Drosophila melanogaster* gene pool sometime this century). There is one notable exception to the low site occupancy, namely a high frequency of insertions at a site near one tip of the X chromosome. This exception appears to prove the rule, for it apparently has been selected because it represses *P*-element activity! The fact that repressors are selected for is further evidence for the parasitic nature of *P*-elements. It would be interesting to know whether this element has retained the ability to transpose, or whether it has lost it and completely gone over to "the other side." Since transposition is costly, mutants that are unable to transpose should be even more favoured by natural selection on the host population.

Evolutionary Dynamics. Though transposable elements can be highly invasive, the long term fate of many of them is extinction. Many genomes have remnants of transposable element families that once were active, but are now moribund. A key factor in these extinctions is likely to be the weakness of purifying selection maintaining function. In a genome containing, say, fifty copies of an element, if one of them is unable to make functional proteins, but is still recognized by proteins made from other elements, then it will be able to transpose as fast as the functional ones. Such nonautonomous elements can therefore accumulate as parasites of the functional elements. Even for the recently invading *P*-elements of *Drosophila melanogaster*, nonautonomous elements typically outnumber complete ones by a ratio of 2:1. Eventually, they may totally replace the autonomous elements, and then the family will be extinct, leaving behind only fossils in the genome.

How, then can transposable elements persist over long evolutionary time spans? The answer appears to be regular horizontal transmission between species, with an average of at least one transfer to a new species before extinction in the old one. There is an ever-growing list of examples of horizontal transmission. We have already mentioned the example of *P*-elements moving from *Drosophila willistoni* to *Drosophila melanogaster*, and recently it has been found that *copia* LTR retroelements have moved in the opposite direction, from *Drosophila melanogaster* to *Drosophila willistoni*. These species ranges have only been overlapping for about 200 years, indicating that horizontal transmission need not be a rare event, at least in some taxa. All else being equal, one expects species for which it is easy to acquire genes by horizontal transmission should have more types of transposable elements than species for which it is more difficult.

Interactions with the Host. As indicated above, most inserts are likely to be harmful to the host organism. However, this need not always be so, and many examples are now known of regulatory regions for host genes that are derived from transposable element insertions. Indeed, in our own genome there are hundreds of thousands of fragments of transposable elements, the vast majority of which have gone to fixation sometime in our tens of millions of years of mammalian ancestry. It is one of the key unanswered questions in transposable element biology what fraction of them (or their equivalents in other genomes) went to fixation because they were beneficial, and what fraction were "effectively" neutral, having been fixed by drift, aided perhaps by hitchhiking. Transposable elements can also cause duplications, deletions, and rearrangements of host genetic material in a bewildering number of ways. As with insertions, the majority of these are likely to be harmful to the host, but some fraction might be beneficial. Transposable elements may thus be important sources of genomic variability, particularly as they produce mutations that are unlikely to arise by other means (e.g., polymerase errors, DNA damage).

As well as particular inserts or rearrangements being useful to the host organism, it is also possible for some transposable element functions (e.g., reverse transcription, cleaving and rearranging DNA, etc.) to be useful to the host. In this case it is possible for a transposable element to be "domesticated." For example, a key component of the vertebrate immune system is that they assemble their immunoglogulin (Ig) and T-cell receptor (TCR) genes in their lymphocyte cells by somatic recombination, and do so slightly differently in different cells, thereby allowing the organism to recognize a tremendous diversity of antigens. This somatic recombination is initiated by proteins encoded by the RAG1 and RAG2 genes cleaving the Ig and TCR genes at inverted repeats. It is thought that RAG1, RAG2, and the inverted repeats they recognize are descendants of an ancient, DNA transposon that has since been domesticated for host benefit.

These potential long term benefits notwithstanding, transposable elements are still best seen as parasites, or selfish genetic elements, because the average effect of their activity on the host organism is harmful. The very structure of transposable elements is incomprehensible under the alternative scenario of them being host adaptations. Why don't they carry genes that benefit the host? Why don't they have mechanisms for excision without integration? Why are the proteins that carry out the transposition reaction encoded by the elements themselves, instead of being stable components of the genome which transpose unrelated fragments of DNA? Similarly for the host: why do they have mechanisms which apparently function to suppress dispersed repetitive DNA, either by silencing it or mutating it? Only by recognizing the conflict in the genome can one make sense of the adaptations of host and parasite.

BIBLIOGRAPHY

Anxloabéhère, D., M. G. Kidwell, and G. Periquet. "Molecular Characteristics of Diverse Populations Are Consistent with the Hypothesis of a Recent Invasion of *Drosophila melanogaster* by Mobile *P* Elements." *Molecular Biology and Evolution* 5 (1988): 252–269.

Biémont, C., F. Lemeunier, M. P. Garcia Guerreiro, J. F. Brookfield, C. Gautier, S. Aulard and E. G. Pasyukova. "Population Dynamics of the Copia, mdg1, mdg3, Gypsy, and P Transposable Elements in a Natural Population of *Drosophila melanogaster*." *Genetical Research* 63 (1994): 197–212.

Jordan, I. K., L. V. Matyunina, and J. F. McDonald. "Evidence for the Recent Horizontal Transfer of Long Terminal Repeat Retrotransposon." *Proceedings of the National Academy of Sciences USA* 96 (1999): 12621–12625.

Lohe, A. R., E. Moriyama, D.-A. Lidholn, and D. L. Hartl. "Horizontal Transmission, Vertical Inactivation, and Stochastic Loss of *Mariner*-Like Transposable Elements." *Molecular Biology and Evolution* 12 (1995): 62–72.

Mackay, T. F. C., R. F. Lyman, and M. S. Jackson. "Effects of *P* Element Insertions on Quantitative Traits in *Drosophila melanogaster*." *Genetics* 130 (1992): 315–332.

Maside, X., S. Assimacopoulos, and B. Charlesworth. "Rates of Movement of Transposable Elements on the Second Chromosome of *Drosophila melanogaster*." *Genetical Research* 75 (2000): 275–284.

Roth, D. B. and N. L. Craig. "VDJ Recombination: A Transposase Goes to Work." *Cell* 94 (1998): 411–414.

Sherratt, D. J. ed. *Mobile Genetic Elements*. Oxford, 1995. A relatively recent collection of reviews on the molecular and evolutionary biology of transposable elements, written at a relatively easy level.

— AUSTIN BURT

TRILOBITES. *See* Arthropods.

TUNICATES. *See* Vertebrates.

U–V

UNITS OF SELECTION. *See* Levels of Selection.

VACCINATION

The successful eradication of smallpox is undoubtedly one of the major achievements of modern medicine, an extinction event for which no tears are shed. It was made possible because of the availability of a safe, cheap, and effective vaccine. However, such a vaccine had been available for 170 years before eradication was achieved. Equally, there are many other childhood infectious diseases that are themselves important causes of mortality for which we already have effective vaccines but for which there is no immediate prospect of eradication. Thus, there is more to controlling diseases through vaccination than developing effective vaccines.

After Jenner's Experiment, the History of Vaccination. After Edward Jenner's 1796 experiment (see Vignette), the next great advances in the development of vaccines were demonstrations in 1880 and 1881 by the French chemist and microbiologist Louis Pasteur (1822–1895) of successful vaccination against cholera in chickens and against anthrax in sheep and goats. But the most dramatic of Pasteur's discoveries came in 1885 with the administration of rabies vaccine to two boys who had been bitten by rabid dogs. Killed vaccines were the next step forward, and by the late 1890s, killed vaccines against typhoid, cholera, and plague became available. The repertoire of live and whole-killed vaccines continued to grow during the first half of the twentieth century. The main technological advance was the introduction of vaccines based on the toxins produced by tetanus and diphtheria. The second half of the twentieth century saw the discovery of cell-culture techniques and the ability to create live-attenuated vaccines. By the end of the twentieth century, vaccination had brought ten important infections under control, setting the stage for the modern era of vaccination.

Types of Vaccines. Figure 1 summarizes the range of applications of vaccination and also the breadth of approaches used to create prophylactic vaccines for infectious diseases. For prophylactic vaccines against infectious diseases, there is a major dichotomy between the passive and active vaccines. Passive vaccines are preparations of preformed antibodies that give temporary protection to the recipient. Active vaccines stimulate the development of the recipient's own immune response to an infection. The resulting protection is typically much longer lasting than that endowed by passive vaccines but is slower to develop.

Passive vaccines. Passive vaccines are used when protection is required immediately, either because the recipient has already been exposed to infection or because the recipient is at particularly high risk of serious disease if exposed. An example of postexposure prophylaxis with a passive vaccine is the administration of rabies immune globulin to people who have been bitten by rabid dogs. This preparation is made from pooled human serum from people known to have high levels of antirabies antibodies. The pooled human serum is treated to inactivate any human immunodeficiency virus (HIV) and hepatitis B viruses that might be present. In the future, it is hoped that pooled human serum will be replaced with recombinant human immunoglobulin, thus avoiding the danger of transmitting infections.

Active vaccines. Active vaccines work by eliciting a new immune response from the recipient, followed, as in an infection, with an immunological "memory." Despite the millennially old observation that after recovering from some infections, individuals tend never to succumb again, the mechanism of this immune memory is still not fully understood and remains one of the most beguiling open questions of modern biology. The prep-

FIGURE 1. Breadth of Applications of Vaccines and of Approaches for Generating Vaccine Antigens.
A. R. McLean. "Mathematical Modelling of the Immunisation of Populations. *Reviews in Medical Virology* 2 (1992): 141–152.

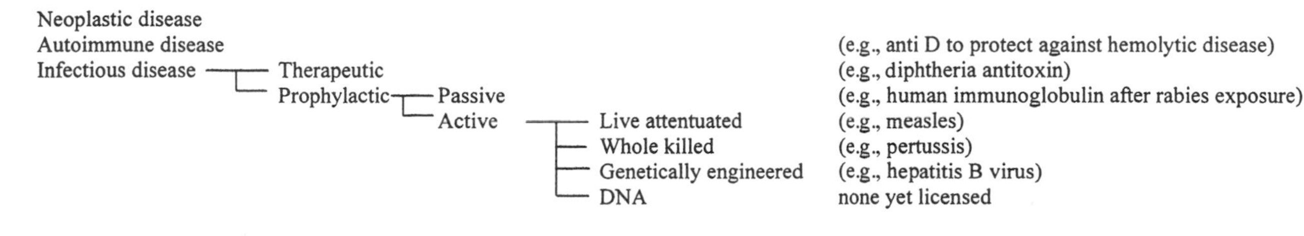

THE DISCOVERY OF VACCINATION

The English physician Edward Jenner (1749–1823) pioneered vaccination, in particular, smallpox inoculation. Jenner lived and worked in the village of Berkeley in rural Gloucestershire. He had trained in London with a leading academic surgeon and, despite his country home, was a member of the intellectual establishment.

In the decades before his birth, it had become widespread practice to purposefully infect children with pus from smallpox scabs to protect them from subsequent infections. This practice, called variolation, had been widely publicized by Lady Mary Montagu after she had seen it take place in Turkey. Jenner himself had been inoculated as a child. He was also aware of the widespread belief among his fellow country dwellers that dairy maids who had suffered from cowpox never subsequently caught smallpox.

Jenner took these two pieces of information and performed a world-changing experiment, the first ever vaccination. In 1796, he inoculated an eight-year-old boy with cowpox by smearing pus from a dairy maid's cowpox lesion into two cuts on his arm. Six weeks later, he inoculated the same boy with smallpox, the boy did not fall ill. This was the first experimental vaccination, and it led to the widespread adoption of the practice and, nearly two centuries later, to the eradication of smallpox.

—ANGELA MCLEAN

arations that elicit this "magical" response are made in a variety of ways. From Mary Montagu's and Jenner's cross-reacting poxy scabs (see Vignette) to the latest high-technology conjugated proteins, the drive to develop safe vaccines uses the whole range of old and new biological techniques.

Live attenuated vaccines. Because the objective of vaccination is to avoid the morbidity associated with natural infection, specially prepared live vaccines need to be used so that they do not cause serious disease in recipients. In the case of smallpox, there was a naturally occurring cross-reacting virus, cowpox, that could be used. The more common situation is that a virus recovered from a human infection is grown in animal cells in vitro for many generations. In ways that are not fully understood, this process can lead to the selection of strains that are attenuated, that is, that cause an infection with few symptoms. Measles, mumps, rubella, and BCG (antituberculosis) vaccine are all live-attenuated vaccines, as are the live versions of polio and influenza vaccines. The advantages of live vaccines include longer-lasting protection, broader protection, and a necessity for fewer doses. One disadvantage is that attenuated vaccines can occasionally cause the disease they are intended to prevent.

Killed vaccines. It is not always possible to generate a suitably attenuated live vaccine, in which case killed vaccines can be used. In many cases, the complete infectious agent is inactivated and administered whole. Rabies virus vaccine and the killed vaccines for influenza, poliovirus, and pertussis all use this strategy. Whole pathogen vaccines have the advantage of eliciting broadly cross-reactive immunity and do not present problems of reversion to virulence that can occur with live vaccines.

Influenza vaccine. The epidemiology of influenza is greatly affected by the virus's ability to change its antigenic properties. The virus has two surface antigens, hemagglutinin (HA) and neuraminidase (NA), which display rapid, immune-driven evolution. Because of this constant antigenic change, the formulation of influenza vaccine is changed every year. Each February, the World Health Organization (WHO) issues recommendations about which influenza strains should be used in vaccines for the coming season. These recommendations are based on surveillance data for currently circulating viruses.

Genetically engineered vaccines. The technologies of genomic manipulation are increasingly used in the manufacture of vaccines. One of the first applications was in the production of a vaccine against hepatitis B. The hepatitis B surface antigen (*HbsAg*) gene is expressed in bakers' yeast, then isolated and used in the vaccine. In this case, the new technology overcame a problem of generating large enough quantities of antigen. Another application of recombinant technology is in the production of conjugated proteins, where the immunogenicity of an antigen is boosted by binding it to a highly immunogenic protein. One of the causes of meningitis is the pathogen *Haemophilus influenzae* type B. A vaccine consisting of the polysaccharide capsule (itself poorly immunogenic in young children) conjugated with tetanus toxoid or another immunogenic protein has been used in a highly successful program controlling this source of meningitis.

DNA vaccines. Vaccines consisting of DNA encoding the desired antigen are the latest advances in vaccine technology. The DNA is taken up by host cells, which then synthesize the antigen. Such vaccines offer particular promise as they elicit cellular as well as humoral immune responses.

What Makes a Good Vaccine? When an individual is successfully vaccinated, not only is there one less person who will be infected, there is also one less person who will be infectious. This confers an element of protection upon those individuals in the same community who have not received the vaccine. This is the basic idea

TABLE 1. Summary of Epidemiological and Demographic Parameters

The table shows the basic reproductive ratio of virus and critical vaccination proportion for eradication.

Infection	*Region*	*Date*	*Mean Age at Infection, A (years)*	*Reciprocal of Average Birth Rate, B (years)*	R_0	p_c (%)
Measles	United Kingdom	1950s	5.0	70	15.0	93
Measles	Senegal	1964	1.8	31	18.0	94
Smallpox	West Africa	1960s	15.0	20	2.3	57
Smallpox	India	1960s	12.5	30	3.4	70

Source: McLean, A. R. "Mathematical Modelling of the Immunization of Populations." *Reviews in Medical Virology* 2 (1992): 141–152.

underlying the principle of herd immunity. The question of how a vaccine will affect the spread of an infection among a population of hosts is essentially an ecological one. A substantial body of work exists exploring the impact of vaccines on the population biology of host–parasite interactions. Much of that work consists of simple mathematical models that predict the outcome of defined levels of intervention.

The basic reproductive ratio $\mathbf{R_0}$. An obvious question to ask is what fraction of the population must be vaccinated to eradicate an infectious disease. For example, why was smallpox eradicated with mass vaccination reaching 80 percent of the population in West and Central Africa, but these same levels of mass vaccination were not sufficient in India and other Asian countries? A single parameter that captures all the different aspects of the host–parasite interaction has proved extremely useful in such discussions. The basic reproductive ratio, or R_0, it is defined as the number of secondary cases caused by one infectious individual introduced into a wholly susceptible population. If $R_0 < 1$, then that infection will die out, whereas if $R_0 > 1$, there is likely to be an epidemic of that infection within that population.

R_0 can be calculated from two easily observable parameters, the average age at infection (A) and the reciprocal of the average birth rate (B). The relationship is:

$$R_0 = \frac{B}{A} + 1 \tag{1}$$

Some illustrative values of R_0 are given in *Table* 1.

The impact of perfect vaccines. The most straightforward situation is that of a vaccine that gives complete, lifelong protection in all recipients. While no such vaccine actually exists, the situation serves as a useful caricature in developing a theoretical description of the impact of vaccines on host–parasite population biology. In considering the community-level impact of vaccines, it is useful to introduce the vaccinated reproductive ratio (R_p), defined as the number of secondary cases caused by one infectious individual introduced into a population where a fraction p have been vaccinated and everyone else is susceptible.

For vaccines that protect completely,

$$R_p = (1 - p)\, R_0 \tag{2}$$

The eradication criterion for an infectious agent is derived by setting $R_p = 1$. Thus, for a perfect vaccine, infection is eradicated when p, the fraction vaccinated, exceeds the critical vaccination proportion, p_c, where

$$p_c = 1 - \frac{1}{R_0} \tag{3}$$

Table 1 lists some numerical values for p_c. Notice that the levels predicted to eradicate smallpox are less than 70 percent, whereas 80 percent vaccine coverage was insufficient to eradicate smallpox from some Asian countries. This discrepancy highlights the extreme simplicity of these one-line models. They serve as useful tools for understanding how demographic and epidemiological processes interact, but for planning control programs, models that incorporate more of the true complexities of disease transmission are needed.

Imperfect vaccines. Of course, vaccines are not perfect. They may completely fail in some people, they may give only partial protection, or their protection may wane over time. These complexities can be incorporated into the framework described above and yield some counterintuitive results. Suppose a vaccine works in only a fraction ε of people and reduces their per exposure infection probability by an amount ψ, and that protection wanes so that the individual is protected for fraction θ of his or her life. It can be shown that, under these circumstances, the equivalent of equation 2 is:

$$R_p = (1 - \varepsilon\psi\theta_p)R_0 \tag{4}$$

This simple equation reveals two aspects of imperfect vaccines that are not otherwise immediately obvious. First, the property of lifetime protection from a vaccine is an extremely important determinant of the community-level impact of a vaccine. For example, a hypothetical anti-HIV vaccine that gave protection that wanes with a half-life of ten years protects for approximately one-fifth of lifetime ($\theta = 0.2$). Equation 4 tells us that such a vaccine (even if perfect in every other way) would only be as good as a vaccine that worked in 20 percent of recipients, but gave complete and lifelong protection. Thus, community-level impact is highly sensitive to the duration of protection. Second, because imperfections in vaccines compound in a multiplicative manner, even quite promising vaccines can have a rather small impact at the community level. For example, for a vaccine that takes in 85 percent of people and endows 90 percent protection that lasts for two-thirds of a lifetime, the combination $\varepsilon\psi\theta = 0.5$. The implication of this is that for any infection for which R_0 is greater than 2, even a campaign that reaches every new recruit cannot eradicate infection.

These calculations show just how good the first live-attenuated vaccines were in that they gave very high efficacy and lifelong immunity.

Vaccine escape. The evolution of vaccine-resistant strains of infectious agents is potentially a huge problem for their control by immunization. Yet, for many infectious diseases, it has been possible to drive them to the verge of extinction (or, for smallpox, beyond) without vaccine escape mutants arising. Mathematical modeling that sets out to ask why this should be so reveals that it is probably a property of the breadth of response endowed by live and whole-killed vaccines. This breadth of response means that the eradication criterion for both the observed strain and any vaccine escape strain will be met before the vaccination campaign gives competitive advantage to any vaccine escape strain. However, another, more worrying possibility is revealed: that vaccine escape mutants could arise many decades after the introduction of a vaccination campaign, and that initial success is no guarantee that eradication can be achieved without vaccine escape mutants arising.

Vaccine escape is a common observation for influenza vaccine and has recently been reported for pertussis and for hepatitis B.

Future Prospects for Vaccination. Vaccination has been one of the major success stories of medical intervention. However, to date, the majority of successful vaccines have been against infectious diseases of childhood that endow lifelong immunity. Important exceptions are hepatitis B and rabies vaccines. The next big challenge is to develop effective vaccines against persistent microbial infections and against parasites. Much attention focuses on the development of vaccines to protect against HIV with its attendant problems of rapid evolution and antigenic variation during a single infection. New technologies that make vaccines to invoke both cellular and humoral immunity hold great promise. The list of new or improved vaccines required is long—one often cited list from the 1990s contains forty items. The proven benefits of safe, cheap, and effective vaccines, intelligently delivered through well-managed logistics, are so great that the development of such a vaccine remains the Holy Grail for every infectious agent.

[*See also* Acquired Immune Deficiency Syndrome, *article on* Dynamics of HIV Transmission and Infection; Basic Reproductive Rate (R_0); Disease, *article on* Infectious Disease; Emerging and Re-Emerging Diseases; Immune System, *article on* Microbial Countermeasures to Evade the Immune System; Influenza; Plagues and Epidemics; Transmission Dynamics.]

BIBLIOGRAPHY

Anderson, R. M., and R. M. May. *Infectious Diseases of Humans: Dynamics and Control.* Oxford, 1991. The canonical text describing the mathematical modeling of infectious disease transmission dynamics. Also includes relevant data review.

Isaacs, D., and E. R. Moxon. "Immunization." In *Oxford Textbook of Medicine*, edited by D. J. Weatherall, J. G. G. Ledingham, and D. A. Warrell. Oxford, 1996. A brief discussion of current immunization practice, with a description of the different vaccines.

McLean, A. R. "Vaccination, Evolution and Changes in the Efficacy of Vaccines: A Theoretical Framework." *Proceeding of the Royal Society B* 261 (1995): 389–393.

McLean, A. R., and S. M. Blower. "Modelling HIV vaccination." *Trends in Microbiology* 3 (1995): 458–463. Covers aspects of the community-level impact of imperfect vaccines.

Moxon, E. R. "Modern Vaccines." In *Current Practise and New Approaches.* London, 1990. A collection of review articles first published in *The Lancet.*

Plotkin, S. A., and W. A. Orenstein. *Vaccines.* Philadelphia, 1999. A comprehensive volume covering all aspects of vaccines.

Reid, R. *Microbes and Men.* London, 1974. A popular science book describing both Mary Montagu's and Edward Jenner's contributions to the practice of vaccination.

— ANGELA MCLEAN

VARIABLE ENVIRONMENTS

Organisms are subject to all sorts of chance events that affect their reproductive success. At one extreme the environment as a whole may fluctuate, with a fluctuation having essentially the same effects on all population members. For example, all population members usually experience the same weather conditions. If rainfall fluctuates from year to year and high reproductive success occurs in rainy years, then whether or not one population member is lucky in terms of its reproductive success will be correlated with whether another population member is lucky. This type of randomness is called environmental stochasticity. For many predators, prey population sizes fluctuate from year to year and provide another source of environmental stochasticity. Superimposed on these large-scale fluctuations are chance events affecting different individuals in the population independently. For example at a given prey density, whether one cougar is lucky in finding prey will usually be independent of whether another cougar in the population is also lucky. This type of randomness is called demographic stochasticity.

The genotype of an individual influences its reproductive strategy. Since different strategies differ in their reproductive success, genotypes differ in the numbers of copies of themselves, or their constituent alleles, that are left in future generations. Natural selection can be expected to produce populations whose genotypes code for strategies that approximately maximize fitness; that is, maximize some appropriate measure of reproductive success. The type of stochasticity determines how fitness should be defined. Suppose first that there is demographic but no environmental stochasticity and consider an individual following a particular strategy. Define the descendants of the individual left next year to comprise any offspring produced this year that survive until then, and the individual itself if it survives. Then the fitness of the strategy is simply the mean number of such descendants, where numbers of descendants are discounted to take account of their relatedness to the focal individual. Here the mean is an average over demographic stochasticity.

In contrast consider the example of an annual plant species that is subject to growing conditions that fluctuate from year to year. For simplicity suppose that growing conditions are either good or bad, each condition being equally likely each year. If a seed germinates in spring and the growing conditions are bad then the new plant dies before maturity. If conditions are good then the plant matures producing a large number of seeds. Those seeds that do not germinate in spring remain dormant, and will germinate in some future spring provided they survive to do so. On average (averaged over growing conditions) a seed leaves more descendants next year if it grows than if it remains dormant. Suppose that the strategy of a plant determines the proportion of its seeds that germinate as opposed to remaining dormant. Consider the strategy of having all seeds germinate. Then a plant following this strategy will leave the highest mean number of descendants next year, where the mean is an average over the different growing conditions. However, the first time that a bad year occurs the line of descent of all plants following the strategy will terminate. This strategy is thus not one that is likely to evolve, and by any reasonable measure it has very low fitness. Contrast this with the strategy of having some seeds grow immediately and some remain dormant until future years. The line of descent of a plant following this strategy is protected, since each time a bad year occurs dormant seeds survive even though those that grow die. This strategy thus has higher fitness (Cohen 1966).

The above example shows that fitness cannot be based on just averaging over environmental stochasticity. The proportionate growth, over a period of several years, in the number of organisms following a strategy, is the product of the growth in each year in the period (cf. Demster 1955). The definition of fitness takes this multiplicative property into account. Specifically, let $r(s)$ be the mean number of descendants left next year when current environmental conditions are s. The mean here is an average over demographic stochasticity for fixed s. Fitness is then the geometric mean of $r(s)$, where this mean is an average over environmental conditions s. This fitness measure is called geometric mean fitness. In mathematical terms it is given by

$$\log \text{(geometric mean fitness)} = \text{mean } \{\log(r(s)\}.$$

(Cohen, 1966; Lewontin and Cohen, 1969).

The above fitness definition is valid in unstructured populations; that is it assumes that all descendants are equivalent. For example in the seed germination problem, all descendants are seeds since the parent plant dies, and it is reasonable to assume all seeds are the same. The population is thus unstructured. In contrast, suppose that we were dealing with a perennial plant such as a tree. Then both the tree itself and its seeds could be descendants, and these are certainly not equivalent. The calculation of fitness is then more complex, being found as the dominant Liapunov exponent of the (random) population projection matrix (Tuljapurkar, 1990).

In the seed germination example, allocating some seeds to dormancy, rather than all seeds to growth, reduced the variance (across environmental conditions) in descendant numbers. Fitness is thus increased even though the mean number of descendants per annum is reduced. In fact fitness can be approximated as follows.

$$\text{Geometric mean fitness} \simeq R - \frac{\sigma^2}{2R}$$

where R is the mean of $r(s)$ and σ^2 is the variance of $r(s)$, with the average in both cases being over environmental conditions, s (Lewontin and Cohen 1969, Gillespie 1977). Thus fitness increases with R but decreases as σ^2 increases. Fitness can be increased by reducing σ^2 in the following two ways (Seger and Brockman 1987).

Reduce the variance in the number of offspring left by each individual at the expense of a reduction in the mean number. Two cases illustrate this principle. (i) Large clutches of eggs may be more affected by adverse conditions than small clutches. Producing a large clutch thus has high R but high σ^2, and the clutch size maximizing fitness is smaller than that maximizing mean descendant number R (Boyce and Perrins, 1987; Haccou and McNamara, 1998). (ii) If some breeding seasons are good and others are bad, then the longer an organism lives the more likely that it will encounter at least one good season during its life. Longevity, with many reproductive bouts in which few offspring are produced, has thus been suggested as a mechanism for increasing fitness by reducing variance in lifetime reproductive success (Schaffer, 1974; Seger and Brockman, 1987).

Diversify the phenotype expressed by a single genotype. The seed germination example illustrates that there may be selection for genotypes that code for a strategy of diversification, with different members taking different action. Diversification does not reduce the variance in reproductive success experienced by an individual genotype member, but reduces the variance in the total reproductive success of all members of that genotype. This spreading of the risk thus reduces the probability of catastrophic failure of the genotype as a whole (Cohen, 1966; Cooper and Caplan, 1982). There is an analogy with portfolio selection in finance, where diversification of investments reduces overall variability of the whole portfolio, and hence guards against large losses. Risk spreading has been extensively discussed in two other biological scenarios, diapause and plant growth. Some insects and other arthropods survive a part of the year in a protected phase called diapause. On emerging from diapause individuals grow to maturity and produce eggs, some of which enter diapause immediately with the rest growing to maturity again. This process can produce several generations within a single breeding season. The partial allocation of eggs to diapause appears to be an adaptation designed to cope with environmental uncertainty. If environmental conditions turn bad during the growth and maturation phase, then all individuals not currently in diapause are likely to die before they can reproduce. The individuals in diapause thus guard against a catastrophic failure for the genotype (Cohen, 1970). Similarly, it has been suggested that the graded allocation to growth and reproduction in some plants may be an adaptation designed to guard against unpredictable adverse growing conditions (King and Roughgarden, 1982; Amir and Cohen, 1990).

There are various ways in which adaptations to fluctuating environments differ from adaptations when there is no environmental stochasticity. The discussion of risk spreading illustrates that it may be optimal to essentially "flip a coin" to determine behavior. When there is no environmental stochasticity such randomized strategies never do strictly better than deterministic strategies. Furthermore, when diversification of phenotypes is adaptive it is not possible to specify the best action for one individual without considering the actions of other members of the genotype. For example, in the seed germination problem, whether it is best for one seed to germinate depends on what proportion of its sibling seeds germinate (cf. McNamara, 1995; McNamara, 1998). This means that in some circumstances it may be best to base actions on past history as well as current circumstances (McNamara, 1997).

A population at evolutionary stability is polymorphic if natural selection maintains two or more genetic types of individual in the population. Environmental fluctuations can stabilize a polymorphic population in the following way. Suppose that the population is composed of two types of individual, labeled 1 and 2. There are also two types of years. In a type 1 year, type 1 individuals are able to reproduce very successfully while type 2 individuals cannot reproduce. The converse holds in type 2 years. Suppose that type 1 individuals predominate in the population. Then in a type 1 year, many type 1 offspring are produced. But these offspring face strong competition since there are so many of them. Thus the advantage to type 1 individuals in type 1 years is reduced. If instead the year is type 2, then since there are few type 2 parents, there are relatively few type 2 offspring and these do not face strong competition. A similar argument holds when type 2 is the predominant type. Thus the rare type has an advantage, and this can stabilize a polymorphic mixture (Frank and Slatkin, 1990).

[*See also* Genetic Polymorphism; Natural Selection, *article on* Natural Selection in Contemporary Human Society; Phenotypic Plasticity.]

BIBLIOGRAPHY

Amir, S., and D. Cohen. "Optimal Reproductive Efforts and the Timing of Reproduction of Annual Plants in Randomly Varying Environments." *Journal of Theoretical Biology* 147 (1990): 17–42.

Boyce, M. S., and C. M. Perrins. "Optimizing Great Tit Clutch Size in a Fluctuating Environment." *Ecology* 68 (1987): 142–153.

Cooper, W. S., and R. H. Caplan. "Adaptive 'Coin-Flipping': A Decision-Theoretic Examination of Natural Selection for Random Individual Variation." *Journal of Theoretical Biology* 94 (1982): 135–151.

Cohen, D. "Optimizing Reproduction in a Randomly Varying Environment." *Journal of Theoretical Biology* 12 (1996): 119–129.

Cohen, D. "A Theoretical Model for the Optimal Timing of Diapause." *American Naturalist* 104 (1970): 389–400.

Dempster, E. R. "Maintenance of Genetic Heterogeneity." *Cold Spring Harbor Symposium on Quantitative Biology* 20 (1955): 25–32.

Frank S. A., and M. Slatkin. "Evolution in a Variable Environment." *American Naturalist* 136 (1990): 244–260.

Gillespie, J. H. "Natural Selection for Variance in Offspring Numbers: A New Evolutionary Principle." *American Naturalist* 111 (1977): 1010–1014.

Haccou, P., and J. M. McNamara. "Effects of Parental Survival on Clutch Size Decisions in Fluctuating Environments." *Evolutionary Ecology* 12 (1998): 459–475.

King, D., and J. Roughgarden. "Graded Allocation Between Vegetative and Reproductive Growth for Annual Plants in Growing Seasons of Random Length." *Theoretical Population Biology* 22 (1982): 1–16.

Lewontin, R. C. and D. Cohen. "On Population Growth in a Randomly Varying Environment." *Proceedings National Academy Science USA* 62 (1969): 1056–1060.

McNamara, J. M. "Implicit Frequency Dependence and Kin Selection in Fluctuating Environments." *Evolutionary Ecology* 9 (1995): 185–203.
McNamara, J. M. "Optimal Life Histories for Structured Populations in Fluctuating Environments." *Theoretical Population Biology* 51 (1997): 94–108.
McNamara, J. M. "Phenotypic Plasticity in Fluctuating Environments: Consequences of the Lack of Individual Optimisation." *Behavioral Ecology* 9 (1998): 642–648.
Schaffer, W. M. "Optimal Reproductive Effort in Fluctuating Environments." *American Naturalist* 108 (1974): 783–790.
Seger J. and H. J. Brockman. "What Is Bet-Hedging?" In *Oxford Surveys in Evolutionary Biology*, eds P. H. Harvey and L. Partridge, pp 182–211. Oxford, 1987.
Tuljapurkar, S. *Population Dynamics in Variable Environments*. Lecture Notes in Biomathematics, Berlin, 1990.

FURTHER READING

Yoshimura, Y., and C. Clark. *Adaptation in a Stochastic Environment*. Lecture Notes in Biomathematics. Berlin, 1993. An edited volume of research papers coving many topics in this area.
Houston, A. I., and J. M. McNamara. *Models of Adaptive Behaviour*. Cambridge, 1999. Chapter 10 provides an extended description of much of the material in this article.

— JOHN M. MCNAMARA

VERTEBRATES

As traditionally defined, Vertebrata includes about 50,000 living species of hagfishes and lampreys, cartilaginous and bony fishes, amphibians, reptiles, birds, and mammals. Most of those creatures are easily recognizable as "animals with backbones," and they occupy all major terrestrial, aquatic, and marine habitats, from the deep ocean floor to several miles above sea level. Vertebrates range in weight from less than a gram to more than 100,000 kilograms, and include humans as well as numerous other organisms that are variously important in our lives. About ten times as many species of vertebrates have existed in the past as are alive today, and the earliest fossil forms are from Early Cambrian marine deposits, about 530 million years ago. If living vertebrates are compared with the approximately 30 other major animal groups, only Arthropoda (horseshoe crabs, spiders, lobsters, centipedes, insects, and their relatives) exhibits comparable diversity in structure and natural history, and only Mollusca (snails, clams, squid, and related forms) even approaches them in terms of size variation and learning capacity.

Vertebrata is classified with Urochordata (tunicates) and Cephalochordata (amphioxus) in phylum Chordata because, at least as embryos, members of all three subphyla possess a stiff supporting rod (the notochord) along the body, as well as a dorsal hollow nerve chord and a muscular tail. Vertebrates are distinguished from other chordates by possession of a segmented vertebral column, an expanded head region, multicellular sense organs, a well formed skull that houses a three-part brain, a duplicate *Hox* gene complex, and a fourth embryonic tissue layer, the neural crest cells. In contrast to cephalochordates, their closest relatives, vertebrates initially differed in having larger body size, higher levels of activity, and predatory rather than filter feeding behavior. Although Agnatha ("jawless fishes") has long been treated as one of two basal-most lineages of living vertebrates, lampreys (Petromyzontoidea) and hagfishes (Mxyinoidea) are probably successively basal to gnathostomes ("jawed vertebrates"), rather than each other's closest relatives. Some zoologists even exclude hagfishes from Vertebrata and name the two of them together as a more inclusive Craniata. As is the case in living hagfishes and lampreys, bone was absent from the first Early Cambrian vertebrates, and that hard tissue type first appears in slightly more recent fossils of heavily armored, jawless fishes called ostracoderms.

Vertebrates are often important in human endeavors, including as domesticated work animals, pets, and food. Comparisons among living and fossils vertebrates have long played a major role in modern biology, from 19th century anatomists grappling with the concept of homology to twenty-first century molecular geneticists deciphering the developmental basis for evolutionary novelties (Hall, 1994; Carroll, 1997; Carroll et al., 2000). Many species of vertebrates are critically endangered as a result of recent environmental changes, and the whale shark, Komodo monitor lizard, giant panda, and other "charismatic megavertebrates" serve as flagship species in global efforts to conserve nature (for an excellent, detailed survey of vertebrate biology, see Pough et al., 2002).

Taxonomic Diversity, Evolutionary Relationships, and Classification. In traditional Linnean taxonomy, organisms are assigned to a ranked hierarchy of seven basic categories (kingdom, phylum, class, order, family, genus, species), each of which may be variously subdivided. Subphylum Vertebrata thus includes the following classes, with only a few of the included categories listed here: Agnatha (more than 100 species of hagfishes and lampreys), Chondrichthyes (almost 900 species of cartilaginous fishes, the sharks, skates and rays, and ratfishes), Osteichthyes (bony fishes, including 24,000 species of subclass Actinopterygii, the ray-finned fishes, and seven species of subclass Sarcopterygii, the fleshy-finned fishes, in the orders Actinistia [coelacanths] and Dipnoi [lungfishes]), Amphibia (4,900 species of salamanders, frogs, and caecilians), Reptilia (7,100 species in the orders Testudinata [turtles], Crocodylia [alligators and their relatives], Spenodontida [tuatara], and Squamata [lizards, snakes, and amphisbaenians]), Aves (9,100 species of birds), and Mammalia (4,500 species of mammals). The Linnean arrangement of vertebrates is thus as follows, for purposes of illustration including here only those subcategories mentioned above:

Kingdom Animalia
  Phylum Chordata
    Subphylum Vertebrata
      Class Agnatha
      Class Chondrichthyes
      Class Osteichthyes
        Subclass Actinopterygii
        Subclass Sarcopterygii
          Order Actinistia
          Order Dipnoi
      Class Amphibia
      Class Reptilia
        Order Testudinata
        Order Crocodylia
        Order Sphenodontida
        Order Squamata
      Class Aves
      Class Mammalia

There are several deficiencies in this traditional classification (de Quieroz and Gauthier, 1994; Greene, 2001). First, although Linnean taxonomy is often claimed to reflect evolution, with respect to vertebrates it actually implies only that all seven classes shared a common ancestor and it reveals almost nothing about the history of their subsequent diversification. Moreover, Linnean classification fails to designate with formal names three of the most significant innovations in vertebrate evolution—the origin of jaws, the well developed limbs that characterize tetrapods, and a shelled or amniotic egg in the common ancestor of reptiles and mammals. Perhaps most importantly, traditional nomenclature recognizes certain paraphyletic groups (those that include a common ancestor but not all of its descendants) and thereby obscures major aspects of the adaptive radiation of vertebrates.

By removing the more than 25,000 species of tetrapods from Osteichthyes and rendering that group paraphyletic (a group of organisms all descended from a common ancestor, but some of whose members are not included), the Linnean system erroneously implies that two ancient lineages of bony fishes, the ray-fins (including almost all of the organisms typically regarded as "fishes") and the fleshy-fins (consisting of coelacanths, lungfishes, and tetrapods), are wildly unequal in numbers of descendant living species. Likewise, by separating Aves and thereby making Reptilia paraphyletic, traditional classification fails to portray crocodylians and birds as each others' closest relatives among living vertebrates. Proponents of Linnean ranked taxonomy argue that by singling out the classes of tetrapods from fishes and birds from the rest of reptiles, an emphasis is placed on extent of evolutionary divergence and possession of special characteristics, such as terrestrial locomotion and flight. Why, however, using this logic, mudskippers that forage on land and bats that fly are not also afforded separate classes, respectively apart from Osteichthyes and Mammalia, is unclear.

An indented phylogenetic classification is based insofar as possible on monophyletic groups (those that include an ancestor and all of its descendants). Unlike traditional Linnean nomenclature, it thus provides formal names for jawed vertebrates, tetrapods, and amniotes, as well as explicitly recognizes the common ancestry of crocodylians and birds. A phylogenetic system also exactly mirrors the genealogy or family tree of vertebrates. The simplified phylogenetic classification that follows includes only a few of the many possible subcategories of vertebrates, for clarity of comparison with the traditional "seven classes" approach described above:

Vertebrata
  Agnatha
  Gnathostomata
    Chondrichthyes
    Osteichthyes
      Actinoptergyii
      Sarcopterygii
        Actinistia
        Choanata
          Dipnoi
          Tetrapoda
            Amphibia
            Amniota
              Mammalia
              Reptilia
                Testudinata
                Lepidosauria
                  Sphenodontida
                  Squamata
                Archosauria
                  Crocodylia
                  Aves

Milestones in Vertebrate Evolution. The first vertebrates were Early Cambrian marine organisms that moved water over their gills with specialized throat muscles, rather than with cilia, as is still the case in amphioxus and other more basal chordates. Additional key innovations of Early Silurian gnathostomes, roughly 450 million years ago, included modification of some gill support elements to form jaws, as well as the origin of true vertebrae, ribs, a complete lateral line sensory system, and paired pectoral and pelvic fins. The evolution of jaws was particularly important in the adaptive radiation of vertebrates, because together with teeth they permit grasping, manipulating, cutting, and even grinding otherwise unavailable food items. Chondrichthians retain many features of those ancestral gnathostomes, and perhaps the initial adaptive radiation of cartilaginous fishes was facilitated by locomotor and feeding advantages over the early jawless vertebrates. Most living chondrichthians are carnivorous, some species even eating large marine mammals, but the gigantic whale shark and manta ray are plankton feeders.

Osteichthyes first appeared in the Late Silurian and a spectacular radiation of bony fishes was well underway by the Middle Devonian, about 425 million years ago. About 73 percent of the Earth's surface is water and, not surprisingly, roughly half of all living osteichthians are ray-finned fishes, the Actinopterygii. Early ray-fins were about 5–25 cm long, and already distinguished from more basal gnathostomes by their highly flexible skull bones and jaws, specialized dentitions, locomotor diversity, and myriad ways of making a living. Although a few sharks and rays inhabit fresh water, among living vertebrates and in terms of numbers of species, actinopterygians rule both that habitat and the marine realm inhabited by most chondrichthians. In terms of diversity, extant ray-finned fishes run the gamut from tiny pet guppies and bizarrely shaped, deep-sea angler fishes to the schooling tuna and migratory salmon many of us use as food. Turning from ray-fins to the other basal lineage of bony fishes, ancestral sarcopterygians were 20–70 cm long, with stout, well-supported fins and massively larger jaws than those of actinopterygians. Fleshy-finned fishes initially diversified in marine and freshwater environments, and descendants of the early sarcopterygians survive today as coelacanths, lungfishes, and tetrapods. Thus, as members of that latter vertebrate branch in the tree of life, and in exactly the same sense that we are primates and we are mammals, humans are fishes.

By the Early Devonian, terrestrial habitats supported diverse assemblages of plants and arthropods, and thus were ripe for invasion by descendants of fleshy-finned fishes. About 0.5–1.2 m long and massively built, the first tetrapods had appeared by the Late Devonian, about 360 million years ago; they conquered land with an internal skeleton and its associated musculature that provided for support and locomotion, ribs and flank muscles that facilitated respiration, and a definitive tongue for capturing, tasting, and manipulating prey. Despite their eventual and spectacular success out of water, those ancestral tetrapods likely lived in aquatic, brackish, or even shallow marine habitats, so perhaps their "terrestrial" innovations first evolved for walking on pond bottoms, breathing air, and feeding in densely vegetated, seasonally dry sloughs.

Early tetrapods diverged into several major lineages, of which Batrachomorpha and Reptilomorpha are represented in modern faunas by amphibians and amniotes, respectively. Among extant tetrapods, Reptilomorpha encompasses about 21,000 living species of mammals and reptiles (collectively Amniota) and is thus now substantially more diverse than Batrachomorpha. Moreover, endothermy (the ability to metabolically maintain an elevated, constant body temperature) independently evolved twice within extant reptilomorphs, in mammals and birds. With respect to Batrachomorpha, although modern amphibians retain ectothermy (a reliance on behavioral rather than metabolic thermoregulation) and some other primitive tetrapod characteristics, they are at best vague icons for the initial conquest of land by vertebrates. Indeed, because of their large size, scaly exteriors, and flattened skulls, the Devonian tetrapod precursors of today's frogs, mice, lizards, birds, and their relatives would have appeared to us as rather more like a clumsy alligator than a salamander.

By the end of the Carboniferous, about 350 million years ago, reptilomorphs had diversified into many terrestrial habitats formerly dominated by batrachomorphs. Among those early reptilomorph lineages, amniotes are distinguished by the amnion, two other embryonic membranes, and the shelled egg, a suite of namesake features that protect embryos from desiccation. The ancestral Late Carboniferous and Early Permian amniotes were small organisms, also characterized by keratinized skin that retarded water loss, skeletal changes that facilitated speed and agility, and jaw specializations that probably enhanced predation; their initial diversification coincided with an extensive adaptive radiation of crawling, jumping, and flying insects. The subsequent evolution of amniotes, however, entailed larger body sizes and divergence into two lineages with extant representatives, the Synapsida (mammals and their extinct stem groups, the reptile-like pelycosaurs and successively more mammallike therapsids) and Reptilia.

A hallmark of the early diversification of Amniota was independent origins of certain feeding specializations. In early reptilomorphs, jaw closing muscles were confined beneath bones on the back sides of the skull, a condition known as anapsid ("no opening"). Most reptiles have a diapsid ("two openings") condition, in which fenestrae ("windows") on each side of the skull permit muscle enlargements and rearrangements that favor rapid snapping for capture of prey. In synapsids ("one opening") a single fenestra enables different muscular specializations and increased food crushing ability. With some impressive exceptions, such as the venom-conducting fangs of some snakes, reptiles generally have teeth that are all approximately the same shape and size (homodont). Synapsids typically exhibit extensively differentiated (heterodont) dentitions, encompassing teeth as different as the elongated canines of extinct saber-toothed cats and huge crushing molars of sea otters.

Synapsids had extensively radiated by the end of the Paleozoic, before some groups of reptiles had even arisen. Although many early synapsids were large animals, some weighing at least 250 kilograms, ancestral mammals were typically only about 100 millimeters long and weighed perhaps 50 grams. Mammalia originated in the Late Mesozoic, more than 65 million years ago, as the large dinosaurs were going extinct. Ranging in size from tiny shrews and bats to multi-ton elephants and whale, mammals are characterized by fur, endothermy, milk production (lactation), and lips for nursing. Although early mammals retained oviparity (egg laying habits), as do the extant platypus and echidnas (Monotremata), they subsequently evolved two radically different styles of viviparity (live bearing habits). Opossums, kangaroos, and other marsupials (Marsupialia) given birth to tiny, poorly formed young after gestation periods as short as a few days; much of their growth and development thus takes place postnatally, in a belly pouch, and is sustained by milk. Other living mammals (the Eutheria) have long gestation periods, nourish their developing fetuses by way of a tissue connection (placenta) with the mother's circulatory system, and give birth to relatively large young.

Turning to the other major amniote lineage, with the exception of birds and as is the case with amphibians, extant reptiles are generally ectothermic. Reptiles other than birds thus have low metabolic needs, averaging roughly 10 percent of that for a comparably sized endotherm, and they can be small; as a consequence, some reptiles survive and even achieve high population densities under conditions so harsh that they would prove impossible for most endotherms. Few birds and mammals weigh less than 3 grams, and many species are in the 100 grams to 1 kilogram range, where as at least 50 percent of ectothermic tetrapods weigh less than 5 grams, and many amphibians and lizards are in the 1 to 10 grams range. Major lineages of reptiles include turtles, renowned for their protective shells; the diverse and often abundant lepidosaurs; the well known archosaurs; and numerous entirely extinct groups, such as marine ichthyosaurs and plesiosaurs. Within Lepidosauria, Squamata provides a fascinating contrast with mammals, in that lizards, amphisbaenians, and snakes have collectively evolved viviparity more than one hundred times (Shine, 1988); parental care and complex sociality, so common in mammals and birds, are relatively rare in squamates.

Archosaurs are among the most diverse, spectacular, and popular of extinct and living vertebrates. Unlike other reptiles, which have generally retained a sprawling posture, and synapsids, which evolved an erect quadrapedal stance, archosaurs are usually at least partially bipedal. With respect to fossils, the giant dinosaurs are unparalleled in terms of public appeal, and no living terrestrial vertebrates are comparable to them in size and natural history. Nevertheless, nearly ten thousand extant species of birds, the surviving members of a dinosaur group called theropods, both captivate people everywhere and inspire intense controversy with respect to their archosaurian heritage. Among living vertebrates, sustained flight, extensive migratory behavior, and song are hallmarks of many birds, but some of the most often admired avian characteristics are also found in other archosaurs. The extinct pterodactyls and other pterosaurs, for example, were powerful fliers, and ranged in size from some as small as sparrows to one that had a wingspan of 13 meters! Parental care, nest building, and acoustic communication among adults and young are typical of all living crocodylians, as they were for at least some extinct dinosaurs, which suggests that those behavioral attributes were present in ancestral archosaurs (Carpenter et al., 1994; Clark et al., 1999). And endothermy, usually exhibited only by mammals and birds among living vertebrates, likely was also characteristic of pterosaurs and at least some extinct dinosaurs.

Repeated Themes in Vertebrate Biology. Vertebrates provide some of the best known case studies in evolutionary biology, as well as countless prospects for further research. Frogs, for example, exemplify adaptive divergence imposed on a suite of homologous specializations, those similarities that reflect their presence in the common ancestor of a group. The more than 4,200 living species of those amphibians all share large hind limbs and other specializations of the trunk, pelvis, and limbs that make them so well designed for jumping. In spite of that anatomical uniformity, various species also burrow in deserts, spend their entire lives in high Andean lakes, or parachute and glide about in tropical rainforest canopies. Although many frogs exhibit primitive tetrapod reproductive biology, involving external fertilization, aquatic eggs and larvae, and metamorphosis to terrestrial froglets, others lay eggs on leaves above a pond or stream, forsake the tadpole stage entirely, or give birth to fully formed young that developed in the female's back skin or stomach. The adaptive radiation of vertebrates also abundantly illustrates homoplasy or convergent evolution, the independent origin of similar attributes in distantly related organisms. Convergence often occurs in concert with similar natural histories, such as locomotion on difficult substrates or predation on particularly challenging prey types. Some lizards, for example, have repeatedly evolved counter-sunk lower jaws, nasal sink traps, and fringed toes, all adaptations for life in and on wind-blown desert sands (Arnold, 1995). Chameleons, frogs, and several groups of salamanders have independently and in different ways solved the problems of extreme tongue projection, as a means of shooting down flying or otherwise difficult to capture insects (Schwenk, 2000).

Vertebrates exemplify several macroevolutionary phenomena, among which one has profound implications for the future of this 500 million year old group of animals (Flannery, 1995). Therapsids, for example, were among the first animals to provide dramatic evidence for the reality of continental drift, when closely related fossils were found in Antarctica and Africa, reflecting the presence of those early mammal relatives on the ancient southern super continent of Gondwanaland. Mass extinctions also have been a recurrent macroevolutionary theme in vertebrate history, as illustrated by the loss of most kinds of dinosaurs at the end of the Mesozoic Era, about 65 million years ago. Megafaunal extinctions have occurred more recently on Australia (about 40,000 years ago), North America (about 12,000 years ago), and Madagascar (about 1,000 years ago), as well as on various small islands; for those land masses and within a relatively short time, the losses encompassed most large mammals as well as some turtles, lizards, and birds. Each of those megafaunal extinction spasms occurred shortly after the arrival of humans, despite the fact that they happened at widely different times and under different ecological regimes, thus implying that we played a causal role. Now, with the Earth's human population at six billion and pollution, deforestation, and other environmental onslaughts ever more pervasive, another major extinction spasm looms. Our most pressing challenge for the twenty-first century is to preserve as much as possible of the spectacular results of evolution, including our fellow vertebrates.

[*See also* Amphibians; Birds; Dinosaurs; Fishes; Mammals; Reptiles.]

BIBLIOGRAPHY

Arnold, E. N. "Identifying the Effects of History on Adaptation: Origins of Different Sand-Diving Techniques in Lizards." *Journal of Zoology* 235 (1995): 351–388.

Carpenter, K., K. F. Hirsch, and J. R. Horner. *Dinosaur Eggs and their Babies*. Cambridge, 1994.

Carroll, R. L. *Patterns and Processes of Vertebrate Evolution*. Cambridge, 1997.

Carroll, S. B., J. K. Grenier, and S. D. Weatherbee. *From DNA to Diversity: Molecular Genetics and the Evolution of Animal Design*. Malden, Mass., 2001.

Clark, J. M., M. A. Norell, and L. M. Chiappe. "An Oviraptorid Skeleton from the Late Cretaceous of Ukhaa Tolgod, Mongolia, Preserved in an Avianlike Brooding Position Over an Oviraptorid Nest." *American Museum Novitates* 3265 (1999): 1–36.

de Quieroz, K., and J. A. Gauthier. "Toward a Phylogenetic System of Biological Nomenclature." *Trends in Ecology and Evolution* 9 (1994): 27–31.

Flannery, T. *The Future Eaters: An Ecological History of the Australian Lands and People*. New York, 1995.

Greene, H. W. "Improving Taxonomy for Us and the Other Fishes." *Nature* 411 (2001): 738.

Hall, B. K., ed. *Homology: The Hierarchical Basis of Comparative Biology*. San Diego, 1994.

Pough, F. H., C. M. Janis, and J. B. Heiser. *Vertebrate Life*. Upper Saddle River, N.J., 2002.

Schwenk, K., ed. *Feeding: Form, Function, and Evolution in Tetrapod Vertebrates*. San Diego, 2000.

Shine, R. "Parental Care in Reptiles." In *Biology of the Reptilia*, edited by C. Gans and R. B. Huey, vol. 16, pp. 275–330. New York, 1988.

— Harry W. Greene

VESTIGIAL ORGANS AND STRUCTURES

Vestigial organs and structures (also called vestigia, rudiments, or remnants) are reduced body parts or organs, often without visible function in the derived bearers, that were fully developed and functioning in earlier members of that phylogenetic lineage. These structures, sometimes described as atrophied or degenerate, are usually small in comparison with their relative size in ancestral generations or in closely related species. Vestigia, etymologically derived from Latin *vestigium* ("footprint, trace, mark"), played an important role in the founding of evolutionary theory because they represented tangible traces of past generations in recent organisms. Today, anatomical vestigia do not represent a major topic in evolutionary research; their presence in all species is taken as given, although the causal mechanisms responsible for both their reduction and their retention are not fully understood.

Vestigial structures are identified by the comparative method. This means that homology (sameness, identity) of the reduced character with a more fully developed, ancestral counterpart must be established not only on the basis of structural and positional criteria but also with regard to the continuous presence of the structure in the lineage leading from the ancestral to the derived form. The latter criterion represents a critical distinction from features called *atavisms*, which occur after prolonged periods of complete absence of a character. Whereas atavisms appear only in exceptional cases and in single individuals, a vestigial structure is present in all representatives of a species. Another distinction should be made with regard to the term *rudiment*. Although often used synonymously with *vestigium*, a rudiment is properly a developmental primordium or anlage, and it coincides with a vestigium only when a primordium is retained in the adult.

Vestigialization affects structures and organs that have reduced or lost their function as the result of an evolutionary change in lifestyle of the derived forms. Classical examples are eye and limb reductions in cave-dwelling or burying organisms, wing reductions associated with ground dwelling, or dental and intestinal reductions attributable to changes of diet. The affected organs can show different stages of reduction in different members of a clade. Thus, we see different degrees of limb reduction in skinks, pelvic bones of different sizes in certain whales and snakes, and various stages of eye regression in vertebrate and nonvertebrate cave organisms, or of wing rudiments in flightless birds and insects. In many cases, the vestigia are present only in fossil forms and are completely lost in extant representatives of a lineage. In other cases, the rudiments are present only in the embryo or juvenile and are completely lost in the adult, such as teeth in the embryos of baleen whales. Extensive collections of vestigial structures from many organisms and from a wide range of organ systems can be found in classical treatises of comparative morphology. Wiedersheim (1893), for example, provides long lists of cases in human anatomy, including the often cited vermiform appendix, the pineal gland, the vomeronasal organ, the coccygeal bone, muscles of the ear, the third molar teeth, remnants of nephric and genital ducts, and many more. The degree to which vestigial organs are developed in the adult can vary extensively both within and among species. Often the structures are relatively more developed in the embryo than in the adult. When their presence is confined to the embryo, the idea of recapitulation is evoked.

The evolutionary mechanisms of vestigialization are thought to be well understood. It is argued that initial behavioral or environmental changes lead to a partial or complete loss of function of an organ, with consecutive selection against that structure, which would have become a burden or hindrance for further evolution. Yet the extent to which vestigialization can be explained in terms of adaptation is debated. Direct selection for vestigialization seems to be rare. Rather, organs often undergo reduction as a byproduct of other adaptive changes, such as limb reduction in association with body elongation or miniaturization. Other explanations invoke indirect selection through energy trade-offs or antagonistic pleiotropy. But nonadaptive scenarios prevail: structures may most often degenerate because of a relaxation of stabilizing selection and the accumulation of selectively neutral mutations.

The specific, developmental realization of vestigialization can be effected by various processes. Mutational changes can affect developmental control, disturb cell division and histological differentiation, or interrupt inductive interactions. Several general modes can be distinguished. Either the entire progression of an embryonic primordium is slowed down or halted, leading to the preservation of a rudiment in the adult; or development of an organ proceeds rather normally up to a certain point at which it is actively destroyed—for example, by hormonal activity, apoptosis, or phagocytosis, as in the tail of anuran tadpoles. A third possibility is that general, heterochronic shifts in larval or embryonic development simply prevent adult structures from forming fully. As a rule of thumb, morphological vestigialization proceeds in the reverse order of the sequence of normal development. For instance, digits of regressing reptilian limbs are deleted in the reverse order of their embryonic formation; eye reduction of cave-dwelling fish and salamanders follows a distal to central progression from cornea, to lens, to bulbus, to nervous components—roughly the opposite order of their formation.

The most interesting question of vestigialization probably concerns not the fact of structural regression itself, but rather the maintenance of remnants that have no obvious function in the derived bearers, or even seem to be under negative selection. An example is the vermiform appendix, which must have caused extensive numbers of prereproductive deaths during hominid evolution. Since it is quite unlikely that structures completely without function would be preserved, certain functions must be assumed to persist. One obvious class of such functions is their intermediate role in development, when a structure that is tightly integrated in developmental interactions cannot be removed without deleterious downstream effects. In other cases, vestigial structures may have acquired new, less obvious functions that differ from the original ones. Hence, a vestigium should not generally be considered without function, or only with respect to its ancestral, adult roles. Finally, vestigia may be maintained because of intact gene networks. Reduction or loss of a phenotypic structure does not necessarily mean that its genetic basis is lost too. Even structural genes that have not been activated for many millions of years, such as those for enamel proteins in birds or lens proteins in moles, are often retained in the genome.

The significance of vestigial structures, these echoes of the past, today lies mostly in the illustration of nonadaptive origins of certain features of organismal construction. They are also of taxonomic interest, and a number of morphological phenomena can be understood only if the processes of vestigialization are known. This is the case with atavisms, which are often based on embryonic vestigia, such as the occasional extra toes in horses. Another case is morphological innovation, which can take its origin from retained rudiments of reduced structures. Experimental enhancements of embryonic vestigia can also uncover suppressed developmental interactions. Finally, certain vestigia in humans are of medical significance, as exemplified by paraovarian cysts, branchial fistulae, or appendicitis. In general, vestigial structures and the processes of their formation shed light on the relationship between development and evolution.

[*See also* Adaptation; Atavisms; Homology; Recapitulation.]

BIBLIOGRAPHY

Darwin, C. *On the Origin of Species*. London, 1859. Chapter 13 illustrates the importance of vestiges and rudiments for the establishment of the theory of evolution.

de Beer, G. R. *Embryos and Ancestors*. London, 1958. A classic in the discussion of the mechanisms of vestigialization.

Fong, D. W., T. C. Kane, et al. "Vestigialization and Loss of Nonfunctional Characters." *Annual Review of Ecology and Systematics*. 26 (1995): 249–268. A comprehensive, modern review.

Griffiths, P. *Adaptive Explanation and the Concept of a Vestige. Trees of Life: Essays in Philosophy of Biology*. Dordrecht, 1992. A theoretical analysis of the relationship between biological function, adaptive explanation, and the concept of vestigialization.

Krummbiegel, I. *Die Rudimentation*. Stuttgart, 1960. Contains numerous examples.

Müller, G. B. "Ancestral Patterns in Bird Limb Development: A New Look at Hampé's Experiment." *Journal of Evolutionary Biology* 2 (1989): 31–47. An example of experimental reactivation of vestigeal structures.

Wiedersheim, R. *Der Bau des Menschen als Zeugniss für seine Vergangenheit*. 2d ed. Freiburg, 1893. Early discussion and comprehensive collection of vestigia in human anatomy. An English edition appeared in 1895.

— GERD B. MÜLLER

VICARIANCE BIOGEOGRAPHY. *See* Biogeography, *article on* Vicariance Biogeography.

VIRULENCE

Evolutionary biologists generally define virulence as the damage to the host caused by infection. The meaning of damage, however, depends on the context. When the topic is the effects of natural selection, damage corresponds to the decrement in evolutionary fitness, specifically inclusive fitness. When evolutionary arguments are applied to other disciplines, adjustments of the definition are useful. In medicine, for example, a lethal infection would normally be considered much more virulent than an infection that caused only infertility.

The virulence of an infection results from the combined effect of the inherent virulence of the parasite and the susceptibility of the host. Parasite virulence refers to the inherent harmfulness of the parasite within a group of similar hosts. Pathogen virulence is used analogously for unicellular or subcellular parasites.

Alternative definitions of virulence have emphasized the ability of a parasite to be productive, transmissible, and damaging, or some combination thereof. Virulence, host resistance, and parasite virulence, however, when defined in terms of damage and used together with other descriptive terms, such as transmissibility, provide the flexibility and clarity necessary for broad-ranging discussions of host–parasite associations and infectious disease.

The Tradeoff Approach to Virulence. For most of the twentieth century, experts in the health sciences concluded that parasites as a rule evolve toward reduced virulence or even mutualism. Although examples of evolution toward reduced virulence have been well documented, this generalization is not based on a sound application of principles of natural selection, which emphasize differential genetic success across generations rather than long-term compatibility.

The modern perspective assumes that high virulence results from high levels of host exploitation to fuel parasite growth, reproduction, or metabolism. Intense exploitation may benefit parasites evolutionarily by increasing their success at competing within hosts and the probability of infecting susceptible hosts upon contact. A trade-off arises because intense exploitation may decrease the probability of contacting susceptibles owing to host death or immobilization. These opposing effects of host exploitation appear to determine in large part the level of virulence favored by natural selection.

Within-host evolution requires genetic variation of parasites within individual hosts. This variation may arise from mutations or the recombination of pathogen genes within the host or from the invasion of the host with genetically distinct organisms. Mutations are a particularly important source of variation for within-host evolution among pathogens with high mutation rates per replicative cycle, continuous replication within hosts, and long durations of infection within each host. The human immunodeficiency virus (HIV), for example, may evolve increased virulence within people as a result of this process, which leads eventually to acquired immune deficiency syndrome (AIDS). [*See articles on* Acquired Immune Deficiency Syndrome.]

Within-host variation in pathogen virulence is also generated by transfer of virulence genes. The virulence of bacteria, for example, can increase upon acquisition of excisable genes, plasmids, or viruses, which coevolve with the bacteria they inhabit, which in turn coevolve with their hosts. The toxin gene in the agent of cholera, *Vibrio cholerae*, for example, is actually a viral gene, the expression of which is regulated by the *V. cholerae* bacterial proteins. As the transmissibility of virulence genes is becoming increasingly recognized, the distinction between community evolution and evolution of any particular pathogen species is becoming blurred.

The hypothesized tendency for virulence to increase when hosts are infected by more than one distinct lineage of the same parasite has remained largely untested. Although this tendency is generally accepted in theoretical arguments, decreased clonality (i.e., increased multiplicity of infection) may sometimes favor reduced virulence. Specifically, if pathogen virulence results from characteristics that benefit the group of pathogens inside of a host at a cost to the individual pathogen, then increased clonality may favor increased virulence by favoring the group success over the success of the individuals within the group. Production of cholera toxin, for example, is costly to *V. cholerae* but probably benefits the entire group of *V. cholerae* in the intestinal tract by shunting out competing species and dispersing *V. cholerae* in a liquid medium. Decreased clonality increases the probably that bacteria producing little or no toxin would be present in the intestinal tract to gain the benefits of toxin action without paying the cost of toxin production. The net effect of increased representation of these mild organisms would be reduced virulence.

In contrast, when the exploitation that causes virulence directly benefits the damaging pathogen, theory predicts that decreased clonality will increase virulence. Unresolved is the general extent to which within-host evolution actually increases virulence above the virulence that would have resulted if transmission opportunities were the only important selective force. Because different transmission modes create different opportunities and constraints, they may favor evolution of differences in virulence independently of within-host competition.

Transmission Modes and Virulence. Vertically transmitted parasites (those transmitted from parents to offspring) tend to be more benign than horizontally transmitted parasites (everything else). Support for this association comes from both cross-species comparisons and experimental manipulations of transmission. The organisms that have been studied span a broad range, including nematode parasites of wasps, viral and protozoal pathogens of mosquitoes, viral pathogens of bacteria, and plasmids that infect bacteria. This association is robustly supported, but vertically transmitted parasites can be highly virulent if they are not entirely vertically transmitted, and tremendous differences in virulence exist among horizontally transmitted parasites as well.

Much of the variation in virulence of horizontally transmitted pathogens is associated with the mode of horizontal transmission. Theory predicts that when pathogens do not rely on the host mobility for transmission, natural selection will favor high levels of host exploitation and hence high virulence. Studies of human pathogens confirm this prediction. Parasites transmitted by arthropod vectors, water, or attendants (e.g., in agricultural settings or hospitals) tend to have little if any dependence on host mobility, because the intervening agent transports the parasites; these parasites tend to be particularly virulent. This argument thus attributes the severity of malaria and sleeping sickness to vectorborne transmission, the severity of cholera, typhoid, and dysentery to waterborne transmission, the severity of hospital-acquired *Staphylococcus aureus* and other bacteria to attendantborne transmission inside of hospitals, and the severity of plant viroids to transmission through pruning and harvesting activities. This argument also explains why parasitoids (species that parasitize as larvae) are so lethal—mobile stages in their life cycles eliminate their dependence on host mobility.

Pathogens that are durable in the external environment also have reduced dependence on the mobility of infected hosts because they can rely on mobility of susceptible hosts to enact transmission. Accordingly, virulence and durability are positively associated among human pathogens of the respiratory tract. Durability in the external environment may thus explain why the tuberculosis bacterium and the smallpox virus are particularly damaging human pathogens and why many pathogens of insects are typically lethal.

The evolutionary trade-offs are different for sexually transmitted pathogens, which tend to have their most damaging effects during the chronic phase of infection. To be competitively successful, such pathogens must have special adaptations for persistence within hosts because opportunities for sexual transmission are generally less frequent than opportunities for other modes of horizontal transmission. But when sexually transmitted pathogens are exposed to a high potential for sexual transmission, selection should favor a buy-now-pay-later strategy, characterized by increased exploitation over the short term. Because of their adaptations for persistence within a host, increased exploitation over the short term may have particularly virulent long-term side effects.

The particular form of the negative effect will depend on the tropisms of the pathogen. HIV, which has tropisms for entering and replicating within particular leukocytes (white blood cells), tends to decimate those leukocytes, making AIDS patients vulnerable to opportunistic infections. Human papillomaviruses, which infect and enhance replication of epithelial cells, cause cervical cancer. *Chlamydia trachomatis*, which infects epithelial cells of the urogenital tract, causes scarring of the oviduct and hence infertility and ectopic pregnancies when the infecting strain is moderately virulent, but pervasive destruction of the reproductive tract when the infecting strain is highly virulent. In accordance with this argument, the strains of *C. trachomatis*, human papillomavirus, and HIV are more virulent in areas where the opportunity for sexual transmission is greater.

Increased density and size of host populations may favor the evolution of increased virulence for several reasons related to transmission. First, parasites may be less dependent on host mobility for transmission in dense populations. Second, highly replicative and hence particularly virulent variants may be best able to capitalize on geometric spread among hosts in large, dense host populations. Third, dense host populations may allow increased transmission during the early stage of each infection, favoring rapidly replicating and hence particularly virulent variants. Finally, highly virulent lineages may be more vulnerable to extinction, the chances of which tend to be lower in larger populations. [*See* Transmission Dynamics.] Although host population size and density are often suggested as determinants of virulence, their net effect on the evolution of virulence as well as the specific effects mentioned above remain largely untested.

Virulence Management: The Domestication of Pathogens. The improving understanding of the evolution of parasite virulence draws attention to the prospects for evolutionary management of virulence. With regard to human diseases, for example, we should be able to intervene in ways that tip the competitive balance in favor of mild strains. One option is to invest in interventions that selectively inhibit transmission of the most virulent variants. For acute infectious diseases, this option involves inhibiting transmission from people who are ill. For diarrheal diseases, waterborne transmission can transport pathogens from a few severely ill people to thousands of susceptibles. Blocking of waterborne transmission should favor mild strains because the modes of transmission that would remain, such as person-to-person or person-to-food-to-person transmission, depend on host mobility and hence host health. Geographic variation in the virulence of the *Shigella* bacteria that cause dysentery and cholera support this hypothesis.

Similarly, by making houses and hospitals mosquito-proof, the sick patients inside should be excluded as sources of infection for mosquitoes, which will instead feed on people who are feeling sufficiently healthy to move outside their dwellings. Because these healthier individuals should tend to be infected with less virulent pathogens, investment in mosquito-proof housing should tend to favor evolutionary reductions in virulence. Although this prediction has not yet been tested, experimental tests are justified ethically and economically by the direct suppressive effects of mosquito-proofing on the prevalence of *Plasmodium* protozoa that cause malaria. Geographic variation in virulence of *Plasmodium* species also suggests that their virulence evolves in responsive to mosquito availability. The effects of *P. vivax*, for example, are mild in northern latitudes, where extreme seasonality of mosquito abundance limits opportunities for transmission. Instead of growing explosively inside humans, *P. vivax* in these regions is prone to generating resting stages, called hypnozooites, which persist in the liver, causing little if any illness. In tropical areas where mosquitoes are more consistently abundant throughout the year, *P. vivax* tends to be more severe. Similar geographic variation occurs throughout the range of *P. falciparum*, although in this case the restricted seasonality of mosquitoes results from seasonality in rainfall more than temperature.

An analogous argument applies to transmission of hospital-acquired pathogens by attendants. Mobility of infected hosts is not necessary for such attendantborne transmission, which has been associated with particularly lethal hospital outbreaks of pathogens such as *Escherichia coli*. Blocking such transmission should therefore favor predominance of benign strains, which are brought in inadvertently from the outside community.

The evidence pertaining to sexually transmitted pathogens suggests that reducing the potential for sexual transmission should shift the competitive balance in favor of mild strains. As is the case for each of the categories of acute infection, a single intervention, such as increasing the reliance on condoms as opposed to birth control pills, should simultaneously favor evolution toward reduced virulence of each sexually transmitted pathogen in the area.

Vaccines could also be developed to favor evolution toward lower virulence. The necessary modification of current vaccination guidelines would be to restrict vaccine antigens to virulence antigens: those molecules that increase the virulence of viable benign organisms. The diphtheria vaccine provides an illustration. The diphtheria toxin, which is responsible for most of the damage caused by diphtheria, was detoxified and then used as the vaccine antigen. In unvaccinated people, the toxin kills host cells, thereby releasing nutrients to the bacterium. In vaccinated people, however, the toxin represents a drain on the bacterium's resources without providing a compensating benefit. This vaccine should therefore favor the *Corynebacterium diphtheriae* that produce no toxin. Accordingly, such benign strains displaced the toxin producers where this vaccine was used.

Evolutionary control of virulence should also help control the evolution of antibiotic resistance. The more benign pathogens become, the less the use of antibiotics, and hence the lower the selective pressure for antibiotic resistance. Geographic comparisons support this idea by revealing that antibiotic resistance is positively correlated with virulence in *V. cholerae* and in *Shigella* species. Where benign variants predominate, antibiotic resistance is rare.

Host Resistance. The other half of virulence involves variation in host resistance. Acquired resistance results from previous exposure to the pathogen. The best studied example is acquired immunological resistance, which results from exposure to antigens similar or identical to those carried by the pathogen. Acquired immunological resistance can reduce the virulence of the infection to such low levels that individuals show no signs of disease.

Genetic polymorphisms in host resistance may be maintained over time by host–parasite coevolution. [*See* Coevolution.] One kind of polymorphism results from the spread of defective alleles, which can provide resistance with little if any negative effects on host physiology when they are present heterozygously. The classic example is the sickle-cell allele, which provides resistance to falciparum malaria in heterozygous form but causes sickle-cell anemia when present homozygously. The allele for cystic fibrosis has been suggested to protect similarly against typhoid fever.

Polymorphisms in immunological alleles also contribute to variation in resistance. Alleles that code for major histocompatibility proteins have been associated with variation in susceptibility to infectious disease. In this case, none of the alleles may be defective, but the proteins from the different alleles may differ at recognizing particular pathogen variants and at presenting the variants to the immune system. Because pathogen composition varies, the relative success of the alternative alleles may vary, thus maintaining variation in resistance within the population.

Understanding the evolutionary basis of such genetic polymorphisms will be increasingly important as the potential for genetic manipulation increases. It would be important to know, for example, whether the allele for Tay-Sachs disease provides resistance to some pathogen before decisions are made to correct the allele. If it increases resistance to tuberculosis, it may be on balance beneficial to heterozygous individuals in areas where *Mycobacterium tuberculosis* is present, especially if the *M. tuberculosis* is often antibiotic resistant.

[*See also Overview Essay on* Darwinian Medicine; Disease, *article on* Infectious Disease; Emerging and Re-Emerging Diseases; Host Behavior, Manipulation of; Immune System, *article on* Structure and Function of the Vertebrate Immune System; Kin Selection; Myxomatosis; Plagues and Epidemics; Symbiosis; Vaccination.]

BIBLIOGRAPHY

Bouma, J. E., and R. E. Lenski. "Evolution of a Bacteria/Plasmid Association." *Nature* 335 (1988): 351–352. The first experimental demonstration that restriction of transmission to the vertical route causes a parasitic entity, in this case a plasmid, to evolve reduced virulence.

Cotter, P. A., and V. J. DiRita. "Bacterial Virulence Gene Regulation: An Evolutionary Perspective." *Annual Review of Microbiology* 54 (2000): 519–565. A description of molecular regulation of bacterial virulence with a focus on *Salmonella*, *Vibrio cholerae*, and *Bordetella*, as variations on the theme.

Dobrindt, U., and J. Reidl. "Pathogenicity Islands and Phage Conversion: Evolutionary Aspects of Bacterial Pathogenesis." *International Journal of Medical Microbiology* 290 (2000): 519–527. A summary of the role of pathogenicity islands and horizontal gene transfer in bacteria.

Ewald, P. W. *Evolution of Infectious Disease*. New York, 1994. An overview of the means by which disease manifestations result from selective pressures acting on parasites, with an emphasis on the associations between transmission modes and virulence, and virulence management.

Ewald, P. W. "Vaccines as Evolutionary Tools: The Virulence Antigen Strategy." In *Concepts in Vaccine Development*, edited by S. H. E. Kaufmann, pp. 1–25. Berlin, 1996. A presentation of the virulence antigen strategy for using vaccines to force pathogens to evolve to lower virulence.

Fine, P. E. F. "Vectors and Vertical Transmission: An Epidemiological Perspective." *Annals of the New York Academy of Science* 266 (1975): 173–194. The first theoretical analysis indicating that vertical transmission should favor lower virulence than horizontal transmission.

Frank, S. A. "Models of Parasite Virulence." *Quarterly Review of Biology* 71 (1996): 37–78. A balanced overview of theoretical models pertaining to the evolution of virulence.

Lenski, R. E., and R. M. May. "The Evolution of Virulence in Parasites and Pathogens: Reconciliation between Two Competing Hypotheses." *Journal of Theoretical Biology* 169 (1994): 253–265. An integration of epidemiological and evolutionary considerations, emphasizing how influences on virulence that are often seen as conflicting may be complementary.

Levin, B. R., and J. J. Bull. "Short-sighted Evolution and the Virulence of Pathogenic Microorganisms." *Trends in Microbiology* 2 (1994): 76–81. Consideration of the extent to which selective pressures on pathogens competing within hosts might lead to levels of virulence that are substantially above levels favored over the entire cycle of infection and transmission.

Levin, S., and D. Pimentel. "Selection of Intermediate Rates of Increase in Parasite–Host Systems." *American Naturalist* 117 (1981): 308–315. A pioneering article using mathematical models to illustrate how virulent competitors can win in competition with benign competitors when both are horizontally transmitted.

May, R. M., and R. M. Anderson. "Epidemiology and Genetics in the Coevolution of Parasites and Hosts." *Proceedings of the Royal Society of London B* 219 (1983): 281–313. Mathematical models of virulence explicitly including genetic bases of fitness differentials.

Messenger, S. L., I. J. Molineux, and J. J. Bull. "Virulence Evolution in a Virus Obeys a Trade-off." *Proceedings of the Royal Society of London B* 266 (1999): 397–404. Demonstration that experimental alterations of the proportions of vertical and horizontal transmission favor different levels of virulence among viruses that infect bacteria.

— PAUL W. EWALD

VIRUSES

Viruses are common intracellular parasites. Their success is such that perhaps every species of cellular life is susceptible to one or more kinds of viral infection, often with detrimental effects. Indeed, it is the association of viruses with disease that gives them their name—the Latin word *virus* means "poisonous fluid." Despite the medical advances of the last century, viruses continue to impose a considerable burden on human health. Current scourges include the human immunodeficiency virus (HIV), which at the current time infects some 16,000 people each day, and the hepatitis C virus (HCV), which has some 175 million sufferers worldwide.

Although viruses differ greatly in their biology, they

have one defining similarity: they cannot replicate in the absence of a cell from a host species, prokaryote, or eukaryote. To some, this dependency means that viruses should not even be classed as "living." Viruses are also noteworthy for their small size; they were discovered at the end of the nineteenth century as particles that were still infectious after being passed through filters small enough to remove bacteria. In modern parlance, this means diameters of only 20–300 nm and genomes that contain at most some 200 genes and usually just ten to fifteen.

The Biodiversity of Viruses. The simplest way of classifying viruses is through the nucleic acid that makes up their genome. Viruses are unique in that their genomes may be either RNA or DNA, rather than just DNA, as is the case for cellular organisms. The majority of viruses have genomes composed entirely of RNA, whereas others have genomes of only DNA. Although fundamental, the division between RNA and DNA genomes is not always clear-cut, because some viruses, like HIV and other retroviruses, have both RNA and DNA genomic stages in their life cycle: the genome that rests in the virus particle is composed of RNA, and a DNA copy is inserted into the host genome.

Within both RNA and DNA viruses further subdivisions can be made according to the number of strands of nucleic acid that make up the genome and their orientation. Most RNA viruses have genomes with a single strand of nucleic acid, usually laid out in a positive polarity so that it corresponds to the messenger RNA (mRNA) of a cellular gene. Other single-stranded RNA viruses have genomes with a negative polarity, so that they are equivalent to the complementary strand of mRNA; a small number have ambisense genomes comprising regions of both positive and negative polarity. A few RNA viruses have double-stranded genomes. DNA viruses, in contrast, all have positive-polarity genomes, and the vast majority are double-stranded. DNA viruses vary greatly in genome size (5–300 kb), whereas RNA viruses are usually much smaller and have a narrower range of length variation (7–30 kb).

Viruses also can be classified according to their structural architecture. All viruses, whatever their genetic-material, possess an outer protein called the capsid. The combination of viral nucleic acid and capsid is known as the nucleocapsid (the entire virus particle is the virion). Most DNA viruses, as well as a large variety of RNA viruses, have nucleocapsids with a regular isosahedral structure with twenty equal triangular sides. Other viruses have nucleocapsids with a simpler "springlike" helical structure. As well as capsids, some viruses possess an outer envelope protein, produced by a combination of viral proteins and lipid membrane acquired from the host cell. By analyzing both genetic and morphological data, distinct families of RNA and DNA viruses have been identified, although the evolutionary relationships among them are often more obscure.

Viruses also vary enormously in the host species and cell types they infect. Most of the viruses described to date infect animal species, although those that prefer plants or bacteria (the bacteriophages) are also well documented. Within multicellular organisms, viruses are able to infect a great variety of cell types, although any particular virus usually specializes in a small number of cell types. The mechanisms that determine this choice are poorly understood. The cell type (tropism) in which the virus replicates is often a function of where and how it enters a cell. For example, viruses that are transmitted by the fecal-oral route (acquired by ingestion, excreted in feces) often replicate in the gastrointestinal tract, a common port of entry. Cell tropism is also a key determinant of the type of disease a virus induces. For example, a number of very different viruses can cause hepatitis. The link between them is that they all show a preference for replication in liver cells and hence cause liver inflammation.

Finally, there are two classes of organisms that share similarities with viruses but that are classified slightly differently—viroids and satellite RNAs. Both are plant pathogens with tiny RNA genomes (usually ~400 bp) that lack protein-coding genes and are the smallest known agents of infectious disease. Viroids are circular, can replicate autonomously, and contain no capsid proteins. Satellite RNAs differ in that they contain capsids but require the assistance of a helper virus to replicate. Similar to satellite RNAs is the hepatitis delta virus (HDV), which infects humans and requires the hepatitis B virus (HBV) to replicate.

The Viral Way of Life. RNA viruses, whatever their genome organization, replicate using an RNA-binding enzyme not found in cellular life forms—the RNA-dependent RNA polymerase. Retroviruses, hepadnaviruses (which include HBV), and caulimoviruses (from plants) differ in that they produce DNA from an RNA template using a virally encoded RNA-dependent DNA polymerase, also known as reverse transcriptase. In contrast, all DNA viruses replicate via DNA-dependent polymerases, usually producing DNA genomic copies, although a few require RNA copies. Apart from parvoviruses and papovaviruses, these DNA polymerases are also encoded in the viral genome. The exceptions, which are too small to carry their own enzyme, parasitize the cell's DNA polymerase.

To establish an infection, a virus first needs to recognize, attach to, and penetrate a host cell using a special receptor molecule on the cell surface. Once inside the cell, the virus moves to the site of replication. In eukaryotic hosts, this can be either the cytoplasm or the nucleus. For example, positive-strand RNA viruses replicate in the cytoplasm because there is no need to par-

BACTERIOPHAGE LAMBDA

Bacteriophages—viruses that infect bacteria—have played a crucial role in the development of molecular biology. Most work has been done on bacteriophage lambda, first isolated from the bacterium *Escherichia coli* in 1951. Lambda is composed of an icosahedral head structure, which contains the linear double-stranded DNA genome (~49 kb in length), and a long tail formation.

As with many bacteriophages, infection with lambda may take one of two courses. In the lysogenic, or temperate, pathway phage, DNA is integrated into the bacterial chromosome, where it is known as a "prophage" and is replicated with the host genome producing a latent infection. Alternatively, a lytic infection may ensue, in which case the bacterial cell is lysed, releasing viral progeny as it does so. Lysed bacteria can be recognized as clear patches or "plaques" on bacteria that are grown on agar media in petri dishes. The switch from a lysogenic to a lytic infection is controlled by the transcription of "late genes," which are repressed in the former but expressed in the latter.

The analysis of bacteriophage lambda infections of *E. coli* has resulted in a number of landmark discoveries, such as that of restriction enzymes—molecular scissors that defend bacteria against phage infection and that are an important tool in molecular biology. The analysis of lambda infections has also provided vital information about the processes of recombination and gene regulation.

—EDWARD HOLMES

asitize the host enzymes active in the cell nucleus, whereas retroviruses replicate in the nucleus as they insert into the host genome. Once the viral genome is replicated and the protein products are translated, the components of the virion are assembled and then released from the cell. Release can occur through cell lysis, a process by which the cell is split open, or by budding, by which the virus leaves the cell, taking part of its membrane with it, this helping to form the viral envelope. The entire replication process is usually very quick, taking only a few hours in most cases, and thousands of viral progeny may be produced from a single cell. Because viruses must compete with cellular genes for the materials they need for replication, many have evolved strategies to give themselves an advantage. For example, some use overlapping reading frames, in which a single gene can be transcribed from different starting points to produce a variety of proteins.

Unfortunately for the host, many viruses have deleterious consequences for the cells in which they replicate. These cytopathic effects may be caused by the toxic properties of viral proteins, or because the cell is deprived of critical resources. The simplest of all cytopathic effects is cell lysis, which results in cell death. Another is cell fusion. In HIV, for example, fusion results in the production of multinucleated giant cells known as syncytia. Many DNA viruses change the properties and structure of host cells, a process known as transformation. The outcome of transformation may be a malignant host cell, often caused by the activation of cellular or viral oncogenes (cancer-causing genes).

Once a virus has successfully infected a host, the course of the infection may vary greatly. Many viruses cause rapid, or acute, infections that last only a few days. These range from entirely asymptomatic to highly virulent. Other viruses infect their hosts for much longer periods. In some of these persistent infections, the virus replicates throughout its stay in the host, whereas in others, such as herpesviruses, replication is absent or occurs only at very low levels so that the infection is latent. Generally, RNA viruses cause acute infections and are often responsible for rapid epidemics, whereas retroviruses and DNA viruses tend to result in persistent infections. Finally, some viruses exist in a noninfectious state within host genomes. These consist of the DNA copies of retroviruses that have integrated into the host genome and (usually) have lost the ability to infect new cells, so that they are passively inherited along with host DNA. Such endogenous retroviruses are a common component of the vertebrate genome and can be considered the ultimate exponents of the parasitic lifestyle.

As well as being able to replicate successfully, viruses must be able to transmit to new hosts and evade immune responses for transmission to take place. There are three basic strategies of viral transmission: direct contact between hosts; vector transmission, usually involving arthropods; and transmission by some other vehicle, such as food, water, aerosols, or even inanimate objects. Because any given virus tends to adopt just one of these strategies, the evolution of the transmission mechanism is not a simple task. Viruses also have adopted a variety of tactics to evade host immune responses. As already noted, some cause latent infections and so stimulate only very weak immune responses. Other viruses, such as influenza A and HIV, adopt a very different approach, producing a vast array of different antigenic types that cannot be controlled simultaneously by immune systems. Additional strategies include infecting cells of the central nervous system, which are only weakly surveyed by the immune system, and infecting infants with immature immune systems.

Mechanisms of Viral Evolution. One of the most important differences between DNA and RNA viruses is the rate at which they evolve. Because replication using DNA polymerases is generally quite accurate, DNA viruses mutate at similar rates to bacteria—on the order of 0.003 mutation per genome, per replication. In contrast, RNA viruses and retroviruses are the fastest evolving of all life-forms. Mutation rates for retroviruses and RNA viruses have been estimated to be about 0.2 and 0.76 mutation per genome replication, respectively, so that a large fraction of the viral progeny will differ from their parental type. Such high rates are caused by the absence of proofreading or repair mechanisms in RNA polymerases. Furthermore, these viruses often have rapid replication rates. HIV, for example, produces approximately 150 generations each year within infected hosts. Together, such rapid mutation and replication result in very high rates of nucleotide substitution, usually in the range of 10^{-3} to 10^{-4} substitutions per nucleotide site, per year, which is approximately one million times faster than the rate seen in mammalian nuclear DNA. Quite clearly, such high rates of nucleotide substitution provide RNA viruses with enormous genetic variability, even within individual patients, in the case of HIV and HCV, which can then be shaped by natural selection. Indeed, such extensive variability is a major factor limiting the development of effective antiviral drugs and vaccines. A consequence of this rampant mutagenesis is that many progeny viruses are defective, containing mutations in critical genes. Although many of these defective viruses do not function correctly, some are able to express enough of their surface proteins to decoy host immune defenses, whereas others—the defective interfering viruses—will compete with competent viruses for the replication apparatus. These very high mutation rates also mean that RNA viruses are susceptible to loss of fitness arising from the accumulation of mutations, a phenomenon known as Müller's ratchet. This is especially true of nonrecombining viruses that live in small populations. It is even possible that rapid mutagenesis could be exploited in the development of new antiviral drugs; if the viral mutation rate could be increased artificially, then all progeny genomes might be so defective that viral extinction occurs.

Besides mutation, viruses are able to undergo recombination, although with greatly differing frequencies. At present, little is known about the mechanics of viral recombination, although it requires at least two viruses to infect a single cell. In DNA viruses, the process is probably similar to that seen in other DNA genomes, involving the breaking and rejoining of DNA strands. Recombination in RNA viruses is likely to be more complex. One favored mechanism is "copy choice," in which the RNA polymerase switches back and forth between viral genomes during replication.

It is also unclear why viruses should vary so much in their ability to recombine. Take, for example, HCV and GBV-C (otherwise known as HGV, or hepatitis G virus). These viruses are close relatives (both members of the Flaviviridae family of RNA viruses), both cause persistent infections, and both are found at relatively high frequencies in human populations. However, there is no evidence that HCV recombines, whereas GBV-C appears to recombine very frequently. Recombination rates also appear to be very high in retroviruses and in RNA viruses with segmented genomes. In the latter, genes are located on independently transcribed genomic segments (effectively chromosomes), which can reassort to produce new genetic combinations. This is best documented in the influenza A virus, where reassortment has led to new combinations of the hemagglutinin and neuraminidase proteins expressed on the viral envelope and which are often associated with major influenza epidemics.

Recombination also may occur among different viruses, even between RNA and DNA viruses, and this may ultimately lead to the generation of entirely new viruses. In particular, recombination may produce new combinations of the key functional genes required by viruses. This process has been termed *modular evolution* because it is possible to think of viral genomes as being composed of a series of functional modules, such as a polymerase module and a capsid module, which could be interchanged. For example, the potyviruses of plants seem to have genomic regions descended from at least four different viral families. Finally, there is evidence that viruses have recombined with cellular genomes and in so doing have captured host genes. This phenomenon is perhaps best described for the *v*-oncogenes found in some retroviruses, and it elegantly demonstrates the parasitic nature of viral life. The capture of host genes might also have a major bearing on how viruses originated.

Origins and Evolutionary History of Viruses. Studies of the evolutionary history of viruses are greatly hampered by the lack of a fossil record. Consequently, phylogenetic analyses can only be conducted using contemporary samples of viruses or those that have been collected and stored from past epidemics. In most cases, such as the influenza A virus, this storage goes back only a little over eighty years. The situation is even more difficult in RNA viruses because their very rapid rates of evolutionary change mean that the molecular footprints of their past history are quickly eroded. As a result, the analysis of viral origins and deep phylogeny is often an exercise in speculation. In particular, although it is generally a simple matter to construct phylogenetic trees for different families of viruses, extensive sequence divergence makes it very difficult to construct trees that can accurately link these families. Although there have been attempts to construct major "supergroups" of viral families, the reality is that all attempts to construct phylogenetic trees of such diverse sequences need to be treated with caution, as there is often no more similarity among them than expected by chance alone. For example, the sequences of RNA-dependent RNA polymerase and reverse transcriptase share only four short motifs, so that it is difficult to determine whether these enzymes have a common ancestry or are similar because of convergent evolution. Further resolution will require information about protein structure that is more stable over very long time periods, although this is difficult to do at present. Recombination between viral families, as suggested in the theory of modular evolution, will also greatly complicate attempts to construct deep viral phylogenies.

Currently, there are two main theories to explain the origin of viruses; first, that they are descendants of escaped host cellular genes that have acquired protective protein coats and the ability to replicate autonomously; second, that they represent the descendants of precellular life forms, with ancestries dating back billions of years, and hence have always existed independently of host genomes.

One of the most attractive aspects of the escaped gene theory is that it can easily explain the evolution of all types of viruses simply by invoking multiple escape events. For example, RNA viruses would be descended from escaped cellular mRNA molecules that either possessed or acquired RNA polymerase activity, and the ancestry of retroviruses, hepadnaviruses, and caulimoviruses would most likely lie in the long terminal repeat retrotransposons (stretches of parasitic DNA with the ability to replicate themselves) that are found widely distributed throughout the genomes of eukaryotes and that also contain reverse transcriptase. Finally, the DNA viruses could be descended from either DNA transposable elements or even the genes found in bacterial plasmids. Evidence for this theory is that phylogenetic trees of retroviruses, hepadnaviruses, and caulimoviruses suggest that they have independent origins, as is the case for trees of DNA viruses constructed using polymerase sequences, which are intermixed with the sequences from cellular species (although these trees are tentative).

The very great genetic divergence among viruses means that direct evidence for the theory that viruses are remnants of ancient precellular life-forms has been less easy to gather. According to this scenario, RNA viruses would be remnants of the earliest time in the history of life on earth—the "RNA world," where the only replicating molecules were composed of RNA and where DNA had yet to evolve. Similarly, DNA viruses would constitute the living representatives of the earliest DNA-based life-forms, and retroviruses would have descended from the first reverse transcriptases, which might have enabled the crucial transition from RNA to DNA replication. It has even been proposed that viroids and satellite RNAs, because of their great simplicity, are remnants of the very earliest types of RNA molecule. Yet it seems equally plausible that these pathogens represent only relatively recently escaped host genes, or perhaps even introns, which have not yet acquired protein coats. In particular, with the exception of HDV, their restricted distribution in plants suggests that they only recently appeared in this group of eukaryotes alone.

There are, however, a number of pieces of circumstantial evidence that suggest viruses might in fact have very ancient origins. First, some RNA molecules—the ribozymes—have catalytic activity necessary for self-replication, and it is not difficult to imagine that these molecules could act as the progenitors of RNA viruses. Second, viruses have a very wide taxonomic distribution, spanning eukaryotes and prokaryotes, suggestive of a long-term interaction with these species, although no single viral family can infect such a wide range of hosts. Finally, there are some examples, especially in DNA viruses, where viruses have clearly cospeciated with their hosts over very long time periods. Although this does not prove the theory of precellular viral origins, it does demonstrate that ancient viral histories can be uncovered in viruses that evolve sufficiently slowly.

Although the two theories for viral origins seem diametrically opposed, they can be partly reconciled by assuming that viruses have been produced throughout the history of life on earth. Hence, it is possible that some viral lineages have indeed existed since the time of the RNA world, whereas others appeared as soon as there were cells to parasitize, and others still may have been produced more recently, when host genomes increased in complexity.

[*See also* Acquired Immune Deficiency Syndrome, *article on* Origins and Phylogeny of HIV; Disease, *article on* Infectious Disease; Emerging and Re-Emerging Diseases; Influenza.]

BIBLIOGRAPHY

Drake, J. W., and J. J. Holland. "Mutation Rates among RNA Viruses." *Proceedings of the National Academy of Sciences USA* 96 (1999): 13910–13913. Key study of mutation rates in RNA viruses.

Eickbush, T. H. "Origin and Evolutionary Relationships of Retroelements." In *The Evolutionary Biology of Viruses*, edited by S. S. Morse, pp. 121–157. New York, 1994. Comprehensive review of the evolution of retroelements.

Eickbush, T. H. "Telomerase and Retrotransposons: Which Came First?" *Science* 277 (1997): 911–912. Update of Eickbush (1994).

Gibbs, M. J., and G. F. Weiller. "Evidence That a Plant Virus Switched Hosts to Infect a Vertebrate and Then Recombined with a Vertebrate-Infecting Virus." *Proceedings of the National Academy of Sciences USA* 96 (1999): 8022–8027. Excellent example of the power of recombination in viral evolution.

Gorbalenya, A. E. "Origin of RNA Viral Genomes: Approaching the Problem by Comparative Sequence Analysis." In *Molecular Ba-*

sis of Virus Evolution, edited by A. Gibbs, C. H. Calisher, and F. García-Arenal, pp. 49–66. Cambridge, 1995. Presents the theory that RNA viruses can be placed into major supergroups and have their origins in the RNA world.

Strauss, E. G., J. H. Strauss, and A. J. Levine. "Virus Evolution." In *Fundamental Virology*, edited by B. N. Fields, D. M. Knipe, and P. M. Howley, 3d ed., pp. 141–159. New York, 1996. A key review of the different theories of viral origins. The book itself is the bible of viral biology.

Webster, R. G., W. J. Bean, and O. T. Gorman. "Evolution of Influenza Viruses: Rapid Evolution and Stasis." In *Molecular Basis of Virus Evolution*, edited by A. Gibbs, C. H. Calisher and F. García-Arenal, pp. 531–543. Cambridge, 1995. Useful review of the evolution of influenza viruses.

Worobey, M., and E. C. Holmes. "Evolutionary Aspects of Recombination in RNA Viruses." *Journal of General Virology* 80 (1999): 2535–2544. Recent review of recombination in RNA viruses.

Zanotto, P. M. de A., M. J. Gibbs, E. A. Gould, and E. C. Holmes. "A Reassessment of the Higher Taxonomy of Viruses Based on RNA Polymerases." *Journal of Virology* 70 (1996): 6083–6096. Analysis of the evolutionary relationships among families of RNA viruses showing that phylogenetic trees cannot be reliably drawn at this level because of extensive sequence divergence.

— EDWARD C. HOLMES

VIVIPARITY AND OVIPARITY

Viviparity, broadly defined, is exactly what the word implies—live birth. Intrinsic to live birth is the retention of the developing embryo in or on the body of a parent, nearly always a female. It is one of many modes of parental care, but one that largely involves physiology, morphology, and development, rather than relying primarily on behavior. The retention of developing young is in contrast to oviparity, or the laying of eggs, which often are fertilized after they are laid. Live bearing is either lecithotrophic, dependent on the yolk for embryonic nutrition, or matrotrophic, in which the mother provides nutrients after the yolk is exhausted. Viviparity may, but need not, include a placenta as an organ for nutrient and gaseous exchange. [*See* Placentation.]

Oviparity is generally assumed to be the ancestral condition in lineages of animals; direct development (laying eggs, but a mode in which the embryos metamorphose before hatching, so that there is not a free-living larval stage in the life cycle) and live bearing are evolutionarily derived reproductive modes. However, not all researchers agree that oviparity is ancestral—some claim that it is derived in some lineages from a viviparous ancestor. It is debated whether members of a lineage, once having evolved viviparity, can regain oviparity. In certain lizard taxa, for example, a parsimonious interpretation based on the phylogenetic relationships of the species indicates that viviparity must have "reversed" during the evolution and speciation of the lineage to oviparity. However, most researchers accept that the evolution of viviparity involves specific modifications of the maternal and embryonic physiologies and morphologies that are unlikely to revert to an ancestral condition.

Evolution of Modes of Development. Viviparity has evolved many times in many groups of animals. Certain flies, scorpions, aphids, and *Peripatus* are examples among invertebrate animals. Within vertebrates, viviparity has developed in all classes except the jawless fish and birds. Several lineages of sharks, teleosts, amphibians, squamate reptiles, and mammals have evolved live-bearing modes of reproduction. Viviparity does not evolve randomly in lineages; it is restricted to only a few in most major groups of vertebrates. For example, no rays among elasmobranchs are viviparous; only teleosts, and few of them, are live bearers among bony fish; few frogs and salamanders, but many caecilians, are viviparous. Viviparity has evolved more than 100 times in reptiles, but only in squamates (snakes, lizards, and amphisbaenians). Among mammals, monotremes are oviparous, although they provide milk to their young after they hatch; marsupials and eutherians are viviparous, with diverse kinds of placentas for nutrient and gaseous exchange.

In most vertebrates, the developing embryos are retained in the oviduct of the female parent. However, there are numerous examples of variation from that theme. Teleost fish lack oviducts, and live-bearing females retain the embryos in the lumen of the ovary or the follicles in which the ova develop. There are several examples of either the male or the female parent attaching the developing eggs to its body and carrying them through part or all of their development. An extreme example is that of sea horses, in which the female deposits her fertilized ova on the abdomen of the male; the abdominal skin then encloses the eggs in a pouch, and there is evidence that the lining of the pouch secretes nutrient material that the embryos absorb. Among amphibians, live-bearing salamanders and caecilians retain the developing embryos in their oviducts. Some species of frogs also are oviductal bearers, but others exhibit numerous "natural experiments" in live-bearing reproduction. Species in different families have evolved mechanisms of depositing fertilized eggs in the skin of the back of the female parent; she maintains them through metamorphosis, so that juveniles emerge. Two Australian species of frogs are (or were; one is presumed extinct) stomach brooders, with the mother swallowing her fertilized eggs and maintaining them in her stomach, so that fully metamorphosed young are regurgitated. The embryos secrete a prostaglandin that inhibits gastric secretion by the mother, so the developing young are not digested. In two South American species in a different family of frogs, the males swallow the eggs and maintain them in their vocal sacs. One species maintains the young through metamorphosis; there is indirect evidence that the lining of the sac secretes material that is ingested by the embryos.

Internal Fertilization. The many viviparous taxa that retain the embryos in their oviducts all have one feature in common: internal fertilization of the ova. Males of the species court the females and deposit sperm into the reproductive tracts of the females. The males develop diverse mechanisms for intromission, depending on the species. Sharks modify their pelvic fin rays as claspers, and teleosts their pelvic rays as gonopodia, that are inserted into the vents (cloacal openings) of the females and transport sperm. One frog retains rudiments of its tail, which is inserted in the vent of the female; other frog species with internal fertilization have the male and the female placing their cloacas in apposition so that sperm move directly from the male to the female. Derived species of salamanders all have a courtship that involves the male depositing a packet of sperm, the spermatophore, on the substrate and luring the female over it, whereupon she grasps the sperm packet with the lips of her cloaca and takes it into her reproductive tract. All male caecilians (members of the third, little known, order of amphibians) have modified the rear part of their cloacas as an intromittent organ that is extruded and inserted in the vent of the female to transport sperm. All amniotes (reptiles, birds, and mammals) have internal fertilization. Reptiles have either paired hemipenes (found in snakes, lizards, and amphisbaenians) or single penises (turtles, crocodilians); some birds have penises (e.g., ducks); mammals have penises that serve as intromittent organs to conduct sperm directly to the female's reproductive tract. However, some of the frogs and most of the salamanders with internal fertilization lay fertilized eggs, rather than retaining them through development; similarly, most reptiles, all birds, and monotremes among mammals lay eggs. Consequently, internal fertilization is necessary but not sufficient for the evolution of oviductal modes of viviparity.

Evolution of Modes of Nutrition. Different modes of nutrition have developed among viviparous species. In many species in several lineages, the developing young in the oviducts or elsewhere in or on the body are dependent on their yolk supply, with no additional nutrients from the parent. They may be born fully metamorphosed or at a larval stage before metamorphosis, whereupon they spend a period actively feeding. Such developing embryos may have gaseous exchange with the parent, though not nutrition per se. This condition is often termed ovoviviparity, although in some lineages it is difficult to distinguish from viviparity, because aspects of nutrient supply are not fully known. Viviparity in the past has been defined in terms of involving a placenta (placental or "true" viviparity) or not (pseudoviviparity). Most workers now prefer a broad definition of viviparity that includes the many modes of live-bearing reproduction, then characterize kinds of viviparity in terms of yolk or other nutrition, kinds of associations of the embryo or fetus with the parent that facilitates gaseous exchange and nutrient uptake a(when present), stage of development of the young at birth, and other variations exhibited by live bearers. Such a broad definition allows examination of trends in the evolution of viviparity within lineages. For example, some closely related species of lizards lay eggs (are oviparous), but the eggs hatch within twenty-four hours; others give birth to juveniles (are viviparous), then shed remnants of the eggshell and the extraembryonic membranes. Several phenomena associated with the evolution of viviparity in these lizards, such as increased time of retention of the eggs, eggshell thinning and loss, maternal supplementation of the yolk with additional nutrients, development of placentas that include the yolk sac and sometimes the other extraembryonic membranes, in association with modifications of the maternal oviduct and the effect of environmental variables such as temperature and moisture on the development of viviparity, can best be examined in the context of nonrestrictive but descriptive terminology. Similarly, the evolution of viviparity in sharks involves diverse levels of maternal nutrition beyond that of the yolk, as well as the development of complex placentas and other absorptive structures in some species. Several studies have examined the contribution of maternal nonyolk nutrients to the newborn young, and recently the structure of the placenta has been intensively investigated. The diverse mechanisms for viviparity in teleost fish are also under investigation, as the maternal-fetal relationship is being explored morphologically, developmentally, and physiologically.

Relatively little is known for most lineages about the selection pressures that resulted in the acquisition of live-bearing modes of reproduction. Furthermore, the morphological modifications that facilitate viviparity are described for relatively few species, and the physiological and endocrinological bases for the evolution of viviparity are known for very few. Research on the evolution of viviparity integrates ecological, behavioral, morphological, physiological, and developmental biology and is currently an exciting area of biology.

BIBLIOGRAPHY

Amoroso, E. C. "The Evolution of Viviparity." *Proceedings of the Royal Society of Medicine* 61 (1968): 1188–1200. A perspective on the evolution of viviparity by one of its primary researchers.

Blackburn, D. G. "Convergent Evolution of Viviparity, Matrotrophy and Specializations for Fetal Nutrition in Reptiles and Other Vertebrates." *American Zoologist* 32 (1992): 313–321.

Blackburn, D. G. "Saltationist and Punctuated Equilibrium Models for the Evolution of Viviparity and Placentation." *Journal of*

Theoretical Biology 174 (1995): 199–216. The Blackburn papers succinctly discuss current views on the patterns of evolution of viviparity.

Blüm, V. *Vertebrate Reproduction: A Textbook*. Berlin, 1986. A textbook with interesting examples of viviparity in vertebrates.

Guillette, L. J., Jr. "The Evolution of Viviparity in Fish, Amphibians, and Reptiles: An Endocrinological Approach." In *Hormones and Reproduction in Fishes, Amphibians, and Reptiles*, edited by D. O. Norris and R. E. Jones, pp. 523–562. New York, 1987. An endocrinological perspective on the evolution of viviparity.

Guillette, L. J., Jr. "The Evolution of Viviparity in Lizards." *BioScience* 43 (1993): 742–751.

Jameson, E. W., Jr. *Vertebrate Reproduction*. New York, 1988.

Packard, G. C., R. P. Elinson, J. Gavaud, L. H. Guillette, J. Lombardi, J. F. Schindler, R. Shine, H. Tyndale-Biscoe, M. H. Wake, F. Xavier, and Z. Yaron. "How Are Reproductive Systems Integrated and How Has Viviparity Evolved?" In *Complex Organismal Functions: Integration* and Evolution in *Vertebrates*, edited by D. B. Wakeand G. Roth, pp. 281–293. Chichester, England, 1989.

Shine, R. "The Evolution of Viviparity in Reptiles: An Ecological Analysis." In *Biology of the Reptilia*, edited by C. Gans and F. Billet, vol. 15, pp. 605–694. New York, 1985.

Shine, R. "A New Hypothesis for the Evolution of Viviparity in Reptiles." *American Naturalist* 145 (1995): 809–823. Shine's papers present an ecological perspective on the evolution of viviparity.

Turner, C. L. "Viviparity in Teleost Fishes." *Scientific Monthly* 65 (1947): 508–518.

Van Tienhoven, A. *Reproductive Physiology of Vertebrates*, 2d ed. Ithaca, N. Y., and London, 1983.

Wake, M. H. "Phylogenesis of Direct Development and Viviparity in Vertebrates." In *Complex Organismal Functions: Integration and Evolution in Vertebrates*, edited by D. B. Wake and G. Roth, pp. 235–250. Chichester, England, 1989. A summary of the many instances of the evolution of viviparity in vertebrates, with a discussion of phylogenetic and historical constraints.

Wake, M. H. "Evolution of Oviductal Gestation in Amphibians." *Journal of Experimental Zoology* 266 (1993): 394–413.

Wourms, J. P. "Viviparity: The Maternal-Fetal Relationship in Fishes." *American Zoologist* 21 (1981): 473–515.

Wourms, J. P., B. D. Grove, and J. Lombardi. "The Maternal-Embryonic Relationships in Viviparous Fishes." In *Fish Physiology: The Physiology of Developing Fish*, part B, *Viviparity and Posthatching Juveniles*, edited by W. S. Hoar and D. J. Randall, vol. 11, pp. 1–134. New York, 1988. Presents summaries of the evolution of viviparity in fish.

— MARVALEE H. WAKE

W–Z

WALLACE, ALFRED RUSSEL

Alfred Russel Wallace (1823–1913), British naturalist whose fame in the history of evolutionary biology rests primarily upon his discovery, independently of Charles Darwin, of the theory of evolution by natural selection. Wallace was a brilliant field observer, a prolific generator of ideas on diverse issues ranging from evolutionary biology to social and political concerns, and a theoretician whose work laid some of the foundations of modern botany and zoology. After a four-year exploration of the Amazon basin of South America (1848–1852), Wallace published *A Narrative of Travels on the Amazon and Rio Negro* (1853), which established his reputation as the foremost analyst of the geographical distribution of animals and plants (biogeography). The recognition that the distribution of closely allied species was often marked by surprisingly precise and abrupt barriers was the most important scientific achievement of his Amazonian travels. However, although Wallace had been committed to some form of general evolutionary theory since 1845, he was not yet prepared to posit an explicit evolutionary mechanism in the *Narrative of Travels*.

Convinced that another voyage of exploration was the most certain means of providing the data required for the theoretical elucidation of what was termed "the species problem" (i.e., how new species originate from preexisting ones), Wallace decided upon an expedition to the Malay Archipelago. He arrived in Singapore on 24 April 1854 to begin eight years of traveling in Java, Borneo, Celebes, New Guinea, Bali, and many smaller islands in the archipelago. From 1854 to 1862, Wallace covered nearly 14,000 miles and collected 125,000 (primarily faunal) specimens. He encountered animals, birds, and insects in bewildering variety and abundance, many of which had not been previously seen by Europeans. Wallace also observed, and lived with, the diverse human inhabitants of those regions.

Wallace's observations on the geographic distribution of species led him to new insights about evolutionary history. He proposed that the species in the western half of the Malay Archipelago are overwhelmingly Indian in origin, whereas the species in the eastern half are predominantly of Australian origin. This was a bold synthesis of evolutionary theory and meticulous observation. The faunal discontinuity separating the Indian from the Australasian segments of the archipelago is called Wallace's Line in his honor. Wallace later generalized these concepts to elaborate a global paradigm for identifying the earth's fundamental biogeographical regions in his magisterial *Geographical Distribution of Animals and Plants* (1876). Details of Wallace's original biogeographic regions—and the precise location of the boundary first suggested by Wallace's Line—have undergone revision as more abundant data have become available and as more recent theories of continental drift have forced a reinterpretation of certain of Wallace's nineteenth-century premises. Nonetheless, his biogeographical synthesis stands as a major development in evolutionary biology.

It was, of course, Wallace's elucidation of the mechanism of evolution that constitutes his greatest scientific legacy from the Malay travels. In February 1858—recalling passages from the English economist Thomas Malthus's *Essay on the Principle of Population* (1798), with its vivid depiction of the competitive struggles for survival among human populations—the principle of natural selection emerged in Wallace's mind as the key mechanism of evolutionary change. He wrote out a draft of his complete theory in the famous essay "On the Tendency of Varieties to Depart Indefinitely from the Original Type" (1858), and mailed it to Darwin in England. A copy of Wallace's essay, along with extracts from an unpublished manuscript on natural selection written by Darwin in 1844, were presented jointly at the historic meeting of the Linnean Society (London) on 1 July 1858, which publicly announced to the world the principle of natural selection. This meeting—a year prior to the publication of Darwin's *Origin of Species* (1859)—ensured that both Wallace and Darwin received recognition and joint priority for their momentous discovery.

Upon his return to England in 1862, Wallace spent the remainder of his long life elucidating the implications of evolutionary theory for a vast array of subjects ranging from biogeography, sexual selection, the phenomenon of organic mimicry (by which one animal species evolves to so closely resemble another animal or even plant species as to be mistaken for it by predators), taxonomy, physical geography and geology, and anthropology. It was Wallace's theories and writings in the last domain, human evolution, that elicited the greatest controversies of his career. Although he remained an ardent selectionist in his overall analysis of evolutionary processes, Wallace regarded natural selection as inadequate to account for the origin and development of certain human characteristics, notably consciousness and the moral sense. Instead, he suggested that other agencies—of a nonmaterial or spiritual nature—had been, and continue to be, instrumental in the origin and future evolution of the human species. Wallace's views on human evolution were inextricably tied to his outspoken advocacy of social and political reforms in the late Victorian period. He remains one of the great figures in the history and development of evolutionary biology precisely because his conception of the evolutionary process integrated so diverse, and (frequently) contentious, a range of subjects. The questions Wallace posed continue to resonate in contemporary debates on the scope, mechanism, and, ultimately, significance of evolution in both scientific and cultural domains.

[*See also* Biogeography, *article on* Island Biogeography; Natural Selection; Speciation.]

BIBLIOGRAPHY

Brooks, J. L. *Just before the Origin: Alfred Russel Wallace's Theory of Evolution.* New York, 1984. One of the most detailed accounts of Wallace's formulation of natural selection, whose utility to the general reader is enhanced by extensive quotations from Wallace's published and unpublished writings in the crucial period 1848–1858. Marred only by Brooks's polemical claim that Darwin owed more to Wallace than he acknowledged, a claim rendered unconvincing by more recent scholarship on the independent paths of Wallace's and Darwin's discovery of natural selection.

Camerini, J. R. "Wallace in the Field." *Osiris* 2.11 (1996): 44–65. Informative account of Wallace's innovative activities as a practicing field naturalist, which situates his scientific achievements within the broader context of the social, political, and economic worlds of Victorian science.

Daws, G., and M. Fujita. *Archipelago: The Islands of Indonesia: From the Nineteenth-century Discoveries of Alfred Russel Wallace to the Fate of Forests and Reefs in the Twenty-first Century.* Berkeley, 1999. A richly illustrated work that gives the reader a vivid sense of Wallace's travels and field work in the Malay Archipelago. Emphasizes his contributions to island biogeography and other tenets of modern biological thought.

Fichman, M. *Alfred Russel Wallace.* Boston, 1981. Useful biography that situates Wallace's scientific achievements within the broader controversies of Victorian culture concerning human evolution.

Knapp, S. *Footsteps in the Forest: Alfred Russel Wallace in the Amazon.* London, 1999.

Marchant, J., ed. *Alfred Russel Wallace: Letters and Reminiscences* (1916). Reprint, New York, 1975. Still indispensable as the standard published edition of much of Wallace's extensive correspondence on a wide range of topics, both scientific and cultural. Marchant's editorial standards, however, no longer satisfy contemporary criteria of scholarly accuracy in all respects. Until a critical edition of the full Wallace correspondence is available, Marchant's remains invaluable for the insights afforded into Wallace's public and private life and thoughts.

Smith, C. H., ed. *Alfred Russel Wallace: An Anthology of His Shorter Writings.* Oxford, 1991. A comprehensive anthology of excerpts and complete selections from Wallace's vast array of contributions to the periodical literature of his era, ranging from evolution and biogeography to sociopolitical and spiritualist concerns. Contains the most complete bibliography to date of Wallace's enormous output of books, articles, essays, and reviews.

ONLINE RESOURCES

"The Alfred Russel Wallace Page." http://www.wku.edu/~smithch/index1.htm. Maintained at Western Kentucky University, Bowling Green. Provides extensive information on current research activities on Wallace, updated bibliographic entries on primary and secondary sources, and online links to electronic versions of a number of texts by and about Wallace. By far the best Internet source on Wallace.

— Martin Fichman

WARFARE

Many anthropologists do not acknowledge an evolutionary background to war, adhering to Margaret Mead's famous dictum, "War is only a cultural invention." Ferguson (1984) is the most outspoken advocate of this position, arguing that war has not been a regular occurrence throughout human history, but most likely became a social institution in Mesopotamia some 8,000 years ago and has been reinvented in many times and places since.

Against this view, the discovery of male coalitional aggression and "lethal male raiding" in free-ranging chimpanzees and battle-type intergroup violence in social carnivores and many primates makes the conventional view of warfare as a singularly human "cultural invention" a few thousand years old an argument about labels. Some aspects of human warfare *are* distinctively human, and some are further restricted to particular times and places in human history. At the same time, questions of phylogenetic continuity and the comparative socioecology of group violence are worthy of study.

An evolutionary perspective prompts questions about the "ultimate" dimension of causality: Why has intergroup violence, or the propensity for warfare, evolved in the first place? Why has it evolved in only so few species? These questions direct attention to the circumstances under which group violence occurs and the fitness effects that participation has on the individuals involved. Such questions should be distinguished from proximate causative factors such as the mechanisms, motives or conditions that led to a particular war. Ultimate and proximate causes are complementary rather than mutually exclusive.

Darwinian Evolution: Sexual Selection and Kin Selection. In general, two types of warfare (generally defined as organized intergroup or intercommunity contest competition) have been distinguished: raiding ("lethal male raiding" or ambush or dawn surprise attack), and battle (confrontation of two opposing lines or phalanxes). When a battle is prearranged, it is called a "pitched battle." In nonstate human societies, raiding is often the most bloody and lethal form of warfare, owing to small but rapidly accumulating casualties and occasional near-genocidal routing.

Lethal male raiding has been explained by Wrangham's

"imbalance-of-power" hypothesis, Tooby and Cosmides's "risk-contract" theory, and Low's and van der Dennen's sexual selection approach. Wrangham has also suggested the distinct possibility that the chimpanzee-hominid common ancestor already had this lethal male raiding pattern in its behavioral repertoire (panid-hominid synapomorphy) some six million years ago.

Battle-type warfare occurs in many primate species and some other group-territorial mammals, such as social carnivores. Battles result mainly from chance encounters by primate groups and from failed raids or surprise attacks and chance encounters in tribal societies, and they occur among standing armies in state societies in which the armies are too big to operate undetected. Turney-High (1949) illuminated the "biomechanics" of the battle line, which develops more or less automatically when two groups meet in an agonistic encounter and every individual organism strives to have its vulnerable flanks protected by its neighbors.

In social carnivores and "female bonded" (or female philopatric) primate species, female participation in these more noisy than bloody battles commonly exceeds male participation. Tournamentlike "ritualized" combat, generally found among tribal societies with fairly dense populations (e.g., New Guinea), is supposed to test the numerical strength of the opponent while leaving room for a more peaceful solution of the conflict by mediators—but the ritual battle can easily develop into a rout and a massacre if a substantial imbalance of power is detected by one of the parties involved.

The question why males are the warriors in raiding-type warfare has been addressed by many, who converge on the hypothesis that raiding-type warfare evolved as a high-risk/high-gain male-coalitional reproductive strategy (or, arguably, even as a parental investment strategy). Reproductive success is the only criterion in the currency of evolution. Male and female organisms have evolved different strategies for the optimization of their reproductive success. For males, females are generally the limiting resource: for human males, women are the highly strategic "good" (always in short supply) that can convert the other resources controlled by the males into offspring.

Wrangham presents the principal adaptive hypothesis for explaining the species distribution of intergroup coalitional killing. This is the imbalance-of-power hypothesis, which suggests that coalitional killing is the expression of a drive for dominance over neighbors. Dominance is not an end in itself but a means to acquiring territory, access to females, and strategic position. Two conditions are proposed to be both necessary and sufficient to account for coalitional killing of neighbors: a state of intergroup hostility, and sufficient imbalances of power between parties that one party can attack the other with impunity. Under these conditions, it is suggested, selection favors the tendency to hunt and kill rivals when the costs are sufficiently low. Wrangham argues that the underlying psychology of male bonding is no different for chimpanzee raiding parties, human urban gangs, pre-state warrior societies, and contemporary armies.

Tooby and Cosmides enumerate some significant implications of their Risk Contract of War: first, men, but not women, will have evolved psychological mechanisms ("Darwinian algorithms") designed for coalitional warfare; and second, sexual access to women will be the primary benefit that men gain from joining male coalitions. This contrasts with most other primate (and social carnivore) species, in which the females have more "vested interests" in the defense of their lineage and the integrity of their group territory.

The rationale for groups to compete *as groups* has been illuminated by scenarios picturing intergroup contest competition in various lights: peaceful coexistence or merging of the groups; peaceful competition between groups, with the losers starving; violent conflict between individuals; scramble competition; and violent group conflict, or warfare. Warfare would be the best alternative for the group that practiced it successfully, assuming it to be within their biological reach. If conflict is inevitable, it makes better evolutionary sense for the troops to determine ownership of the resources as groups, rather than having both the conflict and the decreased inclusive fitness that would accompany a merger. Assuming that different groups tend toward one or another of these strategies in varying degrees, it is easy to see that if imbalances of power are common, warmongers would be the most successful, and could indeed overrun any group attempting to practice one of the other strategies.

Game theory—in particular the well-known hawk-dove game—shows how warfare can be an inevitable outcome of competition for resources, even if it makes everybody worse off. It is the instability of the situation (the logic of the war game) that leads to war. Game theory also underlines the important differences between group and individual costs and benefits. Even though all members of a group may gain or lose from changes in territorial extent, those who engage in the battle may pay costs avoided by those who hang back from dangerous confrontations. This can make group defense a collective action problem. However, benefits that go selectively to fighters (e.g., social status or rewards) "solve" this collective action problem and favor the persistence and recurrence of intergroup violence. Fighters may also acquire reproductive benefits: in many ethnographic settings, a man's quality as a warrior is related to the number of wives he can obtain or his access to nubile women otherwise.

Sociocultural Evolution: Stages in Sociopolitical Complexity. The concept of evolution as an ordering

PEACEFUL SOCIETIES

Van der Dennen's research on the war and peace behavior of some 2,000 "primitive" and historical societies revealed that there were or are more relatively peaceful peoples than is commonly acknowledged in the literature. The research also corroborated the picture of peaceful peoples as cultural or geographical isolates who live in peace by more or less successfully severing all contacts with other peoples. Isolation, splendid or not, seems *prima facie* to be the most prominent condition for peacefulness—so much so, in fact, that Muehlmann virtually identified peaceful peoples with *Rueckzugsvoelker* ("evading/retreating peoples"), a notion that goes back to Spencer.

The importance of isolation and geographical opportunity is nicely illustrated by the example of the Owekeeno and Haihai. The Owekeeno and Haihai belong, together with the Bella Bella, to the Heiltsuk language group of the northwest coast of North America. The Bella Bella were skilled in military strategies. The Owekeeno could easily afford to be peaceful because they were out of the main path of the war canoes, well protected by the easily defendable Wannock River. The unfortunate Haihais, on the other hand, "had no chance to be peaceable as they were forever embroiled in and afflicted by attack from both the northern tribes and the Bella Bellas, having to defend against predatory expeditions directed toward their resource base and to protect themselves from these warring tribes who wanted to practice on them in preparation for more serious expeditions" (S. F. Hilton, "Haihais, Bella Bella, and Owekeeno." In *Handbook of North American Indians*, edited by W. C. Sturtevant and W. Suttles. Vol. 7: Northwest Coast, pp. 312–22, 1990.)

—Johan M. G. van der Dennen

principle in cultural anthropology was proposed about 1840, even before Darwin's *Origin of Species* (1859). This, however, referred to Lamarckian and Spencerian sociocultural evolution, not to Darwinian evolution, which Darwin himself liked to call "descent with modification." The predominant school of social or cultural anthropology to the end of the nineteenth century assumed a linear and progressive course of evolution and history: human societies advance from the simple to the complex, from the "savage" primitive horde, through a barbarian stage, to civilization. A century later, with a much richer ethnographic record, cultural evolutionists used particular aspects of social organization to classify the variation. One widely used typology includes four stages—band, tribe, chiefdom, and state—with concomitant changes in warfare patterns and motives. Scholars differ in the role they see for war in the creation of more complex human societies. Some see it as a prime mover, others as secondary and only reinforcing other trends, and still others as only one of a set of interacting variables.

The evolution of historical war may be succinctly described as the transformation of armed men into manned arms, with the reproductive rewards, as Low suggests, increasingly "unhooked" from warring behavior, although other rewards may still be available. Specialized weapons and complex division of labor can play increasingly important roles. Turney-High argued that social organization has been more important than technology in the history of state-level warfare. Success depends "not upon the adequacy of weapons but the adequacy of team work, organization, and command working along certain simple [tactical] principles." Thus, armies have been characterized by increasing hierarchization of command structure, fighting phalanx-type battles of ever-increasing size, and campaigns for plunder, territorial aggrandizement, and political subordination: power, supremacy, and empire. The fighters in these state-level wars may be people of reduced means (the poor) who have few alternatives, and who may gain valuable societal approval for their actions.

Discipline and coordination in battle fuel the transition from warrior to soldier. The psychology of the warrior gave way in Western history when warfare changed from guerrilla-like raids and ambushes to massive battle formations, but guerrilla-like warfare repeatedly reemerges. Even though war is often a suboptimal solution to political problems, it will in this century probably be waged for scarce resources, particularly fresh water, oil, "security" and "ethnic nationalism" (secessionist, irredentist, and ethnonational wars are, and probably will be, mainly intranational). Thus, the means of production, the means of reproduction, and the means of destruction continue to shape sociocultural evolution.

[*See also* Archaeological Inference; Cooperation; Human Societies, Evolution of; Human Sociobiology and Behavior, *articles on* Human Sociobiology, Evolutionary Psychology, *and* Behavioral Ecology; Tragedy of the Commons.]

BIBLIOGRAPHY

Bigelow, R. *The Dawn Warriors: Man's Evolution towards Peace.* Boston, 1969. The classic work on the origin and evolution of warfare. Bigelow argues that intergroup competition can be a potent and relentless selection force explaining many human traits.

Chagnon, N. A. "Chronic Problems in Understanding Tribal Violence and Warfare." In *Genetics of Criminal and Antisocial Behavior*, edited by G. R. Bock and J. A. Goode, pp. 202–236. New York, 1996. Discusses some of the controversies arising from Chagnon's controversial reports of violence and relentless lethal raiding among the South American Yanomamö.

Cheney, D. L. "Interactions and Relationships between Groups." In *Primate Societies*, edited by B. B. Smuts et al., pp. 267–281.

Chicago, 1987. The first major inventory of more or less violent intergroup conflict in primates, together with the phylogenetic and socio-ecological principles (such as territoriality) governing primate sociality, and intra- and intergroup interactions.

Davie, M. R. *The Evolution of War: A Study of its Role in Early Societies*. New Haven, 1929. Together with Turney-High, Davie belongs to the all-time classics for the students of feuding and warfare patterns in nonstate societies.

Eibl-Eibesfeldt, I. *The Biology of Peace and War*. London, 1979. Presents the views on war and peace practices in human societies of the Lorenzian school of European ethology. In his later work, Eibl-Eibesfeldt emphasizes the roles of indoctrination (and indoctrinability) and group selection in the genesis of band-level and tribal warfare.

Ferguson, R. B. ed. *Warfare, Culture and Environment*. New York, 1984. Reviews the evidence and ideas associated with the (eco)materialist (or cultural materialist) theory of nonstate warfare.

Gat, A. "The Human Motivational Complex: Evolutionary Theory and the Causes of Hunter-Gatherer Fighting. Part 1: Primary Somatic and Reproductive Causes; Part 2: Proximate, Subordinate, and Derivative Causes." *Anthropological Quarterly* 73 (2000): 20–34; 74–88. Argues that fighting is an evolved feature of our nature with interconnected competition over resources, status and prestige, and reproduction as the *root* cause of conflict and fighting in humans as in all other animal species.

Goodall, J. *The Chimpanzees of Gombe: Patterns of Behavior*. Cambridge Mass., 1986/1987. Documents empathy, cruelty, ethnocentrism-cum-xenophobia, cannibalism, and "primitive warfare" in a population of chimpanzees. The discovery of this "lethal male raiding" had a huge impact on many students of nonstate warfare; *Homo sapiens* could no longer be considered the only species killing conspecifics in warlike exploits.

Harris, M. *Culture, People, Nature: An Introduction to General Anthropology*. 3d ed. New York, 1980. This introductory text includes a good discussion of the group selection argument of the cultural materialist theory of band-level and tribal warfare, in which the struggle for scarce resources, preferential female infanticide, warfare, and the so-called male supremacist complex have coevolved as an effective solution to the Malthusian dilemma.

Keeley, L. H. *War Before Civilization: The Myth of the Peaceful Savage*. New York, 1996. Presents the paleoanthropological and archaeological evidence of warfare, and concludes that warfare is documented in the archaeological record of the past 10,000 years in every well-studied region. Acknowledges sociocultural evolution but does not see any special role for "biology."

Low, B. S. "An Evolutionary Perspective on War". In *Behavior, Culture, and Conflict in World Politics*, edited by W. Zimmerman and H. K. Jacobson, pp. 13–56. Ann Arbor, 1993. This evolutionary analysis of "primitive" warfare includes an examination of the roles of natural selection, kin selection, and sexual selection with an emphasis on the reproductive benefits men gain from their status as warriors.

McEachron, D. L., and D. Baer. "A Review of Selected Sociobiological Principles: Application to Hominid Evolution II: The Effects of Intergroup Conflict." *Journal of Social and Biological Structures* 5 (1982): 121–139. Discusses the rationale for groups to compete *as groups*, concluding that the only possible competitive strategy for survival in competition with a group practicing warfare is warfare itself, either defensive or offensive.

Otterbein, K. F. *The Anthropology of Feuding and Warfare: Selected Works by Keith F. Otterbein*. New York, 1994. Investigates the correlations of warfare and feuding in band-level and tribal societies with variables such as territorial expansion, political centralization, patrilocality, polygyny and the existence of so-called fraternal interest groups.

Shaw, R. P., and Y. Wong. *Genetic Seeds of Warfare: Evolution, Nationalism, and Patriotism*. London, 1989. An elaboration and extension of van den Berghe's kin selection theory of ethnocentrism, demonstrating its relevance for the understanding of nonstate- as well as state-level warfare.

Tooby, J., and L. Cosmides. "The Evolution of War and Its Cognitive Foundations." *Proceedings of the Institute for Evolutionary Studies*, 88 (1988): 1–15. Tooby and Cosmides argue that underlying the disposition of males to coalitional violent behavior is an evolved "male coalitional psychology."

van der Dennen, J. M. G. *The Origin of War: The Evolution of a Male-Coalitional Reproductive Strategy*. 2 vols. Groningen, 1995. Reports ongoing investigation of the war and peace behaviors of some 2,000 nonstate and pre-state-level historical societies, as well as intergroup violent interactions in a great number of nonhuman species, with particular emphasis on the evolution of the underlying mechanisms in primates, hominids, and humans. Includes a comprehensive bibliography.

Wrangham, R. W. "Evolution of Coalitionary Killing." *Yearbook of Physical Anthropology* 42 (1999): 1–30. Reviews the evidence of chimpanzee "lethal male raiding" in the communities of Gombe and Taï and presents the imbalance-of-power hypothesis.

Wright, Q. *A Study of War*. Chicago, 1942. In this classic work on all aspects of contemporary ("civilized") warfare, Wright and his collaborators devote a chapter to the statistical analysis of some 650 "primitive" societies, distinguishing social, economic, and political warfare as distinct progressive social-evolutionary stages.

— Johan M. G. van der Dennen

WEISMANN, AUGUST FRIEDRICH LEOPOLD

August Weismann (1834–1914), German biologist, incorporated the first fully developed chromosomal theory of inheritance into the Darwinian framework of evolution by natural selection. Weismann earned his doctorate in medicine in 1856 from the University of Göttingen and practiced medicine, first as a clinical assistant in Rostock, then as a practicing physician in Frankfurt. In 1861, after studying briefly with the zoologist Rudolf Leukart, Weismann accepted a position as personal physician to Archduke Stephan of Austria, who maintained a residence in the Lahn valley. Because the archduke was rarely there, Weismann used the next eighteen months to study the development of dipterans. It was here that he also first read Charles Darwin's *Origin of Species*. His research earned him the Venia legendi at the University of Freiburg and an assistantship in zoology and comparative anatomy on the medical faculty. Despite the onset of a debilitating eye disease and long periods of absence from the university, his colleagues and the Baden government recognized Weismann's scientific achievements by promoting him in 1867 to a Lehrstuhl and gradually transferring his position to the philosophical faculty. With his academic future assured, Weismann married Mary Gruber of Genua. They raised four daughters and a son. Mary Gruber died in 1886. In 1895, Weismann married Wilhelmina Jesse, but the two legally separated in 1901.

Between 1864 and 1874 his overly sensitive eyes forced Weismann to forgo microscopical research; he thus shifted his attention to evolution theory and the study of polymorphism in butterflies. In 1874–1876 he published a collection of papers under the general title of *Studien zur Descendenz-Theorie* (2 vols.), which combined his earlier interest in insect metamorphosis with a strong tendency to view minute external structures in terms of their survival value. A contemporary of the German scientist and philosopher Ernst Haeckel (1834–1919), however, Weismann was also prone to interpret development according to germ layer specificity and a recapitulation of phylogeny.

As his eyes improved after 1874, he extended his investigations to the life cycles of daphnids and to saltwater hydromedusae. These organisms prompted Weismann to interpret the development of gametes as a sequestration of germ cells through the complex life cycles. By 1883, his conclusions evolved into a theory contrasting the continuity of the germ plasm with the mortal soma and led him to challenge the widely held belief in the inheritance of acquired characteristics. Thereafter, Weismann became identified as the day's leading proponent of the neo-Darwinian camp.

In 1887, Weismann argued on a priori grounds that a halving of chromosome number took place by a reduction division during gamete maturation. With the assistance of Chiyomatsu Ishikawa, he endeavored to demonstrate how this might happen during polar body formation. His phylogenetic conception of the chromosome and failure to recognize the details of tetrad formation soon rendered the details, though not the principle, obsolete.

In 1892, Weismann published a full elaboration of his germ plasm theory, embossed with theoretical hereditary and developmental and evolutionary units. The work drew him into heated philosophical arguments both in Germany and England. Accordingly, he adjusted the details of his theory by proposing that a germinal selection process within the body created new and eliminated old structures.

During the first decade of the twentieth century, Weismann published a textbook synthesizing nuclear cytology, embryology, natural history, and natural selection. The text was the first major work to present a comprehensive mechanical explanation of evolution.

In the twentieth century, Weismann's views met with varied reception. His speculations on the architecture of chromosomes suggested a mosaic process of development by subdivision of chromatin material in successive cell divisions. Although this yielded a straightforward explanation of cell differentiation, the "Roux-Weismann hypothesis" of mosaic development was undermined by wide acceptance of Boveri's conception of the chromosome, the Boveri-Sutton correspondences of chromosome behavior in meiosis and Mendelian segregation, Driesch's demonstration of a regulative mode of development, and the rise of classical genetics associated with the study of *Drosophila* and *Zea mays*.

Weismann's germ plasm theory influenced the cytoembryologist E. B. Wilson, author of the important text *The Cell in Development and Inheritance* (London, Macmillan, 1896). Many early geneticists learned of Weismann's theory through this book and came to accept a "nuclear monopoly" on explanations of traits: causes must lie in the nucleus, chromosomes, and ultimately genes, rather than in the cytoplasm or other features of cells and organisms. However, Weismann had treated germ and soma as symmetrical causes: development is both cause and consequence of germinal continuity. Wilson's representation served to emphasize the importance of genetic continuity at the expense of development's role in explaining heredity. Thus, although Weismann's speculative theory of development was rejected, the simplified outlines of his distinction between germinal continuity and somatic discontinuity were largely accepted and updated along Mendelian lines. Ironically, Weismann's germinal theory became enshrined as the basis for partitioning problems of heredity from development rather than unifying them both with selection.

Weismann's ideas on germinal selection and the evolution of development have received renewed attention since the 1980s, as evolutionary developmental biologists, including Leo Buss (1987), sought to reintegrate development with neo-Darwinian evolution. However, after Wilson's representation of the germ–soma doctrine, recent work on the evolution of development has had to reinvent for itself Weismann's views on the symmetry of causal relations between heredity and development.

[*See also* Germ Line and Soma; Sex, Evolution of.]

BIBLIOGRAPHY

Buss, L. *The Evolution of Individuality*. Princeton, 1987.

Churchill, F. B., and H. Risler, eds. *August Weismann. Ausgewählte Briefe und Dokumente*, Freiburg, 1999. This work is annotated in English, contains essays by the editors, and provides a complete bibliography of Weismann's publications and of recent secondary works. The essay in volume 2 by Churchill, "August Weismann: A Developmental Evolutionist," presents a general discussion of Weismann's works and ideas.

Griesemer, J. "Tools for Talking: Human Nature, Weismannism and the Interpretation of Genetic Information." In *Are Genes Us? The Social Consequences of the New Genetics*, edited by Carl Cranor, pp. 69–88. New Brunswick, N.J., 1994. Argues that views on genetic determinism and biologized human nature trace to Weismannism.

Griesemer, J., and W. Wimsatt. "Picturing Weismannism: A Case Study of Conceptual Evolution." In *What the Philosophy of Bi-*

ology Is: Essays for David Hull, edited by M. Ruse, pp. 75–137. Dordrecht, Netherlands, 1989. Traces the history of diagram representations of Weismann's germ–soma distinction from the 1890s to the 1980s.

Weismann, A. *Studies in the Theory of Descent*. Translated by Raphael Meldola, 2 vols. London, 1882. Originally published as *Studien zur Deszendenztheorie*, 2 vols. Leipzig, 1875–1876.

Weismann, A. *Essays Upon Heredity and Kindred Biological Problems*. Edited and translated by Edward B. Poulton, Selmar Schönland, and Arthur E. Shipley, 2d ed., 2 vols. Oxford, 1891–1892. A collection of Weismann's important papers published during the 1880s. The second edition contains a second volume of further essays.

Weismann, A. *The Germplasm: A Theory of Heredity*. Translated by W. Parker and H. Ronnfeldt. New York, 1893. Weismann's major statement about germ plasm. Originally published as *Das Keimplasma. Eine Theorie der Vererbung*. Jena, 1892.

Weismann, A. *Vorträge über Descendenztheorie gehalten an der Universität zu Freiburg im Breisgau*. 2 vols. Jena, 1902. This text appeared in two other editions (1904 and 1913). The second edition appeared in English as *The Evolution Theory*. Translated by J. Arthur Thomson and Margaret Thomson. London.

— James Griesemer and Frederick B. Churchill

WRIGHT, SEWALL

Sewall Wright (1889–1988), leading U.S. population geneticist and evolutionary theorist. Together with R. A. Fisher and J. B. S. Haldane in England, he was responsible for the "modern synthesis," the fusion of Darwinian natural selection with Mendelian inheritance. For many years, these three men dominated the field of population genetics and evolution theory.

Wright was born in Melrose, Massachusetts, but at age two he moved to Galesburg, Illinois, where his father began teaching at Lombard College. Young Sewall was precocious; before starting school, he kept a diary and knew how to extract cube roots. He attended Lombard College, where his interest in genetics was kindled by an inspiring teacher, Wilhelmina Key, and by reading R. C. Punnett's article in the *Encyclopedia Britannica*. After graduation in 1911, he spent a year at the University of Illinois, then moved to Harvard University as a student of William E. Castle. On receiving his doctorate in 1915, he joined the U.S. Department of Agriculture (USDA) as an animal husbandryman. In 1925, he moved to the University of Chicago. Retiring in 1954, he spent his remaining years at the University of Wisconsin.

Wright began his study of guinea pigs at Harvard and continued through his Chicago years. Although best known for population genetics and evolutionary theory, he actually spent most of his time on guinea pigs, where he emphasized the interaction of coat color genes. Using data from the guinea pig colonies at the USDA, he did a masterful study of the effects of inbreeding and crossbreeding. Early in his career he wrote a famous series of papers, "Systems of Mating," which laid the foundation for scientific animal breeding. Wright was particularly skilled with correlations and invented the method of path analysis, which was used extensively by himself and others in developing such widely used concepts as heritability. He also invented the inbreeding coefficient, a measure of the reduction in heterozygosity resulting from inbreeding, and devised a simple algorithm for computing this for any pedigree, however complex. This is now standard and taught in elementary genetic courses.

On moving from the USDA to the University of Chicago, Wright changed his emphasis from animal breeding to evolution. His most famous paper, "Evolution in Mendelian Populations," was published in 1931. This was the beginning of his "shifting balance" theory of evolution, to which he devoted most of the remainder of his life.

To Wright, a major difficulty in evolution was that, because of the complexity of gene interactions, it was often not possible for a population to move from one adapted state, a "fitness peak," to a higher peak, because of maladapted intermediates. To circumvent this, he envisioned a large population broken into numerous partly isolated groups. By random gene-frequency drift, one of these groups might arrive at a better adapted state. It would then export migrants and upgrade the whole population. Then the process could start over.

Starting in the 1930s, Wright's principal opponent was R. A. Fisher, who downplayed the importance of random gene-frequency drift, and argued that evolution is most effective in large populations with mass selection. In the 1960s, Motoo Kimura argued that most molecular evolution took the form of mutation and random drift of selectively neutral alleles. Although Wright also emphasized random drift, it was as an aid to natural selection. He thought of changes small enough to be effectively neutral in a small subpopulation, whereas Kimura thought of changes small enough to be neutral on a global scale.

Wright's theory was very popular among biologists, but was less so with theorists who realized that the delicate balance of selection, migration, and population size might be unusual. Others questioned whether such an elaborate theory was needed; mass selection seemed sufficient. The importance of Wright's theory is still a matter of debate.

Wright was responsible for a number of theoretical developments. His *F* statistics have been especially popular as measures of population subdivision. He introduced the concept of effective population number, or N_e, the size of an idealized population that would have the same amount of random genetic drift as the true population and could be used as a surrogate in equations. Finally, Wright independently discovered the physicist's Fokker–Planck equation to describe the combination of deterministic and stochastic processes in the evolutionary process.

After his retirement, Wright wrote a four-volume treatise, which he completed in his nineteenth year. He died at age ninety-eight, the result of slipping on the ice during one of his regular long walks.

[*See also* Shifting Balance.]

BIBLIOGRAPHY

Fisher, R. A. *The Genetical Theory of Natural Selection.* Oxford, 1930. Variorum edition, J. H. Bennett, ed., Oxford, 1999. This is Fisher's classic work, in many ways the natural successor to Darwin. The 1999 edition has very useful notes, clarifying some points that were confusing in the original edition.

Kimura, Motoo. *The Neutral Theory of Molecular Evolution.* Cambridge, 1983. This is Kimura's definitive summary of his neutral theory.

Provine, William B. *Sewall Wright and Evolutionary Biology.* Chicago, 1986. The definitive Wright biography—a scholarly work, both personal and scientific.

Wright, Sewall. "Systems of Mating." *Genetics* 6 (1921): 111–178. This five-part series laid the foundation for much of the theory of inbreeding and quantitative genetics and had a great influence on scientific animal breeding.

Wright, Sewall. "Evolution in Mendelian Populations." *Genetics* 16 (1931): 97–139. This article is the beginning of Wright's shifting balance theory. It is not, however, the best way to understand Wright's work, for many of his methods have been improved and simplified in later works. See the next reference.

Wright, Sewall. *Evolution and the Genetics of Populations.* 4 vols. Chicago, 1968–1978. These volumes summarize not only Wright's work but also the whole field of population genetics and evolutionary theory as seen by the master.

Wright, Sewall. "Surfaces of Selective Value revisited." *American Naturalist* 131 (1988): 115–123. This is Wright's last paper in which he summarizes his views of the four great contributors, Haldane, Fisher, Kimura, and himself. Written in his last year, it offers a chance to see Wright's mature thoughts.

— JAMES F. CROW

ZEBRA FINCHES

The zebra finch (*Taeniopygia guttata castanotis*) is a small (12-g) estrildine grass finch native to the semi-arid zone of Australia (Figure 1). It is a hardy species that breeds freely in captivity, so it is commonly found in avicultural settings and scientific laboratories throughout the world.

Natural History and Sexual Dimorphism. Zebra finches favor open woodland habitats, requiring surface water and ample grass. Birds form socially monogamous pair bonds that may often persist across breeding seasons. Birds nest in loose colonies in shrubs or trees and favor cryptic or protected (e.g., thorny) sites. Females typically lay two to seven eggs per clutch; clutches are incubated for about two weeks. Hatchlings are highly altricial, but they develop rapidly and leave the nest at about twenty days of age. Parents feed nestlings and recent fledglings on ripe and half-ripe grass seeds, which also constitute the principal diet of adults. Birds reach adulthood by ninety days of age. Little is known about typical dispersal distances, but lack of variation in plumage and morphology throughout the vast range of this species indicates that gene flow between populations is substantial.

FIGURE 1. Male Zebra Finch.

The species-typical plumage color is gray with black-and-white stripes, but atypical colors are common in the avicultural trade. Males possess four plumage markings that females lack (sexual dichromatism). In addition, males' beaks tend to be redder, whereas those of females are more orange, although there is considerable overlap between the sexes in beak color. Males also possess a single, idiosyncratic song type. There is no size difference between the sexes.

Mate Choice and Its Consequences. The sexes have opposing mate preferences for beak color: females prefer males with bright red beaks, and males prefer females with pale orange beaks (beaks do not reflect appreciably in the ultraviolet.) Beak color is not correlated with male viability (in captivity) or with parasite loads (in nature), but it does affect reproductive success via sexual selection. In laboratory experiments, males with bright red beaks reproduce at higher rates. Beak color affects female viability and reproduction: females with pale orange beaks live longer and reproduce at higher rates compared to those with redder beaks. Beak color is heritable, so that offspring of a pair consisting of a male with a bright red beak and a female with a pale orange beak have intermediate beak colors. Thus, there is opposing selection on this sexually dimorphic trait, and the beak color expression of each sex appears not to have not yet reached an evolutionary optimum. The only other sexually dimorphic trait for which mate preferences have been reported is song rate; females prefer males with complex song types and may prefer males that sing at high rates.

In addition to preferences for naturally occurring traits, zebra finches of both sexes have marked preferences for novel traits that are applied in experimental contexts. Females find males banded with any one of several colors—red, yellow, or black—of plastic leg bands (typically used by ornithologists to recognize individual birds) to be attractive, and prefer to mate with them compared to males lacking color bands; males show a narrower range of preferences for novel leg coloration in females; and both sexes find potential mates wearing certain colors (light blue and light green) to be unattractive. In addition, females prefer to mate with males wearing artificial white crests, even though there are no crested species among living estrildines (approximately three hundred species within the family *Ploceidae*). Related species also have color band and crest preferences.

Color band and crest preferences appear to be emergent properties of the central nervous and sensory systems of these birds; that is, they are unpredictable attributes arising from interactions within complex neuro-sensory systems. As such, they are experimentally elicted examples of the "aesthetic" mate preferences (also known as sensory biases) that Charles Darwin hypothesized in 1871 to account for a diverse array of secondary sexual ornamentation in animals.

Females also prefer males wearing symmetrical band color combinations over those that are asymmetrical (different band colors next to each foot). Preferences for leg band symmetry may reflect functional aversions to asymmetry (i.e., preference for low levels of fluctuating asymmetry) or aspects of aesthetic mate preferences.

In breeding experiments designed to explore the potential for evolution of novel attractive traits, birds randomly assigned to wear attractive color bands live longer and have more offspring than birds with unattractive bands. This occurs because attractive birds have differential access to more vigorous mates and contribute less parental investment (differential allocation) than do unattractive birds. Females mated to attractive males deposit more testosterone in egg yolk, which may serve to enhance the competitive ability of their resulting offspring.

Offspring sex ratios also reflect band attractiveness: attractive birds produce more same-sex offspring, whereas unattractive birds produce more opposite-sex offspring. This sex ratio manipulation is effected by selective brood reduction (eviction of nestlings from the nest) several days after hatching.

Socially mated birds engage in extra-pair copulations at variable frequencies. The average rate of extra-pair fertilization (EPF) among zebra finches in nature and in non-color-banded laboratory populations is 5 percent or less. When males are banded with attractive (red) or unattractive (green) bands, the EPF rate increases greatly ($\geq$ 25 percent), in part because females that are socially mated to unattractive males selectively engage in fertile extra-pair copulations with attractive males. The greatly increased frequency of extra-pair relations appears to be an opportunistic response to increased variation in male attractiveness in color-banded populations.

Overall, then, zebra finches behave as if band color were a heritable (genetic) trait. The birds' spontaneous responses to novelty would favor the rapid fixation of novel alleles for attractive traits were such alleles to arise by mutation within populations. The joint occurrence of aesthetic preferences and the dynamic adjustment of reproductive tactics displayed by birds when they encounter novel attractive traits suggest there is an important role to such preferences in speciation processes.

Parental Care and Parent–Offspring Relationships. Both sexes participate in parental care and defense of young, but the male share of care is slightly less, on average, than that of females. The expression of caregiving is variable, however, and the mates of attractively banded individuals compensate for the reduced parental attentiveness of their partners (differential allocation). A functional explanation for the sex-specific begging calls of nestlings has not been reported. It may be the case that parents discriminate against offspring that make ambiguous signals of their sex.

Young birds sexually imprint on the vocalizations made by or visual appearance of their parents. This response is stronger in males, which prefer as mates females that resemble their mother even when maternal traits are highly unusual (e.g., white plumage or artificial grey crests). Female response is less well understood. Females do not imprint on unattractive paternal traits (orange beak color, green leg bands, grey crests), implying that their capacity to recognize attractive traits is relatively independent of their early experience. Both sexes have the ability to recognize kin (siblings and cousins) with which they have no shared experience; presumably this ability involves phenotype matching to traits of kin (parents and same-brood siblings) encountered during development.

[*See also* Fluctuating Asymmetry; Mating Systems, *article on* Animals; Parental Care; Parent–Offspring Conflict; Sexual Selection.]

BIBLIOGRAPHY

Burley, N. "The Organization of Behavior and the Evolution of Sexually Selected Traits." In *Avian Monogamy: The Neglected Mating System*, edited by Patricia Adair Gowaty and Douglas W. Mock, pp. 22–44. Ornithological Monographs #37. Washington, D.C., 1985.

Burley, N. "Sexual Selection for Aesthetic Traits in Species with Biparental Care." *The American Naturalist* 127 (1986): 415–445.

Burley, N. "Sex-Ratio Manipulation in Color-banded Populations of Zebra Finches." *Evolution* 40 (1986): 1191–1206.

Burley, N. "The Differential-Allocation Hypothesis: An Experimental Test." *The American Naturalist* 132 (1998): 611–628.

Burley, N., C. Minor, and C. Strachan. "Social Preference of Zebra Finches for Siblings, Cousins and Non-kin." *Animal Behaviour* 39 (1990): 775–784.

Burley, N. T., P. G. Parker, and K. Lundy. "Sexual Selection and Extra-pair Fertilization in a Socially Monogamous Passerine, the Zebra Finch (*Taeniopygia guttata*)." *Behavioral Ecology* 7 (1996): 218–226.

Burley, N. T., and R. Symanski. "'A Taste for the Beautiful': Latent Aesthetic Mate Preferences for White Crests in Two Species of Australian Grass Finches." *The American Naturalist* 152 (1998): 792–802.

Collins, S. A. "Is Female Preference for Male Repetoires Due to Sensory Bias?" *Proceeding of the Royal Society of London Series B—Biological Sciences* 266 (1999): 2309–2314.

Diego, G., J. Graves, N. Hazon, and A. Wells. "Male Attractiveness and Differential Testosterone Investment in Zebra Finch Eggs." *Science* 286 (1999): 126–128.

Price, D. K., and N. T. Burley. "Constraints on the Evolution of Attractive Traits: Selection in Male and Female Zebra Finches." *The American Naturalist* 144 (1994): 909–934.

Swaddle, J. P. "Reproductive Success and Symmetry in Zebra Finches." *Animal Behaviour* 51 (1996): 203–210.

Vos, D. R. "The Role of Sexual Imprinting for Sex Recognition in Zebra Finches: A Difference between Males and Females." *Animal Behaviour* 50 (1995): 645–653.

Zann, R. A. *The Zebra Finch.* Oxford and New York 1996. A good general reference to this species.

— NANCY TYLER BURLEY

ZOOTYPES

Although an old idea, the term *zootype* was used by Slack et al. (1993) to define a particular spatial pattern of gene expression shared by all animals during embryonic development. Slack et al. suggested that this expression pattern, the zootype, would be most clearly manifested at the phylotypic stage of animal taxa, thereby providing a gene-based definition of an animal.

The zootype concept is based on the observation that certain homeobox-containing genes known to play key patterning roles during development show very similar spatial patterns of expression in embryos of both protostomes and deuterostomes, the two main kinds of triploblastic animals commonly grouped together as bilaterians (Figure 1A). The zootype-defining genes proposed by Slack et al. were orthodenticle and empty spiracles, expressed at the most anterior portion of embryos; five *Hox* (homeobox) genes expressed in the middle portion of the embryo; and even-skipped, which is expressed at the posterior (Figure 1B). The zootype genes have an informational rather than an executive function, so they establish a system of relative positions in the embryos rather than specify the actual structure to be built at a given place.

The phyletic or phylotypic stage of development represents the point during embryonic development at which the basic body plan characteristic of a given phylum is completed. Embryonic development continues after the phylotypic stage is reached, but no major modifications of the body plan take place. Development before and after the phylotypic stage can be very variable in members of the same phylum. It is useful to think of the embryos of members of a phylum converging into the phylotypic stage during early development and diverging from it in late development (Hall, 1992).

Genetic Makeup of Ancestral Body Plan. Under the assumption that animals constitute a monophyletic group, all modern phyla share a single common ancestor, and therefore all current body plans evolved from a single ancestral body plan. One important goal of evolutionary developmental biology is to elucidate how the phylotypic body plans of different phyla originated from that single ancestral body plan. The zootype is a hypothesis about the genetic makeup of that ancestral body plan for animals. The zootype represents a system of positional information, shared by all animals, upon which the diagnostic morphology of different taxa would develop. This argument, however, is problematic. If each taxon shows its characteristic features at the phylotypic stage but all taxa also show the same pattern of expression of key developmental genes (the zootype) at that stage, can these genes be responsible for the differences in body plans (Cohen, 1993)? Probably not. Also essential for the development of different body plans would be other genes providing positional information, for example, the genes responsible for setting the pattern along the dorso-ventral axis, plus the zootype target genes that translate positional into morphological information.

The conservation in the expression of the zootype genes among protostomes and deuterostome embryos simply indicates that the pattern was also present in their last common ancestor, the primitive bilaterian. However, Slack et al. (1993) proposed that "there was once a primordial multicellular ancestor of all existing animals and this ancestor was the first organism to possess the zootype." Because several important animal groups, namely, sponges, cnidarians, and ctenophores, diverged before the proto-deuterostome split, the zootype must also be found in those groups to represent a true defining character (a synapomorphy) of all metazoans. Evidence indicates that the zootype, as defined by Slack et al., does not apply to all living animals. Cnidarians do not seem to contain the six basic HOX genes proposed by the zootype but only four (Finnerty, 1998; Martínez et al., 1998). Furthermore, it is not clear that

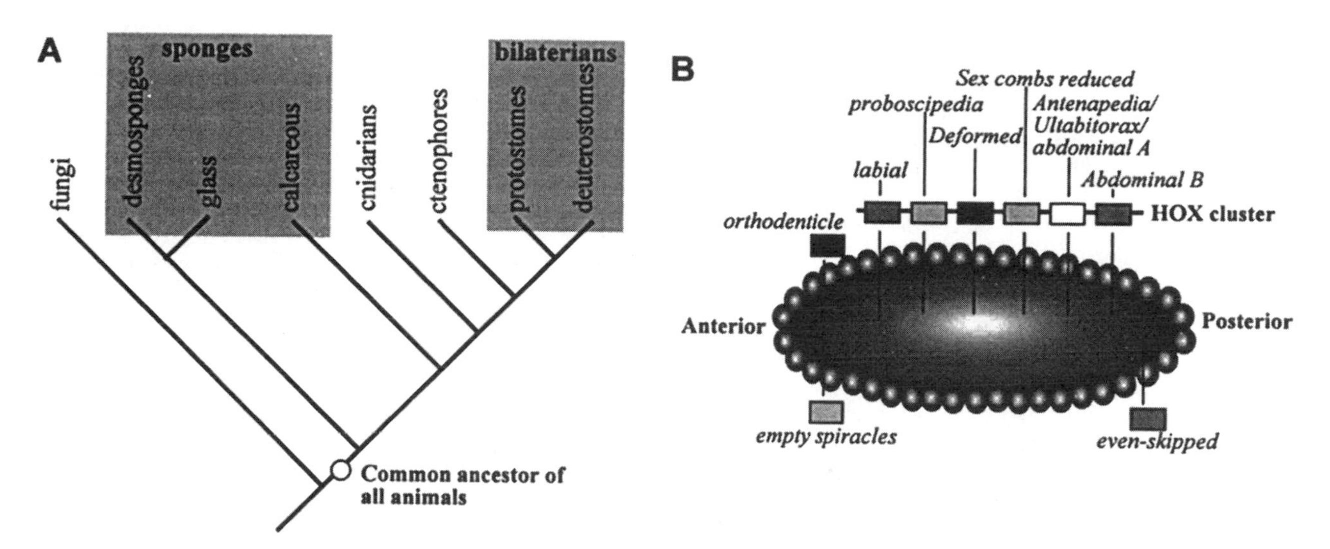

FIGURE 1. Zootypes.
(A) Phylogenetic relationships among the major animal groups. (B) Diagrammatic representation of the presumptive embryo of the common ancestor of all animals. Lines point at the anterior limit of expression of each of the zootype genes, as proposed by Slack et al. (1993). Drawing by Daniel E. Martínez.

cnidarian HOX genes play a role in body axis specification (Mokady et al., 1998; Shenk et al., 1993; Smith et al., 1999). In addition, no HOX gene has been found in sponges.

Could the zootype be redefined to represent a true metazoan synapomorphy? Or better yet, could a unique molecular blueprint of a metazoan exist? The last common ancestor of plants, fungi, and animals was a protist. No one would expect that protist ancestor to contain the blueprint to make a plant or a fungus or an animal. Only the origin of key innovations will explain the protist-to-plant or protist-to-animal transition (Williams, 1992). Should we expect the common ancestor of all animals to contain the animal blueprint? If so, could we deduce that blueprint from the phylotypic body plans of modern animals?

Shared features. Animals presumably have evolved from a colonial choanoflagellate-like creature with a division of labor among cells (e.g., somatic vs. reproductive cells). At a given point during the Precambrian era, this organism acquired certain features that made it an animal. New evidence indicates that sponges may constitute a paraphyletic group and that calcareous sponges (class: Calcarea) are part of a sister group to the rest of the metazoa (Peterson and Eernisse, 2001). This scenario suggests that the last common ancestor of cnidarians, ctenophores, protostomes, and deuterostomes looked like a sponge. The sponge body plan does not include many of the features found in other metazoans: a primary body axis, germ layers, neurons, and so on. Thus, the first metazoans probably did not require many of the genes that we consider essential for bilaterian development, for example, genes for axial spatial organization and germ layer segregation. None of the zootype genes, as described by Slack et al. (1993), have been found in sponges. It is likely that that the key innovations of animal development were not all present in the common ancestor but were acquired later.

Gerhart and Kirschner (1997) hypothesized that the Vendian-Cambrian body plans, recently evolved from a common ancestor, probably shared many features and that the phylotypic body plans were terminal rather than intermediate stages of development. Under this scenario, it is possible to imagine a very simple bilaterian ancestor and its phylotypic stage patterned by the zootype genes. It is more difficult, however, to explain how the ancestral phylotypic body plan could be modified to generate most if not all of the modern body plans while maintaining the ancestral blueprint, the zootype.

The Metazoan "Eve." It is certainly possible to produce a list of developmental genes that are unique to animals. Several years ago, Shenk and Steele (1993) reviewed the presumptive molecular constitution of the metazoan ancestor, the metazoan "Eve." Careful examination of potential animal and nonanimal synapomorphies may help us identify key innovations in the evolution of animals (Hughes and Hughes, 1993). Defining a type to represent all animals, however, may not be possible, simply because the spongelike ancestor of all animals did not possess many of the features that we encounter today in bilaterians.

[*See also* Body Plans; Constraint; Homeobox; *Hox* Genes; Phylotypic Stages.]

BIBLIOGRAPHY

Cohen, J. "Development of the Zootype." *Nature* 363 (1993): 307.

Finnerty, J. R. "Homeoboxes in Sea Anemones and Other Nonbilaterian Animals: Implications for the Evolution of the Hox Cluster and the Zootype." *Current Topics in Developmental Biology* 40 (1998): 211–254.

Gerhart, J., and M. Kirschner. *Cell, Embryos, and Evolution.* Oxford, 1997.

Hall, B. K. *Evolutionary Developmental Biology.* London, 1992.

Hughes, N. C., and S. M. Hughes. "Development of the Zootype." *Nature* 363 (1993): 307.

Martínez, D. E., D. Bridge, L. M. Masuda-Nakagawa, and P. Cartwright. "Cnidarian Homeoboxes and the Zootype." *Nature* 393 (1998): 748–749.

Mokady, O., M. H. Dick, D. Lackschewitz, B. Svhierwater, and L. W. Buss. "Over One-half Billion Years of Head Conservation? Expression of an ems Class Gene in *Hydractinia symbiolongicarpus* (Cnidaria: Hydrozoa)." *Proceedings of the National Academy of Sciences* 95 (1998): 3673–3678.

Peterson, K. J., and D. J. Eernisse. "Animal Phylogeny and the Ancestry of Bilaterians: Inferences from Morphology and 18S rDNA Gene Sequences." *Evolution and Development* 3 (2001): 170–205.

Shenk, M. A., L. Gee, R. E. Steele, and H. R. Bode. "Expression of Cnox-2, a HOM/HOX Gene, is suppressed during head formation in Hydra." *Developmental Biology* 160 (1993): 108–118.

Shenk, M. A., and R. E. Steele. "A Molecular Snapshot of the Metazoan 'Evel.'" *Trends in Biochemical Sciences* 18 (1993): 459–463.

Slack, J. M. W., P. W. H. Holland, and C. F. Graham. "The Zootype and the Phylotypic Stage." *Nature* 361 (1993): 490–492.

Smith, K. M., L. Gee, I. L. Blitz, and H. R. Bode. "CnOtx, a Member of the Otx Gene Family, Has a Role in Cell Movement in Hydra." *Developmental Biology* 212 (1999): 392–404.

Williams, G. C. *Natural Selection. Domains, Levels, and Challenges.* New York, 1992.

— Daniel E. Martínez

Directory of Contributors

Adami, Christoph
Principal Scientist, Jet Propulsion Laboratory, and Faculty Associate, California Institute of Technology
Simulations of Evolution

Adler, Frederick R.
Associate Professor, Department of Mathematics and Department of Biology, University of Utah
Mathematical Models

Ahouse, Jeremy Creighton
Associate Director, Millennium Pharmaceuticals Inc.
Cloning

Aiello, Leslie C.
Professor of Biological Anthropology, University College London
Neogene Climate Change

Alatalo, Rauno V.
Professor of Biology, University of Jyväskylä, Finland
Male-Male Competition

Alcock, John
Professor of Biology, Arizona State University
Social Evolution

Alvarez, Helen
Graduate Student in Anthropology, University of Utah
Life History Theory, *article on* Grandmothers and Human Longevity

Alvarez-Valin, Fernando
Assistant Professor of Biomathematics, Facultad de Ciencias, Universidad de la Republica
Neutral Theory

Ammerman, Albert J.
Professor, Colgate University
Agriculture, *article on* Spread of Agriculture

Andersen, Robert A.
Director, Provasoli-Guillard National Center for Culture of Marine Phytoplankton, Bigelow Laboratory for Ocean Sciences
Algae

Andrews, Paul W.
Graduate Student in Biology, University of New Mexico
Altruism

Antonovics, Janis, FRS
Lewis and Clark Professor of Biology, University of Virginia
Disease, *article on* Infectious Disease

Armelagos, George
Professor of Anthropology, Emory University
Nutrition and Disease

Arnold, Frances H.
Dickinson Professor of Chemical Engineering and Biochemistry, Division of Chemistry, California Institute of Technology
Directed Protein Evolution

Arsuaga, Juan Luis
Professor of Palaeontology, Complutense University of Madrid
Hominid Evolution, *article on* Archaic *Homo sapiens*

Aufderheide, Karl J.
Associate Professor of Biology, Texas A&M University
Cortical Inheritance

Ayala, Francisco J.
Donald Bren Professor of Biological Sciences, University of California, Irvine
Dobzhansky, Theodosius; Malaria

Balding, David
Professor of Statistical Genetics, Imperial College, University of London
Coalescent Theory and Modeling

Baquero, Fernando
Director of the Department of Microbiology, Ramón y Cajal Hospital (INSALUD), Madrid, Spain
Antibiotic Resistance, *article on* Origins, Mechanisms, and Extent of Resistance

Barnes, Ian
Research Scientist, University of Oxford
Ancient DNA

Bascompte, Jordi
Associate Professor of Research, Estación Biológica de Doñana (Spanish Research Council)
Pattern Formation

Bateson, Patrick
Professor of Ethology and Provost of King's College, University of Cambridge, and Biological Secretary of the Royal Society of London
Ethology

Beckerman, Stephen
Associate Professor of Anthropology, Pennsylvania State University
Human Families and Kin Groups

Bennett, Albert F.
Professor of Biological Sciences, Department of Ecology and Evolutionary Biology, University of California, Irvine
Experimental Evolution, *overview article*

Benson, Keith R.
Professor in the History of Science, University of Washington
Epigenesis and Preformationism; Recapitulation

Benton, Michael J.
Professor of Vertebrate Palaeontology, Department of Earth Sciences, University of Bristol
Cope's Rule

Bielawski, Joseph P.
Research Fellow, Department of Biology, University College London
Phylogenetic Inference, *overview article*

Bird, Douglas
Assistant Research Professor, Department of Anthropology, University of Maine
Human Foraging Strategies, *article on* Human Diet and Food Practices

Birkhead, Timothy R.
Professor of Zoology, University of Sheffield
Sperm Competition

Bishop, M. J.
Head of Bioinformatics, HGMP Resource Centre
Overview Essay *on* Genomics and the Dawn of Proteomics; Bioinformatics; Genome Sequencing Projects

Blackmore, Susan
Visiting Lecturer, University of the West of England, Bristol
Meme

Blackwell, Meredith
Boyd Professor, Department of Biological Sciences, Louisiana State University
Fungi

Blumstein, Daniel T.
Assistant Professor of Organismic Biology, Ecology and Evolution, University of California, Los Angeles
Group Living

Bonhoeffer, Sebastian
Professor of Theoretical Biology, Swiss Federal Institute of Technology, Zurich
Transmission Dynamics

Bonsall, Michael B.
Royal Society University Research Fellow, Department of Biological Sciences, Imperial College at Silwood Park
Population Dynamics

Borgia, Gerald
Professor, Department of Biology, University of Maryland
Sexual Selection, *article on* Bowerbirds

Bourguet, Denis
Chargé de Recherches, Institut National de la Recherche Agronomique, La Minière, France
Gene Families

Boyd, Robert
Professor of Anthropology, University of California, Los Angeles
Human Sociobiology and Behavior, *overview article*

Bramble, Dennis M.
Professor of Biology, University of Utah
Human Evolution, *article on* Morphological and Physiological Adaptations

Bronstein, Judith L.
Associate Professor of Ecology and Evolutionary Biology, University of Arizona
Mutualism

Broughton, Jack M.
Associate Professor of Anthropology, University of Utah
Human Foraging Strategies, *overview article*

Brown, Jerram L.
Professor of Biological Sciences, State University of New York at Albany
Alloparental Care

Bryant, Harold N.
Curator of Earth Sciences, Royal Saskatchewan Museum, Regina, Canada
Classification; Nomenclature

Budd, Graham E.
Assistant Professor of Historical Geology and Palaeontology, University of Uppsala, Sweden
Animals

Burley, Nancy Tyler
Professor of Ecology and Evolutionary Biology, University of California, Irvine
Zebra Finches

Burt, Austin
Reader in Evolutionary Genetics, Imperial College at Silwood Park
Selfish Gene; Transposable Elements

Bush, Robin M.
Assistant Professor of Ecology and Evolutionary Biology, University of California, Irvine
Influenza

Bynum, W. F.
Professor of the History of Medicine, Wellcome Trust Centre for the History of Medicine, University College London
Galton, Francis

Cannatella, David
Curator of Herpetology, Texas Natural History Collections, Texas Memorial Museum, and Section of Integrative Biology, University of Texas at Austin
Amphibians

Carey, James R.
Professor of Entomology, University of California, Davis
Demography

Carlson, Bruce M.
Director, Institute of Gerontology, University of Michigan
Development, *article on* Developmental Stages, Processes, and Stability

Carlson, Elof
Distinguished Teaching Professor, Department of Biochemistry, State University of New York at Stony Brook
Muller, Hermann Joseph

Carroll, Robert L.
Strathcona Professor of Zoology, McGill University, Quebec, Canada
Transitional Forms

Cashdan, Elizabeth
Professor of Anthropology, University of Utah
Mate Choice, *article on* Human Mate Choice; Self-Deception

Cavalier-Smith, Thomas, FRS
Professor of Evolutionary Biology and NERC Professorial Fellow, Department of Zoology, University of Oxford
Meiosis

Cavalli-Sforza, L. Luca
Professor of Genetics Emeritus Active, Stanford Medical School
Overview Essay *on* Human Genetic and Linguistic Diversity; Cultural Evolution, *article on* Cultural Transmission

Charlesworth, Brian
Royal Society Research Professor, University of Edinburgh
Fitness; Hitchhiking; Reproductive Value

Charlesworth, Deborah
Professorial Fellow, Faculty of Science and Engineering, Institute of Cell, Animal and Population Biology, NERC Senior Research Fellow (from 1999), Institute of Cell, Animal and Population Biology (ICAPB), Ashworth Lab, University of Edinburgh
Hitchhiking; Sex Chromosomes

Cheney, Dorothy L.
Professor of Biology, University of Pennsylvania
Alarm Calls

Churchill, Frederick B.
Professor Emeritus, Department of the History and Philosophy of Science, Indiana University
Weismann, August Friedrich Leopold

Clark, Andrew G.
Professor of Biology, Pennsylvania State University
Disease, *article on* Hereditary Disease

Clark, David E.
Section Head (Computer-Aided Drug Design), Argenta Discovery Ltd., Harlow, United Kingdom
Engineering Applications

Clarke, Bryan C.
Professor Emeritus of Genetics, University of Nottingham
Cain, Arthur James; Frequency-Dependent Selection

Cochran, Gregory
Independent Scholar, Albuquerque
Natural Selection, *article on* Natural Selection in Contemporary Human Society

Cockburn, Andrew
Professor of Botany and Zoology, Australian National University
Genetic Markers

Cohan, Frederick M.
Professor of Biology, Wesleyan University
Clonal Structure, *overview article*; Clonal Structure, *article on* Population Structure and Clonality of Bacteria

Cohen, Mitchell L.
Director, Division of Bacterial and Mycotic Diseases, National Center for Infectious Diseases, Centers for Disease Control and Prevention
Antibiotic Resistance, *article on* Epidemiological Considerations

Cohn, Martin J.
BBSRC David Phillips Research Fellow, School of Animal and Microbial Sciences, University of Reading
Hox Genes

Comfort, Nathaniel C.
Assistant Professor of History and Deputy Director of the Center for History of Recent Science, The George Washington University
McClintock, Barbara

Conway Morris, Simon
Professor of Evolutionary Palaeobiology, University of Cambridge
Body Plans; Cambrian Explosion

Corne, David Wolfe
Reader in Evolutionary Computation, University of Reading
Genetic Algorithms

Côté, Isabelle M.
Lecturer in Ecology, University of East Anglia, Norwich, United Kingdom
Territoriality

Crandall, Keith A.
Thomas L. Martin Professor of Biology, Department of Zoology, Brigham Young University
Convergent and Parallel Evolution; Reticulate Evolution

Creager, Angela N. H.
Associate Professor of History, Princeton University
Franklin, Rosalind

Crespi, Bernard J.
Professor of Biology, Simon Fraser University, British Columbia, Canada
Eusociality, *article on* Eusocial Insects

Crow, James F.
Professor Emeritus of Genetics, University of Wisconsin
Wright, Sewall

Daly, Martin
Professor of Psychology and Biology, McMaster University, Ontario, Canada
Infanticide

Darnell, Diana Karol
Assistant Professor of Biology, Lake Forest College
Development, *overview article*

Day, Troy
Assistant Professor of Zoology, University of Toronto
Inclusive Fitness

Dean, Michael
Senior Scientist, National Institutes of Health, Frederick, Maryland
Cancer

Delph, Lynda F.
Associate Professor of Biology, Indiana University
Mating Systems, *article on* Plant Mating Systems

Dennett, Daniel
University Professor and Director, Center for Cognitive Studies, Tufts University
Overview Essay *on* The New Replicators

Denton, Michael J.
Senior Research Fellow in Human Genetics, Biochemistry Department, University of Otago, New Zealand
Protein Folding

Desmond, Adrian
Honorary Research Fellow, Department of Biology, University College London
Huxley, Thomas Henry

Donoghue, Michael J.
G. Evelyn Hutchinson Professor of Ecology and Evolutionary Biology, Yale University
Plants

Douglas, Angela E.
Lecturer in Biology, University of York, United Kingdom
Symbiosis

Drayna, Dennis
Senior Fellow, National Institute on Deafness and Other Communication Disorders, National Institutes of Health
Mendelian Genetics; Pedigree Analysis

Dugatkin, Lee Alan
Associate Professor of Biology, University of Louisville
Prisoner's Dilemma Games

Dunsworth, Holly M.
Doctoral Candidate, Department of Anthropology, Pennsylvania State University
Hominid Evolution, *article on* Early *Homo*

Dwyer, Greg
Assistant Professor, Department of Ecology and Evolution, University of Chicago
Myxomatosis

Earle, Timothy
Professor of Anthropology, Northwestern University
Human Societies, Evolution of

Eberhard, William G.
Staff Scientist, Smithsonian Tropical Research Institute, and Profesor Catedratico, Universidad de Costa Rica
Cryptic Female Choice; Genitalic Evolution

Edgecombe, Gregory D.
Senior Research Scientist, Australian Museum
Arthropods

Edwards, A. W. F.
Professor of Biometry, University of Cambridge
Fisher, Ronald Aylmer; Fundamental Theorem of Natural Selection

Eisenberg, John F.
Eminent Scholar Emeritus, University of Florida, Gainesville
Mammals

Embley, T. Martin
Principal Scientist, Department of Zoology, The Natural History Museum, London
Prokaryotes and Eukaryotes

Emlet, Richard B.
Associate Professor, Oregon Institute of Marine Biology and Department of Biology, University of Oregon
Life History Stages

Erwin, Douglas H.
Research Paleobiologist and Curator, Department of Paleobiology, National Museum of Natural History, Washington, D.C.
Metazoans

Ewald, Paul W.
Professor of Biology, Amherst College
Virulence

Eyre-Walker, Adam
Royal Society University Research Fellow, University of Sussex
Codon Usage Bias

Falk, Dean
Professor of Anthropology, Florida State University
Brain Size Evolution

Farber, Paul Lawrence
Distinguished Professor of the History of Science, Oregon State University
Buffon, Georges-Louis Leclerc; Cuvier, Georges; Linnaeus, Carolus

Feldman, Marc
Burnet C. and Mildred Finley Wohlford Professor of Biological Sciences, Stanford University
Cultural Evolution, *article on* Cultural Transmission

Ferrier, David E. K.
Postdoctoral Research Fellow, University of Reading
Homeobox

Fessler, Daniel M. T.
Assistant Professor of Anthropology, University of California, Los Angeles
Emotions and Self-Knowledge

Fichman, Martin
Professor of History of Science, York University, Toronto
Wallace, Alfred Russel

Fleagle, John
Professor of Anatomical Sciences, State University of New York at Stony Brook
Primates, *article on* Primate Classification and Phylogeny

Fogel, David B.
Chief Executive Officer, Natural Selection, Inc.
Artificial Life

Foley, Robert
Director, Leverhulme Centre for Human Evolutionary Studies, University of Cambridge
Environment of Evolutionary Adaptedness; Hominid Evolution, *overview article*

Franciscus, Robert G.
Assistant Professor of Anthropology, University of Iowa
Hominid Evolution, *article on* Neanderthals

Frank, Steven A.
Professor of Biological Sciences, University of California, Irvine
Price, George

Friedman, Thomas B.
Chief, Laboratory of Molecular Genetics, National Institute on Deafness and Other Communication Disorders, National Institutes of Health
Mendelian Genetics; Pedigree Analysis

Futuyma, Douglas J.
Distinguished Professor, Department of Ecology and Evolution, State University of New York at Stony Brook
Speciation

Gage, Matthew J. G.
Royal Society University Research Fellow, School of Biological Sciences, University of East Anglia, Norwich, United Kingdom
Mating Systems, *article on* Animals

Gaston, Kevin J.
Royal Society University Research Fellow, University of Sheffield
Extinction

Gerhardt, H. Carl
Curators' Professor of Biological Sciences, University of Missouri, Columbia
Sexual Dimorphism

Ghiselin, Michael T.
Senior Research Fellow and Chair, Center for the History and Philosophy of Science, California Academy of Sciences
Darwin, Charles; Lyell, Charles

Giambrone, Steve
Associate Professor, Cognitive Evolution Group, University of Louisiana at Lafayette
Consciousness

Gibson, Robert M.
Professor of Biological Sciences, University of Nebraska-Lincoln
Leks

Gifford, Robert
Postgraduate Research Student, Department of Biology, Imperial College at Silwood Park
DNA and RNA

Gifford-Gonzalez, Diane
Professor of Anthropology, University of California, Santa Cruz
Archaeological Inference

Gilbert, Scott F.
Professor of Biology, Swarthmore College
Canalization

Giribet, Gonzalo
Assistant Professor of Biology and Assistant Curator of Invertebrates, Museum of Comparative Zoology, Department of Organismic and Evolutionary Biology, Harvard University
Arthropods

Glazier, Douglas S.
Professor of Biology, Juniata College
Parental Care

Goodall, Jane
Trustee, Jane Goodall Institute for Wildlife Research, Education and Conservation
Overview Essay *on* Culture in Chimpanzees

Goodman, Simon J.
Research Fellow in Genetics, Institute of Zoology, Zoological Society of London
Genetic Polymorphism; Linkage

Gould, Stephen Jay
Alexander Agassiz Professor of Zoology, Curator of Invertebrate Paleontology in the Museum of Comparative Zoology, and Professor of Geology, Harvard University
Overview Essay *on* Macroevolution; Punctuated Equilibrium

Grant, Peter
Professor, Princeton University
Darwin's Finches; Galapagos Islands

Grant, Rosemary
Professor, Princeton University
Darwin's Finches; Galapagos Islands

Grantham, Todd
Associate Professor of Philosophy, College of Charleston
Species Selection

Gray, Michael W.
Professor of Biochemistry and Molecular Biology and Canada Research Chair in Genomics and Genome Evolution, Dalhousie University, Nova Scotia, Canada
Cell Evolution; Endosymbiont Theory

Greene, Erick P.
Associate Professor, Division of Biological Sciences, University of Montana
Development, *article on* Development and Ecology

Greene, Harry W.
Professor and Curator, Department of Ecology and Evolutionary Biology, Cornell University
Vertebrates

Griesemer, James
Professor of Philosophy, University of California, Davis, and Frederick B. Churchill Professor Emeritus of the History and Philosophy of Science, Indiana University
Weismann, August Friedrich Leopold

Hall, Brian K.
George S. Campbell Professor of Biology, Dalhousie University, Nova Scotia, Canada
Atavisms; Dollo's Law; Phylotypic Stages

Hallam, Anthony
Emeritus Professor of Geology, University of Birmingham
Mass Extinctions

Hallerman, Eric M.
Associate Professor of Fisheries and Wildlife Sciences, Virginia Polytechnic Institute and State University
Genetically Modified Organisms

Hallgrímsson, Benedikt
Assistant Professor, Department of Anatomy, University of Puerto Rico, School of Medicine
Phenotypic Stability

Harpending, Henry
Thomas Professor of Anthropology, University of Utah
Modern *Homo sapiens*, *article on* Human Genealogical History; Natural Selection, *article on* Natural Selection in Contemporary Human Society; Race, *article on* Population Genetics Perspectives

Harrison, Richard G.
Professor, Department of Ecology and Evolutionary Biology, Cornell University
Species Concepts

Hart, Michael
Assistant Professor of Biology, Dalhousie University, Nova Scotia, Canada
Germ Line and Soma

Hendrickson, Dean A.
Curator of Ichthyology, Texas Memorial Museum of Science and History, University of Texas at Austin
Fishes

Hewitt, Godfrey M.
Professor of Evolutionary Genetics, University of East Anglia, Norwich, United Kingdom
Hybrid Zones

Hillis, David M.
Alfred W. Roark Centennial Professor, Section of Integrative Biology, University of Texas at Austin
Molecular Systematics

Hiscox, Julian
Lecturer in Virology, University of Reading
Origin of Life, *article on* Origin of Replicators

Holmes, Edward C.
Lecturer in Evolutionary Biology, University of Oxford
Acquired Immune Deficiency Syndrome, *article on* Origins and Phylogeny of HIV; Viruses

Houle, David
Assistant Professor of Biological Science, Florida State University
Quantitative Genetics

Houston, Alasdair I.
Professor of Theoretical Biology, University of Bristol
Optimality Theory, *overview article*; Optimality Theory, *article on* Optimal Foraging

Hrdy, Sarah Blaffer
Professor Emeritus of Anthropology, University of California, Davis
Overview Essay *on* Motherhood

Hull, David L.
Professor Emeritus of Philosophy, Northwestern University
Overview Essay *on* History of Evolutionary Thought; Evolutionary Epistemology; Levels of Selection; Naturalistic Fallacy; Social Darwinism

Hurle, Juan M.
Professor of Anatomy, Universidad de Cantabria, Spain
Apoptosis

Hurst, Gregory
Lecturer in Ecology and Evolution, University College London
Cytoplasmic Genes

Hurst, Laurence D.
Professor of Evolutionary Genetics, University of Bath
Sexes and Mating Types

Irons, William
Professor of Anthropology, Northwestern University
Human Sociobiology and Behavior, *article on* Human Sociobiology

Jablonka, Eva
Associate Professor, The Cohn Institute, Tel Aviv University
Epigenetics; Lamarckism

Johnson, Norman A.
Research Assistant Professor, Department of Entomology, University of Massachusetts-Amherst
Haldane's Rule

Johnston, Mark O.
Associate Professor, Department of Biology, Dalhousie University, Nova Scotia, Canada
Homeosis

Johnstone, Rufus A.
University Lecturer in Zoology, University of Cambridge
Signalling Theory

Jones, Doug
Director, Florida Museum of Natural History
Race, *article on* Sociological Perspectives

Judson, Olivia P.
Research Fellow, Imperial College of Science, Technology and Medicine, London, United Kingdom
Genetic Code

Kaplan, Hillard S.
Professor of Anthropology, University of New Mexico
Life History Theory, *article on* Human Life Histories

Kappeler, Peter M.
Head of Department, Department of Ethology and Ecology, German Primate Center
Primates, *article on* Primate Biogeography

Katz, Laura A.
Assistant Professor in Biological Sciences, Smith College
Protists

Keeling, Patrick
Canadian Institute for Advanced Research Scholar, and Assistant Professor of Botany, University of British Columbia
Cellular Organelles

Keller, Laurent
Professor of Evolutionary Ecology and Head of the Institute of Ecology, University of Lausanne, Switzerland
Kin Selection

King, W. Allan
Professor, Department of Biomedical Sciences, Ontario Veterinary College, University of Guelph, Ontario
Maternal Cytoplasmic Control

Kirkwood, Thomas B. L.
Professor and Head of Gerontology, University of Newcastle upon Tyne
Senescence

Klein, Richard G.
Lecturer in Human Evolution, Stanford University
Modern *Homo sapiens*, *overview article*

Kohn, Alan J.
Professor Emeritus of Zoology, University of Washington
Molluscs

Kotiaho, Janne S.
Research Coordinator, University of Jyväskylä, Finland
Male-Male Competition

Krakauer, David C.
Research Professor, Santa Fe Institute
Genetic Redundancy; Quasi-Species

Kuhn, Steven L.
Associate Professor of Anthropology, University of Arizona
Paleolithic Technology

Laird, Diana J.
Graduate Student, Department of Biological Sciences, Stanford University
Immune System, *article on* Structure and Function of the Vertebrate Immune System

Laland, Kevin N.
Royal Society University Research Fellow, University of Cambridge
Cultural Evolution, *overview article*; Niche Construction

Lamb, Marion J.
Associate Fellow of the Cohn Institute for the History and Philosophy of Science and Ideas, Tel Aviv University, Israel
Epigenetics; Lamarckism

Laporte, Léo F.
Professor Emeritus, Earth Sciences, University of California, Santa Cruz
Simpson, George Gaylord

Layton, Robert
Professor of Anthropology, University of Durham
Art, *overview article*

Lee, Michael S. Y.
Senior Research Scientist, Department of Palaeontology, The South Australian Museum
Reptiles

Lenski, Richard E.
Hannah Professor of Microbial Ecology, Michigan State University
Basic Reproductive Rate (R_0); Experimental Evolution, *article on* A Long-Term Study with *E. coli*; Resistance, Cost of

Leroi, Armand M.
Lecturer, Department of Biological Sciences, Imperial College at Silwood Park, London
Development, *article on* Evolution of Development

Levinson, Orde
Artist and writer, Magdalen College, University of Oxford
Art, *article on* An Adaptive Function?

Li, Liming
Howard Hughes Medical Institute, Department of Molecular Genetics and Cell Biology, University of Chicago
Prions

Li, Wen-Hsiung
Professor of Evolution, University of Chicago
Molecular Evolution

Lieberman, Philip
Fred M. Seed Professor of Cognitive and Linguistic Sciences and Professor of Anthropology, Brown University
Language, *overview article*

Lindquist, Susan
Director of the Whitehead Institute for Biomedical Research, Cambridge, Massachusetts
Prions

Lipsitch, Marc
Assistant Professor of Epidemiology, Harvard School of Public Health
Antibiotic Resistance, *article on* Strategies for Managing Resistance

Lively, C. M.
Professor of Biology, Indiana University
Red Queen Hypothesis

Lonsdorf, Elizabeth Vinson
Doctoral Candidate, Department of Ecology, Evolution and Behavior, University of Minnesota
Overview Essay *on* Culture in Chimpanzees

Low, Bobbi
Professor and Chair of Resource Ecology, School of Natural Resources and Environment, and Research Associate, Population Studies Center, University of Michigan
Tragedy of the Commons

Lutzoni, François
Assistant Professor of Biology, Duke University
Lichens

Lyttle, Terrence W.
Professor of Cell and Molecular Biology, University of Hawaii, Manoa
Meiotic Distortion

Mace, Ruth
Reader in Biological Anthropology, University College London
Demographic Transition; Human Foraging Strategies, *article on* Subsistence Strategies and Subsistence Transitions; Human Sociobiology and Behavior, *article on* Behavioral Ecology

MacLeod, Norman
Keeper of Palaeontology, The Natural History Museum, London
Morphometrics

Maienschein, Jane
Professor of Philosophy and Biology and Director of the Biology and Society Program, Arizona State University
Morgan, T. H.

Majerus, Michael E. N.
Reader in Evolution, University of Cambridge
Genes; Peppered Moth

Marshall, Craig J.
Senior Lecturer in Biochemistry, University of Otago, New Zealand
Protein Folding

Marshall, Diane L.
Professor of Biology, University of New Mexico
Mate Choice, *article on* Mate Choice in Plants

Martínez, Daniel E.
Assistant Professor of Biology and Molecular Biology, Pomona College
Zootypes

Mayden, Richard
W. S. Barnickel Professor of Natural Sciences, Saint Louis University
Biogeography, *article on* Vicariance Biogeography

Mayer, Kimberly M.
Postdoctoral Scholar, California Institute of Technology
Directed Protein Evolution

Maynard Smith, John, FRS
Emeritus Professor of Biology, University of Sussex
Overview Essay *on* The Major Transitions in Evolution; Game Theory

McCabe, Kevin A.
Professor of Economics and Law and IFREE Distinguished Research Scholar, George Mason University
Evolutionary Economics

McLean, Angela
University Lecturer in Biodiversity, Zoology Department, University of Oxford
Vaccination

McNamara, John M.
Professor of Mathematics and Biology, University of Bristol
Optimality Theory, *overview article*; Variable Environments

McNeil, Jeremy N.
Full Professor, Department of Biology, Laval University, Quebec, Canada
Biological Warfare

McVean, Gilean
Royal Society University Research Fellow, University of Oxford
Chromosomes; Introns

Mealey, Linda
Professor of Psychology, College of St. Benedict, and Adjunct Associate Professor of Psychology, University of Queensland, Australia
Human Sociobiology and Behavior, *article on* Evolutionary Psychology

Meinertzhagen, Ian A.
Killam Professor of Neuroscience, Dalhousie University, Nova Scotia, Canada
Cell Lineage

Merino, Ramon
Senior Researcher, Hospital Universitario Martqués de Valdecilla, Santander, Spain
Apoptosis

Michalakis, Yannis
Directeur de Recherches, Centre d'Etudes sur le Polymorphisme des Microorganismes, Montpellier, France
Metapopulation

Mindell, David
Curator of Birds and Associate Professor of Ecology and Evolutionary Biology, University of Michigan
Birds

Mitchell, Melanie
Staff Member, Santa Fe Institute
Complexity Theory

Mitton, Jeffry B.
Professor, Department of Environmental Population and Organismic Biology, University of Colorado at Boulder
Hardy–Weinberg Equation; Heterozygote Advantage, *overview article*

Mock, Douglas W.
Professor of Zoology, University of Oklahoma
Parent–Offspring Conflict; Sibling Rivalry

Mojzsis, Stephen J.
Assistant Professor of Geological Sciences, University of Colorado at Boulder
Origin of Life, *article on* The First Fossils

Møller, Anders Pape
Research Director, University Pierre and Marie Curie, Paris, France
Extrapair Copulations

Moore, Janice
Professor of Biology, Colorado State University
Host Behavior, Manipulation of

Morse, Stephen S.
Director, Center for Public Health Preparedness and Assistant Professor of Epidemiology, Mailman School of Public Health, Columbia University
Emerging and Re-Emerging Diseases

Moxon, E. Richard
Action Research Professor of Paediatrics, Molecular Infectious Diseases Group, Weatherall Institute of Molecular Medicine, John Radcliffe Hospital, Oxford
Immune System, *article on* Microbial Countermeasures to Evade the Immune System

Mueller, Laurence D.
Professor of Ecology and Evolutionary Biology, University of California, Irvine
Density-Dependent Selection

Müller, Gerd B.
Professor of Anatomy, University of Vienna, Austria, and Chairman of the Konrad Lorenz Institute for Evolution and Cognition Research, Altenberg, Austria
Novelty and Key Innovations; Vestigial Organs and Structures

Nelson, Craig E.
Professor of Biology, Indiana University
Creationism

Nelson, C. Riley
Associate Professor, Department of Zoology, Brigham Young University
Insects

Nettle, Daniel
Lecturer in Biological Psychology, The Open University, Milton Keynes, United Kingdom
Language, *article on* Linguistic Diversity

Newman, Mark
Assistant Professor of Physics and Complex Systems, University of Michigan
Complexity Theory

Nielsen, Claus
Associate Professor, Zoological Museum, University of Copenhagen
Protostome-Deuterostome Origins

Nijhout, H. F.
Professor of Biology, Duke University
Gene Regulatory Networks

Nowak, Martin A.
Head, Program in Theoretical Biology, Institute for Advanced Study, Princeton, New Jersey
Quasi-Species

Orel, Vitezslav
Emeritus Head of the Mendelianum, Brno, Czech Republic
Mendel, Gregor

Pagel, Mark
Professor of Evolutionary Biology, University of Reading
Comparative Method; Constraints on Adaptation; Evolution; Phylogenetic Inference, *article on* Methods

Paine, Richard
Associate Professor of Anthropology, University of Utah
Agriculture, *article on* Origins of Agriculture; Disease, *article on* Demography and Human Disease; Population Trends

Patthy, László
Professor of Molecular Biology, Biological Research Center of the Hungarian Academy of Sciences
Proteins

Paul, Diane B.
Professor of Political Science and Director, Program in Science, Technology, and Values, University of Massachusetts at Boston
Eugenics

Pellmyr, Olle
Associate Professor of Biological Sciences, Vanderbilt University
Microevolution

Peterson, Kevin J.
Assistant Professor of Biological Sciences, Dartmouth College
Multicellularity and Specialization

Petrov, Dmitri A.
Assistant Professor, Department of Biological Sciences, Stanford University
Genome Size Evolution

Pfennig, David W.
Associate Professor of Biology, University of North Carolina, Chapel Hill
Kin Recognition

Philippe, Herve
Directeur de Recherches, University Pierre and Marie Curie, Paris, France
Gene Families

Pianka, Eric
Denton A. Cooley Centennial Professor of Zoology, Section of Integrative Biology, University of Texas at Austin
Species Diversity

Pomiankowski, Andrew
Professor of Genetics, University College London
Dosage Compensation; Fluctuating Asymmetry; Genomic Imprinting; Plasmids

Pool, Gail R.
Professor of Anthropology, University of New Brunswick, Canada
Globalization

Povinelli, Daniel J.
Associate Professor and Director, Cognitive Evolution Group, University of Louisiana at Lafayette
Consciousness

Prothero, Donald R.
Professor of Geology, Occidental College
Geological Periods; Paleontology; Stratigraphy

Reeve, Kern H.
Associate Professor of Neurobiology and Behavior, Cornell University
Kin Selection

Reilly, Stephen M.
Associate Professor, Department of Biological Sciences, Ohio University
Neoteny; Paedomorphosis

Reynolds, John D.
Reader in Evolutionary Ecology, University of East Anglia, Norwich, United Kingdom
Mating Systems, *article on* Animals

Rice, Sean H.
Associate Professor, Department of Ecology and Evolutionary Biology, Yale University
Heterochrony

Rich, Stephen M.
Assistant Professor of Genetics, Tufts University
Malaria

Ridley, Mark
Research Associate, Department of Zoology, University of Oxford
Adaptation; Genetic Load; Natural Selection, *overview article*

Rodd, F. Helen
Assistant Professor of Zoology, University of Toronto
Life History Theory, *article on* Guppies

Rodseth, Lars
Associate Professor of Anthropology, University of Utah
Human Evolution, *article on* History of Ideas

Rose, Christopher Stewart
Assistant Professor of Biology, James Madison University
Metamorphosis, Origin and Evolution of

Rowe, Timothy
J. Nalle Gregory Regents Professor of Geology and Director, Vertebrate Paleontology Laboratory, University of Texas at Austin
Dinosaurs

Ruiz, Rosaura
Profesora de Evolución y de Historia de la Biología, Universidad Nacional Autónoma de México
Lamarck, Jean Baptiste Pierre Antoine de Monet

Rupke, Nicolaas A.
Professor of the History of Science and Director of the Institut für Wissenschaftsgeschichte, University of Göttingen, Germany
Owen, Richard

Sachs, Tsvi
Warburg Professor of Plant Physiology, The Hebrew University of Jerusalem, Israel
Developmental Selection

Sanderson, Michael J.
Professor of Evolution and Ecology, University of California, Davis
Homoplasy

Sarkar, Sahotra
Associate Professor of Philosophy and Director of the Program in the History and Philosophy of Science, University of Texas at Austin
Haldane, John Burdon Sanderson

Schlichting, Carl D.
Professor of Ecology and Evolutionary Biology, University of Connecticut
Phenotypic Plasticity

Schluter, Dolph
Professor of Zoology and Canada Research Chair, University of British Columbia
Character Displacement

Schmid-Hempel, Paul
Chair and Head of Experimental Ecology and Professor of Experimental Ecology, ETH Zürich, Switzerland
Marginal Values

Schmidt, Thomas M.
Associate Professor of Microbiology, Michigan State University
Bacteria and Archaea

Schwarcz, Henry
University Professor of Geology Emeritus, McMaster University
Geochronology

Schwenk, Kurt
Professor, University of Connecticut
Constraint

Seeley, Thomas D.
Professor of Biology, Cornell University
Honeybees

Seger, Jon
Professor of Biology, University of Utah
Mathematical Models

Segerstråle, Ullica
Professor of Sociology, Illinois Institute of Technology, Chicago
Neo-Darwinism

Seyfarth, Robert M.
Professor of Psychology, University of Pennsylvania
Alarm Calls

Sheldon, Ben C.
Royal Society University Research Fellow, University of Oxford
Sex Ratios

Sherman, Paul W.
Professor of Animal Behavior, Cornell University
Eusociality, *article on* Eusociality in Mammals; Naked Mole-Rats

Shuster, Stephen M.
Professor of Invertebrate Zoology, Northern Arizona University
Mating Strategies, Alternative

Sibly, Richard M.
Professor, School of Animal and Microbial Sciences, University of Reading
Life History Theory, *overview article*

Sigmund, Karl
Professor of Mathematics, University of Vienna, Austria
Cooperation

Silk, Joan B.
Professor, Biological Anthropology, Primate Behavior, and Evolutionary Biology, University of California, Los Angeles
Human Sociobiology and Behavior, *overview article*

Simms, Steven R.
Professor of Anthropology, Utah State University
Biogeography, *article on* Human Influences on Biogeography

Skelton, Peter W.
Senior Lecturer in Earth Sciences, The Open University, Milton Keynes, United Kingdom
Geology

Smith, Vernon L.
Professor of Economics and Law, George Mason University
Evolutionary Economics

Sniegowski, Paul D.
Assistant Professor, Biology, University of Pennsylvania
Mutation, *overview article*; Mutation, *article on* Evolution of Mutation Rates

Sommer, Ralf J.
Director, Max-Planck Institute for Developmental Biology, Department for Evolutionary Biology, Tübingen, Germany
C. elegans

Spencer, Hamish G.
Associate Professor of Zoology, University of Otago, New Zealand
Eugenics

Stearns, Stephen C.
Edward P. Bass Professor of Ecology and Evolutionary Biology, Yale University
Overview Essay *on* Darwinian Medicine

Stone, Anne C.
Assistant Professor of Anthropology, University of New Mexico
Modern *Homo sapiens, article on* Neanderthal–Modern Human Divergence

Strassmann, Beverly
Associate Professor of Anthropology, University of Michigan
Reproductive Physiology, Human

Sutherland, William J.
Professor of Biology, University of East Anglia, Norwich, United Kingdom
Ideal Free Distribution

Tabashnik, Bruce E.
Professor and Head, Department of Entomology, University of Arizona
Pesticide Resistance

Takahata, Naoyuki
Vice-President, Graduate University for Advanced Studies and Professor of Department of Biosystems Science, Japan
Kimura, Motoo

Tautz, Diethard
Professor of Evolutionary Genetics, University of Cologne, Germany
Regulatory Genes; Segmentation

Taylor, Peter D.
Professor, Mathematics, Biology and Education, Queen's University, Ontario, Canada
Inclusive Fitness

Templeton, Alan R.
Charles Rebstock Professor of Biology, Washington University, St. Louis, Missouri
Coronary Artery Disease

Theriot, Edward
Director, Texas Memorial Museum of Science and History, University of Texas at Austin
Diatoms

Thomas, Mark
Senior Lecturer in Molecular Anthropology, Department of Biology, University College London
Ancient DNA

Thompson, John N.
Professor of Biology, University of California, Santa Cruz
Coevolution

Thomson, Keith S.
Professor of Natural History, University of Oxford
Missing Links

Thorpe, Roger S.
Chair of Animal Ecology, University of Wales, Bangor
Geographic Variation

Tibayrenc, Michel
Director of Research, IRD, Montpellier, France
Clonal Structure, *article on* Population Structure and Clonality of Protozoa

Trivers, Robert L.
Professor of Anthropology and Biological Sciences, Rutgers University
Hamilton, William D.; Self-Deception

Turner, John R. G.
Professor of Evolutionary Biology, University of Leeds
Mimicry

Uptain, Susan M.
Howard Hughes Medical Institute, Department of Molecular Genetics and Cell Biology, University of Chicago
Prions

Valentine, James W.
Professor of Integrative Biology, Emeritus, University of California, Berkeley
Cell-Type Number and Complexity

van der Dennen, Johan M. G.
Senior Researcher, Department of Legal Theory, Section Political Science, University of Groningen, the Netherlands
Warfare

van Schaik, Carel
Professor of Biological Anthropology, Duke University
Primates, *article on* Primate Societies and Social Life

Vrba, Elisabeth S.
Professor of Geology and Geophysics, Peabody Museum Curator of Vertebrate Paleontology and Vertebrate Zoology, Yale University
Exaptation

Wagner, Günter P.
Professor of Biology, Yale University
Homology

Wake, Marvalee H.
Professor of Integrative Biology, University of California, Berkeley
Placentation; Viviparity and Oviparity

Walker, Alan
Distinguished Professor of Anthropology and Biology, Pennsylvania State University
Hominid Evolution, *article on* Early *Homo*

Walsh, Bruce
Associate Professor, Ecology and Evolutionary Biology, University of Arizona
Artificial Selection

Weatherall, Sir David
Director, The Institute of Molecular Medicine, University of Oxford
Heterozygote Advantage, *article on* Sickle-Cell Anemia and Thalassemia

Webster, Andrea J.
Postdoctoral Fellow in Evolutionary Biology, University of Reading
Repetitive DNA

Weir, Bruce
William Neal Reynolds Professor of Statistics and Genetics, North Carolina State University
Forensic DNA

West, Stuart A.
BBSRC David Phillips Research Fellow, University of Edinburgh
Sex, Evolution of; Sex Ratios

Wheeler, Ward
Curator of Invertebrate Zoology, Division of Invertebrate Zoology, American Museum of Natural History
Arthropods

White, Tim
Professor of Integrative Biology and Co-Director of the Laboratory for Human Evolutionary Studies, Museum of Vertebrate Zoology, University of California, Berkeley
Hominid Evolution, *article on Ardipithecus* and *Australopithecines*

Whitlock, Michael C.
Associate Professor of Zoology, University of British Columbia
Genetic Drift; Inbreeding; Metapopulation; Shifting Balance

Whittaker, Robert J.
Reader in Biogeography, School of Geography and the Environment, University of Oxford
Biogeography, *article on* Island Biogeography

Wilcox, Tom
Postdoctoral Associate, Integrative Biology, University of Texas at Austin
Cnidarians

Wilkins, Adam S.
Editor, BioEssays
Sex Determination

Wilkinson, Gerald S.
Professor of Biology, University of Maryland
Reciprocal Altruism

Wills, Christopher J.
Professor, Division of Biology, University of California, San Diego
Plagues and Epidemics

Wilson, David Sloan
Professor, Departments of Biology and Anthropology, Binghamton University
Group Selection

Wilson, Margo
Professor of Psychology, McMaster University
Infanticide

Wodarz, Dominik
Institute for Advanced Study, Princeton, New Jersey
Acquired Immune Deficiency Syndrome, *article on* Dynamics of HIV Transmission and Infection

Wray, Greg
Associate Professor of Biology, Duke University
Echinoderms

Yang, Ziheng
Professor of Statistical Genetics, University College London
Molecular Clock

Yi, Soojin
Research Associate, Department of Ecology and Evolution, University of Chicago
Molecular Evolution

Zeh, David
Associate Professor, Department of Biology and Program in Ecology, Evolution and Conservation Biology, University of Nevada, Reno
Maternal-Fetal Conflict

Zeh, Jeanne
Research Assistant Professor, University of Nevada, Reno
Maternal-Fetal Conflict

Zelditch, Miriam
Associate Research Scientist, Museum of Paleontology, University of Michigan
Heterotopy

Zink, Robert
Breckenridge Chair in Ornithology, University of Minnesota, Twin Cities
Systematic Collections

Zuk, Marlene
Professor of Biology, University of California, Riverside
Assortative Mating; Mate Choice, *overview article*; Sexual Selection, *overview article*

Index

N.B.: Page references in boldface indicate a major discussion.

Vol. 1: pp. E-1–E-94, 1–556;
Vol. 2: pp. 557–1170

D

J

K

O

Vol. 1: pp. E-1–E-94, 1–556;
Vol. 2: pp. 557–1170

T

X

Y

Z